T0140343

Engineering Education 4.0

Sulamith Frerich · Tobias Meisen
Anja Richert · Marcus Petermann
Sabina Jeschke · Uwe Wilkesmann
A. Erman Tekkaya
Editors

Engineering Education 4.0

Excellent Teaching and Learning in Engineering Sciences

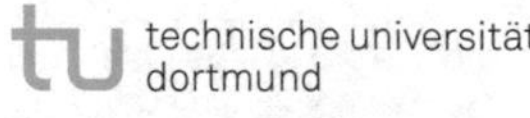

SPONSORED BY THE

Editors
Sulamith Frerich
VVP
Ruhr-Universität Bochum
Bochum
Germany

Tobias Meisen
Anja Richert
Sabina Jeschke
IMA/ZLW & IfU
RWTH Aachen University
Aachen
Germany

Marcus Petermann
Lehrstuhl für Feststoffverfahrenstechnik
Ruhr-Universität Bochum
Bochum
Germany

Uwe Wilkesmann
Zentrum für HochschulBildung
Technische Universität Dortmund
Dortmund
Germany

A. Erman Tekkaya
Institut für Umformtechnik & Leichtbau
Technische Universität Dortmund
Dortmund
Germany

Das Buch wurde gedruckt mit freundlicher Unterstützung der RWTH Aachen University, der Ruhr-Universität Bochum sowie der Technischen Universität Dortmund.
The project ELLI – Excellent Teaching and Learning in Engineering Sciences – with the support codes 01PL11082A, 01PL11082B and 01PL11082C was funded by the Federal Ministry of Education and Research in Germany.

ISBN 978-3-319-83619-5 ISBN 978-3-319-46916-4 (eBook)
DOI 10.1007/978-3-319-46916-4

Mathematics Subject Classification (2010): 68-06, 68Q55, 68T30, 68T37, 97C70, 97D40

Printed on acid-free paper

This Springer imprint is published by Springer Nature
The registered company is Springer International Publishing AG
The registered company address is: Gewerbestrasse 11, 6330 Cham, Switzerland

Foreword

We are very proud to present to you a compilation of scientific papers published during the first funding phase of the cooperative project ELLI—"Excellent Teaching and Learning in Engineering Sciences". This interdisciplinary project, funded by the Federal Ministry of Education and Research in Germany (BMBF), aims to improve study conditions and enhance teaching quality in engineering education. The compilation at hand gives a review on five years of successful cooperation and marks a starting point for the second funding phase of the project, beginning in October 2016. Most publications are peer-reviewed and published in renowned journals, books or conference proceedings of the various disciplinary fields.

The consortium of RWTH Aachen University, Ruhr-Universität Bochum and TU Dortmund University looks back at many years of successful cooperation in developing and establishing innovative concepts in engineering education. The cooperation began in 2010 with the establishment of joint centre for competence and service for teaching and learning in engineering sciences, called TeachING-LeanING.EU and is continued and intensified in the project ELLI.

The combination of research intensive institutes of engineering science with didactical institutions of higher education at all three locations is a unique characteristic of the consortium. Teaching and learning design involves innovative didactic elements, causing a close interaction of didactic concepts and scientific learning. In the field of engineering sciences the consortium includes the following institutes:

- **Institute of Information Management in Mechanical Engineering (IMA) at RWTH Aachen University**
  The research focus at the Institute of Information Management in Mechanical Engineering IMA is to find and establish methods of computer science in all applicable fields of mechanical engineering, such as in the context of virtual laboratories or virtual theatres, industrial robots (ABB, Motoman) and mobile robots (Robotino, NAO). Direct integration allows an experimental evaluation of the developed solutions.

- **Chair of Mechanical Process Engineering (FVT) at Ruhr-Universität Bochum**
  The FVT chair's responsibilities are in research and teaching within process engineering of solid materials. A wide variety of process plants (of pilot plant as well as of bench scale), innovative measurement technologies, and comprehensive chemical analysis facilitate both the development and the optimization of process technologies and help foster the practice-orientedness of both research and teaching.
- **Institute of Forming Technology and Lightweight Construction (IUL) at TU Dortmund University**
  The Institute of Forming Technology and Lightweight Construction (IUL) of TU Dortmund University develops innovative forming technologies and process chains. A focus is on lightweight construction and energy efficiency. Five departments develop and investigate a wide range of forming technologies and keep industrial practice in view.
- **Center for Learning and Knowledge Management (ZLW) at RWTH Aachen University**
  The main focus of the Center for Learning and Knowledge Management ZLW lies on the development, application and professionalization of inter- and transdisciplinary methods in research and teaching. Within the ZLW, interdisciplinary teams address recent scientific problems and questions with regard to the investigation, development and implementation of trend-setting concepts and solutions suitable for complex innovation and organizational development processes on the one hand and learning and knowledge processes on the other.
- **Professional Development Office (IFB) at Ruhr-Universität Bochum**
  Skill enhancement and professional development of our university's teaching staff are the core competencies of the IFB. These objectives are put into practice through a highly diversified training program for didactics in higher education as well as through specific measurements implemented on an individual level, such as mentoring and coaching programs. The overall aim of these activities is to facilitate innovation within teaching and curriculum design, to support and coordinate processes of reform, and to thus foster sustainable organizational development from a holistic perspective.
- **Center for Higher Education (zhb) at TU Dortmund University**
  The Center for Higher Education (Zentrum für Hochschulbildung, zhb) is a central scientific institute unifying three professorships and four service departments. Being part of the professorship for Organizational Studies, Continuing Education and Social Management, the engineering education research group cooperates closely with all zhb service departments. By this, it is ensured that the latest and most innovative research results from the field of Teaching and Learning can be integrated into the training of higher education training.

The project ELLI divides into four core units: Virtual Learning Environments (KB1), Student Mobility and Internationalization (KB2), Student Lifecycle (KB3) and Professional Skills (KB4).

- Within the core unit **Virtual Learning Environments**, remote and virtual labs were established and e-learning as well as mobile-learning solutions were developed across all three sites during the first funding period (October 2011 until September 2016). Those developments were framed by studies analyzing the didactical integration of the respective laboratories into academic teaching. One main task was to technically realize the various laboratories at all three sites. The extensive set-up represent an ideal basis to prepare students for work life in the era of industry 4.0. Requirements for engineering education have changed dramatically following an ongoing integration of information technology and networking of processes and machines in production technology. In the near future traditional engineering jobs will be replaced by operations requiring a profound understanding for digital technologies.
- The requirements of engineering jobs increasingly include acting in globalized development- and production relationships as well as sensitivity for intercultural problems. This is why within the ELLI core unit **Student Mobility and Internationalization** local as well as general barriers for student mobility in engineering education were identified and appropriate measures were established to support student mobility. Studies carried out within the project ELLI and addressing student mobility serve as a basis for strategic acquisition of international students using e.g. video based recruitment campaigns.
- The core unit **Student Lifecycle** focuses especially on the transition between school and university with a special emphasis on barrier-free study conditions. It aims to enthuse young students for technical careers at an early stage. Successful measures, such as career counselling opportunities, reflect a contemporary image of the engineering profession. Offers for seminars and networking events as well as practical inside views strengthen motivation and identification with the studies and accompany the transition from school to graduation, PhD or professional career. The introductory phase of study represents a critical element of the student life cycle where specific measures support the study start.
- The core unit **Professional Skills** has successfully dealt with the integration of professional and interdisciplinary competencies in engineering education. This way, young engineers shall be well prepared and be able to handle professional tasks. Among a variety of professional skills, creativity, interdisciplinary nature, focus on research, problem-based learning as well as competence in diversity were of special interest.

Contributions to this compilation are structured following the core units presented above. Within the core units, articles are in chronological order to let the reader understand the development of the project.

Our deepest thanks go to our scientific staff at RWTH Aachen University, Ruhr-Universität Bochum and TU Dortmund University. With their enthusiasm, interdisciplinary skills and capability to think outside the box, they have made an important contribution to the success of the first project period. Likewise, we would like to thank everyone involved in the development and publication of this compilation. We are looking forward to further successful cooperation.

Contents

Editorial Staff

Kathrin Schönefeld
Thorsten Sommer
Tobias Fischer
Alexandra Koch
Markus Leisten
Laura Rupprecht
Mario Weißkopf
Katharina Schuster
Valerie Stehling
Ute Berbuir
Andreas Kilzer
Tobias Haertel
Tobias R. Ortelt

Editors and Contributors

About the Editors

Jun.-Prof. Dr.-Ing. Sulamith Frerich became assistant professor for "Virtualization of Process Technology—Experimental Experiences in Engineering Education" at Ruhr-University Bochum on August 1st 2013. Her research focuses on simulation and remote operation of test facilities as well as on the resulting underlying thermophysical phenomena and their description with the help of multilevel, coupled models. The research concentrates particularly on the influence of process technological parameters on the end product. Her findings directly influence the development of new teaching concepts that are independent of time and place. Research topics include porous materials, their production and characterization.

Jun.-Prof. Dr.-Ing. Tobias Meisen is the managing director of the Institute of Information Management in Mechanical Engineering (IMA) at RWTH Aachen University since 2014. In July 2012 he earned his doctorate with distinctions, shortly after he became the head of the "production technology" research group. In October 2015 he has been appointed junior professor for "interoperability of simulations in mechanical engineering" at RWTH Aachen University. Tobias Meisen studied computer science with a minor in business administration at RWTH Aachen University. His areas of specialization during his studies were data mining, data exploration and data management. For his research, he was awarded the Young Researcher Award by the DFG excellence cluster "Integrative Production Technology for High-Wage Countries". Furthermore, he was honored with the Borchers Badge for his thesis and he won the 2013 Idea Contest of the Siemens CKI initiative. In his research, he focusses on modern information management in cyber-physical systems (in production and logistic) and industry 4.0 using semantic technologies. His main research areas comprise interoperability of heterogeneous system landscapes (including applications, machinery and products) as well as the conception and the implementation of artificial intelligence in production scenarios.

Jun.-Prof. Dr. phil. Anja Richert is a professor for agile management at the faculty of mechanical engineering at RWTH Aachen University as well as managing director of the Center for Learning and Knowledge Management (ZLW) at the Cybernetic Lab IMA/ZLW & IfU in Aachen. She graduated and did her PhD in communication science, both with distinction. She works at the Cybernetic Cluster since 2003 and from 2008 to 2010 as head of the research department knowledge management. Her fields of specialization are agile management of learning and knowledge processes, researching and developing mixed reality learning concepts as well as the development and testing of data based sociological research designs for complex interdisciplinary research objects. As a member of the DFG-Cluster of excellence for production technology and tailor made fuels from biomass, she is responsible for interdisciplinary scientific communication and foresight processes.

Prof. Dr.-Ing. Marcus Petermann became the head of the Institute for Particle Technology at the Institute for Mechanical Engineering at Ruhr-University Bochum in November 2008. He studied chemical engineering at the Friedrich-Alexander University Erlangen, Nürnberg and in 1999 subsequently earned his doctorate in the field of high pressure process technology with the thesis: "Production of powder coatings by spraying gaseous melts". After transferring to Ruhr-University Bochum to work on his habilitation, he became assistant professor for Particle Technology and Particle Design from 2002 to 2008. His key research areas include high pressure process technology, design of particle systems and the usage of CO2 as a raw material, aerogels, gas cleaning and purification as well measurement technology for high-pressure operation. Marcus Petermann also holds the position as dean of the International Graduate School

Solvation Science at the Clusters of Excellence RESOLV (Ruhr Explores Solvation) as well as being on various committees. Furthermore, Petermann is an appointed member of the Working Party on High Pressure Technology of the European Federation of Chemical Engineering. In 2000 he was awarded the Messer-Innovation-Award as well as the 2003 DECHEMA award to promote up-and-coming teachers in higher education for process technology.

Univ.-Prof. Dr. rer. nat. Sabina Jeschke became head of the Cybernetic Lab IMA/ZLW & IfU of the RWTH Aachen University in June 2009. She studied Physics, Computer Science and Mathematics at the Berlin University of Technology. After research stays at the NASA Ames Research Center/California and the Georgia Institute of Technology/Atlanta, she gained a doctorate on "Mathematics in Virtual Knowledge Environments" in 2004. Following a junior professorship (2005–2007) at the TU Berlin with the construction and direction of its media center, she was head of the Institute of Information Technology Services (IITS) for electrical engineering at the University of Stuttgart from May 2007 to May 2009, where she was also the director of the Central Information Technology Services (RUS) at the same time. Some of the main areas of her research are complex IT-systems (e.g. cloud computing, Internet of Things, green IT & ET, semantic web services), robotics and automation (e.g. heterogeneous and cooperative robotics, cooperative agents, web services for robotics), traffic and mobility (autonomous and semi-autonomous traffic systems, international logistics, car2car & car2X models) and virtual worlds for research alliances (e.g. virtual and remote laboratories, intelligent assistants, semantic coding of specialised information). Sabina Jeschke is vice dean of the Faculty of Mechanical Engineering of the RWTH Aachen University, chairwoman of the board of management of the VDI Aachen and member of the supervisory board of the Körber AG. She is a member and consultant of numerous committees and commissions, alumni of the German National Academic Foundation (Studienstiftung des Deutschen Volkes), IEEE Senior Member and Fellow of the RWTH Aachen University. In July 2014, the Gesellschaft für Informatik (GI) honored her with their award Deutschlands digitale Köpfe (Germany's digital heads). In September 2015 she was awarded the Nikola-Tesla Chain by the International Society of Engineering Pedagogy (IGIP) for her outstanding achievements in the field of engineering pedagogy.

Prof. Dr. Uwe Wilkesmann is Director of the Centre for Higher Education (zhb), TU Dortmund University. He holds a chair for Organization Studies, Management of Continuing Education, and Social Management since 2006. Additionally, he was Adj. Professor at the Knowledge Management and Innovation Research Centre (Department of Industrial and Systems Engineering) at the Hong Kong Polytechnic University between 2008 and 2014 and chaired the German Organization Sociology. Wilkesmann studied philosophy, business administration and social science at the Ruhr-University Bochum and received his doctorate (Ph.D.) in Social Psychology in 1993 from the Ruhr-University Bochum as well as his habilitation (German professor qualification) in 1998. From 2002 to 2003 he was professor for sociology of economics and organizations at the University of Hamburg, from 2004 to 2005 he was appointed as an extraordinary professor for organization studies at the Ruhr-University Bochum, and from 2005 until 2006 he was professor for organizational and educational sociology at the University of Munich. His main areas of research are: (1) Higher Education Research with the focus on higher education institutions as knowledge-intensive organizations. (2) Academic Teaching and Didactics emphasizing the relationship between governance of academic teaching and teaching behavior. (3) Organization Studies where he conducted research regarding the recursive relationship between organizational structures and the individual behavior. He was PI of plenty third party funded research projects and published a lot of articles and books about academic teaching, higher education research, knowledge management, and knowledge transfer for example in Organization Studies, Higher Education, Tertiary Education and Management, VINE Journal of Information and Knowledge Management Systems (Emerald Literati Network Award 2012).

Prof. Dr.-Ing. Dr.-Ing. E.h. A. Erman Tekkaya Since 2007 A. Erman Tekkaya is professor at TU Dortmund University and the head of the Institute of Forming Technology and Lightweight Construction (IUL). In 2011, he initiated the new international master program "Master of Science in Manufacturing Technology" at TU Dortmund University and became executive coordinator of the program. In 2014, he was appointed dean of the faculty of mechanical engineering. Professor Tekkaya studied mechanical engineering at the faculty of mechanical engineering at Middle East Technical University in Ankara and subsequently earned his doctorate with honors at the Institute of Forming Technology at the University of Stuttgart in 1985. His habilitation (Üniversite docenti) at the TC Üniversiteler Arası Kurul, Ankara followed in 1988. From 1986 to 2005 he worked at the Middle East Technical University in Ankara. After his time as assistant and associate professor he became full professor in 1993. Additionally, between 2005 and 2009, Professor Tekkaya managed the manufacturing engineering department at the ATILIM University in Ankara. Until 2013 he was the founding director of the Center of Excellence on Metal Forming. His research interests are metal forming technologies as bulk metal forming, sheet metal forming, bending, and high speed forming as well as the modeling of metal forming processes and material characterization. In recognition of his contributions in the field of metal forming he was awarded the honorary degree Doktor-Ingenieur Ehren halber (Dr.-Ing. E.h.) by the Faculty of Engineering of Friedrich-Alexander-Universität in Erlangen-Nürnberg in 2012. In October 2014, he was awarded the International Prize for Research and Development in Precision Forging of the Japanese Society of Technology of Plasticity for process innovation, process characterization, and international leadership. In 2015, he was also awarded the Steel Innovation Prize for the development of new incremental forming technologies. Prof. A. Erman Tekkaya is a member of numerous national and international committees as well as joint Editor-in-Chief of the Journal of Materials Processing Technology. He is also a visiting professor at the Ohio State University and deputy speaker of the DFG-SFB Transregio 73.

Contributors

Aleksandrova, Gergana IMA/ZLW & IfU, RWTH Aachen University, Deutschland

Antkowiak, Daniela IMA/ZLW & IfU, RWTH Aachen University, Deutschland, Mensch-Maschine-Interaktion in der Medizininformatik

Bach, Ursula IMA/ZLW & IfU, RWTH Aachen University, Deutschland

Baumert, Britta Zentrum für HochschulBildung (zhb) – Bereich Hochschuldidaktik, TU Dortmund, Deutschland, Fachdidaktik und Hochschuldidaktik Theologie

Becker, Christoph Institut für Umformtechnik und Leichtbau (bis Oktober 2015), Technische Universität Dortmund, Deutschland, Ingenieurausbildung, Biegeumformung, Prozesssteuerung und –automatiserung

Benguria, Kari (geb.Wold) College of Communication and Information, University of Kentucky, USA, Education, Instructional Technology, Communication

Berbuir, Ute Stabsstelle interne Fortbildung und Beratung, Ruhr-Universität Bochum, Deutschland, Problembasiertes Lernen (PBL), Forschendes Lernen, Interdisziplinäre Zusammenarbeit

Berens, Tobias Berufsforschungs- und Beratungsinstitut für interdisziplinäre Technikgestaltung e.V., Deutschland, Maschinensicherheit, Ganzheitlicher Arbeitsschutz, Managementsysteme

Bielski, Emanuel Zentrum für HochschulBildung (zhb) – Lehrstuhl für Organisationsforschung, Weiterbildungs- und Sozialmanagement, TU Dortmund, Deutschland, Technikdidaktik, Labordidaktik

Borowski, Esther IMA/ZLW & IfU, RWTH Aachen University, Deutschland

Buescher, Christian IMA/ZLW & IfU, RWTH Aachen University, Deutschland, Virtual Production Intelligence, Semantische Informationsmodellierung in Produktion und Produktionsplanung, Kennzahlenorientierte Evaluation von Fabrikplanungsvorhaben

Chatti, Sami Institut für Umformtechnik und Leichtbau, Technische Universität Dortmund, Deutschland, Ingenieurausbildung, Rohr- und Profilumformung, Leichtbau

Dörfler (geb. Valter), Sarah IMA/ZLW & IfU, RWTH Aachen University, Deutschland, Personalentwicklungsmaßnahmen, Qualifizierung im Themenfeld "professionelle Handlungskompetenz", Moderation von Seminare und Workshops im Bereich Soft-Skills und Hochschullehre

Ewert, Daniel IMA/ZLW & IfU, RWTH Aachen University, Deutschland, selbstoptimierende Produktionssteuerung, kollaborative Robotik für Montage und Intralogistik, Mensch-Maschine-Kollaboration

Frerich, Sulamith Virtualisierung verfahrenstechnischer Prozesse, Ruhr-Universität Bochum, Deutschland, Virtualisierung, Poröse Materialen, Experimentiertechnik in der Lehre

Gies, Sören Institut für Umformtechnik und Leichtbau, Technische Universität Dortmund, Deutschland, Elektromagnetische Umformung, Umformtechnisches Fügen, Inkrementelle Umformung

Guéno, Pierre-Jean Institut für Umformtechnik und Leichtbau, Technische Universität Dortmund, Deutschland, MOOCs, Online-Lernen

An Haack, Alexander IMA/ZLW & IfU, RWTH Aachen University, Deutschland Managementkybernetik, Organisationskulturen, Qualitätsmanagement

Haertel, Tobias Zentrum für HochschulBildung (zhb) – Lehrstuhl für Organisationsforschung, Weiterbildungs- und Sozialmanagement, TU Dortmund, Deutschland, Kreativitätsforschung, Zukunftsforschung, Entrepreneurship

Hauck, Eckart IMA/ZLW & IfU, RWTH Aachen University, Deutschland, Robotik, Wirtschaftskybernetik und technische Kybernetik

Heinze (geb. Heckel), Ute IMA/ZLW & IfU, RWTH Aachen University, Deutschland, Data Analytics, Computerlinguistik und internationale Mobilität

Hoffmann, Max IMA/ZLW & IfU, RWTH Aachen University, Deutschland, Remote Labs, Virtuelle Labore, Virtual Reality

Hosch-Dayican, Bengü Zentrum für HochschulBildung (zhb) – Professorship of Higher Education, TU Dortmund, Deutschland, Hochschulpolitik, Gender, soziale Bewegungen

Isenhardt, Ingrid IMA/ZLW & IfU, RWTH Aachen University, Deutschland, Blended Learning, innovative Ingenieurausbildung, Kommunikation und Sozial Skills für Ingenieure

Jahnke, Isa School of Information Science and Learning Technologies, University of Missouri-Columbia, USA, Sociotechnical Design, Emergent Technologies for Learning, Digital Didactical Designs

Janßen, Daniela IMA/ZLW & IfU, RWTH Aachen University, Deutschland, Blended Learning/eLearning in der Hochschullehre, Virtuelle Lernwelten/Virtual Reality, Human-Computer Interaction

Witt (geb. Janssen), Theresa Stabsstelle interne Fortbildung und Beratung, Ruhr-Universität Bochum, Deutschland, Technikgeschichte, Ingenieurausbildung in Deutschland, Öffentlichkeitsarbeit an Hochschulen

Jeschke, Sabina IMA/ZLW & IfU, RWTH Aachen University, Deutschland, Komplexe IT-Systeme, Robotik und Automatisierung, Virtuelle Welten

Jungmann, Thorsten Fachbereich Ingenieurwissenschaften und Mathematik, Fachhochschule Bielefeld, Deutschland, Technikdidaktik, Engineering Education

Kilzer, Andreas Verfahrenstechnische Transportprozesse, Ruhr-Universität Bochum, Deutschland, Grenzflächenphänomene, Sprühprozesse, Forschendes Lernen

Kruse, Daniel Virtualisierung verfahrenstechnischer Prozesse, Ruhr-Universität Bochum, Deutschland, Remote Labore, Strömungsprozesse, E-Learning

Leisten, Ingo IMA/ZLW & IfU, RWTH Aachen University, Deutschland, Inter- und Transdisziplinarität, Forschungstransfer, kybernetische Kommunikationsund Organisationsentwicklung

Leisyte, Liudvika Zentrum für HochschulBildung (zhb) – Professorship of Higher Education, TU Dortmund, Dortmund, Deutschland, Professional Autonomy, Academic Entrepreneurship, Governance and Management of Higher Education Institutions

Lensing, Karsten Zentrum für HochschulBildung (zhb) – Lehrstuhl für Organisationsforschung, Weiterbildungs- und Sozialmanagement sowie Institut für Umformtechnik und Leichtbau (IUL), TU Dortmund, Deutschland, Engineering Education Research, Kompetenzforschung, Mobile Learning

Lenz, Laura IMA/ZLW & IfU, RWTH Aachen University, Deutschland, Game Design, Virtual Reality, e-Learning

Lieverscheidt, Hille Zentrum für medizinische Lehre, Ruhr-Universität Bochum, Deutschland, Medizindidaktik, Personal- und Organisationsentwicklung

May, Dominik Zentrum für HochschulBildung – Lehrstuhl für Organisationsforschung, Weiterbildungs- und Sozialmanagement, TU Dortmund, Deutschland, Engineering Education Research, Transnationale Lehre, Digitales Lehren und Lernen, Kompetenzforschung

Meisen, Philipp IMA/ZLW & IfU, RWTH Aachen University, Deutschland, Data Integration and Analysis, Time Interval Data Analysis und Visual Analytics

Meisen, Tobias IMA/ZLW & IfU, RWTH Aachen University, Deutschland, Informationsmanagement in cyber-physischen Systemen, Interoperabilität in heterogenen Systemlandschaften, Konzipierung und Entwicklung komplexer lernender Empfehlungssysteme

Meya, Rickmer Institut für Umformtechnik und Leichtbau, Technische Universität Dortmund, Deutschland, Engineering Education, Biegeumformung

Moore, Stephanie Curry School of Education, University of Virginia, USA, Instructional Technology, Online Learning, Engineering and Society

Müller, Kristina Stabsstelle Interne Fortbildung und Beratung, Ruhr-Universität Bochum, Deutschland, Hochschuldidaktik, Beratung, Fortbildung

Ortelt, Tobias R. Institut für Umformtechnik und Leichtbau, Technische Universität Dortmund, Deutschland, Ingenieurausbildung, Prozesssteuerung und –automatiserung, Remote Manufacturing

Ossenberg, Philipp Zentrum für HochschulBildung (zhb) – Lehrstuhl für Organisationsforschung, Weiterbildungs- und Sozialmanagement, TU Dortmund, Deutschland, Ingenieurdidaktik, Forschendes Lernen, Labordidaktik

Pekasch, Sabine Institut für Umformtechnik und Leichtbau, Technische Universität Dortmund, Deutschland, MOOCs, Online-Lernen

Petermann, Marcus Feststoffverfahrenstechnik, Ruhr-Universität Bochum, Deutschland, Feststoffverfahrenstechnik, Hochdrucktechnik, Projektorientierung in der Lehre

Peters, Franz Fakultät für Maschinenbau, Ruhr-Universität Bochum, Deutschland, experimentelle Strömungsmechanik, Gasdynamik, Ingenieurausbildung

Pleul, Christian Institut für Umformtechnik und Leichtbau (IUL), Technische Universität Dortmund, Deutschland, Fertigungstechnik, Engineering Education, Teeoperatives Experimentieren

Plumanns, Lana IMA/ZLW & IfU, RWTH Aachen, Deutschland, Lernen und Lehren in der digitalisierten Welt/Mensch-Maschine Interaktion, Nutzenorientierte Wirtschaftlichkeitsbetrachtungen, Risiko-und Changemanagement

Radtke, Monika Zentrum für HochschulBildung (zhb) – Lehrstuhl für Organisationsforschung, Weiterbildungs- und Sozialmanagement, TU Dortmund, Deutschland, Kompetenzforschung, Tutorenqualifizierung, Labordidaktik

Reinhard, Rudolf IMA/ZLW & IfU, RWTH Aachen University, Deutschland, Datenmodellierung, -visualisierung und -analyse

Richert, Anja IMA/ZLW & IfU, RWTH Aachen University, Deutschland, Agiles Management von Lern- und Wissensprozessen, Erforschung und Entwicklung von Mixed Reality Learning Konzepten, Entwicklung und Erprobung von Data Science gestützten soziotechnischen Forschungsdesigns für komplexe interdisziplinäre Untersuchungsgegenstände

Sadiki, Abdelhakim Institut für Umformtechnik und Leichtbau (bis Juni 2015), Technische Universität Dortmund, Deutschland, Ingenieurausbildung, Prozesssteuerung und –automatisierung

Schilberg, Daniel IMA/ZLW & IfU, RWTH Aachen University, Deutschland, Interoperabilität von heterogenen Systemen, Informationsmodelle, Semantische Annotation von Daten

Schmitz, Daniela Zentrum für HochschulBildung (zhb) – Bereich Hochschuldidaktik, TU Dortmund, Deutschland, Tutorenqualifizierung, Mobile Learning

Schmohr, Martina Stabsstelle interne Fortbildung und Beratung, Ruhr-Universität Bochum, Deutschland, Coaching, Wissenschaftskarriere, Lehre

Schröder, Stefan IMA/ZLW & IfU, RWTH Aachen University, Deutschland, Benchmarking, Text Mining, Inter- und transdisziplinäre Forschungsnetzwerke

Schuster, Katharina IMA/ZLW & IfU, RWTH Aachen University, Deutschland, Mixed Reality Learning, User Experience und Open Innovation im Hochschulwesen

Selvaggio, Alessandro Institut für Umformtechnik und Leichtbau, Technische Universität Dortmund, Deutschland, Ingenieurausbildung, Strangpressen, Prozesssteuerung und –automatisierung

Sigl, Lisa Zentrum für HochschulBildung (zhb) – Lehrstuhl für Hochschuldidaktik und Hochschulforschung, TU Dortmund, Deutschland & Forschungsplattform, Universität Wien, Österreich; Academic Entrepreneurship, Academic Work Cultures, Responsible Research and Innovation

Slemeyer, Andreas Elektro- und Informationstechnik, Technische Hochschule Mittelhessen, Deutschland, Problem- und Projekt-basiertes Lernen, Labordidaktik

Sommer, Thorsten IMA/ZLW & IfU, RWTH Aachen University, Deutschland, Neuroevolution, Artificial Intelligence, Machine Learning, E-Learning

Stehling, Valerie IMA/ZLW & IfU, RWTH Aachen University, Engineering Education, Mentoring, Große Hörerzahlen

Strenger, Natascha Feststoffverfahrenstechnik, Ruhr-Universität Bochum, Deutschland, Internationalisierung in den Ingenieurwissenschaften, Förderung von Auslandsaufenthalten, Hochschulkooperationen

Tekkaya, A. Erman Institut für Umformtechnik und Leichtbau, Technische Universität Dortmund, Deutschland, Umformtechnik, Ingenieurausbildung, Leichtbau

Terkowsky, Claudius Zentrum für HochschulBildung (zhb) – Lehrstuhl für Organisationsforschung, Weiterbildungs- und Sozialmanagement, TU Dortmund, Deutschland, Engineering Education, Labordidaktik, Kreativitätsforschung

Groß (geb. Thöing), Kerstin IMA/ZLW & IfU, RWTH Aachen University, Deutschland, Mediennutzung in der Lehre, Kollaboratives Arbeiten in virtuellen Lernumgebungen, Diversität in Forschung und Lehre

Traphöner, Heinrich Institut für Umformtechnik und Leichtbau, Technische Universität Dortmund, Deutschland, Materialcharakterisierung- und modellierung, Blechumformung, Ebene Torsionsprüfung

Vieritz, Helmut IMA/ZLW & IfU, RWTH Aachen University, Deutschland, barrierefreie IT-Technologien, digitale Medien in der Hochschullehre, Modellierung von Webanwendungen

Vossen, Rene Institut für Unternehmenskybernetik, RWTH Aachen, Deutschland, Wissens- und Innovationsmanagement, Personal- und Organisationsentwicklung, Systemisches Change Management

Wagner, Pia Gemeinsame Arbeitsstelle RUB/IGM, Ruhr-Universität Bochum, Deutschland, Digitalisierung von Arbeit, innovative Betriebsratsarbeit, Interdisziplinarität

Wissemann, Sarah Zentrum für HochschulBildung (zhb) – Lehrstuhl für Organisationsforschung, Weiterbildungs- und Sozialmanagement, TU Dortmund, Deutschland, Lehramt Chemie/Mathematik, Forschungswerkstatt

Zeuch, Mark Persönlicher Referent des Rektors, Ruhr-Universität Bochum, Deutschland, Hochschulkommunikation, Hochschulzugang, Diversity in den Ingenieurwissenschaften

Part I
Virtual Learning Environments

Measurement of the Cognitive Assembly Planning Impact

Christian Büscher, Eckart Hauck, Daniel Schilberg and Sabina Jeschke

Abstract Within highly automated assembly systems, the planning effort forms a large part of production costs. Due to shortening product lifecycles, changing customer demands and therefore an increasing number of ramp-up processes these costs even rise. So assembly systems should reduce these efforts and simultaneously be flexible for quick adaption to changes in products and their variants. A cognitive interaction system in the field of assembly planning systems is developed within the Cluster of Excellence "Integrative production technology for high-wage countries" at RWTH Aachen University which integrates several cognitive capabilities according to human cognition. This approach combines the advantages of automation with the flexibility of humans. In this paper the main principles of the system's core component – the cognitive control unit – are presented to underline its advantages with respect to traditional assembly systems. Based on this, the actual innovation of this paper is the development of key performance indicators.

Keywords Key Performance Indicators · Cognitive Control · Self-Optimization · Assembly Planning

1 Introduction

In this paper, a set of key performance indicators (KPI) is discussed describing the impact of a cognitive interaction system of highly automated assembly systems. The basis is a cognitive interaction system which is designed within a project of the Cluster of Excellence "Integrative production technology for high-wage countries" at RWTH Aachen University with the objective to plan and control an assembly autonomously [1]. The overall objective of the Cluster of Excellence is to ensure the competitive situation of high-wage countries like Germany with respect to high-tech products, particularly in the field of mechanical and plant engineering. Yet these countries are

C. Büscher (✉) · E. Hauck · D. Schilberg · S. Jeschke
Institute of Information Management in Mechanical Engineering IMA,
RWTH Aachen University, Aachen, Germany
e-mail: christian.buescher@ima-zlw-ifu.rwth-aachen.de

Originally published in "ICIRA 2012", Lecture Notes in Computer Science,

DOI 10.1007/978-3-319-46916-4_1

facing increasingly strong competition by low-wage countries. The solution hypothesis derived in the mentioned Cluster of Excellence is seen in the resolution of the so-called Polylemma of Production [1]. The contribution of the project "Cognitive Planning and Control System for Production" is the development of a cognitive interaction system. Cognitive interaction systems in general are characterised by two facts. On the one hand, they comprise cognitive capabilities as mentioned before and on the other hand, they feature an interaction between the technical system and human operators [2]. One of the major challenges of the Polylemma of Production is to increase the efficiency of planning and simultaneously utilise the value stream approach in the domain of assembly. The main results are the implementation of a cognitive control unit (CCU) as the key component of the cognitive interaction system and the construction of an assembly cell on the technical side to practically test the functionality of the CCU.

In this context, assembly tasks are a big challenge for planning systems, especially considering uncertain constraints, as implied in this approach. As a result, classic planning approaches have shown to be of little use due to the huge computational complexity. By calculating the complex planning problems prior to the actual assembly, this problem can be bypassed – but current and temporary changes cannot be taken into account. That is why in this project, a hybrid approach of pre- and re- planning of assembly tasks is followed. While the CCU plans and controls the whole assembly process, the operator only executes assembly steps, which cannot be fulfilled by the robots, and intervenes in case of emergency. In this way, the robot control, which is now based on human decision processes, will lead to a better understanding of the behaviour of the technical system.

The crucial point of the CCU is the reduction of planning costs compared to traditional automated assembly systems. This is reached by means of cognitive capabilities with simultaneously increasing the flexibility during the actual assembly process. To quantify this, a set of four KPIs is developed in this paper. These KPIs point out the influence of implementing a cognitive interaction system for assembly planning to the Polylemma of Production. The new KPIs therefore concentrate on the comparison of production systems with and without cognitive interaction systems while general performance measuring systems itself already exist [3, 4].

2 The Cognitive Control Unit

The main idea of the cognitive control unit is to autonomously plan and control the assembly of a product solely by its CAD description. Hence, it will be possible to decrease the planning effort in advance and to increase the flexibility of manufacturing and assembly systems [5]. Therefore several different approaches, which are suitable for the application on planning problems, are of great interest in the field of artificial intelligence. While generic planners like the ones by Hoffmann [6], Castellini [7] and Hoffmann & Brafman [8] are not able to compute any solution within an acceptable time in the field of assembly planning concerning geometrical analysis, other planners are especially designed for assembly planning, e.g. the widely used Archimedes System [9]. To find optimal plans, it uses AND/OR-graphs and an "assembly by disassembly" strategy. The approach of Thomas follows the same strategy, but uses only geometric information of the final product as input [10]. Nevertheless, both approaches are not adequate enough to deal with uncertainty. Another system developed by Zaeh & Wiesbeck [11] follows an approach which is similar to the CCU apart from the fact that it only plans and does not control the assembly. In this field, the CCU is a sophisticated system on the way to self-optimisation.

The CCU is able to take over tasks from an operator, for example repetitive, dangerous and not too complex operations, as it is capable to process procedural knowledge encoded in production rules and to control multiple robots. As knowledge-based behaviour as well as skill-based behaviour cannot be modelled and simulated by the CCU, it will cooperate with the operator on a rule-based level of cognitive control [12, 13]. The task of the CCU consists of the planning and the controlling of the assembly of a product that is described by its CAD data while the assembly actions are executed by the assembly robots. After receiving an accordant description entered by a human operator, the system plans and executes the assembly autonomously by means of a hybrid planner. With regard to the cooperation between the CCU and an operator it is crucial that the human operator understands the assembly plan developed by the CCU. Furthermore, a robot control which is based on human decision processes will lead to a better understanding regarding the behaviour of the technical system which is referred to as cognitive compatibility [2].

2.1 *Hybrid Planner*

The planning process of the CCU is separated into two assembly-parts to allow fast reaction times: the Offline Planner, executed prior to the assembly, and the Online Planner, executed in a loop during the assembly (Fig. 1). The Offline Planner allows computation times of up to several hours. Its task is to pre-calculate all feasible assembly sequences – from the single parts to the desired product. The output is a graph representing these sequences. This graph is transmitted to the Online Planner whose computation time must not exceed several seconds. Its task is to map repeatedly the current system state to a state contained in the graph during assembly. In a further step, the Online Planner extracts an assembly sequence that transforms the latest state into a goal state containing the finished product. Thus, the proposed procedure follows a hybrid approach [12].

A solution space for the assembly sequence planning problem is derived during the offline planning phase. As mentioned above, an "assembly by disassembly" strategy is applied to generate an assembly graph, which is first generated as an AND/OR-graph and which is then transformed into a state graph that can be efficiently interpreted during online planning [14]. Therefore, a description of the assembled product's geometry and its constituting parts, possibly enriched with additional mating

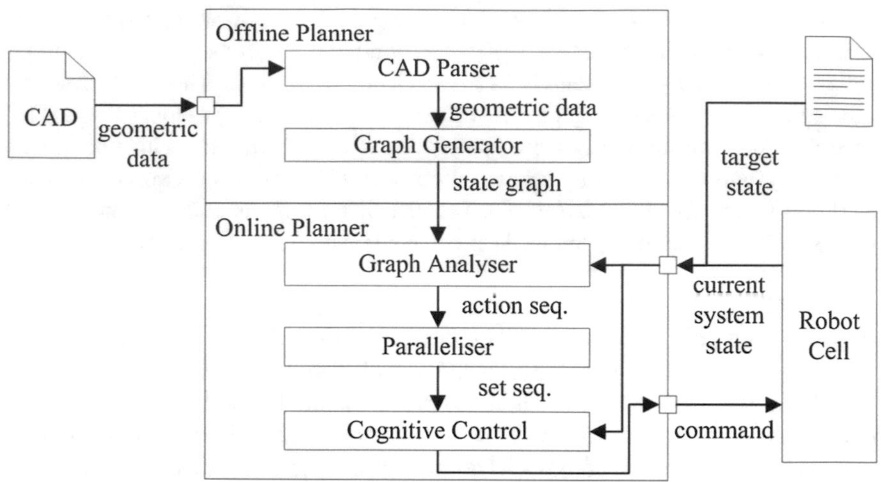

Fig. 1 Hybrid approach of the CCU

directions or mating operation specifications, is used. The geometric data is read by the CCU from a CAD file. The main concept of this strategy is a recursive analysis of all possibilities of an assembly or subassembly [15]. Any assembly or subassembly is separated into two further subassemblies until only single parts remain. All related properties of the product's assembly are stored. Additionally, instances can be used to describe functional aspects. This will be relevant if additional data apart from the part geometries is taken into account by the assembly planner [10].

All data of the separation evaluators is stored in the state graph. The information contains static operation costs and mating operation descriptions for each assembly that might be active during assembly. All transition steps are represented by edges enhanced with the named costs. Each state contains the passive assembly, to which other parts may be added, starting with the empty one at the bottom up to the final product at the top.

The Online Planner derives the assembly plan during the assembly process. It uses the state graph provided by the Offline Planner as well as information about the current robot cell's situation. This approach is similar to the system developed by [11], in which assembly task instructions for a human operator are reactively generated.

The Graph Analyser receives the generated state graph from the offline planning phase and the actual world state describing the current situation of the assembly. Afterwards, the graph analyser maps this world state onto the matching state contained in the state graph. If this node and the "goal-node" are identical, the assembly has been completed and the process ends. Otherwise, the state graph is updated. Dynamic costs in terms of the availability of necessary parts are assigned to the state graphs edges – in addition to the costs, which already have been assigned to the edges during offline planning [12]. After this procedure, the optimal path to the goal-node is calculated using the A* search algorithm, which represents the optimal assembly plan for the given current situation [16]. This path is tested for parallelization and sent to the Cognitive Control component, which executes the assembly in a further step.

The Cognitive Control component receives the assembly sequence, triggers the accordant robot commands and communicates with the human operator so that the latter can operate e.g. in case of unforeseen changes. This component is based on Soar, a cognitive framework for decision finding that aims on modelling the human decision process [17]. Soar contains several production rules which are stored in the knowledge base. Furthermore, human assembly strategies are developed and implemented in the component to generate a higher degree of machine transparency and to enhance cognitive compatibility [13]. Thus, this component implements the cognitive capability of decision making so that the CCU in general is able to optimise its performance according to different delivered target states.

In this section, the background of the cognitive interaction system with regard to the planning algorithm and the possibilities of decision making within the technical system was described. The next section points out how this approach can help to improve the planning process by defining KPIs for cognitive interaction systems.

3 Key Performance Indicators

In order to measure the influence of a cognitive interaction system on the planning process, four key performance indicators were developed. As described in the previous sections, the reduction of planning efforts prior to the assembly is a main objective of cognitive interaction systems like the CCU. This approach enables a faster planning process for assembly and thereby comprises at best an increase of production volume during this phase. This has on the one hand a positive effect on the validity of the data generated and beyond that on the quality of the final production process. In addition, the increased flexibility allows not only a static assembly strategy like traditional automated systems, but the possibility to act adaptively within the framework of the generated plan. The four KPIs which show these advantages within the triple constraint (cost, time and quality) are:

- K_{PE} – planning effort
- K_{APV} – acceleration of production volume growth
- K_{IPV} – increase of production volume
- K_{PQ} – plan quality

The contribution to the the Polylemma of Production technology is determined by the comparison of the KPIs with and without the use of a cognitive interaction system to control an assembly of components of simple geometry like the scenario described in Section 3.2 [18]. All KPIs are defined in a way that the larger the value, the more superior is the cognitive system compared to the traditional one. The turning point where both systems are equal is – depending on the context of the precise KPI – 0 or 1.

3.1 Planning Effort

The first KPI, the planning effort, refers to the phase of mounting and initial programming of the cognitive interaction system within an assembly system. The initial filing and maintenance of the knowledge base in a cognitive interaction system represents significantly more work compared to programming a traditional assembly system, for example by teaching the robot. This effort is too high for a production system that is designed only for one product since a traditional assembly system can be programmed very quickly for a specified manufacturing step and this programming has to be adjusted only marginally during production. However, if the assembly system needs to be able to assemble a wide range of products with small batches, a traditional assembly system has to be repeatedly reprogrammed and optimised. In contrast, an assembly system with a cognitive interaction system can be adapted with little effort on a new product.

The key performance indicator K_{PE} is based on Schilberg [19] and is calculated from the sum over n different products to be assembled by the efforts of the programming of the system respectively the creation of the knowledge base and the optimisation of the assembly:

$$K_{PE} = 1 - \left(\frac{\sum_{i=1}^{n} PE_{i_{cognitive}}}{\sum_{i=1}^{n} PE_{i_{traditional}}} \right) \quad (1)$$

With K_{PE}: Key performance indicator of the planning effort
$PE_{i_{cognitive}}$: Planning effort of an assembly system with a cognitive interaction system
$PE_{i_{traditional}}$: Planning effort of a traditional assembly system.

The interval in which the KPI ranges is $[-\infty, +1]$. The extreme value $-\infty$ of the interval will be reached if the planning effort for $PE_{i_{traditional}}$ is arbitrarily small or if $PE_{i_{cognitive}}$ is an arbitrary large value. The other extreme value of 1 will be reached if $PE_{i_{cognitive}}$ is 0 [18].

By the automated assembly planning within the cognitive interaction system, this system only has to be reprogrammed if the assembly of the product to be manufactured contains steps that were not previously stored in the cognitive interaction system (for example if a new tool is available, which results in new possible operations). By the independent planning of the assembly process, under constraints which have passed by the operator to the cognitive interaction system, no new operation sequences have to be programmed. The adaptive adjustment of the assembly sequence can even ensure an assembly with a not previously known component supply which is impossible in a traditional assembly system. When a new sequence of steps is to be executed, the assembly system has to be reprogrammed and optimised, which represents a significant amount of work. This does not allow flexible responds of the assembly system to changes in product manufacturing or in the assembly sequence.

3.2 Acceleration of Production Volume Growth

The second KPI, acceleration of production volume growth, is determined by comparing the maximum increase in production volume per time unit at the site t i. Therefore, the KPI is calculated as:

$$K_{APV} = \left(\frac{\frac{\partial P_V}{\partial t}|_{cognitive}}{\frac{\partial P_V}{\partial t}|_{traditional}} \right) - 1 \quad (2)$$

With K_{APV}: Key performance indicator for acceleration of production volume growth

$\frac{\partial P_V}{\partial t}|_{cognitive}$: Slope at the inflection point of the production system with cognitive interaction systems

$\frac{\partial P_V}{\partial t}|_{traditional}$: Slope at the inflection point of the traditional production system.

By forming the quotient, the KPI ranges in the interval $[-1, +\infty]$. A value of -1 means that the production volume growth of a production system with cognitive interaction systems is 0, so there is no production. The other extreme value means that the production volume growth with a cognitive interaction system is arbitrarily large respectively the production increase of a traditional production system is 0. This value is never reached because it would mean a discontinuity in the S-curve, which has to be differentiable by definition [18].

In case of a congruence of the inflection points of the ramp-up function f(t) for both production systems, it is sufficient to determine the key performance indicator K APV if one production system dominates the other one. If the two inflection points do not match, it may happen that a production system, which has a steeper gradient but realises this at a significantly later time, possibly has a worse overall production volume. Figure 2 shows such an issue. The points in time and describe the inflection points with the maximum slope of the two curves.

3.3 Increase of Production Volume

In this case, a third KPI, namely the increase of total production volume during ramp-up, should be consulted. It is calculated by integrating over the starting function f(t) in a given period. By the quotient, a direct comparison of production systems with and without cognitive interaction systems can be made.

Therefore, the KPI is calculated as:

$$K_{IPV} = \left(\frac{\int_{t_0}^{t_{S1}} f_{cognitive}(t)dt}{\int_{t_0}^{t_{S2}} f_{traditional}(t)dt} \right) \quad (3)$$

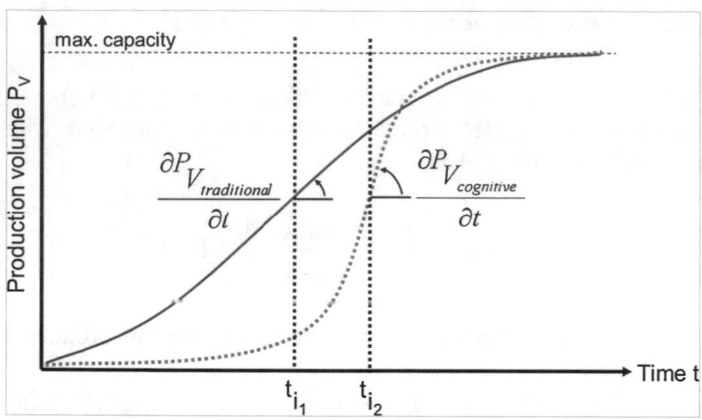

Fig. 2 Example of a ramp-up process where the key performance indicator K_{APV} of a production system is higher, but the total production volume is smaller

With K_{IPV}: Key performance indicator of the increase of production volume
t_0: Start of production
t_{Si}: Time of series production with full capacity.

Under the condition that the integral of f(t) > 0, a cognitive system is superior to a traditional system if the KPI takes a value > 1. The key performance indicator K_{APV} represents therefore the necessary condition for the superiority of a production system with cognitive interaction systems in the assembly, while the key performance indicator K_{IPV} is the sufficient condition for a real improvement [18]. This consideration is only meaningful if both times to volume differ from each other marginally. Otherwise the "faster" production system would always be preferred since time is often the critical variable.

3.4 *Plan Quality*

The fourth KPI is the plan quality, which is borrowed from one criterion for evaluation of planners on the International Planning Competition (IPC) [20]. It is determined from the number of assembly steps required to manufacture a product. In a traditional system, the assembly sequence is either fully optimised and programmed in advance or a heuristic-based optimisation is used by the employee during programming. Depending on the complexity of the product to be assembled, the optimal assembly sequence may not be found in a reasonable time. In the context of this scenario and further scenarios which were analysed within the project, the CCU is able to generate the entire assembly graph in the Offline Planner. At this point no heuristics have to be used but such applications are possible. Thus, this KPI is defined with regard to more complex products where heuristics are relevant.

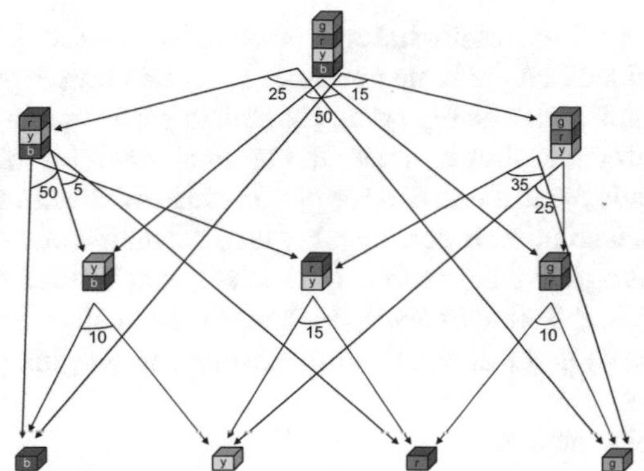

(a) Graph with all possible decompositions with 4 parts and the corresponding edge costs

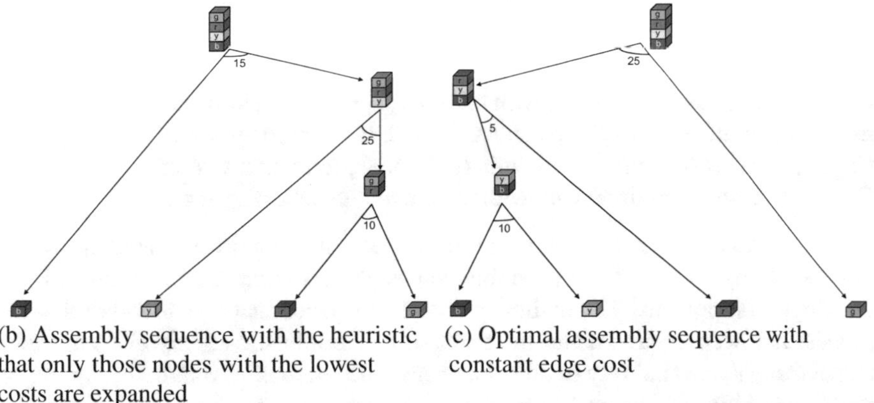

(b) Assembly sequence with the heuristic that only those nodes with the lowest costs are expanded

(c) Optimal assembly sequence with constant edge cost

Fig. 3 Comparison of the plan quality

As described before, in a cognitive interaction system, it may also be necessary to resort to a heuristic to solve the planning problem in a reasonable time in a corresponding product complexity. This planning can be continued during the production, which means that the cognitive interaction system starts with an assembly sequence that was found using a heuristic. Then during the process the system is able to derive a better assembly sequence in parallel by using relaxing heuristics and by conducting a broader search within the AND/OR-graph. If the number of components is below a threshold, all possible decompositions of the AND/OR-graph can be saved and a guaranteed optimal solution of the planning problem can be found [18]. Figure 3a shows the entire AND/OR-graph of a tower of four Lego bricks. In Fig. 3b, an assembly sequence using a heuristic is shown, in which only those nodes of the graph with the lowest costs are expanded.

The plan quality of the traditional and the cognitive assembly system are identical in this case, provided that the same resources for computing capacity and time exist. However, the cognitive assembly system is able to perform a broader search during the actual assembly and thus to create the optimal assembly plan (Fig. 3c). In this example, this would mean a sum of costs of 40 instead of 50 although the first analysis step of the optimal solution is considerably larger than the one of the heuristic. The plan quality of a cognitive interaction system is thus at least as good as a traditional assembly system and is able to achieve a better plan quality which is expressed by a lower sum of costs in the assembly graph through an on-going continuation of the planning.

The KPI is calculated as:

$$K_{PQ} = \sum_{i=0}^{n} C_{i_{traditional}} - \sum_{i=0}^{m} C_{i_{cognitive}} \quad (4)$$

With K_{PQ}: Key performance indicator of the plan quality
n: number of assembly steps with traditional production system
m: number of assembly steps with cognitive interaction system
$C_{i_{traditional}}$: costs on optimal path with traditional production system
$C_{i_{cognitive}}$: costs on optimal path with cognitive interaction system.

Hence, K_{PQ} is never less than zero. In addition, the cognitive interaction system has the ability to adjust the assembly sequence depending on the availability of the single components. The optimal assembly sequence that was found during this process has been created under the terms of a deterministic supply of components respectively the availability of all components and assemblies required. With regard to the possibility of dynamic allocation of the new edge costs within the assembly graph in case of storage of all possible decompositions, the cognitive assembly system is able to adapt the assembly sequence dynamically and to ensure the optimal plan quality in a dynamic environment at any time during assembly.

4 Conclusion

This paper proposes a set of KPIs which can determine the advantages of a cognitive interaction system in contrast to traditional automated systems to improve an assembly system. A precise cognitive interaction system in the domain of assembly planning systems is presented, which is the first self-optimising system in this domain. It comprises several cognitive capabilities implemented in the cognitive control unit (CCU) by a hybrid approach for assembly tasks, which enables robots to decide on their action during assembly autonomously.

To measure the systems' advantages involved, a set of key performance indicators is developed in this paper which can show the impact of this cognitive interaction system. These KPIs concentrate on the main improvements being achieved during assembly and its construction and sequence planning. The interaction of the four KPIs "planning effort", "acceleration of production volume growth", "increase of production volume" and "plan quality" evaluate the improvements in attaining the final production volume and in reducing the planning effort as well as the enhancement of the quality of the derived plan by means of the self-optimising capability of the CCU. These are developed in the context presented in this paper but designed to highlight the impact of cognitive interaction systems on production economics in general.

With respect to future research, there are plans to fill these KPIs with life, while in this paper the theoretical background for the next step has been set. Therefore, industrial applications are to be performed and analysed by comparing the assembly of a product with the use of the CCU on the one hand and with the traditional approach on the other hand. Therein, possible weaknesses can be detected and resolved. This practical testing may then be shifted to other cognitive interaction systems to demonstrate the transferability of the set of KPIs. The challenge of fundamental research like the technological innovations developed within this Cluster of Excellence often comprises the persuasion of industry of the high performance of such solutions and the implementation or the launch of a product out of this. However, companies need reliable predictions on the applicability and economic efficiency. At this, the developed KPIs can play a major role as they provide exactly this required evidence in the examined topic.

References

1. C. Brecher, F. Klocke, R. Schmitt, G. Schuh, *Excellence in Production*. Apprimus Verlag, Aachen, 2007
2. M. Mayer, C. Schlick, D. Ewert, D. Behnen, S. Kuz, B. Odenthal, B. Kausch, Automation of robotic assembly processes on the basis of an architecture of human cognition. Production Engineering Research and Development **5** (4), pp. 423–431
3. A. Pufall, J.C. Fransoo, A.d. Kok, *What determines product ramp-up performance? A review of characteristics based on a case study at Nokia mobile phones, BETA publicaties. Preprints*, vol. WP 228. [Beta, Research School for Operations Management and Logistics], Eindhoven, 2007
4. H. Winkler, M. Heins, P. Nyhuis, A controlling system based on cause-effect relationships for the ramp-up of production systems. Production Engineering Research and Development **1** (1), 2007, pp. 103–111
5. E. Hauck, A. Gramatke, K. Henning, A software architecture for cognitive technical systems suitable for an assembly task in a production environment: Automation control - theory and practice. In: *Two Stage Approaches for Modeling Pollutant Emission of Diesel Engine Based on Kriging Model*, ed. by El Hassane Brahmi, Ghislaine Joly-Blanchard, Lilianne Denis-Vidal, Nassim Boudaoud, Zohra Cherfi, INTECH Open Access Publisher, 2009, pp. 13–28
6. J. Hoffmann, Ff: the fast-forward planning system. The AI Magazine (22), 2001
7. C. Castellini, E. Giunchiglia, A. Tacchella, O. Tacchella, Improvements to satbased conformant planning. In: *Proceedings of the 6th European Conference on Planning (ECP-01)*, ed. by A. Cesta, D. Borrajo. AAAI Press, Palo Alto, California, 2001
8. J. Hoffmann, R. Brafman, Contingent planning via heuristic forward search with implicit belief states. In: *Proceedings of ICAPS'05*. 2005, pp. 71–80

9. S.G. Kaufman, R.H. Wilson, R. Calton, A.L. Ames, *Automated planning and programming of assembly of fully 3d mechanisms, Technical Report*, vol. SAND96-0433. Sandia National Laboratories, 1996
10. U. Thomas, *Automatisierte Programmierung von Robotern für Montageaufgaben, Fortschritte in der Robotik*, vol. 13. Shaker Verlag, Aachen, 2008
11. M.F. Zaeh, M. Wiesbeck, A model for adaptively generating assembly instructions using state-based graph. In: *Manufacturing systems and technologies for the new frontier*, ed. by M. Mitsuishi, K. Ueda, F. Kimura, Springer, London, 2008, pp. 195–198
12. D. Ewert, S. Thelen, R. Kunze, M. Mayer, D. Schilberg, S. Jeschke, A graph based hybrid approach of offline pre-planning and online re-planning for efficient assembly under realtime constraints. In: *ICIRA 2010: 2010 International Conference on Intelligent Robotics and Application, LNAI*, vol. 6425, ed. by H. Liu, H. Ding, Z. Xiong, X. Zhu. Springer, Berlin, 2010, *LNAI*, vol. 6425, pp. 44–55
13. M. Mayer, B. Odenthal, M. Faber, W. Kabuss, B. Kausch, C. Schlick, Simulation of human cognition in self-optimizing assembly systems. In: *Proceedings of the IEA2009 - 17th World Congress on Ergonomics*. 2009
14. Homem de Mello, L.S., A.C. Sanderson, Representations of mechanical assembly sequences. IEEE Transactions on Robotics and Automation **7** (2), 1991, pp. 211–227
15. R.S. Chen, K.Y. Lu, P.H. Tai, Optimizing assembly planning through a three-stage integrated approach. International Journal of Production Economics **88** (3), 2004, pp. 243–256
16. P.E. Hart, N.J. Nilsson, B. Raphael, A formal basis for the heuristic determination of minimum cost paths. IEEE Transactions on Systems Science and **4** (2), 2007, pp. 100–107
17. P. Langley, J.E. Laird, S. Rogers, Cognitive architectures: Research issues and challenges. Journal of Cognitive Systems Research **10** (2), 2009, pp. 141–160
18. E. Hauck, *Ein kognitives Interaktionssystem zur Ansteuerung einer Montagezelle, VDI Reihe 10*, vol. 812. VDI-Verlag, Düsseldorf, 2011
19. D. Schilberg, Architektur eines datenintegrators zur durchgängigen kopplung von verteilten numerischen simulationen. Dissertation, RWTH Aachen, 2010
20. Gerevini, A. E., P. Haslum, Long, D., Saetti, A., Y. Dimopoulos, Deterministic planning in the fifth international planning competition: Pddl3 and experimental evaluation of the planners. In: *Artificial Intelligence*, vol. 173, Elsevier Science, 2009, pp. 619–668

Selfoptimized Assembly Planning for a ROS Based Robot Cell

Daniel Ewert, Daniel Schilberg and Sabina Jeschke

Abstract In this paper, we present a hybrid approach to automatic assembly planning, where all computational intensive tasks are executed once prior to the actual assembly by an Offline Planner component. The result serves as basis of decision-making for the Online Planner component, which adapts planning to the actual situation and unforeseen events. Due to the separation into offline and online planner, this approach allows for detailed planning as well as fast computation during the assembly, therefore enabling appropriate assembly duration even in nondeterministic environments. We present simulation results of the planner and detail the resulting planner's behavior.

Keywords Assembly Planning · Cognitive Production Systems · ROS

1 Introduction

1.1 Motivation

The industry of high-wage countries is confronted with the shifting of production to low-wage countries. To slow down this development, and to answer the trend towards shortening product life-cycles and changing customer demands regarding individualized and variant-rich products, new concepts for the production in high-wage countries have to be created. This challenge is addressed by the Cluster of Excellence "Integrative production technology for high-wage countries" at the RWTH Aachen University. It researches on sustainable technologies and strategies on the basis of the so-called polylemma of production [1]. This polylemma is spread between two dichotomies: First between scale (mass production with limited product range) and scope (small series production of a large variety of products), and second between value and planning orientation. The ICD) "Self-optimizing Production Systems"

D. Ewert (✉) · D. Schilberg · S. Jeschke
IMA/ZLW & IfU, RWTH Aachen University, Dennewartstr. 27, 52068 Aachen, Germany
e-mail: daniel.ewert@ima-zlw-ifu.rwth-aachen.de

Originally published in "ICIRA 2012", © Springer 2012.
Reprint by Springer International Publishing AG 2016,
DOI 10.1007/978-3-319-46916-4_2

focusses on the reduction of the latter dichotomy. It's approach for the reduction of this polylemma is to automate the planning processes that precede the actual production. This results in a reduction of planning costs and ramp-up time and secondly it allows to switch between the production of different products or variants of a product, hence enabling more adaptive production strategies compared to current production. Automatic replanning also allows to react to unforeseen changes within the production system, e.g. malfunction of machines, lack of materials or similar, and to adapt the production in time. In this paper we present the planning components of a cognitive control unit (CCU) which is capable to autonomously plan and execute a product assembly by relying entirely on a CAD description of the desired product.

1.2 Use Case Description

The CCU is developed along a use case scenario for an assembly task in a nondeterministic production environment [2]. This scenario is based on the robot cell depicted in Fig. 1.

Of the two robots of the robot cell, only Robot2 is controlled by the CCU. Robot1 independently delivers parts in unpredictable sequence to the circulating conveyor belt. The parts are then transported into the grasp range of Robot2 who then can decide to pick them up, to immediately install them or to park them in the buffer area.

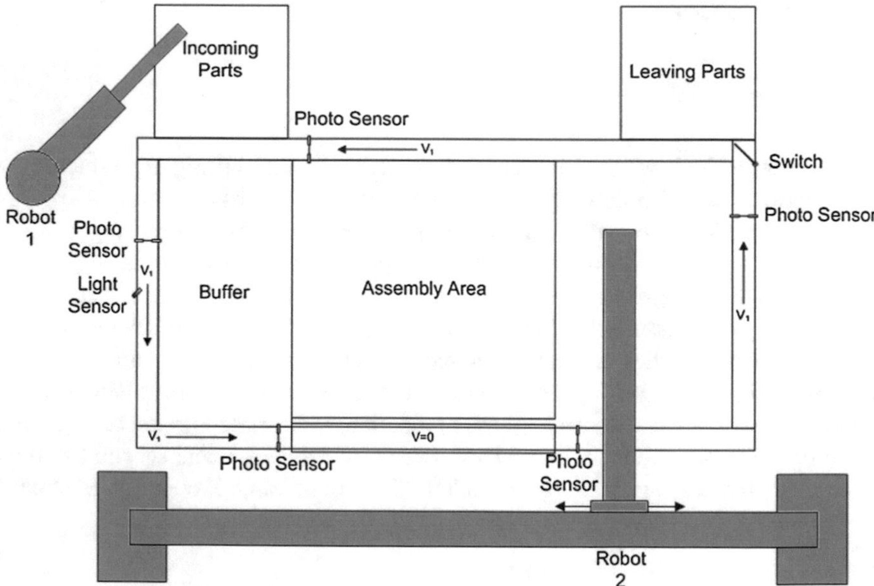

Fig. 1 Schematic of the robot cell

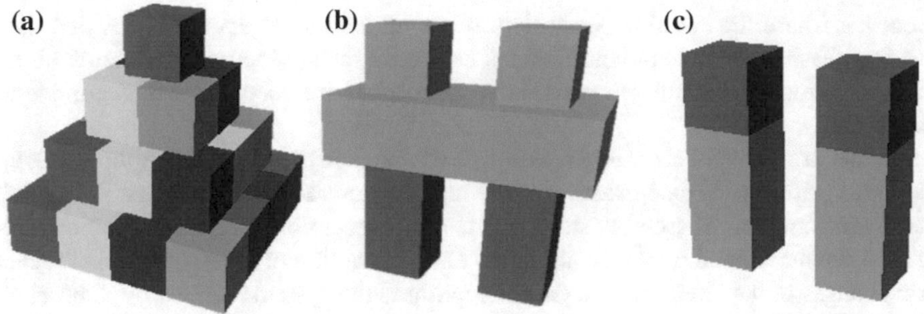

Fig. 2 Toy model products for planner evaluation

The scenario also incorporates human-machine cooperation. In case of failure, or if the robot cannot execute a certain assembly action, the CCU is able to ask a human operator for assistance. To improve the cooperation between the operator and the machine, the operator must be able to understand the behavior and the intentions of the robot [3]. Therefore, machine transparency is a further major aspect in our concept.

The only sources of information to guide the decision making of the CCU are a CAD description of the desired product, the number and types of single parts currently in the buffer and on the conveyor belt and the current state of the assembly within the Assembly Area. The planner is evaluated with the figures described in Fig. 2. The pyramid construct (a) serves here as a benchmark for the computational complexity of our planning approach and has been used in different sizes (base areas of 2×2, 3×3, and 4×4 blocks). Construct (b) and (c) are used to demonstrate the planner's behavior.

2 Related Work

In the field of artificial intelligence planning is of great interest. There exist many different approaches to planning suitable for different applications. Hoffmann developed the FF planner, which is suitable to derive action sequences for given problems in deterministic domains [4]. Other planners are capable to deal with uncertainty [5, 6]. However, all these planners rely on a symbolic representation based on logic. The corresponding representations of geometric relations between objects and their transformations, which are needed for assembly planning, become very complex even for small tasks. As a result, these generic planners fail to compute any solution within acceptable time.

Other planners have been designed especially for assembly planning and work directly on geometric data to derive action sequences. A widely used approach is the Archimedes system by Kaufman et al. [7] that uses And/Or-Graphs and an "Assembly by Disassembly" strategy to find optimal plans. U. Thomas [8] follows this strategy, too, but where the Archimedes system relies on additional operator-provided data to find feasible subassemblies, Thomas uses only the geometric information about the final product as input. However, both approaches are not capable of dealing with uncertainty.

Other products for assembly planning focus on assisting product engineers set up assembly processes. One example is the tool Tecnomatix from Siemens [9], which assists in simulating assembly steps, validates the feasibility of assembly actions etc. All of the mentioned works do not cover online adaption of assembly plans to react on changes in the environment. One exception is the system realized by Zaeh et al. [10], which is used to guide workers through an assembly process. Dependent on the actions executed by the worker, the system adapts its internal planning an suggests new actions to be carried out by the worker. The CCU uses the same technique for plan adaption.

3 Autonomous Assembly Planning

3.1 Hybrid Assembly Planning

The overall task of the CCU is to realize the autonomous assembly in a nondeterministic environment: Parts are delivered to the robot cell in random sequence and the successful outcome of an invoked assembly action cannot be guaranteed. While assembly planning is already hard even for deterministic environments where all parts for the assembly are available or arrive in a given sequence [8], the situation becomes worse for this unpredictable situation. One approach to solve the nondeterministic planning problem would be to plan ahead for all situations: Prior to the assembly all plans for all possible arrival sequences are computed. However, this strategy soon becomes unfeasible: A product consisting of n parts allows for n! different arrival sequences, so a product consisting of 10 parts would already result in the need to compute more than 3.6 million plans. Another approach would be to replan during the assembly every time an unexpected change occurs in the environment. This strategy, however, leads to unacceptable delays within the production process.

Therefore, our approach follows a hybrid strategy. All computational intensive tasks are executed once before the actual assembly. This is done by an Offline Planner component. The results of this step serve as basis of decision-making for the Online Planner component, which adapts planning to the actual situation and unforeseen events. Due to this separation, our approach (see Fig. 3) allows for detailed planning as well as fast computation during the assembly, therefore enabling appropriate assembly duration even in nondeterministic environments. The Offline Planner contains a CAD Parser which derives the geometric properties. The currently supported format is STEP [11]. This data is then processed by the graph generator. The details of this process are explained in Section 3.2. The Online Planner consists of the components Graph Analyzer, Parallelizer and Cognitive Control, which are detailed in Section 3.3.

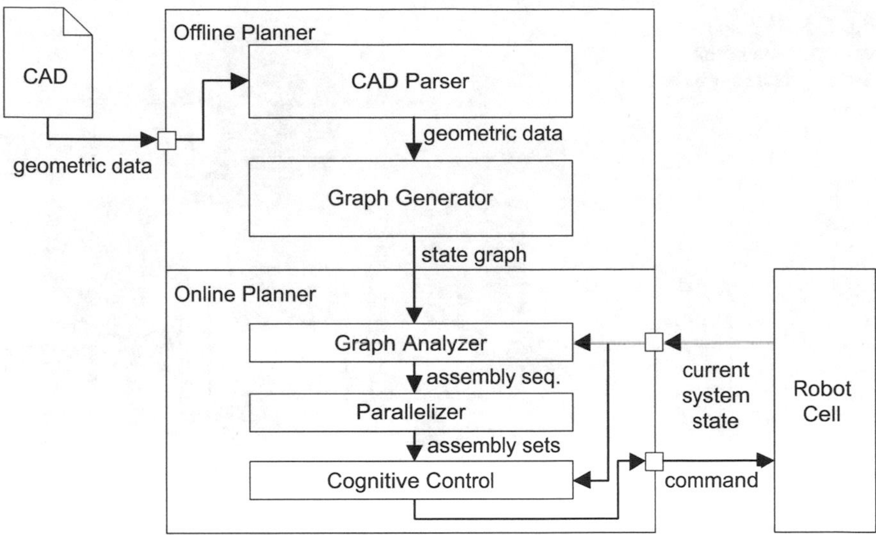

Fig. 3 System overview of the hybrid approach

3.2 *Offline: Graph Generation*

The Offline Planner receives a CAD description of the desired final product. From this input it derives the relations between the single parts of the product via geometrical analysis as described in 0. The results are stored in a connection graph. Assembly sequences are now derived using an assembly-by-disassembly strategy: Based on the connection graph, all possible separations of the product into two parts are computed. The feasibility of those separations is then verified using collision detection techniques. Unfeasible separations are discarded. The remaining separations can then be evaluated regarding certain criteria as stability, accordance to assembly strategies of human operators or similar. The result of this evaluation is stored as a score for each separation. This separation is recursively continued until only single parts remain. The separation steps are stored in an and/or graph [12], which is then converted into a state graph as displayed in Fig. 4 using the method described in Ewert D., D. Schilberg, and S. Jeschke [13]. Here nodes represent subassemblies of the assembly. Edges connecting two such nodes represent the corresponding assembly action which transforms one state into the other. Each action has associated costs, which depend on the type of action, duration, etc. Also, each edge optionally stores information about single additional parts that are needed to transform the outgoing state into the incoming state.

The graph generation process has huge computational requirements for time as well for space. Table 1 shows the properties of resulting state graphs for different products. The results show the extreme growth of the graph regarding the number of parts necessary for the given product. However, as can be seen when comparing the

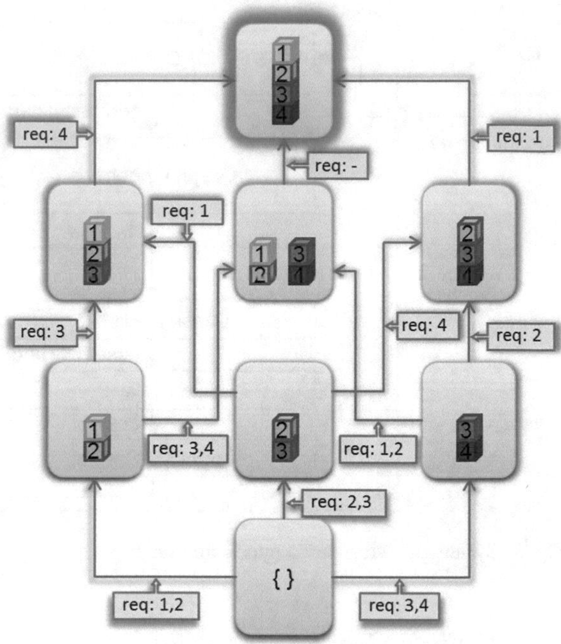

Fig. 4 State graph representation of the assembly of a four blocks tower

Table 1 State graph properties for different products

Product	# Parts	# Nodes of graph	# Edges of graph
Construct (a) (size 2 × 2)	5	17	33
Construct (c)	6	16	24
Construct (b)	14	361	1330
Construct (a) (size 3 × 3)	14	690	2921
Construct (a) (size 4 × 4)	30	141,120	1,038,301

state graphs of both constructs with 14 parts, the shape of a product affects the graph, too: The more possible independent parts are from each other, the more different assembly sequences are feasible. Therefore the graph of the construct (a) with 14 parts has almost twice the size of the state graph resulting from construct (b).

3.3 Online: Graph Analysis

The state graph generated by the Offline Planner is then used by the Online Planner to derive decisions which assembly actions are to be executed given the current situation of an assembly. The Online Planner therefore executes the following process iteratively until the desired product has been assembled: The Graph Analyzer perceives the current situation of the assembly and identifies the corresponding node of the state graph. In earlier publications [13] we suggested an update phase as next step. In this phase all costs of the graphs edges reachable from that node were updated due to the realizability of the respective action. The realizability depends on the availability of the parts to be mounted. Unrealizable actions receive penalty costs which vary depending on how close in the future they would have to be executed. This cost assignment makes the planning algorithm avoid currently unrealizable assemblies. Additionally, due to the weaker penalties for more distanced edges, the algorithm prefers assembly sequences that rely on unavailable parts in the distant future to assemblies that immediately need those parts. Preferring the latter assembly results in reduced waiting periods during the assembly since missing parts have more time to be delivered until they are ultimately needed. Using the A* algorithm [14] the Online Planner now derives the cheapest path connecting the node matching the actual state with a goal node, which presents one variant of the finished product. This path represents the at that time optimal assembly plan for the desired product. The Parellelizer component now identifies in parallel or arbitrary sequence executable plan steps and hands the result to the Cognitive Control for execution. Here the decision which action is actually to be executed is made. The process of parallelization is detailed in [13].

However, updating all edge cost reachable from the node representing the current state is a computational intensive task. To overcome this problem, the edge cost update can be combined with the A* algorithm, so that only edges which are traversed by A* are updated. This extremely reduces the computational time, since only a fraction of the graphs node is examined. So even for large graphs, the Online Planner is able to derive a decision in well under 100 ms in worst case. Figure 5 shows the nodes reachable and examined by the Online Planner during the assembly of a 4×4 construct. Plateaus in the graph depict waiting phases where the assembly cannot continue because crucial parts are not delivered.

3.4 Planner Behaviour

Figure 6 shows the course of the assembly for the construct (c). Newly arrived parts are shown in the third column. They can either be used for direct assembly (first column) or otherwise are stored in a buffer shown in column 2. The right column depicts the plan that is calculated based on the parts located. Here the number of a given block denotes the position where that block is to be placed. In step 0, no parts have been delivered. The planner therefore has no additional information and produces an arbitrary but feasible plan. In step 1 a new green block is delivered, which matches the first plan step. The related assembly action is therefore executed and the new block is directly put to the desired position. In step 2 a new red block is delivered. Given the current state of the assembly and the new red cube, the planner calculates an improved plan which allows to assemble this red block earlier than originally planned: Now it is more feasible to first mount two green blocks on top

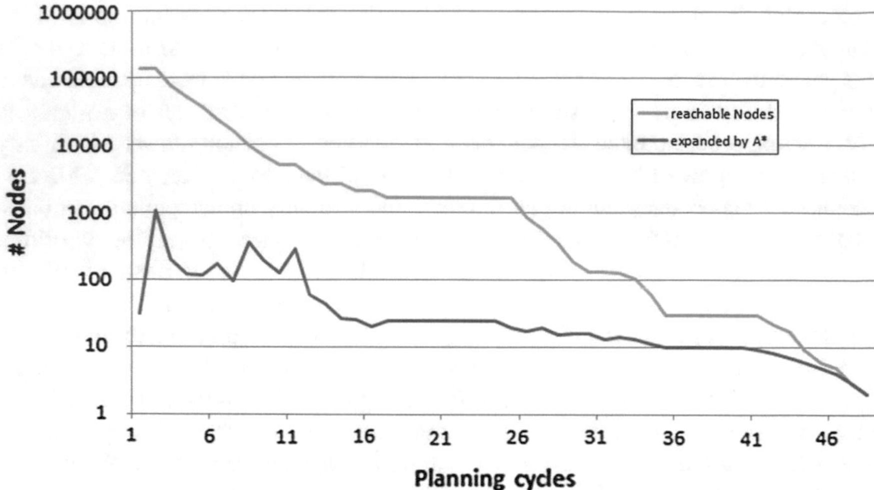

Fig. 5 Number of nodes reachable from the node representing the given situation and number of nodes that are examined by the A*-algorithm. Number of nodes is shown using a logarithmic scale

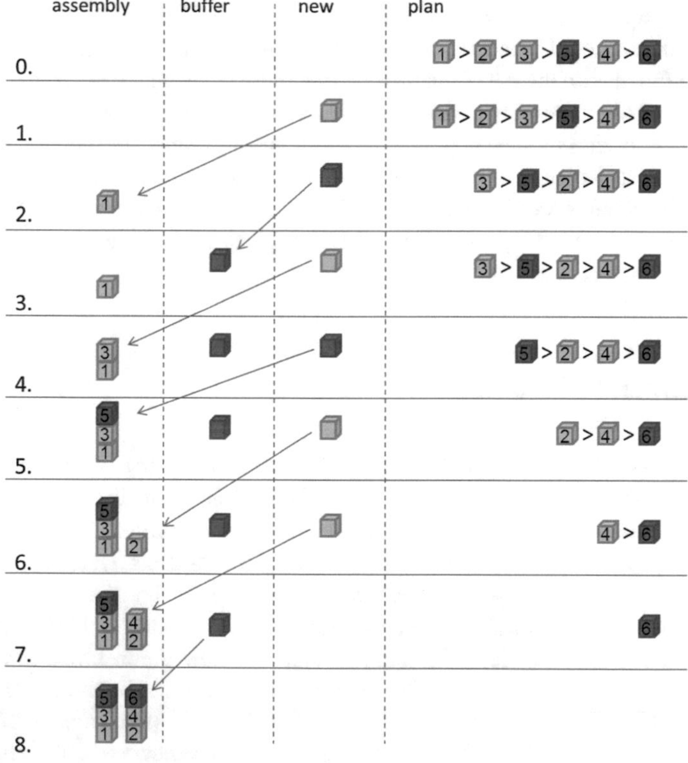

Fig. 6 Exemplary assembly flow for construct (c)

of each other (positions 1 and 3), because then the red block can be assembled, too (position 5). This plan step is executed in step 3 when a second green block becomes available. Now, in step 4, it is possible to mount a red block. From that step on only one feasible assembly sequence is possible, which is then executed.

The described behaviour results in more rapid assemblies compared to simpler planning approaches: A purely reactive planner which would follow a bottom up strategy, would have placed the first two green blocks next to each other (positions 1 and 2). Thus, in step 5 no assembly action would have been possible and the assembly would have to stop until a further green cube would be delivered.

4 Summary

In this paper we presented our hybrid approach for an assembly planner for nondeterministic domains. We described the workflow of the offline planner, which analyses CAD data describing the desired product. The outcome of the offline planner is a state graph which holds all possible (and feasible) assembly sequences. This graph is generated by following an assembly by disassembly strategy: Recursively all possible separations of the final product are computed until only single parts remain. During the actual assembly, this state graph is updated to mirror the current situation of the assembly, specially the availability of newly delivered parts. Using the A* algorithm, the at that time optimal assembly sequence is derived and handed over to the cognitive control unit, which then decides which assembly step gets to be executed. This step is then executed and the outcome of that step is reported back to the planning system. This process is iterated until the product is completed.

5 Outlook

Future work must optimize the described Planners. Using techniques of parallel programming and by incorporating specialized databases which can cope efficiently with large graphs, the planning duration can be improved. Subsequently, the planner will be extended to be able to deal with industrial applications as well as plan and control the production process of a complete production network.

References

1. C. Brecher, F. Klocke, G. Schuh, R. Schmitt, eds., *Excellence in Production*. Apprimus Verlag, Aachen, Germany, 2007
2. C. Brecher, T. Kempf, W. Herfs, Cognitive control technology for a self-optimizing robot based assembly cell. In: *Proceedings of the ASME 2008 International Design Engineering Technical*

Conferences & Computers and Information in Engineering Conference. America Society of Mechanical Engineers, U.S., 2008, pp. 1423–1431
3. B. Kausch, C.M. Schlick, W. Kabuß, B. Odenthal, M.P. Mayer, M. Faber, Simulation of human cognition in self-optimizing assembly systems. In: *Proceedings of 17th World Congress on Ergonomics IEA 2009*. Beijing, China, 2009
4. J. Hoffmann, FF: the fast-forward planning system. The AI Magazine, 2001
5. J. Hoffmann, R. Brafman, Contingent planning via heuristic forward search with implicit belief states. In: *In Proceedings of ICAPS'05*. AAAI, 2005, pp. 71–80
6. C. Castellini, E. Giunchiglia, A. Tacchella, O. Tachella, Improvements to SAT-based conformant planning. In: *Proc. of 6th European Conference on Planning*. 2001
7. S. Kaufman, R. Wilson, R. Jones, T. Calton, A. Ames, Ldrd final report: Automated planning and programming of assembly of fully 3d mechanisms. Technical Report SAND96-0433, Sandia National Laboratories, 1996
8. U. Thomas, *Automatisierte Programmierung von Robotern für Montageaufgaben*, *Fortschritte in der Robotik*, vol. 13. Shaker Verlag, Aachen
9. Tecnomatix, 2011. URL http://www.plm.automation.siemens.com/en_us/products/tecnomatix/index.shtml
10. M. Zäh, M. Wiesbeck, A model for adaptively generating assembly instructions using state-based graphs. In: *Manufacturing Systems and Technologies for the New Frontier*, Springer, London, 2008
11. F. Röhrdanz, H. Mosemann, F. Wahl, HighLAP: a high level system for generating, representing, and evaluating assembly sequences. 1996, pp. 134–141
12. L. Homem de Mello, A. Sanderson, AND/OR graph representation of assembly plans. In: *Proceedings of 1986 AAAI National Conference on Artificial Intelligence*. 1986, pp. 1113–1119
13. D. Ewert, D. Schilberg, S. Jeschke, Selfoptimization in adaptive assembly planning. In: *Proceedings of the 26th International Conference on CAD/CAM Robotics and Factories of the Future*. 2011
14. P. Hart, N. Nilsson, B. Raphael, A formal basis for the heuristic determination of minimum cost paths. IEEE Transactions on Systems Science and Cybernetics **4** (2), 1968, pp. 100–107

RUB-Ingenieurwissenschaften expandieren in die virtuelle Lernwelt

Sulamith Frerich, Eva Heinz and Kristina Müller

1 Einlcitung

Die Gestaltung des akademischen Lehrens und Lernens wird unter anderem durch die Etablierung neuer Medien und Technologien beeinflusst. Ein Ziel des Medieneinsatzes ist die Individualisierung von Lernprozessen. Jede und jeder Studierende soll selbst aktiv werden, um sich „eigenes Wissen" aufzubauen und zu verankern. Eine Möglichkeit, dies künftig stärker zu fördern, ist die Verwendung mobiler Endgeräte zur Vermittlung des Wissens. Besonders förderlich hierfür ist der engere Bezug oder auch Zugriff junger Studierender zu neuen Technologien, der in Untersuchungen zu medialen Nutzungsgewohnheiten vielfach herausgestellt werden [1]. Eine Herausforderung für den Lehr- und Lernalltag von Dozierenden und Studierenden stellt der doppelte Abiturjahrgang dar, der im Wintersemester 2013/14 das Studium in Nordrhein-Westfalen aufnimmt. Bis zum Jahreswechsel 2014/15 richtet die Ruhr-Universität Bochum (RUB) deshalb insgesamt 4.500 zusätzliche Studienplätze ein. Diese werden an zehn Fakultäten in zwölf Studiengängen bereitgestellt, wozu auch die Ingenieurwissenschaften zählen. Zwei der drei ingenieurwissenschaftlichen Fakultäten, Bau- und Umweltingenieurwissenschaften sowie Maschinenbau, stocken das Lehrangebot in drei 1-Fach-B.Sc.-Studiengängen auf und bieten damit einen großen Teil der vereinbarten zusätzlichen Studienplätze an [2].

S. Frerich (✉) · E. Heinz · K. Müller
Ruhr-Universität Bochum, Lehrstuhl für Verfahrenstechnische Transportprozesse (VTP), Bochum, Germany
e-mail: frerich@vtp.rub.de

Originally published in "TeachINGLearnING.EU Discussions",
 DOI 10.1007/978-3-319-46916-4_3

2 Erfolgreiche Ausschreibung zur Ausstattung virtueller und ferngesteuerter Labore

Die Verbundinitiative ELLI – Exzellentes Lehren und Lernen in den Ingenieurwissenschaften – fokussiert in den Kernbereichen „Virtuelle Lernwelten“, „Mobilitätsförderung und Internationalisierung“, „Kreativität und Interdisziplinarität“ sowie in der Verbesserung der Übergänge innerhalb des „Student Lifecycle“ das ingenieurwissenschaftliche Studium mit methodisch-didaktischem Ansatz. Dabei stehen insbesondere die (Weiter-)Entwicklung innovativer Studienkonzepte sowie die Zugänglichkeit der Studienangebote für eine große Studierendenschaft im Mittelpunkt. Beide Laborformen bieten zusätzlich zu den bisher genannten Vorteilen die Möglichkeit, die Vermittlung des Lernstoffs an den einzelnen Versuchsteilnehmer oder die Teilnehmerin anzupassen. Dies unterstützt die eingangs erwähnte individuelle Wissensbildung der Studierenden. An der Ruhr-Universität Bochum wird die Einführung dieser zusätzlichen Laborversuche außerdem als Lösungsansatz für die erwartete deutliche Zunahme der Studierendenzahlen betrachtet, da bereits aktuell die maximale Auslastung der durchgeführten Laborversuche erreicht ist. Dennoch sollen die zusätzlich einzuführenden Laborformen die bisherigen praktischen Versuchserfahrungen nicht ersetzen, sondern in geeigneter Weise ergänzen. So ist eine Aufteilung der Studierenden in verschiedene Gruppen denkbar, die an jeweils unterschiedlichen Kombinationen aus real erfahrbaren und virtuell erlebbar gemachten Versuchen teilnehmen. Dabei ist die zeitliche Verfügbarkeit von ferngesteuerten „Remote Labs“ eine zentral zu lösende Aufgabe. Bereits im Sommer 2012 traten die Projektleiter der ELLI—Initiative an der Ruhr-Universität Bochum an Vertreterinnen und Vertreter aller drei Ingenieurfakultäten heran, um Transparenz über die Vorgehensweise der Ausschreibung zu schaffen und zum Austausch einzuladen. Daraus ging die konkrete Ausschreibung hervor, die im Herbst 2012 von zentraler Stelle aus in Umlauf gebracht wurde. Die einzureichenden Anträge sollten neue Konzepte der Lehrenden vor Ort enthalten, die räumliche und zeitliche Flexibilität schaffen. Neben diesen organisatorischen Rahmenbedingungen rückt die Ausschreibung insbesondere die Kompetenzentwicklung der Studierenden ins Zentrum, indem sie eigenständig ihre Arbeit planen, mit Herausforderungen des beruflichen Ingenieuralltags umgehen und ihren Wissensstand prozessbegleitend überprüfen. Die Beitragsfrist war auf Ende September 2012 festgesetzt. Insgesamt gingen 13 Anträge aus den drei ingenieurwissenschaftlichen Fakultäten ein. Ende Oktober versammelte sich eine unabhängige Jury, bestehend aus den Fachrichtungen Hochschuldidaktik, Bauingenieurwesen, Elektrotechnik und Maschinenbau, um über die Vergabe der Fördermittel zu entscheiden. Die Jury sprach sich für die Förderung von acht Anträgen mit einem Volumen von insgesamt 472 000 Euro aus. Alle drei Fakultäten der Ruhr-Universität Bochum sind nun daran beteiligt, virtuelle und ferngesteuerte Labore in den Ingenieurwissenschaften einzuführen.

3 Auswahl der zu fördernden Laborversuche

Ein wichtiger Aspekt bei der Wahl der Jury war die Darstellung des didaktischen Konzeptes der jeweiligen Anträge. Dabei lag der Schwerpunkt insbesondere auf der geplanten Umsetzung der Lernziele. Die Studierenden sollten in der Lage sein, selbstständig ihr Wissen zu erweitern und ihren individuellen Lernfortschritt zu überprüfen. Zusätzlich war von Interesse, wie der individuelle Lernstand durch die Lehrenden beurteilt werden kann. Positiv bewertet wurden zudem Anträge, die einen möglichst realen Versuchsablauf inklusive Messfehler oder Störfälle skizzierten. Auch sollte ein Betrieb des Experiments in Randbereichen, also beispielsweise mit nichtstationären Versuchswerten möglich sein, um Grenzen des Versuchsbereichs aufzuzeigen. Alle geförderten Vorhaben dienen dazu, das Lehrangebot in den Ingenieurwissenschaften nachhaltig zu erweitern. Sie werden über die Laufzeit der ELLI—Initiative hinaus in den Studienplan integriert. Besonders positiv wurden daher eine möglichst vielseitige Anwendung der Labore, interdisziplinäre Kooperationen oder auch eine Weiterentwicklung bestehender Konzepte beurteilt. Die im Rahmen von ELLI geförderten Vorhaben werden langfristig dazu genutzt, die Lehre praxisnäher und unmittelbarer zu gestalten. Viele von ihnen binden die studentische Perspektive gezielt in die konkrete Ausgestaltung ein. So werden unter anderem Vorhaben zur Prüfung von Elektrofahrzeugen, zur Steuerung von Robotern und zur Überwachung von Kläranlagen umgesetzt.

4 Praxisbeispiel: Virtual Lab und Remote Lab Abwasser (VRL Abwasser)

Exemplarisch für die acht geförderten Anträge wird an dieser Stelle das Virtual Lab und Remote Lab Abwasser (VRL Abwasser) vorgestellt. Der Laborversuch VRL Abwasser ist für drei Lehrveranstaltungen der Fakultät Bau- und Umweltingenieurwissenschaften vorgesehen und wird in Kooperation der Lehrstühle Siedlungswasserwirtschaft und Umwelttechnik sowie Informatik im Bauwesen angeboten. Neben Einsatzmöglichkeiten im Rahmen von Projekt-, Bachelor- und Masterarbeiten in den Studiengängen Bauingenieurwesen (BI) sowie Umwelttechnik und Ressourcenmanagement (UTRM) ist die optimale Steuerung von Anlagen im Besonderen Gegenstand des Moduls „Operations Research und Simulationstechnik" in den Masterstudiengängen BI und UTRM. Entwickelt wurde das Konzept des VRL Abwasser vor allem zur Integration in das Modul „Technische Mikrobiologie". Anhand der Laborversuche sollen die Studierenden online realitätsnahe Überwachungs- und Steuerkonzepte von Kläranlagen und Biogasreaktoren erproben. Die Aufreinigung verschmutzten Abwassers und der Betrieb einer Kläranlage stellen in der Regel 20 % des kommunalen Energieverbrauchs dar, weshalb diese Thematik eine besondere Relevanz in den Studienplänen der Fakultät aufweist. Bis Sommer 2013 sollen daher eine Laborkläranlage und ein Laborbiogasreaktor für diesen Versuch aufgebaut werden, um zum einen die Abbaubarkeit verschiedener im Abwasser enthaltener Substanzen zu untersuchen und zum anderen die mikrobielle Vergärung und Biogasproduktion verschiedener Substrate zu analysieren. Das entsprechende Lehrkonzept ist zweistufig aufgebaut. Zunächst arbeiten die Studierenden mit einer virtuellen Abbildung der Anlage, um verschiedene betriebliche Einstellungen zu testen, bevor sie diese anschließend an den realen Laboranlagen überprüfen. Unter simulierten, realitätsnahen Bedingungen können die Studierenden Erfahrungen im Bereich der Anlagensteuerung machen. Dies ist direkt an der Anlage nicht möglich, da es durch bestimmte Parameterkombinationen zu Problemen bzw. Systemausfällen kommen kann, die gegebenenfalls zu irreparablen Schäden an den sensiblen Anlagen führen. Das Verständnis der einzelnen Komponenten innerhalb der Anlage wird durch deren Visualisierung deutlich verbessert. Die Auswirkung der Änderung einzelner Prozessparameter auf die Anlagen wird dabei abgebildet, sodass verschiedenste Szenarien getestet werden können. Ziel ist es, die betrieblichen Prozessparameter so zu wählen, dass von den Lehrenden vordefinierte Zielwerte, z.B. die zu erreichende Wasserqualität, erreicht werden. Daran anschließend überprüfen die Studierenden an der realen Anlage, ob die gewünschten Ergebnisse mit den gewählten Einstellungen erzielt wurden. Die Studierenden steuern die Anlagen in Kleingruppen an interaktiven Touch Tables. Sie werden angeregt, in der Gruppe Parametervariationen durchzuführen und die Auswirkungen direkt zu diskutieren. Auf diesem Weg sollen die Studierenden in der Lage sein, die vorgegebenen Zielwerte durch Anpassung der Prozessparameter zu erreichen. Durch die Verwendung eines Touch Table erarbeiten sie in Kleingruppen ihre Ergebnisse und schulen dabei ihre Teamfähigkeit und Diskussionsbereitschaft. Die Gruppen speichern ihre gewählten Einstellungen und übermitteln diese zur Bewertung an die Lehrenden. Bei sinnvoll gewählten Betriebsparametern erfolgt eine Freischaltung für den Remote-Versuch in den realen Laboranlagen, andernfalls ist eine weitere Anpassung der Parameter erforderlich. Für einen realistischen Einblick in den ferngesteuerten Versuch wird ein Live-Video der Anlage übertragen.

5 Ausblick

Alle im Rahmen der ELLI—Initiative geförderten Vorhaben werden langfristig dazu genutzt, die Lehre praxisnäher und unmittelbarer zu gestalten, um damit der beschriebenen Herausforderung steigender Studierendenzahlen unter Einbezug neuer Medien und Technologien Rechnung zu tragen. Während der Umsetzung der geförderten Laborversuche erhalten die Antragstellenden Unterstützung vom ELLI-Team, das regelmäßige Vernetzungstreffen während der Umsetzung organisiert.

Literaturverzeichnis

1. Dezernat 8 – Hochschulkommunikation. Maßnahmen zu Lehre und Studium, 08.04.2013. http://www.ruhr-uni-bochum.de/rub2013/massnahmen/lehre-studium/index.html
2. M. Grosch, G. Gidion, *Mediennutzungsgewohnheiten im Wandel: Ergebnisse einer Befragung zur studiumsbezogenen Mediennutzung*. KIT Scientific Publ, Karlsruhe, 2011. http://www.ruhr-uni-bochum.de/elli/download/Mediennutzung%201.pdf

Massive Open Online Courses in Engineering Education

Ute Heckel, Ursula Bach, Anja Richert and Sabina Jeschke

Abstract Though higher engineering education generally lacks students in Germany, some universities are faced with the challenge of dealing with extremely high enrollment numbers due to recent changes in education policy. In the winter term 2011/2012 approx. 1,900 students enrolled for mechanical engineering and industrial engineering and management at RWTH (Rheinisch-Westfälische Technische Hochschule) Aachen University putting the educational skills of teachers to the test. Obviously, new concepts become necessary to find adequate teaching models. Modern information and communication technologies have already become a constant part of everyday life among the new generation of students. But their full potential for higher education has not yet been exploited. Concepts hitherto focused on integrating technologies such as Audience-Response-Systems or mobile applications into face-to-face lectures. Only recently a new approach emerged, bearing the potential of teaching increasingly high numbers of students entirely online and of revolutionizing the higher education landscape: Massive Open Online Courses (MOOCs). They seem to highly motivate their students to actively participate in online courses and to interact with teachers and fellow students using social and technical networks. As demonstrated by initiatives such as the Khan Academy, udacity, edX, or Coursera MOOCs attract enormous numbers of students (In 2011 160,000 students followed the Stanford lecture on Artificial Intelligence by Prof. Thrun and Prof. Norvig with 23,000 earning a certificate). This paper aims to show how MOOCs might help to tackle the challenges of teaching large classes in higher engineering education. As they have attracted large amounts of students especially for engineering topics, they might be adequate for higher engineering education. A variety of MOOCs have emerged so far based on fundamentally different learning principles that cater to the needs of engineering education in different ways. Thus, this paper categorizes MOOCs according to their underlying didactical approaches in a first step. In a second step it is evaluated to what extend the different kinds of MOOCs can be used to implement active and problem-based learning in a large class and for what purposes

U. Heckel (✉) · U. Bach · A. Richert · S. Jeschke
IMA/ZLW & IfU, RWTH Aachen University, Dennewartstr. 27, 52068 Aachen, Germany
e-mail: ute.heckel@ima-zlw-ifu.rwth-aachen.de

Originally published in "Proceedings of ICERI2012. Fifth International Conference of Education, Reserach and Innovation", © 2012. Reprint by Springer International Publishing AG 2016,
DOI 10.1007/978-3-319-46916-4 _4

in engineering education they can be best applied. The results are useful for any university teacher in higher engineering education dealing with large classes.

Keywords Large Classes · Higher Engineering Education · Active Learning · Problem-based Learning · Massive Open Online Courses

1 Introduction

German higher education institutions face the challenge of exceptionally large numbers of students enrolling for undergraduate courses due to recent changes in German education policy.[1] In contrast higher engineering education lacks sufficient graduates mainly due to high drop-out rates caused by challenging curricula, insufficient knowledge of natural sciences and mathematics acquired at school, and too few practical relevance of the learning material [2]. This situation results in the challenging task for higher engineering education institutions to improve the quality of teaching while an increasing number of students head to universities. This is also mirrored in the current situation at RWTH (Rheinisch-Westfälische Technische Hochschule) Aachen University.

The Current Situation – An Extra-Large Lecture in Computer Sciences

In the winter term 2011/2012 approx. 1,900 undergraduate students enrolled for mechanical engineering and industrial engineering and management at RWTH Aachen University [3]. The lecture "Computer Sciences in Mechanical Engineering 1", held by Prof. Sabina Jeschke[2] and her team of the Institute Cluster IMA/ZLW & IfU (Institute of Information Management in Mechanical Engineering (IMA), Centre for Learning and Knowledge Management (ZLW), Assoc. Institute for Management Cybernetics (IfU)), is a compulsory course in the 2nd semester. Thus, the professor and her teaching assistants are faced with approx. 1,900 listeners in the course which is composed of different teaching formats: a lecture, project work, and exercises [4]. This paper focuses on the lecture only.

Teaching at IMA/ZLW & IfU goes along with two corresponding research projects TeachING-LearnING.EU[3] and ELLI[4] (Excellent Teaching and Learning in Engineering Sciences). Both projects explore new ways of teaching in higher engineering

[1]The number of schooling years for the Abitur (German higher education entrance certificate) was reduced from 13 to 12 years. The compulsory military service was suspended as from July 1st, 2011 along with the corresponding civil service. This led to increasing numbers of students heading for universities directly after finishing secondary education and in the future even more students are expected to enter German higher education [1].

[2]Further information on the team can be found at: http://www.ima-zlw-ifu.rwth-aachen.de/institutscluster/mitarbeiter/einzelansicht/team/Sabina-Jeschke.html (last accessed October 1, 2012).

[3]More information on http://www.teaching-learning.eu (last accessed September 24, 2012).

[4]More information on http://www.elli-online.net (last accessed September 24, 2012).

education also focusing on large classes [5]. A study on lectures with more than 1,000 listeners at RWTH showed that involving students into the lecture using interactive methods is even more crucial for successful learning in large classes than in smaller classes. Hence, first experiences where gathered using interactive teaching methods such as Audience-Response-Systems (ARS)[5] where students were asked questions on their understanding of the content. Though first results have yet to be published, it can be stated that ARS resulted in increased motivation and attention of students in the class [6].

While those measures focused on integrating interactive elements into face-to-face lectures, this paper analyzes the opportunities brought about by online education. Thus, a rather new format of online education, the so called Massive Open Online Course (MOOC), is analyzed regarding its potential to improve higher engineering education especially for large classes. First, the special challenges of teaching large classes are briefly described followed by the requirements and pedagogical paradigms in higher engineering education. Finally MOOCs are introduced and categorized before dwelling on their potential for application in higher engineering education.

2 The Challenges of Teaching Large Classes in Higher Education

What Number of Students Makes a Large Class?

When speaking about large classes, it is hard to define a certain number of students as differences between disciplines are predominant. Regardless of the class size, teaching large classes has a certain characteristic as Weimer in one of the first books on teaching large classes points out [7]:

> The focus is on classes in which the possibility of individual relationships between professor and student is precluded, in which not every student who wants to speak in class can be called on, and in which grading essay exams can take up every evening and weekend of the course.

While the lecture focused in this paper is provided for 1,900 students or more, the term of an "extra-large" class might be appropriate according to Stehling et al. [6].

How to Teach Large Classes

Teaching a large class of this size is especially challenging when it comes to integrating active and problem-based learning as required by engineering education (c.f. Section 3). The challenge lies in overcoming anonymity and distance between teachers and students in large classes through creating a group identity and rapport that facilitates discussion, feedback and active learning [8].

---

[5]Audience Response Systems (ARS) were developed for large classes to integrate interactive exchange between the teacher and a large number of students. Questions are answered by students either via smartphone, SMS, or an ARS device [6].

The following list summarizes some of the main tasks for teachers in order to create a favorable learning environment in a large class [8–11]:

- Direct feedback and contact: by enabling personal discourse between the teacher and the students on learning goals, method, and content [12];
- Encourage class participation: by offering group work or making students contribute material;
- Promote active learning: by making lecture outlines available to students before and throughout the class, by using demonstrations or different media to keep students interested, by showing own enthusiasm for the subject, by giving "think breaks" that cut the lecture into 15–20 minute chunks, by designing the lecture around a problem-solving model, or by giving frequent assignments;
- Be organized: by offering a central website with the syllabus and external references.

Most of the methods described above are based on direct in-class interaction between teachers and students. Nevertheless, extra-large classes may still be intimidating for many students. Thus, other approaches e.g. based upon online technology shall be explored.

3 Didactical Approaches in Higher Engineering Education

Shift from Teaching to Learning

Engineering education moves between two poles: hands-on experiences on one side and fundamental knowledge in natural sciences and technology on the other [13]. Student-centered learning approaches were introduced to better cater to those needs by applying activating teaching scenarios and focusing on the teacher as a coach or facilitator rather than an instructor. This is also referred to as the "shift from teaching to learning" first introduced by Barr and Tagg [14]. Thus, students are required to have a certain degree of self-directed learning competences and to take responsibility for their individual learning process. One term often referred to in this context is also "active learning".

Active Learning

Active learning is especially hard to achieve in large classes. It has been tried to use methods such as asking challenging questions, giving students tasks to evocate reasoning and problem solving, or making students repeat learning contents in order to initiate active retrieval and reconstruction of knowledge. As studies have shown, those methods may result in higher attendance and engagement rates and better learning results [15–17]. But still methods that evocate active learning have so far rather focused on smaller classes.

Problem-Based Learning

Learners are perceived as human beings that are intrinsically motivated to solve problems and to acquire the information and the knowledge they need to solve the problems. Therefore, a problem-based learning setting follows a guided design where students are guided through the problem-solving process beginning with a short description of the situation and the problem to be solved along with some introductory material. Their task is to figure out by themselves how the problem can be solved and to finally evaluate the solution they came up with. Thus, problem-based learning pedagogy aims to provide students with the skills of self-assessing their work [17].

Higher engineering education heavily relies on active and problem-based learning which is especially hard to be implemented in extra-large classes. In the next section it will be analyzed whether and how an online technology such as Massive Open Online Courses (MOOCs) may help to put those learning paradigms into practice.

4 Massive Open Online Courses – The Virtual Solution??

4.1 Getting to Know MOOCs

When defining what Massive Open Online Courses (MOOCs) are, a further deep dive into the single attributes *massive*, *open*, *online*, and *course* becomes necessary. When it comes to classify the attributes *massive* and open definitions differ according to the two main underlying didactical designs - connectivist and instructional design which will be described in detail in this section. But regardless of the didactical approach, it seems to be common sense that MOOCs at least refer to *courses* that are taught *online* on the Internet and that are open in the sense of being accessible free-of-charge for anyone [18].

Connectivist MOOCs

According to connectivism[6] as first introduced by Siemens and Downes, learning is defined as being

> "(…) focused on connecting specialized information sets (…) the connections that enable us to learn are more and more important than our current state of knowing. (…) Nurturing and maintaining connections is needed to facilitate continual learning. Ability to see connections between fields, ideas, and concepts is a core skill." [20].

Connectivist MOOCs are based on the principles of connectivism: autonomy, diversity, openness, and interactivity. Students in a connectivist MOOC perform four major activities (c.f. [21]):

---

[6]It has largely been discussed throughout the scientific community as to whether connectivism really states a new learning theory. A brief summary of the criticism of connectivism is provided by Siemens [19].

1. *Aggregate*: Students are asked to pick and choose the content that looks interesting to them and seems to be most appropriate according to their personal learning goals from a wide range of information spilled on the Internet.
2. *Remix*: Students keep track of the information items they accessed by using any tool from lists offline on their computers to online blogs, Twitter, or the like.
3. *Repurpose*: Students describe their own understanding of the material they aggregated and remixed before and thereby create new knowledge based on already existing materials.
4. *Feed Forward*: Students share their thoughts and understanding on the Internet with other course mates and the world at large.

Connectivist MOOCs have been delivered since 2007 with the open courses of David Wiley[7] and Alec Couros mainly targeting the issue of open education itself [22]. But it was not until 2008 when George Siemens and Stephen Downes offered the online course "Connectivism and Connective Knowledge" (CCK08) that the concept of MOOCs became generally known [23]. Many other initiatives followed their example such as e.g. PLENK2010,[8] eduMOOC,[9] mobiMOOC,[10] DS106,[11] or LAK11/12[12] and until today new connectivist MOOCs keep on emerging.

In the context of connectivist MOOCs the attribute *open* rather refers to the fact that learners from different stages of learning from beginners to experts merge together in one common learning community [18]. Connectivist MOOCs so far attracted between 500 and 2,500 participants and set the range for the attribute massive at this margin though not being limited to it.

In this paper the term "cMOOC" (referring to "connectivist MOOC") is used to describe this specific MOOC format.

Instructional MOOCs

Instructional MOOCs often follow a clear teacher-centered, instructional design according to behaviorist and cognitivist learning theory where instructors teach and students listen and accomplish quizzes or assignments. As teaching in this context is more of a one-to-many communication, discussion forums offer each student the possibility to get actively involved. The teachers themselves may answer a certain amount of questions whereas most of the questions are answered by peers around the

[7]In 2007 Wiley first offered the wiki-based course "Open Ed Syllabus". The course material can be accessed at http://opencontent.org/wiki/index.php?title=Intro_Open_Ed_Syllabus (last accessed September 25, 2012).

[8]Personal Learning Environments, Networks and Knowledge 2010 (PLENK2010) http://connect.downes.ca.

[9]eduMOOC: Online Learning Today…and Tomorrow http://edumooc.wikispaces.com (last accessed September 22, 2012).

[10]Learning and training with mobile devices http://mobimooc.wikispaces.com (last accessed September 22, 2012).

[11]Digital Storytelling (DS) http://ds106.us (last accessed September 22, 2012)

[12]Learning and Knowledge Analytics (LAK) http://www.learninganalytics.net (last accessed September 22, 2012).

globe (c.f. [24]). Several MOOCs emerged following the instructional course design such as the Khan Academy,[13] the class on Artificial Intelligence (AI) by Stanford Professors Thrun and Norvig[14], the MITx[15]/edX[16] initiative, Coursera,[17] or udacity[18] just to name the major players in the field [25].

In the context of instructional MOOCs the attribute *massive* clearly refers to an extremely high number of participants (e.g. 160,000 students in AI-class or 680,000 students in Coursera classes). The term *open* refers to the open and free-of-charge access of each learner to materials and content otherwise only being available to tuition paying students [18].

This paper uses the term "iMOOCs" to refer to instructional MOOCs that follow a traditional course structure with pre-defined learning goals, materials provided and presented by a teacher, fixed assignments, and a final exam.

The analyses in this paper are restricted to those two basic types of MOOCs. Having said this, it is nevertheless clear that a variety of other initiatives emerged especially in Germany[19] that are based on the same principles but cannot be taken into consideration in order not to go beyond the scope of this paper.

What Are the Differences Between MOOCs?

Table 1 shows a classification of MOOCs into cMOOCs and iMOOCs according to the different notions of their attributes *massive* (range of number of listeners), *open* (targeted learners), and *online* (form of distribution). The attribute *course* conjointly refers to the course format as already described above.

4.2 *How to Educate Large Classes with MOOCs?*

For analyzing the potentials of cMOOCs the courses "Openness in Education"[20] by Rory McGreal and George Siemens and "Connectivism and Connective Knowledge 2011"[21] by Steven Downes were taken into consideration. The analysis of iMOOCs is based on the courses "Introduction to Statistics" (ST101) by Sebastian Thrun and

---

[13]http://www.khanacademy.org (last accessed September 22, 2012).

[14]https://www.ai-class.com (last accessed October 1, 2012).

[15]http://mitx.mit.edu (last accessed October 1, 2012).

[16]https://www.edx.org (last accessed September 22, 2012)

[17]https://www.coursera.org (last accessed September 22, 2012).

[18]http://www.udacity.com (last accessed September 22, 2012).

[19]German MOOC initiatives: #ocwl11 – Open Course Workplace Learning 2011 http://ocwl11.wissensdialoge.de, OPCO11 http://blog.studiumdigitale.uni-frankfurt.de/opco11, OPCO12 http://opco12.de, iversity http://www.iversity.org, openHPI https://openhpi.de (last accessed September 12, 2012).

[20]The course can be found at http://open.mooc.ca (last accessed September 26, 2012).

[21]The course can be retrieved from http://cck11.mooc.ca (last accessed September 26, 2012).

Table 1 Classification of MOOCs [18, 20, 23]

	connectivist MOOC (cMOOC)	instructional MOOC (iMOOC)
M(assive)	500-2,500 students	up to 160,000 students or more
O(pen)	open to learners from different stages of learning from beginners to experts, no fixed set of learning materials (is gathered by students themselves), no pre-defined outcomes, no formal accreditation	pre-defined classes for different stages of expertise with corresponding learning materials, pre-defined learning outcomes, formal accreditation possible
	open registration, course materials free of charge, accessible and shared	
O(nline)	on the Internet	
C(ourse)	regular courses	

Adam Sherwin, and "Introduction to Computer Science" (CS101) by David Evans provided by udacity.[22]

The analysis focused on the requirements prescribed by higher engineering education and large class pedagogy as discussed before:

- direct feedback and contact,
- encourage class participation
- promote active and problem-based learning, and
- be organized.

Furthermore, especially for online and distance learning, the community of learners is crucial for successful learning and *community building* will, thus, be considered in the following analysis as well. In the following sections each of those elements are matched with the characteristics of cMOOCs and iMOOCs in order to analyze their potential for teaching large classes in higher engineering education.

4.2.1 Direct Feedback and Contact

MOOCs show different ways of integrating direct feedback and contact between teachers and students to facilitate personal discourse. Some MOOCs try to emulate one-to-one tutoring while most of the others provide different ways of direct feedback and contact.

One-to-One Tutoring

Mainly iMOOCs emulate one-to-one tutoring with the teacher being recorded while writing down notes and explaining the content. The student is given the impression of being taught individually by the teacher. As already observed by Bloom in 1984, students tutored one-to-one perform significantly better than students who learn with

[22] Both courses can be retrieved from http://www.udacity.com/courses (last accessed September 26, 2012).

conventional instruction methods [26]. Even though there have been no such studies on the effects of one-to-one tutoring in the context of MOOCs in particular, it is assumed that the emulation of one-to-one tutoring in MOOCs could have positive effects on learning and the feeling of direct contact to the teacher (c.f. [26]).

Feedback

iMOOCs and cMOOCs use discussion boards, forums, or social media tools such as Twitter and Facebook to decrease the feeling of distance between the teacher and students and to encourage active involvement. It is assumed that students are not as hampered to discuss their questions in an online discussion forum as they would be when sitting amidst a class of 1,000 students or more. Nonetheless discussion forums with as many members as in MOOCs are threatened to become difficult to oversee. cMOOCs encourage direct feedback through comments on the learning materials gathered and remixed by each participant. Furthermore, from the teachers point of view student feedback about the course can be gathered in forums or through social media activity helping them to improve their methods [22].

4.2.2 Encourage Class Participation

While in-class courses often implement group work in order to increase participation, MOOCs rely on other mechanisms such as peer learning or the integration of content contributed by students.

Peer Learning

Peer learning happens when "...students learn with and from each other without the immediate intervention of a teacher" [27]. MOOCs heavily rely on peer learning in general since one teacher cannot interact with every single student given the large class size of MOOCs. Especially in iMOOCs the strategies of peer instruction and peer assessment play major roles. Peer instruction [28] aims to actively involve students into the lecture by integrating so called "ConcepTests" that encourage them to discuss the subject of the lecture with their peers. Peer assessment is applied when students give feedback on their peer's work or when they are required to determine and defend their own work [29]. While peer instruction is often used in iMOOCs as a method to structure the course content into demonstrations and question-and-answer sets, peer assessment takes place in forums where students assess posts, questions and answers using badges.[23] Peer learning in cMOOCs takes place through reconstructing and repurposing knowledge by peers.

Students Contribute Course Content

Mainly used by cMOOCs the contribution of course content fosters student participation through connecting students and their individual understanding of the topics

[23] First developed by the Mozilla Foundation, badges are used on the Internet as a "validated indicator of accomplishment, skill, quality or interest" [30]. They can be provided to testify the accomplishment of formal as well as informal knowledge.

[20]. To share their materials students are asked to use the tools already available on the Internet such as blogs, tumblr,[24] Diigo,[25] social network sites, Twitter, and others.

4.2.3 Promote Active and Problem-Based Learning

Active and problem-based learning approaches put the students into the focus and try to incite them to actively take part in their learning process. iMOOCs and cMOOCs cater to those requirements in different ways.

Active Learning

As Koller [31] shows for the case of Coursera courses (iMOOCs), active retrieval practice is used to make students actively engage with the learning material. Based on the findings on the effects of retrieval and reconstruction of knowledge by Karpicke and Blunt [16] students are asked questions assessing comprehension and requiring them to make inferences on the topic. cMOOCs foster the individual's activities of collecting, sharing, and connecting knowledge and thus, achieve to integrate active learning into their courses due to the underlying principle of connectivism.

Problem-Based Learning

As discussed above, engineering courses shall be designed around a problem-solving model which is especially hard to achieve in MOOCs. However, iMOOCs try to implement problem-based learning using examples that depict a problem with high practical relevance and by describing the appropriate problem-solving algorithm subsequently. According to Svinivki and McKeachie [17] this form can rather be classified as "case-based learning" where the students are provided with a problem and a problem-solving algorithm and are asked to critically discuss it. In contrary cMOOCs bear a higher potential to implement problem-based learning. They focus on a certain problem, give first information, and leave students the freedom to come up with own creative solutions. The teacher in this case still cannot guide each of the students individually through the process, but by providing relevant material on the topic problem-solving can be steered to certain directions. Though not having been applied to engineering topics so far, cMOOCs are likely to be adapted to engineering themes.

---

[24]Tumblr is a mirco-blogging platform https://www.tumblr.com (last accessed October 3, 2012).

[25]Diigo is an online bookmarking tool http://www.diigo.com (last accessed October 3, 2012).

4.2.4 Be Organized

iMOOCs are often very clearly organized using a central website where all the information on the syllabus, announcements, discussion forums or chats, and learning materials are aggregated together with all the tools the students need for learning. This provides students with a fixed framework for learning, but is rather time-consuming and expensive to dress-up. cMOOCs on the other side emphasize the effect of self-directed learning. Mostly, they also offer a central website with information on the principles of the course and a syllabus. But, in contrast to iMOOCs they do not offer a fixed set of learning materials. As Siemens also admits, cMOOCs are chaotic on purpose [22]. They constitute an open framework for learning.

4.2.5 Community Building

Community building is a major prerequisite for successful learning. While learning in real-world settings automatically offers a learner community, online learning formats are faced with the challenge of building a virtual community. Svinivki and McKeachie [17] stress the negative effects of student anonymity on motivation for learning:

> Moreover, the sense of distance from the instructor, the loss of interpersonal bonds with the instructor and with other students - these diminish motivation for learning.

Thus, cMOOCs and iMOOCs alike try to use web-based tools such as forums, discussion boards or social media to enhance interaction and to create a group identity and rapport among the course participants. The big advantage arising from the global community of MOOC participants, as Koller pinpoints, is that questions can be answered by peers at any time due to the global reach of the courses leading to a level of service no single teacher is able to provide [31]. Lately, udacity introduced personal meet-ups to encourage face-to-face interaction among students and to add personal and peer teaching in groups to their courses.[26]

4.2.6 Summary

Table 2 summarizes the potentials of cMOOCs and iMOOCs for teaching large classes in higher engineering education.

[26]Meet-ups are organized through a separate platform http://www.meetup.com/Udacity/ (last accessed September 19, 2012).

Table 2 Potentials of cMOOCs and iMOOCs for teaching large classes

	cMOOCs	iMOOCs
Direct Feedback and Contact	– direct interaction among participants by commenting on created course materials and by connecting with other participants through social media tools	– one-to-one tutoring emulated – direct feedback & conversation made available via discussion forums
Encourage Class Participation	– peer learning through reconstructing, repurposing and sharing knowledge by peers – students contribute and share own course content	– peer learning through peer instruction and peer assessment
Promote Active & Problem-Based Learning	– active learning through remixing, reconstructing and sharing content – course content designed around problem-solving model with teachers as guides	– quizzes to engage students in active retrieval and reconstruction of knowledge – course content hard to design around problem-solving model, rather case-based learning provided
Be Organized	– central website with basic syllabus and links to learning resources – widely distributed contents and absent curricula, no common learning goals and methods (learning might be perceived as "chaotic") – open framework for learning provided – use tools already available on the Internet and users are familiar with	– central website with all necessary information and learning resources – fixed framework for learning provided – own tools for communication (forum, chat etc.) have to be developed
Community Building	– virtual community using social media and discussion boards – no direct meet-ups	– virtual community using social media and discussion boards – real-world community through organized meet-ups

5 Conclusion

As shown above the two major types of MOOCs constitute two fundamentally different learning approaches. While connectivist MOOCs focus on a student-centered approach, instructional MOOCs are rather teacher-centered. Both types of MOOCs bear a high potential for being introduced into curricula of higher engineering education. Until recently, iMOOCs have mostly been applied to engineering topics while cMOOCs focus on humanities and science in general. Moreover, iMOOCs lag behind cMOOCs in terms of active and problem-based learning being a basic requirement for good engineering education. Both types of MOOCs require students to have a high level of self-learning competences in order to lead to successful learning. cMOOCs are even more demanding in this respect than iMOOCs due to a total lack of a course outline or guidance throughout the course.

A major weakness of iMOOCs is that they hardly achieve to implement problem-based learning. In contrast, cMOOCs highly encourage problem-based learning. Hence, it is suggested to analyze in a next step whether a combination of both models can be applicable and if so, how an according course design could look like.

Regardless, of this major drawback current iMOOCs can definitely be applied to teach fundamentals in engineering education as is shown by recent courses from Coursera, udacity, or edX. By teaching fundamentals one of the basic pillars engineering competence rests upon can be supported [13]. As statistics show, early drop-outs in higher engineering education are often caused by insufficient basic knowledge in natural sciences and technology. Thus, students struggle in the first semesters to catch up the fundamentals they failed to acquire in secondary education [2]. Hence, by using encouraging methods to teach fundamentals in higher engineering education, the motivation of students to stick to the course might be increased having a positive effect on lowering drop-out rates.

In a next step an integrated concept will be developed for the lecture "Computer Science in Mechanical Engineering 1" that combines the elements mentioned above. Therefore, general advantages and disadvantages of MOOCs will also have to be taken into consideration. The good thing about MOOCs is that just a few technical requirements are necessary such as a computer and Internet access and that they are open to anyone being sufficiently equipped. The work can be shared with others, and students around the world can benefit from each other and outstanding professors. On the other side, technical difficulties can impede the learning process, the learner's preferences for traditional learning formats may not be satisfied, and higher education institutions might not integrate them into curricula as they do not come along formal accreditation (c.f. [32, 33]).

Future Research

There are several topics that have to be dwelled upon as research on MOOCs has just recently emerged. Certainly some interesting questions are to be answered e.g. on the effects of one-to-one tutoring in MOOCs, the consequences of class sizes for activities in discussion forums, peer learning and community building processes in online media, and the implementation of problem-based learning with a MOOC. Moreover, the integration of MOOCs into the higher education system might result in a redefinition of the role of the university. What are the consequences for higher education institutions, teaching and research staff as well as students when a few outstanding professors reach large amounts of students? Questions on the standardization of education will have to be answered as well as on financing models, forms of accreditation, or legal issues.

In addition, the use of technology in education leads to gathering big data on learning patterns and habits of the new generation of learners, the digital natives, being a rich resource for further research. Techniques from data mining, machine learning or natural language processing are adapted to the needs of learning analytics. As indicated by the Horizon Report 2012 [34], learning analytics will be one of the central topics for research on education within the next five years. This will lead to a shift from a hypotheses-driven approach to data-driven research on education and learning. Thus, the foundations to gather data are to be laid already today. MOOCs are one important step in this direction.

References

1. K. KMK. Positiver trend zum studium hält an – kultusministerkonferenz veröffentlicht vorausberechnung, 2012. URL http://www.kmk.org/presse-und-aktuelles/meldung/positiver-trend-zum-studium-haelt-an-kultusministerkonferenz-veroeffentlicht-vorausberechnung.html
2. U. Heublein, C. Hutzsch, J. Schreiber, D. Sommer, G. Besuch, *Ursachen des Studienabbruchs in Bachelor- und in herkömmlichen Studiengängen. Ergebnisse einer bundesweiten Befragung von Exmatrikulierten des Studienjahres 2007/08*. Hochschul-Informations-System GmbH, Hannover, 2009. URL http://www.bildungsserver.de/db/mlesen.html?Id=44422
3. H.D. Hötte, H. Fritz, RWTH aachen university. zahlenspiegel 2011. Tech. rep., RWTH Aachen University, Aachen, 2012
4. S. Jeschke, VisioneerING. future trends in engineering education. In: *Proceedings of IGIP International Conference on Engineering Pedagogy, Villach, Septermber 26–28, 2012*. 2012
5. S. Jeschke, M. Petermann, A.E. Tekkaya, Ingenieurwissenschaftliche ausbildung - ein streifzug durch herausforderungen, methoden und modellprojekte. In: *TeachING-LearnING.EU Fachtagung "Next Generation Engineering Education"*, ed. by U. Bach, S. Jeschke, ZLW/IMA der RWTH Aachen University, Aachen, 2011, pp. 11–22
6. V. Stehling, U. Bach, A. Richert, S. Jeschke, Teaching professional knowledge to XL- classes with the help of digital technologies. In: *ProPEL Conference Proceedings 2012*. 2012
7. Teaching large classes well. new directions for teaching and learning. Jossey-Bass, San Francisco, 1987
8. Schreyer Institute for Teaching Excellence. Teaching large classes - schreyer institute for teaching excellence, 1992. URL http://www.schreyerinstitute.psu.edu/Tools/Large/
9. B. Gross Davis, *Tools for Teaching*. Jossey-Bass, San Francisco, 1993
10. S.M. Ives. A survival handbook for teaching large classes, 2000. URL http://www.teaching.uncc.edu/articles-books/best-practice-articles/large-classes/handbook-large-classes#part1
11. D. Spiller. Maximising learning in large groups: The lecture context, 2011. URL http://www.waikato.ac.nz/tdu/pdf/booklets/3_LargeGroups.pdf
12. P. Arnold, L. Kilian, A. Thillosen, G. Zimmer, *Handbuch E-Learning. Lehren und Lernen mit digitalen Medien*, 2nd edn. Bertelsmann, Bielefeld, 2011
13. F. Klocke, M. Pasthor, Challenge and absolute necessity. In: *TeachING-LearnING.EU Fachtagung "Next Generation Engineering Education"*, ed. by U. Bach, S. Jeschke, ZLW/IMA der RWTH Aachen University, Aachen, 2011, pp. 23–37
14. R.B. Barr, J. Tagg, From teaching to learning - a new paradigm for undergraduate education. Change **27**(6), 1995, pp. 697–710
15. L. Deslauriers, E. Schelew, C. Wieman, Improved learning in a large-enrollment physics class. Science **332**, 2011, pp. 862–864
16. J.D. Karpicke, J.R. Blunt, Retrieval practice produces more learning than elaborative studying with concept mapping. Science **331**(6018), 2011, pp. 772–775
17. M. Svinivki, W.J. McKeachie, *McKeachie's Teaching Tips: Strategies, Research and Theory for College and University Teachers*, 13th edn. Wadsworth, Belmont, 2011
18. O. Rodriguez, MOOCs and the AI-Stanford like courses: Two successful and distinct course formats for massive open online courses. European Journal of Open, Distance and E-Learning, 2012. [Online]. Available: URL http://www.eurodl.org/?article=516. [Accessed: 03.10.2012].
19. G. Siemens. Criticism of connectivism, 2006. URL http://www.connectivism.ca/?p=75
20. G. Siemens. Connectivism: A learning theory for the digital age, 2005. URL http://www.itdl.org/Journal/Jan_05/article01.htm
21. R. McGreal, G. Siemens. Openness in education, 2012. URL http://open.mooc.ca
22. G. Siemens. Designing, developing, and running (massive) open online courses, 2012. URL http://www.slideshare.net/gsiemens
23. A. Fini, The technological dimension of a massive open online course: The case of the CCK08 course tools. International Review of Research in Open and Distance Learning **10**(5), 2009
24. M. Fries. Bildungsprojekt udacity: Hochschulbildung, kostenlos und für alle, 2012. URL http://www.zeit.de/studium/uni-leben/2012-01/udacity-thrun

25. B.B.F. Faviero, Major players in online education market. comparing khan academy, coursera, udacity, & edX missions, offerings. The Tech **132**(34), 2012
26. B.S. Bloom, The 2 sigma problem: The search for methods of group instruction as effective as one-to-one tutoring. Educational Researcher **13**(6), 1984, pp. 4–16
27. D. Boud, R. Cohen, J. Sampson, Peer learning and assessment. Assessment & Evaluation in Higher Education **24** (4), 1999, pp. 413–426
28. E. Mazur, *Peer Instruction: A User's Manual*. Prentice Hall, Englewood Cliffs, NJ, 1997
29. P. Ramsden, *Learning to Teach in Higher Education*, 2nd edn. RoutledgeFalmer, London, New York, 2007
30. K. Carey. A future full of badges, 2012. URL http://chronicle.com/article/A-Future-Full-of-Badges/131455/
31. D. Koller. What we're learning from online education, 2012. URL http://www.ted.com/talks/daphne_koller_what_we_re_learning_from_online_education.html
32. J. Moskaliuk. Bildung zwischen hochschule und web, 2012. URL http://www.wissenmaldrei.de/bildung-zwischen-hochschule-und-web/
33. S. Writers. The world of massive open online courses, 2012. URL http://www.onlinecolleges.net/2012/07/11/the-world-of-massive-open-online-courses/
34. L. Johnson, S. Adams, M. Cummins, *The NMC Horizon Report: 2012 Higher Education Edition*. The New Media Consortium, Austin, Texas, 2012

Using E-Portfolios to Support Experiential Learning and Open the Use of Tele-Operated Laboratories for Mobile Devices

Dominik May, Claudius Terkowsky, Tobias Haertel and Christian Pleul

Abstract The use of laboratories in Engineering Education at universities is an adequate opportunity to implement experiential and research based learning — e.g. in material sciences. Within these laboratories the students have the chance to do own experiments and by that gain own experiences in their learning processes. Recently finished research projects - e.g. like the PeTEX project done by universities in Dortmund (Germany), Palermo (Italy) and Stockholm (Sweden) — implemented an opportunity to do experiential learning by using real laboratory equipment without being physically in the laboratory but having access via the internet. For a couple of reasons this makes sense, for example because limited equipment resources or the high costs of this equipment. Trying to implement experiential learning by the use of these remote laboratories gives the opportunity to the students to do some kind of own research. The question in this context - and for this question we will give a first answer in this paper - is how students could document their own learning process on the one hand and how the teacher can guide the student through this process on the other hand. One possible solution can be seen in the use of e-portfolios. With this e-portfolios the student can document and show to others, what he has been doing. The next step is making the e-portfolio software available for mobile devices so that the student has access from virtually everywhere and every time. With this work in progress paper we show what kind of role e-portfolios can play in the learning process and which kind of scenarios are possible using the software on mobile devices. Furthermore, we show that the combination of experiential learning and the use of e-portfolios offer a great potential to promote the learners' creativity.

Keywords Engineering Education · E-Portfolios · Mobile Learning · Remote Laboratories · Tele-Operated Laboratories

---

D. May (✉) · C. Terkowsky · T. Haertel
Engineering Education Research Group (EERG), Center for Higher Education (zhb), TU Dortmund University, Dortmund, Germany
e-mail: dominik.may@tu-dortmund.de

C. Pleul
Institute of Forming Technology and Lightweight Construction (IUL), TU Dortmund University, Dortmund, Germany

Originally published in "Conference Proceedings of REV2012 - Remote Engineering & Virtual Instrumentation", © IAOE 2012.
Reprint by Springer International Publishing AG 2016,
DOI 10.1007/978-3-319-46916-4_5

1 Introduction

Engineering students once they graduated will work on solving real problems creatively and they will work with real technical equipment - doesn't matter if they go for a career in a company or in the academic sector. But do they get into contact with it during their studies? In most cases we would definitely say, no! Most of their time engineering students are sitting in the lecture hall following the presentation in which the professor explains to them the course's content. In other words: the students try to understand and memorize what they have to know in order to pass the course's exam. This means in many cases that the teacher is showing them results of research activities without giving them the greater context and the research questions which were important at the beginning of the research process. Even if he would like to do so, in many cases there is simply not enough time for it. In "classical" lectures there is only little space and time for the students to understand the big picture of the subject and the inherent research process with its questions, research activities and result interpretation. Furthermore there is seldom enough open space for the students to work creatively with the course content and get in contact with real technical equipment of their future profession [1].

One possibility to change this fact is the use of laboratories in teaching and to implement experiential [2] and research based learning in the teaching and learning process [3]. To bring the students in contact with laboratory equipment means bringing them in contact with the technical equipment of their future profession and giving them the chance to develop central technical competences for the technical part of their future career.

1.1 Constraints and Solutions

A very important factor that hinders the use of laboratories by students in teaching is the cost of such equipment and the organizational aspect of co-locating students, equipment and supervisors. Especially small universities often face the situation that they either cannot afford all the laboratory equipment or that they cannot allow the students to use it because of the risk to damage it. That means in many cases that lab experiments, if the professor tries to integrate them into the lecture, are either only shown via video or that the faculty's staff shows the equipments during guided tours through the laboratory. This is a real dilemma for modern engineering education.

One way out if this dilemma - wanting the students to develop technical competences on the one hand and having them done experiments but not being able to use the equipment on the other hand - are tele-operated (called "remote") and virtual laboratories. With them the laboratory equipment can be used by different universities from different places or very risky experiments can be done completely virtually.

1.2 *PeTEX – Platform for Elearning and Telemetric Experimentation*

Important research on the use of remote laboratories in teaching engineering aspects was done by the universities from Dortmund (Germany), Palermo (Italy), and Stockholm (Sweden) within a project called PeTEX — Platform for e-learning and Teleoperative EXperimentation. The technical part of PeTEX was carried out at TU Dortmund University by the Institute of Forming Technology and Lightweight Construction (IUL, Prof. Tekkaya) and integrated in close co-operation with the Center for Higher Education (former Center for Research on Higher Education & Faculty Development, Prof. Wildt). Within this project comprehensive research in using remote laboratories in teaching was carried out. Therefore a network of three prototypes in the field of manufacturing technology was developed [4–14].

The work presented in this paper is based on the achievements in the PeTEX project, will use its technological infrastructure, and goes ahead to improve the concept by extending the possibilities to use the labs in higher engineering education courses. Fig. 1 shows how the following text is structured.

The overall context for our work is the implementation of research based and experiential learning by using laboratories in higher engineering education at universities. As explained above, our aims are that students get into contact with real technical equipment, understand the greater context of research, and gain technical competences for their future work. In this context we will proceed as follows:

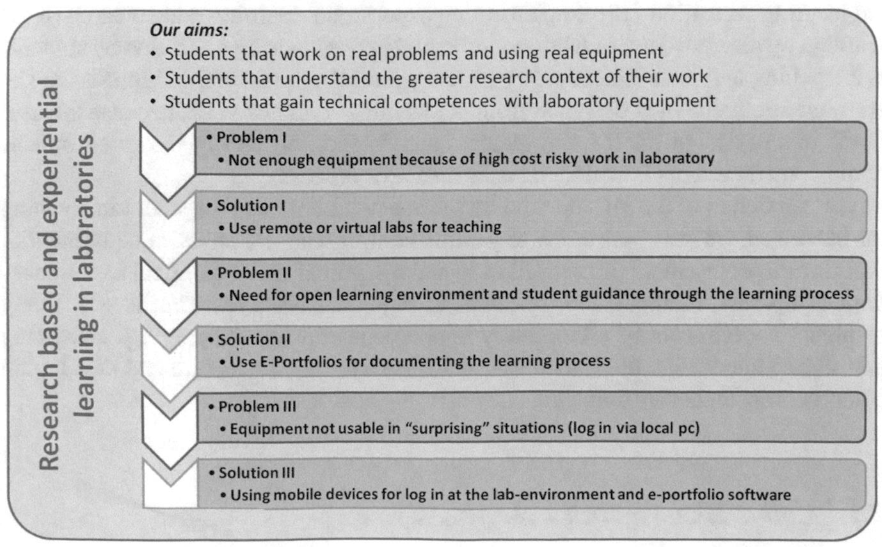

Fig. 1 Text structure for this paper

- Facing the problem that laboratory equipment is expensive and some experiments even can be dangerous for students, we will explain why and how remote laboratories should be integrated in teaching (Section 2).
- Up to today a weakness of such teaching approaches is the need for an open designed learning environment in which the students can act independently on the one hand and can be guided through the learning process on the other hand. We will also show the potential of remote laboratories to foster creativity. We will work out how e-portfolios can help in this context and how they can be used to document and reflect on the learning process (Section 3).
- As a final step we will change over to the topic mobile learning. In order to support the students' learning process as much as possible and giving them the opportunity to use the software virtually every time and from everywhere we will open the software for the use from mobile devices, such as smart phones and tablets. By showing different scenarios we explain how the use of mobile devices can support the learning process significantly and how they can help to promote creativity (Section 4).
- At the end of the paper we will explain shortly what our future steps will be in order to put our plan into action (Section 5).

2 Learning with Remote Laboratories

Learning through experiments in general has become a central part in modern higher engineering education [15]. Implementing experiential learning and research based learning by the active use of laboratories in higher engineering education by students is a teaching and learning concept which supports the constructivist approach. The learning arrangement is designed from the learners' point of view, because the user is the one to design his learning process and "walk" through the learning objects while constructing its knowledge inside an active process.

As mentioned in the introduction by the use of laboratory equipment in teaching the students have the opportunity to get into contact with the physical equipment of their future professional life as well as to make practical and theoretical experiences with equipment, methods and processes of empirical research. That is why doing technical experiments in a laboratory is an adequate way of applying, enhancing and testing knowledge the students have acquired during the lecture and developing central competences by doing so.

2.1 Kolb's Experiential Learning Cycle

The use of laboratories in teaching and learning environments can basically be traced back to understanding of learning explained by Kolb: "Learning is the process whereby knowledge is created through the transformation of experience" [2]. Kolb states that learning involves the acquisition of abstract concepts that can be applied flexibly in a range of situations. In Kolb's theory, the impetus for the development of new concepts is provided by new experiences. Kolb's concept of experience is defined in his experiential learning theory consisting of a four phase cycle in which the learner traces all the foundations of his learning process:

- Concrete Experience: A new experience of situation is faced, or a reinterpretation of an existing experience takes place.
- Reflective Observation: The new experience is analyzed, evaluated, and interpreted. Of particular importance are any inconsistencies between the experience and the understanding of it.
- Abstract Conceptualization: Reflection gives rise to a new idea, or a modification of an existing abstract concept.
- Active Experimentation: Transforming the new abstract concept into operation, the learner interacts to the world around him to check what emerges.

In his four-step learning cycle Kolb explains that at the beginning of each learning process there is a real learner's experience (step 1) which is followed by a reflective observation (step 2). From that point on the learner tries to conceptualize what he has experienced (step 3), starts to experiment actively (step 4) and generates new experiences. This is the start of a new cycle. With every loop - from the simple to the complex - the student enhances his experiences. Thus, the learning cycle transforms learning activities into a helix of experience-based knowledge, skills and competencies.

2.2 Research Based Learning

It is not by coincidence that the research process has quite similar steps, beginning with an experience or a question and ending with real experiments and new research results [16]. That is why research based or experiential teaching and learning in higher education is one adequate way of implementing learner centered teaching. In addition to that Herrington and Oliver worked out the importance of an authentic learning environment for a successful learning process [17]. This authentic learning environment can be offered by teaching and learning activities in laboratories. In these laboratories the students can face a real context and do real activities. By connecting the actions in laboratories in a next step to real problems - e.g. from current research or from the industry - the students are able to go the whole way from the question at the beginning of an experiment to the final use of the results and they can see the relevance of their work.

2.3 *Fostering Creativity*

Going the whole way of a research process corresponds to another important aspect of engineering education: fostering the students' creative potential. Industrial nations are facing tremendous problems. For example, new techniques to tackle climate change, new ideas on how to retain mobility of people or new concepts for energy production without fossil fuels are urgently needed. Engineers play an important role in addressing these challenges. Future prosperity and wealth will depend on their inventions and creativity. In higher education, students' creativity can be fostered in six different facets [18–22]:

1. Self-reflective learning — learners break out of their receptive habit and start to question any information given by the teacher. An internal dialogue takes place and knowledge becomes "constructed" rather than "adopted".
2. Independent learning — teachers stop to determine the way students learn. instead, students start for example to search for relevant literature on their own, they make their own decisions about structuring a text or they even find their own research questions and chose the adequate methods to answer it.
3. Curiosity and motivation — this aspect relates to all measures that contribute to increased motivation, for instance the linking of a theoretical question to a practical example or presenting.
4. Learning by doing — students learn by creating a sort of "product". Depending on the discipline, this might be a presentation, an interview, a questionnaire, a machine, a website, a computer program or similar. Students act like "real" researchers.
5. Multi-perspective thinking — learners overcome the thinking within the limits of their disciplines or prejudiced thinking. They learn to look automatically from different points of view on an issue and they use thinking methods that prevent their brain from being "structurally lazy".
6. Reach for original ideas — learners aim to get original, new ideas and prepare themselves to be as ready-to-receive as possible. Getting original ideas cannot be forced, but by the use of appropriate creative techniques and by creating a suitable environment (that allows making mistakes and expressing unconventional ideas without being laughed out or rejected); the reception of original ideas can be fostered.

A first, small study indicates that especially the facets 2, 5, and 6 might be fostered insufficiently in engineering education [23]. With an appropriate didactical scenario, learning in laboratories provides the potential to foster students' creativity in facet 6, which usually is hard to implement. If students are enabled to evolve their own research questions, to chose a suitable experimentation design and finally to perform the experiment, they will be able to develop some kind of "spirit of research" [23]. This spirit is one important premise for trying to get original ideas (facet 6).

2.4 Active Experimentation Using Tele-operated Equipment

Using remote and virtual laboratories in teaching gives a whole range of opportunities to implement experiential learning in the field of mechanical engineering following the path of research based learning [6]. One example in the context of manufacturing technology, namely forming technology, can be the use of such a special lab concept for material characterization. This could be organized in addition to a normal lecture or to enhance traditional hands-on labs during the phase students prepare themselves for the lab or when they would like to rework some of the test steps while writing the lab report.

Following the approach based on Kolb's experiential learning cycle, students can deal with basic concepts of metal forming during the lecture and test and see what they discussed in class by doing experiments on their own in order to create their own knowledge, using the remote experiential equipment. Another opportunity could be that students are given a real engineering problem related to material behavior. They are asked to work on this problem in small groups by planning and carrying out experiments using the tele-operated equipment. Finally they have to present what they've explored and what they would suggest to deal with the problem [6].

One important aspect in order to support this entire process and especially the step of "active experimentation" an appropriate level of clear interaction and feedback needs to be integrated to the tele-operated experimental setup. In the PeTEX project a complete experimental setup (Fig. 2) has been moved to a new level using innovative engineering design, modern concepts of automation, measurement technology and robotics as shown in Fig. 3.

All aspects have been brought together by developing a clear and interactive user interface providing real time feedback of the running experiment. In Fig. 4 the "window" to the uniaxial tensile test is shown.

When using the live camera stream (1), users can investigate the surrounding test apparatus, e.g. sensors or clamping devices. Afterwards, the learner initiates the preparation of the experiment (2), using the integrated 6-axes robot to select and check an appropriate specimen. To freely configure the experiment, relevant test parameters (3) can be filled in. When the test is started (4) the robot positions the specimen to the fully automated clamping device. The developed innovative concept of the fully automated clamping process and parallel measuring of relevant values is a different story.

Also during the test, a high level of interaction is provided to the user by manipulating the camera view or pausing and continuing the test. Pausing the test — which means the load is not further increased for that moment — causes a reaction by the material. This phenomenon is graphically visible in the real time diagram (6) and also in the real time test data at the header bar (5). Comparisons with prior test data are available by using the data base (7) and the graph (6). After the experiment is finished, learners are provided with data package including all the results for further analysis and investigation.

Fig. 2 Testing

Additionally the entire tele-operated experimental environment was made available with the learning content management system Moodle. There, we cared on the alignment of four, for us elementary, areas for this kind of socio-technical system. This socio-technical alignment for tele-operated laboratory learning consists of the

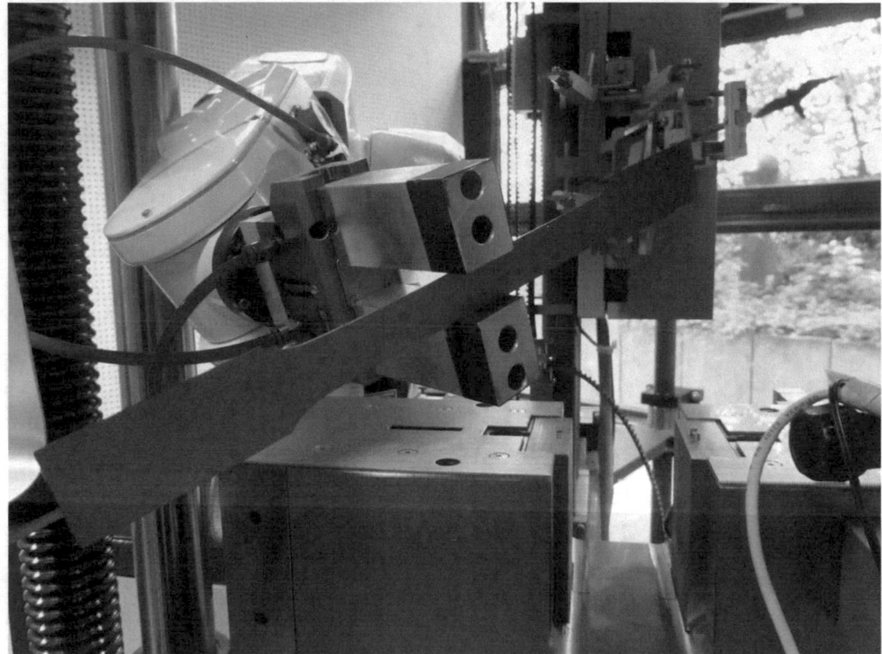

Fig. 3 Robot positioning a specimen

adjustment of the technical, didactical, media and social level. By the implementation into Moodle as shown in Fig. 5, this socio-technical alignment was put into a usable as well as flexible environment.

An often formulated challenge to such open designed learning concepts is that it turns out that a very sophisticated concept is needed to document and evaluate the learners' behavior and achievements during the learning process using the laboratory by the teacher. It is obvious that such a concept requires different systems for the instructor to accompany the learner through the learning process and - above all - to evaluate the achieved learning outcome. The following passages present the future thoughts concerning a concept for the learning process' documentation in context with the use of remote laboratories in combination with e-portfolios.

3 E-Portfolios and Their Use in Experiential Learning

In addition to the open learning concept which is supported by the use of laboratory equipment in general, the use of remote laboratories within the PeTEX project was designed for the usage by a very heterogeneous learner group composed of students and professionals [24, 25]. That means that the software for the learning process'

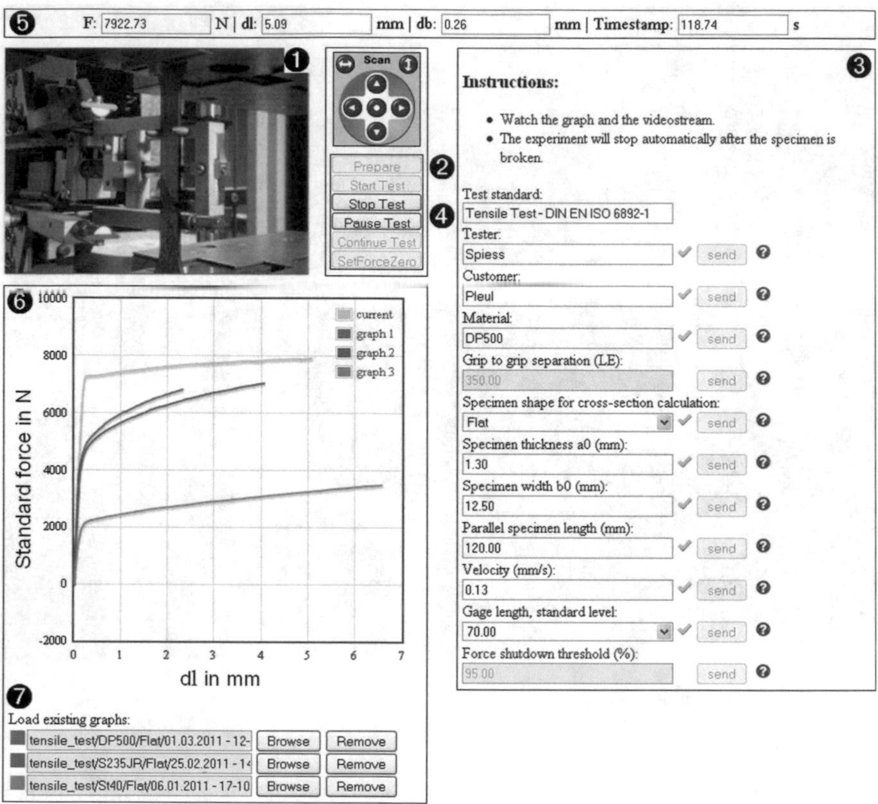

Fig. 4 Interface to the tele-operated experiment

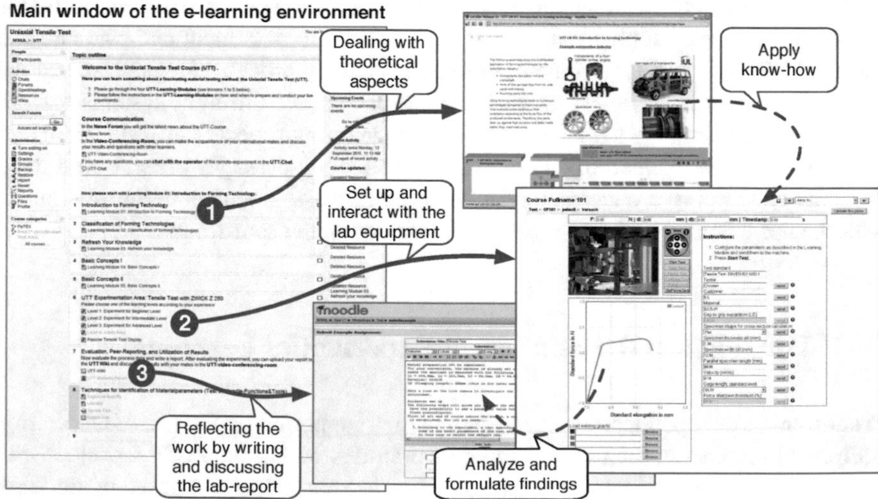

Fig. 5 Experiment environment integrated to Moodle

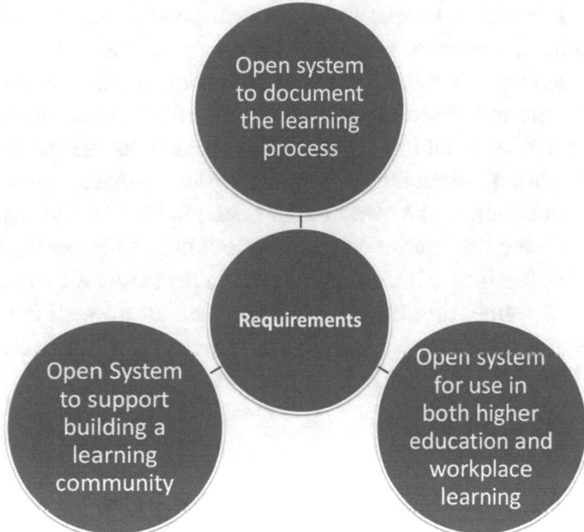

Fig. 6 Requirements for a software system documenting the learning process in remote laboratories

documentation as well must be designed very open in order to prevent system based barriers for different learner types. In this context it is important to keep in mind that the PeTEX system wants to bring higher education and the workplace together and wants to create an international learning community which is not limited to one institute. These 3 aspects – document the learning process, build a learner community and connect the students' work with their future professional work - are requirements the software has to address and accomplish (see Fig. 6).

Software which seems to be adequate and which is frequently discussed in similar contexts is the e-portfolio [26]. E-portfolios are based on the general idea of portfolios, which means to collect different kinds of documents in a folder in order to reflect on your learning process and present it to others [27]. E-portfolios support the same, but they are made online and provide the collection of different kinds of data like texts, tables, photos, videos, and audio [28]. E-portfolio software could be added technically to the Moodle environment - which is already used in the PeTEX context - very easily because an online e-portfolio application especially designed for Moodle already exists — it is called Mahoodle. In the following we will explain why e-portfolios fulfill the three main requirements in the new PeTEX context [7].

3.1 E-Portfolio as a Learning Process Documentation

The user — doesn't matter if in higher education or in professional further education - can arrange all the data he wants to document or show in different ways in order to create his own portfolio just like his personal page in any social network. He can present experiments and its results, show photos from the test set-up, explain his thoughts on the research, and so on. Add to this he can allow other users - other learners or teachers - to see his e-portfolio. By creating such an e-portfolio the learner can document his own learning and research process and start reflecting on the experiments he does during his research based learning process [29]. This reflection is an important aspect because he needs this step in his personal learning circle and especially for students the e-portfolio can give a kind of orientation or checkpoint in the own field of research [2, 29, 30]. By the same way the teacher can evaluate the learner's action by looking at the portfolio, too. Because other persons are able to see the collection in the portfolio it can be said that it is not only a way of documenting the learning process but as well it is a way of communicate it so that a collaborative learning process can be achieved. This leads to the next use of E-Portfolios in the PeTEX context.

3.2 E-Portfolio as a Learning Community Software

Taking the e-portfolio as software for documentation and evaluation is just one use of the system. A constructive enrichment in using the e-portfolios is the community building. Every author of an e-portfolio is able to allow other user to see all or just one part of his portfolio and he can see the others', too. That means that learners, who are doing experiments in the PeTEX system and filling their e-portfolios, can get into contact with each other via the portfolio software. They can see what others are especially interested in, start discussing about it, give comments and help each other in the case of a problem during the experiential learning process. By this way emerges a specialized community on remote laboratories within the PeTEX context. This possibility is also very important for creativity. In order to fostering students' creativity, it is strongly recommended to promote social interaction. With regard to fifth facet of creativity in higher education (multi-perspective thinking, which usually is, like facet 6, hard to implement in engineering education), they should get used to deal with different perspectives. This naturally requires exchange of information, discussions and cooperative problem solving. So learning in laboratories should be designed to promote social interaction as well. Therefore, performing remote experiments should go hand in hand with peer communication. The more heterogeneous the group of users is built up, the more different perspectives are introduced. A critical aspect of the any community is the amount of users. In this context the following use of E-Portfolios can help to build up a growing community with a heterogeneous group of users.

3.3 E-Portfolio as a Bridge Between the University and the Workplace

The PeTEX system is designed for the usage in higher education and in workplace learning. That means in a first step that both user groups can use the e-portfolios for the explained way of use. A further future thought is to use the e-portfolio as a livelong system to document the own competences from the university on and during the whole professional life. This should be explained by an example in three steps:

Step 1 - An engineering student starts working with the PeTEX system at the university. He uses the system in order to document his experiments. During his studies he does different experiments, compares them and collects all his research documentation in his e-portfolio in order to scientifically describe a certain material behavior which was observed (e.g. while pausing the test for a couple of seconds), and reflects on his own way of learning. The teacher is able to evaluate his learning behavior. This can be seen as the main use of e-portfolios at university.

Step 2 - Because the PeTEX system as well addresses workplace learning the e-portfolios can be seen as a bridge from university to professional life. Depending on the concrete use of the e-portfolios by the student he can take his portfolios to present himself to potential employers. They can see what the students did in this field of his studies and if he fits to the company's needs. In this context the e-portfolios can support the process of applying for a job.

Step 3 - Once the former student - now employee at a company - starts working at a company he must not stop working with his portfolio. He still can work on his collection by documenting new experiments as well as gained knowledge and competences in his job. By doing so, the employee doesn't stop reflecting on his learning process. His e-portfolio grows and with every year it becomes a better presentation of his professional life and his competences. Especially the last aspect works perfectly together with the advantages of the PeTEX system, that small and medium sized companies use the system to enhance their technological skills by doing research with the PeTEX hardware. In addition to that they can use the e-portfolios to implement a system for the documentation and measurement of the employees' skills and competences. This could be supported by the lifelong use of e-portfolios.

Summing up all these aspects it can be said, that the use of e-portfolios in the PeTEX context can support the idea of experiential and research based learning even if there are a couple of challenges to meet [29]. The portfolios can be used to document and present the research and learning process, to build up a special focused learning community, and to bring university learning and workplace learning together. Above that, working with e-portfolios fosters the fourth aspect of creativity in higher education: to create something. An e-portfolio is a type of a "product". While working on their e-portfolios, students anticipate that their "product" will be valued by others. Therefore, they will seek to make them more attractive for others, for example by bringing in new aspects or by considering that their ideas must be understood by others as well, which requires a non-contradictory and simple presentation. It will be a future task to integrate the e-portfolio work in the PeTEX learning concept and evaluate the performance.

4 Scenario of using Mobile Devices in Combination with E-Portfolios

Another frequently mentioned new concept in context with higher education is mobile learning. Mobile learning means the use of mobile devices - like cell phones, smart phones or tablet-computers - in the learning process [31]. Only one of the advantages of mobile learning is that not planned time periods can be used for learning and that the learning process can be initiated virtually everywhere [31]. In our context we will focus on the fact that the user carries his mobile device normally at every time and because of that he can use it frequently in order to work with the portfolio software and laboratory equipment.

In addition to that using mobile devices can support the creativity process, because new ideas mainly come spontaneously and having the mobile device with you makes it possibly to at least put down a note with an idea and work on it later or work on it immediately as we will explain in the scenarios. Bringing e-portfolios on tablet-devices for example could be an opportunity to combine the concepts presented in this text with mobile learning. In the following we will present different scenarios how the use of mobile devices can enrich the concept of remote laboratories in higher engineering education. These scenarios differ mainly in terms of individual or collaborative learning processes and self-directed or teacher directed learning processes. These four aspects in different combinations lead to the scenarios, so that it becomes obvious how flexible the use of e-portfolios is in our learning concept. As we wanted an open learning environment for working with the laboratory equipment remotely it seems that the e-portfolio software fulfills this requirement perfectly. Fig. 7 shows the four scenarios that are explained in the following.

4.1 Possible Scenarios

Scenario 1 “Using the software in creative moments” - A first scenario could be that a student is thinking about his experiments while sitting at home and watching TV or while he went out with his friends. He is really struggled by his research work, thinks about his parameters, his results and why his experiments offer the results they showed up. Suddenly he has an idea on a hypothesis and wants to check it by rereading his last experiments in the portfolio or doing a new experiment. Because he can use the software for connecting with the experiential environment by using his tablet computer he doesn’t need to wait until the next day for doing the experiment at the university but he can stay where he is and even can stay sitting on the sofa for

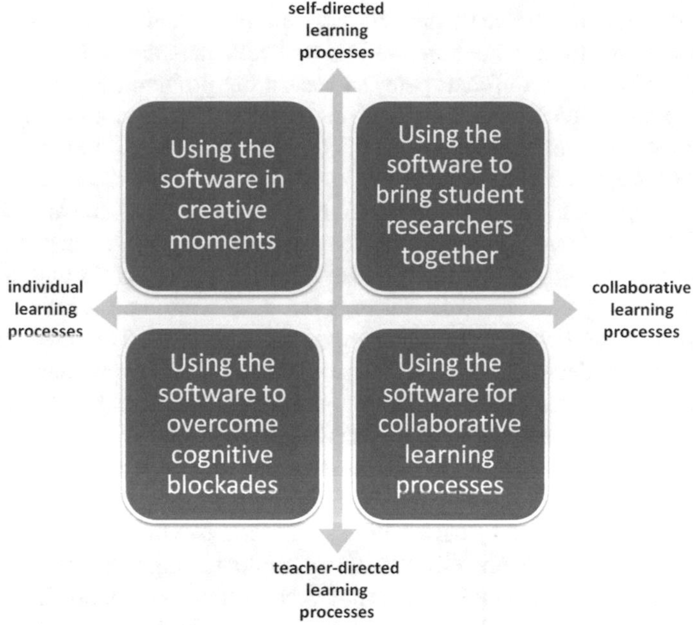

Fig. 7 Four scenarios for using mobile devices in combination with remote laboratories

checking his hypothesis. The new result he can immediately put in his portfolios so that he documents his new step within his research process. Using a simulation in a virtual laboratory instead of the remote experiment can be a method to (pre)-check the hypothesis first and then carry out the real experiment remotely.

Scenario 2 "Using the software to bring student researchers together" - With his mobile device (it doesn't matter if smart-phone or tablet PC) the student can access his personal e-portfolio in which he documents his experiments and his personal competence development in this sector from wherever he wants to. Sitting in the train on the way to or on the way back from the university he could skip through his experiments and look what he found out as different results. At the same moment another student looks on the first student's portfolio. He finds out that his own research had quite the same results even if he used different parameters or - even more challenging - he used the same parameters and material but had different results. Knowing this he contacts the first student via a chat or e-mail, as well using his mobile device, and they can communicate about their common results at this very moment and work together on future experiments.

Scenario 3 "Using the software to overcome cognitive blockades" - A third scenario could foster the students' ability to think in different perspectives about their questions: After performing an experiment that was given to him by the teacher, the student possibly doesn't know why he didn't get the expected results or doesn't know how to interpret the results. He asks himself why the experiment didn't work as it should have, but he can't find the answer. While writing his e-portfolio as documentation for the teacher's evaluation, he could start the "creative-help-app", which helps him to use different perspectives on the problem: Firstly, he is asked to make a (mental) headstand following the question "What else could I do to get the wrong results from experimenting?" If that doesn't help to find the answer, he secondly will be asked to describe his experiential design and his assumptions in a way that a ten-year-old could understand it. If those methods, which are rather close to the problem, still can't help him, the "creative-help-app" will suggest a force-fit technique by showing a picture that doesn't have anything to do with a problem (for example a lady beetle, a daisy chain, a bottle of wine) and asking the student to find relationships between the picture and his experiment. This method helps to leave the well-trodden paths and forces the students to look from completely other perspectives on their problem. It often results in very unconventional or provoking ideas, but rethinking the obviously unsuitable solutions sometimes leads to the one really good idea, that would not have come to mind without making the detour.

Scenario 4 "Using the software for collaborative learning processes" - The fourth scenario could be provoked by the teacher, too. He can give the students - as a kind of homework - to check an explicit hypothesis by implementing adequate experiments. Using the e-portfolios the students can stay in contact without being forces to meet at the university and in combination with mobile devices they can be virtually be anywhere going through a collaborative learning process. Because the e-portfolios software has a connection to the experiential environment the students as a group can do the experiment and discuss the results with regards to the homework's hypothesis in one go and without changing the learning environment.

5 Conclusion, Discussion, and Future Plans

With this paper we explained why the use of laboratories as a place for conducting experiments is important for modern engineering education. The central idea is to engage the students in teaching and learning environments which are connected closely to their future working environment. In addition to that, the aspect of student centered learning environments is very important in higher education because it is central having the students do the things they have to learn by themselves. This is the only way for the students to develop central competences and reach a high level of learning outcomes. Furthermore, we showed the potential of our approach to foster the learners' creativity. See in the following the central advantages of the presented concept:

- As the equipment of laboratories is either very expensive to have it at every university or not always available for the students the use of remote or virtual laboratories is good opportunity to face this dilemma.

- Using the equipment virtually in simulations or remotely from wherever they can help the students to do experiments just as a pre-check on personal hypothesis or even when they are not physically in the laboratory.
- Learning processes that are achieved by the usage of the laboratories can be documented in e-portfolios.
- These portfolios are a good opportunity in order to document the experiments for the personal use or for the evaluation by an instructor. Looking at the portfolios the instructor can either see what kind of experiments the students have done and what they learned from it.
- If the portfolios are not kept hidden from other students but are open for other users to look at them and comment on the achievements, there is the opportunity for a community to evolve working together on the experiments. The e-portfolio software should be made accessible from mobile devices, too. This opens the door to mobile learning, which means that the learning process is not bound to any location. From virtually everywhere and at every time the user can work on their portfolios and communicate with each other.
- With the possibility to promote multi-perspective thinking and a "spirit of research", vital facets of creativity in higher education can be fostered.

As this is a work in progress paper we primarily worked out which opportunities and advantages the use of e-portfolios and mobile devices can add to the learning process. Central research and work on remote laboratories already has been done in the PeTEX project so that the laboratories exist and are usable remotely. The step for the coming year will be to implement the e-portfolio software in the system and make it accessible from mobile devices. Once this has been achieved, first tests with students can be carried out and the system can be evaluated and improved.

As the reader may have recognized we talked only very little about the technical implementation for the ideas. This is not by coincidence or because we forgot to talk about. At this point we are at the beginning of our studies about the use of remote laboratories in combination with e-portfolios and mobile devices. Of course technical problems will arise during the implementation and they may be even quite difficult to solve. But at the end they stay just technical and during the PeTEX project it became obvious that every technical problem will be solved sooner or later. At this moment we concentrate on the concept for our future work and we want to concentrate on the didactic background for this. The question if any of the explained will help the students to learn more and gain real competences during their studies is and will stay our main focus. Not everything which is possible from the technical point of view or even can be technically designed does make sense for the learning process and higher education. So we look at teaching and learning from the students' perspective. That leads us to our concepts explained above first, without asking in detail about how everything can be implemented yet. We want to support better engineering education. That is what our focus lies on.

References

1. H.G. Bruchmüller and A. Haug, "Labordidaktik für Hochschulen - Eine Einführung zum Praxisorientierten Projekt-Labor", Alsbach, Leuchtturm-Verlag, Schriftenreihe report - Band 40, 2001
2. D.A. Kolb, "Experiential learning. Experience as the source of learning and development", Englewood Cliffs, N.J.: Prentice-Hall, 1984
3. L. Huber, J. Hellmer, and F. Schneider, "Forschendes Lernen im Studium - Aktuelle Konzepte und Erfahrungen", Bielefeld, Universitätsverlag Webler, 2009
4. C. Pleul, C. Terkowsky, I. Jahnke, and A. E. Tekkaya, "Platform for e-learning and tele-operative experimentation (PeTEX) - Hollistically integrated laboratory experiments for manufacturing technology in Engineering Education", Proceedings of SEFE Annual Conference, 1st World Engineering Education Flash Week. Lissabon, Portugal, Bernardino, J. and Quadrado, J.C., 2011. 578–585
5. C. Terkowsky, C. Pleul, I. Jahnke, and A. E. Tekkaya, "Tele-operated Laboratories for Production Engineering Education - Platform for E-Learning and Telemetric Experimentation (PeTEX)", International Journal of Online Engineering (iJOE) 7, Special Issue EDUCON 2011, 2011, 37–43
6. C. Pleul, C. Terkowsky, I. Jahnke, and A. E. Tekkaya, "Tele-operated laboratory experiments in engineering education - The uniaxial tensile test for material characterization in forming technology", in "Using Remote Labs in Education. Two Little Ducks in Remote Experimentation", Ed. by Garcıa-Zubıa, J. and Alves, R. G. Deusto Publicaciones, 2011, Chap. 16, 323–348.
7. C. Terkowsky, I. Jahnke, C. Pleul, D. May, T. Jungmann, and A. E. Tekkaya, "PeTEX@Work. Designing Online Engineering Education", in "CSCL at Work - a conceptual framework", ed. by S. P. Goggins, I. Jahnke, and V. Wulf, Springer, in print
8. C. Terkowsky, I. Jahnke, C. Pleul, R. Licari, P. Johannssen, G. Buffa, M. Heiner, L. Fratini, E. Lo Valvo, M. Nicolescu, J. Wildt & A. Erman Tekkaya, "Developing Tele-Operated Laboratories for Manufacturing Engineering Education. Platform for E-Learning and Telemetric Experimentation (PeTEX)", In International Journal of Online Engineering (iJOE). IAOE, Vienna, Vol. 6 Special Issue: REV2010, 2010, 60–70.
9. C. Terkowsky, I. Jahnke, C. Pleul, R. Licari, P. Johannssen, G. Buffa, M. Heiner, L. Fratini, E. Lo Valvo, M. Nicolescu, J. Wildt & A. Erman Tekkaya "Developing Tele-Operated Laboratories for Manufacturing Engineering Education. Platform for E-Learning and Telemetric Experimentation (PeTEX)", In Auer, M.E. & Karlsson, G. (Eds.): REV 2010 International Conference on Remote Engineering and Virtual Instrumentation, Stockholm, Sweden, Conference Proceedings. IAOE, Vienna, 2010, 97–107.
10. C. Terkowsky, I. Jahnke, C. Pleul & A. Erman Tekkaya, "Platform for E-Learning and Telemetric Experimentation (PeTEX) - Tele-Operated Laboratories for Production Engineering Education", In Auer, M.E. , Al-Zoubi, Y & Tovar, E. (Eds.), "Proceedings of the 2011 IEEE Global Engineering Education Conference (EDUCON) - Learning Environments and Ecosystems in Engineering Education". IAOE, Vienna, 2011, 491–497.
11. I. Jahnke, C. Terkowsky, C. Pleul & A. Erman Tekkaya , "Online Learning with Remote-Configured Experiments", In Kerres, M., Ojstersek, N., Schroeder, U. & Hoppe, U. (Eds.): Interaktive Kulturen, DeLFI 2010 - 8. Tagung der Fachgruppe E-Learning der Gesellschaft für Informatik e.V., 2010, 265–277.
12. C. Terkowsky, C. Pleul, I. Jahnke & A. Erman Tekkaya,"PeTEX: Platform for eLearning and Telemetric Experimentation", In Bach, U., Jungmann., T. & Müller, K. (Eds.), "Praxiseinblicke Forschendes Lernen, TeachING.LearnING.EU", Aachen, Dortmund, Bochum, 2011, 28–31.
13. C. Pleul, I. Jahnke, C. Terkowsky, U. Dirksen, M. Heiner, J. Wildt & A. Erman Tekkaya, "Experimental E-Learning - Insights from the European Project PeTEX", 15th International Conference on Technology Supported Learning & Training, Book of Abstracts. ICWE GmbH, Berlin, 2009, 47–50.

14. I. Jahnke, C. Terkowsky, C. Burkhardt, U. Dirksen, M. Heiner, J. Wildt & A. Erman Tekkaya, "Forschendes E-Learning", In "Journal Hochschuldidaktik" 20. Jg. Nr. 2. Oktober 2009, 2009, 30–32.
15. L.D. Feisel, and A.J. Rosa,. "The Role of the Laboratory in Undergraduate Engineering Education", Journal of Engineering Education, 2005, 121–130
16. J. Wildt, "Forschendes Lernen: Lernen im" Format "der Forschung", Journal Hochschuldidaktik, Jg. 20 Heft 2, 2009, 4–7
17. J. Herrington and R. Oliver, "An instructional design framework for authentic learning environments", Educational Technology Research and Development, 48 (3), 2000, 23–48
18. I. Jahnke, I. and T. Haertel, "Kreativitätsförderung in Hochschulen - ein Rahmenkonzept", Das Hochschulwesen, 58, 2010, 88–96
19. I. Jahnke, I., T. Haertel, V. Mattik and K. Lettow, "Was ist eine kreative Leistung Studierender? Medien-gestützte kreativitätsförderliche Lehrbeispiele" In "HDI2010 - Tagungsband der 4. Fachtagung zur" Hochschuldidaktik Informatik"", Ed. by Engbring, D.; Keil, R.; Magenheim, J. and Selke, H., Universitätsverlag Potsdam, 2010, 87–92
20. T. Haertel and I. Jahnke, "Kreativitätsförderung in der Hochschullehre: ein 6-Stufen-Modell für alle Fächer?!" In "Fachbezogene und fachübergreifende Hochschuldidaktik. Blickpunkt Hochschuldidaktik, Band 121", Ed. by Jahnke, I. and Wildt, J., W. Bertelsmann Verl., 2011, 135–146
21. T. Haertel, and I. Jahnke, "Wie kommt die Kreativitätsförderung in die Hochschullehre?" Zeitschrift für Hochschulentwicklung, (6) 3, 2011, 238–245
22. I. Jahnke, T. Haertel and M. Winkler, "Sechs Facetten der Kreativitätsförderung in der Lehre - empirische Erkenntnisse" In "Der Bologna-Prozess aus Sicht der Hochschulforschung, Analysen und Impulse für die Praxis", Ed. by Nickel, S., CHE gemeinnütziges Centrum für Hochschulentwicklung, 2011, 138–152
23. T. Haertel, and C. Terkowsky, "Where have all the inventors gone? The lack of spirit of research in engineering education" In "Proceedings of the 2012 Conference on Modern Materials, Technics and Technologies in Mechanical Engineering" The Ministry of Higher and Secondary Specialized Education (MHSSE) of the Republic of Uzbekistan, Andijan Area, Andijan City, Uzbekistan, 2012, 507–512
24. C. Terkowsky, I. Jahnke, C. Pleul & A. Erman Tekkaya, "Platform for eLearning and Telemetric Experimentation. A Framework for Community-based Learning in the Workplace", Position Paper for the Workshop CSCL at Work, 16th ACM international conference on Supporting group work. Sanibel, Florida, 2010
25. C. Pleul, Claudius T. & I. Jahnke, "PeTEX - platform for e-learning and telemetric experimentation: a holistic approach for tele-operated live experiments in production engineering. GROUP'10", Proceedings of the 16th ACM international conference on Supporting group work. Sanibel, Florida, 2010, 325–326.
26. K. Himpsl and P. Baumgartner, "Evaluation von E-Portfolio-Software", Teil III des BMWF-Abschlussberichts, Einsatz von E-Portfolios an (österreichischen) Hochschulen. www.bildungstechnologie.net. 2009. http://www.bildungstechnologie.net/Members/khim/dokumente/himpsl_baumgartner_evaluation_eportfolio_software_abschlussbericht.pdf/download (accessed 14th July, 2011)
27. A.C. Breuer, "Das Portfolio im Unterricht - Theorie und Praxis im Spiegel des Konstruktivismus", Münster, Waxman Verlag GmbH, 2009
28. R. Reichert, "Das E-Portfolio - Eine mediale Technologie zur Herstellung von Kontrolle und Selbstkontrolle", in "Kontrolle und Selbstkontrolle - Zur Ambivalenz von E-Portfolios in Bildungsprozessen", Ed. by Thorsten Meyer et al. Wiesbaden, VS Verlag für Sozialwissenschaften - Springer Fachmedien Wiesbaden GmbH, 2011
29. G. Reinmann and S. Sippel "Königsweg oder Sackgasse? - E-Portfolios für das forschende Lernen", in "Kontrolle und Selbstkontrolle - Zur Ambivalenz von E-Portfolios in Bildungsprozessen", ed. by Thorsten Meyer et al. Wiesbaden: VS Verlag für Sozialwissenschaften - Springer Fachmedien Wiesbaden GmbH, 2011
30. H. Blom, "Der Dozent als Coach", Neuwied/Kriftel, Luchterhand, 2000
31. A. Kukulska-Hulme and J. Traxler, "Mobile learning: A handbook for educators and trainers", Milton Park and New York, Routlegde and Taylor & Francis Group, 2005

Virtual Production Intelligence—A Contribution to the Digital Factory

Rudolf Reinhard, Christian Büscher, Tobias Meisen, Daniel Schilberg and Sabina Jeschke

Abstract The usage of simulation applications for the planning and the designing of processes in many fields of production technology facilitated the formation of large data pools. With the help of these data pools, the simulated processes can be analyzed with regard to different objective criteria. The considered use cases have their origin in questions arising in various fields of production technology, e.g. manufacturing procedures to the logistics of production plants. The deployed simulation applications commonly focus on the object of investigation. However, simulating and analyzing a process necessitates the usage of various applications, which requires the interchange of data between these applications. The problem of data interchange can be solved by using either a uniform data format or an integration system. Both of these approaches have in common that they store the data, which are interchanged between the deployed applications. The data's storage is necessary with regard to their analysis, which, in turn, is required to obtain an added value of the interchange of data between various applications that is e.g. the determining of optimization potentials. The examination of material flows within a production plant might serve as an example of analyzing gathered data from an appropriate simulated process to determine, for instance, bottle necks in these material flows. The efforts undertaken to support such analysis tools for simulated processes within the field of production engineering are still at the initial stage. A new and contrasting way of implementing the analyses aforementioned consists in focusing on concepts and methods belonging to the subject area of Business Intelligence, which address the gathering of information taken from company processes in order to gain knowledge about these. This paper focusses on the approach mentioned above. With the help of a concrete use case taken from the field of factory planning, requirements on a data-based support for the analysis of the considered planning process are formulated. In a further step, a design for the realization of these requirements is presented. Furthermore, expected challenges are pointed out and discussed.

R. Reinhard (✉) · C. Büscher · T. Meisen · D. Schilberg · S. Jeschke
IMA/ZLW & IfU, RWTH Aachen University, Dennewartstr. 27, 52068 Aachen, Germany
e-mail: rudolf.reinhard@ima-zlw-ifu.rwth-aachen.de

Originally published in "Proceedings of The World Congress on Engineering and Computer Science 2013, WCECS 2013", © Newswood Limited 2013.
Reprint by Springer International Publishing AG 2016,
DOI 10.1007/978-3-319-46916-4_6

Keywords Application integration · Data analysis · Decision support · Digital factory

1 Introduction

Due to the global price competition, the increasing ranges of varieties and customer requirements as well as resulting shorter product lifecycles, production companies in high wage countries face a growing complexity within their production circumstances [1]. Methods and concepts which are used in order to overcome this complexity often fail to address the whole production process. Therefore, solutions are needed, which allow a holistic and integrated view of the relevant processes in order to achieve an increasing product quality, production efficiency and production performance [2].

Within the last years, the usage of simulation applications in the field of production technology became a measure with a growing significance to overcome the complexity mentioned above. Because of the increasing computing performance concerning speed and storage, these simulation applications changed the way of carrying out planning and preparing activities within production. So, instead of engineering a concrete prototype at an early stage of product design, a digital model of this prototype is drafted containing a description of its essential characteristics. In a further step, this model is passed to a simulation application to predict the prototype's characteristics that may have changed after having passed the manufacturing step. The usage of these digital models is subsumed under the notion of virtual production, which "is the simulated networked planning and control of production processes with the aid of digital models. It serves to optimize production systems and allows a flexible adaptation of the process design prior to prototype realization" [3, 4].

Nowadays, various simulation applications exist within the field of virtual production, which allow for the simulated execution of manufacturing processes like heating and rolling. Herein, different file formats and file structures were independently developed to describe digital models. Through this, the simulation of single aspects of production can be examined more easily. Nevertheless, the integrative simulation of complex production processes cannot be executed without large costs and time efforts as the interoperability between heterogeneous simulation applications is commonly not given.

One approach to overcome this challenge is the creation of a new standardized file format, which supports the representation of all considered digital models. However, regarding the variety of possible processes, such an approach results in the creation of a complex, standardized file format. Its comprehension, maintenance and usage, again, require large costs and time efforts. Furthermore, necessary adaptations and extensions take a lot of time until their implementation is finished [5, 6].

Another approach considers the usage of concepts from data and application integration avoiding the definition of a uniform standard. Within this approach, the interoperability between the simulation applications is guaranteed by mapping the aspects of different data formats and structures onto a so called canonical data model

[7, 8]. Newer approaches extend these concepts with regard to semantic technologies by implementing intelligent behavior into such an integrative system. This approach is called Adaptive Application and Data Integration [9, 10].

As a consequence, new possibilities concerning the simulation of whole production processes emerge, which allow the examination of different characteristics of the simulated process, e.g. material or machine behavior. With regard to the analysis of the integrated processes, new questions arise as methods for the analysis of the material or machine behavior mentioned above cannot be transferred to the analysis of the corresponding integrated process. A further challenge comes up as soon as suitable user interfaces are added, which are necessary for the handling of the integrated process and its traceability.

Similar questions emerge whilst the analysis of enterprise data. Applications giving answer to such questions are subsumed under the notion of Business Intelligence. These applications have in common that they identify, aggregate, extract and analyze data within enterprise applications [11, 12].

In this paper, an integrative concept is introduced that transfers the nature of these solutions to the field of application of production engineering. It contains the integration, the analysis and the visualization of data, which have been aggregated along simulated process chains within production engineering. In respect to the concept's application domain and its aim to contribute to the gaining of knowledge about the examined processes, it is called Virtual Production Intelligence. In order to illustrate this approach, in Section 2, a use case scenario from factory planning is taken into consideration. In Section 3, requirements are listed, which arise from the use case scenario described in Section 2. The realization of these requirements makes it necessary to create new concepts, which are presented in Section 4. Section 5 contains a description of expected challenges that come up while realizing the requirements defined in Section 3. This paper concludes with a summary and an outlook on a further use case.

2 Use Case Factory Planning

The notion of virtual production comprises the planning of processes that are characteristic for factory planning. In this chapter, a scenario taken from the field of factory planning is introduced, which follows the concept of Condition Based Factory Planning (CBFP). This concept facilitates an efficient planning process without restricting its flexibility by making use of standardized planning modules [13]. With the help of this scenario, it is pointed out which data are aggregated and which questions are raised concerning the integration, analysis and visualization of data within the planning process of a factory aiming at the support of this planning process. In the following, after having illustrated the use case, the examination of the planning process aforementioned is subsumed under the notion of Virtual Production Intelligence.

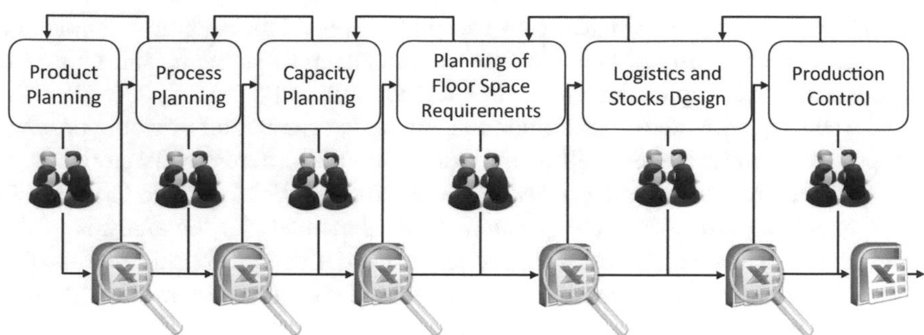

Fig. 1 Exemplary non-linear factory planning process following CBFP

The concept of CBFP is employed to analyze factory planning scenarios with the aim of facilitating the factory planning process by decomposing it into single modules [13]. These modules address various aspects within factory layouting like material flow or logistics. Because of the modular procedure, the characteristical non-linearity of planning processes can be mapped onto each process' modeling.

Within various workshops, requirements concerning the future factory are gathered in collaboration with the customer. For this purpose, table calculation and simulation applications are employed. Subsequent to these workshops, the gathered data are evaluated by one of the factory planners, who participated in the workshops, and suitable planning modules belonging to CBFP are identified. Thereby, different scenarios of the factory's workload are examined to guarantee the future factory's flexibility. Figure 1 illustrates this procedure focusing on the exemplary planning modules Product Planning, Process Planning, Capacity Planning, Planning of Floor Space Requirements, Logistics and Stock Design as well as Production Control.

Although the planning process is supported by the planning modules from CBFP, a significant disadvantage remains as the procedure is vulnerable concerning input errors committed by the user. Furthermore, the automated analysis of gathered data is complicated, due to the lack of a uniform data model.

The support of the planning process is based on a data model, which fulfills the planning modules' requirements. Thereby, the collection of data is performed by making use of familiar applications. Each analytical step, e.g. the calculation of different scenarios of the factory's workload, is computed on a dedicated server by the factory planer during the evaluation phase between two workshops. One advantage of this procedure is the coherent data handling during the entire planning process. Because of this coherent data handling, the design output can be made available for uninvolved and, in particular, new employees as well as for the executives after having finished a planning process. An interactive visualization allows for an explorative analysis of the simulation application's output. Such an integrative solution facilitates the location of possibilities for optimization within the examined processes. As an organizational consequence and a lasting effect, the experiences made during the implementation of optimization processes can be employed with regard to the

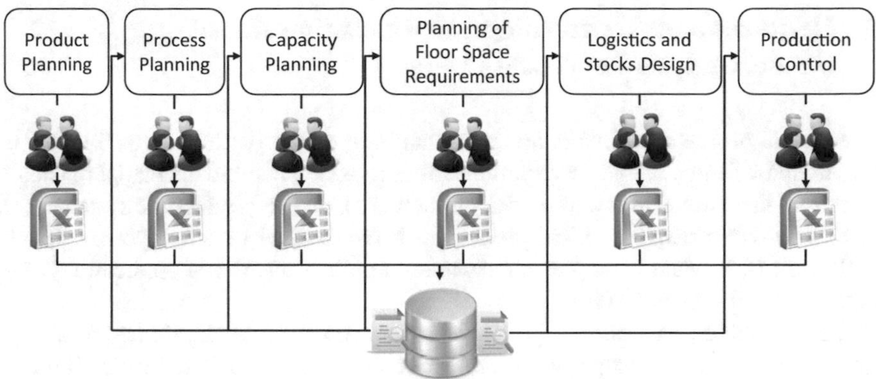

Fig. 2 Exemplary factory planning process following CBFP, supported by an integrative solution

composition of best practices for planning projects within the planning company. The implementation of the integrated solution is depicted in Fig. 2.

As a consequence of the integrated consideration, different fields of employment emerge, for instance the immersive visualization of a digital model of the future factory within a Cave Automatic Virtual Environment (CAVE). This immersive visualization provides the possibility of inspecting the future factory virtually. In doing so, the customer gets the option to feedback the current state of the factory's planning process, which in turn leads to an improved satisfaction regarding the planning's outcome. The creation of the interoperability between the involved applications, on the one hand, and the integrative data handling and analysis, on the other hand, results in the provision of such a solution without generating larger costs and time efforts.

Furthermore, the usage of this solution allows for the examination of different outcomes within a planning problem, e.g. the distribution of machines within a hall and the corresponding planning aspects like logistics or staff security, on a homogeneous data basis. Such a solution might also be adapted to the requirements of another field of application, like marketing, if the attention is directed to the presentation of the planned area rather than to the computing accuracy. Another field of application is the training effort for new employees. Its reduction is an important aim due to high wage costs. In this context, different views adapted to relevant questions of new employees can be used, which comprise the complete and detailed presentation of the current project as well as of past projects.

The scenario described above, which includes methods from factory planning, illustrates how the support of the planning process can be designed. In this scenario, the integration, the analysis and the visualization of data gathered during the planning process is realized. The following chapter comprises a description of requirements that need to be fulfilled when dealing with a system that provides the aforementioned support of the planning process.

3 Requirements on an Integrative Solution Based on an Analysis for Process Data

The virtual production aims at an entire mapping of the product as well as of the production within a model for experimental purposes. Thereby, the mapping should comprise the whole lifecycle of the product and of the production system [14]. Within an enterprise, the virtual production is established by employees, software tools such as Product-Lifecycle-Management applications (PLM applications) and organizational processes [14].

The demanded possibilities for analysis serve the purpose of gaining knowledge by examining already completed planning processes. The term "intelligence" is commonly used to describe activities that are linked to those analyses. Software tools, which support the analysis and the interpretation of business data, are subsumed under the term "Business Intelligence".

As this term can be defined in different ways, at this point, the basic idea of "Business Intelligence" will be pointed out [15–17]. A common feature of the definitions referred to consists in the aggregation of relevant data from different data sources, which are applications within a company, into a central data storage. The transmission of data taken from the application data bases into this central data storage is realized by the well-known Extracting, Transforming and Loading process (ETL). Subsequently, the data are arranged in more dimensional data cubes following a logical order. In doing so, a company's IT is divided into two different categories:

- Operational: This category contains applications customized for e.g. the accounting department, the purchasing department or the production department of a company.
- Analytical: In this case, the category contains applications for the analysis of data arising from the applications mentioned in the operational category.

The fact that operational processes are not influenced by analytical processes can be regarded as an advantage of this division.

Requirements for a system that supports the described planning process in Section 2, in particular the data and application integration, and which additionally follows the idea of Business Intelligence can be subsumed as below:

- Interoperability: Facilitating the interoperability between applications in use.
- Analytical abilities: Systematic analyses providing the recognition of potentials towards optimization and delivering fundamental facts for decision support.
- Alternative representation models: Taylor made visualization for the addressed target group, which provides appropriate analysis facilities based on a uniform data model.

In order to find a solution, which fulfills the requirements mentioned above, a concept formation is needed that addresses the field of application, that is, in this case, the virtual production already mentioned above, as well as the aim of gaining knowledge. This aim is also addressed by the term "Intelligence". The concept formation will take into account approaches, methods and concepts. These will contribute to the achievement of objectives concerning the gaining of knowledge with regard to the processes executed within the considered field of application, which is the virtual production. Therefore the concept formation results in the notion of Virtual Production Intelligence.

This notion will be described in the following chapter.

4 Objectives of the Virtual Production Intelligence

The Virtual Production Intelligence (VPI) is a holistic, integrated concept that is used for the collaborative planning of core processes in the fields of technology (material/machines), product, factory and production planning as well as for the monitoring and control of production and product development:

- Holistic: Addressing all of the product development's sub processes.
- Integrated: Supporting the usage and the combination of already existent approaches instead of creating new and further standards.
- Collaborative: Considering roles, which are part of the planning process, as well as their communication and delivery processes.

The VPI aims at contributing to the realization of the digital factory, which is defined as follows:

Digital factory is the generic term for a comprehensive network of digital models, methods and tools – including simulation and 3D visualization – integrated by a continuous data management system. Its aim is the holistic planning, evaluation and ongoing improvement of all the main structures, processes and resources of the real factory in conjunction with the product [4].

The concept is evaluated by the technical implementation of a web-platform, which will serve as a support tool. This platform will serve for planning and support concerns by providing an integrated and explorative analysis in various fields of application. Figure 3 illustrates how the platform is used in these fields of application by various user groups. Within the figure, the use case "factory planning" is addressed as well other use cases, which will be described in future publications.

5 Challenges

The integrative approach of the VPI concept facilitates the use of various applications, which can, for example, be deployed whilst a planning process without requiring a uniform data format. At the beginning of the use case scenario already described above, different utilization rates of factory capacity are defined. As a consequence, further requirements arise, which concern, for example, the future factory's layout,

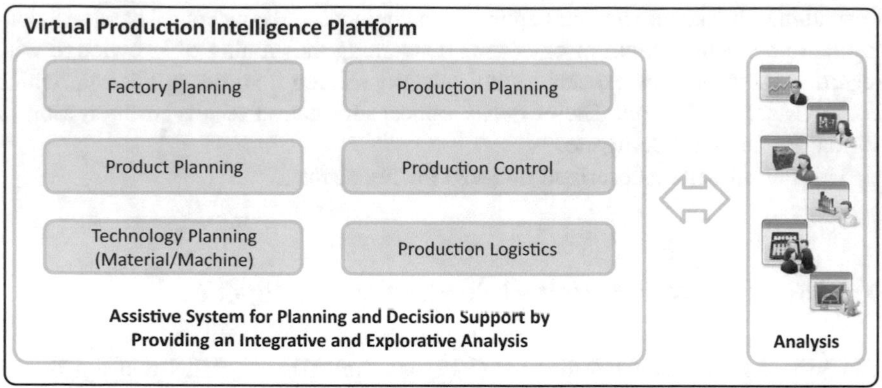

Fig. 3 The concept of the virtual production intelligence' Platform

logistics or stocks. Within the planning process of the factory, data are generated on a large extent. The use of these data depends on the future utilization rates of the factory capacity. In order to analyze the planned processes of the future factory, it is provided that these processes are evaluated beforehand. The planning of these processes will only create an additional value if the identified potentials for optimization are considered in the real process.

Comprehensively, the following questions have to be answered from a planner's point of view:

- Which of the data generated during the process planning are relevant?
- Which key performance indicators are needed with regard to the validation of the considered processes?
- How can the gained knowledge be fed back into the real process?

Regarding the field of information technology, the following questions arise:

- Which data model facilitates the data's analysis?
- Which data analysis methods known from Business Intelligence can serve as role models?
- How to validate the data model's and analysis' functionality appropriately?

Topics that were not considered above will be addressed by the following questions:

- Which simulation model for the considered process is preferred by the user?
- How can the process in consideration be decomposed?
- Which added value may the user expect?

In retrospect, these questions address technical, professional as well as organizational aspects.

6 Summary and Outlook

Within this paper, a concept named "Virtual Production Intelligence" (VPI) has been presented, which describes how the solutions developed within the field of "Business Intelligence" can be adapted properly to the one of virtual production. This concept, which is both holistic and integrated, is used for the collaborative planning, monitoring and control of core processes within production and product development in various fields of application.

Furthermore, the technical implementation of this concept was made a subject of discussion in terms of the Virtual Production Intelligence Platform (VPI-Platform). The platform's implementation is particularly based on concepts and methods established in the field of Cloud Computing. Challenges that might occur during the realization of the platform were taken into account with regard to technical, professional and organizational aspects.

A further scenario, which will point out the VPI's flexibility, will be taken from the field of laser cutting. Thereby, the focus will lie on the problem of analyzing a simulated cutting process in such a way that desired characteristics of the concrete cutting process can be realized. The configuration settings for the cutting machine resulting in the desired cutting quality are a part of the analysis outcome. An additional value for the real cutting process arises after feeding back the analysis outcomes into this process.

References

1. G. Schuh, S. Aghassi, S. Orilski, J. Schubert, M. Bambach, R. Freudenberg, Technology roadmapping for the production in high-wage countries. Prd. Eng. Res. Devel. (Production Engineering), 2011, pp. 463–473
2. C. Brecher, ed. *Integrative Production Technology for High-Wage Countries*. Springer, Berlin and Heidelberg, 2011
3. V. Richtlinie. Digitale fabrik - grundlagen/digital factory - fundamentals
4. V. Richtlinie. Digitale fabrik - digitaler fabrikbetrieb/digital factory - digital factory operations
5. M. Nagl, B. Westfechtel, eds. *Modelle, Werkzeuge und Infrastrukturen zur Unterstützung von Entwicklungsprozessen*. Wiley-VCH, 2003
6. C. Horstmann, *Integration und Flexibilitat der Organisation Durch Informationstechnologie*, 1st edn. Gabler, 2011. 156–162
7. Steve, Canonical data model. Information Management Magazine 2008, 2008
8. D. Schilberg, *Architektur eines Datenintegrators zur durchgängigen Kopplung von verteilten numerischen Simulationen*. VDI-Verlag, Aachen, 2010
9. T. Meisen, P. Meisen, D. Schilberg, S. Jeschke, Application integration of simulation tools considering domain specific knowledge. Proceedings of the 13th International Conference on Enterprise Information Systems, pp. 42–53
10. R. Reinhard, T. Meisen, T. Beer, D. Schilberg, S. Jeschke, A framework enabling data integration for virtual production. In: *Enabling Manufacturing Competitiveness and Economic Sustainability. Proceedings of the 4th International Conference on Changeable, Agile, Reconfigurable and Virtual production (CARV2011), Montreal, Canada, 2–5 October 2011*, ed. by H.A. ElMaraghy. Springer, Berlin, Heidelberg, 2012, pp. 275–280

11. B. Byrne, J. Kling, D. McCarty, G. Sauter, P. Worcester, The value of applying the canonical modeling pattern in soa. IBM Journal, 2008. URL http://www.ibm.com/developerworks/data/library/techarticle/dm-0803sauter/. The information perspective of SOA design, 4
12. M. West, *Developing High Quality Data Models*, 1st edn. Morgan Kaufmann, Burlington, MA, 2011
13. G. Schuh, A. Kampker, C. Wesch-Potente, Condition based factory planning. Prd. Eng. Res. Devel. (Production Engineering) **5**, 2011, pp. 89–94
14. U. Bracht, D. Geckler, S. Wenzel, *Digitale Fabrik: Methoden und Praxisbeispiele*. Springer, Berlin, Heidelberg, 2011
15. H. Luhn, A business intelligence system. IBM Journal, 1958, pp. 314–319
16. H. Kemper, B. Henning, Business intelligence und competitive intelligence. it-basierte managementunterstützung und markt-/wettbewerbsorientierte anwendungen. HMD Praxis der Wirtschaftsinformatik **43** (247), 2006
17. W. Hummeltenberg. 50 jahre bi-systeme, 2009. URL http://www.uni-hamburg.de/fachbereiche-einrichtungen/fb03/iwi-ii/Ausgewaehlte_Publikationen_/W_Hummeltenberg_50_Jahre_Business_Intelligence-Systeme.pdf

Teaching Professional Knowledge to XL-Classes with the Help of Digital Technologies

Valerie Stehling, Ursula Bach, Anja Richert and Sabina Jeschke

Abstract How can the systematic use of digital technologies affect a lecture of 1500 or more students? Moreover, to what extent will it affect the learning outcomes of the students? At RWTH Aachen University, subjects like Mechanical Engineering have to cope with a very high number of students each semester – currently the number lies at approximately 1500 with an estimated increase up to 2000 in the next semester. In order to create an interactive learning environment despite these difficult conditions, the IMA/ZLW&IfU (Institute of Information Management in Mechanical Engineering, Center for Learning and Knowledge Management and Assoc. Institute for Management Cybernetics) of the RWTH Aachen University is currently developing a pilot scheme that includes the application of Audience Response Systems in lectures with such large numbers of student listeners. The implementation of the described system demands a redesign of the lecture with special regards to the content. Questions have to be developed that allow the students to interact with the lecturer as well as each other. This variety of questions ranges from multiple-choice questions to the inquiry of calculation results etc. When giving students the chance to actively take part in a lecture of the described size by answering questions the lecturer asks with the help of technical equipment – which could in the easiest case be their own mobile phones – the lecturer creates a room for interaction. In addition to that he has the chance to get an immediate insight into the perceived knowledge of his or her students. This in turn enables the lecturer to react to obvious knowledge gaps that obstruct successful learning outcomes of the students. An additional benefit hoped for is that the attention of the students – which is a difficult issue for lecturers that face lectures with such a large number of students – might be kept at a higher level than average. The described redeployment of a lecture of the mentioned size is expected to bring about an enhancement of the quality in teaching of professional knowledge. The presumptions made in this paper will be surveyed and thoroughly analysed during and after the realization of the project.

V. Stehling (✉) · U. Bach · A. Richert · S. Jeschke
IMA/ZLW & IfU, RWTH Aachen University,
Dennewartstr. 27, 52068 Aachen, Germany
e-mail: valerie.stehling@ima-zlw-ifu.rwth-aachen.de

Originally published in "The first international ProPEL Conference 2012 - Professions and Professional Learning in Troubling Times: Emerging Practices and Transgressive Knowledges",

DOI 10.1007/978-3-319-46916-4_7

Keywords Teaching Strategies · Audience Response System · Lage Class Managment · Clicker Software

1 Initial Situation

RWTH Aachen University is a highly ranked university especially in the fields of engineering – mechanical engineering, electrical engineering, industrial engineering etc. It therefore attracts a vast amount of students which in the past has lead lecturers to the challenge of having to teach a rising number of students in already large classes each semester. This is even enforced by the German "G8" or "Gy8" concept which reduces studying in school from now 13 to then 12 years. In addition to that the conscription in Germany has been abolished which currently – but temporarily – leads to an even higher enforcement of increasing numbers of students. This means that by 2013 universities in NRW (Northrhine-Westfalia) will face an additional amount of student applications of around 275000. Universities face not only the challenge of having to prepare themselves infrastructurally, the design of their lectures has to be customized as well (Fig. 1).

2 Shifting from Teaching to Learning: A Special Challenge for Large Classes

With the previously pointed out development in mind: How can requested criteria of the Bologna Process like the "student centered approach" and an "emphasis on skills" – in other words a "shift from teaching to learning" be considered when

Fig. 1 Large Class at RWTH Aachen University (http://www.taz.de/!86786/)

planning a lecture with more than 1000 listeners? It is, this has to be said in advance, neither our aim nor any given possibility to split the large lectures into small groups. Lectures in the classical sense serve the purpose of conveying scientific findings and general content that the students need to be able to transform this knowledge into practice in specially designed laboratory or exercise courses. However, current research shows that motivation and attention of students during a classical ex-cathedra lecture decreases drastically already in early stages of the lecture [1].

First outcomes of a current monitoring survey by the ZLW conducted in classes of the engineering sciences show that at RWTH Aachen University most large lectures still have their focus on the teaching aspect rather than promote learning: Students are often passive consumers of knowledge. This of course only goes for the lecture itself – most of these classes have additional exercise courses or labs in which the students get the chance to work together on a subject in small groups. Previous research however shows that one way to gain a certain level of learning in a lecture is to give the students the chance to learn actively – not only in these additional exercise courses but also in the lecture itself [2]. Davis e.g. states that students learn best when they are active participants in the education process [3, 4]. Of course, when talking of lecture sizes beyond a seminar size, this task is a special challenge to every lecturer [5, 6].

Some of the aspects making active learning in a large class a special challenge can be summed up as the following:

- Speaking in front of a large class is often frightening for students. They might compromise themselves when giving a wrong answer.
- Large classes are loud. The level of noise can quickly become very high once the attention of the students starts to diminish. When answering a question in a large class the answer is often overheard because it is too loud in class.
- Even when the lecturer frequently asks questions in his or her class, he or she still cannot involve all the students at the same time [5, 7].

However, finding a solution to this challenge is not that simple. At the IMA/ZLW&IfU, several projects currently deal with the advancement of teaching and learning to an excellent level taking on both perspectives – the teaching perspective as well as the learning perspective (e.g. "ELLI" – Excellence in Teaching and Learning in the Engineering Sciences; "ExAcT" – Center of Excellence in Academic Teaching – for more information see: http://www.ima-zlw-ifu.rwth-aachen.de/en/research/aktuelle_projekte).

TeachINGLearnING.EU (http://www.teaching-learning.eu), a project dealing with teaching and learning concepts especially in the engineering sciences funded by the "Stiftung Mercator" and the "Volkswagen Foundation" and also conducted by (amongst others) the ZLW has – as one of its goals – set up competitions in an "open innovation"-approach. These competitions give students the chance of influencing the learning environments and settings the university offers them. One of these competitions has covered the subject of large class management and how it can be improved. 31 ideas were handed in by students and ten of the proposed ideas concerned the lack of interaction between teachers and students during or after class. Some of these proposals suggested the use of technology to help overcome fears of speaking due to the size of the class or even admitting that one did not understand the subject matter the lecturer has just explained. Findings of Andersen et al. affirm this appraisal. In their pilot study 6 out of 12 students state that in large classes they feel apprehensive of participating [5]. Proposed ideas in the competition mentioned above range from the "IDIOT (I DId nOt get iT)-Buzzers" to classic Audience Response Systems. These offer the possibility to either (in the first case) be able to press a button when a student feels that something has not been explained thoroughly enough or to (in the second case) send questions to the teaching assistant during class that he can either immediately react to in the following lecture.

3 Response Systems – A Possible Teaching Method for XL-Classes?

The described ideas suggested by students show that coherent to current research results and requests for a change in didactics addressed towards university teachers there is a demand for active learning and interaction in class. According to modern age students or in Prensky's words "digital natives" [8], (new) technologies e.g. in the form of Response Systems can be a means to improve the learning outcomes of students.

Jeschke in this context talks about the potential of new media in education [9]. Prensky even states that "our students have changed radically" and that "today's students are no longer the people our educational system was designed to teach" [8]. And this statement leads us to the same conclusion Ketteridge comes to: That "change in education is inevitable as institutions invent and reinvent themselves over time and space" [10].

Using classroom communication systems (CCS) is not a recent innovation – the first popular CCS "Classtalk" was developed in 1985. Writing about the appropriate application, the benefits and downsides of using CCS in class is therefore neither a brand new issue. Of course, since 1985, Response Systems have been improved and advanced as have digital technologies in general. However most research on this topic up to now is dealing with CCS or RS in classes with up to approximately 500 students – so regarding the particular aspect of the as previously described XL-sized classes is definitely interesting (Fig. 2).

Evaluations of best practices from universities worldwide show that the application of Response Systems (RS) in lectures has led to e.g. higher motivation of attendance [11], more attention of the students during class [1] and even higher knowledge acquisition than in conventional (non-interactive) classes [1]. According to Sellar RSs "(…) have been found to be of particular benefit when working with large groups where communication is challenging (…)" [12]. It will be very interesting to see whether this will become evident in classes with up to 2000 students as well.

Fig. 2 Poll Everywhere in use at the MIT (http://polleverywhere.com/how-it-works)

4 Use of Technology Equals Learning??

"Technology doesn't inherently improve learning" [13]. This proposition stated by Beatty does not come out of the blue: Some teachers might be thrilled by the new technology in their classroom but do not use it properly and efficiently due to a lack of pedagogical or didactical conceptualization. Kerres also states that digital media are no "Trojan horses" that can be brought into an organization (or situation such as a class) and unfold their effect "overnight" [14]. Kay and LeSage reinforce these statements by stating that "(…) the integration of an ARS into a classroom does not guarantee improved student learning" but that "it is the implementation of pedagogical strategies in combination with the technology that ultimately influences student success" [15].

According to Kerres [16] universities need for an individual media strategy that ensures an appropriate use and a successful outcome of innovations in teaching by new media. This strategy should at least cover four topics of change:

- Reform of teaching: Which (new) contents of teaching do we want to convey? Reform of teaching methods: Which (new) methods of teaching and learning do we aim for?
- Production of media supported learning environments (inclusive the development of a didactical concept and (if necessary) the development of media) as well as the distribution of the media.
- Designing the personnel and structural conditions for the successful usage of media (HR measures).
- Extension and backup of the infrastructure (hard- and software, installment, attendance, fosterage) [16].

The following chapter will deal with the first two of Kerres' as well as other topics, which are especially relevant for this particular paper.

5 Conceptual Design

There are many best practices to look at when redesigning a lecture to give an appropriate room for the use of Audience Response Systems (ARS) possible. Considering the fact that most of these best practices will have to cope with much less students than RWTH Aachen University, a creative solution needs to be found to ensure that the benefits of the Response System become effective.

The conceptual design of the implementation of a clicker system in the lecture of information technology in the engineering education at RWTH Aachen University with currently approximately 1500 students will be described in the following subsections.

Every lecture is unique due to its specific content, so there is no prototype solution for a conceptual design when introducing RS into large classes. Neither can the conceptual design described in the following sections act as a prototype solution.

Along with the first two topics Kerres [16] sees as indispensable, a few other items need to be considered when planning to introduce Response Systems in this particular large class.

1. Reform of teaching: new contents; methods of teaching and learning
2. Production of media supported learning environments; distribution of the media
3. Financing
4. Motivation for using clickers – student/lecturer approach
5. Roll-Out, Evaluation and Adjustments

These topics will in the following be discussed against the background of the implementation of a Response System in the previously described lecture "Information technology in mechanical engineering" teaching approximately 2000 students.

5.1 Reform of Teaching – Contents and Methods

When planning to introduce a Response System in a lecture, it has to be considered that clicker questions take up a certain amount of time of the lecture. The content of the original lecture thus has to be adjusted to this. And the more clicker questions you as a lecturer ask, the more time for the usual content you will have to give up. According to Beatty, this is a step in the right direction:

> An instructor cannot and should not explicitly address in class every topic, idea, fact, term, and procedure for which students are "responsible". Instead, use class time to build a solid understanding of core concepts, and let pre-class reading and post-class homework provide the rest. [13].

It is not conducive, however, to overload a lecture with clicker questions and turn it into a quizzing lecture. Clicker questions need to be carefully planned and placed. Beatty et al. state that "classroom response systems can be powerful tools

(…). Their efficacy strongly depends on the quality of the questions" [17]. Crews et al. also state that using clickers in the same way every lecture can become monotonous to the students and therefore become counterproductive. "Instructors should be prepared to implement clickers for different purposes throughout the semester (…)" [3]. Examples named here are discussions (pre-, during and post-), quizzes, competitions between groups, student generated questions etc. [3].

Several types of questioning seem appropriate and conducive for the described lecture in the engineering sciences. For example: Since the lecture deals with (amongst other subjects) programming and basics of software engineering it surely is interesting and helpful for further planning processes to get to know your audience better. This can for example be achieved by asking the students in the first lecture which computer languages they already know. Additionally it can be a benefit to ask multiple choice questions and later discuss the answers given. Beatty states that the focus when "quizzing" the audience should not lie on the correctness of the answers but on the reasoning behind it [13]. He also appeals to the right responses when right or wrong answers are given: "How we respond when right or wrong answers are given is crucial. A full spectrum of answers should be drawn out and discussed before we give any indication which (if any) is correct [13]." One very important element here is the "wow factor": By arranging questions cleverly, making mistakes can create an opportunity to learn and considering that mistakes are made anonymously and possibly by a lot of other students too it might therefore lose its negative connotations in class.

Considering that at RWTH Aachen University the use of clickers in class is not a common and well-known teaching method (neither to the students nor to the lecturers), the changes in teaching should not be rushed. Teachers and students need a certain amount of time to learn their "new role" [13] in the lecture: Participants of a more interactive, student-centered approach [2]. This might additionally cause an initial "fear" or "discomfort" on both ends. Especially students might react negatively in the beginning: while the teacher has already had a chance to accustom to his new role in the planning process, students begin with this process at the beginning of the first lecture. Usually, before they see the actual benefits for themselves like a better learning outcome or understanding, they see the "downsides": they need to prepare before the lecture in order to "perform well" in these mini-tests which they might think is the purpose of using Response Systems [13].

On the other hand, lecturers in the described class at RWTH Aachen University will benefit from an expectable rather positive attitude of the students towards technology anchored in their own choice to study mechanical engineering. This is an additional benefit to the assumption that young students today can be described as digital natives or in terms of Wim Veen: "Homo zappiens" [18]. Considering these aspects, we can assume that using this new technology in class will feel natural and intuitional to the students and the playful element [18] participating will have a strong motivational effect on them.

Summing up it is necessary when designing contents and methods for a teaching approach using Response Systems it is highly important to keep a sharp eye on pedagogical as well as learning goals. According to Beatty these learning or pedagogical goals include

- drawing out students' background knowledge and beliefs on a topic,
- making students aware of their own and others' perceptions of a situation,
- discovering points of confusion or misconception,
- distinguishing two related concepts,
- realizing parallels or connections between different ideas,
- elaborating the understanding of a concept,
- exploring the implications of an idea in a new or extended context [13].

These pedagogical goals should be achieved by using the playful element of clickers in the lecture and promoting learning as something that can be fun, too.

5.2 Production of Media

For the lecture, a simple and accessible software is being rented to ensure a very high access rate. The software, "Poll Everywhere" is designed in a way that every student owning any mobile phone, wifi-device or laptop can participate. The screenshot (Fig. 3) below shows the three steps it takes to start a poll in class.

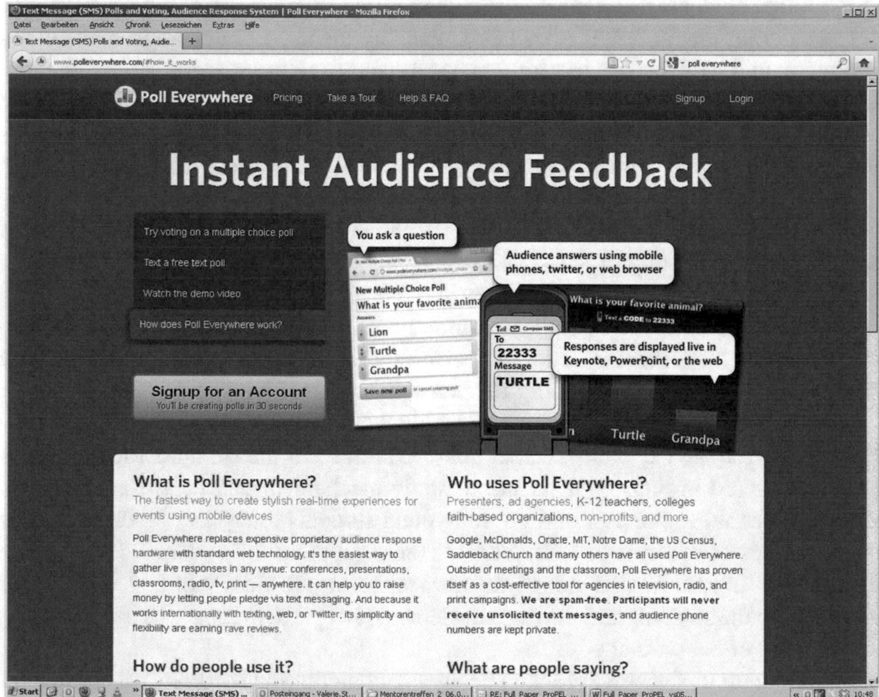

Fig. 3 Screenshot 1: How it works (http://www.polleverywhere.com/how-it-works)

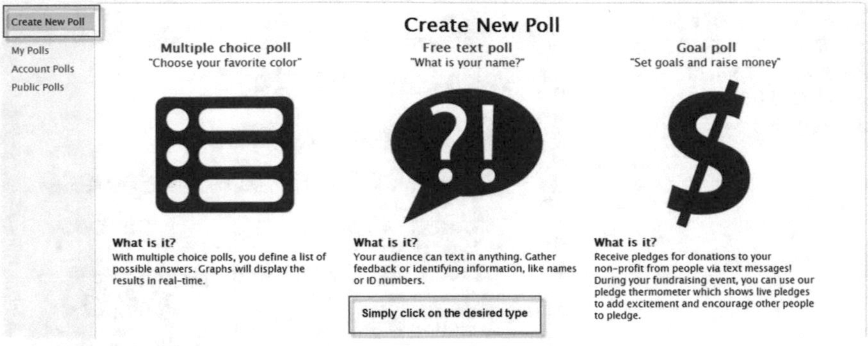

Fig. 4 How to create a new poll [12]

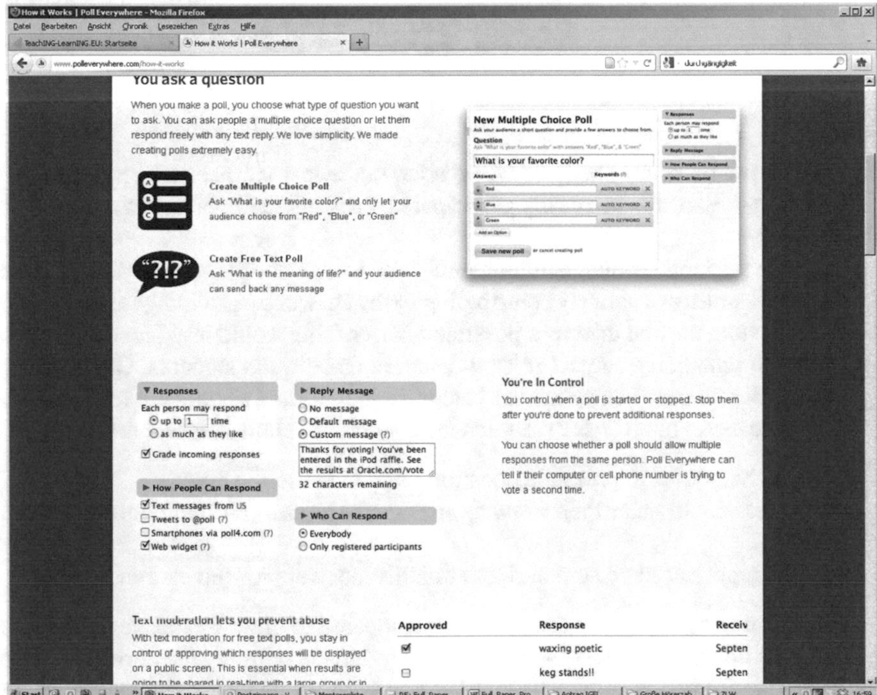

Fig. 5 Installing a poll via Poll Everywhere (http://www.polleverywhere.com/how-it-works)

When you as a lecturer design a poll, you first have to choose which sort of poll you want to use: You can either ask a multiple choice question or let the audience answer a question freely with any text reply (see Fig. 4). If you have decided to ask a multiple choice question, you can type in the question and the possible answers and set the options for participation (see Fig. 5).

Having finished this you can show the students your question and the possible answers including the possible ways of participation (see Fig. 6). Those students

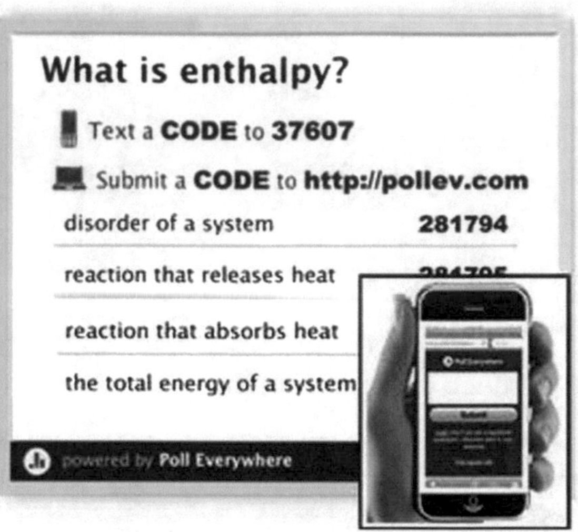

Fig. 6 Example of a poll (http://dukedigitalinitiative.duke.edu/wp-content/uploads/2011/08/pollEverywhere.jpg)

who do not own a smart phone or wifi-device but own a simple mobile phone that supports text-messaging can easily participate by texting their answer to a displayed phone number.

For those students owning neither a wifi-device nor a simple mobile phone, there still is the possibility of actively contributing to the answer by grouping together with their fellow students and discuss a possible answer. This would also serve a positive side effect of enhancing subject-related discussions between students. Once having opened a poll you can then watch the votes rise in real time as they are received. So in the classic sense of a Class or Audience Response System, this technology

- allows an instructor to present a question or problem to the class,
- allows students to enter their answers into their devices (mobile phones, laptops etc.) and
- instantly aggregates and summarizes students' answers for the instructor. [13]

The essential advantage of this software compared to other Response Systems is that students do not have to buy special hardware and accordingly the university does not have to provide devices in class which would be a huge financial burden when lecturing classes of thousands of students. Students are not charged for voting on Poll Everywhere; however, if they vote by text message then standard text messaging charges apply [12].

5.3 *Financing*

By introducing a Response System into a lecture the person responsible for the organization and content of the lecture has to make several expenditures. It will cost time, effort and – of course – money. Usually the lecturer, the institution or the faculty has to pay for the software or application. In the described case the fees are according to class or audience sizes. The bigger expenditure is the amount of time (in terms of staff costs) to adjust or redesign the lecture to the application of clicker questions. As previously described clicker questions need to be carefully designed and should always be strongly tied to a specific learning goal. Current research shows that this is one of the most challenging tasks when introducing clickers into any class and therefore takes up most of the preparation time (Tables 1 and 2).

One advantage of poll everywhere is that due to technical development no clicker hardware or device has to be bought neither by students nor by the university. This is a huge financial improvement – especially in large classes – which also has an impact on student participation in online polls in class – students easily forget to bring their clicker devices to class, because they only use it for class, but the probability of students forgetting to bring their mobile phones is rather small.

At RWTH Aachen University and especially the Faculty of Mechanical Engineering the commitment to advance new concepts in teaching and learning is extraordinarily high. Projects like "The "Students in Focus" Future Strategy" (http://www.rwth-aachen.de/go/id/bbtb) as well as previously described projects like "ELLI", "TeachING-LearnING.EU", "ExAcT" etc. currently put a lot of time and effort into finding and establishing innovative teaching designs. Additional funding for the described careful preparation of the desired implementation this paper deals with has been raised in a special "Exploratory Teaching Space"-Call named "IGEL" (Interactive Large Classes for Excellent Teaching) to enhance excellence in teaching (http://www.cil.rwth-aachen.de/tag/exzellente-lehre/).

5.4 Motivation

Clicker Systems or the implementation of such into a lecture (as every other change process) is as has been pointed out always conjoined by at least financial and temporal aspects. Therefore, the positive aspects of the redesign have to prevail to build up a motivation to take these "burdens" upon oneself (oneself being the lecturer, the institution, faculty or even university). These positive aspects or advantages should be predominant for the lecturer as well as for the students. If you want a Response System to fulfill its "purpose" and increase interactivity by "promoting a two-way-flow of communication between the speaker and the audience" [12], both ends need to be "on board". To point out that an implementation of the described Response System is worth all the cost and effort, the following graphs will sum up the most advantages and disadvantages found by review of literature in the field of Response Systems. These are separated into benefits and disadvantages for the teacher on the one hand and the student on the other.

Evaluations of lectures that are already working with RS in class show a few challenges and downsides aroused by the introduction of RS. These will be summed up in the following chart. It is noticeable, though, that there are a lot more positive aspects on either– the student as well as the teacher – side.

Table 1

benefits of clickers for the student	benefits of clickers for the lecturer
interaction with the lecturer without fear of compromising oneself	identification of knowledge gaps
immediate feedback	identification of shortcomings of the lecture [7]
possibility to actively check their learning outcomes outside of exams	student engagement
be an active participant in class	keeps students focused and involved
anonymity	higher attendance
enhancement of learning	better control of the learning progress
classroom experience more enjoyable	…
…	

Table 2

disadvantages of clickers for the student	disadvantages of clickers for the lecturer
equipment/software functionning	clicker questions take up time pre and during class
equipment accessibility	the implementation itself costs time and money
costs occurring when only option of contributing for the student is a text message	equipment/software functionning
…	diversion by using technical devices in class
	…

One argument that has not been mentioned in this paper before but is often discussed is that students are or might be diverted [1] by using their technical devices – mobile phones, laptops etc. – in class is not steady. One of the results of the monitoring survey conducted in large classes of engineering sciences described earlier in this paper is that most students use their phones or laptops during class for texting and social networks anyway – so why should a lecturer not take advantage of this and use it for educational purposes?

5.5 Roll-Out, Evaluation and Adjustments

Since the lecture in which the Response System is going to be introduced is held in the second semester of the engineering sciences and therefore only takes place in the summer term, the research on this topic is still in progress. First clicker questions will be introduced from the first lecture date of the summer semester 2012.

To secure and check the outcomes of the implementation of a system that enhances a new teaching and learning approach in "XL"-classes and whether previously found results from other universities can be acknowledged, the implementation must be concomitantly evaluated. First evaluations are planned for the middle of the summer semester 2012 – during the pilot study – by questioning both the students and the lecturer to get a first insight on the perception of the new approach. A conclusive and detailed evaluation will follow at the end of the semester.

Considering that all lectures at RWTH Aachen University are being evaluated by the students each semester, we expect to see changes in this evaluation so we can come up with a first rating whether the approach also works for groups of the described size or not.

Results of the evaluations and perceived outcomes (acceptance, estimated learning outcomes, participation, attention etc.) from the pilot study will be published as well.

6 Limitations and Future Prospects

Due to the fact that the paper deals with research still in progress we cannot yet give any valid recommendations or report learnings from the implementation process. These findings will as previously mentioned be published later.

However, by conducting this accompanying study, we expect to find out by analyzing the evaluation sheets whether Response Systems are an appropriate means for advancing interaction and subsequently enhancing student learning and attention in class. The benefits and disadvantages collocated by reviewing previous literature as well as simple estimations in this paper will be revised for the special case of "XL"-classes according to the findings in our evaluations.

References

1. N. Scheele, A. Wessels, W. Effelsberg, Die interaktive vorlesung in der praxis. In: *DeLFI 2004: 2. e-Learning Fachtagung Informatik, Tagug der Fachgruppe e-Learning der Gesellschaft für Informatik e.V. (GI), 6.-8. September 2004 in Paderborn, P-52*. 2004, pp. 283–294
2. C. Fies, J. Marshall, Classroom response system: A review of the literature. Journal of Science, Education and Technology **15**, 2006
3. T.B. Crews, L. Ducate, J.M. Rathel, K. Heid, S.T. Bishoff, Clickers in the classroom: Transforming students into active learners. ECAR Research Bulletin **9**, 2011
4. B. Davis, *Tools for Teaching*. Jossey Bass, San Francisco, 1993
5. R.J. Anderson, R. Anderson, T. Vandegrif, S. Wolfman, K. Yasuhara, Promoting interaction in large classes with computer-mediated feedback. Computer Supported Collaborative Learning Proceedings, 2003
6. U. of Maryland. Large classes: A teaching guide. center for teaching excellence, 2008. URL http://www.cte.umd.edu/library/teachingLargeClass/guide/index.html
7. B.S. Hasler, R. Pfeifer, A. Zbinden, P. Wyss, S. Zaugg, R. Diehl, B. Joho, Annotated lectures: Student-instructor interaction in large-scale global education. Journal of Systemics, Cybernetics and Informatics (5), 2009

8. M. Prensky, Digital natives, digital immigrants. On the Horizon **9** (5), 2001
9. S. Jeschke. Mathematik in virtuellen wissensräumen – iuk-strukturen und it-technologien in lehre und forschung, 2004. URL http://deposit.d-nb.de/cgi-bin/dokserv?idn=971017182&dok_var=d1&dok_ext=pdf&filename=971017182.pdf
10. H. Fry, S. Ketteridge, S. Marshall, The effective academic. In: *Enhancing Teaching in Higher Education*, ed. by P. Hartley, A. Woods, M. Pill, 2005
11. A.R. Trees, M.H. Jackson, The learning environment in clicker classrooms: student processes of learning and involvement in large university-level courses using student response systems. Learning, Media and Technology **32** (1), 2007, pp. 21–40
12. M. Sellar. Poll everywhere, standard review, the charleston advisor, 2010. URL http://docserver.ingentaconnect.com/deliver/connect/charleston/15254011/v12n3/s15.pdf?expires=1330503883&id=67473240&titleid=75002231&accname=RWTH+Aachen+Hochschulbibliothek&checksum=C72E182E1A83A266C60718882EC9BA75
13. I.D. Beatty, Transforming student learning with classroom communication systems. EDUCAUSE, Research Bulletin **2004**, 2004
14. M. Kerres, Wirkung und wirksamkeit neuer medien in der bildung. In: *Eucation Quality Forum*, ed. by R. Keill-Slawik, M. Kerres, Wirkungen und Wirksamkeiten neuer Medien, Waxmann, Münster, 2003
15. R.H. Kay, A. LeSage, A strategic assessment of audience response systems used in higher education. Australian Journal of Educational Technology **25** (2), 2009, pp. 235–249
16. M. Kerres, Zur integration digitaler wissenswerkzeuge in die hochschule. In: *Unbegrenztes Lernen – Lernen über Grenzen? Generierung und Verteilung von Wissen in der Hochschulentwicklung*, ed. by E. Kruse, U. Küchler, M. Kuhl, LIT-Verlag, Münster, 2004
17. I.D. Beatty, W.J. Gerace, W.J. Leonard, R.J. Dufresne, Designing effective questions fpr classroom response system teaching. American Journal of Physics **74** (1), 2006, p. 31
18. W. Veen, B. Vrakking, *Homo Zappiens: Growing Up in a Digital Age*. Continuum, 2006

Intensifying Learner's Experience by Incorporating the Virtual Theatre into Engineering Education

Daniel Ewert, Katharina Schuster, Daniel Johansson, Daniel Schilberg and Sabina Jeschke

Abstract This work introduces the virtual theatre, a platform allowing free exploration of a virtual environment, as an instrument for engineering education. The virtual theatre features three main user components: a head mounted display, a data glove and an omnidirectional floor. These interfaces for perception, navigation and interaction allow for more realistic and intuitive experiences within environments which are inaccessible in the real world. This paper describes the technical properties of the platform as well as studies on human experiences and behavior. It moreover presents current and future applications within the field of engineering education and discusses the underlying didactic principles.

Keywords Virtual Learning Environments · Immersion · Engineering Education

1 Introduction

It is a known and well accepted fact that students need practical experience and practical relevance in higher education, especially in engineering. Practical experience is often gained in laboratory work, a widely-used approach to teach the grass roots of science. Experiments are mostly only conducted for basic subjects like physics or chemistry. A common approach to convey practical relevance to engineering students is by offering internships or excursions.

However, because of the costs, complexity and criticality of industrial sites, it is not possible for a student to explore such installments freely and experiment on her or his own, on one hand to ensure ongoing production and to avoid damage to the machines and on the other hand to protect the student from harm by physical forces or contact to hazardous material. Today, there are still few examples in engineering education where practical experience and practical relevance are truly connected to each other.

D. Ewert (✉) · K. Schuster · D. Johansson · D. Schilberg · S. Jeschke
IMA/ZLW & IfU, RWTH Aachen University,
Dennewartstr. 27, 52068 Aachen, Germany
e-mail: daniel.ewert@ima-zlw-ifu.rwth-aachen.de

Originally published in "Proceedings of the IEEE EDUCON Conference",

DOI 10.1007/978-3-319-46916-4_8

To overcome this, it is possible to rebuild such industrial complexes within virtual environments. In simulations or serious games, the principles behind the industrial site are connected with the respective courses. Until now, the drawback of such virtual learning environments is the artificiality of the experience. Usually, the participant interacts with the virtual environment via a pc workstation. The industrial complex is perceived via a computer monitor, and the viewpoint is moved via keyboard and/or mouse. Interactions with the environment are triggered the same way. Thus, the user has to "translate" the control mode of the computer into the actions of his or her graphical representation, the avatar.

To achieve a more natural experience, new interfaces can be used for visualization, navigation and interaction: For high immersion, the participant needs a seamless 3d surround view of the virtual environment. This can be achieved with modern head mounted displays (HMD). For natural navigation, omnidirectional treadmills can be used, which allow free and unlimited movement by tracking the participant's movement and keeping him in a stationary place at the same time. Finally, data gloves allow interacting with the virtual environment intuitively. These described components – HMD, omnidirectional treadmill, and data glove – are incorporated in the so called virtual theatre [1] (Fig. 1).

Fig. 1 Virtual theatre with user and (simplified) virtual environment

2 Technical Description of the Virtual Theatre

The virtual theatre presents the user with a seamless 3d visualization of a virtual environment. All head movement is instantly reproduced within the simulation, so the user can look around freely. The user can move around within the environment by just walking in the desired direction. Hand gestures can be recognized and allow for manipulation as well as advanced control of the simulation.

2.1 *Hardware Setup*

2.1.1 Omnidirectional Floor

The omnidirectional floor consists of polygons of rigid rollers, with increasing circumferences and a common origo. Each polygon constitutes 16 rollers and together all polygons form 16 sections with an angle of 22.5 degrees. Each section consists of one roller from each polygon, which means from cylinder origo towards the periphery, the rollers are parallel and increasing in length. Rollers are driven from below by a belt drive assembly to minimize distance between sections. In the central part of the floor there is a circular static area where the user can move without enabling the floor. Floor movements here would only cause the feet of the user to be drawn together. As the user moves outside the static boundary, the floor begins to move according to a control algorithm (Fig. 2).

2.1.2 Head Mounted Display

Visual and auditory feedback is received via a zSight [2] head mounted display, providing a 70° stereoscopic field of view with SXGA resolution for each eye, and stereo sound via attached head speakers (see Fig. 3) The HMD weighs 450 grams.

Fig. 2 CAD model of the omnidirectional floor

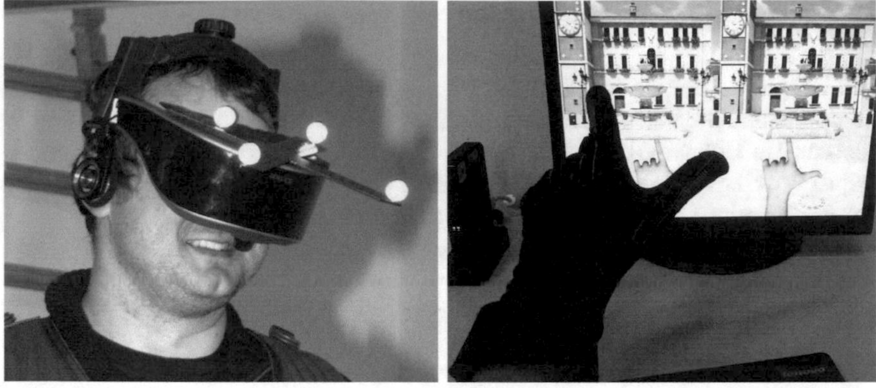

Fig. 3 *Left:* Head mounted display with attached infrared markers; *Right:* Data glove and representation in a virtual environment

2.1.3 Data Glove

To interact with the virtual scenario, a user can utilize hand gestures. These are received by 22 sensors incorporated in a data glove [3]. Currently, only one data glove is incorporated in the virtual theatre, since the added value of a second would glove would be minimal (you can lift and move every virtual object single-handed), while object manipulation and two handed gesture detection would become unnecessarily complex. However, later incorporation of a second glove is generally possible.

2.1.4 Tracking System

To track the movements of a user, the virtual theatre is equipped with 10 infrared cameras. They record the position of designated infrared markers attached to the HMD and the data glove. The position of the HMD serves on the one hand as an input for controlling the omnidirectional floor: The inward speed of the rollers increases linearly with the measured distance of the HMD to the floor center. On the other hand it is used to direct the position and line of vision of the user within the virtual environment. The position of the glove serves for representing the hand within the virtual environment and for triggering emergency shutdown: As soon as the hand drops below 0.5 m, e.g. in the unlikely event of a user falling down, all movement of the omnidirectional floor is immediately stopped.

2.1.5 Integration of the Components

The system architecture of the virtual theatre is depicted in Fig. 4. The fixed components tracking system and omnidirectional floor communicate to a hardware con-

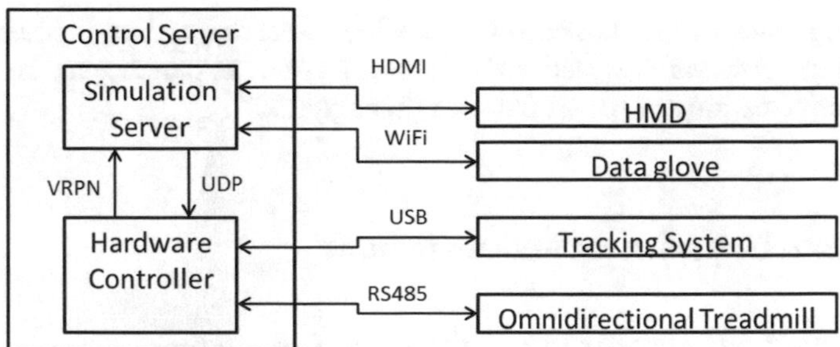

Fig. 4 System architecture of the virtual theatre

troller via cable-based communication channels. However, to allow for unconstrained movement, all hardware that is worn by the user communicates wirelessly.

2.2 *Software Setup*

The theatre's server consists of two personal computers serving different purposes. The hardware control server analyzes the data from the infrared cameras and regulates the omnidirectional floor. For tracking the user's movement it uses Optitrack Tracing Tools [4]. The position information as well as the velocity of the floor is sent to the simulation server via vrpn [5], a device-independent and network-transparent interface to virtual reality peripherals. The simulation server is responsible for high level control and provides the virtual environment. Therefore it makes use of the Vizard Virtual Reality Toolkit [6]. 3D Models for virtual environments are generated with blender [7] and can be loaded into Vizard.

2.3 *Limitations*

Currently the only provided feedback to the user is visual or auditive. Other feedback sources in form of heat or air draft are generally possible but not yet part of the current installation or scenarios. Due to the single center of the treadmill where the user is held, the virtual theatre allows for only one operator. Further spectators can follow the operator's actions on additional screens, but cannot interact directly with the virtual environment.

However, cooperative scenarios, where spectators instruct the user, assist in other ways, or even accompany the user as virtual classically controlled avatars are possible. Nevertheless, the virtual theatre is only applicable for small groups of students. This needs to be kept in mind when a didactical method for the respective courses in engineering education is chosen. An interesting research question is if the students who have the observer perspective also have a learning experience as the theory of observational learning of Albert Bandura [8] suggests.

3 Human Experiences and Behaviour

3.1 Concepual Background of Immersion

In general, all interfaces of simulators have the purpose to create a perfect illusion for the user that he or she actually feels as being in the simulation. Before the virtual theatre is going to be used within engineering education, a better insight in the students' perception of the learning situation and of the learning outcome is needed. Therefore the further development of the hardware and of the scenarios is attended by different psychological studies. Before the experimental design of the studies is explained in detail, a closer look at the conceptual background of the notion of immersion is needed, which appears very often in the context of the work towards more realistic simulators.

The term immersion is frequently used as an objective, gradable measurement to describe how capable technical features are to create perfect illusions. A head mounted display for example is considered a good device to create immersion. Moreover, an HMD which provides the user with a 101° view is more immersive than a display which only provides 100°. Regarding the virtual theatre, a simulator which provides unrestricted movement has a higher grade of immersion than a simulator which doesn't have this characteristic [9].

In gaming circles, the term is used by gamers in order to explain to what extent a game can draw them in and allows them to "lose" themselves in the world of the game. Reading through forums and blogs, it also becomes obvious that within the gaming community, the meaning of immersion is ambiguous and multifaceted, from the ability of a game to "pull you away from your actual world, and swamp you with an actual psychological experience full of emotional turmoil and conflict [10] to "a games ability to draw you in with elements OTHER than it's (sic!) characters and story" [11].

Either way, immersion has a positive connotation and is a quality seal for computer games. Considering the hardware, new interfaces like Nintendo Wii or Kinect for the Xbox console illustrate the craving for more natural control modes in the entertainment sector. However, although there seems to be a broad understanding of immersion amongst hardware developers and in the gaming community, it is still not clear what exactly is meant by immersion and what is causing it from a psychological point of view, especially if the goal is not only entertainment but learning.

In a first general approximation, immersion can be defined as "the subjective impression that one is participating in a comprehensive, realistic experience" [12]. The idea of absorbing and engaging experiences is not a new concept and there are several other concepts that have a relation to immersion. Involvement, spatial presence and flow are considered key concepts to explain such immersive experiences, although they are not clearly distinct and depending on the author overlap and also refer to each other. Involvement is a psychological state experienced as a consequence of focusing one's energy and attention on a coherent set of stimuli, activities or events and it depends on the degree of significance or meaning that the individual attaches to them [13].

Whilst flow describes the involvement in an activity, presence refers to the spatial sense in a mediated environment [14]. Rheinberg et al.'s concept of flow in human-computer-interactions consists of the two subdimensions (1) smooth and automatic running and (2) absorption. The first factor refers to the feeling of utmost concentration and focusing, control over the activity, clarity of the operations, and smooth and automatic cogitations. The second factor refers to the feeling of full involvement, distorted sense of time, optimal challenge, and absent mindedness [15].

It has been shown in many studies that personal characteristics have an influence on immersion. Immersive tendency describes the ability of a person to focus on an activity and to lose track of time and space [13]. Weibel et al. [14] showed that openness to experience, neuroticism, and extraversion of the Big Five personality traits are positively correlated to the tendency to be immersed.

Another personal characteristic that is expected to have an effect on a person's experience in a virtual environment (VE) is affinity for technology. Since the virtual theatre is a very new and innovative technology it is also expected to lead to higher enjoyment during task performance than a laptop if the person generally likes technology. Especially in a simulator which enables free movement like the virtual theatre, cognitive skills like spatial sense might also have an effect on a person's experience but also on his or her performance.

3.2 Experiment Design

The studies on human experience and behaviour in the virtual theatre assess the subjective experience of presence and flow (experience) as well as learning outcome (behaviour). In line with Witmer and Singer [13] the studies carried out in the virtual theatre follow the approach that the strength of presence and flow experienced in a VE varies both as a function of individual differences and the characteristics of the VE as well as the hardware of the simulator. Individual differences, traits, and abilities may enhance or detract from presence and flow experienced in a given VE. Various characteristics of the VE and of the hardware of the simulator may also support and enhance, or detract and interfere with the subjective experience in the learning situation.

Hence, the measures assess individual differences as well as characteristics of the VE and of the virtual theatre that may affect presence, flow and enjoyment. One of the most important questions within an educational framework is whether experiencing

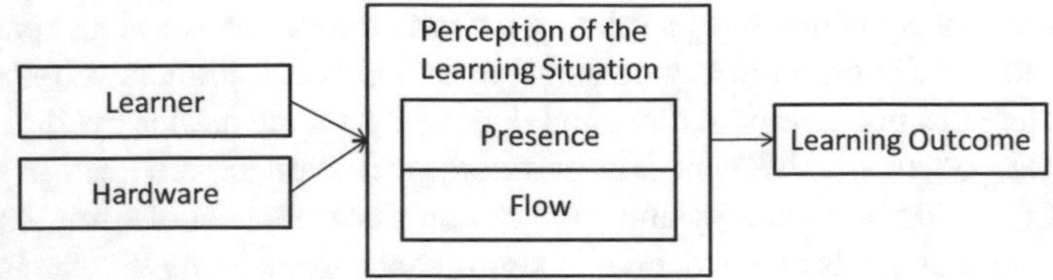

Fig. 5 Expected relationship between learner, hardware, perception of the situation and learning outcome

a VE via a virtual theatre leads to better learning outcomes than e.g. experiencing a VE via a laptop or just an HMD. Therefore the last step of the studies is knowledge retrieval. The expected relationship between learner, hardware, perception of the situation and learning outcome is visualized in Fig. 5.

To measure the strength of the effect of each variable, different study designs will be carried out within the virtual theatre. A first study will measure the immersive quality of the virtual theatre as well as its ability to improve learning. Therefore two groups of students will be compared in a first study, having to use different hardware (independent variable) in the learning situation. The variables virtual environment (a maze) and task (finding different objects in the maze) will be kept constant. The treatment group will have to fulfill the task in the virtual theatre and the control group on a laptop. The retrieval task will also be kept constant for both groups (locating the objects on a map of the empty maze on a tablet pc). Measures will assess the subjects' perception of the learning situation and the accuracy of the location of the objects.

A second study will focus of the immersive quality especially of the omnidirectional floor. Therefore two groups will have to fulfill the same task (looking around freely) in the virtual theatre, in the same virtual environment (Italian piazza). While the treatment group will have to perform physical activity (walking) within the virtual theatre, the control group will only have to make use of the 3d view (independent variable = physical activity). The retrieval task will be the same as in study one (locating the objects on a map of the piazza on a tablet pc). Measures will again assess the subjects' perception of the learning situation and the accuracy of the location of the objects. With a constant validation of the relation between student, hardware and learning outcome, it is possible to predict the level of immersion in the virtual theatre from personality. Thus, a profile could be developed, for which students the virtual theatre would have the best effect and would be most suitable. Moreover, different scenarios could be developed for different profiles.

Further studies on human experiences and behavior in the virtual theatre will include the scenarios developed for engineering education, which are described in the following chapter.

4 Application Scenarios

The virtual theatre offers a lot of possibilities for different application areas. Theoretic background knowledge can immediately be verified in a practical and perceptible way, especially for in reality not accessible situations. At the Institute for Information Management in Mechanical Engineering the virtual theatre will be used for Engineering Education and as part of a school lab of the German Aerospace Center (Forschungszentrum der Bundesrepublik Deutschland für Luft- und Raumfahrt, DLR). The scenarios developed therefore cover the areas of power plants, production facilities, facility planning or space travel. Different target groups are integrated actively in the development of new scenarios. Following the lead user approach [16] different workshops with students and teaching staff are carried out. Their ideas for scenarios and didactical approaches melt into the activities of the research group in charge. After two scenarios are being presented exemplarily, the underlying didactical principles are explained.

4.1 Scenario 1: Mars Mission

The first scenario in development is a Mars scenario where the user can walk on the surface of planet Mars. It consists of a reproduction of the landing area of the rover spirit, based on a height map of the landing are and high-definition pictures of the Mars mission. Within this environment, exact models of the different Mars rovers are placed. Based on the position and point of view of the users, additional information is presented through popup labels (Figs. 6 and 7).

4.2 Scenario 2: Nuclear Power Plant

A second scenario to be developed simulates a nuclear power plant. Students are going to be able to inspect the inner structure of a nuclear reactor and experiment freely within. Conceivable interactions would be to allow for direct manipulation of the fuel and control rods as well as cooling, etc. The consequence of every action is

Fig. 6 Panoramic view taken from the Mars rover (property of NASA)

Fig. 7 Model of the Mars rover opportunity (property of NASA)

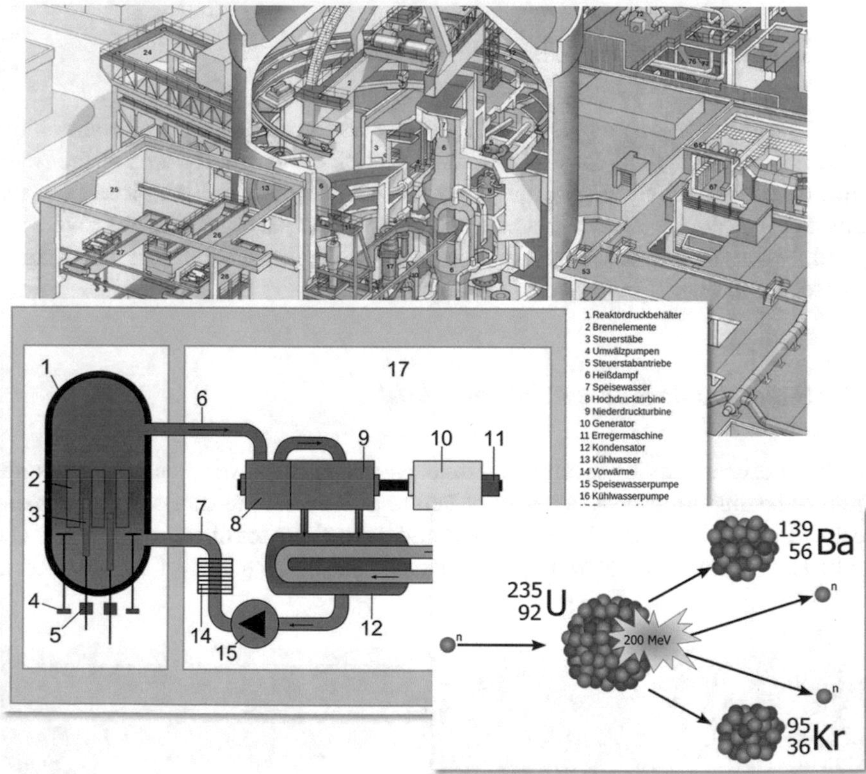

Fig. 8 Schemas of a nuclear power plant

immediately shown in the virtual environment along with the underlying equations. Additionally the varying noise of a contamination meter indicates radiation levels (Fig. 8).

4.3 Didactical Principles

The content of a certain engineering topic does not transfer itself to the students just by presentation. It has to be prepared with a didactical method in order to actively encourage learning processes. Moreover, one must always keep in mind that students not go to university in order to learn content and to accumulate knowledge, but to develop competencies. In both scenarios, different didactical methods and principles come into operation. The fact, that the students can apply theoretical knowledge in practical situations in a computer simulation supported by the repetition of the theory in the pop-ups helps them to sustainably anchor the knowledge to experiences and maybe stories instead of just memorize facts.

Another method which comes into action within the virtual theatre is exploratory learning, where content is not presented sequentially like in a lecture but openly. The main principle is that the learning process is controlled by the students. This offers the chance to activate their interest and also appeals to their motive of curiosity. As a consequence, the students set their own goals (the want to know or be able to do something) and they choose the activities which are necessary in order to reach those goals. Therefore in exploratory learning environments, learning is a non-linear process. Students can walk into dead ends, come back, and repeat certain steps etc. in order to move forward. Self-controlled learning processes are an important premise for students to develop employability, as e.g. demanded in the Bologna Declaration [17].

But there are also inhibiting factors for exploratory learning which have to be considered when putting up a new virtual learning environment. Every student has his or her own subjective mental model of the concept of learning. Vermunt and van Rijswijk [18] analyzed those subjective theories about learning of students and found three different concepts. Within the most common concept (reproductive learning), learning is being understood as a process of copying spoken or written word into the memory. Another subjective theory of learning focuses on the use of knowledge: Here, learning is a needed process, in order to be able to reproduce something subsequently at other times. Within the third concept, which happens rather rarely, learning is the necessity of self-contained examination and construction of knowledge. If students have the subjective theory of learning being a process of copying instead of self-contained construction of knowledge, they mostly also believe that in order to learn something, they need to be instructed properly by an expert [19]. The consequence for the virtual theatre is that the application scenarios need to be adaptable to the different subjective mental models of learning. This could easily be included in the development of scenarios for the virtual theatre by scaling up and down the amount of explicit knowledge repetition in pop-ups or some kind of guiding voice in the background.

In general, computer simulations offer the possibility to change the perspective and therefore making otherwise inaccessible situations accessible to the students. In a future scenario with industrial facilities, the student could take the role of the product in a production machinery in order to fully understand the principle of a value-added process. Other application areas of the same principle are the journey of a container box in logistic processes or a journey through the human body from the perspective of a blood cell, like in the TV series *Once upon a Time…Life* [20].

A promising approach which came up in a workshop with students is game based learning (GBL). The biggest advantage is here that playing or gaming is usually a voluntary activity and not carried out for learning purposes. Nevertheless, through games we learn to deal with quit complex scenarios, like chess [18]. For the virtual theatre, there are different possibilities to turn the developed scenarios into games. First, the student can be given a certain mission which has to be fulfilled or a problem that has to be solved. The student then tries to solve that problem or fulfill the mission with his own skills and knowledge (implicit learning). Only when students can't proceed, they can switch to an explicit mode, where additional information is given to them. This switch can be an inhibiting factor on perceived immersion, so those effects still have to be investigated. Another possibility is to imply different levels to a game or a VE. In an industrial scenario, students first would have to handle a simplified machinery in order to understand its basic principles and then move forward to more complex situations, where certain parts of the machinery get broken and they have to repair the site.

5 Conclusion and Outlook

In this work we introduced the virtual theatre as a new tool for exploratory learning within virtual environments. We described the technical implementation and integration of the hardware as well as the software components. We presented first applications and discussed the impact of increased immersion on the learning situation. Future application scenarios will focus an industrial facilities and machinery. It is considered to combine the virtual theatre with remote laboratories, so that the users movements within the VE are mapped to robotic manipulators. The result of the latter actions are then fed back into the VE and displayed accordingly. First applications here will be based on test benches for metal bending.

References

1. MSEAB Weibull, 2012. URL http://www.mseab.se/The-Virtual-Theatre.htm
2. Sensics. URL http://sensics.com/products/head-mounted-displays/zsight-integrated-sxga-hmd/specifications/
3. Cyber gloves systems. URL http://www.cyberglovesystems.com/products/cyberglove-III/specifications

4. Naturalpoint OptiTrack tracking tools, 2012. URL http://www.naturalpoint.com/optitrack/products/tracking-tools/
5. R.M. Taylor, II, T.C. Hudson, A. Seeger, H. Weber, J. Juliano, A.T. Helser, VRPN: a device-independent, network-transparent VR peripheral system. In: *Proceedings of the ACM Symposium on Virtual Reality Software and Technology*. ACM, New York, NY, USA, 2001, VRST '01, p. 55–61. URL http://doi.acm.org/10.1145/505008.505019
6. WorldViz LLC. vizard virtual reality toolkit, 2012. URL http://www.worldviz.com/products/vizard/index.html
7. T. Roosendaal, S. Selleri, *The Official Blender 2.3 guide: Free 3D creation suite for modeling, animation, and rendering*. No Starch Press, San Francisco, CA, USA, 2004
8. A. Bandura, *Social learning theory*. General Learning Press, Morristown, N.J., 1971
9. D. Johansson, Convergence in mixed reality-virtuality environments. facilitating natural user behavior. Ph.D. thesis, University of Örebro, Sweden, 2012
10. F. Jorgensen. The top 10 most immersive games ever made. URL http://daxgamer.com/2012/03/top-10-immersive-games/. (accessed on November 27th 2012)
11. Xav. Comment on "The top 10 most immersive games ever made", 2012. URL http://daxgamer.com/2012/03/top-10-immersive-games/. (accessed on November 27th 2012)
12. C. Dede, Immersive interfaces for engagement and learning. Science 2 **323** (5910), 2009, pp. 66–69. URL http://www.sciencemag.org/content/323/5910/66. PMID: 19119219
13. B.G. Witmer, M.J. Singer, Measuring presence in virtual environments: A presence questionnaire. Presence: Teleoper. Virtual Environ. **7** (3), 1998, p. 225–240
14. D. Weibel, B. Wissmath, Immersion in computer games: The role of spatial presence and flow. International Journal of Computer Games Technology **2011**, 2011, p. 14. URL http://www.hindawi.com/journals/ijcgt/2011/282345/abs/. Article ID 282345
15. F. Rheinberg, S. Engeser, R. Vollmeyer, Measuring components of flow: the flow-short-scale. Washington, DC, USA, 2002
16. E. von Hippel, *Democratizing Innovation*. MIT Press, Cambridge, 2005
17. European Ministers of Education. The bologna declaration on the european space for higher education, bologna, 1999
18. J.D.H.M. Vermunt, F.A.W.M.V. Rijswijk, Analysis and development of students' skill in self-regulated learning. Higher Education **17** (6), 1988, pp. 647–682
19. M. Kerres, *Mediendidaktik. Konzeption und Entwicklung mediengestützter Lernangebote*. Oldenbourg, München, 2012
20. YouTube. Once upon a time...life - the blood (1 of 3). URL http://www.youtube.com/watch?v=6myuX4ubWRQ. (accessed on December 3 2012)

Entwicklung von Remote-Labs zum erfahrungsbasierten Lernen

Tobias Haertel, Claudius Terkowsky, Dominik May and Christian Pleul

Zusammenfassung In den Ingenieurwissenschaften bietet das Lernen in Laboren ein besonderes Potenzial zum Erwerb auch außerfachlicher Kompetenzen, das in der Regel jedoch kaum genutzt wird. Das Beispiel eines fernsteuerbaren Labors mit einer Lernumgebung, die unterschiedliche Lernpfade einbindet, zeigt, wie erfahrungsbasiertes Lernen in der Hochschule ermöglicht werden kann.

Schlüsselwörter Ingenieurstudium · Kompetenzorientierung · Employability · Labor · Remote-Lab

1 Kompetenzdiskussion im Ingenieurstudium

Das Arbeitsfeld von Ingenieurinnen und Ingenieuren ist vielfältig. Wenn sie in Forschungs- und Entwicklungsabteilungen beschäftigt sind, müssen sie in der Regel mit vielen anderen Abteilungen zusammenarbeiten, z.B. mit der Finanzabteilung, dem Projektmanagement, der Marktforschung, Marketing und Vertrieb, der Service-Abteilung, Qualitätssicherung oder der Rechts- und Umweltabteilung [1]. Im Zuge immer stärkerer Vernetzungen von Unternehmens- und insbesondere Entwicklungsprozessen werden Ingenieurinnen und Ingenieure aber auch zunehmend direkt in diesen Abteilungen beschäftigt. Eine Auswertung von 900 Stellenausschreibungen für Ingenieurinnen und Ingenieure von Siemens Deutschland zeigte, dass nur zu knapp 50 % ein Einsatz in der Forschung und Entwicklung vorgesehen war, gut 50 % der Stellen waren anderen Abteilungen zugeordnet [1].

T. Haertel (✉) · C. Terkowsky · D. May
Engineering Education Research Group (EERG), Center for Higher Education (zhb), TU Dortmund University, Dortmund, Germany
e-mail: tobias.haertel@tu-dortmund.de

C. Pleul
Institute of Forming Technology and Lightweight Construction (IUL), TU Dortmund University, Dortmund, Germany

Originally published in "Themenheft Kompetenzen, Kompetenzorientierung und Employability in der Hochschule", Issue 1, Vol. #8, © 2013.
Reprint by Springer International Publishing AG 2016,
DOI 10.1007/978-3-319-46916-4_9

Ob direkt in anderen Abteilungen beschäftigt oder stark vernetzt mit ihnen zusammenarbeitend – für Ingenieurinnen und Ingenieure in Unternehmen ergibt sich daraus die Notwendigkeit, sich auf andere Sicht- und Denkweisen einlassen zu können, das eigene Fachwissen allgemein verständlich in Dialoge einzubringen und sich jenseits der technischen Probleme auch mit darüber hinausgehenden Zusammenhängen zu befassen.

Die Realität im Ingenieursstudium sieht jedoch anders aus, hier lässt die Dominanz an fachlichen und methodischen Anforderungen traditionell wenig Raum für den Erwerb anderer Kompetenzen. Seit fast 100 Jahren sehen sich die Ingenieurwissenschaften der Forderung ausgesetzt, fachliche Spezialisierung und breiter angelegte Bildungsziele in ein ausgewogeneres Verhältnis zu bringen. Bereits in der Weimarer Republik gab es Bestrebungen, das überspezialisierte Ingenieurstudium mit interdisziplinären Inhalten anzureichern. Schon damals war das Motiv „Employability",[1] das Ziel eine stärkere Verzahnung von Technik und Wirtschaft. Der Reformerfolg blieb jedoch aus, auch die neuen Studiengänge waren geprägt von ihrer technik-fachlichen Spezialisierung [4]. Ebenso wurde in den 1960er Jahren von Wirtschaftsvertretern ein breiterer Blickwinkel im Curriculum gefordert, jedoch auch ohne den gewünschten Erfolg [5].

Zu entsprechenden Ergebnissen kommen auch neuere Studien, die sich mit dem Kompetenzprofil im Ingenieurstudium befassen. Die meiste Aufmerksamkeit erzielten die VDE-Studie Young Professionals [6] und die HIS Absolventenbefragung [7].

Die VDE-Studie kommt zu dem Ergebnis, dass es neben der fachlich hervorragenden Qualität des Ingenieurstudiums Defizite in der Vermittlung von Kommunikations-, Präsentations- und Führungskompetenz gibt. Gleichzeitig wird das Studium mit Blick auf den Nutzen im Beruf als zu theorielastig betrachtet [6]. Wie wichtig solche Kompetenzen sind, zeigt die HIS-Absolventenbefragung: Die befragten Absolventinnen und Absolventen der Elektrotechnik gaben 5 Jahre nach ihrem Examen an, dass Kommunikationsfähigkeit (81 % FH, 79 % Uni), selbstständiges Arbeiten (80 % FH, 81 % Uni), Verantwortungsfähigkeit (69 % FH, 65 % Uni), Organisationsfähigkeit (68 % FH, 69 % Uni) und fachübergreifendes Denken (61 % nur FH) wichtiger sind als spezielles Fachwissen (34 % FH, 37 % Uni) [7]. Die Befragten sahen jedoch gerade bei diesen als wichtig bezeichneten Kompetenzen große Defizite bei der eigenen Beurteilung. Beim Vergleich zweier AbschlussJahrgänge (1997 und 2001), die zum selben Zeitpunkt befragt wurden (2002, also 5 Jahre bzw. 1 Jahr nach dem Examen) zeigten sich kaum Unterschiede bei der Selbsteinschätzung zu Kompetenzdefiziten [7].

Junge (2009) hat in einer bemerkenswerten Literaturanalyse sieben relevante Untersuchungen zum Qualifikationsbedarf von Ingenieurinnen und Ingenieuren zusammengefasst und ausgewertet. So kam er zu einer Gesamtliste mit 60 Kom-

[1] Anders als in vielen anderen Disziplinen wird bei den Ingenieurwissenschaften die Kompetenzdiskussion nicht nur vor dem Hintergrund von Employability geführt, sondern insbesondere auch vor einer gesamtgesellschaftlichen Verantwortung von Ingenieurinnen und Ingenieuren. Zahlreiche Arbeiten der Techniksoziologie belegen die Notwendigkeit der Fähigkeit zum ganzheitlichen, systemischen Denken von allen an der Entwicklung von Technik Beteiligten (s. z.B. [2, 3]) im Sinne einer gesellschaftlich sozial verantwortlichen Technikgestaltung.

petenzen, die er in einem weiteren Schritt auf der Grundlage der Relevanz aus, die den Kompetenzen in den einzelnen Untersuchungen zugeschrieben wurde, zu neun Kompetenzen mit hoher Gesamtrelevanz verdichtete:

„

- breites Grundlagenwissen
- Kenntnisse in EDV
- fachübergreifendes Denken
- die Fähigkeit, wissenschaftliche Ergebnisse/Konzepte praktisch umzusetzen
- die Fähigkeit, Wissenslücken zu erkennen und zu schließen
- selbstständiges Arbeiten
- analytische Fähigkeiten
- Problemlösefähigkeit
- Kooperationsfähigkeit“

“ [8][2]

Auch diese Reduktion der umfangreichen 60 Einzelkompetenzen auf 9 zeichnet immer noch ein anspruchsvoll vielfältiges Bild vom Ingenieurstudium.

2 Verschenkte Chancen beim Kompetenzerwerb im Labor

Werden die gängigen Veranstaltungsformate in ingenieurwissenschaftlichen Studiengängen auf ihr Potenzial zum Kompetenzerwerb jenseits von Fach- und Methodenkompetenzen hin untersucht, bietet sich vor allem das Labor als Ansatzpunkt für die Gestaltung neuer, hochschuldidaktisch fundierter Lehr-/Lernszenarien an [13]. Die Einbindung erfahrungsbasierten und forschungsorientierten Lernens durch den aktiven Einsatz von Laboren in der ingenieurwissenschaftlichen Ausbildung von Studierenden ist ein Lehr-Lernkonzept, das den konstruktivistischen Ansatz sehr gut unterstützen kann. Das Lernen kann im Labor aus Sicht des Lernenden gestaltet werden, sie „bewegen“ sich durch die Lernobjekte und konstruieren im Rahmen eines aktiven Prozesses ihr neues Wissen selbst. Der Umgang mit der Laborausstattung bietet den Studierenden die Möglichkeit, mit den Geräten und Maschinen ihres zukünftigen Arbeitslebens in Kontakt zu kommen und zusätzlich praktische sowie theoretische Erfahrungen beim Experimentieren mit der Ausstattung, den Methoden und den Prozessen der empirischen Forschung zu sammeln [14].

[2]Zur Problemlösefähigkeit wird Kreativität benötigt. Nach [9, 10] und [11] ist Kreativität in der Hochschullehre in sechs Facetten vertreten: (1.) reflektierendes Lernen, (2.) selbstständiges Lernen, (3.) Motivationssteigerung/Forschungsneugier, (4.) kreierendes Lernen, (5.) vielperspektivisches Lernen/neue Denkkultur und (6.) Entwicklung origineller Ideen (ebd.). Eine erste (nicht repräsentative) Auswertung der Modulbeschreibungen aus den Studiengängen Maschinenbau und Elektro- und Informationstechnik von drei Universitäten in Nordrhein-Westfalen gibt jedoch erste Hinweise darauf, dass gerade die für die erfinderische Problemlösung wichtigen Aspekte der Vielperspektivität und der Entwicklung origineller Ideen im Studium von den Lernenden praktisch nicht verlangt wird [12].

Der Einsatz von Laboren in Lehr- und Lernumgebungen kann grundlegend auf das Verständnis des Lernens nach Kolb (1984) zurückgeführt werden: „Learning is the process whereby knowledge is created through the transformation of experience“ [15]. Demnach ist ein Bestandteil des Lernens der Erwerb von abstrahierten Konzepten, die flexibel in unterschiedlichen Situationen angewendet werden können. Nach Kolbs Theorie bieten neue Erfahrungen den Impuls zur Entwicklung neuer Konzepte. Kolbs Zyklus zum erfahrungsbasierten Lernen besteht aus den folgenden vier Phasen als Grundlage des Lernprozesses:

- Konkrete Erfahrungen: Konfrontation mit einer neuen Erfahrungssituation oder Neuinterpretation einer existierenden Erfahrung.
- Reflektierende Beobachtung: Analyse, Evaluation und Interpretation der neuen Erfahrung. Von besonderer Bedeutung sind Unstimmigkeiten zwischen der Erfahrung und dem Verständnis darüber.
- Zusammenfassende Konzeptualisierung: Reflexion führt zu neuen Ideen oder zur Modifikation eines bestehenden Konzeptes.
- Aktives Experimentieren: Transformation des neuen abstrahierten Konzeptes in die Anwendung. Der Lernende interagiert mit seiner Umwelt, um zu erfahren wie diese reagiert.

In diesem vierstufigen Lernzyklus veranschaulicht Kolb, dass der Lernprozess aus einer reale Erfahrung, gemacht vom Lernenden, einer reflektierenden Beobachtung sowie der Konzeptualisierung des Erfahrenen und dem aktiven Experimentieren besteht. Mit jedem Durchlaufen des Zyklus erhöhen Studierende ihre Erfahrungen. Verglichen mit einer aufsteigenden Spirale, wandelt der Lernzyklus die Lernakti vitäten in Erfahrungswissen, Fähigkeiten und Kompetenzen.

Das Labor bietet damit die Chance zum forschungsorientierten Lernen. Es ist kein Zufall, dass Forschungsprozesse nahezu immer dieselben Schritte aufweisen, bestehend aus einer Erfahrung oder einer Frage, realen Experimenten und neuen Forschungsergebnissen [16]. Darüber hinaus haben Herrington und Oliver die Wichtigkeit einer authentischen Lernumgebung für einen erfolgreichen Lernprozess herausgearbeitet [17]. Diese authentische Lernumgebung kann durch Lehr- und Lernaktivitäten in Laborveranstaltungen angeboten werden. In diesen Veranstaltungen setzen sich Studierende mit realen Inhalten in einem realistischen Kontext auseinander und führen reale Handlungen durch. Durch eine weiterführende Verbindung der Laboraktivitäten mit realen Problemstellungen – z. B. der Verknüpfung mit aktuellen Aspekten der Forschung – haben Studierende die Möglichkeit den gesamten Weg zu beschreiten, von der Frage über die experimentelle Erforschung bis hin zum finalen Gebrauch der erarbeiteten Resultate.

In der Praxis allerdings wird das große Potenzial vom Lernen im Labor oft nicht abgerufen. In der eigentlichen Laborveranstaltung werden oft nach schrittweiser Vorgabe experimentelle Daten erhoben. Dabei sind Studierende nur teilweise selbstständiges tätig. Oft werden die einzelnen Arbeitsschritte, z.B. wegen komplizierter Maschinenbedienung oder Sicherheitsvorschriften, nur erklärt und vorgeführt. Die Chance zum forschenden Lernen ist mit diesem in der Praxis oft zu findenden Lehr-Lern-Szenario nicht ausgeschöpft.

3 Experimentierendes Lernen im Remote-Lab

Eine wesentliche Bedingung, um einen umfassenderen Kompetenzerwerb im Labor zu ermöglichen, liegt im Verzicht auf die hohe Einstiegsschwelle (Bestehen des Testats, anschließende Zuweisung einer bestimmten und sehr begrenzten Laborzeit) und die Durchführung bereits vorgegebener Experimente. Stattdessen sollten die Studierenden ihre Forschungsfragen, die sie experimentell beantworten möchten, selbst entwickeln können, oder mit der Lösung realer, praktischer Probleme, wie sie auch in der Berufsausübung auf sie zukämen, beauftragt werden. Ihnen sollte die Möglichkeit gegeben werden, die Lösungswege zur Beantwortung ihrer Fragen selbst zu finden und dabei schließlich zum Experimentieren auf das Labor zugreifen zu können. Jedoch würden bei oft mehreren hundert Studierenden die Labore mit dem freien Zugriff der Studierenden organisatorisch überfordert.

Beides, die Unterstützung der Lernenden bei der selbstständigen Problemlösung und der Durchführung von Experimenten im Labor, lässt sich mit dem im PeTEX-Projekt[3] entwickelten Konzept des Remote-Labors in der Lernumgebung Moodle umsetzen. PeTEX ermöglicht das „aktive Experimentieren“ nach Kolb, indem es in einer integrierten Lernumgebung relevante Informationen, die Möglichkeit zum Experimentieren und von der Suche nach der Problemlösung bis zur Auswertung und Diskussion der Ergebnisse die Gelegenheit zur Kommunikation zur Verfügung stellt [18]. Abbildung 1 zeigt die Laborumgebung (hier zur Bestimmung von charakteristischen Materialeigenschaften), in der Studierende alle Parameter des Experiments einstellen und anschließend das Experiment per Webcam verfolgen können. Die gesammelten Daten stehen unmittelbar danach zur Auswertung zur Verfügung.

Die umfassende Einbindung des tele-operativen realen Versuches erfolgt dabei durch die nahtlose Integration in die Lernplattform Moodle und ist somit direkt in den Lernprozess integrierbar [19]. Der Einsatz von Moodle mit integrierten realen Experimenten bietet aus der didaktischen Perspektive den Vorteil, dass durch die Gestaltung unterschiedlicher Lernpfade die Lehrendenzentrierung mit zunehmender Erfahrung von einer Lernenden-Zentrierung überlagert und schließlich ganz abgelöst wird:

1. Für Anfänger stellt PeTEX einfache Aufgaben zum Kennenlernen der Experimente und der Auswertungsmethoden in vorgeschriebenen Lernpfaden zur Verfügung.
2. Fortgeschrittene Lernende bearbeiten sogenannte Real World Szenarios, die sich mit realen Problemstellungen aus der beruflichen Praxis befassen. Zur Lösung der Aufgaben reicht das innerhalb der Lernplattform angebotene Lernmaterial nicht mehr aus, sondern es müssen Suchbewegungen und strategische Entscheidungen getroffen und durchgeführt werden.

[3]PeTEX - Platform for eLearning and Telemetric Experimentation, gefördert durch die EU, Lifelong Learning Programme (LLLP) / Sub-Programme LEONARDO DA VINCI (ICT), 12/2008 bis 11/2010.

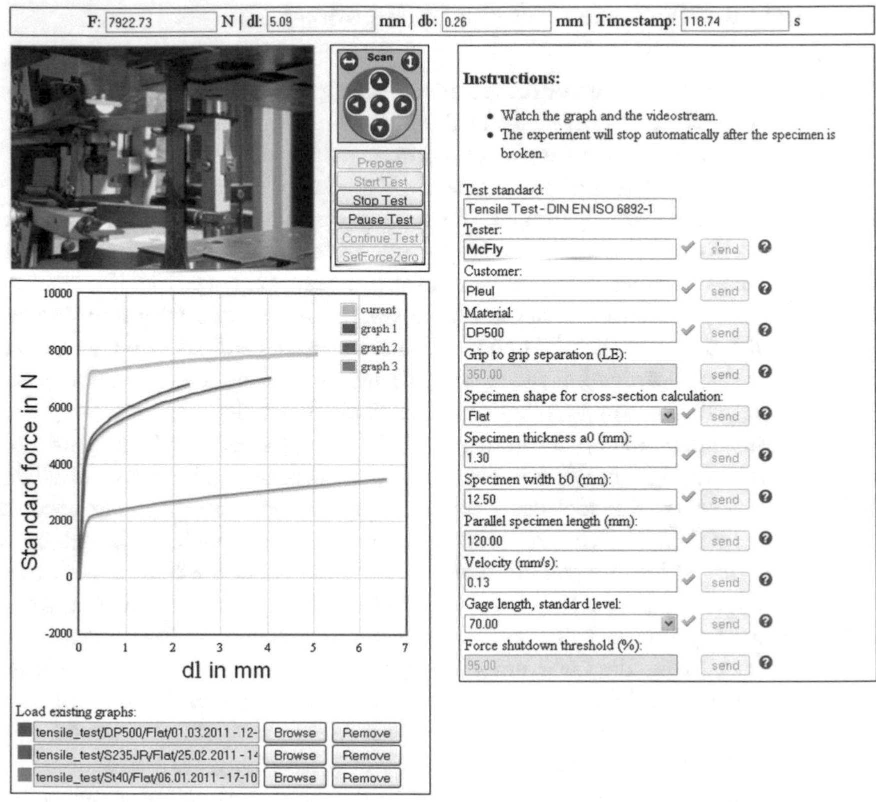

Abb. 1 Remote-Lab-Umgebung PeTEX

3. Für weit fortgeschrittene Lernende wird die Plattform für selbstorganisiertes Forschendes Lernen zur Verfügung gestellt. Die Lernenden müssen hierzu ihre Forschungsfrage und das zugehörige Forschungsdesign selber entwickeln.

Es obliegt den Kursbetreuern, diese drei Szenarien hinsichtlich Aufgabenzusammenstellung, Schwierigkeitsgraden, integrierte Lernobjekte, Einzel- oder Gruppenarbeit, notwendige Unterstützung, usw. entsprechend ihren beabsichtigten kompetenzgenerierenden Lernzielen einerseits, und dem Vorwissen ihrer Lernenden andererseits zielgruppenspezifisch anzupassen.

Auf der technischen Seite hat die Verwendung von realen Experimenten im Modus der Fernansteuerung („remote labs") anstelle von Software-basierten Simulationen („virtual labs") zwar den Nachteil, die Skalierbarkeit–und damit den parallelen Zugriff beliebig vieler User zur gleichen Zeit–nicht wesentlich zu erhöhen, was aber abgewogen werden kann gegenüber den Vorteilen, dass sich erstens verschiedene Einrichtungen die Anschaffung und Nutzung von kostspieligen Versuchseinrichtungen aufteilen können, und dass zweitens reale Experimente den Vorteil haben, reales Verhalten der Versuchsanordnung und der verwendeten Materialproben anzubieten, anstelle von immer gleichen Abläufen des Materialprobenverhaltens bei simulierten Experimenten [12, 13, 20].

4 Fazit

Das PeTEX-Projekt hatte neben der Umsetzung der Remote-Labs in der der Lehre zum Ziel, die Universität als Ausbildungsstätte mit der Wirtschaft als späteren Arbeitgeber für Ingenieure zusammen zu bringen. Aus diesem Verständnis heraus sollte über institutionelle Grenzen hinweg eine, durch die Arbeit am und im Labor verbundene, Community aufgebaut werden, was durch die gemeinsame Lernplattform ermöglicht wurde. Mit einem weiteren Ausbau solcher Angebote kann die beharrlich geforderte Brücke zur Verknüpfung vielfältiger Kompetenzen in der ingenieurwissenschaftlichen Lehre geschlagen werden.

Literaturverzeichnis

1. F.S. Becker. Was heute von Ingenieuren verlangt wird. Markttrends, Erfahrungen von berufsanfngern, Erwartungen von Personalverantwortlichen und Karrieremechanismen: Sonderdruck für den VZEI - Zentralverband Elektrotechnik- und Elektroindustrie e.V. URL http://www.zvei.org/Publikationen/Was%20heute%20von%20Elektroingenieuren%20verlangt%20wird.pdf
2. T. Haertel, J. Weyer, Technikakzeptanz und Hochautomation. Technikfolgenabschätzung - Theorie und Praxis **14** (3), 2005, pp. 238–245
3. T. Haertel, Techniksteuerung durch Normung am Beispiel der Ergonomie von Speditionssoftware: Ergonomienorm oder Ergononienorm? Ph.D. thesis, Duisburg, Essen, 2010
4. W. König, Vom Staatsdiener zum Industrieangestellten: Die Ingenieure in Frankreich und Deutschland 1750–1945. In: *Geschichte des Ingenieurs. Ein Beruf in sechs Jahrtausenden*, ed. by W.K. W. Kaiser, Hanser, München, 2006, pp. 179–232
5. F.S. Becker, Der europäische Hochschulraum: Bekommen wir die Ingenieure, die wir brauchen? In: *Arbeitsmarkt Elektrotechnik Informationstechnik*, ed. by J. Grüneberg, I. Wenke, VDE VERLAG, 2004
6. VDE-Studie Young Professionals, 2007. URL http://www.vde.com/de/Karriere/Ingenieurausbildung/Documents/Studie%20Young%20Prof%202007%20komplett%20DRUCK_1.pdf
7. K.H. Minks, Kompetenzen für den globalen Arbeitsmarkt: Was wird vermittelt? Was wird vermisst? In: *Arbeitsmarkt Elektrotechnik Informationstechnik*, ed. by J. Grüneberg, I. Wenke, VDE VERLAG, 2005
8. H. Junge, Projektstudium als Beitrag zur Steigerung der beruflichen Handlungskompetenz in der wissenschaftlichen Ausbildung von Ingenieure: Dissertation. Ph.D. thesis, TU Dortmund, 2009
9. I. Jahnke, T. Haertel, Kreativitätsförderung in Hochschulen – ein Rahmenkonzept. Das Hochschulwesen **58** (3), 2010, pp. 88–96
10. T. Haertel, I. Jahnke, Kreativitätsförderung in der Hochschullehre: Ein 6-Stufenmodell für alle Fächer?! In: *Fachbergreifende und fachbezogene Hochschullehre*, ed. by I. Jahnke, Wildt. J., Bertelsmann, Bielefeld, 2011, pp. 135–146

11. T. Haertel, I. Jahnke, Wie kommt die kreativitätsförderung in die Hochschullehre? Zeitschrift für Hochschulentwicklung **6** (3), 2011, pp. 283–245
12. T. Haertel, C. Terkowsky, Where have all the inventors gone? the lack of spirit of research in engineering education. In: *Proceedings of the 2012 Conference on Modern Materials, Technics and Technologies in Mechanical Engineering*, ed. by The Ministry of Higher and Secondary Specialized Education (MHSSE) of the Republic of Uzbekistan. 2012, pp. 507–512
13. C. Terkowsky, I. Jahnke, C. Pleul, A.E. Tekkaya. Platform for elearning and telemetric experimentation. a framework for community-based learning in the workplace. position paper for the Workshop cscl at work, 2010
14. H.G. Bruchmüller, A. Haug, *Labordidaktik für Hochschulen Eine Einführung zum Praxisorientierten Projekt-Labor*, *Schriftenreihe report*, vol. 40. Leuchtturm-Verlag, Alsbach, 2001
15. D.A. Kolb, *Experiential learning. Experience as the source of learning and development.* Prentice-Hall, Englewood Cliffs, N.J., 1984
16. J. Wildt, Forschendes Lernen: Lernen im Format der forschung. Journal Hochschuldidaktik **20** (2), 2009, pp. 4–7
17. J. Herrington, R. Oliver, An instructional design framework for authentic learning environments. Educational Technology Research and Development **48** (3), 2000, pp. 23–48
18. C. Pleul, C. Terkowsky, I. Jahnke, A.E. Tekkaya, Platform for e-learning and tele-operative experimentation (petex) – holistically integrated laboratory experiments for manufacturing technology in engineering education. In: *Proceedings of SEFI Annual Conference, 1st World Engineering Education Flash Week*, ed. by J. Bernardino, J.C. Quadrado. Portugal, Lisbon, 2011, pp. 578–585
19. C. Pleul, C. Terkowsky, I. Jahnke, A.E. Tekkaya, Tele-operated laboratory experiments in engineering education the uniaxial tensile test for material characterization in forming technology. In: *Using remote labs in education*, ed. by J. Garca Zuba, G.R. Alves, Publicaciones De La Unive, [Place of publication not identified], 2011, pp. 323–347
20. C. Terkowsky, C. Pleul, I. Jahnke, A.E. Tekkaya, Tele-operated laboratories for online production engineering education. platform for e-learning and telemetric experimentation (petex). International Journal of Online Engineering (iJOE). Special Issue: Educon 2011 **7**, 2011, pp. 37–43

Bringing Remote Labs and Mobile Learning Together

Dominik May, Claudius Terkowsky, Tobias Haertel and Christian Pleul

Abstract Within (remote) laboratories in Engineering Education students have the chance to do own experiments and by that gain own experiences in their learning processes. Apart from technical questions, one of the most intriguing aspects in this context is how students can document their learning process and show to others (teachers and/or other students) what they have achieved. Another aspect concerns the question of learner's mobility during the learning process. If the laboratory can be accessed remotely, why do we constrain learners in their level of liberty by forcing them to sit in front of a fixed computer to use a location-independent environment for experimentation? Therefore, rendering this environment available for mobile devices is the logical consequence. Furthermore, integrating mobile devices into the course's technical environment means to take a whole new approach to the teaching and learning process itself. It is especially a question of embedding mobile devices into the users' workflow (or better "learn flow") rather than a simple question of accessibility. The following article features an example of how remote laboratories can be linked with mobile devices and e-portfolios, thus creating a unique learning environment helping learners to document their personal learning processes and to exchange them with others while at the same time being flexible in means of time and place. This combination of topics has been realized within one subtask of the project "ELLI – Excellent Teaching and Learning in Engineering Education" at TU Dortmund University.

Keywords Engineering Education · E-Portfolios · Mobile Learning · Remote Laboratories · tele-operated Laboratories

D. May (✉) · C. Terkowsky · T. Haertel
Engineering Education Research Group (EERG), Center for Higher Education (zhb), TU Dortmund University, Dortmund, Germany
e-mail: dominik.may@tu-dortmund.de

C. Pleul
Institute of Forming Technology and Lightweight Construction (IUL), TU Dortmund University, Dortmund, Germany

Originally published in "International Journal of Interactive Mobile Technologies (iJIM)", Issue 3, Vol. #7, © IAOE 2013.
Reprint by Springer International Publishing AG 2016,
DOI 10.1007/978-3-319-46916-4_10

1 Introduction to Tele Operated Laboratories as a Place for Learning

Once having graduated, former engineering students will work on solving real problems creatively and they will work with real technical equipment- regardless if they head for a career in a company or in the academic sector. But do these people get into contact with this equipment during their studies? Most of their time engineering students are sitting in the lecture hall following the presentation in which the professor explains to them the course's content. In other words: The students try to understand and memorize what they have to know in order to pass the course's exam. In "classical" lectures there is only little space and time for the students to understand the big picture of the subject and the inherent research process with its questions, research activities and result interpretation. Hence, not few lectures feature results of research activities without providing their greater context or the research questions which were important at the beginning of the research process. Even if a professor would like to do so, in many cases there is simply not enough time for it. One possibility to change this fact could be the use of laboratories in teaching and to implement experiential [1] and research based learning in the teaching and learning process [2]. To bring the students in contact with laboratory equipment means bringing them in contact with the technical equipment of their future profession and giving them the chance to develop central technical competences for the technical part of their future career. In addition to the technical competences, for us the students' work in laboratories offers the opportunity to add aspects of

- systemic thinking
- problem definition
- responsibility
- innovation

The work presented in this paper will base on the achievements of the PeTEX project, will deploy its technological infrastructure and will optimize it. By this we will extend the possibilities innovate the existing concept. The main conception of the further development is the combination of the topics virtual learning environment, mobile learning and creativity. All this work will be carried out as a subtask of the new project ELLI – Excellent Teaching and Learning in Engineering Education which is funded by the German Ministry of Research and Education until 2016.

1.1 Constraints and Solutions for the Use of Laboratories in Education

A very important factor that hinders the use of laboratories by students in teaching is the cost of such equipment and the organizational effort of co-locating students, equipment and supervisors. Especially small universities often face the situation that they neither can afford all the laboratory equipment nor can allow the students to use it by themselves as it might get damaged. This means in many cases that lab experiments, if the professor tries to integrate them into the lecture, are either only shown via video or that the faculty's staff shows the equipment during guided tours through the laboratory. This means a real dilemma for modern engineering education.

One possible way out if this dilemma-wanting the students to develop technical competences on the one hand and having them done experiments but not being able to use the equipment on the other hand-are tele-operated (called "remote") and virtual laboratories. With them the laboratory equipment can be used by different universities from different places or very risky experiments can be done completely virtually.

1.2 PeTEX – Platform For eLearning and Telemetric Experimentation

Important research on the use of remote laboratories in teaching engineering aspects was done by the universities of Dortmund (Germany), Palermo (Italy), and Stockholm (Sweden) within the project PeTEX – Platform for e-learning and Tele-operative Experimentation [3, 4]. The technical part of PeTEX was carried out at TU Dortmund University by its Institute of Forming Technology and Lightweight Construction (IUL, Prof. Tekkaya) and integrated in close co-operation with the Center for Higher Education (former Center for Research on Higher Education & Faculty Development, Prof. Wildt) [5, 6]. Within this project comprehensive research on using remote laboratories in teaching was carried out. Therefore a network of three prototypes in the field of manufacturing technology was developed [7, 8].

Our work's overall context is the implementation of research-based and experiential learning by using laboratories in higher engineering education at universities. As explained above, our aims are that students get into contact with real technical equipment, understand the greater context of research and gain technical competences for their future work. In this context we will proceed as follows:

- Considering that laboratory equipment is expensive and some experiments even can be dangerous for students, we will explain why and how remote laboratories should be integrated in teaching (Section 2).
- Up until today a weakness of such teaching approaches is the need for an open designed learning environment in which the students can act independently while at the same time can be guided through the learning process. We will work out how e-portfolios can help in this context and how they can be used to document and reflect on the learning process (Section 3).
- In a final step we will change over to the topic mobile learning. In order to support the students' learning process as much as possible and to leave them the choice of using the software virtually every time and from everywhere we will open the software for the use from mobile devices such as smart phones and tablet

computers. By showing different scenarios we explain how the use of mobile devices can support the learning process significantly and how they can help to promote creativity (Section 4).
- At the end of the paper we will explain shortly what our future steps will be in order to put our plan into action (Section 5).

2 Learning with Remote Laboratories

Learning through experiments in general has become a central part in modern higher engineering education [9]. Implementing experiential learning and research-based learning by the active use of laboratories in higher engineering education by students is a teaching and learning concept which supports the constructivist approach. The learning arrangement is designed from the learners' point of view because the user is the one to design his learning process and "walk" through the learning objects while constructing his knowledge inside an active process.

As mentioned in the introduction, the students have the opportunity to get into contact with the physical equipment of their future professional life as well as to make practical and theoretical experiences with equipment, methods and processes of empirical research by the use of laboratory equipment in teaching. That is why doing technical experiments in a laboratory is an adequate way of applying, enhancing and testing knowledge the students have acquired during the lecture and developing central competences by doing so.

2.1 Kolb's Experiential Learning Cycle

The use of laboratories in teaching and learning environments can basically be traced back to understanding of learning explained by Kolb: "Learning is the process whereby knowledge is created through the transformation of experience" [1]. Kolb states that learning involves the acquisition of abstract concepts which can be applied flexibly in a range of situations. According to Kolb's theory, the impetus for the development of new concepts is provided by new experiences. Kolb's concept of experience is defined in his experiential learning theory consisting of a four-phase cycle in which the learner traces all the foundations of his learning process:

- Concrete Experience: A new experience of situation is faced or a reinterpretation of an existing experience takes place.
- Reflective Observation: The new experience is analyzed, evaluated and interpreted. Of particular importance are any inconsistencies between the experience and the understanding of it.
- Abstract Conceptualization: Reflection gives rise to a new idea or a modification of an existing abstract concept.

- Active Experimentation: Transforming the new abstract concept into operation, the learner interacts to the world around him to check what emerges.

In his four-step learning cycle Kolb explains that at the beginning of each learning process there is a real learner's experience (step 1) which is followed by a reflective observation (step 2). From that point on the learner tries to conceptualize what he has experienced (step 3), starts to experiment actively (step 4) and generates new experiences. This is the start of a new cycle. With every loop – from the simple to the complex – the student enhances his experiences. Thus, the learning cycle transforms learning activities into a helix of experience-based knowledge, skills and competences.

2.2 The Learning Process in the Light of Research Processes

Since its beginnings university always has been a place not only for learning but for research, too. In order to be more concrete, at university these two processes always were thought and implemented in unison, hence they inspired be each other. For both learning and researching, explicit steps can be defined which describe the process' sequences. For the learning process we already explained the sequences as mentioned above by showing Kolb's learning cycle. For the research process a first approach can be to define the following steps: Make a practical experience, define the research question, implement research activities and interpret the results. If you now look at the learning process as well as the research process and define both as a circle, you will find discover that both can be synchronized. Wildt [10] did this and showed clearly that very similar steps can be identified (see Fig. 1). Surprisingly this fact has never had a severe impact on the way teaching is done at universities. As mentioned above, especially in engineering studies classical lectures are the most often used teaching method. This is the fact although there are some alternatives, which would permit a combination of learning- as well as research processes.

2.3 Research Based Learning

Research based or experiential teaching and learning in higher education is one adequate way of implementing learner centered teaching. In addition to that Herrington and Oliver worked out the importance of an authentic learning environment for a successful learning process [12]. According to them, this authentic learning environment can be achieved by teaching and learning activities in laboratories in which students can face a real context and carry out real activities. By connecting the actions in laboratories in a next step to real problems – e.g. from current research or from

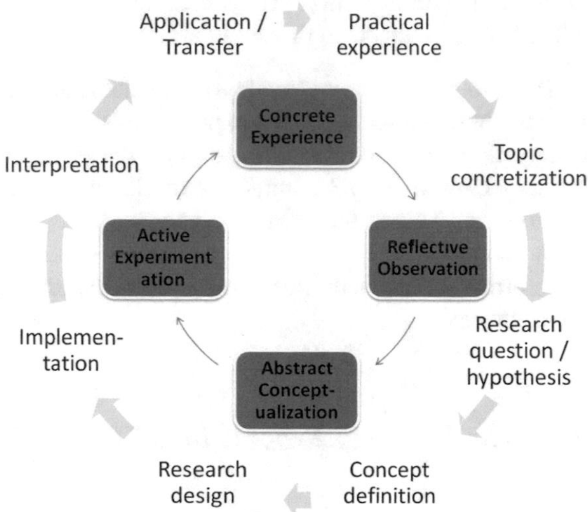

Fig. 1 Synchronized learning and research process [11]

the industry – the students are able to go the whole way from the question at the beginning of an experiment to the final use of the results and can experience the relevance of their work.

2.4 Active Experimentation Using Tele-Operated Equipment

Employing remote and virtual laboratories in teaching provides a vast range of opportunities to implement experiential learning in the field of mechanical engineering following the path of research-based learning [5]. In the following we will explain how a remote laboratory was put into practice. We noticed a discussion within the community whether remote labs can and/or should replace real laboratories. This may not be our concern in this paper as we do not want to advocate for or against one or the other laboratory solution without looking at its circumstances. There are and there always will be situations in which the use of real or remote laboratories makes more or less sense than the other.

One example in the context of manufacturing technology, namely forming technology, can be to use the remote lab concept for the purpose of material characterization. This could be offered in addition to a conventional lecture or in order to enhance traditional hands-on labs during the phase in which students prepare themselves for the lab or when they would like to rework some of the test steps while writing the lab report. Following the approach based on Kolb's experiential learning cycle, students can deal with basic concepts of metal forming during the lecture to test and see what they discussed in class by doing experiments on their own in order to create their own knowledge as well as using the remote experiential equipment. Another opportunity could be that students are given a real engineering problem related to material behavior. They are asked to work on this problem in small groups by planning and carrying out experiments using tele-operated equipment. Finally they have to present their explorations and and suggestions on how they would deal with the problem [5].

Fig. 2 Testing machine

Fig. 3 Robot positioning a specimen

In order to support this entire process and especially the step of "active experimentation" an appropriate level of clear interaction and feedback needs to be integrated to the tele-operated experimental setup. In the PeTEX project a complete experimental setup (Fig. 2) has been moved to a new level using innovative engineering design, modern concepts of automation, measurement technology and robotics as shown in Fig. 3.

All aspects have been brought together by developing a clear and interactive user interface providing real time feedback of the running experiment. In Figure 4 the screen for the uniaxial tensile test is shown. While using the live camera stream (1), users can investigate the surrounding test apparatus, e. g. sensors or clamping devices. Afterwards the learner initiates the preparation of the experiment (2), using the integrated 6-axis robot to select and check an appropriate specimen. To freely configure the experiment, relevant test parameters (3) can be filled in. When the test is started (4) the robot positions the specimen to the fully automated clamping device. The developed innovative concept of the fully automated clamping process and parallel measuring of relevant values is mentioned in [3].

Also during the test, a high level of interaction is provided to the user by manipulating the camera view or pausing and continuing the test. Pausing the test – which means the load is not further increased since that moment – causes a reaction by the material. This phenomenon is graphically visible in the real time diagram (6)

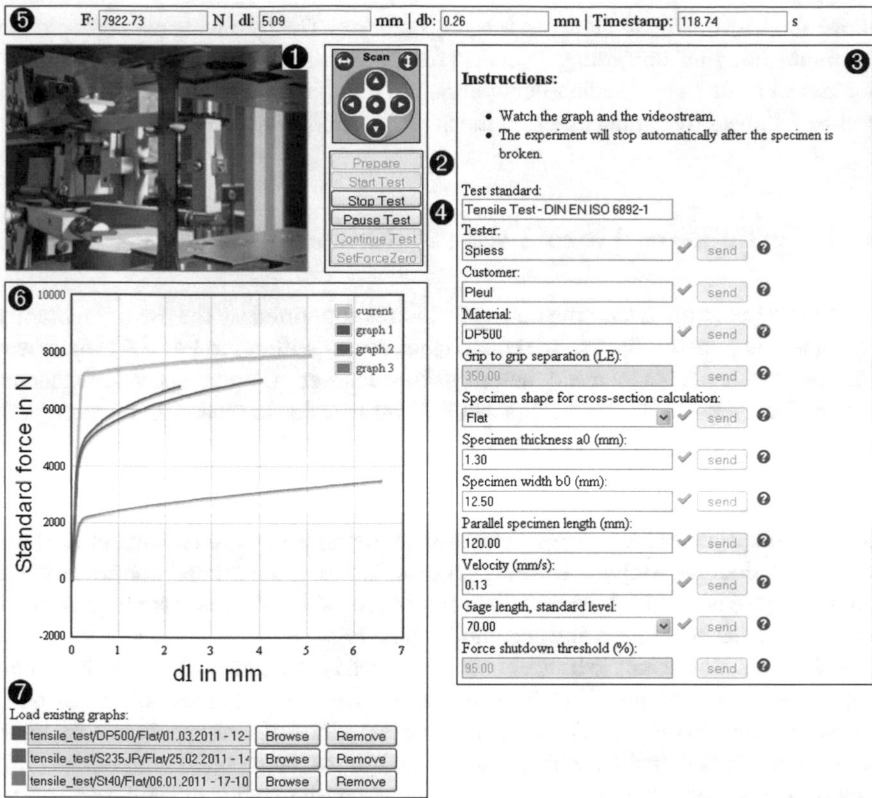

Fig. 4 Interface to the tele-operated experiment [4]

and also in the real time test data at the header bar (5). Comparisons with prior test data are available within the data base (7) and the graph (6). After the experiment is finished, learners are provided with a data package including all results for further analysis and investigation.

Additionally the entire tele-operated experimental environment was made available with the learning content management system Moodle. Within Moodle, we conducted the alignment of four, for us elementary, areas for this kind of socio-technical system. This socio-technical alignment for tele-operated laboratory learning consists of the adjustment of the technical, didactical, media and social level. By integrating the tensile test environment into Moodle, this socio-technical alignment was put into a usable as well as flexible environment.

An often quoted challenge to such open designed learning concepts is that it turns out that a very sophisticated concept is needed for teachers to enable them to document and evaluate the learners' behavior and achievements during the learning process using a virtual laboratory. It is obvious that such a concept requires different systems for the instructor to accompany the learner through the learning process and –

above all – to evaluate the achieved learning outcome. The following passages present the future thoughts concerning a concept for the learning process' documentation in context with the use of remote laboratories in combination with e-portfolios. This will be followed by explanations on the use of mobile devices in this context.

3 E-Portfolios and their Use in Experiential Learning

In addition to the open learning concept which is supported by the use of laboratory equipment in general, the use of remote laboratories within the PeTEX project was designed for the usage by a very heterogeneous learner group composed of students and professionals [13]. This means that the software for the learning process' documentation as well must be designed very open in order to prevent system inherent barriers for different learner types. In this context it should be mentioned that the PeTEX system intends to merge higher education and the workplace as well as to create an international learning community not being limited to one institute. These 3 aspects – documenting the learning process, building a learners' community and connecting the students' work with their future professional work – are requirements the software has to address and accomplish (see Fig. 5).

Software which seems to be adequate and which is frequently discussed in similar contexts is the e-portfolio [14]. E-portfolios are based on the general idea of portfolios, referring to the idea of collecting different kinds of documents in a folder in order to reflect personal learning processes and to exchange them with others [15]. E-portfolios support the same, but they are made online and provide the collection of different kinds of data like texts, tables, photos, videos and audio [16]. E-portfolio

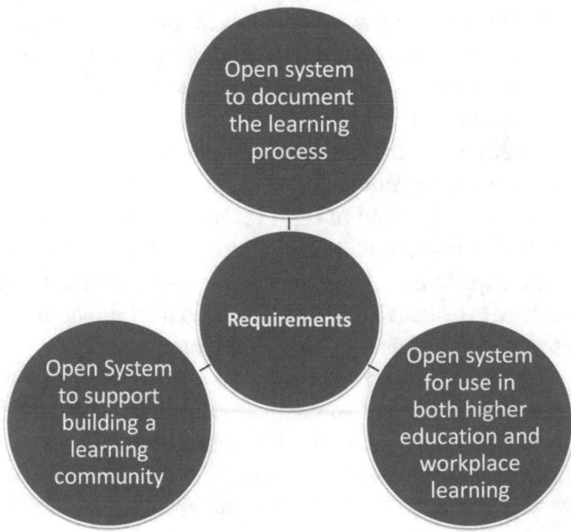

Fig. 5 Requirements for a software system documenting the learning process in remote laboratories

software could be added technically to the Moodle environment – which is already used in the PeTEX context – very easily because an online e-portfolio application especially designed for Moodle already exists by the name of Mahoodle. In the following we will explain why e-portfolios fulfill the three main requirements in the new PeTEX context [6].

3.1 E-Portfolio as a Learning Process Documentation

The user, regardless whether in higher or in professional further education, can arrange all the data he wishes to document or to show in different ways in order to create his own portfolio just like his personal page in any social network. He can present experiments and its results, show photos from the test set-up, explain his thoughts on the research and so on. Additionally, he can permit other users, learners or teachers, to view his e-portfolio. By creating such an e-portfolio the learner can document his own learning and research process and start reflecting on the experiments he does during his research-based learning process [11]. This reflection is an important aspect as it corresponds to his personal learning circle. Especially for students the e-portfolio can serve as a means of orientation or checkpoint in the own field of research [1, 11, 17]. By the same way the teacher can evaluate the learner's action by regarding a learners' portfolio. Because other persons are able to see the collection in the portfolio it can be said that it is not only a way of documenting the learning process but as well a way to communicate it so that a collaborative learning process can be achieved. This leads to the next use of E-Portfolios in the PeTEX context.

3.2 E-Portfolio as a Learning Community Software

Considering the e-portfolio as software for documentation and evaluation is just one application of the system. A constructive enrichment by using e-portfolios is the community building. Every author of an e-portfolio is able to allow other users to view various parts of his portfolio as well as to view others' portfolios. This means that learners who are doing experiments in the PeTEX system and filling their e-portfolios have the chance to get into contact with each other via the portfolio software. They can see what others are especially interested in, start discussing about it, give comments and help each other in case of an emerging problem during the experiential learning process. Along the course of this interaction, a specialized community on remote laboratories evolves within the PeTEX context.

3.3 E-Portfolio as a Bridge Between the University and the Workplace

The PeTEX system is designed for the usage in higher education and workplace learning. This means in a first step that both user groups can use the e-portfolios in the explained way of use. A further future thought is to use the e-portfolio as a livelong system to document own competencies from university level on and during the whole professional life. This should be explained by an example in three steps:

Step 1 - An engineering student starts working with the PeTEX system at the university. He uses the system in order to document his experiments. During his studies he conducts different experiments, compares them and collects all research results and data in his e-portfolio in order to scientifically describe a certain material behavior which was observed (e.g. while pausing the test for a couple of seconds), and reflects on his own way of learning. The teacher is able to evaluate his learning behavior. This can be regarded as the main use of e-portfolios at university.

Step 2 - Because the PeTEX system as well addresses workplace learning the e-portfolios can be viewed as a bridge from university to professional life. Depending on the specific use of e-portfolios by the student he can take his portfolios to present himself to potential employers. Hence, they can see what the students did in this field of his studies and whether he accords with the company's needs. In this context the e-portfolios can support application processes.

Step 3 - Once the former student pursues his career as an employee, there is no need to stop working with his portfolio. He still can work on his collection by documenting new experiments as well as gained knowledge and competencies from his new occupation. By doing so, the employee does not stop reflecting on his learning process. His e-portfolio grows with every year and successively becomes a better representation of his professional life and his competences. Especially the last aspect matches perfectly with the advantage of the PeTEX system as small and medium sized companies use it to enhance their technological skills by doing research with the PeTEX hardware. In addition to that they can use the e-portfolios to implement a system for the documentation and measurement of the employees' skills and competences. This could be supported by the lifelong use of e-portfolios.

Summing up all these aspects it can be concluded that the use of e-portfolios in the PeTEX context can support the idea of experiential and research-based learning even if there are a couple of challenges to meet [11]. The portfolios can be used to document and present the research and learning process, to build up a specially focused learning community as well as to merge university and workplace learning.

4 Scenarios of Using Mobile Devices in Combination with E-Portfolios

Another frequently mentioned new concept in context with higher education is mobile learning. From a technical perspective mobile learning means to integrate mobile devices like cell phones, smart phones or tablet computers in the learning process [17]. Among others, one of the advantages of mobile learning is that previously unplanned time periods can be utilized for learning and that learning processes can be initiated virtually everywhere [17].

In addition to that using mobile devices can support the creativity process because new ideas mainly come spontaneously and the fact of carrying a mobile device throughout the day makes it easily possible to put down a note with an idea and to work on it later or to work on it immediately as we will explain in the scenarios. Bringing e-portfolios on tablet devices for example could be an opportunity to combine the concepts presented in this text with mobile learning. In the following we will present different scenarios how the use of mobile devices can enrich the concept of remote laboratories in higher engineering education. These scenarios differ mainly in terms of individual or collaborative learning processes and self-directed or teacher-directed learning processes. These four aspects in different combinations lead to the scenarios, so that it becomes obvious how flexible the use of e-portfolios is in our learning concept. As we wanted an open learning environment for working with the laboratory equipment remotely it seems that the e-portfolio software fulfills this requirement perfectly. Figure 6 shows the four scenarios that are explained in the following.

4.1 Possible Scenarios

Scenario 1 “Using the software in creative moments” – A first scenario could be that a student is thinking about his experiments while sitting at home and watching TV or while he goes out with his friends. He is really struggled by his research work, thinks about his parameters, his results and why his experiments offered the results that showed up. Suddenly he has an idea on a hypothesis and wants to check it by rereading his last experiments in the portfolio or doing a new experiment. Because he can use the software for connecting with the experiential environment by using his tablet computer there is no need to wait until the next day for doing the experiment over again at the university but he can stay where he is and even can stay sitting on the sofa for checking his hypothesis. The new result can immediately be put in his portfolio so that he documents his new step within his research process. Using a simulation in a virtual laboratory instead of the remote experiment can be a method to (pre-)check the hypothesis first and then to carry out the real experiment remotely [18].

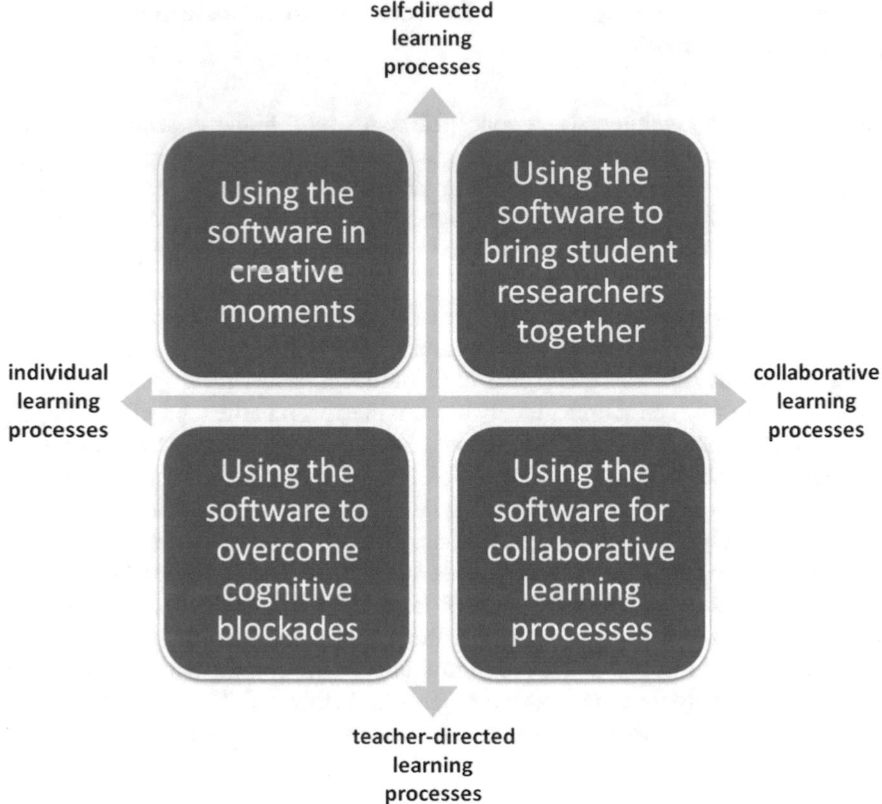

Fig. 6 Four scenarios for using mobile devices in combination with remote laboratories

Scenario 2 "Using the software to bring student researchers together" – With his mobile device (regardless if smart phone or tablet PC) the student can access his personal e-portfolio in which he documents his experiments and personal competency development in this sector from wherever he wants to. Sitting in the train on the way to or on the way back from university he could flip through his experiments and look what he found out as different results. At the same moment another student looks on the first student's portfolio. He finds out that his own research had quite similar results even though he used different parameters or – even more challenging – he used the same parameters and material but had different results. Knowing this he contacts the first student via a chat or e-mail, as well using his mobile device, and they can communicate about their common results at this very moment and work together on future experiments [19].

Scenario 3 "Using the software to overcome cognitive blockades" – A third scenario could foster the students' ability to consider different perspectives about their questions: After having performed an experiment that was given to him by the teacher, the student possibly does not know why he did not get the expected results or does not know how to interpret the results. He asks himself why the experiment did not work as it should have, but he cannot find the answer. While writing his e-portfolio as documentation for the teacher's evaluation, he could start the "creative-help app", which helps him to use different perspectives on the problem: Firstly, he is asked to make a (mental) headstand following the question "What else could I do to get the wrong results from experimenting?" If that does not help to find the answer, he secondly will be asked to describe his experiential design and his assumptions in a way that a ten-year-old could understand it. If those methods, which are rather close to the problem, still cannot help him, the "creative-help app" will suggest a force-fit technique by showing a picture that does not have anything to do with a problem (for example a lady beetle, a daisy chain, a bottle of wine) and asking the student to find relationships between the picture and his experiment. This method helps to leave the well-trodden paths and forces the students to look from completely other perspectives on their problem. It often results in very unconventional or provoking ideas, but rethinking the obviously unsuitable solutions sometimes leads to the one really good idea that would not have appeared to his mind without having made the detour [20].

Scenario 4 "Using the software for collaborative learning processes" – The fourth scenario could be evoked by the teacher. He can give the students – as a kind of homework – the exercise to check an explicit hypothesis by implementing adequate experiments. Using the e-portfolios the students can stay in contact without the necessity to meet at the university and in combination with mobile devices they can be virtually be anywhere going through a collaborative learning process. Because the e-portfolios software has a connection to the experiential environment the students as a group can do the experiment and discuss the results with regards to the homework's hypothesis in one pass and without changing the learning environment [21].

4.2 *Proof of Concept*

In order to put the plans explained above into practice different research has been carried out. Our work on remote laboratories and creativity bases on finished research projects at TU Dortmund University and several other European universities. Based on this research we developed a first proof of concept for the mobile devices running with Android. The software permits the user to do a remotely run experiment by checking the parameters, starting the experimentation process and following the results. The next technical step is to bring the software from the proof of concept status to a level on which it can be tested and improved with students on a larger scale. As a consequence thereof it would have to be further connected with the e-portfolio software and it has to be worked out for iOs devices. These are the steps for the coming year.

5 Conclusion, Discussion and Future Plans

With this paper we explained four different but in our project newly connected aspects:

- We discussed why laboratories are a place for conducting experiments and why they can be important for modern engineering education. The central idea is to engage the students in teaching and learning environments which are connected closely to their future working environment.
- In addition to that, we discussed the aspect of student centered learning environments in general and why they are vitally important in higher education. They are essential for having the students do the things they have to learn by themselves. In our context we underlined this statement by the example of the synchronized learning and research process. This is the only way for students to develop competences and to reach a high level of learning outcomes.
- The learning environment was the next aspect we focused on. If students are learning in laboratories or with the help of laboratory work it is obvious that they need a special learning environment in order to reflect on their learning process. Questions like "How can I document my learning progress for other learners or the teachers?" or "How can I communicate with my classmates during the learning process?" are becoming more and more important. We want to address these questions with the use of e-portfolios as a place to document and to communicate learning processes as well as to merge studies at university with the later workplace environment.
- Finally we explained our plans in context with the use of mobile devices. There are several scenarios thinkable (we concentrated on four of them) in which the use of smart phones and/or tablet PCs extend the opportunities of learning environment substantially. With the use of mobile devices previously impossible learning scenarios become feasible. The fact that these devices are highly portable of course plays a central role in this context. We finished with a technical proof of concept for Android devices.

All the aforementioned aspects are combined in our work. In order to innovate the teaching and learning in engineering education we will design the "Mobile Lab Portfolio". This environment integrates the use of remote labs, e-portfolios and mobile devices. See in the following the central advantages of the presented concept:

- As the equipment of laboratories is either very expensive to purchase and maintain at every university or not always accessible for students the use of remote or virtual laboratories is a good alternative to face this dilemma.
- Using the equipment virtually in simulations or remotely from wherever they can help the students to do experiments just as a pre-check on personal hypothesis or even when they are not physically in the laboratory.
- Learning processes that are achieved by the usage of the laboratories can be documented in e-portfolios.
- These portfolios are a good opportunity in order to document the experiments for personal use or for evaluation by an instructor. By looking at his students'

portfolios the instructor can either see what kind of experiments the students have done and what they learned from it.

- If the portfolios are not kept hidden from other students but are open for other users to look at them and comment on the achievements, there is the opportunity for a community to evolve working together on the experiments. The e-portfolio software should be made accessible for mobile devices, too.
- This paves the way to mobile learning, which means that the learning process is not bound to any location. From virtually everywhere and at every time the user can work on their portfolios and communicate with each other.

As this is a work-in-progress paper the upcoming step for this year will be to implement the e-portfolio software in the PeTEX system and to make it accessible for the students' mobile devices. Once this has been achieved, first tests with students can be carried out and the system can be evaluated and improved.

Of course technical problems will arise during the implementation and they may be even difficult to solve. But in the end most problems stay simply technical and it became obvious in the PeTEX project that every technical problem will be solved sooner or later. At this moment we focus on the concept and first technical steps for our future work. During the whole work we want to concentrate on the didactic background for this. The question if any of the explained aspects will help the students to learn more and gain real competencies during their studies is and will stay our main focus. Not everything which is possible from the technical point of view or even can be technically designed does make sense for the learning process and higher education. Therefore we look at teaching and learning from the students' perspective. That leads us to our concepts explained above first, without asking in detail about how everything can be implemented yet. Our team strives to support a better engineering education – that is what our focus lies on.

References

1. D.A. Kolb, *Experiential learning. Experience as the source of learning and development.* Prentice-Hall, Englewood Cliffs, N.J., 1984
2. L. Huber, J. Hellmer, F. Schneider, *Forschendes Lernen im Studium - Aktuelle Konzepte und Erfahrungen.* Universitätsverlag Webler, Bielefeld, 2009
3. C. Pleul, C. Terkowsky, I. Jahnke, A.E. Tekkaya, Platform for e-learning and tele-operative experimentation (petex) - hollistically integrated laboratory experiments for manufacturing technology in engineering education. In: *Proceedings of SEFE Annual Conference, 1st World Engineering Education Flash Week*, ed. by J. Bernardino, J.C. Quadrado. 2011, pp. 578–585
4. C. Terkowsky, C. Pleul, I. Jahnke, A.E. Tekkaya, Tele-operated laboratories for production engineering education - platform for e-learning and telemetric experimentation (petex)". International Journal of Online Engineering (iJOE) 7, Special Issue EDUCON 2011 , 2011, pp. 37–43
5. J. Garcıa-Zubıa, R.G. Alves, eds., *Using Remote Labs in Education. Two Little Ducks in Remote Experimentation.* Deusto Publicaciones, 2011

6. C. Terkowsky, I. Jahnke, C. Pleul, D. May, T. Jungmann, A.E. Tekkaya, [6] petex@work. designing cscl@work for online engineering education. In: *Computer-Supported Collaborative Learning at the Workplace - CSCL@Work*, ed. by Goggins, S.P., Jahnke, I., V. Wulf, Springer, 2013, pp. 269–292
7. C. Terkowsky, I. Jahnke, C. Pleul, R. Licari, P. Johannssen, G. Buffa, G. Heiner, L. Fratini, E. Lo Valvo, M. Nicolescu, J. Wildt, A.E. Tekkaya, Developing tele-operated laboratories for manufacturing engineering education. platform for e-learning and telemetric experimentation (petex). International Journal of Online Engineering (iJOE). IAOE. Special Issue: REV2010 **6**, 2010, pp. 60–70
8. I. Jahnke, C. Terkowsky, C. Pleul, A.E. Tekkaya, Online learning with remote-configured experiments. In: *Interaktive Kulturen, DeLFI 2010 – 8. Tagung der Fachgruppe E-Learning der Gesellschaft für Informatik e.V.*, ed. by M. Kerres, N. Ojstersek, U. Schroeder, U. Hoppe. 2010, pp. 265–277
9. L.D. Feisel, A.J. Rosa, The role of the laboratory in undergraduate engineering education. Journal of Engineering Education , 2005, pp. 121–130
10. J. Wildt, Forschendes lernen: Lernen im „format“ der forschung. Journal Hochschuldidaktik **20** (2), 2009, pp. 4–7
11. H. Blom, *Der Dozent als Coach.* [16] Luchterhand, Neuwied and Kriftel, 2000
12. J. Herrington, R. Oliver, An instructional design framework for authentic learning environments. Educational Technology Research and Development **48**, 2000, pp. 23–48
13. C. Terkowsky, I. Jahnke, Pleul, C., Tekkaya, A. E., Platform for e-learning and telemetric experimentation (petex) - tele-operated laboratories for production engineering education. In: *Proceedings of the 2011 IEEE Global Engineering Education Conference (EDUCON) – Learning Environments and Ecosystems in Engineering Education*, ed. by M.E. Auer, Y. Al-Zoubi, E. Tovar. IAOE, Vienna, 2011, pp. 491–497
14. A.C. Breuer, *Das Portfolio im Unterricht – Theorie und Praxis im Spiegel des Konstruktivismus.* Waxman Verlag GmbH, Münster, 2009
15. R. Reichert, Das e-portfolio - eine mediale technologie zur herstellung von kontrolle und selbstkontrolle. In: *Kontrolle und Selbstkontrolle – Zur Ambivalenz von E-Portfolios in Bildungsprozessen*, ed. by T.e.a. Meyer, VS Verlag für Sozialwissenschaften – Springer Fachmedien GmbH, Wiesbaden, [14] 2011
16. G. Reinmann, S. Sippel, Königsweg oder sackgasse? – e-portfolios für das forschende lernen. In: *Kontrolle und Selbstkontrolle – Zur Ambivalenz von E-Portfolios in Bildungsprozessen*, ed. by T.e.a. Meyer, VS Verlag für Sozialwissenschaften – Springer Fachmedien GmbH, Wiesbaden, [14] 2011
17. A. Kukulska-Hulme, J. Traxler, *Mobile learning: A handbook for educators and trainers.* Routlegde and Taylor & Francis Group, Milton Park, New York, 2005
18. D. May, C. Terkowsky, T. Haertel, C. Pleul, Using e-portfolios to support experiential learning and open the use of tele-operated laboratories for mobile devices. In: *REV2012 – Remote Engineering & Virtual Instrumentation Conference Proceedings.* IEEE, 2012, pp. 172–180
19. C. Terkowsky, D. May, T. Haertel, C. Pleul, Experiential remote lab learning with e-portfolios - integrating tele-operated experiments into environments for reflective learning. In: *15th International Conference on Interactive Collaborative Learning and 41st International Conference on Engineering Pedagogy.* IEEE, 2012
20. C. Terkowsky, D. May, T. Haertel, C. Pleul, Experiential learning with remote labs and e-portfolios - integrating tele-operated experiments into personal learning environments: Iaoe. International Journal of Online Engineering (iJOE) **9** (1), 2013, pp. 12–20
21. D. May, C. Terkowsky, T. Haertel, C. Pleul, The laboratory in your hand - making remote laboratories accessible through mobile devices. In: *Proceedings of the 2013 IEEE Global Engineering Education Conference (EDUCON).* IEEE, 2013, pp. 335–344

Virtual Production Intelligence – Process Analysis in the Production Planning Phase

Daniel Schilberg, Tobias Meisen and Rudolf Reinhard

Abstract To gain a better and deeper understanding of cause and effect dependencies in complex production processes it is necessary to represent these processes for analysis as good and complete as possible. Virtual production is a main contribution to reach this objective. To use the Virtual Production effectively in this context, a base that allows a holistic, integrated view of information that is provided by IT tools along the production process has to be created. The goal of such an analysis is the possibility to identify optimization potentials in order to increase product quality and production efficiency. The presented work will focus on a simulation based planning phase of a production process as core part of the Virtual Production. An integrative approach which represents the integration, analysis and visualization of data generated along such a simulated production process is introduced. This introduced system is called Virtual Production Intelligence and in addition to the integration possibilities it provides a context-sensitive information analysis to gain more detailed knowledge of production processes.

Keywords Analysis · Digital Factory · Laser Cutting · Production Technology · Virtual Production · Virtual Production Intelligence · VPI

1 Introduction

Considering the individualization and increasing performance of products the complexity of products and production processes in mechanical and automatic processing is constantly growing. This, in turn, results in new challenges concerning the designing as well as the production itself. In order to face these challenges, measures are required to meet the demands which are based on higher complexity. One measure to face this challenge is a more detailed planning of the design and manufacturing of the products by the massive use of simulations and other IT tools which enable

D. Schilberg (✉) · T. Meisen · R. Reinhard
IMA/ZLW & IfU, RWTH Aachen University,
Dennewartstr. 27, 52068 Aachen, Germany
e-mail: daniel.schilberg@ima-zlw-ifu.rwth-aachen.de

Originally published in "Proceedings of The World Congress on Engineering and Computer Science 2013, WCECS 2013", © 2013. Reprint by Springer International Publishing AG 2016, DOI 10.1007/978-3-319-46916-4_11

the user to fulfill the various demands on the product and its manufacturing. To a further improvement of simulations and IT tools it is important not to evaluate them separately but in their usage context: which tool is used to which planning or manufacturing process. It has to be fathomed which information on which effort between the tools are exchanged.

To formulate and execute an appropriate measure, it is necessary to create a basis which allows a holistic, integrative examination of deployed tools in the process. Aim of such an examination is an increasing product quality, efficiency and performance. Due to the rapid development of high-performance computers, the use of simulations in product design and manufacturing processes has already been well-established and enables users to map relations more and more detailed virtually. This has led to a change concerning the way to perform preparatory and manufacturing activities. Instead of an early development of physical existent prototypes, the object of observation is developed as a digital model which represents an abstraction of essential characteristics or practices. The subsequent simulation the digital model is used to derive statements concerning practices and properties of systems to be examined. The use of digital models in production processes is described by the term of virtual production which specifies a "mainstreaming, experimental planning, evaluation and controlling of production processes and plant by means of digital models" [1, 2].

This paper will show an integrative concept which describes an important component to achieve the objective of a virtual production by the integration and visualization of data, produced on simulated processes within the production technology. Taking account of the application domain of production technology and used context-sensitive information analysis, with the aim of an increasing improvement of knowledge concerning the examined processes, this concept is called Virtual Production Intelligence (VPI). The aim of this paper is to present how the VPI contributes to optimize manufacturing processes like laser cutting. The usage of the VPI in a factory planning process is shown in [3].

2 Problem

As a central issue of the virtual production the heterogeneous IT landscape can be identified. As indicated in the introduction, a variety of software tools to support various processes are used. Within these software tools data cannot be exchanged without effort. The automation pyramid offers a good possibility to demonstrate this difficulty. The automation pyramid is depicted in Fig. 1. It shows the different levels of the automation pyramid with the corresponding IT tools and the flow of information between the levels. The level related processes are supported by the mentioned IT tools very good or at least sufficient. At the top level command and control decisions for the company management are supported by Enterprise Resource Planning (ERP) systems. Therefore, these systems allow the decision-makers in the management to monitor any enterprise-wide resources like employees, machinery or materials.

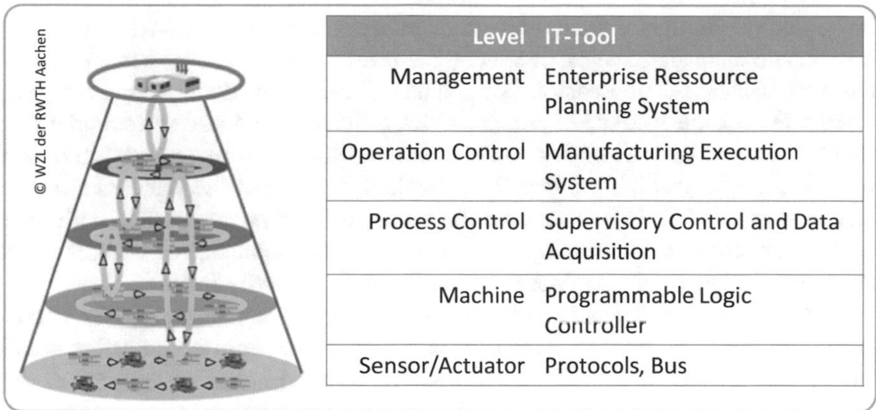

Level	IT-Tool
Management	Enterprise Ressource Planning System
Operation Control	Manufacturing Execution System
Process Control	Supervisory Control and Data Acquisition
Machine	Programmable Logic Controller
Sensor/Actuator	Protocols, Bus

Fig. 1 Automation pyramid [4]

At the lower levels, the Manufacturing Execution Systems (MES), the data acquisition (Supervisory Control and Data Acquisition SCADA) and programmable logic controllers (PLC) are arranged according to the increasing complexity. The field level is the lowest level. Corresponding to protocols the data exchange is organized on this level. The used software tools are developed very well to support the corresponding processes on the appropriate level. The Association of German Engineers (VDI) addressed for the virtual production a unified data management as a way to use data and information across all levels, but this is not realized yet. Without a unified data management the data exchange from a PLC via the SCADA and MES up to the ERP system requires a great effort for conversions and aggregating. The goal is that the ERP can support decisions on the base down to the PLC Data and changes in the ERP system will change the input for the PLC. Currently most companies only exchange data between different levels instead of a flow of information across all levels. This is why a holistic picture of production and manufacturing process is not possible [5].

At present, a continuous flow of information is available only with the application of customized architectures and adapters to overcome the problem of heterogeneity. This involves high costs, why usually small and medium-sized enterprises (SMEs) have no integration of all available data into a system.

There are a high number of different IT tools for virtual production. These enable the simulation of various processes, such as in manufacturing technology, the realistic simulation of heat treatment and rolling process or the digital viewing of complex machinery such as laser cutting machines. At this juncture various independent data formats and structures have developed for a representation of the digital models. Whereas an independent simulation of certain aspects of product and manufacturing planning is possible, the integrative simulation of complex manufacturing processes involves high costs as well as an high expenditure of time because in general an interoperability between heterogeneous IT tools along the automation pyramid is not given. One approach to overcome the heterogeneity is the homogenization with the help of a definition of unified data standards. In this context a transfer of the data formats into a standardization of data by the use of specific adapters as mentioned above. However, this approach is not practical for the considered scenario for two reasons. Firstly, the diversity of possible IT tools that are used lead to a complex data standard. This is why its understanding, care and use are time and cost intensive. Secondly, the compatibility issues for individual versions of the standard are to be addressed (see STEP [6]). Therefore the standard must be compatible with older versions and enhanced constantly to reflect current developments of IT tools and to correspond to the progressive development through research [7, 8].

Another approach, which is chosen as basic in this paper, includes the use of concepts of the data and application integration, which do not require a unified standard. The interoperability of IT applications must be ensured in a different way so that no standard data format is necessary. This is done by mapping the various aspects of the data formats and structures on a so-called integrated data model or canonical data model [8, 9]. In current approaches to these concepts are extended to the use of semantic technologies. The semantic technologies enable a context-sensitive behavior of the integration system. The continuation of this approach enables the so-called adaptive application and data integration [10, 11].

The integration of all data collected in the process in a consolidated data management is only the first step to solving the problem. The major challenge that must be overcome is the further processing of the integrated data along a production process to achieve a combination of IT tools across all levels of the automation pyramid. The question of the analysis of data from heterogeneous sources is addressed in the analysis of corporate data for some time. The applications that enable integration and analysis of data are grouped under the term "Business Intelligence" (BI). BI applications have in common that they provide the identification and collection of data that arise in business processes, as well as their extraction and analysis [12, 13].

The problem in the application of BI on virtual production is that the implementation of the BI integration challenges of heterogeneous data and information conceptually solves in the first place which causes significant problems in the implementation of functional systems. Thus, in concept, for example, a translation of the data into a common data format and context-sensitive annotation is provided. A translation may not be achieved because it is proprietary information which meaning is not known to the annotation. This is also the reason why so many BI integrations have failed so far [14].

The following shows that the previously addressed problems should be solved by the vision of the digital factory. Because this vision is not realized yet, the section heterogeneity of simulations and solution: Virtual Production Intelligence will outline next steps towards the realization of a digital factory. The term "Virtual Production Intelligence" was selected in reference to the problem introduced in the term "business intelligence", which has become popular in early to mid-1990s. It called "business intelligence" methods and processes for a systematic analysis (collection, analysis and presentation) of a company's data in electronic form. Based on gained findings, it aims at improved operative or strategic decisions with respect to various business goals "Intelligence". In this context "Intelligence" does not refer to intelligence in terms of a cognitive size but describes the insights which are provided by collecting and preparing information.

3 Digital Factory

The digital factory is defined by the working group VDI in the VDI guideline [1] as:

> *"the generic term for a comprehensive network of digital models, methods and tools – including simulation and 3D visualization – integrated by a continuous data management system. Its aim is the holistic planning, evaluation and ongoing improvement of all the main structures, processes and resources of the real factory in conjunction with the product."*

According to the VDI guideline 4499 the concept of the digital factory does not include individual aspects of the planning or production but the entire product life cycle (PLC) (Fig. 2). All processes from the onset to the point of decommissioning shall be modeled. Therefore the observation starts with the collection of market requirement, the design stages including all the required documents, project management, prototypes (digital mock-ups), the necessary internal and external logistic processes, planning the assembly and manufacturing, the planning of appropriate manufacturing facilities, installation and commissioning of production facilities, the start-up management (ramp up), series production, sales to maintenance and ends with the recycling or disposal of the product all these points should be part of the

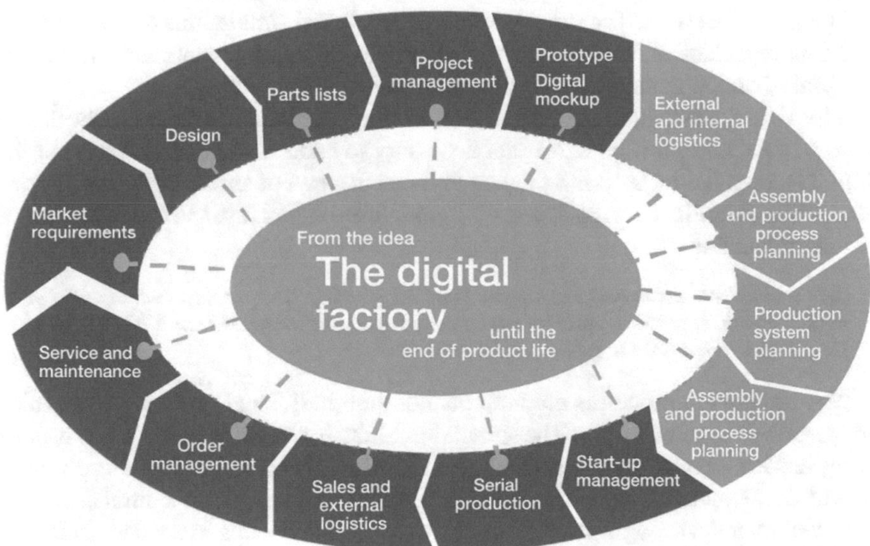

Fig. 2 Product Life Cycle (VDI 4499) and localization of virtual production within the product life cycle in accordance with VDI Directive 4499

Digital Factory. Currently there is no platform which complies with this integration task. But there are already implemented some elements of the digital factory at different levels of the automation pyramid or in phases of the PLC.

Existing PLC Software products help companies to plan, monitor and control the product life cycle in parts. However, these applications are usually only isolated solutions and enable the integration of IT tools that have the same interfaces for data exchange and are provided by the same manufacturer. The detail of the images of individual phases of the product life cycle does not reach this high spatial resolution of special applications to the description of individual phases of the product life cycle or of IT tools that focus on aspects of individual phases. Therefore the recommendation of the VDI to design data management and exchange as homogeneous as possible can only be considered for new developments. Besides there is still no approach about how to implement a standard for such a homogeneous data exchange and how to prevent or avoid the known issues of a standardization process. Therefore even a project that wants to realize the homogenization of the flow of information cannot succeed, because it is not defined what such a condition has to look like. Moreover there is no standard or efforts to standardize as for example the Standard for the Exchange of Product Model Data (STEP) compete with proprietary formats. It must be considered that the proprietary formats were also used to protect the knowledge and skills of the software provider.

With view to a visualization of the digital factory there are tools of Virtual and Augmented Reality which enable users to realize 3D models of factories with or without people as well as to interact with it and to annotate information. A real time-control of a physical existent plant via virtual representation, at which data from the operation in virtual installation are illustrated and further processed for analysis, is right now not possible. The running times of individual simulations do not meet the real-time requirement. With the present techniques, its developments and innovations the goal of digital manufacturing is to be achieved.

The Virtual Production Intelligence serves as a basic building block for the digital factory. To achieve this goal, it is not necessary to address the overall vision of the digital factory, but rather it is sufficient to focus the area of simulation-based virtual production (see Fig. 2). Again, the VDI guideline 4499 is cited to the definition of virtual production:

> *"is the simulated networked planning and control of production processes with the aid of digital models. It serves to optimize production systems and allows a flexible adaptation of the process design prior to prototype realization."*

The production processes are here divided into individual process steps, which are described by simulations. The simulation of the individual process steps is done using modern simulation tools which can represent complex production processes accurately. Despite the high accuracy of individual simulations the central challenge in virtual manufacturing is the sum of individual process steps in a value chain.

The VPI is developed to set the interoperability of IT tools in a first step with distinctly less effort than using tailored solutions mentioned above. In a second step the integrated data is consolidated, analyzed and processed. The VPI is a holistic, integrative approach to support the implementation of collaborative technology and product development. Thereby enabling optimization potentials are identified and made available for the purpose of early identification and elimination of errors in processes. To better understand the terms holistic, integrative and collaborative will be defined as follows:

- Holistic: all parts of the addressed processes will be taken into consideration.
- Integrative: use and integration of existing solutions.
- Collaborative: consideration of all processes addressed in involved roles as well as their communication.

In the next section, the above-mentioned heterogeneities that should be overcome by the use of the VPI, a closer look.

4 Heterogenity

Regarding ISO / IEC 2382-01 [15] interoperability between software applications is realized when the ability exists to communicate, to run programs, or to transfer data between functional units is possible in such a way that the user need no information about the properties of the application. Figure 3 summarizes the heterogeneities, which contribute significantly to the fact that no interoperability is achieved without using customized adapters [16–18].

The syntactic heterogeneity describes the differences in the technical description of data, for example different coding standards such as ASCII or binary encoding, or the use of floating-point numbers as float or double. These two types of heterogeneity can be overcome relatively easy by the use of adapters. Therefore a generic approach should be applied, so that the implemented adapters are reusable. Existing libraries and solutions are available to address the problem of technical heterogeneity. Most modern programming concepts contain methods for implicit type adjustments and controlling explicit conversion of data [16–18].

Overcoming the structural and semantic heterogeneity is the much greater challenge. Structural heterogeneity differences specify the representation of information.

Kind of Heterogeneity	Description/Examples
Syntactical	Presentation of data; e.g. format of numbers, encoding.
Structural	Order, in which data attributes are exported.
Semantical	Meaning of attribute denominations; t = *time* or *temperature* ?

Fig. 3 Types of heterogeneity of simulations

Semantic heterogeneity describes the differences in the importance of domain specific entities and concepts used for their award. e. g. the concept of ambient temperature is used by two simulations, simulation A, uses the concept to define the room temperature of the site where the heating furnace is located. Simulation B uses the concept to define the temperature inside the heating furnace so the temperature in the immediate vicinity of the object to be heated is specified.

In the following section, the VPI is presented, which provides methods to overcome of the mentioned heterogeneity and to facilitate interoperability between applications [16–18].

5 Virtual Production Intelligence

The main objective for the use of the "Virtual Production Intelligence" is to gather results of a simulation process, to analyze and visualize them in order to generate insights that enable a holistic assessment of the individual simulation results and aggregated simulation results. The analysis is based on experts know how and physical and mathematical models. Through an immersive visualization requirements for a "Virtual Production Intelligence" are completely covered.

The integration of result data from a simulation process in a canonical data model (Fig. 4) is the first step to gain knowledge from these data and to realize the extraction of hidden, valid, useful and actionable information. This information includes, for example, the quality of the results of a simulation process or in concrete cases, the causes for the emergence of inconsistencies.

Right now the user who has to identify such aspects has currently limited options to do so. With the realization of an integration solution a uniform view to the data

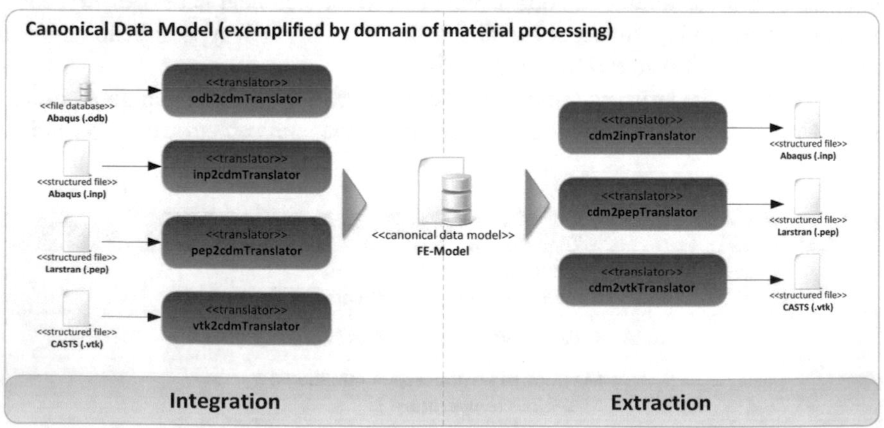

Fig. 4 Canonical Data Model VPI

gets possible. This includes on the one hand the visualization of the entire simulation process in a visualization component and on the other hand, the analysis of the data over the entire simulation process. For this purpose, different exploration methods can be used.

First, the data along the simulation process is integrated into a canonical data model. This is implemented as a relational data model, so that a consistent and consolidated view of the data is possible. Subsequently, the data is analyzed on the analysis level by the user. The user can interact in an immersive environment to explore and analyze the data. With the ability to provide feedback to the analysis component, the user can selectively influence the exploration process and make parameterizations during runtime.

In addition to a retrospective analysis by experts, it is also useful to monitor the data during the simulation process. Such a process monitoring assures compliance with parameter corridors or other boundary conditions. Therefore, if a simulation provides parameter values outside the defined parameter corridors the simulation process will be terminated. Then experts can analyze the current results in order to subsequently perform a specific adaptation of the simulation parameters. A process monitoring could also enable the extraction of point-of-interests (POI) on the basis of features that would be highlighted by the visualization (Fig. 5). The components of the "Virtual Production Intelligence" are shown in [3].

An effective optimization of different structures of production, such as determining the number of process chains and production segment is made possible only by mapping the interdependencies of different planning modules.

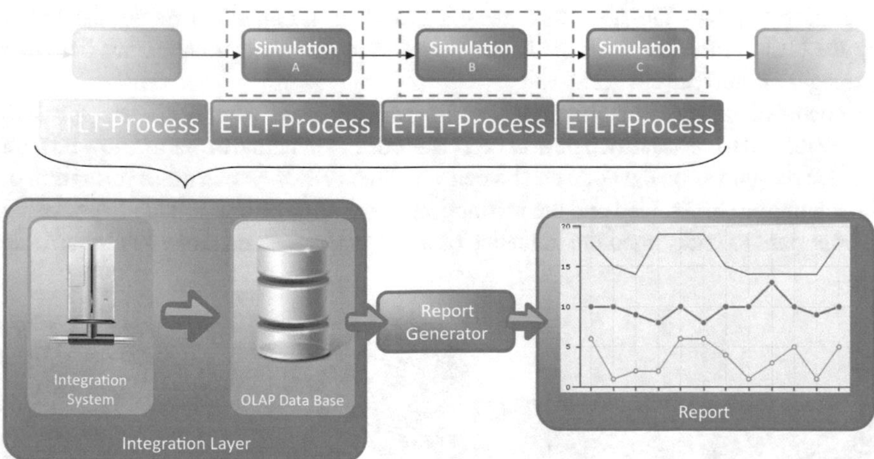

Fig. 5 Extraction of point-of-interests

6 Application Domain Laser Cutting

The VPI is used to identify relevant machine parameters to optimize laser cutting processes concerning different goals like quality or speed. At first a brief look on the process itself is given. Laser cutting is a thermal separation process widely used in shaping and contour cutting applications. Therefore the laser cutting process has some advantages over conventional cutting techniques. The cutting process is very fast and accurate.Because an optical tool is used there is no risk of additional wear.

The ablation process in fusion metal cutting is based on thermodynamics and hydrodynamics. The absorbed laser energy is converted to heat which melts the material. This melt is driven out of the cut by a gas jet that is coming out of the nozzle, coaxially aligned to the laser beam. The VPI is a simulation based tool; therefore not the real process is used for optimization but the simulated. Hence, the analysis results of the VPI strongly correlate to the quality of the simulated laser cutting process. The core of each simulation is the simulation model that is used. The modelling of a laser cutting process requires the modelling of at least three entities at the same time. The optical tool – the laser beam – must be included, the material that should be cut by the laser beam, and the gas jet separating the melt. To gain a good model it is evident that the modeling of the following quantities has to be accomplished as well as their numerical implementation:

- The cutting gas flow
- The radiation propagation
- The ablation of the material

Figure 6 shows the simulation results based on a numerical model developed by the NLD at RWTH Aachen University, Germany, for the ablation ant the beam propagation into the cut kerf.

There are, however, gaps in understanding the dynamics of the process, especially issues related to cut quality. The user of a laser cutting machine needs to know how the surface roughness on cut edge can be better influenced. How can dross formation on the cut bottom be avoided and the inclination of cut edge be controlled? Especially it is important to understand the influence of laser parameters on those quality criteria.

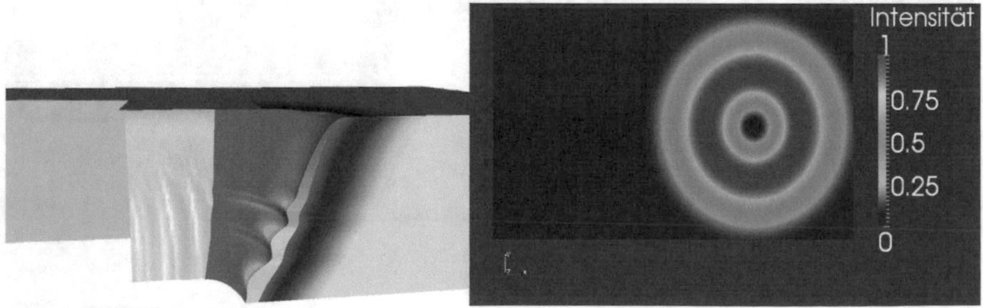

Fig. 6 Ablation simulation for laser cutting and simulated beam propagation into cut kerf

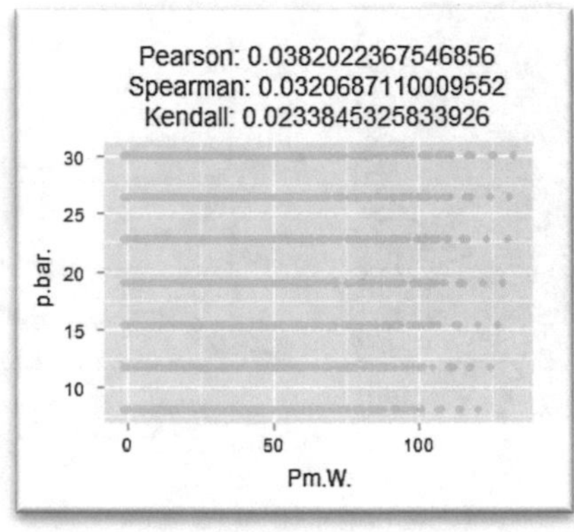

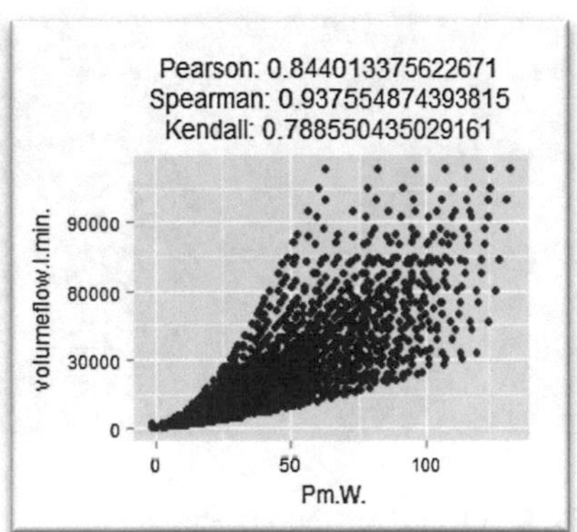

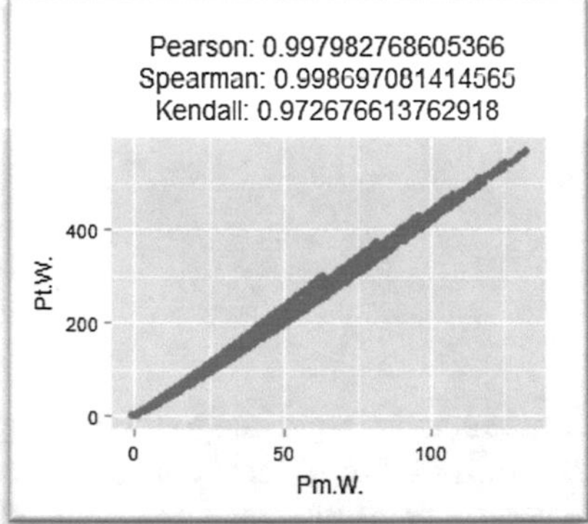

Fig. 7 Methods used by the VPI for reduction of number of parameters

The most important parameters may be the wave length and the modelling of the wave length. Is a gas laser better than a solid state laser? The shape of the beam must be analysed: What is the influence of a Gaussian or a tophat shaped beam? Should the polarization circular or radial? Hence, the goal of the VPI is the reduction of numbers of parameters that are relevant for reaching a certain cutting quality. For that the correlation between chosen output or criterion and parameters or inputs must be determined. The VPI uses different methods to find the correlations (Fig. 7). The VPI uses a sensitivity analysis:

> *"The study of how uncertainty in the output of a model (numerical or otherwise) can be apportioned to different sources of uncertainty in the model input."* [19]

The results of qualitative methods can be visualized by scatter plots and quantitative methods will be used like the computation of rank correlation coefficients between various criteria and parameters.

The VPI is used for the planning of the laser cutting process. It supports the user in three ways. At first by using the VPIs data integration possibilities the user can gather data from various sources and get a consolidated view on these data. In the

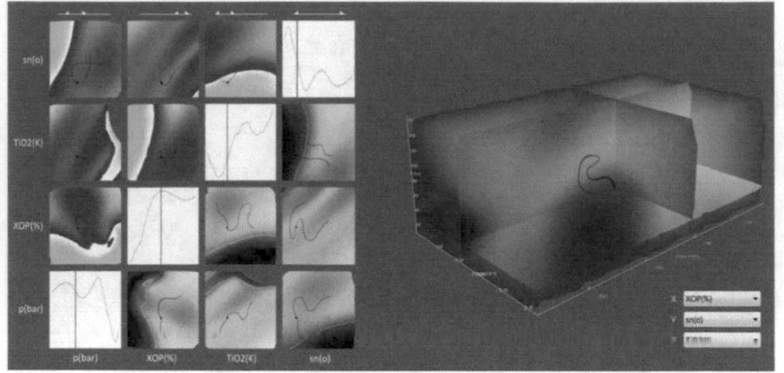

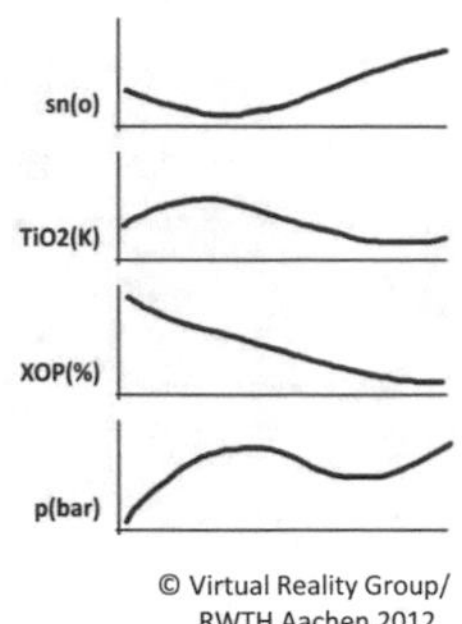

Fig. 8 VPIs explorative visualization

second way the VPI provides an explorative visualization (see Fig. 8) to present the data and to facilitate interaction. The last way is the data analysis to determine where and how to optimize process outcome.

7 Conclusion

In this chapter the Definition of the VPI was given as a holistic and integrative concept for the support of the collaborative planning, monitoring and control of core processes within production and product development in various fields of application. As a role model the idea of the Business Intelligence is used, applied to the domain of Virtual Production. The aim of the VPI is the identification and elimination of error sources in planning processes as well as detecting and taking advantage of enhancement potentials.

With the VPI an essential contribution to the realization of the vision of the digital factory can be achieved. The VPI is an integration platform that enables heterogeneous IT tools in the phase of product and production planning to interoperate with each other. Based on information processing concepts it supports the analysis and evaluation of cause-effect relationships. As product and production planning is the core area of Virtual Production, as part of the digital factory, the contribution is focused on this part. The VPI is the basis to establish interoperability. The functionality of the VPI was presented and illustrated by using the example of factory planning. The use of the VPI allows a significant reduction in engineering effort to create tailored integration and analysis tools, since the VPI is an adaptive solution. Now it is possible to start with a process-oriented and so contextual information processing. Information is now not only based on a single process step, it is related to the overall process, so that the importance and validity of information can be considered.

The future work concerning the VPI in the domain laser cutting will be the determination of machine parameters depending on desired machine states. That will be an optimization problem for a multidimensional function. The solution could be an explorative visualization based on the concept of hyperslices linked with 3D volume visualization. It is important to evaluate what cause-effect relationships can be identified through the exploration process. Furthermore, it must be examined how this information can be presented to the user in an immersive environment, and how can context information understandable and comprehensible be presented. For this purpose, there are various feedback-based techniques in which experts assess results of analysis and optimization. A bidirectional communication is needed, the user gives feedback and this feedback will be used to correct the displayed information. The system will store this feedback to avoid imprecise or erroneous statements.

References

1. VDI Richtlinie 4499, Blatt 1, Digital factory. Tech. rep., 2008
2. VDI Richtlinie 4499, Blatt 2, Digital factory. Tech. rep., 2011
3. D. Schilberg, T. Meisen, R. Reinhard, Virtual production – the connection of the modules through the virtual production intelligence. In: *Lecture Notes in Engineering and Computer Science: Proceedings of The World Congress on Engineering and Computer Science 2013, WCECS 2013, 23-25 October, 2013, San Francisco, USA*. pp. 1047–1052
4. R. Lauber, P. Göhner, *Prozessautomatisierung 1*, 3rd edn. Springer, Berlin, 1999
5. H. Kagermann, W. Wahlster, J. Helbig, *Umsetzungsempfehlungen für das Zukunftsprojekt Industrie 4.0 – Ab-schlussbericht des Arbeitskreises Industrie 4.0*. Forschungsunion im Stifterverband für die Deutsche Wissenschaft, Berlin, 2012
6. DIN EN ISO 10303
7. M. Nagl, B. Westfechtel, *Modelle, Werkzeuge und Infrastrukturen zur Unterstützung von Entwicklungsprozessen*, 1st edn. Symposium (Forschungsbericht (DFG)). Wiley-VCH, 2003
8. C. Horstmann, *Integration und Flexibilitat der Organisation Durch Informationstechnologie*, 1st edn. Gabler Verlag, 2011
9. D. Schilberg, *Architektur eines Datenintegrators zur durchgängigen Kopplung von verteilten numerischen Simulationen*. VDI-Verlag, Aachen, 2010
10. T. Meisen, P. Meisen, D. Schilberg, S. Jeschke, Application integration of simulation tools considering domain specific knowledge. In: *Proceedings of the 13th International Conference on Enterprise Information Systems*. 2011
11. R. Reinhard, T. Meisen, T. Beer, D. Schilberg, S. Jeschke, A framework enabling data integration for virtual production. In: *Enabling Manufacturing Competitiveness and Economic Sustainability – Proceedings of the 4th International Conference on Changeable, Agile, Reconfigurable and Virtual production (CARV2011), Montreal, Canada, 2-5 October 2011*, ed. by A.E. v. Hoda. Berlin Heidelberg, 2012, pp. 275–280
12. B. Byrne, J. Kling, D. McCarty, G. Sauter, P. Worcester, *The Value of Applying the Canonical Modeling Pattern in SOA. IBM (The information perspective of SOA design, 4)*. 2008
13. M. West, *Developing High Quality Data Models*, 1st edn. Morgan Kaufmann, Burlington, MA, 2011
14. W. Yeoh, A. Koronios, Critical success factors for business intelligence systems. Journal of computer information systems **50** (3), 2010, p. 23
15. ISO/IEC 2382-01
16. M. Daconta, L. Obrst, K. Smith, *The Semantic Web: The Future of XML, Web Services, and Knowledge Management*. 2003
17. D. Schilberg, A. Gramatke, K. Henning, Semantic interconnection of distributed numerical simulations via soa. In: *Proceedings World Congress on Engineering and Computer Science 2008*, ed. by I.A. of Engineers. Newswood Limited, Hong Kong, 2008, pp. 894–897

18. D. Schilberg, T. Meisen, R. Reinhard, S. Jeschke, Simulation and interoperability in the planning phase of production processes. In: *ASME 2011 International Mechanical Engineering Congress & Exposition*, ed. by ASME. Denver, 2011
19. A. Saltelli, S. Tarantola, F. Campolongo, M. Ratto, *Sensitivity Analysis in Practice: A Guide to Assessing Scientific Models*. John Wiley & Sons, Ltd., 2004

Virtuelle Produktion - Die Virtual Production Intelligence im Einsatz

Daniel Schilberg, Tobias Meisen and Rudolf Reinhard

Zusammenfassung Die virtuelle Produktion soll einen Beitrag leisten, dass in Hochlohnländern produzierende Industrien weiterhin Konkurrenzfähig sind und sogar Ihren Entwicklungsvorsprung in Hochtechnologien halten und ausbauen können. Um die virtuelle Produktion in diesem Kontext effektiv einsetzen zu können, muss eine Basis geschaffen werden, die eine ganzheitliche, integrative Betrachtung der eingesetzten IT-Werkzeuge im Prozess ermöglicht. Ziel einer solchen Betrachtung soll die Steigerung von Produktqualität, Produktionseffizienz und -leistung sein. In diesem Beitrag wird ein integratives Konzept vorgestellt, das durch die Integration, die Analyse und die Visualisierung von Daten, die entlang simulierter Prozesse innerhalb der Produktionstechnik erzeugt werden, einen Basisbaustein zur Erreichung des Ziels der virtuellen Produktion darstellt. Unter Berücksichtigung der Anwendungsdomäne Produktionstechnik und der eingesetzten kontextsensitiven Informationsanalyse mit der Aufgabe den Erkenntnisgewinn der untersuchten Prozesse zu erhöhen, wird dieses Konzept als Virtual Production Intelligence bezeichnet.

Schlüsselwörter Produktionstechnik · Digitale Fabrik · Datenverarbeitung

1 Einleitung

Der Markt für industriell gefertigte Güter verändert sich immer schneller, so müssen sich Unternehmen der Herausforderung stellen, das einerseits individuelle Kundenanforderungen stetig zunehmen, der für ein Produkt zu erzielende Preis jedoch, trotz des zusätzlichen Aufwands, nur gering steigt. Dies betrifft insbesondere Unternehmen, die in Hochlohnländern agieren, da der globale Wettbewerb für wenig individualisierte Produkte besonders durch die BRICS (Brasilien, Russland, Indien, China, Südafrika) Staaten dominiert wird [1]. Durch die Individualisierung und Leistungssteigerung von Produkten nimmt jedoch die Komplexität von Produkten und Produktionsprozessen in der maschinellen und automatisierten Fertigung stetig zu.

D. Schilberg (✉) · T. Meisen · R. Reinhard
IMA/ZLW & IfU, RWTH Aachen University, Dennewartstr. 27, 52068 Aachen, Germany
e-mail: daniel.schilberg@ima-zlw-ifu.rwth-aachen.de

Originally published in "Exploring Virtuality",
 DOI 10.1007/978-3-319-46916-4_12

Dies wiederum resultiert in neuen Herausforderungen an die Planung von Produkten und die verbundene Planung der Produktfertigung [2]. Um sich diesen Herausforderungen zu stellen werden Maßnahmen benötigt, die den Anforderungen, die aus der höheren Komplexität resultieren, gerecht werden. Eine Maßnahme, um dieses Problem handhabbar zu gestalten, ist eine intensivere Produktdesign- und Produktfertigungsplanung, die durch den massiven Einsatz von Simulationen und weiteren IT-Werkzeugen die Anwender in die Lage versetzt, die an ein Produkt und deren Fertigung gestellten Anforderungen zu erfüllen. Zur weiteren Verbesserung des Einsatzes der Simulationen und IT-Werkzeuge ist es wichtig diese nicht einzeln zu betrachten sondern in ihrem Einsatzkontextes, d.h. welches Werkzeug wird zu welchem Zweck an welcher Stelle des Planungs- oder Fertigungsprozess eingesetzt. Es muss eruiert werden welche Informationen mit welchem Aufwand zwischen den Werkzeugen ausgetauscht werden. Um eine entsprechende Maßnahme zu formulieren und auszuführen, muss eine Basis geschaffen werden, die eine ganzheitliche, integrative Betrachtung der eingesetzten Werkzeuge im Prozess ermöglicht. Ziel einer solchen Betrachtung soll die Steigerung von Produktqualität, Produktionseffizienz und leistung sein [3].

Aufgrund der rapiden Entwicklung der nutzbaren Rechenleistung von Computern ist der Einsatz von Simulationen in der Produktdesign- und Produktfertigungsplanung schon länger etabliert und die Anwender werden immer weiter in die Lage versetzt Zusammenhänge immer detaillierter virtuell abzubilden. Dies hat einen Wechsel hinsichtlich der Art und Weise verursacht, wie Vorbereitungs- und Planungsaktivitäten in der Produktion durchgeführt werden. Anstelle der frühzeitigen Entwicklung von physisch existierenden Prototypen wird der Betrachtungsgegenstand zunächst als digitales Model entwickelt, das eine Abstraktion der wesentlichen Charakteristika oder Verhaltensweisen der Prototypen repräsentiert. In der anschließenden Simulation wird das digitale Model genutzt, um Aussagen über Verhalten und Eigenschaften der zu untersuchenden Systeme und Prozesse abzuleiten. Dieser Einsatz von digitalen Modellen in der Produktion wird durch den Begriff der virtuellen Produktion beschrieben, die eine „durchgängige, experimentierfähige Planung, Evaluation und Steuerung von Produktionsprozessen und anlagen mit Hilfe digitaler Modelle" [4, 5] bezeichnet.

In diesem Beitrag wird ein integratives Konzept vorgestellt, das durch die Integration, die Analyse und die Visualisierung von Daten, die entlang simulierter Prozesse innerhalb der Produktionstechnik erzeugt werden, einen Basisbaustein zur Erreichung des Ziels der virtuellen Produktion darstellt. Unter Berücksichtigung der Anwendungsdomäne Produktionstechnik und der eingesetzten kontextsensitiven Informationsanalyse mit dem Ziel den Erkenntnisgewinn der untersuchten Prozesse zu erhöhen, wird dieses Konzept als Virtual Production Intelligence bezeichnet. Zur Illustration dieses Ansatzes wird zunächst die Problemstellung genauer spezifiziert, danach wird die Vision der Digitalen Fabrik aufgespannt, um mit diesen Kenntnissen ein tieferes Verständnis für die Problematik von Heterogenität von IT-Werkzeugen zu schaffen. Ziel des Beitrags ist die Darstellung wie die Virtual Production Intelligence zur Überwindung der adressierten Herausforderungen beiträgt.

2 Problemstellung

Als Kernproblem der virtuellen Produktion kann die heterogene IT-Landschaft in produzierenden Unternehmen identifiziert werden. Es werden wie in der Einleitung dargestellt unterschiedlichste Softwarewerkzeuge zur Unterstützung verschiedenster Prozesse eingesetzt, wobei Daten und Informationen nicht ohne großen Aufwand zwischen den Softwarewerkzeugen ausgetauscht werden können. Die Automatisierungspyramide bietet eine gute Möglichkeit diese Problematik genauer zu beschreiben. In Abb. 1 sind die Ebenen der Automatisierungspyramide mit den für die Ebene korrespondierenden eingesetzten IT-Werkzeugen sowie der Informationsfluss zwischen den Ebenen dargestellt. Hierdurch wird deutlich, dass es auf jeder Ebene Werkzeuge gibt, die die jeweiligen Prozesse unterstützen. So werden auf der obersten Ebene Steuerungs- und Kontrollentscheidungen für die Unternehmensleitung mit Hilfe von Enterprise Ressource Planning (ERP) Systemen unterstützt. Diese Systeme ermöglichen es den Entscheidern im Management den unternehmensweiten Ressourceneinsatz vom Mitarbeiter über Maschinen bis hin zu Rohstoffen zu überwachen.

Auf den Ebenen darunter sind die Manufacturing Execution Systems (MES), die Betriebsdatenerfassung (Supervisory Control and Data Acquisition SCADA) sowie die Speicherprogrammierbaren Steuerungen (SPS) zu finden und auf der untersten Ebene, der Feldebene, liegt die Datenübertragung auf Basis von entsprechenden Protokollen. Die Softwarewerkzeuge sind auf der jeweiligen Ebene sehr weit entwickelt, um die entsprechenden Prozesse zu unterstützen. Was mit Blick auf die virtuelle Pro-

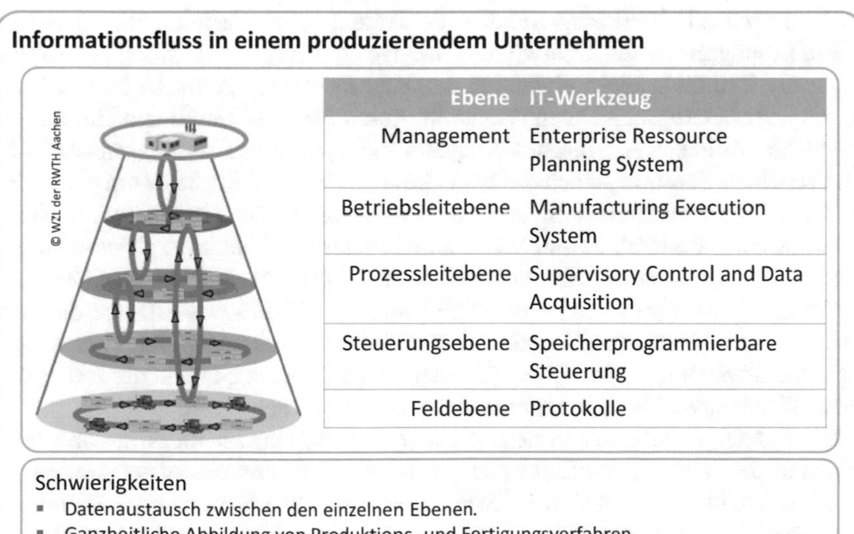

Abb. 1 Automatisierungspyramide [6]

duktion aber nicht realisiert ist, ist das auch vom Verein Deutscher Ingenieure (VDI) adressierte einheitliche Datenmanagement und damit eine Möglichkeit Daten und Informationen über alle Ebenen hinweg verwenden zu können. So ist es in der Regel nur mit sehr großen Aufwänden für Konvertierungen und Aggregieren möglich, die Daten eines SPS über das SCADA und MES bis hinauf zum ERP System zu übertragen, so dass auf Basis von aktuellen Maschinensteuerungsdaten Ressourcenplanungen durchgeführt werden können bzw. sich aus der Ressourcenplanung Steuerungsdaten für die SPS ergeben. Zurzeit gibt es in den meisten Unternehmen nur einen Datenaustausch zwischen einzelnen Ebenen und keinen Informationsfluss über alle Ebenen hinweg. Dadurch ist eine ganzheitliche Abbildung von Produktions- und Fertigungsverfahren nicht möglich [7].

Zurzeit ist ein durchgängiger Informationsfluss nur bei dem Einsatz von maßgeschneiderten Architekturen und Adaptern vorhanden, um die Problematik der Heterogenität zu überwinden. Dies ist mit hohen Kosten verbunden, daher liegt meist bei kleinen und mittleren Unternehmen (KMU) keine Integration aller vorhandenen Daten in ein System vor. Es existiert eine hohe Anzahl unterschiedlicher IT-Werkzeuge für die virtuelle Produktion. Diese ermöglichen die Simulation verschiedenster Prozesse, wie etwa in der Fertigungstechnik die realitätsnahe Simulation von Wärmebehandlungs- und Walzverfahren oder die digitale Betrachtung komplexer Maschinen wie Laserschneidmaschinen. Hierbei haben sich unabhängig voneinander unterschiedliche Datenformate und -strukturen zur Darstellung der digitalen Modelle entwickelt. Während hierdurch die unabhängige Simulation einzelner Aspekte der Produkt- und Produktionsplanung durch einzelne Simulationen möglich ist, ist die integrative Simulation komplexer Produktionsprozesse nicht ohne hohe Kosten- und Zeitaufwand möglich, da in der Regel keine Interoperabilität zwischen den heterogenen IT-Werkzeugen entlang der Automatisierungspyramide gegeben ist.

Ein Lösungsansatz zur Überwindung der Heterogenität ist die Homogenisierung durch die Definition eines einheitlichen Datenstandards, hierdurch ist die Überführung der heterogenen Datenformate in diesen Standard durch den Einsatz von den zuvor erwähnten spezifischen Adaptern möglich. Dieser Lösungsansatz ist für das betrachtete Szenario jedoch aus zwei Gründen nicht praktikabel. Zum einen führt die Vielfalt möglicher IT-Werkzeuge, die eingesetzt werden, zu einem komplexen Datenstandard, wodurch dessen Verständnis, Pflege und Nutzung zeit- und kostenintensiv wird. Zum andren sind Probleme der Kompatibilität zu einzelnen Versionen des Standards zu adressieren (siehe STEP (DIN EN ISO 10303)). So muss der Standard zu älteren Versionen kompatibel sein und ständig weiterentwickelt werden, um aktuelle Entwicklungen von IT-Werkzeugen zu berücksichtigen und der fortschreitenden Weiterentwicklung durch Forschung zu entsprechen [8, 9].

Ein anderer Ansatz, der in dem vorliegenden Beitrag als Basis gewählt wird, beinhaltet die Nutzung von Konzepten der Daten- und Anwendungsintegration, bei denen die Definition eines einheitlichen Standards nicht erforderlich ist. Damit kein Standarddatenformat notwendig ist, muss die Interoperabilität der IT-Anwendungen auf eine andere Art und Weise gewährleistet werden. Dies geschieht durch die Abbildung der Aspekte der verschiedenen Datenformate und -strukturen auf ein sogenanntes integriertes Datenmodell oder kanonisches Datenmodell [9, 10]. In aktuellen

Ansätzen werden diese Konzepte, um den Einsatz semantischer Technologien erweitert. Die semantischen Technologien ermöglichen ein kontextsensitives Verhalten des Integrationssystems. Die Fortführung dieses Ansatzes ermöglicht die sogenannte adaptive Anwendungs- und Datenintegration [11, 12].

Die Integration aller im Prozess erfassten Daten in eine konsolidierte Datenhaltung ist aber nur der erste Schritt zur Lösung der Problemstellung. Die größere Herausforderung, die es zu überwinden gilt, ist die weitere Verarbeitung der integrierten Daten entlang eines Produktionsprozesses, um eine Verknüpfung der IT-Werkzeuge über alle Ebenen der Automatisierungspyramide zu erreichen. Die Fragestellung der Analyse von Daten aus heterogenen Quellen wird seit einiger Zeit bei der Analyse von Unternehmensdaten angegangen. Die Anwendungen, die eine Integration und Analyse der Daten ermöglichen, werden unter der Bezeichnung „Business Intelligence“ (BI) zusammengefasst. Den BI Anwendungen ist gemein, dass sie die Identifikation und das Sammeln von Daten, die in Unternehmensprozessen aufkommen, sowie deren Extraktion und Analyse, bereitstellen [13, 14]. Das Problem bei der Anwendung der BI auf die virtuelle Produktion ist, dass die Umsetzung der BI die Herausforderungen der Integration von heterogenen Daten- und Informationsquellen in erster Linie konzeptionell löst und dies bei der Implementierung funktionsfähiger Systeme erhebliche Probleme verursacht. So wird im Konzept bspw. eine Übersetzung der Daten in ein einheitliches Datenformat und die kontextsensitive Annotation vorgesehen, aber eine Übersetzung kann evtl. nicht erreicht werden, da es sich um proprietäre Daten handelt und für die Annotation die Bedeutung nicht bekannt ist. Dies ist auch der Grund warum so viele BI Integrationen bisher fehlgeschlagen sind [15].

Im Folgenden wird dargestellt, dass mit der Vision der Digitalen Fabrik die zuvor adressierten Probleme gelöst werden sollen. Da die Vision jedoch noch nicht realisiert ist wird in den Kapiteln Heterogenität von Simulationen und Lösungsansatz: Virtual Production Intelligence darauf eingegangen wie die nächsten Schritte zur Digitalen Fabrik realisiert werden können. Der Begriff „Virtual Production Intelligence“ wurde in Anlehnung an den in der Problemstellung eingeführten Begriff „Business Intelligence“ gewählt, der Anfang bis Mitte der 1990er Jahre populär geworden ist. Dabei bezeichnet „Business Intelligence“ Verfahren und Prozesse zur systematischen Analyse (Sammlung, Auswertung und Darstellung) von Daten eines Unternehmens in elektronischer Form. Sie verfolgt das Ziel, auf Basis der gewonnenen Erkenntnisse bessere operative oder strategische Entscheidungen in Hinsicht auf die Unternehmensziele zu treffen. „Intelligence“ bezieht sich in diesem Kontext nicht auf Intelligenz im Sinne einer kognitiven Größe, sondern beschreibt die Erkenntnisse, die durch das Sammeln und Aufbereiten von Informationen gewonnenen werden. Dies entspricht der Verwendung des Wortes „Intelligence“, wie es auch im Kontext für geheimdienstliche Tätigkeiten in der englischen Sprache Verwendung findet (bspw. Central Intelligence Agency – CIA).

3 Digitale Fabrik

Die Digitale Fabrik (Abb. 2) wird durch den VDI Arbeitskreis in der VDI-Richtlinie definiert [4] als

> „der Oberbegriff für ein umfassendes Netzwerk von digitalen Modellen und Methoden, u. a. der Simulation und 3D-Visualisierung. Ihr Zweck ist die ganzheitliche Planung, Realisierung, Steuerung und laufende Verbesserung aller wesentlichen Fabrikprozesse und -ressourcen in Verbindung mit dem Produkt."

Gemäß der VDI-Richtlinie 4499 umfasst das Konzept der Digitalen Fabrik nicht einzelne Aspekte der Planung oder Produktion sondern den gesamten Produktlebenszyklus (Abb. 3). Es sollen alle Prozesse von der Entstehung über den Einsatz bis hin zur Außerdienststellung modelliert werden. Das heißt, die Betrachtung startet bei der Erhebung der Anforderung am Markt, die Entwurfsphasen inkl. aller notwendigen Dokumente, das Projektmanagement, Prototypen (digitale Mockups), die notwendigen internen und externen logistischen Prozesse, die Planung der Montage und Fertigung, die Planung der entsprechenden Fertigungsanlagen, die Montage und Inbetriebnahme der Fertigungsanlagen, das Anlaufmanagement (Ramp Up), die Serien-

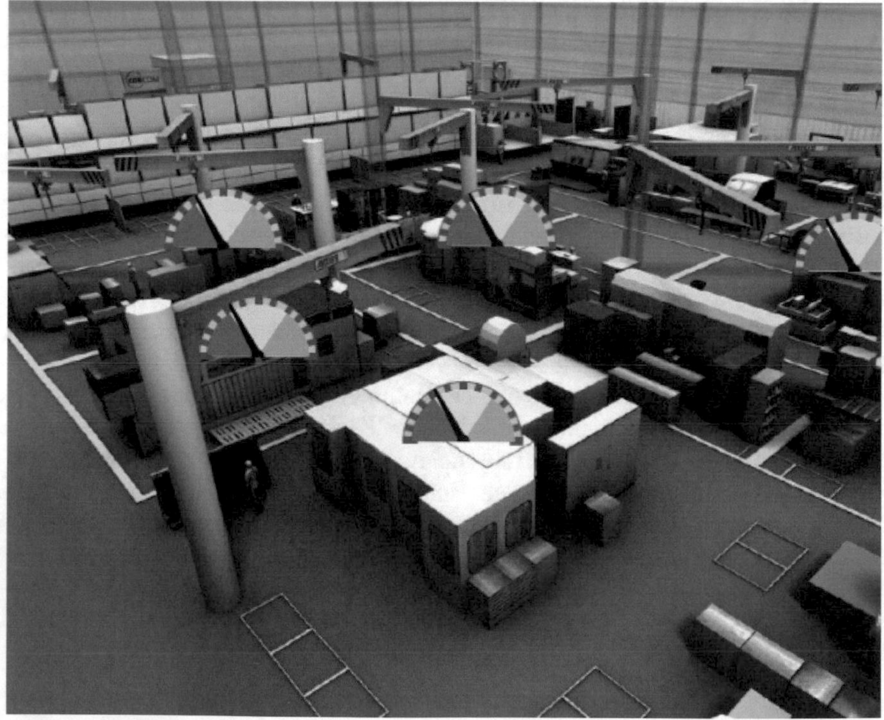

Abb. 2 Digitale Fabrik mit Indikatoren die anzeigen ob ein Prozess läuft © WZL der RWTH Aachen & IMA der RWTH Aachen

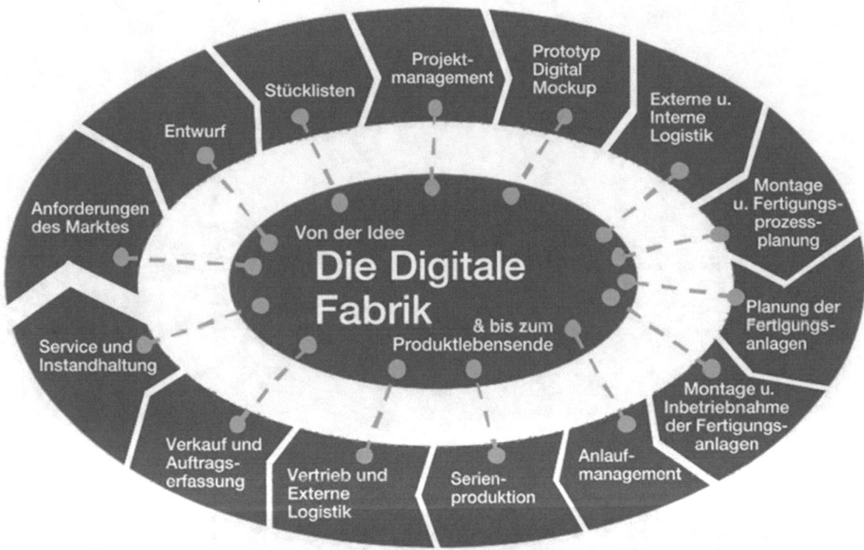

Abb. 3 Produktlebenszyklus (VDI Richtlinie 4499) [4]

produktion, der Vertrieb bis zur Wartung und das Recycling bzw. Entsorgung des Produkts sind damit Teil der Digitalen Fabrik. Zurzeit existiert keine Plattform, die dieser Integrationsaufgabe gerecht wird. Es sind aber schon einige Elemente der Digitalen Fabrik auf unterschiedlichsten Ebenen der Automatisierungspyramide oder in Phasen des Produktlebenszyklus realisiert. So gibt es Produktlebenszyklus Management (PLM) Software, die Unternehmen unterstützt den Produktlebenszyklus zu planen, zu überwachen und in Teilen auch zu steuern. Diese Anwendungen sind jedoch meist Insellösungen und ermöglichen nur die Integration von IT-Werkzeugen, die über die gleichen Schnittstellen zum Datenaustausch verfügen und vom gleichen Hersteller bereitgestellt werden. Die Detailtiefe der Abbildung einzelner Phasen des Produktlebenszyklus erreicht dabei nicht die hohe Ortsauflösung von Spezialanwendungen zur Beschreibung einzelner Phasen des Produktlebenszyklus oder von IT-Werkzeugen die sich auf Teilaspekte einzelner Phasen konzentrieren. Die Empfehlung des VDI, Datenmanagement, und -austausch möglichst homogen zu gestalten, kann daher nur bei Neuentwicklungen berücksichtigt werden. Davon abgesehen existiert bis heute kein Ansatz, wie ein Standard für einen solchen homogenen Datenaustausch umzusetzen ist und wie die angesprochene und wohl bekannten Probleme eines Standardisierungsprozesses verhindert beziehungsweise umgangen werden können. Demnach kann selbst ein Vorhaben, das sich gezielt als Beitrag zur Umsetzung der Vision sieht, nicht zu einer Homogenisierung des Informationsflusses beitragen, da nicht definiert ist, wie ein solcher Zustand auszusehen hat. Dazu kommt, dass es keinen Standard gibt und sich Standardisierungsbemühungen wie bspw. Standard for the Exchange of Product Model Data (STEP) gegen proprietäre Formate behaupten müssen. Hierbei muss berücksichtigt werden, dass

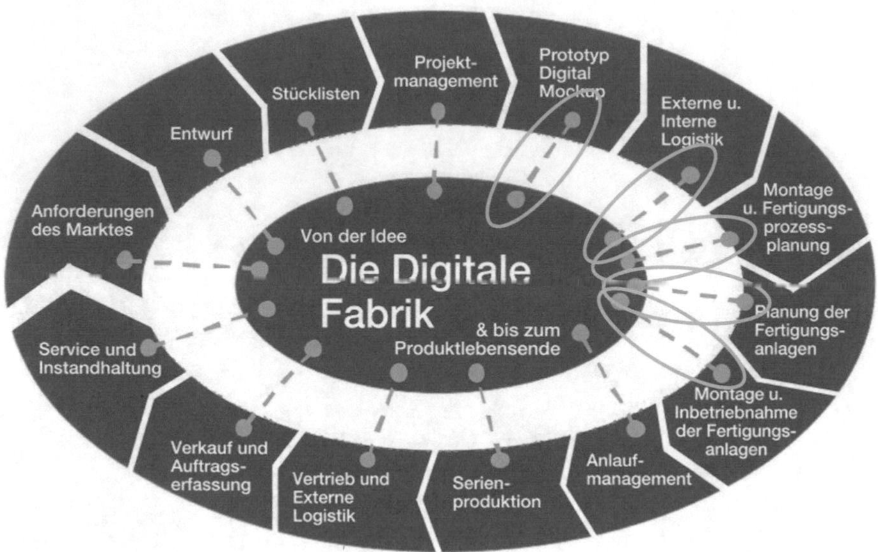

Abb. 4 Verortung der virtuellen Produktion innerhalb des Produktlebenszyklus nach VDI Richtlinie 4499 [4]

die proprietären Formate auch genutzt wurden um das Wissen und Fähigkeiten des Softwareanbieters zu schützen.

Mit Blick auf die Visualisierung der Digitalen Fabrik liegen Werkzeuge der Virtual Reality und Augmented Reality vor, die es den Anwendern ermöglichen 3D-Modelle von Fabrikanlagen mit und ohne Menschen zu realisieren und mit diesen auch zu interagieren und mit Informationen zu annotieren. Es ist jedoch keine Echtzeitsteuerung einer physisch vorhandenen Anlage über eine virtuelle Repräsentation der Anlage möglich, bei der die Daten aus dem Betrieb in der virtuellen Anlage dargestellt und für Analysezwecke weiter verarbeitet werden, da die Laufzeiten einzelner Simulationen die Echtzeitanforderung nicht erfüllen. Mit den vorliegenden Techniken, deren Weiterentwicklungen und Neuentwicklungen soll das Ziel der Digitalen Fabrik erreicht werden.

Die Virtual Production Intelligence dient als Basisbaustein für die Digitale Fabrik. Zur Erreichung dieses Zieles ist es nicht notwendig die gesamt Vision der Digitalen Fabrik zu adressieren sondern es ist vielmehr ausreichend den Bereich der simulationsbasierten virtuellen Produktion zu fokussieren (vgl. Abb. 4). Auch hier wird zur Definition der virtuellen Produktion die VDI-Richtlinie 4499 zitiert:

> „Simulativ durchgeführte vernetzte Planung und Steuerung von Produktionsprozessen mit Hilfe digitaler Modelle. Zweck der virtuellen Produktion ist die Optimierung von Produktionssystemen und flexible Anpassung der Prozessgestaltung vor einer prototypischen Realisierung." [4]

Die Produktionsprozesse werden hierbei in einzelne Prozessschritte zerlegt, die durch Simulationen beschrieben werden. Die Simulation der einzelnen Prozessschritte geschieht unter Verwendung moderner Simulationswerkzeuge, mit deren Hilfe sich selbst komplexe Produktionsverfahren präzise abbilden lassen. Ungeachtet der hohen Genauigkeit einzelner Simulationen besteht die zentrale Herausforderung bei der virtuellen Produktion jedoch in der Zusammenfassung der einzelnen Prozessschritte zu einer Wertschöpfungskette.

Die bereits erwähnte Virtual Production Intelligence (VPI) wird entwickelt, um in einem ersten Schritt die Interoperabilität von heterogenen IT-Werkzeugen herzustellen und zwar mit deutlich geringerem Aufwand bei Einsatz der zuvor erwähnten maßgeschneiderten Lösungen. In einem zweiten Schritt werden die integrierten Daten, die konsolidiert vorliegen, analysiert und weiterverarbeitet. Bei der VPI handelt es sich um ein ganzheitliches, integratives Konzept zur Unterstützung der kollaborativen Durchführung von Technologie- und Produktentwicklung und der Fabrik- und Produktionsplanung. mit dem Ziel die frühzeitige Identifikation und Beseitigung von Fehlerquellen in Prozessen zu ermöglichen wodurch Optimierungspotenziale erkannt und nutzbar gemacht werden. Zum besseren Verständnis werden die Begriffe ganzheitlich, integrativ und kollaborativ folgendermaßen eingegrenzt:

- *Ganzheitlich*: Es werden alle Teile der adressierten Prozesse berücksichtigt.
- *Integrativ*: Nutzung und Zusammenführung vorhandener Lösungsansätze.
- *Kollaborativ*: Berücksichtigung aller in den adressierten Prozessen involvierten Rollen und deren Kommunikation untereinander.

Im nächsten Abschnitt werden die bereits erwähnten Heterogenitäten, die durch den Einsatz der VPI überwunden werden sollen, näher betrachtet.

4 Heterogenität von Simulationen

Nach ISO/IEC 2382-01 liegt Interoperabilität zwischen Softwareanwendungen vor, wenn die Fähigkeit zu kommunizieren, Programme auszuführen, oder Übertragung von Daten zwischen verschiedenen Funktionseinheiten in einer Weise ermöglicht wird, ohne dass der Benutzer Informationen über die Eigenschaften der Anwendung hat. Abbildung 5 fasst die Heterogenitäten zusammen, die wesentlich dazu beitragen, dass keine Interoperabilität ohne maßgeschneiderte Adapter erreicht wird [16–18].

Mit technischer Heterogenität werden die Unterschiede in der Art und Weise wie auf Daten oder Anwendungen von Benutzern oder weiteren Anwendungen zugegriffen wird bezeichnet. Die syntaktische Heterogenität beschreibt die Unterschiede in der Abbildung von Daten, bspw. unterschiedliche Codierungsstandards wie ASCII oder Binär Codierung, oder die Abbildung von Fließkommazahlen als float oder double und ihre interne Repräsentation. Diese beiden Arten der Heterogenität können relativ einfach durch den Einsatz von Adaptern überwunden werden, jedoch ist hier ein möglichst generisches Konzept zu verfolgen, so dass eine weitere Verwendung dieser Adapter ermöglicht wird. Für die technische Heterogenität stehen hierfür eine Vielzahl unterschiedlicher Bibliotheken und Lösungen zur Verfügung. Ebenso verfü-

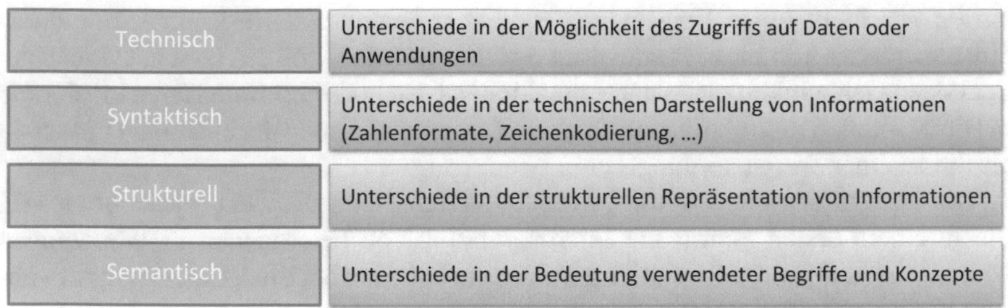

Technisch
Unterschiede in der Möglichkeit des Zugriffs auf Daten oder Anwendungen
Syntaktisch
Unterschiede in der technischen Darstellung von Informationen (Zahlenformate, Zeichenkodierung, ...)
Strukturell
Unterschiede in der strukturellen Repräsentation von Informationen
Semantisch
Unterschiede in der Bedeutung verwendeter Begriffe und Konzepte

Abb. 5 Arten der Heterogenität von Simulationen

gen moderne Programmierkonzepte über implizite Typanpassungen und ermöglichen ebenso die kontrollierte explizite Umwandlung von Daten [16–18].

Die Überwindung der strukturellen und der semantischen Heterogenität stellt die ungleich größere Herausforderung dar. Bei der strukturellen Heterogenität werden Unterschiede in der Repräsentation von Informationen adressiert. Semantische Heterogenität beschreibt die Unterschiede in der Bedeutung der domänenspezifischen Entitäten und der für ihre Auszeichnung verwendeten Begriffe. So können zwei Simulationen den Begriff der Umgebungstemperatur verwenden, bei Simulation A wird damit die Hallentemperatur beschrieben in der ein Aufheizofen steht und in Simulation B wird damit die Temperatur im Ofen in unmittelbarer Umgebung des aufzuheizenden Objekts ausgezeichnet. Im Folgenden wird die VPI vorgestellt, die Methoden bereitstellt um diese Arten von Heterogenität zu überwinden und die notwendige Interoperabilität zwischen den Anwendungen zu gewährleisten [16–18].

5 Lösungsansatz: Virtual Production Intelligence

Die Analyse der Daten erfolgt mit Hilfe von analytischen Konzepten und IT-Systemen, welche die Daten über das eigene Unternehmen, Mitbewerber oder die Marktentwicklung im Hinblick auf den gewünschten Erkenntnisgewinn auswerten.

Die „Virtual Production Intelligence“ hat das Ziel, die in einem Simulationsprozess entstandenen Daten zu sammeln, zu analysieren und zu visualisieren, um Erkenntnisse zu generieren, die eine ganzheitliche Bewertung der einzelnen Simulationsergebnisse und des aggregierten Simulationsergebnisses ermöglichen. Grundlage der Analyse sind Expertenwissen sowie physikalische und mathematische Modelle. Durch eine immersive Visualisierung werden die Anforderungen an eine „Virtual Production Intelligence“ vollständig abgedeckt.

Die Integration von Ergebnissen eines Simulationsprozesses in ein einheitliches Datenmodell ist der erste Schritt, um Erkenntnisse aus diesen Datenbeständen zu gewinnen und die Extraktion von versteckten, validen, nützlichen und handlungsrelevanten Informationen zu realisieren. Diese Informationen umfassen beispielsweise die Qualität der Ergebnisse eines Simulationsprozesses oder in konkreteren Anwen-

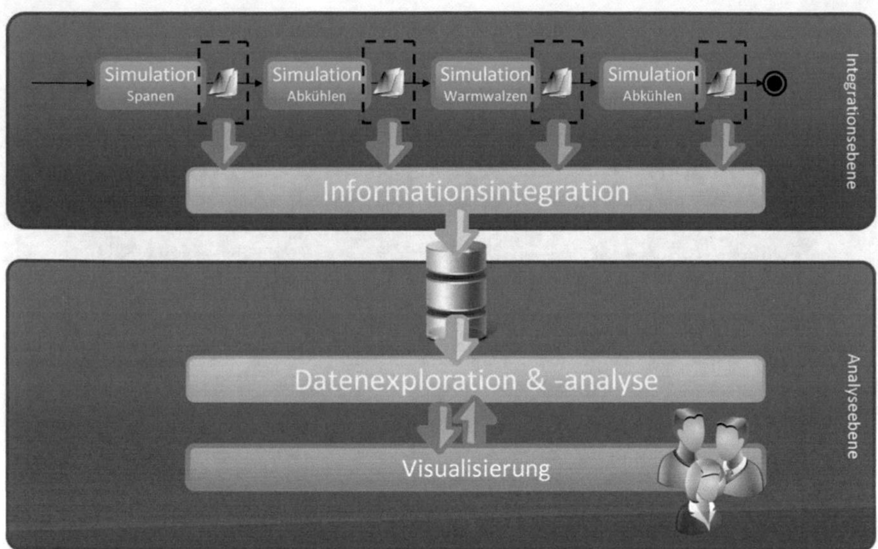

Abb. 6 Datenexploring und -analyse

dungsfällen auch die Ursachen für die Entstehung von Inkonsistenzen. Zur Identifikation solcher Aspekte stehen dem Analysten zurzeit nur begrenzte Möglichkeiten zur Verfügung. Mit der Realisierung einer Integrationslösung aber wird die Möglichkeit einer einheitlichen Betrachtung aller Daten ermöglicht. Dies umfasst zum einen die Visualisierung des gesamten Simulationsprozesses in einer Visualisierungskomponente, zum anderen die Untersuchung und Analyse der Daten über den gesamten Simulationsprozess. Hierzu können unterschiedliche Explorationsverfahren herangezogen werden.

Der beschriebene Sachverhalt der Datenexploration und -analyse ist in Abb. 6 zusammenfassend dargestellt: Zunächst werden die Daten entlang des Simulationsprozesses in ein kanonische Datenmodell integriert, dass in Form eines relationalen Datenmodells umgesetzt wurde, so dass eine einheitliche und konsolidierte Sicht auf die Daten möglich ist. Anschließend werden die Daten in der Analyseebene durch den Anwender unter Verwendung der Visualisierung analysiert. Dabei wird der Anwender mittels Datenexploration und -analyseverfahren unterstützt, die direkt innerhalb der immersiven Umgebung angesteuert werden können. Durch die Möglichkeit zum Feedback an die Analysekomponente kann der Benutzer gezielt den Explorationsprozess beeinflussen und Parametrisierungen von Analysen zur Laufzeit der Ergebnisdarstellung vornehmen.

Neben der nachträglichen Analyse durch Experten ist es ebenso sinnvoll, eine Überwachung der Daten während des Simulationsprozesses zu realisieren, da eine solche Prozessüberwachung beispielsweise die Einhaltung von Parameterkorridoren oder anderen Randbedingungen ermöglicht. Würde ein Simulationswerkzeug Parameterwerte außerhalb der definierten Parameterkorridore liefern, würde dies zu

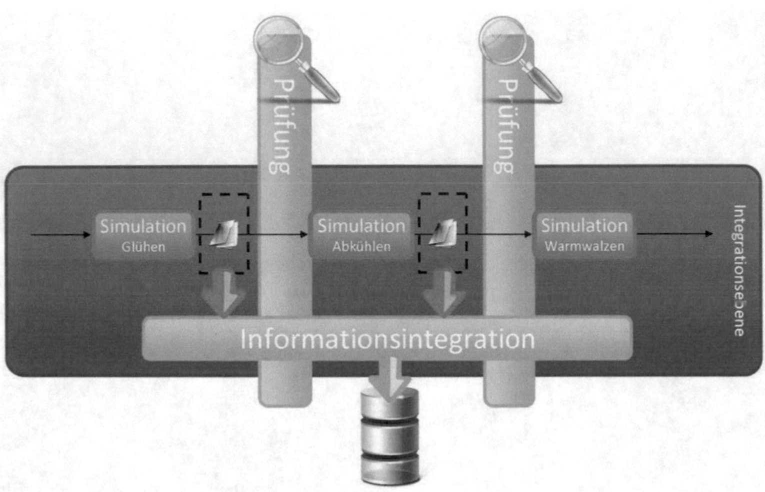

Abb. 7 Prozessüberwachung als querschnittliche Funktion

einem Abbruch des Simulationsprozesses führen. Die bisherigen Ergebnisse könnten dann in der Datenanalyse durch Experten analysiert werden, um anschließend eine gezielte Anpassung der Simulationsparameter durchzuführen. Außerdem wäre die Bewertung von Zwischenergebnissen durch Gütefunktionen denkbar, die nach dem Durchlauf und der Integration der Simulationsergebnisse geprüft werden. Ebenso könnte eine Prozessüberwachung die Extraktion von Point-of-Interests (POI) auf Basis von Funktionen ermöglichen, die anschließend in der Visualisierung hervorgehoben werden würden. Der beschriebene Sachverhalt ist in Abb. 7 zusammenfassend dargestellt.

Abbildung 8 zeigt die Komponenten eines Systems zur Realisierung einer „Virtual Production Intelligence". Die Applikationsebene umfasst die Simulationen, die entlang eines definierten Simulationsprozesses aufgerufen werden. Diese sind über eine Middleware miteinander verbunden, die den Datenaustausch realisiert und für die Sicherstellung der Datenintegration und Datenextraktion innerhalb des Simulationsprozesses verantwortlich ist. Dazu wird ein Integrationsserver bereitgestellt, der über einen serviceorientierten Ansatz Dienste zur Integration und Extraktion zur Verfügung stellt. Der Datenbankserver bildet das zentrale Datenmodell ab und dient als zentraler Datenspeicher für alle im Prozess generierten Daten.

Folgendes Beispiel illustriert die Einsatzmöglichkeit der VPI bei der Unterstützung im Fabrikplanungsprozess. Abbildung 9 fasst die Unterstützung in der Fabrikplanung zusammen.

Die VPI-Plattform wird zur optimierten Fabrikplanung und Prozessketten-Analyse eingesetzt, die Fabrikplanung beruht hierbei auf den Prinzipen des Condition Based Factory Planning (CBFP) [9]. Dieser Ansatz, stellt eine Modularisierung des Planungsprozesses von Fabriken dar, der in Form dezidierter Planungsmodule durchgeführt wird. Die Datenerfassung und Auswertung innerhalb der Planungsphasen

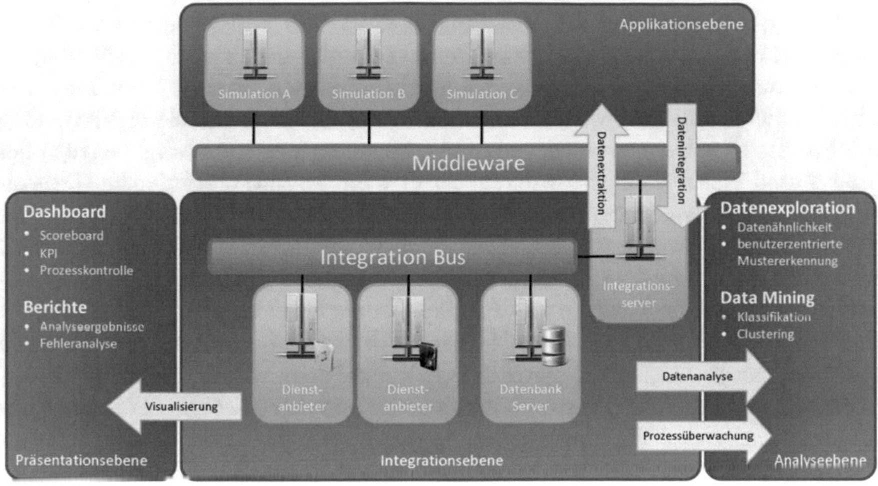

Abb. 8 Komponenten der Virtual Production Intelligence

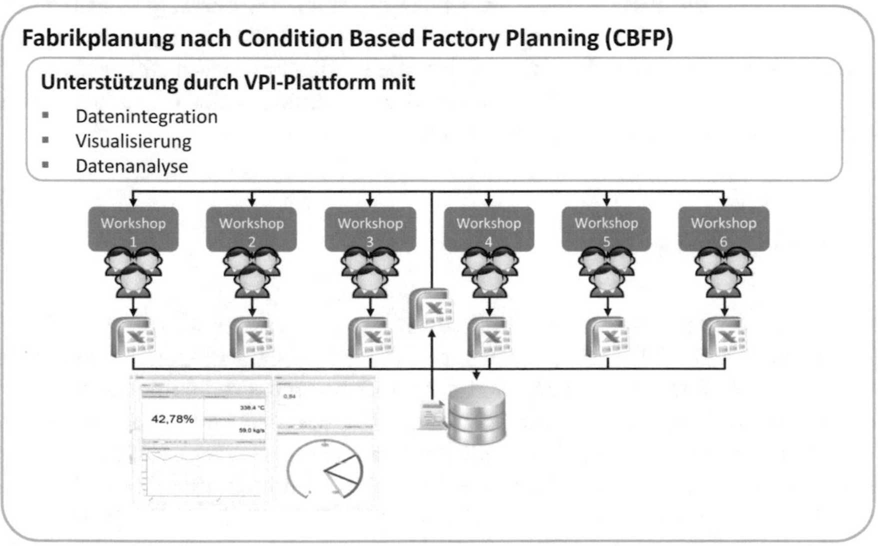

Abb. 9 Unterstützung in der Fabrikplanung

erfolgt durch Nutzung gängiger Office-Anwendungen. Eine Verknüpfung der Planungsmodule untereinander ist daher oft mit einem hohen Übertragungsaufwand hinsichtlich der Datenkonsistenz und -durchgängigkeit verbunden. Insbesondere die Realisation eines Fabrikmodells zur variantenreichen Produktion unter Berücksichtigung verschiedener Szenarien ist nicht ohne weiteres durchführbar, da keine konsolidierte Datenlage bzgl. der einzelnen zu verknüpfenden Planungsmodule vorliegt.

Durch die Anwendung des VPI-Konzeptes auf das CBFP werden vielfältige Möglichkeiten für einen innovativen Produktionsplanungsprozess geschaffen, da eine Zusammenfassung der Planungsmodule in ein gemeinsames Informationsmodell ermöglicht wird. Insbesondere lassen sich durch die zentralisierte Verwaltung von Prozessdaten der verschiedenen Simulationsvorgänge neue Möglichkeiten der Analyse und Visualisierung realisieren. Um eine umfassende und kompakte Darstellung der Ergebnisse und Planungsgrößen der verschiedenen Simulationsmodule zu gewährleisten, wurde aufbauend auf dem VPI-Plattformkonzept eine Web-2.0 Modul entwickelt, die dem Benutzer vielfältige Möglichkeiten der Analyse, Interaktion und Optimierung entlang des Fabrikplanungsprozesses ermöglicht. In Form einer Web-Applikation, welche in jedem modernen Browser ausgeführt werden kann, werden die am Planungsprozess beteiligten Personen mit aktuellen Prozessgrößen aus der Datenbasis versorgt. Somit ist eine Synchronisation der in der Web-Applikation abgebildeten Prozesse und Analysen stets gewährleistet. Der Benutzer hat die Möglichkeit, Anfragen an die Datenbasis zu senden, Manipulationen an den Ein- und Ausgangsgrößen der Simulationen und Änderungen an ihrer Darstellung vorzunehmen. Es wird ermöglicht eine umfassende Prozessketten-Analyse durchzuführen, die insbesondere die Abhängigkeiten mit der Kapazitätsplanung sowie Produktionsstrukturplanung berücksichtigt. Eine wirksame Optimierung verschiedener Produktionsstrukturen, wie etwa die Festlegung der Anzahl der Prozessketten und Produktionssegmente wird erst durch Abbildung der Interdependenzen verschiedener Planungsmodule ermöglicht.

6 Zusammenfassung

Mit der VPI wird ein wesentlicher Beitrag zur Realisierung der Vision der Digitalen Fabrik erreicht. Die VPI ist zum einen eine Integrationsplattform, die es ermöglicht heterogene IT-Werkzeuge im Bereich der Produkt- und Produktionsplanung mit einander zu verknüpfen und zum anderen ein auf dem englischen „Intelligence"-Begriff basierendes Analyse Werkzeug, um Wirkzusammenhänge zu identifizieren und zu bewerten. Da die virtuelle Produktion mit der Produkt- und Produktionsplanung Kernbereich der Digitalen Fabrik ist, wurde im Rahmen des Beitrags auf diesen Teil fokussiert. Basis für die VPI ist die Etablierung von Interoperabilität, Die Funktionsweise der VPI wird dargestellt und mithilfe des Beispiels der Fabrikplanung verdeutlicht. Der Einsatz der VPI ermöglicht eine deutliche Aufwandsreduzierung im Engineering zur Erstellung Maßgeschneiderter Integrations- und Analysewerkzeugen, da mit der VPI eine adaptive Lösung vorliegt. Es ist nun möglich mit einer prozessorientierten und damit kontextsensitiven Informationsverarbeitung zu beginnen. Informationen liegen jetzt nicht nur bezogen auf einen einzelnen Prozessschritt vor, in dessen Zusammenhang sie entstanden sind, sondern stehen im Bezug zu dem gesamt Prozess, so dass die Bedeutung und die Gültigkeit von Informationen intensiver betrachtet werden kann.

Die zukünftigen Arbeiten im Rahmen der VPI werden im Bereich der interaktiven explorationsbasierten Datenanalyse liegen. Dabei ist zu evaluieren, inwiefern sich die durch Explorationsverfahren extrahierten Informationen bewerten lassen. Außerdem ist zu untersuchen, wie diese Informationen dem Benutzer in einer immersiven Umgebung dargestellt werden können und wie sich Zusammenhänge von Informationen verständlich und nachvollziehbar präsentieren lassen. Hierzu bieten sich unterschiedliche feedbackgestützte Techniken an, in denen Experten über Feedbackschnittstellen der Visualisierung Analyseergebnisse bewerten und optimieren. Die Kommunikation verläuft dabei bidirektional, das heißt der Benutzer gibt dem System über eine Schnittstelle Feedback, das wiederum vom Analysesystem verwendet wird, um die dargestellte Information zu korrigieren oder zu präzisieren. Das System versucht dabei, das Feedback zu interpretieren, um zukünftig unpräzise, fehlerhafte Aussagen zu vermeiden.

Literaturverzeichnis

1. A. Chandler, *Scale and Scope. The Dynamics of Industrial Capitalization*. Belknap Press of Harvard University Press, Campbridge, Mass., London, 2004
2. G. Schuh, S. Aghassi, S. Orilski, J. Schubert, M. Bambach, R. Freudenberg, C. Hinke, M. Schiffer, Technology roadmapping for the production in high-wage countries. Prod. Eng. Res. Devel. (Production Engineering) **5** (4), 2011, pp. 463–473
3. C. Brecher, *Integrative Produktionstechnik für Hochlohnländer*. Springer Verlag (Vdi-buch), 2011
4. VDI richtlinie 4499, blatt 1. In: *Digitale Fabrik*, 2008
5. VDI richtlinie 4499, blatt 2. In: *Digitale Fabrik*, 2011
6. R. Lauber, P. Göhner, *Prozessautomatisierung 1*, 3rd edn. Springer, Berlin, 1999
7. H. Kagermann, W. Wahlster, J. Helbig, Umsetzungsempfehlungen für das zukunftsprojekt industrie 4.0 – abschlussbericht des arbeitskreises industrie 4.0. forschungsunion im stifterverband für die deutsche wissenschaft. Tech. rep., Berlin, 2012
8. M. Nagl, B. Westfechtel, Modelle, werkzeuge und infrastrukturen zur unterstützung von entwicklungsprozessen. In: *Symposium (Forschungsbericht (DFG))*, 1st edn., Wiley-VCH, 2003, pp. 331–332
9. C. Horstmann, *Integration und Flexibilitat der Organisation Durch Informationstechnologie*, 1st edn. Gabler Verlag, 2011. S. 156–162
10. D. Schilberg, *Architektur eines Datenintegrators zur durchgängigen Kopplung von verteilten numerischen Simulationen*. VDI-Verlag, Aachen, 2010
11. T. Meisen, P. Meisen, D. Schilberg, S. Jeschke, Application integration of simulation tools considering domain specific knowledge. In: *Proceedings of the 13th International Conference on Enterprise Information Systems (ICEIS 2011), Beijing, China, 8–11 June 2011*. SciTePress, 2011
12. R. Reinhard, T. Meisen, T. Beer, D. Schilberg, S. Jeschke, A framework enabling data integration for virtual production. In: *Enabling Manufacturing Competitiveness and Economic Sustainability – Proceedings of the 4th International Conference on Changeable, Agile, Reconfigurable and Virtual production (CARV2011), Montreal, Canada, 2-5 October 2011*, ed. by H.A. ElMaraghy. Springer, Berlin, Heidelberg, 2011, pp. 275–280
13. B. Byrne, J. Kling, J. McCarty, G. Sauter, P. Worcester. The information perspective of SOA design, part 4: The value of applying the canonical modeling pattern in SOA, 2008
14. M. West, *Developing High Quality Data Models*, 1st edn. Morgan Kaufmann, Burlington, MA, 2011

15. W. Yeoh, A. Koronios, Critical success factors for business intelligence systems. Journal of computer information systems **50** (3), p. 23
16. M.C. Daconta, L.J. Obrst, K.T. Smith, *The Semantic Web a guide to the future of XML, Web services, and knowledge management*. 2003
17. D. Schilberg, A. Gramatke, K. Henning, Semantic interconnection of distributed numerical simulations via SOA. In: *Proceedings of the World Congress on Engineering and Computer Science*. Newswood Limited, Hong Kong, 2008, pp. 894–897
18. D. Schilberg, T. Meisen, R. Reinhard, S. Jeschke, Simulation and interoperability in the planning phase of production processes. In: *Automation, Communication and Cybernetics in Science and Engineering 2011/2012*, ed. by S. Jeschke, I. Isenhardt, F. Hess, K. Henning, Springer, 2013, pp. 1141–1152

Mobile Learning in der Hochschullehre

Daniela Schmitz, Dominik May and Karsten Lensing

1 Einleitung

Im täglichen Leben der Studierenden nehmen mobile Endgeräte eine zunehmend wichtigere Rolle ein. Die Anzahl derer, die weder ein Handy, noch einen MP3-Player oder einen Laptop besitzen, ist verschwindend gering[1]. Seit einigen Jahren steigt auch der Marktanteil von Smartphones und Tablet-PCs immer weiter an [1]. Dabei ist die Nutzung dieser Endgeräte breit gefächert und reicht vom Abrufen des aktuellen Wetters, der Zugverbindungen oder der persönlichen E-Mails von unterwegs bis hin zum Ersatz eines Laptops durch einen Tablet-PC. Gegenüber dieser Entwicklung wird deutlich, dass die aktive Nutzung mobiler Endgeräte in der Lehre an Hochschulen noch in den Kinderschuhen steckt.

Erste wissenschaftliche Publikationen mit Definitionen zu Mobile Learning (kurz m-learning oder auch Wireless Learning, Ubiquitous Learning, Seamless Learning, Nomadic Learning oder auch Pervuasive Learning bzw. Education [2]) lassen sich seit dem Jahr 2000 finden. Anhand der jeweils formulierten Definitionen lassen sich auch die Entwicklungsschritte der Fachdiskussion im Kontext des Mobile Learning nachvollziehen. Während Quinn (2000) davon spricht, dass es sich bei Mobile Learning um „elearning through mobile computational devices: Palms, Windows CE machines, even your digital cell phone“ [3] handelt, zeichnen Sharples et al.

[1]Um eine Datenbasis genau für diese Aussage zu erhalten, findet im Moment eine Umfrage zum Mediennutzungsverhalten im Rahmen des Forschungsprojekts „ELLI“ (http://www.elli-online.net) an den drei Hochschulen RWTH Aachen University, Ruhr-Universität Bochum und TU Dortmund statt. Mit Ergebnissen ist in der zweiten Jahreshälfte von 2013 zu rechnen.

D. Schmitz (✉)
Fakultät für Gesundheit (Department für Pflegewissenschaft), Universität Witten/Herdecke, Witten, Germany
e-mail: Daniela.Schmitz@uni-wh.de

D. May · K. Lensing
Engineering Education Research Group (EERG), Center for Higher Education (zhb), TU Dortmund University, Dortmund, Germany

Originally published in “Journal Hochschuldidaktik” Issue 1-2, Vol. #24,

DOI 10.1007/978-3-319-46916-4_13

(2010) ein deutlich komplexeres Bild von Mobile Learning, indem sie schreiben, dass Mobile Learning auch durch die Mobilität von Lernenden und Wissen sowie durch die Kontextualisierung der Lernprozesse in die Lernenden-Umgebung charakterisiert wird [4].

Deutlich wird an dieser Gegenüberstellung auch, dass es möglich ist, sich dem Mobile Learning von unterschiedlichen Seiten zu nähern. Auf der einen Seite ist eine rein technische Betrachtungsweise möglich, welche die mobilen Endgeräte und ihre Eigenschaften in den Fokus nimmt und somit die technologische Dimension beschreibt. Auf der anderen Seite steht die didaktische Dimension, welche den Lernprozess und dessen Kontextualisierung sowie seine soziale Komponente im Fokus hat [5]. Eine erste, intern am Zentrum für Hochschul-Bildung der TU Dortmund zu Mobile Learning durchgeführte Literaturanalyse betrachtet insgesamt 238 zumeist englischsprachige, wissenschaftliche Publikationen. Die fünf am häufigsten referenzierten Definitionen zu Mobile Learning stellen folgende Aspekte in den Vordergrund (Kursivierung von den Autoren dieses Beitrags ergänzt):

> Mobile Learning devices are defined as handheld devices and [...] should be connected through wireless connections that ensure mobility and flexibility [6].
>
> … provides the potential of personal mobile technologies that could improve lifelong learning programs and continuing adult educational opportunities [7]. What is new in »mobile learning« comes from the possibilities opened up by portable, lightweight devices that are sometimes small enough to fit in a pocket or in the palm of the one's hand [8].
>
> Any sort of learning that happens when the learn er is not at a fixed, predetermined location, or learning that happens when the learner takes advantage of the learning opportunities offered by mobile technologies [9].
>
> Learning will move more and more outside of the classroom and into the learner's environments, both real and virtual […] new learning is highly situated, personal, collaborative and long term; in other words, truly learner-centered learning. […] mobile devices are finding their way into classrooms in children's pockets, and we must ensure that educational practice can include these technologies in productive ways [10].

Aufbauend auf diesen und weiteren Definitionen arbeitet Maske (2011) vier maßgebliche charakteristische Eigenschaften des Mobile Learning heraus, die sich in einem vierdimensionalen Beschreibungscluster für Mobile Learning Lehrveranstaltungen darstellen lassen (s. Abb. 1) [5].

Dieses soll im Folgenden zur Erläuterung der Beispiele herangezogen werden. Das heißt bei der Beschreibung von Mobile Learning Veranstaltungen werden die folgenden Fragen gestellt: „Inwiefern ist die Nutzung mobiler Endgeräte in die Lehrveranstaltung eingeplant?“, „Inwiefern sind die Lernenden ortsunabhängig bei der Teilnahme an der Lehrveranstaltung?“, „Inwiefern sind die Lernprozesse kontextualisiert?“[2] [11] und „Inwiefern sind die Lernszenarien informell?“

---

[2]Nach Göth, Frohberg und Schwabe lässt sich Mobile Learning in vier Kontexte einteilen: (1) in irrelevante Kontexte wie in Bus oder Bahn, wo der Kontext für das Lernen keine besondere Rolle spielt; (2) Lernen in formalisierten Kontexten wie in der Vorlesung oder im Seminar, wo das Lernen in einen institutionalisierten Kontext mit formalen Abläufen eingebunden ist; (3) in sozialen Kontexten, wo das Lernen in Lerngruppen von Bedeutung ist; (4) in physischen Kontexten, wo Lernkontext und Umgebung stimmig sind wie beispielsweise das Lernen im Museum (vgl. [11]).

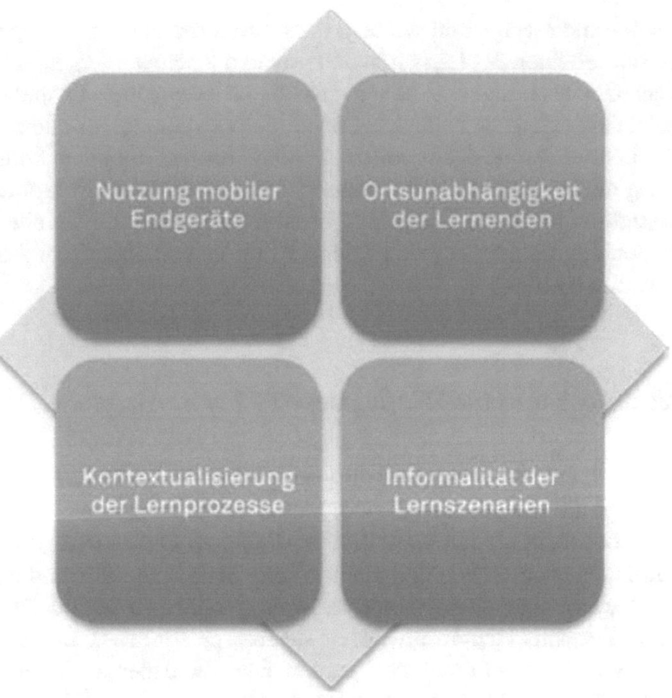

Abb. 1 Beschreibungscluster für Mobile Learning Lehrveranstaltungen

2 Praxisbeispiele Mobile Learning auf Exkursionen

Die folgenden Praxisbeispiele entstanden auf der Basis der Zielvorgabe, Lernprozesse mit mobilen Endgeräten im Kontext der Hochschullehre anzureichern und bei der Realisierung keinen großen technischen Aufwand bzw. möglichst geringe Kosten zu verursachen. Im Folgenden stellen wir zwei schlanke (im Sinne von kostenneutralem und geringem technischen Aufwand), didaktische Einsatzszenarien dar, [12] die als Ergänzung oder Begleitung zur Präsenzlehre einsetzbar sind. Dies sind nur zwei mögliche Lernszenarien aus einem breiten Spektrum, welches durch mobile Lerntechnologie ermöglicht wird.

Worauf kommt es also an, wenn Sie als Lehrende ihre Lehre mit Mobile Learning anreichern möchten? Zunächst sollte geprüft werden, ob ein sinnvoller Einsatz von Mobile Learning zur Erreichung der Lernziele (1) notwendig und (2) überhaupt möglich ist. Weiterhin ist zu klären, ob mobile Lernsequenzen zusätzlich unterstützende oder notwendigerweise zu absolvierende Lernsequenzen sind und welche Mischung von Präsenz-, Online- und Mobile-Lernphasen vorgesehen sind. Als förderliche Faktoren für mobile Lernszenarien bieten sich offene Aufgabenstellungen an, die unter anderem ein Lernen in kleineren Gruppen ermöglichen und auch Raum für informelle Lernprozesse lassen. Bei der Gestaltung mobiler Lernszenarien sollte nicht vergessen werden, dass durch die Nutzung der persönlichen Geräte der Studierenden das Lernen in den privaten Kontext übergeht. Die Nutzung eines eigenen Gerätes bringt eine höhere affektive Bindung mit sich und sollte daher nicht unreflektiert erfolgen. Zudem sollte bei der Gestaltung vor allem der mobile Ansatz der Lerntechnologie ausgereizt werden, um ein mobiles Endgerät nicht für eLearning einzusetzen, welches genauso gut mithilfe eines PCs durchzuführen wäre. Letztendlich sollte die Zielgruppe beachtet werden, da nicht alle Studierenden standardmäßig mit dem mobilen Endgerät immer und überall für jede Lehrveranstaltung lernen (wollen).

2.1 Dokumentation und Reflexion des Lernprozesses

Mithilfe von Textverarbeitungsapplikationen für mobile Endgeräte, die kostenfrei erhältlich und plattformunabhängig sind, halten die Studierenden zu Beginn einer Veranstaltung oder einer Lerneinheit ihre persönlichen Lernziele fest. Dies können neben inhaltlichen auch soziale und methodische Ziele sein. Während einer Exkursion machen sich die Studierenden beispielsweise Notizen zu den Inhalten und ihren Lernzielen und dokumentieren, inwieweit sie ihre persönlichen Lernziele erreicht haben. Zum Abschluss der Lerneinheit werden diese Ziele thematisiert und der Lernprozess wird reflektiert. Mobile Endgeräte fungieren hier als individuelle, mobile Lernprozessunterstützung, zum Beispiel bei Museumsbesuchen oder Betriebsbesichtigungen im Rahmen kleinerer Seminare oder Projektgruppen. Neben Vorgaben zum Lerninhalt können die Studierenden durch die Internetanbindung auch erweiterte Informationen abrufen und ad hoc entstehende Lernbedarfe stillen. Die Realisierung ist mit einer simplen Dokumentenvorlage im txt-Format möglich, welche Vorstrukturierungen für die Notizen oder Satzanfänge für die Lernziele enthält. Diese Vorlage erhalten die Studierenden vom Lehrenden und können diese auf ihre Geräte laden.

2.1.1 Beispielszenario

In der Vorbesprechung am Morgen des Exkursionstages legen die Lernenden ihre Lernziele für den Tag fest. Während des Tages gibt es vor den Pausen die Aufforderung zu dokumentieren und zu prüfen, inwieweit die Lernziele schon erreicht wurden und wie der Lernweg verlaufen ist. In der Nachbesprechung werden die Lernzielerreichung und der individuelle Lernweg reflektiert.

2.1.2 Einordnung in das Mobile Learning Cluster

Der Fokus dieses Beispielszenarios liegt in der Kontextualisierung der individuellen Lernprozesse. Das Lernen findet direkt im relevanten physischen Kontext statt. Durch die Lernform der Exkursion besteht nur eine relative Ortsunabhängigkeit zwischen einzelnen Lernstationen. Das Lernen kann aber nicht außerhalb des Exkursionsortes stattfinden, da der relevante Kontext vor Ort eben das Lernen ermöglicht. Anders sieht die begleitende Dokumentation und Reflexion des Lernprozesses im Rahmen einer klassischen Präsenzlehrveranstaltung aus. Dabei kann das mobile Lernen als dokumentierte Reflexion des Lernens ortsunabhängig und in einem beliebigen Kontext stattfinden. Informelles Lernen didaktisch in einem Lernszenario direkt gestalten zu wollen, erweist sich als schwierig. Informelles Lernen ist interessengesteuert und muss nicht mit den Inhalten der Lehrveranstaltung identisch sein. Sicherlich sind im Rahmen eines Museumsbesuches neben dem Lehrziel der Veranstaltung zahlreiche Möglichkeiten gegeben, informell zu lernen.

2.2 *Generierung von Lernfragen*

Dieser Ansatz bringt im Unterschied zum ersten einen Perspektivwechsel für die Lernenden mit sich. Die Lernenden werden in einem Lernsetting zu einem Perspektivwechsel dazu angeregt, vom Lernenden in die Rolle des Lehrenden zu schlüpfen. Zu vorab definierten Lerneinheiten sollen die Lernenden Lernfragen bzw. Übungsaufgaben mit den richtigen Antworten bzw. Lösungshinweisen generieren (je nach Möglichkeit kann auch schon eine Auswahl an falschen Antworten dazu generiert werden). Diese Lernfragen kommen dann zur inhaltlichen Nachbereitung von Exkursionen oder einer Seminareinheit zum Einsatz. Die Gruppen tauschen ihre Lernfragen aus, erhalten gegenseitig Einblicke in die Inhalte und Interessensschwerpunkte der anderen Gruppen und bereiten so den Lernstoff der Lerneinheit nach.

2.2.1 Beispielszenario

Die Lernenden fertigen zu einem Ausstellungsbereich des Museums Lernfragen an. Inhalte aus diesem Bereich, die die Studierenden für interessant und lehrreich befinden, werden in Frageform verfasst. Neben der richtigen Antwort müssen beispielsweise auch drei falsche Antworten gefunden werden. Diese von den Lernenden erstellten Lernfragen dienen der Nachbereitung der Exkursion vor Ort, indem die einzelnen Lerngruppen ihre Lernfragen austauschen.

2.2.2 Einordnung in das Mobile Learning Cluster

Dieses Lernszenario nutzt mobile Lerntechnologie begleitend und ortsunabhängig für die Erstellung der Lernfragen. Der Kontext des Lernens kann entweder irrelevant sein oder in der Gruppe in einem sozialen Kontext stattfinden. Findet das Lernszenario als Exkursion statt, wäre es ein für das Lernen relevanter physischer Kontext.

Aus diesem Lernszenario kann informelles Lernen entstehen, wenn beispielsweise Lernende ihre Lernfragen interessierten Personen auch außerhalb der Exkursionsgruppe über ein Forum o.ä. zur Verfügung stellen.

3 Teaching Tips und Ausblick

Wenn Sie mobile Lernszenarien umsetzen möchten, sollten Sie auch potenzielle Probleme berücksichtigen. Denn neben technischen und didaktischen Aspekten können zusätzlich weitere unvorhergesehene Aspekte als potenzielle Störfaktoren für das Lernszenario auftauchen:

- Der Faktor „Draußen“: Wettereinflüsse, Verfügbarkeit des Lernangebotes
- Der Faktor Technik: Internetzugang, Akkulaufzeit, Funktion der Geräte und Anwendungen, Nutzung bestehenden Contents oder Contenterstellung, Anbindung an bestehende Technologien wie Lernplattformen
- Der Faktor Lernende: Motivation, Lernbereitschaft mobil zu lernen, Einstellung zu mobilen Endgeräten
- Der Faktor Lehrende: Kompetenzen zur Contenterstellung und -aufbereitung

Resümierend lassen sich aus den bisherigen Überlegungen und vorgestellten Beispielen so genannte Do's and Dont's für Lehrende festhalten. Die folgende Tabelle fasst die Tipps stichpunktartig zusammen:

Do's	Dont's
Informelles Lernen in Kleingruppen ermöglichen	Zu starr vorgegebene Aufgaben
Selbstorganisierte Herangehensweisen bei der Bearbeitung	Ausschließlich mit Leihgeräten für die gesamte Lerngruppe arbeiten
Sinnvolle Einbindung ins Gesamtkonzept der Lehrveranstaltung	Auf Reflexion des Lernwegs und des Technologieeinsatzes verzichten
Notwendigkeit der mobilen Lernunterstützung prüfen	Nur als eLearning auf dem Tablet umsetzen
Abwechslungsreicher Medien einsatz, verschiedene Lerntypen ansprechen	Voraussetzen, dass alle Studierende „Digital Natives“ sind

Das Thema „Mobile Learning“ wird am Zentrum für HochschulBildung im Rahmen des Forschungsprojekts „ELLI – Exzellentes Lehren und Lernen in den Ingenieurwissenschaften“ fokussiert. Neben der Entwicklung und Erprobung von Mobile Learning Anwendungsszenarien geht es auch um deren Einbettung in den Gesamtkontext von virtueller Lehre. Für die hochschuldidaktische Forschung zur Nutzung mobiler Endgeräte in der Lehre sind zurzeit vier konkrete Szenarien in Planung bzw. bereits in der Umsetzung. (1) Neben den mobilen Endgeräten werden im Rahmen von ELLI und in Zusammenarbeit mit dem „Institut für Umformtechnik und Leichtbau“ (IUL) auch Lehr-Lernszenarien mit Remote Laboratories (am IUL sind dies im Speziellen Labore mit entsprechendem Equipment zur Bestimmung und Untersuchung von Materialkennwerten, welche über einen online-Zugang nutzbar sind) erarbeitet. Mobile Learning soll in diesen Kontext integriert werden. Das heißst, dass die Labore zukünftig auch über mobile Endgeräte nutzbar gemacht werden. In Ergänzung dazu wird auch an einem Mobile-Learning-Szenario gearbeitet, welches die Reflexion des Lernens und die Kollaboration der Studierenden mithilfe der mobilen Endgeräte fördert [13]. (2) Ein weiteres Szenario wird im Kontext einer transnationalen online-Lehrveranstaltung zwischen der TU Dortmund und der University of Virginia umgesetzt. Im Rahmen dieser Lehrveranstaltung lernen deutsche und amerikanische Studierende gemeinsam, indem sie die gleiche Lehrveranstaltung online besuchen und dort gemeinsame Lehrprojekte durchführen [14]. In Zukunft soll erforscht werden, welche Vorteile für die Kommunikation und Kollaboration die Nutzung mobiler Endgeräte birgt und welchen Einfluss dies auf die Gestaltung des Lehr-Lernszenarios in der praktischen Umsetzung hat. (3) Für eine Veranstaltung zum Thema „Projektmanagement" wird zurzeit untersucht, wie effektive Kommunikation im Projekt durch die Nutzung mobiler Endgeräte unterstützt werden kann. Für diese Untersuchung wurden Studierende mit entsprechender Hardware ausgestattet und diese erarbeiten in Kooperation mit den Dozierenden ein Lastenheft, welches die notwendigen Ausstattungsmerkmale eines mobilen Endgeräts beschreibt, um Projektkommunikation zu fördern. (4) Ein letztes Szenario zielt auf die Kreativität von Studierenden ab und wird in einer Veranstaltung zu Kreativität selbst umgesetzt. Mithilfe einer speziell entwickelten App sollen Studierende ihre persönlichen, kreativen Momente im alltäglichen Kontext mit Hilfe von Fotos, kurzen Sprachnotizen o.ä. festhalten. Ausgehend davon wird einerseits untersucht, wann Studierende kreativ sind, anderseits sollen die Studierenden durch das Teilen und Reflektieren dieser Momente in einen gemeinsamen, kreativen Prozess kommen.

Die oben beschriebenen Szenarien sind nur Beispiele für eine Unzahl von Möglichkeiten, mobile Endgeräte in der Lehre einzusetzen. Die Forschung dazu hat gerade erst begonnen. Sollten Sie auch eine Idee haben und suchen dafür Unterstützung oder möchten diese intensiv beforschen, so freuen wir uns, wenn Sie sich mit uns in Kontakt setzen. Ansonsten können wir Sie nur ermuntern, Ihre Ideen in diesem Kontext einfach umzusetzen.

Literaturverzeichnis

1. M. Wurm. Mini-Tablets heizen Nachfrage an. http://www.crn.de/hardware/artikel-98610.html
2. D. Frohberg, Mobile-learning. Ph.D. thesis, 2008
3. C. Quinn. mlearning: Mobile, wireless, in yourpocket learning, 2000. URL: http://www.linezine.com/2.1/features/cqmmwiyp.htm
4. M.e.a. Sharples, A theory of learning for the mobile age. In: *Medienbildung in neuen Kulturräumen*, ed. by B. Bachmair, VS Verlag für Sozialwissenschaften, Wiebaden, 2010, pp. 87–99
5. P. Maske, *Mobile Applikationen 1: Interdisziplinäre Entwicklung am Beispiel des Mobile Learning*. Springer Gabler, Wiesbaden, 2011
6. C. Quinn. mlearning: mobile, wireless, in-yourpocketlearning, 2001

7. M. Sharples, *The Design of Personal Mobiletechnologies for Lifelong Learning*, 34th edn. 2000
8. A. Kukulska-Hulme, J. Traxler, *Mobilelearning. A handbook for educators and trainers*. Routledge, London, 2005
9. C. O'Malley, G. Vavoula, J.P. Glew, J. Taylor, M. Sharples, et al. Guidelines for learning/teaching/tutoring in a mobile environment, 2003. https://hal.archives-ouvertes.fr/hal-00696244/document
10. L. Naismith, P. Lonsdale, G. Vavoula, M. Sharples. Report 11: Literature review in mobile technologies and learning, 2004. http://citeseerx.ist.psu.edu/viewdoc/download?doi=10.1.1.459.9648&rep=rep1&type=pdf
11. C. Göth, D. Frohberg, G. Schwabe, Von passivem zu aktivem mobilen Lernen. Zeitschrift für e-learning, Lernkultur und Bildungstechnologie **2** (4), 2007, pp. 12–28
12. D. Schmitz, Mobile Lernprozessunterstützung auf Exkursionen. Hamburger eLMagazin (9), 2012, pp. 23–25. http://www.uni-hamburg.de/eLearning/eCommunity/Hamburger_eLearning_Magazin/eLearningMagazin_09.pdf
13. C. Terkowsky, D. May, T. Haertel, C. Pleul, Experiential learning with remote labs and e-portfolios – integrating tele-operated experiments into personallearning environments. International Journal of Online Engineering (iJOE), **9** (1), 2013, pp. 12–20
14. St. Moore, D. May, K. Wold, Developing cultural competency in engineering through transnationaldistance learning. In: *Transnational DistanceLearning and Building New Markets for Universities*, ed. by R. Hogan, IGI Global, Hershey (PA/USA), 2012, pp. 210–228

Verbesserung der Lernerfahrung durch die Integration des Virtual Theatres in die Ingenieursausbildung

Katharina Schuster, Daniel Ewert, Daniel Johansson, Ursula Bach, René Vossen and Sabina Jeschke

Zusammenfassung In ingenieurwissenschaftlichen Studiengängen zählen praktische Erfahrungen sowie das Verständnis für die praktische Relevanz von Lerninhalten zu den Voraussetzungen eines erfolgreichen Studiums. In Grundlagenfächern wie Physik oder Chemie werden praktische Erfahrungen meist experimentell durch Laborarbeit gesammelt. Ein weiterer verbreiteter Ansatz zur Herstellung praktischer Relevanz sind Praktika oder Exkursionen. Aus Sicherheits- und Kostengründen wie auch aufgrund der Komplexität industrieller Produktionsstätten ist es für Studierende jedoch nicht möglich, solche Industrieanlagen frei zu erkunden und dort eigenständig zu experimentieren.

Um diese Problematik zu umgehen, können Industrieanlagen in virtuellen Umgebungen nachgebaut werden. In Simulationen oder Lernspielen (Serious Games) werden die Eigenschaften und Prinzipien einer Industrieanlage mit entsprechenden Kursinhalten verknüpft und dadurch nachvollziehbar gemacht. In virtuellen Lernumgebungen können Prozesse der Wissensaneignung selbst gesteuert werden. Es können Experimente durchgeführt werden, die unter physisch-realen Bedingungen zu gefährlich oder zu teuer wären. Weiterhin können Orte erschlossen werden, die ansonsten nicht erreichbar wären – sei es weil sie räumlich zu weit entfernt oder zeitlich in der Vergangenheit oder in der Zukunft liegen.

Ein Nachteil des Lernens mit Simulationen besteht in der Künstlichkeit des virtuellen Zugangs. Neben der Art der grafischen Gestaltung einer virtuellen Lernumgebung liegt häufig die Antwort für den Grund auf der Seite der Hardware. Natürliche Nutzerschnittstellen (engl. natural user interface, NUI) für Visualisierung, Navigation und Interaktion können eine authentischere Lernerfahrung als am PC ermöglichen und somit den Wissenstransfer des Gelernten auf spätere Anwendungssituationen erleichtern. In Mixed-Reality Simulatoren wie dem Virtual Theatre werden die Nutzerschnittstellen Head-Mounted Display, omnidirektionaler Boden und Datenhandschuh integriert und ermöglichen so eine uneingeschränkte Kopf- und Fortbewegung zur freien Erschließung virtueller Lernumgebung. Nach einer Beschreibung des technischen Aufbaus des Virtual Theatres beschreibt das Paper einen medienpsychologischen Forschungsansatz, mit dem der Zusammenhang von immersiver Hardware

K. Schuster (✉) · D. Ewert · D. Johansson · U. Bach · R. Vossen · S. Jeschke
IMA/ZLW & IfU, RWTH Aachen University, Dennewartstr. 27, 52068 Aachen, Germany
e-mail: katharina.schuster@ima-zlw-ifu.rwth-aachen.de

Originally published in "TeachING LearnING.EU Discussions", © 2013.
Reprint by Springer International Publishing AG 2016,
DOI 10.1007/978-3-319-46916-4_14

und Lernerfolg gemessen werden kann. Das Paper schließt mit einem Ausblick auf den Einsatz des Virtual Theatres in der Ingenieurausbildung.

Schlüsselwörter Natural User Interfaces · Mixed-Reality · Immersion · Virtual Learning Environments

1 Einleitung

Es ist eine anerkannte Tatsache, dass Studierende der Ingenieurwissenschaften praktische Erfahrungen sowie konkrete Bezüge zur beruflichen Praxis benötigen, um ein ganzheitliches Verständnis ihrer Fachdisziplin erlangen zu können. In Grundlagenfächern wie Physik oder Chemie werden praktische Erfahrungen meist experimentell durch Laborarbeit gesammelt. Ein weiterer verbreiteter Ansatz zur Herstellung praktischer Relevanz sind Praktika oder Exkursionen. Aus Sicherheits- und Kostengründen wie auch aufgrund der Komplexität industrieller Produktionsstätten ist es für Studierende jedoch nicht möglich, solche Industrieanlagen frei zu erkunden und dort eigenständig zu experimentieren. Zum einen mit Blick auf die Produktion, da der stetige Produktionsfluss gewährleistet und Maschinen vor Beschädigung geschützt werden müssen. Zum anderen müssen Studierende selbst vor Verletzungen bewahrt werden, die z. B. durch unsachgemäße Maschinenbedienung sowie den Kontakt mit giftigen oder ätzenden Substanzen zustande kommen könnten. Der große Bedarf von Seiten der Studierenden scheitert folglich schnell am Machbaren.

Um die oben genannte Problematik zu umgehen, können Industrieanlagen in virtuellen Umgebungen nachgebaut werden. In Simulationen oder Lernspielen (Serious Games) können die Eigenschaften und Prinzipien einer Industrieanlage mit entsprechenden Kursinhalten verknüpft und dadurch nachvollziehbar gemacht werden. Ein Nachteil von Simulationen besteht häufig in der Künstlichkeit der Lernerfahrung. Normalerweise interagieren Nutzer/innen mit einer virtuellen Umgebung über einen PC. Am Beispiel der Industrieanlage wird diese auf einem Monitor angezeigt und das Sichtfeld durch eine Tastatur oder eine Maus gesteuert. Interaktionen mit virtuellen Objekten sowie Fortbewegung erfolgen über die gleichen Schnittstellen. Auf diese Weise überträgt der Nutzer oder die Nutzerin den Kontrollmodus des Computers auf die Aktionen seiner grafischen Repräsentation, dem Avatar.

Natürliche Nutzer/innenschnittstellen für Visualisierung, Navigation und Interaktion können eine authentischere Lernerfahrung als am PC ermöglichen und dazu führen, dass Studierende regelrecht in die virtuelle Welt eintauchen können. In diesem Zusammenhang spricht die Fachwelt häufig von Immersion – einem Bewusstseinszustand, auf den später im Rahmen dieses Beitrags noch näher eingegangen wird. Für eine erhöhte Immersion benötigt der Nutzer oder die Nutzerin eine nahtlose 3-D-Sicht

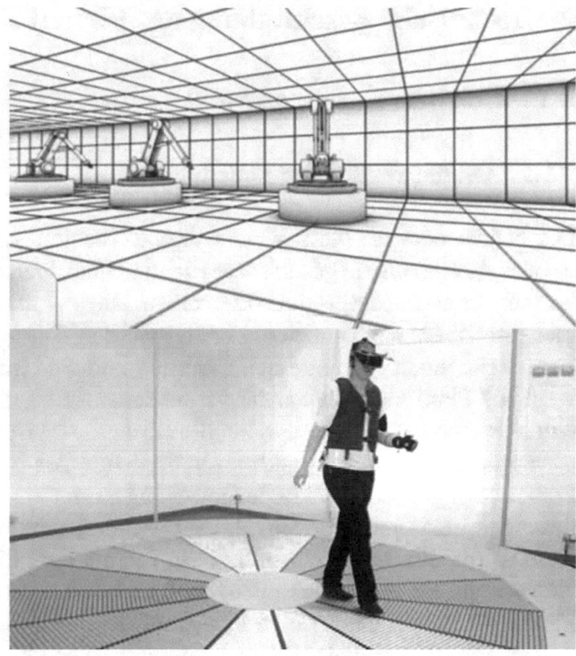

Abb. 1 Das Virtual Theatre mit Nutzerin und (vereinfachter) virtuellen Umgebung

der virtuellen Umwelt. Diese wird häufig durch Head Mounted Displays (HMDs) realisiert. Für eine natürliche Navigation in der virtuellen Umgebung können omnidirektionale Laufbänder verwendet werden, die eine freie und unbegrenzte Bewegung ermöglichen und die nutzende Person doch an einem physisch klar begrenzten Ort lassen. Durch Datenhandschuhe kann der Nutzer bzw. die Nutzerin intuitiv mit der virtuellen Umgebung sowie den Objekten, die sich in ihr befinden, interagieren. Die beschriebenen Komponenten – HMD, omnidirektionales Laufband und der Datenhandschuh – sind im sogenannten Virtual Theatre integriert [1]. Abbildung 1 zeigt die Nutzung des Virtual Theatres sowie ein exemplarisches Anwendungsszenario für die Lehre in den Ingenieurwissenschaften. Im Folgenden werden die technischen Komponenten des Virtual Theatres näher beschrieben. Anschließend wird der Begriff der Immersion erläutert sowie auf weitere Aspekte eingegangen, die die Wahrnehmung und das Verhalten der nutzenden Personen betreffen. Am Ende des Beitrags werden didaktische Prinzipien vorgestellt, die bei der Entwicklung von Lernszenarien für die Ingenieurausbildung berücksichtigt werden müssen sowie exemplarische Szenarien vorgestellt. Der Beitrag schließt mit einem Ausblick auf weiterführende Forschung am und mit dem Virtual Theatre.

2 Technische Beschreibung des Virtual Theatres

2.1 *Hardware Komponenten*

2.1.1 Omnidirektionaler Boden

Der Nutzer oder die Nutzerin des Virtual Theatres kann sich frei bewegen, indem er oder sie einfach die gewünschte Richtung ansteuert. Der omnidirektionale Boden besteht aus 16 trapezförmigen Elementen. Diese sind aus je 16 festen Rollen zusammengesetzt, die nach außen hin breiter werden. Die Elemente haben an der kurzen Grundseite einen gemeinsamen Ursprung und sind im Mittelpunkt über eine unbewegliche Plattform miteinander verbunden. Die Laufrollen werden von unten von einem Riemenantrieb bewegt, um die Abstände zwischen den einzelnen Elementen zu minimieren. Auf der Plattform in der Mitte des Bodens kann sich die nutzende Person bewegen, ohne die Bodenbewegung zu starten. Abbildung 2 zeigt den Aufbau des omnidirektionalen Bodens als CAD-Modell. Ohne die unbewegliche Mittelplattform würde die Bodenbewegung dazu führen, dass die Füße des Nutzers oder der Nutzerin in der Mitte des Bodens aneinander gezogen werden. Wenn die nutzende Person die Mittelplattform verlässt, rotieren die Rollen der Elemente des Bodens gemäß eines Kontrollalgorithmus automatisch in Gegenrichtung zu der Bewegung des Nutzers bzw. der Nutzerin [2].

2.1.2 Head Mounted Display (HMD)

Das Virtual Theatre ermöglicht der nutzenden Person eine nahtlose 3-D-Visualisierung der virtuellen Umgebung. Jede Kopfbewegung wird direkt in der Simulation widergespiegelt, sodass die nutzende Person in seiner oder ihrer Sicht nicht eingeschränkt ist. Das visuelle sowie das auditive Feedback wird über ein zSight HMD realisiert [3], welches ein stereoskopisches 70° Sichtfeld mit SXGA-Auflösung für jedes Auge

Abb. 2 CAD Modell des omnidirektionalen Bodens

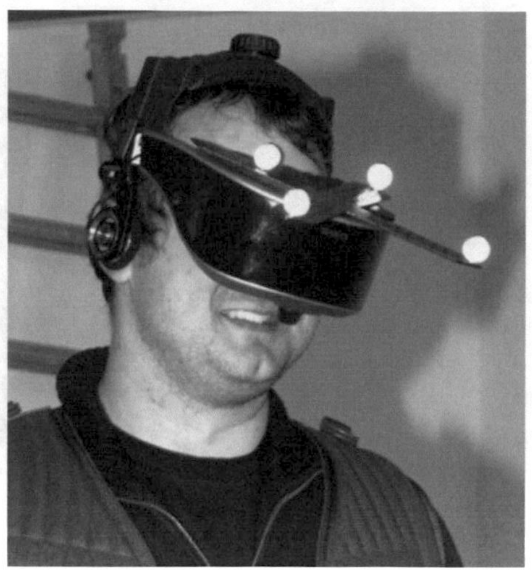

Abb. 3 HMD mit angefügten Infrarot-Markern

erstellt. Der Audiosound erfolgt über an der Brille befestigte Kopfhörer, die Lautstärke kann von der nutzenden Person über ein kleines Rad kontrolliert werden. Das HMD wiegt 450 Gramm. Über einen Drehknopf am Ober- sowie am Hinterkopf kann das HMD auf die individuelle Kopfgröße der Nutzer/innen eingestellt werden. Ein kleiner Drehknopf zwischen den Linsen regelt den Augen-Nasen-Abstand. Über kleine Räder unterhalb der Linsen kann die Auflösung der Mini-Displays an die individuelle Sehstärke angepasst werden. Abbildung 3 zeigt das HMD sowie die an ihm befestigten Infrarot-Marker.

2.1.3 Datenhandschuh

Handgesten können erkannt werden und die Simulation sowohl manipulieren als auch kontrollieren. Um in der virtuellen Umgebung mit Elementen aus der Umwelt interagieren zu können, erhält dic nutzende Person einen Datenhandschuh, der eine Steuerung über natürliche Handbewegungen ermöglicht. Diese werden von 22 Sensoren empfangen, die im Datenhandschuh [4] eingearbeitet sind. Zurzeit ist nur ein Datenhandschuh in das System des Virtual Theatre integriert. Objektmanipulation oder zweihändige Gestenerkennung gestaltet sich derzeit noch als verhältnismäßig zu komplex, wenn man den Entwicklungsaufwand dem Nutzen gegenüberstellt. Dennoch ist die spätere Integration eines zweiten Datenhandschuhs möglich. Abbildung 4 zeigt den Datenhandschuh sowie die Darstellung der Handgesten in einer virtuellen Umgebung.

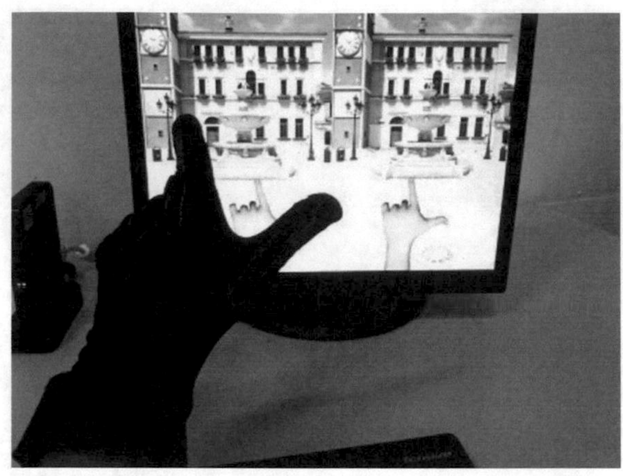

Abb. 4 Datenhandschuh und Darstellung in einer virtuellen Umgebung

2.1.4 Tracking-System

Das Virtual Theatre ist mit zehn Infrarotkameras ausgestattet, um die Bewegungen der nutzenden Person zu registrieren. Die Kameras zeichnen die Positionen der Infrarot-Marker auf, die mit dem HMD und dem Datenhandschuh verbunden sind. Die Position des HMD dient zugleich als Input für die Bewegungssteuerung des omnidirektionalen Bodens. Die nach innen gerichtete Geschwindigkeit der Rollen wird linear mit der Distanz des HMDs zum Zentrum des omnidirektionalen Bodens gesteigert – je weiter der Nutzer oder die Nutzerin vom Zentrum entfernt ist, desto schneller laufen die Rollen. Die Position des HMDs dient zudem der Steuerung der Sicht der nutzenden Person innerhalb der virtuellen Umgebung. Die Position des Datenhandschuhs dient der Repräsentation der Hand innerhalb der virtuellen Szenerie. Außerdem kann über die Position des Handschuhs im Notfall eine Systemabschaltung eingeleitet werden. Sobald die Hand unter eine Grenze von 0,5 Metern bewegt wird, z. B. falls die nutzende Person stürzen sollte, wird jede Bewegung des omnidirektionalen Bodens mit sofortiger Wirkung gestoppt.

2.1.5 Integration der verschiedenen Komponenten

Die Systemarchitektur des Virtual Theatres ist in Abbildung 5 dargestellt. Die fixen Komponenten Trackingsystem und omnidirektionaler Boden kommunizieren mit der Hardware-Steuerung über kabelbasierte Kommunikationskanäle. Um der nutzenden Person uneingeschränkte Bewegung zu ermöglichen, erfolgt jegliche Kommunikation der Hardware, die sich direkt an dem Nutzer oder der Nutzerin befindet, drahtlos (WLAN).

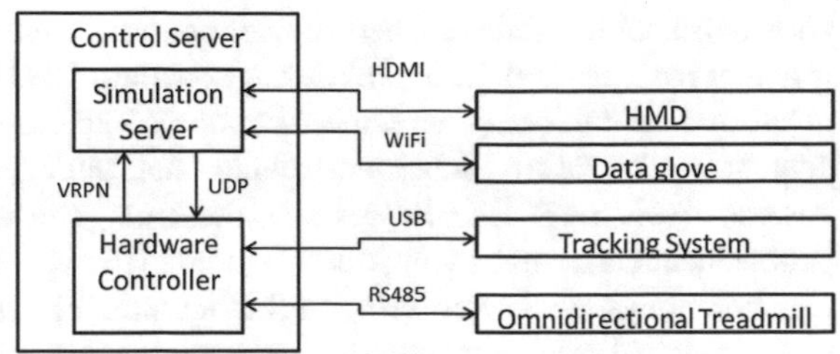

Abb. 5 Systemarchitektur des Virtual Theatres

2.2 *Software Set-up*

Der Server des Virtual Theatres wird von zwei Computern betrieben, die unterschiedlichen Zwecken dienen. Der Hardware-Control-Server analysiert die Daten der Infrarotkameras und regelt den omnidirektionalen Boden. Um die Bewegungen der nutzenden Person auszuwerten, werden OptiTrack Tracking Tools [5] verwendet. Die Information über die Position wie auch die Geschwindigkeit des Bodens wird dem Simulationsserver durch vrpn [6] zugesendet. Hierbei handelt es sich um eine gerätunabhängige und netzwerktransparente Schnittstelle zu Virtual Reality Umgebungen. Der Simulationsserver ist für die High Level Steuerung verantwortlich und stellt die virtuelle Umgebung bereit. Hierfür wird der Toolkit Vizard Virtual Reality [7] verwendet. Die 3-D-Modelle für virtuelle Umfelder werden mit Blender [8] erstellt und können in Vizard geladen werden.

2.3 *Einschränkungen*

Zurzeit kann sich die nutzende Person frei in der virtuellen Welt bewegen und bekommt Feedback in visueller und auditiver Form. Zusätzliche Simulationskomponenten, wie z.B. Temperaturregler oder Luftströmungen sind generell möglich, jedoch nicht Teil der derzeitigen Szenarien. Die parallele Nutzung durch mehrere Personen ist im Virtual Theatre nicht möglich: Der Boden verfügt nur über einen Motor und die Rollen können nicht unabhängig voneinander angesprochen werden. Auch wenn theoretisch mehrere Personen mit Infrarot-Markern ausgestattet und vom Tracking-System verfolgt werden können, so können nur die Daten einer Person verwendet werden, um den Motor entsprechend zu regeln.

Beobachtende können die Aktionen der nutzenden Person zwar an Bildschirmen verfolgen, jedoch nicht direkt mit der virtuellen Umwelt interagieren. Kooperative Szenarien, in denen die Beobachtenden die nutzende Person anweisen und in anderer Weise unterstützen oder ihm bzw. ihr sogar mithilfe eines klassisch gesteuerten Avatars assistieren, sind prinzipiell möglich. Für den Lehreinsatz ist das Virtual Theatre jedoch nur für kleine Kursgrößen geeignet, was u.a. durch die begrenzte Raumkapazität der Halle bedingt ist, in der sich der Simulator befindet.

Dies muss bei der didaktischen Konzeptionierung entsprechender Seminare in den Ingenieurwissenschaften berücksichtigt werden. Die Einzelnutzung ohne den organisatorischen Rahmen eines Kurses ist natürlich auch möglich, jedoch nicht unbedingt kosteneffizient, da aus Sicherheitsgründen stets ein/e geschulte/r Simulationsleiter/in anwesend sein muss. Eine interessante Forschungsfrage lässt sich aus der Theorie des beobachtenden Lernens von Albert Bandura [9] ableiten. So könnte ein zukünftiger Untersuchungsgegenstand sein, ob Studierende, die nur die Beobachtungsperspektive und nicht die der nutzenden Person einnehmen, ähnliche Lernerfolge verzeichnen können, wie tatsächliche Nutzer/innen.

3 Nutzer/innenzentrierte Untersuchungen zu Wahrnehmung und Verhalten

3.1 Konzeptioneller Hintergrund von Immersion

Generell dienen natürliche Nutzer/inn/enschnittstellen immer dem Zweck, die Mensch-Computer-Interaktion zu erleichtern, Illusionen zu erzeugen und bestimmte Situationen so realitätsnah wie möglich zu imitieren. Ziel ist es, bei dem Nutzer oder der Nutzerin den Eindruck zu erwecken, sich tatsächlich in der virtuellen Umgebung zu befinden. Bevor das Virtual Theatre in der Ingenieurausbildung eingesetzt wird, sind wissenschaftliche Erkenntnisse über die Wahrnehmung der Lernsituation und über den tatsächlichen Lernerfolg notwendig. Deswegen wird die weitere Entwicklung der Hardware und der Lernszenarien von unterschiedlichen psychologischen Studien begleitet. Eine Erklärung der experimentellen Gestaltung dieser Studien erfolgt nach einem näheren Blick auf den Begriff der Immersion, der häufig im Kontext der Realisierung immer realistischerer Simulationen verwendet wird.

Bekannte Beispiele realitätsgetreuer Situationsimitationen sind Flug- und Fahrsimulatoren oder Spielkonsolen im Unterhaltungsbereich. Der Begriff der Immersion wird diesbezüglich als ein objektives, steigerbares Maß benutzt, um zu beschreiben, inwieweit technische Eigenschaften dazu beitragen, die Illusionen zu erschaffen. HMDs gelten generell als gute Voraussetzung für Immersion. Ein HMD mit einer 101° Sicht gewährleistet eine intensivere Immersion als ein HMD mit einer Sichtweite von 100°. Ein Simulator wie das Virtual Theatre, in dem man frei herumlaufen kann, erzeugt folglich eine höhere Immersion, als ein Simulator der dies nicht ermöglicht [10].

In Spieler/innenkreisen wird der Begriff für die Erfahrung verwendet, inwiefern man sich in der Welt des Spiels „verlieren" kann. Schon eine oberflächliche Durchsicht verschiedener Gaming Foren und Blogs verdeutlicht, dass die Bedeutung von Immersion mehrdeutig und facettenreich ist. Die Beschreibungen, die diesen Begriff begleiten, reichen von der Fähigkeit eines Spiels „dich von der realen Welt los zu reißen und dich mit tatsächlichen psychischen Erfahrung zu überschwemmen, voll von emotionaler Aufruhr und Konflikten" [11] bis hin zu „der Fähigkeit, den Spieler

(/ die Spielerin) in andere Elemente als den Charakter und die Geschichte des Spiels miteinzubeziehen“ [12]. Der Begriff Immersion besitzt hier eine positive Konnotation und gilt als Qualitätssiegel für Computerspiele. Bezüglich der Hardware verdeutlicht die Entwicklung immer neuer Schnittstellen im Unterhaltungsbereich, wie der Nintendo Wii oder der Kinect für die Xbox Konsole, den Wunsch nach einer natürlicheren Steuerung. Obwohl sich Hardware-Entwickler/innen und die Gaming Community über die Bedeutung von Immersion weitestgehend einig sind, ist der Begriff noch nicht vollständig definiert worden. Weiterhin wurde noch nicht einheitlich geklärt, was Immersion aus psychologischer Sicht ausmacht und welche Effekte sie hat, vor allem wenn das Ziel nicht Unterhaltung sondern Lernen ist.

In einer ersten Annäherung kann Immersion als „der subjektive Eindruck, dass jemand eine umfassende und realistische Erfahrung macht“ [13] definiert werden. Die Idee absorbierender und anregender Erfahrungen ist jedoch nicht neu und es existieren darüber hinaus diverse weitere Begriffskonzepte, die der Idee von Immersion ähneln und in Beziehung zu ihr stehen. Involviertheit, räumliche Präscnz und Flow gelten als Schlüsselkonzepte, um immersive Erfahrungen zu erklären, auch wenn die genauen Definitionen und Bedeutungszuschreibungen je nach Autor/in variieren. Involviertheit beschreibt einen psychischen Zustand, der daraus resultiert, dass jemand all seine Energie und Aufmerksamkeit auf ein Set an Reizen, Aktivitäten oder Ereignissen lenkt. Der Grad an Involviertheit hängt dabei auch von der Wichtigkeit ab, die das Individuum diesen Reizen beimisst [14].

Während Flow die Involviertheit in eine Aktivität beschreibt, bezieht sich Präsenz auf die räumliche Sinneswahrnehmung in einer medialisierten Umgebung [15]. Das auf Csikszentmihalyi [16] zurückgehende Konzept des Flows wurde von Rheinberg et al. auf Mensch-Computer-Interaktionen adaptiert. Hier beschreibt Flow das Gefühl vollkommener Konzentration, Kontrolle über die Aktivität, Klarheit über die Arbeitsschritte, flüssiges und automatisches Denken sowie „glatte“, wie von selbst fließende Handlungsabläufe. Des Weiteren meint Flow das Gefühl der kompletten Involviertheit, ein verändertes Zeitgefühl, das Gefühl der optimalen Herausforderung und einer gewissen Geistesabwesenheit, dem sog. „Verschmelzen“ von Selbst und Tätigkeit [17].

Eine Reihe von Studien zeigt, dass persönliche Charakteristika einen Einfluss auf Immersion haben. Weibel et al. [15] belegen, dass die Eigenschaften Offenheit für neue Erfahrungen, Neurotizismus und Extraversion der sogenannten Big Five Persönlichkeitszüge positiv mit Immersion korrelieren. Ein weiterer Persönlichkeitszug der in Verbindung mit dem Ausmaß der persönlichen Erfahrung in einer virtuellen Umgebung (VU) stehen soll, ist die Affinität zur Technik. Da das Virtual Theatre eine neue und innovative technische Entwicklung ist, wird erwartet, dass die Durchführung einer Aufgabe im Virtual Theatre als anregender empfunden wird als z. B. an einem Laptop, vorausgesetzt die nutzende Person weist generell eine Affinität zu Technik auf. Insbesondere bei einem Simulator, der wie das Virtual Theatre freie Bewegung ermöglicht, wird angenommen, dass kognitive Fähigkeiten wie räumliches Vorstellungsvermögen eine Auswirkung auf die individuelle Wahrnehmung oder Lernleistung haben.

Ein vollständiges Verständnis über das Zusammenspiel individueller Voraussetzungen und den Hardware-Charakteristika sowie über ihren Einfluss auf die von einer Person erlebte Immersion ist von großer Wichtigkeit für weitere Forschung am Virtual Theatre. Ein Einblick in die Gestaltung der Studien von menschlichen Erfahrungen und Verhalten erfolgt im nächsten Abschnitt.

3.2 Studiendesign

Die Studien zu Wahrnehmung und Verhalten der Nutzer/innen im Virtual Theatre bewerten sowohl die subjektive Erfahrung der Präsenz und des Flows als zentrale Kenngrößen für Immersion, als auch den Lernerfolg. In Anlehnung an Witmer und Singer [14] folgen die Studien dem Ansatz, dass Präsenz- und Flow-Erleben in einer virtuellen Umgebung in Abhängigkeit individueller Unterschiede, Charakteristika der VU wie auch der Hardware des Simulators variieren. Individuelle Unterschiede, Merkmale und Fähigkeiten können in einer bestimmten VU das erlebte Präsenz- und Flow-Empfinden steigern oder mindern. Ebenso können sich verschiedene Charakteristika einer VU und der Hardware eines Simulators unterstützend oder beeinträchtigend auf die subjektive Erfahrung in einer Lernsituation auswirken.

Eine der wichtigsten Fragen innerhalb eines Bildungskontextes ist, ob die Erfahrung einer VU durch das Virtual Theatre zu einem besseren Lernergebnis führt als dieselbe Erfahrung über einen Laptop oder nur den Gebrauch eines HMDs. Wissensabfrage ist deshalb eine weitere wichtige Komponente des Studiendesigns. Die erwartete Beziehung zwischen Lernendem bzw. Lernender, Hardware sowie Wahrnehmung der Situation und des Lernerfolges ist in Abb. 6 dargestellt.

Um die Stärke des Effekts jeder Variable zu messen, werden unterschiedliche Studien im Virtual Theatre angesetzt. Eine erste Studie untersucht sowohl die immersive Qualität des Virtual Theatres als auch seine Möglichkeit zur Steigerung der Lernfähigkeit. Hierfür werden zwei randomisierte Studierendengruppen miteinander verglichen, indem sie in einer Lernsituation unterschiedliche Hardware benutzen. Die Hardware stellt die unabhängige Variable (UV) dar. Die Variable der virtuellen Umgebung und die der Aufgabe bleiben konstant: In einem Labyrinth müssen die Versuchspersonen umherlaufen, Objekte finden und sich deren Positionen merken.

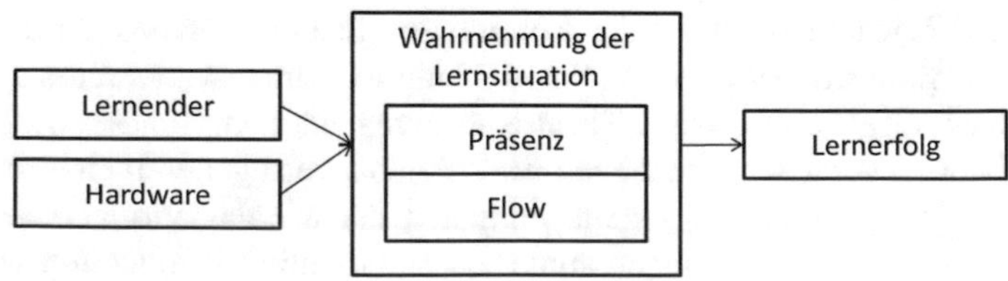

Abb. 6 Erwarteter Zusammenhang zwischen Lernendem bzw. Lernender, Hardware, Wahrnehmung der Situation und Lernerfolg

Die Experimentalgruppe muss die Aufgabe im Virtual Theatre absolvieren und die Kontrollgruppe löst dieselbe Aufgabe am Laptop. Das Wissensabfrageszenario bleibt wiederum für beide Gruppen gleich. Hierfür müssen die Versuchspersonen an einem Tablet PC die Objekte per Drag-and-Drop-Steuerung an ihre korrekten Positionen auf einer Karte des leeren Labyrinths zuordnen. Der Tablet PC wurde gewählt, damit sich das Medium der Wissensabfrage sowohl für die Experimental- als auch für die Kontrollgruppe hinreichend vom Medium der Aufgabe unterscheidet. Die Genauigkeit der Position der Objekte wird vom Computer automatisch erfasst. Die persönliche Wahrnehmung der Lernsituation wird mit entsprechenden Skalen per Fragebogen gemessen. Dies erfolgt im Paper-Pencil-Verfahren, ebenfalls um unterschiedliche Medien bei Befragung und Aufgabe zu verwenden.

Eine zweite Studie konzentriert sich im Speziellen auf die immersive Qualität des omnidirektionalen Bodens. Auch in der zweiten Studie müssen die beiden Gruppen dieselbe Aufgabe in der gleichen Umgebung im Virtual Theatre absolvieren. Hierfür müssen sie auf einer italienischen Piazza frei umherblicken. Während sich die Experimentalgruppe aktiv innerhalb des Virtual Theatres bewegt, sprich: laufen muss, gebraucht die Kontrollgruppe lediglich die 3-D-Sicht. Das aktive Bewegen stellt in diesem Fall die unabhängige Variable dar. Die Aufgabe ist dabei dieselbe wie in der ersten Studie, und die Studierenden müssen sich Positionen von Objekten auf der Piazza merken. Die Positionen werden wieder auf einem Tablet PC auf einer Karte der Piazza markiert. Gemessen werden erneut die persönliche Wahrnehmung der Lernsituation und die Genauigkeit der Position der Objekte.

Durch eine genaue Überprüfung des Zusammenhangs zwischen Nutzer/innen, Hardware und Lernerfolg ist es möglich, den Anteil der Immersion festzustellen, der durch die Persönlichkeit beeinflusst wird. Dadurch kann wiederum ein Profil erstellt werden, für welche Studierenden das Virtual Theatre am besten geeignet ist und am effektivsten wirkt. Außerdem können so maßgeschneiderte Lernszenarien für unterschiedliche Nutzerprofile entwickelt werden. Weitere Studien zu Wahrnehmung und Verhalten der Nutzer/innen im Virtual Theatre werden unter Einbeziehung der Szenarien erfolgen, die parallel speziell für die Ingenieurwissenschaften entwickelt werden. Diese Szenarien sowie die dahinter liegenden didaktischen Prinzipien werden im folgenden Kapitel erläutert.

4 Anwendung in der Lehre

4.1 Rahmenbedingungen

Das Virtual Theatre bietet eine Vielzahl an Möglichkeiten für verschiedene Anwendungsgebiete, wodurch theoretisches Hintergrundwissen in praktischer Weise umgesetzt werden kann. Dies ist insbesondere dann sinnvoll, wenn eine reale Situation zu gefährlich, zu teuer oder generell nicht zugänglich ist. Am Lehrstuhl für Informationsmanagement im Maschinenbau wird das Virtual Theatre für die Lehre sowie als ein Teil eines Schüler/innenlabors des Deutschen Zentrums für Luft- und Raumfahrt (DLR) eingesetzt werden. In der Simulation wird es möglich sein, den Mars Rover Opportunity durch die eigenen Körperbewegungen zu steuern. Ein weiteres Szenario, dass sich vorerst noch in der Planung befindet, behandelt Inhalte und Bedienmöglichkeiten eines Atomkraftwerks.

4.2 *Didaktische Prinzipien*

Es steht außer Frage, dass der Inhalt gewisser Themen in den Ingenieurwissenschaften nicht einfach anhand von Präsentationen vermittelt werden kann. Um den Lernprozess aktiv zu fördern, werden didaktische Methoden benötigt. So können Studierende beispielsweise theoretisches Wissen in praktischen Situationen wie einer Computer-Simulation mithilfe von Pop-up-Fenstern wiederholen. Dies hilft ihnen, das Wissen nachhaltig mit Erfahrungen und Hintergrundgeschichten zu verbinden, anstatt schlichtweg reine Fakten auswendig zu lernen.

Eine andere Methode, die in Verbindung mit dem Virtual Theatre steht, ist das explorative Lernen, wobei Inhalte nicht Stück für Stück wie in Vorlesungen, sondern offen und frei zugänglich präsentiert werden. Das Hauptprinzip dieser Methode ist, dass der Lernprozess von den Studierenden selbst kontrolliert wird und so die Möglichkeit gegeben ist, das Interesse und die Neugier der Studierenden zu wecken. Die Studierenden setzen sich entsprechend ihrer Interessen die Lernziele zu einem bestimmten Thema selbst und wählen eigenständig die Operatoren aus, die nötig sind, um das Lernziel zu erreichen. Beim explorativen Lernen erfolgen Lernprozesse dementsprechend nicht linear, sondern durch individuelle Schwerpunktsetzung. Falls der Lernprozess ins Stocken gerät können die Studierenden die vorangegangenen Lernschritte wiederholen und erst dann fortfahren, wenn diese erfolgreich bewältigt wurden. So bleibt jede/r Studierende stets im eigenen, für sich am besten geeigneten Lerntempo. Wird eine neue virtuelle Lernumgebung erstellt, dürfen mögliche Hemmfaktoren nicht unberücksichtigt bleiben. Dies ist auch deswegen notwendig, da jede/r Studierende über ein eigenes mentales Lernkonzept verfügt. Vermunt und Rijswik [18] haben diese subjektiven Lerntheorien von Studierenden analysiert und drei unterschiedliche Konzepte identifiziert. Das geläufigste Konzept (reproduktives Lernen) beschreibt Lernen als den Prozess der Übertragung von gesprochenem und geschriebenem Wissen ins Gedächtnis. Die zweite subjektive Lerntheorie bezieht sich auf den Gebrauch von Wissen. In diesem Fall ist Lernen ein wichtiger Prozess, um etwas später nachzuvollziehen. Das dritte Konzept, welches eher selten anzutreffen ist, betrachtet Lernen als Notwendigkeit zur selbstgesteuerten Identifikation und Konstruktion von Wissen.

Gesetzt den Fall, dass Studierende Lernen eher als einen Prozess des Nachbildens anstatt einer selbstgesteuerten Konstruktion von Wissen verstehen, wird angenommen, dass gleichzeitig ein großes Bedürfnis nach Instruktion durch Expert/inn/en vorherrscht [19]. Die Konsequenz, die daraus für das Virtual Theatre entsteht ist, dass Anwendungsszenarien auf unterschiedliche subjektive Lernmodelle anwendbar sein müssen. Das dritte Konzept könnte einfach in die Entwicklung von Szenarien eingebaut werden, indem die Abfrage oder Wiederholung von explizitem Wissen durch Pop-up-Fenster oder eine Begleitstimme aus dem Hintergrund erfolgt.

Computersimulationen bieten im Allgemeinen die Möglichkeit des Perspektivwechsels und ermöglichen Studierenden dadurch den Zugang zu normalerweise unzugänglichen Situationen. In einem zukünftigen Szenario, zum Beispiel dem einer Industrieanlage, könnte der/ die Studierende die Abfertigung eines Produkts in einem Produktionsablauf von Anfang bis Ende verfolgen und so wertvolles Verständnis für die Prinzipien eines solchen Prozesses erlangen. Andere Anwendungsbereiche könnten der Weg einer Container Box in einem logistischem Prozess oder auch die Reise durch den Körper des Menschen aus der Perspektive einer Blutzelle darstellen, ähnlich wie in der TV Serie Es war einmal…das Leben [20].

Ein vielversprechender Ansatz, der während eines Workshops mit Studierenden entstanden ist, ist das Game Based Learning (GBL). Computerspielen wird von Studierenden meist als freiwilliger Prozess und nicht als Lernprozess betrachtet, was einen großen Vorteil von GBL darstellt. Nichtsdestotrotz lernen wir durch Spiele mit komplexen Zusammenhängen umzugehen, wie beim Schach [19]. Für das Virtual Theatre gibt es mehrere Möglichkeiten, bereits entwickelte Szenarien in Spiele abzuwandeln. Zuerst kann der oder dem Studierenden eine Mission oder ein Problem aufgetragen werden, das sie bzw. er dann lösen muss. Die oder der Studierende muss versuchen, das Problem oder die Mission nach eigenem Können und Wissen zu lösen (implizites Wissen). Für den Fall, dass die oder der Studierende nicht weiterkommt, hat sie bzw. er die Möglichkeit in einen Erklärungsmodus zu wechseln, in dem zusätzliche Informationen zur Verfügung gestellt werden. Der Wechsel der Modi kann sich jedoch auch hemmend auf die erlebte Immersion auswirken. Dieser Effekt muss in zukünftigen Studien ebenfalls untersucht werden. Eine zusätzliche Möglichkeit ist es, verschiedene Level in das Spiel oder die VU einzuarbeiten. In einer industriellen Szenerie müssten Studierende zum Beispiel anfangs leichtere Maschinen bedienen, um die grundliegenden Prinzipien zu verstehen um dann an komplexeren Situationen arbeiten zu können, in welchen Geräte defekt sind und repariert werden müssen.

Diese Übersicht erhebt keinen Anspruch auf Vollständigkeit. Sie soll aber verdeutlichen, welch großes Potenzial sich durch das Virtual Theatre für die Lehre in den Ingenieurwissenschaften und der Verbesserung der Lernerfahrung ergibt. Im Folgenden werden zwei sich in der Entwicklung befindende Lernszenarien als Anwendungsbeispiele exemplarisch beschrieben.

4.3 Szenario 1: Mars Mission

Beim Marsszenario kann sich die nutzende Person frei auf der Oberfläche des Planeten Mars bewegen. Das Szenario stellt eine Reproduktion der Landebahn des Rover Spirits dar und basiert auf einem Höhenprofil der Landebahn sowie hochauflösenden Fotos der Mars Mission. Innerhalb der Umgebung werden genaue Modelle der verschiedenen Mars Rovers platziert. Abhängig von Position und Sicht des

Abb. 7 Panoramaansicht vom Mars Rover (Eigentum der NASA)

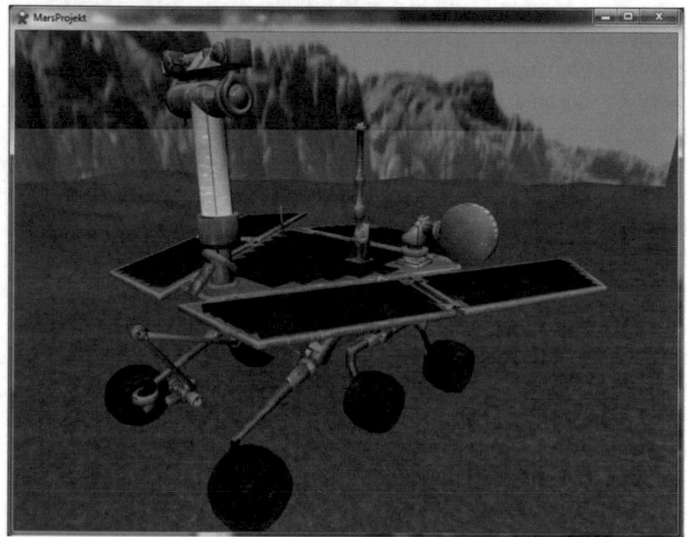

Abb. 8 Model des Mars Rover oppurtinity

Benutzers oder der Benutzerin werden zusätzliche Informationen in Pop-up-Fenstern dargestellt. Abbildung 7 und 8 zeigen Screenshots des Marsszenarios aus Sicht der nutzenden Person sowie aus der Perspektive einer Remote-Steuerung.

4.4 *Szenario 2: Atomkraftwerk*

Das zweite Szenario beinhaltet die Simulation eines Kernkraftwerks. Studierende werden die Möglichkeit bekommen, den inneren Aufbau eines Atomreaktors zu besichtigen und frei in dieser Umgebung zu experimentieren. Naheliegende Interaktionen sind z.B. die direkte Einflussnahme auf Brennstoff und Kontrollstäbe sowie auf die Kühlung. Der Effekt jeder Beeinflussung wird sofort anhand der zugrunde liegenden Gleichungen in der virtuellen Umgebung angezeigt. Zusätzlich verweisen die unterschiedlichen Geräusche des Kontaminationsmessgeräts auf den Anteil der

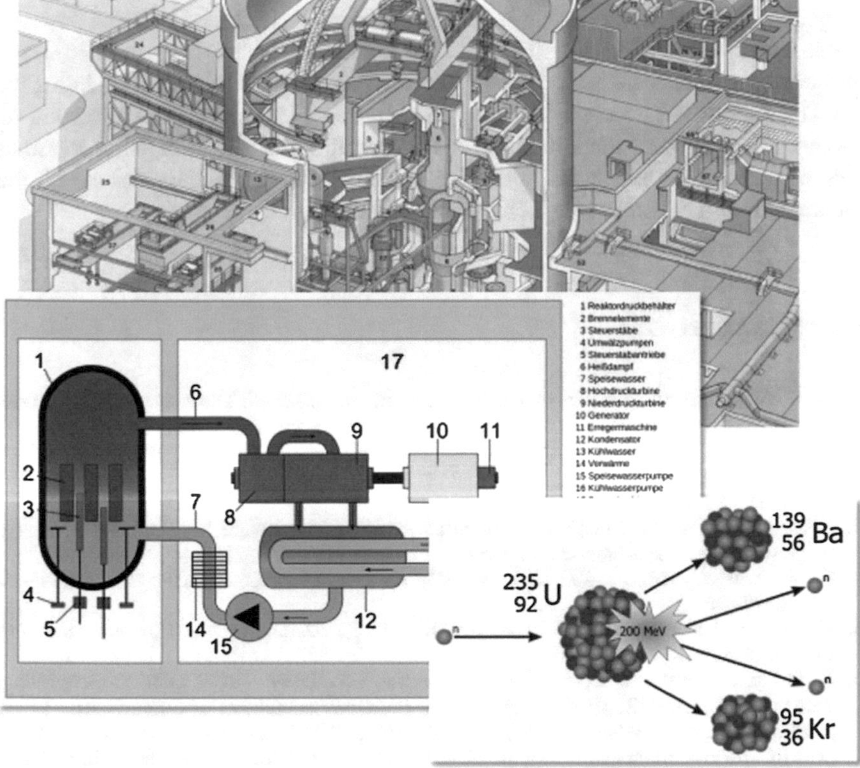

Abb. 9 Schema eines Atomkraftwerks

Verstrahlung. Abbildung 9 zeigt einen Screenshot des Szenarios aus Sicht eines Steuerungsraumes sowie zusätzliche Informationsdarstellungen zur Wissensvermittlung bzw. Vertiefung.

5 Zusammenfassung und Ausblick

In diesem Artikel wurde das Virtual Theatre als ein neues Werkzeug für exploratives Lernen innerhalb virtueller Lernumgebungen vorgestellt. Es wurden die technische Umsetzung und Integration der Hardware sowie der Software beschrieben. Weiterhin wurden erste Anwendungsmöglichkeiten vorgestellt und der Einfluss einer verstärkten Immersion auf Lernsituationen diskutiert. Zukünftige Lernszenarien werden industrielle Anwendungen, Produktionsanlagen und Raumfahrtszenarien fokussieren. Es wird ebenfalls in Betracht gezogen, das Virtual Theatre mit Remote-Laboren zu verbinden und so die Bewegungen der Nutzer/innen innerhalb der virtuellen Umgebung auf robotische Manipulatoren abzubilden. Erste Anwendungen in diesem Bereich werden auf Testanwendungen für Metallbiegung basieren. Zukünftige psychologische Studien werden unterschiedliche Nutzer/innengruppen miteinander vergleichen. So ist es eine interessante Frage, ob Schüler/innen im Virtual Theatre leichter lernen als Studierende. In enger Zusammenarbeit werden die Erkenntnisse aus den Fachrichtungen der Informatik, der Didaktik, den Ingenieurwissenschaften und der Psychologie dazu führen, das gesamte Potenzial des Virtual Theatres für den Einsatz in der Lehre zu erfassen.

Literaturverzeichnis

1. M. Weibull. The virtual theatre and the omnidirectional treadmill: The hardware link between the live and virtual domains, 2012. URL http://www.mseab.se/The-Virtual-Theatre.htm
2. D. Johansson, L. De Vin, Towards convergence in a virtual environment: omnidirectional movement, physical feedback, social interaction and vision. Mechatronic Systems Journal (November Issue), 2011
3. Sensics. Head mounted displays. URL http://sensics.com/products/head-mounted-displays/zsight-integrated-sxga-hmd/specifications/
4. C. Systems. Cyberglove III. URL http://cyberglovesystems.com/products/cyberglove-III/specifications
5. OptiTrack. Naturalpoint OptiTrack tracking tools. URL http://www.naturalpoint.com/optitrack/products/tracking-tools/
6. R. Taylor, T. Hudson, A. Seeger, H. Weber, J. Juliano, A. Helser, VRPN: a device-independent, network-transparent VR peripheral system. In: *Proceedings of the ACM symposium on Virtual reality software and technology, New York, NY, USA*. ACM (VRST'01), 2001, pp. 55–61
7. W. LLC. Vizard virtual reality toolkit. URL http://www.worldviz.com/products/vizard/index.html
8. T. Roosendaal, S. Selleri, *The Official Blender 2.3 Guide: Free 3D Creation Suite for Modeling, Animation, and Rendering*. No Starch Press, San Francisco, CA, USA, 2004
9. A. Bandura, *Sozial-kognitive Lerntheorie*. Klett-Cotta, Stuttgart, 1979
10. D. Johansson, Convergence in mixed reality-virtuality environments: Facilitating natural user behavior. Ph.D. thesis, 2012
11. F. Jorgensen. The top 10 most immersive games ever made, 2012. URL http://daxgamer.com/2012/03/top-10-immersive-games/
12. F.U. Xav. Comment on "the top 10 most immersive games ever made", 2012. URL http://daxgamer.com/2012/03/top-10-immersive-games/
13. C. Dede, Immersive interfaces for engagement and learning. Science 323(5910), 2009, pp. 66–69
14. B.G. Witmer, M.J. Singer, Measuring presence in virtual environments: A presence questionnaire. Presence: Teleoperators and Virtual Environments 7(3), 1998, pp. 225–240
15. D. Weibel, B. Wissmath, Immersion in computer games: The role of spatial presence and flow. International Journal of Computer Games Technology 2011(6), 2011. doi:10.1155/2011/282345
16. M. Csikzentmihalyi, *Beyond Boredom and Anxiety*. Jossey Bass, San Francisco, 1975
17. F. Rheinberg, *Intrinsische Motivation und Flow-Erleben*. Universität Potsdam Arbeitspaper, Potsdam, 2010
18. J.D.H.M. Vermunt, F.A.W.M. van Rijkswijk, Analysis and development of students' skill in selfregulated learning. Higher Education 17(6), 1988, pp. 647–682
19. M. Kerres, *Mediendidaktik: Konzeption und Entwicklung mediengestützter Lernangebote*. Oldenbourg, München, 2012
20. A. Barillé. Es war einmal das Leben...das Blut (Teil 1 von 3), 1991. URL http://www.youtube.com/watch?v=2B-RNnfCmps

Chances and Risks of Using Clicker Software in XL Engineering Classes – From Theory to Practice

Valerie Stehling, Ursula Bach, René Vossen and Sabina Jeschke

Abstract Teaching and learning in XL-classes is a huge challenge to both lecturers as well as students. While lecturers face the difficulty of speaking to a mostly loud and very heterogenic audience, students often lack the opportunity of being an active participant in class. To counteract these difficulties and give the opportunity of immediate feedback, an audience response system has been introduced in the class of information technology in mechanical engineering at RWTH Aachen University. In a previously published paper [1] the theoretical background has been outlined and presumptions have been drawn. The described redeployment of a lecture of the mentioned size was expected to bring about an enhancement of the quality in teaching of professional knowledge. It was also supposed to foster the often desired shift from teaching to learning. Now, after a first trial and evaluation of the method, these presumptions can be tested. In this paper clicker questions from the lecture and their results are the groundwork that allow for a review of the evaluation of the first trial using clicker software in class. Results shall then allow for a comparison of the intended goals with the actual outcomes from the students' point of view. In addition to that, feedback and ideas for improvement through the evaluation of the "further comments" section by the "users" themselves can be gathered. The results of the analysis will then allow for an adjustment and improvement of the concept and may in the future give support to lecturers of other XL-classes that intend to implement an audience response system in their own lecture.

Keywords Large Classes · Clicker Software

1 Initial Situation Outlines

RWTH Aachen University is a highly ranked university (see e.g. [2]) especially in the fields of engineering – mechanical engineering, electrical engineering, industrial engineering etc. It attracts a vast amount of students each semester which in the

---

V. Stehling (✉) · U. Bach · R. Vossen · S. Jeschke
IMA/ZLW & IfU, RWTH Aachen University, Dennewartstr. 27, 52068 Aachen, Germany
e-mail: valerie.stehling@ima-zlw-ifu.rwth-aachen.de

Originally published in "Interdisciplinary Engineering Design Education Conference (IEDEC), 2013", © 2013. Reprint by Springer International Publishing AG 2016, DOI 10.1007/978-3-319-46916-4_15

past has led to lecturers facing the challenge of having to teach a rising number of students each semester [1]. In the lecture “Information Management in Mechanical Engineering I” there are up to 2000 students each semester. This circumstance is partly a result of educational policy that subsequently reduces schooling from 13 to 12 years and thus leads to a vast increase in enrolment.

With such a large number of students it is impossible to give all of them the chance to actively take part in the lecture. Nor is it possible as in smaller seminars to get an idea of the knowledge background of your audience. Audience response systems (ARS) such as hardware clickers or in this case clicker software, however, offer the possibility to interact and thus get a chance of immediate feedback of the students even in large lectures. Therefore, the system was implemented and run in a first trial in the described lecture during the summer semester of 2012. At German universities the use of clicker systems or clicker software is not as widely spread as it is in other countries like e.g. the USA yet. Most clicker literature, however, only deals with implementing clicker systems in classes with up to a few hundred students. Therefore the technical process has been fostered and accompanied by researchers in didactics on the one hand and experts in information technology on the other. This interdisciplinary approach allowed for a multi-angle view on the implementation of the clicker software in anticipation of avoiding typical or common “beginner’s mistakes” beforehand.

2 Inteded Goals

The initial and main goal of the implementation of ARS in the lecture was to give students the chance to actively take part in the lecture instead of taking in all the information as a rather passive consumer. This is especially important when teaching a programming language[1]: just like with any other language you do not automatically learn by listening. It is the exercise that enhances the learning process. Another important goal as previously mentioned was to get an insight in the student learning and possible knowledge gaps. Based on immediate feedback made convenient by the use of modern technology it is possible to react near-term to difficulties in comprehension by repeating difficult subjects, starting a discussion about a controversial point or even changing the focus of the lecture. Rosenberg et al. point out further benefits of the implementation of ARS in class:

> “Research has shown that students in courses using interactive engagement techniques (…) achieve a much greater gain in conceptual understanding than students in traditional lecture courses while also improving their ability to solve quantitative problems.” [3]

Additionally, clicker questions are also known for their motivational benefits. “The questions are designed to motivate students and to unveil potential difficulties with the course topics and material.” [4]. As a positive ‘side-effect’, ARS offer

[1]The language taught in the described lecture is Java.

the opportunity to connect learning with fun. "This approach has two benefits: It continuously actively engages the minds of the students, and it provides frequent and continuous feedback (to both the students and the instructor) about the level of understanding of the subject being discussed." [5].

3 Realization

While using ARS – especially hardware clickers – in class has been a common strategy for many years in countries like the USA, German universities have only discovered the benefits of these systems in the last few years. This means that the launch is expected to be 'virgin soil'. The launch has been carefully planned and realized by an interdisciplinary group of researchers in engineering, information technology and didactics. In a first step, possible questions were sought that matched the aims in teaching and learning of the lecturer. When looking for the appropriate application type, several can be chosen from. Here are some examples of purposes for which ARS can be used:

- Multiple-choice questioning
- Asking for computational results (numbers)
- Asking for coordinates (mark a point)
- Asking for Feedback by Multiple Choice (positive/negative)
- Asking for free text feedback
- Asking to enter questions
- Asking for free text answers
- Sequencing, etc.

Multiple-choice questions again offer several options of application:

- Choice of a right statement/answer
- Choice of a wrong statement/answer
- Choice of the right computational result, etc.

When integrating clicker questions in the topics of the lecture, it has to be considered that they take up a certain amount of time (questioning, polling, discussion) of the lecture – approximately three to five minutes each. So it is highly important to carefully plan when and on which topic they can be applied reasonably. Given the information that previous test results of the lecture have shown that the topic with the highest failure rate was coding, the clicker questions in this particular case were designed specifically for the lectures covering this topic in order to detect difficulties in understanding early on.

Before using the software for the first serious clicker question, the lecturer shortly introduced the software during the lecture by a non-subject-related question. He also explained that the appearance of a question mark in the right hand corner of a slide signalizes a forthcoming question on one of the following slides. This element was

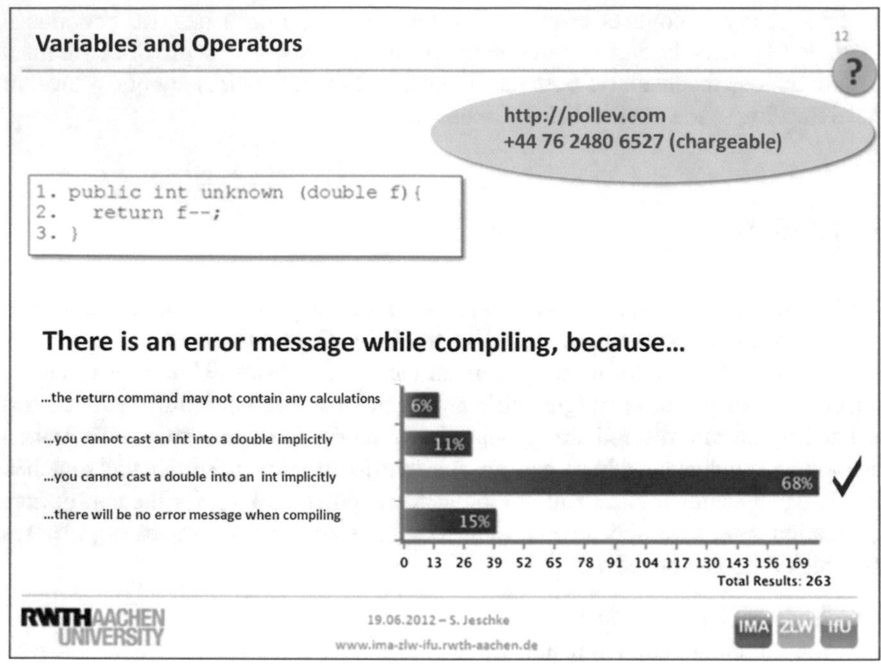

Fig. 1 Screenshot of a clicker question with a clear tendency towards the right answer

chosen to keep the motivation and attention of the students at a high level. The first serious clicker questions were posed to keep track of the difficulties students might have when learning the basics of programming with Java. It was expected to be able to subsequently minimize these difficulties by discussing given answers. Figure 1 shows a screenshot of a clicker question slide and the response frequency in percent from the lecture.

In terms of the subject, the clicker question aims to check and foster the students' basic apprehension of the effects of operators. The question demonstrates decreasing a number variable using the postfix operator for and problems arising from implicit typecasting The slide shows a code. The question posed to the students is a multiple choice question of the right statement: "There is an error message while compiling, because…" The students can choose from four different answers that are supposed to detect the mistake in the code. While the majority (68 %) of the students answers the question correctly, 32% give a wrong answer. With this result the lecturer can seize the opportunity to explain, why the other options are incorrect to eliminate future mistakes.

Figure 2 shows a slightly different result.

On the second question slide the students see another code. This time, the question serves the purpose of showing the effects of different mathematical operators. The question posed to the students is: "Which result does the command calculate (3,2)

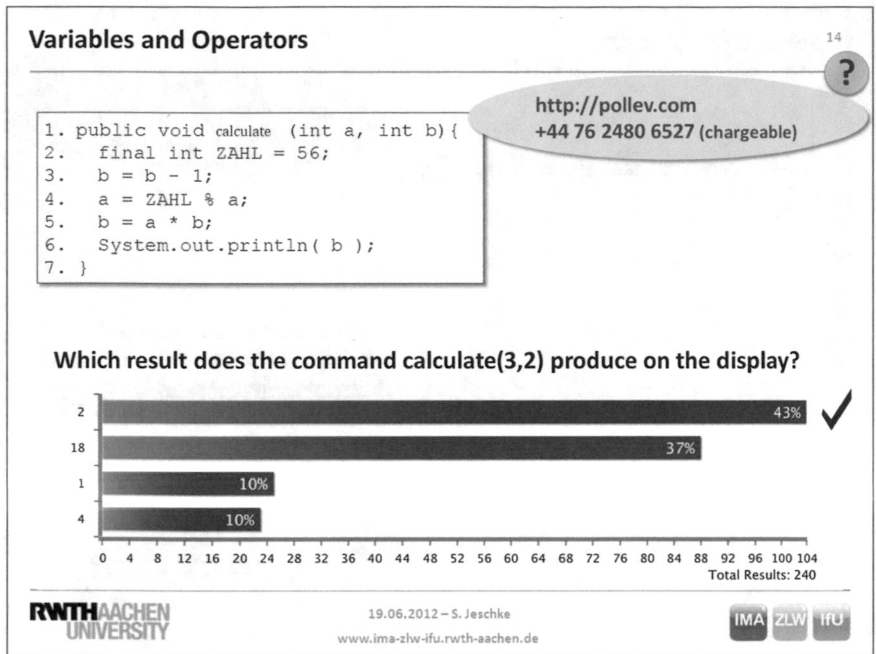

Fig. 2 Screenshot of a clicker question with two very close tendencies, one of them indicating the right answer

produce on the display?". The question thus asks for a choice of the right computational result. The answers to this question appear not to have a clear tendency as in Fig. 1. This indicates that a majority of the students at this point of the lecture do not know the correct answer. This again gives the lecturer the opportunity to discuss and explain the answer. After the discussion and explanation a similar question but with a higher level of complexity is being posed to check possible learnings or difficulties of the students (Fig. 3).

The third question following shortly after the second one is another question of choice of the right computational result. Here, the workflow of a switch statement is demonstrated. The special focus lies on its behavior to execute all subsequent commands once a case has been matched. The question posed to the students is: "Which result does the command calculate (4,11) produce on the display?" The results show that 64 % of the students have chosen the wrong answer. Only 19 % of the students have answered the question correctly. After the display of the result, a positive and estimated side-effect was being observed by the teaching staff: subject-related discussions among the students began to evolve and the students tried to figure out why answer number 2 was the correct one. After that, the lecturer explained again and subsequently went on with the lecture. The following chapter shows and analyzes the results of the first evaluation of the trial implementation of 'poll everywhere' in the lecture by the students.

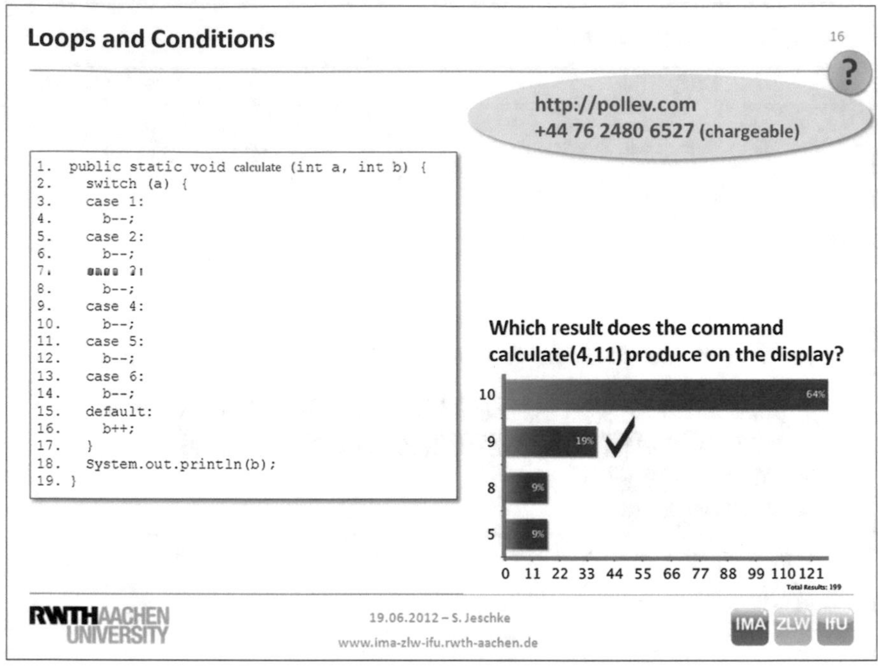

Fig. 3 Screenshot of a clicker question with a clear tendency towards a wrong answer

4 Results of First Evaluation

Worldwide evaluations at universities show that the application of ARS in lectures has led to e.g. higher motivation of attendance [6], more attention of the students during class and even higher knowledge acquisition than in conventional (non-interactive) classes. [7] After several trials the new ARS in the described setting was evaluated by the students near the end of the semester. The evaluation in this particular case was designed to cover the main topics motivation, usability and conceptual design of the questions. The evaluation on the RWTH Aachen University was carried out near the end of the semester, after several trials with the new ARS.

There are about 1800 ARS-related and filled-out evaluation sheets from the lecture. The questions posed were mostly closed-ended questions except for the comments section at the end. They can be divided into several sections which included questions concerning:

- participation,
- impact on comprehension and content,
- motivational aspects,
- rating of the software itself and the methodological launch in the lecture.

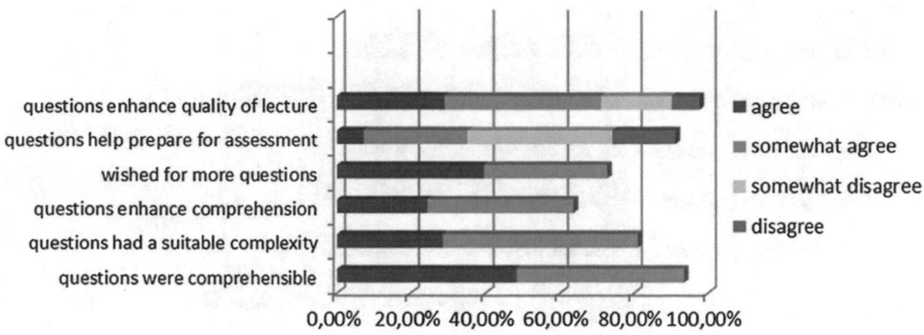

Fig. 4 Results of the evaluation concerning aspects of comprehension and content

The following sections will provide an insight into the most significant results of the evaluation. Every student (with an exception of two) who took part in the evaluation attended the particular lectures in which clicker questions were being posed. 60,3 % of the students actually used the software at least once to answer a question, 35 % of them participated in every single poll. Most of the students (43,3 %) used the software on their smartphone. 27,6 % did not participate at all because they did not have a smartphone or notebook (with them) and they probably did not want to spend extra money by sending a text message.

Figure 4 shows results of the evaluation concerning aspects of subject-related comprehension and content. A vast majority of the students (93,6 %) state that the questions were comprehensible and had a suitable complexity (81,1 %). Around 63 % say that the questions in class enhanced their comprehension which underlines the assumptions as drawn above: that posing adequate clicker questions and discussing right and wrong answers does not only activate students but can also foster and reveal their learning process. Nevertheless, a majority of 56,3 % also state that the questions did not help preparing for assessment. An especially positive outcome is that 70,9 % state that the questions enhance the quality of the lecture which leaves only 26,8 % stating that they don't.

Figure 5 shows the results concerning motivational aspects of the use of ARS in class. A majority (80,1 %) of the students enjoyed participating in polls. As has previously been found in other research, the clicker questions motivated a majority (79,7 %) of the students to be more attentive during the lecture. Only for 29 % of the students, however, see clicker questions as a motivating element for students to attend to the lecture at all. 68,7 % feel motivated to participate in the discussion of the subject and a similar number (69,6 %) state that ARS help feeling more involved in the lecture. A total of 72,8 % even wish for more clicker questions throughout the semester.

Adjacent to the predominance of positive responses there also were some points of criticism. Most of these were related to the technical implementation given that the WiFi connection did not work well as soon as a lot of students wanted to participate at the same time. This explains why only a relatively small number of students

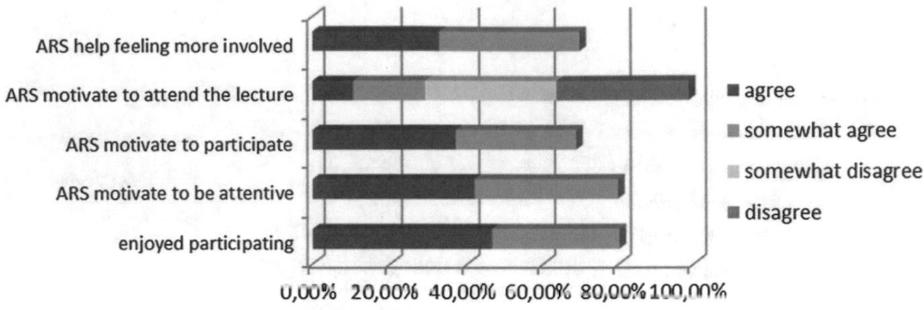

Fig. 5 Results of the evaluation concerning motivational aspects

participated in the polls with numbers around approximately 25 % of the students in the lecture per poll. Although most of the students thought dealing with the software itself was simple (84,4 %), some of them wished for a slightly longer answering period. Other students criticized real time tracking of the polls which has been used for some of the polls as they felt influenced by the already given answers of the other students.

In the comment section of the evaluation ten students specifically complimented the use of ARS. In addition to that there were also suggestions in which part of the lecture clicker questions should be used. Four students said that it would be better if clicker questions were used at the end of the lecture or at the end of a topic. Six students criticized that it was not possible for them to take part, because they had no smartphone or other device to participate.

5 Adjustments and Recommendations

The analysis of the evaluation has shown that most of the anticipated benefits of ARS (for further information see [1]) also apply when dealing with XL classes. Nevertheless, there was also criticism towards some important aspects such as the facility of participation – weak WiFi connection/lack of devices – and the methodology itself – presentation of poll results, frequency of questions, point of time etc. This criticism allows for several recommendations for future use adjustments to be made. The most preeminent points of criticism will here be discussed in terms of possible solutions. The largely positive reaction, however, allows for a recommendation of broadening of the use of ARS in very large classes.

5.1 *Facility of Participation*

The evaluation shows that the main reason for the relatively small number of participants can be ascribed to the weak WiFi connection. This problem can only be solved by the university. In order to avoid the frustration of not being able to participate, it is therefore highly important to address this problem to responsible university staff and give them the chance to improve the conditions.

The problem of participation without an own smartphone or laptop can be solved by having students team up in groups of up to a maximum of five students, each group having at least one functioning device to participate.

One decisive element regarding the software 'poll everywhere' is often underestimated by software designers: the integration of polls in the slides of the lecture – in this case power point slides. This feature allows for the lecturer to avoid discontinuities during the lecture and consequently avoids non-subject-related actions or breaks and the students could be distracted by. A change of the user interface, e.g., takes up time that students will most likely use to chat – and gaining back the attention of 2000 students once one has lost it can be very challenging. It is therefore recommended to choose software that provides the feature outlined above.

5.2 *Methodology*

The solution to the described criticism towards real-time tracking of the polls problem can be a simple one. The software allows for both methods of poll tracking – hidden as well as visible. Both methods have been tested in the trial. Both ways are plausible when linked to specific didactical goals. It is recommended to hide the development of the poll until an acceptable number of answers have been registered, close the poll and show the answers afterwards, unless a specific learning goal is being pursued. Showing the development of a poll in real time, however, can also be useful when e.g. linked to a certain didactical strategy such as peer instruction.

As has been previously stated, the sole implementation of an ARS in class cannot generate an improvement of a lecture – ideally, there is a didactical or conceptual goal behind every single question. Therefore, another suggestion is to combine clicker questions with successful didactical concepts such as just-in-time (JIT) teaching methods or e.g. peer instruction (PI).

> "Just-in-Time Teaching (JiTT) is an ideal complement to PI, as JiTT structures students' reading before class and provides feedback so the instructor can tailor the PI questions to target student difficulties." [8]

> "When PI is used, students are first asked to answer a question individually, and then a histogram of their responses may be displayed to the class. If there is substantial disagreement among responses, students are invited to discuss questions briefly with their neighbors and then revote before the correct answer is revealed. The instructor then displays the new histogram and explains the reasoning behind the correct answer." [9]

In the first ARS-trial of the described lecture solely multiple-choice questions were applied in order to check the apprehension of previously explained methods of e.g. programming or else explain it again. As Crouch et al. state there is, however, a possibility to use open ended questions in lectures with a lot of students:

> "In a course with a large enrollment, it is often easiest for the instructor to poll for answers to multiple-choice questions. However, open-ended questions can also be posed using a variety of strategies. For example, the instructor can pose a question and ask students to write their answers in their notebooks. After giving students time to answer, the instructor lists several answer choices and asks students to select the choice that most closely corresponds to their own. Answer choices can be prepared ahead of time, or the instructor can identify common student answers by walking around the room while students are recording their answers and prepare a list in real time." [10]

Crouch et al. hereby describe one of unlimited options to pose open ended questions. This approach allows for a mix of approved didactical methods with new technology. Technology itself does not replace didactics or in Beatty's words: "Technology doesn't inherently improve learning" [11]. Only when choosing the 'right' strategy can a learning goal be achieved.

6 Future Prospects and Research Fields

The evaluation shows that using clicker software can be an efficient means to engage students – also in a very large class. Students in this particular class found the educational software to be motivating in terms of participation in class and helpful towards understanding. Nevertheless it is not a magic tool that automatically enhances motivation and learning. One result towards learning is that almost half of the students taking part in the evaluation somewhat agree that the questions in the lecture enhanced comprehension of the proposed subject, but did not specifically help when preparing for assessment. There was also the problem of participation – partly students were unable to participate in polls because they either did not own or bring a necessary device such as a phone or a laptop or the weak WiFi-connection did not allow for participation. It will therefore be a task for the second trial to on the one hand give the university the chance to improve the WiFi-connection while on the other to adapt methods that allow e.g. group participation to make sure every student can be involved. Many students wished for more clicker questions throughout the semester. As in the trial clicker questions were only posed in a few of the lectures, there is still latitude for extension.

The second trial shows the advantage of having more time – now that the technical implementation has been completed – to focus even more on subject-specific aspect of the clicker questions. One of the future prospects in the 'big picture' will furthermore be to use results of clicker questions in order to consequently detect knowledge gaps at an early stage. In terms of a holistic approach, the next step will then be to concurrently adapt the supplementary courses and exercises by focusing on these specific most challenging topics and most certainly to subsequently conduct further evaluations.

References

1. V. Stehling, U. Bach, A. Richert, S. Jeschke, Teaching professional knowledge to xl – classes with the help of digital technologies. In: *ProPEL Conference Proceedings 2012*. 2012
2. R.. Accreditation. Master mechanical engineering. http://master-mechanical-engineering.com/content/rankings-accreditation, http://tinyurl.com/aec84xr
3. J.L. Rosenberg, M. Lorenzo, E. Mazur, Peer instruction: Making science engaging. In: *Handbook of College Science Teaching*, ed. by J.J. Mintzes, W.H. Leonard, 2008
4. M. Sievers, W. Reinhardt, D. Kundisch, P. Herrmann. Developing electronic classroom response apps for a wide variety of mobile devices – lessons learned from the pingo project, 2012
5. E. Mazur, Farewell, lecture? Science **323** (5910), 2009, pp. 50–51
6. A.R. Trees, M.H. Jackson, The learning environment in clicker classrooms: student processes of learning and involvement in large university-level courses using student response systems. Learning Media and Technology **32** (1), 2007, pp. 21–40
7. N. Scheele, A. Wessels, W. Effelsberg, Die interaktive vorlesung in der praxis. In: *DeLFI 2004: 2. e-Learning Fachtagung Informatik, Tagung der Fachgruppe e-Learning der Gesellschaft für Informatik e. V. (GI), 6-8. September 2004 in Paderborn, P-52*. 2004, pp. 283–294
8. J. Watkins, E. Mazur, Using jitt with peer instruction. In: *Just in Time Teaching Across the Disciplines, across the academy*, ed. by S. Simkins, M. Maier, 2010, pp. 39–62
9. M.K. Smith, W.B. Wood, W.K. Adams, C. Wieman, J.K. Knight, N. Guild, T.T. Su, Why peer discussion improves student performance on in-class concept questions. The Science Educcation Initiative **323**, 2009, p. 122
10. C.H. Crouch, J. Watkins, A.P. Fagen, E. Mazur, Peer instruction: Engaging students one-on-one, all at once. Reviews in Physics Education Research, 2007, p. 11
11. I.D. Beatty, Transforming student learning with classroom communication systems. EDUCAUSE, Research Bulletin **2004** (3), 2004

Fostering the Creative Attitude with Remote Lab Learning Environments – An Essay on the Spirit of Research in Engineering Education

Claudius Terkowsky and Tobias Haertel

Abstract Creativity has been proclaimed to be one of the most important 21st century skills. Facing tremendous problems, creativity and innovation were seen as key factors of a knowledgebased society able to cope with ongoing and future problems. As Engineers are addressed to play an important role in facing these challenges, the question arises in which way universities could contribute to educate creative engineers. This slightly provoking essay inducts possible boundary conditions and constraints of fostering creativity in engineering education. Moreover, it presents first results from a smallsample pre-study on higher engineering education curricula, conducted in the funded German project "ELLI – Excellent Teaching and Learning in Engineering Education", which suggests a lack of creativity education in the examined curricula. Furthermore, a descriptive analysis of the didactic approach of the finished EU-project "Pe-TEX – Platform for E-Learning and Telemetric Experimentation" provides information about possibilities of fostering the creative attitude in engineering education by means of remote labs. Finally, the essay resumes with future tasks for the ELLI project and open questions addressing relevant future educational and socioeconomic impacts, regarding the role of creativity in engineering education and the professionalization of engineers.

Keywords Fostering Creativity in Higher Engineering Education · Higher Engineering Education Research · Remote Labs · Creativity Supporting Learning Scenarios · Curriculum Development

1 Introduction

Creativity has been proclaimed as one of the key 21st century skills and as a driving force of economic development. With the so-called creative class, comprising different types of creative workers, tackling complex societal problems ranging from

C. Terkowsky (✉) · T. Haertel
Engineering Education Research Group, Center for Higher Education,
TU Dortmund University, Vogelpothsweg 78, 44227 Dortmund, Germany
e-mail: claudius.terkowsky@tu-dortmund.de

Originally published in "International Journal of Online Engineering (iJOE)", Issue 5, Vol. #9, © IAOE 2013. Reprint by Springer International Publishing AG 2016, DOI 10.1007/978-3-319-46916-4_16

solving economic problems through creating innovative technological solutions to devising new ways of social entrepreneurship, the role of creativity will increase dramatically in the years to come. Already today, many of the fastestgrowing jobs and emerging industries rely on workers' creative capacity such as the ability to think unconventionally, inventing new scenarios and producing novel solutions.

To face these demands both engineering education and professional engineering fields have to design and embrace new ways of fostering creativity of engineering students and workers. This raises three paradigmatic questions:

- What means/is creativity in the context of higher engineering education and is there a lack of creativity-fostering education in engineering curricula?
- What is creativity especially in engineering and how could engineering educators foster a "creative attitude" [1] in their engineering courses?
- Could highend remote and virtual labs, as presented in [2] and [3], be powerful instruments for the development of a creative attitude?

Trying to get first answers to these questions, the essay presents a few possibly slightly provoking findings and estimations of a first curriculum survey as pre-study, conducted during the first stage of the nationally funded ELLI project.

Furthermore, the paper discusses the remote lab approach of the finished EU-project "PeTEX – Platform for E-Learning and Telemetric Experimentation" as one successful practice example on fostering creativity.

1.1 What is the Creative Atitude?

Creativity itself can be defined as the interaction among aptitude, process and environment by which an individual or group produces a perceptible product that is both novel and useful as defined within a social context. Creativity generates outcomes that are novel, high-quality and appropriate to the task at hand or some re-definition of that task [4–7].

According to [1] "creativity and progress depend upon asking the right questions at the right time". Reference [8] states that "being creative always requires some amount of deviating from the norm". Moreover, [9] stresses that the function of creativity is twofold: "from the societal viewpoint, the task of creativity is improvement; from the individual viewpoint, the task of creativity is expression". These two perspectives of creativity – improvement and expression – were not extremes of one aspect, they should rather be regarded as singular levels of investigation. "Individual and society interact over time to bring new ideas, products, and solutions into the realm of culture" [9].

Well, what is the creative attitude? Reference [1] defines the creative attitude as "… the desire to go against the mainstream. But such desires are stopped by parents, in school, at work – nearly everywhere [needless to say: in higher education institutions too; CT&TH]. The creative attitude entails posing one's own questions, not answering the questions of others, and it is not always easy to get away with such a point of view".

1.2 The Role of Creativity in Engineering

In 2004, The National Academy of Engineering defined the impact of creativity on engineering in the following way: "Creativity (invention, innovation, thinking outside the box, art) is an indispensable quality for engineering, and given the growing scope of the challenges ahead and the complexity and diversity of the technologies of the 21st century, creativity will grow in importance. (…) Engineering is a profoundly creative process. A most elegant description is that engineering is about design under constraint. The engineer designs devices, components, subsystems, and systems and, to create a successful design, in the sense that it leads directly or indirectly to an improvement in our quality of life, must work within the constraints provided by technical, economic, business, political, social, and ethical issues" [10]. According to [11] engineers shall be able to "demonstrate appropriate levels of independent thought, creativity, and capability in real-world problem solving".

1.3 Creativity and Teaching Engineering

"Although the idea of creativity is attractive to educators, there is a pitfall as well as a promise. From the perspective of educators, creativity is often viewed not as an end, but as a means towards ends such as improving problem-solving ability, engendering motivation, and developing self-regulatory abilities" [12]. Reference [12] proposes three basic aspects of creativity that researchers see as generally comprising the overlap between creativity and education: Respectively, they are

- the use of **creativity** (or insight) **to solve problems** in other subject areas
- **creative ideas for teaching**, and
- teaching for or attempting to **enhance the creativity of learners**

In contrary to that, [13] nominates some teachers' inherent factors that hinder the expression of students' creativity:

- **teachers' prior experiences** during their own school and university years; reproduction of these practices across time, place, person
- prevalence of **teacher-dominated convergent teaching** approaches; personal need for order
- teachers' **need to stick to the plan**; place on the acquisition of facts
- teachers' view that **unexpected student ideas are disruptive**; even soon-to-be teachers generally **prefer expected ideas** over unexpected or unique ideas

- **wrong beliefs**, behaviors and assumptions **about students' motivation** and the role of creativity in the classroom
- **scripted curricula** represent the most extreme form of convergent teaching, separating learning from the development of creative thinking
- **teaching to the test** and increased use of externally mandated, fact-based tests

Teaching to the test points students aware of what is really valued and important: "...the kind of examinations we give really set the objectives for the students, no matter what objectives we may have stated" [14]. "Regardless of how teachers encourage their students to share their creativity, unless teachers also include expectations for creativity in their assignments and assessments, then the message is clear: Creativity really doesn't matter" [13]. Reference [13] resumes that "encouraging creative thinking while learning not only enlivens what is learned but can also deepen student understanding. This is because, in order for students to develop an understanding of what they are learning, they need (...) to come up with their own unique examples, uses, and applications of that information. In order for this to happen, expectations for novel, yet appropriate applications of learning need to be included in classroom assessments of student learning".

1.4 Fostering the Creative Attitude in Higher Engineering Education

Regarding creativity in the field of engineering education, some work has already been done: e.g. [15] presents the creative platform, a concept that focuses on confidence, concentration, motivation and diversified knowledge. But a concrete didactic scenario for engineering education is missing. Such a scenario is delivered e.g. by [16], and [17], combining principles of enhancing creativity with problem-based learning and project-based learning in engineering education.

The research project "ELLI – Excellent Teaching and Learning in Engineering Education", funded by the German Ministry of Research and Education between 2011 and 2016, and its sub-project "KELLI–fostering creativity in engineering education" follows a different strategy to foster and evaluate creativity.

2 "KELLI – Fostering Creativity in Higher Engineering Education"

The sub-project "KELLI – fostering creativity in engineering education" is based on results and outcomes of the already finished German research project "Da Vinci – fostering creativity in higher education". A model of six distinctive facets to define creativity in higher education had been worked out with a comprehensive qualitative

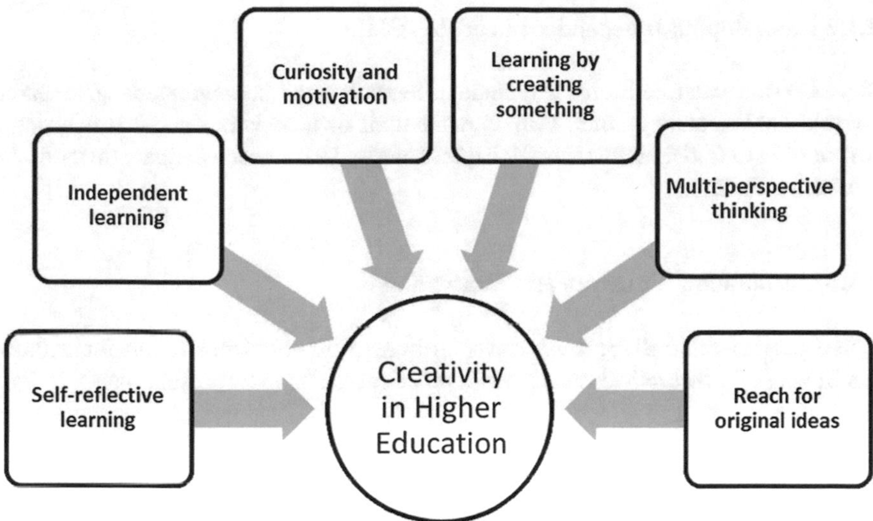

Fig. 1 6 facets of creativity in higher education

approach based on "Grounded Theory" [18], which will be presented in the next paragraph.

These six facets can be deployed to analyze given, as well as to define and stimulate new learning objectives and learning activities.

2.1 Learning Objectives to Foster Creativity in Higher Education

The results of the Da Vinci project show that creativity in higher education (across all disciplines) consists of six different facets [19–22] (see Fig. 1).

2.1.1 Developing Self-Reflective Learning Skills

Learners break out of their receptive habitus and start to question any information given by the teacher. An internal dialogue takes place and knowledge becomes "constructed" rather than "adopted".

2.1.2 Developing Independent Learning Skills

Teachers stop to determine the way students learn. Instead, students start e.g. to search for relevant literature on their own, to make their own decisions about structuring a text or even to find their own research questions and to choose adequate methods for answering them.

2.1.3 Enhancing Curiosity and Motivation

This aspect relates to all measures that contribute to increased motivation, for instance the linking of a theoretical question to a practical example or presenting.

2.1.4 Learning by Doing

Students learn by creating a sort of "product". Depending on the discipline, this might be a presentation, an interview, a questionnaire, a machine, a website, a computer program or similar. Students act like "real" researchers.

2.1.5 Evolving Multi-Perspective Thinking

Learners overcome the thinking within the limits of their disciplines or prejudiced thinking. They learn to look automatically from different points of view on an issue and use thinking methods which prevent their brain from being "structurally lazy" [23].

2.1.6 Reaching for Original Ideas

Learners aim to get original, new ideas and prepare themselves to be as ready-to-receive as possible. Getting original ideas cannot be forced, but by the use of appropriate creative techniques and by creating a suitable environment (allowing to make mistakes and to express unconventional ideas without being laughed down or rejected) the reception of original ideas can be fostered. In the following paragraph these six facets will be used to question the objectives and activities defined in a small sample of engineering curricula.

2.2 *Analysis of Module Descriptions*

In the KELLI subproject, the module descriptions of six engineering education curricula (Manufacturing Engineering and Electrical and Electronic Engineering IT) of three German universities (RWTH Aachen University, Ruhr University Bochum, TU Dortmund University) were analyzed against this background in a first analytical pre-study in order to get to know which aspects of creativity are fostered.

2.3 Results and Discussion

As a result, the creativity-aspects 1 (self-reflective thinking), 3 (curiosity and motivation), and 4 (learning by doing) are highly developed in both courses of all three universities. Apart from one exception these aspects exhibit shares of over 50 %. On the other hand, the aspects 2 (independent learning), 5 (multi-perspective thinking) and 6 (reach for original ideas) can be found only in small proportions at percentages below 50 %, in aspects 5 and 6 even below 10 %, apart from another exception. (see Fig. 2).

To sum up, the pre-analysis of the module descriptions indicates that students were encouraged to think critically and self-reflective in the selected courses. They had to demonstrate levels of motivation and commitment in their courses and were trained to "create" something and to work practically. However, independence, collaborative development of ideas and the exchange with other disciplines for open-minded

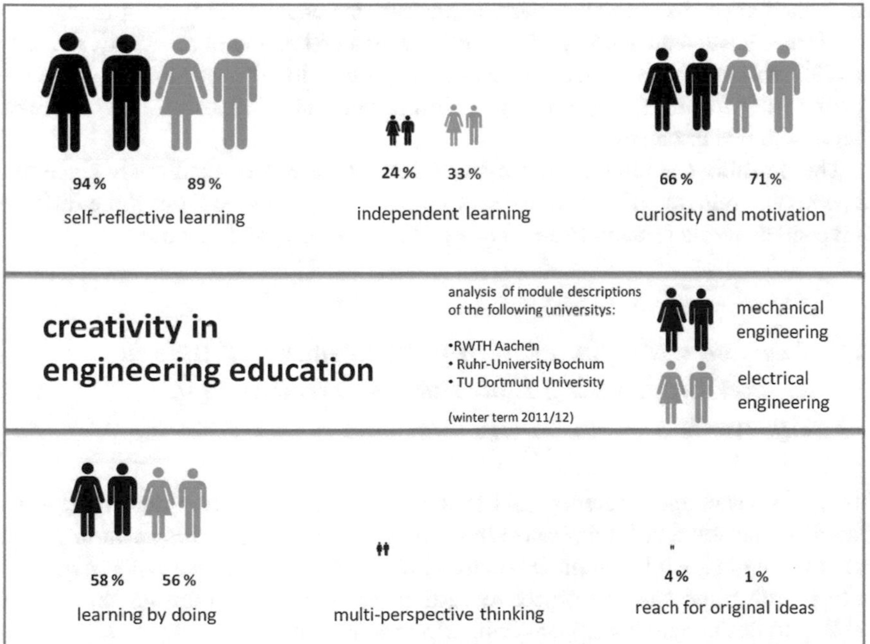

Fig. 2 Creativity in engineering education

discussions, scenarios and experiments, were almost not required or promoted. What emerges is a picture of diligent students who rather work conscientiously on given tasks than on finding new problems, questions and solutions on their own and in discussion with others. The fact that in some of the courses students are not free to pick the topic of their thesis adds to this picture: they can only select from a predefined pool of research questions developed by their respective teachers. In this way many learning tasks may hinder creative cognitive processes relating to research, such as:

- detection of relevant research questions
- deliberation whether an issue is workable
- creation of a feasible research plan, or
- assessment and implementation of eligible methods

Due to this, students are neither encouraged to develop a "bigger picture" of their discipline, nor to build up a "spirit of research" [24–26] as for instance:

- (collaborative) reasoning about current scientific issues in the community
- setting up and discussing new (and sometimes rather risky) theories
- own decision making and seeking collegial advice

If students were to see the "bigger picture", they might get an impression about the value and importance of their work as well as its location and interrelationship in the domain.

Through the findings of this pre-study the question arises whether the current understanding of fostering creativity in engineering education is appropriate.

Moreover, students seem to have a different understanding of creativity. An interdisciplinary survey (n = 320) at TU Dortmund University) shows that students regard "openness", "freedom", "stimulation", "inspiration" and "empowerment" as factors that promote their creativity.

These results of the finished pre-study need to be treated cautiously since they depend on a rather small, arbitrary sample of only six courses from three different universities. Further qualitative and quantitative research will be done.

2.4 *Practice Example: Fostering the Creative Attitude in Higher Engineering Education with a Remote lab Approach*

Just a few years ago reference [27] stated on the basis of a literature review that "hands-on lab adherents emphasize the acquisition of design skills as an important educational goal, while remote laboratory adherents do not evaluate their own technology with respect to this objective". Reference [27] defines design skills as the "ability to design and investigate [which] increases student's ability to solve open-ended problems through the design and construction of new artifacts or processes". This raises the question whether high end remote and virtual labs, as presented in [2] and [3] can be powerful means to foster the creative attitude, and how this could be achieved.

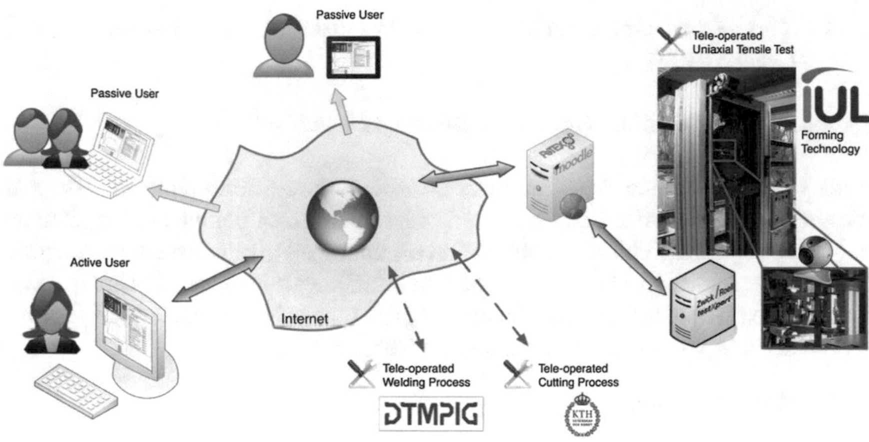

Fig. 3 The PeTEX principle [34]

The didactical concept of the "Platform for E-Learning and Telemetric Experimentation – PeTEX" (see Fig. 3) is one example [28, 29]. The PeTEX system is intended for application in higher education [30, 31] and for workplace learning [32, 33].

The PeTEX system combines a tele-operated experimentation platform (material testing, particularly forming, cutting and joining) with a collaborative learning environment based on Moodle [34, 35]. Established on "experiential learning" [35–37] it provides three different learning levels by deploying three didactic approaches and addressing three problem types [38, 39]. The learning levels correspond to the aforementioned six facets of fostering creativity (see Table 1).

Table 1 Three consecutive learning levels, corresponding to the problem types and the six facets of creativity

Learning Levels	Didactic approach	Problem type	Creativity facet
1. level: Beginner	scripted learning paths	interpolation problems	1. self-reflective learning 2. independent learning skills
2. level: Intermediate	real world scenarios	synthesis problems	3. curiosity and motivation 4. learning by creating something
3. level: Advanced	research-based learning	dialectic problems	5. multi-perspective thinking 6. reach for original ideas

2.4.1 Three Consecutive Problem Levels to Foster Different Facets of Creativity

Beginner Level: Learning with Interpolation Problems

Students in the beginner-level are guided through the learning platform. They are asked to create predefined and expected orders in a given complexity of elements and actions by identifying, assembling and executing all given elements and actions in the right order to solve the task. On the first PeTEX learning level, this is to conduct predefined experiments correctly which comprise interpolation problems. According to [40–42] interpolation problems consist of three elements:

1. a predefined starting point
2. a concrete terminal point
3. a concrete and predefined solution process how to bridge the gap between starting point and terminal point

The challenge of this kind of problem is to correctly fulfill a sufficiently complex task according to the given and scripted path. It deals with recognition of and acting in complexity, e.g. understanding the manual, identifying the relevant units of the real equipment introduced in the manual. The next step is to combine, assemble and connect these elements in the right scripted technical and logical order to fulfill the predefined task as well as to produce the expected results.

The main creativity facets addressed by this kind of task are to break out of the receptive habitus, to create and internalize psycho-motoric operation chains, to gain first experience of basic principles and to start questioning the given information by transforming them into correct action (see facet 1: self-reflective learning).

Moreover, students at the beginner level have the option of repeating the given experiments over and over as long as necessary to pervade the related learning activities. In a further step, students can vary the input data settings to generate a deeper understanding of the underlying process models.

Furthermore, the absence of time and space restrictions in web-based learning facilities allows the students to experiment from almost everywhere at any time. Combined with unconditional repetition and variation of "walk-throughs" [43], learning will be substantiated and enhanced gradually by self-directed "playing" which is usually not offered in common laboratory work settings. This strengthens the students' capabilities for learning in a self-determined manner (see facet 2: independent learning).

Intermediate Level: Learning with Synthesis Problems

At the intermediate level, learners have to apply and transfer their gained knowledge for solving given real-world cases and scenarios. According to [40–42] real world scenarios relate to synthesis problems which consist of three elements:

1. a concrete terminal point
2. a predefined starting point
3. lack of a defined solution process to bridge the gap

The challenge of this problem type is to find, to develop and to deploy a sufficient solution path to a given problem consisting of a presented starting point and an expected terminal point by applying divergent and convergent thinking to find an appropriate solution.

Rewarding and effective problem solving is seen as one main success factor to increase self-efficacy and to intensify interest and enthusiasm [44–46] (see facet 3: curiosity and motivation).

The creative final product is the developed solution gained in iterative loops and mostly "by doing" (see facet 4: learning by doing).

Advanced Level: Learning with Dialectical Problems

Learners at the advanced level have to design own research questions and to develop appropriate experiments. According to [40–42] dialectical problems consist of

1. no predefined starting point
2. no concrete terminal point
3. no predefined solution process

The challenge for learners is to apply developed knowledge, skills and competencies in order to find and define novel and origin problems as research questions, defining a starting point, a final state, and the means for gaining it.

The underlying creative processes can be triggered, animated and intensified by applying changes of perspective. The integration and negotiation among related objectives, views, intentions or theoretical positions, which are involved by the chosen perspectives, may lead to novel scientific questions and more holistic and sustainable solutions for engineering problems. Hence, it is up to the researcher to determine the degree of interdisciplinarity, ranging from different engineering disciplines to the integration of e.g. the humanities, psychology and education, environmental sciences, or liberal arts, considering e.g. participatory design, technology assessment, engineering results assessment, involvement of stake holders and target groups [47, 48], environmental issues, or ethical issues, (see facet 5: multiperspective thinking).

The intended outcome of the advanced level is something like a tangible new product, a prototype, a theory, a process (see facet 6: reach for original ideas).

2.4.2 Dealing with Increasing Complexity in PeTEX for Fostering "the Spirit of Research"

The more the students work with PeTEX, the more freedom they receive to define their own research problems and to find the answers on their own. Furthermore, PeTEX provides collaboration, not only with other students (from other universities and even other countries), but also with lifelong learners. In summary, PeTEX offers an important contribution to foster the "spirit of research" by providing gradually more "openness", "freedom", "stimulation", "inspiration" and "empowerment" to the students; factors which promote their creativity (see Fig. 4 and Fig. 5).

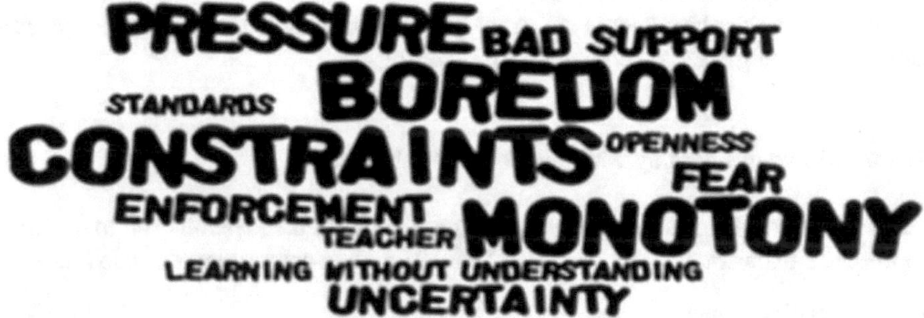

Fig. 4 Students' perspective on factors that hinder creativity

Fig. 5 Students' perspective of factors that foster creativity

3 Conclusions

The essay posed two main questions: What means/is creativity in the context of higher engineering education and is there a lack of creativity fostering education in engineering curricula? To answer these questions the essay presented results of a first curriculum survey as pre-study, conducted during the first stage of the nationally funded ELLI project.

The pre-study confirms an absence of fostering research spirit in the chosen sample of engineering education curricula. Since the results of the pre-study depend on a rather small, arbitrary sample, further qualitative and quantitative research will be undertaken.

Furthermore, the paper discussed the remote lab approach of the finished EU-project "PeTEX – Platform for E-Learning and Telemetric Experimentation" as one successful practice example on fostering creativity.

4 Future Work in ELLI

- Extension of the remote laboratory equipment
- Formative evaluation of creativity fostering teaching and learning practices
- Expansion of the module description analysis with emphasis on creativity
- Implementation of more interviews among university teachers in HEE and employers
- Revision and improvement of the six facets model according to HEE
- "Through the barricades" and "rage against the machine" – HEE workshops for faculty and staff on fostering creativity
- "LabDid 2.0" – HEE workshop on lab didactics for faculty and staff

5 Challenging Open Questions

This contribution resumes with open questions addressing relevant future educational and socio-economic impacts. It remains unclear whether these points also play an important role in the perspective of teachers and, furthermore, parts of the society:

- What wishes and visions do teachers, researchers, industry representatives, professional association representatives have with regard to the education of tomorrow's engineers, their creativity and research spirit?
- Are open experimentation and trying out new ideas, is the search for the unknown new truly important for a society in a globalized world economy?
- Does our economic society indeed need diligent professionals who execute given tasks instead of developing their own initiatives?

- What kind of education will be needed if a society wants to bring up future inventors who are able to cope with challenging future problems?
- How could teachers be trained efficiently and successfully in creativity fostering techniques?
- How could creativity and interdisciplinary knowledge in engineering education courses and curricula be fostered?

These questions should be discussed in a broad social – multidisciplinary – debate and further studies on the impact of teaching creativity in engineering education need to be done.

References

1. R. Schank, P. Childers, *The creative attitude. Learning to ask and answer the right questions*. New York, 1988
2. J. García Zubía, G. R. Alves, eds., *Using Remote Labs in Education. Two Little Ducks in Remote Experimentation. Engineering, no. 8*. University of Deusto, Bilbao, Spain, 2011
3. A. K. M. Azad, M. E. Auer, V. J. Harward, Internet accessible remote laboratories: Scalable e-learning tools for engineering and science disciplines. Engineering Science Reference, 2012
4. J. C. Kaufman, R. J. Sternberg, eds., *The Cambridge Handbook of Creativity*. Cambridge University Press, 2010
5. D. Cropley, A. Cropley, Functional creativity: Products and the generation of effective novelty. In: *The Cambridge Handbook of Creativity*, ed. by J. C. Kaufman, R. J. Sternberg, Cambridge University Press, 2010, pp. 301–320
6. M. Csikszentmihalyi, *Creativity: Flow and the Psychology of Discovery and Inven-tion*. Harper Perennial, New York, 1996
7. D. Gauntlett, *Making is connecting: The social meaning of creativity, from DIY and knitting to YouTube and Web 2.0*. Polity Press, 2011
8. R. J. Sternberg, J. C. Kaufman, Constraints on creativity: Obvious and not so obvious. In: *The Cambridge Handbook of Creativity*, ed. by J. C. Kaufman, R. J. Sternberg, Cambridge University Press, 2010, pp. 467–482
9. S. Moran, The role of creativity in society. In: *The Cambridge Handbook of Creativity*, ed. by J. C. Kaufman, R. J. Sternberg, Cambridge University Press, 2010, pp. 74–90
10. The engineer of 2020: Visions of engineering in the new century, national academy of engineering, 2004, 2004. http://www.nap.edu/catalog.php?record_id=10999#orgs
11. L. D. Feisel, A. J. Rosa, The role of the laboratory in undergraduate engineering education. Journal of Engineering Education, 2005, pp. 121–130
12. J. K. Smith and L. F. Smith, Educational creativity. In: *The Cambridge Handbook of Creativity*, ed. by J. C. Kaufman, R. J. Sternberg, Cambridge University Press, 2010, pp. 250–264
13. R. A. Beghetto, Creativity in the classroom. In: *The Cambridge Handbook of Creativity*, ed. by J. C. Kaufman, R. J. Sternberg, Cambridge University Press, 2010, pp. 447–466
14. J. P. Guilford, Creativity. American Psychologist **5** (9), 1950, pp. 444–454
15. C. Byrge, S. Hansen, The creative platform: A didactic for sharing and using knowledge in interdisciplinary and intercultural groups. SEFI 2008 - Conference Proceedings (9), 2008
16. C. Zhou, J. E. Holgaard, A. Kolmos, J. D. Nielsen, Creativity development for engineering students: Cases of problem and project based learning. In: *Joint International IGIP-SEFI Annual Conference 2010*. 19th–22nd September 2010
17. C. Zhou, Learning engineering knowledge and creativity by solving projects. International Journal of Engineering Pedagogy (iJEP) **2** (1), 2012, pp. 26–31

18. A. Strauss, J. Corbin, *Basics of Qualitative Research: Grounded Theory Procedures and Technics*. Sage Publications Inc., 1990
19. T. Haertel, I. Jahnke, Kreativitätsförderung in der hochschullehre: ein 6-stufen-modell für alle fächer?! In: *Fachbezogene und fachübergreifende Hochschuldidaktik. Blickpunkt Hochschuldidaktik, Band 121*, ed. by H. Hoffmann, J. Wildt, Bertelsmann Verl., Bielefeld, 2011, pp. 135–146
20. T. Haertel, I. Jahnke, Wie kommt die kreativitätsförderung in die hochschullehre? Zeitschrift für Hochschulentwicklung **6** (3), 2011, pp. 238–245
21. I. Jahnke, T. Haertel, M. Winkler, Sechs facetten der kreativitätsförderung in der lehre – empirische erkenntnisse. In: *Der Bologna-Prozess aus Sicht der Hochschulforschung, Analysen und Impulse für die Praxis*, ed. by S. Nickel, Arbeitspapier Nr, CHE gemeinnütziges Centrum für Hochschulentwicklung, 2011, pp. 138–152
22. I. Jahnke, T. Haertel, Kreativitätsförderung in hochschulen - ein rahmenkonzept. Das Hochschulwesen **58**, 2010, pp. 88–96
23. M. Spitzer, *Geist im Netz: Modelle für Lernen; Denken und Handeln*. Spektrum, Heidelberg, 2000
24. T. Haertel, C. Terkowsky, Where have all the inventors gone? the lack of spirit of research in engineering education. In: *Proceedings of the 2012 Conference on Modern Materials, Technics and Technologies in Mechanical Engineering. The Ministry of Higher and Secondary Specialized Education (MHSSE) of the Republic of Uzbekistan*. 2012, pp. 507–512
25. T. Haertel, C. Terkowsky, I. Jahnke, Where have all the inventors gone? is there a lack of spirit of research in engineering education. In: *15th International Conference on Interactive Collaborative Learning and 41st International Conference on Engineering Pedagogy*. 2012
26. C. Terkowsky, T. Haertel, Where have all the inventors gone? the neglected spirit of research in engineering education curricula. In: *Proceedings of the 2012 Conference on Actual Problems of Development of Light Industry in Uzbekistan on the Basis of Innovations. Uzbekistan, Tashkent: The Ministry of Higher and Secondary Specialized Education (MHSSE) of the Re-public of Uzbekistan and The Tashkent Institute of Textile and Light Industry (TITLI)*. 2012, pp. 5–8
27. J. Ma, J. V. Nickerson, Hands-on, simulated, and remote laboratories: A comparative literature review. ACM Computing Surveys **38** (3, Article 7), 2006
28. I. Jahnke, C. Terkowsky, C. Pleul, U. Dirksen, M. Heiner, J. Wildt, A. E. Tekkaya, Experimentierendes lernen entwerfen - elearning mit design-based research. In: *E-Learning. Lernen im digitalen Zeitalter*, ed. by N. Apostolopoulos, H. Hoffmann, V. Mansmann, A. Schwill, Münster, 2009, pp. 279–290
29. I. Jahnke, C. Terkowsky, C. Pleul, A.E. Tekkaya, Online learning with remote-configured experiments. In: *Interaktive Kulturen, DeLFI 2010: 8. Tagung der Fachgruppe E-Learning der Gesellschaft für Informatik e.V.*, ed. by M. Kerres, N. Ojstersek, U. Schroeder, U. Hoppe, 2010, pp. 265–277
30. C. Terkowsky, I. Jahnke, C. Pleul, R. Licari, P. Johannssen, G. Buffa, M. Heiner, L. Fratini, E. Lo Valvo, M. Nicolescu, J. Wildt, A. E. Tekkaya, Developing tele-operated laboratories for manufacturing engineering education. platform for e-learning and telemetric experimentation (petex). International Journal of Online Engineering (iJOE) **6** (Special Issue 1: REV2010; Vienna; IAOE), 2010, pp. 60–70. doi:10.3991/ijoe.v6s1.1378
31. C. Terkowsky, C. Pleul, I. Jahnke, A. E. Tekkaya, Petex: Platform for elearning and telemetric experimentation. In: *Praxiseinblicke Forschendes Lernen, TeachING.LearnING.EU*, ed. by U. Bach, T. Jungmann, K. Müller, Aachen, Dortmund, Bochum, 2011, pp. 28–31
32. C. Terkowsky, I. Jahnke, C. Pleul, D. May, T. Jungmann, A. E. Tekkaya, Petex@work. designing cscl@work for online engineer-ing education. In: *Computer-Supported Collaborative Learning at the Work-place - CSCL@Work, Computer-Supported Collaborative Learning Series*, vol. 14, ed. by S. P. Goggins, I. Jahnke, V. Wulf, Springer, 2013, pp. 269–292
33. C. Terkowsky, I. Jahnke, C. Pleul, *Platform for eLearning and Telemetric Experimentation. A Framework for Community-based Learning in the Work-place". Position Paper for the Workshop on CSCL at Work, 16th ACM international conference on Supporting group work*. Sanibel, Florida, 2010

34. C. Pleul, C. Terkowsky, I. Jahnke, A. E. Tekkaya, Tele-operated laboratory experiments in engineering education - the uniaxial tensile test for material characterization in forming technology. In: *Using Remote Labs in Education. Two Little Ducks in Remote Experimentation. Engineering, no. 8*, ed. by J. García Zubía, G. R. Alves, University of Deusto, Bilbao, Spain, 2011, pp. 323–348
35. C. Terkowsky, C. Pleul, I. Jahnke & A. E. Tekkaya, Tele-operated laboratories for online production engineering education - platform for e-learning and telemetric experimentation (petex). International Journal of Online Engineering (iJOE) **7** (Special Issue: Educon 2011; Vienna; IAOE), 2011. doi:10.3991/ijoe.v7iS1.1725
36. D. May, C. Terkowsky, T. Haertel, C. Pleul, Using e-portfolios to support experiential learning and open the use of tele-operated laboratories for mobile devices. REV2012 - Remote Engineering & Virtual Instrumentation, Bilbao, Spain, Conference Proceedings, 2012, pp. 172–180. doi:10.1109/REV.2012.6293126
37. T. Haertel, C. Terkowsky, D. May, C. Pleul, Entwicklung von remote-labs zum erfahrungsbasierten lernen. In: *Themenheft Kompetenzen, Kompetenzorientierung und Employability in der Hochschule (Teil 2), Zeitschrift für Hochschulentwicklung ZFHE*, vol. 8, ed. by N. Schaper, T. Schlömer, M. Paechter, 2013, pp. 79–87
38. C. Terkowsky, I. Jahnke, C. Pleul, A. E. Tekkaya, Platform for e-learning and telemetric experimentation (petex) – tele-operated laboratories for production engineering education. In: *Proceedings of the 2011 IEEE Global Engineering Education Conference (EDUCON) – Ľearning Environments and Ecosystems in Engineering Education. IAOE*, ed. by M. E. Auer, Y. Al-Zouvi, E. Tovar. 2011, pp. 491–497
39. C. Terkowsky, D. May, T. Haertel, C. Pleul, Experiential learning with remote labs and e-portfolios – integrating tele-operated experiments into personal learning environments. International Journal of Online Engineering (iJOE) **9** (1), 2013, pp. 12–20. doi:10.3991/ijoe.v9i1.2364
40. R. M. Rahn, *Vom Problem zur Lösung*. Heyne Verlag, München, 1990
41. D. Dörner, *Die Logik des Misslingens. Strategisches Denken in komplexen Situationen*. rororo, Reinbek, 2003
42. F. Vester, *Die Kunst vernetzt zu denken: Ideen und Werkzeuge für einen neuen Umgang mit Komplexität. Ein Bericht an den Club of Rome*. DTV, München, 2002
43. C. Terkowsky, T. Haertel, Where have all the inventors gone? fostering creativity in engineering education with remote lab learning environ-ments. In: *Proceedings of the 2013 IEEE Global Engineering Education Conference (EDUCON); S̈ynergy from Classic and Future Engineering Education*. March 13–15; 2013. IEEE; 2013, pp. 345–351
44. S. Lippke, A. U. Wiedemann, J. P. Ziegelmann, T. Reuter, R. Schwarzer, Self-efficacy moderates the mediation of intentions into behavior via plans. American Journal of Health Behavior **33** (5), 2009
45. C. Terkowsky, Vertrauen in elektronische(n) lernumgebungen. In: *Urban Fictions. Die Zukunft des Städtischen*, ed. by M. Faßler, C. Terkowsky, W. Fink, München, 2006, pp. 313–329
46. P. Ilyes, C. Terkowsky, B. Kroll, Das ka-wiki als soziotechnisches system. In: *Wikis in Schule und Hochschule*, ed. by M. Beißwenger, N. Anskeit, A. Storrer, Reihe Ë-Learning, vwh, Boizenburg, 2012, pp. 137–169
47. T. Haertel, J. Weyer, Technikakzeptanz und hochautomation. Technikfolgenabschätzung - Theorie und Praxis **14** (3), 2005, pp. 61–67
48. T. Haertel, *Techniksteuerung durch Normung am Beispiel der Ergonomie von Speditionssoftware: Ergonomienorm oder Ergononienorm?* Duisburg, Essen, Univ., Diss., 2010

Integrating Remote Labs into Personal Learning Environments – Experiential Learning with Tele-Operated Experiments and E-Portfolios

Claudius Terkowsky, Dominik May, Tobias Haertel and Christian Pleul

Abstract The use of laboratories in Higher Engineering Education is an adequate opportunity to implement forms of experiential learning like problem-based or research-based learning into manufacturing technology. The introduction of remote laboratories gives students the opportunity to do self-directed research and by that having their own and unique learning experiences. Recently finished research projects, e.g. the PeTEX project, implemented research-based learning by deploying real laboratory equipment without being physically in the laboratory but by accessing it via the Internet. One essential question in this context is on the one hand how the student can document his/her own learning processes and how the teacher can guide the student through these processes on the other hand. The proposed solution in this paper is a personal learning environment that integrates a remote lab and an e-portfolio system. E-portfolios enable the student to individually and collectively document and reflect what he/she has been doing and to share his/her outcomes with others. The paper outlines the important role that e-portfolios can play as personal learning environments to experience remote laboratory work and to foster creative attitudes.

Keywords Personal Learning Environments · E-Portfolios · Tele-Operated Laboratories · Online Engineering Education · Experiential Learning

1 Introduction

The experience of learning through experiments in general has become an essential part in modern Higher Engineering Education [1]. For the first time during their

---

C. Terkowsky (✉) · D. May · T. Haertel
Engineering Education Research Group (EERG), Center for Higher Education (zhb),
TU Dortmund University, Vogelpothsweg 78, 44227 Dortmund, Germany
e-mail: claudius.terkowsky@tu-dortmund.de

C. Pleul
Institute of Forming Technology and Lightweight Construction (IUL),
TU Dortmund University, Baroper Straße 303, 44227 Dortmund, Germany

Originally published in "International Journal of Online Engineering (iJOE)", Issue 1, Vol. #9, © IAOE 2013. Reprint by Springer International Publishing AG 2016, DOI 10.1007/978-3-319-46916-4_17

education, students can get to know lab equipment and working practices of their future professional world [2]. They can practice experimentation methods and analytical abstraction and are encouraged and sometimes challenged in their scientific and technological self-understanding. This includes, for example, practical implementation of theoretical assumptions, technical engineering or scientific activities through the implementation and evaluation of practical experiments and ideally the critical evaluation of their results and of their own approaches.

However, one of the most important factors that hinder the real use of laboratories by students is the initial and running costs of such equipment. Especially small universities often face the situation that they cannot afford all the laboratory equipment, or that the students are not allowed to use it, because they could damage the test-stands. That means in many cases that experiments are either only shown via video during the lecture or that the faculty's staff demonstratively shows the equipment just during guided tours through the laboratory.

One possible way out of this dilemma — in order to enable students to conduct experiments and to develop technical skills and scientific competencies — are remote and virtual laboratories [3, 4]. With them, laboratory equipment can be shared by separate universities and places, and even more, very risky experiments can be conducted completely virtually. The experience of remote experimentation can be delivered to the learner by technically and didactically integrating the labs into collaborative learning systems like monolithic learning and content management systems (LCMS) or cloud-based personal learning environments (PLE): "A PLE driven approach does not only provide personal spaces, which belong to and are controlled by the user, but also requires a social context by offering means to connect with other personal spaces for effective knowledge sharing and collaborative knowledge creation" [5].

Important research on the use of tele-operated experiments in LCMS-based teaching and learning was done by universities from Dortmund (Germany), Palermo (Italy), and Stockholm (Sweden), within a European project called **PeTEX — Platform for e-learning and Telemetric EXperimentation**. The Dortmund part was carried out by the Institute of Forming Technology and Lightweight Construction (IUL) and the Center for Higher Education. Within this project, fundamental "design-based research" in using tele-operated laboratories for teaching and learning was done [6–10]. A network of three fully functional prototypes in the field of manufacturing technology was developed step by step, formatively evaluated [11–17] and finally demonstrated [18].

The work presented in this paper will be based on the achievements of the PeTEX project, will enhance its technological infrastructure to a personal learning environment by integrating e-portfolio software, and will improve the concept by extending the didactical possibilities with an experiential learning approach.

Further development of an e-portfolio-based personal learning environment (PLE) will be carried out as a subtask of the new project **ELLI – Excellent Teaching and Learning in Engineering Education**. ELLI is funded by the German Ministry of Research and Education until 2016.

2 Pedagogical Foundation

"The most compelling argument for PLE is to develop educational technology which can respond to the way people are using technology for learning and which allows them to themselves shape their own learning spaces, to form and join communities and to create, consume, remix, and share material" [19].

"The development and support for Personal Learning Environments would entail a radical shift, not only in how we use educational technology, but in the organization and ethos of education. Personal Learning Environments provide more responsibility and more independence for learners. They would imply redrawing the balance between institutional learning and learning in the wider world" [20].

2.1 E-Portfolios as Personal Learning Environments

Personal Learning Environments can play an important role to foster and facilitate student-centered learning: "Personal Learning Environments are systems that help learners take control of and manage their own learning. This includes providing support for learners to set their own learning goals, manage their learning; managing both content and process, communicate with others in the process of learning, and thereby achieve learning goals. A PLE may be composed of one or more sub-systems: As such it may be a desktop application, or composed of one or more web-based services" [21].

E-portfolios as one manifestation of personal learning environments are based on the general idea of portfolios. A portfolio gives learners the opportunity to collect and organize different kinds of documents in a folder in order to reflect their learning process, to edit and to present it [22]. E-portfolios support the same processes, but they base on ICT, are accessible online and provide the collection of different kinds of digital data and information like texts, tables, photos, videos, and audio. E-Portfolio-based PLE software, in the presented case Mahara, can be very easily combined with the PeTEX LCMS based on Moodle. For another example of an e-portfolio like system, see [23] which is based on Wiki software. The integrating application Mahoodle combines the properties and functions of the teacher-led LCMS Moodle and the learner-led e-portfolio Mahara into a PLE, which can be deployed as "a facility for an individual [or a group of individuals] to access, aggregate, configure and manipulate digital artifacts of their ongoing learning experiences" [24].

2.2 Kolb's Experiential Learning Cycle

The basic understanding of learning and its use for laboratories in teaching and learning environments can be traced back to [25]: "Learning is the process whereby knowledge is created through the transformation of experience". Kolb states that learning involves the acquisition of abstract concepts that can be applied flexibly in a range of situations. In Kolb's theory, the impetus for the development of new concepts is provided by new experiences. Kolb's concept of experience is defined in his experiential learning theory, consisting of a four phase cycle in which the learner traces all the foundations of his learning process:

- *Concrete Experience*: A new experience of a situation is faced, or a reinterpretation of an existing experience takes place.
- *Reflective Observation*: The new experience is analyzed, evaluated, and interpreted. Of particular importance are any inconsistencies between the experience and the understanding of it.
- *Abstract Conceptualization*: Reflection gives rise to a new idea, or a modification of an existing abstract concept.
- *Active Experimentation*: Transforming the new abstract concept into operation, the learner interacts with the world around him to check what emerges.

In his four-step learning cycle, Kolb explains that at the beginning of each learning process there is a real learner's experience (step 1) which is followed by a reflective observation (step 2). From that point on the learner tries to conceptualize what he has experienced (step 3), starts to experiment actively (step 4), and generates new experiences. This is the start of a new cycle. With every loop — from the simple to the complex — the student enhances his experiences. Thus, the learning activities are transformed by the learning cycle into a helix of experience-based knowledge, skills and competencies. See [7] for a concept to integrate three levels of experience.

2.3 Fostering Creativity

Going the whole way of a research process corresponds to another important aspect of engineering education: fostering the students' creative potential. Industrial nations are facing tremendous problems. For example, new techniques to tackle climate change, new ideas on how to retain mobility of people or new concepts for energy production without fossil fuels are urgently needed. Engineers play an important role in addressing these challenges. Future prosperity and wealth will depend on their inventions and creativity [26–30].

Engineers, who embody the creative inventors and tinkerers more than any other occupation group, carry an important contribution (or even the societal responsibility) to solving current problems. However, engineering education has not been known to be particularly creative or to foster creativity [31].

3 Experiential Learning in the Mode of Research with Tele-Operated Laboratories

Once they graduated, and no matter if they go for a career in a company or in the academic sector, engineering students will mainly work with real technical equipment and they will work on creative solutions for real problems. But will they get the opportunity to have intense experiences with lab equipment during their studies? [1]

One possibility to change this fact is the use of laboratories in teaching, by deploying experiential learning [25] or research-based learning [32]. To bring the students into contact with laboratory equipment means to bring them in contact with the technical equipment of their future profession and to give them the opportunity to develop essential competences for their future career [32].

It is not by coincidence that a research process is quite similar to Kolb's learning cycle theory, beginning with an experience or a question and ending with real experiments and new research results [33].

That is why research-based or experiential learning in higher education is one adequate way of implementing learner centered teaching. In addition to that, [34] had pointed out the importance of an authentic learning environment for a successful learning process.

Only classical telling of knowledge in lectures in combination with theoretical exercises and without giving the real context may not lead to reach higher order learning outcomes, stated by taxonomies like the SOLO taxonomy [35], Bloom's revised taxonomy [36], or the thirteen fundamental objectives of laboratory learning, published and discussed by [1]. But this authentic learning environment can be offered by teaching and learning activities in laboratories where students can face the context of real professional activities. By connecting the actions in laboratories in a next step with real problems — e.g. in current research or the industry — students are able to go the whole way from the question at the beginning of an experiment to the final use of the results which makes them see the relevance of their work. This process requires reflective thinking and independent learning which obviously differ significantly from classical lecture-based courses [37].

Using tele-operated experiments and virtual laboratories gives a whole range of opportunities to implement experiential learning into teaching in the field of Higher Engineering Education. Just one example is its additional use next to normal lectures about forming technology. While students discuss basic aspects of material behavior relevant for forming processes during the lecture, they can simultaneously test and experience what they have discussed by independently doing experiments with the use of tele-operated experimental equipment. Another opportunity is that students receive a real problem in the context of material behavior: in small groups they have to solve the task with the tele-operated lab equipment. Finally, they have to present what they have found out, and what they would suggest for solving the problem [7].

3.1 Active Experimentation Using Tele-Operated Equipment

Using remote and virtual laboratories in teaching gives a whole range of opportunities to implement experiential learning into the field of mechanical engineering following the path of research based learning [7]. One example in the context of manufacturing technology, namely forming technology, will be the use of such a special lab concept for material characterization. This will be organized in addition to a normal lecture or in order to enhance traditional hands-on labs during the phase in which students prepare themselves for the lab. Moreover, the special lab concept helps students to rework some of the test steps while analyzing the data for the lab report.

Following the approach based on Kolb's experiential learning cycle, students can deal with basic concepts of metal forming during the lecture and test and see what they discussed in class by doing experiments on their own. With this they construct their own knowledge using the equipment provided by the remote lab. Another opportunity will be that students have to face a real engineering problem related to material behavior. They are asked to work on this problem in small groups by planning and carrying out experiments using the tele-operated equipment. Finally, they have to present what they have explored and what they would suggest for dealing with the problem [7].

In order to support this entire process and especially the step of "active experimentation", one important aspect is the integration of an appropriate level of interaction and feedback into the tele-operated experimental setup. In the PeTEX project, a complete experimental setup (Fig. 1) has been transformed to a new level by using innovative engineering designs, modern concepts of automation, measurement technology, and robotics, as shown in Fig. 2.

All aspects have been connected by developing a clear, usable, and interactive real time feedback user interface of the running experiment. Fig. 3 shows the "window" of the developed graphical user interface of the uniaxial tensile test.

When using the live camera stream (1), users can investigate the surrounding test apparatus, e.g. sensors or clamping devices. Afterwards, the learner initiates the preparation of the experiment (2) by using the integrated 6-axes robot to select and check an appropriate specimen. Relevant test parameters (3) can be freely set to configure the experiment. When the test is started (4), the robot positions the specimen into the fully automatic clamping device. During the test, a high level of interaction is provided to the user by manipulating the camera view or pausing and continuing the test. Interrupting the test causes a material reaction because the load is not further increased for that moment. This phenomenon is graphically visible in the real time diagram (6) and also in the real time test data at the header bar (5). By using the data base (7) and the graph, comparisons with prior test data are available (6). After the experiment is finished, learners are provided with data package including all the results for further analysis and investigation.

Additionally, within the learning content management system Moodle, the entire tele-operated experimental environment was made available by a developed Moodle module. With Moodle we designed the alignment and the integration of the four necessary structural elements for this kind of socio-technical system. This socio-

Fig. 1 Automated material testing machine

technical alignment for tele-operated laboratory learning consists of the adjustment of the technical, didactical, media and social level. By the implementation into Moodle, as shown in Fig. 4, this socio-technical alignment was put into a usable as well as flexible environment.

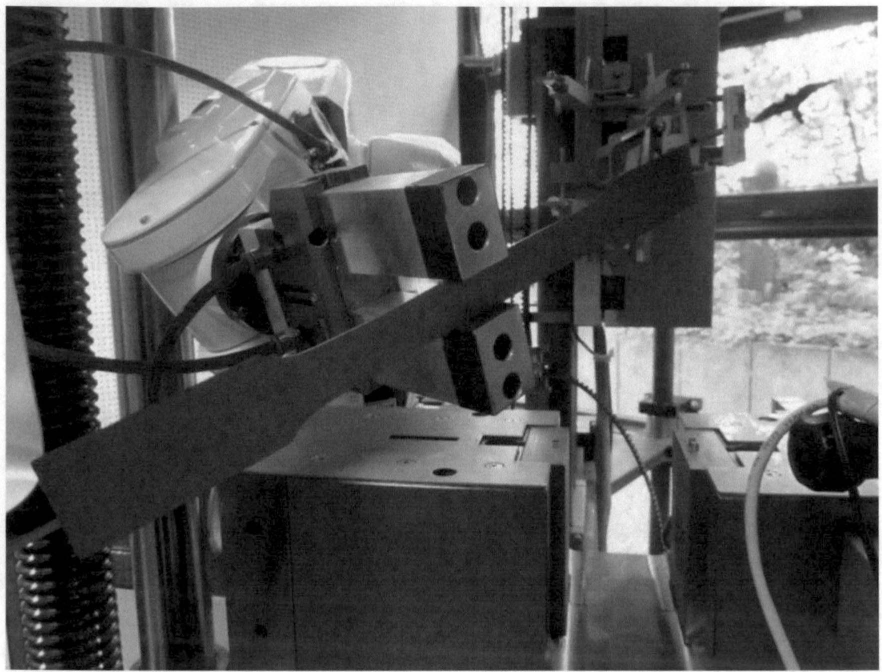

Fig. 2 Robot positioning a specimen

A challenge that is often formulated when talking about such openly designed learning concepts is that the teacher is in need of a very sophisticated concept to document and evaluate the learners' behavior and achievements during the learning processes taking place in the laboratory. It is obvious that such a concept requires different systems for the instructor to accompany the learner through the learning process and, above all, to evaluate the achieved learning outcomes. Software which seems to be adequate and which is frequently discussed in similar contexts is the e-portfolio [38]. The following passages present the concept draft concerning learning process documentation in the context of the combination of remote laboratories and e-portfolios.

3.2 Experiential Learning with E-Portfolios and Tele-Operated Experiments

"ePortfolios are hardly a new idea in the fast developing field of Technology Enhanced Learning" [39]. In the following it will be explained why e-portfolios on the basis of Mahara fulfill the three main requirements in the PeTEX context [40, 41].

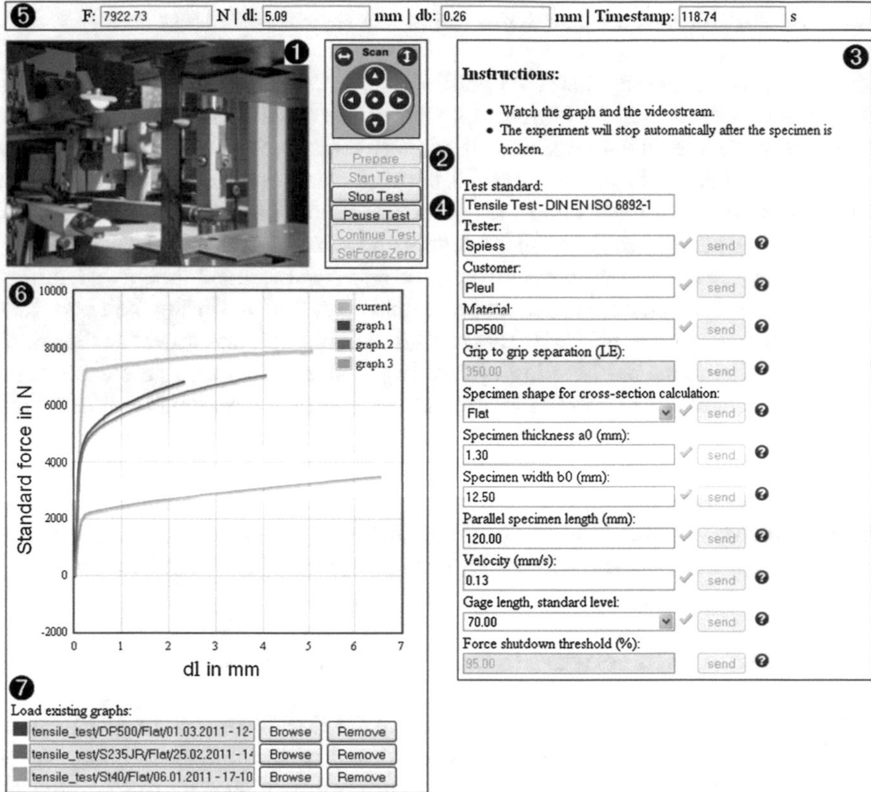

Fig. 3 Interface to the tele-operated experiment

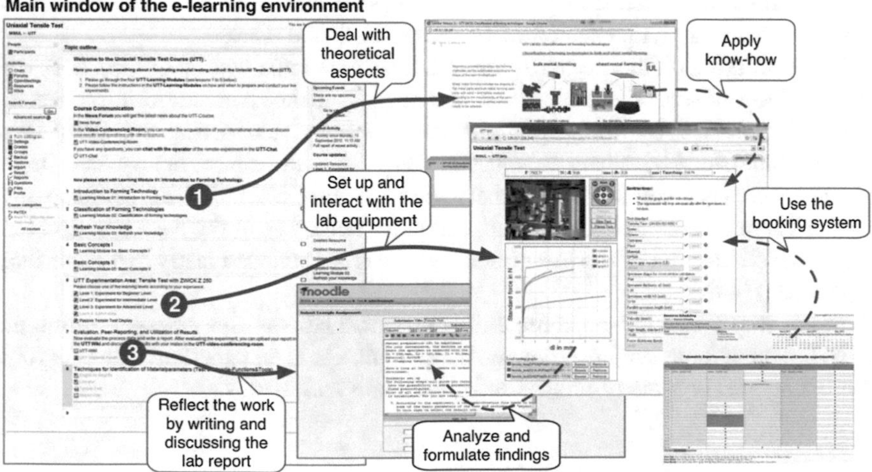

Fig. 4 Experiment environment integrated to Moodle

3.2.1 E-Portfolios as a Documentation of the Learning Process

By creating and designing their own portfolios, users get the opportunity to arrange all data and information they want to document or share with others in different orders. It works just like a personal page in any social network. For example, they can present experiments and their results, show photos from the test set-up, and can explain their research results and thoughts to themselves and others. Furthermore, they can allow other users, like other learners and teachers, to have a look at their e-portfolios. By creating such an e-portfolio, learners can document their own learning and research processes, and start to reflect on their experiments during their research-based learning processes [42, 43]. This reflection is an important aspect as they need this step in their personal learning cycles. Especially for students, the e-portfolio can be an orientation and checkpoint in fields of their own research [44–46]. By the same way, teachers also can evaluate the actions of learners by looking into their e-portfolios. Since other persons are able to see the collection in the portfolio, it can be said that it is not only a way of individually documenting the learning processes, but as well a way of communicating. Thus, a collaborative learning process can be achieved. This leads to the next use of e-portfolios within the PeTEX context.

3.2.2 E-Portfolios as Software to Build up a Learning Community

The deployment of the e-portfolio as software for documentation and evaluation is just one possible use of the system. A constructive enrichment in using the e-portfolios is community building [47, 48]. Every author of an e-portfolio is able to invite other users to look at the entire or just parts of his portfolio, and it works vice versa, too: one can be invited to see other e-portfolios. That means that learners, while conducting experiments in the PeTEX system and filling in their e-portfolios, can get into contact with each other via the portfolio software.

While working on their e-portfolios, students anticipate that their "product" will be valued by others. Therefore, they will seek to make them more attractive for others, for example by bringing in new aspects or by considering that their ideas must be understood by others as well. This requires a non-contradictory and simple presentation. They can see what others are especially interested in, can start discussions about it, can give comments, can help each other in the case of a problem during the conduction of the experiment and its reflection, and lastly can share their experiences [49].

In this way, a specialized community on remote laboratories emerges within the PeTEX context (e.g. see http://www.vrlcom.com.). It is an excellent example for a community on the topic of remote and virtual labs and worlds.

3.2.3 E-Portfolios as a Bridge Between University and the Workplace

The PeTEX system is designed for the usage in higher education and for workplace learning. That means that in a first step, students and workplace learners both can use the e-portfolios in the explained way of use. A further future thought is to use the e-portfolio as a lifelong system. One can document all competences gained from studying at the university, and can continue to document one's challenges, experiences, and advancements during the whole professional life. This should be explained by an example in three steps:

- Step 1 – An engineering student starts working with the PeTEX system at the university. He uses the system in order to document his experiments. During his studies, he does different experiments, collects all documentation of his research in his e-portfolio, and reflects his own learning paths. The teacher is able to evaluate his learning processes, results, and outcomes. This can be seen as the main use of e-portfolios at university.
- Step 2 – Since the PeTEX system addresses workplace learning just as well, e-portfolios can be seen as a bridge from university life to professional life. The student can use his e-portfolios to present himself to potential employers, depending on the concrete thematic design of the e-portfolios. The company can see what the student has acquired during his study in this field, and can decide if he fits to the company's needs. In this context e-portfolios can support the process of applying for a job.
- Step 3 – Once the former student and now employee starts to work in a company, he does not have to stop working with his portfolio. He can still work with his collection and document new experiments as well as gained knowledge and competences in his job. By doing so, the employee will not stop reflecting on his learning processes. His e-portfolio grows and with each year it more and more turns out to be a better presentation of his professional life and his competences. Especially the last aspect works perfectly together with the advantages of the PeTEX system: small and medium sized companies can use the system to skill up their workers by letting them experience research with the PeTEX hardware. In addition to that, they can use the e-portfolios for implementing a system to document and measure skills and competences of their employees.

3.2.4 E-Portfolios as a means of Mobile Learning

Another frequently mentioned new concept in context with higher education is mobile learning. Mobile learning means deploying mobile devices for the support of learning processes — like cell phones, smart phones or tablet-computers [46]. Only one of the advantages of mobile learning is that unplanned periods of time can be used for learning and that learning processes can be virtually initiated everywhere [46]. In our context we will focus on the fact that users actually carry their mobile devices at any time and because of that they can frequently use them in order to work with their e-portfolio software and the related laboratory equipment [49].

3.2.5 E-Portfolios as a Tool in Creative Moments

Consider a student who thinks of his experiments while sitting at home and watching TV or while being at a boys' night out with his friends. However, he is unable to concentrate on soccer and beer because he is really struggling with his research work, is thinking about his parameters, his results and why his experiments offer these results. While he is listening to his friends and how they are ordering the next round of cold beer, he suddenly has an idea for a hypothesis and wants to check it by rereading his last experiments in the e-portfolio, or even by conducting a new sequel of experiments. Since he can use the software for accessing the experimenting environment via his tablet computer, he does not need to wait until the next day or week for doing the experiment at the university. He can just stay where he is and even can stay on the barstool for checking his hypothesis. He can immediately put the new results in his e-portfolio in order to document the new steps within his research process. To celebrate this new step with his friends, he can immediately order a next round of beer remembering that the student is still sitting at the bar.

A researcher from the Australian Labshare project told us during the REV-Conference 2011, that Labshare had been on duty mostly on late Saturday nights and early Sunday mornings. But the reason for that phenomenon had not been investigated, as well as the social situatedness of users.

4 Discussion, Conclusion & Future Work

With this paper we explained why the use of laboratories as a place for conducting experiments is important for modern engineering education. The essential idea is to engage the students in teaching and learning environments which are connected closely with their future working environments. In addition to that, the combination of personal learning environments like e-portfolios and appropriate student-centered approaches gain more and more importance in higher education. This is one essential way for students to reach the high level of learning outcomes, and hereby develop the basis of fundamental competences for their future professional and personal life, as well as attitudes like curiosity, agency, and responsibility. Furthermore, we showed the potential of our approach for fostering learners' creativity. If students are enabled to evolve their own research questions, to choose a suitable experimentation design and finally to perform the experiment, they will be able to develop a kind of "spirit of research" [26, 31]. This spirit is one important premise for developing original ideas.

See in the following the central advantages of the presented concept:

- As the equipment of laboratories is either very expensive to provide at every university or not always available for students, the deployment of remote and virtual laboratories is an impactful means to face this dilemma.
- The use of lab equipment as virtual simulations can help the students to do experiments just as a pre-check on personal hypotheses. Using it remotely from any place they want, it can help them to conduct research even when they are not able to attend the laboratory.
- Learning processes that are achieved by the usage of the laboratories can be documented in e-portfolios.
- These e-portfolios are an adequate opportunity to document experiments for personal use or for the evaluation by an instructor. By examining the portfolios, the instructor can see what kind of experiments the students have done and what they have learned from it.
- If the e-portfolios are not kept hidden for other students but are rather open for other users to take look at them and comment on the achievements, there is an opportunity to evolve a community for collaborative learning and working with experiments [50]. Additionally, the e-portfolio software will be made accessible via mobile devices. This opens new ways of mobile learning, which means that students and some of their learning activities are not bound any longer to specific locations. From virtually anywhere and at any time, the user can use the lab equipment, work on his e-portfolio, and communicate with others [51].
- With the possibility of promoting a "spirit of research", an essential facet of creativity in higher education can be fostered.

Summing up, it can be said that all these aspects of the deployment of e-portfolios in the PeTEX context can support the idea of experiential and research-based learning — even if there are a couple of challenges to overcome [36]. The e-portfolios can be used to document and share the research results and learning processes, to build up an especially focused learning community, and to bring university learning and workplace learning together.

The step for the coming year will be to integrate the e-portfolio software in the system and to make it accessible from mobile devices. Once this will have been achieved, first tests with students can be carried out and the system can be formatively evaluated and improved.

References

1. L.D. Feisel, A.J. Rosa, The role of the laboratory in undergraduate engineering education. Journal of Engineering Education, 2005, pp. 121–130
2. A.G. Bruchmüller, A. Haug, *Labordidaktik für Hochschulen – Eine Einführung zum Praxisorientierten Projekt-Labor*, *Schriftenreihe report – Band*, vol. 40. Leuchtturm-Verlag, Alsbach, 2001

3. J. García Zubía, G.R. Alves, eds., *Using remote labs in education: Two little ducks in remote experimentation*. Publicaciones De La Unive, [Place of publication not identified], 2011
4. A.K.M. Azad, M.E. Auer, V.J. Harward, eds., *Internet accessible remote laboratories: Scalable E-learning tools for engineering and science disciplines*. Engineering Science Reference, Hershey, Pa., 2012
5. M.A. Chatti. Personal environments loosely joined. URL http://mohamedaminechatti.blogspot.de/2007/01/personal-environments-loosely-joined.html
6. I. Jahnke, C. Terkowsky, C. Pleul, Wechselwirkungen hochschuldidaktischer konzepte in fachbezogenen, medien-integrierten lehr-/lern-kulturen: Forschungsbasierte gestaltung. In: *Fachbezogene und fachübergreifende Hochschuldidaktik*, ed. by I. Jahnke, J. Wildt, Blickpunkt Hochschuldidaktik, Bertelsmann, Bielefeld, 2009, pp. 177–189
7. I. Jahnke, C. Terkowsky, Das projekt petex. e-learning und live-experimente verbinden. Journal Hochschuldidaktik **20** (1), 2009, pp. 14–17
8. I. Jahnke, C. Terkowsky, C. Burkhardt, U. Dirksen, M. Heiner, J. Wildt, A.E. Tekkaya, Forschendes e-learning. Journal Hochschuldidaktik **20** (2), 2009, pp. 30–32
9. C. Pleul, C. Terkowsky, I. Jahnke, A.E. Tekkaya, Platform for e-learning and tele-operative experimentation (petex) - hollistacally integrated laboratory experiments for manufacturing technology in engineering education. In: *Proceedings of SEFI Annual Conference, 1st World Engineering Education Flash Week*, ed. by J. Bernardino, J.C. Quadrado. Portugal, Lisbon, 2011, pp. 578–585
10. I. Jahnke, C. Terkowsky, C. Burkhardt, U. Dirksen, M. Heiner, J. Wildt, A.E. Tekkaya, Experimentierendes lernen entwerfen - elearning mit design-based research. In: *E-Learning. Lernen im digitalen Zeitalter*, ed. by N. Apostolopoulos, H. Hoffmann, V. Mansmann, A. Schwill, Waxmann, Münster, 2009, pp. 279–290
11. C. Pleul, C. Terkowsky, I. Jahnke, A.E. Tekkaya, Interactive demonstration of petex platform for e-learning and telemetric experimentation - a holistic approach for tele-operated experiments in production engineering. In: *REV 2011 International Conference on Remote Engineering and Virtual Instrumentation, Brasov, Romania, Conference Proceedings*, ed. by M.E. Auer, D. Ursutiu. IAOE, 2011, pp. 186–190
12. C. Terkowsky, C. Pleul, I. Jahnke, A.E. Tekkaya, Tele-operated laboratories for production engineering education - platform for e-learning and telemetric experimentation (petex). International Journal of Online Engineering (iJOE). Special Issue EDUCON 2011 **7**, 2011, pp. 37–43
13. C. Terkowsky, I. Jahnke, C. Pleul, A.E. Tekkaya, Platform for e-learning and telemetric experimentation (petex) - tele-operated laboratories for production engineering education. In: *Proceedings of the 2011 IEEE Global Engineering Education Conference (EDUCON) – Learning Environments and Ecosystems in Engineering Education*, ed. by M.E. Auer, Y. Al-Zoubi, E. Tovar. IAOE, Vienna, 2011, pp. 491–497
14. C. Terkowsky, I. Jahnke, C. Pleul, R. Licari, P. Johannssen, G. Buffa, M. Heiner, L. Fratini, E. Lo Valvo, M. Nicolescu, J. Wildt, A.E. Tekkaya, Developing tele-operated laboratories for manufacturing engineering education. platform for e-learning and telemetric experimentation (petex). In: *REV 2010 International Conference on Remote Engineering and Virtual Instrumentation, Conference Proceedings, Sweden, Stockholm*, ed. by M.E. Auer, G. Karlsson. IAOE, Vienna, 2010, pp. 97–107
15. C. Terkowsky, I. Jahnke, C. Pleul, Platform for elearning and telemetric experimentation. a framework for community-based learning in the workplace. In: *Workshop CSCL at Work*. Sanibel, Florida, 2010. URL http://www.zhb.tu-dortmund.de/hd/fileadmin/Mitarbeiter/cterkowsky/Group_10_WS._v3x.pdf
16. C. Terkowsky, C. Pleul, I. Jahnke, A.E. Tekkaya, Petex: Platform for elearning and telemetric experimentation. In: *Praxiseinblicke Forschendes Lernen*, ed. by U. Bach, T. Jungmann, K. Müller, TeachING.LearnING.EU, Aachen, Dortmund, Bochum, 2011, pp. 28–31
17. C. Pleul, I. Jahnke, C. Terkowsky, U. Dirksen, M. Heiner, J. Wildt, A.E. Tekkaya, Experimental e-learning – insights from the european project petex. In: *15th International Conference on Technology Supported Learning & Training*. ICWE GmbH, Berlin, 2009, pp. 47–50

18. C. Pleul, C. Terkowsky, I. Jahnke, A.E. Tekkaya, Tele-operated laboratory experiments in engineering education – the uniaxial tensile test for material characterization in forming technology. In: *Using remote labs in education*, ed. by J. García Zubía, G.R. Alves, Publicaciones De La Unive, [Place of publication not identified], 2011, pp. 323–348
19. G. Attwell. Personal learning environments for creating, consuming, remixing and sharing. URL http://www.pontydysgu.org/pontydysgu-and-people/graham-attwell/
20. G. Attwell, Personal learning environments - the future of elearning? eLearning Papers **2**, 2007, p. 5
21. v.M. Harmelen, Personal learning environments. In: *Proceedings of the 6th International Conference on Advanced Learning Technologies (ICALT'06)*. IEEE, 2006
22. A.C. Breuer, *Das Portfolio im Unterricht – Theorie und Praxis im Spiegel des Konstruktivismus*. Waxman Verlag GmbH, Münster, 2009
23. P. Ilyes, C. Terkowsky, B. Kroll, Das ka-wiki als soziotechnisches system. In: *Wikis in Schule und Hochschule*, ed. by M. Beißwenger, N. Anskeit, A. Storrer, E-Learning, vwh, Boizenburg, 2012, pp. 137–169
24. R. Lubensky. The present and future of personal learning environments (ple). URL http://www.deliberations.com.au/2006/12/present-and-future-of-personal-learning.html
25. A. Kolb, *Experiential learning. Experience as the source of learning and development*. Prentice-Hall, Englewood Cliffs, N.J, 1984
26. T. Haertel, C. Terkowsky, I. Jahnke, Where have all the inventors gone? is there a lack of spirit of research in engineering education? In: *Proceedings of ICL 2012 - 15th International Conference on Interactive Collaborative Learning and IGIP 2012 - 41st IGIP International Conference on Engineering Pedagogy*. IEEE, Villach, Austria, 2012
27. I. Jahnke, T. Haertel, Kreativitätsförderung in hochschulen - ein rahmenkonzept. Das Hochschulwesen **58** (3), 2010, pp. 88–96
28. T. Haertel, I. Jahnke, Kreativitätsförderung in der hochschullehre: ein 6-stufen-modell für alle fächer?! In: *Fachbezogene und fachübergreifende Hochschuldidaktik. Blickpunkt Hochschuldidaktik*, vol. 121, ed. by I. Jahnke, J. Wildt, Bertelsmann Verlag, 2011, pp. 135–146
29. T. Haertel, I. Jahnke, Wie kommt die kreativitätsförderung in die hochschullehre? Zeitschrift für Hochschulentwicklung **6** (3), 2011, pp. 238–245
30. I. Jahnke, T. Haertel, M. Winkler, Sechs facetten der kreativitätsförderung in der lehre – empirische erkenntnisse. In: *Der Bologna-Prozess aus Sicht der Hochschulforschung, Analysen und Impulse für die Praxis*, ed. by S. Nickel, CHE gemeinnütziges Centrum für Hochschulentwicklung, 2011, pp. 138–152
31. T. Haertel, C. Terkowsky, Where have all the inventors gone? the lack of spirit of research in engineering education. In: *Proceedings of the 2012 Conference on Modern Materials, Technics and Technologies in Mechanical Engineering*. 2012, pp. 507–512
32. L. Huber, J. Hellmer, F. Schneider, *Forschendes Lernen im Studium - Aktuelle Konzepte und Erfahrungen*. Universitätsverlag Webler, Bielefeld, 2009
33. J. Wildt, Forschendes lernen: Lernen im "format" der forschung. Journal Hochschuldidaktik **20** (2), 2009, pp. 4–7
34. J. Herrington, R. Oliver, An instructional design framework for authentic learning environments. Educational Technology Research and Development **48** (3), 2000, pp. 23–48
35. K. Biggs, C. Tang, *Teaching for quality learning at university. What the student does*, 3rd edn. McGraw-Hill, Maidenhead, 2007
36. L.Q. Anderson, D.R. Krathwohl, *A Taxonomy for Learning, Teaching, and Assessing. A Revision of Bloom's Taxonomy of Educational Objectives*. Addison Wesley Longman, New York, 2011
37. I. Jahnke, T. Haertel, V. Mattik, K. Lettow, Was ist eine kreative leistung studierender? mediengestützte kreativitätsförderliche lehrbeispiele. In: *HDI2010 – Tagungsband der 4. Fachtagung zur Hochschuldidaktik Informatik*, ed. by D. Engbring, R. Keil, J. Magenheim, H. Selke. Universitätsverlag Potsdam, Potsdam, 2010, pp. 87–92

38. K. Himpsl, P. Baumgartner. Evaluation von e-portfolio-software: Teil iii des bmwf-abschlussberichts, einsatz von e-portfolios an (österreichischen) hochschulen. URL http://www.bildungstechnologie.net/Members/khim/dokumente/himpsl_baumgartner_evaluation_eportfolio_software_abschlussbericht.pdf/download
39. G. Attwell, e-portfolios – the dna of the personal learning environment? Journal of e-Learning and Knowledge Society. English Version **3** (2), 2007, p. 1
40. C. Terkowsky, I. Jahnke, C. Pleul, D. May, T. Jungmann, A.E. Tekkaya, Petex@work. designing online engineering education. In: *CSCL at Work - a conceptual framework*, ed. by S.P. Goggins, I. Jahnke, V. Wulf, Springer, New York, 2013
41. I. Jahnke, C. Terkowsky, C. Pleul, A.E. Tekkaya, Online learning with remote-configured experiments. In: *Interaktive Kulturen, DeLFI 2010 – 8. Tagung der Fachgruppe E-Learning der Gesellschaft für Informatik e.V.*, ed. by M. Kerres, N. Ojstersek, U. Schroeder, U. Hoppe. 2010, pp. 265–277
42. G. Reinmann, S. Sippel, Königsweg oder sackgasse? – e-portfolios für das forschende lernen. In: *Kontrolle und Selbstkontrolle – Zur Ambivalenz von E-Portfolios in Bildungsprozessen*, ed. by T. Meyer, et al., Springer Fachmedien, Wiesbaden, 2011
43. R. Reichert, Das e-portfolio - eine mediale technologie zur herstellung von kontrolle und selbstkontrolle. In: *Kontrolle und Selbstkontrolle – Zur Ambivalenz von E-Portfolios in Bildungsprozessen*, ed. by T. Meyer, et al., Springer Fachmedien, Wiesbaden, 2011
44. H. Blom, *Der Dozent als Coach*. Luchterhand, Neuwied, Kriftel, 2000
45. A. Kukulska-Hulme, J. Traxler, *Mobile learning: A handbook for educators and trainers*. Routlegde and Taylor & Francis Group, Milton Park, New York, 2005
46. D. May, C. Terkowsky, T. Haertel, C. Pleul, Using e-portfolios to support experiential learning and open the use of tele-operated laboratories for mobile devices. In: *REV2012 - Remote Engineering & Virtual Instrumentation*, ed. by M.E. Auer, J. García Zubía. 2012, pp. 172–180
47. C. Terkowsky, I. Jahnke, C. Pleul, R. Licari, P. Johannssen, G. Buffa, M. Heiner, L. Fratini, E. Lo Valvo, M. Nicolescu, J. Wildt, A.E. Tekkaya, Developing tele-operated laboratories for manufacturing engineering education. platform for e-learning and telemetric experimentation (petex). International Journal of Online Engineering (iJOE). Special Issue: REV2010 **6**, 2010, pp. 60–70
48. C. Pleul, C. Terkowsky, I. Jahnke, Petex - platform for e-learning and telemetric experimentation: a holistic approach for tele-operated live experiments in production engineering. In: *GROUP '10. Proceedings of the 16th ACM international conference on Supporting group work*. 2010, pp. 325–326
49. C. Terkowsky, D. May, T. Haertel, C. Pleul, Experiential remote lab learning with e-portfolios - integrating tele-operated experiments into environments for reflective learning. In: *Proceedings of ICL 2012 - 15th International Conference on Interactive Collaborative Learning and IGIP 2012 - 41st IGIP International Conference on Engineering Pedagogy*, 2012
50. C. Terkowsky, Vertrauen in elektronische(n) lernumgebungen. In: *Urban Fictions. Die Zukunft des Städtischen*, ed. by M. Faßler, C. Terkowsky, W. Fink, 2006, pp. 313–329
51. M. Faßler, C. Terkowsky, eds., *Urban Fictions. Die Zukunft des Städtischen*. W. Fink, 2006

Virtual Labs and Remote Labs: Practical Experience for Everyone

Sulamith Frerich, Daniel Kruse, Marcus Petermann and Andreas Kilzer

Abstract Laboratory experiences should be available for a great number of students in engineering education, especially at times when the number of students is even more increasing. Virtual Labs and Remote Labs are innovative tools used for improvement. They are either simulating experiments (Virtual Labs) or remotely operated plants (Remote Labs). At RuhrUniversity Bochum, the implementation of eight new labs was supported by the project ELLI (excellent teaching and learning in engineering sciences). Didactical concepts as well as sustainable implementations were among the criteria of the independent jury's decision. After their setup, the management of the variety of labs is the next step. This short paper reports the work in progress in the year 2013, the whole process is to be continued and improved.

Keywords Engineering Education · Student Experiments · Electronic Learning · Civil Engineering · Mechanical Engineering · Electrical Engineering

1 Introduction

Engineers work as trouble shooters; they have to solve problems in professional life and "make things work". Therefore, the ability of transferring theoretical knowledge to practical applications is crucial. Engineering education should take this aspect into account and prepare students for their future tasks. Usually, it combines theoretical knowledge with practical experience in so-called unit operation laboratories, where students run experiments and analyze the corresponding results. However, laboratory capacity is limited. Only a certain amount of people is able to run experiments simultaneously, due to time and safety issues. Thus, the availability of laboratory equipment is limited, especially if the number of students enrolled is increasing. Currently, this situation in engineering education is aggravated in Germany. Recent political decisions altered the duration of school education: Now, it is possible to

S. Frerich (✉) · D. Kruse · M. Petermann · A. Kilzer
Ruhr-Universität Bochum, Lehrstuhl für Verfahrenstechnische Transportprozesse (VTP), Bochum, Germany
e-mail: frerich@vtp.rub.de

Originally published in "IEEE EDUCON", © 2014.
Reprint by Springer International Publishing AG 2016,
DOI 10.1007/978-3-319-46916-4_18

graduate with the general qualification for university entrance after only 12 years in contrast to 13 years, originally. In the meantime, there are students graduating in both systems. In general, the configuration of educational programs is continuously adapted to ideals and moral concepts. Scientific findings influence their contents as well as their methods. Additionally, innovative technologies are implemented [1]. One goal is to individualize learning processes. Students are actively encouraged to learn and deal individually with the subjects taught, as is required for deep-rooted comprehension and transfer of knowledge. It could be achieved by implementing mobile devices, to facilitate the availability of content. The utilization of media by adults of different ages, especially students, has been investigated [2]. In the future, these new tendencies should be implemented sustainably into the curricula of engineering education. The project ELLI (Excellent Teaching and Learning in Engineering Sciences), based on a cooperation between RWTH Aachen University, TU Dortmund and Ruhr University Bochum (RUB), focuses on engineering education improvements. Four different areas including international mobility, interdisciplinary competences and individual progressions from freshman to PhD student work on developing innovative concepts and ideas. One of the main aspects of ELLI is considering new ways of laboratories in lectures and seminars in order to make them available for more students.

2 Implemented Action

Virtual Labs and Remote Labs can be used as innovative tools in providing laboratory experiences. While Virtual Labs use simulation and visualization to create scenarios as realistic as possible, Remote Labs are lab-scaled plants operated remotely. Remote Labs operated under camera surveillance can broadcast events that are actually happening. Therefore, Remote Labs offer an almost realistic experience, since they show parameter influences and phenomena simultaneously. In comparison, Virtual Labs have the big advantage of their safety concepts: Since the whole experiment is simulated, no issues with hazardous materials occur and no treatment of waste is necessary. Both kinds of a laboratory experiment can be alternated according to the individual student or operator's knowledge. Therefore, individually structured learning processes can be created and moderated. In addition, Virtual and Remote Labs can be helpful to make laboratory experiences available to a greater number of students. Although these experiments lack real "hands on"-experience, due to the separation between operator and set up, they can be used as extensions to existing unit operation laboratories. Therefore, the traditional experiments and their virtual equivalents form a great variety. Thus, students choose the experiments they are really interested in, and traditional and new laboratories alternate. At Ruhr University Bochum, a call for projects addressing the local engineering departments was offered in September 2012 to promote the setup of Virtual Labs and Remote Labs. Previously, representatives of all engineering departments were invited for a general exchange of ideas. The resultant call and its content were therefore public and transparent. After a thoroughly consideration of all submissions, eight experiments were selected and supported. The criteria determining which experiments to choose were based on educational concepts, most realistic experimental setup including boundary areas, and their sustainable integration into existing lectures or seminars. The committee who made the selection consisted of independent deputies of all engineering disciplines including students. In total, 472 000 EUR were spent. This sum was split up between the Faculties of Civil and Environmental Engineering, Mechanical Engineering, and Electrical Engineering and Information Technology.

3 Benefits

Virtual and Remote Labs encourage the development of individual competencies. Each student is asked to pursue individually planned experiments while dealing with actual challenges of an engineer's daily life and checking their status of knowledge simultaneously. Therefore, the implementation of didactical concepts was of special importance to the jury's decision. The amplification of student knowledge should be possible at individual terms, but it was necessary to make clear how the respective gain in knowledge would be graded. Submissions implementing realistic experiments including false results or errors in measurement received positive reviews. In addition, experiments run at boundary or unsteady parameters show limitations - essential in realistic scenarios. Finally, all supported Virtual and Remote Labs shall extend the existing tools in engineering education sustainably. Their implementation into curricula is to endure even longer than the project ELLI. Therefore, interdisciplinary concepts integrating different applications in lectures, seminars or summer schools were especially positively evaluated. As a long-term-perspective, implementing Virtual and Remote Labs shall promote the integration of practical experiences and applications in engineering education, and thus improve teaching and learning methods.

4 The Labs in Detail

With a closer look at the individual features, the supported labs can be grouped by three categories [3], as shown in Fig. 1. For a general overview of all labs implemented, see [4]. Since Virtual Labs consist of replications of an environment, they allow several users to access the same tools simultaneously. Depending on the software, the number of users is limited only by the number of licenses used. The simulation of digital processing usually comes close to the real behavior, even with simulated faults and statistical error appearances. One of the examples supported by ELLI is a virtual SPS Programming Environment with access to 3D sensors and actor visualization. It enables students to set up virtual production lines or solve problems in them. The standard SPS programming environment is used, combined with simulated equipment. The whole setup is visualized by additional software and

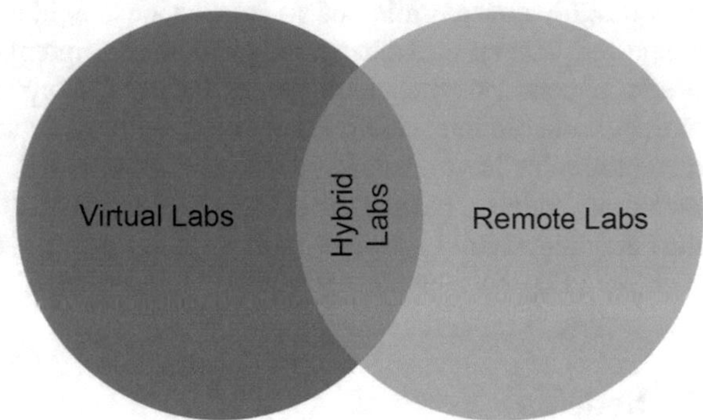

Fig. 1 Lab types supported by the project ELLI in Bochum

allows the building of different settings as well as the customizing of existing ones. Therefore, the content transfer in education is adaptable.

In contrast to Virtual Labs, Remote Labs allow discrete access to real plants for one student within a designated period of time. A remote plant acts and responses in the same way students experience it while actually operating a plant in a unit operations lab. The Remote Labs supported by ELLI cover a large range of fields. All of them are similar concerning required maintenance and safety devices, due to probable hardware failures. In addition, they have to be accessible without hazards to users. Some of them focus on characterizations of substances. As an example, one offers the opportunity to analyze density, conductivity, humidity and acoustic velocity of gases under different pressures. By using specific equipment, this experiment can be used to address the accuracy of properties measured and published. Other Remote Labs focus on devices and sensor functionality. For example, acoustic failure detection by a sound sensor is used to identify abrasion in a gearbox. Again, students can be confronted with measuring techniques and the interpretation of obtained data. Some of the labs supported by ELLI combine the advantages of both Virtual Labs and Remote Labs, since real plants are operated by results from replicated environments [5]. These Hybrid Labs enable the evaluation of simulated results, due to their comparison with measured ones. They even allow a detailed check of the parameters used for a simulation. As an example, students can develop car control units in a virtual environment. In order to check them under realistic circumstances, the virtual environment is implemented in a car simulator, where the interaction between driver and car unit can be tested. (The driving experience is not the main part of this simulator.) The user capacity of this lab is only limited by the time students need to check their results in the car simulator, the part of virtually developing the car control unit can be done by an unlimited number of students at the same time.

5 Management and Accessibility

Although Virtual Labs are available to a great number of students at the same time, depending on software licences, Remote Labs are limited in their accessibility, due to the real plant that is operated remotely. In order to avoid overlapping instructions, Remote Labs need to be operated centralized-controlled. It is necessary to enable exclusive access for one user at each plant any time. Therefore, a booking system is needed to provide specific time slots for everyone, depending on their group affiliation. There are two main groups to represent, students and teachers. While students should be able to gain experience, professors are to get the opportunity of demonstrating theoretical content embedded in applications. At this point, the management system iLab, developed at the Massachusetts Institute of Technology [6], is set up. The open source project enables the attachment of a large number of experiments to a central platform and allows customization if necessary. Its architecture is appropriate to integrate experiments that are based on the software LabView, developed by National Instruments. Therefore, this software was chosen to develop the user and controlling interface of the experiments in Remote or Hybrid Labs.

6 Conclusion and Outlook

At Ruhr University Bochum, Virtual Labs and Remote Labs were chosen as innovative tools to improve engineering education. An independent jury selected eight submissions from all engineering departments, so ELLI supported their setup. During organization, it was found that an early coordination of the involved departments provided enough time and clearance for all partners to communicate expectations and plan the proceedings for the upcoming collaboration. In order to support the usability and fit in the needs of engineering education from a student's point of view, all supported concepts already considered their input. Due to the great variety of parties involved, the implementation of these labs, especially the setup of Remote Labs, is a time-consuming process. It is highly recommended to enable exchanges on a regular basis during this phase, since this will make it much easier to intervene or adjust if necessary. The availability of Remote Labs and Hybrid Labs can be provided through a common management system, if all labs use the same software environment. Since most of the Virtual Labs use specific software, it will be a challenge to integrate all different labs into one management system. The Hybrid Lab deals with a combination of advantages and disadvantages from each lab type, but allows a critical comparison of results. Therefore, its parameter range is comparably high. The required amount of maintenance in order to provide the availability of each experiment is not defined yet, since the respective testing cycles are needed. It is expected that the first labs will be implemented in lectures and seminars in the summer term 2014. During the first cycle, an evaluation for each experiment will be done with the different user groups.

References

1. L. Johnson, S. Adams, and M. Cummins, "NMC Horizon Report: 2012 Higher Education Edition: Deutsche Ausgabe", New Media Consortium, Austin, Texas, USA, 2012
2. M. Grosch, and G. Gidion, "Mediennutzungsgewohnheiten im Wandel – Ergebnisse einer Befragung zur studiumsbezogenen Mediennutzung", Karlsruhe Institue of Technology, KIT Scientific Publishing. http://creativecommons.org/licenses/by-nc-nd/3.0/de/, 2011
3. U. Harms, "Virtual and Remote Labs in Physics education", University of Tuebingen, Tuebingen, Germany, 2000
4. Project ELLI, virtual environments in education. http://www.ruhr-uni-bochum.de/elli/virt.html, Ruhr University Bochum, Germany (11/18/2013)
5. S. Raivo, and S. Seiler, "Learning Situations and Remote Labs in Embedded System Education", Tallinn University of Technology, Tallinn, Estonia, 2013
6. Project iLab: https://wikis.mit.edu/confluence/display/ILAB2/Home, Massachusetts Institute of Technology, Massachusetts, USA (11/18/2013)

Shifting Virtual Reality Education to the Next Level – Experiencing Remote Laboratories Through Mixed Reality

Max Hoffmann, Tobias Meisen and Sabina Jeschke

Abstract Technical universities are more and more focusing on engineering education as a primary discipline. All along with the integration of various innovative fields of application into the curriculum of prospective engineers the need for appropriate educational features into the studies also increases. Unlike exclusively theoretical studies as physics, mathematics or information sciences the education of engineers extensively relies on the integration of practical use-cases into the education process. However, not every university is able to provide technical demonstrators or laboratories for all of the various applications in the field of engineering. Thus, it is the aim of the current paper to propose a method that enables visiting a high variety of engineering laboratories based on Virtual Reality. A Virtual Reality simulator is used to create and emulate remote laboratories that can be located at arbitrary places far away from their Virtual Reality representation. This way, by melting real world demonstrators with virtual environments, we enable a physically and technically accurate simulation of various engineering applications. The proof of concept is performed by the implementation and testing of a laboratory experiment that consists of two six-axis robots performing collaborative tasks.

Keywords Virtual Reality · Mixed Reality · Augmented Virtuality · Virtual Learning Environments · Remote Laboratories · Engineering Education

1 Introduction

In a world that is increasingly based on scientific innovations and technological progress the education of engineering students constantly gains in importance. Furthermore, the field of studies and the variety of possible specializations in engineering classes also increase significantly. In order to qualify graduated engineers to enter

M. Hoffmann (✉) · T. Meisen · S. Jeschke
IMA/ZLW & IfU, RWTH Aachen University, Dennewartstr. 27,
52068 Aachen, Germany
e-mail: max.hoffmann@ima-zlw-ifu.rwth-aachen.de

Originally published in "The International Conference on Computer Science, Computer Engineering, and Education Technologies (CSCEET2014), The Third World Congress on Computing and Information Technology (WCIT)", © APU 2014. Reprint by Springer International Publishing AG 2016, DOI 10.1007/978-3-319-46916-4_19

the working world successfully, the imparting of practical knowledge during studies plays a decisive role in education.

While theoretical knowledge is still transferred using written texts and the spoken word – normally through lecture notes and traditional readings – the experience of practical use-cases in engineering education commonly relies on the visit of laboratory experiments during the studies. However, since computer vision and digitalization techniques have grown to an extensive level, the integration of new media into the curriculum replacing the visit of real laboratories gains in importance [1]. Especially in terms of engineering applications, the use of high quality and realistic visualization techniques as a supplement to the attendance within practical laboratory experiments is of major importance for successfully impart basic concepts of engineering applications to students. Not at least due of the progress in computer science and graphical visualization techniques, the capabilities of visualizing objects of interest embedded into an artificially designed context have grown to an exhaustive amount. In this context, for example physical effects or technical subtleties of engineering applications can be presented in higher detail or in an amplified way in order to emphasize aspects that are not easily observable in reality. These novel potentials can be utilized to explain theoretical knowledge more concrete and tangible and help engineering students to understand the concepts of complex technical applications on different levels and from a practical point of view.

Another major trend that is emerging within the field of education and learning relies on the way of distributing information and knowledge through internet-based media among the students. The high significance of online platforms and social media during the everyday life of a student can be exploited in order to increase communication channels by the establishment of E-Learning-Platforms [2]. The technical possibilities of sharing and representing educational contents, spreading knowledge over the world-wide web and enabling the remote participation of students in engineering classes, open up new opportunities of teaching and learning within universities. In combination with modern visualization techniques these computer-based teaching concepts can for example be implemented through Virtual Reality or Mixed Reality applications [3].

Despite all these technological possibilities, universities are facing more and more obstacles to deal with the high variety of engineering courses and their different technical applications and needs. Thus, each university can only provide a limited amount of experimental or laboratory classes for the different fields of engineering during their curricula. However, the demand for enlarging the variety of engineering education contents is also growing constantly. This leads to a conflict of interests as universities are not supposed to advantage any particular engineering class in comparison to the others. Due to the limited laboratory and teaching capacities, the ability to satisfy the needs for experimental education of engineering students cannot be fulfilled adequately. Furthermore, most of the offered experimental simulations or virtual representations of laboratory environments are lacking the required quality standards in terms of graphical accuracy and interaction capabilities.

The goal of the current paper is to find ways for dealing with these obstacles and proposing possibilities to enable an extensive practical education of engineering classes by answering the following questions that are correlated to the described issues:

1. How can universities address the high variety of engineering related disciplines by ensuring the availability of suitable use-cases, experiments and laboratory exercises for students?
2. How is it possible to integrate concepts of Experiential Learning into engineering education against the background of limited laboratory resources?
3. What is the benefit of integrating Virtual Reality and Mixed Reality applications into engineering education from the technical/physical point-of-view?
4. Is it possible to address educational needs of engineering students in universities by enabling an active manipulation of the virtual laboratory environments?

The present publication intends to answer these questions by introducing a novel method for realizing use-cases for Virtual Reality exercises and laboratory experiments in terms of Remote Laboratories. In this context, the term Remote Laboratory introduces the idea of enabling the visit of distant places for conducting laboratory experiments based on its virtual representation. Hence, it is our aim to enable students to attend specific experimental environments suitable to their field of studies, even if these exercises are not provided at their particular university.

In the next section, the state of the art in teaching with the aid of new media and novel visualization applications is presented. Furthermore, educational advantages of Virtual Reality and Mixed Reality applications are carried out. In Section 3 the technical implementation of our next-generation Virtual Reality simulator is described together with former virtual scenarios that have been implemented using the simulator. Section 4 presents a scenario for enhancing students' learning behavior by the creation of a Remote Laboratory in connection with Mixed Reality approaches. In Section 5 first attempts in evaluating the advantages of such virtual representation of a laboratory experiment are carried out, assessing the learning capabilities of a group of students visiting such Remote Laboratory. Section 6 concludes the outcome of the current paper and specifies the next steps of enhancing the user's experience and creating larger Remote Laboratory environments.

2 State of the Art

New media have gained high significance in university studies and publications in the past decade. These new media – which are mostly based on computer visualization techniques – are continuously replacing traditional books and lecture notes. White boards and projectors are replaced by presentation software, which represents the new standard for visualizing text and pictures [4], with PowerPoint as market leader [5]. This switch from traditional lecture speech to graphical representation has been performed, because this form of presentation enables focusing on the main points of the educational content using illustrative representations and pictorial summaries [6]. Despite the positive, but also critical discussion about an overwhelming usage

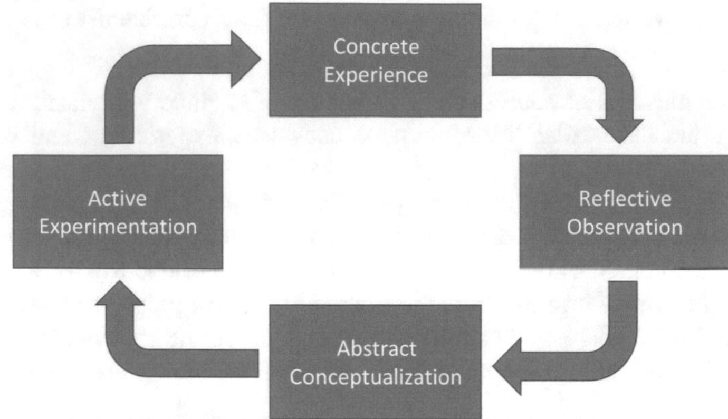

Fig. 1 The Experiential Learning Cycle according to David A. Kolb (1984)

of PowerPoint [7–9] as primary teaching tool, the usage of presentation software in the classroom has grown constantly [10, 11]. In connection with the entry of technological novelties into the classroom it is time to take the next steps from merely presenting pictures using presentation software to the usage of advanced graphical interfaces opening up interaction capabilities for the students involved in the engineering courses.

Presentation software like PowerPoint may be a far reaching advancement for most courses of studies. However, the usage of these meanwhile basic IT tools is also limited to a certain kind of knowledge transfer. Especially for practically oriented study paths like engineering classes active interaction capabilities within courses and exercises are inevitable. In these highly technical studies there is an urgent need for interactive laboratory experiments in order to impart practical and skill-based knowledge tangible to students. Against this background, David A. Kolb's traditional, well-established cycle on Experiential Learning is more up to date than ever [12]. The – almost classical – Learning Cycle is depicted in Fig. 1.

In the picture, we see a never ending process of active experimentation, concrete experience, reflective observation and abstract conceptualization. Starting at active experimentation, there is a need for concrete experience in order to understand abstract concepts. The reflective observation following to the experience helps carrying out an abstract conceptualization that is based on a deeper understanding of the experienced content. Especially the practical part of the learning process in terms of the attendance to experimental courses cannot be replaced by any kind of theoretical knowledge transfer. Other learning related theories, which address the same matters as the experiential learning approach, are action learning, adventure learning, free choice learning, cooperative learning and service learning approaches [13]. In all of these theories, active interaction of the learning person plays an integral part in the learning process.

Due to the high complexity and advanced technological level of the relevant applications in engineering classes, sole static visualizations are not capable to serve as a medium for active experimentation or even concrete experience. In order to address these parts of the learning cycle, novel visualization concepts have to be applied. A key enabler that combines advanced visualization techniques with the experience of a certain scenario is Virtual Reality as shown in the relevant literature [14] and during prior studies in this field of application [15]. In the first step, Virtual Reality cannot serve direct interaction possibilities, but through the use of immersive effects, a Virtual Reality scenario is able to arouse an experienced reality within the perception of the user. This effect can be greatly characterized by the definition of immersion according to Murray [16]: "Immersion is a metaphorical term derived from the physical experience of being submerged in water. We seek the same feeling from a psychologically immersive experience that we do from a plunge in the ocean or swimming pool: the sensation of being surrounded by a completely other reality, as different as water is from air that takes over all of our attention, our whole perceptual apparatus."

In order to address the part of concrete experience during practical education accurately by virtual reality applications, the utilized tools have to fulfill a number of conditions to serve as a suitable complement for laboratory classes. In the following, we will therefore look at existing Virtual Reality applications and discuss if these are capable of bridging the gap of concrete experience to the students.

In terms of existing Virtual Reality technologies, there are already many technical solutions that are primarily focused on the creation of high-quality and complex three-dimensional environments, which are accurate to real-world scenarios in every detail. One example are flight simulators that are capable of tracking the locomotion of a flying vehicle in a virtual scenario [17]. However, these systems are usually not taking into account the position or the head movements of the user. Another Virtual Reality simulator is the well-known Omnimax Theatre, which provides a large angle of view [18], but does not allow any tracking capabilities whatsoever. First attempts to interact with Virtual Reality in a natural way were introduced by head-tracking monitors as conducted by Codella [19] and Deering [20]. These specially designed monitors provide an overall tracking system, but are characterized by a rather limited angle of view [17]. The first mentionable approach to create a Virtual Reality environment with full tracking capabilities of movements and of the head position of the user was introduced by McDowall [21] with the Boom Mounted Display. Despite advanced tracking capabilities these early attempts were characterized by poor resolutions and thus were not capable of a detailed graphical representation of a virtual environment [22].

Thus, in order to enable true user experience in simulated scenarios, Mixed Reality approaches have to be embedded into the Virtual Reality, were reality and virtuality are merged into each other [23]. One far reaching innovation in terms of enabling Virtual Reality and Mixed Reality applications was introduced with the CAVE in 1992 by Cruz-Neira et al. [24]. Hereby, the recursive acronym CAVE stands for Cave Automatic Virtual Environment. By making use of complex visualization techniques combined with various projectors and six projection walls arranged in form of a cube, the developers of the CAVE have redefined the standards in visualizing Virtual Reality scenarios by enabling a new level of immersion.

The CAVE reaches further towards true Virtual Reality which – according to Rheingold [25] – is described as an experience, in which a person is "surrounded by a three-dimensional computer-generated representation, and is able to move around in the virtual world and see it from different angles, to reach into it, grab it and reshape it." These active manipulation activities that can be performed by the user open up various new applications in education by rebuilding industrial use-cases. Thus, by enabling extended interaction capabilities with the scenarios in terms of providing relatively free manipulation of the virtual environment, the immersive effects of the scenarios are enhanced. This effects could lead to the desired impact on the learning behavior of students that consists of the ability to derive abstract conceptualizations on the basis of concrete or practical experience and active experimentation.

However, even the CAVE has got restricted interaction capabilities as the user can only interact in the currently demonstrated perspective of the scenario. Furthermore, natural movement is limited, as locomotion through different scenes of the scenario is usually performed by flying to the next spot. Yet, natural movement as walking, running or jumping through the Virtual Reality is decisive for a highly immersive experience in the virtual environment.

In order to fill this gap of limited interaction and accordingly deeper immersion into the scenario, additional devices and tracking systems for allowing such interaction have to be included into the scenario without losing the high quality of graphical representation of the virtual environment. One promising approach relies in the establishment of the Virtual Theatre that brings together a full-size stereo-vision view and various interaction devices and manipulation capabilities for the user.

3 The Virtual Theatre – Enabler for Extended Immersion

The Virtual Theatre represents a next-level Virtual Reality simulator that allows free locomotion in a virtual environment and active manipulation as well as the control of objects in the visualized scenario. The Virtual Theatre was carried out by a Swedish company [26] and was already described in detail in our previous publications [27] according to the scenarios that have been carried out during our latest research [28, 29]. The centerpiece of the Virtual Theatre is the omnidirectional treadmill, a moving floor that accelerates its centric arranged rollers according to the position of the user (Fig. 2).

The user himself is tracked by an infrared tracking system; hence his head and hand movements are constantly observed and taken into account for a adaptation and manipulation of the virtual environment. The user is wears a Head Mounted Display (HMD) that is equipped with two screens – one for each eye – and enables highly immersive three-dimensional stereo vision for the exploration of the virtual space. The unique characteristic of a HMD is that this kind of devices are capable of measuring the user's head orientation through a perpendicular, which makes it possible to adjust the Virtual Reality according to the head's actual position and orientation. This enriches the immersive experience of the user as he is able to look

Fig. 2 A user experiencing virtual environments in the Virtual Theatre

around and explore the Virtual Reality in a similar way as he does in the real world. An embedded sound system into the HMD completes the plunge into virtuality. For further information about the technical concept of the Virtual Theatre the reader is encouraged to refer to [28], where the hardware and technical setup is explained in detail.

Former scenarios that have been carried out using the Virtual Theatre were well received by the students of engineering classes [29]. One example for the creation of a huge sized virtual environment is our Mars project, where an extensive simulation of a plateau on the surface of the red planet was recreated to enable upcoming astronautics and aerospace students to perform a virtual visit and exploration of the Mars [15]. Another application of the Virtual Theatre was carried out in terms of a study in order to assess the learning behavior and learning efficiency of students while being surrounded in a virtual environment [29]. During the survey, the students were located in a virtual labyrinth, in which they needed to find objects and recognize their location and shape at a later point. Afterwards, the results of their learning efficiency were evaluated and compared to the efficiency using traditional computer screens for performing similar tasks.

However, former studies did not take into account deeper interaction capabilities as an active movement of objects or the remote control of devices for industrial use-cases. However, as mentioned earlier, exactly these interaction capabilities are strongly needed in order to create realistic virtual representation of experimentations, thus to enable Remote Laboratories. Hence, in the next step we present the inclusion extensive interaction capabilities into the use-case and we attempt to enable true immersion based on Mixed Reality concepts.

4 Enabling Remote Laboratories Through Mixed Reality

As part of a network of administrative computers the Virtual Theatre – including its integrated parts and the tracking system – can be expanded by additional hardware in order to enable a more natural user interaction within the virtual scenario.

Taking into account natural user behavior as well as the experience of students using well-established computer game devices, our team decided to carry out a remote control for virtual scenarios based on hardware of the ©Nintendo Wii™ Controller. The new conceptual design of the communication infrastructure for the Virtual Theatre and its surrounding hardware equipment is depicted in Fig. 3.

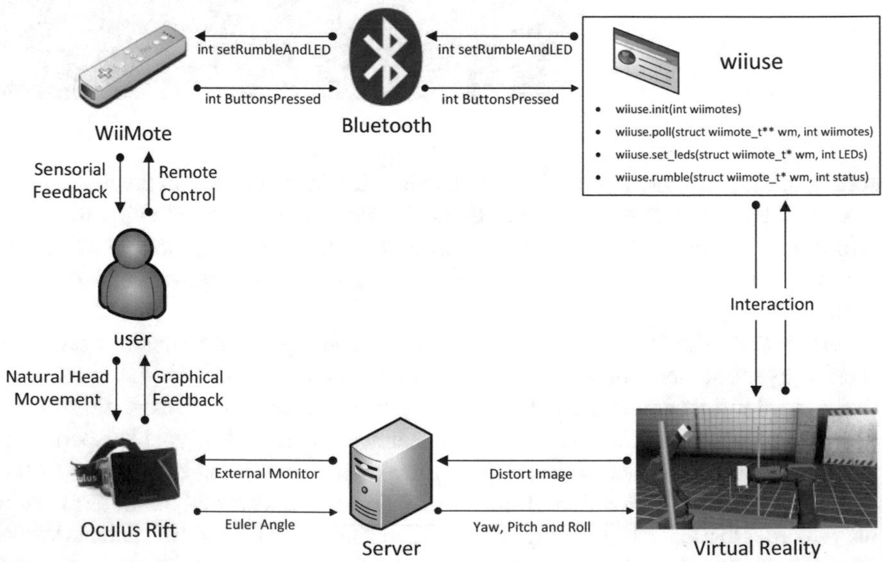

Fig. 3 Communication Infrastructure of the Virtual Theatre with extended interaction capabilities

Fig. 4 Visualization of two six-axis ABB™ robots performing collaborative tasks

In the middle part of the picture's bottom, the central server is visualized. The server deals with the signals of the Head Mounted Display, which is located on the left side, and processes its information for the user's movement, head position and orientation according to the virtual environment that is depicted on the right side at the bottom. The Wii™ remote controller is connected via Bluetooth and sends specific commands to the central server. The server processes these commands to manipulate the virtual environment and visualize the modified scenario in real-time.

A suitable application for including the described interaction device into a virtual scenario is based on an extended laboratory experiment that was virtualized by our team for education purposes. The setup consists of two six-axis robots that are placed on a table in order to perform collaborative tasks. The virtualized model is depicted in Fig. 4.

In reality, the robots are located in the same distance as illustrated in the figure, enabling the ability to perform collaborative, interdepending tasks. As described in [28], the first attempts in carrying out a Remote Laboratory based on these robots consisted in the virtualization of the actual robot movements as well as a real-time alignment of the movements that were performed by the real world robots and the movements of our simulation. As shown in our previous publication, the full setup can be appropriately simulated in real time, i.e. the user inside of the virtual reality simulator can pursue the robot motion without any perceptible time lag. This real-time synchronization between the real world laboratory and its virtual representation enables active remote control of the robot setup as well as various use-cases:

- The actual position of the robots can be tracked and remotely manipulated from arbitrary locations.
- The control of the robot arms can be extended by additional security layers in order to assure save motion of the robots and to avoid collisions.
- The experimenter can easily work with dangerous materials or substances (e. g. chemicals) and is able to operate the robots if these are located at dangerous or non-accessible places.

In order to generate an added value to the remote control of a laboratory environment our research did not only concentrate on the development of remote control devices. We also focused on providing additional features that are enabled through Virtual Reality. One major progress of these efforts relies on the integration of Mixed Reality elements into the laboratory experimental context. In this connection, the term Mixed Reality is characterized as the merging of real and virtual worlds to produce new environments and visualizations, where physical and digital objects co-exist and interact in real-time [30]. Our use-cases including Mixed Reality approaches is able to address several aspects of the application:

- Systematic simulation of experiments with actual machines and components in real time. The exact simulation leads to a co-existence of real world objects in the laboratory and in Virtual Reality with interdepending system states.
- Feedback and manipulation capabilities of the user, which leads to an interaction with both, the objects in the virtual environment and the real components.
- Embedding of real world features into virtual environments by placing cameras into the laboratory environment. This enables the projection of detailed views of experimental insights that are captured by the camera onto a wall in the virtual environment. In terms of the Mixed Reality concept, this effect is also referred to as *Augmented Virtuality*.

Especially the last of the mentioned points of integrating Mixed Reality concept into our Virtual Reality scenarios bears a high potential to enhance the grade of immersion signifycantly. In terms of the user's perception the embedding of Augmented Virtuality – i. e. real pictures into the virtual scenario – is connected to effects of fuzziness between reality and virtuality, which leads to highly immersive impressions for the user. One possible application of integrating a camera in the real world scenario can for example consist in a placement of the camera in a bird's-eye perspective on top of the experimental setup to see the whole scene from a broader point-of-view.

Another possible scenario is to attach the camera onto a robot that is actually moving and thus to observe the scene from the robot point of view. An example for this extended perspective based on this Augmented Virtuality visualization technique is depicted in Fig. 5.

On the bottom of the picture the simulation of the two six-axis robots is shown. The screen that is located above the table shows a video image that is taken from the perspective of the left robot in reality. This image is embedded into the Virtual Reality scenario as a sort video projection and hence represents an Augmented Virtuality element in the simulation.

This enrichment of the virtual scenario is connected to several improvements concerning the technical and the educational application of the scenario:

- The point of view perspective enables a detailed view of the simulated scene. Through the robot's perspective the tasks that have to be performed by the robot arms can be conducted with higher precision due to the overview based on more than one perspective.
- The grade of immersion increases as the user of the simulation is able to see reality objects that are melting with the Virtual Reality scenario in real-time.

Fig. 5 The experimental setup in reality visualized in a Virtual Reality scenario

- For education purposes, multiple cameras can be attached at various locations of the demonstrator, which helps to explain the physical or technical effects of use-cases.

Besides the advantages for the single user that are connected to the embedding of Mixed Reality into the scenario there is also an added value for the remote control of a laboratory experiment from more than one user. Due to the placement of multiple cameras within the surroundings of the experiment, different users can perform collaborative tasks while observing a simulation of the actual system state in real-time, but from different perspectives. This enables a highly precise manipulation of the experimental conditions influenced by different users that can be located at arbitrary places. Especially due to this point, the far reaching benefits of Remote Laboratories as a new class of conducting experiments becomes known.

5 Evaluation of the Technical Implementation and Impact on the Students Learning Behavior

During the previous sections of this paper the added value of Remote Laboratories has been derived in terms of the overall usability and availability of laboratory experiments as well as the impact on the precise conduction of experiments using multiple information channels. In this section we would like to investigate the impact of the utilization of Remote Laboratories not only in terms of the availability of experimental setups in engineering classes, but also its effects on the learning behavior and motivation of students. In terms of the assessment according to this impact, we get back to initial research questions about enhancing the learning environment of students by remote setups or literally: "Is it possible to address the educational needs of engineering students in universities by enabling an active manipulation of the virtual laboratory environments?"

In order to answer this question, different facets of the educational needs of engineering students are taken into account:

- The impact of virtual experience and the grade of sensed immersion into the student's perception in virtual environments to build up a realistic scenario.
- The physical accuracy of simulations embedded into the virtual environment.
- The added value of enhancing the perspective of the user and the personalized view by emphasizing certain physical effects through amplified physical behavior or close-up views.

The first evaluation phase of these effects was performed in-house and based on the personnel and on the student employees of our institution. The visualization accuracy as well as the immersive effects of the simulation into the perception of the user could be verified during the testing phase. Especially the active motion of the six-axis robots through an easily manageable interface while being submerged in the virtual environment had clearly observable impacts on the understanding of robot motion and the need for automation.

In the next steps of the evaluation phase, the investigation of the scenario will be performed by a questionnaire that is carried out for laboratory classes of newcomer students. This evaluation, which will be further concentrating on didactical aspects of the experimentation environment, will take place in the following semester, in which the according students will assess their personal learning success after conducting several experiments with and without the help of the described Mixed Reality-related techniques. During this phase we will examine the effects of the virtual environment on the learning behavior of the engineering students by taken into account the following points:

- The impact of virtual experience and the grade of sensed immersion on the willingness and learning behavior of students in virtual environments.
- Correlations between the learning ability during laboratory experiments, gamification effects and fun in manipulating the laboratory environment in virtuality.
- The effect of hands-on experiments on the learning success of students in comparison to the mere observation of distant experiments that are not accessible in its real environment.

The study will show, if the different aspects like fun in learning, active involvement and free movement in virtual environments as well as the ability to manipulate a virtual representation of a real world demonstrator or – in other words – the reflective observation have a significant impact on the abstract conceptualization of complex engineering applications.

6 Conclusion and Outlook

The need for higher capacities in terms of the practical education of engineering students comprises a major challenge for today's universities. The constantly growing number of students with various educational backgrounds and different experiences as well as the wealth of study opportunities demands for innovative concepts in the organization of a profound engineering education.

In this paper, we have substantiated the idea of conducting real laboratory experiments through a Virtual Reality simulator by enabling Remote Laboratories. These laboratories can serve as an extensive supplement to real experimental setups, because they can be built up at arbitrary places and run simultaneously for multiple users. Various setups for virtual environments can be applied in order to emphasize immersive effects on the user with an expected impact on his learning behavior.

The next steps in connection with the presented scenarios will consist in a quantitative evaluation of the impact of Virtual Reality on the actual learning success of the students by assessing the conceptual knowledge of two different comparison groups, one that visits an actual laboratory experiment without any personal involvement or interaction with actual components, and the other group that visits a Virtual Reality based virtual environment of the laboratory experiment. Furthermore, we will discuss the effect of embedding Mixed Reality components into the Remote Laboratory on the students on the one hand in terms of their qualitative perception of being immersed into the virtual environment and on the other hand in terms of the advantages that are connected to the embedding of camera screens into the virtual scenario for additional perspectives.

On the technical side, the next steps concerning an extension of the Remote Laboratory environment consist in the development of a generic methodology to automate and to control robots of various kinds in virtual environments. In terms of this procedure, aspects of robot security, collision avoidance and inverse kinematics for robot control will be of major importance for an expedient experimentation environment. Next projects will concentrate on the implementation of complex scenarios with multiple robots and interaction devices in order to emphasize the idea of collaborative and concurrent engineering in virtual environments.

References

1. M. Ebner, A. Holzinger, Successful implementation of user-centered game based learning in higher education: An example from civil engineering. Computers & Education **49** (3), 2007, pp. 873–890. URL http://www.sciencedirect.com/science/article/pii/S0360131505001910
2. M.J. Rosenberg, *E-learning: Strategies for delivering knowledge in the digital age*. McGraw-Hill, New York, 2001
3. Z. Pan, A.D. Cheok, H. Yang, J. Zhu, J. Shi, Virtual reality and mixed reality for virtual learning environments. Computers & Graphics **30** (1), 2006, pp. 20–28. URL http://www.sciencedirect.com/science/article/pii/S0097849305002025

4. A. Szabo, N. Hastings, Using it in the undergraduate classroom: Should we replace the blackboard with powerpoint? Computer and Education **35**, 2000
5. R.J. Craig, J.H. Amernic, Powerpoint presentation technology and the dynamics of teaching. Innovative Higher Education **31** (3), 2006, pp. 147–160
6. R.A. Bartsch, K.M. Cobern, Effectiveness of powerpoint presentation in lectures. Computer and Education **41**, 2003, pp. 77–86
7. T. Creed, Powerpoint, no! cyberspace, yes. The Nat. Teach. & Learn. F. **6** (4), 1997
8. D. Cyphert, The problems of powerpoint: Visual aid or visual rhetoric? Business Communication Quarterly **67**, 2004, pp. 80–83
9. P. Norvig, Powerpoint: Shot with its own bullets. The Lancet **362**, 2003, pp. 343–344
10. T. Simons, Does powerpoint make you stupid? Presentations **18** (3), 2005. URL http://global.factiva.com/
11. A.M. Jones, The use and abuse of powerpoint in teaching and learning in the life sciences: A personal view. BEE-j 2, 2003. URL http://www.bioscience.heacademy.ac.uk/journal/vol2/beej-2-3.pdf
12. D.A. Kolb, *Experiential learning: Experience as the source of learning and development*. Financial Times Pren Hall, [S.l.], 2014
13. C.M. Itin, Reasserting the philosophy of experiential education as a vehicle for change in the 21st century. Journal of Experiential Education **22** (2), 1999, pp. 91–98
14. D. Johansson, de Vin, L. J., Towards convergence in a virtual environment: Omnidirectional movement, physical feedback, social interaction and vision. Mechatronic Systems Journal (November 2011), 2011
15. M. Hoffmann, K. Schuster, D. Schilberg, S. Jeschke, Next-generation teaching and learning using the virtual theatre. In: *Proceedings of the 4th Global Conference on Experiential Learning in Virtual Worlds*. Prague, Czech Republic, 2014
16. J.H. Murray, *Hamlet on the Holodeck: The Future of Narrative in Cyberspace*. MIT Press, Cambridge (Mass.), 1997
17. C. Cruz-Neira, D.J. Sandin, T.A. DeFanti, Surround-screen projection-based virtual reality. the design and implementation of the cave. SIGGRAPH '93 Proceedings of the 20th annual conference on Computer graphics and interactive techniques. ACM - New York, 1993, pp. 135–142
18. N. Max, Siggraph '84 call for omnimax films. Computer Graphics **16** (4), 1982, pp. 208–214
19. C. Codella, R. Jalili, L. Koved, B. Lewis, D.T. Ling, J.S. Lipscomb, D. Rabenhorst, C.P. Wang, A. Norton, P. Sweeny, G. Turk, Interactive simulation in a multi-person virtual world. ACM - Human Fact. in Comp. Syst. **CHI 1992 Conf.**, 1992, pp. 329–334
20. M. Deering, High resolution virtual reality. Com. Graph. **26** (2), 1992, pp. 195–201
21. I.E. McDowall, M. Bolas, S. Pieper, S.S. Fisher, J. Humphries, Implementation and integration of a counterbalanced crt-based stereoscopic display for interactive viewpoint control in virtual environment applications. Proc. SPIE **1256** (16), 1990
22. S.R. Ellis, What are virtual environments? IEEE Computer Graphics and Applications **14** (1), 1994, pp. 17–22
23. P. Milgram, A.F. Kishino, Taxonomy of mixed reality visual displays. IEICE Transactions on Information and Systems, 2013, pp. 1321–1329
24. C. Cruz-Neira, D.J. Sandin, T.A. DeFanti, R.V. Kenyon, J.C. Hart, The cave: Audio visual experience automatic virtual environment. Communications of the ACM **35** (6), 1992, pp. 64–72
25. H. Rheingold, *Virtual reality*. Summit Books, New York, 1991
26. MSEAB Weibull. http://www.mseab.se/the-virtual-theatre.htm, 2012
27. D. Ewert, K. Schuster, D. Johansson, D. Schilberg, S. Jeschke, Intensifying learner's experience by incorporating the virtual theatre into engineering education. Proceedings of the 2013 IEEE Global Engineering Education Conference (EDUCON), 2013
28. M. Hoffmann, K. Schuster, D. Schilberg, S. Jeschke, Bridging the gap between students and laboratory experiments. In: *Virtual, Augmented and Mixed Reality*, *Lecture notes in computer science*, vol. Heraklion, Crete, Greece, ed. by R. Shumaker, Springer, Cham, 2014, pp. 39–50

29. K. Schuster, M. Hoffmann, U. Bach, A. Richert, S. Jeschke, Diving in? how users experience virtual environments using the virtual theatre. In: *Virtual, Augmented and Mixed Reality*, *Lecture notes in computer science*, vol. Heraklion, Crete, Greece, ed. by R. Shumaker, Springer, Cham, 2014, pp. 636–646
30. Silva, Adriana de Souza e, D.M. Sutko, *Digital cityscapes: Merging digital and urban playspaces*, *Digital formations*, vol. v. 57. Peter Lang, New York, 2009

Bridging the Gap Between Students and Laboratory Experiments

Max Hoffmann, Katharina Schuster, Daniel Schilberg and Sabina Jeschke

Abstract After having finished studies, graduates need to apply their knowledge to a new environment. In order to professionally prepare students for new situations, virtual reality (VR) simulators can be utilized. During our research, such a simulator is applied in order to enable the visit of remote laboratories, which are designed through advanced computer graphics in order to create simulated representations of real world environments. That way, it is our aim to facilitate the access to practical engineering laboratories. Our goal is to enable a secure visit of elusive or dangerous places for students of technical studies. The first step towards the virtualization of engineering environments, e.g. a nuclear power plant, consists in the development of demonstrators. In the present paper, we describe the elaboration of an industry relevant demonstrator for the advanced teaching of engineering students. Within our approach, we use a virtual reality simulator that is called the "Virtual Theatre".

Keywords Virtual Reality · Virtual Theatre · Remote Laboratories · Immersion

1 Introduction

In terms of modern teaching methods within engineering classes, various different approaches can be utilized to impart knowledge to students. There are traditional teaching techniques, which are still suitable for most of the knowledge transfer. These methods are carried out by the use of written texts or the spoken word. However, due to the increasing number of study paths as well as the specialization of particularly technical oriented classes, there is a need for the integration of new media into the curriculum of most students [1]. Thus, the visualization of educational content in order to explain theory more concrete and tangible has gained in importance. Not least because of the progress in computer science and graphical visualization, the capabilities of visualizing objects of interest within an artificially designed context

M. Hoffmann (✉) · K. Schuster · D. Schilberg · S. Jeschke
IMA/ZLW & IfU, RWTH Aachen University, Dennewartstr. 27,
52068 Aachen, Germany
e-mail: max.hoffmann@ima-zlw-ifu.rwth-aachen.de

Originally published in "16th International Conference on Human-Computer Interaction 2014, HCI 2014", © 2014. Reprint by Springer International Publishing AG 2016, DOI 10.1007/978-3-319-46916-4_20

have grown to an exhaustive amount. However, not only the visualization techniques have emerged, the way of distributing knowledge through teaching media has also grown. One major improvement in reaching students independently to their location are E-Learning Platforms [2]. These technical possibilities of sharing and representing contents open up new opportunities in teaching and learning for students.

Thus, in nearly all courses of studies, new media have gained a high significance in the past decade. These new media are continuously replacing conventional media or in other words traditional, static teaching approaches using books and lecture notes. The new media are mostly based on methods of digital visualization [3], e.g. presentation applications like PowerPoint [4]. This switch from the traditional lecture speech to graphical representations have been performed, because this form of presentation enables focusing on the main points of educational content using illustrative representations and pictorial summaries [5]. Despite the positive [6], but also critical discussion about an overwhelming usage of PowerPoint [7–9] as primary teaching tool [10], the usage of presentation software in the classroom has grown constantly [11].

Applications like PowerPoint may be a far reaching advancement for most courses within university. However, even these IT-based teaching supports are limited to a certain kind of knowledge transfer. Especially practically oriented study paths like engineering courses have an urgent need for interaction possibilities. In these highly technical focused studies, the teaching personnel are facing more and more obstacles in imparting their knowledge tangible. Due to the advanced and complex technology level of the relevant applications [12], progressive methods have to be applied to fulfill the desired teaching goals. In order to make the problem based learning methodologies available [13], novel visualization techniques have to be carried out.

Studies of astronautics or nuclear research can serve as an incisive example for the need of innovative visualization capabilities. During astronomy studies, the teaching personnel will face insurmountable obstacles, if they want to impart practical knowledge about aerospace travelling to the students using theoretical approaches. In order to gain deep, experienced knowledge about real situations an astronaut has to face, realistic scenarios have to be carried out. This can for instance be performed by setting up expensive real-world demonstrators that facilitate practical experiences within aerospace travelling events, e.g. by making use of actual acceleration.

However, there is also a need for a visual representation of the situation. In order to fulfill the requirements of a holistic experience, these visualization techniques need to perform an immersive representation of the virtual world scenario. In this connection, the term immersion is defined according to Murray [14] as follow: "Immersion is a metaphorical term derived from the physical experience of being submerged in water. We seek the same feeling from a psychologically immersive experience that we do from a plunge in the ocean or swimming pool: the sensation of being surrounded by a completely other reality, as different as water is from air that takes over all of our attention, our whole perceptual apparatus."

It is obvious that experience can only be impressive enough to impart experienced knowledge, if the simulation of a virtual situation has an immersive effect on the perception of the user. Our latest research on creating virtual world scenarios has shown that immersion has got a high impact on the learning behavior of students [15]. Following the idea of facilitating the study circumstances for students of astronautics, our first demonstrator was carried out in terms of a Mars scenario [16]. Using novel visualization techniques in connection with realistic physics engines, we have carried out a realistic representation of a plateau located on the red planet.

In our next research phase, we want to go further to increase the interaction capabilities with the virtual environment the user is experiencing. In terms of the Mars representation, there were already few interaction possibilities like triggering of object movements or the navigation of vehicles [16]. However, this sort of interaction is based on rather artificial commands than on natural movements with realistic consequences in the representation of the virtual world scenario.

Hence, in the present paper, we want to introduce a more grounded scenario, which is based on the aforementioned idea of enabling the visit of elusive or dangerous places like an atomic plant. Accordingly, our first step in realizing an overall scenario of a detailed environment like a power plant consists in the development of single laboratory environments. In this context, our aim is to focus especially on the interaction capabilities within this demonstrator.

This target is pursued by carrying out a virtual prototype of an actual laboratory environment, which can be accessed virtually and in real-time by a user in a virtual reality simulator. The realization of these demonstrators is also known as the creation of "remote laboratories". In the present paper, we describe the development, optimization and testing of such a remote laboratory. After a brief introduction into the state-of-the-art of this comparatively new research field in Section 2, our special Virtual Reality simulator, which is used to simulate virtual environments in an immersive way, is described in Section 3. In Section 4, the technical design of the remote laboratory including its information and communication infrastructure is presented. In the Conclusion and Outlook, the next steps in realizing the overall goal of a virtual representation of an engineering environment like an atomic plant are pointed out.

2 State of the Art

In the introduction, we concluded that innovative teaching methodologies have to be adopted to be capable of imparting experienced knowledge to students. Thus, virtual reality teaching and learning approaches will be examined in the following.

Nowadays, an exhaustive number of applications can be found that make use of immersive elements within real-world scenarios. However, the immersive character of all these applications is based on two characteristics of the simulation: The first one is the quality of the three-dimensional representation; the second one is the user's identification with the avatar within the virtual world scenario.

The modeling quality of the three-dimensional representation of a virtual scenario is very important in order to be surrounded by a virtual reality that is realistic or even immersive. However, a high-quality graphical representation of the simulation is not sufficient for an intensive experience. Thus, according to Wolf and Perron [17], the following conditions have to be fulfilled in order to enable an immersive user experience within the scenario: "Three conditions create a sense of immersion in a virtual reality or 3-D computer game: The user's expectation of the game or environment must match the environment's conventions fairly closely. The user's actions must have a non-trivial impact on the environment. The conventions of the world must be consistent, even if they don't match those of the 'metaspace'."

The user's identification with virtual scenario is rather independent from the modeling of the environment. It is also depending on the user's empathy with the "avatar". Generally, an avatar is supposed to represent the user in a game or a virtual scenario. However, to fulfill its purposes according to the user's empathy, the avatar has to supply further characteristics. Accordingly, Bartle defines an avatar as follows: "An avatar is a player's representative in a world. […] It does as it's told, it reports what happens to it, and it acts as a general conduit for the player and the world to interact. It may or may not have some graphical representation, it may or may not have a name. It refers to itself as a separate entity and communicates with the player."

There are already many technical solutions that are primarily focused on the creation of high-quality and complex three-dimensional environments, which are accurate to real-world scenarios in every detail. Flight Simulators, for example, provide vehicle tracking [18]. Thus, the flight virtual reality simulator is capable of tracking the locomotion of a flying vehicle within the virtual world, but does not take into account the head position of the user. Another VR simulator is the Omnimax Theater, which provides a large angle of view [19], but does not enable any tracking capabilities whatsoever. Head-tracked monitors were introduced by Codella et al. [20] and by Deering [21]. These special monitors provide an overall tracking system, but provide a rather limited angle of view [18]. The first attempt to create virtual reality in terms of a complete adjustment of the simulation to the user's position and head movements was introduced with the Boom Mounted Display by McDowall et al. [22]. However, these displays provided only poor resolutions and thus were not capable of a detailed graphical representation of the virtual environment [23].

In order to enable an extensive representation of the aimed remote laboratories, we are looking for representative scenarios that fit to immersive requirements using both a detailed graphical modeling as well as a realistic experience within the simulation. In this context, one highly advanced visualization technology was realized through the development of the Cave in 1991. In this context, the recursive acronym CAVE stands for Cave Automatic Virtual Environment [18] and was first mentioned in 1992 by Cruz-Neira [24]. Interestingly, the naming of the Cave is also inspired by Plato's Republic [25]. In this book, he "discusses inferring reality (ideal forms) form shadows (projections) on the cave wall" [18] within "The Smile of the Cave".

By making use of complex projection techniques combined with various projectors as well as six projection walls arranged in form of a cube, the developers of the Cave have redefined the standards in visualizing virtual reality scenarios. The Cave enables visualization techniques, which provide multi-screen stereo vision while reducing the effect of common tracking and system latency errors. Hence, in terms of resolution, color and flicker-free stereo vision the founders of the Cave have created a new level of immersion and virtual reality.

The Cave, which serves the ideal graphical representation of a virtual world, brings us further towards true Virtual Reality, which – according to Rheingold [26] – is described as an experience, in which a person is "surrounded by a three-dimensional computer-generated representation, and is able to move around in the virtual world and see it from different angles, to reach into it, grab it and reshape it." This enables various educational, but also industrial and technical applications. Hence, in the past the research already focused on the power of visualization in technical applications, e.g. for data visualizations purposes [27] or for the exploration and prototyping of complex systems like the visualization of air traffic simulation systems [28]. Furthermore, the Cave has also been used within medical or for other applications, which require annotations and labeling of objects, e.g. in teaching scenarios [29].

The founders of the Cave choose an even more specific definition of virtual reality: "A virtual reality system is one which provides real-time viewer-centered head-tracking perspective with a large angle of view, interactive control, and binocular display." [18] Cruz-Neira also mentions that – according to Bishop and Fuchs [30] – the competing term "virtual environment (VE)" has a "somewhat grander definition which also correctly encompasses touch, smell and sound." Hence, in order to gain a holistic VR experience, more interaction within the virtual environment is needed.

Though, it is our aim to turn Virtual Reality into a complete representation of a virtual environment by extending the needed interaction capabilities, which are, together with the according hardware, necessary to guarantee the immersion of the user into the virtual reality [31]. However, even the Cave has got restricted interaction capabilities as the user can only interact within the currently demonstrated perspectives. Furthermore, natural movement is very limited, as locomotion through the virtual environment is usually restricted to the currently shown spot of the scenario. Yet, natural movements including walking, running or even jumping through virtual reality are decisive for a highly immersive experience within the virtual environment.

This gap of limited interaction has to be filled by advanced technical devices without losing high-quality graphical representations of the virtual environment. Hence, within this publication, we introduce the Virtual Theatre, which combines the visualization and interaction technique mentioned before. The technical setup and the application of the Virtual Theatre in virtual scenarios are described in the next chapter.

3 The Virtual Theatre – Enabling Virtual Reality in Action

The Virtual Theatre was developed by the MSEAB Weibull Company [32] and was originally carried out for military training purposes. However, as discovered by Ewert et al. [33], the usage of the Virtual Theatre can also be enhanced to meet educational requirements for teaching purposes of engineering students. It consists of four basic elements: The centerpiece, which is referred to as the omnidirectional treadmill, represents the Virtual Theatre's unique characteristics. Besides this moving floor, the Virtual Theatre also consists of a Head Mounted Display, a tracking system and a cyber glove. The interaction of these various technical devices composes a virtual reality simulator that combines the advantages of all conventional attempts to create virtual reality in one setup. This setup will be described in the following.

The Head Mounted Display (HMD) represents the visual perception part of the Virtual Theatre. This technical device consists of two screens that are located in a sort of helmet and enable stereo vision. These two screens – one for each eye of the user – enable a three-dimensional representation of the virtual environment in the perception of the user. HMDs were first mentioned in Fisher [34] and Teitel [35] as devices that use motion in order to create VR. Hence, the characteristic of the HMD consists in the fact that it has a perpendicular aligned to the user and thus adjusts the representation of the virtual environment to him. Each display of the HMD provides a 70°stereoscopic field with an SXGA resolution in order to create a gapless graphical representation of the virtualized scenario [33]. For our specific setup, we are using the Head Mounted Display from zSight [36]. An internal sound system in the HMD ena-bles an acoustic accompaniment for the visualization to complete the immersive scenario.

As already mentioned, the ground part of the Virtual Theatre is the omnidirectional treadmill. This omnidirectional floor represents the navigation component of the Virtual Theatre. The moving floor consists of rigid rollers with increasing circumferences and a common origo [33]. The rotation direction of the rollers is oriented to the middle point of the floor, where a circular static area is located. The rollers are driven by a belt drive system, which is connected to all polygons of the treadmill through a system of coupled shafts and thus ensures the kinematic synchronization of all parts of the moving floor. The omnidirectional treadmill is depicted in Fig. 1.

On the central area that is shown in the upper right corner of Fig. 1, the user is able to stand without moving. As soon as he steps outside of this area, the rollers start moving and accelerate according to the distance of his position to the middle part. If the user returns to the middle area, the rotation of the rollers stops.

The tracking system of the Virtual Theatre is equipped with ten infrared cameras that are evenly distributed around the treadmill in 3 m above the floor. By recording the position of designated infrared markers attached to the HMD and the hand of the user, the system is capable of tracking the user's movements [33]. Due to the unsymmetrical arrangement of the infrared markers the tracking system is not only capable of calculating the position of the user, but is also capable of determining looking directions. That way, the three-dimensional representation of the virtual scenario can be adjusted according to the user's current head position and orientation. Fur-thermore, the infrared tracking system is used in order to adjust the rotation speed of the rollers no only according to the user's distance from the middle point, but also according to the difference of these distances within a discrete time interval. Using these enhanced tracking techniques, the system can deal with situations, in which the user stands without moving while not being located in the middle of the omnidirectional floor.

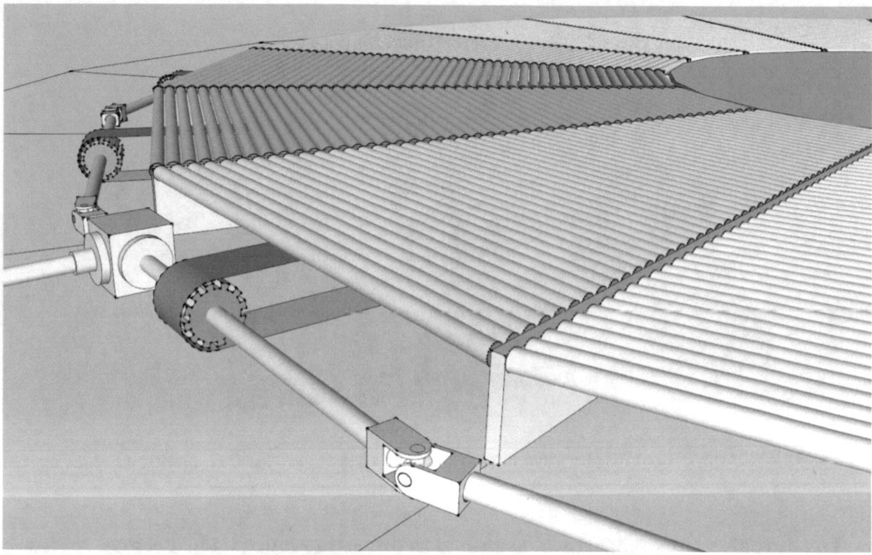

Fig. 1 Technical design of the Virtual Theatre's omnidirectional treadmill

The cyber glove ensures the tactile interaction capabilities. This special hand glove is equipped with 22 sensors, as indicated above, which are capable of determining the user's hand position and gestures [33]. This enables the triggering of gesture based events like the grasping of objects. Additionally, special programmable gestures can be utilized in order to implement specific interaction commands.

After setting up the required hardware of the Virtual Theatre, the user can plunge into different scenarios and can be immersed by virtual reality. After the development of learning and interaction scenarios as described in [16], our main interest here is focused on the development of remote laboratories, which represent the first step towards the realization of a virtual factory. The development, testing and evaluation of our first "Remote Lab" are described in the next chapter.

4 Development of Remote Laboratories in the Virtual Theatre

The described setup of the Virtual Theatre can be used to immerse the user into a virtual reality scenario not only for demonstration purposes, but especially for the application of scenarios, in which a distinctive interaction between the user and the simulation is required. One of these applications consists in the realization of remote laboratories, which represent the first step towards the creation of real-world demon-strators like a factory or an atomic plant into virtual reality.

Fig. 2 Two cooperating ABB IRB 120 six-axis robots

The virtual remote laboratory described in this paper consists in a virtual representation of two cooperating robot arms that are setup within our laboratory environment (see Fig. 2). These robots are located on a table in such a way that they can perform tasks by executing collaborative actions. For our information and communication infrastructure setup, it doesn't matter, if the robots are located in the same laboratory as our Virtual Theatre or in a distant respectively remote laboratory. In this context, our aim was to virtualize a virtual representation of the actual robot movements in the first step. In a second step, we want to control and to navigate the robots.

In order to visualize the movements of the robot arms in virtual reality, first, we had to design the three-dimensional models of the robots. The robot arms, which are installed within our laboratory setup are ABB IRB 120 six-axis robotic arms [37]. For the modeling purposes of the robots, we are using the 3-D optimization and rendering software Blender [38]. After modeling the single sections of the robot, which are con-nected by the joints of the six rotation axes, the full robot arm model had to be merged together using a bone structure. Using PhysX engine, the resulting mesh is capable of moving its joints in connection with the according bones in the same fash-ion as a real robot arm. This realistic modeling principally enables movements of the six-axis robot model in virtual reality according to the movements of the real robot. The virtual environment that contains the embedded robot arms is designed using the WorldViz Vizard Framework [39], a toolkit for setting up virtual reality scenarios.

After the creation of the virtual representation of the robots, an information and communication infrastructure had to be set up in order to enable the exchange of information between the real laboratory and the simulation. The concept of the inter-communication as well as its practical realization is depicted in Fig. 3.

As shown in the figure, the hardware of the remote laboratory setup is connected through an internal network. On the left side of the figure, a user is demonstrated, who operates the movements of the real robot arms manually through a control

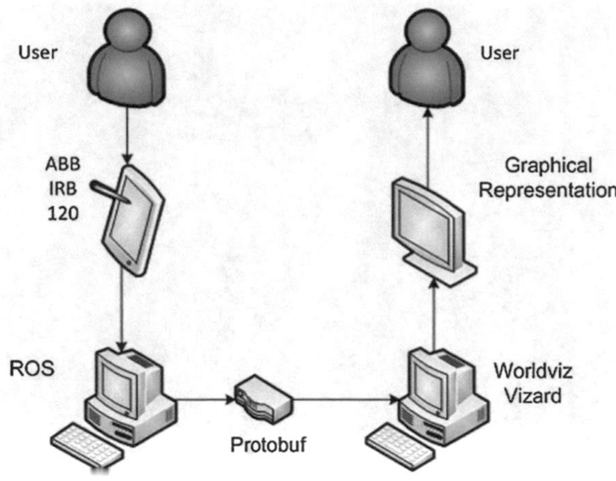

Fig. 3 Information and Communication Infrastructure of the remote laboratory setup

inter-face of the ABB IRB 120 robots. This data is processed by a computer using Linux with embedded Robot Operating System (ROS). The interconnection between the real laboratory and the virtual remote laboratory demonstrator is realized using the Proto-col Buffers (Protobuf) serialization method for structured data. This interface descrip-tion language, which was developed by Google [40], is capable of exchanging data between different applications in a structured form.

After the robots' position data is sent through the network interface, the information is interpreted by the WorldViz Vizard engine to visualize the movements of the actual robots in virtual reality. After first test phases and a technical optimization of the network configuration, the offset time between the robot arm motion in reality and in virtual reality could be reduced to 0.2 seconds. Due to the communication design of the network infrastructure in terms of internet-based communication methods, this value would not increase significantly, if the remote laboratory would be located in a distant place, for example in another city or on the other side of the globe.

The second user, which is depicted in the right upper part of Fig. 3 and who is located in the Virtual Theatre, is immersed by the virtual reality scenario and can observe the positions and motions of the real robots in the virtual environment. In Fig. 4, the full setup of the real and the remote laboratory is illustrated.

In the foreground of the figure, two users are controlling the movements of the actual robots in the real laboratory using manual control panels. In the background on the right side of the picture, the virtual representation of the two ABB IRB 120 robot arms is depicted. The picture on the right side of the wall is generated using two digital projectors, which are capable of creating a 3-D realistic picture by overlapping the pictures of both projections. The picture depicted on top of the robot arms table is a representation of the picture the user in the VR simulator is actually seeing during the simulation. It was artificially inserted into Fig. 4 for demonstration purposes.

This virtual remote laboratory demonstrator shows impressively that it is already possible to create an interconnection between the real world and virtual reality.

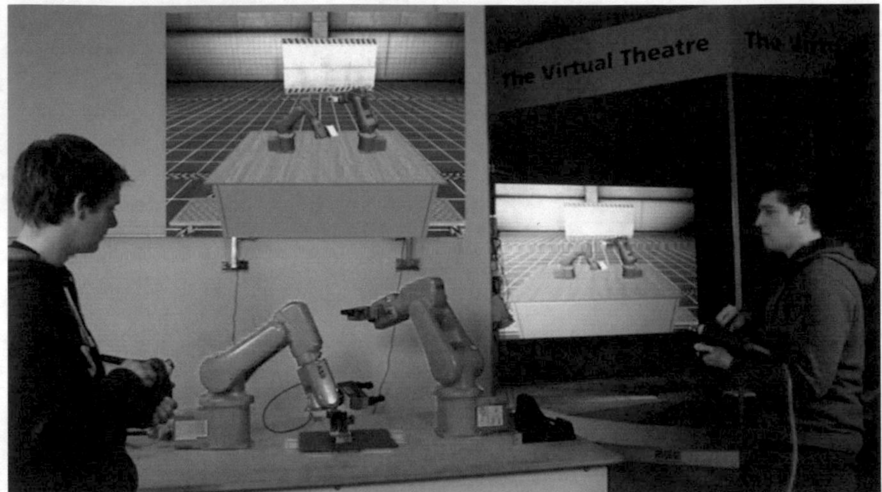

Fig. 4 Manual control of the robots and visual representation in the Virtual Theatre

5 Evaluation

The results of first evaluations within the test mode of our virtual remote laboratory demonstrator have shown that the immersive character of the virtual reality simulation has got a major impact on the learning behavior and especially on the motivation of the users. Within our test design, students were first encouraged to implemented specific movements of an ABB IRB 120 robot using the Python programming language. After this practical phase the students were divided into two groups.

The first group had the chance to watch a demonstration of the six axis robots car-rying out a task using "LEGO" bricks. After seeing the actual movements of the robots within our laboratories, the students were fairly motivated to understand the way of automating the intelligent behavior of the two collaborating robots.

The second group of students had the possibility to take part in a remote laboratory experiment within the Virtual Theatre. After experiencing the robot movements in the simulated virtual environment performing the same task as the real world demonstrator, the students could observe the laboratory experiment they were just experiencing in the Virtual Theatre recorded on video. Their reaction on the video has shown that the immersion was more impressive than the observation of the actual robot's movements performed by the other group. Accordingly, the students of the second comparison group were even more motivated after their walk through the virtual laboratory. The students of the second group were actually aiming at staying in the laboratory until they finished automating the same robot tasks they just saw in virtual reality.

6 Conclusion and Outlook

In this paper, we have described the development of a virtual reality demonstrator for the visualization of remote laboratories. Through the demonstrated visualization techniques in the Virtual Theatre, we have shown that it is possible to impart experi-enced knowledge to any student independent of his current location. This enables new possibilities of experience-based and problem-based learning. As one major goal of our research project "ELLI – Exzellentes Lehren und Lernen in den Ingenieurwissenschaften (Excellent Teaching and Learning within engineering science)", which addresses this type of problem-based learning [13], the implemented demonstrator contributes to our aim of establishing advanced teaching methodologies. The visualization of real-world systems in virtual reality enables the training of problem-solving strategies within a virtual environment as well as on real objects at the same time.

The next steps of our research consist in advancing the existing demonstrator in terms of a bidirectional communication between the Virtual Theatre demonstrator and the remote laboratory. Through this bidirectional communication we want to enable a direct control of the real laboratory from the remote virtual reality demonstrator. First results in the testing phase of this bidirectional communication show that such a remote control will be realized in the near future. In order to enable a secure remote control of the remote laboratory, collision avoidance and other security systems for cooperating robots will be carried out and tested in the laboratory environment. As the overall goal of our project consists in the development of virtual factories in order to enable the visit of an atomic plant or other elusive places, our research efforts will finally focus on the development of a detailed demonstrator for the realistic representation of an industrial environment.

References

1. M. Kerres, *Mediendidaktik: Konzeption und Entwicklung mediengestützer Lernangebote*. Oldenbourg, München, 2012
2. J. Handke, A.M. Schäfer, *E-Learning, E-Teaching and E-Assessment in der Hochschullehre: Eine Anleitung*. Oldenbourg, München, 2012
3. R.J. Craig, J.H. Amernic, Powerpoint presentation technology and the dynamics of teaching. Innovative Higher Education **31** (3), 2006, pp. 147–160
4. A. Szabo, N. Hastings, Using it in the undergraduate classroom: Should we replace the blackboard with powerpoint? Computer and Education **35**, 2000
5. T. Köhler, N. Kahnwald, M. Reitmaier, Lehren und lernen mit multimedia und internet. In: *Medienpsychologie*, ed. by B. Batinic, M. Appel, Springer, Heidelberg, 2008, pp. 477–501
6. R.A. Bartsch, K.M. Cobern, Effectiveness of powerpoint presentation in lectures. Computer and Education **41**, 2003, pp. 77–86
7. T. Creed, Powerpoint, no! cyberspace, yes. The Nat. Teach. & Learn. F. **6** (4), 1997
8. D. Cyphert, The problems of powerpoint: Visual aid or visual rhetoric? Business Communication Quarterly **67**, 2004, pp. 80–83
9. P. Norvig, Powerpoint: Shot with its own bullets. The Lancet **362**, 2003, pp. 343–344

10. T. Simons, Does powerpoint make you stupid? Presentations **18** (3), 2005. URL http://global.factiva.com/
11. A.M. Jones, The use and abuse of powerpoint in teaching and learning in the life sciences: A personal view. BEE-j 2 , 2003. URL http://www.bioscience.heacademy.ac.uk/journal/vol2/beej-2-3.pdf
12. E. André, Was ist eigentlich multimodale mensch-technik interaktion? anpassungen an den faktor mensch. Forschung und Lehre **21** (01/2014), 2014
13. M. Steffen, D. May, J. Deuse, The industrial engineering laboratory: Problem based learning in industrial eng. education at tu dortmund university. EDUCON , 2012
14. J.H. Murray, *Hamlet on the Holodeck: The Future of Narrative in Cyberspace*. MIT Press, Cambridge (Mass.), 1997
15. K. Schuster, D. Ewert, D. Johansson, U. Bach, R. Vossen, S. Jeschke, Verbesserung der lernerfahrung durch die integration des virtual theatres in die ingenieurausbildung. In: *TeachING-LearnING.EU discussions*, ed. by A.E. Tekkaya, S. Jeschke, M. Petermann, D. May, N. Friese, C. Ernst, S. Lenz, K. Müller, K. Schuster, TeachING-LearnING.EU, Aachen, 2013
16. M. Hoffmann, K. Schuster, D. Schilberg, S. Jeschke, Next-generation teaching and learning using the virtual theatre. 4th Global Conference on Experiential Learning in Virtual Worlds, Prague, Czech Republic, 2014
17. Wolf, M. J. P, B. Perron, *The video game theory reader*. Routledge, NY, London, 2003
18. C. Cruz-Neira, D.J. Sandin, T.A. DeFanti, Surround-screen projection-based virtual reality. the design and implementation of the cave. SIGGRAPH '93 Proceedings of the 20th annual conference on Computer graphics and interactive techniques. ACM - New York , 1993, pp. 135–142. doi:10.1145/166117.166134
19. N. Max, Siggraph '84 call for omnimax films. Computer Graphics **16** (4), 1982, pp. 208–214
20. C. Codella, R. Jalili, L. Koved, B. Lewis, D.T. Ling, J.S. Lipscomb, D. Rabenhorst, C.P. Wang, A. Norton, P. Sweeny, G. Turk, Interactive simulation in a multi-person virtual world. ACM - Human Fact. in Comp. Syst. (CHI 1992 Conf.), 1992, pp. 329–334
21. M. Deering, High resolution virtual reality. Com. Graph. **26** (2), 1992, pp. 195–201
22. I.E. McDowall, M. Bolas, S. Pieper, S.S. Fisher, J. Humphries, Implementation and integration of a counterbalanced crt-based stereoscopic display for interactive viewpoint control in virtual environment applications. Proc. SPIE **1256** (16), 1990
23. S.R. Ellis, What are virtual environments? IEEE Computer Graphics and Applications **14** (1), 1994, pp. 17–22. 10.1109/38.250914
24. C. Cruz-Neira, D.J. Sandin, T.A. DeFanti, R.V. Kenyon, J.C. Hart, The cave: Audio visual experience automatic virtual environment. Communications of the ACM **35** (6), 1992, pp. 64–72. doi:10.1145/129888.129892
25. Plato, *The Republic*. The Academy Athens, Athens, 375 B.C.
26. H. Rheingold, *Virtual reality*. Summit Books, New York, 1991
27. C. Nowke, M. Schmidt, van Albada, S. J., J.M. Eppler, R. Bakker, M. Diesrnann, B. Hentschel, T. Kuhlen, Visnest – interactive analysis of neural activity data. 2013 IEEE Symposium on Biological Data Visualization (BioVis), 2013, pp. 65–72
28. S. Pick, F. Wefers, B. Hentschel, T. Kuhlen, Virtual air traffic system simulation – aiding the communication of air traffic effects. 2013 IEEE on Virtual Reality (VR), 2013, pp. 133–134
29. S. Pick, B. Hentschel, M. Wolter, I. Tedjo-Palczynski, T. Kuhlen, Automated positioning of annotations in immersive virtual environments. Proc. of the Joint Virtual Reality Conference of EuroVR - EGVE - VEC , 2010, pp. 1–8
30. G. Bishop, H. Fuchs, et al., Research directions in virtual environments. Computer Graphics **26** (3), 1992, pp. 153–177
31. D. Johansson, *Convergence in Mixed Reality-Virtuality Environments: Facilitating Natural User Behavior*. University of Örebro, Schweden, 2012
32. MSEAB Weibull. http://www.mseab.se/the-virtual-theatre.htm, 2012
33. D. Ewert, K. Schuster, D. Johansson, D. Schilberg, S. Jeschke, Intensifying learner's experience by incorporating the virtual theatre into engineering education. Proceedings of the 2013 IEEE Global Engineering Education Conference (EDUCON), 2013

34. S. Fisher, The ames virtual environment workstation (view). SIGGRAPH '89 (Course #29 Notes), 1989
35. M.A. Teitel, The eyephone: A head-mounted stereo display. Proc. SPIE **1256** (20), 1990, pp. 168–171
36. http://sensics.com/products/head-mounted-displays/zsight-integrated-sxga-hmd/specifications/
37. ABB. http://new.abb.com/products/robotics/industrial-robots/irb-120. Last checked: 27.01.2014
38. Blender. http://www.blender.org/. Last checked: 27.01.2014
39. WorldViz. http://www.worldviz.com/products/vizard. Last checked: 27.01.2014
40. Google. http://code.google.com/p/protobuf/wiki/thirdpartyaddons. Last checked: 27.01.2014

Development of a Tele-Operative Testing Cell as a Remote Lab for Material Characterization

Tobias R. Ortelt, Abdelhakim Sadiki, Christian Pleul, Christoph Becker, Sami Chatti and A. Erman Tekkaya

Abstract Laboratory experiments play a significant role in engineering education. The experience gathered during the labs is one of the most important experiences during studying engineering because there is a strong connection between theory and practical relevance. A tele-operative testing cell for material characterization for forming processes is presented. This testing cell is used as a remote lab so that students can gain their experiences location and time-independent via the internet. In addition, the tele-operative testing cell is also used within the scope of lectures to combine the theory with live experiments in interaction with the students. The main aspects are, on the one hand, the developments in the field of engineering and the implementation of the IT components like iLab and, on the other hand, the integration of the tele-operative testing cell into engineering education.

Keywords Engineering Education · Manufacturing Technology · Forming Technology · Remote Lab · Laboratory Learning

1 Introduction

In times where resources like steel or aluminum become more and more expensive it is important to use full capacity of materials. Parallel to this effect, lightweight construction, particularly in the automotive sector and especially in the field of electric cars, becomes a key feature of efficiency [1]. These two trends can only be implemented with a comprehensive and deep understanding of the material, on the one hand, and the use of simulations like the Finite Element Method (FEM) to exploit full capacity of materials and to develop new forming processes on the other hand. The basis of these two fields is the conventional material characterization in which materials are tested to determine characteristic values like the Young's modulus, which is a value describing the resistance of a material against elastic deformation.

T.R. Ortelt (✉) · A. Sadiki · C. Pleul · C. Becker · S. Chatti · A.E. Tekkaya
Institute of Forming Technology and Lightweight Construction (IUL),
TU Dortmund University, Dortmund, Germany
e-mail: tobias.ortelt@iul.tu-dortmund.de

Originally published in "Proceedings der 2014 International Conference on Interactive Collaborative Learning (ICL)", © 2014. Reprint by Springer International Publishing AG 2016, DOI 10.1007/978-3-319-46916-4_21

Another characteristic value is the yield stress at which the material starts to deform plastically. These two values can be determined by a tensile test. In this test a specimen is typically tensioned until it cracks. This experiment has been implemented into a remote lab called tele-operative testing cell for material characterization and was developed at the Institute of Forming Technology and Lightweight Construction of the TU Dortmund University [2].

Current remote labs are typically used for electrical engineering education and are focused on the programming of processors and controllers or the behavior of circuits. In mechanical engineering virtual labs, such as animated plots of FEM simulations are used, but remote labs for mechanical engineering, and especially in forming technologies, are rarely offered to students.

The main aspect in the development of the tele-operative testing cell was the location and time-independent access for students to forming technology experiments [3]. Providing this access, the students are able to gather their own experience with labs. They can find out if theoretical models and experiments match or if theoretical hypotheses fail in experiments. Regularly, theoretical models are based on a wide process window, but under specific conditions, e.g. a special combination of process parameters, they can fail. For students it is important to know these gaps in theoretical models and with the access to the tele-operative testing cell they can gain this experience via the internet on their pc, notebook or even on their mobile devices. Furthermore, the experiments of the tele-operative testing cell can be integrated into lectures to combine theory and practical relevance in interaction with the students.

The developed tele-operative testing-cell and its integration into engineering education is embedded in the project "ELLI – Excellent Teaching and Learning in Engineering Education", which is a joint research project of RWTH Aachen University, Ruhr University Bochum, and TU Dortmund University with the vision of improving German engineering studies. In the subproject "Virtual Learning Environments" all three partners are dealing with remote labs. At the Ruhr University Bochum a cluster of eight different remote labs is developed [4] and at the RWTH Aachen University a remote lab is combined with Virtual Reality [5].

2 Objectives of the Addressed Tele-Operated Testing Cell

Tele-operation provides the possibility to access equipment over distance via the internet. Generally, this enhances the flexibility of users in terms of time and place when using it [6]. Among other things (e.g. security or training issues), this is one of the most significant aspects to make the potential of innovative lab experiments accessible [7].

But innovation can only be attained when an idea is extended from conceptualization and realization to a usable "product" for the addressed target group. In the context of this paper, this means the integration into learning by considering the student centered learning process [8] and the setup where it should be integrated in.

Therefore, the tele-operated experiment needs to be usable for reaching the addressed learning outcome of the lecture by the students [9].

In connection with the various subject-specific uses of experiments, the application of the scientific procedure should be considered by using experimental verification to quantitatively predict a process parameter. This starts by making reasonable presumptions, testing them creating procedures, and deriving conclusions.

To sum up, for the addressed tele-operated testing cell in forming technology, the following three objectives be stated:

- Technological realization of the tele-operation for sophisticated testing equipment in manufacturing engineering for forming processes.
- Integration into learning processes for engineering students studying the basics of material behavior.
- Consideration of the typical use of the experiment for the discipline of engineering in manufacturing tcchnology.

3 Development of the Tele-Operated Testing Cell

3.1 Available Experiments

The first step in the development of the tele-operative testing cell was the selection of experiments which should be available to the students in the future via internet. This selection was defined by the experiments, on the one hand, and by the machines which provide the experiments, on the other hand.

The focus of this selection was to find experiments which can produce known phenomena of forming technology.

In the previous project PeTEX [10] a universal testing machine was used to guarantee the access to a tensile test [3]. This kind of machine is typically used in non-automatic scenarios in which an operator controls the machine. The operator inserts the specimen, starts the machine, and, in a last step, removes the destructed parts of the specimen.

In the project PeTEX a six-axes-robot took over the handling of the specimen and the machine was controlled by a remote software [11]. This approach, based on the machine for the tensile test and the handling of specimen, was adopted to the tele-operative testing cell.

Furthermore, a new kind of experiment was included in the tele-operative testing cell. A sheet metal testing machine was integrated which provides different experiments for the material characterization of sheet metal. Especially the so-called Nakajima experiment for the construction of Forming Limit Curves (FLC) will be in the focus for the remote use of this machine. FLCs are important for simulations like the FEM for checking the process limits. For the construction of FLCs an optical measuring system is used to record the strain of the specimen during the experiment.

The selected testing machines provide several experiments to cover a wide range of experiments of forming technology. In a tensile test the influence of several parameters and their interaction, like the strain rate (forming speed), the orientation of the specimen or the material itself, with the material behavior can be detected.

In Nakajima experiments the material is tested by different load types. Furthermore, the two described experiments are common in material characterization and most of the students will use these experiments or the gained factors related with process knowledge in their studies or professional life.

3.2 Components of the Tele-Operative Testing Cell

In the following, all relevant components of the tele-operative testing cell are described and Fig. 1 shows a picture of the machines and the IT Components.

Machines and measurement systems:

- Universal testing machine Zwick Z 250 (1): The universal testing machine can conduct tensile and compression tests with a force of up to 250 kN. Measuring systems for force and change of length and width are integrated.
- Sheet metal testing machine BUP 1000 (2): The sheet metal testing machine is used to test and evaluate the ductility of sheet metals and to produce Forming Limit Curves. The experiments can be conducted with forces of up to 1,000 kN and the speed of the punch can reach 50 mm/s.
- Optical measuring system GOM ARAMIS 4M (3): This system is used to record the strain rate during the experiments and it is also used for the calculation of deformation plots.
- Industrial robot KUKA KR30-3 (4): The main task of the robot is the handling of the specimens.

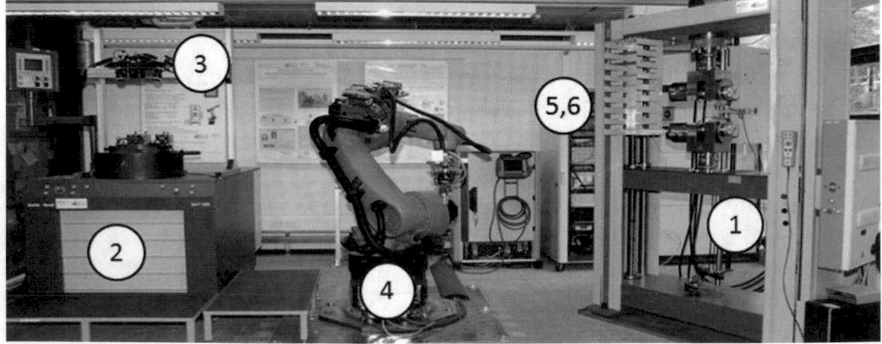

Fig. 1 Fully automated tele-operative testing cell for material characterization

IT components:

- Realtime control system (5): The complete testing cell is controlled by a PXI-System of National Instruments. Hence, the whole software was developed for LabVIEW and controls all machines and the peripheral equipment.
- Safety system (6): A safety system consisting of a programmable logic controller PLC, laser scanners, and protected switches ensuring the shutdown if an error occurs. This system works parallel to the realtime control system and monitors it.
- Camera system: The camera system consists of eight cameras with HD resolution. Most of them can be panned, tilted, and zoomed by the viewers of the experiment.

3.3 Automation of the Machines

All machines of the tele-operative testing cell were originally designed for the non-automatic usage, except for the robot of course. The universal testing machine was delivered with a remote interface which is based on a TCP/IP protocol and is able to control the basic functions of the machine. Basic functions are, for example, the closing of the clamping jaws or the starting of the experiment. Parallel to these basic functions data can be transferred in both ways. It is possible to set parameters like the strain rate, on the one hand, and to record the measuring data stream during the experiment, on the other hand.

The sheet metal testing machine was also delivered with the same kind of remote interface, but here only the software elements were defined. The physical elements like the mechanism to close or open the cap were missing. This means the command to close/open the cap was implemented in the software, but there were no mechanical elements to move the cap. Therefore, a pneumatic mechanism was developed and integrated into the sheet metal testing machine. Furthermore, sensors had to be integrated for safe automation of the machine.

Besides the automation of the machines, the robot and its handling system for the specimen were created. As mentioned above, the tele-operative testing cell should provide at least two different experiments at one time. For each experiment a unique specimen is required. Therefore, two different kinds of grippers were bought or self-developed and integrated. On the one hand, the robot uses commercial 2-finger parallel grippers for the handling of the specimens of the tensile test and, on the other hand, two different pneumatic vacuum grippers for the specimens of the sheet metal tests. These pneumatic grippers were specially developed for the use in the tele-operative testing cell and they use suction grippers to grab the specimens.

Figure 2 shows the pneumatic gripper for the insertion of the specimen into the sheet metal testing machine. The other pneumatic gripper has another geometrical arrangement of the suction grippers. They are arranged spherically so they can catch the deformed specimen after the Nakajima test. The periphery, like the valve manifold to control the grippers or the sensors for checking the vacuum of the suction grippers, is located on the arm of the robot. The control of the robot is also managing the grippers via a PROFIBUS to the valve manifold.

Fig. 2 Developed pneumatic vacuum gripper

3.4 Implementation of the IT Components (Software)

As presented before, several machines are used in the testing cell to provide the user with a wide range of experiments in the field of manufacturing technologies. The specimen handling and the synchronization of several processes during the experiments are the major challenges in the implementation of the remote lab. In this sense, all the scenarios occurring by the conduction of the experiments are defined and considered by the automation and the implementation of the remote lab.

The controlling software, a so-called Experiment Manager, is developed in LabVIEW and is running on a PXI-System with realtime support. This Experiment Manager communicates with the different devices, like the testing machines or the robot, over TCP/IP, PROFIBUS, or digital inputs/outputs and analog channels. The communication type between the devices and the PXI-System depends on the provided automation interface of the device. Some devices do not provide any remote interface, so in that case a communication between the controlling unit of the device and the PXI-System over digital inputs/outputs was realized. Figure 3 shows the machines of the tele-operative testing cell and the connections to the PXI-System.

Due to the various devices used in the lab and the large number of experiments provided by each testing machine, a component-based architecture was designed for the Experiment Manager. All devices are represented as a component with an interface and each component provides the functionalities of a device. Furthermore, the experiments are also implemented as components. In this case, the experiments are abstract components because they do not represent a specific hardware component, furthermore they implement the logic for carrying out the experiment. Moreover, the components from a high level use the components from a low level, for example the tensile test component uses the robot and machine components. Figure 4 shows the components of the controlling software and several relations between the components.

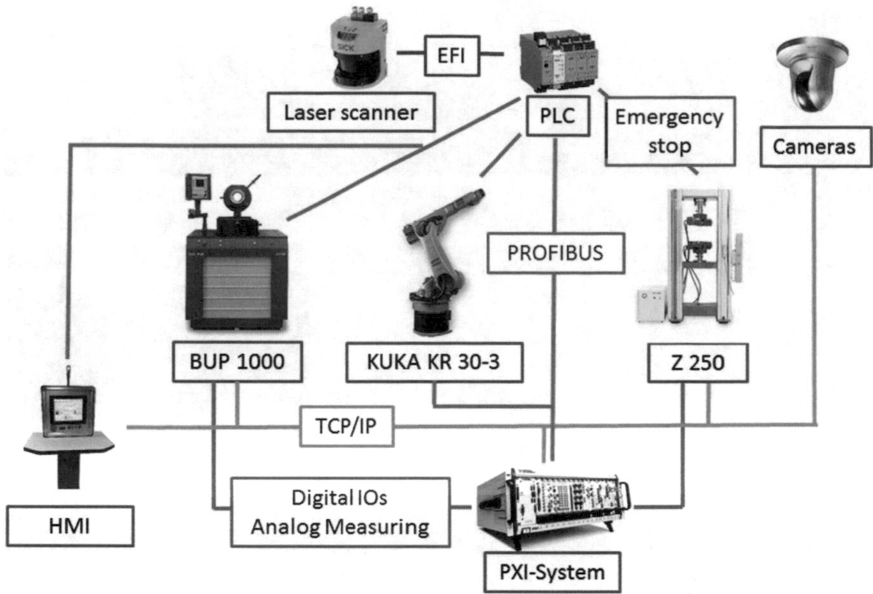

Fig. 3 Different communication interfaces of the tele-operative testing cell

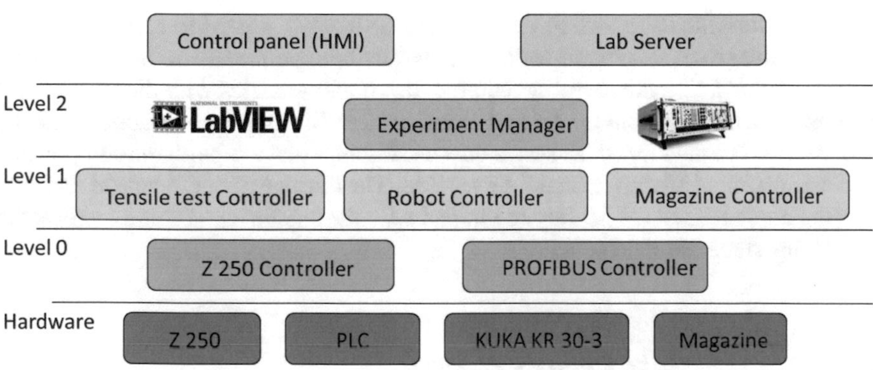

Fig. 4 Component-based platform for controlling the experiments

On the basis of the modularity of the Experiment Manager, new experiments can be simply implemented or existing experiments can be extended by using already implemented components. To allow an external system (e.g. the lab server) to communicate with the Experiment Manager, a component named "Remote" is implemented. After a successful authentication on the Experiment Manager the remote component handles the entire requests coming from the client (lab server). The communication is based on XML and raw strings.

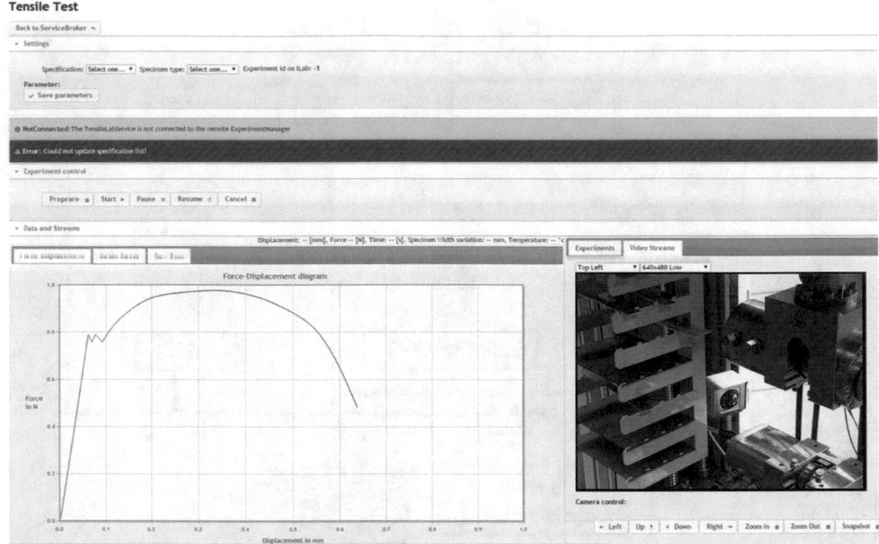

Fig. 5 Graphical user interface of the tensile test

In this work the iLab Shared Architecture (ISA) [12] is used to benefit from the many advantages of this framework. To integrate the experiments in this framework, the development of a lab server is required, this communicates with the Experiment Manager and the lab client for the tensile test to allow the user the interaction with the experiment. The developed lab server uses a C# library that wraps all the components of the Experiment Manager on the server side. The lab client for the tensile test uses also the implemented C# Library to interact with the tensile test. Figure 5 shows the HTML interface of the lab client.

4 Integration into Lecture Course

4.1 Conceptual Aspects

The first application of the tele-operative testing cell was in a 5th semester Bachelor lecture course "Fundamentals of Forming Technology" for 250 mechanical engineering students in October 2013. A scenario to embed a tensile test into the lecture course was designed. Therefore, a key application of engineering experiments has been addressed by the subject-specific, bidirectional use for the development and verification of theoretical models. To start the scenario, one relevant intended learning outcome of this lecture course is the identification of relevant parameters to explain the typical material behavior based on previous theoretical models and to predict a d test the cause and effect. On one hand, this addresses the components of learning outcomes of Model, Data Analysis, and later on Learning from Failure by [13]. On the other hand relevant conceptual understanding at a high quantitative level of multistructural (several but unconnected) aspects followed by the transfer to relational (connected) aspects to develop the understanding qualitatively and apply it further are taken into account [14].

To make the experiment a supporting element for the students' understanding, the integration into the learning process was based on a developed phase model considering an enhanced Kolbs' learning cycle [15, 16]. In connection with previous knowledge, students start with an experience (here an initial experiment was done) and reflect their observation followed by the conceptualization of the phenomena and planning of own experimental examination.

4.2 Step-Wise Procedure

After the connection to students' previous knowledge of theoretical relations, the experimental setup was described in relevant detail followed by a first tele-operative experiment to experience the real characteristic material behavior as shown in Fig. 5. The students were able to observe the whole process by four parallel camera streams, from the beginning, the insertion of the specimen to the machine with the robot, the test execution, and the final removing of the disrupted specimen out of the machine. Figure 6 shows the camera interface. The stream of camera 4 is in this example the active stream and is shown in a large scale. The three other streams show a live preview. Camera 1 is installed on the hand of the robot and camera 3 shows the view of the specimen and its necking and finally its fracture during the experiment. The

Fig. 6 Camera interface of the tele-operative testing cell

position of camera 3 can be detected in the view of camera 4. Camera 2 shows the whole setup of the tele-operative testing cell and camera 4 shows a closer view of the experiment and the universal testing machine. The active camera can be changed at any time by a click on the preview. A video projector showed these camera streams in the lecture hall. A second video projector showed the real time measurement date of the experiment. In this case, it was the stress-strain-diagram.

After the first experiment the results were discussed in interaction with the students and they were asked to apply this analyzed observation to estimate the material behavior under changing boundary conditions. The change was the reduction of the strain rate. The agreement about a possible behavior was arranged by using an interactive poll system eduVote. Three possible answers were given:

- stress will increase,
- stress will remain constant,
- stress will decrease.

Based on the result, a theoretical model was developed which could be used to explain the forecasted material behavior. Now, everything seemed to be fine because of the theoretical model which was verified and is applicable.

4.3 The Unexpected Clash

Finally, the students were asked to test the boundary conditions based on their previous experience to state an explanation according to "How will the specimen restrain if the strain rate is much lower than in the first experiment." Now, students configure the experiment and observe the processing test. At this point, the material starts to behave unexpected and a clash develops. Before explaining the issue, the students had a few minutes to think about an explanation and discuss their ideas with the other students. Afterwards, the lecturer asked for their feedback to develop a conceptual explanation of the observed phenomena. Incorporating the student's feedback, the lecturer discussed the result and explained it theoretically on the basis of the different strain rate.

To support the adopted theoretical model, a third experiment was configured by the students by setting the strain rate to a chosen value. The students chose a very high strain rate. The result of the third experiment matched the expectations and supported the theory.

The developed model shows the integration of the experiment into the learning process for understanding subject-specific concepts. One approach is described showing a typical use of engineering experiments (alternating verifying and generating application) in order to evaluate the advantages of tele-operated real test equipment.

The feedback of the students was continuously positive. They enjoyed this new kind of lecture and the interaction between the lecturer and the audience. In which way this procedure leads to an improved process understanding and to what extent it influences the results of exams will be an aspect on further works.

5 Integration into e-Learning Environment

After the successful first application of the tele-operative testing cell the next step was made in a pilot phase in May 2014. In this pilot phase the tele-operative testing cell was used in its envisaged scenario. The access via internet was made available to students who are studying the "Master of Science in Manufacturing Technology (MMT)". This master program is designed and offered to international students from all over the world. The usage of the tele-operative testing cell was imbedded into an exercise of a lecture course in the field of forming technology and finite element simulation. With the experiments the students were able to calculate factors, like the Young's Modulus, for a subsequent simulation.

In a first step the students registered on the iLab server and booked a timeslot with the service-broker. In the booked timeslot the students were able to access the experiment. They could choose between two materials, steel or aluminum, and its direction of rolling. The specimen was either taken from a direction parallel, in a 45° angle, or perpendicular to the rolling direction when producing the sheet metal. With this kind of specimen the students were able to define or calculate the anisotropy factor which indicates the direction-dependent behavior of a material. This factor is important for sheet metal forming processes like deep drawing. After choosing the specimen the students defined the strain rate. A parameter window for the strain rate was defined previously to guarantee a functional experiment and the safety of the machine and the other equipment. The students started inserting the specimen and gave the start signal of the experiment by clicking on buttons. During the experiment the students could change the camera streams as described above. The average number of executed experiments for each student was two. Subsequent to the experiment the students downloaded the data. The data was delivered in a csv-file with three columns, one for the force and two for the change of the length and the width. With this information and the additional parameters of the specimen, like the initial width and thickness and the measured length, the students calculated the Young's Modulus and the anisotropy factor. In another exercises the students used these calculated parameters for a finite element simulation. After the experiment the students were asked to fill out a feedback paper. The results of this feedback paper were positive, especially from the students, who had never access to this kind of equipment before. Parallel, the students were asked for their used hard- and software for the access to the tele-operative testing cell, but no problems in relation to special hard- or software could be found. Furthermore, it could be observed that few students had no deep experience in analyzing measured data. This shows one aspect of the tele-operative testing cell to combine theoretical knowledge with practical experience, like to handle measured data, which is a typical task in professional life of engineers.

6 Conclusion and Outlook

The described developments of the tele-operative testing cell present the current stage. The first integration into lecture course and the first executed experiments by students via the internet achieved positive feedback. The next stage of the development of the tele-operated testing cell will be initiated in the fall semester 2014/15, when a whole class of engineering students, up to 300, will access the tele-operative testing cell to gather own experience in material behavior.

Parallel to this, new experiments will be integrated. The tensile test will be updated by the integration of an induction generator which can heat a specimen up to 1,000°C in only a few seconds. At this high temperature materials will behave differently to the typical room temperature of 20°C. This high temperature tensile test will become more and more important, because new developed forming processes are hot forming processes [17]. The benefit of the heating by induction in relation to the conventional heating with an oven is, on the one hand, the shorter heating time and, on the other hand, the cameras can film the specimen during the experiment. Two experiments will be developed for the sheet metal testing machine. The first test will be the mentioned Nakajima test and the second experiment will be deep-drawing of a cup. Within these experiments the students will change parameters like clamping force and punch speed and learn more about the process interactions.

References

1. R. Neugebauer, T. Altan, M. Geiger, M. Kleiner, A. Sterzing, Sheet metal forming at elevated temperatures. Cirp Annals-Manufacturing Technology **55** (2), 2006, pp. 793–816. doi:10.1016/j.cirp.2006.10.008
2. T.R. Ortelt, C. Pleul, A. Sadiki, M. Hermes, C. Soyarslan, A.E. Tekkaya, Virtuelle lernwelten in der ingenieurwissenschaftlichen laborausbildung. In: *Exploring Virtuality*. 2012
3. C. Pleul, C. Terkowsky, I. Jahnke, A.E. Tekkaya, *Tele-operated laboratory experiments in engineering education – The uniaxial tensile test for material characterization in forming technology*, Deusto Publicaciones, 2011, vol. 1, book section 16, pp. 323–347
4. S. Frerich, D. Kruse, M. Petermann, A. Kilzer, Virtual labs and remote labs: Practical experience for everyone. In: *IEEE Global Engineering Education Conference, EDUCON*. 2014, pp. 312–314. http://www.scopus.com/inward/record.url?eid=2-s2.0-84903479910&partnerID=40&md5=38575a1f694edb31ae2a4bd6fc311c62
5. M. Hoffmann, K. Schuster, D. Schilberg, S. Jeschke, Bridging the gap between students and laboratory experiments. In: *Lecture Notes in Computer Science (including subseries Lecture Notes in Artificial Intelligence and Lecture Notes in Bioinformatics)*, vol. 8526 LNCS, 2014, pp. 39–50. http://www.scopus.com/inward/record.url?eid=2-s2.0-84903608514&partnerID=40&md5=2e38f4c349d061b5e0a2921a98922c99

6. J.E. Corter, J.V. Nickerson, S.K. Esche, C. Chassapis, S. Im, J. Ma, Constructing reality: A study of remote, hands-on, and simulated laboratories **14** (2), 2007, pp. 1–27. http://doi.acm.org/10.1145/1275511.1275513
7. M. Cooper, J.M.M. Ferreira, Remote laboratories extending access to science and engineering curricular **2** (4), 2009, pp. 342–353. http://portal.acm.org/citation.cfm?id=1683260.1683282, doi:10.1109/TLT.2009.43
8. J. Wildt, *Vom Lehren zum Lernen*, Dr. Josef Raabe Verlags-GmbH, 2006, book section A 3.1, pp. 1–14
9. J.E. Corter, S.K. Esche, C. Chassapis, J. Ma, J.V. Nickerson, Process and learning outcomes from remotely-operated, simulated, and hands-on student laboratories **57** (3), 2011, pp. 2054–2067. http://dx.doi.org/10.1016/j.compedu.2011.04.009, http://ac.els-cdn.com/S036013151100090X/1-s2.0-S036013151100090X-main.pdf?_tid=d6980940-5845-11e4-b076-00000aab0f27&acdnat=1413801950_4e5de0578269382ffcd63c5cfbc61c9a
10. C. Terkowsky, C. Pleul, I. Jahnke, A.E. Tekkaya, Tele-operated laboratories for online production engineering education platform for e-learning and telemetric experimentation (petex). International Journal of Online Engineering **7** (SUPPL.), 2011, pp. 37–43. http://www.scopus.com/inward/record.url?eid=2-s2.0-80054011909&partnerID=40&md5=ffab7781bc312ab4a674acd67f901a75
11. C. Pleul, C. Terkowsky, I. Jahnke, Petex - platform for e-learning and telemetric experimentation: A holistic approach for tele-operated live experiments in production engineering. In: *Proceedings of the 16th ACM International Conference on Supporting Group Work, GROUP'10*. 2010, pp. 325–326. http://www.scopus.com/inward/record.url?eid=2-s2.0-78751697026&partnerID=40&md5=55b70d13ea519cc8a1067b34db8ac738
12. V.J. Harward, J.A. Del Alamo, S.R. Lerman, P.H. Bailey, J. Carpenter, K. DeLong, C. Felknor, J. Hardison, B. Harrison, I. Jabbour, P.D. Long, T. Mao, L. Naamani, J. Northridge, M. Schulz, D. Talavera, C.D. Varadharajan, S. Wang, K. Yehia, R. Zbib, D. Zych, The ilab shared architecture: A web services infrastructure to build communities of internet accessible laboratories. Proceedings of the IEEE **96** (6), 2008, pp. 931–950. http://www.scopus.com/inward/record.url?eid=2-s2.0-60549106226&partnerID=40&md5=83aa07220c41d44ab839ee7e42b7f687
13. L.D. Feisel, A.J. Rosa, The role of the laboratory in undergraduate engineering education. Journal of Engineering Education **94** (1), 2005, pp. 121–130. <Go to ISI>://WOS:000231266300011
14. J.B. Biggs, C. Tang, *Teaching for Quality Learning at University*, 4th edn. Mc Graw Hill, 2011
15. D.A. Kolb, *Experiential learning : experience as the source of learning and development*. Prentice-Hall, Englewood Cliffs, N.J., 1984
16. D.A. Kolb, R.E. Boyatzis, C. Mainemelis, *Experiential Learning Theory: Previous Research and New Directions*, NJ: Lawrence Erlbaum, 2000
17. D. Staupendahl, C. Löbbe, M. Hudovernik, C. Becker, A.E. Tekkaya, *Process Combinations for the Production of Load-Optimized Structural Components*, 2014, p. 638–645

A Web-Based Recommendation System for Engineering Education E-Learning Systems

Thorsten Sommer, Ursula Bach, Anja Richert and Sabina Jeschke

Abstract Today there is a flood of e-learning and e-learning related solutions for engineering education. It is at least a time consuming task for a teacher to find an e-learning system, which matches their requirements. To assist teachers with this information overload, a web-based recommendation system for related e-learning solutions is under development to support teachers in the field of engineering education to find a matching e-learning system within minutes. Because the e-learning market is subject of very fast changes, an agile engineering process is used to ensure the capability to react on these changes. To solve the challenges of this project, an own user-flow visual programming language and an algorithm are under development. A special software stack is chosen to accelerate the development. Instead of classical back-office software to administer and maintain the project, a web-based approach is used – even for a complex editor. The determining of the necessary catalog of related solutions within "real-time" is based on big data technologies, data mining methods and statistically text analysis.

Keywords E-Learning · Recommendation System · Agile Process · Teachers · Professors · Web 2.0 · Software Engineering · Open Source

1 Introduction

To help teachers with their different challenges about finding an e-learning solution, a web-based recommendation system for e-learning systems is under development. This recommendation web-based service enables teachers to choose an engineering education e-learning system, which matches her or his requirements.

T. Sommer (✉) · U. Bach · A. Richert · S. Jeschke
Faculty of Mechanical Engineering, ZLW - Center for Learning and Knowledge Management, IfU - Institute for Management Cybernetics, IMA - Institute of Information Management in Mechanical Engineering, RWTH Aachen University, Aachen, Germany
e-mail: thorsten.sommer@ima-zlw-ifu.rwth-aachen.de

Originally published in "Proceedings of the 6th International Conference on Computer Supported Education (CSEDU)", © SCITEPRESS 2014.
Reprint by Springer International Publishing AG 2016,
DOI 10.1007/978-3-319-46916-4_22

The term "e-learning" is often used in different matters. Therefore, this definition is chosen: "E-learning is an approach to teaching and learning, representing all or part of the educational model applied, that is based on the use of electronic media and devices as tools for improving access to training, communication and interaction and that facilitates the adoption of new ways of understanding and developing learning" [1]. This definition includes any computer- and web-based tool, which is related to the education context.

A variety of e-learning systems and environments [2] are observable and the amount is continuous growing: From classical computer-based training (CBT), web-based training (WBT) [3], wikis and blogs [3], podcasts [4] respectively educasts [3] and game-based learning [3] up to massive open online courses (MOOC) [3, 5].

To illustrate the amount of related and available resources: A simple web-search for "e-learning" ends with over one billion results, a web-search for "e-learning system" with over 480 million results! Nearly 80 unique e-learning systems can be found in short time. This fact demonstrates the problem: The interested teacher must investigate this amount of information to find an e-learning system, which matches their personal requirements. Another problem is: The teacher might not be able to choose a system, which matches their requirements, because it is not trivial to understand all the technologies and differences between the unique systems.

2 Requirements and Challenges

The main precondition of the desired recommendation system is that the related e-learning systems are comparable. Moreover, a catalog of related solutions must exist. The approach to reach the comparable state is to find a set of necessary attributes that describes the characteristics of engineering education e-learning systems. These common e-learning characteristics must base on a broad scientific consensus: To realize this, the input of many experts is acquired.

The collected data about each e-learning system, together with the e-learning characteristics, results in a comparable data sheet about each e-learning system. This data sheet is subject of continuously changes to ensure that the data sheet is up-to-date and represents the current state of each system. Also the e-learning characteristics are subject of changes to cover all related kinds of e-learning.

The traditionally approach to determine the catalog with the solutions uses a lot of resources (time, staff and money): Pay and get every e-learning system, prepare a server environment and install all systems. Then investigate the system (as teacher and student) and fill-up the data sheet. It is possible to speed-up this by using virtualization environments [6] like e.g. a type 1 hypervisor [7] with templates for the required environments, system snapshots and derivation between them.

The new and promising approach to determine the catalog with the available solutions is based on text mining: Crawling and parsing the public vendor and community information about the e-learning systems and store the raw data. Next, the raw text data is able to get analyzed to find out about the textual context. A half-automated algorithm suggests then a value for each characteristic, to assist the employee. Such a process is able to get executed e.g. every quarter to ensure that the data sheets are up-to-date.

To provide a convenient tool to develop and maintain the questionnaire, a new visual user-flow programming language is defined. This language is linking the catalog of solutions, the questions with further explanations for the teachers, the e-learning characteristics and the model of the user-flow together. Compared to existing survey solutions, the visual model and the deep integration are new.

Another challenge is that teachers expect a current, modern, responsive [8] and accessible user interface (UI). This is comprehensible, because it allows any teacher with any device and any handicap to use this web-based recommendation service. A responsive UI saves also time, because there is no need for an additional mobile and tablet websites [8]. Some web frameworks (e.g. Bootstrap[1]) assist the developer in these fields.

For further research (e.g. about e-learning systems and the teachers requirements to these systems), it would be helpful to collect some kind of key performance indicators (KPI) from the recommendation system. The data must be anonymous to keep the teachers privacy. Not only the end results must be logged, also e.g. the reaction time per question and – if present – the cancellation point etc. It is also interesting to capture all single decisions of any anonymous teacher to enable research e.g. in psychology fields.

3 User Flow Language

To enable the scientific assistants to model efficient the user-adapting questionnaire and to provide a convenient tool for maintaining the questionnaire, a new visual programming language [9] is defined.

The language is simple: Different squares – called "function blocks" – are connected by wires. For different purposes, different function blocks are present: Start, end, question, numeric and range blocks. Every function block has no or one input connector and no, one or three output connectors – this depends on the kind of the block. Behind every block, some data is stored: The reference to the common e-learning characteristic, a question, additional explanation text or just a text message – depends on the kind of the block.

Every program must have exact one start block, and at least one end block. The visual program must read from left to right: From the start block at the left, then follow block by block until reaches any of the end blocks. The concrete questionnaire can then deviated out of the visual program by simple traveling through the blocks. There is no special algorithm required to get or generate the questionnaire.

With this visual language, the visual program for the questionnaire has to be built. Explaining Fig. 1 as current example – to clarify the visual language: At the left, the start block was placed. The interviewee gets an introduction message to read. The

[1]http://www.getbootstrap.com.

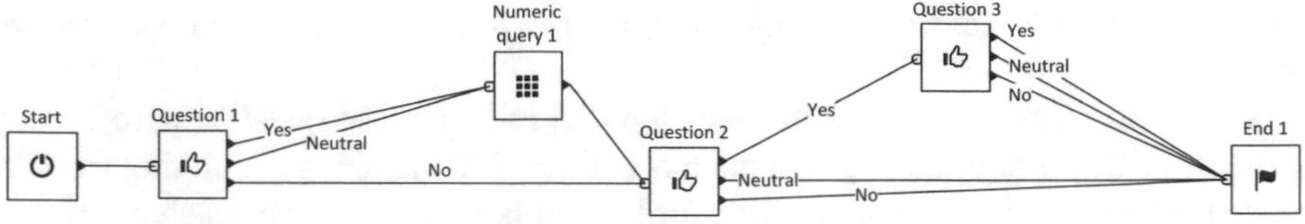

Fig. 1 Example of the user flow visual programming language

flow is directed to “Question 1”, a question block. The interviewee gets a question (the question was defined before by a research assistant) and an additional explanation text. Related to the answer, the flow reaches “Numeric query 1” or “Question 2”. The “Numeric query 1” prompts the teacher to answer a numeric question like e.g. “How many students are in your class?” From this block, the flow also reaches “Question 2”. After “Question 2”, the flow is able to reach “Question 3” or “End 1”, related to the answer. “End 1” is also reached after answering “Question 3”. At the end block, the interviewee is able to read a finished message. By pressing a button, the user submits all their answers to the algorithm.

Each question block has one input and three output connectors for the further flow: A yes-output, a no-output and a neutral-output. This corresponds to the possible answers of the interviewee. “Yes” means that the issue (subject of the question) must be present, “no” means the issue is not present or can be disabled and “neutral” means, that the issue does not matter.

A range block (this kind of block is not part of the example at Fig. 1) contains a text and an explanation text for the interviewee. To represent a range, this block needs two references to the corresponding parts of the common e-learning characteristics. This kind of block has one input and output connector.

While this project grows, the amount of different kind of function blocks will increased as necessary. For any new kind of block, also the algorithm (see Section 5) must be extended to cover the new functionality. New types of blocks are caused by changes at the common characteristics, to cover new requirements on the e-learning market.

For convenient usage, a simple web-based editor is under development. Thereby, it is possible to maintain the questionnaire without any programming skills. The whole life-cycle of the editor and the language is also convenient: There is no installation required, updates are only a server-side deployment and any kind of maintenance just occur on the server-side.

4 Engineering

4.1 Process

To be able to react on new requirements on the whole process (software engineering, determining catalog of solutions, development at the algorithm etc.), agile best practices are chosen [10–12]:

- Use cases: Define a few use cases by drawing diagrams or by writing small so called “user stories” [10].
- Simplicity: Develop just necessary parts and leave anything which is not required [10, 12].
- Fast: Release fast and as many as possible to be able to get feedback from others [10, 12].
- Communication: Get early and continuously feedback from the customer, to ensure the project fits the requirements [10].

Additionally, not manageable challenges are divided into smaller – but manageable – challenges [10, 13]. It is also necessary to choose the right tool for right purpose: Do not use the same tool for anything – there are right purposes for any tool, but not all tools are convenient for all problems.

Even though different changes at the last three months, it was possible to reach the current state after just eight weeks of work with just one person: This is perhaps related to the agile process. The process is promising for the further work and research.

4.2 Client-Side: Web-UI Approach

At least two tools are required for the back-office: The editor for the visual language and the product editor. While the product editor is quite simple, the editor for the visual language is even more complex. Anyhow, a new approach is tested: Instead of developing the website for the teacher’s questionnaire and additional software for the back-office (with Java or .NET), anything will be developed as web-application.

For the web-application (for the back-office tools and also for the teacher’s questionnaire) just HTML5, CSS and JavaScript with Bootstrap[2] and jQuery[3] is used. Thereby, it is also useable on tablets like e.g. iPads – independently of the operating system. Further, also the final client performance is fine.

4.3 Server-Side: Software Stack

On the server-side, a wide range of options are available. The best fit is a “BGMR stack”: FreeBSD,[4] nginx,[5] MongoDB,[6] and Ruby.[7] FreeBSD [14] is very robust,

[2]http://www.getbootstrap.com.

[3]http://www.jquery.com.

[4]http://www.freebsd.org.

[5]http://www.nginx.org.

[6]http://www.mongodb.org.

[7]https://www.ruby-lang.org.

secure, uses small resources and with the concept of "Jails" [15, 16] a powerful virtualization [6] environment is provided.

The web-server nginx [17] is famous as powerful proxy, load-balancer and fasted server for static content [18]. Nginx delivers all necessary static data for this project (images, CSS, JavaScript, fonts etc.) But nginx is the wrong tool to deliver dynamic content, because the handling of external processes like e.g. the common PHP[8] is less efficient [18]. Therefore, non-static requests are passed-through to an application server.

MongoDB [19] is a robust, fast [20, 21] and scalable database. For the most Web 2.0 projects, MongoDB fits perfect, because the database is document-based and uses JSON [22]. If the project grows, the database is also able to scale.

Ruby is an object-oriented programming language [23] that provides a very convenient and fast way to develop web applications: On top of Ruby, the Sinatra[9] framework [24] provides an own web-centric DSL (domain-specific language). This enables a strong focus to web-oriented programming and reduced the necessary overhead a lot. While the project runs on the development environment with Ruby, for the production environment it is designated to run it with JRuby [25] on JavaEE with the JBoss application server [26].

This approach is very promising, because it speeds up the development, keeps the code clear – which makes the maintenance easier – and provides later with JRuby the power and scalability of JavaEE and the JBoss application server [25, 26].

5 Algorithm

To get the expected results out of the teacher's answers, an algorithm is under active development (see Figs. 2, 3, 4 and 5). The current state of the algorithm is constructed and validated with dedicated test data: As developing and testing environment, a normal spreadsheet application is chosen. A table with test data is representing the results of the questionnaire (the teacher's answers).

The term "issue" means in the context of the algorithm an element of the common e-learning characteristics, e.g. a product function, product behavior or a didactic method, but not a numeric or range question (see Section 3 for the differences). As input, the algorithm expects the static catalog of solutions as matrix: Horizontally the columns with the product issues, numeric questions and ranges etc. and vertically the rows with the products. The possible types of columns are corresponding to the available types of function blocks (see Section 3).

Out of the product point of view for the product data: Each issue can obtain the value -100 (the issue is not present), 1 (the issue is present and cannot be disabled) or 2 (the issue is present and can be disabled).

---

[8]http://www.php.net.

[9]http://www.sinatrarb.com.

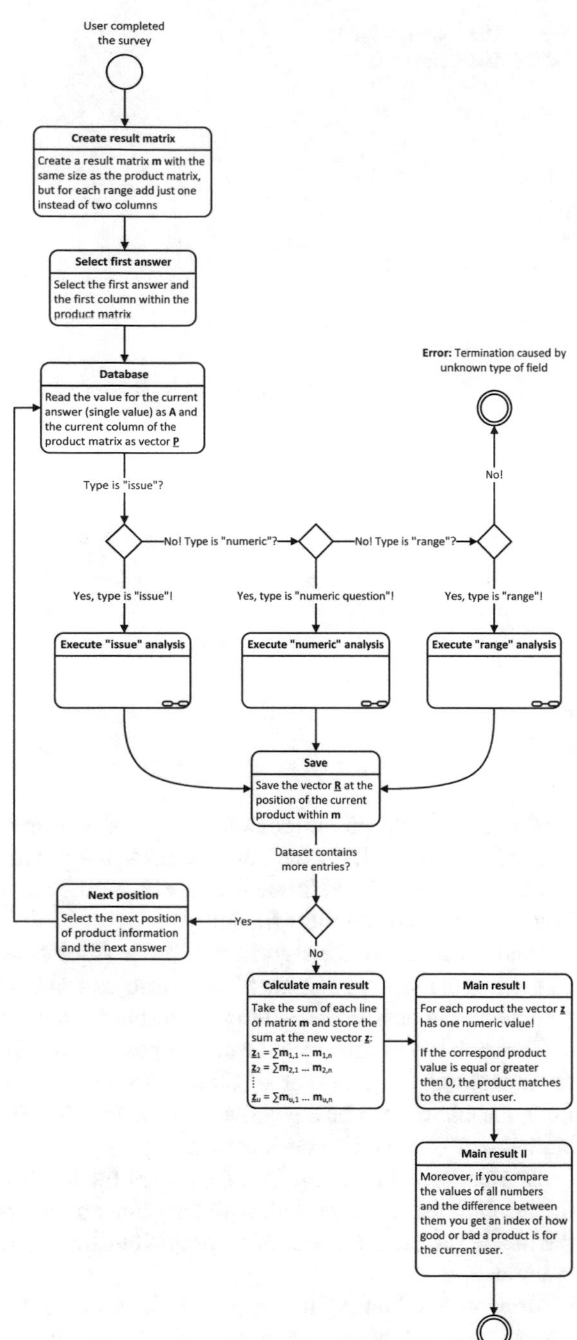

Fig. 2 The recommendation algorithm to suggest e-learning systems

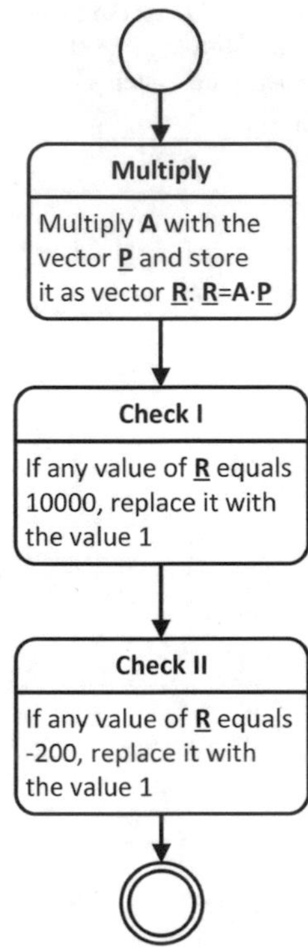

Fig. 3 The sub-algorithm for the "issue" analysis

A range, e.g. the possible amount of students (from/to), is provided as two columns inside the matrix. Moreover, numeric fields are possible for e.g. the price, which are provided as one column inside the matrix. At the moment, numeric fields are able to hold only positive numbers (include 0).

Another input for the algorithm is the teacher's answers, provided as vector. Out of the teacher's point of view: Each issue can obtain the value -100 (issue is not present or can be disabled), 0 (issue does not matter) or 1 (issue must be present).

For each range, the answer can be a positive number (include 0) to represent e.g. the amount of students – or -100 if this does not matter. Finally for each numeric field, the answer can be a positive number (include 0) to represent e.g. the teacher's budget – or -100 if this does not matter.

The start point (see Fig. 2) is the end of the teacher's questionnaire. As first step, the result matrix **m** is created, with the same amount of rows as the product data and the nearly the same amount of columns (but for each range only one instead of two columns).

Important to know: The first question from the questionnaire corresponds to the first answer and this corresponds to the first column within the product matrix and also to the first column within **m**! Furthermore, a teacher receives might not all questions: The questionnaire is dynamic and user-adapting – the teacher gets only necessary questions. Any skipped question results implicit in the answer -100 which means "does not matter".

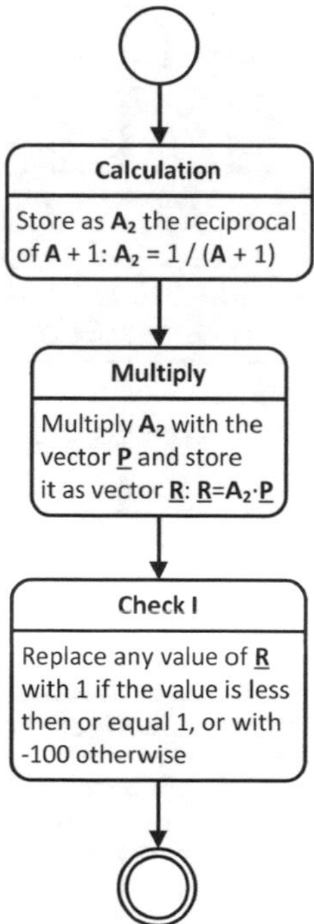

Fig. 4 The sub-algorithm for the "numeric question" analysis

The next steps are repeated for each pair of an answer (the answer as a single value, here called **A**) and corresponding column out of the product matrix as vector $\underline{\mathbf{P}}$.

The algorithm is now branching out by the column type (for now: issue, numeric or range). If an unknown type of column is found, the algorithm gets unexpected terminated. The sub-algorithms for the different types are found at Fig. 3 (type is "issue"), Fig. 4 (type is "numeric question") and Fig. 5 (type is "range").

- In case of "issue", multiply **A** with $\underline{\mathbf{P}}$ and store the result as vector $\underline{\mathbf{R}}$: $\underline{\mathbf{R}} = \mathbf{A} \cdot \underline{\mathbf{P}}$. Check then, if any value of $\underline{\mathbf{R}}$ equals 10000. If so, replace it by 1. If any value of $\underline{\mathbf{R}}$ equals -200, replace it also with 1.
- If the type of the current column is "numeric question", store as $\mathbf{A}_2$ the reciprocal of $\mathbf{A} + 1$: $\mathbf{A}_2 = \frac{1}{\mathbf{A}+1}$. Now multiply $\mathbf{A}_2$ with $\underline{\mathbf{P}}$ and store the result as vector $\underline{\mathbf{R}}$: $\underline{\mathbf{R}} = \mathbf{A}_2 \cdot \underline{\mathbf{P}}$. Replace all values of $\underline{\mathbf{R}}$ with 1 if the value is less or equals 1 or otherwise replace it with -100.
- The "range" type is represented by two columns so instead of one $\underline{\mathbf{P}}$ this type has two vectors $\underline{\mathbf{P}}_{min}$ and $\underline{\mathbf{P}}_{max}$. If **A** equals -100, create the zero vector $\underline{\mathbf{R}}$ with the size of $\underline{\mathbf{P}}_{min}$ and this sub-algorithm is done. Otherwise, store as $\mathbf{A}_2$ the reciprocal

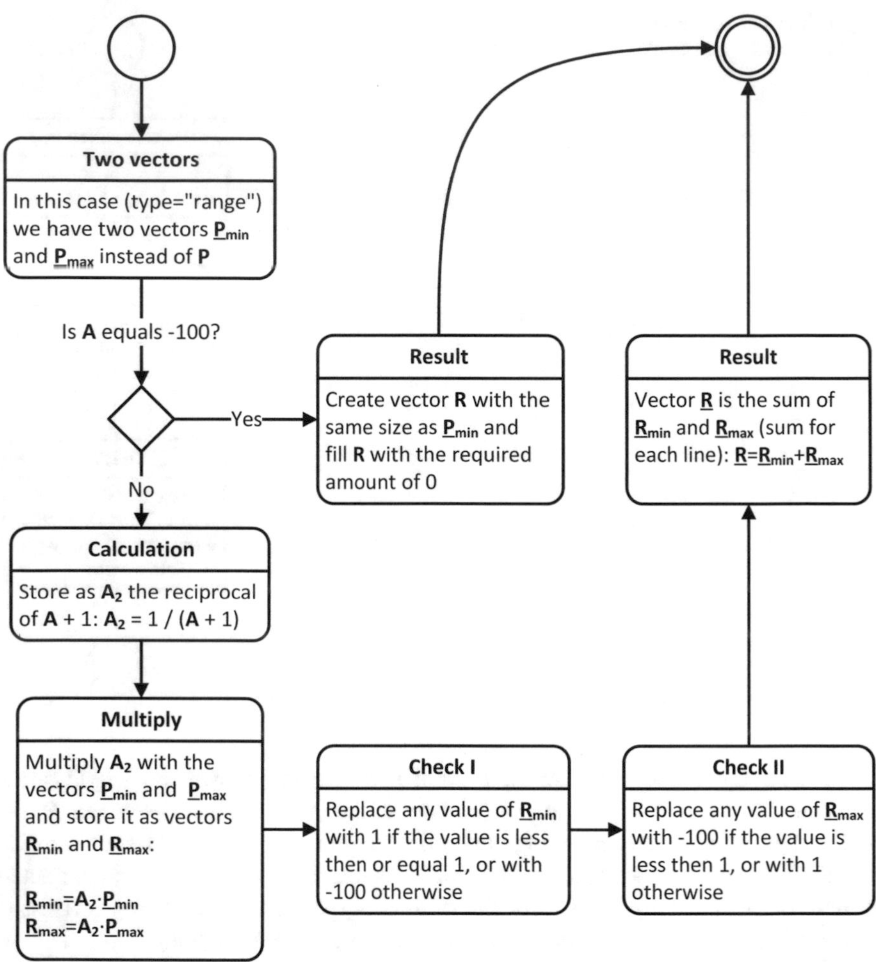

Fig. 5 The sub-algorithm for the “range” analysis

of $\mathbf{A} + 1$: $\mathbf{A}_2 = \frac{1}{\mathbf{A}+1}$. Next, multiply $\mathbf{A}_2$ with $\underline{\mathbf{P}}_{\text{min}}$, $\underline{\mathbf{P}}_{\text{max}}$ and store the result as $\underline{\mathbf{R}}_{\text{min}}$ and $\underline{\mathbf{R}}_{\text{max}}$. Replace now any element of $\underline{\mathbf{R}}_{\text{min}}$ with 1 if the value is less or equals 1, otherwise replace it with -100. Also replace any element of $\underline{\mathbf{R}}_{\text{max}}$ with -100 if the value is less than 1, otherwise replace it with 1. The result vector $\underline{\mathbf{R}}$ is now $\underline{\mathbf{R}} = \underline{\mathbf{R}}_{\text{min}} + \underline{\mathbf{R}}_{\text{max}}$.

After the right sub-algorithm, the result vector $\underline{\mathbf{R}}$ must be stored at the corresponding position in $\mathbf{m}$. If not yet all pairs of answers and product data are executed, then select the next pair and repeat the steps above. If no pairs are left, take the sum of each line of matrix $\mathbf{m}$ and store the sum at the new vector $\underline{\mathbf{z}}$:

$$\mathbf{z}_1 = \sum_{n=1}^{v} \mathbf{m}_{1,n} \tag{1}$$

$$\mathbf{z}_2 = \sum_{n=1}^{v} \mathbf{m}_{2,n}$$

$$\vdots$$

$$\mathbf{z}_u = \sum_{n=1}^{v} \mathbf{m}_{u,n}$$

$$\underline{\mathbf{z}} = \begin{pmatrix} \mathbf{z}_1 \\ \mathbf{z}_2 \\ \vdots \\ \mathbf{z}_u \end{pmatrix} \tag{2}$$

The variable v starts at 1 and grows up to the amount of issues, numeric questions and two times the amount of range questions. The variable u starts at 1 and grows up to the amount of products.

Interpretation of the results: The vector $\underline{\mathbf{z}}$ provides for each product one value. If a value is greater or equals 0, the corresponding product matches to the teacher's requirements. Moreover, if the resulting values are descending sorted, this new ordered list represents the teacher's best matching solution down to the worst solution.[10]

6 Results

The chosen software stack has already proved its power: The development time has been short and the source code is clear, with less overhead and a strong focus to web development. Thereby, the first prototype is a responsive web solution: This saves time, because the mobile devices are covered without an additional mobile app. The agile process made it possible to react on changes to the subject of e-learning and also to common project changes.

The first technical prototype with the visual programming language and the questionnaire is running without errors: It is possible to write a visual program and test the user flow through the resulting questionnaire. It is possible to change the visual program while users are inside the user flow of the questionnaire: The users in front of the changed function blocks are receiving directly these changes, and users behind

[10]If the value is less than 0, the product is of course not a "solution" for this teacher.

the changed blocks are not affected at all. This feature enables to run a long-term system with no or less maintenance impacts.

The visual web-based editor for creating a visual program is convenient and enables also staff without computer science knowledge to create a visual program. The first version of the algorithm is passing all test cases with dedicated test data: The algorithm is working deterministic and the results are correct for any expected input. In case of the comparable characteristics for the related e-learning solutions, which are the precondition for this recommendation system, a first proof of concept exists.

7 Conclusion and Outlook

To meet the precondition of comparable characteristics for related e-learning solutions it is promising to get the input of experts to reach a broad scientific consensus. Because it was possible to build a proof of concept for the characteristics, it is confident to meet this precondition. The recommendation algorithm needs further research to investigate the performance under real conditions with a huge solutions catalog and also huge amount of answers. Therefore, the algorithm must be implemented into the prototype.

The new visual user-flow language enables to build user-adapting questionnaires. This saves processing time for the teachers, because their get just the necessary parts of the whole questionnaire. Nevertheless: The user-flow program is able to cover the complexity from unexperienced to advanced teachers, regarding to the e-learning subject.

In the long-term view: If the web-based recommendation system goes online, the service enables teachers to save time and let them focus to the engineering education. Later, the recommendation system can be expanded to other disciplines, beyond engineering education. After the project reaches a more mature stage, it will be accessible as open source under the 2-clause BSD license to enable others to use and modify it.

References

1. A. Sangrà , D. Vlachopoulos, N. Cabrera, Building an inclusive definition of e-learning: An approach to the conceptual framework. The International Review of Research in Open and Distance Learning **13** (2), 2012, pp. 145–159. http://www.irrodl.org/index.php/irrodl/article/view/1161
2. R.E. Mayer, ELEMENTS OF a SCIENCE OF e-LEARNING. Journal of Educational Computing Research **29** (3), 2003, pp. 297–313. doi:10.2190/YJLG-09F9-XKAX-753D. http://baywood.metapress.com/app/home/contribution.asp?referrer=parent&backto=issue,2,9;journal,78,191;linkingpublicationresults,1:300321,1
3. S. Schoen, M. Ebner, eds., *Lehrbuch fuer Lernen und Lehren mit Technologien*, 2nd edn. 2013. http://www.epubli.de/shop/buch/2000000030471

4. Z. Cebeci, M. Tekdal, Using podcasts as audio learning objects. Interdisciplinary Journal of E-Learning and Learning Objects **2** (1), 2006, pp. 47–57. http://editlib.org/p/44813
5. A. McAuley, B. Stewart, G. Siemens, D. Cormier, *Massive Open Online Courses. Digital Ways of Knowing and Learning. The MOOC Model for Digital Practice.* 2010. http://www.elearnspace.org/Articles/MOOC_Final.pdf
6. M. Rosenblum, The reincarnation of virtual machines. Queue **2** (5), 2004, p. 34–40. doi:10.1145/1016998.1017000. http://doi.acm.org/10.1145/1016998.1017000
7. M. Fenn, M. Murphy, J. Martin, S. Goasguen, An evaluation of KVM for use in cloud computing. In: *Proc. 2nd International Conference on the Virtual Computing Initiative, RTP, NC, USA*. 2008
8. S. Mohorovicic, Implementing responsive web design for enhanced web presence. In: *2013 36th International Convention on Information Communication Technology Electronics Microelectronics (MIPRO)*. 2013, pp. 1206–1210
9. D.D. Hils, Visual languages and computing survey: Data flow visual programming languages. Journal of Visual Languages & Computing **3** (1), 1992, pp. 69–101. doi:10.1016/1045-926X(92)90034-J. http://www.sciencedirect.com/science/article/pii/1045926X9290034J
10. H. Wolf, *Agile Softwareentwicklung: Werte, Konzepte und Methoden*. dpunkt, Heidelberg, 2011
11. M. Fowler, J. Highsmith, The agile manifesto. Software Development **9** (8), 2001, p. 28–35
12. M. Poppendieck, M. Cusumano, Lean software development: A tutorial. IEEE Software **29** (5), 2012, pp. 26–32. doi:10.1109/MS.2012.107
13. J. Kuster, *Handbuch Projektmanagement*. Springer, Berlin [u.a.], 2011
14. B. Niessen, *Der eigene Server mit FreeBSD 9 Konfiguration, Sicherheit und Pflege*. dpunkt verlag, Heidelberg, 2012. http://proquest.safaribooksonline.com/?fpi=9781457170034
15. P.H. Kamp, R.N. Watson, Jails: Confining the omnipotent root. In: *Proceedings of the 2nd International SANE Conference*, vol. 43. 2000, vol. 43, p. 116
16. P.H. Kamp, R. Watson, Building systems to be shared, securely. Queue **2** (5), 2004, p. 42–51. doi:10.1145/1016998.1017001. http://doi.acm.org/10.1145/1016998.1017001
17. C. Nedelcu, *Nginx http server 2nd edition*. Packt Publishing Limited, Birmingham, UK, 2013
18. S. Dabkiewicz, Web server performance analysis , 2012
19. K. Chodorow, *MongoDB: the definitive guide*. O'Reilly, Sebastopol, CA, 2013
20. A. Boicea, F. Radulescu, L. Agapin, MongoDB vs oracle – database comparison. In: *2012 Third International Conference on Emerging Intelligent Data and Web Technologies (EIDWT)*. 2012, pp. 330–335. doi:10.1109/EIDWT.2012.32
21. Z. Wei-ping, L. Ming-xin, C. Huan, Using MongoDB to implement textbook management system instead of MySQL. In: *2011 IEEE 3rd International Conference on Communication Software and Networks (ICCSN)*. 2011, pp. 303–305. doi:10.1109/ICCSN.2011.6013720
22. D. Crockford, The application/json media type for javascript object notation (json), 2006
23. D. Flanagan, Y. Matsumoto, *The Ruby programming language*. O'Reilly, Beijing; Sebastopol, CA, 2008
24. A. Harris, K. Haase, *Sinatra up and running*. O'Reilly Media, Sebastopol, CA, 2012
25. C.O. Nutter, *Using JRuby: bringing Ruby to Java*. Pragmatic Bookshelf, Raleigh, 2011
26. J. Kutner, *Deploying with JRuby: deliver scalable web apps using the JVM*. Pragmatic Bookshelf, Dallas, TX, 2012

A Web-Based Recommendation System for Engineering Education E-Learning Solutions

Thorsten Sommer, Ursula Bach, Anja Richert and Sabina Jeschke

Abstract The e-learning market consists of a wide variety of products, and it is still growing. To find an e-learning solution which fits the particular and situational demands is a very time consuming task, especially for teachers. Moreover, the technical and operational differences between the e-learning solutions are often not easy to understand from the product data and thereby consequences of choices are maybe not clear to the teacher. To solve these problems, a web-based recommendation system for teachers of engineering education is under development. This system is planned to support the decision making process of teachers about the use of an e-learning system. The precondition for setting up a recommendation system is that the desired entries (e.g. products, solutions, music or movies, etc.) are comparable to allow the algorithm to recommend: The current approach is to develop an e-learning scheme and compare the solutions based on this scheme. The determining of the necessary information for each e-learning solution has to be at least a half-automated process to keep the information up-to-date. Since expensive human time is needed to handle the necessary information, some approach with data and text mining as well as text analytics is promising. After the determining phase, each e-learning solution is represented by its data sheet. Apart from the e-learning solutions, also a teacher's requirements have to be comparable to e-learning solutions to allow the algorithm to recommend. That is why a web-based questionnaire is utilized to catch the teachers' requirements. A visual user-flow programming language is under development to provide an adequate environment for the development of the questionnaire and as an interface between the e-learning scheme, questionnaire and user-flow. The next step is to develop a functional prototype for the essential text analysis process to proof the concept. It is also required to analyze the current state of the e-learning scheme further to identify clusters of similar subjects, and to identify all critical properties to provide individual solutions for these.

T. Sommer (✉) · U. Bach · A. Richert · S. Jeschke
IMA/ZLW & IfU, RWTH Aachen University, Dennewartstr. 27, 52068 Aachen, Germany
e-mail: thorsten.sommer@ima-zlw-ifu.rwth-aachen.de

Originally published in "Proceedings of the 9th International Conference on e-Learning (ICEL)", © 2014. Reprint by Springer International Publishing AG 2016, DOI 10.1007/978-3-319-46916-4_23

Keywords E-Learning · Recommendation System · E-Learning Scheme · Text Mining · Visual Programming Language

1 Introduction

To assist teachers of engineering education in selecting an e-learning solution, a web-based recommendation system is under development: It enables teachers to reduce the amount of time finding an e-learning solution that matches their requirements.

The recommendation system requires a catalog with e-learning solutions. Unfortunately there is no common definition for "e-learning". Regarding the recommendation system, any technology aided solution a teacher for engineering education can use to teach or helps students to learn, is an e-learning solution. Therefore, the following definition was chosen: "E-learning is an approach to teaching and learning, representing all or part of the educational model applied, that is based on the use of electronic media and devices as tools for improving access to training, communication and interaction and that facilitates the adoption of new ways of understanding and developing learning" [1]. This definition matches the expectations above and does not exclude any possible e-learning solution. Thus, this recommendation system will be recommend engineering education solutions like e.g. software and web-based solutions [2], hardware solutions [3] as well as game-based solutions [4], etc. Anyhow, this definition is broad but can be restricted if necessary.

A web-search for "e-learning" shows more than one billion results and a book-search for "e-learning" shows more than 60.000 books in English. Obviously, the teacher needs much time to handle this information overload. The consequences of a choice are perhaps not foreseeable: Is the chosen system able to handle the necessary amount of students, is the system prepared for the desired didactic method etc.?

The recommendation system proposed is parted into a public part (the website for teachers, called frontend) and a private administration part (the backend). The precondition for this recommendation system is that all e-learning solutions are comparable with each other, at least in respect to particular aspects: Otherwise it is not possible for algorithms to match the e-learning solutions to the teacher's requirements. The approach to provide a comparability is to find or define a general e-learning scheme. On the teachers' website, the teacher will receive questions out of a questionnaire to catch the teacher's requirements. The teacher's answers are the input for the recommendation algorithm: Together with the product data, the algorithm will be able to rate all solutions and present the teacher a weighted list of e-learning solutions.

To apply the e-learning scheme to the e-learning solutions, it is necessary to determine the required information for any solution. Both is not trivial and contains multiple challenges. This paper is focused to the approaches around the common scheme, the determining of the necessary information and how the teachers' questionnaire can be connected to the scheme.

2 E-Learning Scheme

Recommendation algorithms must be able to compare the desired entries (e.g. products, solutions, music or movies, etc.) among each other and with the input (e.g. answers, choices, requirements). In this case, all desired engineering educational e-learning solutions have to be comparable: It is desired to compare the technical aspects (e.g. hardware and software requirements, known scalability limits, supported data formats) operational behavior (e.g. availability of live interaction, availability of user-roles, possibility to comment, availability of access control) and didactic possibilities (e.g. possibility to declare learning objectives, availability of an interactive whiteboard, possibility of group work). Unfortunately, such a common e-learning scheme does not exist.

However, e-learning standards, specifications and reference models are available, e.g. IEEE P1484.1/D9 [5], SCORM [6], IMS CC [7]: But these are defining e.g. how learning units and content can be exchanged between different systems [6] or defining general e-learning system architectures [5]. Interesting is the "IMS Interoperability Conformance Certification Status" list (IMS Global Learning Consortium, Inc. 2013): These e-learning systems mentioned there are comparable because these systems are implementing the IMS CC standard e.g. at different versions. Nevertheless, this approach is not enough for this e-learning recommendation system: It would be a precondition, that interested teachers know and understand these standards and that all desired e-learning solutions and e-learning related systems are implementing these standard. Therefore, this approach is not sustainable for teachers nor for this recommendation system.

Further to these standards, e-learning classifications are available ([8]: Chapter 1.2): Schulmeister used the degree of interaction and Kolb used the categories "online" and "offline" to classify e-learning solutions. Richert allows also hybrids where an offline e-learning solutions can use additional an internet-connection to allow communication. Baumgartner and Payr define a model to classify the e-learning solutions by learning objective, learning content and teaching methods. None of these classifications compared by Richert [8] and none of the standards above covers the desired technical aspects, operational behavior and also the didactics: Nevertheless, the classifications suggested by Richert are covering parts of the technical aspects and parts of the didactics.

A modern alternative for a recommendation algorithm is to recommend based on e.g. social profiles [9, 10]: For each teacher who want a recommendation, the algorithm catches as much social information as possible and tries to recommend based on this data. Wherever this approach will work for the e-learning context (regarding the teachers' point of view, who is requesting an e-learning solution) is an open question. Especially, wherever the audience (teachers) is using the social networks enough for a good recommendation, is at least a demographic question [11]. Moreover, some aims are not related to the teacher, e.g. the technical aspects maybe related to the institute.

Therefore, the approach is to find or define an e-learning scheme that covers the necessary properties of e-learning solutions. This common e-learning scheme is subject of changes, because the e-learning market is changing [12]. It is preferred that also the teachers' questionnaire is connected to the scheme to ensure that the teachers' answers also comparable to the e-learning solutions: This mean, the e-learning scheme is also an interface between the teachers' answers and the e-learning solutions.

The current development state of the scheme contains more than 40 properties: Some parts of the classifications compared by Richert [8] are already included. However, the distinction by e.g. "online" and "offline" was no longer adequate and was replaced by "offline", "intranet" and "cloud": A "intranet" solution [13] can run on its own server at the institute, which occur also as an online solution to the students however the "cloud" solution runs on a third-party infrastructure and the institute has to pay continuously fees.

It is not possible to discuss more than 40 properties here: The following example properties are illustrating the general principle.

- Property #1: "Commercial solutions versus free solutions"
- Property #2: "Number of students"
- Property #3: "Scheduled content versus static content"
- Property #4: "Usability for teachers"
- Property #5: "Usability for students"

Some desired properties of the scheme are critical, e.g. #4 and #5. How to measure or rate the usability as human without to influence the result [14]? It is supposed, that the necessary degree of usability depends on the teacher's and student's experiences with such solutions. A naive approach (AP1) would be to count how often words or terms e.g. "problem", "complicated" and "does not understand" appears at the community for each e-learning solution to measure how difficult each solution is. However, this approach is unsuitable: Statistically, it is expected that for popular e-learning solutions also proportional more problems are reported.

It would be necessary to scale these measurements to a standardized scale e.g. from 0 to 10. Unfortunately is this not possible, because the population is unknown (e.g. the count of installations or the count of customers). Another approach is required to cover these critical desired properties. Perhaps it is enough to provide for each of these critical properties a suitable treatment: In case of this usability example, a few extra questions in the questionnaire are appropriate to proof or check the teacher's e-learning experience (AP2).

An alternative approach (AP3) uses the collected data from the vendors (see next section) and search for entries that are related to e.g. the usability subject. The questionnaire does not contain questions regarding the critical properties: Instead, the weighted list of results provide for each solution an abstract of related collected content. These abstracts are enabling the teacher to measure e.g. the usability to the own experiences: This is similar to classical customer reviews of online merchants and it is possible to analyze these data [15, 16].

Unfortunately, also this approach solves not all problems as the following thought experiment illustrate: Assume a teacher has completed his questionnaire and reads now the weighted list of results. Some solutions do not provide an abstract about the usability, because there were no records about it. A few solutions are providing abstracts with negative usability opinions, which is not interesting for the current teacher, because the reviews are related to e.g. Microsoft Windows, and the teacher wants to use e.g. a UNIX environment instead. Further, these solutions without abstracts are providing poor usability, which caused the absence of reviews. The current teacher chose an e-learning solution without an abstract, because of the assumption that none critical opinions are better as any critical opinions. At the conclusion, all three approaches have pros and cons but overall is the approach AP2 most convincing to satisfy the teachers.

3 Determining Solutions' Data

The current list of e-learning solutions was created by ourselves by researching public resources (e.g. websites, books, scientific journals). This list contains approximately 80 different solutions; each was selected according to the e-learning definition at Section 1. For each of these solutions, it is necessary to determine the required information to match the scheme: The result is a data sheet for each solution. It is not possible to produce these data sheets manually (purchase, install and evaluate each solution), because even with virtualization environments [17, 18] is this process to slow. It is obvious, that this manual task required many resources (e.g. time, money and staff). One alternative would be a crowdsourcing or community approach: The solutions' data to match the scheme would be mined by the crowd [19] or would be created by the community as user-generated content [20].

Moreover, the resulting data sheets are snapshots of one moment in time. It is required to keep these data sheets up-to-date to ensure that these cover the current state of each solution. Of course, the manual process could also utilize web and book resources instead of buying each solution, etc. Nevertheless, the major cost factor is the required staff costs: It is expected that the utilization of these web and book resources takes a long time. Therefore, an approach with data mining, text mining [21] and text analysis [22, 23] methods is promising: The needed runtime is usually short enough to execute the process continuously e.g. once a quarter. For each desired e-learning solution, the public vendor's information is captured. Additional, it is possible to capture also community information (e.g. discussion boards).

The determining process is split into data retrieval and the information retrieval. The data retrieval part fetches the raw data from the desired sources (e.g. vendors' public websites, communities) as HTML data [24–26]. A preprocessing of the raw data removes e.g. images and stores the data into a database. The information retrieval tries to extract specific information out of the preprocessed data [27]. In the case of this recommendation system, the main advantage is that the kind (vendors' information) and the context (e-learning) of the text data is known [21] and thereby the terminology that can occur is known. This is an important difference to the general text analytics purposes, where the context and kind of the text are unknown: In such a case, additional analytic methods are required to determine the kind and the context of the data.

Where the vendors information are at least not structured text data, the community text data is worse: Noisy text analysis is needed to process such data with issues like e.g. no case-sensitive wording, no punctuation, etc. because it is not possible to detect sentences or nouns [9]. However, a text mining method called "co-occurrences" is promising to get for a word related other words from its neighborhood [21]. It is even possible to get higher-order co-occurrences, if the co-occurrences for each vendor's data is consolidated together [21].

The current approach is to predefine one or more terms for any part of the common e-learning scheme and test these against the collected data. This approach based on the advantage, that the kind and context are known. A few examples to illustrate the principle: Assumed "Platform: Microsoft Windows?" is part of the scheme. The question is, if the solution operates on a Microsoft Windows server environment. If the predefined terms "Windows" and at least one of "not available" or "not ready" occur within one sentence, the process can assume with a certain probability that this e-learning solution is not available for Microsoft Windows environments. The property #1 from above, "Commercial solutions versus free solutions", uses e.g. "dollar" and "euro" and supposed that both are not part of the vendor's information and at least one of "is free", "for free" or "free to use" occurs to assume together that a solution is not commercial. A dictionary with synonyms is not provided: the user who is defining these terms must provide alternatives as above (e.g. "not available", "not ready").

These are simple examples to show the principle, because with real text data, the analysis it not trivial. Use the example "Platform: Microsoft Windows?" once again: With the simple terms from above, the text "A version for Microsoft Windows is may available later." does not match the terms, because the semantic of the text is not known [21]. For this approach, further research and a technically proof of concept are required. With a functional prototype, it is possible to evaluate and compare the performance of this process with a sample (means in this case a test group of humans). In the case of the accuracy, this data mining and text analysis process must reach some degree of accuracy and must avoid serious errors but it is supposed that the sample performs better in such a complicated case: The humans are able to understand the semantic of a text easier. Nevertheless, it is also supposed that the sample is significant slower.

The expected result of this approach is, that at least half automated data sheets are producible with a good accuracy compared to the sample performance. In practice, it is perhaps possible to use natural language processing (NLP) [28] to solve some challenges e.g. understanding the semantic [29]. To work with NLP, programming libraries like e.g. NLTK available to speed up the development [30].

4 Visual User-Flow Language

To capture the teacher's requirements to an e-learning system, a questionnaire that is related to the e-learning scheme is necessary: Each question must be linked to the common scheme to ensure the comparability to the data sheets. To realize this requirement, an own visual programming language [31] was defined to model the user-flow through the questionnaire and to link the related parts of the scheme. The main advantage for a visual language is the visibility of the flow: The observer can directly view the flow through the graph because the flow is represented by e.g. lines or arrows to connect objects.

This language is represented by squares (called "function blocks") connected by arrows. The resulting visual program is linking the questionnaire, the common e-learning scheme and the user-flow together: For each inserted function block, the programmer has to choose the part of the common scheme and provide the related teachers' question.

Figure 1 shows an example to explain and illustrate the language: The program must read from left to right, from the unique start block to any of the end blocks. The starting point directs the teacher to question 1: If the teacher answers with "yes" or "neutral", the teacher is directed to the numeric query 1 - in case of the answer "no", the teacher reaches question 2. After the numeric query 1, the user-flow also reaches question 2. Further, if the teacher answer question 2 with "yes", question 3 follow up by end 1. If the answer for question 2 is "neutral" or "no", end 1 occurs directly. Instead of these example labels (e.g. "Question 1") the used part of the scheme is printed. Therefore, the observer can get a good overview about the program. With a double click, the details behind the function blocks are viewable: This is e.g. the question for the teachers, further explanation, numeric values, etc.

A first technical prototype, and thereby a technical proof of concept, exist: It is possible to program a user-flow with questions and subjects (to simulate the teachers) are able to going through these questions. Moreover, the visual program is changeable even if subjects are currently within the flow. Another thought experiment: Using the example at Fig. 1 once again and assume there are four subjects (A, B, C and D): A is located at the starting block, B is located at "Question 1", C is located at "Question 3" and D is located in the end. Assume that "Question 2" is now changed: If subjects are located before the changed block (subject A and B), these subjects are receiving the changes directly without any interruption. Subject C and D are located behind the changed block: These subjects are not interrupted and effected. The recommendation system is, therefore, convenient to maintain.

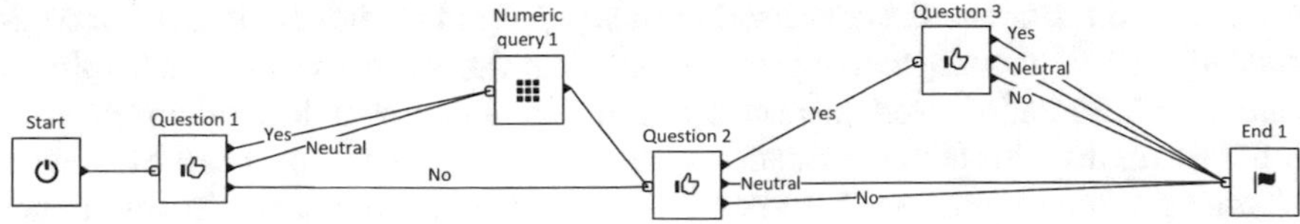

Fig. 1 Example for a visual program

If the common scheme is changed to cover changes on the e-learning market, the visual program must also change to provide related questions. Perhaps, also new function blocks for the language are required by some scheme changes. Therefore, this part of the recommendation system is permanently under development.

5 Results

The current results for this web-based recommendation system are promising: In case of the common e-learning scheme, the e-learning market currently knows multiple standards, specifications and classifications, but none covers the desired properties of technical requirements, operational behavior and didactics. Nevertheless, a common e-learning scheme is necessary to compare e-learning solutions. The approach is to define desired properties for the common e-learning scheme: Currently, approximate 40 plausible properties are part of the scheme. Some of these properties are difficult to measure or to rate because a standardized scale is missing: To define an own scale the underlying population is required - but often not available.

The manual determining of the necessary e-learning solution's data is expensive and time consuming. The scientific field of data mining, text mining, text analysis and NLP provide methods to retrieve information out of the text data. Depending on the text source, the handling of so called "noisy text" is required and e.g. it is not possible to detect nouns and sentences, which is the precondition of many algorithms and methods. Another problem is the semantic of the text data: Negating of the meaning or subjunctive sentences are difficult to analyze, but perhaps it is possible to cover this by using NLP.

Finally, the visual language is under development and allows it to program a user-flow with questions. Teachers are able to going through the questions according to the visual program. The visual program is even changeable if some teachers are currently within the flow: Teachers located before the changed blocks are receiving the changes directly without any interruption. In the case that some teachers are located behind the changed blocks, these teachers are not interrupted and effected. For the maintenance of the recommendation system, this behavior is convenient.

6 Conclusion and Outlook

From many discussions and conversations it was possible to infer that such a recommendation system for engineering education e-learning systems would be helpful to handle the information overload. The precondition of comparable solutions must be fulfilled to realize this recommendation system. The current approach is the common e-learning scheme: Because of the possibility that the mentioned challenges are not solvable, the research for an alternative to recommend is needed. The current over-all development state is promising to achieve this plan.

It is necessary for the determining of the data sheets to develop at least a half automated process to handle the amount of data. To ensure that the data set is up-to-date, this process must be able to run on a continuous basis. The current approach with data mining methods and textual analysis is promising but needs further investigation. After a technical proof of concept, the process must be evaluated: To be able to rate the evaluation results, a comparison with a sample (means here a test group of humans) is needed.

The current development state of the visual user-flow programming language is promising: the integration of the questionnaire, the user-flow and the scheme fulfills the expectations. The visual aspect of the language is also adequate to provide a good overview over the whole program and the flow through the questionnaire. The availability to change the program at runtime without interruption to the teachers is a good foundation for a convenient maintenance. The visual language is continuously under development to cover the current state of the scheme.

At the outlook, there are challenges but it is confident to solve these: Instead of asking the teacher about the own experiences with e-learning, a part of the questionnaire may need to proof the teacher's experiences. More research and different approaches are needed to cover this issue. Also needed is a further analysis to find dependencies between the properties of the scheme to build logical clusters.

The new aspect about the common e-learning scheme would be that different e-learning systems are comparable: It is then possible to compare a web-based solution with a game-based solution or an MOOC solution with a learning management system (LMS) and of course solutions of the same kind. Before this advantage is accessible, the determining of the necessary data for any solution is needed. Further investigation and research are required for the determining of the data, but the first approach is promising to get at least some basic information about any e-learning solution. Unlike a manual process of determining the data, this automated process can run continuously to keep the data up-to-date.

References

1. A. Sangrà, D. Vlachopoulos, N. Cabrera, Building an inclusive definition of e-learning: An approach to the conceptual framework. The International Review of Research in Open and Distance Learning **13** (2), 2012, pp. 145–159
2. S. Schön, M. Ebner, eds., *Lehrbuch für Lernen und Lehren mit Technologien*, 2nd edn. 2013. http://www.l3t.eu
3. K. Hamdan, N. Al-Qirim, M. Asmar, The effect of smart board on students behavior and motivation. In: *2012 International Conference on Innovations in Information Technology (IIT)*. 2012, pp. 162–166
4. D. Short, Teaching scientific concepts using a virtual world—Minecraft. Teaching Science-the Journal of the Australian Science Teachers Association **58** (3), 2012, p. 55
5. J. Tyler, B. Cheikes, *IEEE P1484.1/D9 - Draft Standard for Learning Technology - Learning Technology Systems Architecture (LTSA)*. IEEE Standards Activities Department, 2001
6. V. Devedzic, J. Jovanovic, D. Gasevic, The pragmatics of current e-learning standards. IEEE Internet Computing **11** (3), 2007, pp. 19–27

7. G. Durand, L. Belliveau, B. Craig, Simple learning design 2.0. In: *2010 IEEE 10th International Conference on Advanced Learning Technologies (ICALT)*. 2010, pp. 549–551
8. A. Richert, *Einfluss von Lernbiografien und subjektiven Theorien auf selbst gesteuertes Einzellernen mittels E-Learning am Beispiel Fremdsprachenlernen, Europäische Hochschulschriften/European University Studies/Publications Universitaires Européennes*, vol. 979. Peter Lang Publishing Group, Frankfurt am Main, Berlin, Bern, Bruxelles, New York, Oxford, Wien, 2009
9. M. Michelson, S.A. Macskassy, Discovering users' topics of interest on twitter: a first look. In: *Proceedings of the fourth workshop on Analytics for noisy unstructured text data*. 2010, p. 73–80
10. F. Abel, E. Herder, G. Houben, N. Henze, D. Krause, Cross-system user modeling and personalization on the social web. User Modeling and User-Adapted Interaction **23** (2-3), 2013, p. 169–209
11. M. Duggan, J. Brenner, The demographics of social media users, 2012. Tech. rep., Pew Research Center's Internet & American Life Project, 2013
12. L. Johnson, S. Adams Becker, V. Estrada, A. Freeman, *NMC Horizon Report: 2014 Higher Education Edition*. The New Media Consortium, Austin, Texas, 2014
13. P. Henry, E-learning technology, content and services. Education + Training **43** (4/5), 2001, p. 249–255
14. J. Sauer, K. Seibel, B. Rüttinger, The influence of user expertise and prototype fidelity in usability tests. Applied ergonomics **41** (1), 2010, p. 130–140
15. M. Hu, B. Liu, Mining and summarizing customer reviews. In: *Proceedings of the tenth ACM SIGKDD international conference on Knowledge discovery and data mining*. 2004, pp. 168–177
16. M. Hu, B. Liu, Mining opinion features in customer reviews. In: *AAAI*, vol. 4, 2004, pp. 755–760
17. M. Rosenblum, The reincarnation of virtual machines. Queue **2** (5), 2004, p. 34–40
18. M. Fenn, M. Murphy, J. Martin, S. Goasguen, An evaluation of KVM for use in cloud computing. In: *Proc. 2nd International Conference on the Virtual Computing Initiative, RTP, NC, USA*. 2008
19. A. Doan, R. Ramakrishnan, A. Halevy, Crowdsourcing systems on the world-wide web. Commun. ACM **54** (4), 2011, p. 86–96
20. Nov, O. (2007) What motivates wikipedians? Communications of the ACM, **50** (11), pp. 60–64
21. G. Heyer, U. Quasthoff, T. Wittig, *Text Mining: Wissensrohstoff Text: Konzepte, Algorithmen, Ergebnisse*. W3L-Verl., Herdecke, 2006
22. I. Bierschenk, B. Bierschenk, Perspective text analysis: Tutorial to vertex. Kognitionsvetenskaplig forskning: Cognitive Science Research, 2011
23. T. Nasukawa, T. Nagano, Text analysis and knowledge mining system. IBM Systems Journal **40** (4), 2001, pp. 967–984
24. W. Jicheng, H. Yuan, W. Gangshan, Z. Fuyan, Web mining: knowledge discovery on the web. In: *1999 IEEE International Conference on Systems, Man, and Cybernetics, 1999. IEEE SMC '99 Conference Proceedings*, vol. 2. 1999, vol. 2, pp. 137–141
25. G. Pant, P. Srinivasan, F. Menczer, Crawling the web. In: *Web Dynamics*, Springer, 2004, p. 153–177
26. C. Castillo, Effective web crawling. In: *ACM SIGIR Forum*, vol. 39. 2005, vol. 39, p. 55–56
27. O. Egozi, S. Markovitch, E. Gabrilovich, Concept-based information retrieval using explicit semantic analysis. ACM Transactions on Information Systems (TOIS) **29** (2), 2011, p. 8
28. Y. Chen, Natural language processing in web data mining. In: *2010 IEEE 2nd Symposium on Web Society (SWS)*. 2010, pp. 388–391
29. S. M., S. Vranes, A natural language processing for semantic web services. In: *The International Conference on Computer as a Tool, 2005. EUROCON 2005*, vol. 1. 2005, vol. 1, pp. 229–232
30. M. Lobur, A. Romanyuk, M. Romanyshyn, Using NLTK for educational and scientific purposes. In: *2011 11th International Conference The Experience of Designing and Application of CAD Systems in Microelectronics (CADSM)*. 2011, pp. 426–428
31. D. Hils, Visual languages and computing survey: Data flow visual programming languages. Journal of Visual Languages & Computing **3** (1), 1992, pp. 69–101

Diving in? How Users Experience Virtual Environments using the Virtual Theatre

Katharina Schuster, Max Hoffmann, Ursula Bach, Anja Richert and Sabina Jeschke

Abstract Simulations are used in various fields of education. One approach of improving learning with simulations is the development of natural user interfaces, e.g. driving or flight simulators. The Virtual Theatre enables unrestricted movement through a virtual environment by a Head Mounted Display and an omnidirectional floor. In the experimental study presented (n = 38), the effects of objective hardware characteristics were being tested in two groups. The task was the same: Remembering positions of objects after spotting them in a maze. One group fulfilled the task in the Virtual Theatre, the other group on a laptop. Personal characteristics (gaming experience, locus of control) and perception measures for immersion (spatial presence, flow) were also assessed. Analyses show that the Virtual Theatre indeed leads to more spatial presence and flow, but has a negative effect on the task performance. This contradicts the common assumption that immersion leads to better learning.

Keywords Immersion · Spatial Presence · Flow · Learning · Simulators · Natural User Interfaces

1 Introduction

Simulations are used in various fields of education. By imitating real-world processes, personnel skills can be developed, increased or maintained. Especially if the learning process requires expensive equipment or usually would take place in a hazardous environment, the use of simulations is not only beneficial but absolutely necessary [1, 2]. Apart from the software, the user interfaces of the technological systems applied in the simulation environment can affect the learning process [3]. One approach of improving learning with simulations is the development of natural user interfaces. A common example is the use of flight simulators including authentic user interfaces within pilot training instead of using just the simulation on a regular desktop

K. Schuster (✉) · M. Hoffmann · U. Bach · A. Richert · S. Jeschke
IMA/ZLW & IfU, RWTH Aachen University,
Dennewartstr. 27, 52068 Aachen, Germany
e-mail: katharina.schuster@ima-zlw-ifu.rwth-aachen.de

Originally published in "Proceedings of the HCI International Conference",

DOI 10.1007/978-3-319-46916-4_24

computer. According to the classical memory theory, if the context in which we use our knowledge i.e. in which we have to transfer it to new situations resembles the context in which we learned the information in the first place, our memory works better. Moreover, how well we can retrieve knowledge from our long term memory depends on the quality of how well we encoded the information [4]. In the case of computer-aided learning, encoding information can be considered as a task, which is partitioned in at least two parallel sub-tasks: Dealing with the content and controlling the learning environment with the respected user interfaces [5]. Therefore a lot of research and development activities follow the assumption that if the user can interface with the system in a natural way, more focus can be used for training than for the control itself [6]. However, to assume that hardware or software characteristics automatically lead to better learning outcomes is risky. Not every new approach which is technically feasible improves learning in the sense of task performance. The danger of de-signing complex and expensive virtual learning environments without having a positive impact on learning outcomes is obvious. However, judging the value of a virtual environment simply by its effect on task performance misses out on other factors which support learning. Boosting the students' motivation to deal longer, more steady or more effectively with the given content is also an important goal of virtual learning environments [7, 8]. Apart from learning outcome and motivation, a peak to a different domain reveals a third intended effect of virtual environments. According to the entertainment sector, the extent to which a game or in general a virtual environment can "draw you in" functions as a quality seal [1]. This phenomenon is often referred to as immersion. A figurative definition is given by Murray [9]: "Immersion is a metaphorical term derived from the physical experience of being submerged in water. We seek the same feeling from a psychologically immersive experience that we do from a plunge in the ocean or swimming pool: the sensation of being surrounded by a completely other reality, as different as water is from air that takes over all of our attention our whole perceptual apparatus" [9].

Enabling natural movement as the most basic form of interaction is considered an important hardware quality to create immersion [10]. Manufacturers of hardware that are supposed to enhance immersion claim that "Moving naturally in virtual reality creates an unprecedented sense of immersion that cannot be experienced sitting down" [11]. Almost 20 years ago, this could already be confirmed by Slater [10]. Another basic assumption in the context of virtual learning environments and natural user interfaces is that greater immersion means better learning and potentially higher training transfer [3, 6]. This suggests that immersion would be the precondition for better learning, caused by the qualities of the user interfaces. However, if virtual environments are used in educational contexts, those assumptions need to be confirmed by empirical evidence. The presented study therefore focuses on the following questions:

- Do natural user interfaces create a higher sense of immersion?
- Do natural user interfaces lead to better learning?
- Is immersion a necessary precondition for learning with natural user interfaces in virtual environments?

If assessed in an experimental setting, the construct of immersion needs to be specified. Spatial presence and flow are considered key constructs to explain immersive experiences. In general, flow describes the involvement in an activity [12, 13], whilst spatial presence refers to the spatial sense in a mediated environment [10, 14]. Spatial presence, as indicated in the name of the construct, refers to the spatial component of being immersed, i.e. the spatial relation of oneself to the surrounding environment. If we experience spatial presence in a mediated environment, we shift our primary reference frame from physical to virtual reality [14].

2 Method

2.1 Experimental Setting

The study presented in this paper assesses the relationship between personal characteristics, objective hardware characteristics, subjective experiences and task performance. Their expected relationship is visualized in Fig. 1.

All participants had to solve the same task in the same virtual environment, which was a large-scaled maze in a factory building. Within the maze, 11 different objects were located. The first task for the participants was to navigate through the maze and to imprint the positions of the objects to their memory. For that, they were given eight minutes of time. The second task was to recognize the objects seen before in the maze. The third task was to locate the positions of the objects on a map of the maze. This was done on a self-programmed application on a tablet (Nexus 10) with a drag-and-drop control mode. The view of the maze in the first and second task is pictured in Fig. 2.

For both groups, the participants were given the chance to explore a test scenario (an Italian piazza) freely for about three minutes before the actual task started. This was in order to get used to the respected control mode. All experimenters who conducted the experiments were trained in advance by experienced researchers. First they were being trained the functions of the hardware. In a second step, they took the

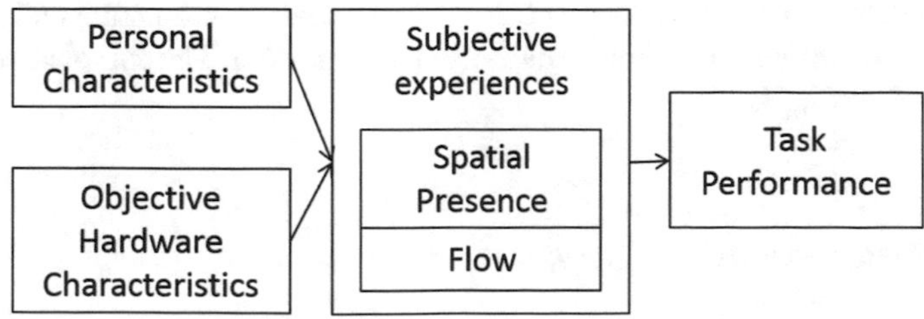

Fig. 1 Expected relationship between personal characteristics, hardware characteristics, subjective experiences and task performance

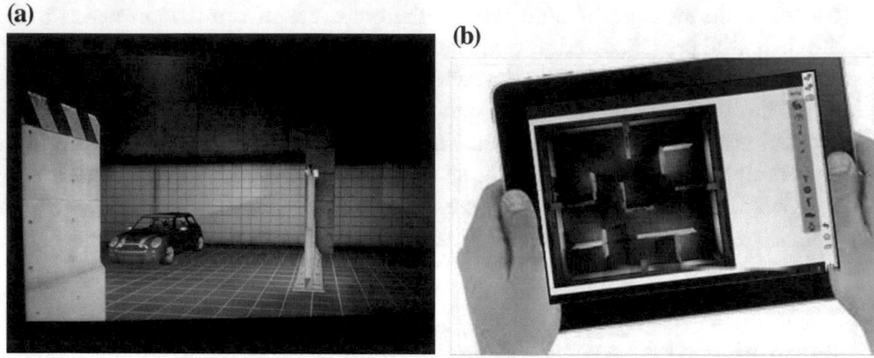

Fig. 2 View of the virtual environment used in the study in the first and in the third task

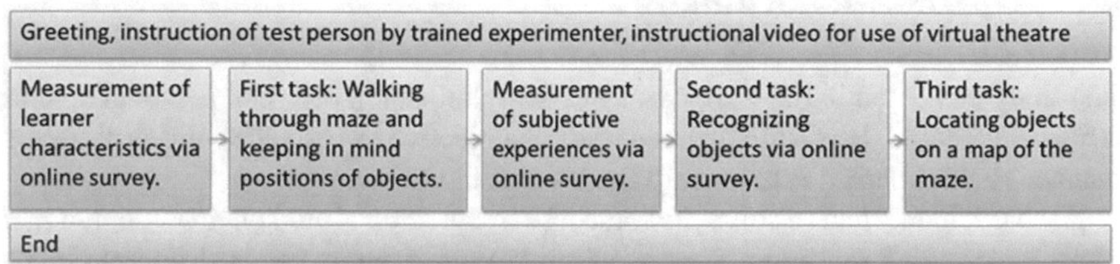

Fig. 3 Procedure of the experiments

observing position in a test run, and thirdly they conducted a test run on their own with the experienced researcher being the observer and giving feedback afterwards. Two groups of test persons were compared, having to use hardware which differed from each other regarding the following characteristics:

- control mode of the field of view,
- control mode of locomotion,
- display and
- body posture of the user.

Due to the composition of the simulator which was applied in the study, the hardware characteristics could only be tested in a certain combination and could not be isolated any further. The whole experiment took one hour. The complete procedure is visualized in Fig. 3.

2.2 Measurements

In this study, spatial presence was measured with elements of the MEC Spatial Presence Questionnaire of Vorderer et al. Several studies conducted by the authors strengthened the postulate of spatial presence being best explained as a two-level

Model. This includes process factors (attention allocation, spatial situation model, self location, possible actions), variables referring to states and actions (higher cognitive involvement, suspension of disbelief), and variables addressing enduring personality factors (i.e. the trait-like constructs domain specific interest, visual spatial imagery, and absorption) [15]. Suspension of disbelief refers to the extent of how much a person pays attention to technical and content-related inconsistencies. The more a person can fade out the action of "looking for errors", the higher the feeling of spatial presence will be according to the theory. In our study, instead of the subscales attention allocation and absorption, we used the Flow-Shortscale of Rheinberg. Flow is the mental state of operation in which a person performing an activity is fully immersed in a feeling of energized focus, full involvement, and enjoyment in the process of the activity. In essence, flow is characterized by complete absorption in what one does, as well as the feeling of smooth and automatic running of all task-relevant thoughts [12, 13].

The perception of a learning situation is highly likely not to be influenced just by objective criteria such as the technical configuration of the learning environment. The strength of spatial presence experienced in a VE is supposed to vary both as a function of individual differences and the characteristics of the VE [3]. A general interest in the topic appeals to a person's curiosity and the motivation to learn something new. If chances to learn or experience something new are low, the motivation to learn decreases [7, 16]. However, we believe that not only interest in a topic but also in the way of presenting it can influence subjective experiences during the learning situation as well as learning outcome. The subscale domain specific interest of the MEC-SPQ refers to the topic of the medium, in our case the virtual environment. Because of the given considerations mentioned above and since interest in mazes didn't seem like a helpful operationalization for domain specific interest, we adapted it to interest in digital games. Additionally, participants were asked to state their gaming frequency, i.e. how many hours they played digital games per week. How well a person learns with the assistance of technology depends not least on whether a person has any experience with the respective technology and if not, feels capable of learning it quickly. The construct "locus of control regarding the use of technology" describes the extent to which a person believes that he or she can control technical devices in everyday life. It is a technology-related personality trait of human-computer interaction and was measured with the short scale from Beier [17]. Based on all theoretical considerations, the general hypothesis of the study was that natural user interfaces should have a positive effect on subjective experiences during the learning situation as well as on learning outcome, in our case operationalized in task performance.

The set of hardware characteristics functioned as the first independent variable in the presented study. Furthermore, the test persons' locus of control regarding the use of technology (second independent variable), interest in digital games (third independent variable) and gaming frequency (fourth independent variable) were measured before the first task. As dependent variables, spatial presence and flow were measured after the first task which had to be fulfilled either in the Virtual Theatre or at the laptop. As dependent measures of task performance, three different variables were analyzed: The number of objects that were correctly recognized in the second task, the third task reaction time and the accuracy of locating the objects on the map in the third task.

2.3 Hardware

Laptop. In the presented study, learning in a Virtual Theatre was compared to a somehow conventional learning with a laptop. The type being used was a Fujitsu Lifebook S761 with a 13,3 inch display and a 1366 × 768 display resolution. The field of view was controlled with a mouse. Locomotion was controlled by WASD-keys, where W/S keys controlled forward and backward while A/D keys controlled left and right. The hardware usually results in a sitting body posture while using the device.

Virtual Theatre. The Virtual Theatre is an innovative platform which enables unrestricted movement through a virtual environment and therefore is used in an upright body posture. The control mode of locomotion is walking naturally. The components of the Virtual Theatre which came to use in the study are pictured in Fig. 4 and moreover described in the following sections. For a more detailed and complete description of the technical system see Ewert et al. [1] and Johansson [6, 18].

Head Mounted Display. In the Virtual Theatre, the field of view is controlled by natural head movement. It presents the user with a seamless 3d visualization of a virtual environment. All head movement is instantly reproduced within the simulation, so the user can look around freely. Visual feedback is received via a

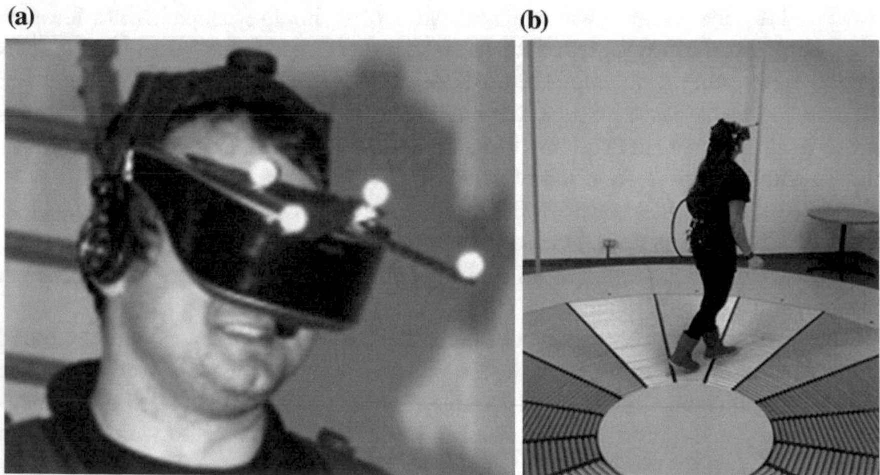

Fig. 4 Head Mounted Display and omnidirectional floor of the Virtual Theatre

zSight [19] Head Mounted Display (HMD), providing a 70° stereoscopic field of view with SXGA resolution for each eye. The HMD weighs 450 grams. It is powered by rechargeable batteries which are located in a vest which the user wears while using the Virtual Theatre. The HMD can be adjusted to the shape of the user's head as well as to his or her eye distance. On a rough scale, the lenses inside the display can be adjusted to short-sightedness and farsightedness.

Omnidirectional floor. The user can move around within the environment by just walking in the desired direction. The omnidirectional floor consists of polygons of rigid rollers with increasing circumferences. Each polygon constitutes 16 rollers and together all polygons form 16 sections with an angle of 22.5 degrees. Each section consists of one roller from each polygon, which means from cylinder origo towards the periphery, the rollers are parallel and increasing in length. Rollers are driven from below by a belt drive assembly to minimize distance between sections. In the central part of the floor there is a circular static area where the user can move without enabling the floor. Floor movements here would only cause the feet of the user to be drawn together. As the user moves outside the static boundary, the floor begins to move according to a control algorithm [18].

Tracking System. To track the movements of a user, the virtual theatre is equipped with 10 infrared cameras. They record the position of designated infrared markers attached to the HMD and an additional hand tracer. The position of the HMD serves on the one hand as an input for controlling the omnidirectional floor: The inward speed of the rollers increases linearly with the measured distance of the HMD to the floor center. On the other hand it is used to direct the position and line of vision of the user within the virtual environment [6, 18]. The position of the hand tracer serves for triggering emergency shutdown: As soon as the hand drops below 0.5 m, e.g. in the event of a user falling down, all movement of the omnidirectional floor is immediately stopped. It should be noted at this point that throughout all studies conducted so far with the Virtual Theatre, this never happened.

2.4 Participants

A total of 38 students between 20 and 33 years ($M = 24.71$; $SD = 3.06$; $n = 13$ female) volunteered to take part in the study. The sample therefore represents a potential user group of virtual environments in higher education. They responded to a call for participation which was hung out at bulletin boards throughout the university but also posted on the front page of the virtual learning platform of the university and on several research and learning related blogs, social media platforms and news feeds. As an incentive and as a sign of appreciation, all participants took part in a drawing for a cordless screwdriver. All participants were healthy and highly interested in participating in the study. They did not report suffering from any physical or mental disorders. To rule out effects due to ametropia, participants were asked in advance to bring their corrective lenses just in case. If participants had been assigned to the Virtual Theatre group, they were asked to wear sturdy shoes.

3 Results

4 Correctional Approach

To gain further information regarding the relationships of the characteristics of the person, the subjective experiences, and the task performance an explorative approach was taken. Thus we calculated correlations between relevant constructs. Results are displayed in Fig. 5.

4.1 Comparing Hardware Conditions

Furthermore hypotheses regarding influences of hardware conditions on subjective experiences and task performance measures were tested with ANOVAs. With regard to the effects of the Virtual Theatre and the laptop on flow, significant differences were found (F (1, 36) = 4.18; $p < .05$). Thus more flow has been experienced in

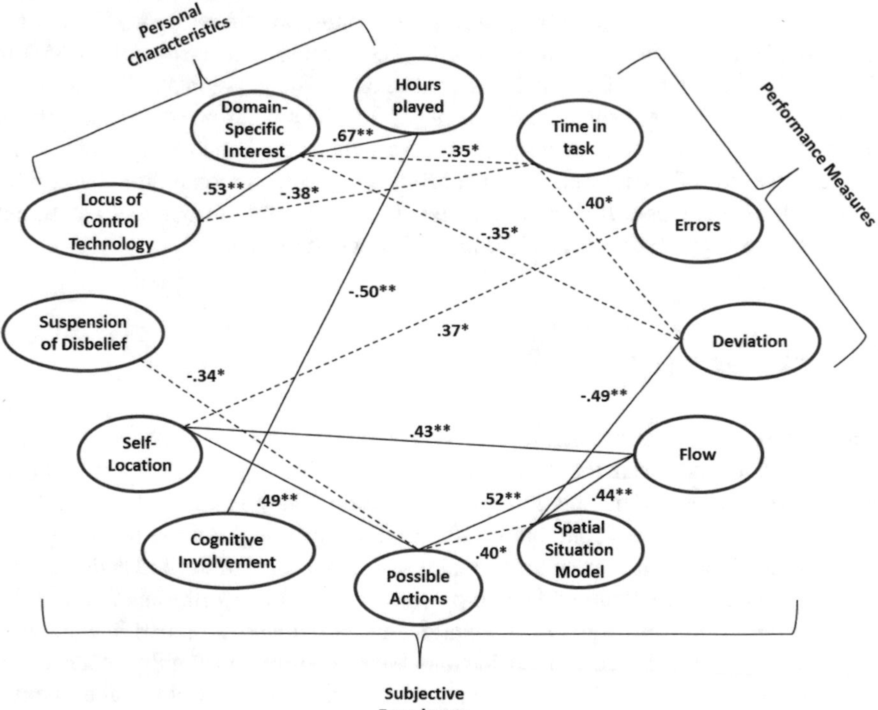

Fig. 5 Correlations between personal characteristics, subjective experiences and task performance

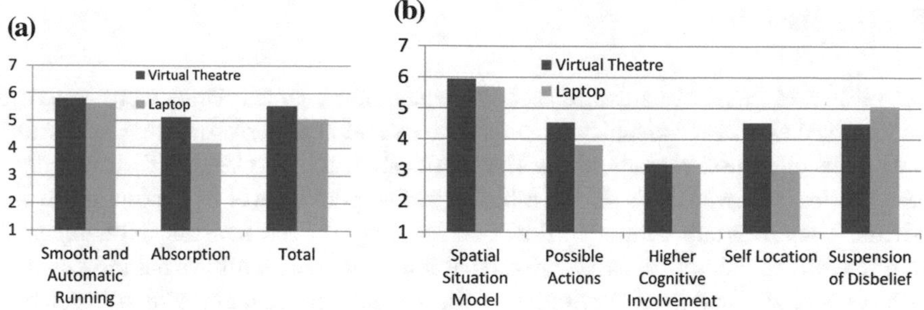

Fig. 6 Effects of Objective Hardware Characteristics on Flow and Spatial Presence

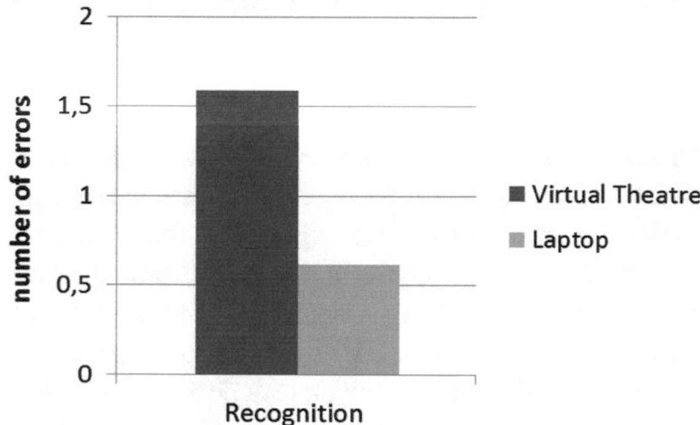

Fig. 7 Effects of Objective Hardware Characteristics on Task Performance, here: Recognition of previously presented objects

the Virtual Theatre (see Fig. 6). Taking a closer look on subscales there is a highly significant difference between conditions in self-reported absorption (F (1, 36) = 10.63; $p < .01$), but not in smooth and automatic running. There are also effects of hardware conditions on spatial presence. Self location in the Virtual Theatre was rated significantly higher (F (1, 36) = 15.79; $p < .001$), which refers to the feeling of actually being in the virtual environment. Similarly, participants in the Virtual Theatre show higher scores on the possible actions subscale of spatial presence (F (1, 36) = 4.90; $p < .05$). There were no further significant effects regarding spatial presence (see Fig. 6).

In addition to that we calculated the effects of hardware on task performance measures. In the recognition task of the objects from the virtual environment, participants in the Virtual Theatre condition made significantly more errors (F (1, 36) = 10.93; $p < .01$), which opposes our hypothesis (see Fig. 7). There were no differences regarding time on task and deviation between the two treatment conditions.

5 Discussion

Two different types of hardware as well as personal characteristics were analyzed regarding their effect on subjective experiences and task performance. Concerning the effects of natural user interfaces, the results show that the Virtual Theatre indeed leads to more flow. This is due to a higher self-reported level of absorption in the Virtual Theatre group. Although flow is an activity related construct, this result is in line with the theoretical assumptions that hardware which allows natural walking can support the feeling of "diving" into the virtual environment, which in general terms is often referred to as immersion.

Next, the effects of the hardware on spatial presence are analyzed in more detail. Students who used the Virtual Theatre reported a higher self-location in the virtual environment which indicates that they had shifted their primary reference frame from the physical to the virtual world. Although the given task didn't require any further nonmental actions but navigating through the virtual environment, students in the Virtual Theatre reported higher on the subscale of possible actions. However, for the other subscales, no differences were measured. Their role will be discussed when looking at the results of the explorative correlations.

According to the results of this study, immersion is not the precondition for better learning in virtual environments with natural user interfaces. Thus, the underlying model of the study (see Fig. 1) needs to be adjusted for further research. The only effect of the Virtual Theatre on task performance was a negative influence on recognition. This result is contradictory to the assumption that immersion leads to better learning. It seems that controlling the hardware was less intuitive than expected. This probably lead to the typical situation for learning with virtual environments: Dividing the available cognitive resources on the two parallel sub-tasks of dealing with the content and controlling the learning environment with the respected user interfaces [5]. Moreover, the combination of an HMD and real physical locomotion could lead to cognitive dissonance. When wearing the HMD, the user can see, where he or she walks in virtuality, but not in physical reality. Therefore the user takes a risk and has to trust in the technology in order to continue his or her actions. Last but not least, walking on the omnidirectional floor is a new experience for users and therefore could result in the fear of falling. All interpretations for the given results are going to be addressed in a follow-up study, where previous participants of the study will be interviewed on their experiences.

When we look at the correlations of the different constructs which were under study we see a slightly different picture: Although not all results are discussed in detail, we want to put the most relevant ones into focus. The positive correlation of self location and errors is in line with the previously discussed results, but therefore also contradictory to the assumption that immersion leads to better learning. However, the negative correlation of the self-reported spatial situation model and deviation supports this assumption. Domain specific interest i.e. interest in digital games correlated negatively with deviation and with time, just like locus of control regarding the use of technology and time. Therefore, in this experimental setting, the experience with virtual environments and the belief of being able to control technology had a bigger influence on those two performance measures than flow and spatial presence, the two indicators for immersive experience. The fact that flow correlated highly significant with the self-reported spatial situation model, possible actions and self location indicates that the constructs are two related facets of immersive experiences. The only correlation between personal characteristics and subjective experiences was between gaming frequency and higher cognitive involvement. In other words, experienced gamers were not challenged enough in this experimental setting.

6 Limitations of the Present Study and Future Research

Finally, some limitations of the present study are considered that should be pursued for future research. One limitation refers to the type of hardware examined. Since the different technical characteristics of the Virtual Theatre can only be tested in a set, it is not possible to isolate single effects. Moreover, the relationship between spatial presence and learning is not clear yet, since one subscale (self location) correlates with worse performance (making more errors) whilst another one (spatial situation model) correlates with better performance (less deviation when locating objects on a map). The other aspect concerns the task chosen for this experiment. Low levels of cognitive involvement in both groups indicate that the whole sample might not have been challenged enough. Since challenge is an important precondition for the motivation for learning, more challenging tasks are going to be tested in the future.

This first exploratory study on the effects of the Virtual Theatre on subjective experiences and learning confirmed a few theoretical assumptions but also contradicted others. In a next step, interviews with participants from both groups are going to be conducted. A deeper insight on the participants' experiences will allow a more differentiated view on the subject of our research. With a constant validation of the relationship between personal characteristics and hardware, it is possible to predict the level of subjective experiences to a certain extent. Thus, a profile could be developed, for whom the Virtual Theatre would have the most additional benefit. Moreover, different scenarios could be developed for different user types.

References

1. D. Ewert, K. Schuster, D. Johansson, D. Schilberg, S. Jeschke, Intensifying learner's experience by incorporating the virtual theatre into engineering education. In: *Proceedings of the 2013 IEEE Global Engineering Education Conference*. Berlin, 2013, pp. 207–212
2. S. Malkawi, O. Al-Araidah, Students' assessment of interactive distance experimentation in nuclear reactor physics laboratory education. European Journal of Engineering Education **38** (5), 2013, pp. 512–518

3. B. Witmer, M. Singer, Measuring Presence in Virtual Environments: A Presence Questionnaire. Presence: Teleoperators and Virtual Environments **7** (3), 1998, pp. 225–240
4. P. Zimbardo, R. Gerrig, *Psychologie*, 7th edn. Springer, Heidelberg, 2003
5. D. Baacke, "Medienkompetenz": Theoretisch erschließend und praktisch folgenreich. Merz – Medien und Erziehung **43** (1), 1999, pp. 7–12
6. D. Johansson, *Convergence in Mixed Reality-Virtuality Environments. Facilitating Natural User Behaviour*. No. 53 In: Örebro Studies in Technology. Örebro University, Örebro, Sweden, 2012
7. A. Hebbel-Seeger, Motiv: Motivation?!– Warum Lernen in virtuellen Welten trotz-dem)funktionieren kann. Zeitschrift für E-Learning – Lernkultur und Bildungstechnologie **7** (1), 2012, pp. 23–35
8. F. Müller, Interesse und Lernen. Report – Zeitschrift für Weiterbildungsforschung **29** (1), 2006, pp. 48–62
9. J. Murray, *Hamlet on the Holodeck: The Future of Narrative in Cyberspace*. MIT Press, Cambridge, 1998
10. M. Slater, M. Usoh, A. Steed, Taking Steps: The Influence of a Walking Technique on Presence in Virtual Reality. ACM Transactions on Computer-Human Interaction **2** (3), 1995, pp. 201–219
11. Virtuix Technologies, 2014. http://www.virtuix.com/
12. M. Csikszentmihalyi, J. LeFevre, Optimal experience in work and leisure. Journal of Personality and Social Psychology **56** (5), 1989, pp. 815–822
13. F. Rheinberg, S. Engeser, R. Vollmeyer, Measuring Components of Flow: the Flow-Shot-Scale. In: *Proceedings of the 1st International Positive Psychology Summit*. Washington DC, USA, 2002
14. M. Slater, Measuring Presence: A Response to the Witmer and Singer Presence Questionnaire. Presence: Teleoperators and Virtual Environments **8** (5), 1999, pp. 560–565
15. P. Vorderer, W. Wirth, F. Gouveia, F. Biocca, T. Saari, F. Jäncke, S. Böcking, H. Schramm, A. Gysbers, T. Hartmann, C. Klimmt, J. Laarni, N. Ravaja, A. Sacau, T. Baumgartner, P. Jäncke. Mec spatial presence questionnaire (mec-spq): Short documentation and instructions for application. report to the european community, project presence: Mec (ist-2001-37661), 2004. http://www.ijk.hmt-hannover.de/presence
16. W. Edelmann, *Lernpsychologie*, 5th edn. Beltz PVU, Weinheim, 1996
17. G. Beier, Kontrollüberzeugung im Umgang mit Technik. Report Psychologie **9**, 1999, pp. 684–693
18. D. Johansson, L. de Vin, Towards Convergence in a Virtual Environment: Omnidirectional Movement, Physical Feedback, Social Interaction and Vision. Mechatronic Systems Journal **2** (1), 2012, pp. 11–22
19. Sensics. http://sensics.com/products/head-mounted-displays/zsight-integrated-sxga-hmd/specifications/

Status Quo of Media Usage and Mobile Learning in Engineering Education

Katharina Schuster, Kerstin Thöing, Dominik May, Karsten Lensing, Michael Grosch, Anja Richert, A. Erman Tekkaya, Marcus Petermann and Sabina Jeschke

Abstract The usage of different kinds of media is part and parcel of teaching and learning processes in higher education. According to today's possibilities of information and communication technologies, mobile devices and app-usage have become indispensable for a big share of the population in everyday life. However, there is little empirical evidence on how students use mobile devices for learning processes in higher education, especially in engineering education. Within the project "Excellent Teaching and Learning in Engineering Sciences (ELLI)", three large technical universities (RWTH Aachen University, Ruhr-University Bochum, Technical University Dortmund) follow different approaches in order to improve the current teaching and learning situation in engineering education. Many of the corresponding measures are media-related. In this context, a broad understanding of media is applied which includes hardware as well as software. Amongst others, research is conducted on the topics of mobile learning, virtual laboratories, virtual collaboration, social media services and e-learning recommendation systems for teaching staff. In order to match the literature and results of the project with the current habits of study related media usage of students, the three universities conducted a survey in cooperation with the Karlsruhe Institute of Technology (KIT). The KIT's questionnaire covers more than

K. Schuster (✉) · K. Thöing · A. Richert · S. Jeschke
IMA/ZLW & IfU, RWTH Aachen University, Dennewartstraße 27, 52068 Aachen, Germany
e-mail: katharina.schuster@ima-zlw-ifu.rwth-aachen.de

D. May
Engineering Education Research Group (EERG), Center for Higher Education (zhb), TU Dortmund University, Dortmund, Germany

K. Lensing
Center of Higher Education, TU Dortmund University, Emil-Figge-Straße 50, 44227 Dortmund, Germany

M. Grosch
Karlsruhe Institute of Technology, Neuer Zirkel 3, 76131 Karlsruhe, Germany

A.E. Tekkaya
Institute of Forming Technology and Lightweight Construction, TU Dortmund University, Emil-Figge-Straße 50, 44227 Dortmund, Germany

M. Petermann
Lehrstuhl für Feststoffverfahrenstechnik, Ruhr-Universität Bochum, Universitätsstraße 150, 44801 Bochum, Germany

Originally published in "Proceedings of the 13th European Conference on e-Learning ECEL 2014", © 2014. Reprint by Springer International Publishing AG 2016, DOI 10.1007/978-3-319-46916-4_25

50 education-related media and IT-Services and has been applied at over 20 universities in 6 countries. For the survey conducted within the ELLI project, the topic of mobile learning was added to the questionnaire. Over 1.500 students were asked about their habits of study related media usage in terms of frequency of use and level of satisfaction. Regarding the topic of mobile learning, the students were asked for the kind of hardware and the kind of apps they use for higher education purposes. The 130 identified apps were clustered regarding subject and function. This paper presents the main results concerning the students' general habits of study related media usage and their mobile learning habits. It concludes with a special focus on the possibilities mobile devices offer to the improvement of engineering education.

Keywords Media Usage Habits · Social Media · Mobile Learning · Engineering Education

1 Introduction

The media usage at universities is highly diverse. Printed learning material, digital documents or complete online courses are just a few examples which illustrate the broad variety of media and IT-services offered to support the students' learning processes. Especially web-based services like search engines, facebook or special tools for online learning induced significant changes in society and in the landscape of higher education during the last years. In contrast to analogue media such as printed books, digital media and IT-services for learning purposes have been referred to as e-learning for the past 20 years. The term can be specified regarding the function of the corresponding service. Reinmann-Rothmeier [16] distinguishes between three lead functions: E-Learning (a) by distributing, (b) by interacting und (c) by collaborating. In the case of e-learning by distributing, the main function of the medium is to distribute information. E-learning by interacting applies to media and IT-services which allow the student to interact with the system without any additional personal help. E-learning by collaborating refers to processes of social problem solving and therefore also refers to the principle of "learning from different perspectives" [16]. Within the first focus of this paper, the research question of which function is needed most from the students' point of view was investigated.

In previous cycles of the KIT's study on habits of study related media usage, Gidion and Grosch [11] found that students prefer services which are linked to face-to-face learning settings, such as printed and electronical script or the online catalogue of the university's library. Media which require active participation are being used less frequently. Virtual teaching and learning environments are also being used not only seldom but also with a low rate of satisfaction [9].

Since one of the research foci of the ELLI project is learning with social media services, another question being investigated within this paper was whether students use services like facebook for studying. Social platforms such as facebook or information services like twitter provide an easy access as well as already implemented possibilities for communication by means of groups, private messages and forums for the students during their studies. Other web based media services for communication and collaboration such as wikis, weblogs and forums were implemented increasingly for the teamwork of students also in the teaching of engineering during the last years [5]. Surveys at other universities and faculties [17] as well as statements of teachers of the engineering sciences [23] show a rather hesitant use of this kind of media by students, which is in line with the results of Gidion and Grosch [9].

The second focus of this paper lies on the use of mobile devices and apps in teaching and learning contexts. The use of such devices and apps has become indispensable for many people's everyday life. According to the Federal Statistical Office of Germany, the number of mobile internet users in Germany increased in 2013 by 43 %. Amongst the 16 to 24 year olds, 81 % use mobile internet [7]. The trend towards mobile computers also applies to the type of technical devices being used: In 2013, 65 % of all German households were in possession of a mobile computer, e.g. a laptop, a notebook or a tablet-PC. The Federal Statistical Office asserts that the share has multiplied almost six-fold compared to surveys in 2003. Most households (33 %) use a mobile device additionally to the stationary PC, followed by almost as many households (32 %) who solely use a mobile device. The number of households, which solely use stationary PCs is decreasing: Whilst in 2003 the share was at 51 %, it dropped 20 % in 2013 [8]. Looking at these statistics, it is a logical consequence that the use of mobile devices is also increasing within the context of higher education. Current literature offers only little empirical data on the question how students are using mobile devices for their learning processes. In addition to that we do not know if, how often and in which contexts the students were asked by teachers to use their mobile devices in learning contexts in class or for the learning process in general.

In recent years the technology shift has changed the way how mobile learning is defined – from PDAs to smartphones and tablets. This is also visible by looking at different definitions over time. Quinn [15] offered a rather technology-centered view in 2001 by saying that mobile learning is „elearning through mobile computational devices: Palms, Windows CE machines, even your digital cellphone“. Sharples et al. [22] have a more user-centered view by describing the learning environment and adding the learners' autonomy regarding the choice over time and place. In order to go even more into detail an extended literature review was conducted in the ELLI-Project. Thus over 100 definitions on mobile learning were found in about 240 different sources. The definitions could be divided into 3 different clusters, either highlighting the mobile devices itself, the flexibility for the learning process or new didactical approaches. The following quotations illustrate this variety and are taken from the most cited sources in the context of Mobile Learning:

> "What is new in 'mobile learning' comes from the possibilities opened up by portable, lightweight devices that are sometimes small enough to fit in a pocket or in the palm of the one's hand" [12].

> "Mobile Learning devices are defined as handheld devices and […] should be connected through wireless connections that ensure mobility and flexibility" [15].

> "[Mobile learning] provides the potential of personal mobile technologies that could improve lifelong learning programs and continuing adult educational opportunities" [21].

Summing up all these perspectives, Crompton et al. (2013) published a description of their literature research regarding the mobile learning evolution over the years. According to them, mobile learning means "learning across multiple contexts, through social and content interactions, using personal electronic devices".

In addition to that, four different but central characteristics of mobile learning can be identified based on literature [13]:

- Use of mobile devises
- Local independence for learning
- Contextualization of the learning process
- Informality of learning

In the following, the methodology of the study is described, including the description of the sample. The results are presented in two sections. The first section describes the ten most and least used media and IT services, as well as their corresponding values of satisfaction. In the second results section, the results on mobile learning are presented. The paper concludes with a broader view on all results presented and depicts further research questions.

2 Methodology and Description of the Sample

The survey used a questionnaire that was developed at Karlsruhe Institute of Technology in 2009 [10]. It was conducted several times before at 20 universities in 6 countries. One of the KIT's long-established survey's aims is to identify the potential of a university, which media and IT sectors still need to be supported further and which are already used by the students. The three universities RWTH Aachen University, Ruhr University Bochum and TU Dortmund University issued the questionnaire in co-operation with the Karlsruhe Institute of Technology in 2013. It focused on the students' habits of study related media usage, in order to expand the empirical database and to match it with the other topics focused in the ELLI project, such as mobile learning, virtual laboratories, virtual collaboration, social media services and e-learning-systems. More detailed, the research questions were:

- Which function of e-learning is needed most and least from a students' point of view?
- Which role do social media services play for study related media usage?
- Which kind of hardware and which kind of apps do students use for studying and what are the consequences for mobile learning activities of universities?

Therefore the survey asked engineering students from the three universities for frequency of use and level of satisfaction for 54 media and IT services as well as hardware devices. To deliver a complete picture of media usage habits, the items related to university internal as well as to external services. Students had to respond to each given item by answering the question "Which of the given devices or services do you use for your studies?" Each item, e.g. e-book reader, had to be responded using a five point scale, ranging from „very frequently" to „never". Due to the special focus on mobile learning, the questionnaire was modified and several items were added to the original KIT questionnaire. Going more into detail with regard to the use of software on mobile devices, the survey asked for apps, which are already used by the students in the context of learning.

The survey was synchronously carried out at the three universities. A total of 1587 engineering students answered the questionnaire between April and May 2013. Most of the students were first- (59.6 %) or second-year (19.4 %) students. 81.6 % were male, 18.4 % were female students. The results of media usage habits in general were conducted with the means of descriptive statistics, calculating the percentage of students who used a medium or an IT-service frequently or very frequently. The same calculation was performed for the least used media and IT-services. In addition, the corresponding satisfaction value was calculated for the most and least used media and IT services. To get deeper insights into the usage of mobile devices, the percentage of how many students used a specific device frequently or very frequently. Moreover, students were asked, which apps they use for study purposes. The given apps were analyzed qualitatively with an explorative approach, i.e. no categories had been conducted previously. After a first round of scanning the data, the apps were clustered by subject. In a second round of analysis, the apps were clustered by function. In a last step, the proportion of the categories was calculated for each cluster.

3 Results

3.1 General Media Usage of Students

The media and IT services most frequently used in the context of academic studies are presented in Fig. 1. The service most frequently used is the Google search engine. This corresponds to the fact that Google is the most-used search-engine in the internet in general [1]. According to different web analytics services, also Wikipedia and Facebook are ranked among the most frequently used sites of the web [2, 14, 20].

As the results show, e-mail services of the university are being used frequently. It seems that communication via e-mail for learning purposes, e.g. to correspond with professors or with fellow students, is common and well established. The frequent use of course accompanying slides (78.7 %) and course accompanying lecture notes (75.3 %) online reflect media usage habits which fall in the category e-learning by distributing [16]. The use of analogue media in the form of course materials printed (68.4 %) is less frequent, but still among the top 10 of all surveyed media and IT services. The fact that google books (77.1 %) and other electronic books (39.5 %) are also amongst the top 10 of study related media, underlines the function of e-learning by distributing as important once more from the students' point of view. 50.4 % of all students stated that they use e-learning opportunities specifically provided by the department frequently or very frequently. This reveals the fact that each e-learning solution provided has to fit the corresponding study field, an approach which is also followed within the ELLI project. Apart from the top 10 of media and IT-services used most in a study related context, the data was also extracted for the least used media and IT-services. The percentage of frequency as well as the corresponding satisfaction rate is shown in Fig. 2.

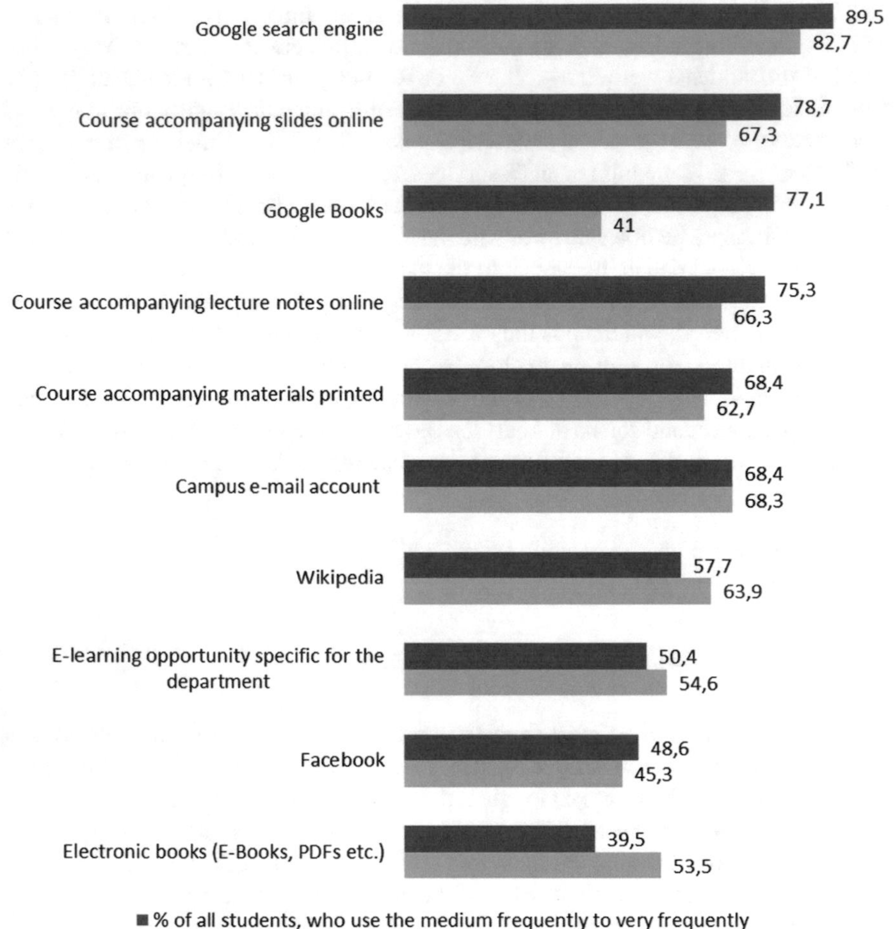

Fig. 1 Most used media and IT-services

A very obvious difference is that the average discrepancy between frequency and satisfaction in this dataset is three times bigger than for the items of the most used media and IT services. Whilst the absolute average discrepancy in the most used media and IT services is 9.7 %, it is 23.1 % in the least used media and IT services. If we look at the services in detail, a first conclusion is that Facebook is the clear leader when it comes to social networks. Whilst Google + is still used by 4.6 % of the students, other providers share only 1.8 % of the lot. This also corresponds to the results of web analytics services mentioned above. The microblogging system twitter is only used by 2.4 % of the students for study purposes.

Although twitter is also amongst the top ten web sites worldwide according to web analytic systems, it has not become common for study purposes yet. Dictionary software is also not used very frequently. This might be a result of frequently updated

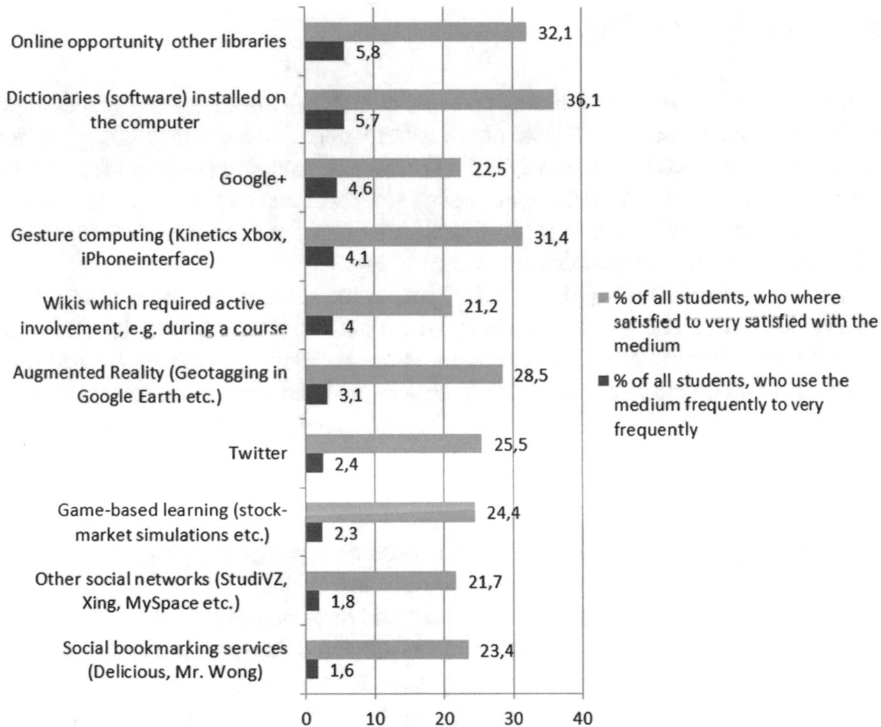

Fig. 2 Least used media and IT-services

and well-established services which are also available online, such as leo.org or duden.de for the German speech community. Social bookmarks don't seem to be needed very often. Only 1.6 % of the students claim to use them frequently or very frequently.

The two second biggest values for the discrepancy between frequency of use and satisfaction occur for two innovative virtual reality technologies which already find their way into higher education: Gesture Computing and Augmented Reality. Both are not used very often (4.1 % and 3.1 %), but the usage leads to high satisfaction. It can be assumed that the seldom usage of such technologies does not happen by choice. It seems more likely that students don't have enough possibilities yet to use the technology, or at least not as a regular part of their studying. The seldom use of the technologies might be the result of lack of supply, not of demand. This could also be the explanation for the seldom use of digital game-based learning. Wikis which require active involvement are not used very frequently, which is in line with previous results of the KIT's survey published by Gidion and Grosch [9]. However, it can not be assumed, that forms of learning with and by user generated content are being avoided in general. An aspect for further investigations is to filter all forms of media and IT services which involve user generated content and to analyse their acceptance on a deeper level.

3.2 Mobile Learning

In the following, all results regarding the topic of mobile learning are presented. Most of the questioned students (87.7 %) own a smartphone. The corresponding value for the tablet PC is much lower with 27.7 %. Therefore if teaching staff decides that the usage of tablet PCs is required for the course, students must be given the opportunity to borrow such a device. The results of hardware usage for study purposes regarding frequency of use and satisfaction are shown in Fig. 3.

The results of the survey show that 63.9 % of the students use their smartphone often or very often in context of learning, whilst only 26 % use their tablet computer often or very often for studying. The latter is not surprising, as tablets are not very distributed among students yet. E-book readers are used even more rarely, as only 4 % use this device often or very often. The last question in this context was, how often the students use mobile apps for learning. The answers reveal that the use of apps explicitly for learning is not very common yet. Only 7.2 % of the students use mobile apps often or very often. This is a fairly low value. Competing conclusions of this result could be that there either are no adequate apps for studying, that existing apps are not usable or that students just do not know them.

Going more into detail the students were asked to name the apps they are already using for study. All in all 13.5 % of the students answered this question with "yes" and named an app which they used for study purposes. In total, the students named 357 apps (139 different ones). Based on the answers a ranking on how often the apps were named was worked out and the top five could be identified. In the following these are explained briefly. Merck PSE is an interactive periodic table of chemical elements and was named 47 times (13 %). Wolfram Alpha, a computational knowledge search engine, was named 26 times (7.3 %). 22 times Schnittkraft-Meister, a game-based learning app to calculate cutting forces, was mentioned. For file hosting and as a cloud service mainly Dropbox is used (4.8 %). From those students who named an app, 4.2 % use Ankidroid, a program to design customized and personalized flashcards and use them for studying.

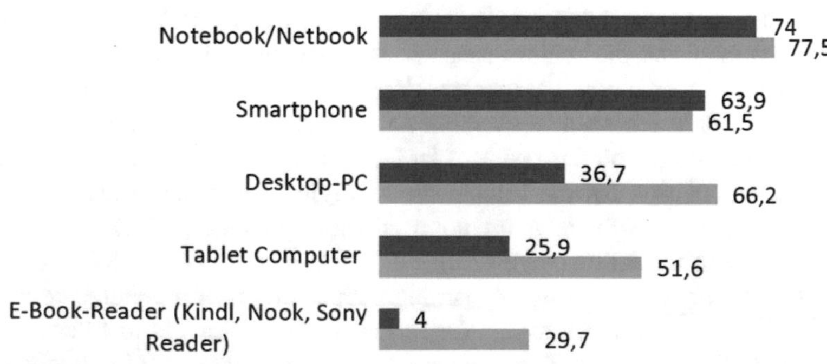

Fig. 3 Frequency and satisfaction of hardware usage for study purposes

Furthermore the different apps were clustered, firstly by function and secondly by subject. In this context again all 357 answers were taken into account and duplications were ignored, in order to calculate the relative distribution among the app naming.

Function oriented cluster: Dividing the apps by the purpose they are used for, 7 different clusters could be identified. 109 (31 %) of the named apps serve as any kind of database. 79 (22 %) apps serve the organization of learning and 47 (13 %) times an app for any kind of application and testing was named. Furthermore the students named 40 (11 %) apps which fall in the category of language dictionaries, 30 (8 %) apps which assist the students to make notes or edit documents and 20 (6 %) apps for cloud computing. 32 (9 %) apps could not be allocated to any cluster. Hence these were summarized under "others". The results are visualized in Fig. 4.

Subject oriented cluster: Looking at the related subjects the named apps could be allocated to six different clusters. The biggest cluster (160; 45 %) is formed by named apps, which had no connection to any special discipline. In this cluster apps like Dropbox or apps used to make notes are summarized. This cluster is followed by mentioned apps for mathematics and chemistry/physics (each 57; 16 %). 49 (14 % of the) times language learning apps, 30 (8 %) times mechanical engineering or logistics apps and finally 4 (1 %) times apps used in context with electrical engineering/informatics were named. These results are visualized in Fig. 5.

The results show that most of the used apps provide a general use without any subject relation. Looking into the list of named apps in this context shows that programs helping to organize the learning process, providing cloud computing and helping to edit documents dominate this kind of use. Hence, this is the area were students currently see the biggest advantages of app usage for study purposes. There are only some exceptions that provide the possibility for concrete knowledge application in a special subject context. Looking at those apps with a special subject connection it is visible that more apps are used in context with basic sciences in comparison to applied sciences. As in this study only engineering students were asked it is clear

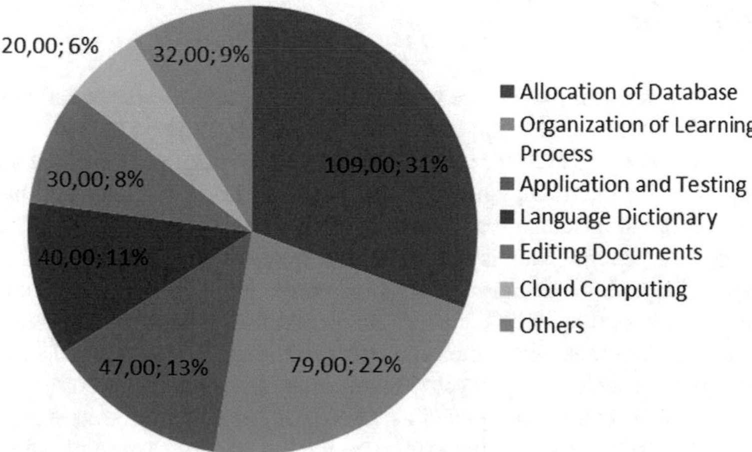

Fig. 4 Apps clustered by function

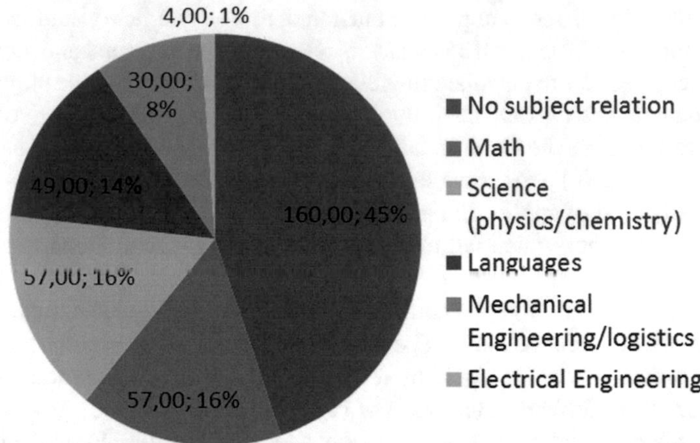

Fig. 5 Apps clustered by subject

that even in engineering classes apps on basic sciences can be used. A second look into the app list shows that these apps in most cases are some kind of database (e.g. Merck PSE) – the biggest among the function oriented clusters. A question for further investigation can be, why only the minority of the mentioned apps could be used for subject related application. Is this a question of app quality or are there simply not that many apps? However the results indicate that are still high potentials lying in the development of specialized apps. Even if there are some apps, those are not used very often. The fact that in general only a little bit more than 13 % of the students could even name apps they are using for study purposes shows that very clearly. Considering furthermore that from all asked students (over 1500) only 47 named the most often mentioned app underlines this finding.

4 Conclusion

To summarize, students mostly use media and IT-services with the function of distribution of information. Facebook as probably the most common example of social media services is being used for study purposes, whilst other social media services like google+ or the microblogging service twitter have not found their way into universities yet. It can be stated that the study related media usage partly reflects the media usage in general. This is shown by the frequent use of popular sites like-Facebook, Wikipedia or Google. Although with the given data, no statements can be made regarding cause and effect, it could be possible that especially in their first years at university, students use their already established mechanisms of media use as a starting point, before they develop other, more demanding and complex forms of e-learning. This can be interpreted as a form of reduction of complexity in the first years of studying. Especially to improve the introductory phase of the studies, which is affected by high dropout rates in engineering education, the relation between media usage and experience needs to be investigated on a deeper level.

E-learning services are being used more frequently than books or printed course materials. As a consequence for the ELLI project, this means that e-learning recommendation systems for teaching staff can be supportive for finding the tool that fits the teaching goals best. This could be of great help, especially for subject-specific e-learning opportunities, as in engineering education. Especially if the e-learning system is supposed to provide more functions than just distributing information, a personal recommendation is of great value. Since the conducted study focused on the students only, it cannot be determined whether the seldom usage of gesture computing or augmented reality is a consequence of the students' choice or of a lack of supply. Assuming that the latter is true, the ELLI project follows the right direction doing research on mixed-reality systems like the Virtual Theatre [18]. However, the application of such innovative hardware requires technical, psychological and subject-specific didactical skills in order to lead to the desired learning outcome [19]. Since "learning from different perspectives" is important for social problem solving [16], it seems to be an important precondition to teach students the competences they need in a globalized and digitalized working world. Why students don't use interactive or collaborative media and IT-services more often still needs to be investigated on a deeper level.

The results on mobile learning show that the use of mobile devices is not as common as it might be expected from general trends [8]. For example only a little bit more than a tenth of the students named apps they are using for study purposes. This could have at least two reasons. On the one hand it is possible that not enough appropriate apps exist, at least those which are usable in a learning context. Another explanation is that the students simply do not know which apps can be used effectively for which purpose. To conclude, the students must be supported in finding adequate apps. Even though the use of mobile apps at this point is not very common, mobile devices offer unique possibilities to support collaborative working processes. Especially the apps the students already named in the survey should stay in focus and it should be investigated why these apps are used for engineering education and what defines their unique selling point. In addition to that more qualitative studies are necessary on that point. In order to go more into detail student interviews should be conducted asking for reasons why they use or do not use their mobile devices for learning.

References

1. Alexa Internet Inc. (2014), "Top Sites in: All Categories > Computers > Internet > Searching > Search Engines", [online], http://www.alexa.com/topsites/category/Top/Computers/Internet/Searching/Search_Engines.
2. Alexa Internet Inc. (2014), "The top 500 sites on the web", [online], http://www.alexa.com/topsites.
3. Atwal, R., Tay, L., Cozza, R., Nguyen, T. H., Tsai, T., Zimmermann, A., and Lu, C. K. (2013), Forecast: PCs, Ultramobiles and Mobile Phones, Worldwide, 2010–2017, 4Q13 Update. Stanford.

4. Bakkers, J. H., Craven, R., Delaney, J., Jeronimo, F., and Montero, I. (2014), European Mobile and Internet Services, 2014: Top 10 Predictions.
5. Dittler, U. (2009), E-Learning 2.0: Von den Hochschulen gehypt, aber von den Studierenden unerwünscht? In: Dittler, U. et al. (Hrsg.). E-Learning: Eine Zwischenbilanz. Kritischer Rückblick als Basis eines Aufbruchs, Waxmann, Münster/New York/München/Berlin, pp. 205–218.
6. Federal Statistical Office (Statistisches Bundesamt) (2012), "Private Haushalte in der Informationsgesellschaft - Nutzung von Informations- und Kommunikationstechnologien", Statistisches Bundesamt, [online], https://www.destatis.de/DE/Publikationen/Thematisch/EinkommenKonsumLebensbedingungen/PrivateHaushalte/PrivateHaushalteIKT2150400127004.pdf;jsessionid=AD21B8CB5822AE878EA61C95630F0569.cae1?_blob=publicationFile.
7. Federal Statistical Office (Statistisches Bundesamt) (2014), "Number of mobile internet users up 43% in 2013". Press Release Nr. 089 (11.03.2014), [online], https://www.destatis.de/EN/PressServices/Press/pr/2014/03/PE14_089_63931.html.
8. Federal Statistical Office (Statistisches Bundesamt, 2013), "Ob als Einzel- oder Zusatz-gerä: Mobile PCs setzen sich durch", Press Release Nr. 386 (18.11.2013), [online], https://www.destatis.de/DE/PresseService/Presse/Pressemitteilungen/2013/11/PD13_386_632.html.
9. Gidion, G. and Grosch, M. (2012), Welche Medien nutzen die Studierenden tatsächlich? Ergebnisse einer Untersuchung zu den Mediennutzungsgewohnheiten von Studierenden. In: Forschung und Lehre 6/12, pp. 450–451.
10. Grosch, M. and Gidion, G. (2011), Mediennutzungsgewohnheiten im Wandel. Ergebnisse einer Befragung zur studiumsbezogenen Mediennutzung, KIT Scientific Publishing, Karlsruhe.
11. Grosch, M. (2012), Mediennutzung im Studium. Eine empirische Untersuchung am Karlsruher Institut für Technologie. Shaker: Aachen.
12. Kukulska-Hulme, A. and J. Traxler, J. (2005), Mobile learning: A handbook for educators and trainers. London: Routledge.
13. Maske, P. (2012), Mobile Applikationen 1: Interdisziplinäre Entwicklung am Beispiel des Mobile Learning. Wiesbaden: Gabler Verlag.
14. Quantcast Corporation (2014), "Rankings - Top sites", [online], https://www.quantcast.com/top-sites.
15. Quinn, C. (2000), mLearning: mobile, wireless, in-your-pocket learning, in LineZine, 2001.
16. Reinmann-Rothmeier, G. (2003), Didaktische Innovation durch Blended Learning. Leitlinien anhand eines Beispiels aus der Hochschule. Bern: Verlag Hans Huber.
17. Schiefner, M., Kerres, M. (2011), Web 2.0 in der Hochschullehre. In: Dittler, U. (Hrsg.). E-Learning: Einsatzkonzepte und Erfolgsfaktoren des Lernen mit interaktiven Medien, Oldenbourg, München, pp. 127–138.
18. Schuster, K., Ewert, D., Johansson, D., Bach, U., Richert, A., Jeschke, S. (2013), Verbesserung der Lernerfahrung durch die Integration des Virtual Theatres in die Ingenieurausbildung. In: Tekkaya, A. Erman; Jeschke, Sabina; Petermann, Marcus; May, Dominik; Friese, Nina; Ernst, Christiane; Lenz, Sandra; Müller, Kristina; Schuster, Katharina (Hg.): TeachING-LearnING.EU discussions. Innovationen für die Zukunft der Lehre in den Ingenieurwissenschaften, pp. 246–260.
19. Schuster, K., Hoffmann, M., Bach, U., Richert, A., Jeschke, S. (2014), Diving in? How users experience Virtual Environments using the Virtual Theatre. In: HCI International 2014 Conference Proceedings, in press.
20. SEOmoz,Inc. (2014), "Moz's list of the top 500 domains and pages on the web", [online], http://moz.com/top500.
21. Sharples, M. (2000), The Design of Personal Mobile technologies for Lifelong Learning, 34th ed.
22. Sharples, M. et al. (2010), A Theory of Learning for the Mobile Age. In: Bachmair, B. (Hrsg.): Medienbildung in neuen Kulturräumen. Wiebaden: VS Verlag für Sozialwissenschaften, pp. 87–99.
23. Thöing, K., Bach, U., Vossen, R., Jeschke, S. (2014), Herausforderungen kooperativen Lernens und Arbeitens im Web 2.0. In: A. Erman Tekkaya et al. (Hrsg.): TeachING-LearnING.EU Tagungsband movING Forward - Engineering Education from vision to mission 18. und 19. Juni 2013, Dortmund, 2014, pp. 191–194.

Enhancing the Learning Success of Engineering Students by Virtual Experiments

Max Hoffmann, Lana Plumanns, Laura Lenz, Katharina Schuster, Tobias Meisen and Sabina Jeschke

Abstract In a world that is characterized by highly specialized industry sectors, the demand for well-educated engineers increases significantly. Thus, the education of engineering students has become a major field of interest for universities. However, not every university is able to provide the required number of industry demonstrators to impart the needed practical knowledge to students. Our aim is to fill this gap by establishing Remote Labs. These laboratory experiments are performed in Virtual Reality environments which represent real laboratories accessible from different places. Following the implementation of such Remote Labs described within our past publications the aim of this contribution is to examine and evaluate possibilities of controlling Remote Labs from arbitrary locations. These control mechanisms are based on the virtualization of two concurrently working six-axis robots in combination with a game pad remote controller. The evaluation of the virtual demonstrator is carried out in terms of a study that is based on practical tests and questionnaires to the measure learning success.

Keywords Virtual Reality · Remote Laboratories · Game-based Learning · Experiential Learning · Virtual Theatre · Immersion

1 Introduction

The current developments within the industry and engineering sciences triggered by the Industry 4.0 pose major challenges for the education of engineering students in universities all over the world. Faster evolving technologies and rapidly changing requirements in industrial environments lead to rising demands in terms of practical education of engineering students. In the course of traditional training methods, the practical education of students is mostly performed by the attendance to laboratory experiments or the visit of factories and production sites. However, in terms of changing circumstances and dynamically performed manufacturing execution the

M. Hoffmann (✉) · L. Plumanns · L. Lenz · K. Schuster · T. Meisen · S. Jeschke
IMA/ZLW & IfU, RWTH Aachen University, Dennewartstr. 27, 52068 Aachen, Germany
e-mail: max.hoffmann@ima-zlw-ifu.rwth-aachen.de

Originally published in "HCI International", © 2015.
Reprint by Springer International Publishing AG 2016,
DOI 10.1007/978-3-319-46916-4_26

scope of laboratory experiments and practical education has to be adopted to these novel requirements as well. It is the aim of this paper to demonstrate novel methods of imparting practical knowledge to students considering the current developments within industrial reality.

One possibility of realizing these practical experiments without neglecting the demands of the Industry 4.0 is to virtualize the experience of visiting laboratory classes or manufacturing sites. In terms of these attempts, Virtual Reality simulations can be carried out in order to create virtual environments that can be adapted according to the current demands and demonstrator configurations. Another application of the described Virtual Reality solutions is to recreate existing laboratory environments from the real world and provide these environments as virtual demonstrators.

This application of Virtual Reality is referred to as Remote Laboratories and can be integrated into the curriculum of students in order to allow engineering students from arbitrary places to visit and experience laboratory environments that are not available at their university or place of study. Prototypical implementations of these Remote Labs have been carried out and examined in previous works of the author [1–3]. In terms of these developments the suitability of creating practical learning environments for engineering students were examined in order to deliver the basis for carrying out virtual experiments of real world demonstrators.

Based on our previous work, it is the aim of the current publications to describe, examine and evaluate ways of direct interaction with real world demonstrators through their virtual representation. Doing so, we extended an existing demonstrator with control mechanisms and implemented remote control solutions for active interaction of a user who is connected to the demonstrator by Virtual Reality tools. In order to evaluate these interaction capabilities the paper is divided into several parts.

In Section 2 we will discuss the state of the art in Game-based Learning in connection with laboratory experiments in the form of Remote Labs. Also, we will point out techniques to examine and create the didactical concepts needed to assess the learning success of students that perform experiments in game-like virtual environments. In Section 3, we will describe in detail the technical solutions that have been carried out and implemented to reach full remote control of distant laboratory environments from arbitrary places. In Section 4, the evaluation of the remote control capabilities takes place in form of a study that have been carried out with students from different universities in Germany. Section 5 summarizes the results and takes a look at further research opportunities in the field of Remote Labs.

2 State of the art

Based on the existing Remote Lab demonstrator that has been carried out and described within our previous publications [1] the different mechanisms for the remote control of these labs are of primary interest in this publication.

Accordingly, the state of the art section of this work deals with evaluation methods that will be selected and implemented to evaluate the learning success of students that are surrounded by virtual environments, thus in terms of a situation comparable to game-based learning/serious gaming scenarios. The evaluation part is realized on the basis of questionnaires that, on the one hand analyzes general suitability of the learning methods for each test person, and on the other hand, assesses the learning success of each individual test person from the technical point of view while taking into account their experience with digital media.

Virtuality-based learning (VBL) is a recent trend not only in engineering education. It is closely related to game-based learning (GBL), which is defined as "[…] a type of game-play that has defined outcomes. Generally, GBL is designed to balance subject matter with game-play and the ability of the player to retain and apply said subject matter to the real world" [4]. What is equal here is the digitalization of a pre-given-subject matter, which has to be learned. The difference is that digitalized places do not necessarily need gamy elements in order to be useful. There is much more about using virtual environments in education. The main advantage is that mistakes can be made without any consequences, that contents are endlessly repeatable plus that it is extremely cost saving. Thus, in terms of Remote Labs, students learn how to use a robot in a virtual environment before actually using it.

Although the advantages of VBL seem to be obvious, the measurement of learning successes presents a major challenge for the parties in charge. It is not only that the learning effect per se needs to be measured, but whether the handling is so unproblematic that users experience a sense of flow [5, 6] (a spontaneous sense of joy while performing a not too easy, not too difficult task), (tele-) presence [7] (the feeling of being enabled to act in this case in a remote lab) and finally immersion [8], the sensation of fully diving into a virtual environment. Obviously, these possible experiences are highly dependent on the user's pre-knowledge (e.g., how to use the WASD plus mouse combination) and his intrinsic technical readiness. The reason is that only users who can forget about the handling of for example a controller can experience a sense of immersion. If they need to look at it and think about the usage again and again, they will constantly be reminded that they are solely performing a virtual task, which is non-existent in reality and might thus attach less importance/meaning to it.

Another big problem in the measurement of subjective virtuality experiences is the question whether to perform the tests quantitatively or qualitatively and which influence the corresponding decision will have on the validity, transparency, causal interrelations and reliability of the results. The usage of the questionnaires on subjective user sentiments and self-assessment is a necessary step since these facts are not objectively observable. The self-assessment questions help to relate the produced results to behavior-parameters, which then lead to tentative conclusions concerning whether there is an interrelation between user preferences/habits and VBL success.

For the pre-assessment of test-persons, the BIG Five questionnaire is named as the most useful way to assess a test person's personality traits. The entailed items cover *neuroticism*, meaning emotional instabilities like fears and sadness, *extraversion*, the willingness to be in the center of attention, *openness to experience*, meaning the willingness to learn, *agreeableness*, the general need to socialize and lastly *conscientiousness*, the willingness to be disciplined [9]. For psychologists, alternative methods to assess personality traits exist; however, in the end, they all come back to the big five although they may be named differently [9].

Another of the most contemporary assessment questionnaires is the MEC-SPQ on general media exposure [10]. The main advantage is that it is highly flexible and may entail eight, six or only four items per scale. It has been used in studies on mobile gaming [11], in the realm of computer gaming [12] and serious games, thus, in the area of game-based learning [13]. So far, it is the only validated and highly consistent measurement instrument on spatial thinking [10].

In addition to this, recent studies by Witte showed that the locus for control of technology (KUT) questionnaire is a validated instrument to measure the performance of test persons while being confronted with technical problems [14]. Burde and Blankertz proved, that there is a correlation between a high score in the KUT and the performance in technical handling [15].

However, besides assessing test persons, an overall system evaluation and technical assessment of all technological devices is of utmost importance. Are software and hardware stabile? Do all components run as desired? Are there any known errors or problems and can the program run 'fluently'? [16]. Secondly, special attention must be paid to the users: how is their first reaction to the virtual robot? Did they spontaneously know what do to? Was there a lot of explanation necessary?

In sum, it must be concluded that the evaluation of virtuality-based learning is partly problematic because of subjective user assessment, talent and perception, which cannot be measured objectively. There is always the risk of users being afraid to truthfully state their abilities or that they even overestimate their capabilities. Our approach addresses this issue by creating an interplay between the estimated technical readiness of individual test persons and their actual real-time learning progress. Accordingly, the risk of falsified results due to inaccurate self-assessment of the test persons can be minimized.

3 Active Interaction for Remote Labs in Virtual Reality Environments

The creation of fully interactive virtual environments is based on the VR techniques that have been utilized by carrying out the technical and virtual environments of the remote labs. To realize a fully capable Remote Lab several steps were performed, i.e.:

1. Virtualization of machines and plants in every detail for three-dimensional representation within virtual environments.
2. Embedding of three-dimensional objects into virtual environments to create a virtual scenario, in which users can move around to exploit objects and the environment.

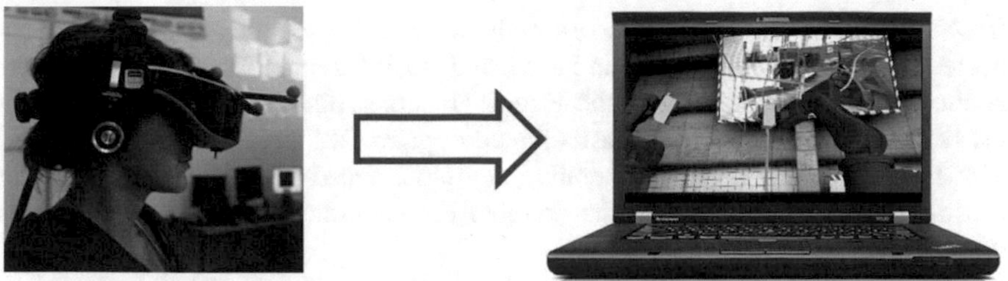

Fig. 1 Remote Lab environment – The user is immersed into the scenario via the Oculus Rift

3. Setup and implementation of an information and communication infrastructure for data exchange between real and virtual laboratory environments.
4. Enabling one-directional communication between the real laboratory environment and its virtual representation in order to reproduce movements of the real world demonstrator within the virtual demonstrator in real-time.
5. Enabling bi-directional communication by embedding control mechanisms and devices for the real laboratory from VR experiments into the scope of the Remote Lab.

The user can interact with the Remote Lab through various interfaces, e.g. the Virtual Theatre described in [1] or other immersive technologies like the Oculus Rift. Fig. 1 shows that a notebook together with a Head Mounted Display is a suitable environment.

The first step of this procedure has already been described by Hoffmann et al. [2]. The virtual demonstrator that is used within the current work consists of two cooperating six-axis robots that are placed on a table in order to perform concurrent tasks. The virtual representation of the robots has been designed using modeling tools for computer graphics and design. The modelling of the robots is performed by integrating a bone structure into the virtual representation whereas the bones of the robot are connected through joints. The meshing of this bone-joint-structure ensures the correct assignment of the single parts in terms of parent and child nodes in order to recreate physically realistic movements of the whole robot, i.e. if the root joint is moved, all subsequent child nodes of the robot (bones and joints) are moved accordingly as well.

The embedding of these robots into a virtual environment is performed by the use of a VR tool for virtual worlds, i.e. WorldViz Vizard as described in [1]. In terms of the modeling, the different components, e.g. the robot table, both robots, the objects to be treated by the robots as well as other elements like avatars or screens are included into this virtual environment to create an immersive scenario for the user experiment.

The information and communication infrastructure (ICT) for Remote Labs has been described in detail in [3] and is an integral part of the virtual laboratory experiment. The ICT consists of the two cooperating robots, which are controlled by two manual control panels and a computer that contains the Robot Operating System

(ROS) environment. Over a network architecture this operating computer is connected to other computers that run the Virtual Reality simulation programs and are connected to VR simulators like the Virtual Theatre as described in [17] or the Oculus Rift in combination with a local client computer [18] as depicted in Fig. 1. The connection between the robot operating computer and the VR simulation systems is established by making use of the Protobuf Protocol interface for the exchange of robot information [19].

The ICT as described allows the one-directional communication between a robot-focused laboratory experiment and a distant representation of this laboratory in terms of a Remote Lab. Using the Protobuf interface standard the angles and joint positions of the robots can be transferred over the network in real-time. Internal tests on the real-time capabilities of such Remote Lab, which allows the observation of distant experiments, determined the maximum lag between reality and virtuality to 0.1–0.2 seconds.

Besides the graphical interface for the visualization of Remote Labs at distant places, e.g. by making use of the Virtual Theatre, there are also interaction devices embedded into the ICT. For our scenario we have chosen a common game pad controller, the Nintendo Wii™, as remote control device for the interaction of the user with the real world demonstrator within the virtual environment. Using this game device, the robots of the simulation and accordingly the real robots can be successfully manipulated. The basic control functions are highlighted in Fig. 2.

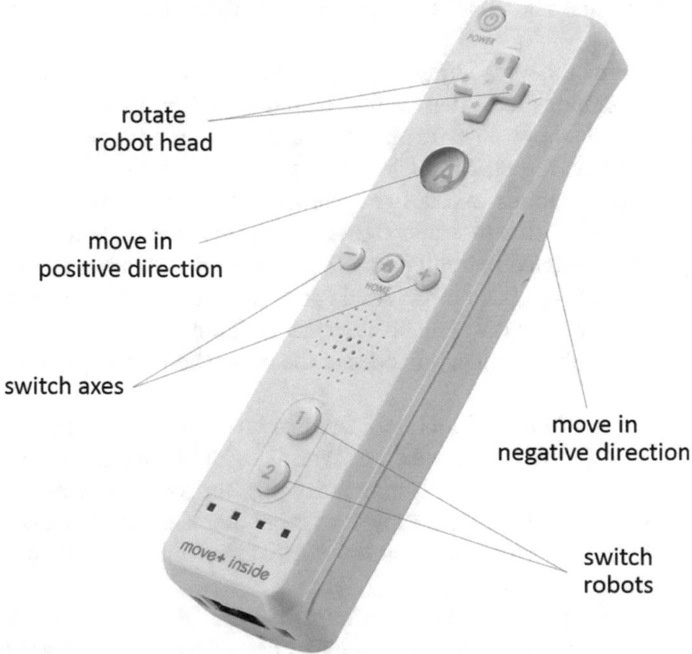

Fig. 2 Nintendo Wii™ controller and basic functions for robot control

There are two control mechanisms that have been carried out for robot control, and which are both based on the usage of the Wii™ gaming controller:

1. Direct kinematics for direct control of the joint angles for each robots.
2. Inverse kinematics for user control of the movement axis (X, Y, Z) whereas the joint angles for the current robot position or moving trajectory are dynamically calculated during the experiments.

In terms of the direct kinematics robot control method, each of the six angles of the selected robot can be individually controlled using the "A" button for positive moving direction and the "B" button on the back of the remote control for negative moving direction. Using the "+" and "–" signs the axes of the robot joints can be subsequently selected. Using the buttons "1" and "2" the according robot can be selected. For direct kinematics the cross on the top of the Wii™ is not used, as the head rotation is represented by the sixth robot joint angle.

Concerning the inverse kinematics the "A" and "B" buttons are used to move the robot claw in positive respectively negative direction of the X, Y or Z axis. The axes are switched again using the "+" and "–" signs on the controller. The "1" and "2" also change the selected robot. The rotation of the robot head for inverse kinematics is implemented using the control cross at the top of the remote control. For the dynamic calculation of the single joint angle values suitable for the goal position or trajectory, an inverse calculation method is used for determining the joint parameters. For our use-case a MATLAB™ Toolbox has been adapted to the needs of the robot demonstrator.

The scenario, in which the described robot control methods are being applied, consists of a setup, where one of the robots has to be moved along a fixed path. This path is represented by a wire. For conducting the experiment an eyelet is attached to

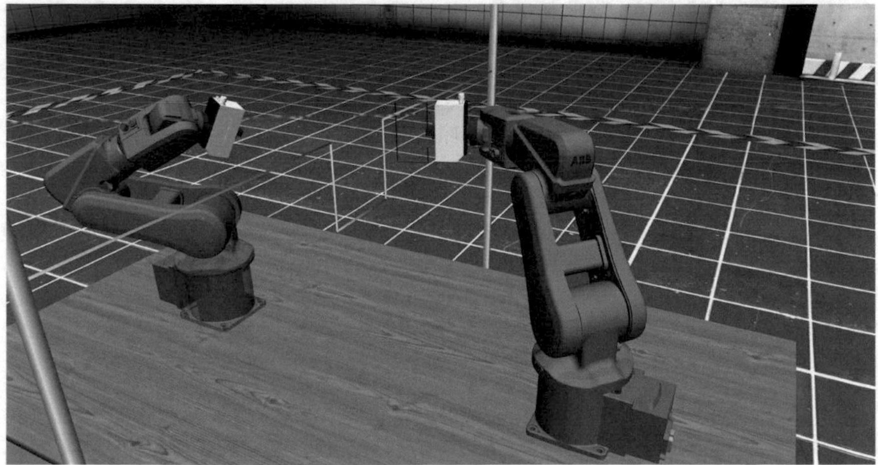

Fig. 3 The eyelet attached to the *right* robot has to be driven along the steel wire in the middle

one of the robots. The task of the controlling person is to move the eyelet attached to the robot along the wire, which forms a certain curve (see Fig. 3).

The described task is performed through the Wii™ remote control either by making use of the direct kinematic mechanism or by making use of inverse kinematics. In order to assess these methods against each other, user studies were performed, which are described in the following chapter.

4 User Studies for Examination and Evaluation of Control Mechanisms for Remote Labs

4.1 Design of Experiment and Expectations

The aim of this study is to examine which of the previously described control mechanisms for six-axis robots is the most beneficial for the implementation in Remote Labs especially with regard to an intuitive control and progress of learning as well as investigating the effects of the sequent comparison of both mechanism.

We expect that students – especially those who are used to gaming – will prefer the inverse control mechanism over the direct one, because it resembles their gaming experience. Furthermore we expect that successful practice experience through the trainings session will enhance the feelings of self-confidence and thus flow.

A representative number of engineering students from different advanced information science courses at multiple sophisticated, technical universities participated in the study in order to evaluate the learning progress of the concurrent methods for remote controlling the robots. The objective of conducting the study is to assess the different methodologies of control mechanisms suitable for engineering students.

The first part of the study is performed in cooperation with the Technical University of Dortmund, where test persons were recruited. The second part of the study is carried out in the course of a lecture with engineering students at the RWTH Aachen University. The study consists of three questionnaires and two practical tests, namely the remote control of the cooperating robots using direct and inverse kinematics. The sequence of the tests (direct and inverse mechanism) is randomized, the participants are accordingly assigned to either Group A or Group B. Participants who are assigned to group A start with the inverse kinematics test whereas participants of group B start with the direct kinematics test. Both user groups conduct both experiments, however in reverse order. The intention of this approach is on the one hand to examine the learning progress of the students during the experiments and on the other hand to equalize the effects of test order. The study is implemented in six steps:

1. Theoretical input in terms of the study design and methods of examination.
2. Pre-questionnaire for general assessment concerning the personal background of the test persons in terms of video game experiences and spatial thinking abilities.

3. First experiment using either inverse kinematics (Group A) or direct kinematics (Group B).
4. Questionnaire for the assessment of the previous test.
5. Second experiment using either direct kinematics (Group A) or inverse kinematics (Group B).
6. Questionnaire for the assessment of the previous test.

The first questionnaire is given to the participants before the experiment and is used for a general classification of the test person. Whereas questions such as the frequency of confrontation with digital games, the frequency of handling a console, whether the participants are active member of a digital sodality and the amount of hours spend on computer games a week are used to assess participants experience of gaming, individuals visual-spatial imagination (in virtual surroundings) are examined by items of the FRS [20] und questions of the subscale DSI of the MEC-Spatial Presence Questionnaire (MEC-SPQ) [10] adapted to computer games. This scale was already used successfully in previous studies and is characterized by fair quality criteria [21].

Besides this scale, items of the KUT [22] are used to assess participants locus of control when confronted with technical problems. Additionally, questions of the BIG Five Inventory [23] are used to assess subjects' personality and psychological biases, to get a broad picture of the participants.

The second and third questionnaire are used to assess the students' technical evaluation of the currently performed tests as well as their experience of learning progress while working. Participants are asked to rate the feasibility, advantages and disadvantages and the control of the just practiced remote mechanism as well as adapted questions concerning the experience of absorption due to the experience of flow [24]. A mental state of operation in which the individual, who is performing the task, is fully involved and immersed by feelings of energized focus [25]. All questions are presented on a seven-point scale, ranging from 1 = *total disagree* to 7 = *total agree*.

4.2 Correlational Approach

To gain further inside of the relationship between the individual gaming experience, such as hours of gaming per week and the evaluation of the control mechanism as well as learning progress during the experiment, the correlation between the pretest data was calculated. Data were analyzed with IBM SPSS statistics software. The results are visualized in Fig. 4 in form of a graph that shows the strength of each correlation.

The correlational approach shows that subjects with more playing hours per week evaluate the inverse kinematics approach better than the direct one, but only if the inverse control mechanism is the initial one. There is no significant correlation

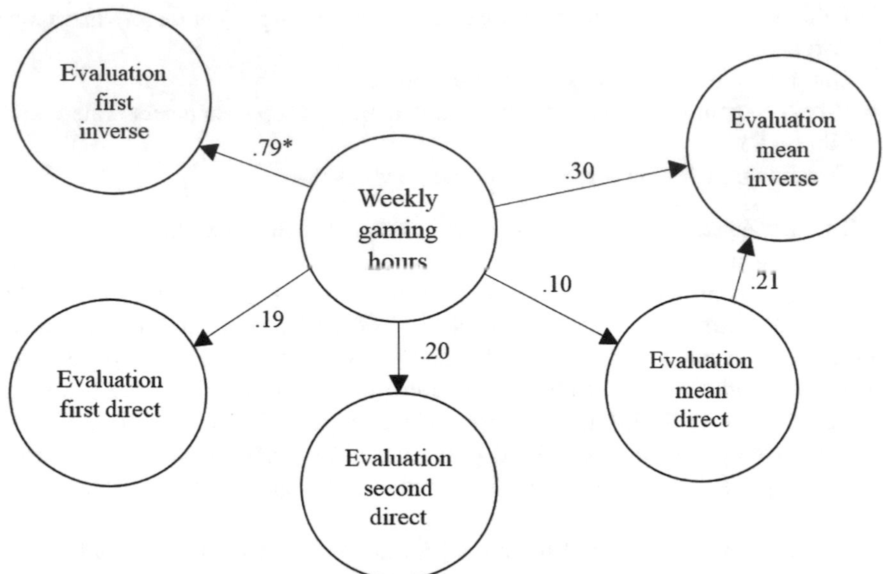

Fig. 4 Correlation between amount of weekly gaming hours and evaluation of remote control

between hours played a week and an appreciation of the direct mechanism, neither as first test nor as second test.

Further differences between the two participant groups were analyzed with a multivariate ANOVA, where each group served as a between-subject factor. The assumption of homogeneity of variances is investigated with Levene's test and shows no significant violations of the assumption for the dependent variable. Inspection of histograms show no significant deviations from normality for the rating of two groups. The analysis shows no main effect of rating due to group assignment, $F(1, 12) = 1.49$, $p = 0.266$, but additional analyses of the within-subject factor task-order show significant differences ($p = <0.05$) between the two tests in both groups (see Table 1).

Statistical analyses reveal that the participants show no significant differences in preference due to both remote mechanisms. In both groups, the second remote mechanism is rated significantly higher in preference than the first one, regardless of group membership. The subscale experienced learning progress is rated above the mean, in particular after the second testing session for the group that starts with the inverse mechanism ($M = 5.33$; $SD = 1.75$) respectively the group that starts with the direct mechanism ($M = 4.45$; $SD = 2.05$). These present findings do not confirm the hypothesis that students prefer the inverse mechanism in statistical terms, despite the fact that the mean values of the inverse mechanism are slightly higher than those of the direct mechanism. However, the results emphasize the importance of learning experience in both groups.

Table 1 Results of the significance analysis of Remote Lab control mechanisms

		Group A	Group B
		Inverse	direct
Test 1	*M*	4.44	4.20
	SD	0.98	1.38
		Direct	Inverse
Test 2	*M*	4.84*	4.84*
	SD	0.99	1.63

* = $p < 0.05$, M = Mean, SD = Standard deviation

Thus, it can be concluded from these results, that the experience of flow and students' valuing of technical mechanism increase over time and are depending on practical experience and learning progress rather than a specific task mechanism per se.

5 Conclusion and Outlook

The aim of this work is to assess different mechanisms for the remote control of laboratory environments at arbitrary places. Based on an existing Remote Lab environment, direct and inverse kinematics control schemes have been carried out and implemented in order to enable the control of two cooperating six-axis robots.

The assessment in terms of the learning success lead to the result that there is not a significant preference for one of the two control mechanisms. However, the inverse kinematics – as expected – has been evaluated slightly better in comparison to the direct specification of joint angles. The study has also shown that the learning effect is equally good using both control methods, hence, both user groups evaluated the second test as preferable to the first one as they gained more self-confidence in controlling the robots during the progress of the study.

During the next steps in enhancing the usability and application of Remote Labs, it is our aim to enable a direct manipulation of the laboratory environment that can be located at arbitrary places. This real laboratory will be moved in real-time and accordingly to the exact digital representation, thus unexpected states of the experiment can be reached in the simulation similarly to the real-world demonstrator. In order to ensure the safety during these remote operations, a collision avoidance system based on the inverse kinematics implementation will be carried. Using this security layer, Remote Labs at arbitrary places can be independently controlled by users from Virtual Reality simulators from various locations. This will enable a holistic coverage of laboratory experiments for universities all over the world.

References

1. M. Hoffmann, K. Schuster, D. Schilberg, T. Meisen, Next-generation teaching and learning using the virtual theatre. In: *At the Edge of the Rift*, ed. by S. Gregory, P. Jerry, N. Taveres Jones, 2014
2. M. Hoffmann, K. Schuster, D. Schilberg, S. Jeschke, Bridging the gap between students and laboratory experiments. In: *Virtual, Augmented and Mixed Reality*, *Lecture notes in computer science*, vol. Heraklion, Crete, Greece, ed. by R. Shumaker, Springer, Cham, 2014, pp. 39–50
3. M. Hoffmann, T. Meisen, S. Jeschke, Shifting virtual reality to the next level: Experiencing remote laboratories through mixed reality. The International Conference on Computer Science, Computer Engineering, and Education Technologies (CSCEET2014), 2014
4. C. Meier, S. Seufert, Game-based learning: Erfahrungen mit und perspektiven für digitale lernspiele in der betrieblichen bildung. In: *Handbuch E-Learning*, ed. by A. Hohenstein, K. Wilbers, Fachverlag Deutscher Wirtschaftsdienst, Köln, 2005
5. M. Csikszentmihalyi, *Finding Flow: The Psychology of Engagement with Everyday Life*. Basic Books, New York, 1997
6. M. Csikszentmihalyi. Creativity: Flow and the psychology of discovery and invention, May 6th, 2013. http://www.books.google.de/books/about/Creativity.html?id=aci_Ea4c6woC&redir_esc=y
7. C. Bracken, P. Skalski. Telepresence and video games. the impact of image quality. http://www.psychnology.org/File/PNJ7(1)/PSYCHNOLOGY_JOURNAL_7_1_BRACKEN.pdf
8. C. Jennett, A.L. Cox, P. Cairns, S. Dhoparee, A. Epps, T. Tijs, A. Walton, Measuring and defining the experience of immersion in games. International Journal of Human-Computer Studies **66** (9), 2008, pp. 641–661. doi:10.1016/j.ijhcs.2008.04.004
9. G. Matthews, I.J. Deary, M.C. Whiteman, *Personality Traits*, 2nd edn. Cambridge University Press, Cambridge, U.K. and New York, 2003
10. P. Vorderer, W. Wirth, F.R. Gouveia, F. Biocca, T. Saari, F. Jäncke, S. Böcking, H. Schramm, A. Gysbers, T. Hartmann, C. Klimmt, J. Laarni, N. Ravaja, A. Sacau, T. Baumgartner, P. Jäncke, Mec spatial presence questionnaire (mec-spq): Short documentation and instructions for application. report to the european community, project presence. MEC (IST-2001-37661), 2001
11. J. Laarni, N. Ravaja, T. Saari, Presence experience in mobile gaming. Proceedings of DiGRA 2005 Conference: Changing Views – Worlds in Play, 2005
12. D. Weibel, B. Wissmath, Immersion in computer games: The role of spatial presence and flow. International Journal of Computer Games Technology **2011** (3), 2011, pp. 1–14. doi:10.1155/2011/282345
13. S. Göbel, *E-learning and games for training, education, health and sports: 7th international conference, Edutainment 2012, and 3rd international conference, GameDays 2012, Darmstadt, Germany, September 18 - 20, 2012; proceedings*, *Lecture notes in computer science*, vol. 7516. Springer, Berlin [u.a.], 2012
14. M. Witte, S.E. Kober, M. Ninaus, C. Neuper, G. Wood, Control beliefs can predict the ability to up-regulate sensorimotor rhythm during neurofeedback training. Frontiers in human neuroscience **7**, 2013, p. 478. doi:10.3389/fnhum.2013.00478
15. W. Burde, B. Blankertz, Is the locus of control of reinforcement a predictor of brain-computer interface performance. Proceedings of the 3rd International Braincomputer Inferface Workshop and Training Course, Graz, 2005, pp. 76–77
16. J. Wakolbinger, P. Kirchner. Netavatar – interaktion mit einem humanoiden roboter, 2004. http://www.hs-augsburg.de/~tr/prj/ss10-IP/07_WK/07_WK_NetAvatar.pdf
17. M. Hoffmann, K. Schuster, D. Schilberg, S. Jeschke, Next-generation teaching and learning using the virtual theatre. In: *4th Global Conference on Experiential Learning in Virtual Worlds Prague, Czech Republic*. 2014
18. OculusVR. https://www.oculus.com/dk2/
19. Google. http://www.code.google.com/p/protobuf/wiki/thirdpartyaddons
20. S. Münzer, C. Hölscher, Entwicklung und validierung eines fragebogens zu räumlichen strategien. Diagnostica **57** (3), 2011, pp. 111–125

21. K. Schuster, M. Hoffmann, U. Bach, A. Richert, S. Jeschke, Diving in? how users experience virtual environments using the virtual theatre. In: *Virtual, Augmented and Mixed Reality*, *Lecture notes in computer science*, vol. Heraklion, Crete, Greece, ed. by R. Shumaker, Springer, Cham, 2014, pp. 636–646
22. G. Beier, Kontrollüberzeugungen im umgang mit technik: Ein persönlichkeitsmerkmal mit relevanz für die gestaltung technischer systeme. Disseration, Humboldt Universität, Berlin, 2004. http://www.dissertation.de
23. B. Rammstedt, O.P. John, Kurzversion des big five inventory (bfi-k). Diagnostica **51** (4), 2005, pp. 195–206
24. F. Rheinberg, S. Engeser, R. Vollmeyer, eds., *Measuring components of flow: the Flow- Short-Scale.* Proceedings of the 1st International Positive Psychology Summit. Washington DC, 2002
25. F. Rheinberg, Vollmeyer, R. Engeser, S., Die erfassung des flow-erlebens. In: *Diagnostik von Motivation und Selbstkonzept*, ed. by J. Stiensmeier-Pelster, F. Rheinberg, pp. 261–279

Using Evernote in Engineering Education – Two Course Concepts to Explore Chances and Barriers

Dominik May, Tobias Haertel and Monika Radtke

Abstract The introduction of mobile devices and fitted software is not that much widespread in German engineering education up to now. Hence, within the ELLI project mobile learning is one focus and the usage of iPads and the software Evernote has been introduced in two different courses. In one of the courses the students were encouraged to use these tools in order to organize and support scientific research processes. In the other one they should track personal procrastination moments in order to analyze and overcome these moments or make use of them for their working processes. In both cases it was evaluated how suitable these instruments were and if they enabled the students to work on their projects anywhere and at any time they decided to do so. The results show that the Evernote software does fit in theses contexts and support the education scenarios. However, there remain barriers and the assumption of merging private and educational contexts cannot be supported by our results.

1 Course Contexts for the Usage of Evernote

Within the research and development project "ELLI-Excellent Teaching and Learning in Engineering Education" at TU Dortmund University (in collaboration with Ruhr University Bochum and RWTH Aachen University; funded by the German Federal Ministry of Education and Research) mobile learning [1] and the usage of mobile devices in educational scenarios is focused. For example a study was done, asking for the influence of mobile devices on higher engineering education at the three involved universities [2, 3]. Moreover in the project's context the scratchpad software Evernote (http://www.evernote.com) has been identified as a tool to be used

D. May (✉) · T. Haertel · M. Radtke
Engineering Education Research Group, Center for Higher Education,
TU Dortmund University, Dortmund, Germany
e-mail: dominik.may@tu-dortmund.de

Originally published in "Proceedings of E-Learn: World Conference on E-Learning in Corporate, Government, Healthcare, and Higher Education 2015", © AACE 2015. Reprint by Springer International Publishing AG 2016, DOI 10.1007/978-3-319-46916-4_27

by students with the help of mobile devices in two different course contexts. With the help of Evernote (just as with other scratchpad tools like e.g. Microsoft OneNote) it is possible to work out and share digital notebooks with various notes and to work on them together without necessarily meeting each other in person.

The first context, in which we used this tool for teaching and learning, is a course called "Procrastination Fighters". In this course the students gained knowledge about different scientific and popular scientific concepts about procrastination in academic settings and strategies how to avoid or reduce it. As a self-experiment, they tested different approaches during the semester and analyzed their impact. Furthermore the students were asked to share every procrastination moment with Evernote: Whenever and wherever they realized that they did something they had not planned in order to avoid doing something else that was planned (e.g. cleaning the bathroom rather than writing a seminar paper), they ought to record this by writing a note at this very moment in Evernote. For this purpose, students without smartphones were equipped with appropriate mobile devices, so that all course members were able to upload their procrastination moments instantaneously. All notes were shared within one notebook, in this way their authors remained anonymous. Each week, these procrastination moments were analyzed in the seminar.

Our second context for the usage of Evernote is the course "Fit for Science". This course is especially designed to learn and practice the most important techniques needed during research processes. All in all the students perform a little research process in groups over one semester: They begin with the definition of a research question, go over to an in-depth literature review, and finally write a scientific paper answering the chosen question on basis of their research. This work is done in groups of 3–4 students, which means that the processes in context with organizing and documenting the work as well as the communication processes can be challenging. Especially involving all group members into the discussion and keeping everyone updated on the current work status can be a serious problem. Therefore the Evernote software was presented to the students exactly for supporting these processes. Our aim was to identify if and how it could help them to handle such issues. Therefor the students were encouraged to organize, document end even present their work throughout the course with the help of Evernote. In addition to that, they were given a tablet PC (iPad in our case), which they could use during the course, too. Hence, we were not only focusing on the usage of Evernote but as well on the usage of tablet PCs throughout the research processes [4].

2 Using Evernote in the Course "Procrastination Fighters"

In the seminar "Procrastination Fighters", the usage of Evernote was a success as well as a failure. Thanks to the user-friendly interface of the Evernote app, no technical barriers or problems appeared; all students were able to upload their notes even without a technical introduction. Hence, on the one hand, a lot of procrastination moments were shared. On the other hand, there could have been recorded more procrastination moments. Although the students praised the comfortable possibility of uploading their notes, they admitted that they did not upload as many procrastination moments as they could have. They argued that their smartphone was not at hand anytime; sometimes it was in the kitchen while the procrastination moment was realized in front of the TV in the living room. Those moments were not shared due to the absence of the mobile device. Students also argued, they sometimes just did not bother to type some words in the Evernote app so they procrastinated uploading the procrastination moments. They stated that taking a picture would have been easier, but even then, they were skeptical that all moments would have been uploaded, because that still was "work" in a moment they wanted to avoid working for university. As a result, a more comfortable way of uploading procrastination moments has to be found for future seminars, although there will remain an insurmountable separation between working for university and moments when students do not want to work for university, even with such a ubiquitous system like smartphones.

3 Using Evernote in the Course "Fit for Science"

As explained in the introduction Evernote and the iPad were important tools to work with during the "Fit for Science" course, too. At the end of the course we asked the students about their feedback and they had to work out reflective papers with focus on their working processes in general as well as on the use of Evernote. As the course was given twice we have the feedback of all in all 23 students. This is not a large number but nevertheless some first work in progress insights definitely can be deduced from their feedback. Furthermore, during the latest course edition one group used the Evernote work chat intensively for their in-group communication and organization, which gave us the chance to examine their chat communication in a more detailed way. Results from both of the research approaches, the students' feedback and the work chat analysis, are briefly explained in the following.

3.1 The Students' Feedback

The results show that Evernote in combination with mobile devices in general can support scientific work but not in all phases. It can be derived from 10 statements that the students merely used the tablet PC and Evernote during the early stages of their research. Especially for the organization at the beginning and for the documentation of the first literature research results the students used that tool. As they were going on in their research process and started to write the paper, Evernote became less important. Some students finally even complained about the fact, that writing longer texts by using a tablet PC is not very comfortable. Therefore they changed over to a classical PC. However, from our perspective this is not a negative result. It is just the other way around. One lesson the students learned was that in most cases technical tools are designed for explicit tasks and can only be used effectively in these contexts. Whereas classical PCs are perfect to work out longer documents, tablet PC have their advantages in collecting information and working when you are on the move. Asking for the tasks the students fulfilled only with the tablet PCs we could work out 8 categories (named here in declining order by the number of naming): Organize and document work processes, search for information, create mind maps (this was an additional task during the first course phase), share and synchronize files in the group, communicate, create documents and time management. A last interesting feedback was that even if most of them carried the tablet PC with them most of the time, they still did not use it in every possible situation. Based on the students' comments, it seldom was the case that they had a quick idea and just noted it in the Evernote notebook. In most of the cases the students still reserved special timeslots at the day, in which they wanted to do work for this course and apart from these slots they simply did not work for it, even if they could do so with such a handy device like a tablet PC. Especially the last point is interesting because the work chat analysis shows a slightly different picture.

3.2 *The Software's Internal Work Chat as Data Source*

As explained above there was one group in the 2015 course edition, which decided to use the program internal work chat as the main communication channel for their group work. Looking at and analyzing that chat communication shows on the one hand when the group was communicating (as the different chat messages are tagged with the corresponding day and time). On the other hand we were able to analyze the chat messages on a content level and cluster them. The examined group transmitted all in all 1078 chat messages during the whole course (beginning May 11th, 2015 and ending at July 20th, 2015). Figure 1 shows the chat messages' distribution by day over the whole course time.

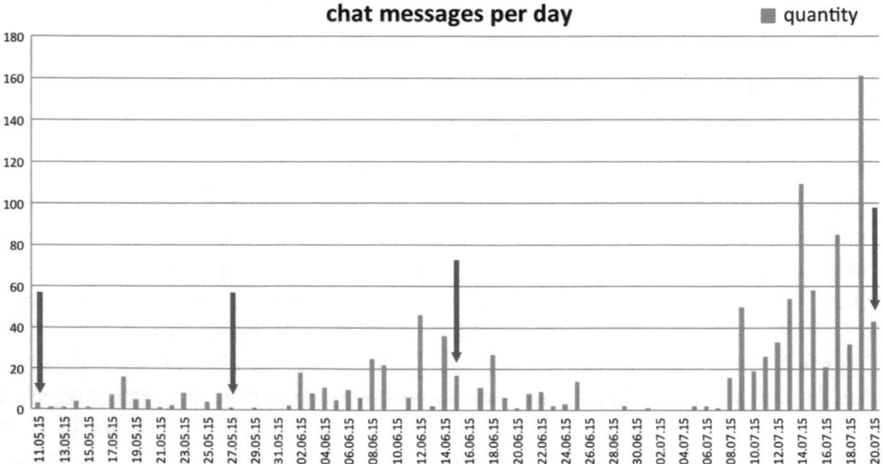

Fig. 1 Quantity of chat messages send out during the course and marked face-to-face meeting days

The red arrows in Fig. 1 mark official face-to-face meetings with all course members. That clearly shows that the third and especially the last course meeting had a significant impact on the chat activity. Looking into the chat conversation content right in advance of these days shows, that the students had to organize their presentations but also discussed content-related questions. The discussed topics will be shown later on in this text more in detail. At first sight, these results seem to be contradictory in comparison to the students' feedback and our argumentation above that especially at the end of the course Evernote seemed to loose importance for the students. However, at this point it must be differentiated between the usage of the scratch pad and the work chat function in the Evernote tool. The scratch pads were mainly used for note taking and information collection particularly at the beginning of the course but not for text production taking place towards the end of the course. That is why the scratch pad function of Evernote did loose its importance for the later working process. This point was clearly stated by the students. Nevertheless, the work process' organization became more and more important towards the end of the course and therefore in the same way the work chat gained importance for the students. This is what the examined work chat shows. In addition to looking at the message distribution over the course time we examined at which time of the day the chat was used. Figure 2 shows the average number of chat messages allocated to daytime.

Figure 2 clearly shows that except for the nighttime the students communicated more or less throughout the whole day. Furthermore it gives an idea and allows a closer look on how the students structure their day. They get up and start interacting around 8 a.m. then they do have lectures and work on other things around 9 a.m. and in the afternoon from 2 p.m. to 7 p.m. Furthermore, they seem to have their lunch and dinner break around noon and from 7 p.m. to 8 p.m. Finally, it can be seen that there is a high level of interaction especially in the evening after 8 p.m. So this figure clearly points out, that course content interaction does not stop after

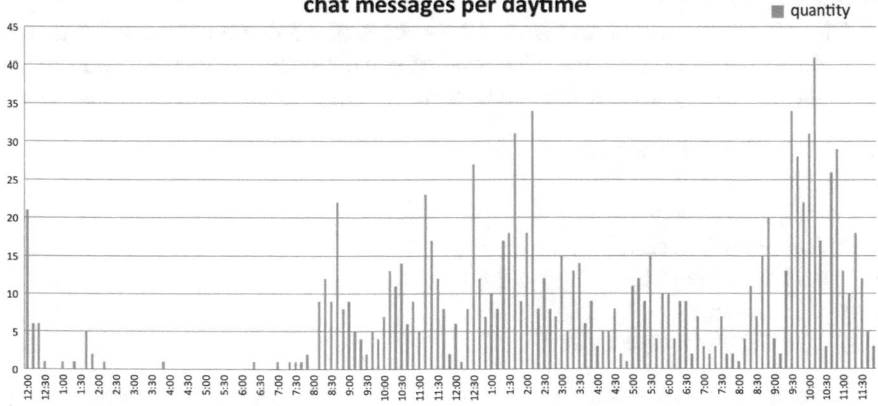

Fig. 2 Quantity of chat messages allocated to daytime

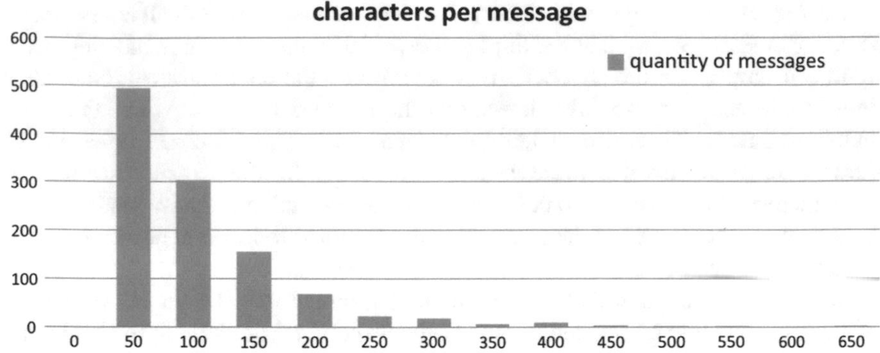

Fig. 3 Quantity of messages with rising number of characters per message

classical working hours. It is even the other way around: Especially in the nighttime before midnight the highest level of interaction could be detected. This supported our idea of introducing Evernote and iPads in order to make the students more flexible in working during the day in the same way make them exempt from necessarily meeting in person for course related interaction.

A third interesting point can be seen in the evaluation of text length when communicating with Evernote. Figure 3 shows the messages in reliance to their number of characters each message contained. The figure clearly shows, that most of the time the system is used to exchange shorter messages up to 50 characters each. Messages with more than 300 characters are seldom exchanged. This goes hand in hand with the insights on longer text production with Evernote. Just like the scratch pads are not made for longer text production the work chat does not serve for texts that normally would be exchanged via e-mail. Nevertheless, the students additionally stated that even with these shorter messages content related discussions were possible. Hence, the software served as an important tool for course related interaction throughout the whole seminar (Fig. 4).

Our last focus was put on the messages' content. Here we analyzed the text of each single message and rated what kind of content could be found. Doing so we were able to identify seven different cluster to which all messages could be allocated:

- content related discussion
- feedback on work results
- making date arrangements
- difficulties with technology
- group internal conflict
- task distribution
- general organizational aspects

From our perspective the main supportive result is that the most important identified cluster are formed by content related discussions and feedback on work results. This is pretty much what we wanted to promote by introducing the tool. One of

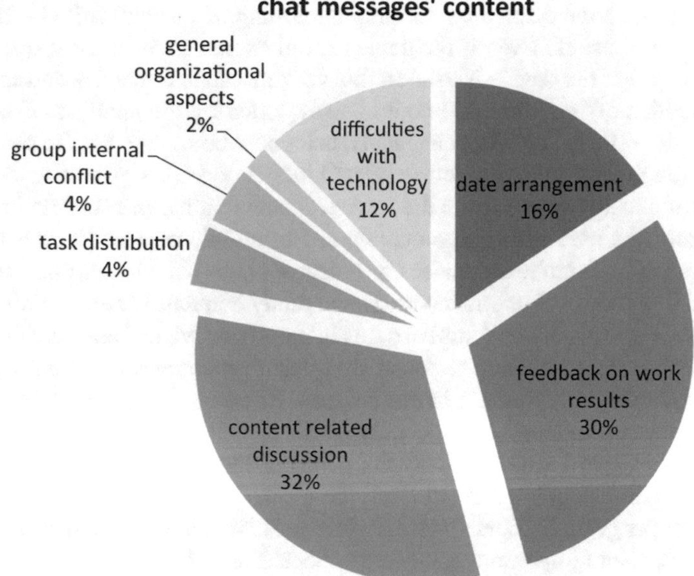

Fig. 4 Cluster build on basis of the messages' content analysis

our aims was supporting the students to make their discussions on the course topics more flexible in terms of location and time by using the Evernote tool. Looking at the posed figures this seems to work out in our explicit context. In addition to that it was interesting to see that there could not be identified any cluster with private or note course relevant messages. Several explanations are possible for this fact. On the one hand it could be argued that the students in their groups did not interact at all on any private or personal level but only with regard to the seminar's topic. A more logical explanation for us would be that neither Evernote nor the iPad itself did find their way into the students' "private live". Hence, the students do differentiate between private interactions (and use their private devices for that purpose) and university interaction. For the latter they used the tools and hardware promoted for the seminar. Nevertheless, at this point these conclusions stay assumptions and need further investigation.

4 Conclusions

As a positive general finding it can be said that mobile devices and scratchpad software can be used in totally different contexts for higher education. However, there still can be seen some significant barriers for the usage, even if the tools are easy to use and the devices are handy. Firstly, the idea of "anywhere", which often goes hand in hand with the discussion on mobile learning, [5] cannot fully be supported by our research results. Even if the device is only some steps away, it doesn't necessarily mean that the device is used in this very moment. Secondly and in contrast to that, the idea of "anytime" [5] could be supported by the analysis of work chat communication. Based on work chat analysis it could be shown that the students use the tool throughout the day for course related interaction. This gives rise to the hope that the introduced tools do make the students' interaction more flexible in terms of time. Thirdly, the assumption that making use of mobile devices (it doesn't matter if these are the students' own devices or borrowed ones) in courses does not necessarily lead to a merge of private contexts and study contexts. There is still a barrier between these two worlds and this barrier is in parts even wished and kept on purpose by the students. Nevertheless, we found significant advantages in introducing tablet PCs, smartphones and Evernote in our courses. However, our research is limited to a small number of students and especially the work chat analysis could only be done for just one group of 4 students. Only this group did use the work chat as their main channel of communication. Nevertheless, the results are promising and the students' feedback on the general course concepts was positive. Hence, this research will be pushed forward and important results are expected for the future.

References

1. A. Kukulska-Hulme, J. Traxler, *Mobile learning: A handbook for educators and trainers*. Routlegde and Taylor & Francis Group, Milton Park and New York, 2005
2. D. May, K. Lensing, A.E. Tekkaya, M. Grosch, U. Berbuir, M. Peterman, What students use - results of a survey on media usage among engineering students. In: *proceedings of 2014 Frontiers in Education Conference "Opening Doors to Innovation and Internationalization in Engineering Education"*, Madrid, Spain, 2014, pp. 92–97
3. K. Schuster, K. Thöing, D. May, K. Lensing, M. Grosch, A. Richert, A.E. Tekkaya, M. Peterman, S. Jeschke, Status quo of media usage and mobile learning in engineering education. In: *proceedings of the 13th European Conference on e-Learning, ECEL 2014*, Copenhagen, Denmark, 2014, pp. 455–463
4. D. May, P. Ossenberg, Organizing, performing and prsenting scientific work in engineering education with the help of mobile devices. In: *'International Journal of Interactive Mobile Technologies Engineering (iJOE)*, Wien, Dresden, New York, 2015, pp. 56–63
5. C. Quinn. mlearning: Mobile, wireless, in your pocket learning: Linezine 2000. URL http://linezine.com/2.1/features/cqmmwiyp.htm

Organizing, Performing and Presenting Scientific Work in Engineering Education with the Help of Mobile Devices

Dominik May and Philipp Ossenberg

Abstract The meaningful use of mobile devices in higher education learning contexts is still underrepresented at German universities. Even if these "new" technologies open up totally new teaching and learning experiences the current use is strongly dominated by simple technology provision. With a special course – designed for engineering students – the authors want to change this and go more into the direction of meaningful interaction and collaboration with the help of mobile devices throughout the learning process. The course is developed to give the students the opportunity to use tablet PCs in context of their studies and simultaneously to improve their ability in the field of scientific working processes. Making use of the online tool Evernote supported this aim. This tool allows users to work out and share digital notebooks and with this to organize as well as document a working process. In this case it was used to support a scientific research process from the beginning to the end. As Evernote can be used with an mobile app it easily can be used on tablet PCs. Hence, it supports the course idea of using mobile devices in perfect way. The course itself is divided into four face-to-face meetings and three working phases. By taking part in the course the students go through their own research project with explicit steps – from having a new idea for research to the results' presentation. The meetings are mainly used in order to introduce tools or techniques for research processes. During the working phases the students do their research, create presentations, a poster, and a scientific report. Based on the internal course evaluation and the students' feedback we observed, that the combination of tablet PCs with the Evernote software is a good opportunity to show how mobile devices can be meaningfully integrated into higher engineering education.

Keywords Mobile Learning · Mobile Devices · Scientific Work · Evernote

D. May (✉) · P. Ossenberg
Engineering Education Research Group, Center for Higher Education,
TU Dortmund University, Dortmund, Germany
e-mail: dominik.may@tu-dortmund.de

Originally published in "International Journal of Interactive Mobile Technologies (iJIM)", © IAOE 2015. Reprint by Springer International Publishing AG 2016, DOI 10.1007/978-3-319-46916-4_28

1 Introduction: Idea and Goal

This paper and the described course concept are dominated by two different topics. On the one hand the use of mobile devices in higher engineering education or better: mobile learning [1] is focused. On the other hand this work deals with the scientific research process and its presence within engineering education. For both topics the authors found potential for future improvements in higher education and the presented work should lead in this direction.

Mobile devices are daily companions for today's students. Most of them own at least a smartphone or even a tablet PC. Virtually everywhere at the university you can see students reading on and texting with their mobile devices. In contrast to that the use of such devices for explicit study purposes is less common in German higher education. That is shown by empiric studies done by the authors over the last two years [2, 3]. The use of smartphones or tablet PCs with the aim to support or even improve the learning experience is seldom. Even if the corresponding pedagogical trend "mobile learning" [1] grew stronger in higher education over the past decades, this process seems to be rather forced by teachers than by the students themselves. This observation was supported by qualitative research in the form of interviews enriching the study results mentioned above.

Another important observation that can be made at many universities – at least in Germany – is, that even if universities are a place for scientific work and research the concept how research works and what it means to work in a scientific manner are only shown to engineering students but their can seldom experience doing it themselves. In many cases the bachelor thesis is the first scientific artifact that is worked out by the students. This of course has a heavily negative impact on the students' ability to work scientifically and write scientific documents.

In the light of those two observations the authors designed and implemented with "Fit for Science" a special course for engineering students in order to improve (1) their ability to work scientifically, (2) their skills to use mobile devices for collaboration in scientific contexts and (3) their scientific writing skills. The course was firstly implemented in summer 2014 and a second edition was given in the winter term 2014/15. In the summer term 2015 the third edition will take place. From one edition to the next one, the evaluation and research results will improve the course concept. All the work is done in context of the project "ELLI — Excellent teaching and learning in engineering education". Within this project two working packages are focusing on mobile learning and on a research workshop for students (Fig. 1). Combining these two views was a first step into design process for the presented course. This paper will explain the underlying research concepts as well a preliminary considerations, the course design itself, the students' feedback and planned improvements for the future.

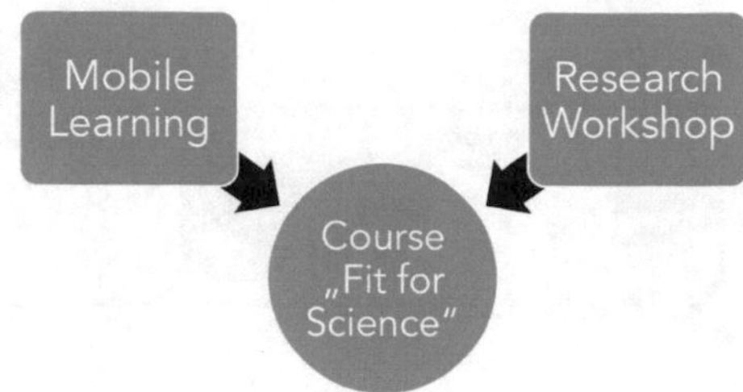

Fig. 1 ELLI working packages contributing to the course concept

2 Research Concept and Previous Observations Important for the Course Design Process

Just as the presented course "Fit for Science" itself is focusing on the process of scientific research the course design process as well is embedded in an underlying research process, too. This process is oriented towards classical research circles, which exist in nearly every science and are more or less similar in its fundamental steps (see e.g. [4]). In our presented case the basic steps are:

- (1) Idea for new research,
- (2) general goal formulation,
- (3) overview over existing work and connections to prior research,
- (4) formulation of concrete research question,
- (5) research action,
- (6) reflection on results, and
- (7) implementation of improvements or start of new research circle.

In order to go through this research steps and work out a course concept at the same time we had to synchronize both processes. Figure 2 shows how this worked out in practice.

2.1 Goal Formulation

As this paper's aim is mainly to present the developed course concept and the students feedback, the focus will lay on these two aspects. They represent the last five steps of our research process. However, the previous considerations during step one to three are important to understand the whole process. Hence they will be explained shortly in the following. The main focus will lay on the effective usage of mobile devices in educational engineering contexts, which reflects the basic idea. Our general goal for our research was to improve engineering education by integrating such devices into educational settings with a main focus on collaborative learning processes. In the following we will give a short overview over existing work. In addition to that we will explain how this work is connected to prior work at our institution.

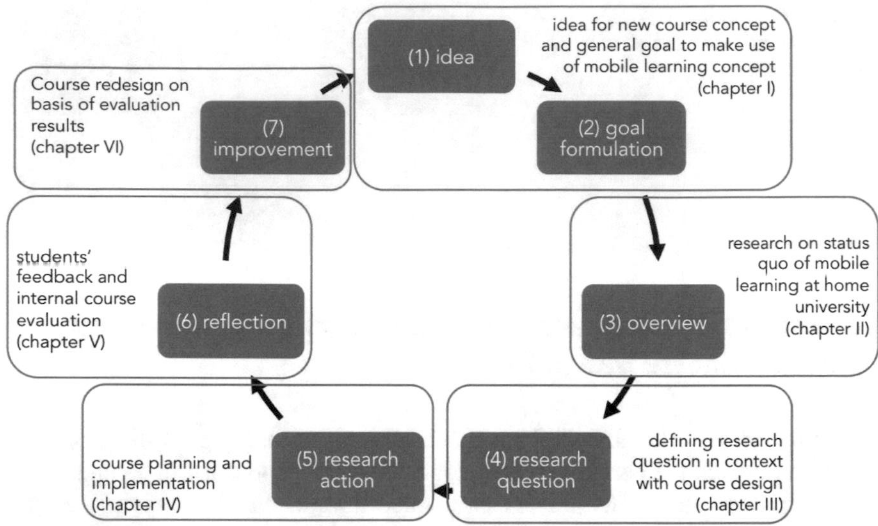

Fig. 2 Synchronized research circle and course design process

2.2 *Overview*

After defining the general goal for the design process we look in the following at existing work in the field of learning with mobile devices and describe the environment in which we implement the course concept.

2.2.1 Use of Mobile Devices in Higher Engineering Education

Using mobile devices is completely normal in the present time. The release of new smartphones or tablet PCs from the company with the apple in its logo is regularly an exceptional period for a whole industry sector. Hence, our idea and goal for the presented work was to bring mobile devices more into educational settings in higher engineering education. Or focus in this context was to put on the support of communicative and co-operative process with the help of mobile devices.

The strong development towards mobile devices also has an impact on universities and higher education. Whereas the universities develop special apps to make the students' life easier (these apps for example contain campus maps, timetables or refectory plans; e.g. RUB App at the Ruhr University Bochum [5]) the field of "mobile learning" as a new learning scenario emerged at the beginning of the nineties' [1]. Since then, this area of research steadily grew so that more and more educators are presently working on the improvement of the students' learning process with the help of mobile devices and conduct connected research. [6] for example recently published a chapter on "iPad didactics" speaking about "[…] new forms of pedagogical concepts for education like transformative learning and complex learning supported by mobile ICT [information and communication technologies] and interactive media and learning (IML)" [6]. In general the company "Apple" seems to be highly interested in the education sector as they are offensively promoting their technology for curriculum design, assessment and teaching [7]. Moreover a growing number of publications on mobile learning can be observed on engineering education conferences. For example on the last "Frontiers in Education Conference 2014" in Madrid more than fifteen papers presented work in context with mobile learning [8]. Most of them deal with the use of mobile devices directly in class or to support the learning process during a course. Nevertheless such approaches remain exceptions in the light of the total amount of engineering courses given all over the world every day.

In addition to that there is only little data on the actual use of mobile devices for study purposes outside of such exceptional approaches. Moreover such approaches seem to be more present in the Scandinavian and the Anglo-American regions. Hence, for us the question was: What is the status quo on mobile learning solutions at German universities in engineering education? It was important for us not only taking the communities' discussion throughout the world into consideration but also having a strong focus on the German educational sector which may show some shortcomings in the implementation of mobile learning in comparison to other regions. Therefore an empirical study with questionnaires was done in 2013 to find out what the students really use [2, 3]. The results show, among others, that firstly nearly every student owns at least a smartphone and around a quarter of them do own a tablet PC, too. That was not surprising at all. What indeed was surprising is that secondly most of the students seldom or never use their mobile device for study purposes. This is clearly shown by the answers on the questions "How often do you use your mobile device (smartphone and tablet) or mobile apps for study purposes?" and "How often have you encouraged by your teacher to use your mobile device for study purposes?":

"The results show that 23.4 % of the students use their smartphone never or very seldom in context of learning, whereas 63.9 % do this often or very often (12.7 % use it from time to time). The same question asking for tablet computers disclosed totally contrary results: 69 % use their tablet never or seldom and only 26 % use it often or very often (in context with studies). […] The last question in this context was, how often the students use mobile apps for learning. The answers reveal that the use of apps explicitly for learning is not very common yet. Only 7.2 % of the students use mobile apps often or very often for learning, but 84.7 % use it seldom or never. […] An additional question asked for the encouragement to use mobile devices during a course. So the students were asked: "Have you been encouraged to use your mobile device during the course? If yes, for what kind of purpose?" The results exposure that only 22.6 % of the students have been asked often or very often to use their devices in context with a course. In most of these cases (65 %) those who were asked to do so, were requested to use them in order to take part at any kind of polling. Polling means in this situation that the teacher asks a question in course-context and the students had to use their device and special software to answer it. The answers are then discussed in the audience. […] Dividing the apps by the purpose they are used for, 7 different clusters could be identified. 109 (31 %) of the named apps serve as any kind of database, 79 (22 %) are used for the organization of learning and 47 (13 %) for any kind of application and testing. 40 (11 %) apps are language dictionaries, 30 (8 %) help to make notes or edit documents and 20 (6 %) are used for cloud computing. Finally 32 (9 %) apps could not be allocated to any cluster. Hence they were summarized under "others" [3]."

Interviews that accompanied the questionnaires supported these observations and made obvious that many students simply do not know how to use their devices meaningfully for learning and co-operation. Hence, a major lack can be seen in the students' inaptitude to work together using their mobile devices for study, or better, for learning purposes. That was one reason to design the course "Fit for Science" with a special focus on mobile devices and their options to support collaborative working and learning processes.

With the Research Workshop at TU Dortmund University we had a perfect setting for this work in order to develop and implement the course. Therefore this setting will be described shortly in the following.

2.2.2 Excursus: Research Workshop for Students

The main idea behind the Research Workshop is to synchronize students learning with the typical steps of a research process [9]. To reach this, we offer various arrangements to the students in a special seminar room, which is furnished to support learning and working in groups as best as possible (Fig. 3). In short extracurricular courses

Fig. 3 Research Workshop at TU Dortmund University

students can improve their key qualifications in "scientific working" or "writing with LaTeX". The course "Fit for Science" is another course example we are offering in the Research Workshop. Last but not least we offer the room itself either for learning or for working on own research projects. During the semester, the seminar room has regular opening hours. Students for example can use experimentation kits like "LEGO Mindstorms" to conduct small experiments, called FLExperiments [10]. To encourage and support the students' investigation we provide small scripts including hypothesis and research questions. During the opening hours students can come in and ask for assistance of a qualified tutor, who can give advice e.g. in fields of "scientific working" or "time management". The Course "Fit for Sciencc" combincs both approaches from above. It takes place in context with the Research Workshop and focuses besides the topic of scientific working the meaningful use of mobile devices.

3 Research Question

Based on the overview on existing research, the observations and the connections within the ELLI projects explained above we were able to formulate the explicit research questions for our work. These were:

- How can tablet PCs support collaboration in research processes?
- How can the results of the first question be applied by designing courses in higher education?

In order to answer these research questions we took into account the general overview explained in II and started with the research action. That means that the course planning and implementation was an essential part for us to answer these questions. This research action was later in validated and supported by the internal course evaluation and the students' final course feedback.

4 Research Action: Methodological Considerations and Course Design Process

4.1 Methodological Considerations

As we wanted to conduct research on the use of mobile devices and simultaneously develop a course concept with the help of such devices we found for the methodological basis the "Design-Based-Research" (DBR) concept as a highly fitting concept to guide us through this process. DBR aims on the development of technology enhanced learning scenarios on the one hand and at the same time it aims on systematic research on these scenarios on the other hand [11, 12]. Hence, the DBR approach is "a systematic but flexible methodology aimed to improve educational practice through iterative analysis, design, development, and implementation, based on collaboration among researchers and practitioners in real-world settings, and leading to contextually sensitive design principles and theories" [13].

Using this methodology gives us the opportunity for being at the same time researchers and teachers. In addition to that DBR bases on a two cyclic phases [11]:

(a) Phase of action: Intervention, design, implementation (see Section 4)
(b) Phase of analysis: reflection of implemented intervention (see Section 5)

That these phases are meant to be cyclical means that after the evaluation a second phase of interaction starts, which is again evaluated in next step.

Having all this in mind we were able go ahead in our own research cycle. "Planning the research action" and "research action" in our case this meant to plan and give the course itself.

In order to give an overview on how we introduced the mobile devices in the course we will firstly explain the course concept in the following. After that we will explain the students feedback, which in our case goes hand in hand with step (6) "reflection on results".

We planned the course "Fit for Science" for the first time for the summer semester 2014. As explained above our overall goal was to design a course concept, which at the same time teaches how scientific research processes are carried out and how to use mobile devices for those research processes in a collaborative manner. In the following we will explain how the work was designed finally. We will begin with the intended learning outcomes, explain afterwards the organizational considerations and finally the instructional activities will be laid out.

4.2 Intended Learning Outcomes

In order to plan the instructional activities it was important to define the intended learning outcomes first. These outcomes are directly deduced from our overall goal. However, they explain the goal more in detail and – that is the most important aspect- they can be used to openly reveal to the students what they are expected to do throughout the course. Following the intended learning outcomes, at the end of the course the students should be able to…

- …conduct own research processes in a group,
- …organize and document scientific work with the help of new collaborative media and mobile devices,
- …present results of scientific work with different media, and
- …write a short scientific paper on basis of common scientific norms with the help of adequate tools.

4.3 Organizational Considerations and Course Schedule

Looking from the general organizational perspective the course can be split into three major thematic parts:

- Organization of scientific work
- Methods of scientific work with focus on literature search
- Presentation of scientific work and results

These major parts are designed so that the students were able to carry out an own research process. Furthermore looking at the detailed schedule the course consists of four full-day classroom sessions at the university. For these sessions all participants come together. This is accompanied by three working phases in between the meetings. During these working phases most of the course's work is carried out. Hence, the meetings mainly serve for presenting achieved results and preparing the students for the next working phase.

As the general course goal indicates the research process with its different parts is the overall orientation for the course design. Hence, the "Fit for Science" course schedule itself reflects a simplified research process with the following six steps:

- Starting for new research on specific topic,
- Gaining an overview and searching for existing work,
- Confining research question,
- Carrying out detailed literature research,
- Answering research question on basis of literature, and finally,
- Presenting the results in written and oral form.

Figure 4 shows the course concept with all its different meetings and steps. It furthermore explains how on the one hand the three major phases are distributed among the four in-class sessions and the three working phases. On the other hand it explains how this course schedule is synchronized with the research steps shown above.

For the whole course the student group is split into smaller teams of three to four students. Each team has to perform its own research and go through the whole process once. As the course concept heavily bases on the use of tablet PCs the students either could use their own ones or borrow one from the university for the whole course time. Most of the instructional activities are designed so that they can be carried out with the help of tablet PCs. Based on our own experiences we chose the online tool "Evernote" [14] as the fitting tool to the support the students' collaboration among each other and organize the working process. Evernote is an online workspace that allows writing notes, collecting search results, finding data and presenting work in one digital notebook, which can be shared among the team members. Additionally we chose the online tool "Mindmeister" [15], which serves for working out digital mind-maps, which also can be shared among different team members. In the following we will briefly explain each course parts and its meeting. Furthermore the instructional activities and the contextualized use of tablet PCs will be laid out.

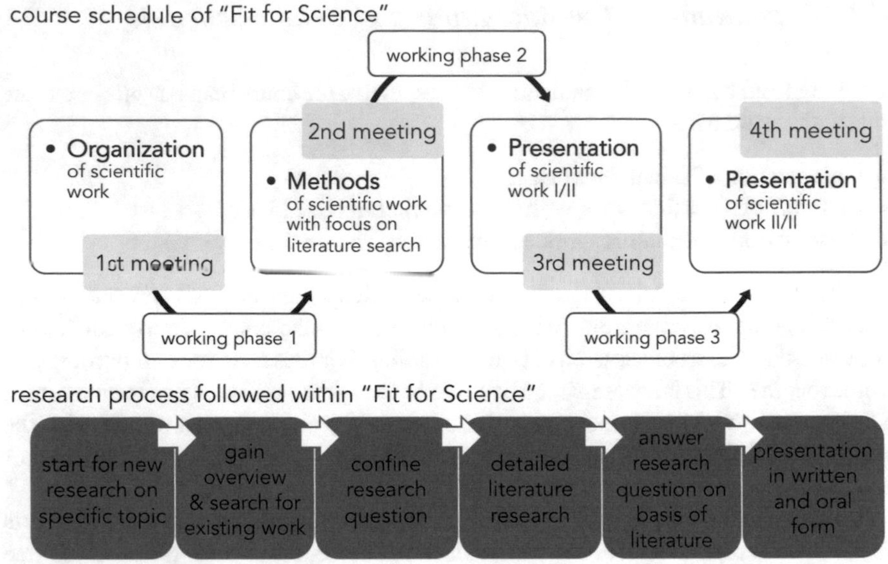

Fig. 4 Course schedule and followed research process

4.4 Course Parts and Instructional Activities

4.4.1 First Part: Organization of Scientific Work

The first major course part equals with the first meeting. In this first part the focus lies on the organization of the research processes. Within this step the course concept with its intended learning outcomes is presented to the students. In addition to that they have to organize their working process in the group, and make themselves familiar with basics of scientific work in general. All of this is done during the first meeting by a mixture of several instructional methods. For example the general process of scientific work is on the one hand presented to the students but on the other hand additionally worked out more in depth by themselves with the help of several group tasks. In order to not being too theoretically at this early point the students group already choose their topic of interest for the following research process. As the topic in this case is just a means to an end and the process itself is more in focus, the groups can choose their own favorite topic. We as lecturers just give possible topics in order to help if the students can not decide. For example in the 2014 summer term's edition of this course the groups finally worked on the topics "e-mobility", "fuel cell technology", "industry 4.0", and "resource efficiency in production processes" and during the following winter term's edition the research work was focused on "e-mobility" and "mobility concepts in mega cities". At this point the topics are defined very broadly as one of the students task for later on is to narrow the topic so that a concrete research question can be formulated and an effective literature research can be carried out. The task for the first working phase is to gain an overview of the topic and search for existing work. During the phase they additionally develop first ideas for the research question.

In order to support the group work the tools Evernote and Mindmeister are introduced at this point and have to be used by the students from the first meeting on. Both are easily usable with tablet PCs so that their use perfectly fits to our course goals. The advantage of those tools for the working processes is that with both of them work artifacts can be jointly produced even if the groups do not physically meet. Hence, the idea of mobility and flexibility that lies beyond the concept of mobile learning is supported here [16]. To be more concrete, the students have to use Evernote to work out a shared notebook, in which their work processes are organized and documented. Moreover they have to use Mindmeister in order to work out a mind map on their topic of research. This mind map should reflect the overview and the search results on existing work in context with their topic. The development of the mind map is the main task for the first working phase.

4.4.2 Second Part: Methods of Scientific Work with Focus on Literature Search

At the beginning of the second phase (and so at the beginning of the second face-to-face meeting) the students present their mind map to the rest of the class. With this result they can start in the following research steps. Guided by several detailed methodological steps the students define their explicit research question and present it to the group. At this point it is important that the lecturer and the rest of the course carefully scrutinize the formulated research question. As this question will guide the students during the following research it is necessary that the question is very detailed. The students have to try to answer this question only on basis of a literature search of some weeks and have to write a scientific paper of only seven pages on this later on. Hence, a very board question does not help in this context This must be understood by the students, as they tend to formulate broad questions. In a next step basic concepts and explicit methods for literature research come into focus. This includes concepts to find relevant texts and to organize the emerging literature database. For both aspects additional tools like online library databases or “Citavi” (a tool for literature administration and knowledge organization [17]) are presented and the students learn their usage guided by explicit tasks.

An additional but very essential topic of the second meeting is the underlying idea and importance of using, respecting and listing references during scientific work. As the students start with their scientific paper after this meeting different concepts of referencing sources in papers are discussed, too. Moreover the text type “scientific paper” with its important parts and basic rules for writing such a text are discussed. With this knowledge and skills the research groups start into the second working phase, in which they firstly conduct a detailed literature research in order to answer their research question and secondly start with the writing process.

4.4.3 Third Part: Presentation of Scientific Work

In the final part of "Fit for Science", which covers the last two meetings, the presentation of scientific work comes into focus. In order to show the diversity of presenting methods different options are discussed as well as used by the students. This starts with the presentation of their current working process and first research results using the Evernote tool. We explicitly ask the student to use this in order to have them experiencing several options to present working results apart from the classical PowerPoint presentation. Evernote has a special presentation mode, which allows the students to easily present the working process using the tablet PC. After that the course focuses on the presentation of scientific work with the help of the formats presentation slides and scientific posters, as those two formats are very common in the scientific community. During several assignments the students acquire characteristics of good talks, presentations and posters. That means that all these publication forms are discussed in course regarding questions like "What are good and bad examples?", "What are the differences in the use of texts, slides and posters?" and "What is the impact of the different aspects for the design?".

At this point the students start into the third working phase, which is mainly used for finally answering the research question based on the found literature. The results should be prepared for the final meeting in three different ways. Firstly a short presentation with slides should be given. This should be enriched with a scientific poster summing up the results. Additionally and as mentioned above the students have to work out a scientific paper and hand it in. A feedback on their presentation and the poster is given by the whole group and by the course instructors. The scientific papers is only rated by the instructors. With this phase of presenting the work the course ends after the fourth meeting.

5 Reflection: Course Evaluation and Students' Feedback

16 students took part in this first edition and 7 students joined the second edition (for the third edition, which will take place, during the publication process of this paper, again 16 signed up). In order to enrich our own course valuation and to receive a proper student feedback on the course, they had to write a final reflective paper and fill out questionnaires ($n1 = 13$, $n2 = 7$; mainly asking the students to rate statements using a five point scale from "I fully agree" to " I do not agree"). The feedback mainly based on two central approaches: On a general feedback and on the digital instructional resources used during the course.

5.1 General Feedback on the Course

The general feedback from the students was very positive. They stated that they liked the course in its current form and did improve their abilities to conduct own research processes. As explained in the introduction, working scientifically and writing scientific papers is seldom taught in general engineering curricula. The students recognize this, too. One question in the questionnaire directly addressed this observation. In that context 7 students agreed and even 12 students fully agreed with the statement that they visit the course to learn something about methodological concepts of scientific work (students from both editions were taken into account). Hence, they are thankful for such opportunities. This is even more visible in additional written student statements like "In general I learned a lot in this course and the atmosphere was continuous comfortable".

The most frequently mentioned point of criticism is that the students did not like the given software Evernote (parts of them would prefer MS OneNote). Another point of interest is the average time the students use the tablet PCs. The statements lead to the conclusion that they quite seldom use their tablet PCs in context of actual learning processes but instead for communication and online research 10 statements support this conclusion, whereas 6 students use their tablet PC frequently.

Asking for the most important course topic especially the part about literature research and reference managements systems was voted very positive. We additionally were very pleased that 14 students full agreed with the statement: "I was encouraged to think about practical application of theoretical knowledge".

5.2 *Feedback on Instructional Resources*

The other aspect is about the usage of instructional resources within the course. Hence, we analyzed the reflective papers in a first step with the intention to find out which programs students mentioned having used during the course. With this we want to find out which activities were done with desktop computers and the tablet PCs on the one hand and on the other hand we want to identify additional programs or apps they used. Therefore we counted the mentioned programmed and identified categories corresponding with explicit activities. At this point we did not distinguish between activities done with the help of a desktop computer or a tablet PC. Altogether we have analyzed 18 reflective papers.

The coding of the reflective papers led to 6 definable categories and 1 category of "others". The final categories are (sorted in descending order):

- Scratch Pad (e.g. Evernote)
- File Hosting Service (e.g. Dropbox)
- Instant Messaging (e.g. Facebook messenger)
- Classic Document Processing (e.g.Word)
- Mind Mapping Tool (e.g. Mindmeister)
- Reference Management (e.g. Citavi)

It is not surprising that Scratch Pad is the most often mentioned category as this category is deeply connected to the requirements we defined by designing the course, e.g. by promoting Evernote as the program for collaborative work processes. The category "others" includes for example apps like YouTube but also Doodle (tool to find a date for appointments). All groups used a File Hosting Service like Google Drive or Dropbox especially when writing the scientific paper. Some groups used instant massaging to communicate efficiently. Especially during the second edition quite a number of students used the Evernote Work-Chat to communicate, as the app designers introduced this opportunity just before the course started. Mind Mapping Tools are only mentioned in context with the first course phase. Even if all the students mentioned to have used a reference management system like Citavi at least once in context with the course the scientific papers clearly showed that not all groups have used that properly. This was one aspect we especially focused on during the second course edition and it will be an important aspect we will have to keep our focus on during the next editions.

However our question not only was, which programs the students used in general but even more important was to find out how they finally used the tablet PCs. Even if some of the students said, that they had to change to a normal PC for explicit tasks (for example for longer writing processes for the paper), they were positively surprised by the options tablet PCs and connected apps offer. By analyzing the reflective paper in a second step we identified categories of activities students did with their tablet PCs in context with the course activities. Finally we were able to clearly define the following eight different categories (again sorted in descending order; supplemented by a category "others" and a category for "leisure activities"):

- Organize and document work processes (e.g. with Evernote)
- Search for information (e.g. with library databases)
- View and edit PDF files (e.g. with PDF viewer)
- Create mind maps (e.g. with Mindmeister)
- Share and synchronize files (e.g. Dropbox)
- Communicate (e.g. with Skype)
- Create documents (e.g. with http://www.titanpad.com)
- Time management (e.g. calendar app)

Because of the special course design, it was not surprising that the most frequently mentioned activity is to organize and document work processes with Evernote. Following the course requirements (use information from library databases to write a scientific paper) the another frequently mentioned activity was to find course relevant information in databases and on websites by using the tablet PCs. Additionally most of the students used the tablet PCs to view and edit PDF files, e.g. to find and analyze information or make notes on slides during lectures (some of the answers indicated that this usage was not limited to our course context but the students even used the borrowed tablets in other courses for this purpose). Unfortunately the students created mind maps only because they had to present one in the second meeting. However they have created this mind maps by using the tablet PCs and with the advised application. The ability to share and synchronize files between desktop computers and tablet PC using explicit tools like Dropbox is another good opportunity the students see and made use of. With this they always have relevant documents for the working processes at the hand, no matter which device they are using at the very moment. Students reported that the tablet PCs was seldom used for pure communication purposes and in this context mainly replaced by the smartphone. Those who did use the tablet PC for communication mainly did this with Skype or any mailing application. Although most of the students finished up writing the paper at a desktop PC and not using the tablet PC for creating longer documents, 5 of them students even did this task with the help of their mobile device. In this case they made use of online writing tools like http://www.titanpad.com, which allows several users to synchronously write in one document at the same time. Last but not least 4 students reported that they use their tablet PC to organize their time and used a special calendar app for that.

At this point we will finish with the overview on the course evaluation and will change over to the future plans with this concept.

6 Improvement: Conclusion and Future Plans

In context with this research we formulated the following research questions:

- How can tablet PCs support collaboration in research processes?
- How can the results of the first question be applied by designing courses in higher education?

In order to answer these two questions we designed and implemented a special course concept, which heavily relied on the use of mobile devices in combination with several tool; mainly Evernote. From our perspective this combination is a very good opportunity in order to meaningfully integrate mobile devices into the teaching and learning process. The evaluation findings support our assumption. The students successfully used this tool and the device in order to organize and document their research process, which in short means they did collaborate with it even if they had to get used to it first. It is obvious that tablet PCs itself do not support collaboration but they need to be integrated in context with adequate application. This is what we productively did with this course.

The course has been implemented only twice up to this moment. By the time of the publication a third round will be running. For the coming courses some adjustments will be done, which are based on the evaluation findings. For example the scientific paper showed clearly that most of the students really struggle with the writing process. When questioned, most of the students said that this was the first longer paper they had to write after finishing school. Hence, a special focus will be laid on the scientific writing process.

Nevertheless the course will become a part of the Research Workshop course program. Hence, it will be taught two times a year and continuously improved with every edition. With this the evaluation and the research methodologies will be further developed, too. That means that we will be able to get more and more meaningful data on the use of mobile devices and advance the research findings on that. In addition to that this course will also be a place in which new ideas in context with the use of mobile devices in teaching can be tested. Successful approaches will be then transferred to other engineering classes.

References

1. A. Kukulska-Hulme and J. Traxler, *Mobile learning: A handbook for educators and trainers*. Milton Park and New York: Routlegde and Taylor & Francis Group, 2005.
2. K. Schuster, K. Thöing, May, D., Lensing, K., M. Grosch, A. Richert, A.E. Tekkaya, M. Petermann, S. Jeschke, "Status Quo of Media Usage and Mobile Learning in Engineering Education, Presentation on the ECEL Conference 2014, Copenhagen, Denmark, not published yet".
3. D. May, K. Lensing, A. E. Tekkaya, M. Grosch, U. Berbuir, M. Peterman, "What students use - Results of a Survey on Media Usage among Engineering Students," in *Proceedings of 2014 Frontiers in Education Conference "Opening Doors to Innovation and Internationalization in Engineering Education,"* 2014.
4. T. Jungmanm, "Forschendes Lernen im Logistikstuidum: Systemaitische Entwicklung, Implementierung und emprirische Evaluation einers hochschuldidaktischen Modells am Beispiele des Projektmanagemnts," doctoral dissertation, TU Dortmund University, Dortmund, 2011.
5. RUB mobile. Available: http://www.ruhr-uni-bochum.de/mobile/ (2014, Nov. 16).
6. I. Jahnke, S. Kumar, "iPad-Didactics - Didactical Designs for iPad-classrooms: Experiences from Danish Schools and a Swedish University," in *The New Landscape of Mobile Learning: Redesigning Education in an App-based World*, C. Miller and A. Doering, Eds.: Routledge publisher.
7. Apple Inc, *Curriculum, Assessment, and Teaching Tools for iPad*. Available: https://www.apple.com/education/docs/Curriculum_Assessment_Teaching_Tools_for_iPad_8-14.pdf (2014, Nov. 16).
8. IEEE, Ed. *Proceedings of 2014 Frontiers in Education Conference "Opening Doors to Innovation and Internationalization in Engineering Education,"* 2014.
9. T. Jungmann, P. Ossenberg, "Research Workshop in Engineering Education: Draft of New Learning," in *Proceedings of the 2014 IEEE Global Engineering Education Conference (EDUCON)*. 2014, pp. 83–87.
10. P. Ossenberg, T. Jungmann, "Experimentation in a Research Workshop: A Peer-Learning Approach as a First Step to Scientific Competence," *International Journal of Engineering Pedagogy (iJEP)*, vol. 3, no. 3, pp. 27–31, 2013.
11. I. Jahnke, C. Terkowsky, C. Burkhardt, U. Dirksen, M. Heiner, J. Wildt, A.E. Tekkaya, "Experimentierendes Lernen entwerfen - eLearning mit Design-based Research," in *E-Learning: Lernen im digitalen Zeitalter*, N. Apostolopoulos, H. Hoffmann and Mansmann V. & Schwill, A, Eds, Münster: Waxmann, 2009, pp. 279–290.
12. T. Reeves, J. Herrington, R. Oliver, "Design research: A socially responsible approach to instructional technology research in higher education," *Journal of Computing in Higher Education*, vol. 16, pp. 97–116, 2005.
13. F. Wang, M.J. Hannafin, "Design-based research and technology-enhanced learning environments," *Educational Technology Research and Development*, vol. 53, pp. 5–23, 2005.
14. Evernote, *The workspace for your life's work*. Available: http://www.evernote.com (2014, Nov. 16).
15. Mindmeister, *Mind Mapping Software - Create Mind Maps Online*. Available: http://www.mindmeister.com (2014, Nov. 16).
16. C. Quinn, "mLearning: mobile, wireless, in-your-pocket learning," *LineZine*, 2000.
17. Citavi, *Organize your knowledge. reference management, knowledge organization, and task planning*. Available: URL http://www.citavi.com (2014, Nov. 16).

Using Remote Laboratories for Transnational Online Learning Environments in Engineering Education

Dominik May, Tobias Ortelt and Erman Tekkaya

Abstract Working in transnational contexts will be the normal case for the future, especially for engineers. In general this competence can be gained by going abroad during the course of studies. However, not every student can afford going abroad. E-learning technologies can help connect students in different countries and have them work together. Especially for the engineering context, remote laboratories are a very promising technology. With the help of these laboratories, students can perform real live online experiments from their devices at home. Connecting students via the internet and combine this with online experimentation leads to international working groups. The presented course follows such a concept. The online course was developed to prepare students for an international work environment and for a stay in Germany. The use of remote laboratories plays an important role. The course concept and findings are explained on the basis of the first course edition in 2014.

1 Introduction

Being able to work in transnational and intercultural professional contexts already is and will be even more important in the future. That is why intercultural competence is seen as one of the key competencies for acting and communicating successfully in the professional world (seen as one source representing many others [1]). However, this competence means more than just showing foreign language skills or having explicit knowledge of different cultures. Without going too much into detail at this point, being intercultural competent means, in short, to be able to effectively interact in intercultural situations and respond adequately to these situations [2]. This is, con-

D. May (✉)
Engineering Education Research Group, Center for Higher Education,
TU Dortmund University, Dortmund, Germany
e-mail: dominik.may@tu-dortmund.de

T. Ortelt · E. Tekkaya
Institute of Forming Technology and Lightweight Construction (IUL),
TU Dortmund University, Dortmund, Germany

Originally published in "Proceedings of E-Learn: World Conference on E-Learning in Corporate, Government, Healthcare, and Higher Education 2015", © AACE 2015.
Reprint by Springer International Publishing AG 2016,
DOI 10.1007/978-3-319-46916-4_29

sequently, one of the competence areas students have to develop during their period of studies. However, the development of intercultural competences is merely understood as a requirement the students have to fulfill by going abroad and studying or working for some time in a foreign country. There exist a lot of opportunities for that: Going abroad for one or two semesters in the context of a special exchange program, studying a whole course of studies abroad, or doing an internship at an international company. In 2014, for example, more than 205,000 international students came to Germany in order to study at a German university and more than 115,000 German students went abroad for the same purpose and for gaining international experiences [3]. Nevertheless, the figures from UNESCO show that even if many students are going abroad, the percentage in comparison to the total number of students studying in Germany is rather small. A survey from 2013 shows that only 26 % of the German students could show any international experience in study contexts [4]. For engineering students, this figure is with 19 % (universities) and 16 % (technical universities) even smaller [4]. This reveals that engineering students are the student group that shows the lowest rate of mobility in study context. These figures are contrasting the requirements of the professional world, which students will enter later on.

However, for years now, the developments in the E-Learning sector have opened up totally new opportunities for international collaboration and transcultural interactions among students in educational contexts. From our perspective, especially the opportunity to bring international students together in transnational course scenarios with the help of web conferencing tools are perfectly fit in order to have them interacting with each other and with this, developing intercultural competences as needed in the professional world (for such concepts see, e.g., [5, 6]). One advantage of these approaches is that the students do not necessarily have to leave their home country for those experiences. This turns out as the critical point as there are many barriers to going abroad, such as financial or time constraints or even very personal aspects. In the following, we will present an online course concept, that has been designed for international students allowing them to take part from their home country. We will focus on the aspect that we did not only make use of web conferencing tools but also of remote laboratories. This technology describes really existing laboratory equipment, which can be used remotely via the Internet. From our perspective, this is a perfect technology to build up transnational engineering classes, in which the teaching of intercultural and professional engineering competence can be combined.

2 The Course Concept "Engineering and Mobility in a Globalized World"

The described course has been developed as a preparational course especially for international engineering students coming to Germany for the international master program "Master of Science in Manufacturing Technology" (MMT) given at the Faculty of Mechanical Engineering at TU Dortmund University and coordinated by the Institute of Forming Technology and Lightweight Construction (IUL) (official MMT website http://www.mmt.mb.tu-dortmund.de). The students take part in this course before they come to Germany. This contextualization opened up two different perspectives on the course concept and its aims. On the one hand, the students should be prepared for their stay in Germany. That includes the preparation for the university, the study program as well as cultural aspects. On the other hand the perspective to prepare the students for working in international working environments is equally important. Finally, the course was designed as a three-week online course and was given for the first time in August/September 2014. As mentioned above, the students took part in this course from their home country before coming to Germany (for more detailed information on the students group, see the explanations later on in the findings).

2.1 Course Description

The chosen core topic of this course was the question of future mobility as this is a classical engineering topic. Furthermore, it can be discussed from very different technical as well as cultural perspectives. Throughout the course different mobility concepts for public transport in big cities, found all over the world, were discussed and the students should bring in their very national perspectives for rating these concepts with regard to the applicability in cities of their countries. As our focus lies on the use of the remote laboratories, we will only briefly explain the whole course and its activities (for more detailed information about the course see [7]).

As to course design, we firstly worked out the learning objectives as the basis of our instructional activities on them. These objectives heavily rely on the circumstances explained in the introduction and the two perspectives on the course explained above: Preparing the students for both their stay in Germany and working internationally in engineering contexts later on during their career. In addition to that, it was important for us that they do not only work together in transnational teams but also start thinking about their home country, the role of engineers in the very own cultural context, and comparing this to others. These reflection processes are an important first step to gain intercultural competence. Only on the basis of such self-reflective processes is it possible to compare one‘s own position to others, work out differences as well as similarities, and, with this, start an intercultural exchange process [8]. In order to give this course a strong notion of engineering, we decided to include real engineering experimentation in this class. This was connected to the topic of mobility, too. These considerations finally led to eight overall learning outcomes for the course: After taking part in this course the students should be able to…

1. …describe their destination on the basis of internet research
2. …describe their own concept of engineering in comparison to that of others
3. …being able to reflect on the international differences in engineering
4. …use a basic model for describing technology from different perspectives (technical, cultural, organizational)
5. …run experiments like the tensile test and use the gained data for engineering work by connecting theory and practice

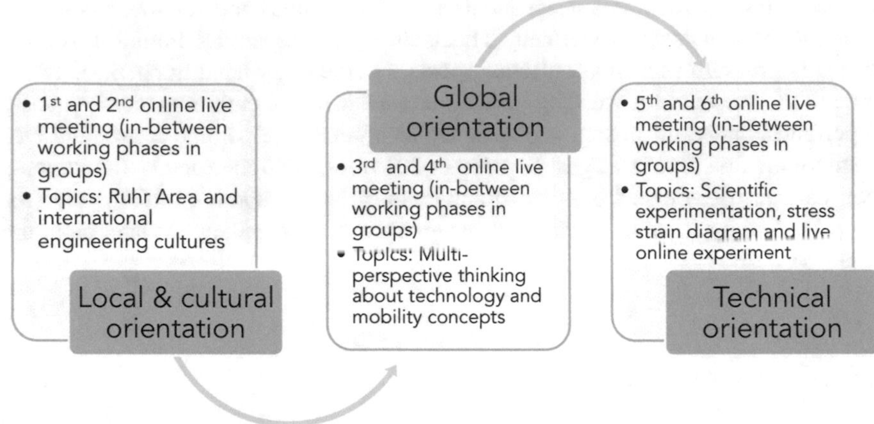

Fig. 1 MMT-Pre-Course concept in three phases

6. …explain a general stress-strain-diagram by identifying important points as well as explaining their relevance for forming technologies
7. …communicate successfully with international engineering colleagues using internet technology
8. …organize themselves in a working process and work successfully together in multinational teams in order to fulfill explicit tasks

Based on this, the preparational course is divided into three main phases: Local and cultural orientation, Global orientation, and Technical orientation (see Fig. 1). Each of the three phases consists of two online meetings, in which we meet with the students in an online room (in our case we used Adobe Connect). Between these meetings the students have to work in groups on explicit tasks in context of the current course subjects. During the first phase the focus lies on the orientation in Germany and, even more in detail, in Dortmund (as a city lying in the economical area of the Ruhr Area). Therefore, the students have to do some online research on this region, its history, its current economical strengths, the current areas of research in the scientific sector, and the living conditions. By changing over to the topic of engineering cultures in Germany and in the rest of the world, the second course phase starts. In this context it is important to have a look at the profession of engineers in different countries and work out the difference, especially with regard the countries the students live in. Having made this step to international comparisons, we start to talk about mobility and different mobility concepts for public transport. For this purpose, the students have to inform themselves about the concepts, look at them not only from a technical but also a social and organizational perspective and finally work out a statement on the applicability of the concept in their home country. With this, we make the step to the third and last phase. Here, the engineering experimentation and trough this the use of remote laboratories come into focus.

2.2 The Remote Laboratory at the IUL

For the experimentation, we make use of the remote laboratory developed at the IUL in the context of the research and development project "ELLI – Excellent Teaching and Learning in Engineering Education", sponsored by the German Federal Ministry for Education and Research (the following description is partly taken from [7]).

With the developed tele-operative testing cell, the students are able to conduct their experiments via Internet from their own devices [9]. In the context of this shown online course, they were able to use a lab, which they had never visited. The students can modify the parameters like the strain rate or the kind of specimen. In a next step, they can start and observe the experiment by four different cameras and by real time data streams. After the test, they can download the data and do their calculations to determine material properties like the Young Modules, for example. The iLab-technology and its components are used to provide the Internet access to the tele-operative testing cell.

The remote laboratory at the IUL gives the students and teachers the opportunity to conduct experiments especially in the field of manufacturing technologies, mainly for material characterization. Figure 2 (right) shows the laboratory with two testing machines for sheet metal forming and tensile tests. In addition to that, the lab contains an industrial robot with several grippers for the specimen handling and the needed equipment for the experiments' automation and control. The tensile test is the first implemented experiment in this remote lab. This test is one of the common and efficient tests to get the material properties of the tested specimen [10]. The determined properties describe the behavior of the material. Furthermore, the properties can be used in forming applications like FEM simulations (e.g., simulation of forming processes or production processes). This is why it is a very basic but as well an important test in the context of manufacturing technology. In order to perform the online experimentation a user-friendly Graphical User Interface (GUI) for the tensile test is implemented and integrated (see Fig. 2 left). This GUI is divided in four regions: parameter settings, experimentation actions, real time data, and experiment video stream.

Fig. 2 Remote Laboratory at IUL: Lab Client (*left*) and testing cell (*right*)

2.3 Online Experimentation in the Course Context

The third course phase is the technical orientation, in which we make use of the above-described remote laboratory. Here, we connect the topic of lightweight construction and vehicle design with the question of material properties (for example, the strain rate). On the basis of these properties, decisions on different materials for different technical applications and component design statements can be taken. The work with the laboratory equipment is divided into two stages. In the first stage the students are asked to work out a short overview of stress-strain diagrams. This can be seen as a preparation for the following experimentation. These diagrams and the connected experimentation are basic works in engineering in order to gain material properties. For that first technical task the students are given two simple but differing stress strain diagrams and on the basis of their previous knowledge and additional Internet research they should answer questions like "What do the diagrams show?", "How are they worked out?", "What are important areas?" or "Which material properties can be gained through the connected data and how?"

This preparational task is not only important for the students to start into the experiments but also us as the students taking part in this course did their bachelor degree in their home countries. Hence, they are on different levels talking about their skills and expertise with experimentation in general and the explicit knowledge needed for understanding the stress-strain-diagram. With this first task we can make sure that everybody at least shows enough understanding of the theory in order to understand the experiments.

After that task the students start into the real experimentation with the remote technology. For this, they are given an explicit component design task, which is to design a security component for a vehicle: little pins that prevent the engine from entering the passenger compartment during a frontal crash. For this component, the students have steel and aluminum as different material options and they are asked to perform the live online experimentation for those two materials. For the experimentation itself, the students are free to do this whenever they are able to do it in their group. The equipment is online available and ready for usage for three and a half days. Hence, the students have to develop a working plan in their group when to do the experimentation and arrange the experimentation by booking a time slot. The groups can arrange the working process with the help of Adobe Connect so that they jointly do the experimentation. That means that one of the group members books the laboratory equipment, logs in, shares its screen via Adobe Connect, and does the experiment. With this technique all of the group members can see and discuss what is happening.

On the basis of the experimented results, they have to work out a short report talking about the differences between the two materials and in how far this has an impact on the production process. Talking about the actions that have to be performed that means that the students need to work out the stress strain diagram, calculate the material properties and the component dimensions, and finally make a statement on which material would be better for the component. One of the basic expected statements is that safety requirements often are contradictory to lightweight construction goals. Using lighter material often (not always) leads to bigger component dimensions in order to reach the same safety requirements. Before the last meeting the reports have to be handed in and they are discussed in the sixth and last online meeting. Moreover, the last meeting is used to do a final feedback discussion on the class, the course concept, the learning outcomes, and the students' expectations coming to Germany after the course.

3 Findings on Basis of the First Course Edition

As mentioned above, the first edition was given in August/September 2014. A total of 12 students from 10 different countries took part, speaking 10 different mother tongues (the students came, e.g., from India, China, Pakistan, Nigeria, Brazil, Turkey, to name just a few). Because of the different time zones, we split the course into two groups. One group consisted of participants from Far East countries (from Iran to China) and another group of the westernmost participant from Brazil and the students from the near east (Lebanon was the easternmost located country in this group). 11 of the participants were male, one female. 7 of them were between 21 and 24 years old, which 5 of them were between 25 and 29. All of them either learned English during earlier education or because it is a second official language in their country. That meant for us that the language barrier could be seen as rather small. Nevertheless we tried to use multiple communication channels in order to prevent misunderstandings. The tasks, for example, were always given orally but as well in written form so that those students having problems with the language could reread everything. Most of the participating students already had experiences in international contexts: 6 had friends from multiple cultures, 3 already studied abroad, 3 were living outside of their home country and 5 were already multilingual. However, 3 students stated not to have any intercultural experiences.

From the final student feedback the students handed in by filling out a short reflective survey, it can be stated that all of the students were satisfied with the course and its online concept. They did appreciate the opportunity to learn something about Germany, Dortmund, and their future classmates as well as about their future teachers in advance of their stay (for example, they built up a Facebook group in order to stay connected). Moreover, they generally appreciated to work in international teams as they, too, think that this ability will be important in the future. Being asked to assess their level of interaction with their internationally scattered classmates, they would rate this interaction on average with a 4 (scale from 1: no interaction to 5: high level of interaction). For the students, this transnational online concept fitted perfectly as most of them started their journey just some days after finishing this course. When asked about their favorite task during the course, 8 out of the 12 students named the online experimentation. As they had no contact to such equipment before, they were really interested in (and partly even impressed by) these learning opportunities for students. This, of course, is a supporting result for our approach to use this technology for building up transnational student working groups. From our perspective, we can state that the technical equipment worked fine. Except for some minor problems the equipment worked solidly and 13 experiments could be performed. Even if some students accessed the laboratory environment from countries with a comparatively weak Internet connectivity in general (according to personal statements from some of the participants), this did not turn out to be a big problem for the experimentation. In contrast to that, we had to cope with weak Internet connectivity throughout the course. That literally meant that students sometimes lost connection right in the moment they were talking. This risk has to be taken into account when designing activities and the teachers have to be prepared to step in these moments in order to keep the course running. Such technical challenges were rated as the most negative aspect of the course, even though the students did not feel too much annoyed by it.

We teachers also draw a positive conclusion on the basis of this first course edition. For us it was proved, that online conferencing tools and especially the remote laboratory environment can serve as a connection point to bring internationally scattered students together and have them work together in groups on engineering tasks. Moreover, we argue that the task of reflecting upon international engineering cultures and mobility concepts helped the students to get a better idea of what it means to be an engineer and work in intercultural contexts as it will become more and more normal. Nevertheless, there are aspects about the course we will change on the basis of the students' feedback. This will be a task for this year, as the second edition of this class will be given in September 2015. With this second edition, the research on the course concept will be carried on and will become more detailed.

References

1. U. Weithner, Vorwort. In: *Das Andere lehren. Handbuch zur Lehre interkultureller Handlungskompetenz*, ed. by O. E, Waxmann, New York, Mnchen, Berlin, 2010, pp. 7–8
2. Bertelsmann Stiftung, *Interkulturelle Kompetenz - Schlsselkompetenz des 21. Jahrhunderts?: Thesenpapier der Bertelsmann Stiftung auf Basis der Interkulturellen-Kompetenz-Modelle von Dr. Darla K. Deardorf.* Gtersloh, 2006
3. UNESCO. Global flow of tertiary-level students: Interactive online source, 07.01.2014. URL http://www.uis.unesco.org/education/Pages/international-student-flow-viz.aspx
4. HIS., ed. *7. Fachkonferenz "go out! studieren weltweit" zur Auslandsmobilitt deutscher Studierender: Ausgewhlte Ergeb-nisse der 4. Befragung deutscher Studierender zu studienbezogenen Aufenthalten in anderen Lndern 2013.* 2013
5. D. May, K. Wold, S. Moore, Using interactive online role-playing simulations to develop global competency and to prepare engineering students for a globalised world. European Journal of Engineering Education, 2014
6. G. Schuh, T. Potente, R. Varandani, C. Witthohn, ipodia-innovative, international, interactive higher education. In: *TeachING-LearnING.EU discussions - Innovationen fr die Zukunft der Lehre in den Ingenieurwissenschaften*, ed. by A.E. Tekkaya, S. Jeschke, M. Petermann, D. May, N. Friese, C. Ernst, S. Lenz, K. Mller, K. Schuster. 2013
7. D. May, A. Sadiki, C. Peul, A.E. Tekkaya, Teaching and learning globally connected using live online classes for preparing international engineering students for transnational collaboration and for studying in germany. In: *Proceedings of International Conference on Remote Engineering and Virtual Instrumentation (REV)*, ed. by IAOE. Wien, 2015, pp. 114–122
8. D.K. Deardorff, The identification and assessment of intercultural competence as a student outcome of internationalization at institutions of higher education in the united states: Unpublished dissertation. Ph.D. thesis, Raleigh, North Carolina, 2004
9. T.R. Ortelt, A. Sadiki, C. Pleul, C. Becker, S. Chatti, A.E. Tekkaya, Development of a teleoperative testing cell as a remote lab for material characterization. In: *International Conference on Interactive Collaborative Learning (ICL).* 2014
10. A.E. Tekkaya, Metal forming. In: *Handbook of Mechanical Engineering*, ed. by K.H. Grote, E.K. Antonsson, Springer, 2009, pp. 554–606

Transnational Connected Learning and Experimentation – Using Live Online Classes and Remote Labs for Preparing International Engineering Students for an International Working World

Dominik May, Abdelhakim Sadiki, Christian Pleul and A. Erman Tekkaya

Abstract Students, who are leaving their home country for taking part in an international study program, face several challenges. Not only the new course of studies can be very challenging but also their whole living conditions may change significantly. This can be a severe clash especially for students who are moving to a country with a totally different cultural background in comparison to their home countries. Moreover, it can profoundly complicate the first weeks at the new university. Knowing about the difficulties the Institute of Forming Technology and Lightweight Construction (IUL) at TU Dortmund University in Germany developed a preparational online course for those international students, who are coming to the IUL for their Master of Science program in Manufacturing Technology (MMT; a special international master program). In context of this course the use of the IUL's remote laboratory equipment was a key aspect. The course itself was implemented and delivered for the first time in 2014. By now a second updated edition was delivered in 2015. It was designed to prepare the students as best as possible for their new studies at a German university and at the same time prepare them for transnational collaboration. Hence, this course is a good example for a meaningful integration of remote laboratories into an innovative online course concept. On the one hand making use of remote laboratories and its practical integration in online courses helps to connect the international students and on the other hand it brings them into the situation to interact in context of a typical engineering situation, the experiment. The paper presents the course itself and experiences from its first and second implementation.

Keywords Intercultural Competences · Online Teaching and Learning · Remote Laboratories · Transnational Teaching

D. May (✉)
Engineering Education Research Group, Center for Higher Education,
TU Dortmund University, Dortmund, Germany
e-mail: dominik.may@tu-dortmund.de

A. Sadiki · C. Pleul · A. E. Tekkaya
Institute of Forming Technology and Lightweight Construction (IUL),
TU Dortmund University, Dortmund, Germany

Originally published in "Proceedings of International Conference on Remote Engineering and Virtual Instrumentation (REV 2015)", © 2015.
Reprint by Springer International Publishing AG 2016,
DOI 10.1007/978-3-319-46916-4_30

1 Introduction and Course Idea

The world is more and more globally connected. That counts for the economical as well as for the educational sector. Whereas producing companies distribute their production processes all over the globe, students increasingly seek to gain international experiences by studying abroad or doing internships in foreign countries in order to prepare themselves for working in international environments later on. Programs and opportunities to study abroad are nearly uncountable. In 2014 for example 206,986 international students studied in Germany and 117,576 German students went abroad to gain new international experiences [1]. A special case in this context are international study programs that offer the opportunity to students to not only going abroad for one or two semester but studying a full bachelor or master program at an university outside their home country. These programs are mainly taught in English so that the language barrier could be rather small.

The engineering faculty at TU Dortmund University in Germany offers such an international Master of Science program in manufacturing technology (MMT). To attract engineering students from around the world, the course offers a compact 2-year English taught program. MMT consists of theoretical fundamentals in machining, materials and forming technology in connection with comprehensively applied hands-on science studies. Students are given the opportunities to carry out their hands-on experiments side-by-side with researchers in highly equipped labs. The included one term thesis should be done with leading companies in the sector manufacturing technology. Based on the experience gained with prior MMT cohorts, foreign students often need some time to overcome some of the typical difficulties, like cultural and academic habits, when starting. The basic idea for the presented course concept was to work out an online course the students could go through before they come to Dortmund. Two different perspectives mainly drive the course idea and its concept.

The first perspective is the students' preparation for their stay in Germany. As the students, who sign up for this international master program, come from all over the world they all have differing cultural backgrounds and are used to different educational systems as well as teaching methods. Hence, on the one hand the idea was to bring them into contact with the German culture and the educational concepts they will be facing during their future studies. On the other hand, we wanted to bring them into contact with their future classmates. From former student cohorts we know that especially students from the same home countries tend to build a closed peer group, which hinders international collaboration during their stay in Germany.

That leads to the second perspective on the course idea: All of the students will work as engineers after their master program, either at scientific institutions or in industry. For both options it can be said that transnational collaboration in the working context is more important than ever. Hence, the students are expected to develop real intercultural competences during their studies. Without going to much into detail about the notion of intercultural competence itself, the following should serve as a working definition: "Intercultural competence describes the ability to effectively and

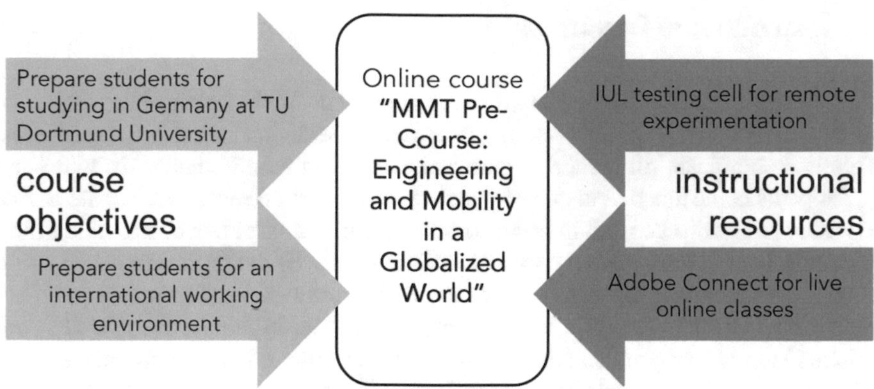

Fig. 1 Course objectives and instructional resources for the MMT Pre-Course design

adequately interact in intercultural situations on basis of explicit attitudes as well as the special ability to act and reflect" (own translation) [2]. This definition states clearly that intercultural competence can only be shown in intercultural situations. Consequently, this means that its development can only happen in corresponding learning situations. The developed course should serve as such a situation in the run up to their international experience in Germany.

Both considerations from above inspired the course conception. Summing up, its main goal was to prepare the students for two different contexts: Their stay in Germany and their future working environment. This all should be done in a combined use of modern online communication tools and the comprehensive remote laboratory equipment at the IUL (see Fig. 4; for details on the tools and instructional resources see part 2).

Finally, the course was developed with respect to the guiding theme of mobility and called "MMT Pre-Course: Engineering and Mobility in a Globalized World". Giving the whole course the context of mobility opened up the opportunity to talk on the one hand on core engineering topics like production and material sciences and make use of the remote laboratory equipment at the IUL. On the other hand mobility was identified as an ideal topic to tackle a future global challenge. Furthermore, mobility can be discussed on basis of many differing international perspectives. Hence meaningful discussions in terms of multi-perspective and multi-cultural thinking can be expected discussing mobility with international students. For more details on the course concept and the connected activities see part 3.

The course is delivered for the first time in August/September 2014 and a second time in August/September 2015, each time in advance of the students' stay for their master program in Germany. That means that all of the students are in their home country at this time. Therefore we take advantage of online tools in order to deliver this course fully Internet based and having the students taking part from their home. In the following the instructional resources and the laboratory equipment will be explained. After that the course itself will be explained more in detail.

2 Instructional Resources

As the participants for the presented course concept are globally distributed and take part via the Internet, a special focus has to be put on the digital instructional recourses. This is a significant difference to classical classroom based courses, in which all participants come together in one physical existing room. Hence, an adequate online environment has to be used in order to build up meaningful teaching and learning activities. In addition to that a special focus for this class is put on the laboratory work part. The students are expected to carry out real experiments and therefore remotely use the IUL's laboratory equipment. Remote laboratories in general describe physical existing laboratory equipment that can be accessed and used via the Internet. Hence, this technology fits perfectly in the described online course concept. Both, the online learning environment and the remote laboratory, will be explained in the following.

2.1 Using Adobe Connect for Live Online Learning Experiences

The core instructional resource for this course was Adobe Connect [3]. This technology is a classical online meeting tool and gives the opportunity to run live class sessions just as in a real classroom. The only difference is that all the participants are not present in a physically existing room but enter an online room by using their

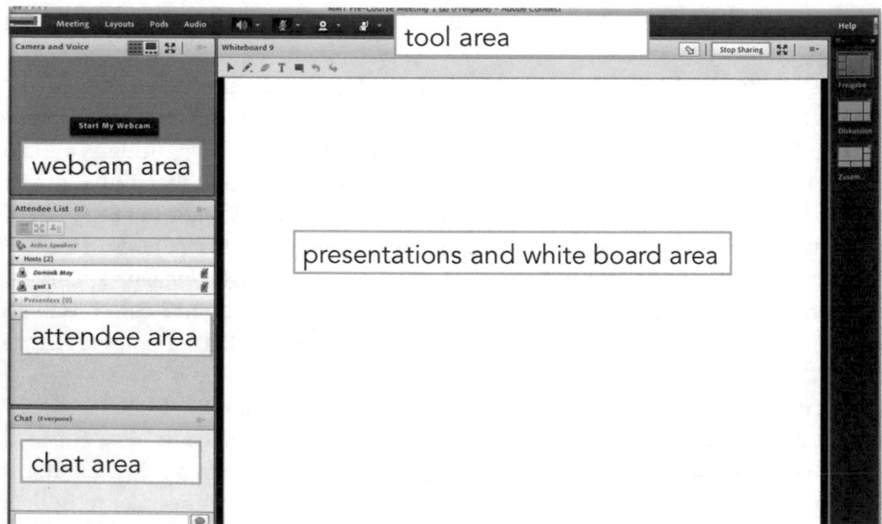

Fig. 2 Desktop in the Adobe Connect environment

personal computer, a headset, a webcam and, of course, their internet connection. Figure 2 shows the environment, which the instructor and the students mainly see on their desktop during class.

As it can be seen in Fig. 2 there are different areas in the environment. At the very top left corner, there is the webcam area, where every student can transmit its webcam picture. The tool automatically recognizes the speaking person, so that only her or his picture is transmitted at this time. Below that a list of the current meeting attendees is displayed and the participants can also identify the different attendee roles, e.g. the instructor, participant. In the left bottom corner a chat area can be used in order to communicate in written from. Normally, in the chat area short messages are posted, for example to ask short questions without interrupting an ongoing presentation or to note that the audio quality is temporally low. The desktop's biggest part is the presentation and white board area. This can be used as a classical whiteboard for note making or quick sketching. If a presentation is shown to the participants it is also displayed in this area. This design can be changed and enhanced with other applications, just as it is needed during class.

Moreover, Adobe Connect has many opportunities in terms of instructional tools, depending on which activities should be performed during class. In this course the desktop sharing option was one of the most frequently used application. That means that one of the meeting attendees is able to share his or her desktop with all the others and with this e.g. give a presentation. The shared desktop is also displayed in the white board area.

Another important application for this class was the break out room option. Normally, all participants are in one virtual room so that everybody can speak to all the others or, to put it another way, all the participants can take part in one discussion. The break out room application gives the opportunity to distribute the students to individual virtual rooms. In these rooms they can discuss in smaller groups without

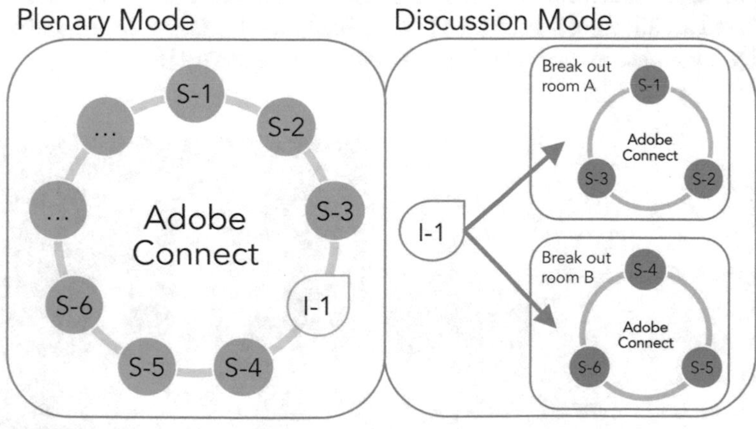

Fig. 3 Two mainly used modes during the MMT pre course

hearing the other groups talking. After a period of time the instructor can bring everybody back to the main room and ask them to explain what they recently discussed in the smaller groups. Figure 3 displays the two main course situations.

2.2 The IUL Testing Cell for Live Online Experimentation

Another very important technology for this course was the remote laboratory at the IUL [4]. With this it was possible to make the testing machine at TU Dortmund accessible via the Internet. With the help of this technology the students could sit at home in front of their computer, start real experiments, monitor the experiments and download the generated results. For the experimentation we made use of the iLab technology from the Massachusetts Institute of Technology [5].

At the IUL an own remote laboratory on manufacturing technology has been developed. This laboratory gives the students and teachers the opportunity to conduct experiments in the field of manufacturing technologies especially for material characterization. Figure 4 (right) shows the laboratory with two testing machines for sheet metal forming and tensile tests. In addition to that, the lab contains an industrial robot with several grippers for the specimen handling and the needed equipment for the experiments' automation and control. The tensile test is the first implemented experiment in this lab. This test is one of the common and efficient tests to get the material properties of the tested specimen [6]. The determined properties describe the behavior of such material. Furthermore, the properties can be used in forming applications like FEM-Simulations (e.g. simulation of forming processes or production processes). This is why it is a very basic but as well an important test in the context of manufacturing technology.

Due to the global requirements, such as the share of the experiments with other Universities, managements of users, user groups and reservation of timeslots, the iLab shared architecture is used here [7]. Basically the lab is developed so that it can be easily integrated in other platforms like the weblabdeusto [8].

Fig. 4 Remote laboratory at IUL: Lab Client (*left*) and testing cell (*right*) [4]

As mentioned above, a user friendly Graphical User Interface (GUI) for the tensile test is implemented and integrated in iLab (see Fig. 4 left). This GUI is divided in four regions. The first region consists of a field of parameters, which can be set by typing a numerical value or selecting a value from a list. In the second region, different actions can be performed "Setting Parameters", "Start", "Pause", "Resume" and "Cancel". With the help of these actions an interaction with the experiment is guaranteed. The captured experiment real time data are shown as numerical value in the third area. This data are the acting force on the specimen, the displacement and the width variation of the specimen. Furthermore to illustrate the data, it is displayed in form of diagrams, too. In the fourth area (last), a video live stream of the experiment is shown. The user can change the observation perspective by selecting another camera. This is very helpful to give the user an all-around perspective of what is happening in the experiment.

3 The MMT-Pre-Course Concept

After having explained the instructional resources and its functionality, the course itself comes into focus. All this technology would be useless if a meaningful integration into an educational setting would be lacking. Hence, in the following the course design concept, the learning objectives and a brief overview of the course activities will be given.

3.1 Constructive Alignment as a Backbone for the Course Design

Constructive Alignment is a fundamental concept in higher education [9]. Within this concept the intended learning outcome, the teaching and learning activities, and finally the examination are in focus (see Fig. 5). As the name indicates, all three of these course parts have to be aligned in order to design a well-prepared course; meaning they have to be designed so that they show clear conjunctions. Following this model, a course design process ranges in a triangle between these three components and takes them into account – beginning with the intended learning outcome. The intended learning outcomes are the goal for the whole course. These outcomes define what the students should be able to do after having successfully passed the course. Based on them the teaching and learning activities should be planed and designed. That means that the course designers have to take great care in order to design activities that give the students the opportunity to achieve the course goals. For example, if the course goal is to be able to perform experiments and with this to gain material characteristics after the course, there is no other way than letting the students do their own experiments. No oral lecture would fully bring them into the

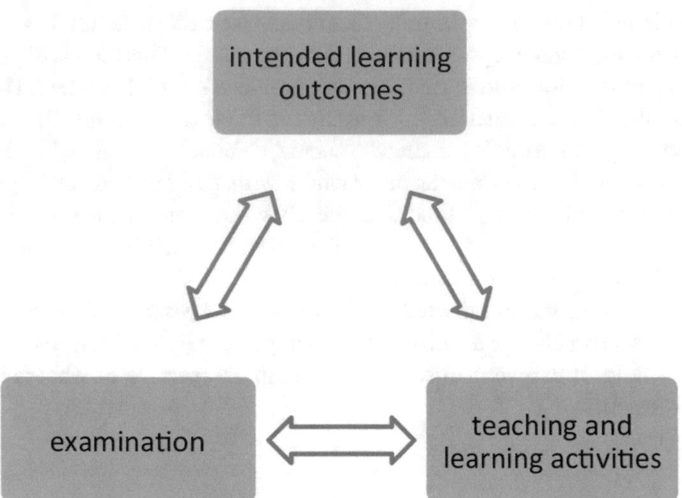

Fig. 5 Constructive alignment as a backbone for the course design process [9]

situation to gain such a competence. Finally, the examination has to measure exactly these outcomes. Even if the constructive alignment does not force the designer to begin at any of the three points it proved helpful to begin with the learning outcomes as a guiding factor for the following process. It is not necessary to go on with the activities and finish with the exam as both are in such a close interconnection so that it will be a back and forth process anyway.

In the following the intended learning outcomes and the teaching and learning activities for the MMT-Pre-Course will be explained. As this course does not finish with any exam or something similar this part of the constructive alignment will be left out.

3.2 Learning Objectives

The learning objectives for this course are heavily influenced by the course objectives displayed in Fig. 1. Hence, a particular emphasis is put on the development of intercultural competence. As intercultural competence has something to do with interaction with others and with self-reflection [2], both of these activities are included in the course concept. In addition to that, bringing the students into contact with their destination's culture and with their future classmates is another focus. Moreover, the fact that the course should prepare the students for an engineering master program profoundly affects the intended learning outcomes. Experimentation is an important part of engineering curricula. Therefore it should be part of this course, too. Hence, the students should do core engineering work, in this case by executing own experiments (a tensile test to be even more specific). By making use of remote laboratories recent technical developments in production engineering education are added to the course concept. In this context seems to be important to us to talk about technologies not only form a technical but from different perspectives. Talking about the technical as well as cultural and organizational aspects of technologies broadens the students' view and gives the opportunity to compare different perspectives on an intercultural level. This approach subsequently leads back to development of intercultural competence.

All these considerations lead to the following intended learning outcomes. After the course the students should be able to…

1. … describe their destination on the basis of internet research
2. … describe their own concept of engineering in comparison to others
3. … reflect on the international differences in engineering
4. … use a basic model for describing technology from different perspectives (technical, cultural, organizational)
5. … run experiments like the tensile test and use the gained data for engineering work by connecting theory and practice
6. … explain a general stress-strain-diagram by identifying important points as well as explaining their relevance for forming technologies
7. … communicate successfully with international engineering colleagues using internet technology
8. … organize themselves in a working process and work successfully together in multinational teams in order to fulfill explicit tasks

These learning outcomes can be divided into 4 different groups. Whereas outcome number 1 focuses on Dortmund and the students' destination, the outcomes 2–4 express the necessity of being a reflective engineer who is able to put its own profession and work in context with other fields and finally in a global context. Outcome 5 and 6 focus on central engineering aspects and define the objectives that are connected to the work with the laboratory equipment. Finally, outcomes 7 and 8 have to be seen in a broader context. They express the goals in context with international co-operation with the help of Internet technology. All of these intended outcomes were presented to the students during the first meeting, so that they knew what was expected from them during the coming course.

Based on these intended learning outcomes the course activities are designed. They will be outlined in the following.

3.3 Course Activities

The first course edition in 2014 was delivered during a period of three weeks. In the second edition from 2015 we extended this to four weeks as the experiences from the previous year showed, that more time was needed in order to successfully achieve

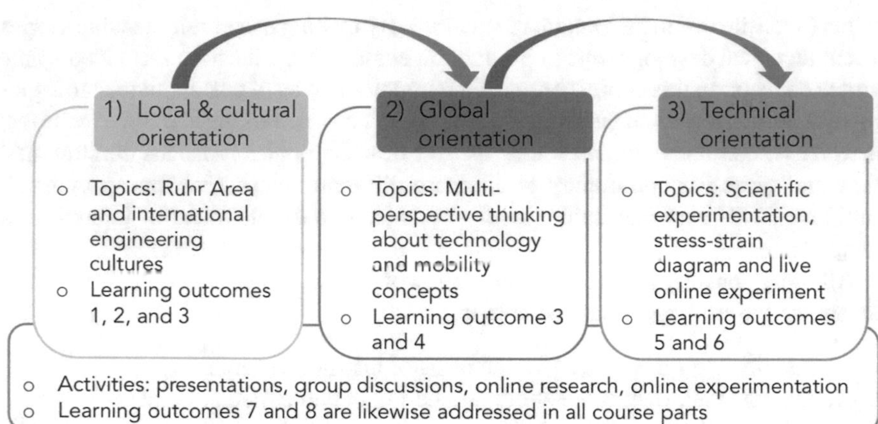

Fig. 6 MMT-Pre-Course concept in three phases

the course goals. However, each course week covers two live online meetings. In these live sessions all participants come together in an Adobe Connect online room (see Fig. 2). The sessions mainly consist of presentations by the students, discussion groups or explanations by the lecturer in order to introduce a new topic or the next step in class. Even more important are the working phases between the live sessions. In these phases the students do own research on various topics or carry out preparational work for the coming session, mainly in form of group work. Therefore the whole student group is split into smaller working groups. The students are mixed so that every group is a blend of students from different countries with different cultural backgrounds. In these groups they work on the given tasks. Speaking frankly, this is the time were the international students actual have to co-operate with each other with the help of various Internet technologies.

Based on the different topics the course can be divided into three main parts: Local and cultural orientation, global orientation and technical orientation. These parts reflect the different intended learning outcomes. Fig. 6 shows the course concept in total. The details will be explained in the following.

3.3.1 Local and Cultural Orientation (First Part)

The first course part is dominated by getting to know each other, the course instructor and the course concept itself. However, not too much time is planned for that in order to get into proper interaction with each other as soon as possible. Therefore, in this very early phase of the course a focus is put on the personal orientation in the students' future destination. During this local orientation a short presentation about the Ruhr Area itself and its location in Germany, the city of Dortmund and the TU Dortmund University is given. This presentation is designed so that the students gain a broad overview over the area and are able to start with this knowledge in a first working phase. Divided into three groups the students have to do an online research about their destination for their future master program, led by several guiding questions. These questions lead the students from the local orientation to the cultural orientation, too. The groups have to focus on different aspects during their research. One group has to find out more about regional information and the historical role of engineering in the Ruhr Area. Another group has to focus on current strengths in industry, science and research. The third group concentrates on quality of life and future plans for the region. The task connected to this online research is to find online available information to work out a presentation on these topics. In addition to that, all of the groups are asked to compare their results to their home countries and answer if they find any major similarities or differences. The results are presented and discussed in class during one of the early live sessions. With this the students simultaneously orient themselves in their destination and put that into context with their own cultural background. Furthermore, they are introduced into the home countries of their future classmates.

After the local orientation and the presentations the following meeting is dominated by the several discussions on the engineering profession itself. Therefore, we make use of the option to split the students into smaller groups and send them to several breakout rooms for more intimate discussions. After some minutes they come back and share their discussion results with the whole group. This phase is sequenced into four steps. In the first step the groups discuss the role and the status of engineers in their home countries. Furthermore, in the second step they are asked to describe typical tasks of engineers from their perspectives. Based on that, they work out a list of competences engineers need to fulfill the described tasks. In a third step the students discuss on the necessity of international experiences for the job of an engineer and furthermore they are asked to describe global technical challenges especially engineers face in the future. Again, they finally work out a list of competences that are needed for facing these global challenges.

With these activities learning outcome No. 1 to 3 are in focus. The last two discussion topics explained above are meant as a transitional period from the cultural orientation into the following course part: the global orientation.

3.3.2 Global Orientation (Second Part)

The global orientation is characterized by two different working phases. On the one hand the students have to work on a concept to describe technologies by looking at technical, cultural and organizational aspects. Furthermore, they should develop a concept on how technology progress and society development are connected processes and determine each other. This activity is supported by out of class readings (e.g. [10, 11]) and by in-class discussions based on those readings. The important aspect in this phase is to bring the students into the position to discuss on technology from different and even non-technical perspectives. From our viewpoint this is absolutely necessary in order to develop intercultural competences. From our perspective the pure technical understanding of technology is not enough for future engineers, as we consider a meaningful exchange on the implication of technology in different cultural contexts will gain in importance in the future. Even if the retraction on pure technical considerations may be the favored aspect for many engineers about their profession, this won't be enough in order to act successfully in intercultural situations.

In a second step the students have to use the discussed models to work out a presentation on different future mobility concepts from all over the world: The *Land Airbus* developed in China, the *Personal Rapid Transit System* used at Heathrow Airport in Great Britain and the *Car Sharing* system, for example introduced in Berlin, Germany. For this task the students again work in the three smaller groups and in these groups they have to do an in-depth research one of these technologies. Based on that research and the previous readings they have to work out their presentation and answer the following questions:

- What are technical, organizational and cultural issues of that technology?
- What are relevant social groups that might have an interest in the technology or might have an impact on the technology's development?
- What are advantages and challenges of the technology?
- Would that explicit technology be applicable in each of your home countries? If yes/no, why?

Especially the last question indicates that an international comparison of these concepts is an important part of this task. Hence, the students were forced to think about their home country, make a statement on the compatibility of such a technology in their country and compare that to others. The presentations are given during one of the live class meetings and normally lead to highly interesting discussion points. For example one student commented in 2014 that public transport in general is problematic in his country for security reasons. Even if a technology like the Land Airbus would work perfectly and may be a solution for decreasing the traffic in the city center, nobody would use it because of being frightened of being robbed. Comments like that are the backbone of the connected discussions and point out that every technology has cultural as well as organizational aspects, which differ from country to country.

As shown in Fig. 6 this course part references back to outcome No. 3 but mainly addresses learning outcome No. 4. Talking about mobility concepts at the same time is the door opener for the technical orientation in the last part.

3.3.3 Technical Orientation (Third Part)

For the technical orientation the remote laboratories come into focus. In the presented course they are connected to the topic of vehicle design, or better, lightweight construction in vehicle design. Therefore the students finally have to do a tensile test with two different materials using the laboratory equipment at the IUL. Before that, they are asked to work out a short overview on stress strain diagrams in general.

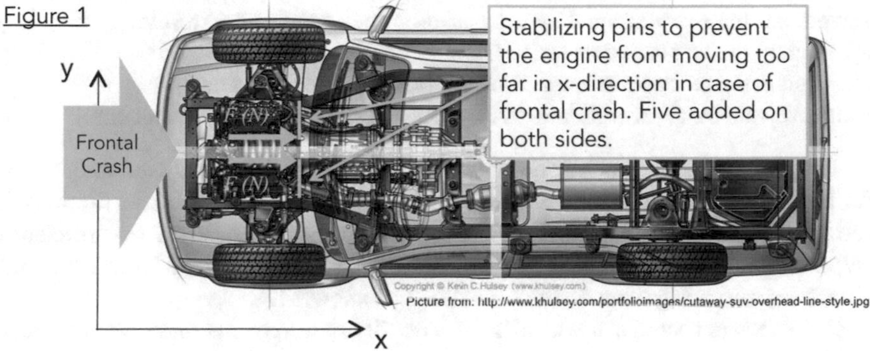

Fig. 7 Context for component design task [12]

These diagrams and the connected experimentation are a basic work in engineering in order to gain material properties. These properties can then be used for design tasks. For that first technical task the students are given two simple but differing stress strain diagrams and based on their previous knowledge and additional internet research they have to answer the following questions:

- What do the diagrams show?
- What is the difference between the two diagrams?
- How are they worked out?
- What are important areas?
- Which material properties can be gained through the connected data and how?

After that, the students are introduced into the iLab environment and the experimentation software. Furthermore they are given an explicit component design task. Within this task they have to design a security component for a vehicle: Little pins, which prevent the engine from entering into the passenger cabin during a frontal crash (see Fig. 7). For designing those pins the students have to compare two materials options (steel and aluminum). Therefore, they are asked to execute the live online experimentation with both materials remotely using the IUL's testing cell. For the experimentation itself the students are free to do this whenever they are able to do it in their group. The remote lab is online available and ready to use for three half days. Hence, the working groups have to develop a working plan when to do the experimentation and arrange the experimentation by booking a time slot. In order to jointly do the experiment in their group the students are asked to use Adobe Connect. That means that one of the group members books the laboratory equipment, logs in at iLab, shares its screen via Adobe Connect and does the experiment. With this technique all of the group members can see and discuss what is happening.

Based on the results the students have to work out a short report talking about the differences between the two materials and in how far this has an impact on the production process. Summing up, that means that the students have to

- work out the stress strain diagram based on the experiments' results,
- calculate the respective material properties,
- calculate the component dimension for bot materials and finally
- make a statement on which material would be better for the component.

This preparational work is of highly importance for the following experimentation. As the participating students have bachelor degree from different institutions in completely different countries it is necessary to check if all of them are sufficiently prepared for understanding the experimentation and the connected knowledge and, if necessary, to bring them on the same level of knowledge.

One of the expected statements is that security requirements often are contradictory to lightweight construction goals. Using lighter material often (not always) leads to bigger component dimensions, if the same security requirements should be met. The last live meeting is used to do a final feedback discussion on the class, the course concept, the learning outcomes and the students' expectations coming to Germany after the course.

The last course part mainly addresses learning outcomes No. 5 and 6. The learning outcomes No. 7 and 8 cannot be allocated to one of the three course phases, as there is constant interaction between the participants. Especially during the working phases, in which the students have to do online research, work out presentation or do the experimentation, they learned a lot about international collaboration over the Internet. After having explained the course in the following the experiences will be explained. This will be enhanced by some insights into the students' feedback.

4 Experiences and Students' Feedback

As mentioned above, the course was delivered in 2014 for the first time. A second edition was delivered in 2015, so that we can look at experiences from two editions by now. Hence, these still are the first hands-on experiences the instructor team made with the instructional activities and laboratory's usage in a transnational course concept. Therefore, this chapter is divided into three main parts. Firstly, a short overview on the participants of each edition is given. Secondly, the experiences that were made with the laboratory equipment and its usage will be explained. Thirdly, the students' feedback will be explained.

4.1 Course Participants

In 2014 edition of the course all in all 12 students from 10 different countries took part, speaking 10 different mother tongues (the students came e.g. from India, China, Pakistan, Nigeria, Brazil, Turkey, to name just a few; see also Fig. 8). These different countries meant practical challenges for the course design as they the students lived in very different time zones. In order to have synchronous online course parts, which were seen as essential for the course success, this fact made it necessary to split the participants into two groups from the beginning on. One group consisted of participants from Far East countries (from Iran to China) and another group of the most western located participant from Brazil and the students from the near east (Lebanon was the most eastern located country in this group). Even if it would have been better to put all participants in one group in order to have a more intercultural mixed group, there was seen no other option to face the time-zone problem. 11 of the participants were male, one female. 7 of them were between 21 and 24 years old, whereas 5 of them were between 25 and 29. All of them either learned English during earlier education or because it is a second official language in their country. In terms of earlier intercultural experiences only for stated that they did not have any. Most if them commented that they have parents from different cultures, already studied abroad, worked with international colleagues or had other experiences with different cultures.

In the 2015 course edition at the beginning 16 students (5 between 20 and 22, 7 between 23 and 25, 2 between 26–28 and 2 between 29–31) took part. Due to several reasons this number dropped to 14 participants. The students came from 6 different countries: Mexico, Turkey, Iran, Pakistan, India, and Nepal (Fig. 8). All of them (except for 4 students) stated that they already gained intercultural experiences through friends or personal experiences abroad.

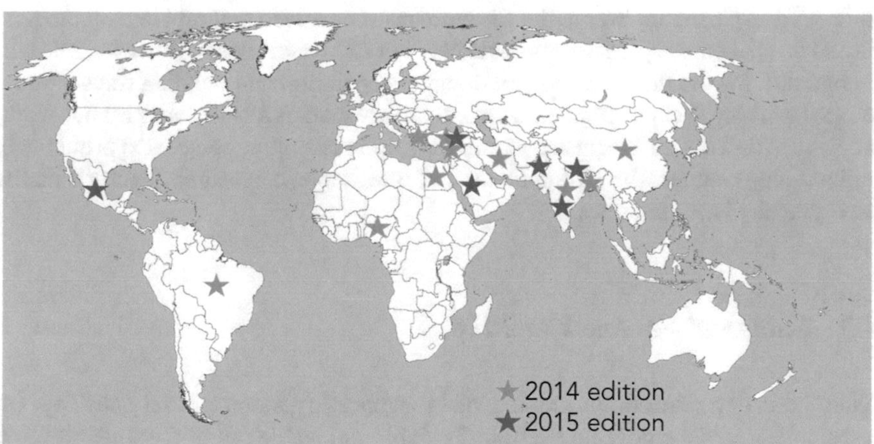

Fig. 8 Course participants and their home countries [13]

4.2 The Remote Laboratory Experience with International Distributed Student Groups

In the 2014 MMT-Pre-Course the students performed a total of 13 tests. Each student performed at least one experiment. Some of the experiments were reserved but not performed or canceled by the conduction. Due to security issues (bugs in the safety network of the machines) the control unit interrupted some of the tests. But these issues have been resolved within few hours, so that the students could restart there interrupted experiments. Some of the students complained that the captured real time data are not displayed correctly in the diagrams, after some investigations we could found the problem. This was related to the bandwidth and to the process power of the used device. The problem was fixed by separating the data transmission from the display process.

During the 2015 edition severe problems occurred on one of the experimentation days. The case is not completely explained yet, but it seemed to be problem between the browser, which is used to log into the iLab experimentation software, and Adobe Connect. It seemed to us that the desktop sharing application of Adobe Connect overlaid the browser window in a way so that the students were not able to click several buttons in the graphical user interface for the experimentation. Even if we saw the students clicking on these buttons, nothing happened. Hence, the students were not able to prepare, start, and finish the experiment. This problem was the reason for canceling one of the experimentation sessions completely and retrying it the next day. Changing over to the Chrome browser solved that problem on the next day so that the course participants could carry out 21 experiments. However, this will be one task for the future. As international co-operation and building transnational student working groups in context with laboratory work is a main focus for our work, we have to find out an adequate solution for jointly executing remote experiments in online groups, either with an additional tool like Adobe connect or directly with the used graphical user interface.

4.3 Students' Feedback in 2014

After the 2014 course edition the students' feedback in general was very positive. On a scale from 1 (not enjoyable at all) to 5 (highly enjoyable) all of the students rated the MMT-Pre-Course with a 4 or 5. They appreciated very much the opportunity to get into contact with their future classmates and to learn something about the Ruhr Region. In addition to that especially the online experimentation part was rated as highly interesting as most of them never had worked with such equipment before. In order to get a more detailed impression of the students' feedback, they filled out a short questionnaire after the class. Answering the question "What was your favorite part of the course?" 8 out of 12 students stated, that the experimentation with the tensile test was their favorite part. Others indicated that the the component design task was the most interesting and one student stated that the international collaboration in general was the best.

In general most of the students indicated that they were satisfied with the course outcomes. Their principal personal goal (in addition to the intended learning outcomes posed by us) was to get to know their future classmates and their destination, which was achieved. Others commented that this course helped them to "switch their minds" back into study modus as they finished their bachelor degree some time ago. A critical comment on the course was that even more about the daily life in Dortmund could have been learned. Asked for the major benefits they experienced while working with students from other countries, the students mentioned the insight in different cultures and countries in the context of technology application. Furthermore, they found it important to make first experiences in working with students distributed all around the world. For example having the different time zones in mind and its impact on working processes was new to some of them.

As a major challenge for the collaboration different English language accents and the weak Internet connectivity of some participants was identified. That even had some negative influences on the live online classes. Whereas some participants obviously had a strong connectivity and regularly could take part in all meetings, some of the students had bigger problems with that. Some of the students experienced the problem of repetitively loosing the Internet connection, so that they had to reconnect and rejoin the online meeting several times. In one case one student even missed a whole class as his home city was undergoing an electricity shutdown for a couple of hours. A bad Internet connectivity also had negative impacts on the audio transmission so that some participants were hard to understand or sometimes even could barely follow the course, as they did not understand the speaking person.

Finally the students were asked about their personal interaction with their international classmates. On a scale from 1 (no interaction) to 5 (high level of interaction) the students rated their personal level of interaction on an average with 4. In additional comments on this question the students stated that they regularly met between the classes in order to fulfill the given tasks and that everybody's point of view were considered. Some even founded new and course bound Facebook groups.

4.4 Students' Feedback and Course Evaluation in 2015

Based on the feedback and additional focus interviews the course concept was slightly edited for the 2015 edition. As already indicated above, the course concept – e.g. the intended learning outcomes, the three course parts and the used instructional resources – basically remained unchanged. However, we expanded the course time from 3 to 4 weeks and a major change can be seen in a more in-depth evaluation process, which was carried out during the 2015 edition. This evaluation was based on a newly at TU Dortmund developed holistic model for online course evaluation with a special focus on the online experimentation. This concept is composed of several online questionnaires done by the students at several points of measurement throughout the course. Explaining the whole model of evaluation here goes beyond the scope of this paper. The full concept and more detailed research results will be explained in additional papers and contributions to the respective conferences in the future. However, some of the results will be shown in the following.

11 students took part in the final questionnaire, which was meant to receive general feedback on the course concept. The results from the 2014 course edition could be confirmed in 2015. 82 % of the students rated the course as "Highly enjoyable". Moreover, 64 % of them were highly satisfied with the course outcome (36 % answered "satisfied"). Especially the group work and the online experimentation (despite the technical problems) again was favored by the students, as three of the students' comments show: "*I really enjoyed team working in this class.*" "*Interaction with other students was very much helpful to improve my knowledge and I also learned a lot about remote experimentation.*" "*Get to know most of my future course mates and doing the online experiment was great.*"

The latter two statements are supported by the fact that 73 % of the students rated the remote experimentation as the most interesting course part.

In another part of the applied evaluation model we had a closer look on the impact the remote experimentation has on the students' level of proficiency in context with several learning objectives for engineering instructional laboratories. Therefore we asked the students to assess if they think that their perceived level or proficiency in context with fifteen aspects of laboratory work (see items Table 1) has changed during the experimentation task. The asked items mainly base on the work of [14, 15].

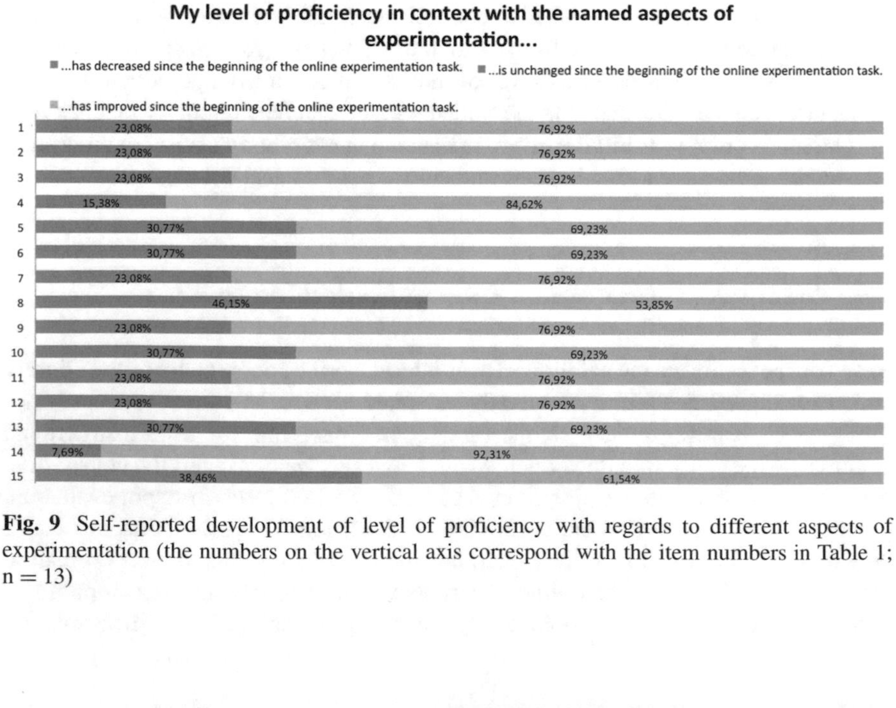

My level of proficiency in context with the named aspects of experimentation...

...has decreased since the beginning of the online experimentation task.
...is unchanged since the beginning of the online experimentation task.
...has improved since the beginning of the online experimentation task.

1 23,08% 76,92%
2 23,08% 76,92%
3 23,08% 76,92%
4 15,38% 84,62%
5 30,77% 69,23%
6 30,77% 69,23%
7 23,08% 76,92%
8 46,15% 53,85%
9 23,08% 76,92%
10 30,77% 69,23%
11 23,08% 76,92%
12 23,08% 76,92%
13 30,77% 69,23%
14 7,69% 92,31%
15 38,46% 61,54%

Fig. 9 Self-reported development of level of proficiency with regards to different aspects of experimentation (the numbers on the vertical axis correspond with the item numbers in Table 1; n = 13)

Table 1 Used items and questions to evaluate the level of proficiency in context with experimentation

Items
1. … handling laboratory equipment, measurement tools and software for experimentation
2. … identifying strengths and weaknesses of engineering specific theoretical models as a predicator for real material behavior
3. … planning and executing common engineering experiments
4. … converting raw data from experimentation to a technical meaningful form
5. … applying appropriate methods of analysis to raw data
6. … designing technical components or systems on Basis of experiments results
7. … recognizing whether or not experiment results or conclusions based on them "make sense"
8. … improving experimentation processes on basis of experiment results, that do not "make sense"
9. … relating laboratory work to the bigger picture and recognizing the applicability of scientific principles to specific real world problems in order to solve them creatively
10. … choosing, operating and modifying engineering equipment
11. … handling technological risks and engineering practices in responsible way
12. … presenting experimentation results to technical and non-technical audiences in written form
13. … presenting experimentation results to technical and non-technical audiences in oral form
14. … working effectively in a team
15. … applying professional ethical standards in terms of objectivity and honesty in context with data handling
Question: Please state if your level of proficiency in context with the above named aspect of experimentation…
• **… has decreased since the beginning of the online experimentation task during the course** • **… is unchanged since the beginning of the online experimentation task during the course** • **… has improved by doing the online experimentation task during the course**

All in all 13 students took part in the laboratory evaluation. Figure 9 displays the results. For us these results are more than encouraging, as in none of them a perceived decrease is reported and in the vast amount of the items the students stated a perceived improvement of their personal proficiency. In 9 out of these 15 aspects over 75 % of the course participants had this positive impression.

5 Current Limitations and Future Work

The experiences displayed in this paper base on two course edition delivered so far. In the future a third edition of the course will be taught. Changes to the concept will be implemented with respect to each year's feedback results. Even if such a course concept is highly innovative, there remain explicit limitations so far. In each of the first and second edition only a small group of students took part. It is not answered yet, if such a concept, which heavily relies on interaction and discussions, can be scaled up to a larger number of students. However, it will be a task for the future to scale up such concepts in order to reach even more students and having them participate in such transnational course concepts.

Another question will be, if the students, who took part in class, behave – once they are in Germany – in a different way than those who did not take part. This question goes in the same direction like the measurement of the learning outcomes. At the end of the 2014 course edition it was not measured in any way, if and in how far the students reached the intended learning outcomes. Only on basis of the internal evaluation it could be assumed that this happened. Nevertheless, this measurement task is important for designing effective instructional course concepts. Hence, for the 2015 course edition a holistic evaluation concept was designed and applied. The data displayed in IV D. only shows a small portion of the evaluation results. More detailed results on the development of intercultural competences, the achievement of the intended learning outcomes, and even on the IUL's remote lab's effective integration will be broadly discussed in additional papers. This paper mainly served for explaining the course and its design process. Nevertheless, the evaluation model will be improved on basis of the 2015 experiences and applied to the future course edition and even to other courses, in which the remote lab plays an important role.

However, once again we would like to emphasize the positive experiences the students and we as instructors made during the first two course editions. All of the participants would recommend the course to their classmates and future MMT students. Moreover, it is successfully proofed that remote laboratories can be used for online courses with students coming from all over the world, even if they are only connected via the Internet. Hence, from our perspective this approach in general opens up the opportunity to include remote laboratories in international educational online contexts in manufacturing technology. With this the design of totally new instructional concepts against the background of a globalized industrial and educational world will be possible. We just have to take advantage of these opportunities.

References

1. UNESCO. "Global flow of tertiary-level students", interactive online source. URL http://www.uis.unesco.org/education/Pages/international-student-flow-viz.aspx
2. Bertelsmann Stiftung, *Interkulturelle Kompetenz - Schlüsselkompetenz des 21. Jahrhunderts?: Thesenpapier der Bertelsmann Stiftung auf Basis der Interkulturellen-Kompetenz-Modelle von Dr. Darla K. Deardorf*. Gütersloh, 2006
3. Adobe. Adobe connect. URL http://www.adobe.com/de/products/adobeconnect.html
4. T.R. Ortelt, A. Sadiki, C. Pleul, C. Becker, S. Chatti, A.E. Tekkaya, Development of a tele-operative testing cell as a remote lab for material characterization. In: *International Conference on Interactive Collaborative Learning (ICL)*. 2014
5. iLab Service Broker. URL http://ilab.mit.edu/iLabServiceBroker/
6. A.E. Tekkaya, Metal forming. In: *Handbook of Mechanical Engineering*, ed. by K.H. Grote, E.K. Antonsson, Springer, 2009, pp. 554–606

7. V.J. Harward, J.A. Del Alamo, S.R. Lerman, P.H. Bailey, J. Carpenter, K. DeLong, et al., The ilab shared architecture: A web services infrastructure to build communities of internet accessible laboratories. In: *Proceedings of the IEEE*, vol. 96. 2008, vol. 96, pp. 931–950
8. M. Kaluz, P. Orduña, J. García-Zubia, M. Fikar, L. Cirka, Sharing control laboratories by remote laboratory management system weblab-deusto. In: *Proceedings of conference: 10th IFAC Symposium Advances in Control Education*. 2013, pp. 345–350
9. K. Biggs, C. Tang, *Teaching for quality learning at university. What the student does*, 3rd edn. McGraw-Hill, Maidenhead, 2007
10. K.A. Neeley, Toward an integrated view of technology. In: *Technology and Democracy–A Sociotechnical System Approach*, ed. by K.A. Neeley, Cognella, San Diego, 2010, pp. 37–45
11. T.J. Pinch, W.E. Bijker, The social construction of facts and artifacts: Or how the sociology of science and the sociology of technology might benefit each other. In: *The Social Construction of Technological Systems: New Directions in the Sociology and History of Technology*, ed. by W.E. Bijker, T.P. Hughes, T.J. Pinch, The MIT Press, Cambridge, Massachusetts, 1987, pp. 17–50
12. 2014. URL http://www.khulsey.com/portfolioimages/cutaway-suv-overhead-line-style.jpg
13. 2014. URL http://www.naportals.com/wp-content/uploads/2014/02/A_large_blank_world_map_with_oceans_marked_in_blue.png
14. L.D. Feisel, A.J. Rosa, The role of the laboratory in undergraduate engineering education. Journal of Engineering Education, 2005, pp. 121–130
15. Laboratory educators in Natural Sciences and Engineering. Pre-lab self evaluation form. URL http://www.owlnet.rice.edu/~labgroup/assessment/selfeval.html

Bitmap-Based On-Line Analytical Processing of Time Interval Data

Philipp Meisen, Tobias Meisen, Diane Keng, Marco Recchioni and Sabina Jeschke

Abstract On-line analytical processing is in the focus of research over the last couple decades. Several papers dealing with summarizability problems, cube computations, query languages, fact-dimension relationships or different types of hierarchies have been published. Nowadays, analyzing time interval data became ubiquitous. Nevertheless, the use of established, reliable, and proven technologies like OLAP is desirable in this respect. In this paper, we present an OLAP system capable to process time interval data. The system is based on bitmaps, enabling performant selection and fast aggregation. Moreover, we introduce a two-step aggregation technique, which enables the calculation of relevant measures in the context of time interval data. We evaluate the performance of our system using different bitmap implementations and a real-world data set. To our knowledge, there are no other systems available enabling OLAP and providing correct results considering the summarizability of time interval data.

Keywords Time Interval Data · Time Series · Bitmap · On-Line Analytical Processing · Two-Step Aggregation

1 Introduction

Time interval data are recorded in various situations, e.g. task executions, gesture tracking, or behavioral observations. Several disciplines like artificial intelligence, music, medicine, ergonomics, and cognitive science analyze time interval data to detect patterns (frequent sequences) or association-rules [1]. Lately, some attention on assessing the similarity between sets of time interval data (so-called event sequences) arose [2]. Nevertheless, very limited attention has been given on the aggregation of time interval data in the context of on-line analytical processing (OLAP).

P. Meisen (✉) · T. Meisen · D. Keng · M. Recchioni · S. Jeschke
IMA/ZLW & IfU, RWTH Aachen University, Dennewartstr. 27, 52068 Aachen, Germany
e-mail: philipp.meisen@ima-zlw-ifu.rwth-aachen.de

Originally published in "12th International Conference on Information Technology: New Generations (ITNG 2015)", © 2015.
Reprint by Springer International Publishing AG 2016,
DOI 10.1007/978-3-319-46916-4_31

OLAP is generally described as a user-driven method to analyze multidimensional data. An important technique used within such applications is the aggregation of data along defined members of different hierarchies belonging to different dimensions [3]. In addition to the aggregating performance, the filtering capabilities (i.e. handling high selectivity) of the underlying database system are of great importance regarding the performance of an OLAP system. Since the beginning of multidimensional models (MDM), the research community has introduced three different approaches used to handle data within a MDM; multidimensional (MOLAP), relational (ROLAP), and hybrid approaches (HOLAP). The different advantages and disadvantages are frequently discussed in numerous publications. Likewise, other issues concerning MDM have been addressed over the last years; non-strict, non-onto or non-covering hierarchies [4, 5], many-to-many relationships and summarizability problems [6, 7], low granularities leading to high cardinality [3], computation of data cubes [8], and query languages [9].

When dealing with time interval data we are faced with several of these issues, e.g.

- every interval defines a many-to-many relationship between the facts of the interval and the time dimension which leads to summarizability problems,
- real-world data sets contain non-onto and non-covering hierarchies,
- the time dimension may use a lowest granularity leading to millions of granules (e.g. using milliseconds), and
- the querying of data for aggregated values needs temporal operators, e.g. as defined by Allen [10].

Our contributions: In this paper, we present an OLAP system for time interval data. The heart of our system is a bitmap-based implementation, enabling fast selection and aggregation. Additionally, we will introduce our two-step aggregation technique, which enables the calculation of relevant measures in the context of time interval data. To our knowledge, there are no other OLAP systems available providing sufficient solutions for the raised issues.

This paper is organized as follows: First, we provide a short recap of the time interval data analysis model (TIDAMODEL) presented in detail in [11] Next, we present the problems occurring while aggregating time interval data from a semantic and temporal point of view and we introduce our two-step aggregation technique. In the following section, we describe our bitmap-based implementation used to enable OLAP of time interval data supporting the presented aggregation methods. Finally, we provide performance measures of the system before the paper is concluded.

2 Background: Modeling Time Interval Data

A characteristic of time interval data is the start and end time-value specifying the validity range of other descriptive values associated to the interval (cf. Fig. 1). Sets

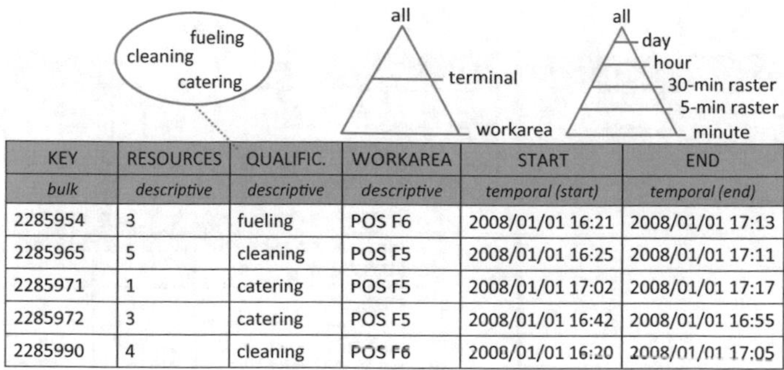

KEY	RESOURCES	QUALIFIC.	WORKAREA	START	END
bulk	*descriptive*	*descriptive*	*descriptive*	*temporal (start)*	*temporal (end)*
2285954	3	fueling	POS F6	2008/01/01 16:21	2008/01/01 17:13
2285965	5	cleaning	POS F5	2008/01/01 16:25	2008/01/01 17:11
2285971	1	catering	POS F5	2008/01/01 17:02	2008/01/01 17:17
2285972	3	catering	POS F5	2008/01/01 16:42	2008/01/01 16:55
2285990	4	cleaning	POS F6	2008/01/01 16:20	2008/01/01 17:05

Fig. 1 Illustration of our running example TIDAMODEL, the specified interval is read to be half-open, i.e. [start, end)

of time interval data can be represented in many ways. When modeling time interval data in the context of analytics, it is necessary to define not only the schema of a time interval data record but also the MDM. A MDM is defined by the different dimensions, its hierarchies, levels and members. In the case of time interval data, the time axis and the time dimension are of special interest. In this paper, we use the TIDAMODEL introduced in [11]. The model is defined by a 5 tuple (P, Σ, τ, M, Δ) in which P denotes the time interval database, Σ the set of descriptors, τ the time axis, M the set of measures, and Δ the set of dimensions. The time interval database P contains the raw time interval data records and a schema definition of the contained data. The schema associates each field of the record (which might contain complex data structures) to one of the following categories; temporal, descriptive, or bulk. The model explicitly allows the definition of non-strict, non-onto, and non-covering hierarchies.

Our running example within this paper is illustrated in Fig. 1. The figure shows five raw time interval records of a database P. Each record of the database is defined to provide values[1] for the fields KEY, RESOURCES, QUALIFICATION, WORKAREA, START, and END which are categorized in the specified groups. The example does not define the set of values, dimensions or mapping functions for every descriptor, which is normally necessary. Instead, it exemplifies the valid values for the QUALIFICATION descriptor (i.e. fueling, cleaning, and catering) as well as two hierarchies - one for the WORKAREA descriptor and the other one for the time dimension. The time granularity within our example is defined to be minutes. Without introducing the term descriptor any further, it should be mentioned that a descriptive value of the raw record is mapped to one or many descriptor values. This enables the system to support generally many-to-many relationships between facts and dimensions. In our running example we use the identity function to map a descriptive value to a descriptor value.

[1]The values do not have to be provided, but if not the system uses a fallback strategy to determine a value, which can be defined by the mapping function (e.g. unknown, or not set).

MOLAP

	POS F5	POS F6
catering	AGGR(1, 3)	
cleaning	AGGR(5)	AGGR(4)
fueling		AGGR(5)

ROLAP

id	qualific.
1	catering
2	cleaning
3	fueling

id	qualific.
1	POS F5
2	POS F6

key	wa_id	qual_id	res
2285954	3	2	3
2285965	2	1	5
2285971	1	1	1
2285972	1	1	3
2285990	2	2	4

Fig. 2 MOLAP and ROLAP schemas of the running example excluding the START and END fields

3 Aggregating Facts

Aggregation is the predominant operation in OLAP systems [3]. The facts of the systems are aggregated along the defined members of the different hierarchies defined by the MDM. Typically, the OLAP system stores the collected facts in a cube (MOLAP) or retrieves facts from a relational database (ROLAP). The former aggregates the facts of each measure, while the data is loaded into the cube. The latter applies the aggregation of the measure when retrieving the data (e.g. using group by) or is loaded with pre-aggregated data generated during integration [12]. In Fig. 2, a two-dimensional MOLAP cube and a ROLAP star schema are illustrated while taking into consideration that the raw records from our running example are used, excluding the START and END fields.

Whenever a user selects data (e.g. by roll-up, slice or dice operations), the collected facts are aggregated by the system according to the selected pre-defined measures or the aggregation specified in the multidimensional expression (MDX). The latter aggregation method is independent of the used technology. The use of aggregations within an MDX is mostly complicated, it is not applicable to end-users and quite error-prone. Considering the running example and the models shown in Fig. 2: if a user were to select all qualifications, the work-area *POS F5* and a measure defined as SUM(RESOURCES), the calculation of the system would result in 9. More complex queries, like retrieving the average amount of resources per work-area,[2] are not provided easily via a user-interface and therefore have to be formalized using MDX queries directly.

[2]This could be achieved by calculating the SUM per work-area and calculate the AVG of the retrieved summed values.

3.1 Aggregation and Many-to-Many Relationships

A many-to-many relationship occurs if a fact is associated to multiple members instead of exactly one. In the case of time interval data, each interval defines facts which are related to multiple members of the time dimension. Generally, modern OLAP systems like Microsoft Analysis Services, Oracle Database OLAP, IBM Cognos, or icCube support many-to-many relationships. Nevertheless, the (implicitly) adapted MDM increases the integration effort dramatically, especially when a low granularity (e.g. minutes) is used for the time dimension [7]. Additionally, the performance of the OLAP system decreases tremendously [13] and some systems are not capable of handling the amount of data necessary to be created in memory when querying time interval data [11].

The main difference between the aggregations of facts being related to dimensions by a many-to-many relationship is the selection of the facts to be aggregated. In the case of a many-to-many relationship, the system ensures that facts of the same record are not counted multiple times which would lead to summarizability problems. To achieve this, the OLAP system typically uses a unique identifier for each record to identify and remove duplicates. After the selection of facts is performed, the aggregation is done as described for the default case of one-to-many-relationships.

3.2 Aggregation of Time Interval Data

Figure 3 illustrates our running example in a Gantt-chart. The WORKAREAs are illustrated as swim-lanes, whereby different QUALIFICATIONs are color-coded. The figure shows the already mentioned many-to-many relationship, i.e. each fact of an interval is associated to multiple members of the time dimension (e.g. the descriptor value 1 of the time interval starting at 17:02 is associated to 15 granules 17:02, 17:03, ..., 17:16). Additionally, the figure shows the values of three measures; *needed resources*, *time spent (min)* and *tasks finished*. The values are aggregated to the *5-min raster* members of the time hierarchy (i.e. [16:20, 16:25), ..., [17:15, 17:20)).

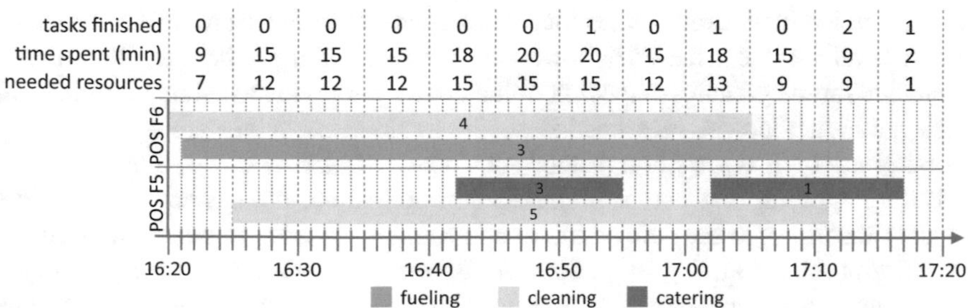

Fig. 3 Running example illustrated in a Gantt-chart with three measures

Defining these three measures within a modern OLAP system is currently not possible while only using multidimensional modeling. Heavily adapting and adding temporal logic to the integration process are some of the required steps to perform, accepting the negative impacts mentioned previously. To resolve these problems, we suggest a new aggregation technique, as well as the definition of new temporal aggregation operators like *count started*, *count finished* and *count running*. We introduce our aggregation technique initially by explaining each of the three measures. Thereafter, we define our technique in detail.

To calculate the measure *needed resources* for a specific member of a level of a time hierarchy, it is necessary to calculate an intermediate result for each member of the lowest granularity. The intermediate result of the measure of a granule is calculated by summing up all the RESOURCES facts. Looking at our running example shown in Fig. 3 and the granule 16:42, the intermediate result is calculated by summing 5, 3, 3 and 4, i.e. 15. The intermediate result for the granule 16:41 is 12, calculated by 5 + 3 + 4. In a second phase, the maximum of all the retrieved intermediate results is picked as a result of the measure for the member of the higher level. In our example, the result of the measure for the member [16:40, 16:44] of the 5 min raster is thereby given by the maximum of 12 (16:40), 12 (16:41), 15 (16:42), 15 (16:43) and 15 (16:44).

The measure *time spent (min)* is also calculated using two steps. In the first step, the number of associated facts to a granule is counted. In the second step, the calculated counts are summed up. Looking at the example and the granule, 16:20 results in a count of 1, whereby 16:21 results in 2. Adding up all values for the member [16:20, 16:24] of the *5 min raster* results in 9, i.e. 1 (16:20) + 2 (16:21) + 2 (16:22) + 2 (16:23) + 2 (16:24).

The third measure *tasks finished* uses an aggregator based on temporal logic. In the first step, the temporal aggregation method *count finished* is applied on the lowest granularity counting the intervals finishing at the specified granule. In the second step, the retrieved values are summed up.

3.3 Two-Step-Aggregation Technique

As exemplified in the previous section, the calculations for measures in the context of time interval data are achieved by a two-step aggregation. In the first step, the value is calculated for the granules of the time-dimension by applying a default or a temporal aggregation method. In the second step, these calculated values are aggregated by applying a default aggregation method such as *count*, *sum*, *average*, or *mean*.

Before applying the two-step aggregation technique, the system has to select the data according to the defined filter and grouping criteria. To overcome summarizability problems arising by the mentioned reasons pertaining to many-to-many relationships, it is necessary that the system is capable of associating each fact to its interval. We present our bitmap-based solution considering the filtering in Section 4.

We denote the set of selected intervals by Θ and the set of selected intervals for a specific group g by Θ_g. We denote the fact of a record ρ for the granule t of the time-dimension by $\phi(\rho, t)$. The set of all time granules selected by the query using a specified hierarchy level of the time-dimension or a time-window is denoted by T. Let the first aggregation method be denoted by AGGR1 and the second by AGGR2. The result of the first aggregation for a specific interval set Θ_g and a granule $t \in T$ is defined by

$$i_t := \text{AGGR1}(\{\phi(\rho, t) \mid \rho \in \Theta_g\}). \tag{1}$$

The result of a measure mg for a specific group g and set of granules T is defined by

$$m_g := \text{AGGR2}(\{i_t \mid t \in T\}). \tag{2}$$

Figure 4 illustrates the defined entities used when calculating a measure applying the two-step aggregation technique. The illustration is based on our running example, showing the application of the technique for Query 1 (Table 1).

As shown in the figure, a fact function may return *null* values. This indicates that no interval covers the specified granule. The *null* values might be important for the first aggregation function used. In some cases (e.g. *sum*), a *null* value can be understood as 0, in others (e.g. *average*) it might provide important information (e.g. *NaN* as result).

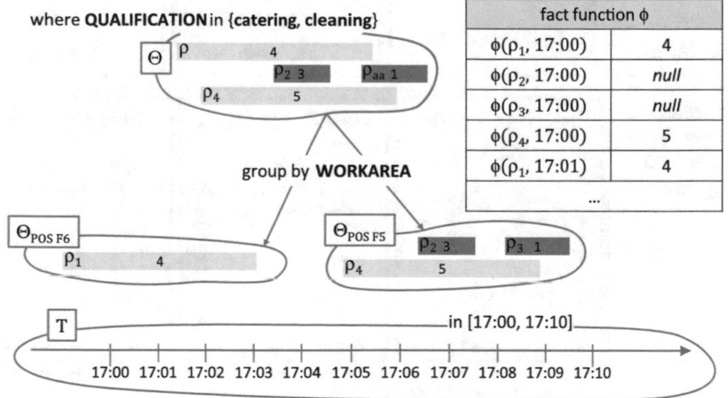

Fig. 4 Entities defined for two-step aggregation technique

Table 1 Query 1. Example query used in Fig. 4

`select`	`MAX (SUM (RESOURCES)) as 'needed Res.'`
`on`	`RASTER.5MIN`
`in`	`[17:00, 17:10]`
`where`	`QUALIFICATION in {catering,cleaning}`
`group by`	`WORKAREA`

4 Bitmap-Based OLAP of Time Interval Data

As mentioned in the previous sections, aggregation is the predominant operation in an OLAP system. Nevertheless, selecting the needed data for aggregation and supporting incremental calculations are important aspects to increase aggregation performance. Additionally, the need to identify each interval individually is a significant requirement to ensure correct summarizability.

Bitmaps have been used in several implementations and publications to achieve fast data selection in mostly read-only environments. Depending on the application context of the bitmaps, it has been shown that the selection of the encoding and compression scheme of a bitmap is crucial, considering the performance gained and storage needed. Important criteria to select the best encoding and compression scheme are the types of queries [14], the order of the data [15], and the complexity considering the logic operations used within queries [16]. Furthermore, in specific scenarios implementations using bitmaps outperform popular commercial database management systems (DBMS) by a factor larger than ten [17].

Figure 5 illustrates the components of our bitmap-based OLAP system enabling fast selection and aggregation of time interval data. The illustration shows the dif-

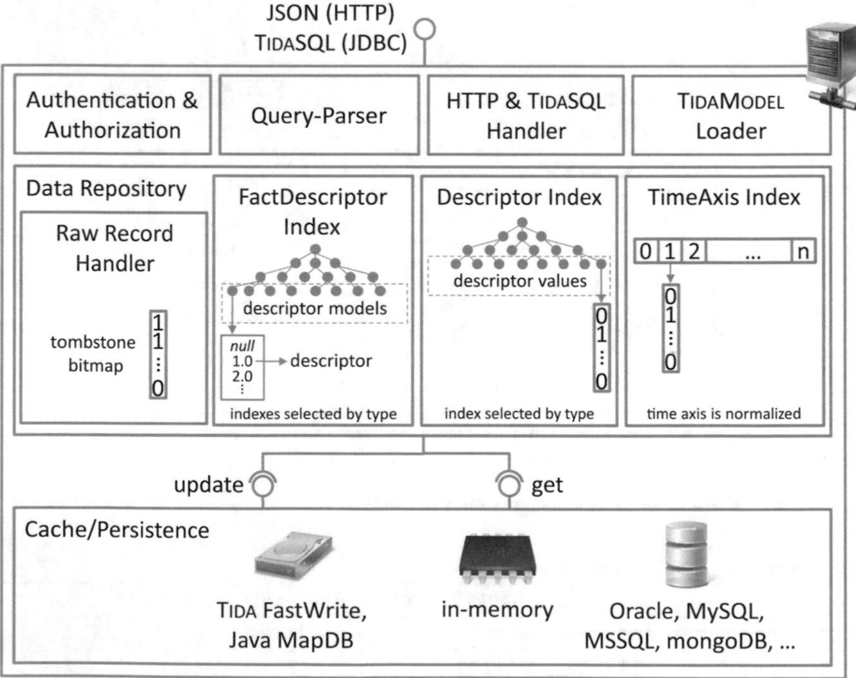

Fig. 5 Components of the TIDASERVER focusing on the bitmap-based Data Repository and Cache/Persistence components

ferent components focusing on the Data Repository and the Cache Persistence components. Before we introduce the application of the two-step aggregation technique presented in Section 3 and the selection of data needed for the aggregation based on the bitmap-based implementation, we briefly introduce the different components of the Data Repository.

4.1 Raw Record Handler: Inserts, Deletes and Updates

Inserts, deletes and updates[3] of time interval data records are handled by the *Raw Record Handler* to receive newly or updated raw time interval data records. The handler is responsible of managing the so called tombstone bitmap which identifies if a record is still valid (i.e. value of 1) or not (i.e. value of 0).

When a new record is received by the handler, it creates a unique identifier for that record. Additionally, the record is processed using the mapping functions defined by the TIDAMODEL to which the record belongs. Afterwards, the different bitmaps using the *Descriptor* and *TimeAxis Indexes* are updated. Finally, the tombstone bitmap is updated and set to 1 for the record.

The deletion of a record is achieved by setting the value within the tombstone bitmap to 0. A clean-up process picks up the dead stored records and performs a clean-up in the background. An update of a time interval record is performed by deleting the record and inserting the updated record as a new instance.

4.2 Descriptor, TimeAxis and FactDescriptor Index

The different indexes used within our running example are illustrated in Fig. 6. In the example, it is assumed that each descriptor uses a fact-function invariant of the record (RESOURCES uses the identity function, WORKAREA and QUALIFICATION use a fact-function returning always 1).

The *Descriptor Index* is used to select the different bitmaps for the different descriptors. The system selects the best fitting index (from the ones known by the system) for the identifiers of the descriptors considering performance (e.g. TroveInt[4] is used for int values).

The *TimeAxis Index* is implemented using an array-like structure. Each granule of the time-axis is represented by an integer. The bitmaps for a specific granule can be reached by using the integer as index. Besides the bitmap, the index provides access to a *FactDescriptor Index* containing information about the facts associated to the granule of the time-axis. The *FactDescriptor Index* is used to retrieve the

[3]The insertion process is introduced in detail in [11]. We refer the interested reader to that paper for more insights.

[4]http://trove.starlight-systems.com/.

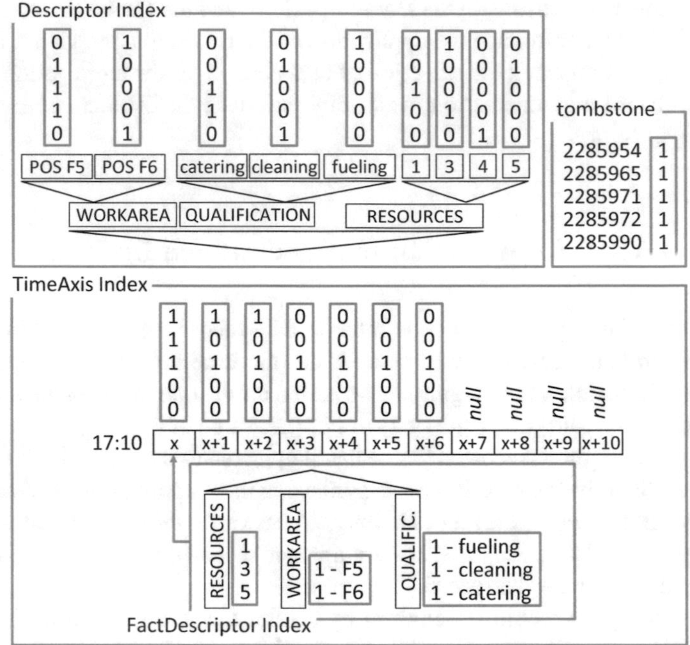

Fig. 6 The indexes of the running example: TimeAxis Index is only illustrated for [17:10, 17:20] and only one FactDescriptor Index is shown

different facts for a granule. Facts are retrieved from a descriptor by its fact-function. The fact-function can thereby be defined to be invariant of the record or not. As defined in our example, the RESOURCES descriptor is an invariant descriptor, i.e. the value of the descriptor is provided by the descriptor itself. An example for a variant descriptor could be a temperature value measured over a time interval, assuming that the descriptive value (e.g. 120.54 °F) is mapped to a descriptor value categorizing the descriptive value (e.g. 120.54 °F is mapped to 'high'). When aggregating the facts of the temperature descriptor, the facts have to be retrieved directly from the record. The descriptive value 'high' does not define the exact temperature needed for aggregation. The *FactDescriptor Index* sorts all the facts of the different descriptors in ascending order. Variant descriptors are thereby moved to the top of the list (with a fact-value of *null*). Whenever the system has to retrieve a fact for a specific descriptor, it uses the index to retrieve a sorted list. If the list contains *null* values indicating variant facts, the fact of each record is retrieved. Otherwise the values of the list are used directly (cf. Fig. 7).

4.3 Bitmap-Based Time Interval Data Selection

Whenever a (new) query is fired against the system, the *Query Parser* processes the query and receives the filtering, grouping and time dependent criteria. In addition, the parser determines the measures to be calculated and which level of the time hierarchy is to be used. Using this information, the system creates a set of bitmaps containing

- the resulting bitmap of the filtering criteria (determined by applying bitwise and $*$, or $+$, not $\neg$, or xor $\bigoplus$ operations)
- the bitmaps for each defined group, and
- the bitmaps for each selected time-granule.

The resulting set of bitmaps of the previously introduced Query 1 is shown in Formula 3, in which bit_x denotes the bitmap of the descriptor value x or the bitmap of the time-granule x.

$$\begin{aligned} r_1 = \{&\text{tombstone} * (\text{bit}_\text{catering} + \text{bit}_\text{cleaning}), \\ &\{\text{bit}_{17:00}, \ldots, \text{bit}_{17:10}\}, \{\text{bit}_\text{POS F5}, \text{bit}_\text{POS F6}\}\} \end{aligned} \tag{3}$$

To create the different time-series for each time-granule within each group, the algorithm iterates over each bitmap of a time-granule and each bitmap of a group. Within each iteration the algorithm combines the filtered bitmap with both the current time-granule bitmap and the group bitmap using the and operation. For our example, the resulting bitmap for time-granule 17:10 and the group POS F5 is shown in Formula 4.

$$\text{bit}_\text{POS F5, 17:10} = (0, 1, 1, 0, 0)^T \tag{4}$$

If a level of a hierarchy is used within the grouping or the filter criteria of the query, the algorithm resolves the granules for the specific descriptor and combines the bitmaps according to the hierarchy definition (using the or-operation).

4.4 Bitmap-Based Aggregation

After the selection is performed the aggregation algorithm is applied. The input for the aggregation is a set of bitmaps (cf. Formula 3) created by the selection algorithm. The following aggregation algorithm is based on the two-step aggregation technique presented in Section 3.

The first aggregation is performed on every bitmap of the sequence. The implementation, and thereby the performance, of the aggregation method depends on the type of aggregation. A *count* aggregation can be performed most efficiently by almost any bitmap implementation [16]. Temporal aggregations need additional information

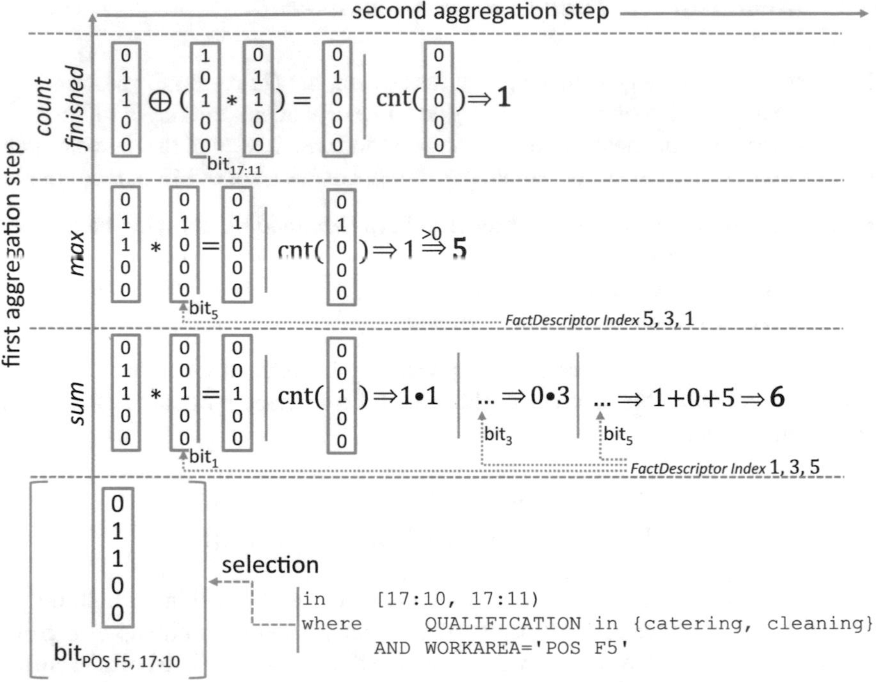

Fig. 7 Example of the first aggregation step of *sum*, *max* and *count finished* based on bitmaps

about the next or previous time-granule to be performed.[5] Other aggregations such as *sum*, *max* or *min* typically need the fact-values to be applied. These values are retrieved using the *FactDescriptor Index* to determine the sorted descriptors associated with the time-granule. If the facts of the descriptors are variant, the algorithm creates an array of facts retrieved from the records and applies the aggregation technique to the array. Otherwise, if the facts are invariant, the records are not needed and the result of the aggregation can be determined using bitmaps only. The algorithm iterates over the descriptors (descending if the aggregation is *max*, otherwise ascending) and combines the bitmap of the descriptor with the bitmap of the time-granule using the *and*-operation. For a *min* or *max* aggregation, the count of the resulting bitmap is used to determine the value (a value larger than 0 indicates that the algorithm can stop and return the invariant fact, otherwise the algorithm moves on to the next descriptor). In the case of a *sum* aggregation, the algorithm multiplies the count with the invariant value and sums all the results. Figure 7 illustrates selected aggregation methods useful as a first aggregation and the bitmap based calculation.

[5]The aggregation combines the previous (count started) or following (count finished) bitmap with the actual one using the xor-operation and counts the result. We refer to the example shown in Fig. 7.

The second step aggregation is performed on the set of numeric values returned during the first step and mathematically straight forward. The aggregation collects the calculated values from the first step and applies the mathematical function to the set of numbers, e.g. *sum*, *min*, *max* or *count*.

5 Performance

To assess the performance of our implementation, we ran several tests on an Intel Core i7-3632QM with a CPU clock rate of 2.20 GHz, 8 GB of main memory, a SSD, and running 64-bit Windows 7 Enterprise. For our Java implementation, we used a 64-bit JRE 1.6.45, with XMX 4,096 MB and XMS 512 MB. We tested the runtime performance of our implementation using two different bitmap implementations, i.e. word-align (EWAH)[6] [16] and roaring bitmap [19]. We used a real-world data set containing 1,122,097 records collected over one year. The records have an average interval length of 48 minutes and three descriptive values (person (cardinality: 713), task-type (cardinality: 4), and work area (cardinality: 31)). The used time-granule was minutes (i.e. time cardinality: 525,600). Each test ran 100 times and the average CPU time is used as result.

The results are shown in Fig. 8. As illustrated both implementations achieve almost the same performance when counting. Roaring bitmaps perform way better considering logical operations, which are used excessively when aggregating (e.g. *min* and *sum*). Additionally, the setter method of roaring bitmaps executes almost constantly per record compared to word-aligned bitmaps when the size of the bitmap increases. The tests show that using roaring bitmaps can increase the performance by a factor of 25. With respect to results based on JavaEWAH shown in [11] using roaring bitmaps is promising.

6 Conclusions and Future Work

In this paper, we presented a bitmap-based OLAP system for time interval data. The system is capable of creating aggregations by loading the raw time interval data into the system, persisting the raw data (if required), and calculating measures needed to analyze time interval data using our introduced two-step aggregation technique. Additionally, we presented temporal aggregation methods like count finished, which are supported by the bitmap-based implementation as well.

In future work, we plan to use the implementation presented here in order to enable on-line analytical mining (OLAM) on time interval data. One of the challenges in that field is the remapping from aggregated time-series patterns back to the time-

[6]We use the JavaEWAH library, which is considered to offer the best query time for all Java world-align distributions [18].

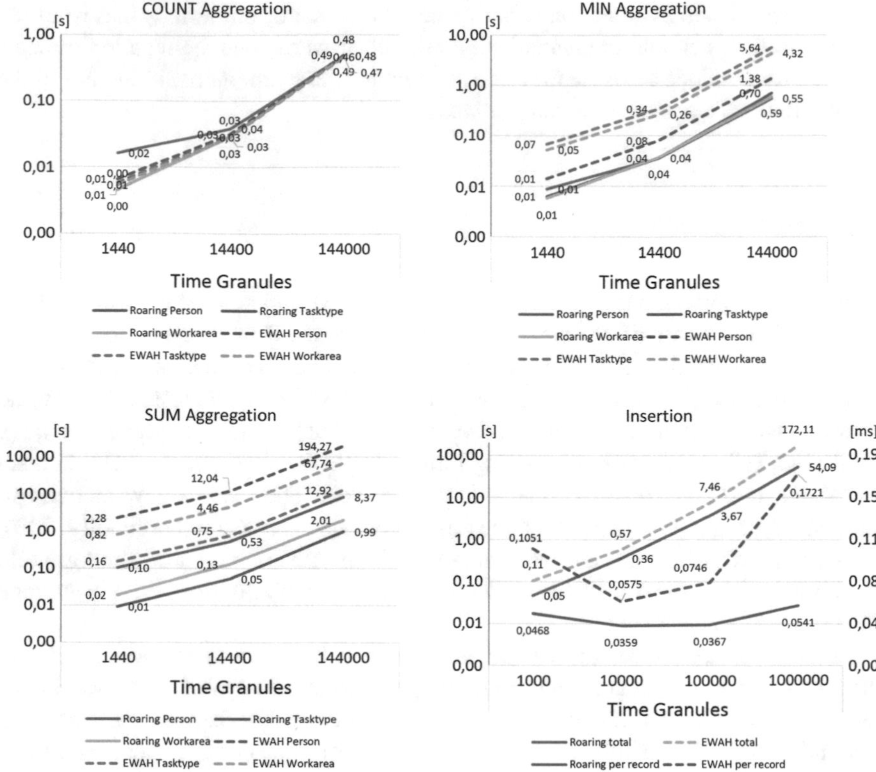

Fig. 8 Results of the performance tests using our implementation based on word-align and roaring bitmaps using a real-world data set

intervals. We also plan on defining a simplified query language to replace the mostly complex MDX queries. Additionally, we plan to present in detail the capabilities of distributed calculations and load balancing available through our implementation.

References

1. P. Papapetrou, G. Kollios, S. Sclaroff, D. Gunopulos, Mining frequent arrangements of temporal intervals. Knowledge and Information Systems **21** (2), 2009, pp. 133–171
2. A. Kotsifakos, P. Papapetrou, V. Athitsos, Ibsm: Interval-based sequence matching. In: *Proceedings of the 13th SIAM International Conference on Data Mining, Austin, Texas, USA*. 2013
3. S. Agarwal, R. Agrawal, P. Deshpande, A. Gupta, J. Naughton, R. Ramakrishnan, S. Sarawagim, On the computation of multidimensional aggregates. In: *Proceedings of the 22nd VLDB*. 1996, pp. 506–521
4. T. Pedersen, Aspects of data modeling and query processing for complex multidimensional data. Ph.D. thesis, Aalborg University, 2000. Ph.D. thesis

5. S. Banerjee, K. Davis, Modeling data warehouse schema evolution over extended heirarchy semantics. Journal on Data Semantics XIII, Lecture Notes in Computer Science **5530**, 2009, pp. 72–96
6. J. Mazón, J. Lichtenbörger, J. Trujillo, Solving summarizability problems in fact-dimension relationships for multidimensional models. In: *Proceedings of DOLAP '08*. 2008, pp. 57–64
7. J. Mazón, J. Lichtenbörger, J. Trujillo, A survey on summarizability issues in multidimensional modeling. Data & Knowledge Engineering **68**, 2009, pp. 1452–1469
8. A. Vaisman, R. A. Mendelzon, A temporal query language for olap: Implementation and a case study. In: *Proceedings of the 8th Biennial Workshop on Data Bases and Programming Languages (DBPL 2001), Frascati, Italy, September 8–10*. 2001
9. Y. Yuan, X. Lin, Q. Liu, W. Wang, J. Xu Yu, Q. Zhang, Efficient computation of the skyline cube. In: *Proceedings of the VLDB '05 Proceedings of the 31st international conference on Very large data bases*. 2005, pp. 241–252
10. J. Allen, Maintaining knowledge about temporal intervals. Commun. ACM **26** (11), 1983, pp. 832–843
11. P. Meisen, T. Meisen, M. Recchioni, D. Schilberg, S. Jeschke, Modeling and processing of time interval data for data-driven decision support. In: *Proceedings of the SMC2014, San Diego, USA*. 2014
12. S. Chaudhuri, U. Dayal, An overview of data warehousing and olap technology. SIGMOD Rec. **26** (1), 1997, pp. 65–74
13. M. Russo, A. Ferrrari. The many-to-many revolution 2.0: Advanced dimensional modeling with microsoft sql server analysis service. URL http://www.sqlbi.com/wp-content/uploads/The_Many-to-Many_Revolution_2.0.pdf. Last Checked: October, 2011
14. C. Chan, Y. Ioannidis, An efficient bitmap encoding scheme for selection queries. In: *SIGMOD 1999, Proceedings ACM SIGMOD International Conference on Management of Data, Philadephia, Pennsylvania, USA*. 1999
15. K. Wu, A. Shoshani, K. Stockinger, Analyses of multi-level and multi-component compressed bitmap indexes. ACM Trans. Database Syst. **35** (1), 2008. Article 2
16. O. Kaser, D. Lemire. Compressed bitmap indexes: beyond unions and intersections, 2014. Eprint arXiv:1402.4466
17. K. Wu, S. Ahern, E. Bethel, J. Chen, H. Childs, E. Cormier-Michel, Fastbit: interactively searching massive data. Journal of Physics Conference Series 08/2009 **180** (1), 2009
18. G. Guzun, C. Guadalupe, D. Chiu, J. Sawin, A tunable compression framework for bitmap indices. In: *Proceedings of the 30th International Conference on Data Engineering (ICDE 2014)*. 2014
19. S. Chambi, D. Lemire, O. Kaser, R. Godin. Better bitmap performance with roaring bitmaps, 2014. Eprint arXiv:1402.6407

TIDAQL: A Query Language Enabling On-line Analytical Processing of Time Interval Data

Philipp Meisen, Diane Keng, Tobias Meisen, Marco Recchioni and Sabina Jeschke

Abstract Nowadays, time interval data is ubiquitous. The requirement of analyzing such data using known techniques like on-line analytical processing arises more and more frequently. Nevertheless, the usage of approved multidimensional models and established systems is not sufficient, because of modeling, querying and processing limitations. Even though recent research and requests from various types of industry indicate that the handling and analyzing of time interval data is an important task, a definition of a query language to enable on-line analytical processing and a suitable implementation are, to the best of our knowledge, neither introduced nor realized. In this paper, we present a query language based on requirements stated by business analysts from different domains that enables the analysis of time interval data in an on-line analytical manner. In addition, we introduce our query processing, established using a bitmap-based implementation. Finally, we present a performance analysis and discuss the language, the processing as well as the results critically.

Keywords Time Interval Data · Query Language · On-Line Analytical Processing · Distributed Query Processing

1 Introduction

Nowadays, time interval data is recorded, collected and generated in various situations and different areas. Some examples are the resource utilization in production environments, deployment of personnel in service sectors, or courses of diseases in healthcare. Thereby, time interval data is used to represent observations, utilizations or measures over a period of time. Put in simple terms, time interval data is defined by two time values (i.e. start and end), as well as descriptive values associated to the interval: like labels, numbers, or more complex data structures. Figure 1 illustrates a sample database of five records.

P. Meisen (✉) · D. Keng · T. Meisen · M. Recchioni · S. Jeschke
IMA/ZLW & IfU, RWTH Aachen University,
Dennewartstr. 27, 52068 Aachen, Germany
e-mail: philipp.meisen@ima-zlw-ifu.rwth-aachen.de

Originally published in "17th International Conference on Enterprise Information Systems (ICEIS 2015)", © 2015. Reprint by Springer International Publishing AG 2016, DOI 10.1007/978-3-319-46916-4_32

key	resources	type	location	start	end
2285954	3	cleaning	POS F6	2015/01/01 16:21	2015/01/01 17:13
2285965	5	maintenance	POS F5	2015/01/01 16:25	2015/01/01 17:10
2285971	1	maintenance	POS F5	2015/01/01 17:02	2015/01/01 17:17
2285972	3	room service	POS F5	2015/01/01 16:42	2015/01/01 16:55
2285990	4	miscellaneous	POS F6	2015/01/01 16:20	2015/01/01 17:05

Fig. 1 A sample time interval database with intervals defined by [start, end), an id, and three descriptive values

For several years, business intelligence and analytical tools have been used by managers and business analysts, inter alia, for data-driven decision support on a tactical and strategic level. An important technology used within this field, is on-line analytical processing (OLAP). OLAP enables the user to interact with the stored data by querying for answers. This is achieved by selecting dimensions, applying different operations to selections (e.g. roll-up, drill-down, or drill-across), or comparing results. The heart of every OLAP system is a multidimensional data model (MDM), which defines the different dimensions, hierarchies, levels, and members [1].

The need of handling and analyzing time interval data using established, reliable, and proven technologies like OLAP is desirable in this respect and an essential acceptance factor. Nevertheless, the MDM needed to model time interval data has to be based on many-to-many relationships which have been shown to lead to summarizability problems. Several solutions solving these problems on different modeling levels have been introduced over the last years, leading to increased integration effort, enormous storage needs, almost always inacceptable query performances, memory issues, and often complex multidimensional expressions [2, 3]. Additionally, these solutions are, considering real-world scenarios, only applicable to many-to-many relationships having a small cardinality which is mostly not the case when dealing with time interval data. As a result, the usage of MDM and available OLAP systems is not sufficient, even though the operations (e.g. roll-up, drill-down, slice, or dice) available through such systems are desired.

Enabling such OLAP like operations in the context of time interval data, requires the provision of extended filtering and grouping capabilities. The former is achieved by matching descriptive values against known filter criteria logically connected using operators like *and*, *or*, or *not*, as well as a support of temporal relations like *starts-with*, *during*, *overlapping*, or *within* [4]. The latter is applied by known aggregation operators like *max*, *min*, *sum*, or *count*, as well as temporal aggregation operators like *count started* or *count finished* [5].

The application of the *count* aggregation operator for time interval data is exemplified in Fig. 2. The color code identifies the different types of a time interval (e.g. cleaning, maintenance, room service, miscellaneous). Furthermore, the swim-lanes show the location. The figure illustrates the count of intervals for each type over one day across all locations (e.g. POS F5 and POS F6) using a granularity of minutes (i.e. 1,440 aggregations are calculated).

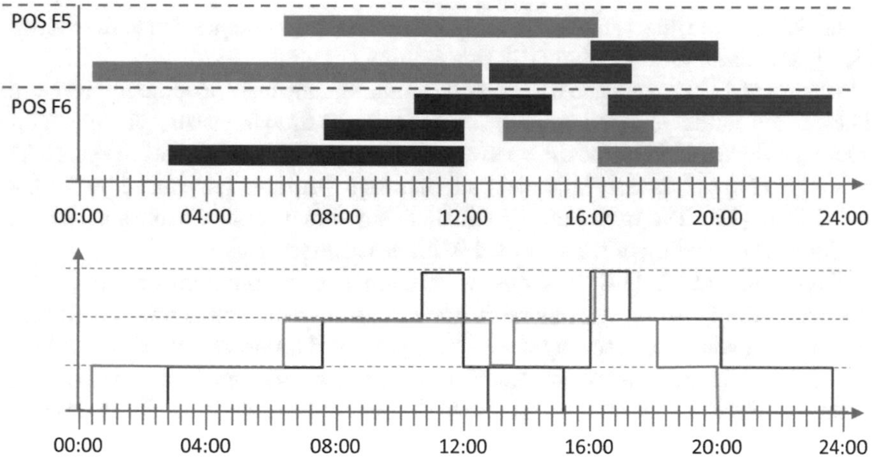

Fig. 2 On top the time interval data (10 records) shown in a Gantt-Chart, on the bottom the aggregated time-series

In this paper, we present a query language allowing to analyze time interval data in an OLAP manner. Our query language includes a data definition (DDL), a data control (DCL), and a data manipulation language (DML). The former is based on the time interval data model introduced by Meisen et al. [6], whereby the latter supports the two-step aggregation technique mentioned in Meisen et al. [5]. Furthermore, we outline our query processing which is based on a bitmap-based implementation and supports distributed computing.

This paper is organized as follows: In Section 2, we discuss related work done in the field of time interval data, in particular this section provides a concise overview of research dealing with the analyses of time interval data. We provide an overview of time interval models, discuss related work done in the field of OLAP, and present query languages. In Section 3, we introduce our query language and processing. The section presents among other things how a model is defined and loaded, how temporal operators are applied, how the two-step aggregation is supported, how groups are defined, and how filters are used. We introduce implementation issues and empirically evaluate the performance regarding the query processing in Section 4. We conclude with a summary and directions for future work in Section 5.

2 Related Work

When defining a query language, it is important to have an underlying model, defining the foundation for the language (e.g. the relational model for SQL, different interval-based models for e.g. IXSQL or TSQL2, the multidimensional model for MDX, or the graph model for Cypher). Over the last years several models have been introduced in the field of time intervals, e.g. for temporal databases [7], sequential pattern mining [8, 9], association rule mining [10], or matching [11].

Chen et al. [12] introduced the problem of mining time interval sequential patterns. The defined model is based on events used to derive time intervals, whereby a time interval is determined by the time between two successive time-points of events. The definition is based on the sequential pattern mining problem introduced by Agrawal and Srikant [13]. The model does not include any dimensional definitions, nor does it address the labeling of time intervals with descriptive values.

Papapetrou et al. [14] presented a solution for the problem of "discovering frequent arrangements of temporal intervals". An e-sequence is an ordered set of events. An event is defined by a start value, an end value and a label. Additionally, an e-sequence database is defined as a set of e sequences. The definition of an event given by Papapetrou et al. is close to the underlying definition within this paper (cf. Fig. 1). Nevertheless, facts, descriptive values, and dimensions are not considered.

Mörchen [15] introduced the TSKR model defining tones, chords, and phrases for time intervals. Roughly speaking, the tones represent the duration of intervals, the chords the temporal coincidence of tones, and the phrases represent the partial order of chords. The main purpose of the model presented by Mörchen is to overcome limitations of Allen's [4] temporal model considering robustness and ambiguousness when performing sequential pattern mining. The model neither defines dimensions, considers multiple labels, nor recognizes facts.

Summarized, models presented in the field of sequential pattern mining, association rule mining or matching do generally not define dimensions and are focused on generalized interval data, or support only non-labelled data. Thus, these models are not suitable considering OLAP of time interval data, but are a guidance to the right direction.

Within the research community of temporal databases different interval-based models have been defined [7]. The provided definitions can be categorized in weak and strong models. A weak model is one, in which the intervals are used to group time-points, whereas the intervals of the latter carry semantic meaning. Thus, a weak interval-based model is not of further interest from an analytical point of view, because it can be easily transformed into a point-based model. Nevertheless, a strong model and the involved meaning of the different operators – especially aggregation operators – are of high interest from an analytical view. Strong interval-based models presented in the field of temporal databases lack to define dimensions, but present important preliminary work.

In the field of OLAP, several systems capable of analyzing sequences of data have been introduced over the last years. Chui et al. [16] introduced S-OLAP for analyzing sequence data. Liu and Rundensteiner [17] analyzed event sequences using hierarchical patterns, enabling OLAP on data streams of time point events. Bebel et al. [18] presented an OLAP like system enabling time point-based sequential data to be analyzed. Nevertheless, the system neither support time intervals, nor temporal operators.

Recently, Koncilia et al. [19] presented I OLAP, an OLAP system to analyze interval data. They claim to be the first proposing a model for processing interval data.

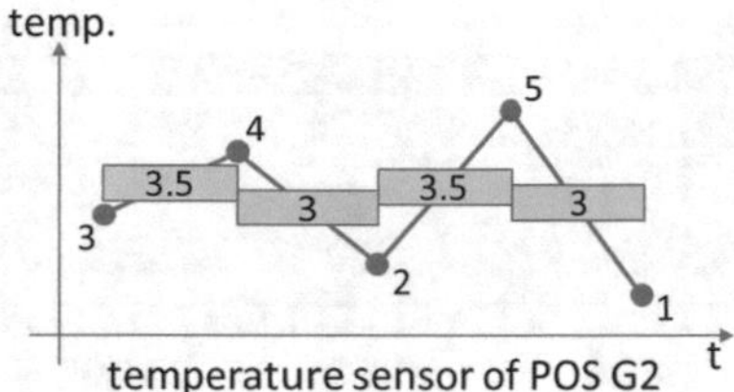

Fig. 3 Illustration of the model introduced by Koncilia et al. [19]. The intervals (rectangles) are created for each two consecutive events (dots). The facts are calculated using the average function as the *compute value function*

The definition is based on the interval definition of Chen et al. [12] which defines the intervals as the gap between sequential events. However, Koncilia et al. assume that the intervals of a specific event-type (e.g. temperature) for a set of specific descriptive values (e.g. POS G2) are non-overlapping and consecutive. Considering the sample data shown in Fig. 1, the assumption of non-overlapping intervals is not valid in general (cf. record 2,285,965 and 2,285,971). Figure 3 illustrates the model of Koncilia et al. showing five temperature events for POS G2 and the intervals determined for the events. Koncillia et al. mention the support of dimensions, hierarchies, levels, and members, but lack to specify what types of hierarchies are supported and how e.g. non-strict relations are handled.

Also recently, Meisen et al. [6] introduced the TIDAMODEL "enabling the usage of time interval data for data-driven decision support". The presented model is defined by a 5-tuple (P, Σ, τ, M, δ) in which P denotes the time interval database, Σ the set of descriptors, τ the time axis, M the set of measures, and δ the set of dimensions. The time interval database P contains the raw time interval data records and a schema definition of the contained data. The schema associates each field of the record (which might contain complex data structures) to one of the following categories: temporal, descriptive, or bulk. Each descriptor of the set Σ is defined by its values (more specific its value type), a mapping- and a fact-function. The mapping-function is used to map the descriptive values of the raw record to one or multiple descriptor values. The mapping to multiple descriptor values allows the definition of non-strict fact-dimension relationships. Additionally, the model defines the time axis to be finite and discrete, i.e. it has a start, an end, and a specified granularity (e.g. minutes). The set of dimensions δ can contain a time dimension (using a rooted plane tree for the definition of each hierarchy) and a dimension for each descriptor (using a directed acyclic graph for a hierarchy's definition). Figure 4 illustrates the modeled sample database of Fig. 1 using the TIDAMODEL. The figure shows the five intervals, as well as the values of the descriptors location (cf. swim-lane) and type (cf. legend). Dimensions are not shown. The used mapping function for all descriptors is the identity function. The used granularity for the time dimension is minutes.

Another important aspect when dealing with time interval data in the context of OLAP, is the aggregation of data and the provision of temporal aggregation operators. Kline and Snodgrass [20] introduced temporal aggregates, for which several

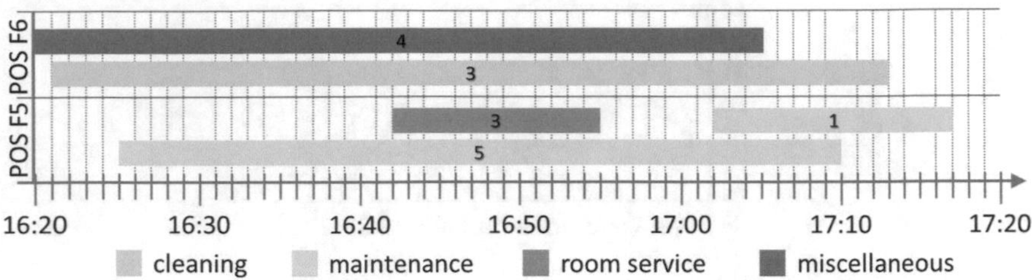

Fig. 4 Data of the sample database shown in Fig. 1 modeled using the TIDAMODEL [6]

start	end	res.	loc.
16:21	16:31	4	POS1
16:21	16:38	3	POS1
16:22	16:32	2	POS2
16:25	16:38	5	POS2

Fig. 5 Two-step aggregation technique presented by Meisen et al. [5]

enhanced algorithms were presented over the past years. Nevertheless, the solutions are focused on one specific aggregation operator (e.g. SUM), do not support multiple filter criteria, or do not consider data gaps. Koncilia et al. [19] address shortly how aggregations are performed using the introduced *compute value functions* and *fact creating functions*. Temporal operators are neither defined nor mentioned. Koncilia et al. point out that some queries need special attention when aggregating the values along time, but a more precise problem statement is not given. Meisen et al. [5] introduce a two-step aggregation technique for time interval data. The first one aggregates the facts along the intervals of a time granule and the second one aggregates the values of the first step depending on the selected hierarchy level of the time dimension. Figure 5 illustrates the two-step aggregation technique. In the illustration, the technique is used to determine the needed resources within the interval [16:30, 16:34]. Within the first step, the sum of the resources for each granule is determined and within the second step the maximum of the determined values is calculated, i.e. 14. Additionally, they introduce temporal aggregation operators like *started* or *finished count*.

The definition of a query language based on a model and operators (i.e. like aggregations), is common practice. Regarding time-series, multiple query languages and enhancements of those have been introduced [21]. In the field of temporal databases time interval-based query languages like IXSQL, TSQL2, or ATSQL have been defined [7] and within the analytical field, MDX [22] is a widely used language to query MDMs. Considering models dealing with time interval data in the context of analytics, Koncilia et al. [19] published the only work the authors are aware of that mentions a query language. Nevertheless, the query language is neither formally defined nor further introduced.

Summarized, it can be stated that recent research and requests from industry indicate that the handling of time interval data in an analytical context is an important task. Thus, a query language is required capable of covering the arising requirements. Koncilia et al. [19] and Meisen et al. [5, 6] introduced two different models useful for OLAP of time interval data. Different temporal aggregation operators, as well as standard aggregation operators, are also presented by Meisen [5]. Nevertheless, a definition of a query language useful for OLAP and an implementation of the processing are, to the best of our knowledge, not formally introduced.

3 The Tida Query Language

In this section, we introduce our time interval data analysis query language (TIDAQL). The language was designed for a specific purpose; to query time interval data from an analytical point of view. The language is based on aspects of the previously discussed TIDAMODEL. Nevertheless, the language should be applicable to any time interval database system which is capable of analyzing time interval data. Nevertheless, some adaptions might be necessary or some features might not be supported by any system.

3.1 Requirements

The requirements concerning the query language and its processing were specified during several workshops with over 70 international business analysts from different domains (i.e. aviation industry, logistics providers, service providers, as well as language and gesture research). We aligned the results of the workshop with an extended literature research. Table 1 summarizes selected results.

3.2 Data Control Language

The definition of the DCL is straight forward to the DCL known from other query languages e.g. SQL. As defined by requirement [DCL1], the language must encompass authorization features. Hence, the language contains commands like **ADD**, **DROP**, **MODIFY**, **GRANT**, **REVOKE**, **ASSIGN** and **REMOVE**. In our implementation, the execution of a DCL command always issues a direct commit, i.e. a roll back

Table 1 Summary of the requirements concerning the time interval analysis query language (selected results)

Requirement	Description
Data Control Language (DCL)	
[DCL1]: authorization aspects	It is expected that the language encompasses authorization features, e.g. user deletion, role creation, granting and revoking permissions
[DCL2]: permissions grantable on global and model level	Permissions must be grantable on a model and a global level. It is expected that the user can have the permission to add data to one model but not to another. For simplicity, it should be possible to grant or revoke several permissions at once
Data Definition Language (DDL)	
[DDL1]: loading and unloading	The language has to offer a construct to load new and unload models. The newly loaded model has to be available without any restart of the system. An unloaded model has to be unavailable after the query is processed. However, queries currently in process must still be executed
[DDL2]: non-onto, non-covering, non-strict hierarchies	Each descriptor dimension must support hierarchies which might be non-onto, non-covering, and / or non-strict [23]
[DDL3]: raster levels	A raster level is a level of the time dimension. For example: the *5-minute raster*-level defines members like [00:00, 00:05) …[23:55, 00:00). Several raster levels can form a hierarchy (e.g. 5-min → 30-min → 60-min → half-day → day)
Data Manipulation Language (DML)	
[DML1]: raw data records	The language must provide a construct to select the raw time interval data records
[DML2]: time series by time-windows	The language must support the specification of a time-window for which time-series of different measures can be retrieved
[DML3]: temporal operators	It must be possible to use temporal operators for filtering as e.g. defined by Allen [4]. Depending on the type of selection (i.e. raw records or time-series) the available temporal operators may differ
[DML4]: The two-step aggregation technique	Meisen et al. [5] present a two-step aggregation technique which has to be supported by the language. Both aggregation operators (see Fig. 5) must be specified by a query selecting time-series, no predefined measure should be necessary

(continued)

Table 1 (continued)

Requirement	Description
[DML5]: complete time series	A time-series is selected by specifying a time-window (e.g. [01.01.2015, 02.01.2015) and a level (e.g. minutes). The resulting time-series must contain a value for each member of the selected level, even if no time interval covers the specified member. The value might be N/A or null to indicate missing information
[DML6]: insert, update and delete	The language must offer constructs to insert, update and delete time interval data records
[DML7]: open, half-open, or closed intervals	The system should be capable of interpreting intervals defined as open, e.g. (0, 5), closed, e.g. [0, 5], or half-opened, e.g. (0, 5]
[DML8]: meta-information	It is desired that the language supports a construct to receive meta-information from the system, e.g. actual version, available users, or loaded models
[DML9]: bulk load	It is desired, that the language provides a construct to enable a type of bulk load, i.e. increased insert performance

is not supported. Figure 6 shows the syntax diagram of the commands. Because of simplicity, a value is not further specified and might be a permission, a username, a password, or a role.

To fulfill the [DCL2] requirement, we define a permission that consists of a scope-prefix and the permission itself. We define two permission-scopes GLOBAL and MODEL. Thus, a permission of the GLOBAL scope is defined by

```
GLOBAL.<permission>
```

(e.g. GLOBAL.manageUser). Instead, a permission of the MODEL scope is defined by

```
MODEL.<model>.<permission>
```

(e.g. MODEL.myModel.query).

For query processing, we use the Apache Shiro authentication framework (http://shiro.apache.org/). Shiro offers annotation driven access control. Thus, the permission to e.g. execute a DML query is performed by annotating the processing query method.

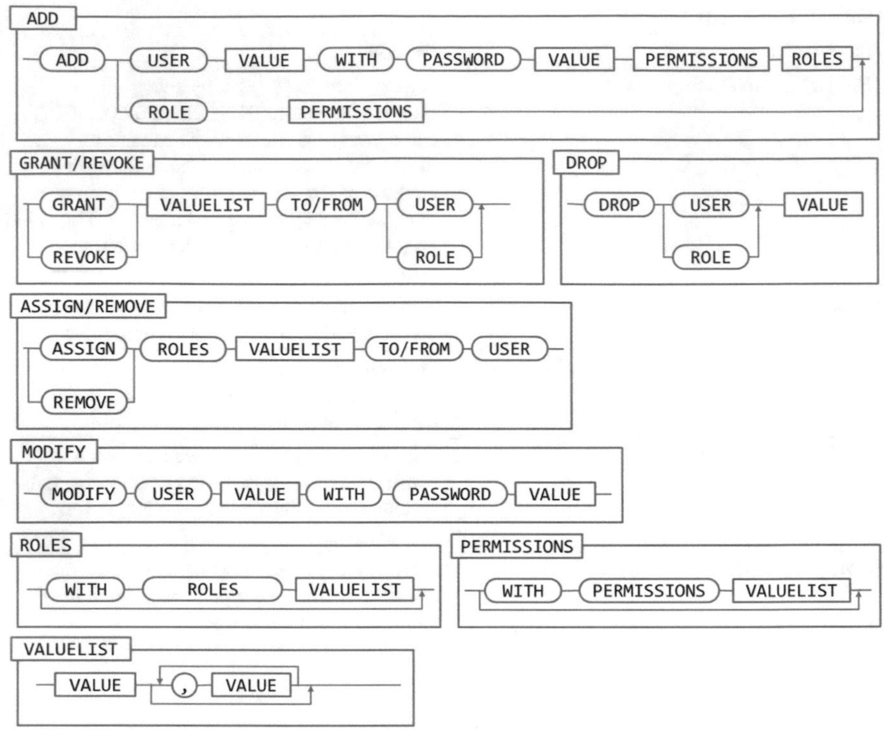

Fig. 6 Commands of the DCL query language

3.3 Data Dafinition Language

The DDL is used to define, add, or remove the models known by the system. [DDL1] requires a command within the DDL which enables the user to load or unload a model. The syntax diagram of the **LOAD** and **UNLOAD** command is shown in Fig. 7. A model can be loaded by using a model identifier already known to the system (e.g. if the model was unloaded), or by specifying a location from which the system can retrieve a model definition to be loaded. Additionally, properties can be defined (e.g. the *autoload* property can be set, to automatically load a model when the system is started). In the following subsection, we present an XML used to define a TIDAMODEL.

3.3.1 The XML TIDAMODEL Definition

As mentioned in Section 2, the TIDAMODEL is defined by a 5-tuple (P, Σ, τ, M, δ). The time interval database P contains the raw record inserted using the API or the **INSERT** command introduced later in Section 3.4.1. From a modelling perspective

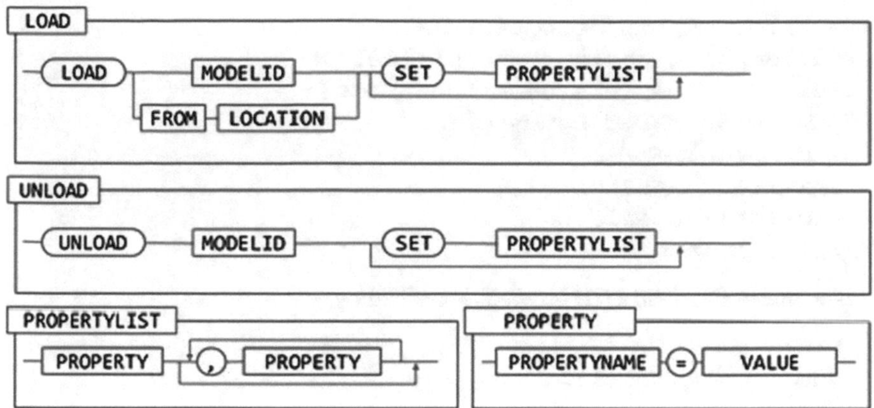

Fig. 7 Commands of the DDL query language

it is important for the system to retrieve the descriptive and temporal values from the raw record. Thus, it is essential to define the descriptors Σ and the time axis τ within the XML definition. Below, an excerpt of an XML file defining the descriptors of our sample database shown in Fig. 1 is presented:

```
<model id="myModel">
 <descriptors>
  <string id="LOC" name="location" />
  <string id="TYPE" name="type" />
  <int id="RES" null="true" />
 </descriptors>
</model>
```

The excerpts shows that a descriptor is defined by a tag specifying the type (i.e. the descriptor implementation to be used), an id-attribute, and an optional name-attribute. Additionally, it is possible to define if the descriptor allows *null* values (default) or not. To support more complex data structures (and one's own mapping functions), it is possible to specify one's own descriptor-implementations:

```
<descriptors>
 <ownImpl:list id="D4" />
</descriptors>
```

Our implementation scans the class-path automatically, looking for descriptor implementations. An added implementation must provide an XSLT file, placed into the same package and named as the concrete implementation of the descriptor-class. The XSLT file is used to create the instance of the own implementation using a Spring Bean configuration (http://spring.io/).

```
<!- File: my/own/desc/List.xslt ->
<xsl:template match="ownImpl:list">
 <xsl:call-template name="beanDesc">
  <xsl:with-param name="class">
    my.own.desc.List
  </xsl:with-param>
 </xsl:call-template>
</xsl:template>
```

The time axis of the T*IDA*M*ODEL* is defined by:

```
<model id="myModel">
 <time>
  <timeline start="20.01.1981"
            end="20.01.2061"
            granularity="MINUTE" />
 </time>
</model>
```

The time axis may also be defined using integers, i.e. [0, 1000]. Our implementation includes two default mappers applicable to map different types of temporal raw record value to a defined time axis. Nevertheless, sometimes it is necessary to use different time-mappers (e.g. if the raw data contains proprietary temporal values) which can be achieved using the same mechanism as described previously for descriptors.

Due to the explicit time semantics, the measures M defined within the TIDAMODEL are different than the ones typically known from an OLAP definition. The model defines three categories for measures, i.e. *implicit time measures*, *descriptor bound measures*, and *complex measures*. The categories determine when which data is provided during the calculation process of the measures. Our implementation offers several aggregation operators useful to specify a measure, i.e. *count*, *average*, *min*, *max*, *sum*, *mean*, *median*, or *mode*. In addition, we implemented two temporal aggregation operators *started count* and *finished count*, as suggested by Meisen et al. [5]. We introduce the definition and usage of measures in Section 3.4.2.

The TIDAMODEL also defines the set of dimensions δ. The definition differs between descriptor dimensions and a time dimension, whereby every dimension consists of hierarchies, levels, and members. It should be mentioned that, from a modelling point of view, each descriptor dimension fulfills the requirements formalized in [DDL2] and that the time dimension supports raster-levels as requested in [DDL3]. The definition of a dimension for a specific descriptor or the time dimension can be placed within the XML definition of a model using:

```
<model id="myModel">
 <dimensions>
  <dimension id="DIMLOC" descId="LOC">
   <hierarchy id="LOC">
    <level id="HOTEL">
     <member id="DREAM" rollUp="*" />
     <member id="STAR" rollUp="*" />
     <member id="ADV" reg="TENT"
         rollUp-"*" />
    </level>
    <level id="ROOMS">
     <member id="POSF" reg="POS F"
         rollUp="DREAM" />
     <member id="POSG" reg="POS G"
         rollUp="DREAM" />
    </level>
    <level id="STARROOMS">
     <member id="POSA" reg="POS A"
         rollUp="STAR" />
    </level>
   </hierarchy>
  </dimension>
 </dimensions>
</model>
```

Figure 8 illustrates the descriptor dimension defined by the previously shown XML excerpt. The circled nodes are leaves which are associated with de-scriptor values known by the model (using regular expressions). Additionally, it is possible to add dimensions for analytical processes to an already defined model, i.e. to use it only for a specific session or query. The used mechanism to achieve that is similar to the loading of a model and will not further be introduced.

The definition of a time dimension is straight forward to the one of a descriptor dimension. Nevertheless, we added some features in order to ease the definition. Thus, it is possible to define a hierarchy by using pre-defined levels (e.g. templates

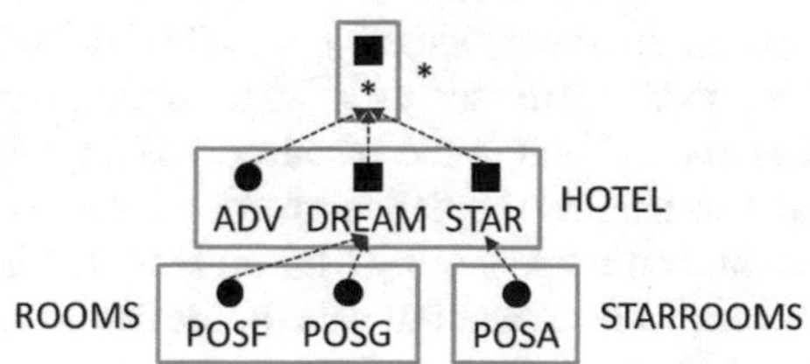

Fig. 8 Illustration of the dimension created with our web-based dimension-modeler as defined by the XML

like 5 min raster, day, or year) and by defining the level to roll up to, regarding the hierarchy. The following XML excerpt exemplifies the definition:

```
<model id="myModel">
 <dimensions>
  <timedimension id="DIMTIME">
   <hierarchy id="TIME5TOYEAR">
    <level id="YEAR" template="YEAR"
       rollUp="*" />
    <level id="DAY" template="DAY"
       rollUp="YEAR" />
    <level id="60R" template="60RASTER"
        rollUp="DAY" />
    <level id="5R" template="5RASTER"
        rollUp="60R" />
    <level id="LG" template="LOWGRAN"
        rollUp="5R" />
   </hierarchy>
  </timedimension>
 </dimensions>
</model>
```

A defined model is published to the server using the **LOAD** command. The following subsection introduces the command, focusing on the loading of a model from a specified location.

3.3.2 Processing the Load Command

The loading of a model can be triggered from different applications, drivers, or platforms. Thus, it is necessary to support different loaders to resolve a specified location. In the following, some examples illustrate the issue. When firing a **LOAD** query from a web-application, it is necessary that the model defi nition was uploaded to the server, prior to executing the query. While running on an application server, it might be required to load the model from a database instead of loading it from the file-system. Thus, we added a resource-loader which can be specified for each context of a query. Within a servlet, the loader resolves the specified location against the upload-directory, whereby our JDBC driver implementation is capable of sending a client's file to the server using the data stream of the active connection. After retrieving and validating the resource, the implementation uses a model-handler to bind and instantiate the defined model. As already mentioned, the bitmap-based implementation presented by Meisen et al. [5] is used. The implementation instantiates several indexes and bitmaps for the defined model. After the instantiation, the model is marked to be up and running by the model-handler and accepts DML queries.

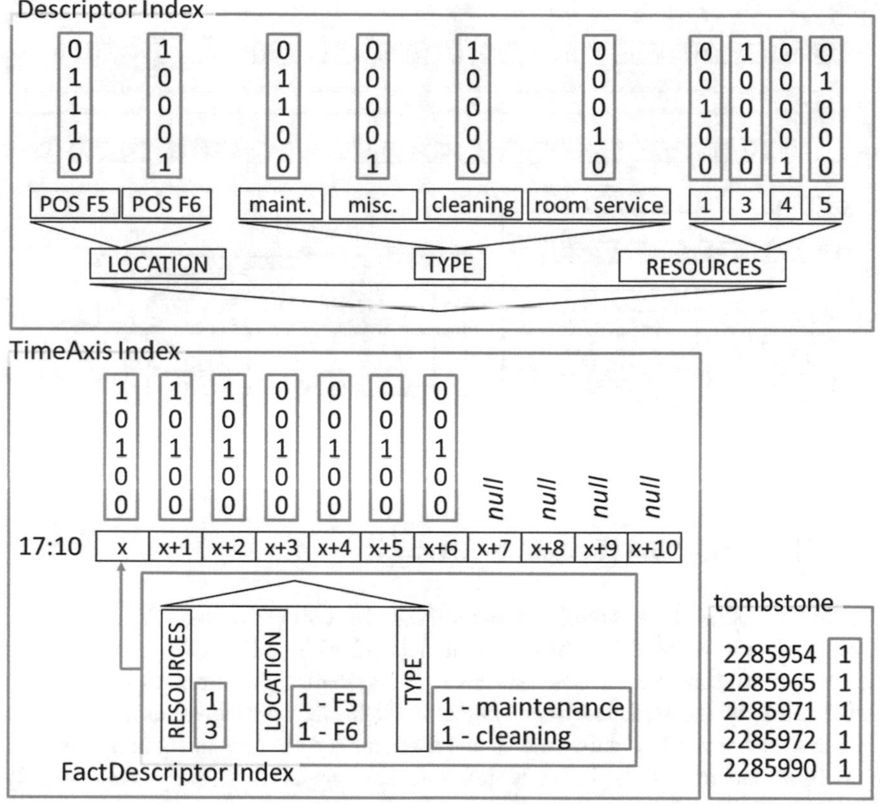

Fig. 9 Example of a loaded model [5] filled with the data shown in Fig. 1

Figure 9 exemplifies the initialized bitmap-based indexes filled with the data from the database of Fig. 1.

3.4 Data Manipulation Language

Considering the requirements, it can be stated that the DML must contain commands to **INSERT**, **UPDATE**, and **DELETE** records. In addition, it is necessary to provide **SELECT** commands to retrieve the time interval data records, as well as results retrieved from aggregation (i.e. time-series). Furthermore, a **GET** command to retrieve meta-information of the system is needed.

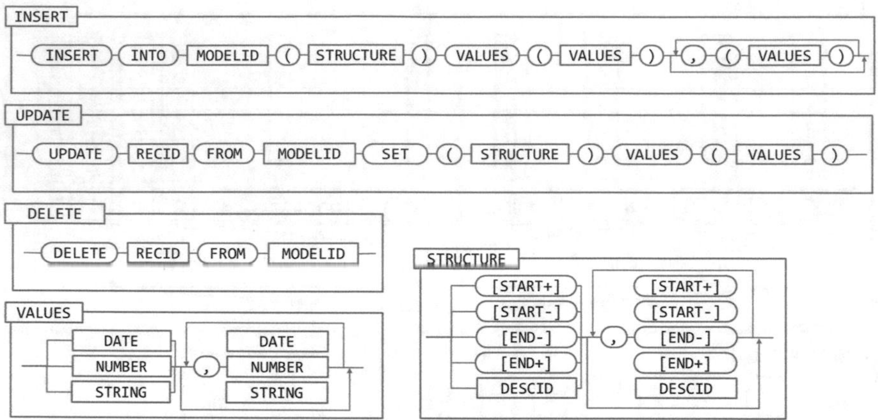

Fig. 10 Syntax diagrams of the commands **INSERT**, **UPDATE** and **DELETE**

3.4.1 INSERT, UPDATE, and DELETE

Figure 10 illustrated the three commands **INSERT**, **UPDATE**, and **DELETE** using syntax diagrams which fulfill the requirement [DML6]. The **INSERT** command adds one or several time interval data records to the system. First, it parses the structure of the data to be inserted. The query-parser validates the correctness of the structure, i.e. the structure must contain exactly one field marked as *start* and exactly one field marked as *end* even though the syntax diagram suggest differently. Additionally, the parser verifies if a descriptor (referred by its id) really exists within the model. Finally, it reads the values and invokes the processor by passing the structure, as well as the values. The processor iterates over the defined values, validates those against the defined structure, uses the mapping functions of the descriptors to receive the descriptor values, and calls the mapping function of the time-axis. The result is a so-called *processed record* which is used to update the indexes. The persistence layer of the implementation ensures that the raw record and the indexes get persisted. Finally, the tombstone bitmap is updated which ensures that the data is available within the system.

A deletion is performed by setting the tombstone bitmap for the specified id to 0. This indicates that the data of the record is not valid and thus the data will not be considered by any query processors anymore. The internally scheduled clean-up process removes the deleted records and releases the space.

An update is performed by deleting the record with the specified identifier and inserting the record as described above.

To support bulk load, as desired by [DML9], an additional statement is introduced. The statement **SET BULK TRUE** is used to enable the bulk load, whereby **SET BULK FALSE** stops the bulk loading process. When enabling the bulk load, the system waits until all currently running **INSERT**, **UPDATE**, or **DELETE** queries of other sessions are performed. New queries of that type are rejected across all sessions during the waiting and processing phase. When all queries are handled, the system responds to the bulk-enabling query and expects an insert-like statement, whereby the system directly starts to parse the incoming data stream. As soon as the structure is known, all incoming values are inserted. The indexes are only updated in memory. If and only if the memory capacity reaches a specified threshold, the persistence-layer is triggered. In this case, the current data in memory is flushed and persisted using the configured persistence-layer (e.g. using the file-system, a relational database, or any other NoSQL database). The memory is also flushed and persisted whenever a bulk load is finished.

3.4.2 SELECT Raw Records and Time-Series

The **SELECT** command is addressed by the requirements [DML1], [DML2], [DML3], [DML4], [DML5], and [DML7]. Figure 11 illustrates the select statements to select records and time-series. Because of space limitations, we removed more detailed syntax diagrams for the **LOGICAL** and **GROUP EXPRESSION**. The non-terminal **MEASURES** is specified later in this subsection when introducing the **SELECT TIMESERIES** in detail.

As illustrated, the intervals can be defined as open, half-open or closed (cf. [DML7]). The processing of the intervals is possible, thanks to the discrete time-axis used by the model. Using a discrete time-axis with a specific granularity makes it easy to determine the previous or following granule. Thus, every half-open or open interval can be transformed into a closed interval using the previous or following granule. Hence, the result of the parsing always contains a closed interval which is used during further query processing.

As illustrated in Fig. 11, the **SELECT RECORDS** statement allows to retrieve records satisfying a logical expression based on descriptor values (e.g. **LOC = "POS F5" OR (TYPE = "cleaning" AND DIMLOC.LOC.HOTEL = "DREAM")**) and/or fulfilling a temporal relation (cf. [DML3]). Our query language supports ten different temporal relations following Allen [4]: **EQUALTO**, **BEFORE**, **AFTER**, **MEETING**, **DURING**, **CONTAINING**, **STARTINGWITH**, **FINISHINGWITH**,

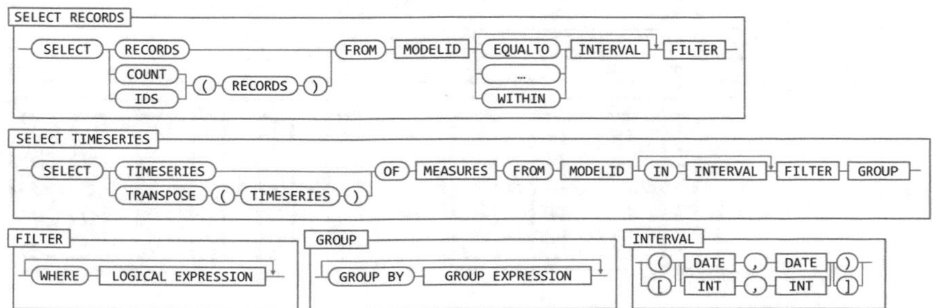

Fig. 11 Syntax diagrams of the **SELECT RECORDS** and **SELECT TIMESERIES** commands

OVERLAPPING, and **WITHIN**. The interested reader may notice that Allen introduced thirteen temporal relationships. We removed some inverse relationships (i.e. inverse of meet, overlaps, starts, and finishes). When using a temporal relation-ship within a query, the user is capable of defining one of the intervals used for comparison. Thus, the removed inverse relationships are not needed, instead the user just modifies the self-defined interval. In addition, we added the **WITHIN** relationship which is a combination of several relationships and allows an easy selection of all records within the user-defined interval (i.e. at least one time-granule is contained within the user-defined interval).

When processing a **SELECT RECORD** query, the processor initially evaluates the filter expression and retrieves a single bitmap specifying all records fulfilling the filter's logic [5]. In a second phase, the implementation determines a bitmap of records satisfying the specified temporal relationship. The two bitmaps are combined using the and-operator to retrieve the resulting records. Depending on the requested information (i.e. count, identifiers, or raw records (cf. [DML1])), the implementation creates the response using bitmap-based operations (i.e. count and identifiers) or retrieving the raw records from the persistence layer. Figure 12 depicts the evaluation of selected temporal relationships using bitmaps and the database shown in Fig. 1.

The **SELECT TIMESERIES** statement specifies a logical expression equal to the one exemplified in the **SELECT RECORDS** statement. In addition, the statement specifies a **GROUP EXPRESSION** which defines the groups to create the time-series for (e.g. **GROUP BY DIMLOC.LOC.ROOMS**). Furthermore, the measures

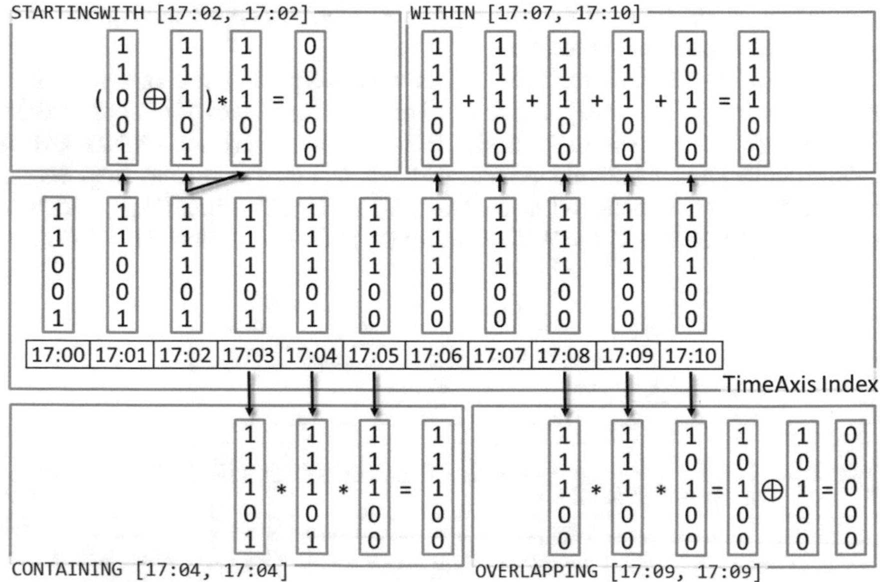

Fig. 12 Examples of the processing of temporal relationships using bitmaps (and the sample database of Fig. 1)

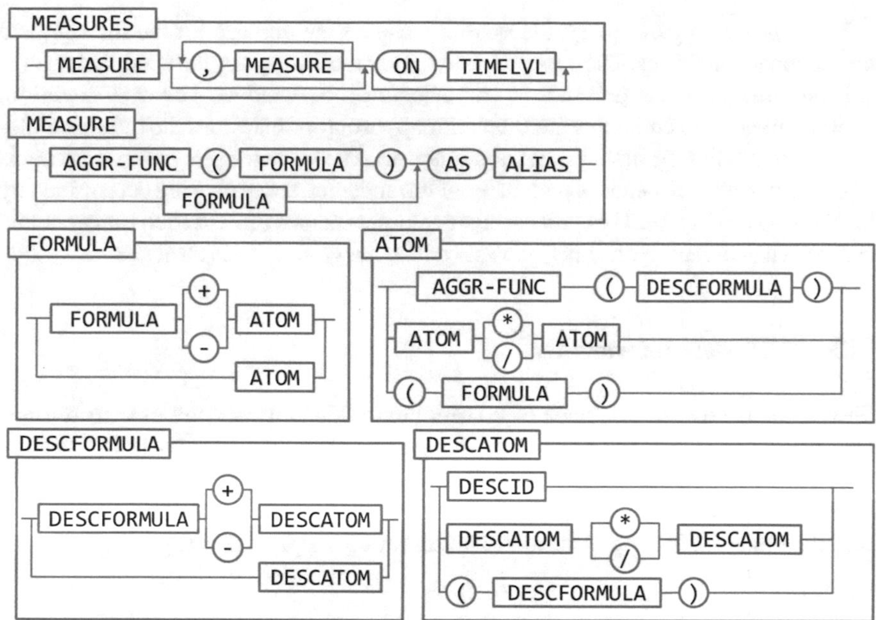

Fig. 13 Syntax diagrams of the MEASURES definition

to be calculated for the time-series and the time-window (cf. [DML2]) are specified. It is also possible to specify several comma-separated measures. Figure 13 shows the syntax used to specify measures (cf. **MEASURES** in Figure 11).

A simple (considering the measures) example of a **SELECT TIMESERIES** query is as follows:

```
SELECT TRANSPOSE(TIMESERIES)
  OF MAX(SUM(RESOURCES)) AS "needed Res"
  ON DIMTIME.TIME5TOYEAR.5RASTER
  FROM myModel
  IN [01.01.2015, 02.01.2015)
  WHERE DIMLOC.LOC.HOTEL="DREAM"
  GROUP BY TYPE
```

As required by [DML4], a measure can be defined using the two-step aggregation technique. The first aggregation (in the example SUM) is specified for a specific descriptor and the second optional aggregation function (in the example MAX) aggregates the values across the stated level of the time-dimension.

When processing the query, the system retrieves the bitmaps for the filtering and the grouping conditions. The system iterates over the bitmaps of the specified groups and the bitmaps of the granules of the selected time-window. For each iteration, the implementation combines the filter-bitmap, group-bitmap, and the time-granule-bitmap and applies the first aggregation function. The second aggregation function is applied whenever all values of a member of the specified time-level are determined by the first step (cf. Fig. 6). This processing technique ensures that for each time-granule a value is calculated, even if no interval covers the granule (cf. [DML5]).

3.4.3 GET Meta-Information

[DML8] demands the existence of a command which enables the user to retrieve meta-information, like the version of the system. This requirement is fulfilled by adding a **GET** command to the query language. A statement like **GET VERSION**, **GET USERS**, or **GET MODELS** enables the user to retrieve information provided from the system. Filtering is currently not required and thus, not supported.

4 Implementation Issues

This section introduces selected implementation aspects of the language and its query processing. First, we introduce processing implementations for the most frequently used query-type **SELECT TIMESERIES** and show performance results for the different algorithms. In addition, we present considerations of analysts using the language to analyze time interval data and address possible enhancements.

4.1 SELECT TIMESERIES Processing

In Section 3, we outlined the query processing based on the TIDAMODEL and its bitmap-based implementation (cf. Sections 3.3.2 and 3.4.2). For a detailed description of the bitmap-based implementation we refer to Meisen et al. [5]. In this section, we introduce three additional algorithms which are capable to process the most frequently used **SELECT TIMESERIES** queries, introduced in Section 3.4.2.

Prior to explaining the algorithms, it should be stated, that we did not implement any algorithm based on AGGREGATIONTREEs [20], MERGESORT, or other related aggregation algorithms defined within the research field of temporal databases. Such algorithms are optimized to handle single aggregation operators (e.g. count, sum, min, or max). Thus, the implementation would not be a generic solution usable for any query. Nevertheless, such algorithms might be useful to increase query performance for specific, often used measures. It might be reasonable to add a language feature, which allows to define a special handling (e.g. using an AGGREGATIONTREE) for a specific measure.

Next, we introduce our naive implementation. All three presented algorithm do not support queries using group by, multiple measures, nor multi-threading scenarios. To support these features, commonly used techniques (e.g. iterations and locks) could be used.

```
01 TimeSeries naive(Query q, Set r) {
02 TimeSeries ts = new TimeSeries(q);
03 // filter time def. by IN [a, b]
04 r = filter(r, q.time());
05 // filter records def. by WHERE
06 r = filter(r, q.where());
07 // it. ranges def. by IN and ON
08 for (TimeRange i : q.time()) {
09 // filter records for the range
10 r' = filter(r, i);
11 // det. measures def. by OF
12 ts.set(i, calc(i, r', q.meas());
13 }
14 return ts;
15 }
```

The algorithm filters the records of the database, which fulfill the defined criteria of the **IN** (row 04) and **WHERE** clause (row 06). Next, it calculates the measure for each defined range (row 10). The calculation of each measure depends mainly on its type (i.e. measure of lowest granularity (e.g. query #1 in Table 2), measure of a level (e.g. query #2), or two-step measure (e.g. query #3)). Because of space limitations, we state the complexity of the **calc**-method instead of presenting it. The complexity is **O(k·n)**, with k being the number of granules covered by the **TimeRange** and n being the number of records.

The other algorithms we implemented are based on INTERVALTREEs (INTTREE) as introduced by Kriegel [24]. The first one (A) – of the two INTTREE – based implementations – uses the tree to retrieve the relevant records considering

Table 2 The shortened queries used for testing

#	Query
1	**OF** COUNT(TASKTYPE **IN** [01.JAN, 01.FEB)
	WHERE WA.LOC.TYPE='Gate'
2	**OF** SUM(TASKTYPE) **ON** TIME.DEF.DAY
	IN [01.JAN, 01.FEB) **WHERE** WORKAREA='SEN13'
3	**OF** MAX(COUNT(WORKAREA)) **ON** TIME.DEF.DAY
	IN [01.JAN, 01.FEB) **WHERE** TASKTYPE='short'

the **IN**-clause (row 05 of the naive algorithm). Further, the algorithm proceeds as the naive algorithm.

The second implementation (B) differs from the first one, by created a new INTTREE for every query.

```
01 TimeSeries iTreeB(Query q, Set r) {
02 TimeSeries ts = new TimeSeries(q);
03 // filter records def. by WHERE
04 IntervalTree iTree =
05 createAndFilter(r, q.in(),
06 q.where());
07 // it. ranges def. by IN and ON
08 for (TimeRange i : q.time()) {
09 // use iTree to filter by i
10 r' = filter(iTree, i);
11 // det. measures def. by OF
12 ts.set(i, calc(i, r', q.meas());
13 }
14 return ts;
15 }
```

As shown, the algorithm filters the records according to the **IN**- and **WHERE**-clause and creates an INTTREE for the filtered records (row 04). When iterating over the defined ranges, the created iTree is used to retrieve the relevant records for each range (row 08).

4.2 Performance

We ran several tests on an Intel Core i7-4810MQ with a CPU clock rate of 2.80 GHz, 32 GB of main memory, an SSD, and running 64-bit Windows 8.1 Pro. As Java implementation, we used a 64-bit JRE 1.6.45, with XMX 4,096 MB and XMS 512 MB. We tested the parser (implemented using ANTLR v4) and processing considering correctness. In addition, we measured the runtime performance of the processor for the three introduced algorithms (cf. Section 4.2), whereby the data and structures of all algorithms were held in memory to obtain CPU time comparability. We used a real-world data set containing 1,122,097 records collected over one year. The records have an average interval length of 48 minutes and three descriptive values: person (cardinality: 713), task-type (cardinality: 4), and work area (cardinality: 31). The used time-granule was minutes (i.e. time cardinality: 525,600). We tested the performance using the **SELECT TIMESERIES** queries shown in Table 2. Each query specifies a different type of query (i.e. different measure, usage of groups, or filters) and was fired 100 times against differently sized sub-sets of the real-world data set (i.e. 10, 100, 1,000, 10,000, 100,000, and 1,000,000 records).

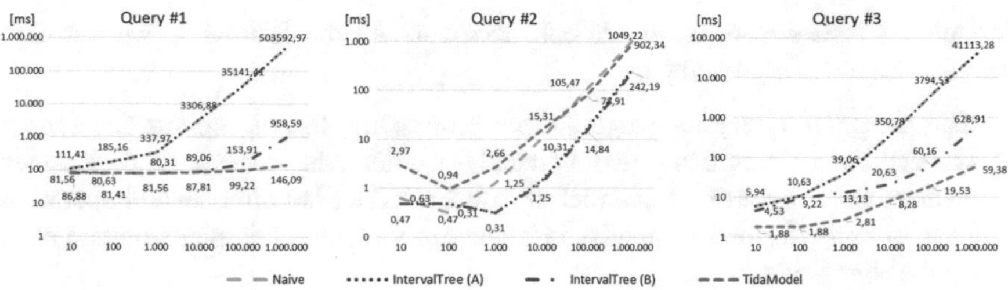

Fig. 14 The measured average CPU-time performance (out of 100 runs per query)

Table 3 Statistics of the test results

in DB	Number of records			Selectivity		
	Selected by query			Selected / in DB		
	#1	#2	#3	#1	#2	#3
10^1	1	0	0	0.1000	0.0000	0.0000
10^2	5	0	7	0.0500	0.0000	0.0700
10^3	12	2	46	0.0120	0.0020	0.0460
10^4	147	9	480	0.0147	0.0009	0.0480
10^5	1.489	121	5.148	0.0149	0.0012	0.0515
10^6	15.378	1.261	51.584	0.0154	0.0013	0.0516

The results of the runtime performance tests are shown in Fig. 14. As illustrated, the bitmap-based implementation performs better than the naive and INTTREE algorithms when processing query #1 and #3. Regarding query #2 the INTTREE-based implementations perform best. As stated in Table 3, the most important criterion to determine the performance is the selectivity. Regarding a low selectivity the INTTREE-based algorithm (B) performs best.

Nevertheless, considering persistency and reading of records from disc the algorithm might perform worse. We would also like to state briefly, that other factors (e.g. kind of aggregation operators used) influence the performance of the bitmap algorithm, so that it outperforms the INTTREE-based implementation, even if a low selectivity is given.

4.3 *Considerations*

The query language and processing introduced in this paper, is currently used within different projects by analysts and non-experts of different domains to analyze time-interval data. In the majority of cases, the introduced language and the processing is capable of satisfying the user's needs. Nevertheless, there are limitations, issues, and preferable enhancements. In the following, we introduce selected requests/improvements:

1. The presented query language and its processing do not support any type of transactions. A record inserted, updated, or deleted is processed by the system as an atomic operation. Nevertheless, roll-backs needed after several operations have to be performed manually. This generally increases implementation effort on the client-side.
2. The presented XML definition of dimensions (cf. Section 3.3.1) uses regular expressions to associate a member of a level to a descriptor value. Regular expressions are sometimes difficult to be formalized (especially for number ranges). An alternative, more user-friendly expression language is desired.
3. The **UPDATE** and **DELETE** commands (cf. Section 3.4.1) need the user to specify a record identifier. The identifier can be retrieved from the resultset of an **INSERT**-statement or using the **SELECT RECORDS** command. Nevertheless, users requested to update or delete records by specifying criteria based on the records' descriptive values.
4. When a model is modified, it has to be loaded to the system as new, the data of the old model has to be inserted and the old model has to be deleted. Users desire a language extension, allowing to update models. Nevertheless, the implications of such a model update could be enormous.

5 Conclusion

In this paper, we presented a query language useful to analyze time interval data in an on-line analytical manner. The language covers the requirements formalized by several business analyst from different domains, dealing with time interval data on a daily basis. We also introduced four different implementations useful to process the most frequently used type of query (i.e. **SELECT TIMESERIES**).

An important task for future studies is to confirm, or define new models and present novel implementations solving the problem of analyzing time interval data. In addition, future work should focus on distributed and incremental query processing (e.g. when rolling-up a level). The mentioned considerations (cf. Section 4.3) of our introduced language and its implementation should be investigated. Another interesting area considering time-interval data is on-line analytical mining (OLAM). Future work should study the possibilities of analyzing aggregated time series to discover knowledge about the underlying intervals. Finally, an enhancement of the processing of the two-step aggregation technique should be considered. Depending on the selected aggregations an optimized processing strategy might be reasonable.

References

1. E. Codd, S. Codd, C. Salley, *Providing OLAP (On-Line Analytical Processing) to User-Analysts: An IT Mandate*. 1993. E. F. Codd and Associates (sponsored by Arbor Software Corp.)
2. J. Mazón, J. Lichtenbörger, T. J., Solving summarizability problems in fact-dimension relationships for multidimensional models. In: *11th Int. Workshop on Data Warehousing and OLAP (DOLAP '08). Napa Valley, California, USA, 26.–30. October*. 2008, pp. 57–64
3. R. Kimball, M. Ross, *The data warehouse toolkit: The definitive guide to dimensional modeling*, 3rd edn. Wiley Computer Publishing, 2013
4. J. Allen, Maintaining knowledge about temporal intervals. Communication ACM **26** (11), 1983, pp. 832–843
5. P. Meisen, D. Keng, T. Meisen, M. Recchioni, S. Jeschke, Bitmap-based on-line analytical processing of time interval data. In: *12th Int. Conf. on Information Technology. Las Vegas, Nevada, USA, 13.–15. April*. 2015
6. P. Meisen, T. Meisen, M. Recchioni, D. Schilberg, S. Jeschke, Modeling and processing of time interval data for data-driven decision support. In: *IEEE Int. Conf. on Systems, Man, and Cybernetics, San Diego, California, USA, 04.–08. October*. 2014
7. M. Böhlen, B. R., J. C. S., Point-versus interval-based temporal data models. In: *14th Int. Conf. on Data Engineering, Orlando, Florida, USA, 23.-27. Feburary*. 1998, pp. 192–200
8. P. Papapetrou, G. Kollios, S. S., G. D., Mining frequent arrangements of temporal intervals, knowledge and information systems **21** (2), 2009, pp. 133–171
9. F. Mörchen, Temporal pattern mining in symbolic time point and time interval data. In: *IEEE Symp. on Computational Intelligence and Data Mining (CIDM 2009), Nashville, Tennessee, USA, 30. March–2. April*. 2009
10. F. Höppner, F. Klawonn, Finding informative rules in interval sequences. In: *IDA2001. LNCS*, vol. 2189, ed. by F. Hoffmann, N. Adams, D. Fisher, G. Guimarães, D. Hand, Springer, Heidelberg, 2001, pp. 123–132
11. A. Kotsifakos, P. Papapetrou, V. Athitsos, Ibsm: Interval-based sequence matching, 13th siam int. conf. on data mining (sdm13), austin, texas, usa, 02.–04. may. 2013
12. Y. Chen, M. Chiang, M. Ko, Discovering time-interval sequential patterns in sequence databases. Expert Systems with Applications **25** (3), 2003, pp. 343–354
13. R. Agrawal, R. Srikant, Mining sequential patterns. In: *Int. Conf. Data Engineering, Taipei, Taiwan*. 1995, pp. 3–14
14. P. Papapetrou, G. Kollios, S. S., D. Gunopulos, Discovering frequent arrangements of temporal intervals. In: *5th IEEE Int. Conf. on Data Mining (ICDM'05), IEEE Press*. 2005, pp. 354–361
15. F. Mörchen, A better tool than allen's relations for expressing temporal knowledge in interval data. In: *12th ACM SIGKDD Int. Conf. on Knowledge Discovery and Data Mining, Philadelphia, Pennsylvania, USA*. 2006
16. C. Chui, B. Kao, E. Lo, D. Cheung, S-olap: An olap system for analyzing sequence data. In: *ACM SIGMOD International Conference on Man-agement of Data, Indianapolis, Indiana, USA*. 2010
17. M. Liu, E. Rundensteiner, K. Greenfield, C. Gupta, S. Wang, I. Ari, A. Mehta, E-cube: multi-dimensional event sequence analysis using hierarchical pattern query sharing. In: *ACM SIGMOD International Conference on Management of Data, Athens, Greece*. 2011
18. B. Bebel, M. Morzy, T. Morzy, Z. Królikowski, R. Wrembel, Olap-like analysis of time point-based sequential data. In: *Advances in Conceptual Modeling*, ed. by S. Castano, P. Vassiliadis, L. Lakshmanan, M. Lee, 2012. 978-3-642-33998-1
19. C. Koncilia, T. Morzy, R. Wrembel, E. J., *Interval OLAP: Analyzing Interval Data, Data Warehousing and Knowledge Discovery (DaWaK 2014)*, vol. 8646. Springer Int., 2014
20. N. Kline, R. Snodgrass, Computing temporal aggregates. In: *11th Int. Conf. on Data Engineering (ICDE 1995), Taipei, China, 06.–10. March*. 1995, pp. 222–231
21. D. Rafiei, A. Mendelzon, Querying time series data based on similarity. IEEE Transactions on Knowledge and Data Engineering **12** (5), 2000

22. G. Spofford, S. Harinath, C. Webb, D.H. Huang, F. Civardi, *MDX-Solutions: With Microsoft SQL Server Analysis Services 2005 and Hyperion Essbase*. John Wiley & Sons, 2006
23. T. Pedersen, Aspects of data modeling and query processing for complex multidimensional data. Ph.D. thesis, Aalborg Universitetsforlag, Aalborg, Department of Computer Science, Aalborg Univ., 2000. No. 4
24. H. Kriegel, M. Pötke, T. Seidl, Object-relational indexing for general interval relationships. In: *7th Int. Symposium on Spatial and Temporal Databases (SSTD 2001), Los Angeles, California, 12.–15. July*. 2001, pp. 522–542

Investigating Mixed-Reality Teaching and Learning Environments for Future Demands: The Trainers' Perspective

Lana Plumanns, Thorsten Sommer, Katharina Schuster, Anja Richert and Sabina Jeschke

Abstract The first three industrial revolutions were characterized by the invention of water and steam engine, centralized electric power infrastructure and mass production as well as digital computing and communications technology. The current developments caused by the fourth revolution, also known as "Industry 4.0", pose major challenges to almost every kind of work, workplace, and the employees. Due to the concepts of cyber-physical systems, Internet of Things and the increasing globalization, remote work is a fast-growing trend in the workplace, and educational strategies within virtual worlds become more important. Especially methods as teaching and learning within virtual worlds are expected to have an enormous impact on advanced education in the future. However, it is not trivial to transfer a reliable educational method from real to the virtual worlds. Therefore, it is important to adapt, check and change even small didactic elements to guarantee a sustainable learning success. As there is a lot of ongoing research about using virtual worlds for the training of hazardous situations, it has to be figured out which potential those environments bear for the everyday education of academic staff and which competencies and educational support trainers need to have respectively can give in those worlds. The used approach for this study was to investigate the trainers' didactic perspective on mixed-reality teaching and learning. A total of ten trainers from different areas in Germany took part in this study. Every participant pursued both roles: the teaching and the learning part in a virtual learning environment. In order to assess the learning success and important key factors the experiment yields data from the participants' behavior, their answers to a semi-structured interview and video analysis, recorded from the virtual world. Resulting data were analyzed by using different qualitative as well as quantitative methods. The findings of this explorative research suggest the potential for learning in virtual worlds and give inside into influencing variables. The online gaming experience and the age of participants can be shown to be related

L. Plumanns (✉) · T. Sommer · K. Schuster · A. Richert · S. Jeschke
IMA/ZLW & IfU, RWTH Aachen University, Dennewartstr. 27, 52068 Aachen, Germany
e-mail: lana.plumanns@ima-zlw-ifu.rwth-aachen.de

Originally published in "Proceedings of the 18th International Academic Conference", © 2015. Reprint by Springer International Publishing AG 2016, DOI 10.1007/978-3-319-46916-4_33

to participants' performance in the virtual world. It looks like the barriers for the affected trainers are low regarding utilization of virtual worlds. Together with the mentioned advantages and possible usages, the potential of these setups is shown.

Keywords Education · Mixed-Reality · Teaching · Virtual World

1 Introduction

Recent examinations of 702 today's occupations show how many million tasks and areas are affected by the ongoing digitalization [1]. While some occupations will be ceased, others will change, and new ones will occur. Responsible for these change are today's concepts like e.g. "Industry 4.0" [2–5] or Internet of Things (IoT). The ongoing globalization will not end and therefore, employees have to follow the trend. Occupations like e.g. teachers and trainers change through the trend to massive and remote teaching as e.g. massive open online courses (MOOCs) [6]. Powered by serious games [7] and gamification concepts, virtual worlds push into the teaching and training activities. Further, the produced data by the usage of these technologies is an enabler for learning analytics [8] and general analysis-driven methods.

Regardless of these possibilities, a teacher and trainer must be able to reflect those options. The potential usage of media and technology depends on the learning subject [9]. Thus, for some learning subjects virtual worlds are suitable. Today, virtual worlds are used to train uncommon scenarios e.g. major incidents [10] or can be used to teach invisible processes e.g. the basics of a calculator. Besides incidents and inaccessible or non-existing places, also the training of dangerous activities is a possibility for such virtual worlds [11]. Researchers identified effects and advantages of virtual worlds as method for trainings, e.g. in some cases an increase of team performance of about 50 % or the fact that in case of 62 %, the usage of a virtual worlds had the same effectiveness as traditional methods [10].

Further, remote work is constantly pushing forward. Working remotely whether from home, a coffee bar or another place is booming. Research suggests that more than half of today's office-based employees will regularly be working remotely within the next decade, thanks to technological advances in the workplace [12]. Advantages are among others more efficient agreements due to avoided travel time and a reduction of costs and the enhanced comfortability for the user [13].

Due to current technological capacities it is possible to control machines and even whole factories remotely, so that no instructor has to be in place [14]. This technical development obviously shows the future requirements for such employees: While in the past an engineer was responsible for handling a specific local machine, tomorrow's engineer can control multiple factories remotely across the world. Hence, the remote collaboration is an important part of future companies and engineers might be confronted with e.g. intercultural issues. Prospectively, the future engineers must be aware of all processes, which are running at a factory instead of controlling one single process step locally. Further, the engineer has to understand and know all kind of machines at a factory and must be empowered to know their limits in order to control the whole factory.

Therefore, with the advent of Industry 4.0, a large market arises in the field of virtual training and settings for collaboration and schooling. But not every approach that is technically feasible improves users' learning outcomes; hence the danger of designing expensive virtual learning environments without having a positive effect on the users' learning is obvious. Thus, in order to ensure sustainable learning outcomes in virtual learning environments, people who provide professional skills in the physical world have to be involved. Based on previous studies about students' perspective on virtual education [5], this study analyses the trainers' perspective of teaching and training using a virtual world for educational purposes with immersive hardware.

Before such technologies can be used in everyday training, teaching and learning with groups, further research regarding the transfer of common methods is necessary. Is it suitable for a trainer to moderate a group of students in a virtual world just like in the real world, even if the simulated area is huge? Is it possible to transfer well-known methods such as e.g. think-pair-share into a virtual reality setting? Currently, answers to questions like this are unknown and object of further research activities.

Following the high expectations regarding learning and working within virtual worlds, this research assessed trainers' behavior in and opinion about virtual learning environments. Hence, this experiment yields data from the participants' spatial behaviour and movements, their answers to questions regarding education and their experience within mixed-reality virtual learning environments to answer the question: Is today's society ready for remote training by using these technical possibilities? This study tries to give a first answer to this question by investigating the trainers' perspective inside the mixed-reality with virtual learning environments by a threefold purpose. First, an overview of the experimental setup is given; then the experimental study shows the challenges that tomorrow's trainers have to face and variables that might affect their performance. Finally, some technical and conceptual limitations are shown to guide further research in this field.

2 Setup and Virtual Environment

In order to get inside of technological details of this investigation, this section gives an elaborate description of the setup, the used environment, and the technical conditions. Hence, today's minimal technical requirements to provide an immersive virtual learning environment are shown. As the essential structure of this study, only one location was used in which two persons were participating for each pass. Thus, no headset was required for the verbal communication. A head mounted display (Oculus Rift DK2) served as immersive hardware to enter the virtual learning environment. Head-mounted displays had been used successfully in previous studies and the usage of these displays within virtual worlds is connected to strengthen sensations of immersion, flow, and spatial presence [15]. Due to users' attention allocation

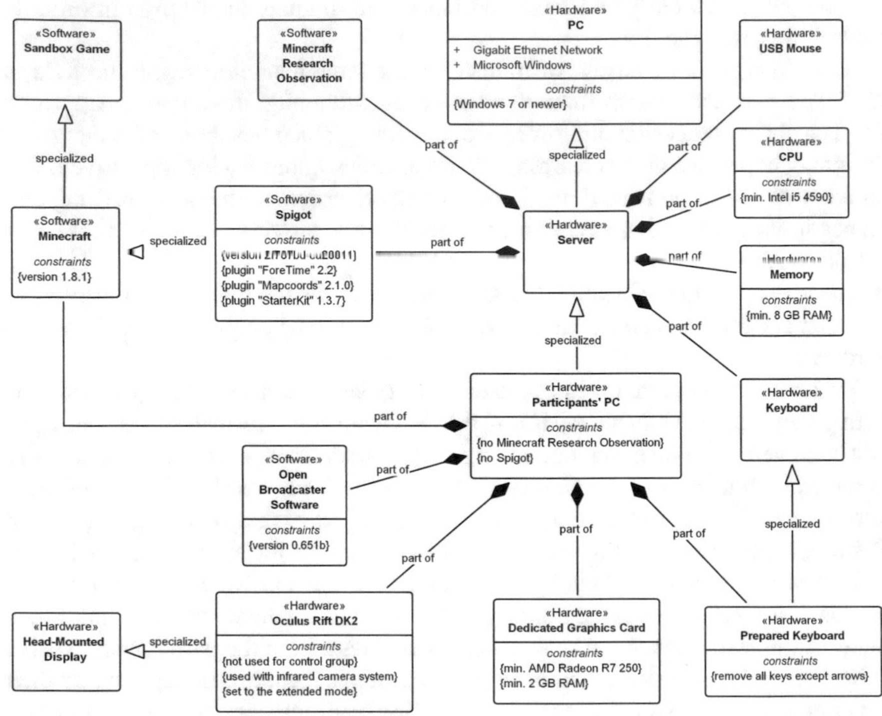

Fig. 1 The UML (Unified Modeling Language) model about the hard- and software requirements for this study (Source: Self-created model)

processes while wearing these displays it is possible to blend the physical with the virtual world and hence design virtual interaction and knowledge transfer as naturally as possible.

The hardware requirements for this investigation are almost standard, as the model in Fig. 1 shows. A few components, however, need further consideration: Since this study was investigating trainers perspective on virtual education with the aid of immersive hardware, specifically the Oculus Rift, the participant's PC needed a dedicated graphics card in order to work with the head-mounted display. Further, for the convenient control of the participant's avatar in the virtual learning environments with equipped head-mounted display, a prepared keyboard was used. Except the arrow keys, all other keys were dismounted to prevent participants from pressing wrong or undesired keys. This modification is necessary because the participants cannot see the physical world while the head-mounted display is equipped. Additionally, a standard-sized USB mouse was used to enable a 360-degree turn, while sitting on in front of the PC.

The head-mounted display was used with the infrared camera system and was set to the extended mode. Thus, it behaves like an additional display. However, this setup is not the optimal setting regarding lowest system response time; it is very stable and durable. In comparison, the direct mode offers a lower response time, which should prevent the simulator sickness better, but it appears to be not very stable within forgone pretests. For the optimal immersion into the virtual world, the system was calibrated regarding participants' individual interpupillary distance (IPD) and their body height.

As setting for the virtual learning environment, the open-world and sandbox game Minecraft was chosen [16]. This program has been used successfully for learning and teaching purposes beforehand and allows participants among other things to freely explore a virtual environment [17]. With Minecraft, the creation and manipulation of virtual worlds are efficiently possible. For this study, two virtual worlds respectively virtual learning environments were necessary: The first world comprised a tutorial to teach the basics of Minecraft, which was the participants' trainee part. The second world served as the task that the participants should solve as a trainer. To prevent simulator sickness [18] as good as possible, both worlds are designed to be usable without jumping and climbing.

Figure 1 also shows, that "Spigot" was used on the server-side to provide the worlds. Its configuration was also important to make the worlds useful for scientific studies and virtual learning environments. The difficulty was changed to be friendly, which means that eventually present non-player characters do not attack the participants during the experiment. Further, the appearance of any non-player characters and the participants' possibility to attack each other were disabled.

3 Method

3.1 Participants

Ten professional trainers aged between 24 and 60 years ($M = 40.7$; $SD = 13.2$; $n = 2$ female) participated. Participants were recruited from different areas in Germany and active in various domains of personal development to cover a broad range of professions.

3.2 Assessment of Participants' Objective Behaviour

A screen capture tool recorded participants' behavior within the virtual learning environment. The assessment of these objective data is important to clarify whether the basic competencies for learning within virtual learning environments are given and to track participants' performance during the experiment [19]. A calibration with subsequent validation procedure was conducted prior to the experiment. To assess participants attention allocation processes, their gaze fixations, collaboration behavior, communication as well as fluency and speed of movement during the

Fig. 2 Screenshot of the teaching scenario (left side); participant entering the virtual learning environment (right side) (Source: Own figure)

experiment was assessed with the OBS (Fig. 2). These assessments were invisible to participants. Participants' movement fluency and speed allowed assumptions about their habituation progress, whereas the kind of communication and the progress of solving the subtasks reflected the efficiency as a trainer.

Participants' way of movements within the virtual learning environment was also assessed and extracted using Minecraft Research Observation tool to investigate whether they were e.g. rather following or autonomous and leading in the trainer part. These psychomotor skills can be classified into seven different categories, ranging from simple to complex: (1) perception, (2) readiness to act, (3) imitation, (4) habitual movement patterns, (5) complex overt response, (6) adapting the movement pattern to reach an aim and (7) creation of new movement patterns [20].

3.3 Semi-structured Interview

In order to assess trainers opinion and experience within the virtual learning environment, a semi-structured interview was conducted after the experiment. The quantitative questions were assessed anonymously on a laptop and covered items as questions regarding simulator sickness, sensations within the virtual world, potential use of these worlds as a training method and benefit of these (rated on a 5-point Likert scale with 1 = not helpful at all to 5 = very helpful). The qualitative questions of the expert interview covered among other things questions like the trainers experience within the virtual learning world, advantages, and disadvantages of this kind of mixed-reality education, further fields of application. Data of the interview were recorded for further transcription (see analysis section).

3.4 Procedure

Upon arrival, all participants signed the informed consent and filled in their demographic data (among other age, gender, previous gaming experience and used gaming modus, specific field of expertise). Each experiment started with a short introduction into the virtual learning environment to get the trainer habituated to the environment and the hardware. During the experiment, the participants' spatial behavior within the virtual learning environment was assessed with a screen capture tool and documented by a scientific researcher. In order to focus on modern engineering education, an engineering task was given to the participants, who entered the virtual environment by wearing a head-mounted display, displayed in Fig. 2 (right side).

In order to get a deeper insight into the trainers' perspective of this new way of education as well as the discrepancy between learning and teaching within the virtual environment, every trainer pursued two roles: first the trainee and then the trainer part. The participants entered the virtual environment in groups of two. During this virtual meeting, both participants (the trainer and the confederate) were represented by avatars. In the first part of the experiment (trainee-part), the participant was instructed by the research director to restore a broken electrical circuits based on the "Redstone" system [21]. This part of the experiment also enabled the trainer to get comfortable with the virtual learning environment and the handling of the unknown hardware. In the second part of the experiment, a similar problem was stated, but this time the participant had to instruct the confederate about how to solve the stated problem, without anticipating the problem-solving process (the trainer part).

This second world provides a small area with a house and a lighting system. The issue was similar to the first world, but, in this case, the electrical circuit was bigger, and the light was partially working. The participants must repair the circuit to activate all the lightings. This task can be divided into several sub-tasks: (1) Get an overview of the electrical setting of the building. (2) Find out, at which spots the electrical circuit is broken. (3) Remember the necessary steps to repair the electrical circuit. (4) Find the spot within the building from where the success of the problem solving can be controlled [5].

Fig. 3 Screenshot of the OBS recordings with the trainees view (right side) and the trainers view (left side) (Source: Own figure)

Starting with the briefing of the handling of the simplified keyboard and the mouse, the participants had to make sure that the other person (the trainee) feels comfortable within the virtual learning environment and solve the stated problem without further instructions. During the experiment, both screens (trainers and trainees) were recorded and gathered by the mentioned video capture tool for assessing the trainer's and the trainee's behavior in the virtual learning environment (Fig. 3).

4 Analysis

The quantitative data were analyzed by using IBM SPSS, version 22 (http://www.spss.de). Three independent scientific raters coded all qualitative data and scored from 1 = low to 6 = high. Since the interrater variability was high the scores of the three rates were collapsed into one score for each variable and analysed by using SPSS, results of Spearman correlation (two-sided) are displayed in the parentheses below.

5 Results

The screening of the gathered video data indicates that age and online-gaming experience were shown to be related to participants' spatial coordination within the virtual learning environment ($r_{age} = -0.60$, $p < 0.05$; $r_{gaming} = 0.78$, $p < 0.01$). Older participants, and those who had no gaming experience showed initial difficulties with the hand-cursor coordination in the habituation phase, which was indicated by more questions, slower and less fluid movements and spatial problems when it came to entering the buildings or reading instructions that were hanged on a wall. Furthermore, these participants were shown to stay longer in the more simple psychomotor categories as proposed by Simpson [20]. However, most difficulties diminished after around the first minute (*M* = 43.4 seconds; *SD* = 23.8 seconds) by all participants.

This study showed that participants who reported higher sensations of immersion, got used to the virtual world faster, as seen by their objective behavior within the virtual learning environment (e.g. number of gaze fixations ($r_{gaze} = 0.92, p < 0.01$), spatial coordination ($r_{spatial} = 0.94, p < 0.01$), general task performance (efficiency) as trainee, respectively trainer ($r_{trainee} = 0.91, p < 0.01$); ($r_{trainer} = 0.92, p < 0.01$)). However, whether this is a consequence of participants precondition and mixed-reality devices could not be analyzed in this study and deserves further investigation. All participants were able to solve the stated problem as a trainee and were capable of instructing another person verbally within the second virtual learning environment. Thereby, it was shown that participants who documented their course of action out loud as trainee performed more efficiently; hence they finished the sub-goals faster and transferred their knowledge into their role as a trainer. This behavior was particularly seen by trainers with extended working experience. However, the amount of

Table 1 Inter-correlation matrix with Cronbach's alpha at the diagonals (Source: Own analysis). *Correlation is significant by level 0.05 (two-sided); **Correlation is significant by level 0.01 (two-sided)

	1	2	3	4	5	6	7	8
1. Age	1							
2. Gaming	−0.65**	1						
3. Gaze	−0.63*	0.86**	0.91					
4. Spatial	−0.60*	0.78**	0.98**	0.92				
5. Work	0.62	−0.25	0.02	0.06	1			
6. Immersion	−0.82*	0.74	0.92**	0.94**	−0.35	0.87		
7. Efficiency Trainee	−0.72*	0.84**	0.98**	0.95**	0.04	0.91**	0.94	
8. Efficiency Trainer	−0.50	0.77*	0.87*	0.85**	0.31	0.92**	0.93**	0.90

work experience as trainer does not predict efficiency within the virtual world solely (Table 1).

Other factors that seemed to influence the success of the trainee phase were among others the self-confidence appearance to the trainer within the virtual learning environment and the time they spend on reading the instructions. No participant got sick due to the simulation. The experimental session took around 35 minutes per participant. After the experiment, the participants were asked to evaluate their previous learning experience in the virtual learning environment anonymously. The participants' overall conclusion was very positive.

It was shown that even the barriers for the affected trainers are low regarding utilization of virtual learning environments for teaching. The benefit of virtual worlds in teaching got a mean score of 4.5 ($SD = 1$), rated on a 5-point Likert scale (with 1 = not helpful at all to 5 = very helpful). In reply to the question concerning the potential use of these worlds as a training method, all trainers answered with the highest rating (5 = very helpful). The additional expert interview yield more insight into the gathered data and the trainer's perspective. The training was rated as very adaptive, and the participants pronounced the feeling of immersion into the virtual world. The trainers' particularly mentioned the possibility to represent and adapt specific learning content and their feeling of deep and conscious learning as well as the fast familiarization with the virtual learning environment. The speed of movement and the visualization of the environment within the scenarios were stated as pleasant just like the navigation after the habituation phase. The level of difficulty was rated as appropriate for the purpose, and the setup of the virtual learning environment was rated as immediately intuitive for those with gaming experience ($n = 8$).

As possible fields of further application the trainer called among other things fields like emergency services, schooling of security staff and training of techniques, which are too hazardous for the training in field, as well as everyday schooling for higher education and development. As a particular advantage of learning in virtual worlds, the resource efficiency and flexibility, as well as the targeting of many senses at once and the consequential deep learning, were emphasized. Also, the possibility to change or adapt single parameters for training or learning success were mentioned as benefits of virtual learning environment as well as the exploration of environments or settings that are hard to visualize on plane surfaces.

The trainers emphasized the chance to visualize learning success immediately. Also, they told that it is forward-looking, to develop academic and personal needs with the aid of gamification of learning content. To the question of potential difficulties with virtual worlds for trainees and trainers, initial problems with the usage and the acceptance of technology were mentioned as well as a partial negative delay when looking into depth. However, the benefits exceeded the possible adverse effects. When asked about further suggestions, the trainers emphasized the importance of virtual learning environment for learning and future work forms and mentioned their interest in the progressive interlocking of economy and research due to the ongoing digitalization.

6 Conclusion and Outlook

This research gives an insight into trainers view on virtual education. Often, teenagers' and young adults' opinions are positive regarding modern technologies. However, the average age of the participants was relatively high ($M = 40.7$ years; $SD = 13.2$ years). Nevertheless, their opinion regarding training in virtual worlds with immersive hardware was positive. Though, a representative study would be interesting to get an average result from the trainers' and therefore, the teachers' cohorts. A shortcoming of this study was the cross-sectional design, a longitudinal investigation could be useful to give additional insight. Furthermore, a larger sample size would be needed in order to verify the current findings.

After the initial minute, the participants had no major problems with the hand-cursor coordination and everyone was able to show the necessary spatial movement patterns to complete the task. The possibility to verbal communicate within in the virtual learning environment was perceived as beneficial in the trainers self-report for both parts, the trainer, and the trainee part. Thus, the process of pointing toward e.g. an issue was possible by using the verbal communication as well as the avatar's gaze direction and arms. Furthermore, due to real-time communication any issues could be resolved directly. The trainers mentioned as an advantage of this setup that multiple senses are targeted at once. It would be interesting to investigate effects regarding learning, caused by targeting multiple senses in longitudinal studies.

The experience with this study shows possible improvements for further studies: The so-called "spawn point" is the point, where participants enter the world. Although the spawn point was predefined, the virtual worlds do not guarantee a particular point. Instead, a probable spawn area is used. The consequence was that the most participants enter at the right position, but some enter the world e.g. on the roof of the building. For further studies, the usage of an appropriate plugin is planned which allows the definition of a single static spawn point where also the point of view (the angel of the head) is pre-defined.

It is going to be a new challenge for trainers and users to teach, learn and work in the virtual world. Next to nowadays-required competencies, tomorrow's trainers need technical expertise to support users in case of technical problems, malfunctions, and digital literacy to tutor and collaborate appropriately within virtual environments. These requirements represent a major challenge for trainers especially when it comes to groups of users instead of a single one. Also, trainers must be aware if this technology is the right method for a particular subject and if they can transfer their didactic methods into these worlds.

For the successful usage of mixed-reality virtual learning environments for everyday teaching, training and learning for suitable topics, the corresponding industry must provide efficient programs for the creation of such settings. The creation and preparation of suitable worlds must be able and time-efficient as the creation of today's presentations. Some concepts are already promising: For example, some police restricts in Germany got already a suitable solution [22, 23]: They can utilize a pre-defined world with an editor that is customized for the police training to easily setup suitable scenarios. Sandbox games like Minecraft are another approach, which affords the changeability and openness of a suitable tool. However, often, these games are limited regarding related learning topics like e.g. physics, mathematics and training of major incidents, which is often caused by missing mechanical systems at the virtual worlds.

Regardless of the tooling, the virtual reality hardware needs more research regarding usage for training and teaching purposes: Currently, it is not suitable for the students to take or read notes while they are wearing the virtual-reality headsets. One promising approach is the usage of today's speech recognition to take notes. Reading of notes is potentially possible with the right tooling. Another approach is the upcoming augmented reality hardware. They combine the physical world with the virtual reality, respectively can blend the physical world with virtual elements. Because the students can also perceive the reality, notes can be written and read. Nevertheless, for both technologies are topics and scenarios suitable.

Due to new technological developments, the integration of activating learning elements in virtual environments is possible, even as working remotely and controlling machines from afar, offering training with the aid of avatars and many more. This study showed the potential of nowadays-recent progress regarding the ongoing digitalization, but it is necessary to take care that everyone can be involved in this developments. Therefore, continuing this kind of research is an important contribution towards tomorrow's proliferous and digitalized world.

References

1. Frey, C. B. & Osborne, M. A. (2013). *The future of employment: how susceptible are jobs to computerisation?*
2. Federal Ministry for Economic Affairs and Energy. (2015). *Industrie 4.0 und Digitale Wirtschaft: Impulse für Wachstum, Beschäftigung und Innovation*. Berlin.
3. Geissbauer, R., Schrauf, S., Koch, V., & Kuge, S. (2014). *Industrie 4.0: Chancen und Herausforderung der vierten industriellen Revolution.*

4. Jeschke, S. (2013, December). *Everything 4.0 - Drivers and Challenges of Cyber Physical Systems*. Forschungsdialog Rheinland, Wuppertal. Retrieved from http://www.ima-zlw-ifu.rwth-aachen.de/keynotes/Forschungsdialog4Dez2013.pdf
5. Schuster, K., Groß, K., Vossen, R., & Richert, A. (2015). Preparing for Industry 4.0 - Collaborative Virtual Learning Environments in Engineering Education. In D. Guralnick (Ed.), *The International Conference on E-Learning in the Workplace Conference Proceedings*.
6. Sursock, A. (2015). *Trends 2015: Learning and Teaching in European Universities*. Brussels, Belgium.
7. Moreno-Ger, P., Martinez-Ortiz, I., Freire, M., Manero, B., & Fernandez-Manjon, B. (2014). Serious games: A journey from research to application. In *Frontiers in Education Conference Proceedings* (pp. 1–4).
8. Baker, R. S. (2014). Educational Data Mining: An Advance for Intelligent Systems in Education. *IEEE Intelligent Systems*, *29*(3), 78–82. doi:10.1109/MIS.2014.42
9. Tesar, M., Stöckelmayr, K., Pucher, R., Ebner, M., Metscher, J., & Vohle, F. (2013). Multimediale und interaktive Materialien: Gestaltung von Materialien zum Lernen und Lehren. In M. Ebner & S. Schön (Eds.), *Lehrbuch für Lernen und Lehren mit Technologien* (2nd ed.).
10. LeRoy Heinrichs, W., Youngblood, P., Harter, P. M., & Dev, P. (2008). Simulation for team training and assessment: case studies of online training with virtual worlds. *World journal of surgery*, *32*(2), 161–170. doi:10.1007/s00268-007-9354-2
11. Encarnação, J. L. (2008). *Serious Games*, SS 2008, Darmstadt.
12. Sawers, P. (2012). *60% of UK Employees Working Remotely Within a Decade*. Retrieved from http://thenextweb.com/uk/2012/02/22/home-sweet-home-60-of-uk-employees-could-be-working-remotely-within-a-decade/
13. Ubell, R. (2010). *Virtual Teamwork: Mastering the Art and Practice of Online Learning and Corporate Collaboration*. New York: Wiley.
14. Höpner, A. (2012). *Steuerungstechnik: Die ferngesteuerte Fabrik*. Retrieved from http://www.handelsblatt.com/technik/forschung-innovation/steuerungstechnik-die-ferngesteuerte-fabrik/6913260-all.html
15. Schuster, K. (in press). *Einfluss natürlicher Benutzerschnittstellen zur Steuerung des Sichtfeldes und der Fortbewegung auf Rezeptionsprozesse in virtuellen Lernumgebungen* (Dissertation). RWTH Aachen University, Aachen.
16. Short, D. (2012). Teaching scientific concepts using a virtual world - Minecraft. *Teaching Science*, (3), 55–58.
17. Schifter, C., & Cipollone, M. (2013). Minecraft as a teaching tool: One case study. In R. McBride & M. Searson (Eds.), *Proceedings of Society for Information Technology & Teacher Education International Conference* (pp. 2951–2955). Association for the Advancement of Computing in Education (AACE).
18. Höntzsch, S., Katzky, U., Bredl, K., Kappe, F., & Krause, D. (2013). Simulationen und simulierte Welten: Lernen in immersiven Lernumgebungen. In M. Ebner & S. Schön (Eds.), *Lehrbuch für Lernen und Lehren mit Technologien* (2nd ed.).
19. Wilson, K. A., Bedwell, W. L., Lazzara, E. H., Salas, E., Burke, C. S., Estock, J. L., ... Conkey, C. (2008). Relationships Between Game Attributes and Learning Outcomes: Review and Research Proposals. *Simulation & Gaming*, 40(2), 217–266. doi:10.1177/1046878108321866
20. Simpson, E. J. (1972). *The classification of educational objectives in the psychomotor domain*. Washington, DC: Gryphon House.
21. Dezuanni, M., O'Mara, J., & Beavis, C. (2015). 'Redstone is like electricity': Children's performative representations in and around Minecraft. *E-Learning and Digital Media*, (12(2)), 147–163. doi:10.1177/2042753014568176
22. Herkersdorf, M. (2013, October). *VIRTUELL-INTERAKTIVES TRAINING (ViPOL) - eine bundesweit einmalige Lösung der Polizei BW*. Retrieved from http://www.pfa.nrw.de/PTI_Internet/pti-intern.dhpol.local/TagSem/Seminar/Nr48_13/07_Herkersdorf_Internet/TriCAT_ViPol_15102013.pdf
23. Lecon, C., & Herkersdorf, M. (2014). Virtual Blended Learning virtual 3D worlds and their integration in teaching scenarios. In *Computer Science Education (ICCSE), 2014 9th International Conference on* (pp. 153–158).

The Challenge of Specimen Handling in Remote Laboratories for Engineering Education

Abdelhakim Sadiki, Tobias R. Ortelt, Christian Pleul, Christoph Becker, Sami Chatti and A. Erman Tekkaya

Abstract The robot controlled specimen handling for experiments in the field of material characterization for forming technology is presented. The testing cell consists of testing machines, an industrial robot, and other necessary components for the automation and conduction of experiments. First, a methodology is introduced how the key sequence of the robot tasks are identified, planned, and simulated. Afterwards, the design process of the testing cell is described. Finally, the implementation of the methodology and the integration of the robot tasks in a remote laboratory are presented.

Keywords Robot Control · Remote Laboratory · Automation · Simulation · Specimen Handling

1 Introduction

To describe, understand, and develop forming processes, the characteristic material behavior has to be understood. The mechanical properties of materials can be quantified by several material characterization procedures. Common tests in this field are the tensile, cupping, and compression tests as well as the test to determine the forming limit curve (FLC) [1]. These experiments are especially important for students in mechanical engineering because they will be able to connect theoretical models and reality much better by interpreting material properties and the influence of test parameters by gathering experiences in lab sessions [1, 2]. The conduction of such experiments requires highly specialized testing machines and equipment. Some of the main issues in carrying out meaningful experiments are the safety conditions and the knowledge about the use of these machines. Due to time and personal arrangement, the student may not have the opportunities or just restricted access to free experimentation. Furthermore, remote laboratories have a great potential for teaching [3–5].

A. Sadiki (✉) · T.R. Ortelt · C. Pleul · C. Becker · S. Chatti · A.E. Tekkaya
TU Dortmund University, Institute of Forming Technology and Lightweight Construction (IUL), Dortmund, Germany
e-mail: Abdelhakim.Sadiki@tu-dortmund.de

Originally published in "REV 2015" © 2015.
Reprint by Springer International Publishing AG 2016,
DOI 10.1007/978-3-319-46916-4_34

In order to provide the access to such kind of experiments, a remote laboratory was implemented at the Institute of Forming Technology and Lightweight Construction at TU Dortmund University [6, 7]. For this laboratory, the iLab Shared Architecture (ISA) [8] is used. In contrast to electrical engineering experiments, which are commonly implemented in remote laboratories, experiments in the field of manufacturing technology require a high effort and precise handling of the specimen. This is a main challenge during the implementation and automation of such remote laboratories [9]. For this purpose, a handling system is required. In the developed remote laboratory, a system is used which consists of a six axis industrial robot KUKA KR 30-3, several grippers for the various geometries of the specimens, a collision detection unit (mounted on the flange of the robot), and a gripper switching unit. The handling process consists of several tasks. These tasks have to be previously identified and carefully analyzed in order to provide a reliable and safe mode of operation [5].

2 Methodology

The presented method comprises three steps:

A Identification of the specimen handling tasks
B Layout design of the testing cell
C Simulation of the robot tasks

With the help of this method, a testing cell for material characterization, with an industrial robot as a specimen handling system, was designed. Furthermore, the robot tasks can be analyzed and implemented.

2.1 Identification of the Specimen Handling Tasks

When conducting an experiment, the operator performs the following tasks:

- Measure the specimen (required for setting up the machine parameters)
- Lubricate the specimen if necessary
- Enter the test parameters into the machine
- Move the machine into initial position
- Open the clamping device
- Place the specimen into the machine
- Close the clamping device
- Start the test
- After the test, retrieve the specimen and remove it

The automation of the process requires a precise synchronization of the mentioned tasks. However, not each task has to be addressed by the robot itself (e.g. the robot must wait before it places the specimen into the machine until the machine is

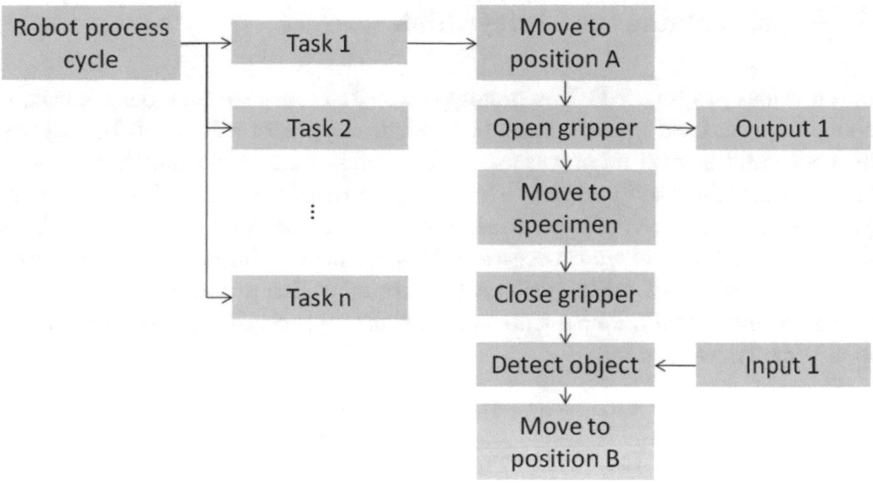

Fig. 1 Flowchart of a robot process cycle [10]

ready). For this reason, an exchange of information between machine and robot is necessary. This automatic sequence control was implemented on a master controller in LabVIEW. The tasks of the robot must be extracted from the tasks listed above and defined. Each robot task consists of actions. These actions can be motion, computational, or I/O operations. Figure 1 shows a schematic representation of the robot process cycle.

2.2 *Layout Design of the Testing Cell*

The first step during the design and implementation of the testing cell is the positioning of the components relative to the main coordinate system. Here, two main types of conditions need to be considered. The first one depends on the robot tasks and the second one on the measurements and the available space for the testing cell (e.g. electrical and pneumatic terminals). An effective method to address this task is the modeling of the testing cell and kinematic properties of components with a CAD software [11]. With the help of this model, it is possible to roughly visualize the robot tasks and to verify the robot accessibility to the components.

2.3 *Simulation of the Robot Tasks*

Based on the CAD model of the testing cell, the simulation of the robot tasks should be carried out for two different reasons. On the one hand, the tasks can be analyzed thoroughly, collisions can be eliminated, and the movements of the robot can be optimized. On the other hand, the robot code can be generated and easily updated.

3 Implementation and Integration

In a previous project, PeTEX, a prototype was developed to carry out a tensile or compression test and was integrated in an e-learning platform [12–14]. To deal with the specimens, a small robot was used. However, its range is not sufficient to serve the other machines and the robot has no gripper changing unit with the necessary grippers for other types of specimens. For these reasons, a new conceptual design for the automation process is necessary. This evolution is furthermore necessary to extend the number of experimental setups for education purposes. This broadens even more the portfolio for students to choose the appropriate experiment according to the problem worked on.

3.1 Material Characterization Testing Cell

As mentioned before in the introduction, fundamental and common experiments (tensile, compression, cupping, and FLC tests) will be automated and provided to students and teachers over the internet. For that purpose, several material testing machines and necessary components are used. In the following, the relevant components are presented:

- Universal testing machine Zwick Z 250: This machine can conduct tensile, compression, and bending tests with a force of up to 250 kN.
- Sheet metal testing machine BUP 1000: Cupping and FLC tests can be done with this machine.
- GOM ARAMIS 4M: This is an optical measuring system to measure the strain of the specimen during a sheet metal test.
- Lubrication unit: This device is used to lubricate the specimen before testing it (e.g. by a cupping test).
- Control Unit: The control unit consists of a PXI Real Time System of National Instruments (master controller), a safety PLC to guarantee a safe execution of the experiment, and communication devices (like switches and routers).
- Robot: This robot is an industrial robot with six axis of freedom.
- Close the clamping device
- Cameras: In order to provide a comprehensive viewing perspective to the user (e.g. on the robot flange), several cameras are placed in different places in the testing cell.
- Magazine: To store the specimens for the tests.

Figure 2 shows the testing cell with the machines, industrial robot and other necessary components. Additionally to the handling system, different magazines are designed and constructed. Two types of magazines were developed and are used for the remote laboratory: static and automated ones. In a static magazine, the specimens are stored for the tensile test in multiple positions (usually horizontally) and different

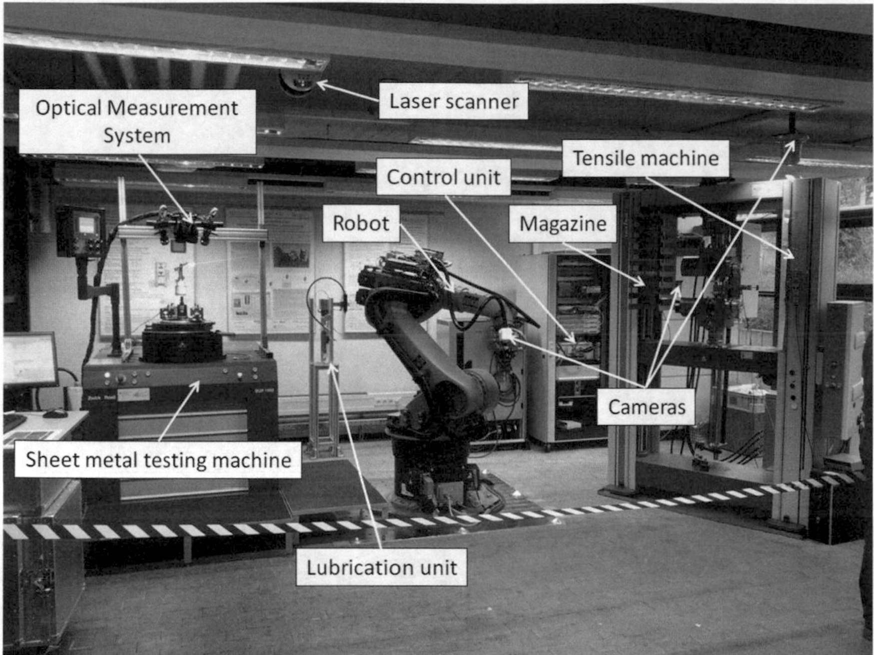

Fig. 2 Material Characterization testing cell

levels (see Fig. 3). If the specimen does not fulfill the static requirements, it is placed vertically, because the horizontal placement would lead to its deformation (bending). Moreover, a very important restriction to consider for the retrieval of a specimen is that the robot can only pick up a specimen if the prior positions of the specimen are empty. The robot program to retrieve the specimen from this magazine has two parameters, the first parameter indicates the row in which the specimen is placed and the second parameter indicates the position of the specimen in the row. These two parameters are determined by the master controller after the user selects the specimen type.

The automated magazine has the same principle to store the specimen. The difference here is that the specimens are delivered with a conveyor system to a specific positon and every specimen can be transported without restrictions. Moreover, multiple specimen types can be stored in this magazine. The delivery position is always the same. This has the advantage that the robot has always the same grasping position. However, the robot needs information about the geometry type of the specimen, because this magazine can store specimens for more than one experiment and with different geometries.

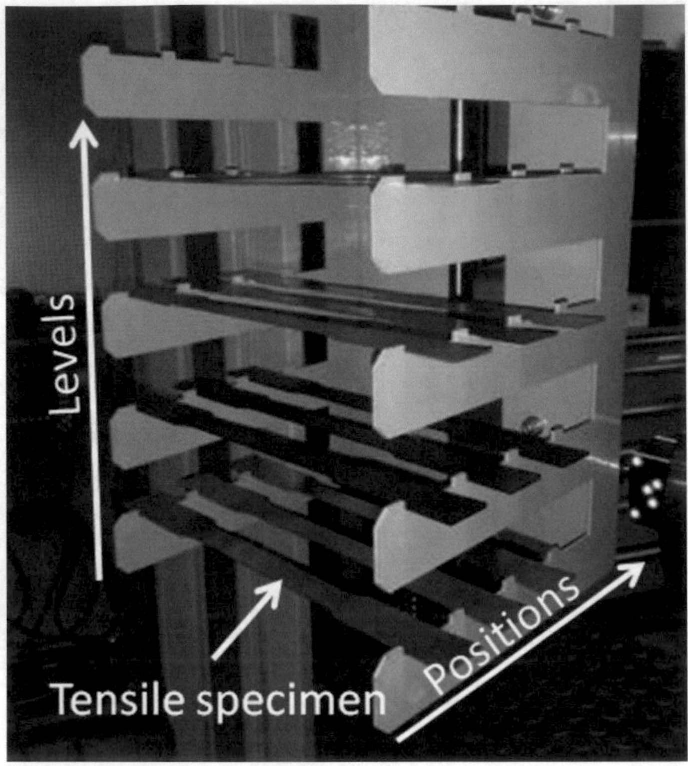

Fig. 3 Storage of the tensile specimens in the magazine

3.2 Components of the Robot

As described above, a six axis robot is used as handling system in the tele-operative testing cell in interaction with different additional devices and different grippers. This kind of robot can handle loads up to 30 kg (the specimens used for the experiments are around 100 gram) and its working envelope is approx. 27.2 m^3. The flexibility and especially the expandability was the main aspect during the selection of the robot. Furthermore, the size and the wide work envelope of the robot are in accordance with the positions of the testing machines (see Fig. 7). The robot is fixed on a basement, which consists of a cement fundament and a steel disc. The steel disc provides several mounting points for the robot to change the position in different directions. The robot allows the handling of the specimens in interaction with different grippers. Due to the fact that there are several kinds of specimens with totally variable geometries, different grippers are needed. These grippers are, on the one hand, typical two finger parallel grippers, which are available on the market and were modified for the use in the tele-operative testing cell, and, on the other hand, pneumatic suction grippers. The two finger grippers are used to deal with the specimens for the tensile or the

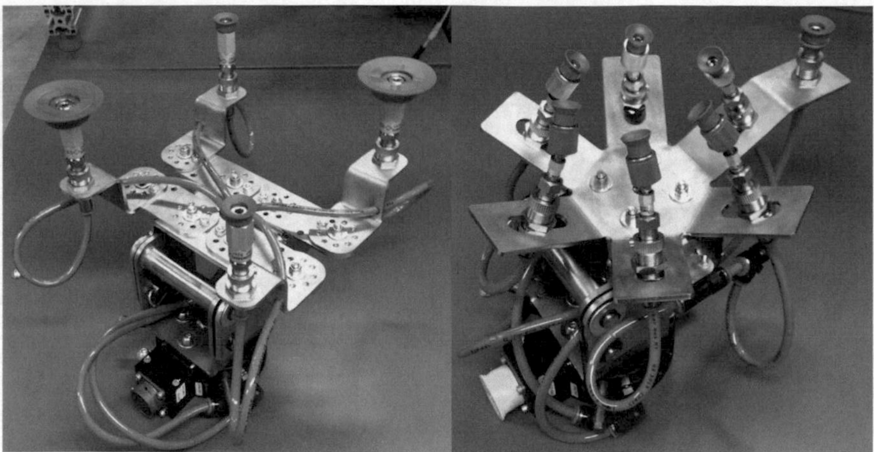

Fig. 4 Pneumatic suction grippers

compression test. These grippers were modified by milling to fulfill the requirements regarding the geometries of the specimens and the available space in the machine for the specimens. A second type of grippers was designed with two different pneumatic suction systems. These grippers were developed for the handling of specimens for the sheet metal testing machine (see Fig. 4). One gripper is designed to insert the specimen into the machine and the other gripper to pick up and draw the deformed specimen out of the machine. Therefore, special requirements regarding the geometry were identified and these requirements were the most relevant for the design process. For example the specimen is plain before the FLC testing and deformed (spherical dome) after the test. Hence, the suction nozzles are designed spherical and are also provided with spring elements to guarantee a good grip or small distance to the spherical surface of the specimen. It is obvious that a gripper change system is also needed to switch between the different grippers in the automatic mode of the tele-operative testing cell. This system can handle loads up to 25 kg so that the requirements regarding the weight of the grippers, and the specimens are fulfilled. Furthermore, the system forwards all the signals from, and into the gripper. Another advantage of this system is the quick change process of the grippers. Furthermore, an additional system is installed between the change system and the flange to protect the robot from collisions and overloads.

This anti-collision and overload protection with automatic reset ensures the interoperability of the different systems like mounting holes for example (see Fig. 5). In the case of a collision or an overload, an emergency shutdown of the robot and the other systems of the tele-operative testing cell is executed. The automatic reset allows a return to the zero position when the gripper moves away from the collision object. The collision protector operates also with pneumatics. The operating pressure can be modified by a regulator, which is installed next to the valve manifold on the arm of the robot. By modifying the operation pressure, the sensitivity for collisions can be changed.

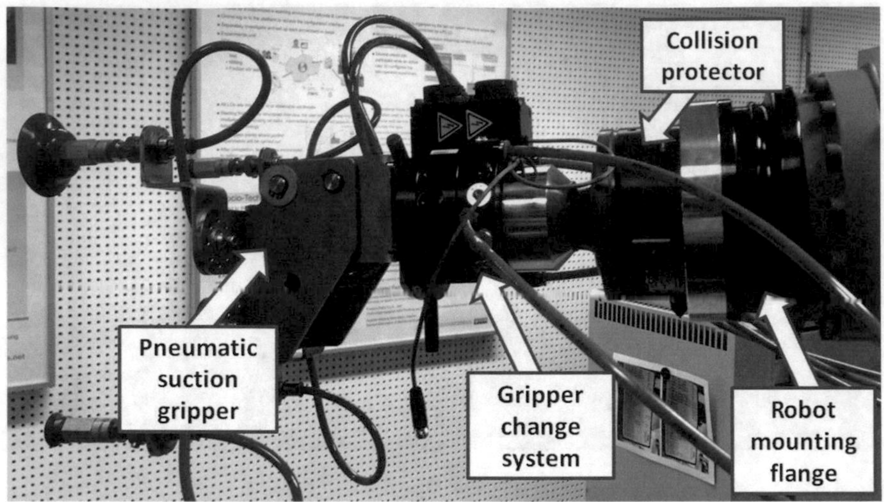

Fig. 5 Devices installed on the flange of the robot

Figure 6 shows the valve manifold and five regulators on the arm of the robot. These regulators provide the different systems with different pressure levels. As mentioned above, one regulator controls the operating pressure of the collision protector and other regulators control the pressure level of the vacuum generators for the pneumatic suction grippers. The valve manifold provides also digital inputs and outputs. Currently, the digital inputs are used only to receive signals from sensors of the grippers and in the future to identify the different grippers. For example two sensors are installed on the two finger gripper to detect the position of the fingers. With these two sensors, it is possible to check if the gripper is open or closed and, additionally, if a specimen is picked. Vacuum sensors to check the vacuum of the grippers during the handling process are also installed. These vacuum sensors are used to detect whether a specimen is gripper or not.

The valve manifold combines different kinds of valves and vacuum generators. The valves, typical 3/2-way valves, are used to control the grippers and the opening-closing mechanisms of the changing system. The vacuum generators also have an ejector pulse to separate the specimen from the vacuum suction cups after the handling process. The valve manifold is connected via PROFIBUS to the robot and operates as a slave in the network.

3.3 *Layout Design and Simulation*

As a first step, by the modeling of the testing cell and the simulation of the presented scenarios before (see Section 2), a 2D layout was created, in which the components

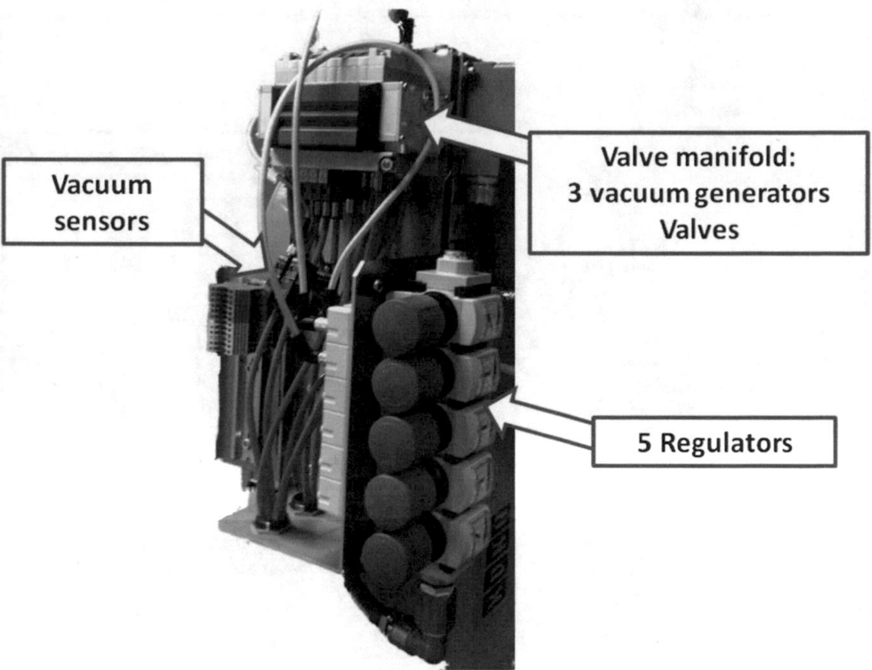

Fig. 6 Valve manifold and regulators on the arm of the robot

were placed. Figure 7 shows the top view of the testing cell. Regarding the boundaries of the testing cell (e.g. dimensions of the room), a rough and fast positioning of the components could be carried out. In the second step, a 3D environment was created in CATIA. In order to visualize the movements of the machines and the robot, the kinematic properties of the components were defined and set in the 3D environment. Now, the reachability of the robot could be tested and the fine positioning of the machines could be done.

After the final positions of the components were found, the components were exported in a compatible format (STL) for KUKA.Sim Pro then imported into the simulation software and positioned. Since the behavior of the objects could not be exported for the simulation, these were again defined to simulate the scenarios. Thereafter, the robot tasks were visualized and analyzed for collision and time efficiency. Figure 8 shows a simulated robot task during the conduction of a tensile test.

3.4 *Implementation of the Robot Tasks*

As mentioned before, the master controller (implemented in LabVIEW) manages and controls the experiments. During the conduction of an experiment, the controller

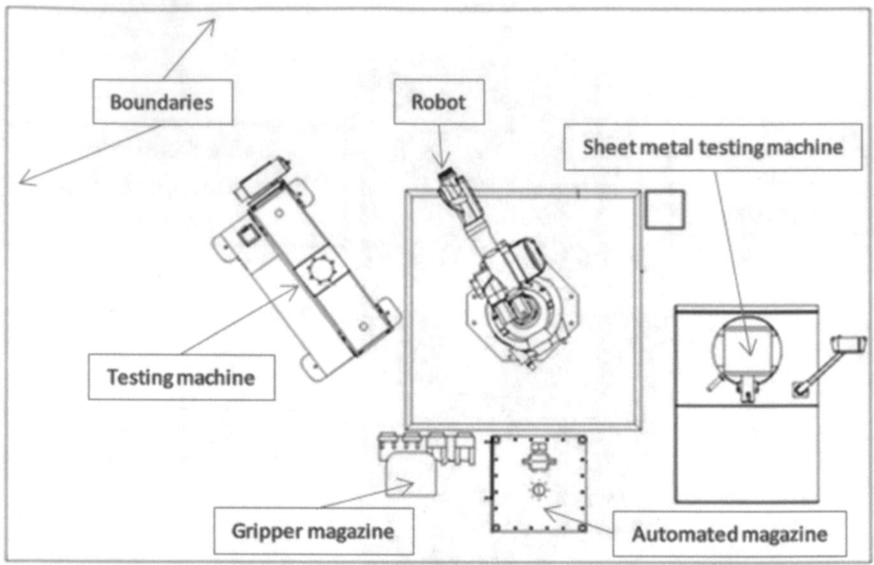

Fig. 7 2D layout of the testing cell (top view)0

checks, as a first step, the parameters and defines the specimen position of the selected specimen type, then the parameters are sent to the machine and the determined position of the specimen is sent to the robot. After this, the controller synchronizes the tasks of the robot with the conducted experiment on the testing machine. This means, for example, that the robot retrieves the specimen from the magazine and feeds it to the machine before the test is started. Due to the many active components and their supported communication interfaces, multiple technologies/protocols are used. Figure 9 shows the communication infrastructure of the testing cell. The robot provides a digital communication interface which is used in industrial applications. For the communication between the controller, the robot, and other components (e.g. Safety PLC), the PROFIBUS DP (Process Field Bus Decentralized Peripheries) is used here as a protocol. In order to ensure a smooth and secure control of the robot, the automatic external interface defines multiple inputs/outputs [15] so that it is possible to run a program stored on the robot to stop and to get the state of the robot (e.g. whether a program is running or if the robot is in its initial position). This interface is extended to three bytes as inputs and two bytes as outputs. This is necessary because some programs of the robot need parameters (e.g. refer to a specimen position) or they return information (e.g. if the robot has grasped a specimen). Figure 10 illustrates the data transmission layer between the master controller and the robot. In order to test the implemented robot controller in LabVIEW, which provides the communication to the robot, a test VI (Virtual Instrument) was implemented, in which the signals of the robot can be read and changed and data (numbers) sent and received. Figure 11 shows the test VI with all the controls, inputs, and indicators.

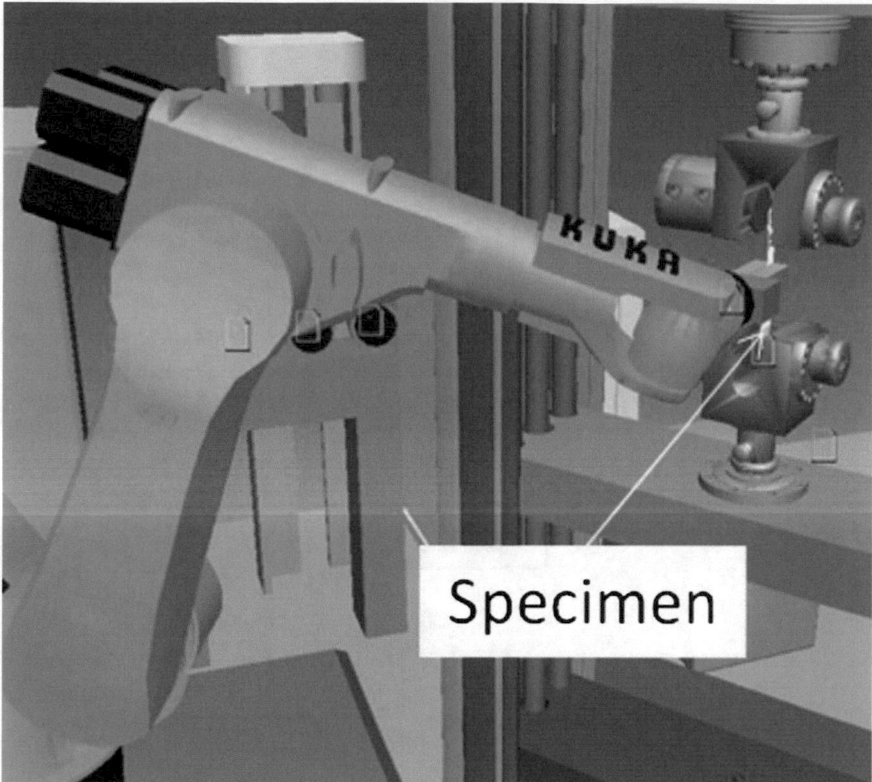

Fig. 8 Simulation of a robot task

3.5 Integration of the Robot in the Automated Tensile Test

As part of the introduced project, the tensile test was automated. A controller was developed in LabVIEW for the testing machine to control and conduct parametrized tensile tests on the machine. With its help, several parameters could be set, such as strain rate in the elastic-plastic area and the initial gauge length l0. After the test is started, the measured data, like the force acting on the specimen, displacement and width variation of the specimen, are captured by the controller. For the attempt to carry out the tensile test in a fully automated manner, a specimen handling is required. The specimen has to be placed into the machine and, after the test, it should be retrieved to allow the conduction of a new experiment. The implemented and tested robot controller as shown before was used to execute all the needed and defined specimen handling tasks. The tasks of the robot were programmed by teaching the robot manually. This method had proven to be the best option for the requirements as it is very accurate in contrast to the code generated by the simulation program. Additionally to teaching the robot, some programs were parameterized and extended by custom code. This was necessary, because for example the program which retrieves

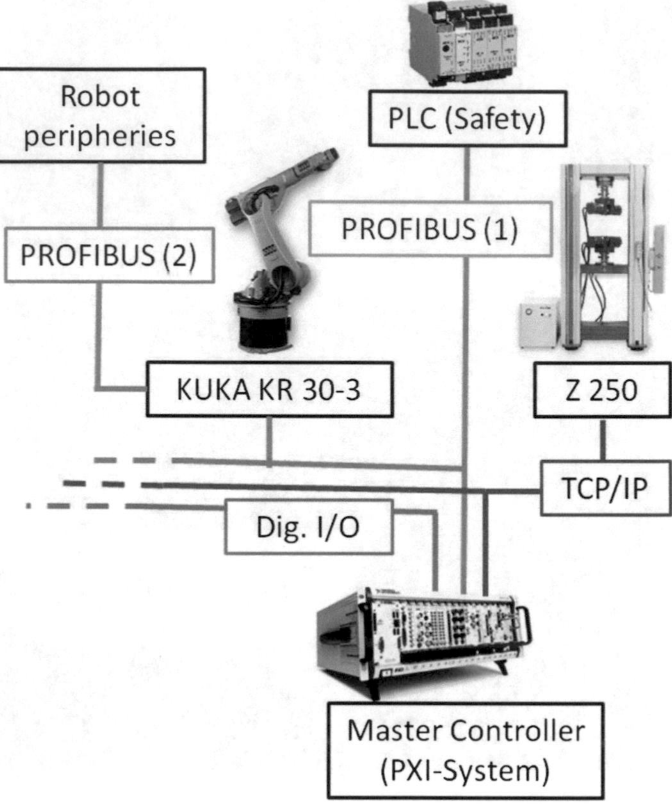

Fig. 9 Communication infrastructure of the automated testing cell

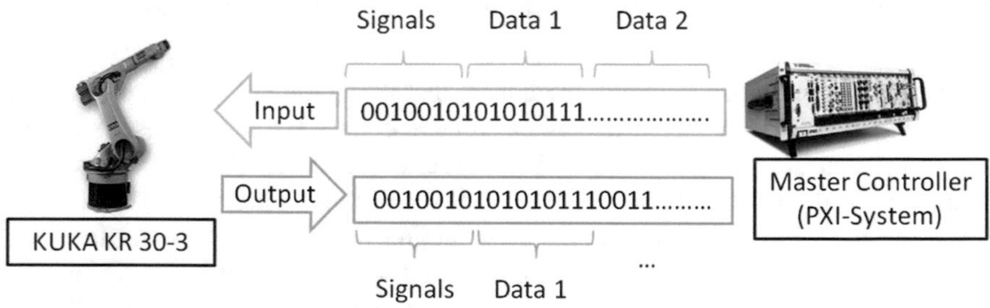

Fig. 10 Data transmission between the controller and the robot

the specimen from the magazine needs to know the position of the specimen. The position is read from the controller before the robot starts moving and grasps the specimen. As a final step, by the integration of the robot in the automated tensile test, the two controllers for the testing machine and the robot were tied together in a tensile controller. Figure 12 shows a prototype for conducting a fully automated and parametrized tensile test.

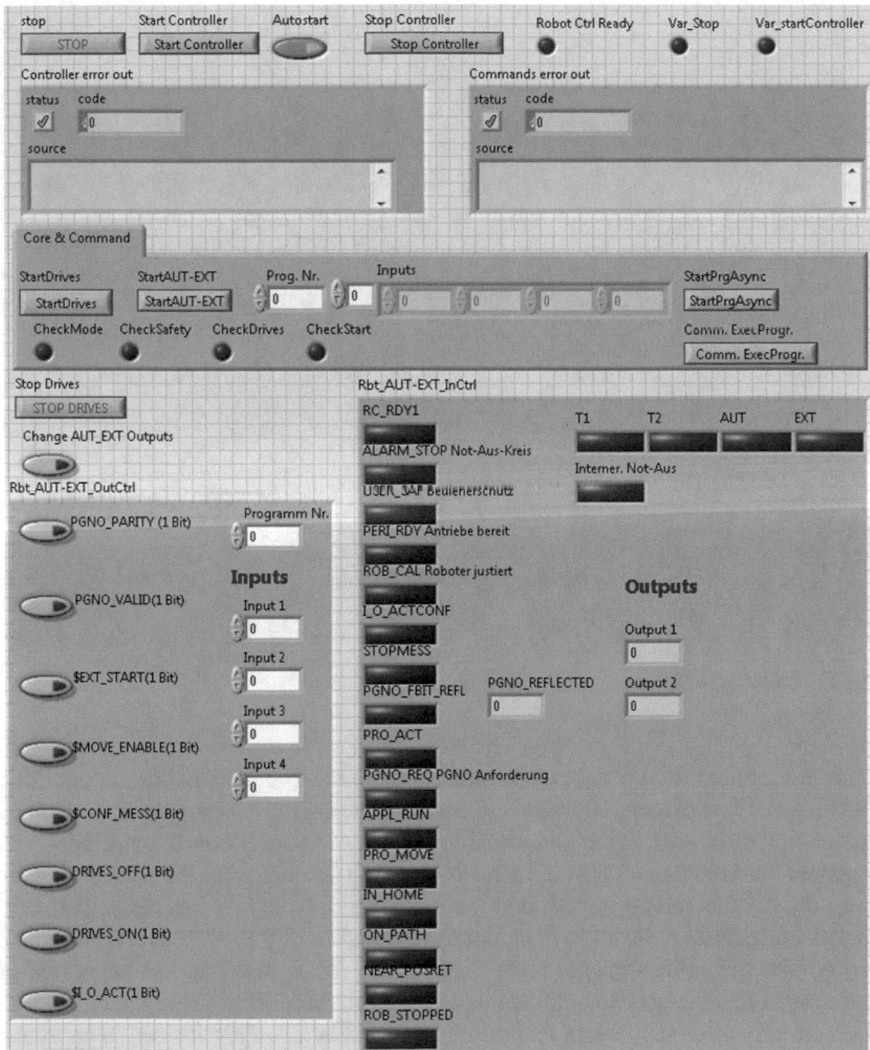

Fig. 11 Test VI to control the robot and to execute stored programs on the robot

3.6 *Integration into a Learning Scenario*

Since material characterization is an essential part in manufacturing processes of forming technologies, students deal with the aspect during lectures [16], different labs [17, 18] or seminars in various facets. The described highly sophisticated teleoperated testing cell was integrated into a fundamental lecture of forming technology. For this, a student centered approach was used to activate the students and make

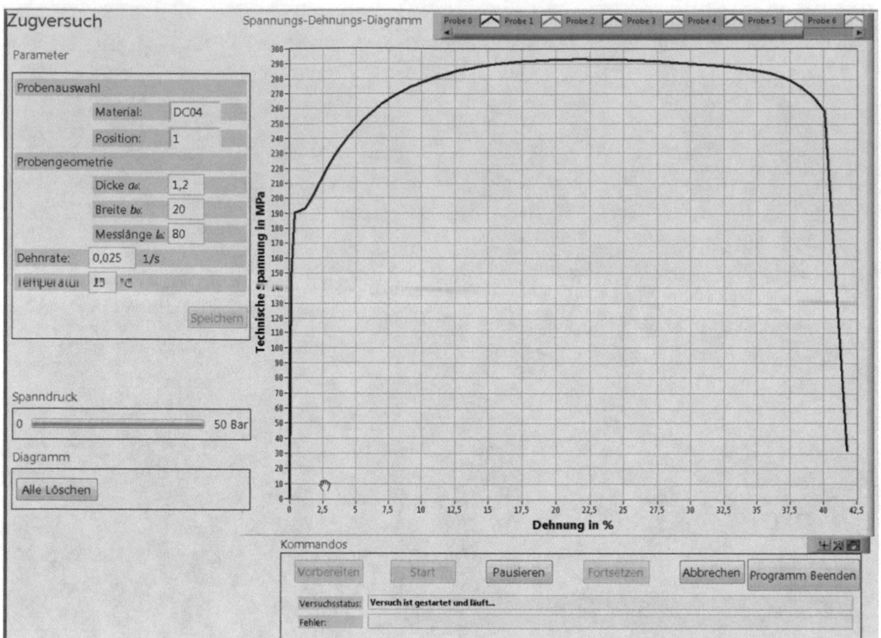

Fig. 12 Prototype of the tensile test with integrated robot controller

them think about the concept of material behavior while exchanging ideas with a classmate. For the integration of the interactive real-time experiment, a tensile test was carried out based on the understanding of the previously covered concepts. The observed phenomenon was then justified by voting on questions. After resolving the question, the theoretical background was developed. Based on that, the experiment was reconfigured on the basis of students' hypothesis of the expected behavior and carried out again. Depending on the final results, the hypothesis was supported or disproved and a further explanation was developed. According to the students' feedback directly after the lecture and during the evaluation, they found the approach interesting and motivating to deeply think about the concepts dealt with. Furthermore, they appreciated being activate during a lecture by discussing with class mates about an observed phenomenon.

4 Conclusion

Based on the presented scenarios for performing an experiment, the robot tasks to carry out a tensile is concretely defined and described in detail. After this, a CAD model of the testing cell is modeled in CATIA to find out valid positions of the components. In order to visualize the robot task and to verify the motions of the robot, a simulation in KUKA.Sim Pro is made. The implemented communication architecture between the robot and the LabVIEW controller is explained. Finally, the programming method used to program the robot tasks and a prototype of an automated tensile test is presented. The implemented robot controller in this work will be used for further automation of other experiments for the remote laboratory at the IUL.

References

1. A. E. Tekkaya, "Metal Forming," *In Handbook of Mechanical Engineering. Ed. by K.-H. Grote, E. K. Antonsson*, Springer, 2009. Chap. 7.2, pp. 554–606.
2. M. Merklein, "Fließkurven," *In Handbuch Umformen. Edition — Handbuch Fertigungstechnik. Ed. by H. Hoffmann, R. Neugebauer, G. Spur*, München: Carl Hanser Verlag, 2012, pp. 66–76.
3. L. Gomes, and S. Bogosyan, "Current Trends in Remote Laboratories," *In IEEE Transactions on Industrial Electronics 56.12, pp. 4744–4756*, 2009.
4. J.E. Corter, J.V. Nickerson, S. K. Esche, C. Chassapis, S. Im, and J. Ma, "Constructing reality: A study of remote, hands-on, and simulated laboratories," *In ACM Transactions on Computer-Human Interaction 14.2 (2 2007)*, pp. 1–27, 2007.
5. J. Ma, J. V. Nickerson, "Hands-on, simulated, and remote laboratories: A comparative literature review," *In ACM Computing Surveys 38.3 (7 2006)*, pp. 1–24, 2006.
6. T. R. Ortelt, A. Sadiki, C. Pleul, C. Becker, S. Chatti, and A.E. Tekkaya, "Development of a Tele-Operative Testing Cell as a Remote Lab for Material Characterization". *Proceedings of 2014 in International Conference on Interactive Collaborative Learning (ICL)*, pp. 977–982, 2014.
7. T. R. Ortelt, C. Pleul, A. Sadiki, M. Hermes, C. Soyarslan, and A. E. Tekkaya, "Virtuelle Lernwelten in der ingenieur-wissenschaftlichen Laborausbildung," *in Exploring Virtuality, Aachen*, 2012.
8. V. J. Harward, J. A. Del Alamo, S. R. Lerman, P. H. Bailey, J. Carpenter, K. DeLong, et al., "The iLab shared architecture: A web services infrastructure to build communities of internet accessible laboratories," in *Proceedings of the IEEE, vol. 96, pp. 931–950*, 2008.
9. C. Pleul, C. Terkowsky, I. Jahnke, and A. E. Tekkaya, "Teleoperated laboratory experiments in engineering education – The uniaxial tensile test for material characterization in forming technology," *in Using Remote Labs in Education. vol. 1, J. Garcia- Zubia and G. R. Alves, Eds.: Deusto Publicaciones*, pp. 323–347, 2011.
10. P. Freedman, and R. Alami, "Repetitive Sequencing: from robot cell tasks to robot cell cycles," *Conference on Robotics and Automation*, 1962–1967, 1990.
11. I. Davila-Rios, L. M. Torres-Trevino, and I. Lopez-Juarez, "On the Implementation of a Robotic Welding Process Using 3D Simulation Environment," *in Electronics, Robotics and Automotive Mechanics Conference, CERMA*, 2008, pp. 283–287.
12. C. Terkowsky, C. Pleul, I. Jahnke, and A. E. Tekkaya, "Teleoperated laboratories for online production engineering education platform for e-learning and telemetric experimentation (PeTEX)," *International Journal of Online Engineering*, vol. 7, pp. 37–43, 2011.
13. C. Pleul, C. Terkowsky, I. Jahnke, U. Dirksen, M. Heiner, J. Wildt, and A. E. Tekkaya, "Experimental e-learning – insights from the European project petex," *in Book of Abstracts: ONLINE EDUCA BERLIN 2009 – 15th International Conference on Technology Supported Learning and Training*, pp. 47–50, Berlin, 2009.
14. C. Pleul, C. Terkowsky, I. Jahnke, and A. E. Tekkaya, "Platform for e-learning and teleoperative experimentation (PeTEX) – Holistically integrated laboratory experiments for manufacturing technology in engineering education," *In Proceedings of SEFI Annual Conference. 1st World Engineering Education Flash Week. (Lissabon, Portugal, Sept. 27–Oct.*

4, 2011). Ed. by J. Bernardino, J. C. Quadrado, pp. 578–585. url: http://www.sefi.be/wp-content/papers2011/T12/104.pdf (visited on 11/17/2012).

15. KUKA Roboter GmbH, "Roboterprogrammierung für Experten," *in Arbeitsheft MP4 05.10.03*, 2010.
16. C. Pleul, M. Hermes, C. Becker, and A.E. Tekkaya, "ProLab@Ing – Projekt-Labor in der modernen Ingenieurausbildung". *TeachING-LearnING.EU innovations. Flexible Fonds zur Förderung innovativer Lehre in den Ingenieurwissenschaften. Hrsg. von Petermann, M., Jeschke, S., Tekkaya, A. E., Müller, K., Schuster, K. und May, D., pp. 16–21*, 2012.
17. C. Pleul, A. Sadiki, M. Hermes, S. Chatti, and A.E. Tekkaya, "miniLABs – Focused lab sessions in manufacturing technology related to forming processes". *International Journal of Engineering Pedagogy (iJEP) 3 (Special Issue: EDUCON2013 2013). invited contribution for best paper section, S. 52-56*, 2013.
18. C. Pleul, D. Staupendahl, M. Hermes, S. Chatti, and A.E. Tekkaya, "Problem-based Laboratory Learning in Engineering Education – PBLL@EE". *TeachING-LearnING.EU discussions. Innovation für die Zukunft der Lehre in der Ingenieurwissenschaften. Hrsg. von Tekkaya, A. E., Jeschke, S., Petermann, M., May, D., Friese, N., Ernst, C., Lenz, S., Müller, K. und Schuster, K. 2013. chap. V. Experimente und Labore, pp. 193–198.*

Development of a Cupping Test in Remote Laboratories for Engineering Education

Alessandro Selvaggio, Abdelhakim Sadiki, Tobias R. Ortelt, Rickmer Meya, Christoph Becker, Sami Chatti and A. Erman Tekkaya

Abstract A remote controlled cupping test for sheet metal as material characterization for forming technology is presented. This formability test is included in a tele-operative testing cell consisting of an additional testing machine, an industrial robot, and other necessary components for the automation and execution of experiments. First, a methodology is introduced explaining how the remote cupping test is realized. Afterwards, the integration of the cupping test in a remote laboratory is presented.

Keywords Cupping Test · Robot Control · Remote Laboratory · Automation · Simulation · Specimen Handling

1 Introduction

In the field of forming technology the mechanical properties of materials can be quantified by several material characterization experiments. Common tests in this field are the tensile, compression, and cupping tests as well as the test to determine the forming limit curve (FLC) [1]. These experimental tests are also important for students in mechanical engineering because they will be able to connect theoretical models and reality much better by interpreting material properties and the influence of test parameters by gathering own experiences in lab sessions. The conduction of such experiments requires highly specialized testing machines and equipment. Some of the main issues in carrying out meaningful experiments are the safety conditions and the knowledge about the use of these machines. Due to time and personal arrangement, the student may not have the opportunities or just restricted access to free experimentation. Furthermore, remote laboratories have a great potential for teaching [2, 3]. In order to provide the access to such kind of experiments, a remote

A. Selvaggio (✉) · A. Sadiki · T.R. Ortelt · R. Meya ·
C. Becker · S. Chatti · A.E. Tekkaya
Institute of Forming Technology and Lightweight Construction (IUL),
TU Dortmund University, Dortmund, Germany
e-mail: alessandro.selvaggio@iul.tu-dortmund.de

Originally published in "13th International Conference on Remote Engineering and Virtual Instrumentation (REV)", © 2015. Reprint by Springer International Publishing AG 2016, DOI 10.1007/978-3-319-46916-4_35

laboratory was implemented at the Institute of Forming Technology and Lightweight Construction at TU Dortmund University [4]. For this laboratory, the iLab Shared Architecture (ISA) [5] is used.

Exemplary the ISA is used in a partnership between the Massachusetts Institute of Technology (MIT) and the Obafemi Awolowo University in Nigeria, the Makerere University in Uganda and the University of Dar-es-Salaam in Tanzania in coordination with the Maricopa Advanced Technology Education Center (MATEC). This association has focused its work on the development of iLabs around the National Instruments Educational Laboratory Virtual Instrumentation Suite (ELVIS) platform. The ELVIS is a low-cost, small-footprint unit that contains most of the common test instruments found in a typical electrical engineering lab. Students using this iLab will be able to perform a variety of measurements on an analog electronics system such as a multi-stage audio filter [6].

2 Material Characterization Testing Cell

With the developed tele-operative testing cell for the material characterization fundamental and common experiments (tensile, compression, cupping, and FLC tests) will be automated and provided to students and teachers over the internet. For that purpose, several material testing machines and necessary components are used.

- Universal testing machine Zwick Z 250 to conduct tensile, compression, and bending tests with a force of up to 250 kN.
- Sheet metal testing machine BUP 1000 for Cupping and FLC tests.
- GOM ARAMIS 4M as an optical measuring system to measure the strain of the specimen during a sheet metal test.
- Lubrication unit to lubricate the specimen before testing it (e.g. by a cupping test).
- Realtime Control Unit.
- An industrial robot with six axis of freedom for the handling.
- Cameras in order to provide a comprehensive viewing perspective to the user.
- Magazine to store the specimens for the tests.

Figure 1 shows the testing cell with the machines, industrial robot, and other necessary components as described above.

To enable the automation, the communication of all elements must be ensured. In Fig. 2, the respective components and the status is displayed. Using the Overview it can be checked if all components are ready for operation or a malfunction has to be fixed before starting an experiment.

The robot allows the handling of the specimens for the Zwick Z250 as well as for the BUP1000. Due to the fact that there are several kinds of specimens with totally variable geometries, different grippers are needed. These grippers are, on the one hand, typical two finger parallel grippers, and, on the other hand, pneumatic suction grippers.

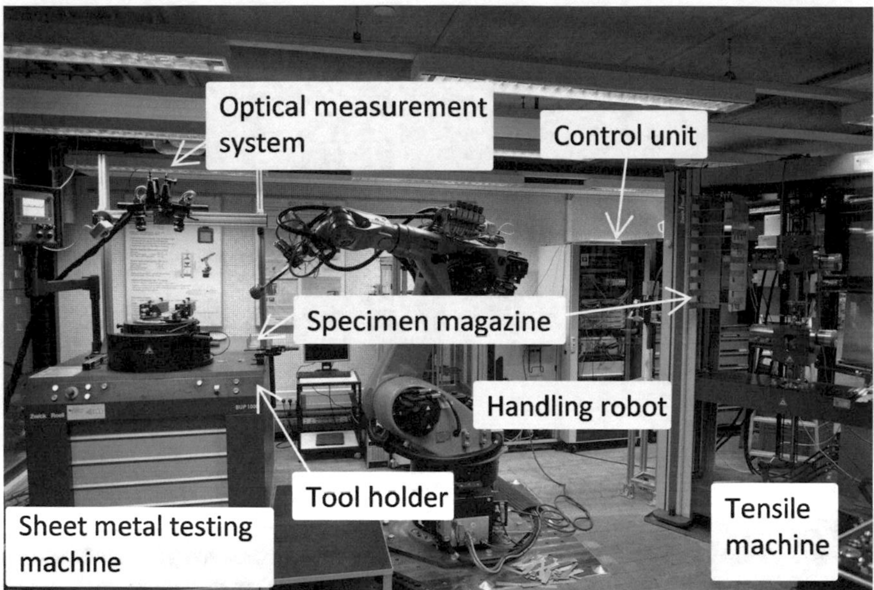

Fig. 1 Material characterization testing cell

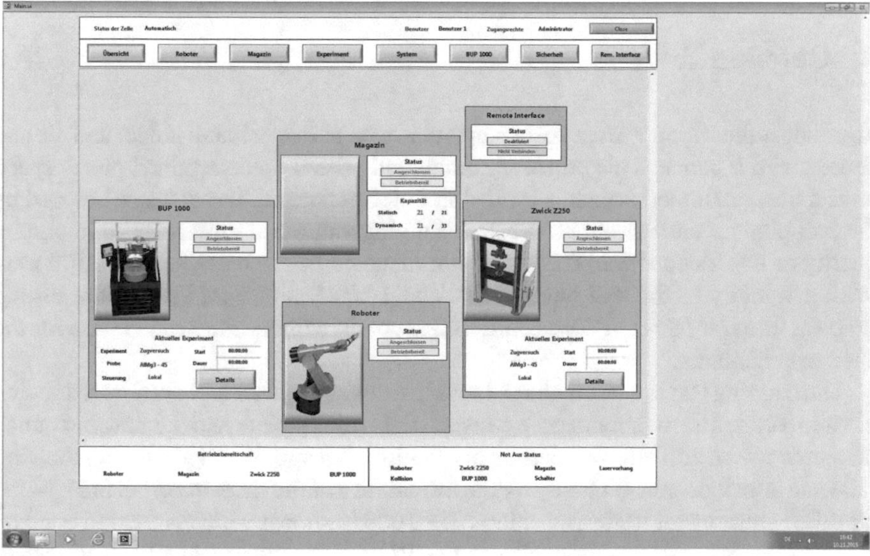

Fig. 2 Overview of the components for cupping test

The two finger grippers are used to deal with the specimens for the tensile or the compression test. A second type of grippers was designed with a pneumatic suction system. These gripper are needed for the handling of specimens for the sheet metal testing machine (see Fig. 3).

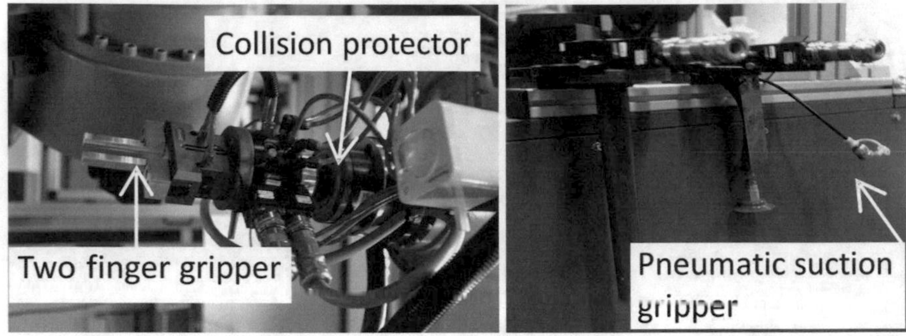

Fig. 3 Devices installed on the flange of the robot. Two finger gripper (left) and pneumatic suction gripper (right)

It is obvious that a gripper change system is also needed to switch between the different grippers in the automatic mode of the tele-operative testing cell.

Furthermore, an additional system is installed between the change system and the flange to protect the robot from collisions and overloads. If this additional system detects a collision the automated movement is stopped.

3 Cupping Test

To conduct the cupping test a specimen is clamped between blank holder and die and dented with a punch. This process is continued with a predetermined punch speed until a fine, continuous crack is formed in the sheet metal. The distance covered by the ball plunger until the crack is called cupping value and represents an important quality of the tested sheet. The maximum distance of the ball plunger is 100 mm. With a velocity of the ball plunger between 1 mm/s and 5 mm/s the pure testing time for one test between 20 seconds 100 seconds without the needed time for the specimen handling.

The cupping test is a test method, in which a circular plate is shaped by a drawing die into a cup. The maximum ratio between the circular plate and the drawing punch diameter, which still allows a perfect production of a cup, is called limiting drawing ratio and is an indicator of quality for the formability of the sheet material. In Fig. 4 the modified sheet metal testing machine BUP 1000 is shown which is used to conduct remote cupping tests. Additionally to the BUP 1000 video cameras, a specimen holder, a tool holder, a handling robot, and some control units are needed.

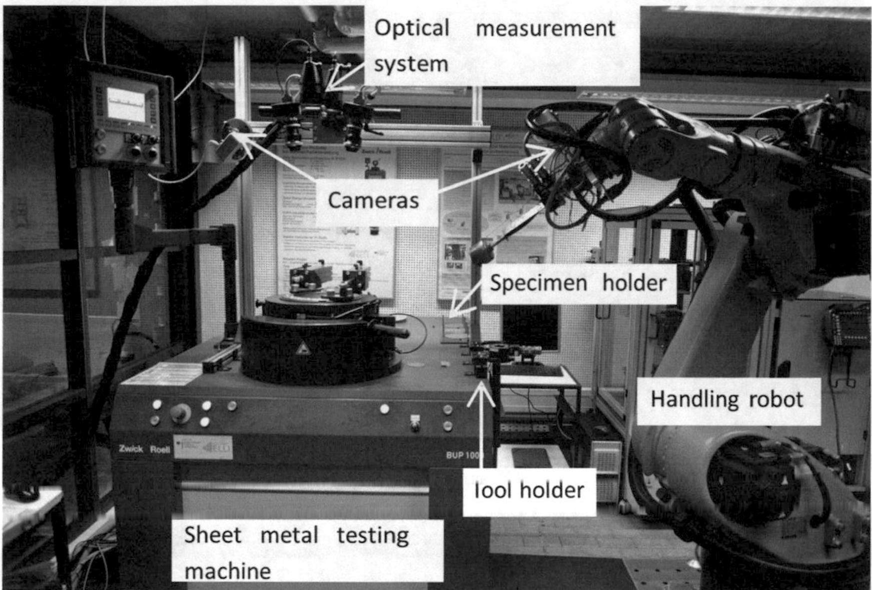

Fig. 4 Tele-operated sheet metal testing machine BUP 1000

4 Automation of the Cupping Test

In order to realize a teleoperated cupping test an automated control is needed. Here the cupping test especially in the view of automation can be divided into several little tasks [7]:

- Specimen handling
  - Getting specimen
  - Lubrication
  - Inserting Specimen
  - Removing the specimen (cup)
- Remote control of BUP 1000
  - Opening/Closing of the bayonet lock
  - Opening/Closing of the head
  - Starting and stopping the experiment
  - Crack identification

To conduct a remote controlled cupping test a human-machine-interface (HMI) was developed (see Fig. 5). With this interface the user can perform predefined functions without danger of damaging the setup. The depicted HMI is used to control the whole testing cell including additional machines like the tensile test machine.

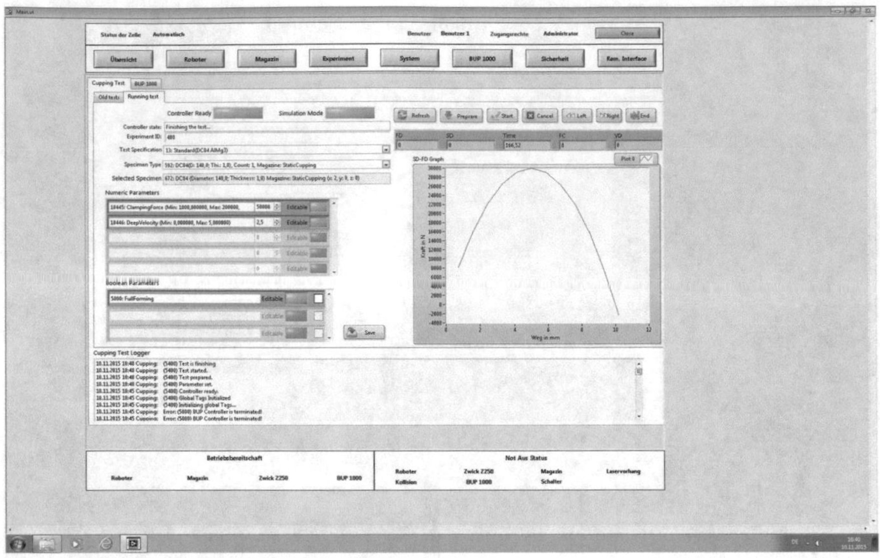

Fig. 5 Human-machine-interface (HMI) to conduct a remote cupping test

To perform a cupping using the HMI only few steps are required. In the following the basic steps are described. First of all the test specification has to be choosen. In our mask shown in Fig. 5 point 1 we have predefined specifications for materials like DC04 or AlMg3. Beside these specifications also other specifications can be defined. Afterwards, the specimen type has to be selected (Fig. 5 point 2). To define the specimen type the user has to enter the tap magazine (Fig. 6). For the respective positions in the magazine, the values of the corresponding specimens have to be filled in. Within the specification of the specimen the thickness is essential for the automation. If the thickness of the defined specimens don't correlate with the real thickness it would cause an error during the automation. The defined specimens can be identified later on by an individual ID which is created automatically in this step.

In Fig. 7 the real magazine is shown. This realized magazine is capable to manage specimens with the same diameter and has actually a capacity of 300 specimens. Variations are only possible by changing the thickness or the material of the specimen. Taking a look on the real magazine it is obvious that the defined thickness of the specimen is essential for the automation of the process. As the specimens are stored in a staple wrong defined thicknesses would result to a wrong calculated height of the staple and thus causing a collision of the handling robot.

In the next step the test parameters has to be defined. The user has to set up the clamping force of the head and the velocity of the ram. After that it can be choosen whether to draw a full or a half cup (Fig. 5 point 4). After saving the defined parameters the defined experiment can be initialized (Fig. 5 point 5 and point 6). In the initialization step the parameters are transferred to the BUP1000 and the gripper is changed if needed. After that the test specimen is picked up from the magazine

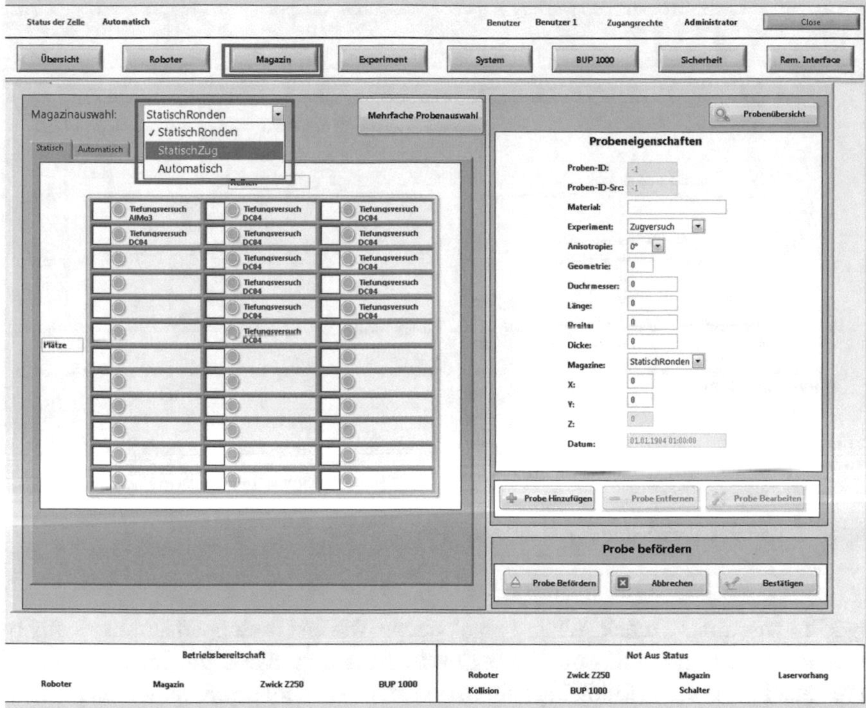

Fig. 6 Virtual magazine

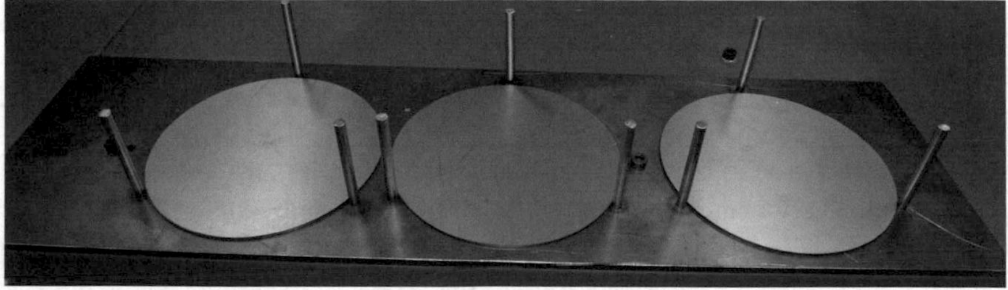

Fig. 7 Real magazine

and inserted into the testing machine. Thereby the specimen is also removed from the virtual magazine shown in Fig. 6. Finally the test can be started. In this step the measured ram force over the ram position is shown in a graph while the process can be observed by several installed cameras.

To prevent problems during the automatization a crack identification algorithm is integrated into the implemented software. The implemented algorithm continuously calculates the gradient of the force curve. Normally, if no crack occurs, the gradient changes only slowly. If the gradient of the force curve changes too fast the

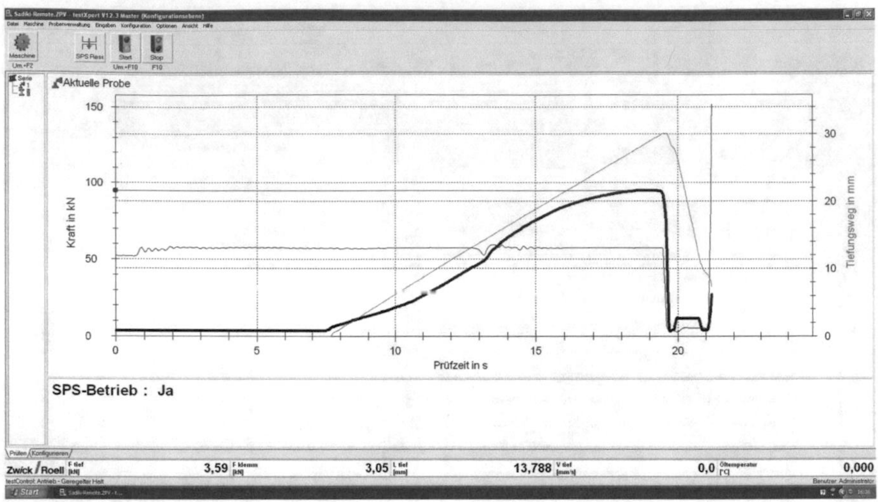

Fig. 8 Determined crack during a sheet metal test

algorithm assumes a crack. When a crack occurs the test is stopped automatically by the algorithm. Without this monitoring there is the risk, that a specimen is clamped in the machine and the following automation steps will not be continued. In Fig. 8 an example of a crack is depicted. In this case a crack was determined by the algorithm causing a stop of the experiment.

After the test the cup is removed by the robot and can be analyzed with the help of an extra camera (see Fig. 9). Here the user has the possibility to rotate the cup to be able to watch the cup from all sides.

In Fig. 10 the appereance of the test specimens before and after the test are shown. On the left side of the picture the initial specimen of cold rolled DC04 is shown. The two specimens in the center show results of a half cup. Here the left cup is without defects while the right one has a crack. The last sample on the right side is the result of a full drawn cup. In this sample the earing can be observed. The reason herefor is a greater resistance to deformation in rolling direction than in an angle of 45. In the deep drawing of a sheet therefore ears remain in the rolling direction and at right angles thereto. The earing can be prevented through a special rolling process or heat treatment.

In order to improve the students knowledge old experimental results can be compared with each other within the developed HMI. Hereby all conducted tests were stored in a database. If a user wants to compare old results they are loaded from the database and displayed on the HMI (see Fig. 11).

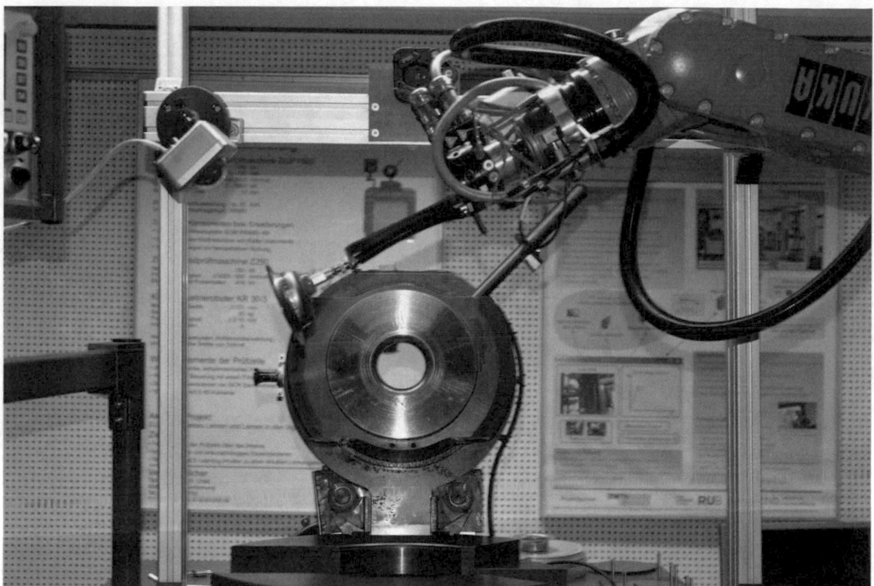

Fig. 9 Additional camera to analyze the results

Fig. 10 Appearence of the test specimens before and after the test

5 Integration into a Learning Scenario

Since material characterization is an essential part in forming processes, students deal with the aspect during lectures [8], different labs [9] or seminars in various facets. The described teleoperated testing cell was integrated into a fundamental lecture of forming technology in form of a live experiment. For this, a student centered approach was used to activate the students and make them thinking about material behavior.

For the integration of the interactive real-time experiment, the remote cupping test was realized. Students have the opportunity to experiment freely in predefined timeslots. They can carry out experiments with their own parameters and evaluate the results.

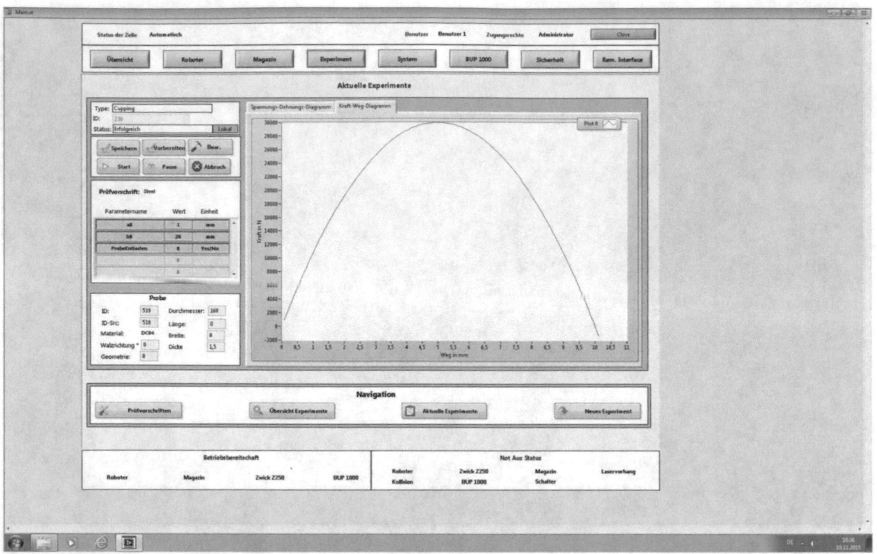

Fig. 11 Old results loaded from the database

The effects can be derived theoretically and afterwards shown practically or vice versa. It is very useful for lectures because it shows a real process with demonstrative effects. Here for example the reasons for earing and the influence of the rolling direction can be asked. Usually remote labs are used in electrical engineering so that this remote laboratory is a new aspect in lecture of forming technology. By integrating this concept into the lecture, many students can see and understand the process and the material behavior without being at the machine itself. The main advantage of developing a tele-operative control instead of showing a video is that the students can really influence the process. They have to think what parameters could be useful and can see their effect in the process directly. In addition they have several cameras to observe the process from different views as if they were right in front of it. The students can make their own experiences, which could lead to an open-minded student for new creative ideas. In this way they can deepen their theoretical knowledge by doing real experiments.

The process will also be integrated into remote-labs, so that students can do experiments with the machine and investigate the phenome at home.

After the successful first application of the remote cupping test the next step will be made in a pilot phase in January 2016. In this pilot phase the remote cupping test will be integrated into a lecture called “Umformende Fertigungstechnologien”. This lecture is part of the degree program mechanical engineering and industrial engineering and management. The access via internet will be made available for the lecturer conducting the experiment.

In a further step the students will be registered on a server where also timeslots can be booked. Within the booked timeslot the students will be able to access freely the experiment. They will be able to choose between two materials, steel or aluminum, and its direction of rolling. The specimen was either taken from a direction parallel or in a 45 angle to the rolling direction when producing the sheet metal. Also they have to define all described parameters needed to start the experiment like ram speed, clamping force or the lubrication.

It is a functional experiment and the safety of the machine and the other equipment. During the experiment the students could change the camera as described above.

Furthermore, it can be observed if students have experience in analyzing measured data. Here they can combine theoretical knowledge with practical experience, like to handle measured data, which is a typical task in professional life of engineers.

6 Conclusion

In this paper a developed teleoperated cupping test in the field of material characterization is presented. The cupping test is included in a testing cell consisting of an additional testing machine, an industrial robot, and other necessary components for the automation and conduction of experiments. The testing cell can be controlled via the internet from all over the world and is used to improve the quality of lectures. First, a methodology is introduced showing how the remote cupping test is realized. Afterwards, the integration of the cupping test in a remote laboratory is presented. The first integration into a lecture course will be in January 2016.

References

1. A. E. Tekkaya, Metal forming. In: *Handbook of Mechanical Engineering*, ed. by K.-H. Grote, E. K. Antonsson, Springer, 2009, pp. 554–606
2. L. Gomes, S. Bogosyan, Current trends in remote laboratories. In: *IEEE Transactions on Industrial Electronics 56.12*. 2009, pp. 4744–4756
3. J.E. Corter, J.V. Nickerson, S. K. Esche, C. Chassapis, S. Im, J. Ma, Constructing reality: A study of remote, hands-on, and simulated laboratories. In: *ACM Transactions on Computer-Human Interaction 14.2 (2 2007)*. 2007, pp. 1–27
4. T. R. Ortelt, A. Sadiki, C. Pleul, C. Becker, S. Chatti, A. E. Tekkaya, Development of a teleoperative testing cell as a remote lab for material characterization. In: *Proceedings of 2014 in International Conference on Interactive Collaborative Learning (ICL)*. 2014, pp. 977–982
5. V. J. Harward, J. A. Del Alamo, S. R. Lerman, P. H. Bailey, J. Carpenter, K. DeLong, et al., The ilab shared architecture: A web services infrastructure to build communities of internet accessible laboratories. In: *Proceedings of the IEEE, vol. 96*. 2008, pp. 931–950
6. A. Jiwaji, J. Hardison, K. P. Ayodele, S. Stevens, A. Mwanbela, V. J. Harward, J. A. Del Alamo, B. Harrison, S. Gikandi, Collaborative development of remote electronics laboratories. In: *2009 ASEE Annual Conference & Exposition*. June 14–17, 2009.

7. A. Sadiki, T. R. Ortelt, C. Pleul, C. Becker, S. Chatti, A. E. Tekkaya, The challenge of specimen handling in remote laboratories for engineering education. In: *12th International Conference on Remote Engineering and Virtual Instrumentation (REV)*. 2012
8. C. Pleul, M. Hermes, C. Becker, A. E. Tekkaya, Prolab@ing – projekt-labor in der modernen ingenieurausbildung. In: *TeachING-LearnING.EU innovations. Flexible Fonds zur Förderung innovativer Lehre in den Ingenieurwissenschaften*, ed. by M. Petermann, S. Jeschke, A. E. Tekkaya, K. Müller, K. Schuster, D. May, 2012, pp. 16–21
9. C. Pleul, D. Staupendahl, M. Hermes, S. Chatti, A. E. Tekkaya, Problem-based laboratory learning in engineering education – pbll@ee. In: *TeachING-LearnING.EU discussions. Innovation für die Zukunft der Lehre in der Ingenieurwissenschaften*, ed. by A. E. Tekkaya, S. Jeschke, M. Petermann, D. May, N. Friese, C. Ernst, S. Lenz, K. Müller, K. Schuster, 2013, pp. 193–198

Preparing for Industry 4.0 – Collaborative Virtual Learning Environments in Engineering Education

Katharina Schuster, Kerstin Groß, René Vossen, Anja Richert and Sabina Jeschke

Abstract In consideration of future employment domains, engineering students should be prepared to meet the demands of society 4.0 and industry 4.0 – resulting from a fourth industrial revolution. Based on the technological concept of cyber-physical systems and the internet of things, it facilitates – among others - the vision of the smart factory. The vision of "industry 4.0" is characterized by highly individualized and at the same time cross-linked production processes. Physical reality and virtuality increasingly melt together and international teams collaborate across the globe within immersive virtual environments. In the context of the development from purely document based management systems to complex virtual learning environments (VLEs), a shift towards more interactive and collaborative components within higher educational e-learning can be noticed, but is still far from being called the state of the art. As a result, engineering education is faced with a large potential field of research, which ranges from the technical development and didactical conception of new VLEs to the investigation of students' acceptance or the proof of concept of the VLEs in terms of learning efficiency. This paper presents two corresponding qualitative studies: In a series of focus groups, it was investigated which kinds of VLEs students prefer in a higher education context. Building upon the results of the focus groups, a collaborative VLE was created within the open world game Minecraft. First screenings of the video material of the study indicate a connection between communicational behavior and successful collaborative problem solving in virtual environments.

Keywords Engineering Education · Minecraft · Oculus Rift · Virtual Collaboration · Virtual Learning Environments

K. Schuster (✉) · K. Groß · R. Vossen · A. Richert · S. Jeschke
IMA/ZLW & IfU, RWTH Aachen University, Dennewartstr. 27, 52068 Aachen, Germany
e-mail: katharina.schuster@ima-zlw-ifu.rwth-aachen.de

Originally published in "Proceedings of the International Conference on E-Learning in the Workplace (ICELW 2015)", © 2015. Reprint by Springer International Publishing AG 2016, DOI 10.1007/978-3-319-46916-4_36

1 Introduction: Today's Learning and Working in Preparation for Industry 4.0

Today's portfolio of e-learning solutions is as diverse as never before. Different kinds of media services, software for teaching and learning as well as innovative hardware solutions not only become a bigger part in higher education and the workplace but increasingly adapt to the massive changes our working world is going through. A common and frequently cited example is the use of learning management software, based amongst others on the open-source management system Moodle [1]. Today, Moodle counts 53.738 registered installations with 68.7 million users of 226 countries in 7.7 million courses [2]. Platforms like Moodle have different functions, which can also be viewed as e-learning solutions themselves: With chats, forums or messenger systems, students or workers can communicate in synchronous or in asynchronous ways. Wikis enable cooperative text production and different kinds of assessment modes or quizzes give teachers the chance to test the students whenever they want and as many times they want during the semester. Here, one of the biggest advantages is that the tests are rated automatically, which makes the frequent testing also suitable for large groups. Being tested frequently, the students get instant feedback about their current state of knowledge. Digitally supported learning brings direct individual advantages in terms of self-awareness of the content of the lecture.

The digitalization of education also means that learning becomes more collaborative [3]. The key word "user generated content" describes the fact that in times of web 2.0, content rarely is produced by just one single provider of content, but is generated by several users instead. Transferred to the context of higher education, the students' role changes. Whereas back in the days, when the teacher was more or less the only source who provided information, today students can get basically any information they want from the internet, but can also contribute actively within forums, wikis or blogs. The potential is there to switch the students' role from rather passive users of information to creators of knowledge in networked structures – with all accompanying advantages and disadvantages. With the goal in mind not only to boost the students' knowledge and to support them to strengthen their personality over the years, but also to develop crucial competences for the working world they are about to step in, various types of collaboration have to be trained and tested in learning scenarios.

In a first step, one can differentiate between cooperative and collaborative learning. In cooperative learning scenarios, each group member is given a sub-task e.g. reading and interpreting different parts of scientific literature, technical reports etc. The individually produced results, e.g. a presentation, are simply being added up. Therefore, the main result mostly doesn't represent the state of knowledge of each individual group member. It is more a question of how to divide the work in an efficient, but not necessarily in an effective way.

Collaboration instead focuses on the creation of a new knowledge baseline, which is built through interlinked and co-referenced work during the learning process [4]. Especially in engineering, collaborative learning in virtual environments is highly important in the context of a dynamic and digitalized working world. This can be realized by analyzing a defective machine, coming up with a logistics concept for a virtual factory or designing a virtual car. The last example points out the importance for engineers to link their own specific technical expertise with expertise of other domains. Working in interdisciplinary teams situated all over the world is standard practice. The increasing digitalization of economy and society links knowledge over borders of time, space and systems. In times of industry 4.0, physical reality melts with virtuality [5]. For almost decades now, e.g. finite element models, data models, analytical models or CAD-models of machine elements have been produced with software. The data is used, provided and linked within socio-technical working systems via clouds, ubiquitous computing, product-lifecycle management and product data management. Thus, in engineering, human work processes are increasingly being transferred to virtual spaces of an internationally networked world.

2 Collaborative Learning in Virtual Environments

In higher engineering education computer-supported cooperative and collaborative learning (CSCL) have long been established as methods which support self-driven and work-related learning processes. By further technical development as well as new requirements of the changing working world such common methods can be lifted to a new level. Virtual learning platforms like moodle can systematically be linked to virtual or teleoperative laboratories. Every student gets the opportunity to experiment with physically real equipment without the necessity to be physically present at the location of the machine [6]. With special booking systems, expensive equipment for teaching and training processes can be used more efficient, since students from different time zones (e.g. USA and Germany) can log in at different schedules. Thus, it is possible to introduce students to learning settings, which would otherwise be too dangerous (e.g. an atomic power plant), too hard to access (e.g. the surface of mars) or too big a risk for ongoing production (e.g. in a factory) [7].

Moreover, in massive open online courses (MOOCs), each student can learn at his own speed. Serious games offer the possibilities to learn in a playful manner, in single-player or in multi-player mode. Innovative virtual knowledge spaces therefore offer all kinds of possibilities for learning and working in times of industry 4.0. In order to use the new technologies for engineering education in a proper way, deeper insights in reception, cognition and communication in virtual environments are necessary. Simply providing the technical infrastructure doesn't automatically guarantee successful collaboration. Therefore, the analysis of key factors for successful collaboration in virtual environments is an important field of research in the context of the working world of the future. Linking the different fields of this research is a core point for its success.

In the project "Excellent Teaching and Learning in Engineering Sciences" the three large german universities RWTH Aachen University, Ruhr-Universität Bochum and Technical University Dortmund focus on the development of virtual and remote laboratories as well as non-experimental collaborative learning spaces. In order to show students the "bigger picture" of the engineering profession, but also in the light of increasing numbers of students in engineering, the necessity of experimental equipment is obvious.

When working on the development of virtual or remote laboratories, the focus is clearly laid upon the final product and its future way of use. From a different point of view, looking at the current media use of students can help to predict the steps that still need to be done, if one day collaboration in virtual learning environments is supposed to prepare for industry 4.0 on a large scale. But are today's students ready for innovative teaching methods? Current studies of digital media usage show a rather passive usership. The majority of students hardly uses media services which require an increased work load by generating content (e.g. wikis or blogs), as a long-term study on media use of students shows [8]. A study with german engineering students (n = 1587) focused on the frequency of usage of different kinds of media services. The results show that interactive and collaborative media services are not used very often by the majority of the sample. This conclusion also counts for blogs or tools for collaborative text production such as wikis. In other words: Absorbing content is still more popular than generating it [9].

But the same study reveals another important aspect. Although not many students have been in contact with innovative teaching formats such as serious games or virtual courses in real-time; those who have experience with such formats are highly satisfied with it. These results are of high relevance for the development of virtual collaboration spaces, but also for companies who wish to use them. Developers need to know which the crucial features of a virtual environment are that really solve students' problems and are not just "nice to have". Moreover, universities or companies who invest in virtual learning environments aim for some kind of return on invest, which is not likely to come if the VLE isn't used. However, there is still little evidence on the motifs of students for using, or better for not using media services which require active participation. Although today's students all grew up in a digital society, user profiles are highly diverse. Providers of virtual collaboration spaces such as universities or companies need deeper insights in actual user preferences of specific target groups. Which level of graphical precision is required to understand complex processes, which level of gamification is preferred or which kinds of narrative scenarios would motivate this user group to deal with the content longer or more often still needs to be answered. Therefore, in order to investigate the described quantitative research in the field of collaboration in virtual environments, a qualitative research design was chosen, which will be explained in more detail in the following chapter.

3 Experimental Setting for the Analysis of Collaboration in Virtual Environments

3.1 *Lead User Workshops with Future Engineers*

Since today's students are going to be working within industry 4.0 contexts it is important to integrate them in the research process on VLEs which are supposed to prepare for the corresponding requirements. The approach of user-centered design is well-known in the field of software development, but also under the label of open innovation in the case of new product development [10]. Within idea competitions or lead user workshops, the approach has also proven to be helpful in the development of new teaching methods or new formats of virtual learning in context of the Bologna Process [11]. As representatives of future user groups of such innovative learning and working spaces, students are questioned within focus groups. Two workshops with 23 students from Germany and one workshop with 13 participants of a European study program were conducted (Fig. 1).

The two major aims of the workshop series were to collect the students' requirements but also their retentions on VLEs. The students were asked the following questions:

Fig. 1 Lead user workshops with students on the topic of virtual learning environments

- Which scenarios would you like to experience within VLEs?
- Which didactical method would you prefer, e.g. game based learning, free exploration etc.?

The students first had to work on the questions in small teams and then presented the results to the whole group. Each idea had to be written down on a prompt card. For a deeper insight into the workshop results, the cards were analyzed with qualitative analysis [12]. The contributions to the two questions were therefore clustered into topics. Afterwards, the quantity of contributions in each category was counted. The topics with the most contributions were considered the ones of greatest interest or greatest concern of the students.

The results of the analysis show that students equally prefer realistic (e.g. factory simulations) and fictional scenarios (e.g. traveling through a factory from the product's perspective). In case of fictional scenarios, the main principle is to exceed the limits of time, space and physics. The students like to be immersed by the virtual environment and to interact with it intuitively and naturally. The possibility to get instant feedback is valued very positive by the students. To combine learning with playing in terms of game based learning is welcomed by the students, but not necessary to enjoy the learning process within the VLE or to consider the VLE useful. The students had no major retentions to VLEs in general, but a few contributions referred to the concern that too many unnecessary features of the VLE could distract from the actual task and the content that should be learned or practiced.

Although surely not being the main contribution to learning success of students, the insights in students' preferences on VLEs delivered important information for the didactical design of future learning environments.

3.2 Work in Progress: Study on Collaboration in Virtual Environments

In line with the preference of students to be immersed in a VLE, a previous experimental study showed that students who used natural user interfaces for interacting with the virtual environment in individual learning scenarios experienced more immersion than students who solved the task on a laptop. Immersion generally referred to as "diving into the virtual environment" had been operationalized with the constructs of spatial presence and flow. An interaction of experienced flow and errors in task performance revealed the complexity of working in virtual environments: Being immersed by the environment unfortunately can also mean that one is absorbed more by the exploration of the environment than by solving the given task [13]. This finding stands in conflict with the user preferences found in the lead student workshops.

However, the nature of collaboration might help to compensate this problem. Since more people are involved in solving a given problem, more attention can be spent on problem-related details. Moreover, as mentioned in the introduction, the prediction of the working world of the future under the label of industry 4.0 specifi-

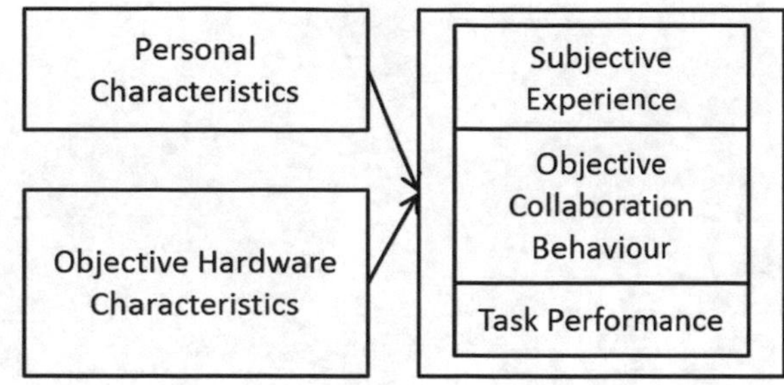

Fig. 2 Expected relationship between personal characteristics, hardware characteristics, subjective experiences, objective collaboration behaviour and task performance

cally emphasizes the importance of controlling complex, geographically distributed industrial processes [5]. Collaborating in teams with diverse professional and cultural backgrounds is an important precondition for the success of such processes.

To understand the complex interactions of different human factors in situations of virtual collaboration, a current experimental study focuses on preconditions for successful collaborative problem solving in virtual environments. This study assesses the relationship between personal characteristics, objective hardware characteristics, subjective experiences, objective collaboration behavior and task performance. Their expected relationship is visualized in Fig. 2.

The virtual environment is based on the results of the lead user workshops. Therefore the VLE had to be immersive, interactive, give instant feedback on the task performance and have elements of gamification in it. Since in the context of higher education personal and financial resources are mostly small for the development of virtual learning environments, the open-world game Minecraft was chosen as the setting for the learning environment. Minecraft has already been used for teaching and learning settings in the USA and the UK [14] and provides many features which are crucial for virtual collaboration:

- Quick construction of simple learning settings without any programming skills,
- Possibility to build more complex technical environments with the use of blue prints, available mostly for free within the Minecraft community [15],
- Simple and easy to learn modes of interaction without sophisticated gaming skills,
- Possibility to move around freely and to explore the scenario as a user actively.

To link the study to an industry 4.0 scenario, pairs of students are given a task with an engineering background. All participants have to solve the same problem in the same virtual environment, which is to restore electricity in a virtual building. From no perspective within the VLE the whole electrical setting can be viewed completely, which leads to the necessity for the students to actively communicate with each other. The students only know the target state, but not the steps how to get there. The task of restoring electricity within the building can be divided into the following sub-tasks, which have to be encountered by the students without further instructions:

- Get overview of the complete electrical setting of the building,
- Find out, at which spots the electrical circuit is broken,

Fig. 3 Screenshot of the VLE, implemented in Minecraft

- Remember the necessary steps to repair the electrical circuit,
- Find the spot within the building from where the success of the problem solving can be controlled.

The process of the collaborative problem solving has to be organized by the students themselves. Before they start as a pair, each student has to run through a tutorial individually. A screenshot of the virtual environment is pictured in Fig. 3. To analyze the possible interaction of immersion and task performance, the effect of natural user interfaces is integrated in the research design. The research plan consists of two groups. In both groups, the students work on laptops. Both groups use a simplified keyboard, where all keys except the arrows have been removed. With the arrow keys, the participants control horizontal movement. With a mouse, participants in both groups interact with the VLE. By clicking on the keys of the mouse, they can select different kinds of tools or resources they need to solve the problem. The experimental group fulfills the collaborative task wearing a head mounted display (Oculus Rift, DK 2). The field of view is therefore controlled by natural head movement. The control group controls the field of view by twisting the mouse.

Subjective experiences focus on immersion, operationalized with the constructs of spatial presence and flow. In this study, spatial presence is measured with elements of the MEC Spatial Presence Questionnaire of Vorderer et al. [16]. Flow as the mental state of operation in which a person performing an activity is fully immersed in a feeling of energized focus, full involvement, and enjoyment in the process of the activity, is measured with the shortscale of Rheinberg. According to this instrument, in essence, flow is characterized by complete absorption in what one does, as well as the feeling of smooth and automatic running of all task-relevant thoughts [17].

Additionally to the questionnaire, the participants of the study are being interviewed about their experiences. Within the interviews, the experiences are linked with personal characteristics of the participants, e.g. their gaming experience. The subjects are also asked about the experienced quality of the collaboration itself, more precisely their strategies of problem solving, communication and task management. Since diffusion of responsibility within teams has proven to be an inhibiting factor for the success of group work [4, 18] this aspect is also part of the interview. The behavior of the users is being captured by video camera, screen casts and spatial tracking systems. The task performance is measured in time needed for solving the problem.

A total of 8 students between 24 and 34 years ($M = 26$; $SD = 3.28$; $n = 5$ female) volunteered to take part in the pre-study. First screenings of the video material and the screen casts indicate a connection of problem-related speech-acts and task performance. Students who explicitly verbalize what they do and what they think the other one should do, get quickly to the point when they identify the necessary sub-tasks. For some students, especially those using the oculus rift, the tendency to "chit-chat" about the virtual environment from a meta-perspective and about its immersive effects was noted. For the analysis of the interviews it will be necessary to link this fact to the corresponding task performance. On the one hand, a strong interest in such "meta-information" can mean a positive effect regarding motivational aspects, but on the other hand it can mean some sort of distraction from the actual problem solving task, as it was indicated in previous studies on learning in virtual environments [13].

Comparing subjective experiences of the quality of collaboration with the actual task performance of the subjects will be another important aspect of the data analysis. However, corresponding conclusions will always have their limitations, since one of the greatest advantages of virtual learning environments is that people can use them at their own preferred speed.

4 Conclusion

The qualitative research on collaboration in virtual environments gives deeper insights into the relationship of personal preferences for VLEs, subjective experiences within them and actual task performance. To focus on virtual collaboration delivers important research results for universities and companies who wish to use virtual environments in the future with the vision of industry 4.0 in mind. The preliminary results of the video analysis of the pre-study indicate the importance of communication within virtual environments. If performance within VLEs was to be enhanced even from today's point of view, to train staff in virtual communication skills would be a promising point to start.

The approach of user-centered design helped to get deeper insights into specific design preferences of VLEs from a user group, who grew up in a digitalized society. However, the preferences can also be due to age instead of cohort. Especially in the light of demographic change and aging populations in Europe, it will be crucial to continue the kind of studies which have been presented in this paper. For the effective virtual collaboration within diverse teams, another research area covers collaboration of pairs with different skill levels: of novices and experts, of young and old co-workers or of IT-close vs. IT-distant people. Who can work best with whom, and in what kind of virtual environment will be an important aspect for effective human resources planning in companies.

Should research one day prove actual financial benefits of virtual collaboration, e.g. by reducing travel costs, this way of working will soon become established. Companies who know how to collaborate in virtual environments efficiently will have a strong competitive advantage compared to those who don't. Continuing this kind of research therefore is an important contribution towards a globalized, connected and digitalized working world in terms of industry 4.0.

References

1. Haerdle, B. (2013). Die Digitalisierung der Lehre. Wirtschaft und Wissenschaft, vol. 4, 2013, p. 12.
2. Moodle: Moodle Statistics. Online: http://moodle.net/stats/ [accessed on 12.03.2015].
3. Johnson, L. et al. (2014). NMC Horizon Report: 2014 Higher Education Edition. The New Media Consortium, Austin, Texas.
4. Niegemann, H. M. et al. (2008). Kompendium multimediales Lernen. Springer, Berlin Heidelberg, p. 338 f.
5. Federal Ministry of Education and Research: Industrie 4.0. Online: http://www.pt-it.pt-dlr.de/de/3069.php [accessed on 12.03.2015].
6. Terkowsky, C., May, D., Haertel, T., & Pleul, C. (2012, September). Experiential remote lab learning with e-portfolios: Integrating tele-operated experiments into environments for reflective learning. In *2012 15th International Conference on Interactive Collaborative Learning (ICL)*, pp. 1–7. IEEE.
7. Ewert, D. et al. (2013). Intensifying learner's experience by incorporating the virtual theatre into engineering education. In *Proceedings of the IEEE EDUCON Conference*, pp. 13–15. Berlin.
8. Dahlstrom, E., Walker, J.D. and Charles Dziuban, mit einem Vorwort von Morgan, G. (2013). ECAR Study of Undergraduate Students and Information Technology (Research Report). EDUCAUSE Center for Analysis and Research. Louisville, CO. Online: http://www.educause.edu/ecar [accessed on 12.03.2015].
9. Gidion, G., Grosch, M. (2012). Welche Medien nutzen die Studierenden tatsächlich? Ergebnisse einer Umfrage zu den Mediennutzungsgewohnheiten von Studierenden. In: Forschung & Lehre. Deutscher Hochschulverband (Hrsg.), 6/12. Online: http://www.forschung-und-lehre.de/wordpress/Archiv/2012/ful_06-2012.pdf [accessed on 12.03.2015].
10. von Hippel, E. (2005): Democratizing Innovation. Cambridge, MIT Press.
11. Koch, J. et al. (2011). Open Innovation - Kunden als Partner Wie Hochschulen von Unternehmen lernen können, in Wissenschaftsmanagement, vol. 17, no. 1, pp. 31–35. Lemmens, Bonn.
12. Gläser, J.; Laudel, G. (2010): Experteninterviews und qualitative Inhaltsanalyse. Wiesbaden: VS Verlag für Sozialwissenschaften.
13. Schuster, K. et al. (2014). Diving in? How Users Experience Virtual Environments Using the Virtual Theatre, in Proceedings of the 3rd International Conference on Design, User Experience, and Usability (DUXU 2014), Heraklion, Crete, 22.-27 of June 2014, pp. 636–646.
14. The Minecraft Teacher. Online: http://minecraftteacher.tumblr.com/ [accessed on 12.03.2015].

15. Elterpro: Realistisches Atomkraftwerk mit Kontrollzentrale. Video on Youtube. Online: https://www.youtube.com/watch?v=l72VcTI4D88 [accessed on 12.03.2015].
16. Vorderer, P.; Wirth, W.; Gouveia, F. R.; Biocca, F.; Saari, T.; Jäncke, F.; Böcking, S.; Schramm, H.; Gysbers, A.; Hartmann, T.; Klimmt, C.; Laarni, J.; Ravaja, N.; Sacau, A.; Baumgartner, T.; Jäncke, P. (2004): MEC Spatial Presence Questionnaire (MEC-SPQ): Short Documentation and Instructions for Application. Report to the European Community, Project Presence: MEC (IST-2001-37661). Online: http://www.ijk.hmt-hannover.de/presence [accessed on 14.10.2012].
17. Rheinberg, F.; Vollmeyer, R. Engeser, S.; (2003): Die Erfassung des Flow-Erlebens. In: Stiensmeier-Pelster, J.; Rheinberg, F. (Hrsg.): Diagnostik von Motivation und Selbstkonzept. Test und Trends. Göttingen: Hogrefe, pp. 261–279.
18. Niegemann, H. M. et al. (2008). Kompendium multimediales Lernen. Springer-Verlag, Berlin Heidelberg. S. 338 f.

New Perspectives for Engineering Education – About the Potential of Mixed Reality for Learning and Teaching Processes

Katharina Schuster, Anja Richert and Sabina Jeschke

Abstract The majority of mixed reality scenarios have been mainly the subject of game engines. 'Mixed Reality' describes the combination of virtual environments and natural user interfaces. Here, the user's field of view is controlled by his natural head movements via a head mounted display. Data gloves e.g. allow direct interaction with virtual objects and omnidirectional treadmills enable unrestricted navigation through a virtual environment by natural walking movements. To evaluate perspectives and potential for the use of mixed reality settings within engineering education an experimental study has been carried out, focusing on the impact of spatial presence and flow on cognitive processes. To assess the effects of natural user interfaces on cognitive processes, a two-group-plan (treatment and control group) was established. The mixed reality simulator was used as main stimulus of the treatment group whereas the control group used a laptop as interaction device. The learning environment was kept constant over both groups. The data were collected and interpreted with quantitative methods. Constraints of data collection exist since the influence of the hardware can only be evaluated within a set of independent variables, which consists of a combination of different user interfaces to a mixed reality simulator. Thereby not all of the disruptive factors could be eliminated. In this paper the study and the detailed results are described, which showed advantages especially regarding affective and motivational factors of virtual environments for cognitive processes. In particular, the depth of the resulting spatial presence and the phenomenon of flow are discussed. The paper closes with a discussion of the question, to what extend such innovative technologies establish new possibilities for educational sciences and pedagogics, especially focusing on engineering education and the field of virtual experiments.

Keywords Immersion · Spatial Presence · Flow · Learning · Natural User Interfaces · Engineering Education

K. Schuster (✉) · A. Richert · S. Jeschke
IMA/ZLW & IfU, RWTH Aachen University, Dennewartstr. 27,
52068 Aachen, Germany
e-mail: katharina.schuster@ima-zlw-ifu.rwth-aachen.de

Originally published in "Proceedings of the ASSE's 122nd Annual Conference & Exposition", © ASSE 2015. Reprint by Springer International Publishing AG 2016, DOI 10.1007/978-3-319-46916-4_37

1 Introduction – New Perspectives for Engineering Education Through Mixed Reality

A main goal of engineering education is the development of professional skills, so that graduates can apply their knowledge in their later working environment. A proper knowledge transfer is an important precondition for engineers to act competently and to solve different kinds of problems. However, due to the increasing number of study paths as well as the specialization of particularly technical oriented classes, there is a need for the integration of new media into the curriculum of most students [1]. Thus, the visualization of educational content in order to explain theory more concrete and tangible has gained importance. To prepare students adequately for new situations in their work life, virtual reality (VR) can be an effective instrument for learning and teaching processes. By imitating real-world processes, professional skills can be developed, increased or maintained. Especially if the learning process requires expensive equipment or usually would take place in a hazardous environment, the use of simulations is not only advantageous but necessary [2, 3]. Apart from the virtual learning environment (VLE), the hardware of the given user interfaces applied in the simulation environment can affect the learning process [4]. One approach of improving learning with simulations is the development of natural user interfaces.

Classical memory theory claims that if the context in which knowledge is applied resembles the context in which the information has been learned initially, the memory works better. Moreover, how well we can retrieve knowledge from our long term memory depends on the quality of how well we encoded the information in the first place [5]. Sweller's Cognitive Load Theory postulates that learning is as a task, which is partitioned in at least two parallel sub-tasks: Dealing with the content and controlling the learning environment with the respected user interfaces [6]. Therefore a lot of research and development activities follow the assumption that if the user can interface with the system in a natural way, more cognitive resources are available for dealing with the actual content related exercise, which would increase the efficiency of VLEs [7].

However, to assume that hardware or software characteristics automatically lead to better learning outcomes is risky. Not every new approach which is technically feasible improves learning in the sense of task performance. The danger of designing complex and expensive VLEs without having a positive impact on learning outcomes is obvious. However, judging the value of a VLE simply by its effect on task performance misses out on other factors which support learning. Boosting the students' motivation to deal longer, more steady or more effectively with the given content is also an important goal of VLEs in engineering education [8, 9]. Apart from learning outcome and motivation, a peak to a different domain reveals a third intended effect of virtual environments. According to the entertainment sector, the extent to which a game or in general a virtual environment can "draw you in" functions as a quality seal [2]. This phenomenon is often referred to as immersion [10].

Enabling natural movement as the most basic form of interaction is considered an important hardware quality to create immersion [11]. Manufacturers of hardware that are supposed to enhance immersion claim that "Moving naturally in virtual reality creates an unprecedented sense of immersion that cannot be experienced sitting down" [12]. Almost 20 years ago, this could already be confirmed by Slater [11]. Another basic assumption in the context of VLEs and natural user interfaces is that greater immersion means better learning and potentially higher training transfer [4, 7]. This suggests that immersion would be the precondition for better learning, caused by the qualities of the user interfaces. However, if virtual environments are used in educational contexts, those assumptions need to be confirmed by empirical evidence. The presented study therefore focuses on the following questions:

- Do natural user interfaces create a higher sense of immersion?
- Do natural user interfaces lead to better task performance?
- In what way do immersion and task performance interact in mixed reality learning environments?

If assessed in an experimental setting, the construct of immersion needs to be specified. Spatial presence and flow are considered key constructs to explain immersive experiences. In general, flow describes the involvement in an activity [13, 14], whilst spatial presence refers to the spatial sense in a mediated environment [11, 15]. Spatial presence, as indicated in the name of the construct, refers to the spatial component of being immersed, i.e. the spatial relation of oneself to the surrounding environment. If we experience spatial presence in a mediated environment, we shift our primary reference frame from physical to virtual reality [15].

2 Experimental Analysis of the Potential of Mixed Reality for Learning and Teaching Processes

2.1 *Study Design – Focusing on Spatial Abilities*

The study presented in this paper assesses the relationship between objective hardware characteristics, subjective experiences and task performance. Their expected relationship is visualized in Fig. 1.

All participants had to solve the same task in the same virtual environment, which was a large-scaled maze in a factory building. Within the maze, 11 different objects were located. The first task for the participants was to navigate through the maze

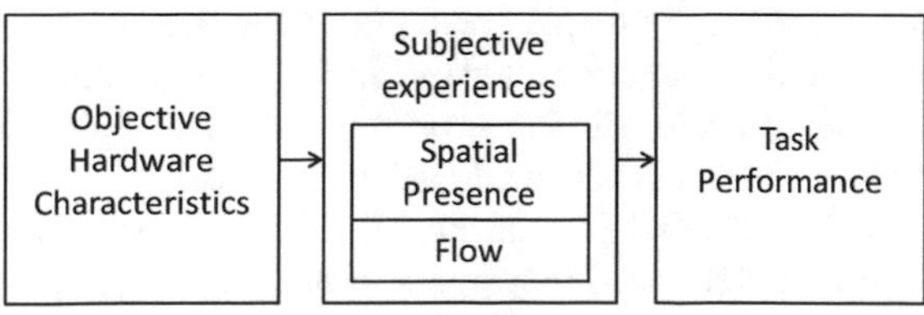

Fig. 1 Expected relationship between hardware characteristics, subjective experiences and task performance

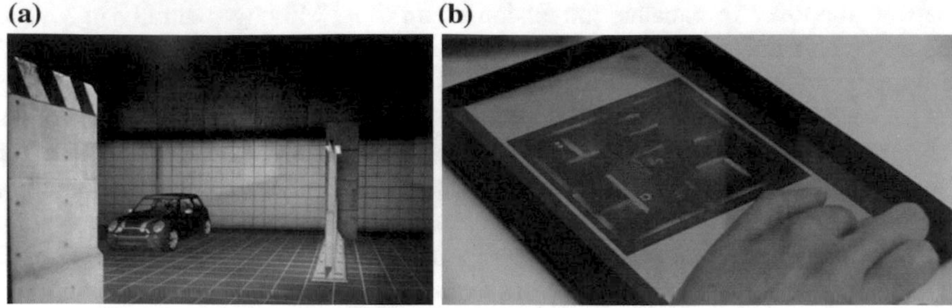

Fig. 2 View of the virtual environment used in the study in the first and in the third task

and to imprint the positions of the objects to their memory. For that, they were given eight minutes of time. The second task was to recognize the objects seen before in the maze. The third task was to locate the positions of the objects on a map of the maze. This was done with a self-programmed application on a tablet (Nexus 10) with a drag-and-drop control mode. The view of the maze in the first and second task is pictured in Fig. 2.

For both groups, the participants were given the chance to explore a test scenario (an italian piazza) freely for about three minutes before the actual task started. This was in order to get used to the respected control mode. All experimenters who conducted the experiments were trained in advance by experienced researchers. First they were being trained the functions of the hardware. In a second step, they took an observing position in a test run, and thirdly they conducted a test run on their own with the experienced researcher being the observer and giving feedback afterwards. Two groups of test persons were compared. The mixed reality simulator was used as main stimulus of the treatment group whereas the control group used a laptop as interaction device. The different hardware being used differed regarding the following characteristics:

- control mode of the field of view,
- control mode of locomotion,
- display and
- body posture of the user.

In the presented study, learning in a Mixed-Reality-Simulator was compared to a somehow conventional learning with a laptop. The technical equipment is described in more detail in the following:

Laptop. The type being used was a Fujitsu Lifebook S761 with a 13,3 inch display and a 1366 x 768 display resolution. The field of view was controlled with a mouse. Locomotion was controlled by WASD-keys, where W/S keys controlled forward and backward while A/D keys controlled left and right. The hardware usually results in a sitting body posture while using the device.

Mixed Reality Simulator. The Virtual Theatre is a mixed reality simulator which enables unrestricted movement through a virtual environment and therefore is used

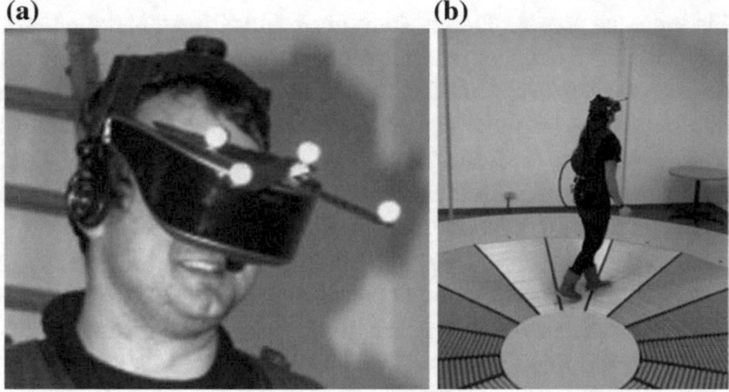

Fig. 3 Head Mounted Display and omnidirectional floor of the Virtual Theatre

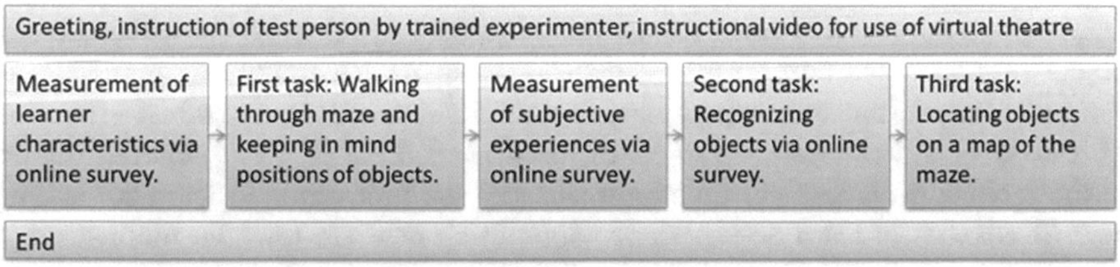

Greeting, instruction of test person by trained experimenter, instructional video for use of virtual theatre

Measurement of learner characteristics via online survey. → First task: Walking through maze and keeping in mind positions of objects. → Measurement of subjective experiences via online survey. → Second task: Recognizing objects via online survey. → Third task: Locating objects on a map of the maze.

End

Fig. 4 Procedure of the experiments

in an upright body posture. The user can move around within the environment by just walking in the desired direction. Therefore the control mode of locomotion is walking naturally. To track the movements of a user, the virtual theatre is equipped with 10 infrared cameras. They record the position of designated infrared markers attached to the HMD and an additional hand tracer. The components of the Virtual Theatre which came to use in the study are pictured in Fig. 3. For a more detailed and complete description of the technical system see Ewert et al. [2] and Johansson [7, 16].

Due to the composition of the simulator which was applied in the study, the hardware characteristics could only be tested in a certain combination and could not be isolated any further. The whole experiment took one hour. The complete procedure is visualized in Fig. 4.

2.2 *Variables and Measurements*

In this study, spatial presence was measured with elements of the MEC Spatial Presence Questionnaire of Vorderer et al. Several studies conducted by the authors strengthened the postulate of spatial presence being best explained as a two-level

Model. This includes process factors (attention allocation, spatial situation model, self location, possible actions), variables referring to states and actions (higher cognitive involvement, suspension of disbelief), and variables addressing enduring personality factors (i.e. the trait-like constructs domain specific interest, visual spatial imagery, and absorption) [15]. Suspension of disbelief refers to the extent of how much a person pays attention to technical and content-related inconsistencies. The more a person can fade out the action of "looking for errors", the higher the feeling of spatial presence will be according to the theory. In the presented study, instead of the subscales attention allocation and absorption, the Flow-Shortscale of Rheinberg was used. In this scale, flow is operationalized as the mental state of operation in which a person performing an activity is fully immersed in a feeling of energized focus, full involvement, and enjoyment in the process of the activity. In essence, flow is characterized by complete absorption in what one does, as well as the feeling of smooth and automatic running of all task-relevant thoughts [13, 14].

The perception of a learning situation is highly likely not to be influenced just by objective criteria such as the technical configuration of the learning environment. The strength of spatial presence experienced in a VE is supposed to vary both as a function of individual differences and the characteristics of the VE [4]. A general interest in the topic appeals to a person's curiosity and the motivation to learn something new. If chances to learn or experience something new are low, the motivation to learn decreases [8, 17]. However, not only interest in a topic but also in the way of presenting it can influence subjective experiences during the learning situation as well as learning outcome. The subscale domain specific interest of the MEC-SPQ refers to the topic of the medium, in this case the virtual environment. Because of the given considerations mentioned above and since interest in mazes didn't seem like a helpful operationalization for domain specific interest, it was adapted to interest in digital games. Based on all theoretical considerations, the general hypothesis of the study was that natural user interfaces should have a positive effect on subjective experiences during the learning situation as well as on learning outcome, in this case operationalized in task performance.

The set of hardware characteristics functioned as the first independent variable in the presented study. Furthermore, interest in digital games (second independent variable) was measured before the first task. As dependent variables, spatial presence and flow were measured after the first task which had to be fulfilled either in the Virtual Theatre or on the laptop. As dependent measures of task performance, three different variables were analyzed: The number of objects that were correctly recognized in the second task, the third task reaction time and the accuracy of locating the objects on the map in the third task.

2.3 Sample and Results

A total of 38 students between 20 and 33 years ($M = 24.71$; $SD = 3.06$; $n = 13$ female) volunteered to take part in the study. The sample therefore represents a potential user

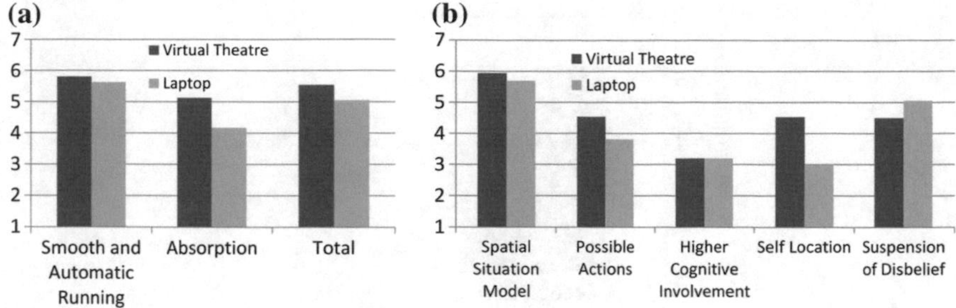

Fig. 5 Effects of Objective Hardware Characteristics on Flow and Spatial Presence

group of virtual environments in engineering education. They responded to a call for participation which was hung out at bulletin boards throughout the university but also posted on the front page of the virtual learning platform of the university and on several research and learning related blogs, social media platforms and news feeds. As an incentive and as a sign of appreciation, all participants took part in a drawing for a cordless screwdriver. All participants were healthy and highly interested in participating in the study. They did not report suffering from any physical or mental disorders. To rule out effects due to ametropia, participants were asked in advance to bring their corrective lenses just in case. If participants had been assigned to the mixed reality group, they were asked to wear sturdy shoes.

Hypotheses regarding influences of hardware conditions on subjective experiences and task performance measures were tested with analyses of variance (ANOVAs). With regard to the effects of the Virtual Theatre and the laptop on flow, significant differences were found (F (1, 36) = 4.18; $p < .05$). Thus more flow has been experienced in the Virtual Theatre (see Fig. 5). Taking a closer look on subscales there is a highly significant difference between conditions in self-reported absorption (F (1, 36) = 10.63; $p < .01$), but not in smooth and automatic running. There are also effects of hardware conditions on spatial presence. Self Location in the Virtual Theatre was rated significantly higher (F (1, 36) = 15.79; $p < .001$), which refers to the feeling of actually being in the virtual environment. Similarly, participants in the Virtual Theatre showed higher scores on the possible actions subscale of spatial presence (F (1, 36) = 4.90; $p < .05$). There were no further significant effects regarding spatial presence (see Fig. 5).

In addition to that the effects of hardware on task performance measures were calculated. In the recognition task of the objects from the virtual environment, participants in the Virtual Theatre condition made significantly more errors (F (1, 36) = 10.93; $p < .01$), which opposes the hypothesis that the use of natural user interfaces leads to better learning outcomes (see Fig. 6). There were no differences regarding time on task and deviation between the two treatment conditions.

Regarding the question, in what way immersion and task performance interact in mixed reality learning environments, a significant interaction between flow and hardware characteristics was found in the case of the task performance indicators "deviation" (F (1, 36) = 9.53; $p < .01$, see Fig. 7) and "total duration" (F (11, 36) =

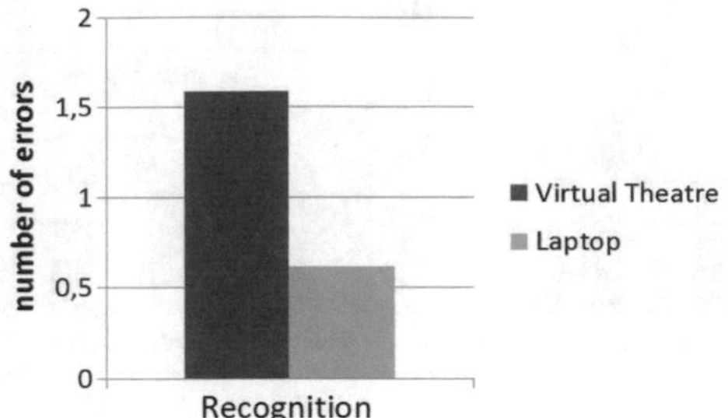

Fig. 6 Effects of Objective Hardware Characteristics on Task Performance, here: Recognition of previously presented objects

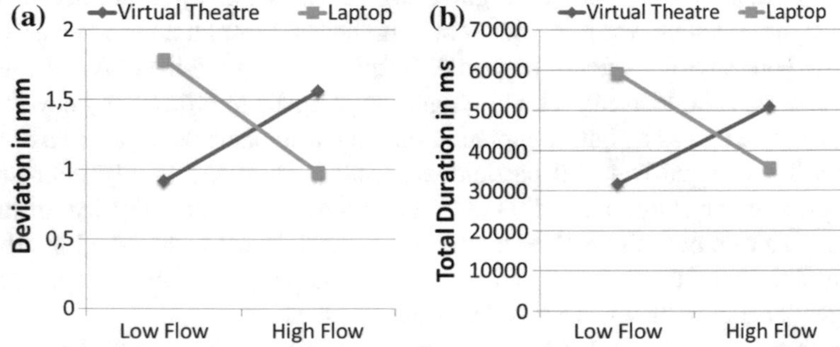

Fig. 7 Interaction between flow and user interface according to deviation and total duration (shown in percentile split)

4.65; $p < .05$, see Fig. 7). In other words, high values of experienced flow subside with better task performance, if a laptop has been used for learning, but with worse task performance if a mixed reality simulator has been used.

3 Discussion

The results show that mixed reality learning environments indeed lead to more flow. This is due to a higher self-reported level of absorption in the experimental group. Although flow is an activity related construct, this result is in line with the theoretical assumptions that hardware which allows natural walking can support the feeling of "diving" into the virtual environment, which in general terms is often referred to as immersion.

Next, the effects of the hardware on spatial presence are analyzed in more detail. Students who used the Virtual Theatre reported a higher self-location in the virtual environment which indicates that they had shifted their primary reference frame from the physical to the virtual world. Although the given task didn't require any further non-mental actions but navigating through the virtual environment, students in the

Virtual Theatre reported higher on the subscale of possible actions. However, for the other subscales, no differences were measured.

The interaction of flow and hardware characteristics in the case of the task performance indicators "deviation" and "total duration" leads to the assumption that in a mixed reality simulator, the subtasks of learning "switch roles": Controlling the mixed reality environment the phenomenon of "diving in" becomes the main task, while dealing with the actual content-related task moves gets less attention [6]. Although the given data set gives no insight to the question, if this is a conscious or a subconscious decision, this is a result of high importance in engineering education. Especially if the trend of developing immersive learning environments continues, the possibility for students to get used to the control mode is absolutely necessary in order not to inhibit the learning process.

According to the results of this study, immersion is not the precondition for better learning in virtual environments with natural user interfaces. Thus, the underlying model of the study (see Fig. 1) needs to be adjusted for further research. The only effect of the Virtual Theatre on task performance was a negative influence on recognition. This result is contradictory to the assumption that immersion leads to better learning. It seems that controlling the hardware was less intuitive than expected. This probably led to the typical situation for learning with virtual environments: Dividing the available cognitive resources on the two parallel sub-tasks of dealing with the content and controlling the learning environment with the respected user interfaces [6]. Moreover, the combination of an HMD and real physical locomotion could lead to cognitive dissonance. When wearing the HMD, the user can see where he or she walks in virtuality, but not in physical reality. Therefore the user takes a risk and has to trust in the technology in order to continue his or her actions. Last but not least, walking on the omnidirectional floor is a new experience for users and therefore could result in the fear of falling. All interpretations for the given results are going to be addressed in a follow-up study, where previous participants of the study will be interviewed on their experiences.

Finally, some limitations of the present study are considered that should be pursued for future research. One limitation refers to the type of hardware examined. Since the different technical characteristics of the Virtual Theatre can only be tested in a set, it is not possible to isolate single effects. The other aspect concerns the task chosen for this experiment. Low levels of cognitive involvement in both groups indicate that the whole sample might not have been challenged enough. Since challenge is an important precondition for the motivation for learning, more challenging tasks are going to be tested in the future.

This exploratory study on the effects of mixed reality learning environments on subjective experiences and task performance confirmed a few theoretical assumptions but also contradicted others. In a next step, interviews with participants from both groups are going to be conducted. A deeper insight on the participants' experiences will allow a more differentiated view on the subject of our research. Since the digitalization and virtualization of engineering education will play an increasingly important part in the future, it is of high value to know the exact advantages and disadvantages of immerging teaching and learning technologies, before they come into practice.

References

1. M. Kerres, *Mediendidaktik: Konzeption und Entwicklung mediengestützter Lernangebote*, 3rd edn. Oldenbourg Verlag, München, 2012
2. D. Ewert, K. Schuster, D. Johansson, D. Schilberg, S. Jeschke, Intensifying learner's experience by incorporating the virtual theatre into engineering education. In: *Proceedings of the 2013 IEEE Global Engineering Education Conference*. Berlin, 2013, pp. 207–212
3. S. Malkawi, O. Al-Araidah, Students' assessment of interactive distance experimentation in nuclear reactor physics laboratory education. European Journal of Engineering Education **38** (5), 2013, pp. 512–518
4. B. Witmer, M. Singer, Measuring presence in virtual environments: A presence questionnaire. Presence: Teleoperators and Virtual Environments **7** (3), 1998, pp. 225–240
5. P. Zimbardo, R. Gerrig, *Psychologie*, 7th edn. Springer, Heidelberg, 2003
6. J. Sweller, Element Interactivity and Intrinsic, Extraneous, and Germane Cognitive Load. Educational Psychology Review **22** (2), 2010, pp. 123–138
7. D. Johansson, *Convergence in Mixed Reality-Virtuality Environments. Facilitating Natural User Behaviour*. No. 53 In: Örebro Studies in Technology. Örebro University, Örebro, Sweden, 2012
8. A. Hebbel-Seeger, Motiv: Motivation?!– Warum Lernen in virtuellen Welten trotz-dem)funktionieren kann. Zeitschrift für E-Learning – Lernkultur und Bildungstechnologie **7** (1), 2012, pp. 23–35
9. F. Müller, Intresse und Lernen. Report – Zeitschrift für Weiterbildungsforschung **29** (1), 2006, pp. 48–62
10. J. Murray, *Hamlet on the Holodeck: The Future of Narrative in Cyberspace*. MIT Press, Cambridge, 1998
11. M. Slater, M. Usoh, A. Steed, Taking Steps: The Influence of a Walking Technique on Presence in Virtual Reality. ACM Transactions on Computer-Human Interaction **2** (3), 1995, pp. 201–219
12. Virtuix Technologies, 2015. URL http://www.virtuix.com/
13. M. Csikszentmihalyi, J. LeFevre, Optimal experience in work and leisure. Journal of Personality and Social Psychology **56** (5), 1989, pp. 815–822
14. F. Rheinberg, S. Engeser, R. Vollmeyer, Measuring Components of Flow: the Flow-Shot-Scale. In: *Proceedings of the 1st International Positive Psychology Summit*. Washington DC, USA, 2002
15. M. Slater, Measuring Presence: A Response to the Witmer and Singer Presence Questionnaire. Presence: Teleoperators and Virtual Environments **8** (5), 1999, pp. 560–565
16. D. Johansson, L. de Vin, Towards Convergence in a Virtual Environment: Omnidirectional Movement, Physical Feedback, Social Interaction and Vision. Mechatronic Systems Journal **2** (1), 2012, pp. 11–22
17. W. Edelmann, *Lernpsychologie*, 5th edn. Beltz PVU, Weinheim, 1996

Please Vote Now! Evaluation of Audience Response Systems – First Results from a Flipped Classroom Setting

Valerie Stehling, Katharina Schuster, Anja Richert and Ingrid Isenhardt

Abstract Many University lecturers in Germany face the challenge of teaching very large classes, sometimes including 1000 or even more students. They often have to cope with a very high level of noise, bad room conditions, an extremely low level of participation as well as interaction and feedback. Some lecturers therefore try to overcome these challenges by using technology in their classroom. Previous research has already focused on evaluating the use of audience response systems (ARS) in a traditional but very large engineering lecture. This sort of technology has proven to be an effective tool in order to e.g. increase student motivation, give them additional support in the learning process and on the other hand give the lecturer feedback about the students' learning progress as well as possible crucial points of the lecture. This paper, however, goes one step further. It analyzes the use of ARS in a flipped classroom setting of a large engineering lecture for first-year-students. After having completed almost two thirds of the flipped classroom lecture, students were being questioned about their experiences and opinions about the use of ARS in this particular educational setting. The standardized questionnaire included questions issuing e.g. comprehension, motivation, frequency, enjoyment, interaction, involvement as well as usability aspects. First results show that e.g. the majority of the students feel that clicker questions foster their comprehension, motivate them to be attentive and increase the quality of the lecture. When comparing the results to findings from previous research in a traditional lecture, however, one thing becomes apparent: The evaluation of the use of ARS in the in a flipped classroom setting has turned out to be slightly less positive than that of the traditional lecture. This finding will be particularly discussed and may even call for further research in the designated field of interest. In a first step, the lecture itself will be described considering content, background and general settings. Subsequently, the survey instrument and methodology will be presented. In a third step, the results of the survey will be presented and discussed. Finally, further research fields will be identified.

Keywords Large Classes · Clicker Questions · Flipped Classroom

---

V. Stehling (✉) · K. Schuster · A. Richert · I. Isenhardt
IMA/ZLW & IfU, RWTH Aachen University,
Dennewartstr. 27, 52068 Aachen, Germany
e-mail: valerie.stehling@ima-zlw-ifu.rwth-aachen.de

Originally published in "Proceedings of the 10th International Conference on e-Learning (ICEL 2015)", © 2015. Reprint by Springer International Publishing AG 2016, DOI 10.1007/978-3-319-46916-4_38

1 Introduction

Teaching engineering to a vast amount of students can be an extremely challenging task. Teaching these engineering students why communication and organizational aspects are extremely important for almost every field in their future careers is an even harder goal to be achieved. At the Center for Learning and Knowledge Management (ZLW) of RWTH Aachen University, the teaching method of flipping the classroom has been combined with the clicker technology in order to master this challenge. Previous research has already shown that the use of clicker questions in a traditional large lecture at the same institution with the same target group – students of mechanical engineering – is e.g. motivating the students to participate and be more attentive in class [1, 2]. The research at hand shall answer the question how the mix of the two teaching methods affects the class and how the results of the class evaluation turn out compared to previous evaluations of simply using clickers in the traditional classroom.

2 Lecture "Communication and Organization Development" ("KOE")

2.1 Content and Genesis

The lecture focused on in this paper is a lecture for first year mechanical engineering students. It is called "Communication and Organization Development" and will in the following be referred to as "KOE", from the German term "Kommunikation und OrganisationsEntwicklung". As has already been hinted at, the KOE lecture pursues one decisive goal. It aims to motivate young engineers to think interdisciplinary and learn how communication and organizational aspects are the key to success in student team work as well as professional life. To reach the targeted goal, many consecutive approaches have been introduced and tested in the lecture over the past years. In the early years, the lecture has been held as a traditional ex-cathedra lecture. It did, however, already include an extracurricular lab tutorial as well as online assessment at this stage. Shortly after, a learning management system in order to exchange data and provide online learning material has been added to the lecture. These data included slides as well as first video recordings of the lecture and served as helpful means for the students to prepare for and postprocess the lecture. In the following years it has proved as good practice to invite experts from the industry as well as alumni to the lecture. These experts were asked to give talks and enrich the lecture by combining their own experiences and practical examples with basic theoretical information about the topic that the specific class or module focused on. Altogether, the lecture consists of twelve modules [3]. Each of these modules covers one specific topic such as listed below.

1. Introduction
2. Basics of communicating in organizations
3. Basics of problem solving in teams
4. Basics of organizational development
5. Professional/cooperative processes in organizations
6. Learning and knowledge management
7. Global division of work
8. Intercultural cooperation
9. Organizational models and management approaches
10. Virtual Production
11. Innovation management
12. Change management

As a consecutive step, the lecture has been professionally videotaped and processed in small chunks of content in order to support student learning by allowing them to pause and afterwards easily reconnect and reenter the class. This adjustment of the lecture has been combined with the application of an audience response system in the lecture. The software-based system enables the students' participation and subsequently also the student-teacher interaction in class. The current and still prevailing adjustment to the lecture has been developed and introduced in 2012. It adds the method of flipping the classroom by giving an online lecture and using lecture time for discussion. This large group discussion is regularly fostered by the use of the previously described audience response system and shows the current educational setting of the KOE lecture which will be described in the following Chapter.

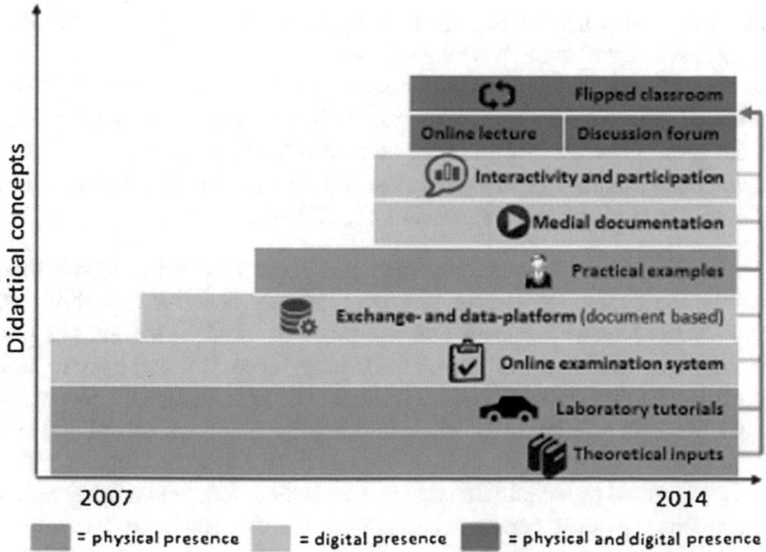

Fig. 1 Didactical concept development of the "KOE" lecture from 2007 to 2014 [3]

Figure 1 shows the development and advancement of the KOE teaching approach and consecutive settings from 2007 until 2014 [3].

2.2 Current Setting of the KOE Lecture

The KOE lecture has been held for many years at RWTH Aachen University. In earlier years the lecture has been extremely challenging for both students as well as lecturers, due to the high number of students (up to 1500), a resulting high level of noise and an extremely low level of student-teacher interaction. Therefore the lecture has been gradually redesigned over the years and is now being held as a "flipped classroom". The teaching model of "flipping the classroom" is a lecture design originated in the field of blended learning. The blended learning approach combines classical classroom lectures with various e-learning designs. In the case of the flipped classroom model that means that the lecture itself is made available to the students in the form of online videos. In addition to that there are classroom sessions, where students bring in the questions aroused by watching the lecture videos. In contrast to the traditional lecture models, the classroom roles are therefore "flipped" by making the students active participants asking questions and thereby defining the content of the classroom sessions. This active participation is the one decisive element that leads to the success and popularity of the method. "Students learn best when they are active participants in the education process" [4, 5]. When talking of lecture sizes beyond a seminar size, however, this task is a special challenge to every lecturer [6, 7] as well as their students:

- Speaking in front of a large class is often frightening for students. They might fear being exposed when giving a wrong answer.
- Large classes are loud. The level of noise can quickly become very high once the attention of the students starts to decline. When answering a question in a large class the answer is often overheard because it is too loud in class.
- Even when the lecturer frequently asks questions in his/her class, he/she still cannot involve all the students at the same time [1, 2, 6, 8].

As a result, it has become common practice at many universities around the world that lecturers use technology as a means to overcome the described possible obstacles and to still gain as much feedback and insight in the student learning as possible. One technological tool that has proved to be very successful and easy to handle in the described teaching scenario is the application of an audience response system – "ARS" or often also referred to as "clickers". A software-based model of these ARS which is accessible via smartphone, laptop or any device that can connect to the internet has been acquired especially for large lecture interaction. When using technology in class there are, however, certain aspects to be carefully considered. Beatty's statement from 2004 gets to the heart of the dilemma: "Technology doesn't inherently improve learning" [9]. "Kerres also states that digital media are no 'Trojan horses' that can be brought into an organization (or situation such as a class) and unfold their effect 'overnight'. Kay and LeSage reinforce these statements by stating that '(…) the integration of an ARS into a classroom does not guarantee improved student learning' but that 'it is the implementation of pedagogical strategies in combination with the technology that ultimately influences student success'" [1, 9–11].

Along with the application of the ARS in the flipped classroom a didactical concept has been developed in order to reach the highest possible outcome of the method

Table 1 Benefits of using ARS in class [1]

Benefits of clickers for the student	Benefits of clickers for the lecturer
Interaction with the lecturer without fear of compromising oneself	Identification of knowledge gaps
Immediate feedback	identification of shortcomings of the lecture [11]
Possibility to actively check their learning outcomes outside of exams	Student engagement
Be an active participant in class	Keeps students focused and involved
Anonymity	Higher attendance
Enhancement of learning	Better control of the learning progress
Classroom experience more enjoyable	…
…	

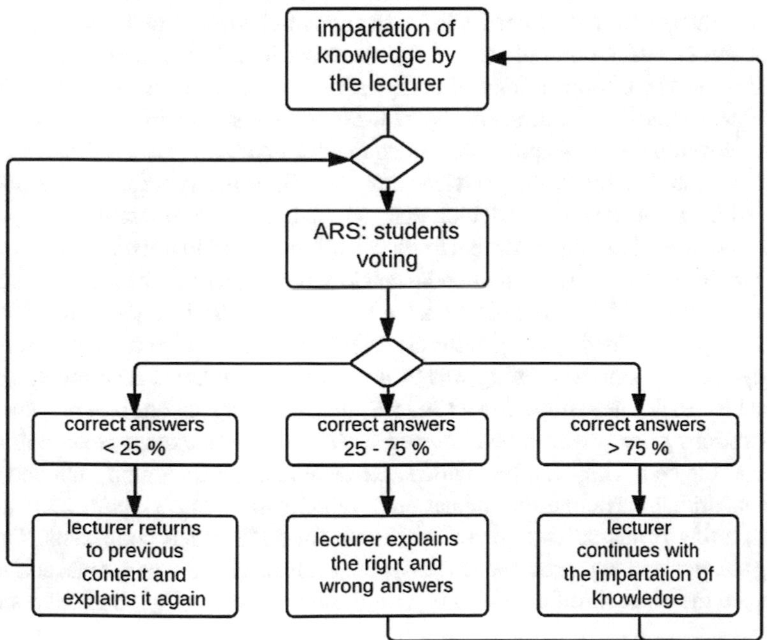

Fig. 2 Didactical concept of the "KOE" lecture [3]

for both. Major benefits such as fearless interaction through anonymity, immediate feedback etc. have been put together in previous research and are shown in Table 1.

The didactical concept provides three different procedures depending on the poll results. It starts with the lecturer explaining specific concepts, posing a poll question afterwards and giving the students time to submit their answers. If the number of correct answers is at less than 25 %, the lecturer will explain the previously conveyed content again, pose the poll question again until it shows significant improvement and then proceed. If the number of correct answers ranges from 25 to 75 %, the lecturer will take up these answers and explain the correct answer. Given that the content is mostly understood by the students (more than 75 % correct answers), the lecturer will move on to the next topic [3].

Figure 2 shows the didactical concept of the application of the audience response system in the flipped KOE lecture.

3 Survey Instrument and Methodology

Previous research in this particular field of study has focused on the introduction of clicker questions in a traditional but also large lecture with the same target group: first-year engineering students [1, 2]. The topic of the lecture then, however, was an introduction into computer programming for mechanical engineers. The paper at hand, as has been previously mentioned, concentrates on the evaluation of the use of clicker questions in an interdisciplinary flipped classroom setting. Of course, a comparison of evaluations of the same lecture but in two different settings would have allowed for a more intensive analysis of the one lecture itself. But despite the different topics and settings of the lectures, the questionnaire handed out to the students covered the same questions. Since the object of analysis – effects of clicker questions on student motivation and learning – was the same as before, it was possible to collect the same sort of data which makes both surveys comparable.

Worldwide evaluations show that the application of ARS in university lectures has led to e.g. higher motivation of attendance [12], more attention of the students during class and even higher knowledge acquisition than in traditional, non-interactive classes [13]. After having frequently used clicker questions from the beginning of the flipped classroom, the setting was evaluated by the students near the end of the semester in 2014. The sample covers 367 responses of first year engineering students. The questions posed were mostly closed-ended questions except for a comments section at the end. They can be divided into several sections which included questions concerning participation, impact on comprehension and content, motivational aspects, rating of the software itself and room for open ended comments. The following chapter will elaborate the results of the evaluation and – in a consecutive step – compare the most significant results with these of a previously conducted survey.

4 Results

4.1 Survey Results

The application of clicker questions in the flipped KOE lecture has – judging by the results of the survey – at large been received well by the students. Approximately 75 % of those questioned thought the application to be a reasonable approach. Around 70 % had a lot of fun answering the clicker questions and almost half of the questioned students felt that the clicker questions were motivating them to be attentive in class. The results of the survey can be summed up in five categories of the questionnaire: participation, motivation, questions, software as well as open ended questions.

4.1.1 Participation

97.8 % of the questioned students were in class when clicker questions were being posed. Of the 367 respondents, 33 % indicate to have participated in every single one of the polls. 44.7 % state that they participated only sometimes and a number of 21.5 % never entered a poll at all. To participate in a voting process, more than half of the respondents – 57.5 % – used their smart phone. Of the students who did not take part in the voting process, 34.5 % state that they do not own a suitable device for participation. A number of 25.5 % indicate that a lack of interest made them miss out on the clicker questions.

4.1.2 Comprehension and Content

85.6 % agree or somewhat agree, that the questions posed were comprehensible. 72.8 % of the respondents state that the questions posed were at a suitable or rather suitable complexity. Another 42.6 % agree or somewhat agree that the questions fostered their comprehension and were therefore helpful. 37.6 %, however, somewhat or even completely disagree. A 47.7 % stated that they would have wished for more clicker questions during the lecture and another 53.6 % even wished for more lectures with clicker questions in general. 76.3 % thought the application of clickers in class to be very useful.

4.1.3 Motivational Aspects

66.2 % of the students enjoyed or rather enjoyed participating in the polls. Around half of the respondents (54.2 %) agreed or somewhat agreed, that the application of ARS lead to a higher motivation to be attentive during the lecture. 38.5 %, however disagreed or rather disagreed.

Approximately half of the respondents (52.5 %) were motivated to participate in the polls. Another half of the questioned students (48.8 %) disagreed that the application of polls contributed to them attending class and a number of 67 % stated that the application of clicker questions enhanced the quality of the lecture.

4.1.4 Software

Around half of the questioned students stated that the software operated well while 37.9 % disagreed. The majority, a total of 85.3 % agreed that the software and terms of use were introduced and explained adequately by the lecturer. 55.9 % thought the software to be easily manageable, another 26.4 % rather agreed to that.

Figures 3 and 4 sum up the most significant results of the survey.

4.1.5 Open Ended Questions

Most comments in the open-ended comments section at the end of the survey are related to the internet connection and the ability to participate. Apparently the WiFi connection in class often broke down before all students were able to participate. Some students stated that the questions posed in the classroom sometimes were too

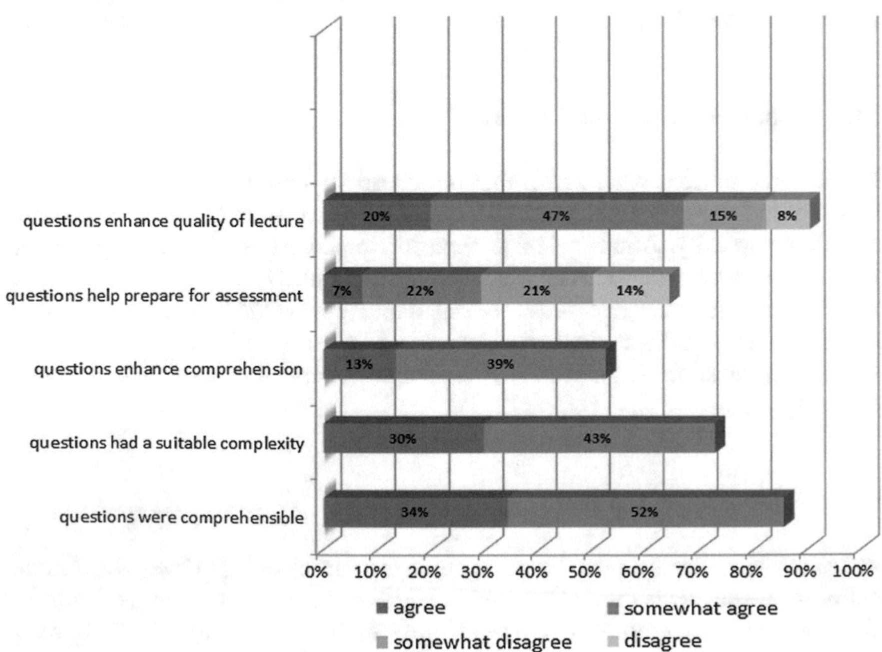

Fig. 3 Results of the flipped classroom clicker evaluation

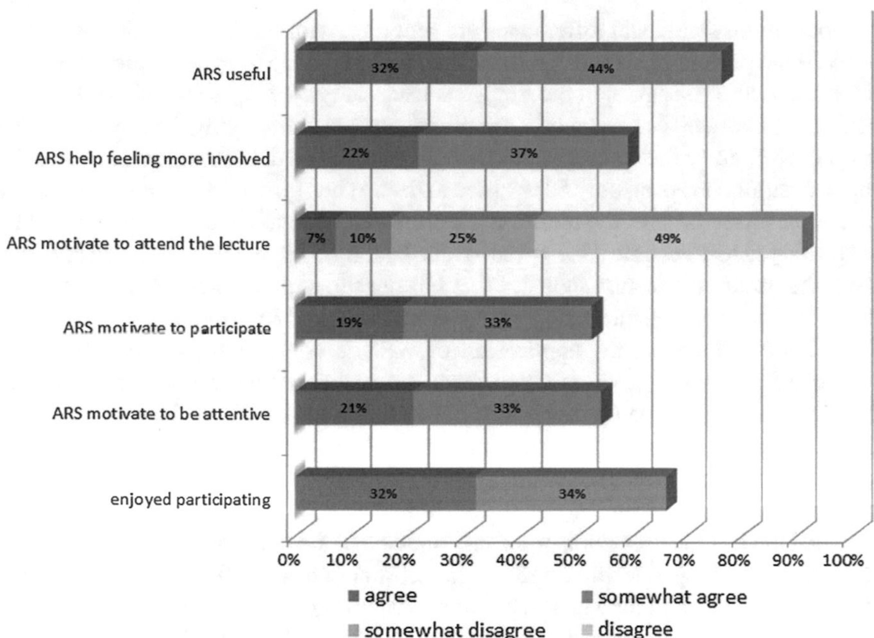

Fig. 4 Results of the flipped classroom clicker evaluation

easy and that they would have wished for a higher complexity factor. Other comments concern motivational state that the clicker questions motivated students to focus on the content again after having been diverted.

4.2 *Flipped Classroom Versus Traditional Lecture*

When comparing the results of both surveys – the first conducted in a traditional lecture and the second conducted in a flipped classroom setting – the first impression is that in both settings the application was rather successful. Only by taking a closer look at the results, smaller differences as well as some significant differences in the rating are noticeable. In order to identify similarities and differences, the most significant results elaborated in the previous chapter will be picked up again and compared to those of the earlier conducted survey. For the purpose of a higher clarity, the numbers in percentage will be rounded and the classifications "agree" and "rather agree" as well as "disagree" and "rather disagree" will be aggregated.

In the traditional lecture, 71 % of the students agree or rather agree that the clicker questions enhance the quality of the lecture. In the flipped classroom setting, the number of agreement lies slightly below at 67 %.

Some more significant differences are recorded with the following items. When asked if the polls help prepare for assessment, a total of 35 % agree in the traditional lecture and 56 % disagree. In the flipped setting, only 29 % agree but also only 35 % of the respondents disagree while the others abstain from voting. In the traditional setting, 63 % agree that the application of polls enhanced their comprehension. In the flipped setting, however, only 52 % agree to that. When queried about the complexity of the questions, 82 % of the traditional lecture respondents state that it was suitable. In the flipped lecture, only 73 % agree to that. Finally, 93 % of the questioned students from the traditional lecture indicate that the questions posed were comprehensible while in the flipped lecture a slightly smaller number of 86 % agree.

In both evaluations, the application of ARS is considered a useful approach, although the percentage of approval from the traditional lecture is slightly higher (83 %) than in the flipped setting (76 %). The following questions deal with motivational effects of the application of clickers in the classroom. 70 % of the traditional lecture state that ARS help them feel more involved in the lecture. In the flipped classroom evaluation, only 59 % agree to that. When asked if the ARS motivate students to attend the lecture, only 29 % of the respondents of the traditional lecture compared to 17 % of the flipped lecture, agree. The most significant differences between the two surveys appear in the last three items concerning motivational aspects towards participation, attention and enjoyment. In the traditional lecture, 69 % state that the use of clickers in the classroom motivates them to actively participate in the lecture.

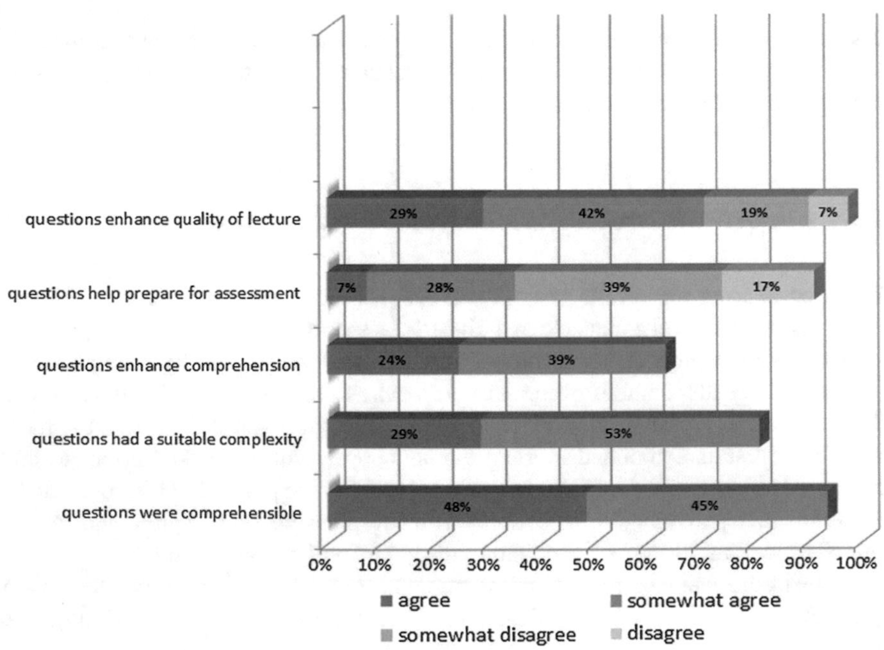

Fig. 5 Results of the traditional lecture clicker evaluation

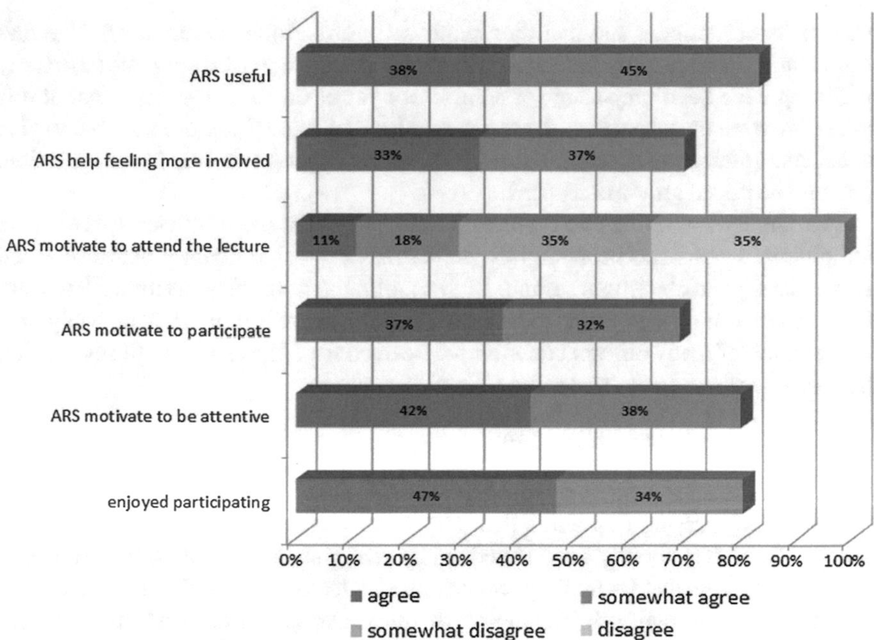

Fig. 6 Results of the traditional lecture clicker evaluation

In the flipped classroom, however, only 52 % agree to that. Considering one key factor for the lecturer of the class, that ARS motivate the students to be attentive and focus on the lecture, the following results are recorded: 80 % of the traditional lecture agree while only 54 % of the flipped classroom students support this view.

The last item in focus deals with the fun factor of the polls. 81 % of the students in the traditional lecture enjoyed participating while only 66 % of the flipped lecture students agree to that.

Figures 5 and 6 sum up the most significant results of the clicker evaluation of the traditional lecture.

5 Discussion and Further Research Fields

To sum up the results of the survey, audience response systems (ARS) are altogether received well with first-year engineering students. Most students appreciate the chance to actively participate in a lecture. In addition to that it also encourages them to be more attentive in order to be prepared for the questions coming up. This again helps the lecturer to stay focused and not be distracted by the high level of noise. The two major points of disagreement in both surveys were that ARS help to prepare for assessment and that they motivate to attend the lecture at all. The first result can be explained by the fact that students at the point of filling out the survey might not have been preparing for assessment yet or on the other hand that at this point they were unsure of how the test would look like and therefore not able to give an adequate statement about this just yet. In order to get an answer to this question, it should be posed after assessment.

The comparison of the two evaluations shows that the use of clicker questions in a traditional lecture has been slightly, sometimes even significantly better received than in the flipped classroom setting. At this particular point reasons for such a result can only be estimated and need to be verified by further research. Most possibly they are a mixture of many different conditions. Derived from the results of the evaluation these can be summarized under the following captions:

Complexity of/interest in the topic

One significant comment reinforces the impression that one gets when speaking to a lot of engineering first-years that attend the lecture. It reads: "mathematics is more fun". As has already been stated, engineering students in their first semester often do not understand why they should attend a lecture that at first sight does not directly connect to their original subject of study. Compared to many other courses in the curriculum of mechanical engineering, the subject of communication and organization management is felt to have a lower complexity and seems more easily intelligible to many students, which is why they often underestimate its importance. This, in turn, can lead to students feeling insufficiently challenged and sometimes even bored and frustrated.

Level of interaction

The flipped classroom setting combines several methods of interaction with poll questions just being one of many other interactive elements. This might lead to a slight reduction of the impact of the audience response system. Compared to that, in the traditional lecture, the poll questions were the only interactive element of the lecture which might have been a cause for the slightly better evaluation. The students attending the "KOE" lecture are usually in their first semester and mostly haven't attended any university lectures yet, they are rather used to traditional and obligatory courses from school. The method of flipping the classroom is yet a rather uncommon one in Germany and might also lead to an overstimulation-induced shutdown with some students. This should be subject of further research.

Technical difficulties

Apparently, the WiFi connection broke down several times during class times due to the high number of participating students in the poll. The resulting network traffic peaks led to connection difficulties or even errors. Most comments in the free comments section refer to this circumstance. Presumably this circumstance caused frustration with several students who tried to participate but couldn't.

Another interesting object of further research would be to query lecturers who use ARS in their lectures in order to find out whether or to what extent the described benefits of ARS are perceptible. Further research can be dedicated to the question if and – if yes – how the results of the polls are used for curricular optimization and didactical adjustments.

References

1. V. Stehling, U. Bach, A. Richert, S. Jeschke. Teaching professional knowledge to XL-classes with the helph of digital technologies, 2012
2. V. Stehling, U. Bach, R. Vossen, S. Jeschke. Chances and risks of using clicker software in XL engineering classes - from theory to practice, 2013
3. L. Koettgen, S. Schröder, E. Borowski, A. Richert, I. Isenhardt. Flipped classroom on top - excellent teaching through a method-mix, 2014
4. T.B. Crews, L. Ducate, J. Rathel, K. Heid, S. Bishoff, Clickers in the classroom: Transforming students into active learners. ECAR Research Bulletin (9), 2011
5. B. Davis, *Tools for Teaching*. Jossey Bass, San Francisco, 1993
6. R.J. Anderson, R. Anderson, T. Vandegrift, S. Wolfman, K. Yasuhara, Promoting interaction in large classes with computer-mediated feedback. In: *Designing for Change in Networked Learning Environments*, ed. by B. Wasson, S. Ludvigsen, U. Hoppe, no. 2 In: Computer-Supported Collaborative Learning, Springer Netherlands, 2003, pp. 119–123
7. University of Maryland. Large classes: A teaching guide. center for teaching excellence, 2008. URL http://www.cte.umd.edu/library/teachingLargeClass/guide/index.html
8. B. Hasler, R. Pfeifer, A. Zbinden, P. Wyss, S. Zaugg, R. Diehl, B. Joho, Annotated lectures. student-instructor interaction in large-scale global education. Journal of Systemics, Cybernetics and Informatics **7**, 2009
9. I.D. Beatty, Transforming student learning with classroom communication systems. EDU-CAUSE, Research Bulletin **2004**, 2005
10. R.H. Kay, A. LeSage, A strategic assessment of audience response systems used in higher education. Australian Journal of Educational Technology **25** (2), 2009, pp. 235–249
11. M. Kerres, Wirkungen und wirksamkeit neuer medien in der bildung. In: *Education Quality Forum. Wirkungen und Wirksamkeit neuer Medien*, ed. by R. Keill-Slawik, M. Kerres, Waxmann, Münster, 2003
12. A. Trees, M. Jackson, The learning environment in clicker classrooms: student processes of learning and involvement in large university-level courses using student response systems. Learning, Media and Technology **32** (1), 2007, pp. 21–40
13. A.W. Nicolai Scheele, Die interaktive vorlesung in der praxis, 2004, pp. 283–294

Development of a Tele-Operative Control for the Incremental Tube Forming Process and Its Integration into a Learning Environment

Rickmer Meya, Tobias R. Ortelt, Alessandro Selvaggio, Sami Chatti, Christoph Becker and A. Erman Tekkaya

Abstract A deficient access to experimental equipment leads to the usage of remote labs to improve engineering education and open experiments for every student location- and time-independent. The usage of a tele-operative controlled industrial bending process in lecture combines theoretical learning contents with practical experiences. Lecturers can make experiments in interaction with the students, who are able to assist in choosing the process values. The chosen and presented bending process is the incremental tube forming process that uses in contrast to many ordinary bending processes targeted the superposition of stresses. By superposing of stresses, in this process for example a tube bending and a tube spinning process, several fundamental process characteristics can be observed and integrated into lectures to visualize the theoretical fundamentals behind. Incremental tube forming combines the tube spinning process, which affects the diameter of the tube all along the tube and creates a compressive stress, and a bending process. The understanding of superposition of stresses and the process phenomena are ambitious, so that experimental experience is very useful. By using a tele-operative control, the experiment is location- and time-independent available for lecturers and students all over the world. They can interact with the process like stopping it, influencing it during the process or laying it up. The possibilities for a usage in learning environments are described and pointed out.

Keywords Remote Lab · Experiment · Incremental Tube Forming

1 Introduction

At first the setup of the machine is described. Therefore, the important process characteristics are described to understand which phenomena will be shown with this process. Then the setting of the tele operative control and important programming

---

R. Meya (✉) · T.R. Ortelt · A. Selvaggio · S. Chatti · C. Becker · A.E. Tekkaya
TU Dortmund University, Institute of Forming Technology and Lightweight Construction (IUL), Dortmund, Germany
e-mail: rickmer.meya@iul.tu-dortmund.de

Originally published in "IEEE Global Engineering Education Conference (EDUCON)", © 2016. Reprint by Springer International Publishing AG 2016,DOI 10.1007/978-3-319-46916-4_39

aspects in LabVIEW are shown. LabVIEW is used because it is a powerful programming environment with good measuring applications and good graphical design possibilities.

After describing the most important program features the integration concept into the learning environment is presented. The main focus of this tele-operative control is to use it in lectures but it can also be used in labs, e-learning or online-courses. A theoretical explanation will be presented. This model can also be simplified for other precognitions, like schools for example.

2 Structure of the Tele-Operative Control

2.1 *Important Process Characteristics*

In the incremental forming process, a tube bending and a tube spinning process are superposed [1]. By using this superposition, a reduction of bending forces and springback can be achieved [2]. In Fig. 1 the setup of the incremental tube forming machine is shown. The process has been developed at the Institute of Forming Technology and Lightweight Construction at the TU Dortmund University. The ready-made machine with its Siemens control is available for engineering education and its main elements

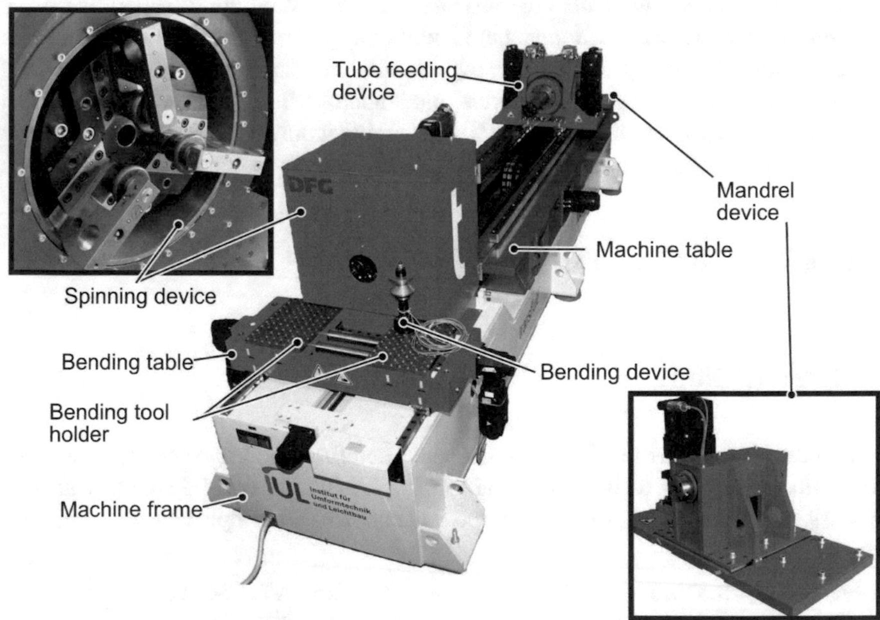

Fig. 1 Overview of the incremental forming machine [2]

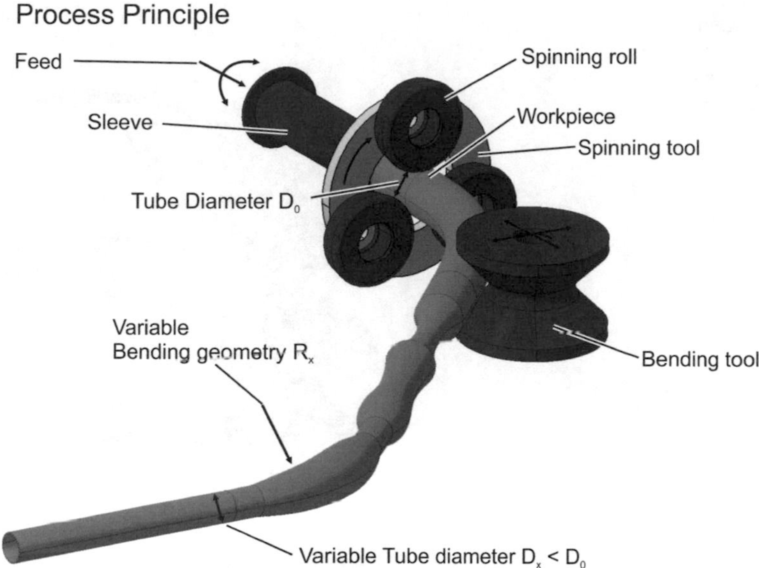

Fig. 2 Process principle of the incremental forming process [4]

are described in the following (Fig. 2). For feeding the tube to the bending unit an unbent tube is fixed in a feed unit. The feed unit pushes the unbent tube through a sleeve so that it is routed. These sleeves are changeable so that different raw diameters can be used in the process. Behind the sleeve there are spinning rolls to vary the diameter of the tube. These rolls rotate at a high adjustable rotational speed so that a spinning process is superposed. The radial infeed is also adjustable and leads to a decrease of the tube diameter. It is possible to change these parameters in the running process. The usage of a mandrel supports the bending process and gives the opportunity to decrease the wall thickness because it bears the tube from the inside. After the spinning rolls there are two bending units with adjustable levers. The bending rolls are path controlled and bend the tube vertical to the feed direction. The rolls are designed that they can bend many different tube diameters. In addition, they are easily changeable if another diameter is desired. By usage of the spinning and the bending there is a superposition of stresses in the forming zone [3].

The phenomena in this process is that the superposition of two stresses reduces the bending force and the springback. The bending force is recorded through the whole process by strain gauges that are mounted in two directions at the bending tool (Fig. 3).

LabVIEW uses the data and creates a graph to show the bending force over the time directly at the control panel. Many other graphs can be created by evaluation of the movement of the different axis which are recorded at the machine. It is then possible to record the bending force over the movement from the bending device for

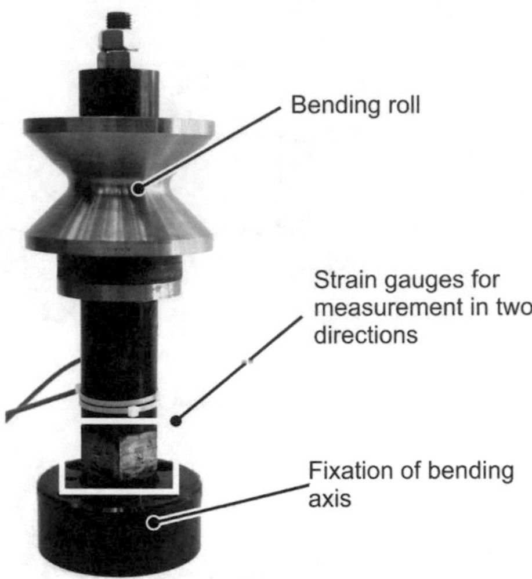

Fig. 3 Strain gauges in wheatstone bridge at the bending axis [2]

example. Out of this, the reduction of the springback can be estimated. Alternatively, the springback can be measured after the process by the usage of camera systems.

For the representation of the process phenomena different process steps will be performed. At first the process will be carried out without the spinning rolls in engagement. By feeding the tube and moving the bending axis simultaneously, the tube will be bent to a certain radius under load. The required bending force will be logged. By multiplying the required bending force with the lever arm of force the required bending moment can be evaluated. In the next steps the pushing rolls are additionally used with adjustable feed and rotational speed. By adjusting these parameters up to a certain maximum and recording the bending force, the effect of the superposition can be shown. The bending force and therefore the bending moment will decrease apparently. The springback will act similar. By increasing the rotational speed and increasing the feed of the pushing rolls the springback will be reduced [5]. This phenomena are shown schematically in Fig. 4.

2.2 Setup of the Tele-Operative Control

The tele-operative control is integrated into the existing environment of the incremental tube forming process. With the help of LabVIEW the connection between the user and the machine control is done. A tele-operative accessible user interface will be designed, so that the lecturers in interaction with the students can interact and change important process values. These parameters are implemented in a safe process window, so that the machine environment will be always safe and the tube will not exhibit failures normally, if the process is remote controlled.

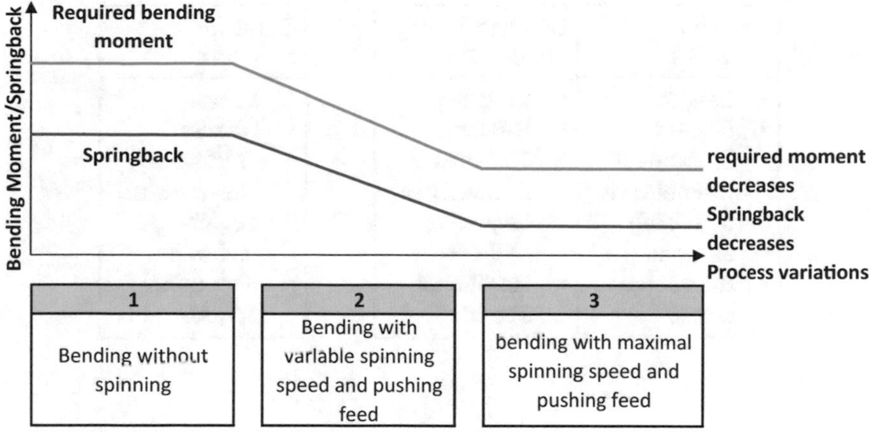

Fig. 4 Schematic effects shown with the process

For the programming of the tele-operative control the LabVIEW program has several demands to meet, which are described afterwards:

1. Creating the geometry data
    a. Interact with the user tele-operatively
    b. Only accept data which is safe for the process
    c. Visualize the tube in 2D and 3D
2. Machine data
    a. Creating the machine data out of the geometry
    b. Visualize the movements of the machine axis
3. Transfer to machine
    a. The machine data has to be transferred from the LabVIEW Computer to the control of the machine
4. Measurements
    a. Measurement of bending force
    b. Measurement of springback
    c. Measurement of axis movement
    d. Analysis of the input data
    e. Analysis of springback
5. Tools
    a. Tools helping students to understand the process phenomena

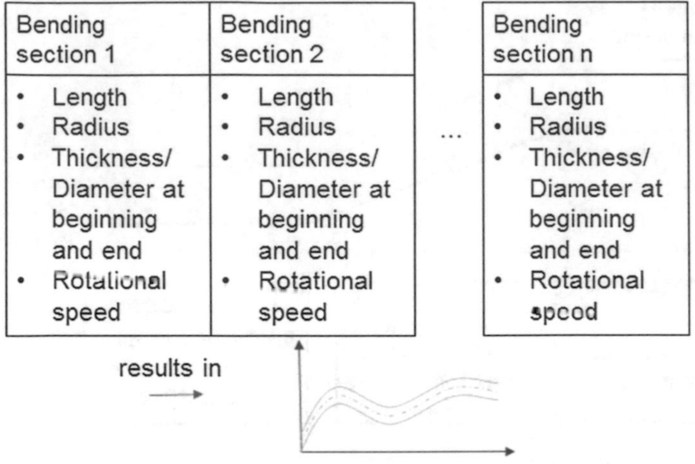

Fig. 5 Different bending sections for defining a variable geometry

6. Environment
    a. Physical connection to machine control
    b. Camera implementation
    c. Implementation of a cutting tool
7. Data management
    a. The data should be available for saving after the experiment
    b. Students should be able to take their results home

For starting the experiment, a description how the bent tube will look like under load has to be done. A mostly tool independent and nearly free geometry can be defined (Fig. 5). Therefore, the following geometry parameters are described for every bending sections:

- Radius of the bending section
- Length of the bending section
- Diameter at the beginning of the bending section
- Diameter at the end of the bending section
- Wall thickness of the section
- Rotational speed

The input data for a tube with four different sections in the whole geometry is shown in Fig. 6, but is arbitrary and easily extendible. In the first column the length of the bending section is described. Then the diameter at the beginning and the diameter at the end is described and the diameter gradient is calculated linearly in between. In the next three columns the bending radius (+/− for different directions, and "0" for no bending), the wall thickness ("0" for undefined thickness) and the rotational speed are shown. By using this data, the program calculates all points of the tube.

Geometry	Length	Diameter 1	Diameter 2
0	100	40	37
0	100	40	37
	200	37	37
	100	37	37

Bending radius	Wall thickness	Rotational speed
100	1	100
-100	1	100
200	1	100
0	1	100

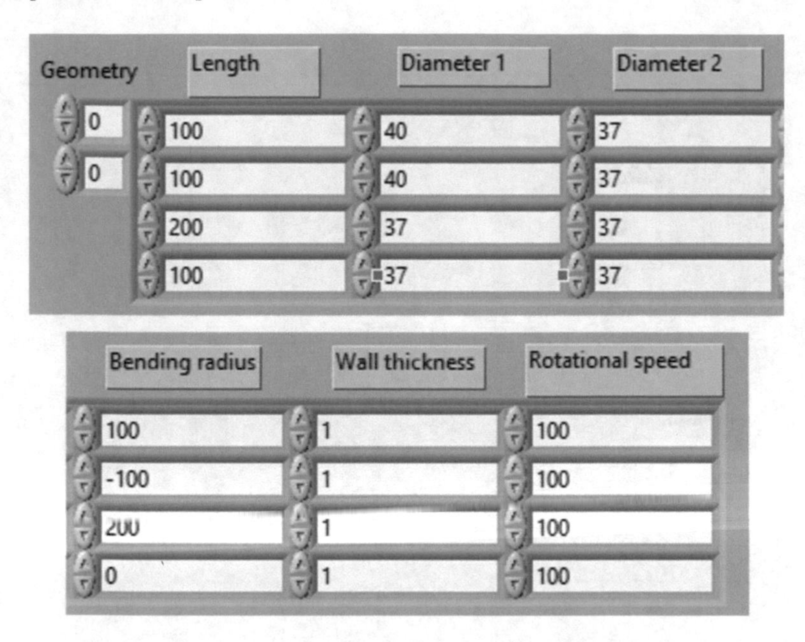

Fig. 6 Input data for an example of a tube with four different bending sections

If all bending sections are described and the machine parameters are set, the LabVIEW program will put them all together to a complex geometry. After this LabVIEW shows a preview of the designed tube, so that the user can check if the geometry is correct (Fig. 7). In case of a wrong geometry, the parameters can be adjusted.

In the next step superordinate machine parameters have to be set:

- Maximum rational speed
- Bending lever
- Geometry of bending unit
- Current position of the feeding unit
- Accuracy of the geometry definition
- Step of feeding

These parameters cannot be changed by the lecturers because wrong inputs can destroy the machine or increase the calculating time extensive.

After describing the geometry LabVIEW creates the machine parameters that are required for the process and transfers them to the control. The student is then able to see how the process will look like and how the axis will move. The following Fig. 8 shows one step of feeding. At every step of feeding the positions of all axis are derived and shown in a graph. By putting all these steps into one video, the student can see how all axis will move.

As an outlook it should be possible to visualize the whole process in 3D, so that students can check what the machine will do, without making a real experiment

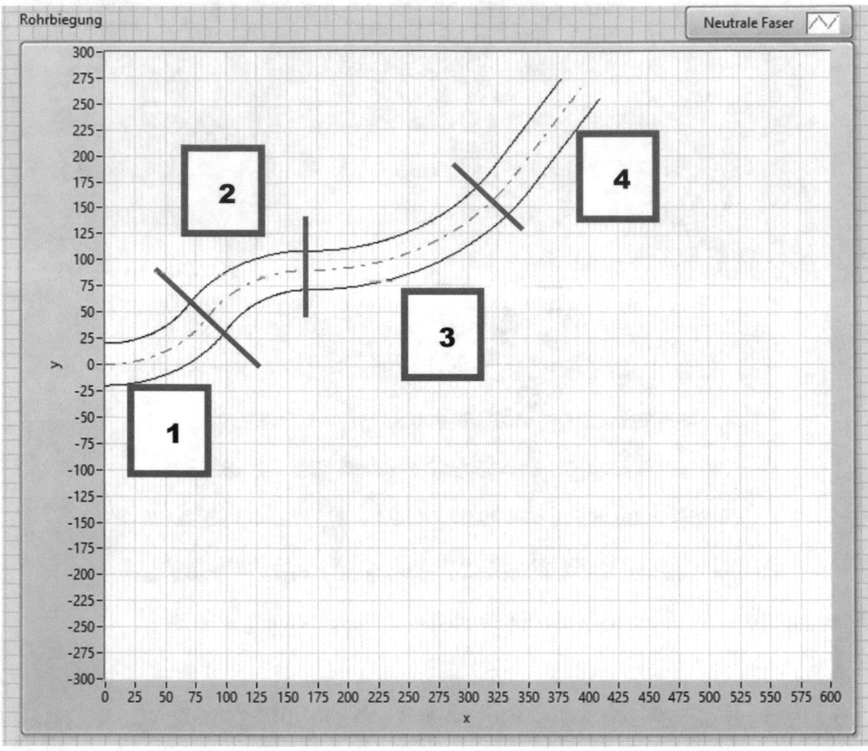

Fig. 7 Drawn geometry in LabVIEW for the four defined bending sections

(Fig. 9). Afterwards the students can transfer the NC data to the control and start the process. Students can then compare the simulation of the process movement with the experiments in reality. At this state of work springback for example is not calculated in this program, so that every shown radius is the loaded radius.

For the usage of the process as tele-operative control cameras will be integrated so that the students can observe the process from different and adjustable angles. They will be able to adjust the camera focus such as zooming or tilting. In addition, they can stop and start the process and adjust it whenever they wish to. Important process parameters as for example the bending force are logged and monitored in the LabVIEW interface in real time and can be saved as a text file. For the implementation of springback measurements a camera system can be integrated. This system compares the resulting tube geometry with the geometry the user has entered. Another way of comparing the different springback is by measuring the movements of the axis in LabVIEW over the bending force. In Fig. 10 the development of the bending force over the bending axis position is shown for the two process variations.

The one with the higher required bending force is the development for the bending without pushing. At $x = 0$ the bending axis is in contact with the tube, but does not bend the tube. Then the tube begins to be bend, first elastic and then plastic up to the

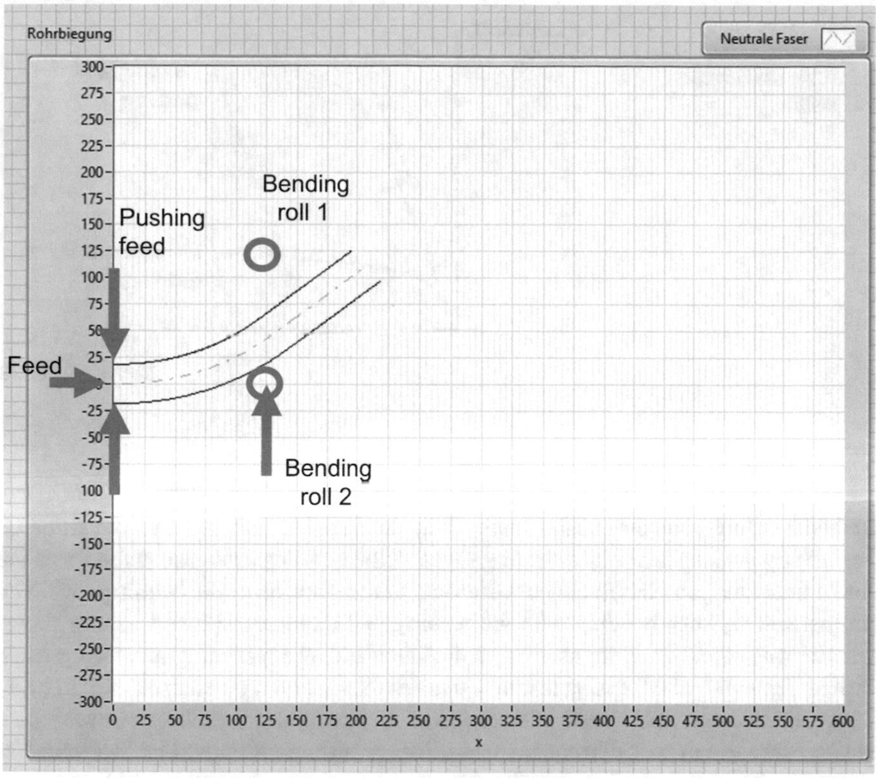

Fig. 8 One step in the calculation and visualization of the machine data

Fig. 9 Picture of the process in which the 3D movement could be integrated [2]

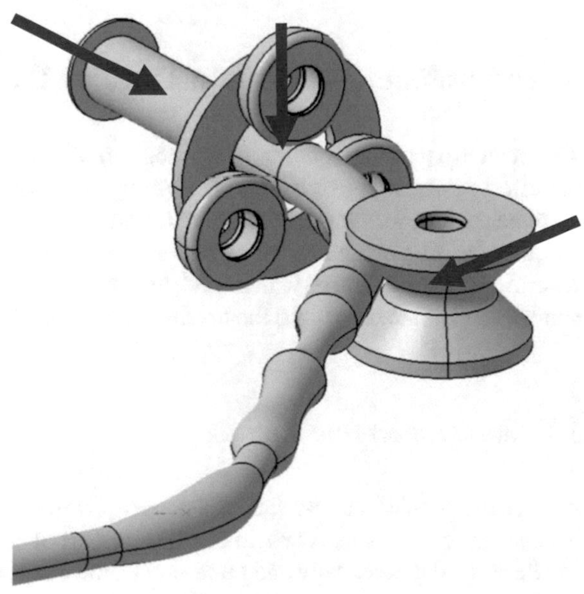

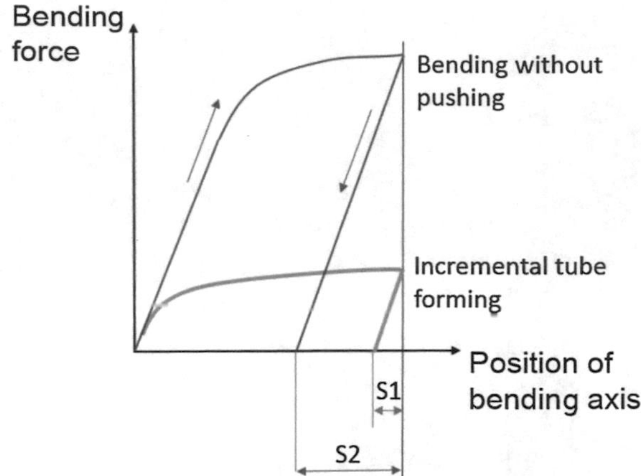

Fig. 10 Schematic development of the bending force over the bending axisposition

wished bending geometry under load. Then the bending axis releases the force on the tube by moving back. The strain gauges at the bending axis record the force the whole time and LabVIEW connects this data to the position of the bending axis. The comparison between bending without pushing and incremental tube forming shows that the way of the bending axis *S1* while releasing and recording a bending force with incremental tube forming is much smaller than without pushing *S2*. So this way is a qualitative description of the reduction of the springback, while a camera system can measure the exact difference. In addition, a cutting unit could be integrated at one of the bending tables so that the tube can be cut and does not collide with the environment.

3 Integration Concepts into Teaching Environment

The main aspect of integrating this process into a learning environment is to show the effect of stress superposition in forming processes and its characteristics. The incremental tube forming process serves as an example and is in principle transferable on other forming processes with stress superposition. Superposition of stresses has several interesting effects which lead to innovative processes. This tele-operative control can be integrated into the teaching environment by many ways:

3.1 Integration into Lecture

At first the process can be integrated into lectures in the theory of plasticity for instance in form of a live experiment. The effects of superposition of two processes can be derived theoretically and afterwards shown practically or vice versa. It is very useful for lectures because it is an interesting process with demonstrative effects. It improves the creativity of the students by showing a creative process with superposition of stresses. Therefore, students can make their own experiences, which could lead to an open-minded student for new process ideas. In addition, it is interesting to show real industrial bending machines with high potential for lightweight structures. Usually remote labs are used in electrical engineering or for material characterizations so that this tele-operative control is a new aspect in lecture. By integrating this concept into lecture, many students can see and understand the process without being at the machine itself. The main advantage of developing a tele-operative experiment instead of showing a video is that the students can really influence the process. They have to think what parameters could be useful and can see their effect in the process directly. In addition, they have several cameras to observe the process from different camera angles as if they were right in front of it. Furthermore, the experiments are very safe because students do not come in direct contact with the machine.

The following table shows how the experimental procedure could look like, while the bending force and the springback are measured for every experiment.

The experimental procedure is made so that students can see the effects of the incremental tube forming. At first they can try two conventional bending processes without the tube spinning. They can observe how the required bending force is different by comparing different bending radiuses. They learn how to calculate the bending moment out of the bending force in a real process.

In the experiment-steps two and three of Table 1 the incremental tube forming process is integrated. Students can choose different diameter reductions and rotational speeds and observe their influence on the bending force and springback. In the last experiment they can use the maximum speed and reduction to get the most extreme result. By giving them the opportunity to choose the parameters by themselves they are more interested in the process and they can see how an experimental procedure could look like.

Table 1 Conceivable experimental procedure for the incremental tube forming in lecture

Steps	Number of experiments	Experimental setting and observation
Step 1	1–2	Bending without pushing with two different bending radius. Observation: Required bending moment can be calculated, springback can be recorded
Step 2	2–3	Incremental tube forming: • With individual reduction of diameter • With individual rotational speed of the pushing rolls Observation: The required bending moment and the springback reduce
Step 3	1	Incremental tube forming with the highest diameter reduction and the fastest rotational speed the machine and the tube can keep up too. Observation: The required bending moment and the springback are at their low point

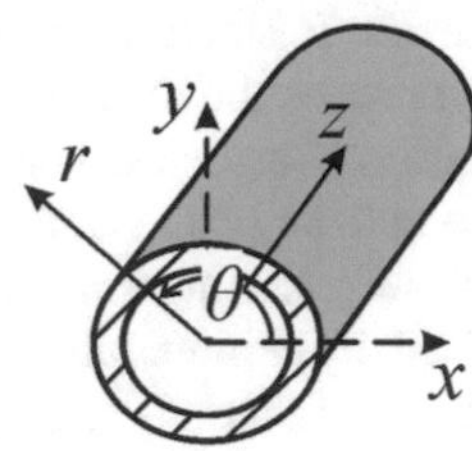

Fig. 11 Coordinate system of a tube section [2]

After observing these experiments, a theoretical explanation will be done. Therefore, different degrees of difficulty can be used. For freshmen other theoretical explanations are used than for almost completed engineers. In this paper an easy approach is presented. For integration into lecture some basic theories are described before the experiment. The fundamental terms for the theory of plasticity should be known and understood. The process phenomena can be shown in a simple experiment like the Hoogenboom's experiments at first. In the Hoogenboom's experiment a wire is fixed hanging and a weight is attached so that the wire elongates plastically [7]. In a second experiment step with the same experimental setting the same weight is attached to the wire, but a torque is applied on the wire by rotating the weight. Therefore, by superposition of stresses out of the torque and the pulling, the wire elongates. The v. Mises yield criterion can be used as explanation for this phenome and also for the process phenomes occurring by the incremental tube forming which are described in the following section.

In Fig. 11 the tube and its coordinate system for the theoretical explanation [2] is shown. The first assumption is that the modulus of resistance is constant, because the diameter reductions are negligible compared to the effects of superposition of stresses. Second assumption is that the yield stress is constant all over the process. So strain hardening is not considered.

At first the yield criterion by von Mises is shown:

$$k_f = \sqrt{\frac{1}{2} \cdot [(\sigma_{\theta\theta} - \sigma_{zz})^2 + (\sigma_{zz} - \sigma_{rr})^2 + (\sigma_{rr} - \sigma_{\theta\theta})^2]} \tag{1}$$

Under assumption that is a plane stress follows:

$$k_f = \sqrt{\sigma_{\theta\theta}^2 - \sigma_{\theta\theta} \cdot \sigma_{zz} + \sigma_{zz}^2} \tag{2}$$

With this equation the phenomena can be described so that students learn that there is a clear effect on the process by superposition of stresses.

There are two borderline cases which can be achieved with this process to plastify the material. On the one hand there is the bending process without the pushing process. On the other hand, the pushing process could be carried out without the bending process. The effect of these borderline cases on the yield stress is shown in Table 2.

Table 2 Two borderline cases and their influence on the yield stress

Bending without pushing	Pushing without bending
$\sigma_{\theta\theta} = 0$	$\sigma_{zz} = 00$
$k_f = \sigma_{zz}$	$k_f = \sigma_{\theta\theta}$

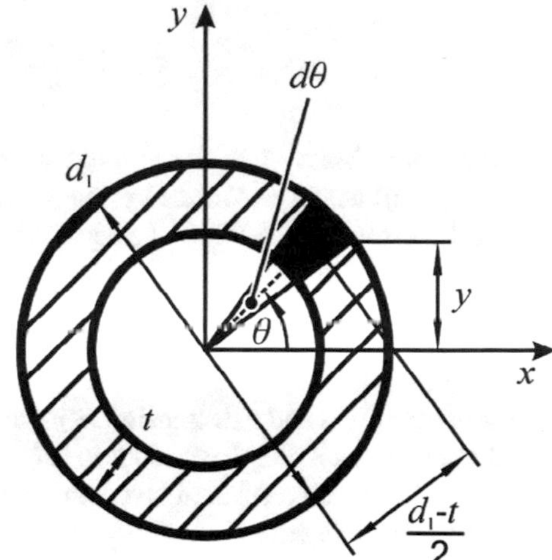

Fig. 12 Increment of area at one tube section [2]

The incremental tube forming process is therefore a process which combines both stresses. Students have observed how the bending force and the springback behave. So they need a theoretical explanation how the bending moment and the springback is described.

The bending moment can be calculated out of the bending stress σ_{zz}, the wall thickness t, and the tube diameter d_1 under the assumption of thin walled tubes:

$$M_B = \int_{dA} \sigma_{zz}(\theta) \cdot \frac{d_1 - t}{2} \cdot sin(\theta)dA \tag{3}$$

$$= \frac{d_1 - t}{2} \cdot \int_{\theta=0}^{2\pi} \sigma_{zz}(\theta) \cdot sin(\theta)dA \tag{4}$$

The increment of the area as shown in Fig. 12 can be described as

$$dA = \frac{1}{2} \cdot t \cdot (d_1 - t)d\theta \tag{5}$$

By usage of (3) and (4) you get the equation for the bending moment

$$M_B = \frac{(d_1 - t)^2}{4} \cdot t \cdot \int_{\theta=0}^{2\pi} \sigma_{zz}(\theta) \cdot sin(\theta)d\theta \tag{6}$$

In (5) it is obvious that MB is dependent on the σ_{zz} and will increase if σ_{zz} increases. By looking at the yield criterion students learn an increase of $\sigma_{\theta\theta}$ reduces σ_{zz} and therefore M_B and vice versa. They learn that the superposition has a huge influence on the process. For describing the springback two equations are needed. At first the definition of springback S by usage of the unloaded bending radius R_U and the loaded bending radius R_L is shown:

$$S = \frac{R_U - R_L}{R_L} \cdot 100\% \tag{7}$$

Students will observe that the difference between unloaded radius R_U and loaded radius R_L will heavily decrease by usage of the incremental tube forming.

For the forming area following equation can be found [2]

$$\frac{1}{R_U} = \frac{1}{R_L} - \frac{M_B}{E \cdot I} \tag{8}$$

By usage of (7) and a decreasing M_B the difference between R_U and R_L will get smaller and so the springback reduces [2].

To visualize the correlation between the stresses, moments and forces the tool in Fig. 13 will be created.

This tool calculates the other two parameters of $M_B, \sigma_{zz}, \sigma_{\theta\theta}$ by one of them given. It needs a preset k_f and the geometry of the tube cross section. By adjusting one of the scroll bars, students can see how the parameters influence them mutual. After this theoretical explanation the students will be able to understand why superposition of stresses influence the process and can transfer it to other processes which use superposition of stresses.

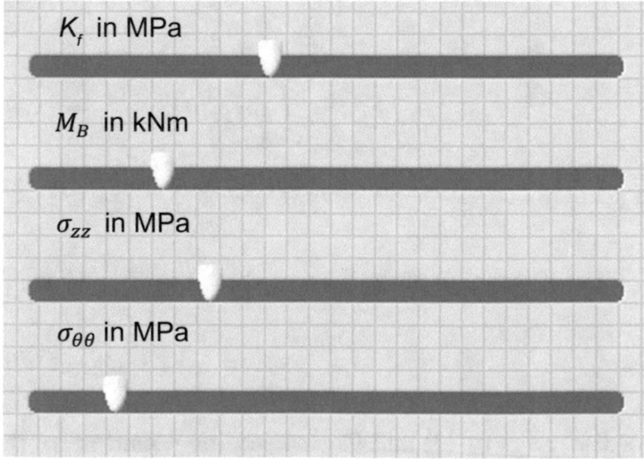

Fig. 13 Tool for the correlation between stresses, moments and forces

3.2 Integration into Online Courses

As an outlook the tele-operative control of the bending machine can also be used in an online course for foreigners that plan to study in Germany. They can get an overview what machines are the subject of investigation and learn how German engineering education looks like. The task could be to observe what effects the superposition of stresses has. Simultaneously they get a theoretical description of the process phenomena as described the chapter before. So students could learn about superposition of stresses by themselves room and time independent.

3.3 Integration into Labs

The process can also be integrated into labs, so that students can do experiments with the machine and investigate the process phenome at home. In difference to the pre-courses the whole setting could be more ambitious. The lab could also be as a teamwork. A plan of an integration into labs could look like this:

1. Literature research about superposition of stresses
2. Create a meaningful experimental program
3. Realization of the experimental program and observation of the phenomena
4. Attempt to explain the phenomena on basis of the observation and the literature research
5. Creating a lab report for the incremental tube forming
6. Present the experimental result as a team and discussion

3.4 Integration into e-Learning Environment

In the same way the tele-operative control of the bending process with superposition can be integrated in a e-learning environment so that students from all over the world can interact and learn from it. It can be integrated in a platform with many experiments which are using stress superposition or other interesting forming processes.

4 Conclusion and Outlook

The development of a tele-operative control of the incremental bending process will lead to an improvement of engineering education. The effect of superposition of stresses is shown and can be understood. In this process a reduction of the required bending moment and springback by superposition of two processes is clearly shown and will be accessible to students. The fact that LabVIEW interacts with a conventional control of a machine opens this integration into lecture for other processes that have superposition of stress. In addition, by showing this combination of stresses and interesting process the creativity of students is stimulated. Similar processes with distinguishable effects can be integrated into lecture and e-learning environment so that many students have time and room independent access to the experiments.

References

1. C. Becker, A. E. Tekkaya, M. Kleiner, Fundamentals of the incremental tube forming process. CIRP Annals – Manufacturing Technology 63/1, 2014, pp. 253–256.
2. C. Becker, 2014, Inkrementelles Rohrumformen von hochfesten Werkstoffen, Aachen, Shaker (Dortmunder Umformtechnik, 79)
3. C. Becker, K. Isik, A. Bayraktar, S. Chatti, M. Hermes, C. Soyarslan, A. E. Tekkaya, Numerical Investigation of the Incremental Tube Forming Process, Key Engineering Materials 554–557, 2013, pp. 664–670.
4. C. Becker, G. Quintana, M. Hermes, B. Cavallini, A. E. Tekkaya, Prediction of surface roughness due to spinning in the incremental tube forming process, Production Engineering-Research and Development 7 (2–3), 2013, pp. 153–166
5. C. Becker, D. Staupendahl, M. Hermes, S. Chatti, A. E. Tekkaya, Incremental Tube Forming and Torque Superposed Spatial Bending – A View on Process Parameters, Steel Research International, Special Issue, 2012, pp. 415–41
6. Ortelt, T.R., Sadiki, A., Pleul C., Becker, C., Chatti, S., A. E. Tekkaya, "Development of a Tele-Operative Testing Cell as a Remote Lab for Material Characterization" Proceedings of 2014 International Conference on Interactive Collaborative Learning (ICL), December 2014; Dubai, UAE
7. A. E. Tekkaya, C. Becker, T. R. Ortelt, G. Grzancic, "Utilizing Stress Superposition in Metal Forming," in 7th JSTP ISPF International Seminar on Precision Forging, Nagoya, Japan, 2015, pp. 1–6

Integration of new Concepts and Features into Forming Technology Lectures

Tobias R. Ortelt, Sören Gies, Heinrich Traphöner, Sami Chatti and A. Erman Tekkaya

Abstract New concepts and features are shown for forming technology lectures. On the one hand, technological and electronic developments as audience and response system and the integration of a remote lab is presented. On the other hand, didactical approaches focus on the aim to combine theory with practical relevance. The implemented developments were evaluated by the students and the influence to their grade points in exams are presented. Additionally, new concepts in the current winter semester 2015/2016 are shown.

Keywords Lecture · Forming Technology · Remote lab

1 Introduction

Today's lectures are not much different from lectures 20 years ago and slides are still the most common medium. Although the overhead projectors were replaced by digital projectors and the slides are presented with PowerPoint or a similar software tool. However, the concept is still the same: A lecturer is giving a talk and the students are listening. Often they have to make notes because the presented slides are also the script of the course.

The lecture "Umformende Fertigungstechnologien (UFT) - Fundamentals of Forming Technology" at the Institute of Forming Technology and Lightweight Construction (IUL) of the TU Dortmund University is a basic lecture in bachelor courses in mechanical engineering in the 5th semester. In the last few years this lecture was redesigned and new concepts were integrated. On the one hand, these new concepts were based on technological and electronic improvements. On the other hand, new didactical approaches were embedded. The combination of theory and practical relevance was the main aim of almost all developments.

T.R. Ortelt (✉) · S. Gies · H. Traphöner · S. Chatti · A.E. Tekkaya
Institute of Forming Technology and Lightweight Construction (IUL) Dortmund,
TU Dortmund University, Dortmund, Germany
e-mail: tobias.ortelt@iul.tu-dortmund.de

Originally published in "IEEE Global Engineering Education Conference (EDUCON)",

DOI 10.1007/978-3-319-46916-4_40

2 Technological and Electronic Developments

2.1 Audience Response System

Almost all technological and electronic developments are connected with new opportunities in this field. 10 years ago, only few students brought electrical devices like PDAs or heavy, uncomfortable notebooks to lectures. Today, almost all students have their own mobile devices, like smartphones, notebooks or tablets with them all the time. Some lectures think these devices are distracting students and are disturbing their lectures or lessons. Only few lectures realize the upcoming new possibilities. In the UFT lecture, mobile devices are used in interaction with an app or with a website as an audience response system. We use the system eduVote [1] in lectures. The biggest effort of this software tool is the aspect that no other hardware is needed to conduct a poll. On the presentation notebook an software add-in for PowerPoint is used to setup a slide with a question and possible answers like A/B/C/D or Yes/No and to start the poll. After the lecturer has released the question, the students can login with their devices to answer the question. During the poll, the votes are counted and displayed on the slide. The lecturer finishes the poll after enough votes and by one click, a chart with the percentage of answers is shown on the slide. A big benefit of this system is the possibility for the students to pick the answer anonymous. If a lecturer asks a question to students in the lecture hall two aspects can be observed:

- Frequently, the same few students, normally the top students of the course, will answer the question.
- If the lecturer asks the audience to raise the hands to vote a lot of students are not answering.

These two aspects are based on the same problem. The students are afraid to give a wrong answer. This fear can be explained with two points:

- The students are afraid to give a wrong answer because the lecturer may recognized the person later in the exams or especially in oral exams.
- The students are afraid to give a wrong answer in front of the whole class because they do not want to be the "losers".

The observed answer rate of the students is above 50 %. This is much better than the rate for answering by raising hands. Figure 1 shows the typical setup of a slide with a multiple choice question. A button (Start, Stop, Show chart, Reset) controls the audience response system and an interactive field is counting the votes during the polling.

The usage of this audience response system pursues two aims. On the one hand, the students are engaged through their active participation in the lecture and they have to use their latest knowledge to find the right answer. On the other hand, the lecturer gets an immediately feedback of the students and can recognize if they have understood the content Additionally, the integration of questions and audience

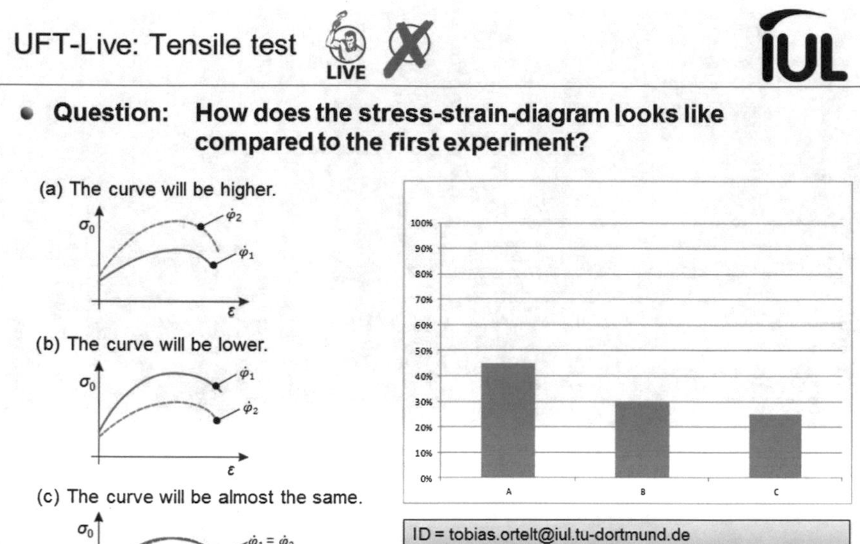

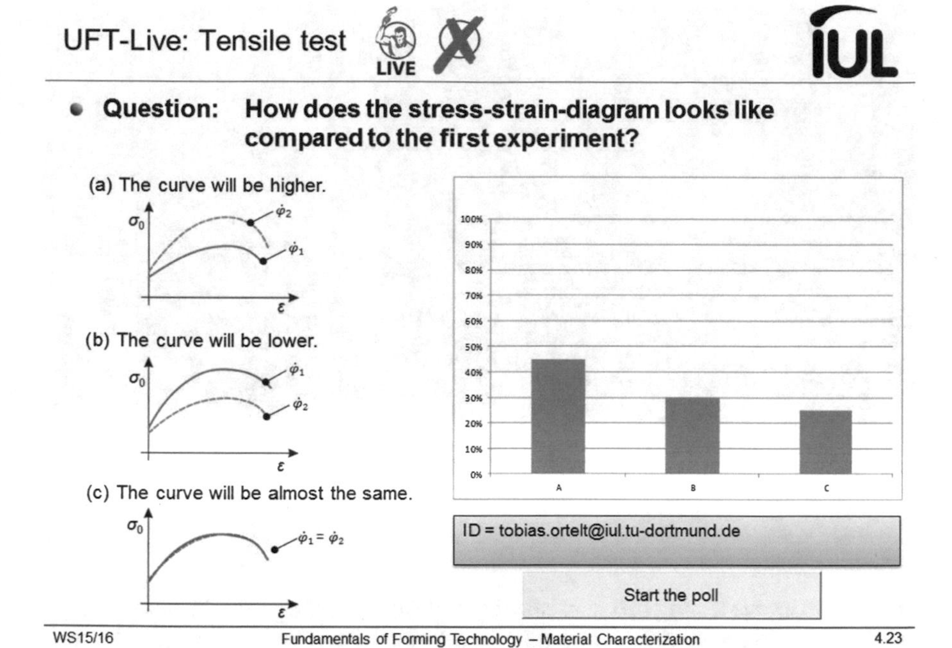

Fig. 1 Typical setup of a eduVote slide (Translated into English)

response systems can lead to better exam grads because of the interaction during the lecture [2]. It should be mentioned that the design of these questions is very tricky. The best questions or especially the right answer should generate a "wow effect". This "wow effect" can be reached by disprove of misconceptions of the students or with a gap between theory and practical relevance. In the lecture UFT these questions are often connected to live experiments. With this approach the theoretical knowledge is directly compared to practical relevance by conducting an experiment. These experiments can be done with small setups directly in the lecture hall (see III.C Connection of theory and practical relevance) or with remote labs.

2.2 Remote Lab

To reach a wow effect in a special way a remote lab is integrated to the lecture. A remote lab is a laboratory, which can be accessed via internet. At the IUL a tele-operative testing cell for material characterization [3] was developed within the research project "ELLI – Excellent Teaching and Learning in Engineering Science". The tele-operative testing cell provides different experiments in the field of material characterization for forming technology via internet. Therefore, a web site in interaction with the iLab platform is used to guarantee the remote-access.

Fig. 2 Tele-operative testing cell

The tele-operative testing cell consists of two different testing machines, a robot and several IT components like real-time control system and camera system. Figure 2 shows the machines of the tele-operative testing cell.

In the current stage of development two different experiments can be controlled via the internet. On the one hand, a uniaxial tensile test can be carried out with a universal testing machine. On the other hand, a deep drawing test can be carried out with a sheet metal testing machine. Both tests or experiments can be configured by setting parameters in the preparation stage. At the tensile test different parameters like material (steal or aluminum) or strain rate (forming speed) can be chosen. The described remote experiments are norm experiments for forming technology and are typically used to determine material properties, like the Young's Modulus with the tensile test. Figure 3 shows the concept and the different stages of the integration of a remote lab to a lecture.

The integration of the remote experiment into a lecture can be divided into different stages. Obviously, the first stage is the classical theory presented by slides. After this theory block the experiment is introduced to the students. The used machines and the influencing parameters will be mentioned. The first set of parameters is specified in interaction with the students and the experiment is started remotely. The students can observe the handling of the specimen and the movements of the robot by different cameras. A camera, next to the specimen and the measuring devices, provides a good view during the experiment. In a tensile test the expected necking and the following crack of the specimen can be observed. The destroyed specimen is removed by a robot. Additionally to the camera views the experiment can be understood by observing real-time measuring data. Figure 4 shows the ilab interface to control the tensile test. In this case, three different sets of parameters (different specimen temperatures: black curve 20 °C, red curve 350 °C, green curve 750 °C) were tested.

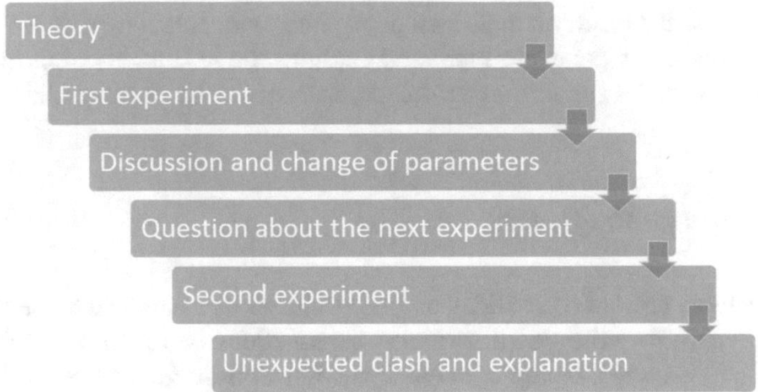

Fig. 3 TIntegration of the remote experiment into lectures

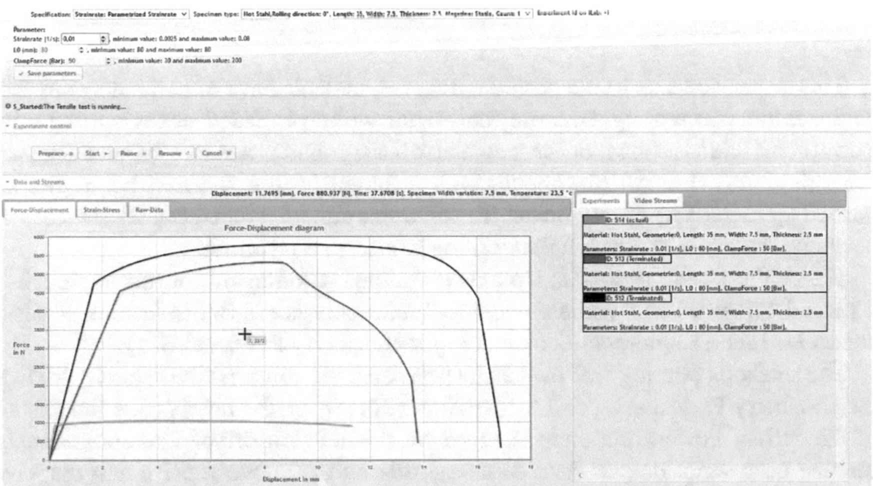

Fig. 4 iLab interface to control and observe the experiment

After conducting and observing the first experiment the result is discussed by the students in a first step and then discussed in interaction with the lecturer. To change the result of the next experiment the parameters are changed. In the shown example in Fig. 4 the temperature of the specimen was changed. Therefore, an induction heater was integrated into the tele-operative testing cell to heat the specimen up to 1000 °C in only few seconds. Following the setting of the parameters a multiple choice question is asked by the lecturer about the estimated result of the second experiment. Then the experiment is started and observed by the students. Moreover, these experiments should be constituted to create an unexpected clash. In this example for forming technology the material behave unexpected. After this wow effect the students discuss the result on their own or together with colleagues. In the next step the lecturer discusses the experiment and explains the results. To support this new adopted theory model sometimes a third experiment can be conducted.

2.3 Use of two Projectors

Usually lectures present their slides on one projector and they often use the blackboard and chalk to write down formulas or equations or to draw graphs. In the described UFT lecture a laptop with touchpad-functions in interaction with a digital pen is used. The normal process (on the projector) looks like:

1. PowerPoint Slide
2. Digital writing on the laptop
3. Next PowerPoint Slide

This sequence leads to the problem, that the students have to make notes as fast as the lecturer is writing, because the digital writing is faded out when the next PowerPoint Slide is shown or the new blank background is needed. When a lecturer is writing on a classical blackboard with different boards, the first board can be pushed up to get new empty board. To solve this situation an electrical device was developed to mirror the last digital writing to a second projector.

Beside the second projector, the only additional requirement for this approach is a second computer. Since this second computer only needs to display image files, we decided for a single-board computer type Raspberry PI Model B.

The notebook running the PowerPoint presentation can access the network drive of the Raspberry PI. Using a previously defined hot key on the notebook a screenshot of the current PowerPoint slide is saved on the network drive. The continuously running bash-script detects the new image file on the network drive and displays the screenshot using the previously defined image viewer on the second projector. At this moment, both projectors show the same content. If the lecturer skips to the next PowerPoint slide on the notebook the image on the second project remains unchanged until a new screenshot is generated by pressing the hot key combination. This way the students have enough time to take their notes from the duplicated slide shown on the second projector (Fig. 5).

In a next step, a software update for the Raspberry PI should be developed. This new software will be able to check, if a new slide or a new writing is presented. With this approach the lecturer does not need to press any hot keys to copy the picture from the first projector to the second projector. Moreover, this approach is more stable because the problem of forgetting to press the hot key is solved.

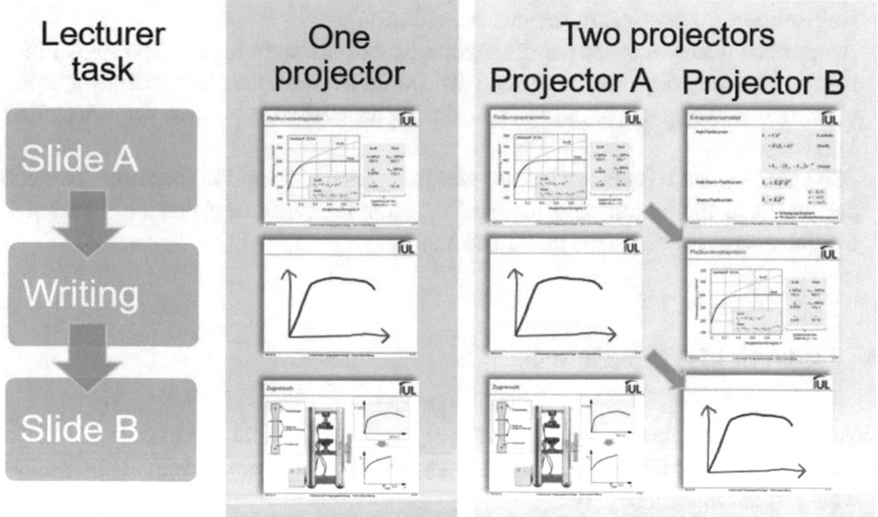

Fig. 5 Old and new concept (using two projectors)

2.4 *Online Learning*

A further aspect was the availability of the slides and other digital files online. For years, a classical FTP-Server was used. In the first lecture of a semester the login and the password was shown and the students were able to download the PDF-files from the homepage. This procedure was very user-unfriendly and linear. To provide digital files in a more user-friendly format a Learning Management Software (LMS) was needed. The TU Dortmund University uses moodle as a Learning Management Software [4]. Therefore, a course in moodle was generated for the described lecture. The main aim of moodle was the download option for PDF files. Furthermore, new functions of moodle were used for the lecture:

- Communication
  To communicate with the participants a newsfeed is integrated to moodle. Information like "PDF of the upcoming lecture is available" or "Please register for the field trip" can be posted online. Furthermore, all participants of the course are registered with their email address. Therefore, urgent information like "The lecture tomorrow takes place in a different lecture hall" can be send immediately via email to the students.
- Calendar
  A calendar for the lecture was used to schedule the dates for lectures and tutorials. This was very useful if a lecture has to be rescheduled or if there was a change of the lecture place.

- Registration for post-exam review or field trips
  A interface was developed for the registration of different topics. The registration for post-exam review was integrated. The students were able to register for a time slot. Hereby, the spread of the students during the post-exam review was controlled.
- Feedback
  A feedback system for every lecture was integrated to moodle. Therefore, students can evaluate the lecture and its level of complexity. A free text field was also available to give comments about the lecture.

3 Didactical Approaches

Next to the approaches in the technological and electronical field some didactical approaches were implemented and are described below. These didactical approaches can be divided into three parts:

3.1 Additional Learning Materials

In evaluations of the course, the students considered some lectures or some topics as "complex" or "tricky". They also breathed the wish to have a written script for these topics. The 2nd and 3rd lecture handle with the topic "Theory of Plasticity". In these two lectures the basics of forming technology like definition of strain and stress, the Yield criterion or the flow rule are described. This basic knowledge is needed for the next lectures or even for the entire course. Therefore, two research associates wrote a script with 80 pages additionally to the PowerPoint slides (about 120 slides). Mainly the slides are described in formulated text and some background information is added. This written script can be seen as a book chapter for "Theory of Plasticity".

3.2 Bonus Exam

Another aspect in the evaluation of the course was the unknown requirements in the exams after the course. Also, the students complained about the wide field of possible topics in the exams. Furthermore, in the exams the students showed gaps in the knowledge of forming technology basics. To fulfil these two wishes of the students and to solve the problem with the basic knowledge an additional bonus exam in the middle of the course was added. The date for this bonus exam is before the Christmas break at the end of December. In this bonus exam the basic knowledge (first five lectures) of forming technology is tested.

To create attraction of this optional bonus exam the students can get scores for the real exam after the whole lecture. Up to 15 % of the score of the real exam can be earned with the bonus exam. This bonus score is only available in the first attended exam after the current lecture. For instance, a student gets 10 bonus scores with the additional exam in December 2015. These 10 bonus scores can be used in the first try of the exam in spring 2016 or in fall 2016. With the start of a new lecture in winter semester 2016/2017 all bonus scores of all students are erased. The students can attend the additional exam again to earn bonus scores for the next two exams. The evaluation of the bonus exam and the influence on the grades of the students will be shown below.

3.3 Connection of Theory and Practical Relevance

To combine theory with practical relevance live experiments were integrated to the lecture. On the one hand, like described above, is the integration of remote labs. This approach is very cost-intensive because of the development but allows the lecturer to do very specialized experiments during a lecture via the internet. Additionally, the integration of the remote lab is time-intensive in a 90 minutes lecture. The equipment and the experimental setup must be described before the experiment and the different steps (two experiments or discussion with the students) need also some minutes. One the other hand, quick live experiments were integrated to the lecture. These experiments have a lower level of technical complexity compared to the remote lab. A third approach is the so called "lecture in the lab", which takes place in the experimental hall of the IUL.

3.3.1 Remote Lab

The integration of the remote lab is described above in detail. It should be mentioned that the remote lab will also be integrated to online tutorials. Students will be able to conduct experiments via internet during their exam preparation.

3.3.2 Quick Live Experiments During the Lecture

Quick live experiments were integrated to several lectures with different topics to combine theory with practical relevance. The quick live experiments are up to few minutes long and are carried out live in the lecture hall on the platform by the lecturer or a research associate. The principle is the same like the integration of the tele-operative testing cell, but on another level of complexity. As example, the so-called Hoogenboom Experiment [5] was done to show the influence of stress superposition. These little experiments, which can also be done at home by the students, lead to a wow effect or new knowledge.

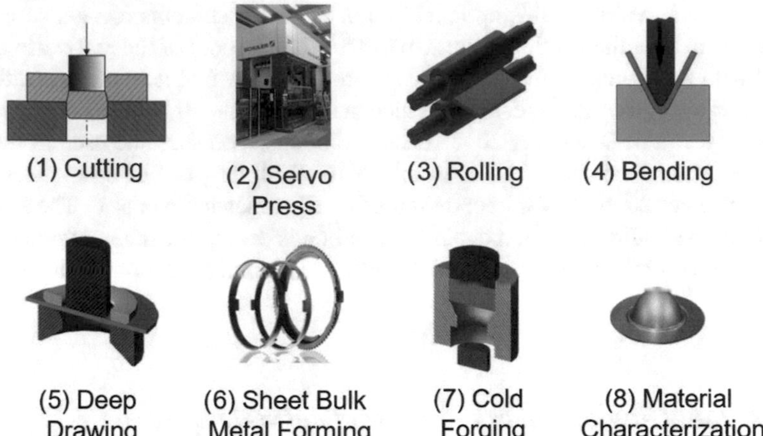

Fig. 6 Stages during the lab lecture

3.3.3 Lecture in the lab

Furthermore, on a special date all students were invited to join a lecture in the IUL laboratory. At this so-called "Hallenvorlesung – lecture in the lab", different machines or processes can be watched live. These presentations or experiments are connected to the theoretical lectures. The experiments are shown in Fig. 6.

Each experiment or machine is shown in an eight-minute presentation and ends with a comprehension question, which is relevant for the exam.

4 Evaluation of the new Concepts

At the TU Dortmund University every lecture is evaluated by the students at the end of the semester by a survey. In this evaluation the students put a score on different questions or statements. The answers are given by numbers and with different definitions, like "I totally agree" to "I totally disagree" or from "much" to "little". Some questions are scored by marks form "1 – excellent" to "5 – insufficient". The evaluation was done by 80 students in winter semester 2012/2013, by 100 students in winter semester 2013/2014 and by 77 students in winter semester 2014/2015. The evaluation is done before the exam of the course. Additionally, to the ticking of the questions, the students were able to write free comments on the lectures.

All the shown approaches or changes in the concept to improve the lecture were not implemented contemporary but step by step. The first integration of the different approaches or concepts is shown below:

- Polling system: winter semester 2012/2013
- Remote Lab: winter semester 2013/2014

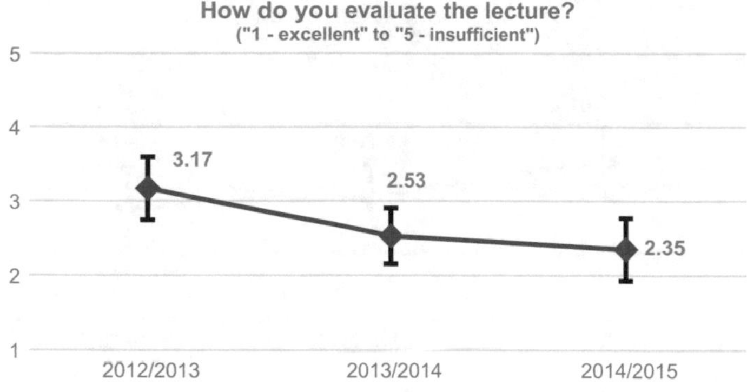

Fig. 7 Evaluation of the entire lecture

- Bonus cxam: winter semester 2013/2014
- Moodle as LMS: winter semester 2014/2015
- Written script: winter semester 2014/2015

4.1 Evaluation of the Entire Lecture

Figure 7 shows the line graph of the evaluation of the entire lecture and the given grades by the students. We assume that the integration of the remote lab and the using of the polling system in this context improved the lecture for the students. The written comments of the students strength this assumption. The students wrote "Remote Lab was good" or "Bonus exam is nice".

4.2 Motivation for the Lecture and Workload of the Students

Figure 8 shows the motivation of the students to work on their own for the last three years. It can be observed, that the students are a little more motivated to work on their own, but a score of 2, 72 with "2, 5 – I am neutral" is not good for a basic lecture. An explanation can be given by the fact that the students believe they have a high workload for this lecture.

4.3 Workload of the Students

The shown lecture UFT is a four credit points lecture. One credit point is equal to a workload of 30 hours. Therefore, four credit points are equal to a workload of

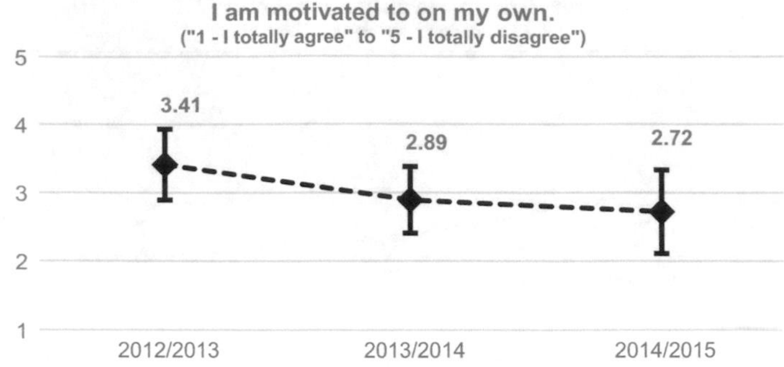

Fig. 8 Motivation of the students to work on their own

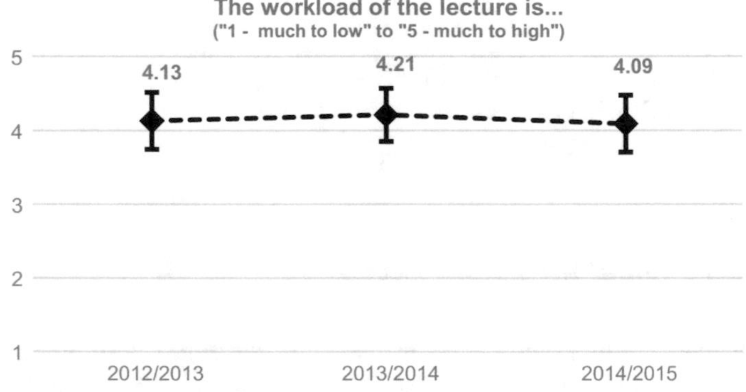

Fig. 9 Workload of the lecture

120 hours for one semester. The UFT course has 14 lectures and 5 tutorials with each 90 minutes and this is totalized to 28.5 hours per semester. For the preparation 25 % equal to 30 workhours can be defined. Therefore, the available workload (for the lecture) is equal to 61.5 hours (120 h - 28.5 h – 30 h = 61.5 h) for the entire lecture or divided by 14 lectures and 5 tutorials to about 3 hours per lecture.

Figure 9 shows the rated workload by the students from "1 – much to low" to "5 – much to high". The line graph shows that the students think the workload of the lecture is too high.

Figure 10 shows the workload of the students per week in hours. For all semesters more than 50 % of the students work less than 1 hour per week for this lecture. As described above a workload of 3 hours per week is available based on the credit points. To solve this dilemma and to improve the recapitulation of the lectures for the students the lectures are recorded in the current semester. With this approach the students should be motivated to recapture the lectures on their own at home.

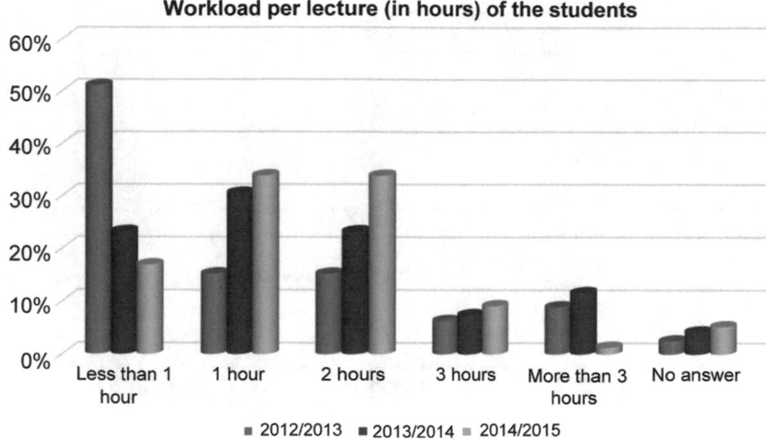

Fig. 10 Workload per week (in hours) for the lecture

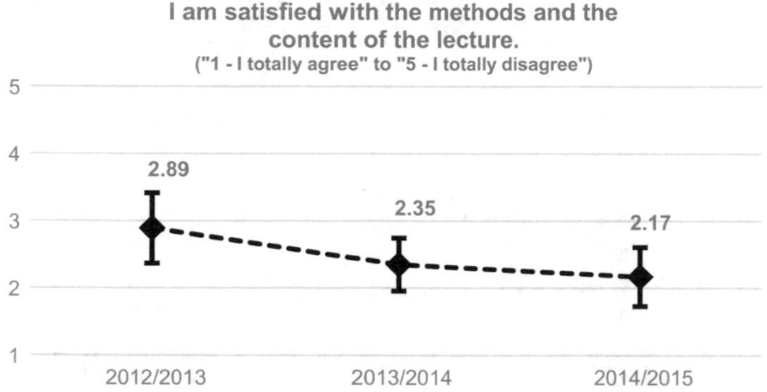

Fig. 11 Satisfaction of the students with the methods and the content of the lecture

4.4 *Satisfication with the Methodic and the Conten of the Lecture and the Lecturer Himself*

Independently from the workload or the motivation, the students are satisfied with the lecture and with the lecturer himself. Figure 11 shows the satisfaction of the students with the methods and the content of the lecture. The score raised from 2.89 in winter semester 2012/2013 to 2.17 in winter semester 2014/2015.

Figure 12 shows that the students are satisfied with the lecturer. The score was at 2.78 in winter semester 2012/2013 and is now at 2.07 at the winter semester 2014/2015. We connect this good results with the available feedback and with the changes of the lecture and the lecturer himself. In some comments of the evaluation the students wrote "write more clear with the digital pen" and "the lecturer is too fast". These two examples lead to the effect that the second projector is used to enable

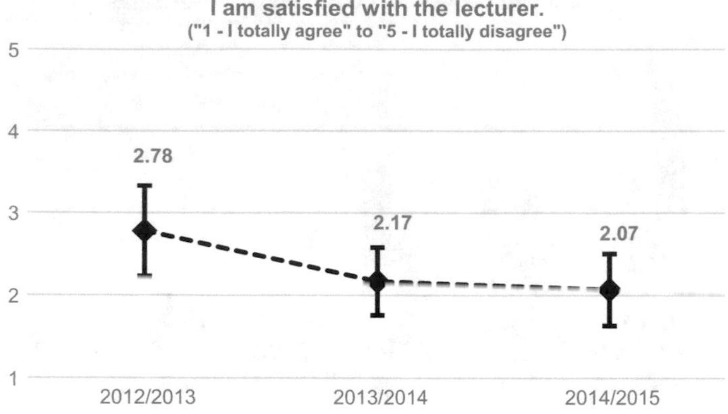

Fig. 12 Satisfaction with the lecturer

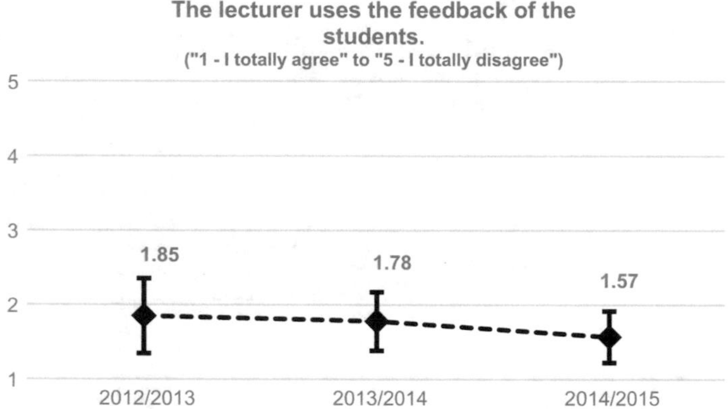

Fig. 13 Using of student feedback by the lecturer

the students to make notes during the lecture. A further proof of this assumption is shown in Fig. 13. The students rate the outcome of the feedback of the students by the lecturer himself. A score of 1.57 in winter semester 2014/2015 reflects the attempt of the lecturer and the involved research associates to improve the lecture.

4.5 Suggestions of the Evaluation

In many cases the new approaches or the approaches combined led to a better evaluation based on the scores of winter semester 2012/2013. In winter semester 2012/2013 the grade of the entire lecture was 3.17 and is in winter semester 2014/2015, after the implementation of the different approaches, at the level of good (2.35). Unfortunately, the motivation of the students to work on their own is still on a low level. New

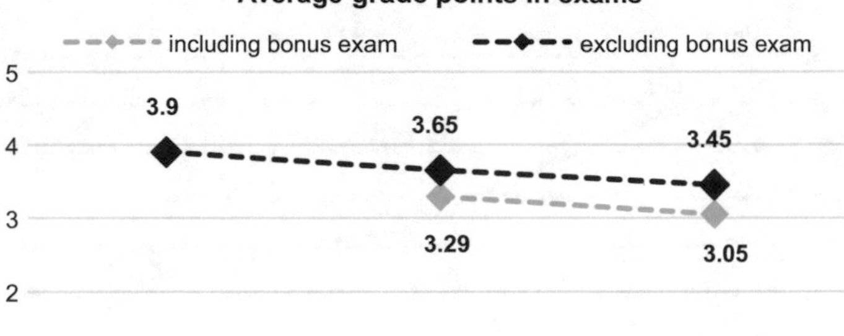

Fig. 14 Average grade points in exams

approaches in the current semester aim at solving this problem. A very nice output of the evaluation is the fact that the students honour the work of the lecturer and the involved research associates to improve the lecture.

5 Effects on the Grades of the Exams

Next to the evaluation of the lecture the results of the exams are very important for the students and for the lecturer. The approach to offer a bonus exam after the first lectures is from our point of view the most influencing factor on the results.

Figure 14 shows the average grade points in the exams of the last three years. Sadly, the results are not as good as desired. In winter semester 2012/2013 the average grade point was 3.9 and is now at 3.45 in winter semester 2014/2015. A nice effect is the fact that the results are even better without the extra points from the bonus exam. This effect is clearly shown in Fig. 15. Even without the bonus exam the failure rate from up to 34 % was dropped to 22 % in winter semester 2013/2014 respectively to 16.4 % in winter semester 2014/2015. With the extra points from the bonus exam the failure rate was decreased to 9.4 %.

6 Developments in the Current Semester (2015/2016)

For students, basics lectures in engineering costs a lot of time for the repetition of each lecture, as well as for the preparation for the exam. Slides and self-made notes are the common basis for them to learn. Many details and explanations, which are told by the lecturer during the lecture to enhance the understanding of the contents of each lecture will be lost afterwards.

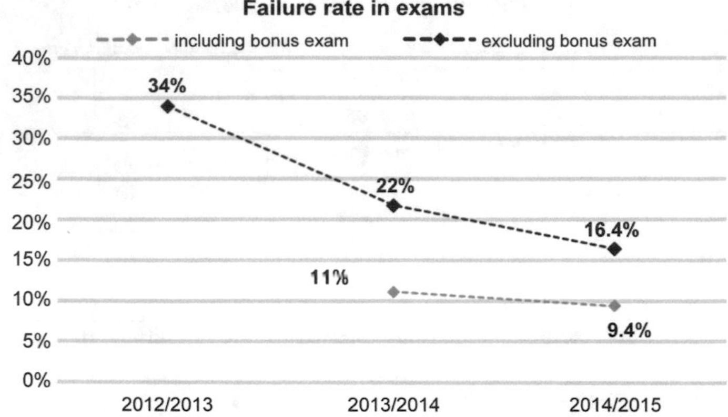

Fig. 15 Failure rate in exams

Since the current semester all lectures are recorded. The recording is based on Audio (microphone) and Video (pictures of the notebook respectively the projector). With this approach the students can rewatch every lecture at home and they have access to the lecturers expertise at a later time. The videos can be downloaded from the moodle platform few days after the lecture.

As the basic equipment, the Camtasia Studio software is used for a digital recording of the notebook screen during the presentation on the touch notebook. Each slide, as well as notes, videos and sketches are recorded. A wireless microphone receiver is used to insert the audio of the lecture hall sound system in the recording. By these means, the slides with all the notes and the spoken word of the lecturer are available. The lecturer itself is not shown in the video.

In addition to this recording, a webcam is orientated to a defined position in the lecture hall, where the teacher can show live demonstrations or live experiments, which will later be integrated in the recording. Some rules are especially important for the lecturers for a successful recording of the lecture. These must be followed during the whole recording.

1. Before answering, any question of a student the question must be repeated by the lecturer for the audio recording, because only his microphone sound is recorded.
2. Instead of the usual way to point on the projector canvas with a laser pointer to areas that are described, the lecturer has to point them directly on the touchpad.
3. Each demonstration in the lecture hall must take place at the certain place where the webcam is focused.

If the lecturer differs from this concept, the information will be lost for the video. For video editing each lecture is divided thematically in 10 to 20 minute long chapters, which can be viewed and downloaded independently. This enables to repeat topics individually, without having to see the full lecture. In addition, the concentration of students on a video of 10 minutes is higher than on a video that lasts 90 minutes.

7 Satisfication with the Methodic and the Conten of the Lecture and the Lecturer Himself

All shown approaches led to an improvement of the lecture, but the redesign of this lecture is not yet completely finished. For example, audio and video recording of the slides are carried out in the current run. With this media students can watch the lecture at home again. This step of improvement maybe leads to a Massive Open Online Course for "Fundamentals of Forming Technology" in the future.

References

1. eduvote, 2015. http://www.eduvote.de/en/
2. R.E. Mayer, A. Stull, K. DeLeeuw, K. Almeroth, B. Bimber, D. Chun, M. Bulger, J. Campbell, A. Knight, H. Zhang, Clickers in college classrooms: Fostering learning with questioning methods in large lecture classes. Contemporary Educational Psychology **34** (1), 2009, pp. 51–57. doi:10.1016/j.cedpsych.2008.04.002
3. T.R. Ortelt, A. Sadiki, C. Pleul, C. Becker, S. Chatti, A.E. Tekkaya, Development of a tele-operative testing cell as a remote lab for material characterization. In: *Proceedings of 2014 International Conference on Interactive Collaborative Learning, ICL 2014*. 2014, pp. 977–982. doi:10.1109/ICL.2014.7017910
4. Moodle overview, 2015. http://www.moodle.com/moodle-lms/
5. A.E. Tekkaya, C. Becker, T.R. Ortelt, G. Grzancic, *Utilizing Stress superposition in metal forming*, 2015, p. 1–6

Concepts of the International Manufacturing Remote Lab (MINTReLab) – Combination of a MOOC and a Remote Lab for a Manufacturing Technology Online Course

Tobias R. Ortelt, Sabine Pekasch, Karsten Lensing, Pierre-Jean Guéno, Dominik May and A. Erman Tekkaya

Abstract A concept for a Manufacturing Technology Online Course, which combines the MOOC approach and the use of a remote laboratory is presented. Hence, this online course is a combination of a MOOC-concept and a physically existing but tele-operatively usable experimentation laboratory for real engineering experimentation. Developing this online course can be either used as an advertising for mechanical engineering education in Germany and for the "Master of Science in Manufacturing Technology (MMT)" or as one part of the preparation process for international students coming to Dortmund in advance of their stay in Germany. For doing so the course offers not only core engineering content around the topic of manufacturing technology but also explains the German educational system at higher education institutions and presents cultural information about Germany as a place to live.

Keywords MOOC · Remote Lab · Online Course · Online Learning

1 Introduction

In the recent years Massive Open Online Courses (MOOC) have been constantly popular in the United States of America and lately they are getting more and more popular in Europe as well, especially in Germany. Certainly there is one major difference between the US and the European MOOC offers. In the United States a MOOC is often an element of a corresponding business model, meaning education institutions earn money by every student participating in a course and attending for

T.R. Ortelt (✉) · S. Pekasch · P.-J. Guéno · A.E. Tekkaya
Institute of Forming Technology and Lightweight Construction (IUL) Dortmund,
TU Dortmund University, Dortmund, Germany
e-mail: tobias.ortelt@iul.tu-dortmund.de

K. Lensing · D. May
Engineering Education Research Group (EERG), Center for Higher Education (zhb),
TU Dortmund University, Dortmund, Germany
e-mail: Dominik.may@tu-dortmund.de

Originally published in "IEEE Global Engineering Education Conference (EDUCON)",

DOI 10.1007/978-3-319-46916-4_41

an exam. In contrast the German education system, meaning universities and similar higher education institutions, provides education free for everyone and is only partially organized with an economical objective of earning money. Due to that fact, students do not need to pay any fee for their education while studying – what, as a general rule, also applies to international students.

Over the last few years an increasing number of international students has been welcomed at TU Dortmund University. The number of international students has raised from 2.618 to 3.603, considering winter semester 2010/2011 until winter semester 2014/2015. So there has been a rise of nearly 1000 international students in the last four years, which can be described as a significant percentage increase of almost 38 %.

Going more into detail the number of international students enrolled at the engineering faculty, unfortunately, is rather low. In winter semester 2006/2007 20 % of the mechanical engineering students originally came from foreign countries. Until winter semester 2012/2013 the share even declined to 11 %, but then had a rebound resulting in 15 % international students enrolled at the last winter semester. Having in mind that the world's economy and its labor market is growing more and more together and is becoming a fully international market, these numbers must be alarming. In our opinion, future engineers need to be trained within international work situations right from the start of their educational programs, what implies the necessity of a diverse, heterogeneous and multinational student body, especially at the engineering faculty [1]. So, which type of course concept could attract international students to study engineering in Dortmund?

Knowing that one important barrier for international students not coming to Germany is the language, Faculty of Mechanical Engineering at TU Dortmund University successfully launched a master study program in the field of manufacturing and production engineering. The "Master of Science in Manufacturing Technology (MMT)" started in 2011 and is accredited by ASIIN.

In addition to that, and in order to attract even more students from all over the world for studying in Germany and especially for studying at TU Dortmund University, a special online course has been designed and will be further worked out in detail until the coming summer. The course is a combination of the MOOC concept and a teleoperative remote lab for material characterization. Accompanying to the theoretical and lab inputs, there will be content about living and studying in Germany especially at the TU Dortmund University.

2 Technological and Electronic Development

The concept of the so-called "International Manufacturing Remote Lab (MINTReLab)" was created to combine the advantages of a MOOC with the advantages of a remote lab. At best, such a combination fosters a learning experience characterized by multinational web-based collaboration and practical engineering laboratory work at an anywhere-/anytime-basis, enabling young engineering students to have a first glance at their potential future working conditions in a globalized world. In the following we will explain the MOOC approach, the remote laboratory, and finally the combination of both in context with the online course.

2.1 The MOOC Approach

The Acronym MOOC stands for a Massive Open Online Courses. This concept describes a recent manifestation in the evolution of distant education that take advantage of the current possibilities offered by the modern information technologies (IT): simultaneity, interactivity, multimedia and broad access.

What could be regarded as the first occurrence of a massive open e-learning concept, is the MIT OpenCourseWare (OCW) launched in 2000 as "a collection of more than 2000 course syllabi, lecture notes, assignments, and exams [...] provided free of charge" [2]. The OCW lead to the creation of the OpenCourseWare Consortium in 2005 and the MIT began to "introduce course materials designed specifically for use by independent learners, which [...] include complete sets of content, increased focus on problem-solving, and additional self-assessment opportunities" [2]. The term MOOC was coined in 2008 to describe the course Connectivism and Connective Knowledge (CCK08) developed in Canada by Georges Siemens of the Athabasca University and Stephen Downes of the National Research Council. In the following years leading American Universities have been creating their own platforms such as Udacity (for-profit, Stanford), Coursera (for-profit, Stanford) and EdX (non-profit, MIT, Harvard) and the New York Times described 2012 as "The Year of the MOOC" [3]. This development set off new academic research interests about the stakes and challenges of e-learning [4, 5]. The main issues are the effectiveness of the learning effect on IT platforms, the fostering of motivation and high dropout rates.

In their article of the Proceedings of the 10th European Conference on Technology Enhanced Learning, Bakki et al. reviewed four different approaches to design an effective MOOC [6]:

- Pedagogical strategies approach with, for example, regular motivational messages [7]
- Personalization and/or adaptivity approach.
- Gamification approach [8]
- Technological approach using all the current IT possibilities

These solutions were consciously and conscientiously implemented in the concept of the MINTRelab MOOC. Hence, such as participants in current and famous American MOOCs, MINTRelab's ones can benefit of a large variety of advantages. Firstly, the design of the courses, which mixes video, interactive content and quizzes, as well as the structure of the lessons developed by professional educationalists improve the learning experience. Participants can also study at their own pace and repeat a single lesson or videos until they really understand it. Using online forums and social media networks, they can share experience with peers and, thus, improve even more.

2.2 The Remote Laboratory

A remote lab, as we define it here, can be described as a system of web-based architectures enabling access to a real world physical experiment – providing learners to ad-hoc manipulate process parameters from distance through Internet technology. The use of these technologies is not entirely new in higher engineering education, for example there have been several comparative studies exploring the assets and drawbacks of similar approaches since the late 1990's [9, 10]. Consensus is, that the learning outcome and motivation of the participants attending these remote experiments is strongly reliant to the instructional content design on the one hand, as it applies to e-learning approaches in general, and to the extent of embeddedness and usability regarding the surrounding learning environment on the other hand [11]. This should be kept in mind while developing the course content, as well as while designing the necessary hardware and software infrastructures. In recent years there has also been a strong focus on creating standardized platforms to make these labs also available to the student body of other universities and thereby create a global network of different remote experiments [12]. This has to be seen as an excellent opportunity to evolve a practical engineering education on a transnational level [13].

2.3 Combination of MOOCs and Remote Labs

Similar Approaches of integrating remote labs into a MOOC have been quite promising, even if the integration of telemetric laboratories in the field of mechanical engineering, in terms of complexity, seems to be a considerable elaborate process [14]. The MINTReLab benefits from experiences made by giving transnational online classes and using the remote lab in several course contexts [15, 16]. Like this transnational online class the developed MINTReLab's main audience are is the body of international students. Hence, the concept of MINTReLab basically combines three different educational topics: MOOCs, remote labs, and international students as primary target group (see Fig. 1).

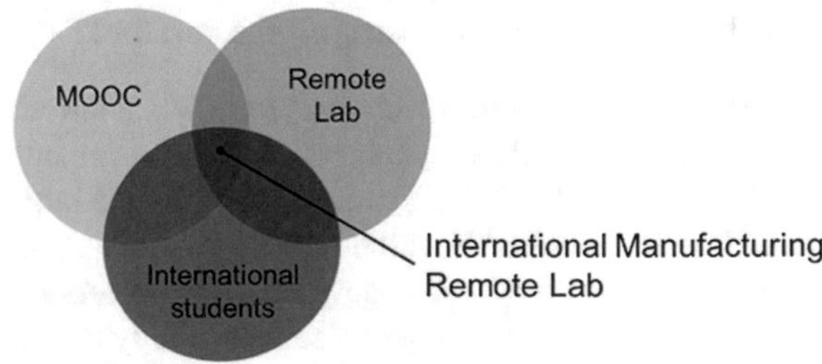

Fig. 1 Concepts of the International Manufacturing Remote Lab

3 MINTReLab Objectives

Looking from TU Dortmund's perspective, there are four main objectives pursued by developing the MINTReLab. They will be described in the following:

3.1 *Increase of International Students*

As mentioned above one main objective is to attract international students to German degree programs, especially to engineering degree programs at the TU Dortmund University. There is not only a profit for the university, but also for the students who can get in exchange with each other. In consideration of the more and more connected word economy, there is a need for special competences like collaboration, language and social skills. Teams, who work on a certain topic, will have members from all over the world and all these people will have different social and academic backgrounds.

3.2 *International Distribution of Learning Opportunities*

One of the specifics of German mechanical engineering studies is the practice oriented work in laboratories. All students have to do hands on labs during their studies. For us laboratory experience is very usefull to combine theory and practical relevance. A deep knowledge of processes is needed to fulfil the German idea of "Industrie 4.0", where products and their process need to have special properties [17]. If you want to control processes to create products with a batch size of 1 the first product has to be correct. This is only possible, if engineers of the future, so the current students all over the world, have a deep knowledge in theory and practical relevance. A further aspect of "Industrie 4.0" is added with the tele-operative testing cell. Students of the MINTReLab will control real machines in near real-time via internet.

3.3 *Preparation of Future Incoming Students*

Nevertheless, it is not the case that no international students are coming to Germany at the moment, even if the number could be higher. From the author's perspective there is a lack of opportunities for these student in order to prepare themselves for their stay in Germany. The MINTReLab is integrating social videos in the MOOC, which present Germany and the TU Dortmund University as places to study and which are thus preparing the learners for a future life in a foreign country.

3.4 Usage and Exploration of Didactical Potentials of MOOCs

The development of a MOOC with manufacturing engineering as the topic was and currently is kind of unique unique in Germany. Thus, the main objective from the pedagogical view is to explore the potentials that offers a MOOC for engineering education. Especially the integration of learning videos in the course is on the foreground because they are the "main media" in a MOOC. The question is how a "good" learning video should look like that learning successfully can takes place. Besides the videos different kinds of activities, such as tests, exercises and forums, will be integrated in the course not only to improve the interaction between the learning material and the learners but also between the learners themselves.

4 Structure of the MINTReLab

In the following we will explain the MINTReLab's working consortium at TU Dortmund University and content development's status quo.

4.1 Consurtium of the MINTReLab

Five different institutions of the Faculty of Mechanical Engineering of the TU Dortmund University and experts for engineering education are working on the development of the MINTReLab course. The following list shows the involved chairs, departments or institutes:

- Chair of Polymer Technology
- Department of Materials Test Engineering
- Institute of Mechanics
- Institute of Machining Technology
- Institute of Forming Technology and Lightweight Construction
- Center for Higher Education

The list above shows the consortium's wide expertise in mechanical engineering and that it is intentionally supplemented by an institution focusing on the pedagogical development aspects.

A first step in the corporate development of this MOOC was to find a similarity to combine these different expertise to one topic. The uniaxial tensile test was chosen as the connecting element. Every institution is working with this standard test, certainly from a different view. Next to this main section "Learning" of the MOOC there are two more sections (see Fig. 2). On the one hand there is the section "Lab", which

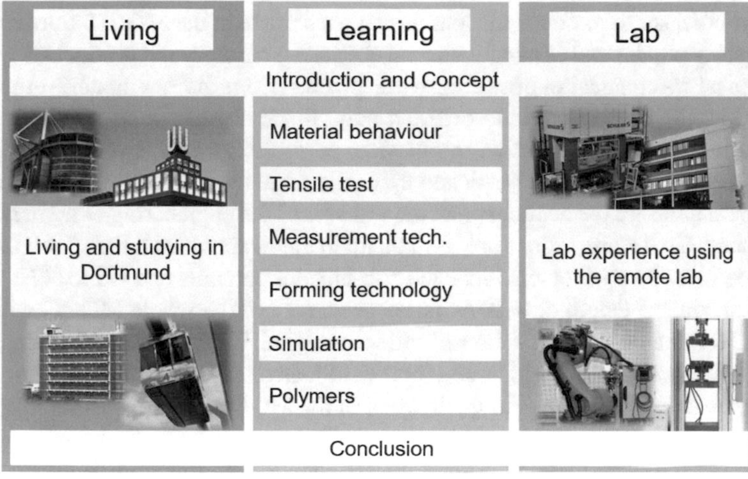

Fig. 2 Structure of the MINTReLab MOOC

consists of the different experiments in this online course. On the other hand, there is the section "Living", which focusses on the living in Germany and in Dortmund especially.

4.2 *Section – Learning*

The section "Learning" of the MOOC represents the MOOC's theoretical base. Besides an introduction and concluding chapter, there are six chapters with professional engineering content (see Table 1).

The eight chapters should be attended in chronological order because they are subsequently organized. In the introductive chapter the course's main content and the learning objectives will be introduced. Also the remote lab will be briefly presented

Table 1 Chapters of the section "learning"

Chapter No.	Chapter Title
I	Introduction
II	Material behaviour
III	Tensile test
IV	Measurement technology
V	Forming technology
VI	Simulation
VII	Polymers
VIII	Conclusion

as a sort of teaser to motivate the learners to participate in the MOOC. Furthermore, a short story will be told as a motivational element. This story is a production process of a typical mechanical engineering work piece – from the raw material up to the finished product. The goal is to create a naturalistic context in which the learners can pull together the theoretical input with engineering processes in real life. In every chapter this story will be picked up to illustrate the relevance of the particular theoretical input. In the next chapters the transfer of the engineering content is in the foreground. Therefore, videos are used as main elements of MOOCs. Each Chapter has three to four sub-chapters and each sub-chapter contains one video. The videos have a maximum length of 8-10 Minutes each, so that the whole MOOC will have about 230 minutes (around 4 hours) video material. With the additional materials and exercises the learners will have a total workload of about 8 hours. Assuming that the learners spend about one hour per week in the MOOC, the course can fully done in about 8 weeks.

In addition to the videos each sub-chapter contains the following content:

- Tests: There are three to five multiple choice questions for the learners' self-assessment. Therewith they can check whether they understand the theoretical input in the video or not.
- Additional materials: For further information there are additional materials in sort of PDF's, bibliographical references or references to further websites or videos. The reading and studying of these materials are optional. At the conclusion of every chapter there is a big exercise to work on for the learners.
- Exercises: In this context the learners either have to operate the tensile test for themselves in the remote lab as a practical task or they have to solve a theoretical task in sort of a text. However, in each of the exercises the learners have to upload their answers in the MOOC by generate a threat in the appropriate forum.

As mentioned above the learners have the possibility to upload their task solution in the particular forum. With this we want to provoke an intensive interaction between the learners in our MOOC. This interaction will be initiated by additional request in the exercises itself. In addition to that the MINTReLab Team as well as the staff of the involved chairs, departments or institutes will give feedback at one time or another, too. This will heavily depend on the number of participants taken part in the MOOC. Based on first experiences we will have a closer look at this aspect and check in how far this interaction has to be automated.

In the concluding chapter the whole course will be summarized with the special view of the story. Furthermore the particular professional perspectives of the participating chairs, departments or institutes will be compared. After the conclusion of the professional content another social video about degree programs at the TU Dortmund University follows.

Fig. 3 Tele-operative testing cell

4.3 Section "Lab"

Remote labs enable students to connect theory with practical relevance via the internet. They can conduct experiments time- and location-independent. Remote labs are popular in the field of engineering education because the needed hardware is often cheap in relation to the costs for engineering machines. However, a tele-operative testing cell as a remote lab for material characterization was developed in the project "ELLI – Excellent teaching and learning in engineering science" at the Institute of Forming Technology and Lightweight Construction at TU Dortmund University [18]. The applied concepts of the integration of the remote lab are based on the project "PeTEX" [19]. At the tele-operative testing cell an industrial robot is able to feed two different machines with specimens [20]. Figure 3 shows the tele-operative testing cell.

The remote lab uses the iLab structure [21] to provide access to the different experiments. At the current stage of development two different experiments in the field of material characterization for forming technology are available for students via the internet. In this online course the uniaxial tensile test will be conducted by the students from all over the world.

4.4 Section "Living"

As mentioned above the section "Living" implies an additional social aspect. It shall present Germany, North Rhine-Westphalia (federal state) and the TU Dortmund University as locations to live and to study in. Another decisive issue is the preparation

Table 2 Chapters of the section "Living"

Episode No.	Location	Content
I	Abroad	Germany as an attractive place to study engineering
II	Entry to Germany	Prerequisites to study in Germany
III	North Rhine-Westphalia	NRW as a key region for economics and engineering
IV	North Rhine-Westphalia	Industrial heritage trail NRW, final stop Dortmund
V	Dortmund	Dortmund as an innovative city for living, studying and working
VI	TU Dortmund University	TU Dortmund University and the Mechanica Engineering Faculty

of the future incoming students to the life and to the studies in Germany, while advising them about preconditions and potential difficulties (Table 2).

Therefore, this section consists of six episodes, which start as a wide shot and then focusing step by step on the TU Dortmund University and its Faculty of Mechanical Engineering. These short video implement different content type, such as shorts animations, motion pictures, interviews and graphics, not only with a common graphic charter and a common pattern, but also with a common protagonist: the viewers themselves. The voice-over keeps addressing the viewers in order to guide them along the path to their studies at TU Dortmund University.

TU Dortmund University, representatives of its International Office and peers - foreign students, who already study in Dortmund - count among the interviewees. They serve as experts, explaining important points and convincing the future incoming student, that the TU Dortmund University complies with their wishes.

According to the theme of the episode, each video presents facts and stakes at a particular level. For example, Germany's prominence in the fields of large scale. More funny features of Germany are also investigated, such as common stereotypes on the country and its inhabitants. This combination advertises for Dortmund and should foster the international spreading of Dortmund's degrees in the field of Engineering, as well as the increasing of foreign students at the TU Dortmund University.

5 Conclusion and Outlook

The shown concepts will connect a MOOC with a remote lab to inspire students from all over the world for studies in Germany. The concepts will be implemented in the MOOC "MINTReLab - International Manufacturing Remote Lab". With this, the presented course concept combines two current topics of online education and puts them into practice in context with manufacturing technology education. At the moment the different video episodes are under development and over the summer

2016 the videos will be worked out. By the end of September the MOOC will be ready and first students can take part. More details about the final course development and our experiences are to be expected in future publication.

References

1. D. May and A. E. Tekkaya, "The globally competent engineer: What different stakeholders say about educating engineers for a globalized world," in *Interactive Collaborative Learning (ICL), 2014 International Conference on,* 2014, pp. 924–930.
2. C. d'Oliveira, S. Carson, K. James, and J. Lazarus, "MIT OpenCourseWare: Unlocking Knowledge, Empowering Minds," *Science,* vol. 329, pp. 525–526, 2010-07-30 00:00:00 2010.
3. L. Pappano, "The Year of the MOOC," in *The New York Times,* ed, 2012.
4. M. Chadaj, C. Allison, and G. Baxter, "MOOCS with attitudes: Insights from a practitioner based investigation," in *Frontiers in Education Conference (FIE), 2014 IEEE,* 2014, pp. 1–9.
5. J. Reich, "Rebooting MOOC Research," *Science,* vol. 347, pp. 34–35, 2015-01-02 00:00:00 2015.
6. A. Bakki, L. Oubahssi, C. Cherkaoui, and S. George, "Motivation and Engagement in MOOCS: How to Increase Learning Motivation by Adapting Pedagogical Scenarios?," in *Design for Teaching and Learning in a Networked World: 10th European Conference on Technology Enhanced Learning, EC-TEL 2015, Toledo, Spain, September 15-18, 2015, Proceedings,* G. Conole, T. Klobucar, C. Rensing, J. Konert, and É. Lavoué, Eds., ed Cham: Springer International Publishing, 2015, pp. 556–559.
7. J. J. Williams, "Applying cognitive science to online learning," *Available at SSRN,* 2013.
8. M. Romero and M. Usart, "Serious Games Integration in an Entrepreneurship Massive Online Open Course (MOOC)," in *Serious Games Development and Applications: 4th International Conference, SGDA 2013, Trondheim, Norway, September 25-27, 2013. Proceedings,* M. Ma, M. F. Oliveira, S. Petersen, J. B. Hauge, Eds., ed Berlin, Heidelberg: Springer Berlin Heidelberg, 2013, pp. 212–225.
9. J. E. Corter, J. V. Nickerson, S. K. Esche and C. Chassapis, "Remote versus hands-on labs: a comparative study," in *Frontiers in Education, 2014. FIE 2004. 34th Annua,* 2004, pp. F1G-17-21, Vol. 2.
10. J. Ma and J. V. Nickerson, "Hands-on, simulated, and remote laboratories: A comparative literature review," *ACM Computing Surveys (CSUR),* vol. 38, p. 7, 2006.
11. N. E. Cagiltay, E. Aydin, C. C. Aydin, A. Kara, and M. Alexandru, "Seven principles of instructional content design for a remote laboratory: a case study on erl," *Education, IEEE Transactions on,* vol. 54, pp. 320–327, 2011.
12. E. Sancristobal, S. Martin, R. Gil, P. Orduna, M. Tawfik, A. Pesquera, *et al.*, "State of art, initiatives and new challenges for Virtual and Remote Labs," in *Advanced Learning Technologies (ICALT), 2012 IEEE 12th International Conference on,* 2012, pp. 714–715.
13. V. Chennam Vijay, M. Lees, P. Chima, C. Chapman, and P. Raju, "Knowledge based educational framework for enhancing practical skills in engineering distance learners," in *Global Engineering Education Conference (EDUCON), 2015 IEEE,* 2015, pp. 124–131
14. G. Diaz, F. Garcia Loro, M. Castro, M. Tawfik, E. Sancristobal, and S. Monteso, "Remote electronics lab within a MOOC: Design and preliminary results," in *Experiment@ International Conference (exp. at'13), 2013 2nd,* 2013, pp. 89–93.
15. D. May, A. Sadiki, C. Pleul and A. E. Tekkaya, "Teaching, and learning globally connected using live online classes for preparing international engineering students for transnational collaboration and for studying in germany," in *Proceedings of 2015 12th International Conference on Remote Engineering and Virtual Instrumentation, REV 2015,* 2015, pp. 118–126.

16. D. May, T. R. Ortelt, A. E. Tekkaya, "Using remote laboratories for transnational online learning environments in engineering education," in *E-Learn: World Conference on E-Learning in Corporate, Government, Healthcare, and Higher Education*, 2015, pp. 632–637.
17. A. E. Tekkaya, J. M. Allwood, P. F. Bariani, S. Bruschi, J. Cao, S. Gramlich, *et al.*, "Metal forming beyond shaping: Predicting and setting product properties," *CIRP Annals - Manufacturing Technology,* vol. 64, pp. 629–653, 2015.
18. T. R. Ortelt, A. Sadiki, C. Pleul, C. Becker, S. Chatti, A. E. Tekkaya, "Development of a tele-operative testing cell as a remote lab for material characterization," in *Proceedings of 2014 International Conference on Interactive Collaborative Learning, ICL 2014,* 2015, pp. 977–982
19. C. Terkowsky, I. Jahnke, C. Pleul, R. Licari, P. Johansson, G. Buffa, *et al.*, "Developing tele-operated laboratories for manufacturing engineering education," *Platform for E-Learning and Telemetric Experimentation (PeTEX) International Journal of Online Engineering (iJOE) Vol,* vol. 6, pp. 60–70, 2010.
20. A. Sadiki, T. R. Ortelt, C. Pleul, C. Becker, S. Chatti, A. E. Tekkaya, "The challenge of specimen handling in remote laboratories for engineering education," in *Proceedings of 2015 12th International Conference on Remote Engineering and Virtual Instrumentation, REV 2015,* 2015, pp. 180–185.
21. V. J. Harward, J. A. Del Alamo, S. R. Lerman, P. H. Bailey, J. Carpenter, K. DeLong, *et al.*, "The ilab shared architecture: A web services infrastructure to build communities of internet accessible laboratories," *Proceedings of the IEEE,* vol. 96, pp. 931–950, 2008.

Part II
Mobility and Internationalization

International Student Mobility in Engineering Education

Ute Heckel, Ursula Bach, Anja Richert, Sabina Jeschke and Marcus Petermann

Abstract Engineering students are, compared to their counterparts in other disciplines, less mobile resulting in limited intercultural skills. Globalization requires professionals being excellent in their fields and being able to work on a global scale at the same time. So far, engineering education has put too little stress on integrating intercultural competences into curricula. This paper shows new approaches to incorporate international experiences into higher engineering education. First, it analyzes the current situation of international student mobility in Germany, before emphasizing the general motivation for international student exchange especially in engineering science. A consortium of three excellent German engineering universities was put up to introduce new measures for increasing student mobility as is described subsequently. This paper represents work in progress. Thus, further results will be published continuously.

Keywords International Student Mobility · Inbound · Outbound · Engineering Education · Curricula Development · Intercultural Competences

1 Introduction

In Germany international student mobility in engineering science lies below average compared to other disciplines. Even the Bologna reform of introducing joint degrees throughout Europe did not result in higher numbers of engineering students going abroad (referred to as "outbound mobility") [1]. Also, the number of foreign engineering students coming to Germany (referred to as "inbound mobility") especially

U. Heckel (✉) · U. Bach · A. Richert · S. Jeschke · M. Petermann
IMA/ZLW & IfU, RWTH Aachen University, Dennewartstr. 27,
52068 Aachen, Germany
e-mail: ute.heckel@ima-zlw-ifu.rwth-aachen.de

U. Bach
e-mail: ursula.bach@ima-zlw-ifu.rwth-aachen.de

A. Richert
e-mail: anja.richert@ima-zlw-ifu.rwth-aachen.de

S. Jeschke
e-mail: sabina.jeschke@ima-zlw-ifu.rwth-aachen.de

Originally published in "Proceedings of the 2012 IEEE Global Engineering Education Conference (EDUCON)", © 2012. Reprint by Springer International Publishing AG 2016, DOI 10.1007/978-3-319-46916-4_42

from industrialized countries such as the US, the UK, Japan, or France is below expectations [2].

In order to educate excellent engineers, intercultural competences are essential. Engineering is not limited by national borders. Not only big multinationals but also small and medium sized enterprises, where most engineers are employed in Germany, compete on global markets generating the need for intercultural competent employees. Thus, excellent universities, that aim to educate excellent engineers, have to face the challenge of integrating international experiences into engineering curricula.

A consortium of three excellent German universities, RWTH Aachen University (RWTH), Ruhr University Bochum (RUB), and Technical University Dortmund (TUD), was put up to develop new approaches for improving higher engineering education. First, the partners worked together to build the "Competence and Service Center for Teaching and Learning in Engineering Science – TeachING-LearnING.EU" [3]. Within the project ELLI ("Exzellentes Lehren und Lernen in den Ingenieurwissenschaften" Excellent Teaching and Learning in Engineering Science) the same partners discuss key issues of engineering education also focusing on international student mobility.

After introducing basic terms and definitions (Section 2), the paper gives an overview over international student mobility numbers in Germany (Section 3) and introduces the basic motivation of students, higher education institutions (HEI) and industry to promote international student exchange in engineering education (Section 4). Key measures proposed within the project ELLI to foster student exchange in the field of engineering education are presented (Section 5) together with their expected impact (Section 6). The paper represents work in progress that will be put into practice during the next five years. Concrete results will be published continuously.

2 Terms and Definitions

When talking about mobility and mobile students some basic terms have to be defined. According to the definitions of [2, 4, 5] the following terms are introduced:

- National student: student with German nationality.
- Foreign student: student with a nationality different from Germany.
- Bildungsinlaender students: foreign students who gained their higher education entrance qualification at a German school or who passed a Gifted Students Test (Begabtenprüfung) or an Aptitude Test (Eignungsprüfung) in Germany.
- Bildungsauslaender students: foreign students who gained their higher education entrance qualification at a foreign school and/or complemented their foreign school qualifications by attending a German Studienkolleg (preparatory course for higher education admission).
- Mobile students: students who cross national borders for the purpose or in the context of their studies.
- Inbound mobile students: students who move to a country for the purpose of study or study related activity.

- Outbound mobile students: students who leave their country to another for the purpose of study or a traineeship in the context of study.
- Inbound mobility: refers to the number of inbound mobile students.
- Outbound mobility: refers to the number of outbound mobile students.
- Country of origin: the country where the student moves from.
- Country of destination: the country which the student moves to.

Using those definitions, the next chapter introduces statistics showing the current mobility activities of students in Germany.

3 Statistical Background

The Bologna reform focused on structural changes to increase student mobility throughout Europe as a central goal. In 2009 the European Ministers of Education and Research set the aim that until 2020 20 % of European students should have lived and worked abroad. On the conference in Budapest and Vienna in 2010 it became clear that this goal will not be accomplished [6]. This chapter gives an overview over current inbound and outbound mobility data in Germany in general and on university level mainly based on the survey carried out by the Hochschul-Informations-System (HIS) in co-operation with the German Academic Exchange Service (DAAD) [2].

3.1 *Inbound Mobility at German HEI*

In 2010 11.5 % of students in Germany were foreign students (c.f. Fig. 1). In 2009 the most common countries of origin were China (10 %), Turkey (9.9 %) and Russia (5.2 %). 16.5 % of foreign students were enrolled in engineering studies with mechanical engineering (5.9 %) and electrical engineering (4.5 %) being the most popular engineering disciplines.

In terms of subject preferences most Bildungsauslaender students are inclined to choose courses in language and cultural studies and show less interest in engineering as is shown in Fig. 2.

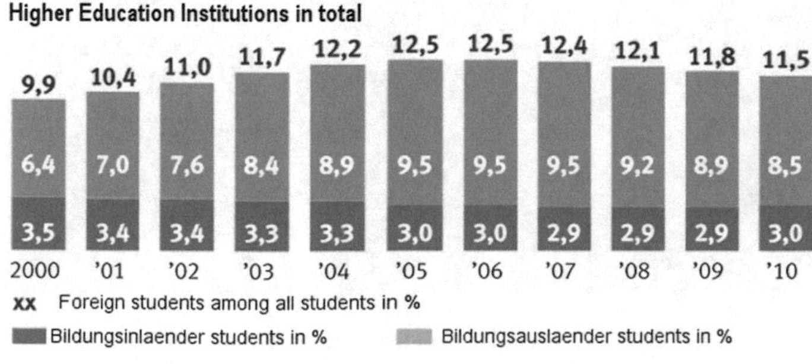

Fig. 1 Foreign students as a percentage of all students in German HEI [2]

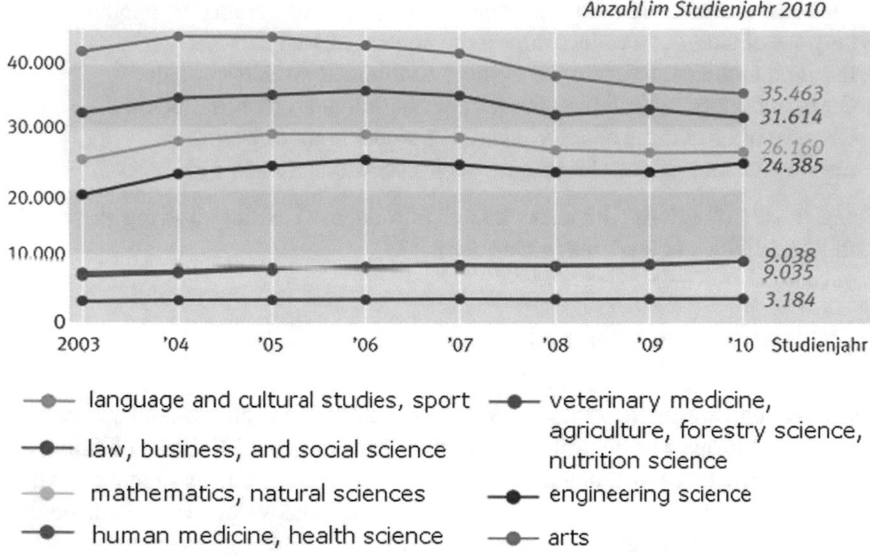

Fig. 2 Bildungsauslaender students at universities, by subject group [2]

3.2 *Outbound Mobility at German HEI*

In 2008 4.2 % of national German students pursued their studies abroad. Depending on the degree different countries of destination were most popular: 40 % of German bachelor students pursuing their studies abroad choose The Netherlands as their country of destination, 37 % of German doctoral students abroad were enrolled in Switzerland and 61 % of students with other degrees (such as diploma or master) took up studies in Austria. As is shown in Fig. 3 the neighboring countries seem to be most attractive to German students with Austria and the Netherlands being the most visited countries as well as the UK and the US as Anglophone countries.

Destinations	Quantity	Amount
Austria	20.019	19,6
The Netherlands	18.972	18,6
Great Britain	12.895	12,6
Switzerland	11.005	10,8
USA	9.679	9,5
France	6.071	5,9
Australia	3.418	3,3
Sweden	3.400	3,3

Fig. 3 German students abroad in 2008, by country of destination in % [2]

Compared to other countries German students are rather mobile. In France only 2.9 % of national students went abroad, in the UK only 1.2 % and in the US 0.3 %. In other countries such as Ireland (10.4 %), Slovakia (11.8 %), or Iceland (21.8 %) national students choose their course of studies more often abroad than in Germany. The reason for this lies in the higher education system of those countries not being able to provide enough capacities to enroll all of their national students. Thus, those students are more often and to a higher degree mobile than students in other nations [5]. China, India, and South Korea are currently the countries with the highest outbound mobility rates [7].

Mostly national students from language and cultural studies envisage a mobility period in their course of studies (12 %). The lowest rate of outbound mobility is encountered among engineering students (4 %). Thus, in terms of subject preferences inbound and outbound mobile students show similar priorities (c.f. Fig. 2).

Regarding the development within the last 17 years, mobility rates developed differently throughout the disciplines as is shown in Table 1. Sincc 1994 mobility rates in language and cultural studies stayed constantly at a high level (1994: 12 %, 2000: 13 %, 2009: 12 %) whereas students in e.g. engineering science became a lot more mobile over the same timespan where mobility rates nearly doubled between 1994 (2 %) and 2009 (4 %) [7].

Regarding the disciplines different ways of foreign exchange seem to be commonly chosen by students. Students of language and cultural studies often chose to pursue their studies abroad whereas engineering students prefer to take up internships in the respective country of destination as is shown in table II [7] (Table 2).

Table 1 Rate of mobile students, by discipline in % [7]

Disciplines	Year					
	1994	1997	2000	2003	2006	2009
Engineerings	2	3	4	4	3	4
Language and cultural studies	12	12	13	12	12	12
Maths/nat. s.	4	5	4	5	5	5
Medicine/health	4	5	5	7	6	5
Law and business studies	5	8	9	8	9	11
Social/education s. psychology studies	2	4	4	5	6	8

Table 2 Rate of mobile students with study-related activities, by discipline in % [7]

Disciplines	Studies		Internship		Language course		Other	
	06	09	06	09	06	09	06	09
Engineering s.	3	4	6	7	2	1	1	1
Language and cultural studies	12	12	9	8	7	6	4	4
Maths/nat. s.	5	5	5	5	2	1	2	2
Medicine/health	6	5	18	16	3	2	3	2
Law and business studies	9	11	9	7	5	3	1	1
Social/education s. psychology	6	8	7	7	3	3	2	2

3.3 International Mobility Rates on Institutional Level

The picture is also mirrored in statistics on institutional level. The following numbers represent international student mobility at RWTH being also emblematic for RUB and TUD. In the winter term of 09/10 32,943 students were enrolled in study programs at RWTH Aachen University with the faculties of mechanical engineering (8,721 students), civil engineering (1,340 students) and electrical engineering and information technology (3,106 students) being one of the biggest faculties of RWTH. In 2009 624 German students took part in international exchange programs to pursue a part of their studies abroad. The most popular countries were Spain, France and the UK [8].

More than 5,000 foreign students contribute to an international profile of RWTH. In the winter term (wt) 09/10 5,164 foreign students were enrolled in courses at RWTH which amounts to 15.7 % of all students (c.f. Fig. 4).

Despite of mobility programs initiated by the Bologna reform numbers remained constantly at the same level between 1999 and 2010. The biggest community of foreign students is stemming from China, followed by Turkey and Luxemburg [8]. Fewer than 100 students from countries such as the US, UK, France, and Japan pursued their studies at RWTH in 2009. The most popular faculties among foreign students are the faculties of mathematics, computer science and natural sciences (21 %), the faculty of mechanical engineering (20 %) and the faculty of electrical engineering and information technology (14.3 %). RWTH ranks sixth among the most popular universities for foreign students in Germany [8].

Those numbers show that there exists a discrepancy between inbound and outbound mobility with a lot more foreign students being enrolled in German universities than German students attending courses abroad. Furthermore, only a low percent-

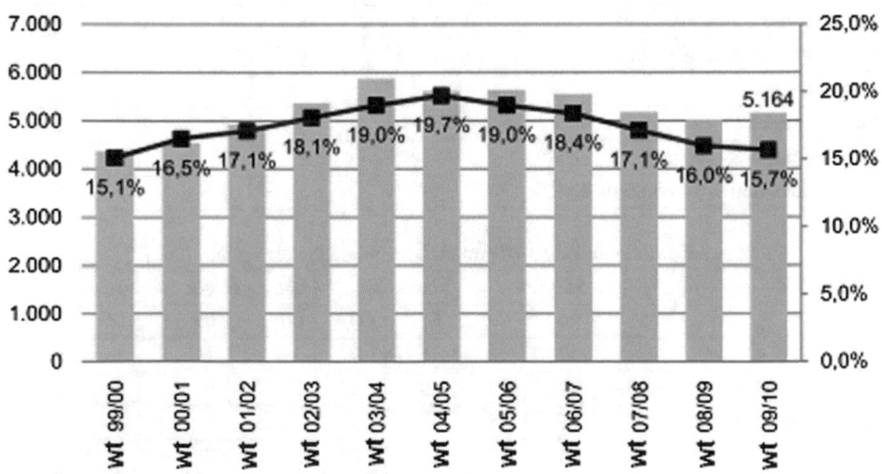

Fig. 4 International students at RWTH in winter terms 99/00 – 09/10 [8]

age of foreign students from the industrialized world take part in German university programs. The following chapter gives an overview over basic reasons for these phenomena.

4 Motivation for Mobility

The reasons for small numbers in outbound mobility are multifaceted (c.f. Table 3). Compared to former German higher education degrees such as the diploma, new curricula developed throughout the Bologna reform (i.e. the bachelor degree) seem to be less flexible leaving the students not enough time to spend a semester or more time abroad. 35 % of bachelor students and 30 % of master students at universities state that they consider a temporary stay abroad as a loss of time [2]. Furthermore, insufficient mutual recognition of studies and qualifications present major problems. 21 % of bachelor students and 13 % of master students at universities report problems in the recognition process of their records accomplished abroad. Also financial problems, little support from higher education institutions (HEI), and mandatory industry internships during semester breaks seem to hamper the students's opportunities to integrate stays abroad into their courses of studies.

The factors motivating German students to take part in foreign exchange programs are manifold. The prospect of developing personal and professional skills seems to be the main motivator. But also other positive experiences were reported by German students abroad as is shown in Table 4 [2].

German engineering education is highly regarded in emerging countries resulting in increased inbound mobility rates from China, or India whereas inbound mobility from Western countries stays rather low (c.f. Section 3). Engineering students from the industrialized world seem to be attracted to strong research nations and English speaking countries such as the US, the UK, Canada, or Australia. From a research perspective it is crucial to attract students from industrialized nations in order to intensify connections between German and other leading research universities in the industrialized world, to boost research excellence, and to broaden networks for research and education.

Table 3 Rate of mobile students with study-related activities, by discipline in % [2]

Study-related problems	Bachelor university	Master university
Loss of time	35	30
Little support from HEI	24	21
Financing problems	22	23
Limited compatibility with study specifications	26	20
Difficulties finding an appartment	18	20
Problems with recognition of records	21	13

Table 4 Experiences of German students during study-related visits abroad, by type of degree in % [2]

Experiences	Bachelor university	Master university
Managed to deal with the foreign mentality	81	84
Gathered new experiences	86	81
Feeling of being integrated	72	68
Accomplished all planed records	63	72
Took part in all classes as intended	63	69
Successful communication in foreign language	70	63
Very good support	63	68
Transfer of important discipline-specific knowledge	47	50
Learned a lot about the future profession	40	43

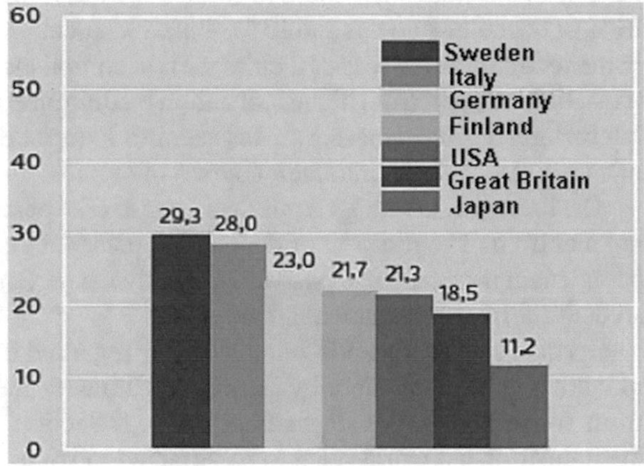

Fig. 5 Female graduates in engineering science in 2004 in % [11]

From an industry perspective intercultural skills play a major role in engineers's professional profiles. Due to globalization, companies do no longer operate on local but on global markets. Therefore, students have to be prepared to face the challenges of globalized businesses. Thus, international and intercultural skills will have to be integrated into engineering education [9].

Shortages of skilled workers and managerial staff in Germany increase the demand for highly skilled foreign workers. Especially the potential of female engineers is not yet reached. With currently 10 % female engineers, Germany lags behind other countries such as Portugal (24 %) or France (16 %) [10]. According to OECD statistics Germany also lies behind other western countries such as Sweden or Italy considering the number of female engineering graduates from higher education institutions as is shown in Fig. 5 [11].

Thus, attracting female students for higher engineering education in Germany might also bear the potential to solve the problem of current labor shortages.

5 Measures for Increasing Student Mobility

Within the project ELLI (c.f. Section 1) RWTH, RUB, and TUD develop new measures to increase both inbound and outbound mobility rates. The following chapter describes those measures in detail.

5.1 Measures for Increasing Outbound Mobility

The project ELLI aims to create a favorable framework enabling national students of each of the three universities to integrate at least one stay abroad during their course of studies. Therefore, the following measures will be put into practice:

- Measure 1: Establishment of a program for advancing international mobility,
- Measure 2: Development of an interactive mobility map,
- Measure 3: Integration of mobility periods in curricula.

5.1.1 Measure 1: Establishment of a Program for Advancing International Mobility

In a first step specific barriers hampering international student mobility at RWTH, RUB, and TUD are analyzed. Therefore, a survey is carried out among engineering students attending courses in undergraduate and graduate studies. As a result the introduction of a common international course record book will be envisaged. It aims at easier and faster recognition of study records accomplished abroad and will be developed in close co-operation with the international partners of the consortium (c.f. measure 2: interactive mobility map). The international course record book will be based on agreements concerning common learning outcomes and their evaluation at all partnering universities. But also new concepts and measures for integrating international mobility periods into the course of studies based on present strategies at the universities are discussed together with the international offices of each of the partnering universities. Furthermore, the program for advancing international mobility includes the development of an indicator system based on the already existing indicator system developed within the project TeachING-LearnING.EU (c.f. Section 1). Thus, quality of international study periods as well as compatibility of records between universities can be measured. The basis of the indicator system will be the annual number of international exchanges indicating the acceptance of the measure among students and its impact on the overall international student mobility.

The program will be implemented at the participating universities for two years. Moreover, professors will be supported in encouraging the best and most committed students for international study terms. Students planning their stay abroad will be supported by a comprehensive discipline-specific counseling service.

5.1.2 Measure 2: Development of an Interactive Mobility Map

Mostly international exchange is organized through programs offered by funding organizations such as DAAD or others. But, successful exchange always relies on personal and informal contacts between professors enabling students to be exchanged between partnering institutions. Often students are not aware of the contacts of their home institution to other higher education institutions abroad. In order to increase transparent information transfer and the opportunities for students to find a suitable institution abroad, the development of an interactive mobility map is envisaged.

The contacts and networks of the professors bear a tremendous potential for exchange. In order to bring out this potential, the networks are analyzed especially focusing on common strategies and contents in teaching and research. Following a bottom-up approach, an interactive mobility map of Europe and/or the world is developed that integrates personal contacts of professors as well as general contacts of each participating university. Already existing mobility maps such as moveonnet.eu [12] are adapted according to the purposes of the consortium. The interactive mobility map will be made available online supplemented by a mobile App. Additionally students will be able to upload discipline-specific experience reports as well as to integrate further metadata through geo-tagging. The map will be maintained by the ELLI consortium to guarantee discipline-specific contents in teaching and learning in engineering science. The interactive mobility map will be evaluated regularly using surveys among its users.

5.1.3 Measure 3: Integration of Mobility Periods in Curricula

Mutual degree recognition and accreditation are central topics of the Bologna process and the development of the European Higher Education Area. The aim is to facilitate the recognition of study records accomplished at a foreign research institution. Therefore, common quality standards for the recognition process as well as harmonized curricula have to be developed [13].

Due to different contents taught in the institutions of higher engineering education across the borders, recognition of foreign study records is still difficult. The measure aims at simplified recognition procedures to promote study-related stays abroad combined with industry internships. Therefore, curricula have to be made transparent between the participating universities and their international partners. In this regard double degree programs have a strong potential to enhance mobility periods. Common strategies to integrate international mobility periods into curricula will be developed together with the departments and faculties of the three German universities and their international partners as well as partners from industry. The aim is to design an individual plan listing all necessary study records or industry internships for each student. The evaluation of the measure integrates a survey during the students's stay abroad as well as after returning home.

5.2 *Measures for Increasing Inbound Mobility*

As shown above (c.f. Section 3) the majority of foreign students taking up their studies in Germany come from China, Turkey and Russia whereas German students often tend to choose their closest neighbors such as The Netherlands, Austria, or Switzerland and the English speaking countries such as the UK, or the US. In order to build fruitful co-operations for research and teaching, a balanced network of partnering universities has to be drawn up.

The project ELLI aims to create the necessary framework to increase numbers of foreign students mainly stemming from the industrialized world such as the US, UK, Japan, or France in order to balance the inbound/outbound ratio and to build connections to strong research nations. The measures are built on experiences gathered from former projects and initiatives. First experiences from joint projects between Drexel University Philadelphia and RUB show that courses provided during semester breaks are widely accepted. From the spring/summer term 2011 on, eleven students from the US take part at lectures and laboratory courses at RUB. Due to high standards in laboratory equipment, the demand from US students for German laboratory courses is very high. The courses are offered by German and American lecturers as block seminars in English and German. Furthermore, the project Undergraduate Research Opportunities Program International (UROP) at RWTH invites students from renowned universities in the US and Canada to pursue a 10-weeks research stay in Aachen during the summer break.

Based on those experiences, specific barriers hampering foreign students to integrate a stay in Germany into their curricula are analyzed in close co-operation with the network of international partners (c.f. interactive mobility map). Taking those findings into account, the three partners RWTH, RUB and TUD will develop curricula enabling foreign students to integrate a stay abroad easily into their course of studies. Furthermore, experiences gathered within the project RISE (Research Internships in Science and Engineering) where short term research internships of undergraduate students from the US, the UK, and Canada at German higher education institutions are funded by DAAD [14] as well as experiences from the International Summer Program (ISP) at TUD [15] for student exchanges during summer breaks will be laid down in the concept.

For example the effect of study courses being taught in English as well as harmonized curricula during term breaks will be evaluated. Therefore, concepts are developed for selected co-operations between German and international universities and put into practice over a period of four years. The evaluation is based on attendance rates and surveys among course participants. Long-term effects such as retention rates and successful career development of participants are explored.

6 Expected Impact

All of the proposed measures aim to increase inbound and outbound student mobility rates in engineering education. Intercultural competences already play a major role in nowadays professional profiles of engineers but will do so even more in the future. Due to globalized markets and companies acting on a global scale, engineers will have to cope with intercultural settings. Intercultural competences can only be gathered through international exchange. Approaches focusing on language courses or intercultural seminars "at home" do not prove to show sufficient effects on the participants's intercultural problem solving competences [2]. Therefore, "true" exchange programs and measures inciting international student mobility will have to be put into practice. Programs do not start from scratch and will be adapted to current exchange programs at RWTH, RUB, and TUD. But also programs and measures are developed taking the special needs and circumstances of engineering education into account. Thus, the prominent integration of intercultural competences into the curricula of engineers will be at the center of the introduced measures within the project ELLI.

References

1. U. Heublein, *Internationale Mobilität im Studium 2009. Wiederholungsuntersuchung zu studienbezogenen Aufenthalten deutscher Studierender in anderen Ländern*, Hannover:Hochschul-Informations-System (HIS), 2009.
2. DAAD (ed.), *Wissenschaft weltoffen 2011. Daten und Fakten zur Internationalität von Studium und Forschung in Deutschland*, Bielefeld:HIS, DAAD, 2011.
3. U. Bach, Th. Jungmann, K. Müller, "Projektbeschreibung TeachING-LearnING.EU." in *Journal Hochschuldidaktik*, 2/2010, HDZ der TU Dortmund (eds.), Dortmund, 2010.
4. M. Kelo, U. Teichler, B. Wächter (eds.). (2011, November 17). *Eurodata. Student mobility in European higher education* [Online]. Available: http://ec.europa.eu/education/erasmus/doc/publ/eurodata_en.pdf.
5. M. Leidel. (2011, November 17). *Statistische Erfassung der Mobilität von Studierenden. Nationale und internationale Datenquellen und Indikatoren* [Online]. Available: http://www.destatis.de/jetspeed/portal/cms/Sites/destatis/Internet/DE/Content/Publikationen/Querschnittsveroeffentlichungen/WirtschaftStatistik/BildungForschungKultur/ErfassungMobilitaet102004,property=file.pdf
6. H.-W. Rückert, "Studentische Mobilität. Mehr Bewegung bitte!" *in fundiert. Das Wissenschaftsmagazin der Freien Universität Berlin*, 02/2010, Berlin, pp. 50–55.
7. BMBF (ed.), *Internationalisieurng des Studiums - Ausländische Studierende in Deutschland - Deutsche Studierende im Ausland. Ergebnisse der 19. Sozialerhebung des Deutschen Studentenwerks durchgeführt von HIS Hochschul-Informations-System*, Bonn, Berlin: BMBF, 2010.
8. RWTH Aachen (ed.). (2011, November 17). *RWTH International - Internationalisierungsreport* 2010 [Online]. Available: http://www.international.rwth-aachen.de/global/show_document.asp?id=aaaaaaaaaabyddv.
9. S. Jeschke, M. Petermann, E. Tekkaya, *Ingenieurwissenschaftliche Ausbildung - ein Streifzug durch Herausforderungen, Methoden, Modellprojekte, Aachen*, 2011.
10. U. Pfenning, O. Renn, U. Mack. (2011, November 19) *Zur Zukunft technischer und naturwissenschaftlicher Berufe. Strategien gegen den Nachwuchskräftemangel* [Online]. Available: http://www.think-ing.de/index.php?media=966.

11. OECD Online Education Database. (2011, November 19). Available: http://www.oecd.org/document/54/0,3746,en_2649_37455_38082166_1_1_1_37455,00.html.
12. Moveonnet.eu (2011, Novmeber 18) *Moveonnet. Higher Education Worldwide* [Online]. Available: http://www.moveonnet.eu.
13. HRK (eds.). (2011, November 18) *Projekt nexus - Konzept und gute Praxis für Studium und Lehre* [Online] Available: http://www.hrk.de/de/projekte_und_initiativen/5913.php.
14. DAAD (eds.). (2011, November 18). *RISE - Research Internships in Science and Engineering* [Online]. Available: http://www.daad.de/rise/de.
15. TU Dortmund (ed.). (2011. November 18) *International Summer Program* [Online]. Available: http://www.aaa.uni-dortmund.de/cms/en/International_Students/International_Summer_Program__ISP_/index.html.

Motivationen und Hindernisse für die Auslandsmobilität von Studierenden in MINT-Fächern – eine vergleichende Studie an der RWTH Aachen University

Ute Heinze, Ursula Bach, René Vossen and Sabina Jeschke

Zusammenfassung 64 von 1.000 deutschen Studierenden absolvierten im Jahr 2010, laut DAAD, einen studienbezogenen Auslandsaufenthalt. Die Mobilitätsrate der verschiedenen Fachdisziplinen variiert jedoch stark. Besonders unter den Studierenden der Ingenieurwissenschaften ist die Mobilitätsrate mit 4 % vergleichsweise gering. Während aktuelle Statistiken lediglich den Istzustand abbilden, gibt es wenige Daten zu den Gründen hinter der bei Ingenieuren geringer ausfallenden Auslandsmobilität. Daher wurde vom IMA/ZLW & IfU in enger Kooperation mit dem International Office der RWTH im November 2012 eine weitreichende Online Umfrage unter sämtlichen Studierenden der Universität durchgeführt. Die Studie "GoING abroad – Auslandsmobilität an der RWTH Aachen University" konzen-trierte sich auf mobilitätsfördernde sowie -hemmende Faktoren. Dabei konnte einerseits festgestellt werden, dass die Ingenieurstudierenden an der RWTH vergleichsweise häufig ins Ausland gehen. Motivationsgründe sind hauptsächlich die Verbesserung von Fremdsprachenkenntnissen oder das Kennenlernen einer neuen Kultur. Hindernisse werden vor allem bei der Anerkennung der im Ausland erbrachten Studienleistungen berichtet.

Schlüsselwörter Auslandsmobilität, Deutsche Studierende im Ausland, MINT-Wissenschaften, Interkulturelle Kompetenzen, Internationalisierung

---

U. Heinze (✉) · U. Bach · R. Vossen · S. Jeschke
IMA/ZLW & IfU RWTH Aachen University,
Dennewartstr. 27, 52068 Aachen, Germany
e-mail: ute.heinze@ima-zlw-ifu.rwth-aachen.de

U. Bach
e-mail: ursula.bach@ima-zlw-ifu.rwth-aachen.de

R. Vossen
e-mail: rene.vossen@ima-zlw-ifu.rwth-aachen.de

S. Jeschke
e-mail: sabina.jeschke@ima-zlw-ifu.rwth-aachen.de

Originally published in "TeachING-LearnING.EU-Sammelband",

DOI 10.1007/978-3-319-46916-4_43

1 Einleitung

Durch die Globalisierung müssen Unternehmen vermehrt international denken und handeln. Interkulturelle Erfahrungen und Kompetenzen prägen zunehmend auch das Berufsfeld und -profil von Wissenschaftlerinnen und Wissenschaftlern der Natur- und Ingenieurwissenschaften sowie der Informatik. Diesen globalen Entwicklungen und Herausforderungen müssen auch Hochschulen durch eine adäquate Ausbildung begegnen. Aus diesen Gründen ist es wichtig, dass sich Studierende während des Studiums die Fähigkeit aneignen, mit Menschen aus anderen Kulturen zusammenzuarbeiten um diese Qualifikation später sicher auf dem globalen Arbeitsmarkt anzuwenden zu können [1].

Aktuelle Statistiken, auf die im Folgenden noch detaillierter eingegangen wird, zeigen, dass Studierende der mathematisch-naturwissenschaftlichen, informatischen und vor allem technisch-ingenieurwissenschaftlichen Fächer (kurz MINT-Fächer) weitaus weniger auslandsmobil sind als Studierende bspw. aus den Rechts-, Wirtschafts- oder Sozialwissenschaften [2–5]. Über die Gründe ist laut Heublein et al. [6] bisher wenig bekannt, sodass lediglich hypothetische Annahmen dazu getroffen werden können. Es wird vermutet, dass die Umstellung der Studiengänge auf das Bachelor-/Mastersystem und die damit einhergehende Zeitproblematik, einen Auslandsaufenthalt in das Studium zu integrieren, einen nicht unwesentlichen Einfluss auf die Auslandsmobilität ausüben [7].

Um jedoch die Beweggründe der Studierenden insbesondere aus MINT-Fächern für oder gegen studienrelevante Auslandsaufenthalte näher zu beleuchten, wurde im Rahmen des Projekts Exzellentes Lehren und Lernen in den Ingenieurwissenschaften (ELLI) im November 2012 eine Studierendenbefragung an der Rheinisch-Westfälischen Technischen Hochschule (RWTH) Aachen University unter der Leitung des Lehrstuhls für Informatik im Maschinenbau (IMA), des Zentrums für Lern- und Wissensmanagement (ZLW) und des Instituts für Unternehmenskybernetik (IfU) in Kooperation mit dem International Office der RWTH durchgeführt. Die Intention der Studie war die Ermittlung der Motivationen und Hindernisse für Auslandsmobilität insbesondere der Studierenden der MINT-Fächer an der RWTH Aachen University. Berücksichtigt wurden finanzielle Belange, Anerkennungsprozesse, universitäre Informations- und Beratungsangebote sowie persönliche Gründe für die Entscheidung für oder gegen einen studienbezogenen Auslandsaufenthalt.

Nach einer kurzen Beschreibung vorangehender Studien zur Auslandsmobilität deutscher Studierender in MINT-Fächern wird zunächst auf die angewandte Methodik der RWTH-internen Studie näher eingegangen, bevor die erzielten Ergebnisse dargestellt werden.

2 Studien zur Auslandsmobilität von Studierenden der MINT-Fächer

Laut der Datengrundlage des Deutschen Akademischen Austauschdienstes (DAAD) und des Instituts für Hochschulforschung (HIS) [2] sowie des Statistischen Bundesamtes [3] absolvierten 2010 insgesamt 64 von 1.000 an deutschen Hochschulen eingeschriebenen Studierenden einen studienrelevanten Auslandsaufenthalt. Diese Zahl beinhaltet sowohl zeitweilig an einer ausländischen Hochschule immatrikulierte Studierende, als auch jene, die einen Studienabschluss im Ausland anstrebten.

Zahlen zu den Mobilitätsraten der verschiedenen Fachdisziplinen sind bisher hauptsächlich nur für das europäische Austauschprogramm ERASMUS erfasst worden. Demnach verweilen im Studienjahr 2010/2011 von insgesamt 30.274 deutschen ERASMUS-Austauschstudierenden ca. 40 % (12.174) von ihnen aus den Rechts-, Wirtschafts- und Sozialwissenschaften im Ausland, gefolgt von ca. 25 % (7.531) aus den Geisteswissenschaften und der Kunst, ca. 12 % (3.726) aus den Ingenieurwissenschaften sowie ca. 9 % (2.956) aus naturwissenschaftlichen, mathematischen und informatischen Studienfächern [2]. Auch andere Studien wie bspw. die 19. Sozialerhebung des Deutschen Studentenwerks [4] kommen zu einem ähnlichen Ergebnis und bestätigen eine Tendenz zu geringer Auslandsmobilität unter Studierenden der MINT-Fächer.

Einige wenige Studien untersuchen die Motivation und Hindernisse von Studierenden bei der Planung und Durchführung eines Auslandsaufenthaltes. Laut Analysen des DAAD [5], der HIS [8] oder des nordrhein-westfälischen Ministeriums für Innovation, Wissenschaft und Forschung [9] werden die folgenden Hindernisse häufig von Studierenden angeführt:

- Probleme bei der Anerkennung von Studienleistungen,
- Zeitverlust im Studium,
- Probleme bei der Vereinbarkeit des Auslandsaufenthaltes mit den an der Heimathochschule geforderten Leistungen sowie
- finanzielle Probleme.

Wie bereits Heublein et al. [6] feststellen, sind weitergehende Studien zur Motivation der Studierenden bei der Planung und Umsetzung von Auslandsaufenthalten rar. Deshalb zielt die im Folgenden beschriebene Studie darauf ab, motivierende und hemmende Faktoren für die Durchführung eines studienrelevanten Auslandsaufenthaltes von Studierenden an der RWTH Aachen University im Vergleich zwischen MINT-Studierenden und Nicht-MINT Studierenden näher zu beleuchten.

3 Methodik

3.1 *Online-Fragebogen*

Die Befragung „Going abroad - Auslandsmobilität an der RWTH Aachen University" richtete sich an alle Studierenden mit deutscher oder ausländischer Staatsangehörigkeit, die ihre Hochschulzugangsberechtigung an einer deutschen Ausbildungseinrichtung erworben haben, sogenannte „Bildungsinländer". Insgesamt wurden n = 33.003 Studierende gebeten, an der Studie teilzunehmen. Mit N = 3.218 vollständig auswertbaren Fragebögen beträgt die Rücklaufquote 9,75 %.

Um möglichst viele Studierende zu erreichen, wurde die Studie als Online-Umfrage implementiert und über die Zeitspanne von vier Wochen vom 5.–30. November 2012 mithilfe einer E-Mail-Benachrichtigung sowie einer Erinnerung nach zwei Wochen verbreitet. Die Stichprobe wurde zudem durch Filterfragen in verschiedene Gruppen mit entsprechend individuell zugeschnittenen Fragen eingeteilt, um so möglichst präzise auf verschiedene Planungs- und Durchführungsphasen eines Auslandsaufenthaltes eingehen zu können.

3.2 *Stichprobenbeschreibung*

3.2.1 Demografische Daten

Die erhobene Stichprobe zeigt auf, dass mit 61,5 % der Großteil der befragten Studierenden männlich ist. 3.064 Teilnehmende stammen aus Deutschland (ca. 95 %), 16 aus China (ca. 0,5 %), 15 aus der Türkei (ca. 0,5 %) und 10 aus Russland (0,3 %). Die Hochschulzugangsberechtigung erlangten 97,4 % der Befragten in Deutschland.

3.2.2 Studienbezogene Daten

Im Vergleich zu der Verteilung der Gesamtanzahl der an der RWTH eingeschriebenen Studierenden im Wintersemester 2012/2013 [10] verteilten sich die Teilnehmenden der Studie in einem ähnlichen Verhältnis auf die jeweiligen Fakultäten wie die folgende Abb. 1 zeigt. Somit kann keine Verschiebung innerhalb der Daten zugunsten einer bestimmten Fakultät beobachtet werden, weswegen die Ergebnisse der Studie einen repräsentativen Schnitt der Studierenden abbilden.

Die Befragungsteilnehmenden strebten unterschiedliche akademische Grade an. Die drei am häufigsten vertretenen Gruppen in Bezug auf den angestrebten akademischen Abschluss waren Masterstudierende mit 44,3 %, Bachelorstudierende mit 26,5 % und Promovenden mit 11,7 %.

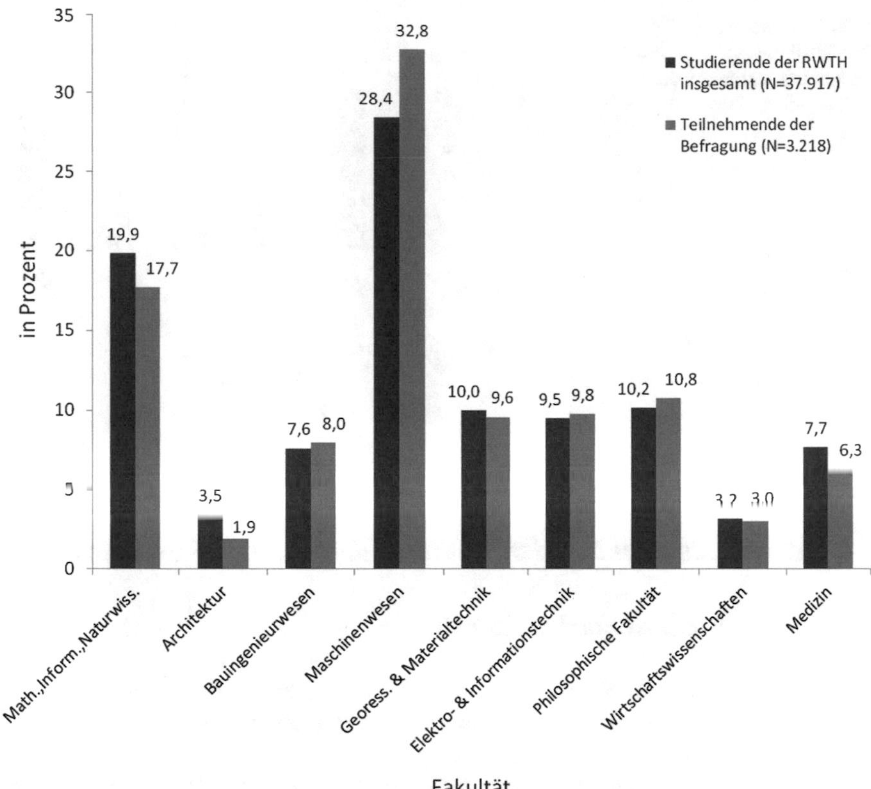

Abb. 1 Verteilung der Befragten über die Fakultäten

4 Ergebnisse der Studie

Um die Beweggründe bei der Planung und Organisation eines Auslandsaufenthaltes von Studierenden der MINT-Fächer im Vergleich zu Studierenden der Nicht-MINT Fächer zu analysieren, wird sich die weitere Auswertung der Ergebnisse auf die Motivationen und Hindernisse bei der Organisation eines Auslandsaufenthaltes der Studierenden der Fakultäten Mathematik, Informatik, Naturwissenschaften, Bauingenieurwesen, Maschinenwesen, Georessourcen und Materialtechnik sowie Elektrotechnik und Informations-technik konzentrieren und sie gegen jene der Studierenden der Fakultäten für Architektur, Philosophie, Wirtschaftswissenschaften und Medizin abgrenzen. Insgesamt nahmen N = 2.507 Studierende der MINT-Fächer an der Umfrage teil, was einem Anteil von 12,3 % der Gesamtstudierenden der MINT-Fächer der RWTH Aachen University entspricht.

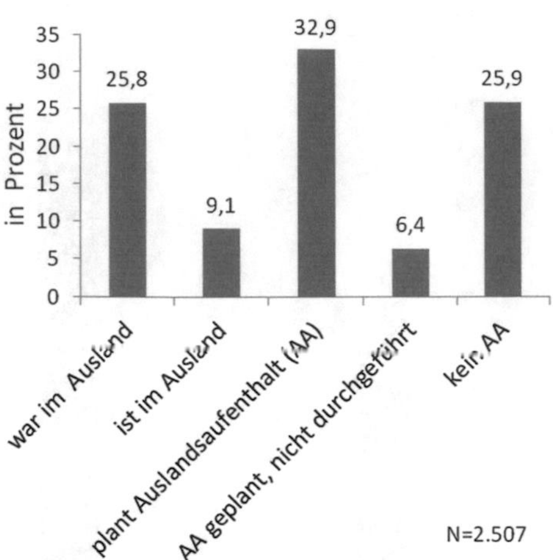

Abb. 2 Planungs- und Durchfü]hrungsphasen des Auslandsaufenthaltes der Befragten

4.1 Mobilitätsverhalten der Studierenden in MINT-Fächern

4.1.1 Stadium des Auslandsaufenthaltes

Die Teilnehmenden der Studie befanden sich in unterschiedlichen Planungs- und Durchführungsphasen ihres Auslandsaufenthaltes bzw. hatten überhaupt keinen Auslandsaufenthalt vorgesehen, wie Abb. 2 zeigt.

Demnach waren insgesamt 34,9 % der befragten MINT-Studierenden zum Zeitpunkt der Umfrage im Ausland bzw. hatten bereits Auslandserfahrungen im Studium gesammelt. 32,9 % planen einen Auslandsaufenthalt und insgesamt 32,3 % planen keinen Auslandsaufenthalt ein bzw. traten einen zuvor geplanten Auslandsaufenthalt nicht an. Somit ergibt sich hier fast eine Gleichverteilung auf die unterschiedlichen Planungs- und Durchführungsstadien.

4.1.2 Mobilität der MINT-Studierenden im Vergleich

Vergleicht man die Mobilitätsrate der Studierenden der MINT-Fächer mit Studierenden der anderen Fakultäten der RWTH Aachen University, bringt die Studie keine spezifischen Erkenntnisse hervor. Die Mobilität der Befragten der Fakultäten Mathematik, Informatik und Naturwissenschaften, Bauingenieurwesen, Maschinenwesen, Georessourcen und Materialtechnik sowie Elektrotechnik und Informationstechnik („MINT-Studierender") wurden mit jener der Studierenden der Fakultät für Architektur, der Philosophischen, Wirtschaftswissenschaftlichen und Medizinischen Fakultäten („kein MINT-Studierender") verglichen. Dabei bemaß sich die Mobilität danach, ob die Teilnehmenden während ihres Studiums bereits einen Auslandsaufenthalt absolviert hatten oder sich während der Umfrage im Ausland befanden. Diese Befragten wurden der Variable „war/ist im Ausland", alle anderen der Variable „war

Tabelle 1 Kontigenztabelle für das Mobilitätsverhalten von MINT- und nicht-MINT-Studierenden

	war/ist im Ausland	war (bisher) nicht im Ausland	Gesamt
MINT-Studierender	873	1634	2507
kein MINT-Studierender	234	477	711
Gesamt	1107	2111	3218

(bisher) nicht im Ausland" zugeordnet. Die folgende Tabelle 1 verdeutlicht die Häufigkeitsverteilung der getesteten Variablen.

Der Chi-Quadrat-Test nach Pearson wurde auf die zwei nominal skalierten Variablen „MINT-Studierender" und „kein MINT-Studierender" sowie „war (bisher) nicht im Ausland" und „war/ist im Ausland" durchgeführt. Folgende Werte wurden berechnet:

$$\chi^2(1, N = 3218) = .90;\ p = .05 \rightarrow \text{nicht signifikant}$$

Die Ergebnisse können die Nullhypothese nicht bestätigen. Es kann keine signifikante Aussage über die Mobilitätsaffinität hinsichtlich verschiedener Fakultäten getroffen werden. Es kann also durch die vorliegende Untersuchung nicht festgestellt werden, dass Studierende der MINT-Fächer weniger mobil als Studierende anderer Fakultäten sind. Somit kann kein Zusammenhang zwischen der Zugehörigkeit zu den MINT-Fakultäten und einem Auslandsaufenthalt identifiziert werden.

Dies steht im deutlichen Gegensatz zu den Zahlen bisheriger Studien (siehe Kapitel 1), die über eine gravierend geringere Mobilität der Studierenden in MINT-Fächern berichten. Über Gründe hierfür kann allerdings nur gemutmaßt werden, da diese Befragung darüber keinen Aufschluss gibt. Es ist zu vermuten, dass sich einerseits die Effekte der besonderen Förderung der Auslandsmobilität von MINT-Studierenden an der RWTH zeigen. Vor allem ein breit gestreutes Informationsnetzwerk an Auslandsstudienbeauftragten an den Fakultäten hilft dabei, Studierende aktiv bei der Planung und Durchführung ihres Auslandsaufenthaltes zu unterstützen. Andererseits ist jedoch auch ein Selektionseffekt durch eine nicht konsistente Randomisierung der Stichprobe nicht auszuschließen. Eventuell nahmen eher mobilitätsaffine Studierende an der Befragung teil und beeinflussten die Ergebnisse entsprechend.

4.2 Motivation für studienrelevante Auslandsaufenthalte

Die Studienteilnehmenden wurden dazu aufgefordert, einer Reihe von möglichen Motivationsgründen auf einer 6-stufigen Skala einen Wert von 1 (gar kein Motivationsgrund) bis 6 (großer Motivationsgrund) zuzuordnen. Anschließend wurden die Mittelwerte der Bewertungen errechnet, um somit eine Aussage über die Motivation für studienrelevante Auslandsaufenthalte treffen zu können. Die folgende Abb. 3 zeigt einerseits die abgefragten Kriterien und andererseits die jeweils dazugehörigen Bewertungen der MINT-Studierenden sowie der Studierenden der anderen Fakultäten („Nicht MINT-Studierende").

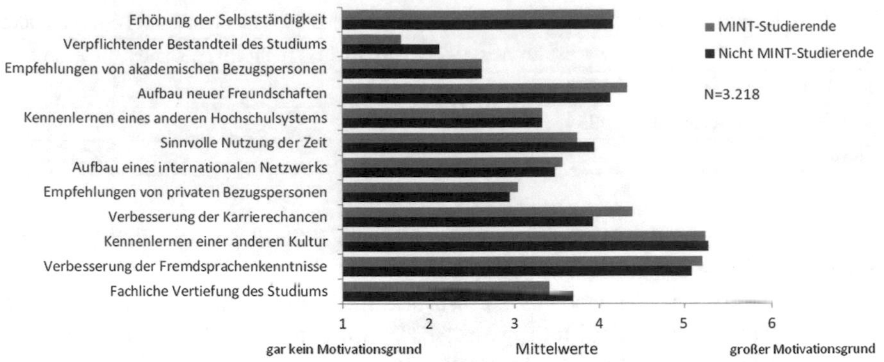

Abb. 3 Motivation für studienrelevante Auslandsaufenthalte für MINT- und Nicht MINT-Studierende

Disziplinübergreifend wurden solche Faktoren wie die Verbesserung der Fremdsprachenkenntnisse, das Kennenlernen einer neuen Kultur, der Aufbau neuer Freundschaften sowie die Erhöhung der Selbstständigkeit als besonders große Motivationsgründe bewertet. Diese hauptsächlich der Gruppe der Soft Skills zuzuordnenden Faktoren wirken offenbar stärker auf die Motivation als andere Faktoren wie bspw. die fachliche Vertiefung des Studiums oder das Kennenlernen eines anderen Hochschulsystems. Die Bedeutung der Empfehlungen von privaten oder akademischen Bezugspersonen wurde disziplinübergreifend als gering eingeschätzt. Am wenigsten wurde eine Verpflichtung zu einem Auslandsaufenthalt als Motivationsgrund angegeben, was jedoch darauf zurückzuführen ist, dass in den wenigsten Curricula bisher Auslandsaufenthalte als verpflichtende Bestandteile vorgesehen sind.

Unterschiede zwischen den Fachdisziplinen wurden in Bezug auf die Wertung folgender motivierender Faktoren festgestellt: sinnvolle Nutzung der Zeit, fachliche Vertiefung des Studiums und die Verbesserung der Karrierechancen. Der Aspekt der sinnvollen Nutzung ihrer Zeit bei einem Auslandsaufenthalt wird von Studierenden der MINT-Fächer im Vergleich zu anderen Studierenden als geringerer Motivationsgrund eingeschätzt. Eine ähnliche Wertung wurde auch für die fachliche Vertiefung des Studiums gegeben. Erstellt man eine Rangliste der zwölf oben dargestellten Motivationsgründe und sortiert sie aufsteigend nach ihrer Bewertung, so befindet sich das Kriterium „fachliche Vertiefung des Studiums" auf Rang 8 bei MINT und auf Rang 7 bei nicht MINT-Studierenden. Somit zeigt sich, dass dieses Kriterium für Studierende aller Fachrichtungen gleichermaßen weniger als Motivationsgrund für die Durchführung eines Auslandsaufenthaltes dient. Die Verbesserung der Karrierechancen hingegen motiviert Studierende der MINT-Fächer wesentlich stärker dazu, ins Ausland zu gehen, als andere Studierende. Bei MINT-Studierenden werden verbesserte Karrierechancen als dritthäufigstes Kriterium genannt, während es bei nicht MINT-Studierenden nur auf Rang 6 liegt.

4.3 Hindernisse für studienrelevante Auslandsaufenthalte

Bei der Betrachtung der Hindernisse zeigt sich ein ähnlich homogenes Bild, wie in der folgenden Abb. 4 dargestellt. Auch hier wurden die Befragten gebeten, ihre Bewertung der aufgeführten Hindernisse auf einer 6-stufigen Skala von 1 (gar kein Hindernis) bis 6 (großes Hindernis) abzugeben.

Als die wichtigsten Hindernisse für Auslandsaufenthalte während des Studiums werden vor allem Zeitdruck während des Studiums, finanzielle Gründe und zu wenig Austauschplätze genannt. Ebenso werden unklare Zuständigkeiten bei den für die Organisation eines Austausches relevanten Stellen von allen Studierenden gleichermaßen als hemmend eingestuft. Am wenigsten fühlen sich Studierende durch einen nicht erkennbaren Nutzen für ihren Karriereweg, die Angst vor der Herausforderung oder dem Unbekannten sowie Sprachbarrieren an der Durchführung eines Auslandsaufenthaltes gehindert.

Unterschiede in den Bewertungen zwischen den Disziplinen sind bei den Hindernissen hingegen deutlich erkennbar. Vor allem Probleme bei der Anerkennung und ein mit dem Auslandsaufenthalt einhergehender zu großer zeitlicher Aufwand werden von Studierenden der MINT-Fächer vermehrt als Hindernis angegeben. Auch uneinheitliche Semesterzeiträume werden von MINT-Studierenden als hinderlicher bewertet als von anderen Studierenden. Als wesentlich weniger hinderlich schätzen Studierende der MINT-Fächer im Vergleich zu ihren Kommilitoninnen und Kommilitonen anderer Studienrichtungen gesundheitliche sowie finanzielle Gründe oder die Bindung an Freunde und Familie ein.

Damit werden die zuvor von anderen Studien ermittelten Ergebnisse (siehe Kapitel 1 *Studien zur Auslandsmobilität von Studierenden der MINT-Fächer*) auch durch diese Studie bestätigt.

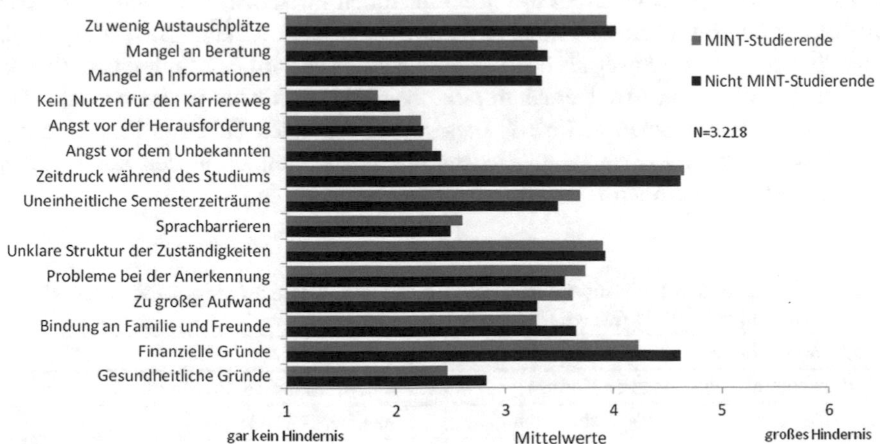

Abb. 4 Vergleich der Hindernisse für Auslandsaufenthalte zwischen MINT- und Nicht MINT-Studierenden

5 Zusammenfassung

Die vorgestellte Studie untersuchte motivierende und hemmende Faktoren, die Studierende der MINT-Fächer im Vergleich zu den Studierenden der Nicht-MINT-Fächer dazu veranlassen, einen Auslandsaufenthalt in ihr Studium zu integrieren oder nicht. Vergleicht man dabei die Top 5 der Motivationen und Hindernisse für MINT-Studierende ergibt sich folgende Tabelle 2.

Neben den in der Praxis stark nachgefragten Soft Skills wie der Beherrschung von Fremdsprachen oder einer eigenständigen Arbeitsweise wird der Verbesserung der Karrierechancen eine höhere Priorität zugeordnet als bspw. der fachlichen Vertiefung des Studiums, die erst auf Platz 8 der Motivationsgründe rangiert (siehe auch Kapitel 3.2). Diese Priorisierung scheint eine Tendenz zur Karriereorientierung der MINT-Studierenden anzudeuten: Für den Fall, dass Studierende der MINT-Fächer eine Verbesserung ihrer Karrierechancen durch einen Auslandsaufenthalt als gering einschätzen und sich zusätzlich bestimmter Hindernisse wie hohem Zeitdruck oder fehlender finanzieller Unterstützung ausgesetzt sehen, werden sie vermutlich keinen Auslandsaufenthalt organisieren. Hinzu kommt die besondere soziale Situation der Studierenden der MINT-Fächer. Wie die 18. Sozialerhebung des Deutschen Studentenwerks [11] zeigt, immatrikulieren sich häufig Studierende der „unteren sozialen Herkunftsgruppen" in Bezug auf allgemein und berufsbildende Abschlüsse und die berufliche Stellung der Eltern für ein Studium der Ingenieurwissenschaften– einem der großen MINT-Fächer. Somit liegt die Vermutung nahe, dass mit der sozialen Herkunft auch eine entsprechende finanzielle Unterstützung der Studierenden durch das Elternhaus einhergeht. Finanzielle Probleme wiederum zählen zu den Haupthindernissen bei der Organisation eines Auslandsaufenthaltes.

Wie auch schon von Heublein et al. [6] angemerkt, scheint sich die Entwicklung der Auslandsmobilität derzeit in einer Übergangsphase zu befinden. Trotz vermehrter Förderbemühungen stagnierten die Mobilitätsraten zwischen 2000 und 2010. Die Bedeutung eines Auslandsaufenthaltes für die zukünftige Berufstätigkeit wird jedoch von Studierenden durchweg als hoch eingeschätzt. Dies wird nicht zuletzt durch eine erhöhte Nachfrage an Absolventen mit fremdsprachlichen Kompetenzen von global agierenden Unternehmen gefördert. Diese begünstigenden Faktoren lassen darauf hoffen, dass sich die Rate an auslandsmobilen Studierenden in den kommenden Jahren weiter steigern wird.

Tabelle 2 Top 5 der Motivationen und Hindernisse bei Auslandsaufenthalten unter MINT-Studierenden an der RWTH Aachen University

Top 5 Motivationen	Top 5 Hindernisse
1. Kennenlernen einer anderen Kultur	1. Zeitdruck während des Studiums
2. Verbesserung der Fremdsprachenkenntnisse	2. Finanzielle Gründe
3. Verbesserung der Karrierechancen	3. Zu wenig Austauschplätze
4. Aufbau neuer Freundschaften	4. Unklare Struktur der Zuständigkeiten
5. Erhöhung der Selbstständigkeit	5. Probleme bei der Anerkennung

Literaturverzeichnis

1. S. Jeschke, M. Petermann, A.E. Tekkaya, Ingenieurwissenschaftliche Ausbildung– ein Streifzug durch Herausforderungen, Methoden und Modellprojekte. In: *TeachING-LearnING.EU Fachtagung "Next Generation Engineering Education"*, ed. by U. Bach, S. Jeschke, ZLW/IMA der RWTH Aachen University, 2011, pp. 11–22
2. DAAD, HIS-Institute für Hochschulforschung, eds., *Wissenschaft weltoffen 2012. Daten und Fakten zur Internationalität von Studium und Forschung in Deutschland.* W. Bertelsmann Verlag, Bielefeld, 2012
3. Statistisches Bundesamt, ed., *Deutsche Studierende im Ausland. Statistischer Überblick 2000–2010.* Statistisches Bundesamt, Wiesbaden, 2012
4. W. Isserstedt, M. Kandulla, *Internationalisierung des Studiums – Ausländische Studierende in Deutschland – Deutsche Studierende im Ausland. Ergebnisse der 19. Sozialerhebung des Deutschen Studentenwerks durchgeführt durch HIS Hochschul-Informations-System.* BMBF, Bonn, Berlin, 2010
5. Deutscher Akademischer Austauschdienst (DAAD), ed., *Wissenschaft weltoffen 2011. Daten und Fakten zur Internationalität von Studium und Forschung in Deutschland.* HIS, Bielefeld, 2011
6. U. Heublein, J. Schreiber, C. Hutzsch, Entwicklung der Auslandsmobilität deutscher Studierender. Tech. rep., HIS, 2011
7. W. Schmitz. Die Scheu des Ingenieurstudenten vor dem Ausland, 2007. http://vdi-nachrichten.com
8. N. Netz, D. Orr, C. Gwosc, B. Huß, What deters students from studying abroad? Evidence from Austria, Switzerland, Germany, the Netherlands and Poland. In: *HIS: Discussion Paper*. HIS, Bielefeld, 2012
9. S. Rieck, O. Märker, Ergebnisbericht. http://www.besser-studieren.NRW.de. Tech. rep., Ministerium für Innovation, Wissenschaft und Forschung des Landes Nordrhein-Westfalen, Bonn, 2012
10. H.D. Hötte, H. Fritz, RWTH Aachen University. Zahlenspiegel 2011. Tech. rep., RWTH Aachen University, Aachen, 2012
11. W. Isserstedt, E. Middendorf, G. Fabian, A. Wolter, *Die wirtschaftliche und soziale Lage der Studierenden in der Bundesrepublik Deutschland 2006. 18. Sozialerhebung des Deutschen Studentenwerks durchgeführt durch HIS Hochschul-Informations-System.* BMBF, Bonn, Berlin, 2007

International Exchange in Higher Engineering Education – A Representative Survey on International Mobility of Engineering Students

Ute Heinze, Ursula Bach, René Vossen and Sabina Jeschke

Abstract The attitude of German engineering students towards spending a certain period of time abroad is highly ambivalent. Although German engineering students assume that international experience is a key competence and a career-enhancing qualification, outbound mobility (referring to the number of students who leave Germany to another country for the purpose of study or a traineeship) in higher engineering education lies below average compared to other disciplines. One reason frequently listed by students is the intensive workload in engineering sciences hampering their opportunities to go abroad as well as difficult recognition procedures of study credits accomplished abroad. Statistics show that the changes brought about by the Bologna Reform (initiated in 1999 to align European higher education and to establish a common credit transfer system) did not result in increased student mobility throughout Europe. It rather resulted in less flexible curricula leaving students not enough time for spending a certain part of their studies abroad. This paper analyzes the current situation of international student mobility in German higher engineering education especially focusing on outbound mobility. It is based on the results of a survey carried out among approx. 35.000 students at one of Germany's most renowned technical universities. The study deals with the motivation and the obstacles faced by engineering students aiming at integrating international exchange periods into their curricula. Therefore, an anonymous wide ranging survey was distributed among all students at the above mentioned university. Topics such as financial issues, the recognition of credit points, and career advice services influencing their decision in or not in favor of an international exchange are investigated among other aspects. The results

U. Heinze (✉) · U. Bach · R. Vossen · S. Jeschke
IMA/ZLW & IfU, RWTH Aachen University,
Dennewartstr. 27, 52068 Aachen, Germany
e-mail: ute.heinze@ima-zlw-ifu.rwth-aachen.de

U. Bach
e-mail: ursula.bach@ima-zlw-ifu.rwth-aachen.de

R. Vossen
e-mail: rene.vossen@ima-zlw-ifu.rwth-aachen.de

S. Jeschke
e-mail: sabina.jeschke@ima-zlw-ifu.rwth-aachen.de

Originally published in "120th ASEE Annual Conference 2013 Proceedings",

DOI 10.1007/978-3-319-46916-4_44

of the survey are finally summarized together with derived measures to increase the participation of engineering students in international exchange programs.

Keywords International Student Mobility · Inbound · Outbound · Engineering Education · Curricula Development · Intercultural Competences

1 Introduction

Nowadays, intercultural competences and social skills are inevitable for a successful engineering career because they play a significant role in professional profiles of engineers and will do so even more in the future. Those competences can only be gathered through international exchange. Due to the progressing globalization companies do no longer operate only on local but on global markets. Thus, especially engineering should not be limited by national borders. Therefore students have to be prepared to face the challenges connected with globalized markets [1].

Nevertheless, the number of students in engineering science who leave Germany to another country for the purpose of study or traineeship (referred to as "outbound mobility" [2–4]) is distinctly below average compared to other disciplines [5]. Statistics show that the goals of the Bologna Reform, whose key aim was the unification of European higher education to boost international mobility by establishing a common credit transfer system, were not obtained satisfactorily [2]. The reform focused on structural changes to increase student mobility throughout Europe as a central goal. On the Ministerial Conference 2009 in Leuven and Louvain-la-Neuve the European Ministers of Education and Research set the aim that until 2020 20 % of European students should have lived and worked abroad [6, 7]. On the Bologna conferences in Budapest and Vienna in 2010 it became clear that this goal will not be accomplished [8, 9].

Especially engineering students tend to be less mobile as several studies indicate [2, 10, 11]. In order to take a deeper look at the motivational factors and obstacles engineering students face when planning international exchange periods, a survey was carried out among approx. 33,000 students at Rheinisch-Westfälische Technische Hochschule (RWTH) Aachen University in November 2012.

The survey was performed by the institute cluster IMA/ZLW & IfU – Institute of Information Management in Mechanical Engineering (IMA), Center for Learning and Knowledge Management (ZLW), Associated Institute for Management Cybernetics e.V. (IfU) in cooperation with RWTH's International Office. Its aim was to investigate motivational factors and obstacles students face in different planning and realization stages of a foreign exchange. The anonymous and wide ranging survey covers topics such as financial issues, the recognition of credits, and career advice services among other personal factors influencing a decision on international exchange.

The present paper introduces current studies related to the topic of international mobility of engineering students, and describes the method as well as the results of the RWTH survey and puts them into relation to major studies representing the current state of the art in the research on international student mobility: the study on international mobility and study-related exchanges of German students carried out by Heublein and Hutzsch [5] of Hochschulinformations GmbH (HIS) in 2009 and 2011, the annual report "Wissenschaft weltoffen" performed by the German Academic Exchange Service (DAAD) [2], and the 19th Social Survey of the Deutsches Studentenwerk (German National Association for Student Affairs) on the economic and social conditions of student life in Germany in 2009 [11].

2 Current Studies on International Mobility of Engineering Students

Several studies take a deeper look at outbound mobility of German students. According to data from the German Academic Exchange Service (DAAD) and the Higher Education Information System Institute (HIS) [12] as well as the German Federal Statistical Office [13] 64 out of 1,000 students from German higher education institutions pursued a study-related exchange in 2010. These numbers include those students who reside abroad on a temporary basis as well as those who work towards a degree from a foreign higher education institution. Compared to the mobility numbers of the past decade, an increase in outbound mobile students can be observed as is shown in Fig. 1.

When looking at mobility rates for different subject groups, data are available for example from the European exchange program ERASMUS, as shown in Fig. 2.

Out of 30,274 German ERASMUS exchange students, approx. 40 % (12,174) came from the social sciences, business and economy, or law, approx. 25 % (7,531) from the humanities and liberal arts, approx. 12 % (3,726) from engineering sciences and approx. 9 % (2,956) from natural sciences, mathematics and computer science in the study term 2010/2011 [12]. Apparently, students from the subject groups of

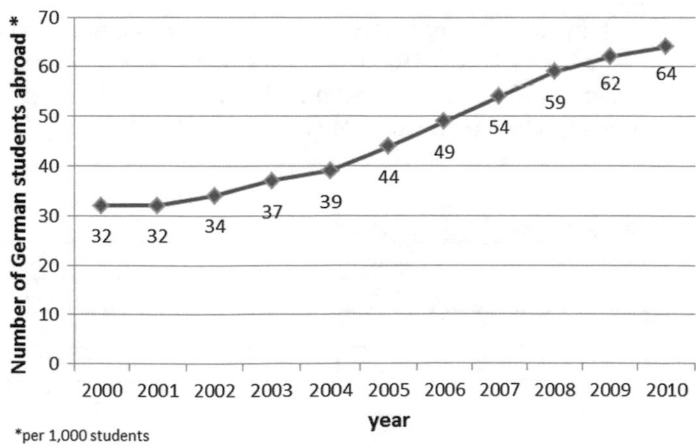

Fig. 1 Development of mobility rate of German students from 2000-2010, per 1,000 students [13]

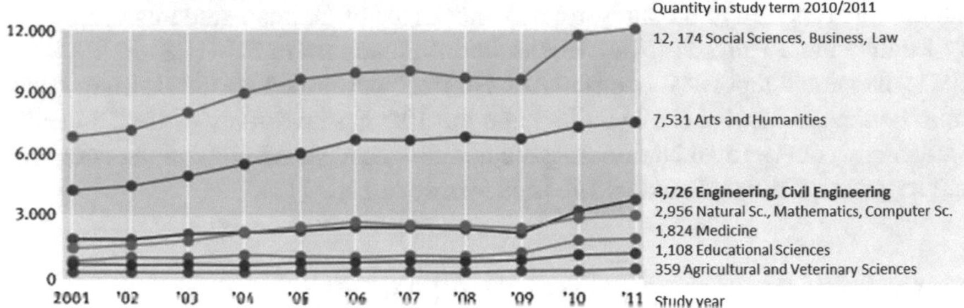

Fig. 2 German ERASMUS students by subject group [12]

science, technology, engineering and mathematics (STEM fields) lag behind their counterparts from other subjects in terms of study-related international mobility. Certainly, the data presented here might be biased due to the mere focus only on one single program, the ERASMUS program. It may be the case that the program itself mainly attracts students from other subjects than the STEM fields through its funding scheme or program design. EU official statistics show that approx. 60 % of ERASMUS funded students in 2010 and 2011 were women [14]. As in Germany the majority of students in STEM fields are male, they might participate in ERASMUS exchange to a fewer degree and thus do not appear in the official statistics which results in generally lower exchange numbers for students in STEM fields.

Nevertheless, other studies such as the 19th Social Survey of the Deutsches Studentenwerk (German National Association for Student Affairs) on the economic and social conditions of student life in Germany in 2009 [11] or the study of Heublein and Hutzsch [5] show similar results and tendencies.

Even though the mobility rates among students from STEM fields are comparatively low, their development over the last years is steeper than in other disciplines as Table 1 shows.

Since 1994 mobility rates among students of language and cultural studies stayed constantly at a high level (1994: 12 %, 2000: 13 %, 2009: 12 %) whereas students in engineering science became a lot more mobile over the same timespan where mobility rates nearly doubled between 1994 (2 %) and 2009 (4 %) [11].

Table 1 Rate of mobile students, by discipline in % [11]

Disciplines	Year					
	1994	*1997*	*2000*	*2003*	*2006*	*2009*
Engineering science	2	3	4	4	3	4
Language and cultural studies	12	12	13	12	12	12
Maths/natural science	4	5	4	5	5	5
Medicine/health	4	5	5	7	6	5
Law and business studies	5	8	9	8	9	11
Social/education science, psychology	2	4	4	5	6	8

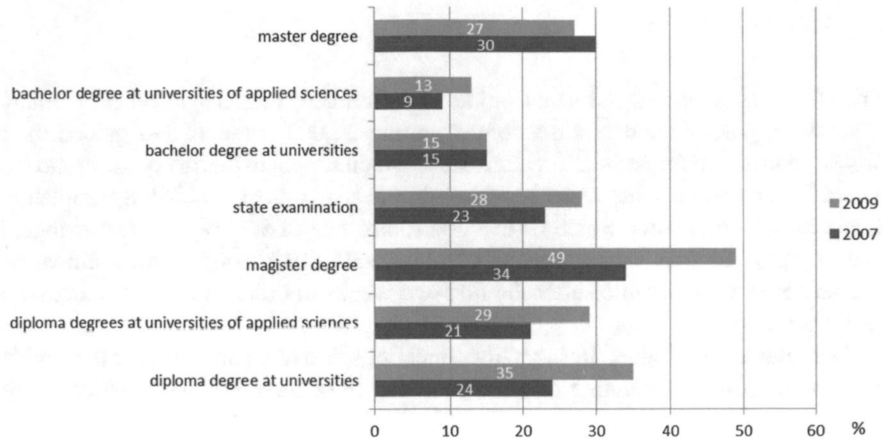

Fig. 3 Study related international exchange periods of German students according to degree program, in % [5]

Heublein and Hutzsch [5] also discovered differences in mobility levels according to degree programs as Fig. 3 shows.

Students enrolled in the traditional German degree programs that were in place before the Bologna Reform such as the magister and diploma degrees show a higher mobility rate than bachelor students. They argue that this is mostly due to the higher average age and number of study semesters as well as the lacking enrollment of new students in the traditional programs. Those numbers suggest that new degree programs seem to hamper international mobility.

Just a few studies dwell on the motivational factors and obstacles students face when planning and accomplishing international exchange phases. According to numbers from DAAD [2], HIS [5] and the Ministry of Innovation, Science and Research of the German state of North-Rhine Westphalia [15] the following obstacles are often reported:

- problems in the recognition process of credits accomplished abroad,
- time lost in the overall course of studies,
- problems with harmonizing the international exchange phase with the study requirements of the home institution, and
- financial problems.

As already Heublein et al. [16] argue further studies on student motivation for international exchange hardly exist. Therefore, the present survey focuses the motivational factors and obstacles for international exchange periods at RWTH Aachen University in further detail.

3 Method

The survey was carried out as an online questionnaire in order to reach as many students as possible and was distributed among 33,003 students that gained their higher education entrance qualification at a German school in Germany or abroad (so called 'Bildungsinlaender students'). The sample accounts to $N = 3{,}218$ completely answered questionnaires which results in a return rate of 9.75 %. It was distributed over a period of four weeks from November 5-30, 2012 using email notification and an email reminder once after the first two weeks in order to remind students to participate.

The sample was divided through filter questions into five parts in order to provide students in different planning or realization stages of their exchange periods with tailored questions.

- part 1: students who were abroad and were back in Germany at the time of the survey
- part 2: students who were abroad at the time of the survey
- part 3: students who were still in Germany but planning an exchange period at the time of the survey
- part 4: students who had planned an exchange period before, but finally did not realize it
- part 5: students who were not planning any exchange period at all

Furthermore, the questions were divided into nine thematic blocks:

- block 1: demographic data
- block 2: study related data
- block 3: information on the exchange
- block 4: motivation and evaluation of exchange
- block 5: obstacles
- block 6: experiences and problems
- block 7: financing of exchange
- block 8: recognition of study credits
- block 9: used sources of information

The questions of blocks 1, 2, and 8 were mainly composed of single or multiple choice questions, drop-down lists or entry fields. Question blocks 3, 4, 5, 6, and 7 also used the above mentioned question types added by questions where participants were asked to rate their answer tendency according to a specific statement on a scale ranging from 1 "does not apply at all" through to 6 "applies fully".

4 Description of the Sample

4.1 Demographic Data

A majority of 61.5 % of students that took part in the survey were male with an average age of 23.44 years. 3,064 participants were German (approx. 95 %), 16 Chinese (approx. 0.5 %), 15 Turkish (approx. 0.5 %) and 10 Russian (approx. 0.3 %). 97.4 % of the participants gained their higher education entrance qualification in Germany.

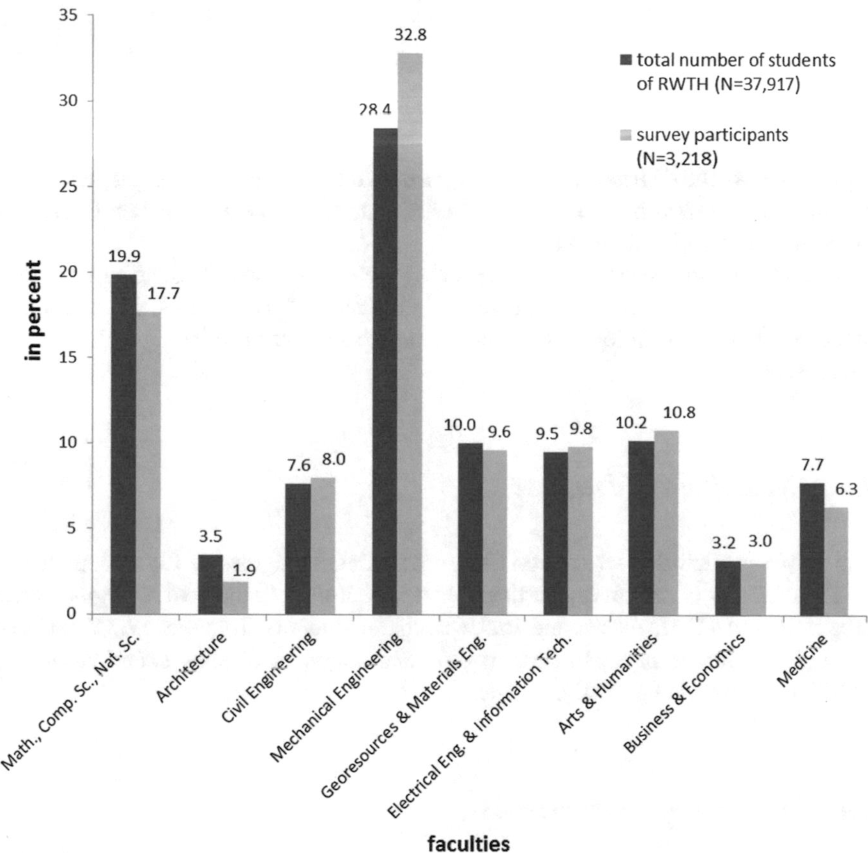

Fig. 4 Distribution of participants among university faculties, in % [17]

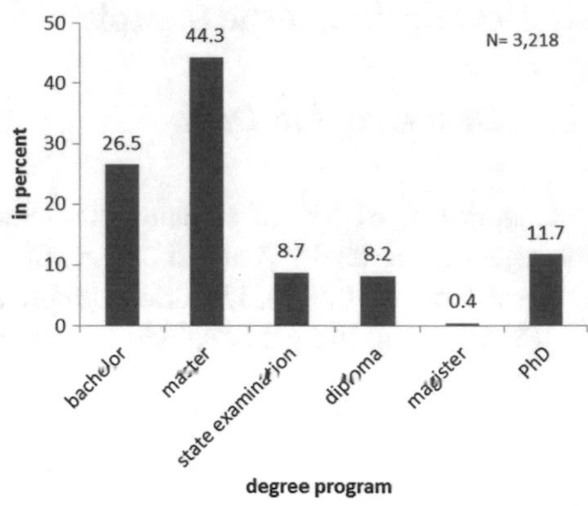

Fig. 5 Repartition of degrees among survey participants, in %

4.2 Study Related Data

Figure 4 shows the distribution of survey participants among the different faculties of the university comparing the total number of students at RWTH Aachen University in the winter term 2012/2013 [17].

It shows that the participants of the survey were almost evenly distributed among faculties compared to the general distribution of all RWTH students. Thus, any bias in the distribution of survey participants cannot be observed producing well balanced data resource.

4.3 Repartition of Degrees

The survey participants worked on different academic degrees as Fig. 5 depicts.

The majority of the survey participants were enrolled in the post Bologna degree programs with 44.3 % master and 26.5 % bachelor students. Together 17.3 % pursued a traditional degree program such as diploma, magister, or state examination and 11.7 % were enrolled as PhD students.

4.4 International Experiences

A clear majority of 50.7 % of all survey participants had never gathered any experiences abroad before starting their higher education programs. While 27.8 % gathered international experiences through one exchange, 20.4 % pursued two or more exchanges before entering university as is demonstrated in Fig. 6.

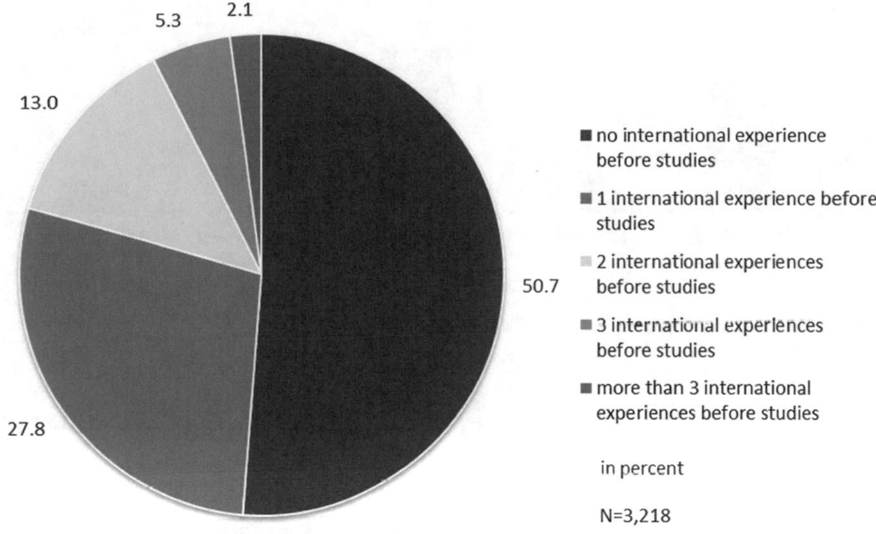

Fig. 6 International experiences before studies, in %

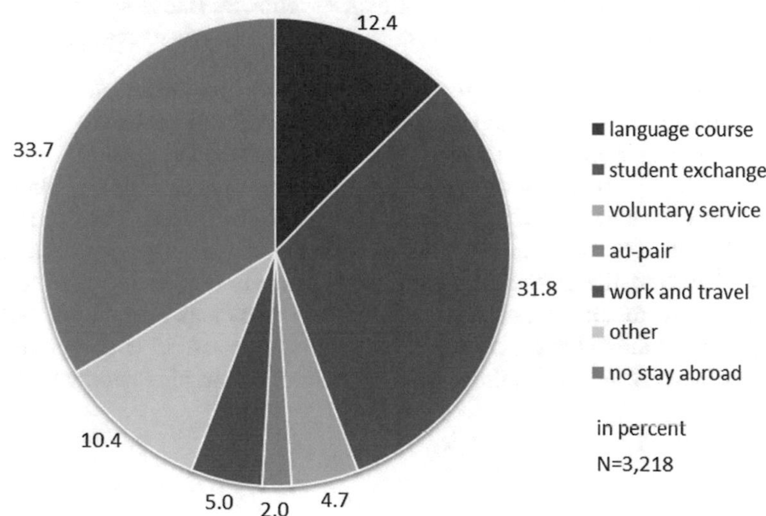

Fig. 7 Types of international experiences of survey participants before studies, in %

With 66.3 % two thirds of the survey participants gathered international experiences before studying, mostly through student exchanges at high school with 31.8 % or language courses with 12.4 % as is shown in Fig. 7.

The survey participants were in different planning and realization stages of their exchange or were not planning to go abroad at all as Fig. 8 shows.

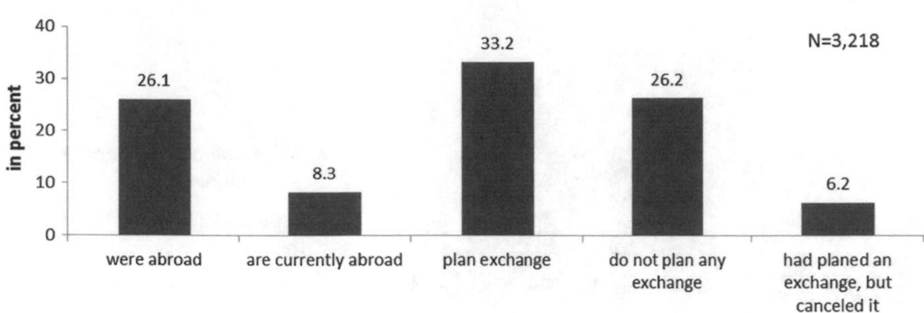

Fig. 8 Planning and realization stages of survey participants, in %

The sample can be divided into three groups in terms of international experiences gathered:

- group A: students with international experiences (those who were abroad and are currently abroad)
- group B: students currently planning an exchange
- group C: students without any international experiences (those who had planned an exchange, but cancelled it and those who do not plan any exchange)

With 34.4 % more than one third of the sample can be attributed to group A. 33.2 % of students belong to group B and thus actually plan to go abroad during their studies. All together 32.4 % of the survey participants belong to group C and have not gathered any international experiences whereas 6.2 % had originally planned to undertake an exchange and 26.2 % did not plan to go abroad at all. Thus, the sample is well balanced in terms of international experiences of the survey participants.

Those characteristics of the sample can also be found in similar studies. According to Heublein and Hutzsch [5], approx. one third of all students plan an international exchange during their studies, 20 % are not decided yet, and 35 % do not envisage going abroad at all. A third of those who went abroad even plan another exchange period.

5 Results of the Survey

5.1 Mobility According to Disciplines

When comparing mobility of engineers with students of other disciplines, the survey did not discover any specific differences. The participants from the engineering faculties of Mathematics, Computer Science and Natural Sciences, Civil and Mechanical Engineering, Georesources and Material Engineering as well as Electrical Engineering and Information Technology ('engineer') were compared to those of the faculties

Table 2 Contingency table for mobility & engineering students, in frequency of answers

	was/is abroad	**has not been abroad**	**total**
engineer	873	1634	2507
no engineer	234	477	711
total	1107	2111	3218

of Architecture, Arts and Humanities, Business and Economics and Medicine ('no engineer').

Those participants who had already been abroad or were abroad at the time of the survey were attributed to the variable 'was/is abroad', all others to the variable 'has not been abroad'. Table 2 shows the frequency distribution among the tested variables.

A Pearson's chi-squared test was performed on the two nominal variables 'engineer' or 'no engineer' and 'has not been abroad' or 'was/is abroad'. The null hypothesis (H_0) on the relationship between the two variables was tested and the following results calculated (χ^2: chi-squared value, p: probability value):

$$\chi^2(1, \mathrm{N} = 3,218) = .90; \mathrm{p} = .05$$

The results fail to reject the null hypothesis, which means that there is no significant difference in mobility affinity between the disciplines. Thus, engineering students are not less mobile than other students that took part in the RWTH survey.

These results are contrary to the numbers introduced above showing that engineering students are mostly less mobile than students from other disciplines such as the social sciences, business and economics, or arts and humanities [12]. While those numbers rely on studies with a much broader data background, the present survey only shows the picture of RWTH Aachen University. The reasons for these deviant results can only be speculated. Most certainly, the bigger engineering faculties in terms of student numbers dispose of better structures than the smaller faculties such as the arts and humanities faculty by providing i.e. explicit coordinators for international relations that encourage their students to undertake international exchanges [18]. Furthermore, RWTH has installed double degree programs [19] with partner universities in China, Japan and France explicitly for engineers and students from STEM fields which is also reflected in the higher exchange numbers among those students. Moreover, RWTH has built a unique reputation also on international level due to the fact that it succeeded for the second time within the Excellence Initiative of the German federal and state governments, and thus consolidated its leading position among German universities [20]. This led to more university partnerships especially in the highly requested target regions such as the US, Spain or Sweden (c.f. *5.3 Countries of destination*). Together with the English-taught master's degree courses in computer science, mechanical and electrical engineering as well as geophysics not only the incoming numbers of international students might have been increased

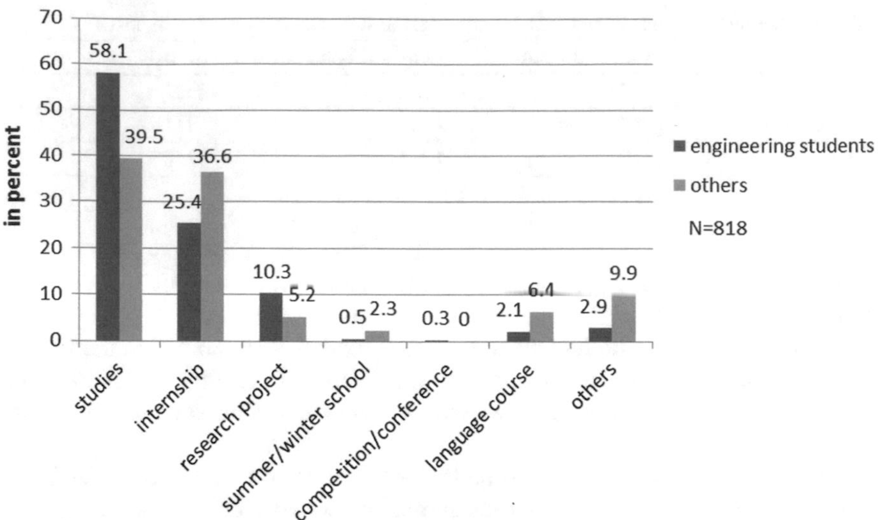

Fig. 9 Prioritized types of exchange by engineering students and others, in %

but also national students that are more aware of the positive effects of international experiences for their later careers might have been encouraged to go abroad. Last but not least, exchange numbers among the RWTH engineering students might have increased due to close links of the respective faculties to industry partners resulting in a wide range of industry internships abroad.

Apart from those measures a certain selection effect within the survey due to imperfect randomization of the sample cannot be ruled out. It may be the case that rather mobility-prone students took part in the survey and thus influenced the results accordingly.

5.2 *Types and Duration of Exchange*

Different types of foreign exchange seem to be commonly chosen by students from the different disciplines. As the following figures show the mobility preferences of engineering students differ significantly from other students. When asked to rate the three most important types of exchange, they answered as follows (Fig. 9).

Hence, studies and internships are highly and almost evenly prioritized among all students whereas engineering students prefer to pursue their studies more than an internship abroad.

This is contrary to the numbers of the Deutsches Studentenwerk. According to them, students of language and cultural studies often chose to pursue their studies abroad whereas engineering students prefer to take up internships in the respective country of destination as is shown in Table 3 [11].

Table 3 Rate of mobile students with study-related activities, by discipline in % [11]

Disciplines	Studies		Internship		Language course		Other	
	06	*09*	*06*	*09*	*06*	*09*	*06*	*09*
Engineering s.	3	4	6	7	2	1	1	1
Language and cultural studies	12	12	9	8	7	6	4	4
Maths/ nat. s.	5	5	5	5	2	1	2	2
Medicine/health	6	5	18	16	3	2	3	2
Law and business studies	9	11	9	7	5	3	1	1
Social/education s. psychology	6	8	7	7	3	3	2	2

The difference in results compared to the RWTH survey may be attributed to a wide ranging network of partnering universities rather than contacts to industry partners. Presumably, students tend to take advantage of already existing networks rather than organizing an industry internship themselves. Regarding the duration of the exchange most of the students that were abroad before and that were abroad at the time of the survey chose a period of 4-6 months (43.5 %) which perfectly fits into one study term. 19.9 % of all participants pursued a 3-month exchange or shorter. 19.3 % preferred to stay abroad for 10-12 months. On average students went abroad in their 7th semester (M = 7.04, SD = 3.02) (Fig. 10).

This also corresponds with the numbers of Heublein and Hutzsch [5]. While engineering students tend to pursue short-term stays abroad such as internships, students in the humanities often organize long-term studies. The average international exchange lasted for approximately 6 months. Only 8 % of students stayed longer than 12 months abroad. While studies lasted 6 months, internships lasted 3 months on average.

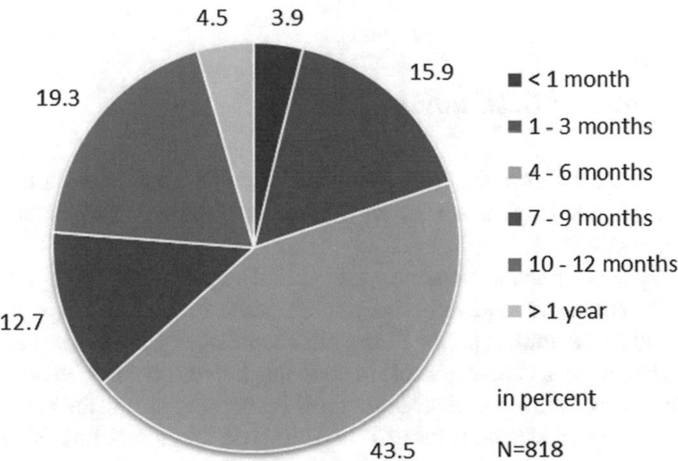

Fig. 10 Duration of international exchanges, in %

Table 4 Most popular countries of destination among survey participants

rank	engineers	others
1	US	Spain
2	Spain	France
3	Sweden	US
4	UK	UK
5	France	Italy

Table 5 German students abroad in 2008, by country of destination, in % [2]

Destinations	Quantity	Amount
Austria	20.019	19,6
The Netherlands	18.972	18,6
Great Britain	12.895	12,6
Switzerland	11.005	10,8
USA	9.679	9,5
France	6.071	5,9
Australia	3.418	3,3
Sweden	3.400	3,3

Just very few curricula integrate mandatory exchange periods. The majority of 96.2 % of all RWTH survey participants answered that exchange periods were not mandatory. When asked whether those periods shall be made mandatory the majority of 58.3 % declined it. Thus, the students seem to appreciate that international exchange periods have to be self-organized and can be integrated at a time of their choice into their studies. 44.2 % of participants took semesters off in order to go abroad and accepted a related extension of the overall study time (60.6 %).

5.3 Countries of Destination

The following table shows the five most popular countries of destination for international exchange among the survey participants separately for engineers and other fellow students (Table 4).

Hence, the most popular countries are almost evenly distributed among disciplines. While engineers tend to prefer the US more than the European countries, students of other disciplines prefer European countries slightly more than the US.

Also the language plays a major role in choosing a country of destination. 61.6 % of the participants used the respective language of the country during their studies. Thus, the languages learned at school such as the European languages English, French, or Spanish or the ones being close to German such as Swedish are preferred.

The tendency of a clear preference of European countries is also mirrored by the numbers of DAAD [2], while different countries of destination were most popular in different disciplines: 40 % of German bachelor students went to the Netherlands, 37 % of German PhD students abroad enrolled in Switzerland and 61 % of students with other degrees (such as diploma or master) took up studies in Austria. As is shown in Table 5 the neighboring countries seem to be most attractive to German students with Austria and the Netherlands being the most visited countries as well as the UK and the US as Anglophone countries.

Also Heublein and Hutzsch [5] show that the most popular countries of destination of German students such as the UK, France, and Spain were situated in Western Europe. Only 12 % went to the US or Canada, 11 % to Eastern Europe, 11 % to Asia, and 7 % to Latin America or Africa.

5.4 *Experiences of Exchange*

When asking those who are currently abroad or who have already finished their international exchange on the experiences their gathered the following picture emerges. Across disciplines positive experiences were gathered in developing language skills, getting to know a new culture and friends and an increased autonomy.

While non-engineers seem to rate the introduction to another culture and an increase in autonomy as more relevant than engineers, they seem to have gathered better experiences with good counseling services abroad or the improvement of career options.

Heublein and Hutzsch [5] also confirm that students report to gather rather positive experiences abroad regardless of the country of destination. 81 % of students were able to deal with the new culture without any problems and report of having had the feeling of being integrated into society. This highly corresponds with good language skills and can also be attributed to a good preparation leaving students with appropriate expectations for international exchange (Fig. 11).

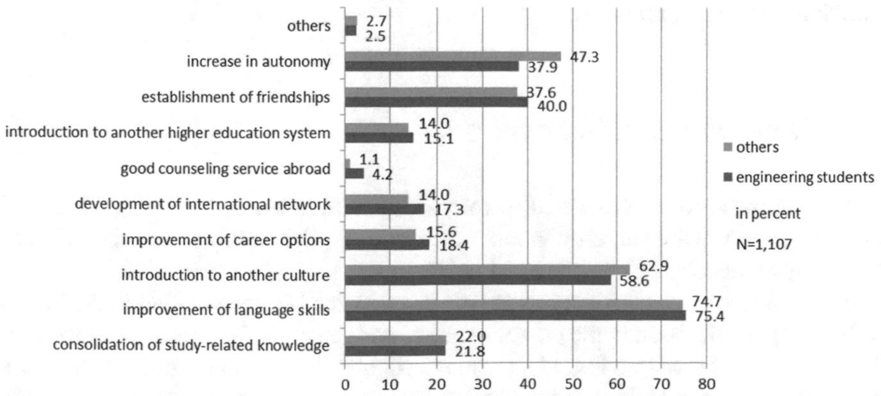

Fig. 11 Experiences during exchange, by discipline in % (multiple answers possible)

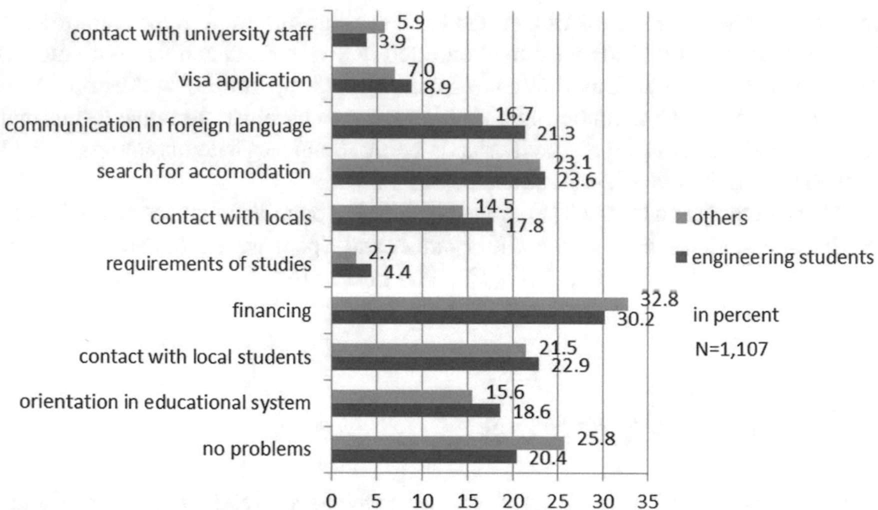

Fig. 12 Problems during exchange, by discipline in %

5.5 *Problems of Exchange*

Most problems were reported concerning financing and the search for accommodation. While engineering students seem to have fewer problems with financing their studies abroad, they report more problems with communicating in the respective foreign language, in contacting locals or in dealing with the requirements of studies. 20-25 % of students reported to having had no problems at all (Fig. 12).

Heublein and Hutzsch [5] report that approximately one quarter of the students had difficulties in financing their studies as well as criticized a lack of support by their sending university. Fewer problems are reported concerning the requirements of studies abroad and the recognition process. Only 16 % referred to problems in the search for an accommodation.

5.6 *Motivation for Exchange*

The participants were also asked to rate a pre-defined set of motivational factors on a range between 1 (no motivational factor) to 6 (high motivational factor). The following tendencies can be observed in Fig. 13.

As the figure shows there seems to be a good match between motivational factors influencing the decision in the planning phase and the actually gathered experiences during the stay abroad (c.f. Fig. 11). There is a slight difference between the estimation of improved career options before and during the exchange period. While in the planning phase students tend to estimate a high influence of international experiences

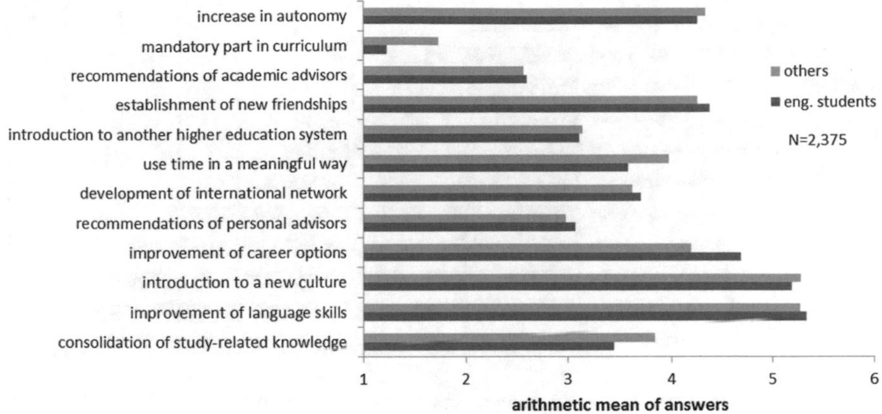

Fig. 13 Motivational factors for planning an international exchange, by discipline

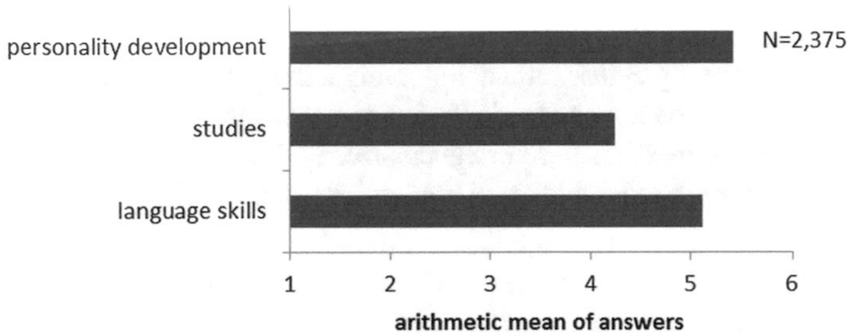

Fig. 14 Evaluation of positive effects of exchange

on their professional career, they rate it rather low during their stay abroad. The most important motivational factors across disciplines seem to be the improvement of language skills and the introduction to a new culture along with the establishment of new friendships and an increase in autonomy. This corresponds with the positive experiences students gather when being abroad (c.f. *5.4 Experiences of exchange*).

Differences between engineers and students of other subjects can be observed at the factors ‘mandatory part in curriculum’, ‘use time in a meaningful way’, ‘improvement of career options’ and ‘consolidation of study-related knowledge’. While exchange periods are often mandatory in curricula of non-engineers, it is not common in engineering sciences. Engineers tend to be more motivated by improved career options and less motivated through using time in meaningful way, or by consolidating their study-related knowledge than other disciplines.

When asked to evaluate their exchange participants scaled its effects on their personal development and the improvement of their foreign language skills higher than on their studies as is shown in Fig. 14.

Heublein and Hutzsch [5] also point to motivational factors for international exchange. Most of the students focus on improving social as well as language

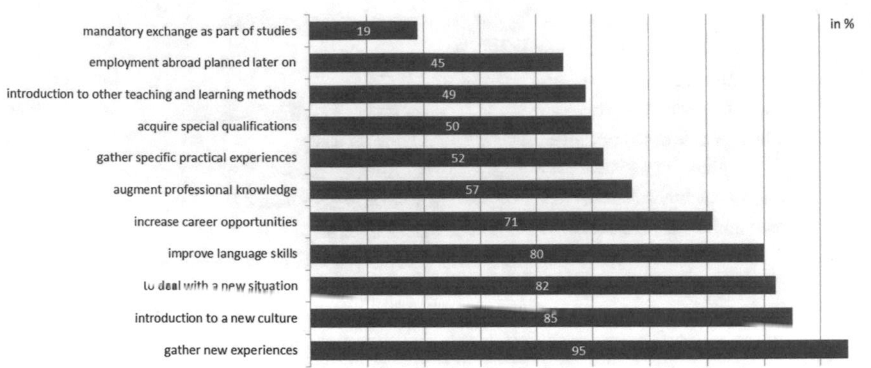

Fig. 15 Reasons for study-related exchange, answers on a scale from '1 = not important at all' to '5 = very important', in % [21]

skills and on gathering experiences in another culture. 71 % perceive an international exchange as being career-enhancing. Only half of the students aim to improve their discipline-specific knowledge. 45 % plan to work abroad in their profession and thus try to gather international experiences already during their studies. Figure 15 summarizes the motivational factors of students to go abroad.

5.7 *Obstacles for Exchange*

The most important obstacles across disciplines are time pressure, financial problems, and too few exchange opportunities. Financial problems seem to be more severe for other students while engineering students rated those problems less relevant. Participants did not see any special problems imposed by the fear of the unknown or the challenge of going abroad. They also rated the fact of an international exchange being a problem for their career as rather low, as is summarized in Fig. 16.

According to Heublein and Hutzsch [5] approx. 10 % of all students fail to realize an international exchange period regardless of their specific course of studies. The most evident reasons were problems with financing the exchange (49 %), a lack of support by their home university (45 %), low compatibility with the requirements of their studies (43 %), or a loss of time (39 %). Only 33 % refer to problems in the recognition process.

Apparently, the participants of the RWTH survey seem to feel an extraordinarily high time pressure. This may be due to strict curricula leaving not enough time to integrate an international exchange.

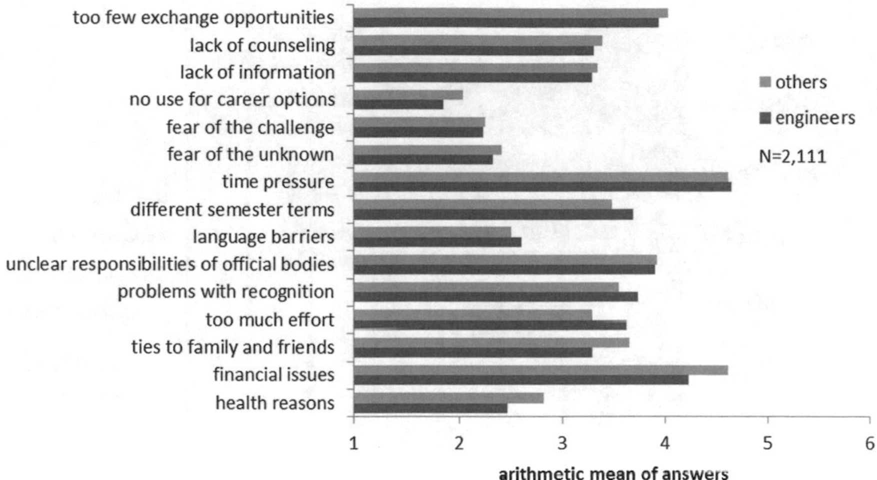

Fig. 16 Obstacles for international exchange, by discipline

5.8 *Financing of Exchange*

Most of the students take advantage of their own savings, their parents' financial support, or scholarships to go abroad. Engineering students tend to use scholarships to finance their exchange more than other students while the others rely more on private savings and family support. Only 10.3 % receive state-funded grants that support their studies in Germany (international BAföG) also abroad. Generally students are very well informed about possible ways of financing their international exchange, but use only few of their opportunities as Fig. 17 demonstrates.

5.9 *Recognition of Study Credits*

Most of the students report not to have had any problems in the recognition process at all. Of those problems that occurred, most frequently the duration of the recognition process was criticized along with different syllabi and problems with the conversion of acquired credits. Obviously, engineering students face bigger problems in all of the stages of the recognition process as demonstrated in Fig. 18. Thus, special measures to tackle those problems for engineers are necessary.

Heublein and Hutzsch as well as the Deutsches Studentenwerk confirm these tendencies. According to Heublein and Hutzsch5 only a fifth of all students face problems within the recognition process. Whereas the Deutsches Studentenwerk [11] points out that students in the planning phase of an international exchange tend to be hampered by the prospect of having problems in the recognition process more than the problem really occurs later on.

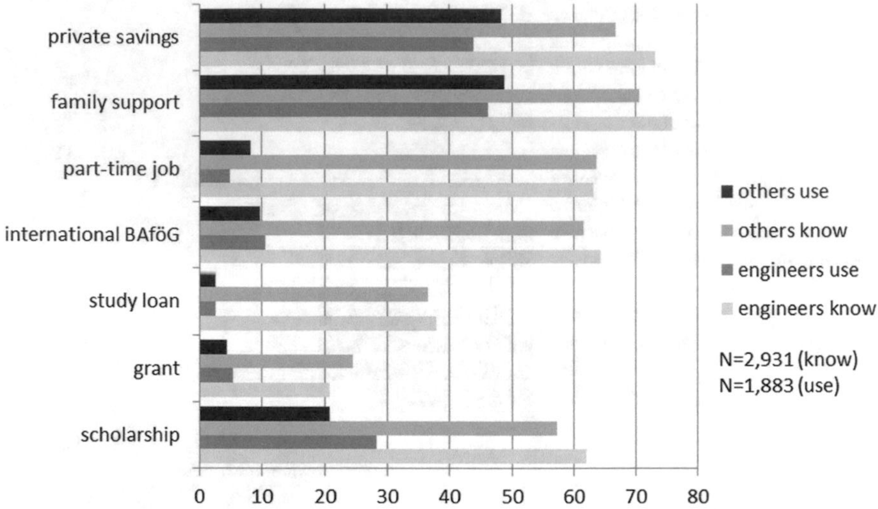

Fig. 17 Level of information and use of financing for exchange, by discipline in %

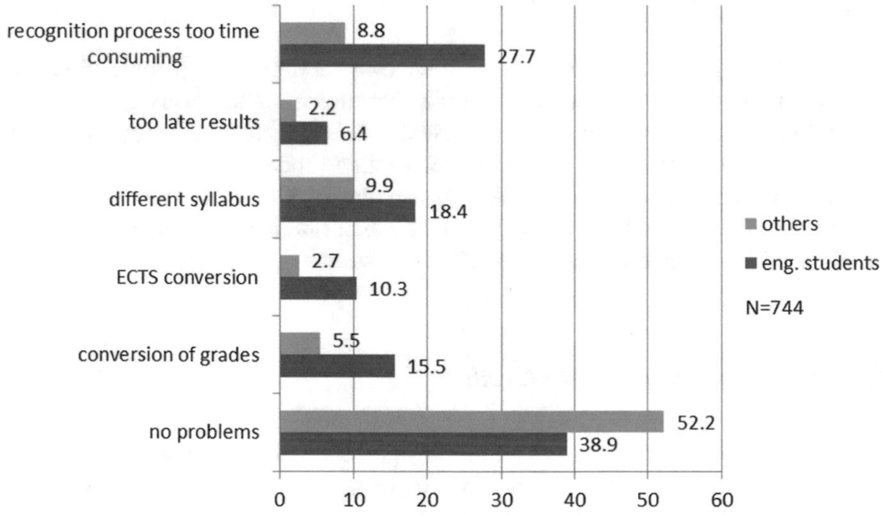

Fig. 18 Problems in recognition process, by discipline in %

6 Summary and Conclusion

The survey among approx. 33,000 RWTH students has led to the following key findings.

- The survey could not confirm the tendency of low mobility among engineering students compared to fellow students from other subject groups at RWTH Aachen University.

- Regarding the types of exchange engineering students of RWTH tend to prefer studies rather than internships abroad which may be attributed to an extended network of partnering higher education institutions rather than industry partnerships abroad. Most students go abroad for 4-6 months which corresponds to one study term and obviously fits best into study curricula.
- There are only a few slight differences in the preferred countries of destination. While engineering students tend to prefer the US more than the European countries, their fellow students from other disciplines rather prefer to stay within European borders to gather international experiences. The foreign language nevertheless plays an important role in the decision for the country of destination in so far as the majority of students communicate in the respective language of the country.
- Most of the students who are currently abroad or who have already finished an international exchange report having gathered rather positive experiences. They rated the improvement of their language skills and the introduction to another culture as most relevant.
- Most problems among those with international experiences occurred with financing the exchange and with finding an appropriate accommodation abroad. Especially engineering students face problems with communicating in the foreign language and with contacting local students.
- Regardless of the different planning and realization stages of an international exchange, students are mostly motivated by the prospect of improving their language skills, by getting to know a new culture, by enhancing their career options, by building new friendships, and by increasing their autonomy. The development of discipline-specific knowledge or other factors such as recommendations by counselors or getting to know another higher education systems seem to be less central.
- The major obstacles students face in any planning or realization phase are financial problems, time pressure during their studies, and too few exchange possibilities especially.
- For financing their stay abroad most students rely on private savings, their family's support, or scholarships – while the latter is especially important for engineering students. The minority of students finances international exchange through state funded grants.
- The majority of survey participants report no problems in the recognition process of study credits. Nevertheless, those problems reported such as the long duration of the recognition process and problems with differing syllabi were most severely encountered among engineering students.

Those key findings suggest the following conclusions and requirements for further research.

There are several deviations in the results of the survey compared to prior studies such as the fact that engineering students at RWTH are not less mobile than other students or face fewer problems in the recognition process of study credits. This might be partly due to the specific situation at RWTH and its specialized programs for engineering students that already tackle those challenges successfully. Nevertheless, further investigations on the specific reasons and comparisons to other universities will become necessary. An in depth-analysis will have to measure what components of those programs and to what extent they successfully encourage outbound mobility among engineering students in order to facilitate their transfer to other higher education institutions.

It is striking that students can only integrate an international exchange when it is financed by their families due to lacking financial support by official bodies. This bears the danger of privileging one specific group of students and discriminating mainly students from lower social classes. As Finger [22] argues the social background of students is especially important when it comes to the decision to go abroad. Once students decided to go abroad the influence of the social background on the choice of country of destination and the duration of mobile periods decreases. Thus, in order to support wide ranging exchange programs more opportunities and financing options shall be offered. Further investigation on the social background especially of engineering students shall be undertaken in order to analyze the special needs of this group of students.

The survey has also shown that students seem to be mainly motivated to go abroad by improving their social competences and language skills. Thus, information and counseling should focus on those issues but should also point out the positive effects on an international exchange on the development of discipline-specific skills and the professional career.

References

1. Jeschke, S., Petermann, M. & Tekkaya, A. E. Ingenieurwissenschaftliche Ausbildung - Ein Streifzug durch Herausforderungen, Methoden und Modellprojekte. *TeachING-LearnING.EU Fachtagung 'Next Generation Engineering Education'* (Bach, U. & Jeschke, S.) 11–22 (ZLW/IMA der RWTH Aachen University, 2011).
2. DAAD. *Wissenschaft weltoffen 2011. Daten und Fakten zur Internationalität von Studium und Forschung in Deutschland.* (HIS, 2011). at <http://www.wissenschaft-weltoffen.de/>
3. *EURODATA - Student mobility in European higher education.* (Lemmens Verlags- & Mediengesellschaft, 2006). at <http://ec.europa.eu/education/erasmus/doc/publ/eurodata_en.pdf>
4. Leidel, M. *Statistische Erfassung der Mobilität von Studierenden. Nationale und internationale Datenquellen und Indikatoren.* 1167–1180 (Statistisches Bundesamt, 2004). at <https://www.destatis.de/DE/Publikationen/WirtschaftStatistik/BildungForschungKultur/ErfassungMobilitaet102004.pdf?__blob=publicationFile>
5. Heublein, U. & Hutzsch, C. *Internationale Mobilität im Studium 2009. Wiederholungsuntersuchung zu studienbezogenen Aufenthalten deutscher Studierender in anderen Ländern.* (Hochschul-Informations-System GmbH, 2009).
6. European Commission. The Bologna Process 2020 - The European Higher Education Area in the new decade. Communiqué of the Conference of European Ministers Responsible for Higher Education, Leuven and Louvain-la-Neuve, 28–29 April 2009. (2009). at <http://www.ond.vlaanderen.be/hogeronderwijs/bologna/conference/documents/leuven_louvain-la-neuve_communiqu%C3%A9_april_2009.pdf>
7. European Commission. *The European Higher Education Area in 2012: Bologna Process Implementation Report.* (European Commission, Education, Audiovisual and Culture Executive Agency, 2012).

8. Westerheijden, D. F. *et al. The Bologna Process Independent Assessment. The first decade of working on the European Higher Education Area. Executive summary, overview and conclusions.* (2010). at <http://www.ond.vlaanderen.be/hogeronderwijs/bologna/2010_conference/documents/IndependentAssessment_executive_summary_overview_conclusions.pdf>
9. Grigat, F. Bologna als Mobilitätsbremse. Ergebnisse von aktuellen Studien. *Forschung und Lehre* 9/11, (2011).
10. Heublein, U., Hutzsch, C., Schreiber, J. & Sommer, D. *Internationale Mobilität im Studium 2009. Ergebnisse einer Wiederholungsbefragung zu studienbezogenen Auslandsaufenthalten deutscher Studierender in anderen Ländern.* (HIS, 2011). at <http://go-out.de/imperia/md/content/go-out/mobilit__t_2009_171111_din-a4.pdf>
11. Isserstedt, W. & Kandulla, M. Internationalisierung des Studiums - Ausländische Studierende in Deutschland - Deutsche Studierende im Ausland. Ergebnisse der 19. Sozialerhebung des Deutschen Studentenwerks durchgeführt durch HIS Hochschul-Informations-System. (BMBF, 2010). at <http://www.studentenwerke.de/pdf/Internationalisierungbericht.pdf>
12. *Wissenschaft weltoffen 2012. Daten und Fakten zur Internationalität von Studium und Forschung in Deutschland.* (W. Bertelsmann Verlag, 2012).
13. *Deutsche Studierende im Ausland. Statistischer Überblick 2000–2010.* (Statistisches Bundesamt, 2012). at <https://www.destatis.de/DE/Publikationen/Thematisch/BildungForschungKultur/Hochschulen/StudierendeAusland5217101127004.pdf?__blob=publicationFile>
14. European Commission. *ERASMUS - Facts, Figures & Trends. The European Union support for student and staff exchanges and university cooperation in 2010–11.* (2012). at <http://ec.europa.eu/education/pub/pdf/higher/erasmus1011_en.pdf>
15. Rieck, S. & Märker, O. *Ergebnisbericht.* www.besser-studieren.NRW.de. (Ministerium für Innovation, Wissenschaft und Forschung des Landes Nordrhein-Westfalen, 2012).
16. Heublein, U., Schreiber, J. & Hutzsch, C. *Entwicklung der Auslandsmobilität deutscher Studierender.* (HIS, 2011).
17. Hötte, H.-D. & Fritz, H. *RWTH Aachen University. Zahlenspiegel 2011.* (RWTH Aachen University, 2012).
18. Naderer, H., Piel, B., Hötte, H.-D. & Gawlik, M. *RWTH International. Internationalisierungsreport 2012.* (RWTH Aachen University, 2012).
19. RWTH Aachen University. Dopplelabschlussprogramme - RWTH AACHEN UNIVERSITY Fakultät für Maschinenwesen. (2012). at <http://www.maschinenbau.rwth-aachen.de/cms/Fakultaeten/Maschinenbau/Studium/Internationales/Austauschprogramme/~djcy/Dopplelabschlussprogramme/>
20. RWTH Aachen University. RWTH Aachen University Launches the Second Phase of the Excellence Initiative - RWTH AACHEN UNIVERSITY. (2012). at <http://www.rwth-aachen.de/cms/root/Die_RWTH/Exzellenzinitiative/~eob/Exzellenzinitiative/lidx/1/>
21. Heublein, U. *Internationale Mobilität im Studium 2009. Wiederholungsuntersuchung zu studienbezogenen Aufenthalten deutscher Studierender in anderen Ländern.* (Higher Education Information System (HIS), 2009).
22. Finger, C. *The Social Selectivity of International Mobility among German University Students. A Multi-Level Analysis of the Impact of the Bologna Process.* (Social Science Research Center Berlin (WZB), 2011). at <http://bibliothek.wzb.eu/pdf/2011/i11-503.pdf>

Student Exchange Programs in Engineering Sciences Between USA and Germany

Natascha Strenger, Marcus Petermann and Sulamith Frerich

Abstract In the Unites States of America (USA), as well as in Germany, the international mobility of students in engineering sciences is rather low in comparison to students of other disciplines. While the USA are the most attractive host country for German engineering students, American students' interest in coming to Germany is not equally great. Facing different teaching languages and semester times, summer school formats seem to be a good solution for German universities to attract students from the USA. However, summer schools might not always be the most suitable way of providing possibilities for the development of international skills. At the Ruhr University Bochum, the project ELLI works on the further development of student exchange programs in engineering education. This short paper represents work in progress. It gives an overview of the context of student exchange in and between the USA and Germany and describes the results of one case study that was made in Bochum.

Keywords Cultural Differences · Engineering Education · Internationalization · Student Exchange · Stay Abroad

1 Introduction

In the light of an increasing complexity of global production processes, internationalization in engineering education has become a subject of central importance. Facing a future employment in an international environment, it is necessary that engineering students acquire intercultural and language competences during their studies, preferably during a stay abroad. The idea is that students acquire intercultural and language competences by living in alternate cultural surroundings. Those competences are presumably gained by interacting with the people in the host country, ideally being forced to communicate in another language and getting to know

N. Strenger (✉) · M. Petermann · S. Frerich
Ruhr-Universität Bochum, Project ELLI – Excellent Teaching and Learning in Engineering Sciences, Bochum, Germany
e-mail: strenger@fvt.rub.de

Originally published in "IEEE EDUCON, April 2nd–4th, 2014",

DOI 10.1007/978-3-319-46916-4_45

cultural differences [1]. In order to provide engineering students with those competences assumed to be relevant for their future career, educational institutions try to include periods abroad in their curricula. Engineering universities and faculties in countries all over the world have been establishing various types of student exchange programs with one or several partner universities during the past decades.

2 Student Exchange in and Between the USA and Germany

2.1 Comparatively Low Mobility in Engineering Sciences in the USA and in Germany

If exchange in and between the USA and Germany is examined more closely, several common denominators become apparent with regard to outgoing mobility in engineering education: In Germany, just as in the USA, students in engineering sciences do generally participate less in study abroad programs than students of other disciplines. In the USA, the underrepresentation of engineering students in staying abroad programs even led to the development of several initiatives to promote outgoing mobility among them, such as RISE (Research Internships in Science and Engineering) [2]. In Germany, a representative student survey conducted in the beginning of 2013 by the Higher Education Information System (HIS) listed students in engineering sciences as less mobile than all other subject groups [3]. Hindering factors for spending a semester abroad on both sides are the financial efforts as well as (an assumed) prolonged time to graduation. Among engineering students in the USA, also the language barrier is considered to be an impediment to stays abroad [4].

2.2 Student Exchange Between the USA and Germany: Imbalanced Appeal to Exchange Students

While the general mobility among engineering students is rather low in both countries, it is interesting as well to examine the type of relationship between the two countries when it comes to academic exchange. In this equation, the power to draw interested students from abroad to one's own national universities becomes a factor that cannot be neglected: In the opinion of the German Association for Electrical, Electronic and Information Technologies (VDE), higher education in engineering sciences in Germany used to be attractive for students from all over the world, but it has nowadays lost much of its appeal to other nations. Among those countries whose universities attract most international students are the United States, next to Great Britain and Australia. The reasons for this are presumably to be found in those universities' good marketing and in the worldwide dominance of the English language [5]. The attractiveness of US-American universities becomes clearly perceivable also in German higher engineering education: Among German engineering students, the USA are the most sought-after country for temporary, study-related visits abroad [6]. For US-American students in general, the preferred student-exchange country is Great Britain by far, followed by other countries with the same mother tongue, such as Australia and New Zealand [7]. So in Germany, the overall mobility among engineering students might be comparably low, but to those who do go abroad, US-American universities are a very appealing destination.

2.3 Attracting US-American Students with Summer School Formats

Consequently, to promote outgoing mobility, German engineering faculties face the challenge of attracting enough US-American students to provide their own students with the exchange opportunities they are most interested in. Curricula in German engineering education, however, consist mostly of courses in German, which makes it rather difficult to attract students from an English-speaking country who clearly prefer other English-speaking destinations. At this point, summer school formats present themselves as a possible approach on the German side for initiating academic exchange with the USA. For American engineering students, "compact" experiences abroad seem to have become a successful alternative with regard to many of the general obstacles to outgoing mobility [8]. Such programs usually consist of several weeks spent at faculties abroad in countries in Europe or Asia for example. They are sometimes accompanied by additional language classes but otherwise taught in English. Besides solving the language problem, the shorter period of time might be less expensive than spending an entire semester in the country [4].

3 Case Study: Engineering Exchange at Ruhr University Bochum

3.1 ELLI and Internationalisation in Engineering Education

The project ELLI – Excellent Teaching and Learning in Engineering Sciences – aims at improving the conditions of teaching and learning in engineering education. It has four key areas, (including virtual learning environments, interdisciplinary competences and the transitions within the student lifecycle,) internationalization and mobility being one of them. ELLI is funded by the German Federal Ministry of Education and Research and based on cooperation between RWTH Aachen University, TU Dortmund and the Ruhr University Bochum (RUB). At RUB, the key area internationalization has started to investigate the conditions of exchange programs for incomings students in 2013, focusing on a first case study of an exchange between the Ruhr University and Drexel University in Philadelphia, USA. Starting from this point, the program aims at exploring the challenges to designing optimal forms of student exchange that fit the students' needs as well as institutional requirements.

3.2 *Exchange at the RUB Faculty of Mechanical Engineering*

At the Faculty of Mechanical Engineering of the Ruhr University Bochum, a summer school program was designed especially for incoming students from a partner university in the USA: Beginning in 2011, students from Drexel University in Philadelphia came to the RUB and as a result, German students spent the following spring term at Drexel University. The program has been repeated annually since and proved to be a success with regard to the students' general interest in participation. Until now, more than 50 students studied abroad, so the general goal of increasing mobility between universities was achieved. Nevertheless, several aspects in the design of the program made a restructuring for 2014 seem adequate. First of all, different requirements of German and American students resulted in a rather complex and work-intensive organization of the entire program: While German students take part in the regular semester in Philadelphia, their counterparts receive a special program of lectures. They come to RUB for a period of approximately 8 weeks between July and September. Usually, there are no lectures at RUB during the summer break and classes in English are generally limited. Therefore, in order to provide the US-American guests with classes that fit into their curricula at home, special courses in Statistics, Fluid Dynamics and Unit Operations were offered in English. This program was split between German lecturers and a professor from Philadelphia who came to Germany together with the students. Apart from the lectures in engineering sciences, the students attended German classes at beginners' level. They were accommodated within the housing facilities the university offers to exchange students via its International Office. So with regard to the framework of the program, similar services were required by American students in Germany as by German students in the US.

3.3 *Difficult Conditions for Academic and Intercultural Contact*

However, the organizational format of the program presented further challenges with regard to the student's intercultural interaction: RUB students visiting Drexel University have the chance to melt in with the cultural and academic surroundings of their host country and university as they are attending the regular courses there; American students are lacking similar opportunities in Bochum. The special courses prevent them largely from getting into contact with German students: They do not participate in regular classes and German students have no lectures during the summer months, so they rarely meet in an educational context. In their German classes, the US-students will not encounter German students either. Also, the relatively short time the American students spent in Bochum naturally makes them want to seize the opportunity of getting to know the surroundings of Bochum and the Ruhr Area. Trips are usually organized in groups of American students and not always accompanied by German guides. Overall, a packed schedule and the special course program prevent them from getting in touch with German students either on campus or during free-time activities. The students themselves partially perceive this situation as a rather disappointing element in an otherwise very successful program, as a survey among them has revealed this year.

3.4 The Students' View

In the summer of 2013, a qualitative survey was done among the students visiting from Philadelphia (a group of 14 students, all of whom participated). Following previous talks with the organizers of the program and several individual students, a questionnaire was designed and handed to the students at the end of their last day of class. This questionnaire contained questions about the program and its organization in general and also asked for previous interest and experience in or with Germany and the German language. Furthermore, it contained open questions asking for the students opinion on what could be done by the organizing universities for improvement. All participants rated the overall program (the classes and lectures, the organization and services such as housing, excursions) as very good and most confirmed that they would recommend it to fellow students in the US or participate again themselves if they had the chance. With regard to previous experiences in Germany, some students stated that they had visited the country before – in a high school exchange, or during a private visit of family and friends living in Germany. A few had also taken German classes before or were able to acquire some knowledge through German-speaking relatives back in the USA. So for those students it can be said that they had already established some sort of cultural interest in Germany before they took part in the summer school. The open questions, however, revealed that several students would wish to get more into contact with German students. Some stated that they would like to participate in classes and programs together with Germans; others expressed their wish to take part in German everyday life or have a German student 'buddy' to introduce them to it. This first qualitative evaluation has shown that the students, albeit very satisfied with the program itself, are missing intercultural and academic contact with German students, in which they would be interested. The fact that the students were not specifically asked about this topic but for general recommendations, underlines the importance of their statements.

4 Conclusion and Outlook

In the quest for higher student mobility and more international exchange in engineering education, academic institutions have to take into account the aims of such exchange when designing programs that fit into the institutional framework. If one aim is considered to be the development of intercultural experience as a skill necessary for engineers that face future employment in a globalizing economy, it should be ensured that such experiences are actually made. Summer school formats offer a comprised international experience that also accommodates the different semester-times of two countries and therefore seem to be a good strategy to intensify German-US-American student exchange in engineering sciences. Nevertheless, the closer examination of one such program has so far indicated that a true immersion of US-students into the German host culture might not necessarily take place in this case. For this reason, combined with a high organizational effort that comes along with the isolated summer courses, the program coordinators at both universities are currently pondering possibilities of integrating US-American students into the regular summer semester. The fact that the program in Bochum is designed for the accommodation of only one culturally homogenous group might distinguish it from other summer schools. Without knowing if other programs are facing similar challenges, so far no general recommendations can be made. It will be necessary to keep evaluating the exchange with regard not only to organizational aspects, but also to the actual intercultural and international skills which engineering students can possibly acquire through it. The ELLI-Project in Bochum is planning on continuing this evaluation, intensifying research on the topic and possibly taking into consideration comparative studies of similar programs in the future.

References

1. Grandin, John M. and Hedderich, Norbert (2013): "Intercultural Competence in Engineering - Global Competence for Engineers", The Sage Handbook of Intercultural Competence. Darla K. Deardoff (Ed.). 362–374.
2. Daniel Obst, Rajika Bhandari and Sharon Witherell (2007): "Meeting America's Global Education Challenge: Current Trends in U.S. Study Abroad & The Impact of Strategic Diversity Initiatives.", Institute of International Education (IIE).
3. Heublein, Ulrich (2013): "Gehen oder Bleiben? Internationale Mobilität im Studium. Erste Ergebnisse der 4. Befragung deutscher Studierender zu studienbezogenen Aufenthalten in anderen Ländern 2013." Hannover:Hochschul-Informations-System (HIS).
4. Parkinson, Alan (2007) "Engineering Study Abroad Programs: Formats, Challenges, Best Practices," Online Journal for Global Engineering Education: Vol. 2: Iss. 2, Article 2. Available at: http://digitalcommons.uri.edu/ojgee/vol2/iss2/2
5. German Association for Electrical, Electronic and Information Technologies (VDE) (2013): "Die Internationalisierung der Ingenieurausbildung in Deutschland" Availabe at: http://www.vde.com/de/Karriere/Ingenieurausbildung/Seiten/Internationalisierung.aspx#seitenanfang
6. German Academic Exchange Service (DAAD) (ed.) (2013), Wissenschaft weltoffen 2013. Facts and Figures on the International Nature of Studies and Research in Germany, Bielefeld:HIS, DAAD.

7. Raisa Belyavina, Jing Li and Rajika Bhandari (2013): "New Frontiers: U.S. Students Pursuing Degrees Abroad: A 2-year analysis of key destinations and fields of study", Institute of International Education (IIE).
8. Schubert, Thomas F. Jr. and Jacobitz, Frank G. (2013): "Compact International Experiences: Expanding Student International Awareness Through Short-Term Study Abroad Courses With Substantial Engineering Technical Content", Online Journal for Global Engineering Education: Vol. 7: Iss. 1, Article 1. Available at: http://digitalcommons.uri.edu/ojgee/vol7/iss1/1

Higher Education Institutions as Key Actors in the Global Competition for Engineering Talent – Germany in International Comparison

Natascha Strenger, Sulamith Frerich and Marcus Petermann

Keywords Globalization of EE · Internationalization · Attracting Talent

1 Introduction

In order to prevent the predicted shortage of highly qualified labour in technical professions, Germany and other industrialized nations show increasing interest in strategies of controlled immigration, one being the recruitment of foreign students: Many European countries are seizing the opportunity to attract international students to support their labour market after graduating from their universities [1]. In Germany, the legal conditions for working in the country as foreigners have become more relaxed during the last decade. Especially for international students from third countries, the latest developments in legal regulations have simplified a transition from German universities to companies in international comparison [2]. Additionally, fundamental changes were undertaken in the German higher education system during the past 20 years to make it more attractive for international students. Since the General Agreement on Trade in Services (GATS) and the implementation of bachelor's and master's degrees throughout the European Union via the Bologna Process, the recognition of foreign degrees has become much easier. By offering international study programs in technical subjects and natural sciences, German universities are trying to attract international students who have already obtained a bachelor's degree in their home country and who might decide to stay in Germany after graduating. This paper describes the latest legal developments regarding foreign students in Germany. Furthermore, it considers the political and economic influence on the German

N. Strenger (✉) · S. Frerich · M. Petermann
Ruhr-Universität Bochum, Project ELLI – Excellent Teaching and Learning in Engineering Sciences, Bochum, Germany
e-mail: strenger@fvt.rub.de

Originally published in "SEFI Annual Conference, June 29-July 2, 2015",

DOI 10.1007/978-3-319-46916-4_46

higher education system. With a special focus on academic migration in engineering education, it aims to analyse the position of Germany within the global movements of educational mobility. It deals with the implementation of international study programs in technical disciplines and evaluates whether the German endeavours to attract international talent are successful.

2 Meeting the Labour Shortage with Strategies for Highly Qualified Immigration

Germany and its fellow industrialized nations are predicted to be facing a shortage of highly qualified labour in the not too distant future. One of the reasons for this shortage of skills is seen to be the upcoming demographic change, as more people are retiring while less are born and growing up to enter the future labour market. At the same time, continuing technological progress and economic growth are about to call for an adequately increasing number of highly qualified workers [3, 4]. Another problem for Germany is the emigration of highly educated, often young people who move to Switzerland, Austria or the United States of America. Again, this loss could be absorbed by attracting highly qualified immigrants [5]. On a global scale, when it comes to recruiting international talent, the member states of the European Union find themselves in competition not only with each other, but also with highly industrialized, technology-oriented third countries. Especially in comparison with the USA, the European Union's share of highly qualified immigration is rather small [6].

3 Attracting and Retaining International Students

As part of their approaches to increase the immigration of highly educated employees, Germany and further EU countries are starting to recognize the potential of keeping foreign students in the country to join their workforces after graduation [7]. In the much sought-after natural sciences and technological disciplines, the EU is currently working on strategies to connect higher education institutions and work markets. Another goal is to attract international researchers to improve its position in the global competition, especially against the United States [1]. From the viewpoint of employers, foreign students after graduation are ideally not only equipped with a university degree, but also with cultural and language skills acquired in the host country. Preferably, they have gained first work experience during their studies and are directly available for the host country's work market after their graduation [8, 9]. On a macroeconomic level, the costs and returns of educating international students at domestic universities are another important aspect: As German public universities do not charge tuition fees, the money invested in the higher education of foreigner scan only be regained if those students stay in Germany at least a few years to generate tax incomes [10].

4 Legal Developments

The latest developments in the legal framework for highly qualified foreigners and international students in Germany represent a strong increase of political interest in attracting these target groups.After the guest-worker policy in the course of the economic growth during the 1950s and 1960s, a general recruitment ban for workers from third countries has been valid in Germany since 1973. In 2000, however, German migration politics began to focus on covering economic demands by passing special regulations for highly qualified foreigners. A famous example was the Green Card initiative which was directed at recruiting foreign IT professionals [3]. In October 2007, the access regulation for university graduates was passed by the German Minister for Labour and Social Affairs, which facilitated the conditions for foreign students and certain immigrants from the new EU member states. For employees in the fields of mechanical and automotive engineering from the new member states (who did not yet benefit from the freedom of movement for workers within the EU) and for third country residents who graduated from a German university, the regulation allows for the priority review to be dropped [11, 12]. This means that for job applicants from these two target groups, the German Federal Employment Agency does no longer check whether there are German employees that might fill the vacant position. However, the employment has to be according to the foreign applicant's qualifications and his or her salary has to be at least as high as that of German workers in the same field. Apart from the abolition of the priority examination, the options for extending the residence permit have been systematically improved for foreign graduates of German universities. After successfully completing a degree programme, international graduates from a German higher education institution used to be granted a one-year period in order to find an adequate employment in Germany [5]. With the most recent changes in the Residence Act in August 2012, international graduates can now prolong this period by 6 months, adding up to a total of 18 months to search for a job in Germany. Compared to other EU countries, the legal possibilities for international students to enter the German labour market are rather favourable. Especially with regard to the period of time granted for the job search, since fellow EU nations such as Finland or Austria allow foreign students only six months to apply for a residence permit. However, the German restrictions are comparatively strict when it comes to the consideration of the kind of work, as it has to match the graduate's qualifications. In Finland, for example, international students can apply for a residence permit with any sort of employment [2].

5 Internationalization of the German Higher Education System

When it comes to attracting international students as highly qualified workers for the German labour market, higher education institutions are in a key position [8]. Along with the development of legal regulations, the German higher education system has undergone a series of structural changes during the past 20 years. The General Agreement on Trade in Services (GATS) in 1995 liberalized the global exchange of services, which also affected the higher education sector. With the nowadays almost complete implementation of bachelor's and master's degrees in consequence of the Bologna reform, German higher education institutions increasingly opened up to the global competition for talent. In the 1990s, a political discussion had started in Germany about the international attractiveness of its universities. One of the gravest disadvantages of the former system was that the courses foreign students had taken at their home universities often could not be satisfyingly transferred into the German system if they came to absolve part of their studies in Germany. The amendment to the Framework Act for Higher Education in 1998 marked the starting point of fundamental changes to the system: It opened the possibility for introducing the two-stage model of study programmes including bachelor's and master's degrees according to the Anglo-American example. For a simplified transfer of courses and exams taken at different universities, a credit point system was introduced [13].

6 Implementation of International Programmes

Facing the above mentioned competition for international talent, the implementation of new, international study programmes was another part of the higher education reform. New courses of studies were supposed to be explicitly designed for the requirements of foreign students, considering teaching language and offers for orientation and support during their studies in Germany [14]. Such internationally-oriented degree programmes were supported by the German government and enthusiastically implemented by the higher education institutions, especially on the master level. Currently, the German Academic Exchange Service (DAAD) has more than 800 international master and 140 bachelor programmes in its database [15]. In this database, no less than 245 international master programmes appear when the discipline "engineering" is selected. The largest share of them is taught entirely in English, while several others use German as a second teaching language. Right from the beginning, the engineering disciplines were strongly represented in the process of implementing study programmes for international students. A reason for this is considered to be a wish on the part of these subjects to increase their student numbers, which were and still are relatively low among German students if compared to other disciplines. Another beneficial factor for the implementation was that especially the technical disciplines in Germany capitalize on a high international reputation [16]. The impact of these changes in the higher education system with regard to attracting international talent as a potential answer to the predicted labour shortage will be explored in the following sections.

7 Germany's Position Within Global Student Mobility

The legal enhancements described above were directed at international students and at highly qualified foreign workers in shortage occupations alike. However, it seems to be the target group of international students that is most successfully attracted by the improvements in German migration politics and in the German higher education system: Latest OECD findings show that the largest share of highly qualified immigration in Germany during the last years has been due to international students [9]. In international comparison, Germany ranks among the top three destination countries for global student mobility after the USA and Great Britain [17]. The total number of Bildungsausländer among foreign students in Germany (non-nationals who acquired their entrance qualification for higher education abroad) has doubled since 1997, to reach a total number of about 200,000 in 2014. Their share of all students enrolled in Germany has increased during this time from 7.0 % up to 9.5 %. Consequently, the number of foreign graduates has risen as well. Half of the approximately 30,000 foreign students who graduated in Germany in 2013 stated that they wished to stay to work in Germany at least for several years [18]. An investigation into the student numbers from third countries who studied in Germany between 2005 and 2013 showed that, in 2014, actually 54.1 % (about 100,000 people) are still living in the country, while 26,700 of them are already working [2].

8 Specifics of International Students in German Engineering Disciplines

With a share of 25 %, engineering sciences belong to the most popular disciplines chosen by foreign students in Germany [19]. In addition, Germany is one of the countries that attract the largest share of all globally mobile engineering students worldwide [20]. Furthermore, the percentage of engineering students among all foreign students in Germany is higher (24.8 %) than the percentage of engineering students among all German students (19.8 %) and the technical disciplines show the greatest difference profile of all subject groups [8]. With regard to the characteristics of international students in Germany, it is note worthy that students from developing countries tend to study technical disciplines or natural sciences while students from industrialized countries prefer cultural or language studies. Moreover, a gender-specific choice of subjects can be observed: Similar to the distribution among German students, male foreign students in Germany choose mostly engineering disciplines (36 % of all foreign students from developing countries, 27 % from industrialized nations), while female students prefer philological or cultural studies (36 % from developing countries, 44 % from industrialized countries) [13]. Foreign students in German engineering disciplines are comparatively often matriculated in master's programmes, while international students of other disciplines have larger shares in bachelor's (humanities) or PhD programmes (natural sciences). Regarding their national background, the largest groups of foreign engineering students are from East-Asian or Asian countries (each with a share of 40 %) [19]. A disadvantage of this distribution of nationalities among engineering students might be that especially students from Asian regions rather do not tend to stay in Germany after studying here, at least not in a long-term perspective [2].

9 International Reputation of German Engineering Education

When it comes to the specifics of student migration in the engineering disciplines, it is note worthy as well that the German reputation in this field plays an important role in attracting foreign students: While 61 % of all international students state that the reputation Germany as a high tech country has influenced their decision for studying abroad, 85 % of the foreign students in engineering name this as a reason [19]. These findings correlate with the previously mentioned expectations of the higher education institutions which played a role in the implementation of international study programmes in engineering. A good impression of the strong reputation of a German engineering qualification worldwide is given in a statement from British experts on international connection in higher education who compare it with those from the world's top-ranking universities: "Occasionally, such arrangements [for mutual recognition of degrees obtained abroad] are superfluous simply because the qualifications obtained are so well-known and respected: every international employer knows, or thinks they know, the worth of a degree from Harvard Law School, a doctorate from Oxford or Cambridge, a German engineering qualification, or a degree from the Ècole Polytechnique." [20].

10 Conclusions and Outlook

This paper has taken a closer look at the developments in legal regulations for highly qualified foreigners and international students and at the internationalization of higher education in Germany. It can surely be stated that those changes are rather favourable for attracting international talent. Relatively relaxed conditions and abroad offer of international study programmes invite large numbers of students from third countries not only to study in Germany, but to stay on to work there after graduation. Especially in the engineering disciplines, Germany benefits from its reputation for technical subjects. However, in order to attract international talent for the German labour market, the legal conditions for working in the country as a foreigner after graduating from a German university could be further simplified. With regard to international students in engineering disciplines, more diversity regarding gender and national background could be pursued. While there is a beneficial international reputation, the higher education institutions in Germany hold the key position in the global competition for talent. It is their task to secure and further promote their reputation by assuring the quality of its educational offers to students from all over the world [21, 22]. For future investigations, it would be interesting to consider in greater detail the student's motives for academic migration. Further analyses might be directed at intentions of working in the host country after graduation and define whether such objectives affect the selection of destination countries for studying abroad.

References

1. Bhandari, R., Belyavina, R., Gutierrez, R. (2011), Student mobility and the internationalization of higher education, Institute of International Education, New York.
2. BAMF – Bundesamt für Migration und Flüchtlinge (2011), Hoch qualifizierte Migrantinnen und Migranten. Deckung des Fachkräftebedarfs durch Zuwanderung, Integrationsaspekte und Kosten der Nichtintegration, Nürnberg.
3. Hays AG, Institut für Beschäftigung und Employability (IBE) (2008), Internationale Rekrutierung - Realität oder Rhetorik? Available online: http://www.hays.de/mediastore/pressebereich/Studien/pdf/Hays_Rekrutierungs_Studie_2008.pdf?nid=41e4cce6-e533-4aac-b763-54bab0a00db9
4. Sachverständigenrat deutscher Stiftungen für Migration und Integration (2012), Migrationsland 2011. Jahresgutachten 2011 mit Migrationsbarometer. Available online: http://oezoguz.de/wp-content/uploads/2011/04/Integrationsbarometer_2011_des_SVR.pdf
5. Angenendt, S., Parkes, R. (2010), Wanderer, kommst du nach Europa? Strategien zur Anwerbung Hochqualifizierter in der EU, Internationale Politik (IP), Stiftung Wissenschaft und Politik, pp. 74–78.
6. Chaloff, J., Lemaitre, G. (2009), Managing highly-skilled labour migration: A comparative analysis of migration policies and challenges in OECD countries, OECD Social, Employment and Migration Working Papers (79).
7. DAAD, Hochschulrektorenkonferenz (HRK) (2013), Internationalität an deutschen Hochschulen. Vierte Erhebung von Profildaten 2013.
8. OECD (2013): Zuwanderung ausländischer Arbeitskräfte, Paris.
9. Nationales MINT Forum (Hg.) (2014), Empfehlungen zur Internationalisierung des Studiums in den MINT-Fächern, Herbert Utz Verlag, München (Empfehlungendes Nationalen MINT-Forums, 2).
10. BAMF – Bundesamt für Migration und Flüchtlinge (2008), Migrationsbericht 2007. Nürnberg.
11. Welte, H.-P. (2008), Arbeitsmigration und Studium von Ausländern. Praxishandbuch zum Zuwanderungsrecht mit Aktionsprogramm zur Sicherung der Fachkräftebasis, Walhalla-Fachverlag, Regensburg.
12. Sachverständigenrat deutscher Stiftungen für Migration und Integration (SVR) (2012), Mobile Talente? Ein Vergleich der Bleibeabsichten internationaler Studierender in fünf Staaten der Europäischen Union. Available online: http://www.svr-migration.de/wp-content/uploads/2012/04/Studie_SVRFB_Mobile_Talente.pdf
13. Hanganu, E. (2015), Bleibequoten von internationalen Studierenden im Zielstaaten-Vergleich. Available online: http://www.bamf.de/SharedDocs/Anlagen/DE/Downloads/Infothek/Forschung/Studien/artikel-auswertung-zuabsolventenstudiefb23.pdf?__blob=publicationFile
14. Schnitzer, K. (1999), Wirtschaftliche und soziale Lage der ausländischen Studierenden in Deutschland (1999). Ergebnisse der 15. Sozialerhebung des Deutschen Studentenwerks (DSW) durchgeführt durch HIS Hochschul Informations-System. BMBF, HIS, DSW. Bonn.

15. Kultusministerkonferenz (KMK) (1997), Stärkung der internationalen Wettbewerbsfähigkeit des Studienstandortes Deutschland. Bericht der Kultusministerkonferenz an die Ministerpräsidentenkonferenz zu den Umsetzungsmaßnahmen, Available online: http://www.kmk.org/fileadmin/veroeffentlichungen_beschluesse/1997/1997_10_24-Staerkung-Wettbewerb-Studienstandort-Deutschl.pdf
16. DAAD (2015), online database of international study programmes in Germany: https://www.daad.de/deutschland/studienangebote/international-programs/en/
17. Rotter, C. (2005), Internationalisierung von Studiengängen. Typen, Praxis, Empirische Befunde. Available online: http://www-brs.ub.ruhrunibochum.de/netahtml/HSS/Diss/RotterCarolin/
18. Stemmer, P. (2013), Studien- und Lebenssituation ausländischer Studierender an deutschen Hochschulen. Analyse - Handlungsfelder - strategische Entscheidungsmöglichkeiten, *Studien zum sozialen Dasein der Person*, Vol. 14, Nomos, Baden-Baden.
19. DAAD; Deutsches Zentrum für Hochschul- und Wissenschaftsforschung (DZHW) (2014), Wissenschaft Weltoffen 2014. Daten und Fakten zur Internationalität von Studium und Forschung in Deutschland, Bielefeld.
20. Apolinarski, B., Poskowsky, J., Kandulla, M., Naumann, H. (2013), Ausländische Studierende in Deutschland 2012. Ergebnisse der 20. Sozialerhebung des Deutschen Studentenwerks durchgeführt vom Deutschen Zentrum für Hochschul- und Wissenschaftsforschung (DZHW), Berlin.
21. Macready, C., Tucker, C. (2011), Who goes where and why? An Overview and Analysis of Global Educational Mobility, Institute of International Education, New York.
22. Heidel, U. (2006), Hochschulmarketing, DAAD, *Die internationale Hochschule*, Bielefeld.

Part III
Student Lifecycle

Access to UML Diagrams with the HUTN

Helmut Vieritz, Daniel Schilberg and Sabina Jeschke

Abstract Modern software development includes the usage of UML for (model-driven) analysis and design, customer communication etc. Since UML is a graphical notation, alternative forms of representation are needed to avoid barriers for developers and other users with low vision. Here, Human-usable Textual Notation (HUTN) is tested and evaluated in a user interface modeling concept to provide accessible model-driven software design.

Keywords Unified Modeling Language (UML) · Human-usable Textual Notation (HUTN) · Accessibility · Modeling

1 Introduction

Software development is an interesting work for visually impaired programmers and developers. It is mostly text reading and writing based on programming languages as Java or C# and therefore accessible for Screenreaders, Braille devices and other assistive technology. However, writing and testing source code is only one activity among others in the development process. Especially in professional environments, analysis and modeling gain more and more importance with the growth of the project. Beside programming languages, other formal notations are used for the communication between software architect, developer, customer and user. In the last decade, the Unified Modeling Language (UML) [1] became the Lingua Franka in design and modeling of software artifacts. It facilitates the communication for the involved people with formalized and easy-to-understand diagrams such as the use case diagrams.

H. Vieritz (✉) · D. Schilberg · S. Jeschke
IMA/ZLW & IfU, RWTH Aachen University, Dennewartstr. 27, 52068 Aachen, Germany
e-mail: helmut.vieritz@ima-zlw-ifu.rwth-aachen.de

D. Schilberg
e-mail: daniel.schilberg@ima-zlw-ifu.rwth-aachen.de

S. Jeschke
e-mail: Jeschke.office@ima-zlw-ifu.rwth-aachen.de

Originally published in "ASSETS 2012: The 14th International ACM SIGACCESS Conference on Computers and Accessibility, Colorado, USA",

DOI 10.1007/978-3-319-46916-4_47

Within UML, a graphical notation and a generic metamodel are combined for object-oriented software modeling. Behavior description is supported by use case or activity diagrams etc. Component or class diagrams allow the description of software structures. Especially UML class diagrams are widely supported by development tools as the Eclipse Modeling Framework (EMF), Rational Rose, Magic Draw etc. All these reasons help to understand why UML became the new standard for software design.

But on the other side, the UML graphical notation creates new barriers for some visually impaired people. Diagrams mix up text and graphical information and can be very detailed. Essential information such as object relations is only given by graphical layout. Therefore, without the support of alternatives, the use of UML diagrams is very restricted for software developers with low vision. Thus, a requirement exists to provide accessible presentation alternatives for UML diagrams.

In the next chapters, the state of the art is discussed first. Then, a concept for a text-based notation is used as an alternative for UML diagrams. The results of the evaluation are discussed and finally, a conclusion and outlook is given.

2 Related Research

The TeDUB project and Accessible UML [2] provides access to UML based on the XMI format. At first, diagrams are interpreted and then the user can navigate them with a diagram navigator using a joystick or keyboard navigation. Thus, diagrams need an extra transformation for accessible usage. Direct editing of diagrams is not possible.

Different approaches exist to translate visual graphics into haptic presentation. Typically, they are restricted to very simple diagrams. An interesting approach was elaborated by Kurze [3] when haptic presentation was much more powerful if the mental representation by the user is considered.

3 HUTN-Based UML Notation

3.1 Concept

The presented concept of textual notation for graphic models is part of the INAMOSYS approach in user interface (UI) modeling. INAMOSYS uses UML Activity and class diagrams for the task and presentation modeling of accessible Web applications and product automation systems [4]. Activity diagrams describe navigation; user action and workflow (e.g. see Fig. 1). Class diagrams describe the abstract structure of the UI.

The alternative notation of the task and presentation models aims to have direct access to the design process for users with low vision. Direct access means that

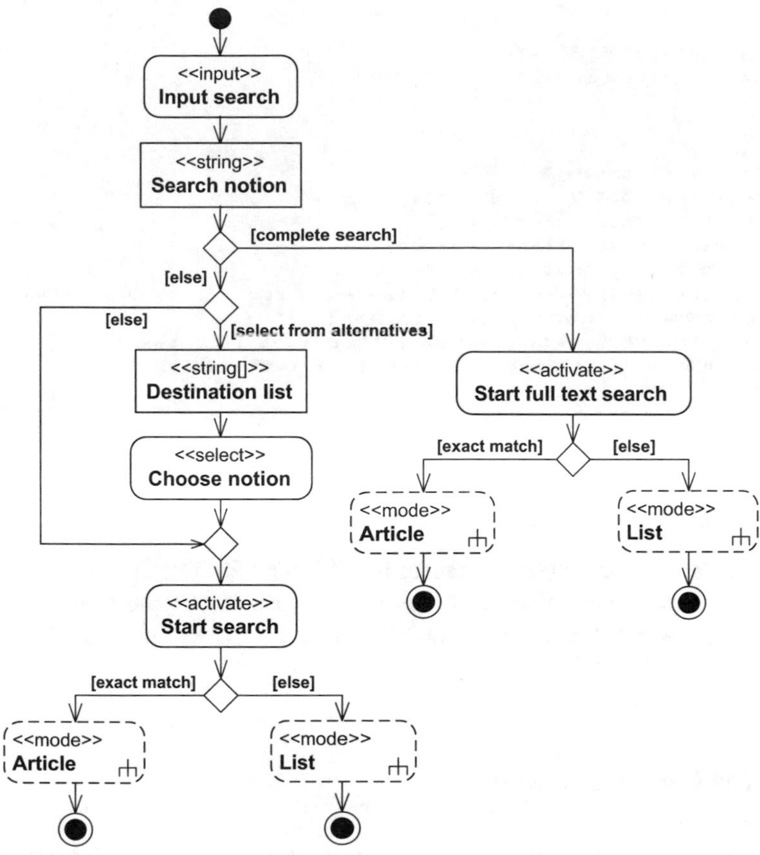

Fig. 1 Example of a UML Activity Diagram

the model presentation and manipulation does not need additional transformations. Therefore, the Human-Usable Textual Notation (HUTN) [5] was used for UML diagram manipulation. As UML and XML Metadata Interchange (XMI), HUTN is based on the same metamodel Meta Object Facility (MOF) [6] and was developed in 2004 by the OMG. It is intended to provide an easy and quick way for changing details in diagrams by developers. HUTN provides short notations for modeling artifacts as objects, relations, attributes or methods. Until now, tool support such as the Epsilon-plugin [7] is very rare and a UML adaption for HUTN is still missing. Therefore, usage of HUTN needs the translation from UML to HUTN with the MOF meta-model.

For our evaluation, HUTN was adapted for the INAMOSYS task and presentation modeling with activity and class diagrams. The models were generated with the EMF tools for the Eclipse platform including Epsilon for the HUTN access. Listing 1 shows a part of the corresponding HUTN presentation for Fig. 1.

```
@Spec {
  metamodel "Task_Model" {
    nsUri: "Task_Model"
  }
}
package {
  Activity "Submode_Search" {
    activity: Activity "Submode_Search"
    Rel1: Workflow "Workflow1" {
      Rel2: input "[Input search]" {
        name: "[Input search]"
      }, Structured Activity Node "<<Search>>[Search notion]" {
        name: "<<Search>>[Search notion]"
      }, activate "<<activate>>[start full text search]" {
        name: "<<activate>>[start full text search]"
      }
      ...
    }
  }
}
```

Listing 1 Part of HUTN-notated Task Model

Model notation is semantic identical in UML and HUTN. The textual notation does not describe graphical layout as the size and position of symbols, boxes, arrows and lines. However, the power of modeling is not affected since layout is a non-semantical aspect of UML modeling.

3.2 Evaluation and Results

Evaluation method was a heuristic test by two modeling experts. The quality and functionality of the diagrams were tested with screen reader (JAWS 10.0) including the following issues:

- User can identify the metadata of the diagram (title, type, author etc.)
- User has orientation and overview during reading the diagram and navigation is possible
- Diagram elements can be identified
- User understands the meaning of the text elements
- User can manipulate the models

Table 1 summarizes some positive and negative results of the evaluation.

Generally, HUTN provides a good access to UML models. Since it has low redundant information, the user must know the syntax properly.

Table 1 Evaluation Summary

+	Direct manipulation of UML diagrams
+	Simple and generic principles for notation
+	Short, well-readable notation
+	Avoidance of redundant information
−	Complex nesting of elements is hard to read
−	Closing of elements may be ambiguous
−	Additional data for automatic model transformation as ID attributes cause more workload for human reading

4 Conclusions and Outlook

In the presented concept, HUTN was used for the text-based presentation of UML activity and class diagrams and evaluated with screen readers. The results have shown that HUTN presentation is much shorter and easier to understand than earlier approaches based e.g. on the XML-conform XMI format. An accessible presentation and manipulation of UML models is possible. Nevertheless, HUTN has some weaknesses regarding the orientation in long documents or the deep nesting of elements. HUTN files are presentable with standard assistive technologies as Braille devices or screen readers. Until now, tool support is very rare and does not corresponding with the advantages for people with low vision. Due to the used tools, the model support is restricted to a subset of MOF. Further work is necessary to provide a complete HUTN-support for UML models. Even, the publicity is not accord with the chances for inclusion of people with low vision.

References

1. O.M.G. (OMG). Unified modeling language (version 2.4.1), 2010. http://www.omg.org/spec/UML/2.4.1/
2. C. Schlieder. Technical drawings understanding for the blind (tebub), 2005. http://www.kinf.wiai.uni-bamberg.de/research/projects/TeBUB/
3. M. Kurze, Methoden zur computergenerierten darstellung räumlicher gegenstände für blinde auf taktilen medien. Ph.D. thesis, FU Berlin, 1999
4. H. Vieritz, User-centered design of accessible web and automation systems. USAB 2011, 2011, pp. 367–378
5. OMG. Human-usable textual notation (hutn), 2004. http://www.omg.org/spec/HUTN/
6. OMG. Meta object facility, 2012. http://www.omg.org/mof/
7. E. Community. Eclipse-epsilon. http://www.eclipse.org/epsilon/

Innovative XXL-Lehre: Das Beispiel „Kommunikation und Organisationsentwicklung“ an der RWTH Aachen

Daniela Janßen, Stefan Schröder and Ingrid Isenhardt

Zusammenfassung In den letzten zehn Jahren haben tief greifende Veränderungen in der Hochschullehre stattgefunden. Unterschiedliche Strömungen, z. B. die Vermittlung überfachlicher Kompetenzen, ein verändertes Mediennutzungsverhalten der heutigen Schulabsolventen und -absolventinnen sowie steigende Studierendenzahlen haben Einfluss auf die Gestaltung der Hochschullehre genommen. Aufgrund der veränderten Umstände ist eine Anpassung der Lehrmethoden und -konzepte erforderlich. Studierende rücken dabei ins Zentrum des Lehr- und Lernprozesses. Besonders für Lehrveranstaltungen mit großen Hörerzahlen ist dies eine besondere Herausforderung. Am Beispiel der Lehrveranstaltung „Kommunikation und Organisationsentwicklung“ des Zentrums für Lern- und Wissensmanagement (ZLW) der RWTH Aachen University wird aufgezeigt, wie die Vorlesung ein volständiges Redesign erfahren hat, um den veränderten Rahmenbedingungen und Anforderungen gerecht zu werden.

Schlüsselwörter Überfachliche Kompetenzen · Engineering Education · Audience Response System · Best-practice

1 Einleitung

In diesem Beitrag wird ein Best-Practice Beispiel vorgestellt, das vor dem Hintergrund verschiedenster Einflussfaktoren im gesamten universitären Lehrprozess innovative Lehrmethoden und -konzepte umsetzt. Studierende rücken dabei ins Zentrum ihres Lernprozesses, insbesondere bei Lehrveranstaltungen mit großen Hör-

D. Janßen (✉) · S. Schröder · I. Isenhardt
IMA/ZLW & IfU, RWTH Aachen University, Dennewartstr. 27, 52068 Aachen, Germany
e-mail: daniela.janssen@ima-zlw-ifu.rwth-aachen.de

Originally published in “TeachING LeranING.EU discussions: Innovationen für die Zukunft der Lehre in den Ingenieurwissenschaften”, © 2013. Reprint by Springer International Publishing AG 2016, DOI 10.1007/978-3-319-46916-4_48

erzahlen. Am Beispiel der Lehrveranstaltung „Kommunikation und Organisationsentwicklung“ des Zentrums für Lern- und Wissensmanagement (ZLW) der RWTH Aachen University wird aufgezeigt, wie die Vorlesung ein vollständiges Redesign erfahren hat, um den veränderten Rahmenbedingungen und Anforderungen aller am Wertschöpfungsprozess „Lehre“ beteiligten Stakeholdern gerecht zu werden.

2 Veränderte Einflussfaktoren im Kontext der Hochschullehre

In den letzten zehn Jahren haben tiefgreifende Veränderungen in der Hochschullehre stattgefunden. Unterschiedliche Strömungen haben Einfluss auf die Gestaltung der Hochschullehre genommen. Zu diesen zählen u.a. der Fokus auf die Vermittlung überfachlicher Kompetenzen, das veränderte Mediennutzungsverhalten [1] der heutigen Schulabsolventen und -absolventinnen sowie die Vielzahl von Studierenden an deutschen Hochschulen. Diese wichtigsten Einflussfaktoren werden im Folgenden näher erläutert.

2.1 Überfachliche Kompetenzen

Die Vermittlung von überfachlichen Kompetenzen spielt in der Hochschullehre im Allgemeinen und in den Ingenieurstudiengängen im Speziellen eine zentrale Rolle. Nicht nur Fachwissen wird heutzutage von Ingenieuren und Ingenieurinnen erwartet – dieses wird weitgehend als selbstverständlicher Output der Universität vorausgesetzt–, sondern vor allem auch überfachliche Kompetenzen wie Methoden, Selbst-, Organisations- und Sozialkompetenzen [2].

Das ZLW hat bereits seit langem aus seinem interdisziplinären und ganzheitlichen Ansatz heraus bei der Konzeption und Durchführung der Lehrveranstaltungen den Anspruch, Lehre so zu gestalten, dass zusätzlich die Vermittlung überfachlicher Kompetenzen im Fokus steht. Unterstützt bzw. implementiert wird dies im Curriculum durch den Bologna Prozess [3]. Ein Kernziel der Reform ist es, die Beschäftigungsfähigkeit („Employability“) von Hochschulabsolventen und Hochschulabsolventinnen zu fördern [4]. Studierende sollen demnach so ausgebildet werden, dass sie dem Arbeitsmarkt schnell, effizient und adäquat ausgebildet zur Verfügung stehen [5].

Zudem setzt sich der Paradigmenwechsel „shift from teaching to learning“ [6], also die zunehmend studierendenzentrierte und beteiligungsorientierte Gestaltung der Lehre, immer mehr in der Praxis durch: „It is characterized by innovative methods of teaching which aim to promote learning in communication with teachers and other learners and which take students seriously as active participants in their own learning,fostering transferable skills such as problem-solving, critical thinking and reflective thinking“ [7]. Die Gestaltung der Lehre orientiert sich nun vielmehr an „Learning Outcomes“, also an den Lernergebnissen von Studierenden und deren Erreichung als an der „Content-Orientierung“, d.h. der Vermittlung von Inhalten.

2.2 *Digital Natives – der Student von heute?!*

Für eine studierendenorientierte Gestaltung von Lehrveranstaltungen ist es für Lehrende von zentraler Bedeutung, die Zielgruppe – die Studierenden – zu kennen und sie aktiv in die Lehrveranstaltung einzubinden [8]. Die heutige Generation von Jugendlichen wird häufig als „Digital Natives“ [9] bezeichnet. Digital Natives sind „[…] Menschen […], die nach 1980 direkt in das digitale Zeitalter hineingeboren wurden, […]. Sie sind durchweg vernetzt und mit den neuen digitalen Medien und Möglichkeiten bestens vertraut.“ [10]. Hier wird die zentrale Rolle von neuen und sozialen Medien wie beispielsweise Blogs, Foren oder sozialen Netzwerken für die Studierenden deutlich. Neue technologische Entwicklungen begünstigen diesen Trend zusätzlich. Neue Medien wie Audience Response Systeme (ARS) bieten an dieser Stelle sowohl die technologischen als auch die didaktischen Möglichkeiten, um Studierende in Lehrveranstaltungen einzubeziehen und mit ihnen in Interaktion zu treten [11]. Vor diesem Hintergrund ist heutzutage von einem anderen Mediennutzungsverhalten der Studierenden auszugehen. Sie bringen einerseits andere Voraussetzungen mit als Studierende, die vor den 1980er Jahren geboren sind der 1990er Jahre und haben gleichzeitig andere Ansprüche an die Gestaltung der Lehre in Hinblick auf den Einsatz von neuen und sozialen Medien [1].

Der Bedarf nach Einsatz neuer Medien im Kontext der Hochschullehre spiegelt sich im Ideenwettbewerb von TeachING-LearnING.EU (siehe Teaching-Learning-.EU[1]) wider. Studierende der Ingenieurwissenschaften können durch die Einreichung eigener Ideen aktiv an der Verbesserung der Lehre und des Lernprozesses ingenieurwissenschaftlicher Studiengänge mitwirken. Die eingereichten Ideen zeigen die Forderung nach neuen technologischen Möglichkeiten wie sozialen Netzwerken, der Aufzeichnung von Lehrveranstaltungen, der Nutzung von Lehrportalen zur Bereitstellung von Lernmaterialien und Skripten. Die Vermittlung überfachlicher Kompetenzen ist nicht nur ein Ziel von Bologna, sondern wird ebenso in den eingereichten Ideen der Studierenden gewünscht.

2.3 *Steigende Studierendenzahlen*

Der steigende Bedarf an Ingenieuren und Ingenieurinnen [12] und damit einhergehend die steigende Anzahl von Studienanfängern und Studienanfängerinnen stellt ingenieurwissenschaftliche Studiengänge vor die Herausforderung, in Lehrveranstal-

[1] http://www.teaching-learning.eu/aktuelles/ideenwettbewerbe.html.

tungen mit einer großen Zahl an Studierenden umgehen zu müssen bzw. auch zu können. Dieses Problem wird zusätzlich durch die Verkürzung der Schulzeit um ein Jahr im Zuge der Schulreform verschärft. 2013 steht der doppelte Abiturjahrgang an, weswegen die Zahl der Einschreibungen an deutschen Universitäten deutlich steigen wird. Der Wegfall der Wehrpflicht begünstigt diese Entwicklung zusätzlich. Dies hat zur Folge, dass neue Interaktionsmöglichkeiten für die Studierenden im Rahmen „klassischer" Lehrveranstaltungen mit großen Hörerzahlen geschaffen und implementiert werden. Hierzu bedarf es ebenso geeigneter technischer Lösungen, wie entsprechend neu aufbereiteter Inhalte.

Zusammengefasst ergeben sich für das Lernverhalten, aber auch das Lehrkonzept neue Anforderungen sowie Möglichkeiten, die mit steigenden Hörerzahlen, wachsenden (überfachlichen) Anforderungen vonseiten der Wirtschaft und der Vielfalt medientechnischer Lösungen umgehen müssen, wie zum Beispiel: Die Ansprache der Studierenden über verschiedene Kanäle und der angepasste Zugang zum Lehrstoff:

- Interaktive Angebote, die individuelle Lernprozesse, -wege und -geschwindigkeiten zulassen
- Mitgestaltung der Inhalte mit praxis- und projektbasierter Ausrichtung neben reiner Wissensvermittlung

Im Folgenden wird unser Versuch, diesen Bedarfen und Ansprüchen in der Lehrveranstaltung „Kommunikation und Organisationsentwicklung" durch Neukonzipierung gerecht zu werden, vorgestellt.

3 Redesign der Vorlesung „Kommunikation und Organisationsentwicklung"

3.1 Umsetzung des neuen Lehrkonzepts

Die Erstsemester-Pflichtveranstaltung „Kommunikation und Organisationsentwicklung" (KOE) wird unter Verantwortung der promovierten Soziologin und habilitierten Maschinenbauerin Ingrid Isenhardt organisiert und durchgeführt. Insgesamt besuchen ca. 1.200 Studierende, größtenteils des Maschinenbaus, die Veranstaltung. Neben der wöchentlich stattfindenden Vorlesung findet zusätzlich eine Laborübung statt, die den Studierenden die Simulation eines Unternehmens und dadurch die direkte Erprobung in einer möglichst authentischen Umgebung ermöglicht. Unter den eingangs dargelegten veränderten Einflussfaktoren und der Prämisse der Lernerfolgsmaximierung sowie der nachhaltigen Vermittlung von praxisrelevanten und überfachlichen Inhalten wurde die Vorlesung kontinuierlich weiterentwickelt und im Wintersemester (WS) 2012/2013 einem kompletten Redesign unterzogen. Dabei wurden neue lern- und lehrdidaktische Elemente implementiert bzw. bestehende weiter ausgebaut. Die Veranstaltungselemente sind in Abb. 1 visualisiert und werden folgend erläutert (im WS 2012/2013 neu eingesetzte Elemente sind mit einem * gekennzeichnet).

3.2 *Theorieinput & Vorlesungsmodule*

Insgesamt besteht die Vorlesung aus zwölf Modulen, von den Grundlagen der Kommunikation- und Organisationsentwicklung, Lern- und Wissensmanagementkonzepten und interkulturellen Aspekten weltweiter Arbeitsteilung über Management- sowie systemtheoretische Ansätze bis hin zu Praxisbeiträgen verschiedener Expertinnen und Experten aus Wirtschaft und Wissenschaft (siehe Expertenvorträge in Abb. 1). Das gesamte Lehrkonzept folgt dabei einer streng linearen Struktur (siehe Abb. 2). So knüpfen beispielsweise die Praxisvorträge an bereits vermittelte Theorieinhalte an, greifen diese auf und schaffen so den idealtypischen Transfer von Theorie und Praxis.

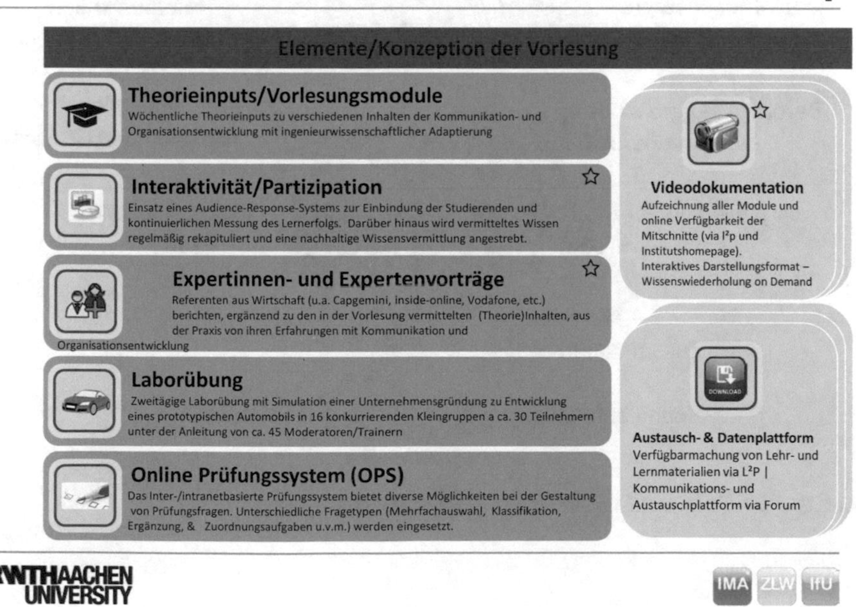

Abb. 1 Struktur der Vorlesung „Kommunikation und Organisationsentwicklung" der RWTH Aachen University [Eigene Darstellung]

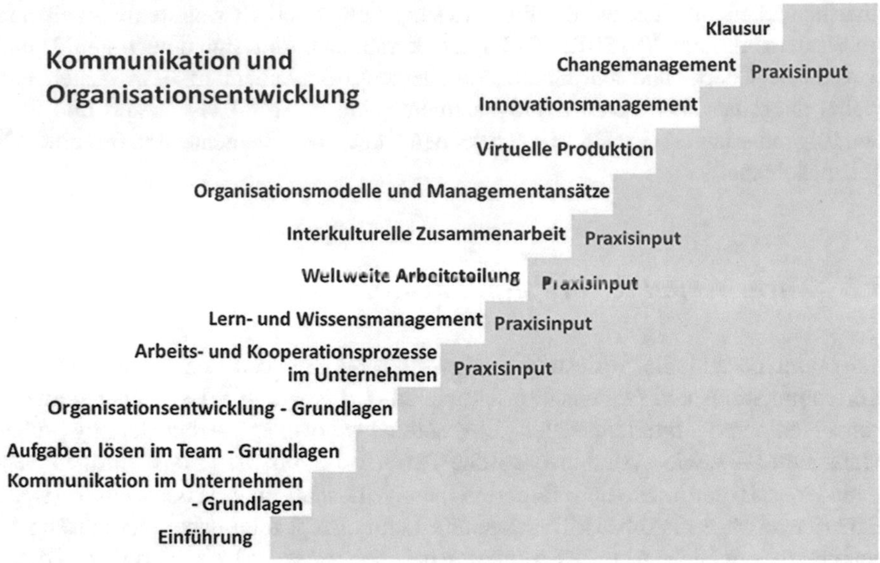

Kommunikation und Organisationsentwicklung

Klausur
Changemanagement Praxisinput
Innovationsmanagement
Virtuelle Produktion
Organisationsmodelle und Managementansätze
Interkulturelle Zusammenarbeit Praxisinput
Weltweite Arbeitsteilung Praxisinput
Lern- und Wissensmanagement Praxisinput
Arbeits- und Kooperationsprozesse im Unternehmen Praxisinput
Organisationsentwicklung - Grundlagen
Aufgaben lösen im Team - Grundlagen
Kommunikation im Unternehmen - Grundlagen
Einführung

Abb. 2 KOE Vorlesungstreppe [Eigene Darstellung]

Wissen- und Innovationsmanagement

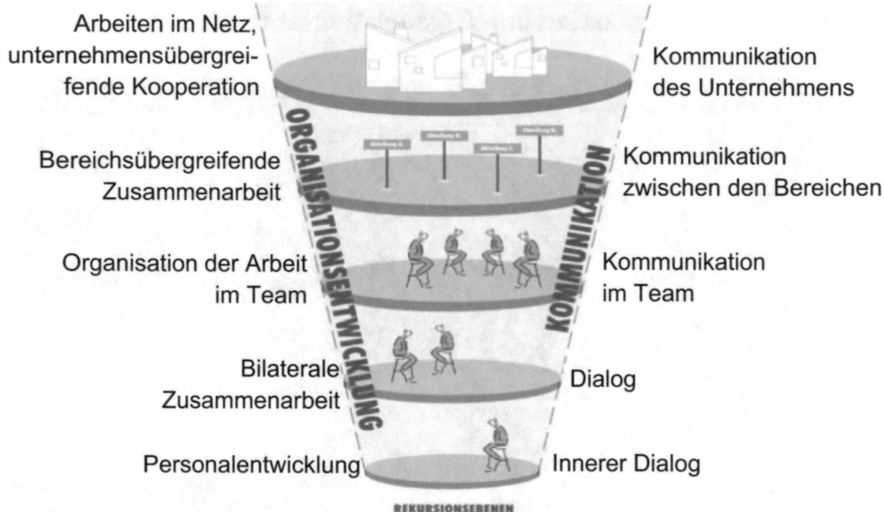

Abb. 3 KOE Trichter/Rekursionsebenen [Eigene Darstellung nach 13]

Am Ende eines jeden Moduls werden die Vorlesungsinhalte auf den modulrelevanten Rekursionsebenen reflektiert und resümiert. Hierzu wird der KOE-Trichter eingesetzt (siehe Abb. 3), der auf sowohl die systemische Sichtweise (auf der linken Seite des Trichters) als auch die unterschiedlichen Unternehmensebenen aufgreift (auf der rechten Seite des Trichters). Dabei werden Kommunikation und Organisationsentwicklung als entscheidende Entwicklungsstränge der Gestaltung der Interaktion von Mensch, Organisation und Technik auf unterschiedlichen organisationalen Ebenen (z. B. innerhalb von Abteilungen oder in Projektteams) reflektiert.

3.3 Interaktivität durch Einsatz eines „Audience Response System"

Um den Bedarfen der Studierenden, beispielsweise hinsichtlich interessensgerechter Informationsvermittlung gerecht zu werden, wird seit dem WS 2012/2013 ein Audience Response System (ARS) eingesetzt (Online Abstimmungstool zum Einbeziehen des Auditoriums). Den Studierenden wird dadurch mehrfach (ca. 2-3 Mal) ermöglicht, aktiv in die Vorlesung einzugreifen und z. B. (Teil-)Inhalte auszuwählen, Wissen gezielt zu rekapitulieren, zu interpretieren und zu reflektieren oder sich auf mögliche Klausurfragen vorzubereiten (siehe Abb. 4). Dabei werden die Taxonomien von Lernzielen, wie z. B. wissen, verstehen, anwenden und analysieren [13, 14], durch beispielsweise Wissens- und Verständnisfragen, sowie Anwendungsbeispiele nahezu ganzheitlich adressiert, was zur Folge hat, dass Problemlösungskompetenzen der Studierenden gefördert werden. Die Teilnahmebarrieren für die Studierenden sind denkbar gering. Benötigt werden ein internetfähiges Endgerät (Smartphone, Laptop) und eine funktionierende Internetverbindung (z. B. eduroam).

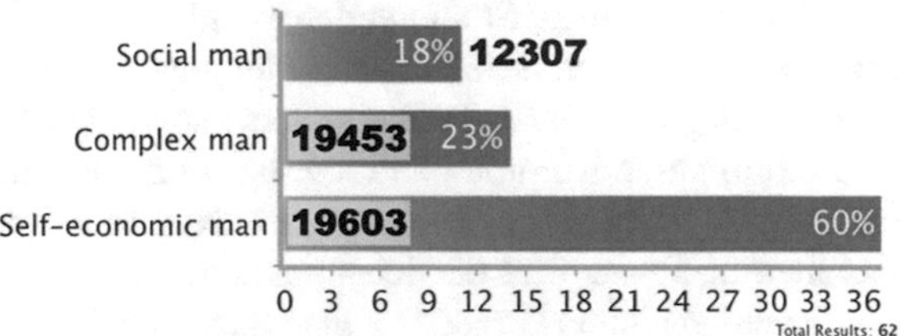

Abb. 4 Exemplarische ARS Darstellung [Eigene Darstellung | polleverywhere.com]

3.4 Expertinnen- und Expertenvorträge

Wie eingangs bereits erwähnt, strebt das Redesign der Vorlesung einen erhöhten Relevanz- und Praxisbezug für die Studierenden an und erhöht gleichzeitig die Fähigkeit, das Arbeits- und Berufsleben in seiner Komplexität zu verstehen. So geben mehrere renommierte Gastredner aus der Industrie (darunter u.a. Vertreter der Unternehmen Vodafone Group, Inside Unternehmensgruppe, Capgemini und der p3 Ingenieurgesellschaft) Einblicke in ihre Unternehmen und Berufserfahrungen. Inhaltlich thematisieren die Experten beispielsweise ihre Erfahrungen im Rahmen internationaler Projektarbeit (Thema weltweite Arbeitsteilung & kulturelle Diversität bei Offshore Projekten mit/in Indien) und interdisziplinärer/fachkultureller Herausforderungen (Thema interkulturelle Zusammenarbeit am Beispiel des Projektes Airbus). Dabei schlagen diese gleichsam die Brücke zwischen bereits vermittelten theoretischen Inhalten und der Relevanz von Kommunikation und Organisationsentwicklung im täglichen Arbeitsprozess. Darüber hinaus vermitteln die Experten wichtige essenziell-notwendige überfachliche Kompetenzen, die weit über die rein fachliche Qualifikation der ingenieurwissenschaftlichen Studierenden hinausgehen, beispielsweise Sozial-, Selbst- und Methodenkompetenz. Um dem Spannungsfeld zwischen der Vermittlung von wissenschaftlichen Grundlagen und praxisbezogener Lehrform gerecht zu werden [15], wurde auf eine adäquate Inhalts- und Grundlagenverteilung geachtet (nicht zulasten unzureichender wissenschaftlicher Grundlagen).

3.5 Laborübung

Beim Thema Kommunikation und Organisationsentwicklung versteht es sich von selbst, dass nicht alle Vorlesungsinhalte und Lernziele zur Wissensvermittlung in Großvorlesungen geeignet sind bzw. nicht internalisiert werden können. Daher findet vorlesungsbegleitend einmal im Semester an 1,5 Tagen zusätzlich eine Laborübung statt, in der es den Studierenden möglich ist, zuvor theoretisch vermittelte Grundlagen praktisch in Form einer Unternehmenssimulation (simulation-based-learning) anzuwenden und sich in Kommunikations- und Arbeitsprozessen zu erproben. In Gruppen von bis zu 30 Studierenden gründen sie ein fiktives Unternehmen der Automobilbranche mit verschiedenen Abteilungen (u.a. Marketing/Vertrieb, Konstruktion & Design), entwickeln Zielsysteme sowie Unternehmensstrategien, definieren und koordinieren Kommunikationswege und konstruieren ein innovatives Automobil unter der Anleitung von 40 erfahrenen und aufwendig geschulten Coachs. Das Arbeiten in Teams und die Kommunikation zwischen den Abteilungen zeigt den Studierenden binnen kürzester Zeit authentisch die Relevanz der Vorlesung (bzw. der vermittelten Inhalte) sowie den Bezug zur Arbeitswelt und schafft gleichzeitig die Sensibilisierung für die Notwendigkeit strukturierter Kommunikation und Organisationsprozesse.

3.6 *Online Prüfungssystem (OPS)*

Um die Prüfung von über 1.000 Studierenden ressourceneffizient und gleichzeitig lernzielorientiert zu gestalten, erfolgt die Abnahme des Prüfungsnachweises (insgesamt ca. sechs Kohorten à 120 Minuten) über ein digitales Online Prüfungssystem. Die von den Lehrinhalten abgeleiteten Klausurfragen werden in unterschiedlichen Taxonomiestufen (Schwierigkeitsstufen) geclustert und über ein randomisiertes Verfahren den Studierenden zugeteilt [13]. Dabei werden gleichermaßen Wissen abgefragt, als auch Transferleistung eingefordert. Die Aus- und Bewertung der Prüfungen erfolgt im Anschluss automatisiert und elektronisch.

3.7 *(Interaktive) Videodokumentation*

Wie bereits die Ergebnisse des Ideenwettbewerbs (siehe Teachling-Learning.EU, www.teaching-learning.eu) gezeigt haben, besteht von Studierendenseite ein großer Bedarf an digital aufbereiteten Lehrinhalten. Aus diesem Grund und durch das Redesign ermöglicht, wurde im WS 2012/2013 die komplette Vorlesung neu aufgezeichnet und sukzessive den Studierenden über das Lehr- und Lernportal der RWTH Aachen (L^2P) und den Institutsserver zur Verfügung gestellt.

Wie Abb. 5 zeigt, handelt es sich dabei um eine interaktive Videodokumentation. Den Lernenden wird durch die Aufbereitung der Inhalte eine gezielte Wissensrekapitulierung ermöglicht. Außerdem werden sowohl die Folien inkl. Freihandzeichnungen und Animationen als auch ein Mitschnitt der Ausführungen der Dozentin bzw. Expertinnen und Experten dargestellt. Gleichsam wird so der Grundgedanke des „Corporate Learnings" aufgegriffen. Der bzw. die Lehrende fungiert als Vermittler oder Vermittlerin und stellt Inhalte über unterschiedliche Kanäle zur Verfügung[16].

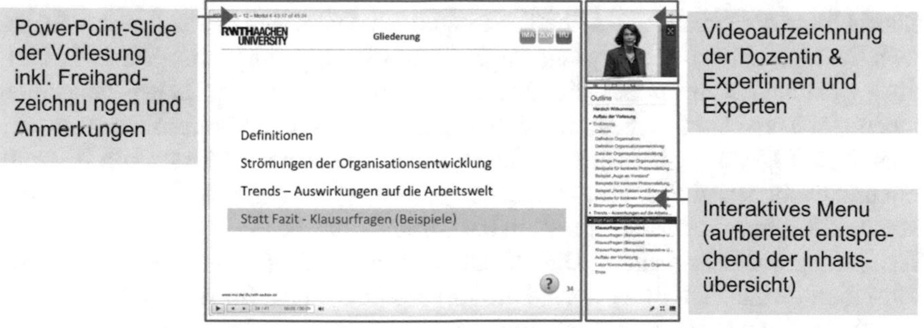

Abb. 5 Oberfläche Videomitschnitt der Vorlesung [Eigene Darstellung]

Den Lernenden wird so ermöglicht (on Demand), für sie relevantes Wissen zu wiederholen und zu verstetigen sowie sich gezielt auf die lehrveranstaltungsabschließende Prüfung (Klausur) vorzubereiten (in Kürze stellen wir auf unserer Institutshomepage http://www.ima-zlw-ifu.rwth-aachen.de einen ausgewählten Vorlesungsvideomitschnitt exemplarisch für Interessierte zur Verfügung).

3.8 Austausch- und Datenplattform

Als Reaktion auf die veränderten Einflussfaktoren (z. B. neues Nutzungsverhalten) werden den Studierenden bereits seit dem Jahr 2007 Inhalte (Präsentationen, weiterführende Literatur, Informationen zu Terminen etc.) über das Lehr- und Lernportal der RWTH Aachen „L^2P" digital zur Verfügung gestellt. Darüber hinaus haben diese die Möglichkeit, sich in einem Diskussionsforum untereinander und mit den Lehrenden sowie Betreuerinnen und Betreuern auszutauschen. Diese mediale, den Lern- und Lehrprozess unterstützende Plattform ist passwortgeschützt und ausschließlich den für die Veranstaltung angemeldeten Studierenden vorbehalten. Andere können Vorlesungsinformationen, Impressionen sowie Ansprechpartner und -partnerinnen über die Institutshomepage einsehen.

4 Ausblick

Im Zuge des Redesigns der Vorlesung „Kommunikation und Organisationsentwicklung" wurden erstmalig eLearning Elemente wie die Nutzung des Lehr- und Lernportals (L^2P) zur Bereitstellung von Lernmaterialien sowie die aufbereiteten Videomitschnitte der Vorlesung integriert. Damit wurde die Vorlesung entsprechend der Bedarfe seitens der Studierenden und der zuvor dargestellten veränderten Einflussfaktoren didaktisch und methodisch angepasst. Die ersten Evaluationsergebnisse nach dem Redesign der Vorlesung zeigen, dass die Neuerungen im WS 2012/2013 einen positiven Einfluss auf die studentische Beurteilung der Lehrveranstaltung haben. Im Durschnitt verbesserte sich die Bewertung der Veranstaltung um eine halbe Note (0,5 | artih. Mittelwert). Beispielsweise wurde der Einsatz von Hilfsmitteln und Demonstrationen im Rahmen der Vorlesung mit der Note 1,7 (arith. Mittelwert, n=382) bewertet (Steigerung um 0,4). Ebenso wurde ein positiver Trend in der Vermittlung der Lehrinhalte („Die Dozentin kann den Stoff verständlich erklären") konstatiert (arith. Mittelwert von 2,4 im Vorsemester auf 2,1 im WS 2012/2013). Als Gründe hierfür werden die neu implementierten Elemente, wie der Einsatz des ARS-Systems, der digital aufbereitete Videovorlesungsmitschnitt sowie die Praxisbeiträge genannt.

Für eine kontinuierliche und bedarfsgerechte Verbesserung der Lehre besteht weiterhin Optimierungspotenzial der KOE-Vorlesung. Neben Überlegungen, die bislang im ersten Semester stattfindende Lehrveranstaltung zu einem späteren Zeitpunkt im

Curriculum (ggf. ab dem vierten Semester) zu positionieren und dadurch eine erhöhte thematische Sensibilisierung der Studierenden zu erreichen, ist ein langfristiges Ziel, die Vorlesung nach der Methode des „just-in-time teaching“ [17] zu konzipieren und (Inhalte) aufzubereiten bzw. anzupassen. Danach werden Präsenzveranstaltungen nicht mehr primär dazu genutzt, Lehrinhalte zu vermitteln, sondern vielmehr wird angestrebt, die Studierenden in der Vorlesung zur Interaktion zu motivieren sowie Fragen zum und Probleme mit dem Lehrstoff zu diskutieren. Vor jeder Präsenzveranstaltung werden den Studierenden hierzu Aufgaben und Fragen online zur Verfügung gestellt. Lehrende können so die Ergebnisse der Studierenden vor der nächsten Vorlesung „just in time“ sehen und entsprechend ihre Vorlesung aufbauen. Mit dieser Methode erhalten Lehrende zum einen Feedback über den Wissens- und Lernstand der Studierenden, womit sie auf die Bedarfe der Studierenden eingehen können. Zum anderen werden neue, studierendenaffine Medien eingesetzt. Des Weiteren erlernen Studierende überfachliche Kompetenzen, indem sie eigenständig Fragen formulieren sowie Aufgaben selbstorganisiert bearbeiten müssen. Die stetige Weiterentwicklung der Lehrveranstaltung „Kommunikation und Organisationsentwicklung“ dient primär dem Ziel, für große Hörerzahlen ein noch bedarfs- und zielgruppenorientierteres Lehrangebot zu schaffen.

Literaturverzeichnis

1. M. Grosch, G. Gidion, Mediennutzungsgewohnheiten im Wandel: Ergebnisse einer Befragung zur studiumsbezogenen Mediennutzung. KIT, 2011
2. F. Pankow, *Die Studienreform zum Erfolg machen! Erwartungen der Wirtschaft an Hochschulabsolventen*. Deutscher Industrie- und Handelskammertag, Berlin, 2008
3. K. Schuster, U. Bach, A. Richert, S. Jeschke, Openbologna – a strategic instrument for integrating students in curriculum developement. engineering education **6** (2), 2011, pp. 47–56
4. The bologna declaration, 1999
5. T. Jungmann, K. Müller, K. Schuster, Shift from teaching to learning. Anforderungen an die Ingenieurausbildung in Deutschland. Journal Hochschuldidaktik (02), 2010, pp. 6–8
6. J. Wildt, The shift from teaching to learning – Thesen zum Wandel der Lernkultur in modularisierten Studienstrukturen. Fraktion Bündnis 90/Die Grünen im Landtag NRW, Unterwegs zu einem europäischen Bildungssystem, 2003, pp. 14–18
7. Angele Attard, Emma Di Iorio, Koen Geven, Robert Santa. Student-centred learning – toolkit for students, staff and higher education institutions
8. M. Petermann, I. Isenhardt, E. Tekkaya. Learning by doing – Wie steigern wir den Praxisbezug im Ingenieurstudium?, vortrag 6, 2012
9. R. Schulmeister, Gibt es eine Netgeneration?, 2009
10. J.G. Palfrey, U. Gasser, *Born digital: Understanding the first generation of digital natives*. Basic Books, New York, 2008
11. V. Stehling, U. Bach, A. Richert, S. Jeschke, Teaching professional knowledge to xl-classes with the help of digital technologies. In: *ProPEL Conference Proceedings*, vol. 2012. 2012, vol. 2012
12. Gute Jobaussichten für Ingenieure trotz wirtschaftlicher Eintrübung | Pressemitteilung VDI Verein deutscher Ingenieure, 11.12.2013, 2013. http://www.presseportal.de/pm/16368/2362459/gute-jobaussichten-fuer-ingenieure-trotz-wirtschaftlicher-eintruebung-hoher-bedarf-an-ingenieuren

13. L.W. Anderson, D.R. Krathwohl, *A taxonomy for learning, teaching, and assessing: A revision of Bloom's taxonomy of educational objectives*. Longman, New York, 2001
14. B.S. Bloom, *Taxonomie von Lernzielen im kognitiven Bereich*. Beltz, Weinheim, 1972
15. Wegner, Elisabeth Julia Erika, M. Nückles, Die Wirkung hochschuldidaktischer Weiterbildung auf den Umgang mit widersprüchlichen Handlungsanforderungen. Zeitschrift für Hochschulentwicklung **6** (3), 2011, pp. 171–188
16. S. Mader. On-demand-lernen: Komprimiertes Wissen, hier und jetzt, 2013. http://blog.getabstract.com/on-demand-lernen-komprimiertes-wissen-hier-und-jetzt/?lang=de
17. E. Mazur, J. Watkins, Just-in-time teaching and peer instruction http://mazur.harvard.edu/sentFiles/Mazur_263828.pdf

Simulation-Based Learning for Conveying Soft-Skills to XL-Classes

Daniela Janßen, Sarah Valter, Esther Borowski, René Vossen and Sabina Jeschke

Abstract Soft skills have become more important in higher education in order to prepare students for employability in later career. In XL-Classes, the theoretical conveyance of soft skills to students presents a special challenge. One approach for the application of theoretically imparted knowledge in practice is the concept of simulation-based learning. Simulations have been used for a long time in a variety of disciplines, particularly in high-risk areas such as medicine, aviation and space industries, using virtual environments to prepare professionals for real life situations. The term 'simulation-based learning' is particularly used in medical education. Approaches of simulation-based learning are increasingly used by other disciplines in the context of higher education and the education of students. Based on a definition of the term 'simulation-based learning' a concept to convey soft skills in higher education courses is developed. A practical implementation of the concept is demonstrated in the paper by using it in the XL-Class "Communication and Organizational Development" for students in the bachelor programme Mechanical Engineering at RWTH Aachen University. Here, the foundation of an enterprise in the automotive industry is simulated within 1.5 days. Key skills such as team building, time management and project management are applied, experienced and trained in the simulation. Overall, 600 students pass an organizational development process in which they establish a fictional automotive company with various departments, develop target systems as well as business strategies and construct an innovative car prototype. The basic knowledge for the realization of this task is mediated via microteaching units. Therefore the developed concept transfers soft skills knowledge to students by experiencing and training them in a simulated environment.

Keywords Simulation-based Learning · XL-Class · Higher Education · Soft Skills

D. Janßen (✉) · S. Valter · E. Borowski · R. Vossen · S. Jeschke
IMA/ZLW & IfU, RWTH Aachen University, Dennewartstr. 27, 52068 Aachen, Germany
e-mail: daniela.janssen@ima-zlw-ifu.rwth-aachen.de

Originally published in ?International Conference on Education and New Developments 2013 - Book of Proceedings?, © 2015.
Reprint by Springer International Publishing AG 2016,
DOI 10.1007/978-3-319-46916-4_49

1 Introduction

In order to promote students' employability in later career, soft skills have become more important in higher education. The transfer of soft skills especially within engineering degree programs is indispensable [9]. Professional and certified knowledge is not the only required output of higher education, soft skills like method-, self-, organizational- and social competences are also expected from graduating students nowadays [6]. As a consequence, the higher education system has to face this challenge. The already emerging change in the design of teaching focuses on 'learning outcomes', the students' learning results and the way of achieving these results. The more student-centered and involvement-oriented design of teaching, the so called 'shift from teaching to learning' [7], is preferred over 'content-oriented' approaches. This is a major challenge for large-audience classes, the so-called 'XL-Classes', especially when soft skills have to be learned and applied in practice. In this case, classical teaching no longer fulfills the learning outcomes, as the application of theoretically imparted knowledge falls short to convey any practical relevance for future professional life. One approach for the application of theoretically imparted knowledge in practice is the concept of simulation-based learning (SBL). Simulations in general have been used for a long time in a variety of disciplines, particularly in high-risk areas such as aviation, space industries and medicine, using virtual environments to prepare professionals for real life situations [1]. Their application in medical education is described below resulting in a particular definition of the term 'simulation-based learning' and in the development of a concept for university education, using the example of a laboratory training, called 'KOE-Labor', at the Center for Learning and Knowledge Management (ZLW) at RWTH Aachen University.

2 A Definition of Simulation-Based Learning

The term 'simulation-based learning' is particularly used in the context of medical education and can be outlined in its use in this area. The transfer of practical skills, as the most common learning objective [5], has become an important part of medical education. SBL emerges as an appropriate teaching approach for practical and soft skills training in medical education, since research in medical education shows that the development of communication skills should not be separated from the curriculum, as it is only effective in combination with clinical knowledge (Barlow et al., 1999; Benbassat & Baumal 2002 quoted from: [3]). SBL offers a practical and helpful tool for deepening medical students' knowledge, skills and attitudes, while protecting patients from incorrect diagnosis or treatment [8]. It also addresses the communication competency in doctor-patient-interaction, which is a crucial soft skill for a clear imparting of diagnosis to patients and relatives.

Abdulmohsen [1] defines simulation as a generic term that refers to an artificial representation of a real world process to achieve educational goals through experimental learning: "Experiential learning …is an active learning process during which the learner constructs knowledge by linking new information and new experiences with previous knowledge and understanding" ([1]: 36). Simulations in general are considered as situations, in which a certain set of conditions is artificially generated in order to create real-life experiences [1]. Concluding "simulation based medical education can be defined as any educational activity that utilizes simulation aides to replicate clinical scenarios" ([1]: 36). Regarding the implementation of SBL in university didactic, the following definition is derived: SBL is a didactic approach for the active training of practical skills in a constructed situation. The active learning process creates experience-based learning outcomes, which prepare the learners specifically for their future performance in real situations in a professional context.

3 A Concept of Simulation-Based Learning in Higher Education

The outlined approaches of SBL in medical education can be transferred to other disciplines in higher education. Based on the definition of SBL stated above, a didactical concept for the transfer of soft skills in higher education is derived.

The didactical concept is developed for its use in an environment close to reality with face to face-situations. Initial point of the simulation is a complex problem or an overall task, which is structured in separate stages and embedded in a realistic background story of a professional context. A class of students, which is divided in subgroups, experiences a realistic situation in time lapse. The students act as agents in the simulation and are challenged to solve problems in their subgroup in order to prevail against the competition. To solve tasks, students must activate and apply theoretical knowledge from the lecture. The separate stages should be consecutively structured, so that students have to repetitively apply skills and competences acquired in the simulation to solve the problem successfully. Therefore a continuous learning process is initiated that leads to an even higher learning success.

The different simulation stages and problem elements stimulate communication and team processes, which promote soft skills like communication-, teamwork- and problem-solving-skills. The features professional context, team training, skill acquisition, deliberate practice and feedback, identified by McGaghie et al. [5] are the key features of simulation based learning. As an element of simulation, the team development process makes an important contribution towards the 'learning outcomes'. The initial divided subgroups meet each other during different stages of the simulation. Through the extension of the groups at different stages, the whole class of students is united again at the end of the simulation. One key element of the team development process is the definition of each role within the team, which should be made transparent to all participants. In order to create turbulences within teams, agents are purposely substituted because by mixing up students to new and different teams, new situations evolve and the ability to work in a team is fostered. Like in the real world, the subgroups receive predefined and standardized instructions with a limited number of information and a limited time frame to carry out their task. Strategies for problem solving should be independently developed by the subgroups. Each task stage should be followed by a period of reflecting, modeled after Kolb's Learning Cycle [4]. Therefore debriefing after a task as well as feedback, where the groups reflect upon their group performance in the last stage, plays a crucial role in the simulation [5].

Overall, coaches play a central role in SBL. They assess the progress, diagnose problems, provide feedback and evaluate overall results. For a successful implementation of simulations, coaches need to undergo intensive training. To experience the learning process at firsthand, coaches receive a detailed speeded up training on learning contents and objectives in a separate training for them.

Simulation-based education provides opportunities to train soft skills in a risk-free environment by giving and receiving direct feedback and learning from mistakes [5]. Traditional teaching, such as frontal lectures, should not be replaced, but rather complemented by SBL in order to integrate soft skills and their practical application in teaching.

4 The Practical Example 'KOE Labor'

The concept of SBL as stated above has been transferred to other disciplines in higher education by the ZLW. This best practice example of SBL in higher education takes place in the education of mechanical engineers at RWTH Aachen University. Within their first semester, students attend the compulsory lecture 'communication and organizational development' (KOE) for engineers. This lecture is mainly designed as a traditional classroom lecture with almost 1200 students. As mentioned above, the practical application of soft skills is a challenge for XL-Classes. Therefore, in addition to the lecture, a simulation-based laboratory tutorial of 1.5 days for small groups of about 35 students is implemented, where they practice the application of the lecture content 'communication and organizational processes'. This laboratory format is based on a simulated company start-up in the automotive industry in face-to-face situations with other students. In the scenario of this competition organized by the chamber of commerce and industry (CCI), 600 students per date go through a process of organizational development by founding a fictitious company with various departments, developing target systems as well as business strategies and constructing a prototype of an innovative automobile. The learning environment is characterized by a fictitious but nonetheless realistic situation and certain challenging tasks, like real-world problems for example communicational challenges, new and unexpected information and demands, limited information. The basic knowledge for the realization of tasks is mediated via microteaching units [2] (Fig. 1).

On the first day of the laboratory, lecture content is repeated and deepened by the practical application. The students have the task to develop an innovative concept for a one-day introductory tour for first semester students, which they have to present at

Fig. 1 Students during the laboratory

the end of the day. Collaboration in teams fosters their teamwork ability, the ability to co-ordinate tasks, to delegate assignments, to set goals and develop their presentation skills through the presentation of the concept in front of the other students.

On the second day, the students found their own company in three stages: company development; internal set up of departments and coordination with other departments; planning and implementation of an automobile prototype which serves as a discipline specific content. In the first phase, the students develop a general and organizational concept for their company. This includes a model, an organizational structure and identity as well as the process of communication. The departments design, construction, body construction and marketing departments are established in the process. In the second phase, the previously founded departments define their tasks and objectives in order to be able to coordinate communication channels and the cooperation of the separate departments. The third and last stage of the company foundation is about planning and constructing the prototype. The planning phase is essential for the future construction of the prototype. In this stage, the plan for the construction is specified, as the students develop project-, time- and action-plans which determine who performs which task when and how the departments should communicate and cooperate with each other. The subsequent constructing phase shows how effective

Fig. 2 Students during teamwork

the previous phases were. During this phase, they develop a marketing concept that is presented together with the prototype afterwards in front of the competitors (Fig. 2).

At the end of the simulation, all prototypes are presented at the market place and are judged according to the criteria of technical innovation - regarding engine, equipment, sustainability, quality of workmanship and design - as well as the presentation according to the criteria of professionalism and creativity and the communication and organizational processes. These processes are of an essential importance for the detailed phases of the business formation. Communication skills, team building, time and project management, decision making and organizational skills are promoted. During the simulation, the students are supported by specially trained coaches in an advisory role, who bring the students out of the simulation into a meta-level reflecting the process in terms of communicative, collaborative and organizational aspects. Supported by the respective coach, central learnings are recorded and visualized on posters for future stages. During the reflection process the coaches provide professional feedback concerning the performance of the group in the last stage. This feedback provides specific recommendations of actions for the next phases and reflects the self-perception of the group through the extrinsic perception of the coach. Apart from the direct effect of the feedback the coach indirectly influences the students as a role model by giving them professional feedback according to predefined rules. This also fosters a learning effect on students' feedback ability. Reflection and feedback of ongoing group processes foster a gradual competence gain from stage to stage. The simulation sensitizes the students' necessity of structured communicational and organizational processes. The relevance of these processes in the real working environment and the relevance of the lecture content become apparent for the students. Even the processes become shapeable by the individual experience in small groups. The developed concept transfers soft skills knowledge and practical application to the students through experience and trains them in a simulated environment. Based on experience and the run through business simulation, awareness for relevance of soft skills in daily routine is fostered.

5 Outlook

The concept of SBL is especially suited for the integration of practical competencies and soft skills in lectures in the field of higher education. In the case of lectures with large-audience, SBL serves as a concept to impart and apply practical skills. The evaluations of the KOE-Labor since 2007 show that the integration of SBL in the first semester is to be regarded as critical. To be able to draw on basic knowledge and existing experiences, fundamental course contents as well as practical knowledge from traineeships and scientific activities must be existent. Therefore, while designing a lecture based on SBL, the integration in the curriculum should be considered to be imbedded later on in the curriculum, when students have completed the orientation period. Simulation-based education should be scheduled and executed throughout the curriculum in order to increase the practical relevance sustainably, and combined with other educational methods such as problem-based learning, practice modules, laboratory work and others [5]. For the purpose of constructive alignments, more attention should be paid to the interconnection with the lecture as well as with the examinations at the end of the semester. Also an assessment concept, which examines the soft skills directly during the simulation, should be developed.

According to the subject, increasing the fidelity of the settings is reasonable. For example this can be done in virtual or remote labs. Virtual and remote experiments allow experiments that are not possible in reality due to increased risks and financial reasons. Furthermore, through the experimental approach to abstract topics experiments provide access to "hands-on" experience.

References

1. Abdulmohsen, H. (2010). Simulation-based medical teaching and learning. *Journal of family and Community Medicine, 17*, 35–40.
2. Brall, S., Hees, F. (2007). Effektives Lernen mit Kurzlerneinheiten. *Kompetenzentwicklung in realen und virtuellen Arbeitssystemen.*

3. Jünger, J., & Köllner, V. (2003). Integration eines Kommunikationstrainings in die klinische Lehre. Beispiele aus den Reformstudiengängen der Universitäten Heidelberg und Dresden. *Psychother Psych Med, 53* (2), 56–64.
4. Kolb, A. Y., Kolb, D. A. (May 15, 2005). The Kolb Learning Style Inventory – Version 3.1 2005 Technical Specifications. Retrieved May 2, 2013, from http://www.whitewater-rescue.com/support/pagepics/lsitechmanual.pdf
5. McGaghie, W., & Issenberg, S. B., Petrusa, E. R., Scalese, R. J. (2010). A critical review of simulation-based medical education research: 2003-2009. *Medical Education, 44*, 50–63.
6. Pankow, F. (2008). *Die Studienreform zum Erfolg machen! Erwartungen der Wirtschaft an Hochschulabsolventen*. Berlin: DIHK – Deutscher Industrie- und Handelskammertag e.V.
7. Wildt, J. (2003). The Shift from Teaching to Learning Thesen zum Wandel der Lernkultur in modularisierten Studienstrukturen. In Fraktion Bündnis 90/ Die Grünen im Landtag NRW (Eds.), *Unterwegs zu einem europäischen Bildungssystem*, (14–18). Düsseldorf.
8. Ziv, A., & Wolpe, P. R., Small, S. D., Glick, S. (2003). Simulation-based Medical Education: An Ethical Imperative. *Academic Medicine, 78*(8), 783–788.
9. VDI-Ingenieurstudie (2008), from http://ipih.de/system/files/upload/ipih-archive/VDI_Ingenieurstudie_Berichtsband.pdf.

On-Professional Competences in Engineering Education for XL-Classes

Stefan Schröder, Daniela Janßen, Ingo Leisten, René Vossen and Ingrid Isenhardt

Abstract Far reaching changes in university higher education have taken place in the last ten years. Different factors, e.g. necessity of on-professional competences in engineering education, rising or vast student numbers and new technical possibilities, have influenced the academic teaching and learning process. Therefore interdependence between requirements and didactical-educational possibilities is given. Because of changed circumstances an adaption of teaching methods and concepts is required. At the same time Bologna arrogates students to be placed in the centre of the teaching and learning process and claims on-professional competences for today's students. Especially for XL-Classes this is a specific challenge. One of the questions ensuing is how to increase learning success by the use of specific didactical methods? With a research approach connecting different proven didactical concepts and considering the previously shown conditions, the concept of the lecture "communication and organizational development" (KOE) at RWTH Aachen University has been redesigned. This lecture, organized by the Institute Cluster IMA/ZLW & IfU at RWTH Aachen University, is mainly frequented by up to nearly 1.300 students of the faculty of mechanical engineering and inherent part of the bachelor-curriculum. The following practical example prospects the multi-angulation of didactical concepts and shows up innovative educational teaching.

Keywords Engineering Education · Audience Response System · On-professional Competences · Best-practice

1 Introduction

Innovative teaching techniques and concepts have been developed in the last years against the background of different factors of influence in the teaching process of higher education. This paper presents a best-practice example, which implements and combines different didactical concepts in higher education. The combination

S. Schröder (✉) · D. Janßen · I. Leisten · R. Vossen · I. Isenhardt
IMA/ZLW & IfU, RWTH Aachen University, Dennewartstr. 27, 52068 Aachen, Germany
e-mail: stefan.schroeder@ima-zlw-ifu.rwth-aachen.de

Originally published in "2013 Frontiers in Education Conference Proceedings",

DOI 10.1007/978-3-319-46916-4_50

of different concepts is necessary as set out in section 2 – Challenges in context of higher education.

In the course of years multifarious didactical concepts have been created. Research reveals different human learning types and success, depending on learning environments [1]. It follows and carved out that integration of learning methods and styles is essential.

Bologna claims the design of student-centered education formats and conveyance of on-professional competences [2], which is challenging especially in large-audience courses, the so-called "XL Classes". On-professional competences are understood as competences which enable students to deal with their theoretical specialized knowledge and prepare them for working life. These competences comprise e.g. methodological-, social-, self-reflecting- and media-skills.

By means of the lecture "communication and organizational development", organized by the Center for Learning and Knowledge Management (ZLW) of the RWTH Aachen University, it is highlighted, how and against which backdrop the lecture was subjected to a whole redesign to cope with the changed influence-factors, technical possibilities and student-requirements. This redesign is mentioned in section 3 – Redesign of the lecture 'KOE'.

2 Challenges in Context of Higher-Education

During the last decade far reaching changes in higher education teaching took place. Different factors have influenced the design of higher education, for example conveying on-professional competences in higher education, the high number of students at German universities and (new) technical opportunities. These most important challenging factors will be elaborated in the following.

2.1 *The Challenge 'XL Classes'*

In Germany, many Universities face the challenge of a vast number of students each semester, especially in engineering education degrees at RWTH Aachen University [3]. This challenge is tightened by the reduction of school time by one year due to a change in German educational policy. In addition to that the demand for engineers in Germany is growing [4]. More than ten years after implementing the Bologna Declaration, the higher education system in Germany has changed significantly [2]. Since then students and their learning process represent the core of the teaching and learning process. This change, also known as a 'shift from teaching to learning' [5], is a more student-centered and involvement-oriented design of teaching. 'It is characterized by innovative methods of teaching which aim to promote learning in communication with teachers and other learners and which take students seriously as active participants in their own learning, fostering transferable skills such as problem-solving, critical thinking and reflective thinking' [6]. Instead of being 'content-oriented', which means focusing on the transmission of content, the design of teaching now focuses on 'learning outcomes', hence the students' learning results and the way of achieving these results. This is, however, a major challenge to lecturers of XL-Classes, as they predominantly focus on the transmission of content due to their difficulties of involving a large number of students at the same time. Therefore the question arises how a lecture can be student-oriented in XL-Classes.

2.2 *Conveyance and Promotion of On-Professional Competences*

Due to the huge number of students attending lectures- XL-Classes have the problem of a content-oriented focus. Therefore the conveyance of on-professional competences to students in XL-Classes is often in a theoretical way. Indeed the transmission of on-professional competences plays an important role in higher education in general and particularly in engineering degree programs [7]. Not only professional knowledge is required to be the output of higher education, but also soft skills like method-, self-, organizational - and social competences are expected from today's students [8].

Against this background the required competences are considered at the redesign. From its multidisciplinary and holistic approach, the ZLW aims at designing and conducting lectures also including the transmission of on-professional competences. Thereby the ZLW tries to accomplish one major goal of the Bologna Process, which is "to create a European space for higher education in order to enhance the employability [...]" [2]. Students should be educated in a way which makes them available for the employment market in a fast, efficient and adequately educated manner [9]. This means students should be enabled to convert learned knowledge in higher education in later working life [10].

2.3 *Combination of Didactical Concepts*

Didactical concepts are the tools of today's lecturers. Those concepts and methods are combinable and applicable in various ways. Due to the didactical diversity of methods a student-centered alignment of lectures is possible. The students learning process is further supported through didactical multi-angulation [11].

Additionally, the students obtain skills in dealing with various methods through methodical diversity (e.g. with presentation techniques). Böss-Ostendorf and Senft confirm that useful methodical diversity is a factor of success for university education, because of the multiplicity access to teaching content [12]. The reasons for integration and combination of different didactical concepts are mentioned by Flechsig [1]:

- Various learning styles and types of students with different learning success
- Diversity of study motivation and interest
- Variety of competences and fields of knowledge
- Variety of context in which learning is placed

Any didactical method aims at enforcing learning and knowledge permanently [13]. As a result a sensible combination of didactical methods and concepts to increase learning-success for students is necessary. As a consequence concepts, explained in the following, have been integrated in the lecture 'KOE'.

3 Redesign of the Lecture 'KOE'

The compulsory lecture "communication and organization development" (KOE) is held every winter term. Almost 1300 students, mostly engineering students, participate in this weekly lecture. In addition a laboratory session takes place, in which students experience a simulated company situation in an authentic environment [14]. Based on the changed influence factors, which are mentioned before, and with the premise to maximize learning success and also among the maxim of sustainable teaching of practice-oriented and on-professional contents, the lecture is continuously refined and was subject to a complete redesign in the winter term 2012/2013. Existing elements were further developed and new didactic elements were implemented. The lecture's elements (shown in Fig. 1) will be explained in the following.

Didactical multi-angulation of the lecture 'KOE'	
A	**Theoretical inputs and lecture modules**
B	**Practice oriented theoretical inputs through expert-lectures with professional practice**
C	**Interaction in XL-Classes through Audience Response Systems (ARS)**
D	**Learning on demand through medial preparation**
E	**Teaching and learning online portal/platform**
F	**Simulation based learning through organizational simulation**
G	**Online examination system**

Fig. 1 Didactical elements 'KOE' lecture at RWTH Aachen University

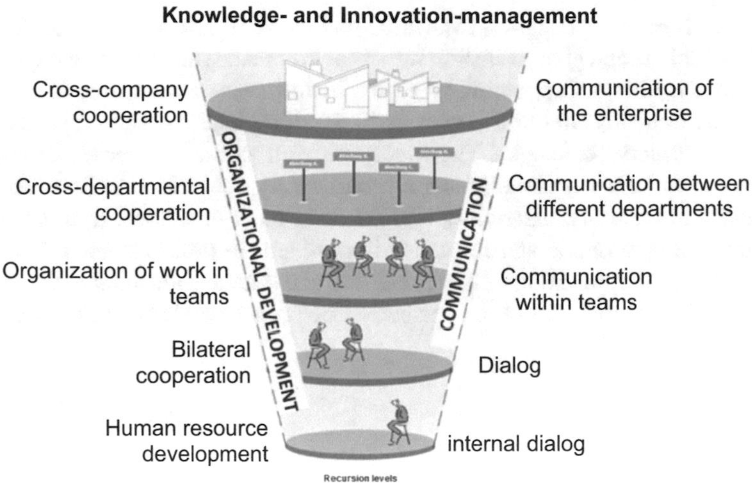

Fig. 2 'KOE' funnel/recursion-layers [14]

3.1 Theory Input and Lecture Modules

The lecture consists of twelve modules, including the basics of communication and organization development, learning- and knowledge management concepts and intercultural aspects of global work division management as well as system-theoretical approaches and practical inputs by experts from industry. The whole concept of teaching follows a linear structure. Practical lectures tie in with earlier taught theoretical contents. Thus an ideal transfer from theory to praxis is established. At the end of every module contents are reflected and summed up. For this purpose the 'KOE funnel' is applied (Fig. 2). Both the systemic view and the different levels of a company are included. Thereby communication and organization are regarded as the most important requirements for the development of interaction between humans, technique and organization on the different organizational levels, for example within departments or project teams.

3.2 Expert Lectures

As already stated the lectures redesign aims for an increasing relation to practice applications and does not only focus on technical expertise. Additionally it also aims at increasing the ability of understanding the complexity of working life. The inclusion of on-professional competences in studies is not only an issue since the Bologna reform [15]. Furthermore it is important to connect the studies with professional practice, not only from the student's perspective, but also from the perspective of company representatives [16]. For this reason, it is more important to connect teaching with practical insights [15]. Based on this, guest speakers with professional industrial background give an insight into their companies and working experiences (e.g. Vodafone Group, Capgemini, p3 group, inside group of companies). For example: Lecture contents are the experiences in handling international projects as well as interdisciplinary challenges. The experts as well as the lecturer try to establish the connection between already taught theoretically contents and the meaning of communication and organization development in daily processes. In this way theories are linked to praxis and relevance is outlined. In order to solve the area of conflict between the teaching of basic sciences and practical teaching at the expense of theoretical essentials [17] the distribution of proper contents and basics is kept reasonable.

3.3 Interactivity Through the Application of "Audience Response Systems"

As outlined by Prensky, that "our students have changed radically" and that "today's students are no longer the people our educational system was designed to teach" [18] (new) technologies e.g. in the form of 'Audience Response Systems' (ARS) may improve the learning outcomes of students [19]. Due to the application of an ARS, students are further involved in the education process.

The ARS is a valuable didactical element as already mentioned by Brinker/ Schumacher in 2009 [20]. Capabilities to participate in a lecture are quite diverse for students, which means that they can choose particular contents, recap their knowledge, interpret, reflect and prepare for examinations. Taxonomies of learning goals, like knowledge, understanding, application and analyzing [21, 22] are addressed holistically by the use of knowledge and comprehension checks. As a result problem-solving skills can be improved. The barriers for participation are marginal. Only an end device with access to the internet is required.

To satisfy the student's demands and with the objective to enable interactions between lecturer and students [23], an ARS is used since the winter term 2012/2013. The implementation of the described system demands a redesign of the lecture with special regards to the content. Questions have to be developed that allow the students to interact with the lecturer as well as with each other. This variety of questions ranges from multiple-choice questions to the inquiry of calculation results etc. [19]. Furthermore iteration-loops, to repeat misunderstood content, can be taken into account.

Figure 3 illustrates the didactical concept of an Audience Response System, which is treated in the 'KOE'. Two fields are shaded grey, as this case is not applied in the lecture. If there are less than 30% correct answers, an explanation on the right answer is given. Provided that the content is understood (>80% correct answers), the lecturer switches to the next topic.

Evaluations of the success of the implementation of ARS in large university lectures show that the application of this "tool" has led to e.g. higher motivation of

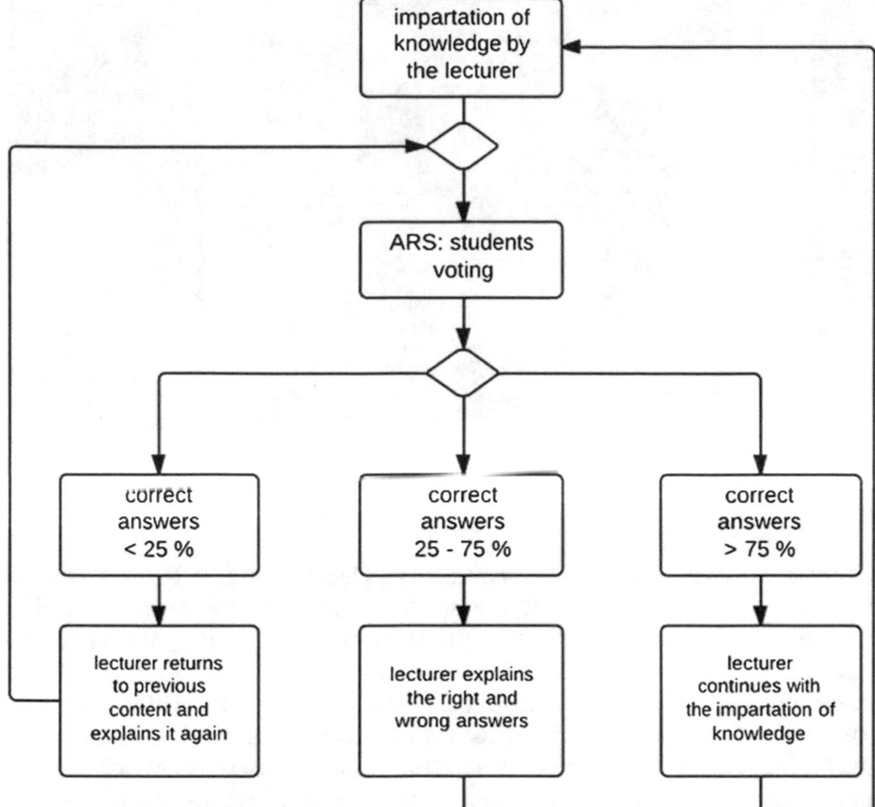

Fig. 3 Didactical concept of using the ARS [23]

attendance, more attention of the students during class and even higher knowledge acquisition than in conventional (non-interactive) classes [19]. If and how the new didactical implementation of the media Audience Response System influences the learning process of students in the lecture 'KOE', has to be evaluated during the next terms.

3.4 Video Documentation

In the winter term 2012/2013 the whole lecture was recorded by a professional camera crew, edited and afterwards provided to the students via the online teaching and learning platform L^{2}P of RWTH Aachen University to support the learning process in auto-didactical-phases of the students. As shown in Fig. 4 it is a matter of an interactive video documentation. For each module the chapters can be retrieved individually.

Fig. 4 Surface of the video recording of the lecture

Knowledge is thereby no longer appropriated as a lecturer-reserve, but accessible for students on demand. This form of making content available for students resembles the concept of learning on demand [24] and results in time saving potential for student learning [25]. Through this video documentation, the students can recapitulate content without any limit and at any time. Furthermore the presentation papers and manual sketches as well as animations and explanations of the lecturer are shown (see Fig. 4). The moderator works as a mediator and presents contents using different channels based on corporate learning. This way offers the possibility to review important contents and especially the preparation of exam papers could be done in a more efficient way. The result is an approach to the teaching for "individual learning", because students can decide when, where and how fast they organize their learning process [26].

3.5 Exchange and Dataplatforms

As a reaction to the various influencing factors like the new technical possibilities, the teaching and learning platform L^2P is used since the end of 2007. The learning platform is password secured and is just accessible for registered students.

The platform is implemented in order to provide important contents like teaching materials (e.g. lecture notes, videos and presentations of experts) and (further) literature for the students on a web based server to support time-consuming and complex teaching and learning processes [27, 28]. Additional L^2P can be used to conduct surveys and simulate electronic tests.

Moreover the students have the possibility to discuss certain topics in an online forum talking to professors, mates or coaches and to obtain important clues to further arrangements. Thus discussion and dealing with lecture-content is supported and promoted. The aim is to influence the learning process to the effect that knowledge is understood by the students and can be applied. According to Palloff & Pratt the use of L^2P in the lecture 'KOE' can be described as a "web-enhanced course" [29]. That implies physical as well as virtual components and thus copes with the European university-tradition in independent preparation of subject-specific content [30].

3.6 Laboratory Tutorial

XL-Classes are faced with the challenge of a theoretical conveyance of on-professional competences to students. During the lecture 'KOE' not all contents can be taught within XL-Classes [31]. Therefore parallel to the lectures a laboratory tutorial of one and a half days takes place. The laboratory tutorial is based on the concept of 'simulation-based learning' which is a research or training method that tries to create a realistic experience in a controlled environment [32]. According to this, simulation replaces or boosts real experiences [33] and thus offers enormous advantages in mediating knowledge long-acting [33, 34]. In groups up to 40 students a foundation of a fictitious automotive company with different branches is simulated. Target systems and various strategies are developed and communication ways are defined and coordinated. An innovative vehicle is constructed under the guidance of 40 professional coaches. The basic knowledge for the realization of this task is mediated via microteaching units. In the next step, earlier learned theoretically basics are applied practically during a company simulation in order to obtain first practical experiences in organizational communication and working processes. Students try to construct an abstract concept in a theoretical framework to finally put their findings in an active experiment. The teamwork and the communication between the different branches demonstrate the importance of the lecture as well as the relation to working life. Additional to the contents of the lecture 'KOE', key skills such as team building, time management and project management are applied, experienced and trained in the simulation.

3.7 Online Examination System

In order to examine around 1300 students efficiently and content orientated, a digital online examination system (called OPS = Online-Prüfungssystem | developed externally especially for 'KOE') is used and applied since 2007 [35, 36]. Examination questions are derived from the lectures contents and distinguished in different taxonomy levels (degrees of difficulty). Thereby the whole taxonomy-spectrum by Bloom 1956 [24] on a cognitive base is addressed to the students. The taxonomy levels are: knowledge, comprehension, usage, analysis, synthesis and evaluation. For the preparation of exam questions it is important to take into account, that the taxonomy levels are hierarchically arranged [37]. Hence the mediated competences for each taxonomy level must gain a specific manifestation before they can be applied in the next taxonomy level [38]. For example, without knowledge, usage of taught content (transfer capacity) is impossible [38].

Supported by a holistic handling of the mentioned difficulty degrees, the OPS enables to retrieve the student levels of awareness. With the help of the OPS knowledge and transfer capacity can be interrogated and evaluated electronically within a few minutes, so that no further staff is required.

4 Conclusion & Vision

Each semester a university and lecture wide evaluation is stated. First evaluation results indicate that the adaption of the lecture concept and the implementation of new technical and didactic elements have a positive impact on the student evaluation of the course, especially the use of technical aids and demonstrations as part of the lecture as well as the teaching of contents by the lecturer. The average evaluation grade rises in the order of 0.5 (arithmetic average; scale from 1 (very good) to 6 (very bad)). For example the usage of devices and demonstrations in the lecture are evaluated with 1.7 (arithmetic average). The mediation of contents also has a positive trend. Lecture records as well as implementations of ARS Systems are inter alia reasons for that.

Altogether the multi-angulation of various didactical concepts can be outlined as a success for the lecture 'KOE'. This educed concept states that students are centered in educational learning process. In a next step the transferability of the redesigned concept of the 'KOE' lecture must be elaborated. Therefore indicators for a survey will be operationalized and developed.

Based on the already mentioned scientific studies of combining various methodical concepts to increase learning-success and the first positive evaluation after the 'KOE' redesign, more innovative teaching and learning methods will be developed, taken up and implemented for this purpose in the next few years.

Furthermore there is still potential for a continuous and appropriate optimization of the lecture 'KOE'. A long-term goal is to implement additional to the existing concepts the method of 'just in time teaching' to optimize and adjust lecture contents [39].

Accordingly presence lectures are not used anymore for conveying contents, but rather to motivate the students to participate and interact in the presence lecture and to discuss questions and problems with the lecture content. Therefore questions and tasks are provided online before every presence lecture. So the lecturer can see the student's results before every lecture 'just in time' and develop the lecture according to the level of awareness. This method is used to obtain feedback related to the student knowledge. This can be used in order to precisely react to the students demands on certain contents. In addition, students get to know on-professional competences while formulating their own questions as well as working on tasks independently. The continuous development of the lecture 'communication and organizational' primarily serves the goal of creating demand-oriented and target group-oriented courses for XL-Classes.

References

1. K.H. Flechsig, *Kleines Handbuch didaktischer Modelle*. Neuland, 1996
2. Europäischen Bildungsminister. The bologna declaration, 1999. Page 4
3. RWTH Aachen University. Zahlenspiegel 2011, 2012
4. VDI. Ingenieurmonitor: Der arbeitsmarkt für ingenieure im februar 2013: Klassifikation der berufe 2010, 2013. No. 3
5. J. Wildt, The shift from teaching to learning - thesen zum wandel der lernkultur in modularisierten studienstrukturen. Fraktion Bündnis 90/ Die Grünen im Landtag NRW. Unterwegs zu einem europäischen Bildungssystem. Reform von Studium und Lehre an den nordrhein-westfälischen Hochschulen im internationalen Kontext, 2003, pp. 14–18
6. The European Students' Union, *Student-centered learning: Toolkit for students, staff and higher education institutions*. Brüssel, 2010. Page 5
7. VDI. Vdi Ingenieurstudie: Studie der vdi wissensforum gmbh, 2007
8. F. Pankow, *Die Studienreform zum Erfolg machen! Erwartungen der Wirtschaft an Hochschulabsolventen*. Berlin, 2008
9. T. Jungmann, K. Müller, K. Schuster, Shift from teaching to learning: Anforderungen an die ingenieurausbildung in deutschland. Journal Hochschuldidaktik 21 (2), 2010, pp. 6–8
10. N.V. Paetz, F. Ceylan, J. Fiehn, S. Schworm, C. Harteis, *Kompetenz in der Hochschuldidaktik: Ergebnisse einer Delphi-Studie über die Zukunft der Hochschullehre*. Wiesbaden, 2011
11. G. Macke, U. Hanke, P. Viehmann, *Hochschuldidaktik: Lehren, vortragen, prüfen*. Freiburg, 2008
12. A. Böss-Ostendorf, H. Senft. Einführung in die hochschul-lehre: Ein didaktik-coach, 2010
13. M. Kerres, *Mediendidaktik: Konzeption und Entwicklung mediengestützter Lernangebote*. München, 2012
14. I. Leisten, S. Brall, F. Hees, eds. *Everyone wants them – we enable them: communicative Engineers*. Kuala Lumpur, 2008
15. M. Oechsle, G. Hessler, Praxis einbeziehen – berufsorientierung und studium. HDS Journal – Perspektiven guter Lehre (2), 2010, pp. 11–22
16. M. Winter, Praxis des studierens und praxisbezug im studium: Ausgewählte befunde der hochschulforschung zum „neuen" und „alten" studieren. In: *Nach Bologna: Praktika im Studium - Pflicht oder Kür?*, ed. by W. Schubarth, K. Speck, A. Seidel, Universitätsverlag Potsdam, Potsdam, 2011, pp. 7–44
17. E. Wegner, M. Nückles, Die wirkung hochschuldidaktischer weiterbildung auf den umgang mit widersprüchlichen handlungsanforderungen. Zeitschrift für Hochschulentwicklung **6** (3), 2011, pp. 171–188
18. M. Prensky, Digital natives, digital immigrants. On the horizon **9** (5), 2001, pp. 1–6
19. V. Stehling, U. Bach, A. Richert, S. Jeschke, eds. *Teaching professional knowledge to XL-classes with the help of digital technologies*. 2012
20. T. Brinker, E.M. Schumacher, Geben sie regelmäßig rückmeldung. lehridee – Ideen und Konzepte für das Lernen und Lehren, 2007, pp. 1–7
21. L.W. Anderson, B.S. Bloom, *A taxonomy for learning, teaching, and assessing: A revision of Bloom's Taxonomy of educational objectives*. New York, 2011
22. B.S. Bloom, *Taxonomie von Lernzielen im kognitiven Bereich*. Weinheim, 1976
23. M. Möhrle, Qualitätsverbesserung interaktiver lehre durch das lead-learner-konzept. In: *Hochschuldidaktik und Hochschulökonomie – Neue Konzepte und Erfahrungen*, ed. by H. Albach, P. Mertens, Gabler, Wiesbaden, 1994, pp. 41–52

24. T. Reglin, C. Speck, Zur kosten-nutzen-analyse von elearning. VBM-Verband der Bayerischen Metall-und Elektro-Industrie eV & Christof Prechtl (Hrsg.), Leitfaden E-Learning, 2003, pp. 221–235
25. S. Seufert, D. Euler, *Nachhaltigkeit von eLearning-Innovationen – Ergebnisse einer Delphi-Studie*. St. Gallen, 2004
26. O.K. Ferstl, K. Schmitz, Integrierte lernumgebungen für virtuelle hochschulen. Wirtschaftsinformatik **43** (1), 2001, pp. 13–22
27. J. Handke, A.M. Schäfer, *E-Learning, E-Teaching und E-Assessment in der Hochschullehre: eine Anleitung*. München, 2012
28. A.L. Dyckhoff, P. Rohde, P. Stalljohann, An integrated web-based exercise module. In: *Proceedings of the 11th IASTED International Conference on Computers and Advanced Technology in Education, September 2008*, ed. by V. Uskov. Acta Press, 2008
29. R.M. Palloff, K. Pratt, *Lessons from the cyberspace classroom: The realities of online teaching*. John Wiley & Sons, 2001
30. R. Schulmeister, Zur didaktik des einsatzes von lernplattformen. Lernplattformen. Web-based Training. Empka-Akademie, 2005, pp. 11–19
31. I. Leisten, S. Brall, F. Hees, Fostering entrepreneurship in engineering education at rwth aachen university. In: *Proceedings of the International Conference on Global Cooperation in Engineering Education: Innovative Technologies, Studies and Professional Development, 4-6 October, 2008, Kaunas, Lithuania*. 2008
32. P. Mack. Understanding simulation-based learning, 2009
33. D.M. Gaba, The future vision of simulation in health care. Quality & safety in health care **13 Suppl 1**, 2004, pp. i2–10
34. J. Castronova, Discovery learning for the 21st century: what is it and how does it compare to traditional learning in effectiveness in the 21st century. Literature Reviews, Action Research Exchange **1** (2), 2002
35. F. Hees, A. Hermanns, A. Huson, Prüfungserstellung mit total quality management (tqm). In: *Prüfungen auf die Agenda! Hochschuldidaktische Perspektiven auf Reformen im Prüfungswesen, Blickpunkt Hochschuldidaktik*, vol. 118, W.Bertelsmann Verlag, 2008, pp. 129–141
36. P. Blum, *Ein inter-, intranetbasiertes System zur Erstellung, Durchführung und automatisierten Bewertung von dynamischen Leistungstests in der medizinischen Lehre*. Inside-Verl. für Neue Medien, Aachen, 2005
37. W. Sitte, *Beiträge zur Didaktik des 'Geographie und Wirtschaftskunde'-Unterrichts, Materialien zur Didaktik der Geographie und Wirtschaftskunde*, vol. Bd. 16, 4th edn. Inst. für Geographie und Regionalforschung, Wien, 2006
38. F. Bruckmann, O. Reis, M. Scheidler, eds., *Kompetenzorientierte Lehre in der Theologie: Konkretion - Reflexion - Perspektiven*, vol. 3. Berlin, 2011
39. E. Mazur, J. Watkins. Just-in-time teaching and peer instruction, 2014. URL http://mazur.harvard.edu/sentFiles/Mazur_263828.pdf

Die Studieneingangsphase als Weichensteller im Ingenieurstudium – Herausforderungen begegnen

Natascha Strenger, Theresa Janssen, Gergana Aleksandrova and Alexander an Haack

1 Einleitung

ELLI ist ein Verbundprojekt an den drei Standorten der RWTH Aachen University, der Ruhr-Universität Bochum sowie der Technischen Universität Dortmund und wird im Rahmen des Qualitätspakts Lehre gefördert. ELLI zielt auf die Verbesserung der Studienbedingungen und die Weiterentwicklung der Lehrqualität in der Ingenieurausbildung. Die Projektarbeit konzentriert sich dabei auf vier Kernbereiche: 1. Virtuelle Lernwelten und E-Learning, 2. Mobilität und Internationalisierung, 3. Student Lifecycle und 4. Professionelle Handlungskompetenzen.

Den Herausforderungen der Studieneingangsphase widmet ELLI sich an den Standorten Aachen und Bochum im Bereich des Student Lifecycle: Im Hinblick auf eine Verringerung von Studienabbruchquoten (vgl. [1]) werden die Übergangsphasen des Ingenieurstudiums mit beratenden Angeboten begleitet. Neben dem Studieneinstieg zählen hierzu auch die Übergänge im gestuften Bachelor-/Master-Reformmodell sowie zur Promotion bzw. der Einstieg ins Berufsleben. Auch der geringen Auslandsmobilität unter Ingenieurstudierenden will ELLI im Kernbereich Mobilität und Internationalisierung frühzeitig entgegenwirken, indem Maßnahmen in der Studieneingangsphase Anwendung finden. Im Folgenden werden beispielhaft einige der bisher von ELLI umgesetzten und geplanten Maßnahmen beschrieben.

N. Strenger (✉) · T. Janssen
Ruhr-Universität Bochum Project ELLI – Excellent Teaching and Learning in Engineering Sciences, Bochum, Germany
e-mail: strenger@fvt.rub.de

G. Aleksandrova · A. an Haack
RWTH Aachen University, Aachen, Germany

Originally published in “Nürnberg, HD-MINT Symposium Tagungsband”,

DOI 10.1007/978-3-319-46916-4_51

2 Am Standort Bochum: ALLES ING! – Role-Model Konzepte für einen erfolgreichen Studieneinstieg

Durch ELLI konnte am Standort Bochum die Initiative ALLES ING! umgesetzt werden: In dieser verfolgen die drei ingenieurwissenschaftlichen Fakultäten der Ruhr-Universität Bochum ihre gemeinsamen Ziele, das Interesse und die Identifikation mit den Technikwissenschaften zu stärken. Durch eine attraktive Darstellung der Ingenieurwissenschaften und der „Menschen hinter der Technik", sollen interessierte Schüler/innen in ihrer Entscheidung für ein technisches Studium bestärkt und Studienanfänger von vielfältigen Vorbildern durchs Studium begleitet werden. Dieses Konzept der Role-Models, die Wege in und durch das ingenieurwissenschaftliche Studium aufzeigen, setzt ALLES ING! auf zwei Ebenen um: In einem Internetportal, welches sich verschiedenster medialer Präsentationsformen wie Kurzvideos, Interviews oder Twitter-Einträgen bedient, um Ingenieurstudierende im Studienalltag, beim Auslandsaufenthalt, mit ihrem Forschungsthema oder beim Berufseinstieg zu porträtieren (vgl. [2]). Begleitet wird das Konzept durch eine Reihe von Veranstaltungen, bei denen technikinteressierte Schüler in Kontakt mit Studierenden, Alumni und Wissenschaftlern kommen. Das in ALLES ING! medial umgesetzte Konzept der Role-Models kommt in den von ELLI konzipierten Beratungsformaten am Standort Bochum jeweils thematisch begleitend zum Einsatz.

3 Am Standort Bochum:

3.1 ALLES ING! – Role-Model Konzepte für einen erfolgreichen Studieneinstieg

Durch ELLI konnte am Standort Bochum die Initiative ALLES ING! umgesetzt werden: In dieser verfolgen die drei ingenieurwissenschaftlichen Fakultäten der Ruhr-Universität Bochum ihre gemeinsamen Ziele, das Interesse und die Identifikation mit den Technikwissenschaften zu stärken. Durch eine attraktive Darstellung der Ingenieurwissenschaften und der „Menschen hinter der Technik", sollen interessierte Schüler/innen in ihrer Entscheidung für ein technisches Studium bestärkt und Studienanfänger von vielfältigen Vorbildern durchs Studium begleitet werden. Dieses Konzept der Role-Models, die Wege in und durch das ingenieurwissenschaftliche Studium aufzeigen, setzt ALLES ING! auf zwei Ebenen um: In einem Internetportal, welches sich verschiedenster medialer Präsentationsformen wie Kurzvideos, Interviews oder Twitter-Einträgen bedient, um Ingenieurstudierende im Studienalltag, beim Auslandsaufenthalt, mit ihrem Forschungsthema oder beim Berufseinstieg zu porträtieren (vgl. [2]). Begleitet wird das Konzept durch eine Reihe von Veranstaltungen, bei denen technikinteressierte Schüler in Kontakt mit Studierenden, Alumni und Wissenschaftlern kommen. Das in ALLES ING! medial umgesetzte Konzept der

Role-Models kommt in den von ELLI konzipierten Beratungsformaten am Standort Bochum jeweils thematisch begleitend zum Einsatz.

3.2 Beratungsangebote speziell für Schüler/innen im Übergang Schule-Hochschule

Im Austausch mit den Mitarbeiter/innen der Ingenieurwissenschaften aus dem Themenfeld des Übergangs Schule-Hochschule wurden von ELLI die Ziele definiert, nicht-traditionelle Studierende für ein Ingenieurstudium zu begeistern und besonders den noch immer geringen Frauenanteil zu erhöhen (vgl. [3]). Im Rahmen des Schülerinnen-Projektworkshops im Oktober 2012 begründete das Projekt in Zusammenarbeit mit der Fakultät für Elektrotechnik und Informationstechnik das Format der Tea-Time, in der weibliche Studieninteressierte in den Austausch mit Vorbildern aus der Technik kamen. Von der jungen Studentin bis zur erfolgreichen Professorin stellten sich Ingenieurinnen ihren Fragen und hielten anschließend ihren Werdegang und ihre Botschaft an den Nachwuchs in sehr persönlichen Videostatements fest. Zum Tag der offenen Tür der Ruhr-Universität im März 2013 – wurde gemeinsam mit den Fakultäten ein Informationsangebot für weibliche Studieninteressierte umgesetzt. Die 7 besten Gründe IngenieurIn zu werden bildeten eine erste Chance für Schülerinnen, in den Austausch mit Studierenden und Studienfachberater/innen zu kommen. Informationsveranstaltungen wie der Markt der Studienmöglichkeiten und die Immatrikulationszeit werden im Projekt zusammen mit den Mitarbeiter/innen der Fakultäten durch die Planung des gemeinsamen Auftrittes und die Erstellung gemeinsamer Poster und Flyer gestaltet.

3.3 GoING Abroad – Frühzeitige Beratung zu Auslandsaufenthalten im Ingenieurstudium

Um der vergleichsweise geringen Auslandsmobilität unter ingenieurwissenschaftlichen Studierenden entgegenzuwirken (vgl. [4]) und das Interesse für Austauschprogramme schon möglichst früh im Studium zu wecken, setzt auch der Kernbereich Mobilitätsförderung und Internationalisierung auf Beratungsformate, die die Studierenden bereits in den ersten Semestern in Kontakt zu wichtigen Ansprechpartnern und auslandserfahrenen Kommilitonen bringen. Hierzu wurde gemeinsam mit dem International Office, den Auslandkoordinatoren der Fakultäten sowie Studierendenvertretern das GoING Abroad-Programm entwickelt: Zentrales Element ist eine Informations- und Beratungsveranstaltung, welche jedes Semester stattfindet und bei der auslandserfahrene Kommilitonen sowie Fachberater Interessierten für alle Fragen rund um das Thema Ausland zur Verfügung stehen. Ein Novum ist hierbei die fachspezifische Herangehensweise (vgl. [5]): Als Referenten werden gezielt Studierende der ingenieurwissenschaftlichen Fächer gewonnen und es werden Fakultätskooperationen sowie die für Ingenieurstudierende besonders attraktiven, berufsbezogenen Praktika im Ausland vorgestellt. Auch im Kernbereich Mobilität und Internationalisierung kommt das ALLES ING!-Konzept der Role-Models zum Einsatz: In der eigenen Rubrik Globetrotter berichten Ingenieurstudierende im Ausland von ihren Erfahrungen. Die frühe Thematisierung von Auslandsaufenthalten im Studium, gekoppelt mit der Präsentation von Vorbildern, wirkt den Herausforderungen einer schwierigen Organisation und der Angst vor Studienzeitverlängerung entgegen, die von deutschen Studierenden häufig als Mobilitätshemmnisse angeführt werden.

3.4 Beratung und Unterstützung im Studienverlauf und Aufzeigen von Forschungs- und Berufsperspektiven

Im Hinblick auf eine nachhaltige Sicherung des Studienerfolges liegt die Herausforderung innerhalb des Student Lifecycle darin, die Studierenden von Anfang an für das Studium zu begeistern und ihnen praktische Einblicke in Wissenschaft und Wirtschaft zu ermöglichen. Die Studierenden sollen dazu motiviert werden, ihr Studium aktiv selbst zu gestalten und durch den Austausch mit Kommiliton/inn/en anderer Semester Tipps zu erhalten.

Den Austausch mit den Studierenden und gleichzeitig eine stärkere Vernetzung der ingenieurwissenschaftlichen Studiengänge untereinander initiiert ELLI in mehrmonatigen Abständen bei großen Fachschaftstreffen mit den sieben Fachschaften der Ingenieurstudiengänge in Bochum. An der Tagesordnung sind hierbei regelmäßig Themen wie studentische Beteiligung an der Fachschaftsarbeit, Studienorganisation und -struktur. Auch konkrete Veranstaltungen werden in diesem Rahmen gemeinsam geplant; darunter die Begrüßung und Beratung der neuen Studierenden bei der Immatrikulation zum Wintersemester oder die fakultätsübergreifende Beteiligung der Fachschaften bei der VDE Techniknacht Ruhr im Oktober 2013.

Weiterhin zeigt ELLI Perspektiven für Forschungs- und Abschlussarbeiten sowie den Berufseinstieg auf. Für höhere Semester werden die Beratungsveranstaltungen Mit dem Master in die Praxis sowie Wege zum Dr.Ing. angeboten. Für angehende und Jungingenieure kommen auch hier mediale Formate im Rahmen der ALLES ING! Initiative zum Einsatz: Quer durch alle Ausbildungsstufen sind hier Studienarbeiten, Dissertationen, Forschungsprojekte und Abschlussarbeiten zu finden. Die Videoreihe Young Professionals besucht RUB-Alumni an ihrem Arbeitsplatz und zeichnet ihren individuellen Weg in und durchs Studium bis hin zum Beruf nach und stellt gleichzeitig potenzielle Arbeitgeber für Ingenieure vor.

4 Am Standort Aachen:

4.1 StartINGs meet Alumni – Vernetzung von Studierendengenerationen

Die Fachschaften sind zu Beginn des Studiums eine häufige Anlaufstelle für die Studienanfänger aller Fachrichtungen. Insbesondere nach dem Doppelabiturjahrgang in NRW dieses Jahr sind die Herausforderungen „hoch zwei". Es wird befürchtet, dass durch die zunehmende Studentenanzahl nicht nur unzureichend Lern- und Wohnplätze zur Verfügung stehen, sondern auch die zentralen Beratungsstellen der Hochschulen überlastet werden. An dieser Stelle knüpfen die ersten Maßnahmen zur Gestaltung der Studienanfangsphase an, die im Rahmen von ELLI an der RWTH Aachen eingeleitet wurden.

Neben bereits bestehenden Angeboten der Fachschaften wie z. B. der Gestaltung der Einführungswoche werden zur Vorbereitung auf das Studium und als weitführendes Beratungsangebot gemeinsame Treffen ingenieurwissenschaftlicher Fakultäten geplant. Hierbei wird zum einen der intensive Austausch zwischen den Ingenieuren aller Fachrichtungen an der RWTH Aachen unterstützt. Zum anderen wird der Kontakt der Studienanfänger mit Studierenden fortgeschrittener Semester gefördert. Im Vorfeld werden spezifische Fragestellungen der Studienanfänger/innen erhoben und systematisiert. Auf Basis der aufgefassten Daten werden die Orientierungstage in den folgenden Jahren angepasst, um die Anforderungen der Studierenden optimal zu erfüllen. Ziel ist es, den Start ins Studium möglichst flexibel und einfach für die Studierenden zu gestalten. Weiterhin sollen gemeinsame semesterübergreifende Treffen der angehenden Ingenieure und Ingenieurinnen zwei Mal im Jahr durchgeführt werden. In Form eines Stammtisches werden die jungen Studierenden mit höheren Semestern und Alumni an einen Tisch gebracht. Dabei werden zum einen organisatorische Fragen besprochen und geklärt und zum anderen Erfahrungen zwischen den Studierenden unterschiedlicher Semester und Fachdisziplinen ausgetauscht. Durch diese gemeinsamen Treffen werden der interdisziplinäre Austausch unterschiedlicher ING-Disziplinen sowie die Steigerung der Varietät angestrebt.

Die jeweilige Ausgestaltung und Weiterentwicklung der Maßnahmen wird gemeinsam mit aktiven Mitgliedern sämtlicher Fachschaften und Alumni-Vereine ausgearbeitet. Die Organisation sowie die Durchführung eines solchen Treffens übernehmen die Fachschaften vor Ort. Einerseits werden die StartINGs durch Studierende älterer Semester bei der Gestaltung des Studienplans sowie bei alltäglichen Herausforderungen unterstützt. Andererseits wird die Vernetzung zwischen Studenten, Alumni und Unternehmen stärker gefördert. Durch das Netzwerkformat der geplanten Treffen erhalten die Studierenden die Gelegenheit, in zwangloser Atmosphäre Kontakte zu knüpfen. Ehemalige können über den Stammtisch hinaus als Mentor/in beratend fungieren. Somit können die Studierenden auf die Erfahrungen von Absolventen zurückgreifen, typische ingenieurwissenschaftliche Arbeitsfelder kennenlernen und eventuelle künftige Praktikumsplätze ergattern.

4.2 learnING by doING – Vermittlung von Praxisbezügen

Eine Vielzahl der Studienabbrecher ingenieurwissenschaftlicher Fächer konnte aufgrund mangelnder Lern- und Lehrsituationen nicht die notwendige Bindung zu ihrem Studienfach entwickeln. Gründe hierfür sind unter anderem mangelnde studienbindende Faktoren sowie mangelnde soziale, fachliche und akademische Integration (vgl. [6]). Um dieser Problematik entgegenzuwirken, werden Maßnahmen geplant, die das Technikverständnis fördern, Praxisbezüge vermitteln und somit das fachliche Vertrauen bestärken. Hierbei werden Blockveranstaltungen ausgearbeitet, in denen Studierende praxisnahe Studienaufgaben in Lerngruppen gemeinsam erledigen können. Im Rahmen eintägiger Seminare für Studierende der Ingenieurwissenschaften werden auf Basis eines erfahrungsbasierten Lernzyklus (vgl. [7]) Theorieinputs zunächst praxisbezogen vermittelt und anschließend durch aktives Experimentieren erprobt und dadurch für die Lernenden konkret erlebbar gemacht. Hierbei werden die Seminarinhalte so gestaltet, dass diese den Anforderungen der technisch interessierten Studienanfänger und Studienanfängerinnen gerecht werden. Sie werden vor die Herausforderung gestellt, unterschiedliche kleine Aufgaben in einem Organisationsentwicklungsprozess und in Zusammenarbeit mit den anderen Teilnehmern zu bewältigen und das eigene Handeln im Team zu reflektieren. Die Studierenden erhalten Grundlagen des Projekt- und Zeitmanagements vermittelt und werden in die Lage versetzt, diese anschließend im Alltag intuitiv und leicht einsetzen zu können. Neben der Theorievermittlung stehen das praktische Erleben und Erproben, aber auch der Anschluss an Kommilitonen anderer technischer Fächer im Vordergrund. Daher erhalten die Studierenden die Gelegenheit, sich untereinander besser kennenzulernen, in Arbeitsgruppen zu organisieren und nicht nur technische sondern auch soziale und kulturelle Inhalte ungezwungen zu erlernen. Geplant sind die Seminare als weiterführendes Unterstützungsangebot für Studierende aller technischer Fachrichtungen und Semester. Der Fokus liegt allerdings auf den Bedürfnissen und Anforderungen von Studienanfängern und Studienanfängerinnen an der RWTH Aachen.

5 Evaluation und Ausblick

Die Arbeit mit Role-Models und die möglichst frühzeitige Vernetzung der Studierenden aus verschiedenen Studienphasen untereinander sowie mit den für ihre Bedürfnisse wichtigen Ansprechpartnern erweisen sich als höchst effektiv. Sie bieten ein niedrigschwelliges Kontaktangebot, fördern die Identifikation mit dem ingenieurwissenschaftlichen Studium und erhöhen die Attraktivität individueller Profilbildungsmöglichkeiten, wie zum Beispiel von studienbezogenen Auslandsaufenthalten oder Themen für Abschlussarbeiten. Erfahrungswerte von Gleichaltrigen erhalten eine große Akzeptanz und Erstsemester erhalten in der Studieneingangsphase konkretere Vorstellungen vom Studienverlauf. Mit einer neuen, durch vielfältige mediale Konzepte unterstützten, Ansprache können langfristig auch mehr Nicht-traditionelle Studieninteressenten erreicht werden.

Die langfristige Implementierung von erfolgreichen Maßnahmen wie diesen an den beteiligten Standorten stellt eine Herausforderung dar, welcher ELLI durch die Einbindung aller relevanten hochschulischen Akteure sowie der Studierendenvertreter begegnen will. Die von ELLI eingesetzten Maßnahmen für die Studieneingangsphase werden regelmäßig evaluiert, indem Teilnehmer- und Besucherzahlen erhoben sowie quantitative Befragungen durchgeführt werden. Die qualitative Rückkopplung mit der Zielgruppe wird durch regelmäßige Treffen mit Studierendenvertretern gesichert, welche ein wertvoller Projektbestandteil sind. Besonders positive Erfahrungen wurden auch in der Zusammenarbeit mit den Mitarbeiter/inne/n der ingenieurwissenschaftlichen Fakultäten gesammelt, die sehr motiviert und engagiert an neuen Ideen für die Studieneingangsphase mitwirken.

Literaturverzeichnis

1. Koppel, Oliver. 2012: Ingenieure auf einen Blick: Erwerbstätigkeit, Innovation, Wertschöpfung. URL https://www.vdi.de/fileadmin/vdi_de/redakteur_dateien/bag_dateien/Beruf_und_Arbeitsmarkt/2012_-_Ingenieure_auf_einen_Blick.pdf
2. I. Bertozzi, M. Klick, J. Lippmann. Alles ING! Technik an der RUB, 09.02.2016. URL http://www.ing.rub.de/
3. M. Winde, J. Schröder. Hochschulbildungsreport 2020: Jahresbericht 2015. URL http://www.stifterverband.de/bildungsinitiative/hochschul-bildungs-report_2015.pdf
4. S. Burckhart, U. Heublein, J. Kercher, J. Mergner, J. Richter, *Wissenschaft Weltoffen 2013: Facts and Figures on the International Nature of Studies and Research in Germany // Daten und Fakten zur Internationalität von Studium und Forschung in Deutschland*. Bertelsmann W, Bielefeld, 2013. URL http://www.wissenschaftweltoffen.de/publikation/wiwe_2013_verlinkt.pdf
5. DAAD Deutscher akademischer Austauschdienst. Eine Initiative von:Informations- und Werbekampagne „Go out! Studieren weltweit“: Wirtschaftsrelevante Stipendienprogramme des DAAD für Studierende und Graduierte deutscher Hochschulen. URL http://www.go-out.de/imperia/md/content/go-out/konferenzmaterial_2011-05-19_neu.pdf
6. K. Gensch, C. Kliegl. Studienabbruch in MINT-Fächern – welche Gegenmaßnahmen können Hochschulen ergreifen? URL http://www.ihf.bayern.de/uploads/media/IHF_kompakt_Mai_2012.pdf
7. D.A. Kolb, *Experiential learning: Experience as the source of learning and development*. Prentice-Hall, Upper Saddle River, NJ, 1984. URL http://academic.regis.edu/ed205/kolb.pdf

„Ist digital normal?“ – Untersuchung des Mediennutzungsverhaltens Studierender in der ingenieurswissenschaftlichen Lehre

Kerstin Thöing, Ursula Bach, René Vossen and Sabina Jeschke

Zusammenfassung More than ten years ago, Marc Prensky coined the digital age by his concept "Digital Native" [1]. Replies from opposing points of view were published directly afterwards, predominantly by Rolf Schulmeister in the German-speaking area. Overall, neither the homogenous characterization of a "digital generation" nor the attribution of certain characteristics like multitasking ability could be maintained [2]. Regarding habits of media usage and media acceptance, younger generations appear to be equally ambivalent: the same students who nowadays have their finger on the pulse of current technical affairs, spending most of their time surfing on Facebook and the internet with smartphones or Tablet PCs, often show a reserve towards the media offer of their higher education institutions. After the discrepancy between habits of media usage and media acceptance in educational and private context was identified by higher education institutions, the number of evaluations relating to this topic has steadily increased [3, 4]. The results of these surveys show almost consistently that e.g. web 2.0 resources are hardly ever used [5]. Regarding the evaluation of these results it has to be considered that, for example, almost every survey had an interdisciplinary approach and did not focus on specific subjects. It can be assumed that students with a background in engineering and affinity to technology show other preferences than students from disciplines in arts and humanities. Consequently, important data, especially for the adjustment and sustainable development of media offers for engineering education, is missing. Taking these aspects into consideration, the Centre for Learning and Knowledge Management at RWTH Aachen University and the Karlsruhe Institute of Technology are going to conduct a survey regarding habits of media usage among engineering students during the winter semester 2012/2013. The outcomes will serve as aid to orientation in strategic media development for engineering education.

Schlüsselwörter media usage · web 2.0. · evaluation · engineering sciences · academic teaching

K. Thöing (✉) · U. Bach · R. Vossen · S. Jeschke
IMA/ZLW & IfU, RWTH Aachen University, Dennewartstraße 27, 52068 Aachen, Germany
e-mail: kerstin.thoeing@ima-zlw-ifu.rwth-aachen.de

Originally published in "Tagungsband 7. IGIP Regionalkonferenz 2012",

 DOI 10.1007/978-3-319-46916-4_52

1 Einleitung

Der Einsatz digitaler Medien zur Unterstützung von Lehr- und Lernprozessen, zusammengefasst unter dem Begriff E-Learning [6], ist ein wesentlicher Teil moderner Hochschuldidaktik. Neben der inzwischen konventionellen Verwendung von digitalen Medien für die Präsentation und Distribution von Lehrinhalten bezieht E-Learning in der jüngeren Entwicklung zusätzlich die Studierenden durch deren aktive Generierung von Inhalten in offene Lernsysteme ein und wird unter diesem Aspekt zum E-Learning 2.0 [6]. Die hochschuldidaktische Ausgestaltung mit digitalen Medien ergibt sich unter anderem im Zuge des „Shift from Teaching to Learning", bei dem die Studierenden und ihre Lernprozesse durch das Medienangebot ihrer Hochschule unterstützt und gefördert werden [7].

Um in diesem Sinne einen adäquaten Medieneinsatz innerhalb eines Fachbereichs, einer Hochschule und im gesamten deutschen Hochschulraum strategisch und didaktisch sinnvoll planen zu können, werden Erkenntnisse aus mehreren Fachdisziplinen, wie der Medienpädagogik, Kommunikationswissenschaft, Mediensoziologie, Medienwissenschaft und Medienpsychologie herangezogen. Ein weiterer Untersuchungsbereich ist die Mediennutzung Studierender. Hier wird erforscht, ob und wie häufig Studierende das interne Medienangebot ihrer Hochschule sowie externe Medien, wie Wikipedia oder Google, in ihrer Freizeit und insbesondere für das Studium nutzen. Faktoren werden identifiziert, die diese Nutzung beeinflussen, wie Zufriedenheit mit der Qualität des Angebotes, curriculare Einbindung der Medien oder Erfahrung im Umgang mit Medien. Erkenntnisse aus diesen Untersuchungen unterstützen bei der Entwicklung des Medienangebotes, indem sie beispielsweise Bereiche des E-Learnings aufdecken, in denen künftig gezielt gefördert und finanziert werden sollte, etwa wenn die Nutzungshäufigkeit eines bestimmten Angebotes zunimmt oder hinter den Erwartungen zurück bleibt.

Der folgende Beitrag gibt einen Überblick über die bisherige Entwicklung und den aktuellen Stand der Untersuchungen zum Mediennutzungsverhalten Studierender. Dabei geht er insbesondere auf das Mediennutzungsverhalten Studierender der Ingenieurwissenschaften ein und stellt Forschungsbedarfe für weitere Untersuchungen fest. Der anschließende Ausblick weist auf eine Untersuchung des Mediennutzungsverhaltens ingenieurwissenschaftlicher Studierender hin, mit dem die RWTH Aachen University diesem Forschungsbedarf nachkommt.

2 Der Studierende als Nutzer digitaler Medien im Fokus

Die ausführliche Auseinandersetzung mit dem Mediennutzungsverhalten Studierender begann 2001 mit dem „Digital Native"-Konzept von Marc Prensky [1]. Prensky bezeichnet die Studierendenjahrgänge der jüngeren Generation, die mit digitalen Medien und dem Internet aufwachsen, als „Digital Natives". Er formulierte die Metapher der „native speakers of the digital language of computers, video games and the Internet“ [1], die sowohl in Fachkreisen, als auch in der Öffentlichkeit kontrovers diskutiert wurde. Prensky berief sich auf seine Beobachtung, dass die Digital Natives einen offensichtlich selbstverständlichen und souveränen Umgang mit den neuen technischen Möglichkeiten pflegen als ihre Lehrenden, die er „Digital Immigrants“ nennt. Eine seiner Thesen besagt: „Digital Immigrant instructors, who speak an outdated language (that of the pre-digital age), are struggling to teach a population that speaks an entirely new language“ [1]. Prensky forderte für die Digital Natives angepasste Lehrkonzepte mit neuen Methoden und Lehrinhalten. Digital Natives präferieren seiner Meinung nach einen spielerischen Zugang zu den Lerninhalten und arbeiten am liebsten in kollaborativen Szenarien mit ausgeprägten Feedbackstrukturen und kleineren Lehreinheiten [1]. Seine Thesen wurden bald von zahlreichen weiteren Autoren, unter anderen Wim Veen, Don Tapscott, Horst Opaschowski, Claudia de Witt, Neil Howe, Diana und James Oblinger weiterentwickelt. Sie fanden für die jüngere Studierendengeneration weitere Bezeichnungen, wie „Born digital“ [8], „Generation NeXt“ [9] oder „Net Generation“ [10].

Forciert durch die vielfältigen Möglichkeiten und fortschreitende Entwicklung der digitalen Medien wurden in den folgenden Jahren entsprechende didaktische Angebote entwickelt, die den Bedürfnissen der Digital Natives entgegenkommen sollten. Gerade E-Learning 2.0-Angebote mit ihren Möglichkeiten zum vernetzten Arbeiten schienen in dieser Hinsicht besonders geeignet zu sein.

3 Zur Untersuchung des Mediennutzungsverhaltens von Studierenden

Rolf Schulmeister war einer der ersten, der das Konzept des Digital Natives wissenschaftlich tiefer gehend überprüfte. Er verglich 2007 bis 2009 mehr als 45 empirischen Studien zu Mediennutzung und Nutzermotiven Studierender und Jugendlicher [2]. Seine Ergebnisse zeigen, dass weder das homogenisierte Bild einer „digitalen Studierendengeneration“ noch die generalisierte Zuschreibung der genannten charakteristischen Eigenschaften und Präferenzen haltbar sind. Schulmeister stellte fest, dass Studierende vielmehr dazu tendieren, digitale Medien passivrezeptiv zu nutzen und in kollaborativen Arbeitsszenarien der E-Learning 2.0-Dienste kaum aktiv partizipieren, wobei die Gründe für dieses Verhalten, wie möglicherweise eine noch zu geringe Qualität des Angebotes, nicht näher untersucht wurden. Er widersprach in seiner Metaanalyse weitere Thesen Prenskys, wie beispielsweise das Vorhandensein einer „digitalen Sprache“ der jüngeren Generation [2]. Das Digital Native-Konzept und viele der Prenskys Annahmen weiterführenden Konzepte anderer Autoren wurden damit wissenschaftlich zumindest infrage gestellt und eignen sich nicht mehr als fundierte Empfehlungsgrundlage für die Weiterentwicklung der mediengestützten Hochschuldidaktik.

Seit 2008 werden an Hochschulen daher vermehrt dezidiertere Untersuchungen zum Mediennutzungsverhalten Studierender durchgeführt [3, 4]. Besonders zu erwähnen sind die HISBUS [11, 12]- und ECAR-Studien [13]. Die nationalen HISBUS-Studien erreichten mit 4.400 Studierenden 40% der Grundgesamtheit und wählten ihr Panel so aus, dass sie als repräsentative Stichprobe gelten können. Die ECAR-Studien werden in den USA jährlich mit bisher annähernd 26.100 Studierenden an über 100 Hochschulen durchgeführt, umfassen eine umfangreiche Literaturrecherche, eine quantitative Erhebung des detaillierten Mediennutzungsverhaltens sowie eine qualitative Befragung von Fokusgruppen und einen Langzeitvergleich der erhobenen Daten.

Die Ergebnisse des aktuellen Stands der Untersuchungen zum Mediennutzungsverhalten Studierender werden im Folgenden zusammengefasst und diskutiert.

4 Ergebnisse der bisherigen Untersuchungen des Mediennutzungsverhaltens von Studierenden

In der Metabetrachtung zeigen die Ergebnisse der Erhebungen zum Mediennutzungsverhalten von Studierenden nahezu einheitlich, dass häufig Medien genutzt werden, die bei möglichst geringem Aufwand einen schnellen und ausreichenden Informationsmehrwert erzielen. So sind benutzerfreundliche Informationsdienste wie Google-Websuche, Wikipedia, Online-Wörterbücher und digitale Unterrichtsmaterialien unter den Studierenden sehr beliebt [5].

Lernangebote wie Lernplattformen werden wenig genutzt [5]. Erfordern Lernangebote eine aktive Partizipation, wie bei Wikis, Blogs und interaktiver Lernsoftware, verringert sich die Nutzung durch die Studierenden auf ein Minimum [5].

„Since we began this study we have found in both the quantitative and the qualitative data that students say convenience is the most valuable benefit of IT in courses“, so eine ECAR-Studie [15]. Es zeigt sich jedoch, dass digitale Medien im Unterricht dann wieder verstärkt gewünscht werden, wenn die Studierenden den Einsatz und die Vorteile ausreichend erfahren konnten. Dementsprechend nutzen insbesondere Studierende höherer Fachsemester E-Learning-Angebote, nachdem ihnen die Vorteile durch ihre Professoren nahe gebracht wurden [13]. In diesem Zusammenhang ist interessant, dass die meisten Studierenden, entgegen der Digital Native-Hypothese, zu Beginn ihres Studiums trotz ihres bisherigen intensiven Umgangs mit den digitalen Medien nur über geringe Medienkompetenzen, wie beispielsweise Informationskompetenz, verfügen [16]. Weiterhin zeigen sich Erstsemester gegenüber dem Einsatz von digitalen Medien im Unterricht am wenigsten aufgeschlossen [17]. Die Erfahrungen von Erstsemestern mit E-Learning-Angeboten sind gering, die Möglichkeiten dieser Medien oft unbekannt [4]. Mit dem Erwerb der Fähigkeiten durch die curricular vorgegebene Nutzung des Angebots an digitalen Medien im Studium steigt gleichfalls das Interesse [13].

Insgesamt bevorzugten Studierende bisher einen moderaten Einsatz von E-Learning und sahen diesen nicht als Ablösung, sondern als Ergänzung zur konventionellen Lehre. Sie wünschten sich einen direkten Kontakt mit ihren Lehrenden und Kommilitonen [17] und präferierten die Präsenzlehre aufgrund besserer Kommunikation, Kooperation, Betreuung und Erwerb von Fähigkeiten [18].

Grosch fasst angesichts der Ergebnisse zur Mediennutzung Studierender treffend zusammen, dass eine „tendenzielle Stagnation expliziten E-Learnings und umfassende Durchsetzung der Mediennutzung im Studium zeitgleich“ stattfindet [19].

Eine gesonderte oder vergleichende Auswertung von Daten speziell zum studiumsbezogenen Mediennutzungsverhalten von Studierenden der Ingenieurwissenschaften findet sich kaum unter der Gesamtheit der Untersuchungsauswertungen. Die ECAR-Studien sowie die Erhebung des Karlsruher Instituts für Technologie (KIT) bieten die bisher umfangreichste Datenbasis und zeigen, im Vergleich zu den Ergebnissen interdisziplinärer Befragungen, deutlich andere Tendenzen bei der Mediennutzung ingenieurwissenschaftlicher Studierender.

Kvavik und Caruso stellen in der ECAR-Studie [17] von 2005 fest, dass Studierende ingenieurwissenschaftlicher Fächer in der Einzelbetrachtung ein anderes Mediennutzungsverhalten aufzeigen als die Studierenden anderer Fächer. Als Grund vermuten sie die unterschiedlich weit fortgeschrittene curriculare Einbindung von E-Learning-Medien. In den ingenieurwissenschaftlichen Fächern, so die Studie, wurde diese Einbindung durch die Forderungen nach dem Erwerb von Medienkompetenzen bei den Studierenden seitens der Regierung und der Fachverbänden forciert [17]. Studierende der Ingenieurwissenschaften besitzen und nutzen im Vergleich zu anderen Disziplinen das vielfältigste Medienangebot. Die Studierenden der Ingenieurwissenschaften schreiben sich selbst im Vergleich mit Studierenden anderer Fächer die höchsten Werte im Bereich Medienkompetenz zu, insbesondere in Bezug auf Computer- und Software-Nutzung. Interviews mit Studierenden der Ingenieurwissenschaften zeigen, dass die Studierenden besonders zufrieden mit der Quantität und Qualität des Medieneinsatzes in ihren Fächern sind. Sie betonten in der Befragung, dass Kurse, die zum Einsatz digitaler Medien ermutigen und diesen forcieren, die Aneignung von Medienkompetenzen unterstützen. Die Studierenden halten sich dazu befähigt, sich schnell in neue Medien einzuarbeiten, wünschen sich allerdings unterstützend adäquate Einweisungen und Übungen begleitend zum Einsatz neuer Medien.

Insgesamt, und hier unterscheiden sich die Befunde im Vergleich zu denen der anderen Fachkulturen, wünscht sich die Mehrzahl der ingenieurwissenschaftlichen Studierenden einen umfangreichen oder sogar ausschließlichen Einsatz digitaler Medien im Studium [17].

Diese Ergebnisse werden auch in neueren ECAR-Studien bestätigt: „Over the years we have found that certain majors are associated with the IT that respondents use, which makes sense because required technologies vary by major“ [15].

Während die ECAR-Studien nicht zwischen den ingenieurwissenschaftlichen Fachdisziplinen, wie beispielsweise Maschinenbau, Bauingenieurwesen oder Elektrotechnik, differenzieren, so betrachtet die KIT-Studie das disziplinär unterschiedliche studiumsbezogene Mediennutzungsverhalten an der durchführenden Universität. Grosch unterscheidet medienaffine und medienaverte Studiengänge, wobei die Mediennutzung ingenieurwissenschaftlicher Studiengänge divergiert: Maschinenbauer nutzen insgesamt mehr und unterschiedlichere Medien als Bauingenieure oder Bioingenieure. Er zeigt, dass zum Beispiel die Nutzungshäufigkeit von Lernsoftware nach Studienfächern signifikant variiert, wobei im Maschinenbau und in der Elektrotechnik die mit Abstand häufigste Nutzung stattfindet [19]. Im Maschinenbau verwenden die Studierenden zudem Wikis mit überdurchschnittlicher Häufigkeit, gleichzeitig aber unterdurchschnittlicher Zufriedenheit, während im Studiengang Bauingenieurwesen Wikis verhältnismäßig selten, dann allerdings mit hoher Zufriedenheit genutzt werden. Auch im Gebrauch von Vorlesungsaufzeichnungen in Form von Videos, Audiodateien oder Folien ergeben sich deutliche Unterschiede nach Studienfächern. Sie werden beispielsweise häufig im Studiengang Wirtschaftsingenieurwesen genutzt, im Bauingenieurwesen dagegen ist die Nutzung gering [19].

Grosch vermutet die Ursache für die unterschiedliche Intensität der Mediennutzung in unterschiedlichen curricularen Strukturen und spricht sich für Anschlusserhebungen zur Überprüfung dieser Annahme aus.

5 Diskussion der Ergebnisse der bisherigen Untersuchungen

Zunächst ist bei der Interpretation der Ergebnisse der Studien das Untersuchungsdesign zu beachten, welche Stichproben gewählt und wie die Daten erhoben wurden, bevor generelle Aussagen getroffen werden. Einige Studien haben ausschließlich Erstsemester befragt [20], um insbesondere die Charakterisierung der jüngsten Studierenden als Digital Natives zu untersuchen. In Hinblick auf die studiumsbezogene Mediennutzung kann diese Gruppe keine Ergebnisse beitragen, da sie innerhalb der ersten Studienwochen noch keinen näheren Kontakt mit dem gesamten Medienangebot ihrer Hochschule hatten. Zudem ändert sich, wie oben ausgeführt, die Einstellung zur studiumsbezogenen Mediennutzung mit zunehmend positiver Erfahrung im Verlauf des Studiums.

Die Art der Datenerhebung kann die Auswahl der Stichprobe ebenfalls beeinflussen: Mit einer rein online zur Verfügung gestellten Umfrage könnten eher diejenigen Studierenden erreicht werden, die sich gerne und regelmäßig im Internet aufhalten, während eine reine Pen-und-Paper-Umfrage an der Hochschule online mobile Studierende außer Acht lassen könnte.

Eine weitere Einschränkung bei der Ergebnisauswertung ist der Umstand, dass der Großteil der Untersuchungen zum Mediennutzungsverhalten Studierender nicht zwischen studiumsbezogener und privater Mediennutzung unterscheidet. So gibt eine undifferenzierte Aussage über den Gebrauch von Social Media keinen Aufschluss darüber, ob und inwieweit solche Communities in Bezug auf das Studium genutzt werden [4].

Zudem muss auch die technische Entwicklung und didaktische Einbindung der im akademischen Bereich eingesetzten E-Learning-Dienste beachtet und kritisch hinterfragt werden, ob die Qualität insbesondere des E-Learning 2.0-Angebotes zu Anfang und teilweise bis heute hoch genug war, um eine Nutzung attraktiv zu machen. Hier ist weitere Forschung nötig, um die Ansprüche Studierender an einzelne Angebote sowie fördernde und hemmende Faktoren der Nutzung zu identifizieren.

Weiterhin wurden die meisten Untersuchungen, wie beispielsweise die KIT-Studie, an den jeweiligen Hochschulen und nicht institutionsübergreifend durchgeführt. Die studiumsbezogene Mediennutzung ist stark vom Angebot der Hochschule abhängig. Ist ein bestimmtes digitales Medium gar nicht, nur gering oder in unzureichender Qualität an einer Hochschule präsent, so wirkt sich dies auf die Ergebnisse zur Nutzung durch die Studierenden aus. Aus diesem Grund sind Vergleiche bis hin zu Validierungen mit externen Erkenntnissen anderer Hochschulen erforderlich, sollen gültige und zuverlässige Aussagen getroffen werden.

Zuletzt, und schließlich für die Erforschung der studiumsbezogenen Mediennutzung ingenieurwissenschaftlicher Studierender relevant, ist für die meisten Untersuchungen festzustellen, dass die Daten nicht fachspezifisch ausgewertet wurden. Wie oben ausgeführt, zeigen Studierenden der Ingenieurwissenschaften ein signifikant abweichendes Mediennutzungsverhalten und wahrscheinlich auch Varianzen innerhalb der einzelnen Disziplinen, die für die mediale Ausgestaltung der ingenieurwissenschaftlichen Fachrichtungen in weiteren Untersuchungen erforscht werden sollten.

6 Ausblick

Die bisherigen Untersuchungsergebnisse lassen bereits, mit Einschränkungen, fundierte Aussagen zum Mediennutzungsverhalten Studierender zu. Doch trotz der vorhandenen Forschungsaktivität in dem Bereich der studentischen Mediennutzung mangelt es an Vergleichsdaten, vor allem zu einzelnen Studienfächern und -disziplinen. Speziell für die Anpassung und nachhaltige Entwicklung von Medienangeboten für das Ingenieurstudium müssen weitere Daten erhoben werden.

Unter diesen Gesichtspunkten führt das Zentrum für Lern- und Wissensmanagement der RWTH Aachen University in Zusammenarbeit mit dem Karlsruher Institut für Technologie im Sommersemester 2013 im Rahmen des Projektes ELLI - Exzellentes Lehren und Lernen in den Ingenieurwissenschaften – eine Umfrage zu den Mediennutzungsgewohnheiten von Studierenden der Ingenieurwissenschaften durch. Maßnahmen und Lösungsansätze zur Verbesserung der Studienbedingungen und der Qualität der Lehre in den Ingenieurwissenschaften konsequent zielgruppenorientiert zu gestalten ist zentraler Anspruch des Projektes ELLI.

Die Umfrage erhebt Daten zur studiumsbezogenen Nutzungshäufigkeit, -zufriedenheit und -akzeptanz von etwa 50 Mediendiensten und vergleicht 40 verschiedene Umgebungsvariablen. Dabei identifiziert sie mögliche Einflussfaktoren auf die Mediennutzung im Studium, wie Freizeitnutzung von Medien, Lernverhalten, soziodemografische Größen sowie Zusammenhänge der Mediennutzung mit der Qualität des Studiums im Allgemeinen. An der Mediennutzungserhebung sind inzwischen über 25 Hochschulen in elf Ländern beteiligt, so dass nationale und internationale Vergleichsdaten zur Mediennutzung im ingenieurwissenschaftlichen Studium zur Verfügung stehen.

Die aus der Untersuchung resultierenden Erkenntnisse werden für die gezielte Medienentwicklung in der ingenieurwissenschaftlichen Lehre Orientierung geben. Erste Ergebnisse sind zu Anfang des Wintersemesters 2013/2014 zu erwarten.

Literaturverzeichnis

1. Prensky, M.: Digital Natives, Digital Immigrants Part 1. On the Horizon, Oktober 2001. Vol. 9 No. 5, S. 1-6. Online abrufbar unter http://www.emeraldinsight.com/journals.htm?issn=1074-8121&volume=9&issue=5&articleid=1532742&show=pdf
2. Schulmeister, R.: Gibt es eine „Net Generation"? Erweiterte Version 3.0. Hamburg, 2009. Online abrufbar unter http://www.zhw.uni-hamburg.de/uploads/schulmeister_net-generation_v3.pdf
3. Rohs, M.: Studierendenbefragung E-Learning 2008 Ergebnisbericht. Universität Zürich, 2009. Online abrufbar unter http://www.scribd.com/doc/23681534/Studierendenbefragung-E-Learning-2008
4. Ebner, M., Schiefner, M., Nagler, W.: Has the Net-Generation Arrived at the University? - oder Studierende von heute, Digital Natives? In: Campus 2008. Offener Bildungsraum Hochschule - Freiheiten und Notwendigkeiten. Zauchner, S., Baumgartner, P., Blaschitz, E., Weissenbäck, A. (Hrsg.). Münster, 2008. S. 113-123. Online abrufbar unter http://www.pedocs.de/volltexte/2010/2977/pdf/Zauchner_Baumgartner_etal_2008_Offener_Bildungsraum_HS_D_A.pdf
5. Gidion, G., Grosch, M.: Welche Medien nutzen die Studierenden tatsächlich? Ergebnisse einer Umfrage zu den Mediennutzungsgewohnheiten von Studierenden. In: Forschung & Lehre. Deutscher Hochschulverband (Hrsg.), 6/12. Online abrufbar unter http://www.forschung-und-lehre.de/wordpress/Archiv/2012/ful_06-2012.pdf
6. Downes, S.: E-Learning 2.0. E-Learn Magazine - Education and Technology in Perspective, Oktober 2005. Online unter http://elearnmag.acm.org/featured.cfm?aid=1104968
7. Welbers, U., Gaus, O. (Hrsg.): The Shift from Teaching to Learning. Konstruktionsbedingungen eines Ideals. Bielefeld, 2005.
8. Palfrey, J., Gasser, U.: Born Digital: Understanding the First Generation of Digital Natives. New York, 2008.
9. Taylor, M. L.: Generation NeXt Comes to College: 2006 Updates and Emerging Issues. A Collection of Papers on Self-Study and Institutional Improvement, 2006. Vol 2. S. 48-55. Online abrufbar unter http://www.taylorprograms.org/images/Gen_NeXt_article_HLC_06.pdf
10. Oblinger, D. G., Oblinger J. L. (Hsrg.): Educating the Net Generation. Educause 2005. Online abrufbar unter http://www.educause.edu/research-and-publications/books/educating-net-generation
11. Kleimann, B., Weber, S., Willige, J.: E-Learning aus Sicht der Studierenden: HISBUS-Kurzbericht Nr. 10, 2005. Online abrufbar unter http://www.hisbus.de/results/pdf/2005_hisbus10_e-learning.pdf
12. Kleimann, B., Özkilic, M., Göcks, M.: Studieren im Web 2.0 - HISBUS-Kurzbericht Nr. 21, 2008. Online abrufbar unter https://hisbus.his.de/hisbus/docs/hisbus21.pdf
13. Kvavik, R. B., Caruso, J. B., Morgan, G.: ECAR Study of Students and Information Technology 2004: Convenience, Connection and Control. Online abrufbar unter http://www.educause.edu/library/resources/ecar-study-students-and-information-technology-2004-convenience-connection-and-control
14. Jadin, T., Richter, C., Zöserl, E.: Formelle und informelle Lernsituationen aus Sicht österreichischer Studierender. In: Zauchner, S., Baumgartner, P., Blaschitz, E., Weissenbäck, A.

(Hrsg.): Offener Bildungsraum Hochschule: Herausforderungen und Notwendigkeiten. Münster: Waxmann 2008. S. 169-180 Online abrufbar unter http://www.pedocs.de/volltexte/2010/2977/pdf/Zauchner_Baumgartner_etal_2008_Offener_Bildungsraum_HS_D_A.pdf

15. Smith, S. D., Borreson Caruso, J.: The ECAR Study of Undergraduate Students and Information Technology 2010. Vol. 6, 2010. Online abrufbar unter http://net.educause.edu/ir/library/pdf/ERS1006/RS/ERS1006W.pdf
16. Bundesministerium für Bildung und Forschung (Hrgs.): Kompetenzen in einer digital geprägten Kultur. Medienbildung für die Persönlichkeitsentwicklung, für die gesellschaftliche Teilhabe und für die Entwicklung von Ausbildungs- und Erwerbsfähigkeit, 2010. Online abrufbar unter http://www.bmbf.de/pub/kompetenzen_in_digitaler_kultur.pdf
17. Kvavik, R. B., Caruso, J. B.: ECAR Study of Students and Information Technology 2005: Convenience, Connection, Control and Learning. Online abrufbar unter http://net.educause.edu/ir/library/pdf/ers0506/rs/ers0506w.pdf
18. Paechter, M., Fritz, B., Maier B., Manhal, S.: eSTUDY - eLearning im Studium: Wie beurteilen und nutzen Studierende eLearning? Projektbericht, Juni 2007. Online abrufbar unter http://www.e-science.at/dokumente/eSTUDY_Endbericht.pdf
19. Grosch, M., Gidion, G.: Mediennutzungsgewohnheiten im Wandel. Ergebnisse einer Befragung zur studiumsbezogenen Mediennutzung. Karlsruhe, 2011. Online abrufbar unter http://uvka.ubka.uni-karlsruhe.de/shop/download/1000022524
20. Kennedy, G. E.: First year students' experience with technology: Are they really digital natives? In: Australasian Journal of Educational Technology, 24(1), 2008, S.108-122

Early Accessibility Evaluation in Web Application Development

Helmut Vieritz, Daniel Schilberg and Sabina Jeschke

Abstract Existing accessibility guidelines are mainly focused on runtime behavior and do not provide recommendations and evaluation for conceptual design of Web applications. Our approach aims to support more abstract principles for analysis and design of accessible Web applications. Combined with a prototype evaluation, it provides early integration of accessibility requirements into the process of Web application development. The approach is based on a model-driven user interface design method. Analysis of tasks and workflow is used to design a prototype which is evaluated with a simple screening technique to get fast and efficient results on selected accessibility requirements. The longtime objective of this work is a general concept for software development which bridges the gap between user requirements and developers needs in the field of accessibility.

Keywords Accessibility · Evaluation · Web Development · User-Centered Design

1 Background

Accessibility of Web-based user interfaces (UI) is an important research topic on human-computer interaction (HCI). Accessibility and usability [1] have as non-functional requirements an important impact on software architecture and design. As other non-functional needs e.g. scalability and reliability, they have to be considered early in the design process to realize them successfully. Common existing accessibility recommendations as the Web Content Accessibility Guidelines (WCAG 2.0 [2]) of the World Wide Web Consortium (W3C) are focused on runtime behavior

H. Vieritz (✉) · D. Schilberg · S. Jeschke
IMA/ZLW & IfU, RWTH Aachen University, Dennewartstr. 27, 52068 Aachen, Germany
e-mail: helmut.vieritz@ima-zlw-ifu.rwth-aachen.de

D. Schilberg
e-mail: daniel.schilberg@ima-zlw-ifu.rwth-aachen.de

S. Jeschke
e-mail: Jeschke.office@ima-zlw-ifu.rwth-aachen.de

Originally published in "HCI (7): Universal Access in Human-Computer Interaction. User and Context Diversity - 7th International Conference, UAHCI 2013, Held as Part of HCI International 2013", © 2013. Reprint by Springer International Publishing AG 2016, DOI 10.1007/978-3-319-46916-4_53

and require implementation to be evaluated. They do not match the needs in early development phases as software analysis and design.

Typically, prototyping or mockups help to overcome this lack of rules in analysis and design. Use case description provides the required information regarding user task objectives and activities. Based on task analysis, user-centered design (UCD) allows focusing on UI interaction. Here, an analysis and modeling approach is described to implement and evaluate a part of the WCAG 2.0 recommendations in early UI prototypes. Research questions are:

1. How can accessibility requirements be integrated early in analysis and design of Web application development?
2. What kind of technique is useful to evaluate the prototypes regarding accessibility?
3. What kind of technology can be used to design and implement accessible UI prototypes?

The next section gives an overview regarding the related research. The overall concept of the approach and the research questions are presented in Section 3. A case study in Section 4 details the approach and finally in Section 5, the conclusions and the outlook complete the discussion.

2 Related Research

User-centered Design (UCD) is a common method to ameliorate the usability. Accessibility in UCD is described by Henry [3] focusing on integration of users with disabilities in analysis, design and evaluation. Early usability evaluation is discussed in [4]. The approach is based on the Model-driven Architecture (MDA) and a usability framework. The specifics of accessibility are not focused.

Approaches for Model-driven Design (MDD) in UI development are User Interface Markup Language (UIML) [5], User Interface Description Language (UIDL) [6], User Interface Extensible Markup Language (UsiXML) [7], useML [8], Unified Modeling Language for Interactive Applications (UMLi) [9], Task Modeling Language (TaskML) [10] and Dialog Modeling Language (DiaMODL) [10]. Three UI models are known as essential [11] – the task, dialog and (abstract) presentation model.

Only few publications address the integration of accessibility in MDD of UIs. The Dante project [12] uses annotations in UI modeling to improve the navigation capabilities for visually impaired users. The authors have discussed particular aspects of UCD and MDD for accessible UIs in former publications (e.g. [13, 14]. As well, more details of our UI modeling process can be found in [13, 14].

3 The Approach

In short, accessibility means that all required information is perceivable, operable, understandable and robust (usable with different UI technologies) [2]. Information is required if the user needs it to accomplished his or her working tasks. Thus, analysis and identification of workflow-related information is the key to early accessibility integration. Analysis starts with use case description including possible scenarios of HCI. Important information is the identification and temporal order of main activities at the one side and of user actions and workflow at the other. Complex tasks need different modes of HCI and navigation between them. Modes are consistent subsets of user actions e.g. edit the shipping address in online shops [4]. They are related to a user main activity respectively scenario – e.g. carry out online purchases.

The presented approach is based on a general UI modeling concept which is discussed in former publications [13, 14]. The objective of this concept is the integration of accessibility requirements during the development process of an e.g. Web application. Figure 1 gives an overview. The concept separates four different structure levels. Two are the (macro)-levels of modes and the change of modes (navigation). A mode is particular context of HCI (e.g. a single Web site) or a view in a desktop application (e.g. an editor or a file browser) [15]. The other two are the basic user actions and their temporal order (workflow). For analysis and design, three essential UI models [11] are required – the task model, the dialog model and the presentation model. The task model describes user's point of view on the HCI and represents the more technology-independent part of all UI models. The presentation model describes the (abstract) structure and behavior of the UI technology and represents the system view on the HCI. The dialog model combines both worlds in one view and specifies the interaction by describing the information interchange between user and system.

In this publication, the focus is set on earliest accessibility integration in analysis, design and implementation of the two macro-level navigation and modes. Based

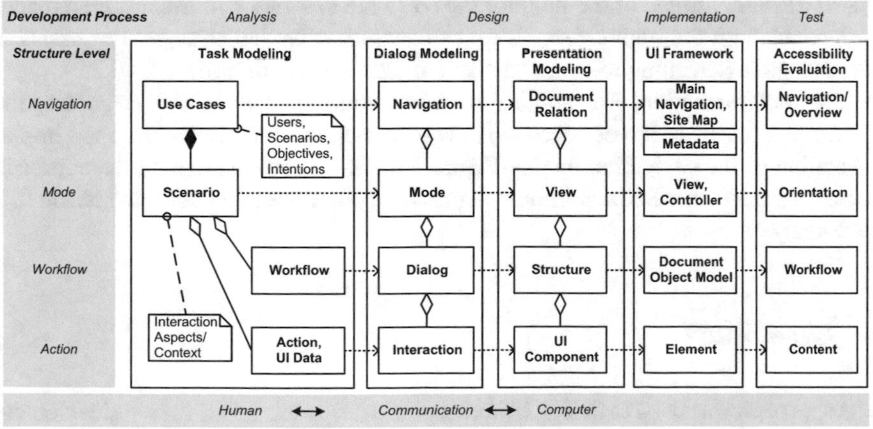

Fig. 1 Development process at a glance

Table 1 Relevant WCAG 2.0 guidelines

No.	Guideline
1.3.3.	Sensory Characteristics
2.1.1.	Keyboard
2.1.2.	No Keyboard Trap
2.4.3.	Focus Order
2.4.4.	Link Purpose
2.4.5.	Multiple Ways
2.4.6.	Headings and Labels
2.4.7.	Focus Visible
2.4.8.	Location
3.2.3.	Consisted Navigation

on task analysis and the modeling of HCI modes, a simple UI prototype can be implemented and evaluated for some accessibility criteria (first research question). Related aspects include user's orientation, over-view and navigation in HCI with Assistive Technology (AT) e.g. Screen readers. Affected criteria of the WCAG [2] are given in Table 1. More abstract accessibility requirements were derived from guidelines:

4. Clearly structured activities – a particular Web site is focused on only one main activity which can be identified by the title-element. Main activities are mapped one by one to a UI mode.
5. Serial order of modes corresponds with user's expectation to support assistive technology as screen readers etc.
6. Navigation is accessible and corresponds with user's mental representation.

To answer the second research question, a simple and easy evaluation method is the screening technique. Most screening techniques are based on interaction with the UI with limited sensory or physical abilities (see [3] for more details) e.g. low vision glasses. Instead of the monitor the software designer can use a screen reader to interact with the application. Tests can include adaptive strategies or assistive technologies. Screening tests are best-suited during early design.

For implementation, Web application frameworks were evaluated. Regarding the third question, Java Server Faces (JSF) was chosen since it allows the easy implementation of the navigation model. Figure 2 shows the seven steps of the approach in an overview. In the next section, the process is discussed in detail and tested in a case study.

4 Case Study

The approach was used to design a prototypical user-centered UI for a finite element (FE) integration system in virtual production. In complex processes different physical

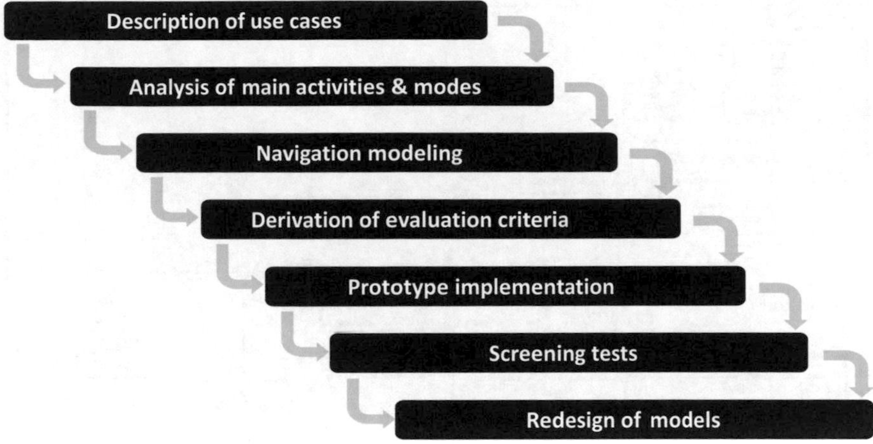

Fig. 2 Process overview

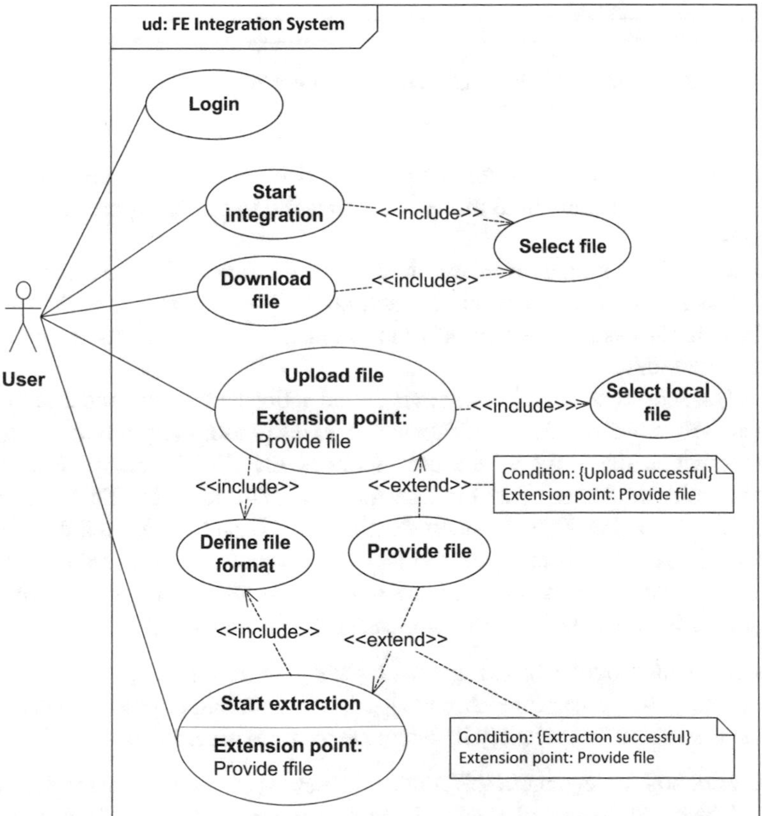

Fig. 3 Use cases for an Information Integration System

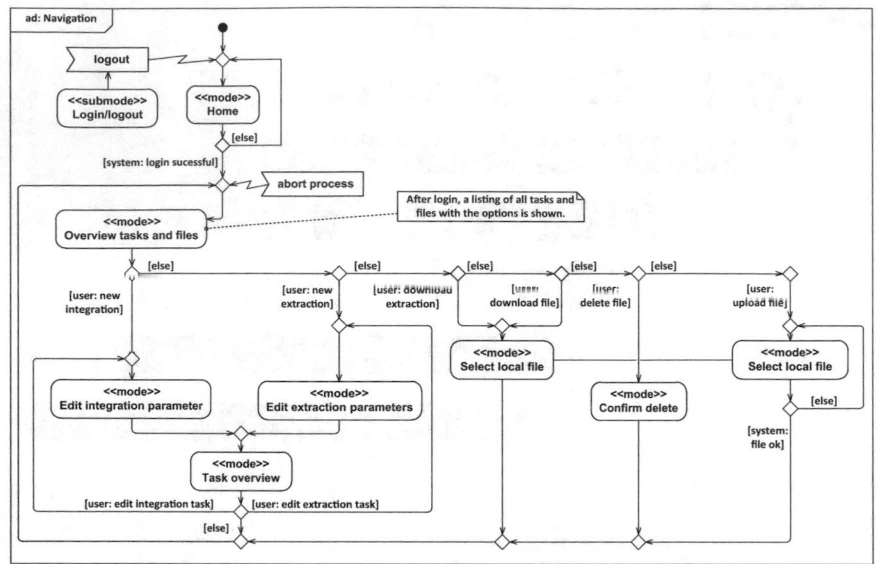

Fig. 4 Navigation Model for an information integration system

simulations are needed to analyze and parameterize the whole process. Information integration provides inter-simulation communication to combine particular physical processes.

The usage of the system was described by the customer with a use case diagram (Fig. 3) and an additional description of activities. Typical use cases of the customer were login to the system, start an information integration or extraction, select, load and download files.

Based on scenarios, the related activities and actions were analyzed and modeled within activity diagrams. The first step was to identify user main activities, to group similar activities and to model the use modes of the UI. The grouping of similar activities avoids the design of use modes which are almost identical. This step reduces the number of modes. Fewer modes simplify the UI and reduce learning efforts. Alternatively, structure layering techniques etc. can be used for customers without technical background when they e.g. are not familiar with UML activity diagrams. Evaluation criteria are derived from use case analysis including:

7. Can the main activities be recognized by using a screen reader?
8. Is the navigation accessible? Are navigation options and targets always clear?
9. Is the navigation from site to site adequate to the expected work-flow?

The temporal arrangement of modes is described in the navigation model (Fig. 4). The model is notated as a UML activity diagram. Modes and sub-modes are assigned with UML stereotypes.

For the implementation of the navigation model, Java Server Faces (JSF) was chosen. It supports the easy model-to-text transformation into the faces-config.xml file which describes the application navigation. Information from the model was used to implement the particular JSF views which represent the sites of the Web application. The prototype does not contain a particular navigation system. Thus, the changes from site to site were implemented as hyperlinks. Additionally, the sites contain some meta-information as the title which corresponds to the mode name in the navigation model.

A screening test with the simple prototype was done by three developers with screen readers. Since the developers were not familiar with the AT, a longer warm-up was necessary. The test has shown that:

- User clearly identify main activity
- User identify next expected activity
- Last activity was sometimes not clear for the user
- Users could evaluate the accessibility to main navigation, home mode etc.
- User overlooks the main steps of workflow within the application

Additionally, the mentioned WCAG criteria were checked. For some criteria (2.4.3, 2.4.7, 3.2.3) a complete testing was possible. Test results were used to fix errors in the prototype. Complete testing for the other criteria (1.3.3, 2.1.1, 2.1.2, 2.4.4, 2.4.5, 2.4.6 and 2.4.8) requires additionally the implementation of content. Therefore, testing had to be finished in later development phases. It was remarked by the test persons that the short-time feedback of the test helps to understand better the accessibility requirements.

5 Conclusions

The presented concept evaluates early accessibility evaluation in Web application design and development. Accessibility requirements regarding navigation, orientation and overview are combined with UCD and MDD. The concept is based on task analysis and modeling. The case study has shown that a limited early accessibility evaluation is possible. WCAG testing can be finished for some criteria. The majority of criteria is content-related and need more implementation. Even, some general aspects as overview, orientation and (macro) navigation can be evaluated early. Limitations of the approach are:

- Only subset of accessibility criteria is testable
- Evaluation focused on AT for visual impairments
- Need for more comprehensive evaluation

Further research would be necessary to test more criteria related to others impairments and disabilities e.g. a systematic approach based on the International Classification of Functioning, Disability and Health (ICF) of the World Health Organization (WHO). Systematic accessibility guidelines for the design and conception of Web applications (or software in general) are still not available. They are required for a holistic integration of accessibility in development processes. Last but not least, the approach is focused on early accessibility integration. Further work is necessary to cover the whole development process in a systematic way.

Acknowledgements The research leading to these results has received funding from the Federal Ministry of Education and Research (BMBF) for the project Excellence in Teaching and Learning in Engineering Sciences (ELLI).

References

1. E. Folmer, J. van Gurp, J. Bosch, Software architecture analysis of usability. In: *EHCI-DSVIS 2004*, *LNCS*, vol. 3425. Springer, 2004, *LNCS*, vol. 3425, p. 38–58
2. World Wide Web Consortium 2008. Web content accessibility guidelines 2.0, 2008. URL http://www.w3.org/TR/WCAG20/. Last visited: 11/09/2011
3. S. Henry. Just ask, 2007. Lulu.com
4. S. Abrahao, E. Insfran, Early usability evaluation in model driven architecture environments. In: *Sixth International Conference on Quality Software, 2006. QSIC 2006*. IEEE, 2006, pp. 287–294
5. U. community 2009. User interface markup language (UIML), 2009. URL http://uiml.org/. Last visited: 11/09/2011
6. U. community 2008. User interface description language (UIDL), 2008. URL http://www.uidl.net/. Last visited: 11/09/2011
7. U. community 2010. User interface extensible markup language (UsiXML), 2010. URL http://itea.defimedia.be/. Last visited: 11/09/2011
8. A. Reuther, *useML - systematische Entwicklung von Maschinenbediensystemen mit XML*. Technische Universität Kaiserslautern, 2003
9. P. Da Silva, *Object Modelling of Interactive Systems: The UMLi Approach*. University of Manchester, 2002
10. H. Trætteberg, *Model-based User Interface Design*. Norwegian University of Science and Technology, Trondheim, 2002
11. K. Luyten, *Dynamic User Interface Generation for Mobile and Embedded Systems with Model-Based User Interface Development*. Transnational University Limburg, 2004
12. Y. Yesilada, S. Harper, C. Goble, R. Stevens, Dante annotation and transformation of web pages for visually impaired users. In: *Proceedings of the 13th international World Wide Web conference on Alternate track papers & posters (WWW Alt 2004)*. ACM Press, 2004, p. 490–491
13. P. Göhner, S. Kunz, S. Jeschke, H. Vieritz, O. Pfeiffer, Integrated accessibility models of user interfaces for IT and automation systems. In: *Proceedings of the Conference on Computer Applications in Industry and Engineering (CAINE-2008), Honolulu, Hawaii, USA, 12–14 November 2008*. 2008-11, pp. 280–285
14. H. Vieritz, F. Yazdi, D. Schilberg, P. Göhner, S. Jeschke, User-centered design of accessible web and automation systems. In: *Information Quality in e-Health. Proceedings of the 7th Conference of the Workgroup Human-Computer Interaction and Usability Engineering of the Austrian Computer Society, USAB 2011, Graz, Austria, November 25–26, 2011, Lecture Notes in Computer Science*, vol. 7058, ed. by A. Holzinger, K.M. Simonic. Springer Berlin Heidelberg, 2011, *Lecture Notes in Computer Science*, vol. 7058, pp. 367–378. URL http://www.springerlink.com/content/47724v115vt7h320/
15. H. Thimbleby, *Press on: Principles of Interaction Programming*. MIT Press, Cambridge, MA, 2010

ALLES ING! Count me in! – Attracting Human Talents in Providing Open Access to Universities with Focusing on Individual Opportunities in Engineering Sciences

Theresa Janssen, Mark Zeuch, Marcus Petermann and Andreas Kilzer

Abstract In higher engineering education of the 21st century, it is necessary to "go with the times" and make use of many and varied ways when it comes to attracting young talents for technical studies. At the Ruhr University Bochum, the Project ELLI decides to start an offensive with the engineering faculties: They give an insight into "Who we are and what we are doing" in the framework of their new, joint initiative "ALLES ING!". In this endeavor, the faculties for Civil and Environmental Engineering, Mechanical Engineering and Electrical Engineering and Information Technology present themselves with a shared "business card" to the public beyond, as well as within the university. This paper documents the concept and development of "ALLES ING!" as a means of attracting human talent for engineering education, as well as the challenges and opportunities that still lie ahead.

Keywords Engineering Education · Engineering Students · Social Network Services · Knowledge Transfer

1 Introduction

In Germany the number of graduates in the engineering discipline is stagnating although the data of first-year students has slightly increased. Among others by dynamic and demographic development there is an increasing demand of technically educated academics on the labor market: "Zwar ist der Bedarf an Ingenieuren [...] vom höchsten Stand im September 2008 (68.800) auf 21.200 im Januar 2010 gesunken. Aber seither werden Ingenieure wieder verstärkt gesucht. Das Ganze hat neben der wirtschaftlichen Entwicklung noch eine demographische Komponente: 30.000 Absolventen verlassen zurzeit [in Deutschland] im Jahr die Ingenieurwissenschaftlichen Fakultäten, aber 36.000 Ingenieure gehen innerhalb eines Jahres in Rente" [1].

T. Janssen (✉) · M. Zeuch · M. Petermann · A. Kilzer
Ruhr-Universität Bochum, Bochum, Germany
e-mail: janssen@fvt.rub.de

Originally published in "Global Engineering Education Conference (EDUCON), IEEE 3-5 April 2014 Istanbul DOI: 10.1109/EDUCON.2014.6826225",

DOI 10.1007/978-3-319-46916-4_54

The resulting general skills shortage and the high drop-out rates during the first semesters of engineering studies underline the need to discuss this topic. "In den Studienbereichen Maschinenbau, Elektrotechnik und Bauingenieurwesen beendet an Universitäten mehr als jede(r) Zweite das Bachelorstudium ohne Abschluss. In Mathematik und Informatik erreicht der Studienabbruch ähnliche Größenordnung" [2]. The university drop-out is frequently caused by lacks of social, technical and academic integration [3]. Therefore universities should open up for new potential target groups and to new ways of communication in the long run, thus meeting the students' changing demands for information [4]. For in an interconnected media society such us ours, they are regularly confronted with a variety of information, competing for their attention [5].

Especially technical fields of study are confronted with the challenge of presenting matter-of-fact technical contents in a vivid way and from a personal perspective, in order to speak the language of those interested and increasing possibilities of identification with the subject of study, providing technology with a "human face". The variety of the engineering job profile is challenging because there are difficulties in giving a precise scope of duties: "Jobs are defined by their field of activity but college and university students commonly do not have a clear vision of the engineer's scope of duty." [6].

In higher engineering education of the 21st century, it is necessary to "go with the times" and make use of many and varied ways when it comes to attracting young talents for technical studies.

2 ELLI – Excellent Teaching and Learning in Engineering Science

The project ELLI meets this challenge. ELLI – Excellent Teaching and Learning in Engineering Science – is a common and financially supported initiative of the German Federal Government and the individual states for enhancing studying conditions and the quality of teaching.

ELLI is a cooperative project of the Ruhr University Bochum, the RWTH Aachen University and the Technische Universität Dortmund and is financed within a common initiative of the German Federal Government and the individual states. The project work is focused on four main areas: 1. Virtual Learning Environments, 2. Internationalization and Mobility, 3. Student Lifecycle and 4. Additional Skills [7].

At the Ruhr University Bochum the project ELLI started a new initiative to meet the challenge particularly with regard to the beginning of the engineer's education and is especially pursuing the goal of attracting non-traditional students for technical fields of study [8]. Role models from technology and science are applied in order to establish a higher identification with technical topics and fields of work. The initiative concentrates on the demonstration of a multifaceted mixture of personalities, experiences and knowledge, which describes university's everyday life of engineering students as well as scientific assistants of the departments in science, technology and administration. Due to an attractive depiction of the engineering discipline and the "people behind the technology", already interested college students shall be encouraged and beginners be accompanied by manifold examples.

Within the new common initiative "ALLES ING!" ELLI answers the slogan "Who we are and what we do". For this initiative the three engineering faculties Civil and Environmental Engineering, Mechanical Engineering and Electrical Engineering and Information Technology have joined forces to show up with a common visiting card to present the project to both the interested general public and within the university and the faculties themselves.

The medially implemented Role-Models-Concept is accompanied by the advisory services at Ruhr University designed by ELLI. Research and professional fields for engineers are introduced as part of these advisory services and the students are supported during their studies.

3 ALLES ING! – New Ways for Engineering Sciences

ALLES ING! makes use of the varied possibilities of the Smartphone age and places emphasis on a variety of media. In a new internet portal, which makes use of manifold ways of presentation, students and employees in engineering science report on their typical working day, a stay abroad or their research topic, or also on the start of their professional career.

On their profile pages, they get into a direct exchange with potential students and other people interested via videos, interviews, or short messages. A series of events called "ALLES ING! Live" accompanies the web offer, which is especially directed to an interested public from the region and relates technology to socially relevant topics [9].

The website went online on the 21th of March and there are now – 6 months later – approximately 150 visitors each week, clicking through the categories, research topics or watching short videos. Thus, the people in engineering science at RUB are really getting to know each other.

The object is to inspire interested college and university students to reflect themselves and to enable them to make their decisions relating to their study choice considering their own interests and strengths. Capacity for team work and communication are essential because although our students are self-confident, independent, creative and heterogeneous especially at the beginning of their studies there is a lack of orientation relating organization and requirements of the studies. Therefore they require a lot of motivation to get into the technical basics, the teamwork with other students as well as the exchange and support between the students of different semesters.

ELLI offers advice formats for the transitions in the engineering studies to enable the student to organize their studies self-determined right from the beginning.

Within the annual "Schülerinnen-Projektworkshops" with its Tea-Time ELLI also gets involved before the start of the studies. In this program, 55 female college students from the region get in contact with examples from the technology field. They get the possibility to ask questions and exchange experiences with these examples from young female student to successful female professors. In 2012 the role models captured insights into their professional career and a personal message for the young academics in video statements. These are to be found on the ALLES ING! website.

On the open house presentation ELLI and the engineering faculties offer information for interested female students: "The seven best reasons to become an engineer" are a first opportunity for college students to talk to female students and expert advisers. In 2013, approximately 40 female college students were pleased about the welcome and intensive exchange with the students.

The project supports more informative meetings such as "Market of study opportunities" and the period of matriculation before the semester start. Moreover, ELLI demonstrates perspectives for research work and final theses as well as the career entry.

During the informative meetings "Mit dem Master in die Praxis" and "Wege zum Dr.-Ing." students get into an exchange with supervisors for doctoral theses and RUB-Alumni as well as RUB doctoral candidates and thus they obtain very individual information. The annual peer-to-peer-consultation carried out by young professionals and doctoral candidates obtain an especially positive evaluation, which attracts approximately 60 students, respectively.

In a further step ELLI would like to unite the occupational history into the industry and science under the slogan "Karrierewege in den Ingenieurwissenschaften" and thereby present RUB-Alumnis‘ career advices and prospects.

Students find topics for final thesis on the common website of the initiative. ELLI motivates to exchange experiences of this important obstacle in the engineering studies to make it more transparent in organization and diversity of topics. There are already wanted posters of investigation projects online which arouse interest of other universities in Germany. This sector shall be accompanied by an online advisor for final thesis: step by step to success!

This interactive advisor tool offers online tips for writing final thesis as well as organization and the search of topics.

Advice formats and websites provide a low-threshold contact-offer, promote the identification with technical studies and increase the attractiveness of individual profile formation.

Peers 'experiences data achieve a huge acceptance and first-year students gain a more precise idea of their study pathway. With a new designation supported by manifold medial concepts, more non-traditionally interested students can be attracted for technical studies. The long term implementation of successful measures such as those mentioned above rises to a challenge which is confronted by ELLI due to the integration of all relevant protagonists of the university.

The measures used for the beginning of the studies are regularly evaluated by the collection of participants‘ and visitors' data as well as by quantitative opinion surveys. The qualitative feedback of the target group is assured by regular meetings with students' representatives, who are a valuable part of the project. There are very positive experiences in the collaboration with assistants of the engineering faculties, too. They are highly motivated and dedicated to participate in the realization of new ideas for the period of beginning the studies.

References

1. Friedrichsen, Heike: Ingenieur-Gehalt: Krisensicher und überdurchschnittlich hoch. Available at: www.academics.de/wissenschaft/ingenieurgehalt_krisensicher_und_ueberdurchschnittlich_hoch (Stand 30.10.2013).
2. Hafner, Theo: Nachfrage nach MINT-Studium steigt - Potenziale bei den Frauen noch nicht ausgeschöpft. Publikation HIS:Forum Hochschule 11|2013, www.his.de (Stand 30.10.2013).
3. Gensch, K./Kliegl, C. (2011): Studienabbruch - was können die Hochschulen dagegen tun?, Bayerisches Staatsinstitut für Hochschulforschung Hochschulplanung, München.
4. Verein Deutscher Ingenieure e. V. (Hrsg.) (2012): Ingenieure auf einen Blick. Erwerbstätigkeit, Innovation, Wertschöpfung. Online available at: https://www.vdi.de/fileadmin/vdi_de/redakteur/dps_bilder/SK/2012/2012_-Ingenieure_auf_einen_Blick.pdf (Stand: 29.05.2013).
5. Stifterverband für die deutsche Wissenschaft/Mc Kinsey & Company (2012): *Hochschul-Bildungs-Report 2020*, Edition Stifterverband, Essen.
6. Deutsche Akademie der Technikwissenschaften/Verein Deutscher Ingenieure e. V. (Hrsg.) (2009): Nachwuchsbarometer Technikwissenschaften. Available at: http://www.bmbf.de/pubRD/nachwuchsbarometer_technikwissenschaften.pdf (Stand: 29.05.2013), S. 61.
7. www.elli-online.net.
8. The Ruhr University Bochum lies in the Ruhr Area of the Federal State of North Rhine Westphalia (with about 17.8 Million inhabitants). With ten of Germany's biggest universities, the focus of the area lies on education and science. In the 1960s and 1970s, crucial reforms in engineering education were developed here. Today, the region is marked by a strong structural change.
9. www.ing.rub.de.

What Students Use – Results of a Survey on Media Use Among Engineering Students

Dominik May, Karsten Lensing, A. Erman Tekkaya, Michael Grosch, Ute Berbuir and Marcus Petermann

Abstract Nowadays, university students are facing a large number of highly diverse media, including conventional books as well as online-based mobile applications - all used to support learning. Especially the internet with connected social media services or e-learning possibilities induced significant changes in society and in the landscape of higher education during the last years and still do so. The four universities RWTH Aachen University, Ruhr-University Bochum, TU Dortmund University, and the Karlsruhe Institute of Technology conducted an exploratory student survey on media and information use, in order to expand the empirical database on that topic. A special focus was laid on mobile learning. In this context the survey asked for the hardware and software the students are using and for moments in which they already got in contact with any kind of mobile learning – e.g. by using special apps for learning or because they were asked by their teachers to use a mobile device. The results of the survey elucidate that the use of online media and especially social media as well as mobile devices in higher education are to be promoted in future. Furthermore, it reveals demands for action in the field of media competency concerning students and teachers.

Keywords Media Use · Engineering Education · Social Media · Academic Teaching · Mobile Learning

---

D. May (✉) · K. Lensing
Engineering Education Research Group, Center for Higher Education,
TU Dortmund University, Dortmund, Germany
e-mail: Dominik.may@tu-dortmund.de

A.E. Tekkaya
Institute of Forming Technology and Lightweight Construction (IUL),
TU Dortmund University, Dortmund, Germany

M. Grosch
Institute of Education and Vocational Training, Karlsruhe Institute of Technology, Kalsruhe, Germany

U. Berbuir · M. Petermann
Center for Higher Education and Institute of Forming Technology
and Lightweight Construction TU Dortmund University, TU Dortmund University,
Dortmund, Germany

Originally published in "Proceedings of 2014 Frontiers in Education Conference "Opening Doors to Innovation and Internationalization in Engineering Education"",

DOI 10.1007/978-3-319-46916-4_55

1 Introduction

The media use at universities – generally spoken – is highly diverse. From conventional books over digital documents to fully online submitted courses the institutions offer nearly everything. In contrast to that, the empirical insights on what students use and how often they use it is little. The project "Excellent Teaching and Learning in Engineering Education" (ELLI) – executed by the three German universities RWTH Aachen University, Ruhr University Bochum and TU Dortmund University – focuses on these changes. The project's aims are, among others, to show current possibilities in the e-learning context and to reveal new solutions e.g. for mobile learning scenarios – always in context with engineering education. Therefore the three universities – in cooperation with the Karlsruhe Institute of Technology – issued a questionnaire focusing this topic. With this survey engineering students from the three universities were asked about their media use.

Presenting the survey's results this paper is divided into three main parts. Within the first part the questionnaire's approach and methodological basis is laid out in short. The second part will cover general aspects, which have been explored through the survey. Hence, the collaborative use of online and social media will be focused. The third part is motivated by an explicit working package in the ELLI project. This part talks about the results concerning the use of mobile devices in learning – in general called mobile learning [1]. Mobile devices and apps have become indispensable for many people today. This is one reason why the use of mobile devices is more and more diffusing into higher education. Current literature offers only little empirical data on the question how students are using mobile devices for their learning processes. In addition to that, we do not know if, how often and in which situations the students were asked by teachers to use their mobile devices in learning contexts in class or for the learning process. Therefore the survey concentrated on mobile devices in particular and asked for the kind of hardware and the kind of apps the students are already using or they were encouraged to use by teachers. The paper will present the survey's results on the one hand and will draw conclusions for the future of higher engineering education research on the other hand.

2 General Methological Aspects

The survey's purpose was to get knowledge on media services the students are actually using, how satisfied they are with using these services and which advantages they see in using the services for their study. A special focus was set on mobile media and mobile devices.

The survey used a questionnaire that was developed at Karlsruhe Institute of Technology in 2009 [2]. Surveys with the help of this questionnaire were carried out several times and applied by 20 universities in six countries. It is based on the Media Acceptance Model MAM [3]. In this model media are understood as technologies supporting and extending human communication. Media acceptance is considered to be a special form of technology acceptance. It is seen as an indicator of the quality of a medium from a subjective (here: the student's) point of view. Due to the special focus on mobile technologies, the questionnaire was modified and several question batteries were added regarding this topic. The questionnaire contained items regarding frequency of use and satisfaction of use concerning 54 university-internal as well as external media services. Students had to answer using a five point scale (e.g. "very satisfied" – "very unsatisfied").

The survey was synchronously carried out and the questionnaires were distributed among engineering students at the three universities working together in the ELLI project: RWTH Aachen University (paper & pencil after lectures), University of Bochum (online) and TU Dortmund University (online). As two survey methods were mixed, it is difficult to define an overall return rate. Whereas the paper & pencil method led to a return ratc of nearly 100 %, the online survey were answered and send back only by 4 % of the asked students. By doing so, a total of 1587 samples from students were collected from April to May 2013. (gender aspects: 81.6 % male, 18.4 % female; age range: 2.9 % 18 years or younger, 14.7 % 19 years, 33.7 % 20 years, 15.3 % 21 years, 33.4 % 22 or older)

3 Results Concerning Collaborative Use of Online Media

Collaborative online media – such as wikis, weblogs and forums – were more and more implemented for the students' teamwork during the last years [4]. Surveys at other universities and faculties [5] as well as teacher statements of the engineering sciences [6] proved a rather hesitant use of this kind of media by students. Therefore the use of collaborative online media was of particular interest for the interpretation of the media use survey in the ELLI project. 87.3 % of the respondents stated, that they seldom or never actively use created Wikis, e.g. for a seminar. Furthermore, 3.4 % answered to use Wikis often, 0.6 % do this very often and 8.7 % were undecided. Newsgroups or internet forums are never or very seldom used by 55.9 % of the asked students. Only 3.5 % utilize this kind of collaboration very often (24.4 % undecided and 16.2 % often).

In the context of collaborative online media the students' information literacy was also requested. Here the question was if they trust third-party contents and if they accept them without checking the background. Moreover, a corresponding question was if they validate provided online information (Full question: Do you trust internet pages' content, which is controlled by other users? e.g. Wikipedia). 19.3 % of the engineering sciences students are less skeptical and have full trust in such homepages, on which contents are created or controlled by other users. Further 37.6 % of the asked students still do agree (even if not fully), 26.1 % are undecided and 17.1 % do rather or fully disagree Fig. 1.

In addition to that another question in this context asked for the validation of online information quality. Only slightly more than a third (37.2 %) always validates the quality of the provided information. 34.7 % check their online sources more often, 17.7 are undecided and 10.4 % check the quality seldom or never Fig. 2.

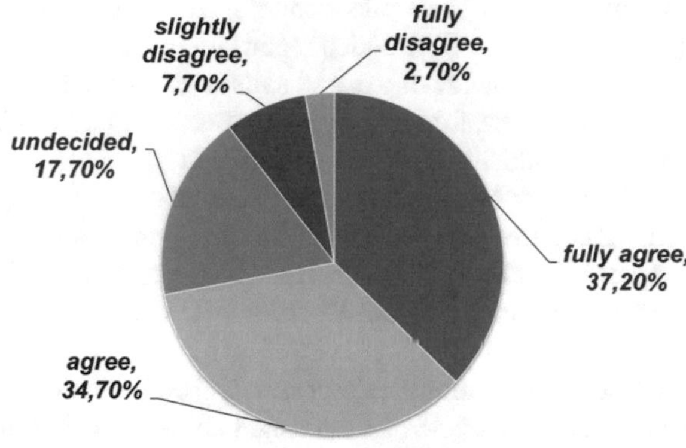

Fig. 1 Answers on question: Do you validate the quality of found information?

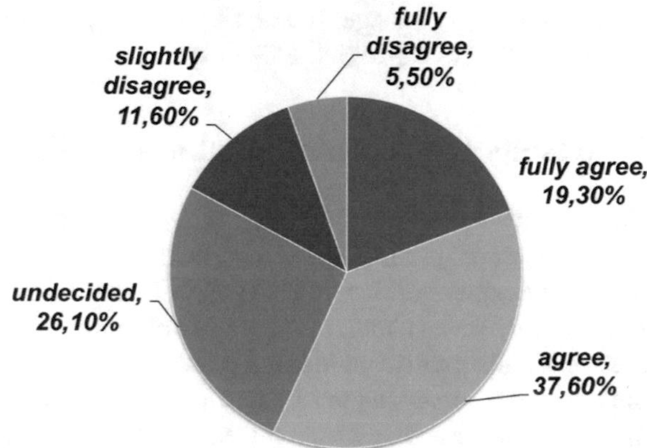

Fig. 2 Answers on question: Do you trust internet pages' content, which is controlled by other users?

3.1 Social Media use

Social platforms such as Facebook or information services like Twitter provide an easy access as well as already implemented possibilities for communication by means of groups, private messages and forums for the students during their studies. About one third of the students (30.2 %) stated to use Facebook for study purposes very often and 18.4 % do this at least often. Nevertheless, a high level of satisfaction with the use was indicated by only 12.7 %. On the other hand 20.6 % never use Facebook for their studies. The remaining 30.8 % of the asked students use it at least occasionally (16.8 % undecided, 14.0 % seldom) Fig. 3. Similar figures are reached with regard to Twitter, because 91.5 % of the survey's participants never use this application in context with their studies and only 1.1 % communicate very frequently via Twitter in this context (1.3 % often, 2.2 % undecided, 3.8 % seldom) Fig. 3. Google+ is being utilized by only a few students (2,7 %) for study purposes (85,7 % never use it). Services like StudiVZ, Xing, MySpace etc. are being used even less. Only 1.0 % of the asked students use these networks for their studies very often and the vast amount (90 %) never do this.

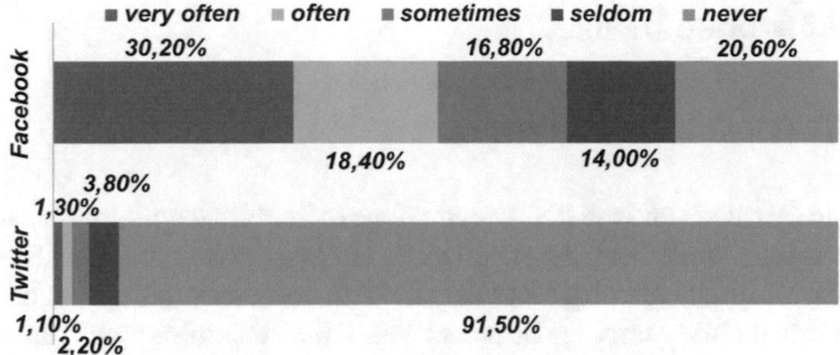

Fig. 3 Answers on question on frequency of use of Facebook and Twitter in context with studies

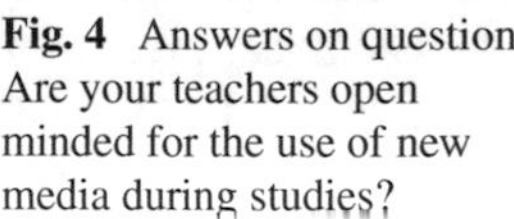

Fig. 4 Answers on question: Are your teachers open minded for the use of new media during studies?

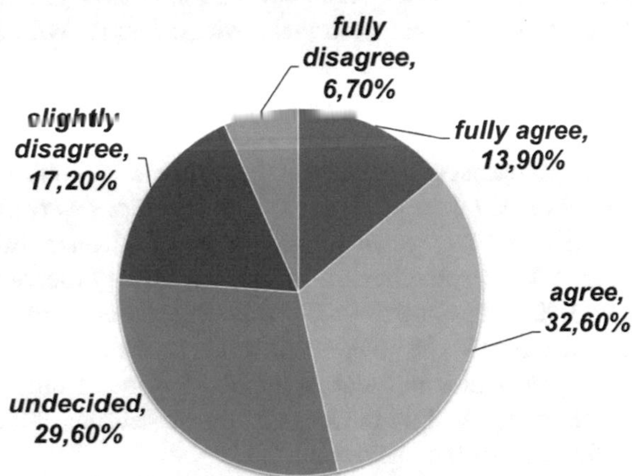

3.2 *Teachers' Position Towards Media use in Teaching*

Talking about media usage in context of learning can be seen on the one hand from the learner's perspectives. Another important perspective are the teachers and their openness for media us in the university context. Therefore the students of the ELLI-media use study were asked to rate the position of their teachers regarding the use of new media. Only 13.9 % of the students fully agreed with the statement, that their teachers are open for the use of digital media in the respective course of studies. 32.6 % are more appealing to it, 29.6 % are likely undecided, 17.2 % have the opinion that their teachers are not very open for the use of digital media and 6.7 % fully disagreed (Fig. 4).

After having explained some of the survey's results concerning (social) media usage at the university, in the following the focus will change to mobile devices. Mobile devises can be used in order to carry out mobile learning. Hence in the following mobile learning will be explained in a excursus first and then the survey's results will laid out.

4 Use of Mobile Devices

4.1 Excursus: Mobile Learning

Referring to Gartner's and the IDC's predictions on mobile technologies and internet services, usable mobile learning experiences are just around the corner [7, 8]. But what exactly is mobile learning? At this point it is necessary to define what is meant with the term mobile learning. In recent years the technology shift has changed the way we define Mobile Learning (PDAs to smartphones and tablets). That is also visible by looking at different definitions throughout the years. [9] e.g. offered a rather technology-centered view in 2000 by saying, that mobile learning is "elearning through mobile computational devices: Palms, Windows CE machines, even your digital cellphone" [10]. Found a more user-centered view by describing the learning environment and adding the learners' autonomy in the choice over time and place.

In order to go even more into detail an extended literature review was conducted in the ELLI-Project. More than 100 definitions on mobile learning in about 240 different sources were taken into account. The definitions were divided into 3 different clusters. These clusters represent different perspectives on mobile learning by highlighting the mobile devices, the flexibility for the learning process or new didactical approaches. The following quotations illustrate this variety and are taken from the most often cited sources in the context of Mobile Learning.

"What is new in "mobile learning" comes from the possibilities opened up by portable, lightweight devices that are sometimes small enough to fit in a pocket or in the palm of the one's hand." [1].

"Mobile Learning devices are defined as handheld devices and [...] should be connected through wireless connections that ensure mobility and flexibility." [9].

"[Mobile learning] provides the potential of personal mobile technologies that could improve lifelong learning programs and continuing adult educational opportunities." [11].

"Summing up all this perspectives [12] just released an up to date description of their literature research regarding the mobile learning evolution over the years. Following them mobile learning means learning across multiple contexts, through social and content interactions, using personal electronic devices."

In addition to the definition of mobile learning, it is possible to identify four different but central characteristics based on literature [13]:

- Use of mobile devises
- Local independence for learning
- Contextualization of the learning process
- Informality of learning

After having defined what mobile learning is, the survey and its results come back into focus. They survey in this context did not asked directly for the term mobile learning itself – as it might be unknown among students anyways. Instead the students were asked about the devices and software they use. The results with regard to hardware, apps and use at the university will be explained in the following.

4.2 *Survey Results on Mobile Devices and Mobile Learning*

Looking at today's campuses it is obvious to everybody that the number of students without any mobile device is extremely small. In order to verify this impression the survey asked firstly for the hardware and the operating system the students are using.

4.2.1 Hardware and Operating System

The questions concerning their personal mobile phone answered all in all 1345 students. 165 of them said that they do not possess a smartphone but a classical cellphone instead (12.3 %). Consequently the majority (87.7 %) of the students own a smartphone. The corresponding value for tablet computer is much lower. Only 27.7 % of the asked students (438) did reply that they have an own tablet pc. These values indicate that if educational developers want to plan the use of tablet pcs in the course, they cannot rely on the assumption that most of the students do posses such a device (even if this figure might have changed slightly over the last year). That means on the one hand that a course concept basing on the use mobile devices, at least has to be open to the use of smartphones instead of tablet pcs. On the other hand it means, that if it is necessary to use a tablet computer (for whatever reason) there must be given an opportunity to the students to borrow such a device.

Within the ELLI-project one work package is designing a special app that offers the opportunity to control laboratory equipment remotely via mobile devices (for more information on that see e.g. [14]). Hence it is important to know, on what kind of operating systems the students' mobile devices run. The published values of market shares between iOs and Android do not help in this context, as [15] found out for the US. Such figures are calculated on basis of sold devices but do not take the actual use of thc devices into account. Therefore the students were asked which operating system their personally used devices provide. For tablets 438 and for smartphones 1180 students answered this question. The results are shown in Fig. 5.

The results show that – as expected – Android and iOs make up the vast majority of the market. Whereas Smartphones are mainly run by Android (62 % in comparison to 29 %), tablets are mostly run by iOs (45.2 % in comparison to 40 %). That means that the majority of the students' tablets are iPads. Such information is important to those who want to design new software as up to today software has to be programmed separately for both systems.

After explaining the results with regard to hardware and operating systems, in the following the results with special focus on learning with mobile devices will be explained.

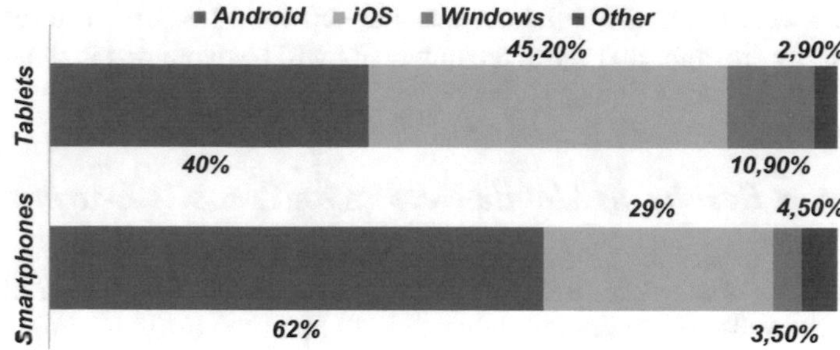

Fig. 5 Distribution of operating systems among the students' devices

4.2.2 Mobile Devices in Context with Learning

Asking for the hardware and operating system is only one perspective. Even more interesting is the question if and how often students already use mobile devices or mobile apps for their learning process. So the survey asked for the frequency of use in this context (possible answers ranged on a five-step scale from never to very often).

Frequency of Use Fig. 6

The results show that 23.4 % of the students use their smartphone never or very seldom in context of learning, whereas 63.9 % do this often or very often (12.7 % use it from time to time). The same question asking for tablet computers disclosed totally contrary results: 69 % use their tablet never or seldom and only 26 % use it often or very often (in context with studies). E-book reader are used even more rarely, as 93.1 % use such devices never or seldom and just 3.9 % often or very often. 27.3 %

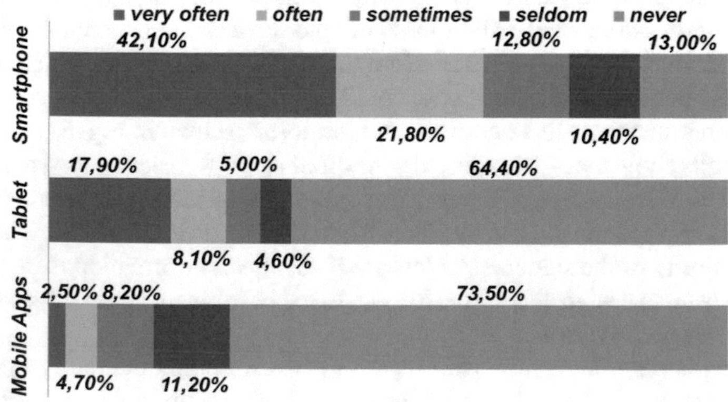

Fig. 6 Frequency of use of mobile devices and apps in context with studies

of the students said that they use a mobile internet-connection – via smartphone or tablet – never or seldom and 59.8 % use it often or very often. The last question in this context was, how often the students use mobile apps for learning. The answers reveal that the use of apps explicitly for learning is not very common yet. Only 7.2 % of the students use mobile apps often or very often for learning, but 84.7 % use it seldom or never. This drives e.g. the question, if this can be explained due to the fact that there are no adequate apps, that existing apps are not usable or that the students just do not know them.

Value of Acceptance

Based on the results explained above and bringing them into correlation with the level of satisfaction with the devices (not explicitly constitute in this paper), it was possible to calculate a value of acceptance (value could range from 0 to 4). The survey showed the following results: Smartphone 2.8; tablet computer 2.4, e-book-reader 1.3; mobile apps 1.4; mobile internet connection 2.8. Considering these values separately does not make sense, but comparing them with each other disclose that mobile apps are less accepted than using the smartphone and tablet computer. But even those devices exhibit a lower level in comparison to classical notebooks (3.1).

Encouragement for Use in Learning Contexts

An additional question asked for the encouragement to use mobile devices during a course. So the students were asked: "Have you been encouraged to use your mobile device during the course? If yes, for what kind of purpose?" The results exposure that only 22.6 % of the students have been asked often or very often to use their devices in context with a course. In most of these cases (65 %) those who were asked to do so, were requested to use them in order to take part at any kind of polling. Polling means in this situation that the teacher asks a question in course-context and the students had to use their device and special software to answer it. The answers are then discussed in the audience. Going more into detail with regard to the use of software on mobile devices, the survey asked for apps which are already used by the students in the context of learning. The results will be discussed in the following.

4.2.3 App Use

One part of the survey asked for software use on mobile devices. So the students named apps they are using in context with learning. 214 students (13.5 %) answered this question and named all in all 357 apps, or more specific 139 different apps. Based on these answers it was possible to cluster the apps, firstly by subject and secondly by kind of use. In this context all 357 answers are taken into account.

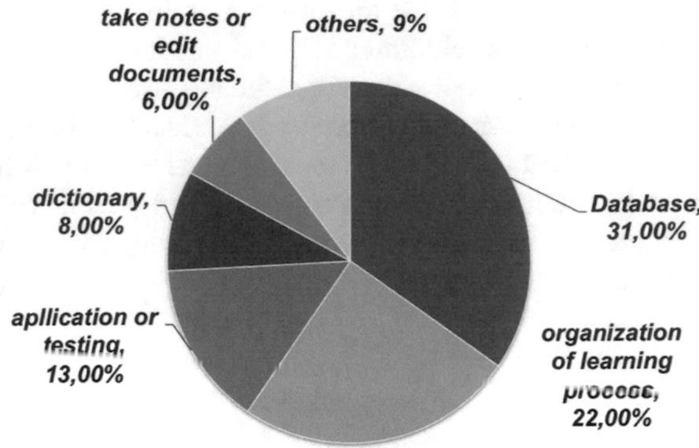

Fig. 7 Use oriented clusters for named apps

Subject Oriented Cluster:

Looking at the related subjects the named apps could be allocated to six different clusters. The biggest cluster (160; 45 %) is formed by apps, which had no connection to any special discipline. In this cluster apps like dropbox or apps used to make notes are summarized. This cluster is followed by apps for mathematics and chemistry/physics (each 57; 16 %). 49 (14 % of the) apps are used for language learning, 30 (8 %) for mechanical engineering or logistics and finally 4 (1 %) are used in context with electrical engineering/informatics.

Use Oriented Cluster Fig. 7:

Dividing the apps by the purpose they are used for, 7 different clusters could be identified. 109 (31 %) of the named apps serve as any kind of database, 79 (22 %) are used for the organization of learning and 47 (13 %) for any kind of application and testing. 40 (11 %) apps are language dictionaries, 30 (8 %) help to make notes or edit documents and 20 (6 %) are used for cloud computing. Finally 32 (9 %) apps could not be allocated to any cluster. Hence they were summarized under “others”.

Top 5 Apps:

Finally the top 5 apps were identified. In the following these apps are explained briefly. Merck PSE is an interactive periodic table of chemical elements and was named 47 times (13 %). Wolfram Alpha, a computational knowledge search engine, was named 26 times (7.3 %). 22 times Schnittkraft-Meister, a game-based learning app to calculate cutting forces, was mentioned. For file hosting and as a cloud service mainly Dropbox is used (4.8 %). From those students who named any app, 4.2 % use Ankidroid in order to design customized and personalized flashcards.

5 Discussion

The results of the ELLI survey correspond with the results of various surveys on the subject of media use at other universities and faculties. It can be assumed, on closer consideration of some factors (e.g. how students rate the quality of certain offers) and framework conditions (e.g. if contents are only provided via certain media), that there exist indeed disparities within different subjects and disciplines, even within some courses. Nevertheless there are some limitations, which must be stated.

Most of the asked students were first- (59.6 %) or second-year (19.4 %) students. Some of the requested media could be still unknown for them. If they already knew and used the media offer at their university, they know and use it only recently. That leads to the question, if the frequency of use would decrease with better knowledge about the facilities of the offer. A wider random sample including more students from different academic years would have been better within this aspect. In addition to that the survey did not inquire if the different media the questions asked for are present at the different universities. So it is not very clear if certain media are not used because the students do not want to use or because they are not able to use them in context with their studies. Questions like these must be answered in a second study or with the help of interviews for example. The conclusions, which instead can be drawn are, that social media like Twitter or Facebook are not that present in teaching as it may be the perception – for whichever reason. Another limitation of this study can be seen in wording. The definition of satisfaction for example is very subjective. Indications like "e.g. would you recommend this media to your fellow student" had been given a common understanding about this term. A next point is, that the survey didn't asked, with one exception, for active (with own input) or passive (only reading) use. Especially regard to collaborative use of online media this would be a very interesting question for additional research. If most of the students prefer a passive use, more motivation for an active use has to be given. Closing this discussion, it must be stated, that the present survey is an exploratory survey with several limitations in terms in generalizing. There are a lot of open questions to answer and a lot of possible connecting factors for further research.

After having discussed shortly the survey's limitations, in the following some of the results will be highlighted again in order to show prospective future work and connected further research questions.

6 Future Work

The overall evaluation point in context with collaborative online media and social media is, that those media are being used very scarcely for study purposes. In public perception the use of social media in private life is very present. Following the survey's results social media have not that much found their way into teaching up to know. Only one third for example do use Facebook in context with their studies.

It would be interesting to examine how those use it. Do they for example use it in order to arrange meetings or do they really discuss content from lectures with others on Facebook? Here it becomes clearly recognizable that further empirical surveys are needed in order to detect the reasons for this and to uncover criteria for the future implementation of both media types during the course of study. Various studies, including the yearly conducted "ECAR Study of Undergraduate Students and Information Technology" [16], explained with their results that an open position and a distinct media competency of the teachers have a positive effect on the media use and the competent handling with digital media by their students. The survey showed that around 45 % (so nearly half of them) of the teachers are quite open minded to new medias in teaching. In the first step this is good result. Further activities in this context should seek to promote new media use in teaching contexts even more. A following research question is "Which kind of new media teachers prefer and how do they use it?".

In particular, the results regarding media competency of the students prove a need for action: The majority of the students take contents, which are being provided, online without a prior quality control. Especially with regard to collaborative media, a support of the media education of the students is recommended by the results. This support should come from the teachers since they design the learning scenarios with digital media and supervise the media-assisted learning process. The teachers have to run through an academic media education for this purpose. Regarding the survey's results a better education in the field of media-assisted teaching also encourage the openness to the implementation of media in future.

In addition to these more general results, the survey results on mobile learning in particular as well show that the use of mobile devices is not as common as it might be expected. For example only a little bit more than a tenth of the students could name apps in context with learning. That can have at least two reasons. On the one hand it might be possible that there are not enough appropriate apps that are in fact usable in learning context. On the other hand it can be the case that the students simply do not know which apps can be used effectively for what. Moreover the students must be supported in finding adequate apps. Even though the use of mobile apps at this point is not very common, mobile devices offer unique possibilities to support collaborative working processes. Especially the apps the students already named in the survey should stay in focus and it should be investigated why these apps are used for learning and what makes them successful.

Moreover additional qualitative studies are necessary on these points. In order to go more into detail student interviews should be conducted asking for reasons why students use or do not use their mobile devices for learning. It might be for example a topic of privacy. If the mobile device is accepted as very personal, students could have problems using them in formal and more official learning processes. Such topics should be investigated in the future in order to use new technical developments like personal mobile devices in the best possible way for learning.

References

1. A. Kukulska-Hulme, J. Traxler, *Mobile learning: A handbook for educators and trainers*. Routledge, 2005
2. M. Grosch, G. Gidion. Mediennutzungsgewohnheiten im wandel. ergebnisse einer befragung zur studiumsbezogenen mediennutzung, 2011. URL http://digbib.ubka.uni-karlsruhe.de/volltexte/1000022524
3. M. Grosch, *Mediennutzung im Studium. Eine empirische Untersuchung am Karlsruher Institut für Technologie*. Shaker, Aachen, 2012
4. U.e.a. Dittler, ed., *E-Learning: Eine Zwischenbilanz. Kritischer Rückblick als Basis eines Aufbruchs*. Waxmann, Münster, New York, München, Berlin, 2009
5. M. Schiefner, M. Kerres, Web 2.0 in der hochschullehre. In: *E-Learning: Einsatzkonzepte und Erfolgsfaktoren des Lernen mit interaktiven Medien*, ed. by U. Dittler, Oldenbourg, München, 2011, pp. 127–138
6. K. Thöing, U. Bach, R. Vossen, S. Jeschke, Herausforderungen kooperativen lernens und arbeitens im web 2.0. In: *TeachING-LearnING.EU Tagungsband movING Forward – Engineering Education from vision to mission18. und 19. Juni 2013*, ed. by A. Erman Tekkaya et al. 2014, pp. 191–194
7. R. Atwal, L. Tay, R. Cozza, T.H. Nguyen, T. Tsai, A. Zimmermann, C.K. Lu, *Forcast: PCs, Ultramobiles and Mobile Phones, Worldwide, 2010–2017, 4Q13 Update*. Stamford, 2013
8. J.H. Bakkers, R. Craven, J. Delaney, F. Jeronimo, I. Montero, *European Mobile and Internet Services, 2014: Top 10 Predictions*. 2014
9. C. QUINN. mlearning: mobile, wireless, in-your-pocket learning, 2000
10. M.e.a. Sharples, A theory of learning for the mobile age. In: *Medienbildung in neuen Kulturräumen.*, ed. by B. Bachmair, VS Verlag für Sozialwissenschaften, Wiesbaden, o. J., pp. 87–99
11. M. Sharples, *The Design of Personal Mobile technologies for Lifelong Learning*, 34th edn. 2000
12. H. Crompton, A historical overview of mobile learning: Toward learner-centered education. In: *Handbook of mobile learning*, ed. by Z.L. Berge, Routledge, London, 2013, pp. 3–14
13. P. Maske, *Mobile Applikationen 1: Interdisziplinäre Entwicklung am Beispiel des Mobile Learning*. Gabler Verlag, Wiesbaden, 2012
14. D. May, C. Terkowsky, T. Haertel, C. Pleul, The laboratory in your hand - making remote laboratories accessible through mobile devices. In: *Proceedings of the 2013 IEEE Global Engineering Education Conference (EDUCON)*. IEEE, 2013
15. C. Arthur, Why an 80% market share might only represent half of smartphone users. The Guardian, 07.11.2013. URL http://www.theguardian.com/technology/2013/nov/07/android-market-share-smartphone-users-google-apple
16. E. Dahlstrom. Ecar study of undergraduate students and information technology, 2012. URL http://www.educause.edu/library/resources/ecar-study-undergraduate-students-and-information-technology-2012

Inspiration für die Studienentscheidung. Mit Role-Models Orientierung gewinnen!

Natascha Strenger, Theresa Janssen, Gergana Aleksandrova and Daniela Antkowiak

1 Allgemeine Informationen

ELLI ist ein Verbundprojekt an den drei Standorten der RWTH Aachen University, der Ruhr-Universität Bochum sowie der Technischen Universität Dortmund und wird im Rahmen des Qualitätspakts Lehre gefördert. ELLI zielt auf die Verbesserung der Studienbedingungen und die Weiterentwicklung der Lehrqualität in der Ingenieurausbildung. Die Projektarbeit konzentriert sich dabei auf vier Kernbereiche: 1. Virtuelle Lernwelten, 2. Mobilität und Internationalisierung, 3. Student Lifecycle und 4. Professionelle Handlungskompetenz. Den Herausforderungen der Studieneingangsphase widmet sich ELLI an den Standorten Aachen und Bochum im Bereich des Student Lifecycle: Im Hinblick auf eine Verringerung von Studienabbruchquoten (vgl. [1]) werden die Übergangsphasen des Ingenieurstudiums mit beratenden Angeboten begleitet. Neben dem Studieneinstieg zählen hierzu auch die Übergänge im gestuften Bachelor-/Master-Reformmodell sowie zur Promotion bzw. Einstieg ins Berufsleben. Auch der geringen Auslandsmobilität unter Ingenieurstudierenden will ELLI im Kernbereich Mobilität und Internationalisierung frühzeitig entgegenwirken, indem entsprechende Maßnahmen in der Studieneingangsphase Anwendung finden. Im Folgenden werden beispielhaft einige der bisher von ELLI umgesetzten und geplanten Maßnahmen beschrieben.

N. Strenger (✉) · T. Janssen
Project ELLI – Excellent Teaching and Learning in Engineering Sciences,
Ruhr-Universität Bochum, Bochum, Germany
e-mail: strenger@fvt.rub.de

G. Aleksandrova · D. Antkowiak
Project ELLI – Excellent Teaching and Learning in Engineering Sciences,
RWTH Aachen University, Aachen, Germany

Originally published in "Tagungsband zum Mosbacher Tag der Lehre, Qualifizierung von Studierenden im Student-Life-Cycle" © 2014.
Reprint by Springer International Publishing AG 2016,
DOI 10.1007/978-3-319-46916-4_56

2 Am Standort Bochum

2.1 ALLES ING! – Wer wir sind und was wir machen!

Durch ELLI konnte am Standort Bochum die Initiative ALLES ING! umgesetzt werden: In dieser verfolgen die drei ingenieurwissenschaftlichen Fakultäten der Ruhr-Universität Bochum ihr gemeinsames Ziel, das Interesse und die Identifikation mit den Technikwissenschaften zu stärken. Durch eine attraktive Darstellung der Ingenieurwissenschaften und der „Menschen hinter der Technik" sollen interessierte Schüler/innen in ihrer Entscheidung für ein technisches Studium bestärkt und Studienanfänger von vielfältigen Vorbildern durchs Studium begleitet werden. Dieses Konzept der Role-Models, die Wege in und durch das ingenieurwissenschaftliche Studium aufzeigen, setzt ALLES ING! auf zwei Ebenen um: In einem Internetportal, welches sich verschiedenster medialer Präsentationsformen wie Kurzvideos, Interviews oder Twitter-Einträgen bedient. Ziel ist es dabei, Ingenieurstudierende im Studienalltag, beim Auslandsaufenthalt, mit ihrem Forschungsthema oder beim Berufseinstieg zu porträtieren.[1] Begleitet wird das Konzept durch eine Reihe von Veranstaltungen, bei denen technikinteressierte Schüler/innen in Kontakt mit Studierenden, Alumni und Wissenschaftler/inne/n kommen. Das in ALLES ING! medial umgesetzte Konzept der Role-Models kommt in den von ELLI konzipierten Beratungsformaten am Standort Bochum jeweils thematisch begleitend zum Einsatz.

2.2 Beratungsangebote speziell für Schüler/innen im Übergang Schule-Hochschule

Gemeinsam mit den Mitarbeiter/innen der Ingenieurwissenschaften aus dem Themenfeld des Übergangs „Schule-Hochschule" und ELLI wurden die Ziele definiert, nicht-traditionelle Studierende für ein Ingenieurstudium zu begeistern und besonders den noch immer geringen Frauenanteil zu erhöhen (vgl. [2]). Hierfür begeistern einmal die virtuellen Porträts der RUB-Ingenieurinnen, in denen von der jungen Studentin bis zur erfolgreichen Professorin Ingenieurinnen mit ihrem individuellen Lebensweg für ein technisches Studium begeistern. In Kontakt kommen Studieninteressierte mit den Ingenieurinnen im Rahmen des Tags der offenen Tür, bei dem erstmalig seit Frühjahr 2013 die drei ingenieurwissenschaftlichen Fakultäten gemeinsam und mit Role-Models beraten. Das von ELLI initiierte Format wurde bereits in das feste Programm der Fakultäten integriert und 2014 selbstständig durchgeführt. Um dem häufigen Studienabbruch erfolgreich zu begegnen, besteht das zweite Ziel für die Ingenieurwissenschaften in der richtigen Informierung über Studiengänge und Berufsperspektiven. Hierfür bringt das Projekt die Zielgruppe der Interessierten bzw. Erstsemester in Kontakt mit Studierenden bei gemeinsamen Treffen, die regelmäßig

[1] vgl. http://www.ing.rub.de

im Rahmen der Fachschaftsarbeit stattfinden und durch die aktive Ansprache mittels gemeinsamer Auftritte im Rahmen von Informationsveranstaltungen und der Immatrikulationsphase. Die Vernetzung der Studierenden untereinander schafft den Raum für ein Peer-Mentoring, das die Erwartungen realistisch, in der schwierigen ersten Phase (hoher Studienabbruch) im Studium allerdings auch motivierend gestaltet. Berufsperspektiven für angehende und Jung-Ingenieure kommen auch hier in Form von medialen Formaten im Rahmen der ALLES ING!-Initiative zum Einsatz: Quer durch alle Ausbildungsstufen sind hier Studienarbeiten, Dissertationen, Forschungsprojekte und Abschlussarbeiten zu finden. Die Videoreihe Young Professionals besucht RUB-Alumni an ihrem Arbeitsplatz und zeichnet ihren individuellen Weg in und durchs Studium bis hin zum Beruf nach und stellt gleichzeitig potenzielle Arbeitgeber für Ingenieure vor. Bereits Studierenden zeigt ELLI Perspektiven für Forschungs- und Abschlussarbeiten sowie den Berufseinstieg für Semester im Rahmen der Beratungsveranstaltungen Mit dem Master in die Praxis sowie Wege zum Dr. Ing. Hier kommen Studierende in direkten Kontakt mit RUB-Promovierenden und RUB-Alumni und erfahren im World Café, wer auf welche Weise das Studium und den Berufseinstieg geschafft hat. Außerdem wird die Veranstaltungsreihe professional durch Career Service und Research School begleitet.

2.3 GoING Abroad – Frühzeitige Beratung zu Auslandsaufenthalten im Ingenieurstudium

Unter der Zielsetzung einer Erhöhung der Auslandsmobilität unter Studierenden ingenieurwissenschaftlicher Fächer (vgl. [3]) hat das Projekt ELLI aufbauend auf einer qualitativen Bedarfserhebung das fachspezifische Informations- und Beratungskonzept GoING Abroad entwickelt. Gerade für Studierende der ersten Semester stellt eine Vielzahl an Ansprechpartnern und Förderprogrammen auf Fakultäts- sowie Universitätsebene häufig eine erste Hürde bei der Planung von Auslandsaufenthalten dar. Gesprächskreise mit studentischen Vertretern in den Ingenieurfakultäten der Ruhr-Universität Bochum haben darüber hinaus die Ergebnisse quantitativer Studien bestätigt, dass Studierende bei der Planung eines Auslandsaufenthaltes besonderen Wert auf die Erfahrungen und Empfehlungen von Kommiliton/inne/en desselben Fachbereichs legen. Gemeinsam mit den Auslandskoordinatoren der Fakultäten und dem International Office der Ruhr-Universität wurde von ELLI ein Veranstaltungsformat entwickelt, welches gezielt auf diese Bedarfe eingeht: Zentrales Element des GoING Abroad Programms ist eine Informations- und Beratungsveranstaltung, welche seit April 2013 einmal pro Semester stattfindet und bei der auslandserfahrene Kommilitonen sowie Fachberater Interessierten Fragen rund um das Thema Ausland beantworten. Als Referenten werden gezielt Studierende der ingenieurwissenschaftlichen Fächer gewonnen und es werden Fakultätskooperationen sowie die für Ingenieurstudierende besonders attraktiven, berufsbezogenen Praktika im Ausland vorgestellt. Die Veranstaltung findet im

November 2014 bereits zum fünften Mal statt und die Veranstaltungsbewertungen haben ergeben, dass die durchschnittlich 30 bis 40 Teilnehmer der Fakultäten für Bauingenieurwesen und Maschinenbau sich größtenteils in den ersten Studiensemestern befinden. Auch im Kernbereich Mobilität und Internationalisierung kommt außerdem das ALLES ING!-Konzept der Role-Models zum Einsatz: In der eigenen Rubrik Globetrotter berichten Ingenieurstudierende im Ausland von ihren Erfahrungen. Die Evaluationen der Maßnahmen bestätigen, dass eine frühe Thematisierung von Auslandsaufenthalten im Studium, gekoppelt mit der Präsentation von Vorbildern, den Herausforderungen einer schwierigen Organisation und der Angst vor Studienzeitverlängerung aufseiten der Studierenden entgegenwirken kann.

2.4 Wissenschaftliches Schreiben als Querschnittsaufgabe in Ingenieurstudium und -beruf

Den Austausch mit den Studierenden und gleichzeitig eine stärkere Vernetzung der ingenieurwissenschaftlichen Studiengänge untereinander initiiert ELLI in regelmäßig stattfindenden großen Fachschaftstreffen mit den sieben Fachschaftsräten der Ingenieurstudiengänge in Bochum. An der Tagesordnung sind regelmäßig Themen wie studentische Beteiligung an der Fachschaftsarbeit, Berufsorientierung im Studium, Studienorganisation und -struktur. Auch konkrete Veranstaltungen werden in diesem Rahmen gemeinsam geplant – das Projekt ELLI nutzt dieses Forum um aktuelle Themen mit den Projektzielen rück zu koppeln. Im Februar 2014 ging es schwerpunktmäßig um das Thema „Abschlussarbeiten im Studium“ bzw. „Schreiben im Ingenieurstudium“. Seit Sommer 2013 veröffentlich das Projekt im Newsbereich „Forschung“ auf der gemeinsame Plattform ALLES ING! Forschungsthemen – von der Projektarbeit bis zur Promotion – aus den Ingenieurwissenschaften. Diese geben bereits einen Einblick, wie mögliche Themenstellungen aussehen können, zeigen, woran die jeweiligen Lehrstühle forschen und vermitteln einen Eindruck, was ganz praktisch auf die/den Einzelne/n zu kommt. Doch gestaltet sich auch die Organisation von schriftlichen Arbeiten in den Studiengängen höchst unterschiedlich, Anforderungen sind nicht transparent und stellen die Studierende vor große Herausforderungen. Dem begegnet der von ELLI entwickelte virtuelle Ratgeber für Abschlussarbeiten mit verbindlichen Aussagen der Studienfachberater/innen sowie Studiendekanen aus den Fakultäten Bau- und Umweltingenieurwesen und Maschinenbau zu Themen wie: Themenfindung – Art der Arbeit – Ansprechpartner – Literatur finden u.a.m. Neben der Organisation steht im bisher einmaligen Pilotprojekt „Vorbereitung auf eine Semesterarbeit“ ganz zentral das wissenschaftliche Schreiben im Fokus. Die auf sechs Termine gekürzte zusätzliche Seminarveranstaltung mit Beginn im Wintersemester 2014/15 richtet sich an Studierende im fünften Semester (B.A.) – die Semesterarbeit wird laut Studienplan des Maschinenbaus im fünften Semester geschrieben – und hier besonders an diejenigen, die sich unsicher in Bezug auf ihre wissenschaftlichen Schreibkompetenzen fühlen. Gemeinsam mit dem Studiendekan

Prof. Dr.-Ing. Franz Peters werden Merkmale wissenschaftlichen Schreibens in den Ingenieurwissenschaften besprochen, konkrete Schreibübungen durchgeführt wie bspw. Versuchsbeschreibungen und das Peer-Feedback gestärkt. Die Evaluation der Veranstaltung umfasst dabei auch einen strategischen Austausch der Fakultätsmitarbeiter/innen und der zentralen Einrichtung Schreibzentrum organisiert durch das Projekt ELLI im Frühjahr 2015.

3 Am Standort Aachen

3.1 StartINGs meet Alumni – Vernetzung von Studierendengenerationen

Insbesondere kurz nach Beginn des Studiums ist die Abbruchquote in den Ingenieurwissenschaften hoch. Trotz guter schulischer Voraussetzungen und technischen Interesses können viele der Beginner das Pensum und die fachliche Tiefe im Studium schwer abschätzen und entwickeln meist nicht die notwendige Bindung zu ihrem Studienfach. Präzisiert sind die Gründe hierfür zum einen mangelnde studienbindende Faktoren, aber auch mangelnde soziale, fachliche und akademische Integration (vgl. [4]). Zur besseren Gestaltung der Studieneingangsphase wurde das Format „Start-INGs meet Alumni“ am Standort Aachen eingeleitet. Die Maßnahme zielt darauf ab, den Kontaktaufbau zu Hochschulabsolventen zu fördern und somit eine frühzeitige Integration und Identifikation mit dem Studium und den möglichen beruflichen Zukunftsperspektiven zu erreichen. Im Rahmen von regelmäßig stattfindenden Treffen kommen interessierte junge Studenten mit Ingenieuren diverser Fachrichtungen in Kontakt und können diese in informeller Atmosphäre oder in deren Berufsalltag kennenlernen und porträtieren. Dabei werden verschiedenste mediale Präsentationsformen eingesetzt wie Kurzvideos oder Interviews. Die virtuellen Porträts fördern den intensiven Austausch von persönlichen aber auch fachlichen Erfahrungen unterschiedlicher Generationen, Fachdisziplinen und Hintergründe und sollen somit auch ein breites Publikum für ein technisches Studium begeistern. Die Kurzvideos thematisieren dabei unterschiedliche Herausforderungen in der Studieneingangsphase wie beispielsweise „Wie finde ich eine Lerngruppe“, persönliche Erfahrungen im Umgang mit der Angst vor Klausuren bis hin zu hilfreichen Tipps rund um die Freizeitgestaltung. Mit dieser Maßnahme werden ein interdisziplinärer Austausch unterschiedlicher ING-Disziplinen erreicht sowie für ein erfolgreiches Studium fachübergreifende Kompetenzen vermittelt. Die Studierenden können auf die Erfahrungen von Absolventen zurückgreifen, typischen ingenieurwissenschaftlichen Arbeitsfeldern begegnen und eventuelle künftige Praktikumsplätze ergattern.

3.2 Careerblockveranstaltung – Vermittlung von Praxisbezügen

Ein weiterer Hauptgrund für einen Studienabbruch ist der fehlende Praxisbezug der Lehrveranstaltungen, was sich durch sinkende Motivation und Studienengagement bemerkbar macht. Um dieser Problematik entgegenzuwirken, werden Maßnahmen geplant, die das Technikverständnis fördern, Praxisbezüge vermitteln und das fachliche Vertrauen bestärken. Neben der Konstruktion virtueller Labore, die eine anwendungsbezogene und experimentelle Lehre gewährleisten, werden verschiedene Vorträge von Praktikern im Rahmen von Lernveranstaltungen in den Ingenieurwissenschaften angeboten. Diese sogenannten Praxisvorträge vermitteln den Studierenden erste Einblicke in die Aufgaben und Herausforderungen des Arbeitsalltags von Absolventen mit ingenieurtechnischem Hintergrund aus verschiedensten Bereichen. Dabei lernen die teilnehmenden Studenten spezifische Arbeitsgebiete und Einsatzorte von Ingenieuren sowie herausfordernde Problemstellungen aus dem Berufsalltag kennen. Die geschilderten Problemstellungen veranschaulichen neben maschinenbaunahen Herausforderungen auch solche aus dem Bereich Management und Mitarbeiterinteraktion. Die Veranstaltung trägt ebenfalls zur Kontaktpflege zwischen RWTH Studenten und Absolventen der Ingenieurwissenschaften bei und bietet den interessierten Studenten die Möglichkeit, Kontakte zu potenziellen Arbeitgebern frühzeitig zu knüpfen. Die eingeladenen Referenten bekommen andererseits die Gelegenheit, potenzielle Werkstudenten, Praktikanten oder künftige Arbeitnehmer zu akquirieren.

3.3 Fachsprachkurse – Verständnisschwierigkeiten vermeiden

Aktuelle Studien belegen, dass für Studierende mit Migrationshintergrund durch sprachliche Barrieren, insbesondere zur jeweiligen Fachsprache, ein deutlicher Mehraufwand im Studium entsteht. Diese Barriere tritt insbesondere in der Eingangsphase auf, wodurch gerade in den ersten zwei Semestern viele Studienabbrecher mit Migrationshintergrund zu verzeichnen sind. Die Heterogenität der Studierendenschaft wird unter anderem durch kulturelle Unterschiede, verschiedene Vorbildung aufgrund differierender Bildungswege oder „first-generation-students" begründet. Das sind nur einige Faktoren, die dazu führen, dass studienbegleitende Maßnahmen zur Förderung gerade der Studieneingangsphase mehr und mehr an Bedeutung gewinnen. Durch ein im Rahmen von ELLI entwickeltes Angebot von fachspezifischen Sprachkursen sollen die Barrieren zur spezifischen Fachsprache für Studierende mit Migrationshintergrund gesenkt werden. In Zusammenarbeit mit Vertreterinnen und Vertretern der Fakultäten, den Fachschaften und Anbietern von Sprachkursen wurde ein Sprachkurskonzept entwickelt, das die ingenieurwissenschaftliche Fachsprache fokussiert und eine Brücke zwischen den diversen sprachlichen Hintergründen und der praktizierten wissenschaftlichen Lehre schlagen soll. Vor dem Hintergrund von studienfachspezifischem Aufgabenmaterial – bereitgestellt von den im Bachelorstudiengang „Maschinenbau" lehrenden Instituten – wurden eigens für den Fachsprachkurs Lehreinheiten zu den thematischen Bereichen allgemeiner Wortschatz, Mathematik, Mechanik, Physik, Geometrie, Materialien, Werkstoffe, Chemie, Computer und Programme sowie Werkzeuge und Maschinen entwickelt. Diese zielen sowohl auf das Hörverstehen als auch auf Schrift- und Aussprachekompetenzen ab. Vertieft werden neben Phonetik und Rechtschreibung ebenso grammatikalische Grundlagen. Außerdem werden universitätsspezifische Kommunikationssituationen wie das Gespräch in einer Sprechstunde, einer Klausureinsicht beim Prüfungsamt oder Referats- und Vortragssituationen eingeübt. Zudem findet ein Blended Learning Angebot über die elektronische Plattform der RWTH Aachen University (L2P) statt. Das Kursmaterial wird hier eingestellt und erlaubt eine individuelle Wiederholung und Nachbereitung des behandelten Lernstoffs. Im L2P können auch Fragen an den Kursleiter gestellt werden oder über den Modus der Gruppendiskussion ein direkter Austausch zwischen den Kursteilnehmern stattfinden. Für den Kurs gilt eine Höchstteilnehmerzahl von 20 Personen und die Zugangsvoraussetzung von einem B2 – C1 Sprachniveau. Dadurch werden die intensive Kommunikation und dynamische Lernsituation zwischen den Teilnehmern gewährleistet. Der durch ELLI initiierte Kurs wird im kommenden Wintersemester 2014/2015 erstmalig als Intensivkurs durchgeführt werden. Der Unterricht findet zweimal wöchentlich à 90 Minuten im Zeitraum von Oktober bis Dezember statt. Durch das Angebot eines Intensivkurses soll einer möglichen Fluktuation der Teilnehmer aufgrund der anstehenden Prüfungsphase entgegengewirkt werden.

4 Evaluation und Ausblick

Die Arbeit mit Role-Models und die möglichst frühzeitige Vernetzung der Studierenden aus verschiedenen Studienphasen untereinander sowie mit den für ihre Bedürfnisse wichtigen Ansprechpartnern erweisen sich als höchst effektiv. Sie bieten ein niedrigschwelliges Kontaktangebot, fördern die Identifikation mit dem ingenieurwissenschaftlichen Studium und erhöhen die Attraktivität individueller Profilbildungsmöglichkeiten, wie zum Beispiel von studienbezogenen Auslandsaufenthalten oder Themen für Abschlussarbeiten. Erfahrungswerte von Gleichaltrigen erhalten eine große Akzeptanz und Erstsemester konkretere Vorstellungen vom Studienverlauf. Mit einer neuen, durch vielfältige mediale Konzepte unterstützten, Ansprache können langfristig auch mehr nicht-traditionelle Studieninteressenten erreicht werden. Die langfristige Implementierung von erfolgreichen Maßnahmen wie diesen an den beteiligten Standorten stellt eine Herausforderung dar, welcher ELLI durch die Einbindung aller relevanten hochschulischen Akteure sowie der Studierendenvertreter begegnen will. Die von ELLI eingesetzten Maßnahmen für die Studieneingangsphase werden regelmäßig evaluiert, indem Teilnehmer- und Besucherzahlen erhoben sowie quantitative Befragungen durchgeführt werden. Die qualitative Rückkopplung mit der Zielgruppe wird durch regelmäßige Treffen mit Studierendenvertretern gesichert, welche ein wertvoller Projektbestandteil sind. Besonders positive Erfahrungen wurden auch in der Zusammenarbeit mit den Mitarbeitenden der ingenieurwissenschaftlichen Fakultäten gesammelt, die sehr motiviert und engagiert an neuen Ideen für die Studieneingangsphase mitwirken.

Literaturverzeichnis

1. VDI. Ingenieure auf einen Blick: Erwerbstätigkeit, Innovation, Wertschöpfung, 2012. URL https://www.vdi.de/fileadmin/vdi_de/redakteur_dateien/bag_dateien/Beruf_und_Arbeitsmarkt/2012_-_Ingenieure_auf_einen_Blick.pdf
2. Stifterverband für die deutsche Wissenschaft/Mc Kinsey & Company, *Hochschulbildungs-Report 2020*. Edition Stifterverband, Essen, 2012
3. DAAD/HIS. Wissenschaft weltoffen. Daten und Fakten zur Internationalität von Studium und Forschung in Deutschland, 2013. URL http://www.wissenschaftweltoffen.de/publikation/wiwe_2013_verlinkt.pdf
4. K. Gensch, C. Kliegl. Studienabbruch – was können die Hochschulen dagegen tun?, 2011. URL http://www.ihf.bayern.de/uploads/media/IHF_kompakt_Mai_2012.pdf

Herausforderungen kooperativen Lernens und Arbeitens im Web 2.0

Kerstin Thöing, Ursula Bach, René Vossen and Sabina Jeschke

Zusammenfassung Hochschuldidaktische E-Learning-Angebote werden zunehmend mit interaktiven Web 2.0-Komponenten kombiniert. Aus dem „Web 2.0" wird ein „E-Learning 2.0" (vgl. Downes, S.: E-Learning 2.0. E-Learn Magazine – Education and Technology in Perspective, Oktober 2005. Online unter http://elearnmag.acm.org/featured.cfm?aid=1104968). Der vielfältige Einsatz und die Gestaltungsmöglichkeiten dieser Medien bringen für Lehrende und Lernende neue Erfahrungen mit sich, die Gegenstands des Workshops „Herausforderungen kooperativen Lernens und Arbeitens im Web 2.0" im Rahmen der TeachING-LearnING.EU Tagung waren. Die Teilnehmender, größtenteils Dozenten mit mehrjähriger Lehrerfahrung, berichteten über ihre Erlebnisse mit Web 2.0-Anwendungen in der Lehre und bezogen dabei die Perspektive ihrer Studierenden mit ein. Aus diesen Erfahrungswerten heraus wurden fördernde und hemmende Faktoren für mediengestütztes gemeinschaftliches Lernen identifiziert. In der Diskussion wurde deutlich, dass an eine mediengestützte Lehre im ingenieurwissenschaftlichen Bereich besondere Ansprüche gestellt werden, die es künftig zu berücksichtigen gilt.

Digitale Medien fungieren bei Studierenden in unterschiedlicher Ausprägung als Alltagsritual, Lebensmodell und Experimentierfeld, soziale Anschlussstelle, aber auch als Accessoire und Requisite der Selbstdarstellung.[1] Dies bezieht sich vor allem auf die Freizeitnutzung. In der Lehre hingegen bevorzugen Studierende einen moderaten Einsatz neuer Technologien und stellen Qualität im Sinne des didaktischen Nutzens vor Quantität.[2]

[1]Vgl. Smith, S. D., Borreson Caruso, J.: The ECAR Study of Undergraduate Students and Information Technology 2010. Vol. 6, 2010. Online abrufbar unter http://net.educause.edu/ir/library/pdf/ERS1006/RS/ERS1006W.pdf.

[2]Vgl. Kleimann, B., Özkilic, M., Göcks, M.: Studieren im Web 2.0 – HISBUS-Kurzbericht Nr. 21, 2008. Online abrufbar unter https://hisbus.his.de/hisbus/docs/hisbus21.pdf.

K. Thöing (✉) · U. Bach · R. Vossen · S. Jeschke
IMA/ZLW & IfU, RWTH Aachen University, Dennewartstr. 27, 52068 Aachen, Germany
e-mail: kerstin.thoeing@ima-zlw-ifu.rwth-aachen.de

Originally published in "Tagungsband zum Mosbacher Tag der Lehre, Qualifizierung von Studierenden im Student-Life-Cycle", © 2014.
Reprint by Springer International Publishing AG 2016,
DOI 10.1007/978-3-319-46916-4_57

Diskussionsforen, Arbeitsplattformen, gemeinsam gepflegte Wikis und die Einbindung von Social-Media-Formaten bieten je nach didaktischem Einsatz Vorteile gegenüber der klassischen Vor-Ort-Lehre, stellen gleichzeitig jedoch Lehrende und Lernende gleichermaßen vor neue, bisher unbekannte Herausforderungen. Wie reagiert ein Lehrender beispielsweise auf die Herabsetzung eines Studierenden durch Kommilitonen in einem anonym eingerichteten Lehr-Forum? Was soll er tun, wenn Studierende den Arbeitsbereich dazu nutzen, um offen Kritik an der Lehrveranstaltung zu äußern?

In der virtuellen Welt sind die bisherigen Handlungsoptionen für Lehrende und Lernende neu zu überdenken. Ist es als Gegenmaßnahme in solchen Fällen angebracht, die Anonymität aufzuheben und damit den freien Meinungsaustausch eventuell einzuschränken? Mit welchen Optionen sollte oder muss eine Web 2.0-Komponente ausgestattet werden, um ein für die jeweilige Veranstaltung angemessenes und zweckmäßiges Lernumfeld zur Verfügung zu stellen? Und was ist überhaupt ein angemessenes und zweckmäßiges Lernumfeld im Web 2.0?

Solche und viele weitere, praxisorientierte Aspekte diskutierten die Teilnehmenden des Workshops „Herausforderungen kooperativen Lernens und Arbeitens im Web 2.0".

Um möglichst viele Blickwinkel auf dieses Thema einbringen zu können, wurde für den Workshop die Moderationsmethode Dynamic Facilitation[3] gewählt. Ziel der Methode ist weniger eine konkrete Lösungssuche, als das Thema aus verschiedenen Blickwinkeln zu reflektieren und somit intensiver aufzuschließen. Sie ermöglicht dem kommunikativen Prozess eine Eigendynamik, wodurch Raum für spontane Richtungswechsel in der Diskussion und neue Einsichten gegeben wird. Das Thema des Workshops wurde dabei aus den Perspektiven „Lehren im Web 2.0", „Lernen im Web 2.0" und „Web 2.0 in der ingenieurwissenschaftlichen Lehre" betrachtet. Weitere Facetten ermöglichte die Unterteilung der Themenbereiche in

- „Herausforderungen/Fragen": Sammlung von Aussagen, die zu lösende Probleme beschreiben,
- „Bedenken/Einwände": Sammlung jener Befürchtungen, die zu den bereits bestehenden Lösungen formuliert wurden,
- „Lösungen/Ideen": Sammlung von Lösungen, die sich nicht zwingend auf bereits formulierte Probleme und Fragestellungen beziehen mussten, sowie
- „Informationen/Sichtweisen": Sammlung aller weiteren Fakten und Informationen, die im Verlauf der Diskussion geäußert wurden.

Auf diese Weise wurde eine komplexe Betrachtung aller Aspekte ermöglicht, ohne das Thema des Workshops zu verlassen. Ergebnis der einstündigen Diskussion war eine Vielzahl an praxisrelevanten Faktoren zum Themenfeld „Kooperatives Lehren und Lernen im Web 2.0", die die langjährigen Erfahrungen der Teilnehmenden in der Lehre sowie ihre Erfolgserlebnisse, aber auch Befürchtungen im Umgang mit den neuen Medien widerspiegeln.

---

[3] Vgl. Junge, N.: Change Management – Möglichkeiten der Kommunikation, 2009.

Herausforderungen beim Einsatz von Web 2.0-Anwendungen sahen die Teilnehmenden vor allem in der Frage, wie sich diese Medien didaktisch sinnvoll einbinden lassen. Zudem sahen viele in dem zusätzlichen administrativen Aufwand, etwa zur Kontrolle der von den Studierenden erstellten Inhalte, eine Mehrbelastung der Lehrtätigkeit. Das gleiche kann auch für die Studierenden gelten: Das neue Medienangebot kann zusätzliche, statt Arbeit zu erleichtern oder effektiver zu gestalten – zusätzliche provozieren, wodurch das Leistungspotenzial der in der virtuellen Welt kooperierenden Gruppe beeinflusst wird. Diskutiert wurde, wie viel Kontrolle der studentischen Nutzung möglich, förderlich oder rechtlich zulässig ist. Gerade im Bereich Datenschutz gibt es viele offene Fragen, darunter, ob die Web 2.0-Angebote geschlossen, offen oder öffentlich zur Verfügung gestellt werden sollten.

Zur Planung eines sinnvollen Einsatzes von Web 2.0-Anwendungen sollten, so die Workshopteilnehmer, zunächst verschiedene Aspekte wie Gruppengröße, Eignung des Themas, Angemessenheit des Aufwandes für den Lehrenden und zu erwartender Umfang des zu erstellenden Inhaltes beachtet werden. Um den administrativen Aufwand gering zu halten, erwogen die Teilnehmenden eine selbstverantwortliche Kontrolle der Studierenden, beispielsweise durch eine Bewertung der Beitragsqualität untereinander. Auch die Vorbildfunktion des Lehrenden wurde hervorgehoben: Der Lehrende sollte die angemessene und regelmäßige Nutzung vorleben. In Bezug auf die Häufigkeit der Nutzung konnten einige Teilnehmende beobachten, dass die Aktivität der Studierenden im gemeinschaftlichen virtuellen Arbeitsraum zum Ende des Semesters hin zunimmt.

Im Verlauf der Diskussion nahmen die Teilnehmenden auch immer wieder die Perspektive ihrer Studierenden ein. Für diese ergeben sich ebenfalls viele neue Fragen in Bezug auf das „Lernen im Web 2.0": Wie können bestimmte, im Arbeits- und Lernprozess erstellte virtuelle Inhalte mit möglichst geringem Aufwand extrahiert und archiviert werden? Wie kann gleichzeitig Anonymität und die Sichtbarkeit des eigenen Beitrages für die individuelle Bewertung gewährleistet werden? Wie funktioniert überhaupt Gruppenarbeit im Web 2.0? Bestehen Unterschiede gegenüber der bisherigen Zusammenarbeit vor Ort an der Hochschule? Darüber hinaus formulierten die Teilnehmenden Bedenken gegenüber dem Medieneinsatz von studentischer Seite, wie zusätzlicher Zeitaufwand, nicht aufbereitete Inhalte und eine zunehmende Anzahl an Log-in-Daten.

In der ingenieurwissenschaftlichen Lehre identifizierten die Workshopteilnehmer ganz fachspezifische Herausforderungen im Umgang mit Web 2.0 in der Lehre. Lehrende müssen zum Beispiel die Möglichkeit haben, nichttextuelle Informationen, wie technische Zeichnungen, in Foren oder Social-Media-Formaten integrieren zu können. Zudem sollten Dateiformate leicht austauschbar sei – eine Standardisierung der Formate steht der Ermöglichung individueller Programmwahl gegenüber. Zwar ist eine bildliche Darstellung häufig möglich, die nachträgliche Editierung jedoch nicht.

Insbesondere im Studium der Ingenieurwissenschaften muss eine Web 2.0-Anwendung mit großen Hörerzahlen kompatibel sein. Bei einer großen Anzahl an Studierenden, die in der virtuellen Umgebung gemeinsam lernen soll, ergeben sich komplexe Anforderungen an die mediengestützte Lehre: Die kooperative Arbeit muss anders organisiert und verwaltet werden, als in virtuellen Lernräumen mit kleiner Teilnehmerzahl. Andererseits bietet der Web 2.0-Medieneinsatz gerade bei großen Hörerzahlen Vorteile und neue Möglichkeiten, wie etwa Feedback für jeden einzelnen Lehrenden durch entsprechend aufgebaute Anwendungen.

Am Ende des Workshops hatten die Teilnehmenden sechs Metaplanwände mit ihren Kommentaren, Einwänden sowie Ideen gefüllt und zeigten sich überrascht über die Vielzahl an neuen Aspekten, die ihnen bisher wenig oder noch nicht bewusst waren. Vor allem die praxisrelevanten Anregungen waren ihnen wichtig, ebenso die im Workshop gemachte Erfahrung, dass sich andere Lehrende mit ähnlichen Problemen in der Web 2.0-gestützten Lehre konfrontiert sehen. Für die Zukunft bleiben noch einige Lösungen für Fragen und Probleme in diesem Bereich sowie für die fachspezifischen Anforderungen in der ingenieurwissenschaftlichen Lehre, zu finden. Doch letztendlich war die Stimmung gegenüber dem Medieneinsatz positiv: Die Teilnehmenden zweifelten kaum die Sinnhaftigkeit und den Mehrwert eines kooperativen virtuellen Werkzeuges sowohl für Lehrende als auch für Lernende grundsätzlich an, sodass – trotz auch kritischer Einwände – stets der konstruktive Austausch darüber, wie den Herausforderungen zu begegnen sei, im Vordergrund stand.

Retaining Talent, Addressing Diverse Requirements: Academic Writing for Engineering Students

Theresa Janssen, Natascha Strenger, Sulamith Frerich and Franz Peters

Abstract In the light of the anticipated and much-discussed shortage of skilled labor and university graduates in the STEM fields, attracting students for technological studies is an important strategic aim of German universities and engineering faculties. Currently, the so-called "doppelter Abiturjahrgang" (a result of reducing school years from 13 to 12) plays into their hands by releasing twice the amount of high school graduates into the study and work market in Germany. Still, dropout rates among students of engineering sciences are higher than in other subject fields and, consequently, universities and employers are losing potential engineering graduates despite the fact that the total number of enrollments increased. Among the factors considered to be responsible for university dropout are lacks of social and academic integration. Furthermore, formats such as mass lectures that are often part of undergraduate studies in engineering prevent teaching approaches that deal with individual requirements of students. However, a more individual approach is what the increasing diversity among university students would call for. Against the backdrop of an increasingly heterogeneous student body, academic writing presents one of the biggest challenges in engineering studies. In order to support students during their first semesters but also in writing their final theses, the project ELLI is currently implementing a multi-level concept to foster academic writing in engineering studies. At the Ruhr University Bochum, quantitative and qualitative research was done among students and faculty members of the three engineering faculties, based on which a guidebook for writing final theses was developed. Moreover, a seminar for academic writing for undergraduate engineering students was designed, which will take place for the first time in the winter term of 2014.

Keywords Academic Writing · Diversity · Engineering Education · Scientific Writing Skills

---

T. Janssen (✉)
Ruhr-Universität Bochum, Project ELLI – Excellent Teaching and Learning in Engineering Sciences, Bochum, Germany
e-mail: janssen@fvt.rub.de

Originally published in "Global Engineering Education Conference (EDUCON), 18-20 March 2015 IEEE, Tallinn DOI: 10.1109/EDUCON.2015.7095945",

DOI 10.1007/978-3-319-46916-4_58

1 Introduction

In order to retain their technological performance, Western industrialized nations are increasingly depending on academic qualifications for various professional fields. In Germany, universities play a significant role when it comes to maintaining the competitiveness of the German economic system, as they are responsible for the education of a highly-qualified workforce. Fortunately, the country can currently rely on an increasing number of students, which is partly caused by singular effects. Among these are the so-called "doppelter Abiturjahrgang" (twice the number of high school graduates due to a reduction of school years from 13 to 12) and the abolishment of compulsory military service. The attractiveness of German higher education for foreign students appears to be growing as well, for their numbers have been continuously increasing during the past years. Nevertheless, future prospects indicate that, especially due to demographic changes, the number of students will shrink again during the next years [1]. Therefore, in order to ensure the supply of new talent especially in the technical sciences, educational reserves need to be mobilized on a broader range. Educational participation of people with a migration background or a qualification from professional training should be increased and the transitions from school to university and into later working life made more transparent [2]. Opportunities for further education and life-long-learning need to be extended and, for those who are already studying, it is essential to understand and approach the causes for premature drop-out.

However, the situation at German universities is becoming increasingly critical, especially at the engineering faculties. As Prof. Dr.-Ing. Franz Peters, dean of the Faculty for Mechanical Engineering at the Ruhr University Bochum, states: "We are currently not educating enough master students." The reasons for this circumstance are presumably to be found way earlier in the course of studies: While employers report a shortage of highly-skilled labor and graduates from these disciplines, the number of drop-outs during the first semesters of engineering studies is dangerously high. According to several statistics, more than half of the students of Mechanical Engineering, Electrical Engineering and Civil Engineering at German universities leave their studies without completing their degrees. Similar numbers are valid for Mathematics and Information Technologies [3–6]. A lack of bonding factors such as social, disciplinary and academic integration is considered to be among the reasons for premature drop-out [7]. Consequently, higher education institutions are supposed to take action by opening up for new target groups and ways of communication, catering to the changing needs for information of potential students [8]. Still, university everyday life looks very different from this idea: mass lectures, especially during the first semesters, confront university teachers with the challenge of being responsive to the individual needs of their students and simultaneously maintaining a high educational standard [9]. Students who are sufficiently prepared for university by taking their A-levels and melting right into the labor market after obtaining their university degree, have long since become exceptional, rather than the rule [10].

Today's students in the STEM disciplines display a broad variety of cultural and social backgrounds – a pleasant result of successfully implemented programs that promote equal opportunities for first generation students and the internationalization of higher education [10]. Anyway, these new target groups have to face a number of significant challenges with regard to the curriculum during the early stages of their engineering studies. For international students for example, these comprise linguistic barriers and a lack of contact to their German counterparts and higher education institutions should adjust their curricular structures and offers accordingly [10, 11].

At the Ruhr University Bochum, the Project ELLI is currently developing strategies to deal with the challenges in the initial stages of engineering studies.

2 Project ELLI

The project ELLI – Excellent Teaching and Learning in Engineering Sciences - aims at improving the conditions of teaching and learning in engineering education. ELLI is a common and financially supported initiative of the German Federal Government and the individual states for enhancing studying conditions and the quality of teaching. It is a cooperative project of the Ruhr University Bochum, the RWTH Aachen University and the TU Dortmund University. The project work is focused on four main areas: 1. Virtual Learning Environments, 2. Internationalization and Mobility, 3. Student Lifecycle and 4. Additional Skills [12]. In all of these working fields, the project applies an approach of constructive cooperation with the faculty members and students at the three engineering faculties of the Ruhr Universität. ELLI accompanies the Faculties' Change management by regularly offering strategy workshops and establishing a sustainable communication structure within and between the engineering faculties, as well as with central institutions of the University, such as Career Service or International Office.

The working field Student Lifecycle deals with the transitional steps characterizing a student lifecycle as well as the initial and PhD phases. With a view to the reduction of dropout rates and an increasingly heterogeneous student body, transitional steps within engineering study programs at the RUB are accompanied by advisory services. Stages considered are the transition phase from school to university, the early stage of studies as well as the transition to PhD studies or the start of professional life as an engineer. The overall objective of all measures in this field is to allow a higher number of undergraduate students to gain access to an engineering degree. Furthermore, study-related barriers for students with professional qualifications shall be overcome.

With regard to fostering academic writing skills in engineering studies, ELLI is currently implementing a multi-level concept for Scientific Writing in Engineering Studies at the Ruhr University Bochum in order to provide students with the necessary support, especially during the early stages of their studies. The aim of the project is to increase the significance of academic writing as a cross-cutting skill for engineering studies, also with regard to the professional life of engineers, and to offer students an introduction to German academic writing culture [13]. The project partner, TU Dortmund University, could already achieve the first successes in this area: Their Forschungswerkstatt (research workshop) frequently offers seminars regarding scientific writing for students in the technical disciplines [14].

3 A Demand-Oriented Approach: Focus Groups with Engineering Students Styling

At the Ruhr University Bochum, the project ELLI has been regularly initiating focus groups (at least 4 times p. a.), containing representatives of the student councils of the seven different study programs in engineering sciences. The participants are studying Civil Engineering, Environmental Engineering and Resources Management, Mechanical Engineering, Sales Engineering and Product Management, Electrical Engineering and Information Technology, as well as IT Security. The composition of the group is usually gender balanced and the attending students represent various semesters of their study programs. The focus topics for each meeting are prepared by the project manager of this key area in ELLI and the relevance of these topics for engineering education is discussed with the focus group by means of three central questions: 1. How would you, as members of the student councils, rate the importance of this topic for your studies (students' demand)? 2. How do you evaluate the structural implementation of the topic within the faculties (contact persons, consulting offers)? 3. What does this topic require, according to you, in order to become more easily accessible for students and employees of the engineering faculties (strategies, measures)?

In February and April 2014, two of these focus groups centered on the topic of "academic writing in engineering studies". In order to get a more differentiated picture of the student's opinions and problems, the discussion was subdivided in the focus areas "significance of scientific writing in engineering studies" and "organization and realization of the final thesis" (respectively Bachelor, Master). One of the key findings on the latter topic was that the organization of final theses was dealt with differently; not only between the engineering faculties, but also between the various chairs of each faculty, thus, it can be difficult to find the right contact person and relevant information for finding a topic and writing one's thesis. The students generally consider the final thesis to be of great significance, and they see it as a means of presenting their scientific profile to potential employers. But despite this important role the final thesis plays for their studies, the students do not feel sufficiently guided and informed with regard to this topic and clearly express a demand for orientation.

4 Conduction of a Survey on the Significance of Academic Writing

In order to complement the focus group's findings on the key area "significance of academic writing in engineering studies", a quantitative survey was conducted among the members of the student councils and the tutorial groups. With an especially high number of students in Mechanical Engineering (62 %), a total of 81 students participated in this survey. Questions centered on how important the students consider scientific writing for their studies, to what extent they make use of the counselling services of the faculties in this respect and – in the form of free text answers – which services they are still missing during their studies with regard to academic writing and especially the writing of their final theses.

The following graph displays the distribution of reference group among the three engineering faculties of the Ruhr University (see Fig. 1).

When asked who they would consult for advice (while multiple answers were possible), 46 % of the participants selected "employees of a chair". As these employees are usually also their supervisors, the result shows that the students chose the right contact person in this case. Further answers that were often selected were professors with 41 % and other social contacts such as family and friends with about 28 %. (The absolute numbers of the selected answers are shown in Fig. 2.)

Not unexpectedly, the free text answers dealing with "Would you like to receive support during the organization/writing of your final thesis?" and "Do you wish to be supported in academic writing?" were far more differentiated. Nevertheless, the request for more information on the writing of final theses in a company or abroad turned out to be a clear favourite among the students: About 85 % of the answers to question one described a wish for further support with these topics, which shows the students' demand for practical, yet discipline-specific experiences outside their studies. Three other categories seem to be almost equally important to the students: A wish for more support from the supervising staff, a more transparent structuring of the

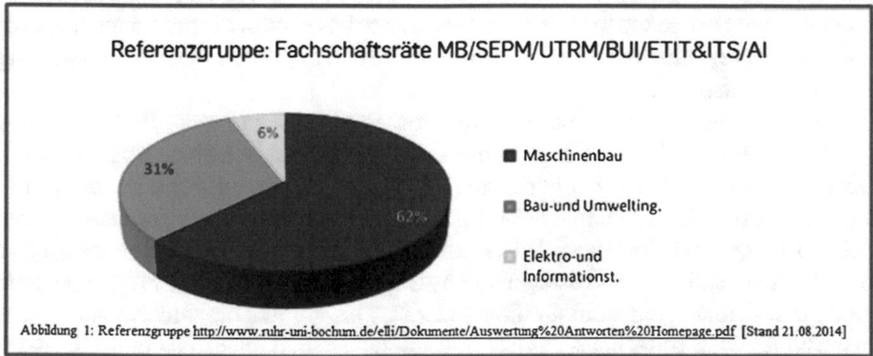

Fig. 1 Application 1: Referenzgruppe

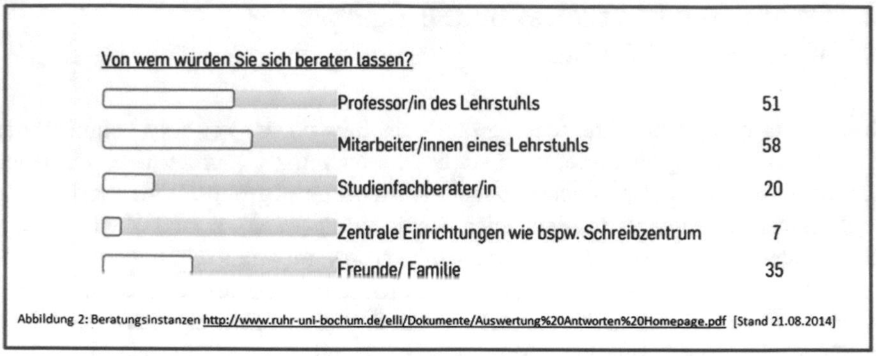

Von wem würden Sie sich beraten lassen?

Professor/in des Lehrstuhls	51
Mitarbeiter/innen eines Lehrstuhls	58
Studienfachberater/in	20
Zentrale Einrichtungen wie bspw. Schreibzentrum	7
Freunde/ Familie	35

Abbildung 2: Beratungsinstanzen http://www.ruhr-uni-bochum.de/elli/Dokumente/Auswertung%20Antworten%20Homepage.pdf [Stand 21.08.2014]

Fig. 2 Application 2: Beratungsinstanzen

courses offered by the Writing Center, a central institution of the university which offers discipline-specific courses for academic writing and the request for sample theses. The findings of this survey shall soon be presented to the faculty members of Mechanical Engineering within a session of the faculty council in order to provide professors, scientific employees and thus potential supervisors of final theses insight into the student's situation. Results will be presented by the project ELLI and a special focus will be on the student's requests for information on practical research with companies or cooperation with partner universities abroad. Furthermore, the wish for sample theses shall be addressed by collecting links to all the different guidelines of the various chairs of the faculty on a common homepage [15].

5 Expert Interviews with Student Advisors

In addition to the focus groups and the quantitative survey, a third step was to conduct qualitative interviews with advisors and deans of studies of the engineering programs. These experts were asked to formulate their advice based on six different testimonials which show the profiles of students who are at the end of their studies and/or planning their final thesis.

Thus, they describe possible studying profiles and ways to write the final thesis, based on their experience as experts in the field of counseling students in the engineering disciplines: How should the student Katharina proceed if she wants to write her final thesis with a company? Is it possible for Christian and Thomas to write their Bachelor thesis together? The manifold experiences and recommendations of the experts are collected and categorized by the project ELLI, in order to merge them within one virtual "guidebook for final theses". This guidebook, which is accessible on the project's website, helps engineering students to find their first counseling with regard to writing papers and final theses in a manner that is flexible in time and

Fig. 3 Application 3: Ratgeber Abschlussarbeiten

space and prepares them for further, personal counseling with the student deans or academic supervisors (Fig. 3).

6 Demand for Orientation: Development of the "Guidebook for Final Theses"

Based on the focus groups, the student survey and the expert interviews explained above, a guidebook for writing (final) theses in engineering studies was developed in February 2014 and has been continuously updated ever since. The guidebook can be found on the website of the project ELLI [16]. The regular feedback of students (gained from the focus groups that continue to take place about four times per year) helps to keep the website up to date and tailored to the student's needs. In order to make the guidebook widely accessible, it is furthermore linked to the courses of the e-learning system of the Ruhr University. Moreover, also the student counselors and deans of studies in the engineering faculties promote the offer on the websites of their respective courses of studies. Regarding its content, the guidebook focuses on the following nine key issues: How do I find a topic for my thesis? – What sort of thesis suits me best? – What do I have to keep in mind for the organization of my thesis? – Who is my contact person and who will rate my work? – How and where do I find the right literature for my research? – What does academic writing mean? – What is different about writing a thesis abroad? – What is different about writing a thesis with a company? – How significant is my final thesis for my future professional life? The example "how do I find a topic?" makes transparent which ways students have to go in order to find a scientific topic for their thesis. The student counselors recommend seeking direct contact to the chairs of the faculty in order to gain a deeper insight into the research area of this chair. The students can then their own research interests against the key areas of the chair. The special situation in engineering sciences – where, in comparison to other disciplines, the students are supervised individually by the scientific staff of the chair – highly requires personal contacts and a strong interest in the research topic. Thus, a first competitive situation arises here. Consequently, the students might not be able to find a topic in their most wanted research field. In this situation, the experts in the guidebook wish to widen the view of the students and suggest: "Detach yourself a little from your favorite topic!" and "It is not only essential, what you are doing, but how you are doing it!"

In addition to the expert's recommendations, the project worked closely together with central institutions of the university and the faculties in order to develop further guidelines for the students. So, for example, documents for a successful literary research were developed together with the departmental library [17] and checklists for completing a reasonable design of one's project in cooperation with the Project Office for Civil and Environmental engineering [18]:

- Literature meets technology
- Ten steps to academic writing
- Ten steps to finding (the right) literature
- As well as individual guidelines of the faculties' chairs

7 Academic Writing in Engineering Studies: A Pilot Scheme for the Faculty of Mechanical Engineering

While the guidebook for final theses is already established and available for students seeking advice regarding the first steps in organization and planning of their written semester and final theses, the Faculty of Mechanical Engineering at the Ruhr University in cooperation with the project ELLI will also start a pilot project for fostering academic writing in engineering studies in the winter term of 2014/15. A series of six seminars will be offered to students in the third semester of their studies in Mechanical Engineering, Sales Engineering and Product Management, or Environmental Engineering and Resources Management. With their seminar series: Preparation for the Term Paper, the lecturers – dean of studies of the Faculty of Mechanical Engineering, Prof. Dr.-Ing. Franz Peters and specialist in German Studies M.A. Theresa Janssen – want to address especially those students who feel insecure with regard to their competencies in academic writing. The target group of students who are in their third semester will soon have to write their first paper. The term paper in the fourth semester usually confronts engineering students with one of their first big challenges in the course of their studies. It also represents their only opportunity to practice their academic writing skills for the next big project, the bachelor thesis. Furthermore, the grade of the term paper already has an influence on the later application for a bachelor thesis. An exchange between the above mentioned lecturers had revealed a demand for linguistic integration of the heterogeneous student body, which is now about to be addressed for the first time through this basic seminar for academic writing in engineering sciences. In the course of the seminar, a general introduction to academic writing will be followed by a description of the specific requirements (regarding literature, citation, etc.) in engineering sciences. During the subsequent seminar sessions, writing exercises will be conducted by means of experiment descriptions. The texts are written manually during the seminar and the feedback system includes peer-reviews and an appreciative guiding by the two lecturers that is both, discipline-specific and linguistically sound. The description of currently relevant experiments shall be practiced, while a special emphasis is put on the exactness which will be essential for the writing of future research papers and theses. Thus, the seminar presents a first opportunity for students to check their own capabilities against the requirements of academic writing in the course of their studies. These first trials of students who are willing to improve their scientific writing skills are not graded directly and the students do not receive credit points for their participation. Instead, in order to reward their efforts, their participation in the seminar will contribute positively to the evaluation of their future term paper. In order to further define the strategies for fostering academic writing in engineering sciences, evaluations of the seminar as well as the experiences from developing the guidebook and the student's survey will soon be discussed in a strategic talk with the managing staff of the faculty, the dean of studies, the center for academic writing of the university and the project ELLI.

8 Conclusions and Outlook

Against the background of increasingly diverse requirements of today's engineering students in Germany, the project ELLI is trying to provide individual support during the transitional phases of the study programs in technical disciplines. In the course of an engineering study program, academic writing seems to be one of the greatest challenges. At the Ruhr University Bochum, the qualitative focus groups with student representatives and a quantitative survey have underlined the students' demand for guidelines and more orientation regarding academic writing and especially for writing their final thesis. In 2014, several measures have been introduced by ELLI in order to meet this demand, which are starting to bear fruits already. The guidebook for writing final theses was well-received by students and faculty members alike and it served to raise awareness for the significance of the topic. The seminar on academic writing will have to be evaluated after its pilot run in the winter term of 2014/15 and, if it proves to be equally well-received, the Faculty of Mechanical Engineering will consider strategies to make it available to all its students.

References

1. Bildung und Qualifikation als Grundlage der technologischen Leistungsfähigkeit Deutschlands 2014, Alexander Cordes, Andre Donk, Christian Kerst, Michael Leszensky, Tanja Meister
2. Cordes, Alexander; Donk, André; Kerst Christian; Leszczensky, Michael, Meister, Tanja: Bildung und Qualifikation als Grundlage der technologischen Leistungsfähigkeit Deutschlands 2014, http://www.e-fi.de/fileadmin/Innovationsstudien_2014/StuDIS_1_2014.pdf (Stand 20.11.2014).
3. Hafner, Theo: Nachfrage nach MINT-Studium steigt - Potenziale bei den Frauen noch nicht ausgeschöpft. Publikation HIS:Forum Hochschule 11|2013, https://idw-online.de/de/news548080 (Stand 11.09.2014)
4. Lübke, Friederike: Haltet die Ingenieure, 03|2013, http://www.zeit.de (Stand 11.09.2014).
5. Heine, Christoph , Egeln, Jürgen, Kerst, Christian, Müller, Elisabeth, Park, Sang-Min: Bestimmungsgründe für die Wahl von ingenieur- und naturwissenschaftlichen Studiengängen, Zentrum für Europäische Wirtschaftsförderung (ZEW) (ftp://ftp.zew.de/pub/zew-docs/docus/dokumentation0602.pdf Stand 11.09.2014)
6. und Hafner, Theo: Nachfrage nach MINT-Studium steigt - Potenziale bei den Frauen noch nicht ausgeschöpft. Publikation HIS:Forum Hochschule 11|2013, http://www.his.de (Stand 30.10.2013)
7. Gensch, K./Kliegl, C. (2011): Studienabbruch - was können die Hochschulen dagegen tun?, Bayerisches Staatsinstitut für Hochschulforschung Hochschulplanung, München
8. Verein Deutscher Ingenieure e. V. (Hrsg.) (2012): Ingenieure auf einen Blick. Erwerbstätigkeit, Innovation, Wertschöpfung. Online unter: https://www.vdi.de/fileadmin/vdi_de/redakteur/dps_bilder/SK/2012/2012_-_Ingenieure_auf_einen_Blick.pdf (Stand: 29.05.2013).
9. Hetz, Pascal: Nachhaltige Hochschulstrategien für mehr MINT-Absolventen, Essen 2011.
10. Hochschul-Bildungs-Report 2020
11. Die Relevanz bestätigt sich außerdem in der Regelmäßigkeit der Thematisierung von Schreibkompetenzen im Rahmen der wissenschaftlichen Beiträge zur EDUCON vgl.: Kramberg-Walker, Carol: The Need to provide Writing, Support for academic Engineers (1993); Olds, Barbara et al.: Writing in Engineering and Technology Courses (1993); Damron, Rebecca L., High, Karen A.: Innovation in Linking and Thinking: Critical Thinking and Writing Skills of First-Year Engineering Students in a Learning Community (2008).
12. http://www.elli-online.net
13. Kümmerling, Franziska: Studienqualitätsmonitor 2013: Studierende identifizieren sich mit ihrer Hochschule, https://www.idw-online.de/de/news602747 (Stand 11.09.2014).
14. http://www.zhb.tu-dortmund.de/hd/fowe_wissenschaftliches-arbeiten/ (Stand 11.09.2014).
15. Die vollständigen Ergebnisse der Befragung, der Fragebogen sowie der Ratgeber finden sich auf der Projektseite: http://www.rub.de/elli/abschlussarbeit
16. http://www.rub.de/elli/abschlussarbeit
17. http://www.ub.ruhr-uni-bochum.de/fachbib/verbund-ic/
18. http://pbu.rub.de/

Are Virtual Learning Environments Appropriate for Dyscalculic Students? – A Theoretical Approach on Design Optimization of Virtual Worlds Used in Mixed-Reality Simulators

Laura Lenz, Katharina Schuster, Anja Richert and Sabina Jeschke

Abstract In Germany, there are more than four million people (almost 6 % of the entire population) living with dyscalculia, a disorder which alludes numbers as well as general arithmetic and is closely related to dyslexia [1]. The estimated number of unreported cases is probably even higher. Medical researchers talk about a "forestalled elite" since these people are commonly not less intelligent than non-handicapped individuals. Still, they rarely make it to a university-entrance diploma; they get lost on the way because of missing standby facilities offered in primary and continuative schools [2]. They require special needs and attention in order to learn and show their de facto potential. This paper deals with the dyscalculic-friendliness of learning environments provided by Mixed-Reality Simulators. After a presentation of the scientific state of the art on the specific needs of affected students, it will be elaborated in how far virtual environments used in the education of mechanical engineering students can sufficiently not only meet those needs but support them in their study.

Keywords Dyscalculia · Dyslexics and Dyscalculics in Academia · Virtual Learning Environments · Learning Content Adaptation · Mixed Reality Simulator · Dybuster Calcularis

1 Introduction

In autumn 2013/2014, more than 2.5 million students were enrolled at German universities – with a growing tendency [3]. The German Student Union claims that around 8.000 students of these are afflicted with dyslexia or dyscalculia. The affected students often suffer silently since many do not know about the opportunities their

L. Lenz (✉) · K. Schuster · A. Richert · S. Jeschke
IMA/ZLW & IfU, RWTH Aachen University,
Dennewartstr. 27, 52068 Aachen, Germany
e-mail: laura.lenz@ima-zlw-ifu.rwth-aachen.de

Originally published in "Proceedings of the 7th IEEE Consumer Electronics Society Games, Entertainment, Media Conference (GEM 2015)",

DOI 10.1007/978-3-319-46916-4_59

university offers for them [4]. Instead of informing their professor in advance, they fail exams because of numerous arithmetic errors although they are as well prepared as their fellow examinees. Institutions like the AStA Representatives for Disabled and Chronically Ill Students at RWTH Aachen University, which deal with all issues regarding studying with a handicap, provide manifold administrative prospects to assist their students throughout their daily study routine [5]. However, administrative acts are not where integrational processes and the aspiration for diversity and equal opportunities for students should come to a stop. Instead, the significance of particular (but often expensive) needs, such as a very personal educational mentoring, must be part of an inclusive, barrier-free curriculum design – independent of study field. At the IMA/ZLW & IfU institute cluster, which is part of RWTH University's Faculty of Mechanical Engineering, major seminal, didactic and curriculum related decisions on how a contemporary and attractive syllabus could be created are being made. Thus, it is inter alia their task to also include handicapped students into the designated curricula as far as possible. For all students, in order to make mechanical engineering 'touchable' instead of 'dry' as well as often extremely complex theory, virtual learning environments (VLEs) have been part of engineering education for several years.

For this sake, the RWTH acquired a worldwide unique mixed reality simulator. The so called 'Virtual Theatre' combines a head-mounted display (HMD) with an omnidirectional treadmill, so that students can physically walk through virtual worlds. But do these digital settings and their set-up sufficiently serve the special needs of dyscalculics? What happens if the possibility to physically move through VLEs is contingent? Can VLEs serve as a less-costly alternative for these target groups to learn, as they are available at any time, customizable and randomly often repeatable? And if they are not appropriate, which parts of their design and setting should and could be adapted and how? These are the core questions this paper seeks to answer.

To do so, the authors will proceed as follows: First, it will be summarized what exactly dyscalculia is. Second, the state of the art of academic provisions in connection with dyslexic and dyscalculic students will be elaborated. Third, in order to stress the specific needs of dyscalculics in education, the scientifically proven, dyscalculic-friendly Swiss concept of 'Dybuster Calcularis', a special learning software for this audience, will be examined. It shall serve as one example out of many, equally functioning software tools for the handicapped. Fifth, the relevance, assumptions and limits of this work will be outlined. Sixth, the researched hardware, the 'Virtual theatre' will be explained in technical terms. Seventh, it will be investigated in how far the VLEs used at IMA/ZLW & IfU suit the requirements for studying with dyscalculia as proposed by 'Dybuster'. In here, possible design adjustment suggestions concerning how to make the analyzed learning tools more optimal will be made. The conclusion will include a tentative outlook concerning how the academic education of dyscalculics in Germany could and must evolve.

2 What is Dyscalculia

Dyscalculia is a difficulty in learning arithmetic correlations as well as fundamental mathematical cohesions in general. It is similar to dyslexia, but adverts to numbers rather than letters. Dyscalculics are 'blind' for numbers; they cannot understand or manipulate them. Like dyslexia, dyscalculia is not related to a high or low IQ. Besides arithmetic issues, patients do often have difficulties with time, measurement and spatial thinking. Their frustration level is relatively low, wherefore they become easily frustrated if they cannot complete tasks correctly and feel embarrassed. Dyscalculia often comes in combination with ADHD and concerns 3–6 % of the population [6]. If it is caused by a brain injury, the correct designation is acalculia. In contrast, dyscalculia has a developmental origin. In daily life, developmental dyscalculics furthermore face difficulties with recognizing the largest out of many numbers, budgeting, basic calculations, differentiating between left and right, mental visualization, the estimation of distances, the ability to complete mentally exhaustive tasks and the recollection of names as well as designations.

For a dyscalculic, the ideal learning environment contains as many as possible sensual stimuli (for example audible and visual) and is user- and context-adaptive [6]. The reason is that the three task-specific modules (verbal, symbolic and analogue magnitude, summarized in the so called 'triple-code model') are located in various parts of the brain and must thus be stimulated differently. In the context of number processing, a high overlap between these modules leads to an increased arithmetic understanding, which is less present in the minds of dyscalculics. However, in the brain of any person, the three modules develop hierarchically over time, wherefore each learner reacts variably to diverging mental attraction [6]. Still, the too little overlaps in a dyscalculic mind make individualization of learning crucial.

3 Provisions for Dyscalculics in German Academic Education

In terms of dyscalculic-friendly curriculum adaptation, there are currently no non-administrative provisions taken at German universities. However, there are some common guidelines to support handicapped persons; the measure is nationwide known as disadvantage-compensation and does inter alia concern dyslexics and dyscalculics. The disadvantage- compensation includes arrangements such as the non-consideration of spelling, punctuation and grammar mistakes in (take-home) exams, an extended exam-duration, the usage of a notebook with auto-correct instead of having to write with pen and paper, an oral instead of a written exam or several evaluation/correction loops before having to hand in an assignment. These provisions are taken by numerous well-known German universities [7]. Thus, the status quo demonstrates that the prevailing measures are exclusively of administrative, not syllabus-adaptive nature. Universities expect their dyslexic students to master teaching and learning contents, but do only make compromises when it comes to examinations. However, the actual problem starts much earlier – when dyslexic students must apprehend and internalize study-related information.

4 Learning Requirements of Dyscalculic Students: An Overview

After it has been clarified how a dyscalculic brain functions and that there a no curriculum-related measures taken by Germany universities yet, it must be examined which special learning requirements evolve hereof, which were the results of multitudinous studies [8, 9]. First of all, it must be ascertained to which of the three representational mental modules the respective dyscalculic predominantly reacts. This can be captured by a controlling algorithm embedded into a Bayes-Net, which will store, categorize and analyze individual learning performance as well as it recognizes the interconnectedness of skills and adapt the provided tasks accordingly. A dynamic, independent of age Bayesian model is part of most learning software tools for dyscalculics and targets the modelling of the intelligent mind [6]. Thus the selection of learning paths should be non-linear and flexible. Common learning software fulfills these requirements by offering the opportunities 'to stay', meaning to continue with the actual task, to 'go back', to return to the former task and to 'go forward', to proceed with the next (assembled) task. These opportunities benefit the adaptability of content in terms of individual needs, the locality, the configuration based on "[…] nodes and neighbors […]" [6] and generality, meaning the common applicability on any structure, model and content.

5 Dyscalculic, Adaptive Learning Through Gamification: An Example

In order to make the general theoretical requirements explicit, one example for dyscalculic learning software shall be given. Besides applications like 'Cool Math Games', 'ETA Cuinenaire' and 'NUMBER SENSE', 'Dybuster Calcularis' is one of the most commonly known tools [10]. In general, the mentioned software tools function equally and are relevant for persons of any age group since patients react to similar stimuli based on their module group as outlined in Section 2. 'Dybuster Calcularis' was nonetheless chosen here because it was created due to public funding by the ETH Zurich and the University of Zurich, Switzerland, scientifically proven and verified in two scientific studies and is nowadays used by more than 35.000 dyscalculics [11]. In addition, the parties in charge display their methods in a transparent way on their webpage and in numerous publications. However, what must be internalized is that Dyscalculia can solely be treated, not cured.

'Dybuster Calcularis' stimulates the interplay between the representational verbal, symbolic and analogue magnitude brain modules. It does so by offering alternative representations to existing arithmetic problems, which suit the individual mental peculiarity of the learning dyscalculic. This process, which is called transcoding, provides three principles of number understanding: cardinality, the (in-) finite numbering of elements, ordinality, the indexing of elements and relativity, the interrelatedness of elements [6] (Fig. 1).

In the 'Landing Game', a basic level of 'Dybuster Calcularis', the user must position the displayed number (83). The completion of this task does not only give him a feeling for the correct ordering of numbers fosters the understanding of ordinality (which number is how much bigger/smaller?) and relativity (where between 80 and 90 is 83?). The numbers themselves are shown in an analogue magnitude (a block). Important is also that the numbers do have different colors in order for the patient not to exchange them (Fig. 2).

On an advanced level, in the Plus-Minus game, analogue magnitude (eight blue and five green blocks of diverging form)) are combined with verbal (numbers) and symbolic (colored points) elements. The addition of 85 + 8 is hereby displayed on several modes in order to not only apply to one, but several mental schemes. Thus, 'Dybuster Calcularis' advances the connection building between the three mental modes of arithmetic thinking and does thereby accomplish improvement in the realm

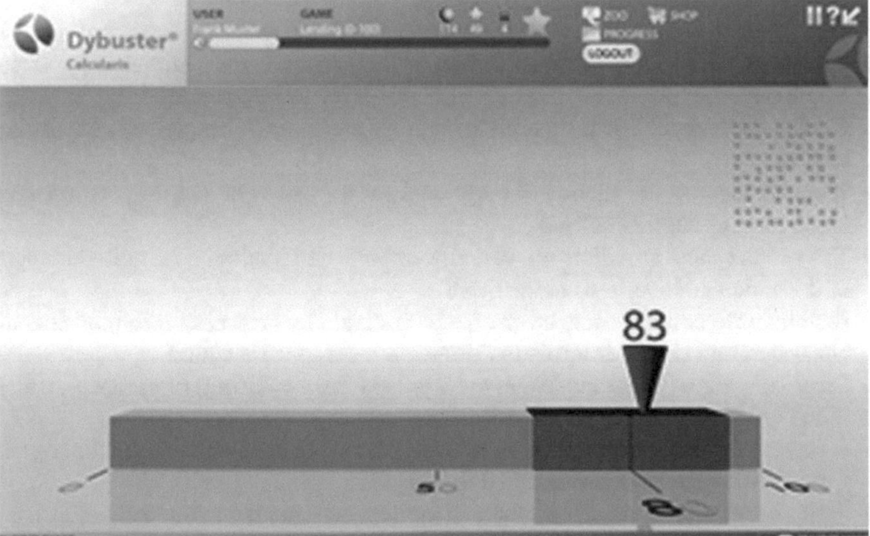

Fig. 1 Dybuster Calcularis – the Landing Game

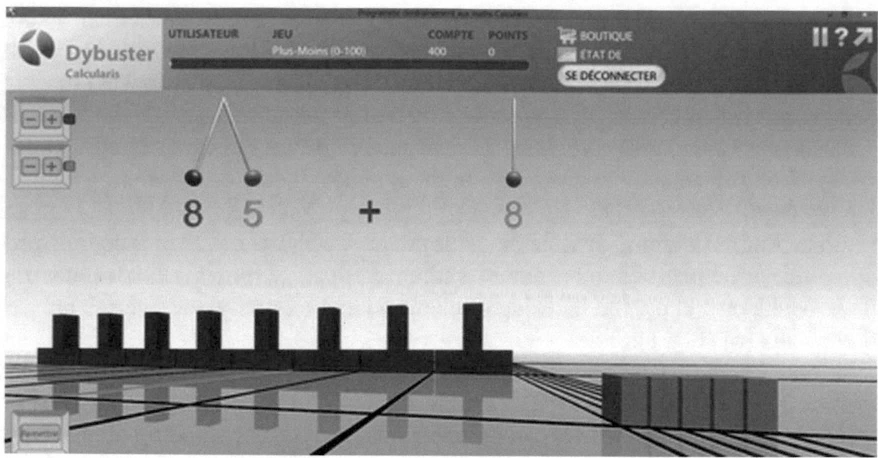

Fig. 2 The Plus-Minus game

of mathematical thinking and understanding. If tasks are completed correctly, the user may adapt the background of the game as a kind of reward.

6 In-Between Summary: What Must be Kept in Mind

For now, as an interim-summary, it must be kept in mind which the special requirements of a dyscalculic for successful learning are:

- The visualization of what must be learned. Dyscalculics are attracted by visual stimuli.
- Ideally, there should be a supportive auditory element to complement the visualization.
- If applicable, the stimulation of additional senses would be helpful, for example in the form of haptic feedback.
- The content must stimulate the verbal, symbolic and analogue magnitude modes of the brain and ideally combine those
- A stimulus to communicate an error, which should by no means be a punishment, but a short and simple information, for example a specific sound.
- Clear separation (for example due to complementary coloring) of numbers, which sound or look equal to a dyscalculic
- A Bayes-Net to not only analyze and store errors, but to categorize and interrelate them for the sake of self-controlled feedback.
- Forms in combination with numbers. If content is presented by the help of a visual depiction, it can easier be kept in mind [8, 11].

- The possibility to repeat new and old contents whenever desired (although this is crucial for any student). Ideally, there should be a mix of both to not only internalize what has already been learned, but to interconnect it with the new.
- There must be an opportunity to obtain a reward in order to increase (long term) motivation.

All in all, this makes 10 requirements, which must be checked in the 'Virtual Theatre' scenarios created by universities who want to design their virtual learning environments towards dyscalculic-friendly support.

7 Relevance, Assumption and Limits

Before starting with the analysis of the 'Virtual Theatre' scenarios, some relevant assumptions and limits of this work must be outlined. One important assumption of this paper is that the dyscalculic-friendly learning software 'Dybuster Calcularis' is indeed helpful for the learning process of dyscalculics. The scientific results they present to verify its impact must be taken for genuine because it is beyond the limits of a theoretical approach on a dyscalculic-friendly syllabus design to reappraise the medical results. Moreover 'Dybuster' is solely one example for many similarly functioning software tools and not the only solution for the problems of dyscalculics.

Another important assumption is that for the investigation of the learning environments' importance for the learning process of dyscalculics, one must suppose that the user has already had some training with 'Dybuster' or comparable dyslexia software. There must be tie points to this basic software in order to guarantee a continuous learning process based on equal principles. This is what this work seeks to do, identifying these connection points. It is beyond the limits of a university to (re-)teach arithmetic basics. Instead, one main ambition should be to make existing contents dyscalculic-friendly instead of creating entirely new learning scenarios.

As for additional limits, this paper shall solely serve as a theoretical fundament for further research on dyscalculia in academia. In the future, practical quantitative and qualitative research on and with affected students must be conducted in order to further prove and qualify the results. Thus, this work provides a subjective, tentative, self-critical assessment concerning in how far the special needs of dyscalculics are currently being dealt with throughout their learning process.

Lastly, it must be stated that RWTH Aachen University also uses digital games like 'Minecraft' for the education of engineers. However, this paper shall exclusively deal with virtual environments which were either created by RWTH itself or which they are enabled to adapt. It appears less meaningful do make fundamental design suggestions for learning games the university cannot adjust anyways.

8 Hardware Description: The Virtual Theatre – A Mixed-Reality Simulator

To proceed with the evaluated hardware device, mixed reality simulators like the 'Virtual Theatre' combine the natural cut surface of a Head Mounted Display (HMD) with an omnidirectional floor. It can be connected to a hand tracker, which does however only support the position tracking. This is otherwise being done by ten infrared cameras. The omnidirectional floor consists of 16 trapezoidal elements. These are equipped with rolls, which have a common provenance in the middle of the 16 trapezes. In the middle, the user can stand still. As she leaves it, the motor and the rolls start moving. They allow for natural walking and will rotate faster as the user comes closer to the edge in order to not let her run out of the surrounding [12].

The 'Virtual Theatre' makes a full 3D visualization of a virtual environment possible. The movements are captured in real time so that natural ways of acting and moving become possible. An infrared marker, which is attached to the HMD tracks the head movements. Although this is (not yet) possible, hand tracking will in the long run enable the user to actively interact with the virtual environment presented in the 'Virtual Theatre'. For now, the users can hold a cross-like hand tracker in their hand, which is also vested with infrared markers. If this hand tracker undercuts the height of 0.5 meters, the simulation and motors stop, for example, if the user stumbles. Any interaction between user and 'Virtual Theatre' happens wirelessly (Fig. 3).

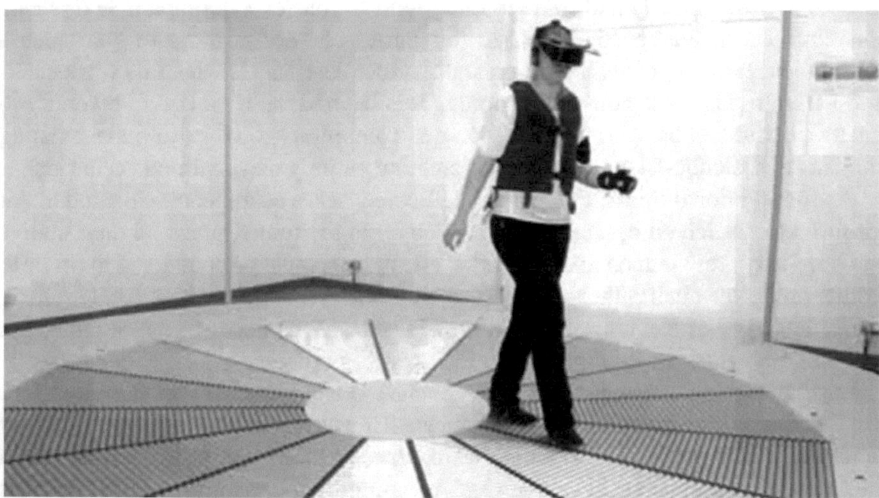

Fig. 3 A test-person walking through a virtual world on the 'Holodeck'

One of the purposes of the 'Virtual Theatre' is its usage for the education of mechanical engineering students. It is supposed to depict repeatable, danger free scenarios in which young engineers may test their knowledge in a realistic way. Obviously, this also saves material costs [13]. Currently, there are four scenarios available for the 'Virtual Theatre', the test scenario 'Piazza', which depicts a typical Italian marketplace. Second there is the 'Maze', in which users need to find items such as a ball or a rubber duck. The task is to be able to locate the found items on a map afterwards in order to investigate the user's spatial abilities. Third there is another test scenario, the 'Gallery'. Users walk through a museum-like room with famous drawings in it. Fourth, there is the 'Mars' scenario, which was designed on the corresponding basis map material of the NASA. The 'Mars' is used as an explorative landscape, a space-museum or an obstacle parcours by the National Aeronautics and Space Research Center of the Federal Republic of Germany (DLR) School Lab for primary and secondary school pupils inter alia located at RWTH [14]. There are several space shuttles and crafts located on it, which are either drivable by the user or move freely. They are closely located to signs displaying their actual real life name/designation.

All scenarios are available for both computers and the 'Virtual Theatre'. Potentials and opportunities for dyscalculics shall already tentatively be traced by evaluating the status quo. It will be outlined whether and how the 'Virtual Theatre' gives rise to the inclusion of students with special needs from the genesis onwards and to which extent it could therefore be groundbreaking for dyscalculic academic education if executed carefully and correctly.

Before starting with the analysis of the 'Mars', 'Maze' 'Gallery' scenarios, it must be clarified, why only these two were chosen. The reason is that actually, only three of the four scenarios contain textual and or numerical elements. In order to analyze whether the virtual learning environments used at RWTH Aachen University are appropriate for dyscalculics, it does not make sense to investigate scenarios, which do not contain text or any other potential handicap-related content and do therefore not require any spatial thinking or arithmetic comprehension skills. The 'Piazza' shall be excluded because it depicts 'solely' a quadrangular place. It is trivial to talk about spatial thinking in a quadrangle. Nevertheless, it must be underlined that generally, any mathematical content could be included into any scenario. Additionally, the interaction with objects inside scenarios is generally an option due to a physics-engine, but not fully activated and declared yet.

9 Analysis: The 'Virtual Theatre' Scenarios 'Mars', 'Maze' & 'Gallery'

Ultimately, it is of utmost importance to present the reader of this paper what the analyzed scenarios look like. For this purpose, the authors will first give some impressions on the virtual environments. The three will be analyzed separately because they apply to varying modules and learning channels. At the end of the analysis, there will be a score concerning in how far each scenario fulfills the ten previously mention dyscalculic requirements and whether it offers any extra potential.

Fig. 4 A part of 'Virtual Theatre' scenario 'Mars' (1)

Fig. 5 A part of the 'Virtual Theatre' scenario 'Mars' (2)

Fig. 6 Gaze inside the Maze. The user just found a car

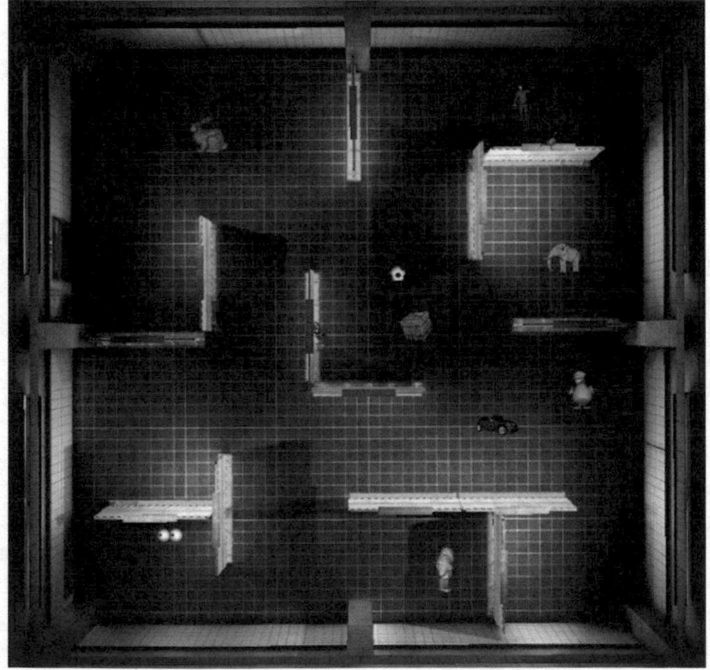

Fig. 7 The 'Maze' from above with the all items to be found in it

Fig. 8 The 'Gallery' scenario with Da Vinci's 'Mona Lisa'

On 'Mars', it is only partly possible to interact with the presented shuttles. They either stand next to their name tag or move independently. Only one specific rover can be maneuvered by the user by Q, O, P, R control, but exclusively on laptops. Inside the 'Virtual Theatre', a Wii-controller can be attached to manipulate the VLE [15]. Apart from this, users may explore the virtual space and can learn about the shuttle's name. The aim is to enthuse pupils about engineering and inspire them. The crafts are empathized by their original paradigms and users can investigate their outer appearance and construction. In the 'Maze', users walk around to find hidden subjects in it, but without interacting with these. Similarly, users can walk around the virtual 'Gallery' and learn about the artworks and respective artists. Whenever the user virtually stands and looks right at the drawing in front of him, title and painter appear above it. The idea is to create an association between text and drawing in order to support the learning efficiency by not only saving information on a cognitive, visual level, but due to the physical movement in and around it.

To begin with the requirement of visual elements/stimuli, this need is clearly being served on 'Mars', the 'Maze' and in the 'Gallery'. Not only that the numerous presented crafts and pictures can be investigated from all sides. Especially on 'Mars', although both scenarios are in 3D, the shuttles also move (either by themselves or by the user's manipulation) and hereby become credibly imaginable and alive. Thus, instead of seeing machines and drawings on book pages, they can be experienced and are therefore more attractive for dyscalculics; they can easier remember them. In addition to the depiction of the machines and artworks, a continuative positive effect is their name tags located nearby, either on a sign or (in case of the 'Gallery') as lettering on top of it. Independent from the character's color scheme, one can expect a positive connection between display (which is already targeting the 'translation into forms' criterion) and appellation. The 'Maze' is of special interest here because the finding of hidden items fosters spatial thinking as recent RWTH studies show [13]. As spatial thinking is a weakness of dyscalculics, the 'Maze' may serve as an especially stimulating environment here since it caters on numerous senses as it will be examined below.

As for the audible stimulus, there is none available (yet) in all scenarios. Generally, it is possible to attach headphones to the HMD, but this is currently not being done. Moreover, the roles of the 'Virtual Theatre' are relatively noisy, so the headphones would have to block external noises and foster the internal at the same time in order to create an immersive experience. Since 'Dybuster Calcularis' solely suggests the repetition of one certain tone in one specific context, it does not necessarily have to be an actual close to reality machine sound (in the case of 'Mars' or the 'Maze'), but a random one, affiliating one single item.

To continue with the stimulation of additional senses of dyscalculic users besides the auditory and the visual, 'Dybuster Calcularis' proposes the subjoining of additional tenors in order to create a more intense, immersive feeling. This is not crucial, but still to an extent desirable. It is not really one of the five senses (at best groping), but presumably no disadvantage to offer test persons the opportunity to physically access the virtual learning environments to herein move naturally and freely. Besides this, one cannot talk about supplementary stimulation like for example (virtual/digital) haptic feedback. There is prevailingly no way the user can actively manipulate his surroundings (besides being able to move one shuttle by a portable keyboard), hear what happens inside of (in this case) the space craft or how the duck quacks (an item in the 'Maze').

Proceeding with a fixed color scheme, the letter color (brown-grey on 'Mars' and white in the 'Gallery') does not vary per machine or drawing, which is according to 'Dybuster Calcularis' highly problematic for the absorption process of a dyscalculic student. For her, it would make a major difference if each item's name tag had its own color and one respective sound. Although not implemented, the individual and adjustable coloring of numbers, letters and words is a change that can be undertaken with a marginal programming effort. The maze shall be left out here because it does not entail any signs, numbers or letters.

Currently, all scenarios are non-competitive, meaning that there are no errors to be made. However, if students would for example have to learn the space crafts' or artworks' technical terms or names and either allocate them from a selection of the same or type the names in directly, an analogical feedback tune could be created. It must be claimed that this ability is heavily dependent on the auditory feedback criterion and asks for its successful completion and implementation to be realized itself.

In two scenarios (the 'Mars' and the 'Gallery', but not in the 'Maze'), there are some technical appellations of crafts and drawings, which might be mixed up by dyscalculic students and cause confusion and disorientation depending how severe their handicap is and if it strikes dyslexic traits, too. One example could be the 'Surveyor' and the 'Sojourner' as depicted in Figs. 4 and 5. 'Surveyor' and 'Sojourner' sound very akin. According to the color scheme of (oppositional) colors as proposed by 'Dybuster Calcularis', these two are not similar enough to demand fully complementary dyeing, but must by all means differ explicitly. As a side remark, there is a lot of potential for dyslexic students in the 'Gallery'. In accordance with 'Dybuster Ortograph', a dyslexia learning software, it is especially appealing to this target group if they first see the content and then its description. However, the analysis of dyslexia-friendliness goes beyond the limits of this paper [11, 16].

Currently, as there is no audible stimulus integrated into the 'Mars', 'Maze' and 'Gallery' scenarios, there is in consequence no auditory feedback (whether positive or negative) hearable. Furthermore, the engineering contents are still under construction as it was elaborated above. Therefore, there is no Bayes-Net (yet) to store, analyze, interrelate and categorize possible errors of the dyscalculic exploring the virtual environment and for example locating items. This is why one can on the one hand not talk about a rudimental fulfillment of this demand and on the other hand not about a growing new potential (Figs. 6, 7 and 8).

In contrast, there is a translation of mechanical/artistic learning contents into 3D forms, meaning that the theoretical construction of a space shuttle or a picture frame are transferred into an experience the 'Virtual Theatre' user can actually undergo. In 'Dybuster Calcularis', this is being done in a more simplistic way, namely by depicting arithmetic elements as geometrical figures. Still, this fact underlines a dyscalculic's affection for imagination through accurate, 'touchable' forming, about which one can due to the credible digital realizations of shuttles and artworks talk here.

As for the fore-last criterion, the ability to arbitrarily repeat contents, the 'Virtual Theatre' appears to be a suitable technology. The repetition does not only reduce costs and dangers. The simulation always looks exactly the same and subject matters can theoretically be adjusted as the learning progress increases. At the same time, this leaves enough space for the adaptation in line with specific dyscalculic demands like individual focus points. If the VLEs are being adjusted slightly, dyscalculic students will have the opportunity to test their knowledge freely without being hesitant about breaking material and thus being ashamed in front of others, which will supposedly be a very positive experience for (not only) a dyscalculic.

Lastly, there are no tasks to pass or fail in all scenarios, which is why there are no rewards available (yet), which would be motivating for especially dyscalculics. However, in the long run, there appears to be a major potential here. An imaginable reward could be the random modification of the environment by the user (like in 'Dybuster Calcularis') or additional points for a final exam if the user completes certain assessment-relevant tasks in the virtual world.

From the analysis of the RWTH's 'Virtual Theatre' scenarios, the following tables can be conducted (additional potential shall be covered under the 'Other' section, but not be part of the dyscalculic-friendliness calculation) (Tables 1, 2 and 3).

Table 1 Evaluation of the 'Mars' scenario

MARS	
Requirement	Application
Visual stimulus	✓
Audible stimulus	x
Additional senses	✓
Fixed color scheme	x
Error-feedback	x
Separation of similar words	x
Bayes-Net	x
Translation into forms	✓
Repetition	✓
Rewards	x
Other	✓ *Active manipulation*
Total	**4/10 (+1)**

Table 2 Evaluation of the 'Maze' scenario

MAZE	
Requirement	Application
Visual stimulus	✓
Audible stimulus	x
Additional senses	✓
Fixed color scheme	x
Error-feedback	x
Separation of similar words	x
Bayes-net	x
Translation into forms	✓
Repetition	✓
Rewards	x
Other	✓ *Spatial thinking*
Total	**5/10 (+1)**

Table 3 Evaluation of the 'Gallery' scenario

GALLERY	
Requirement	Application
Visual stimulus	✓
Audible stimulus	x
Additional senses	✓
Fixed color scheme	x
Error-feedback	x
Separation of similar words	x
Bayes-net	x
Translation into forms	x
Repetition	✓
Rewards	x
Other	✓ *Dyslexic-friendly*
Total	**3/10 (+1)**

10 Potentials for Dyscalculics

It has been shown that for dyscalculics, the 'Virtual Theatre' offers a lot of not fully fathomed possibilities. From the technical point of view, it is capable to provide free and unbound navigation, which means that the user is enabled to move the way she usually does and can thus immerse without further effort. It also benefits orientation and spatial capabilities as well as content-related visualization. The 'Virtual Theatre' might have some weaknesses in comparison to other mixed-reality simulators like the 'Cave', for example in terms of the visual stimulus quality, but offers in the long run more interactive ways to actively manipulate virtual environments. Due to this interactivity, any representational model, verbal, symbolic and analogue magnitude can be satisfied by attracting up to three senses (audible, visual, haptic). As for the status quo, the verbal is already served by the names of painters and drawings displayed in the 'Gallery'. Although it is currently 'only' letters, any scenario could be adapted by inserting arithmetic elements. Scenarios like the 'Mars' go far beyond the depiction of names and symbols. The 'Virtual Theatre' provides actual 3D machines (space shuttles) to manipulate, which is likely to increase the understanding of technical properties. Analogue magnitudes can be found in the 'Maze' where a specific number of items must be found and mapped.

For the didactical advantages, the 'Virtual Theatre' offers not only possibilities to arbitrarily repeat any content, but also self-determined learning (SDL). In SDL, students can freely decide when, how, where and what to learn, which leads to an increased learning success [17]. This does not only benefit non-handicapped students, which have a greater tendency to adapt to learning requirements and stimulated learning channels, but dyscalculics, who need to mainly focus on one specific channel, at least at the beginning. In the long run, module-overlaps must be fostered. Thus, if the sensual stimuli through the 'Virtual Theatre' are further refined, it will serve the representational mode of any dyscalculic and can thus serve as a means to transfer any learning content.

11 Conclusion

All in all, it has been shown that the 'Virtual Theatre' at RWTH Aachen University has a lot of potential to attract the learning 'channels' of dyscalculics. However, it is not at all fathomed. Currently, the creation of a surrounding, which enables the user to almost behave naturally in a virtual world which is accompanied with didactically valuable elements, for dyscalculics but also for non-handicapped students, is still in its initial stage. Nonetheless, the implementation of dyscalculic-equitable, academic learning environments is only one of the many measures which have to be taken in order to make the German education landscape more suitable for them. Ideally, they should be initiated in primary school and continued throughout continuative training of any kind. The question is whether other universities will in the long run offer equally interactive and multi-sensory learning conditions for marginal groups, since acquisitions like the 'Virtual Theatre' are not only costly, but require a lot of programming and didactic expertise. Until comparable learning conditions can be found in more academic institutions, large-scale cooperation is and will be key. What is crucial is that the facilitation of a dyscalculic's every-day study life is not only of administrative, but content-related nature.

Back to the micro-level, the non-handicapped studentship as well as the general public should be better informed about dyscalculics and their mental potential. They should internalize that this marginal group will contribute to technological and societal developments once they are enabled to be a fully-fledged part of academic education. Thus, there is a threefold interplay between public information and the thereof resulting acquaintance with dyscalculia, the early advancement and inclusion of dyscalculic pupils in schools and finally, ties into higher education needed. Only then, the "forestalled elite" will make it to where they belong, just like any other fringe group – into the middle of a creative, tolerant and forward thinking society 4.0.

References

1. Die Zeit Online. Ziffern ohne sinn, 2013. URL http://www.zeit.de/2013/29/dyskalkulie-zahlenblind-teilleistungsstoerung. Retrieved online, May 21st, 2015
2. Die Zeit Online. Verhinderte elite, 2003. URL http://www.zeit.de/2003/42/C-Legasthenie-Schule. Retrieved online, May 21st, 2015
3. Spiegel Online – Unispiegel. Studenten in deutschland: So viele gab's noch nie, 2014. URL http://www.spiegel.de/unispiegel/studium/studentenzahl-2-7-millionen-studieren-an-deutschen-hochschulen-a-1005107. Retrieved online, May 21st, 2015

4. Deutsches Studentenwerk. Nachteilsausgleich: Antragsverfahren und nachweise. URL http://www.studentenwerke.de/de/content/nachteilsausgleich-antragsverfahren-und-nachweise. Retrieved online, May 21st, 2015
5. Allgemeiner Studierendenausschuss der RWTH Aachen. Gleichstellung. URL https://www.asta.rwth-aachen.de/de/startseite. Retrieved online, May 21st, 2015
6. T. Käser, A.G. Busetto, G.M. Baschera, J. Kohn, K. Kucian, M. von Aster, M. Gross. Modelling and optimizing the process of learning mathematics, 2012. URL http://link.springer.com/chapter/10.1007%2F978-3-642-30950-2_50. Retrieved online, May 21st, 2015
7. Universitüt Würzburg. Informationen zum nachteilsausgleich für studierende mit behinderung und chronischer erkrankung. URL http://www.behindertenbeauftragter.uni-wuerzburg.de/fileadmin/32500250/_temp_/Broschuere_Nachteilsausgleich.pdf. Retrieved online, May 21st, 2015
8. L. Breiman. Bagging predictors, 1996. URL http://statistics.berkeley.edu/sites/default/files/tech-reports/421.pdf. Retrieved online, 21st, 2015
9. J. Haffner, K. Baro, P. Parzer, F. Resch. Heidelberger rechentest, 2005. URL http://www.testzentrale.de/programm/heidelberger-rechentest.html. Retrieved online, May 21st, 2015
10. Dyscalculia.org (n.d.). Best tools. URL http://www.dyscalculia.org/math-tools. Retrieved online, May 21st , 2015
11. Dybuster. Dybuster ortograph & dybuster calcularis. URL http://www.dybuster.com/orthograph. Retrieved online, May 21st, 2015
12. K. Schuster, M. Hoffmann, U. Bach, A. Richert, S. Jeschke, Diving in? how users experience virtual environments using the virtual theatre. In: *Proceedings of the 3rd International Conference on Design, User Experience, and Usability (DUXU 2014), Heraklion, Crete, 22-27 June 2014, Lecture Notes in Computer Science Springer*, vol. 8518. Springer, 2014, *Lecture Notes in Computer Science Springer*, vol. 8518, pp. 636–646
13. K. Schuster, D. Ewert, D. Johansson, U. Bach, R. Vossen, S. Jeschke, Verbesserung der lernerfahrung durch die integration des virtual theatres in die ingenieurausbildung. In: *Innovationen für die Zukunft der Lehre in den Ingenieurwissenschaften*, ed. by A.E. Tekkaya, S. Jeschke, M. Petermann, TeachING-LearnING.EU discussions, TeachING-LearnING.EU, 2013, pp. 246–260
14. M. Hoffmann, K. Schuster, D. Schilberg, S. Jeschke, Bridging the gap between students and laboratory experiments. In: *Virtual, Augmented and Mixed Reality. 6th International Conference, VAMR 2014, Held as Part of HCI International 2014, Heraklion, Crete, Greece, June 22-27, 2014: proceedings, Heraklion, Crete, Greece*, ed. by R. Shumaker. Springer, Cham, 2014, Lecture notes in computer science, 8525-8526, pp. 39–50
15. M. Hoffmann, K. Schuster, D. Schilberg, S. Jeschke, Next-generation teaching and learning using the virtual theatre. In: *4th Global Conference on Experiential Learning in Virtual Worlds Prague, Czech Republic*. 2014
16. A. Schabman, C. Klicpera, *Lehasthenie – LRS: Modelle, Diagnose, Therapie und Förderung*. UTB GmbH, Stuttgart, 2013
17. F. Peschel. Unterricht in der evaluation, 2006

Access all Areas: Designing a Hands-On Robotics Course for Visually Impaired High School Students

Valerie Stehling, Katharina Schuster, Anja Richert and Sabina Jeschke

Abstract In recent years, student laboratories have been established as effective extracurricular learning areas for the promotion of educational processes in STEM fields. They provide various stimuli and potentials for enhancements and supplements in secondary school education [1]. Most courses, however, do not offer full accessibility to all students. Those who e.g. suffer from visual impairment or even sightlessness find themselves not being able to participate in all tasks of the courses. On this account, the Center for Learning and Knowledge Management and Institute of Information Management in Mechanical Engineering at RWTH Aachen University have redesigned one of their robotics laboratory courses as a first step towards accessibility. This paper presents the work in progress of developing a barrier-free course design for visually impaired students. First feedback discussions with the training staff shows that even little changes can sometimes have a huge impact.

Keywords School Laboratories · Barrier Free · LEGO Mindstorms · Visual Impairment · High-School Students

1 Introduction

Extracurricular school laboratories have proven to be an effective way to let students playfully experience the fundamentals of robotics, computer science or other technology-related topics. In combination with a hands-on approach, e.g. by working with LEGO MindStorms, they get a chance to learn on a cognitive, emotional and haptic level. While this is a widespread approach these days, not every pupil, however, is able to participate in courses like these due to a lack of accessibility. Ludi e.g. states that "awareness of potential career paths and access to adequate preparation remain barriers to students who are visually impaired" [2]. Due to their impairment or lack of sight it is rather impossible for them to fully participate in a programming process or when building a robot using e.g. LEGO MindStorms sets.

V. Stehling (✉) · K. Schuster · A. Richert · S. Jeschke
IMA/ZLW & IfU RWTH Aachen University, Dennewartstr. 27, 52068 Aachen, Germany
e-mail: valerie.stehling@ima-zlw-ifu.rwth-aachen.de

Originally published in "Proceedings of the 17th International Conference on Human-Computer-Interaction (HCI 2015)", © Springer 2015.
Reprint by Springer International Publishing AG 2016,
DOI 10.1007/978-3-319-46916-4_60

To overcome this sort of discrimination, the Center for Learning and Knowledge Management and Institute of Information Management in Mechanical Engineering of RWTH Aachen University have teamed up with a group of experts in order to develop a special barrier-free course design. This group of interdisciplinary researchers and practitioners – psychologists, school and university teachers, experts in the field of accessibility as well as robotics etc. – took the original course design from an existing robotics course for high school students and transformed it into an accessible course design by applying specific changes. Applying solely technical adjustments to the course, however, cannot be fully sufficient in the development of a new and adequate course design. Therefore, all changes applied to the course went hand in hand with an adjustment of teaching and learning strategies.

When designing a programming course for pupils with handicaps in a first step these strategies as well as required tools have to be thoroughly identified. The resulting new course design allows students with a handicap such as impairment of sight to access the same courses and benefit from the same experiences as their fellow pupils. This paper will present the original course design followed by results from the expert design workshops in terms of technical and didactical adjustments to the course. Finally it will present first indications through feedback discussions on the achievements made in first courses.

2 Original Course Design: "Roborescue" and "Rattlesnake"

In the original robotics course design high school students are given the chance to get an insight in building and programming robots using LEGO Mindstorms sets in a school laboratory. The main focus of the course lies on the construction and programming of various robot models with LEGO Mindstorms construction kits. By using the graphical programming interface NXT-G, which is also suitable for non-professionals, students find an easy access into the world of programming [3].

In order to prepare and motivate students for a future career in robotics, they can try their hands at building, programming and testing robots in a highly interactive and playful environment. The course allows them to experience the fascination of robotics by letting the students create either a "rescue robot" [4] that can search for virtual victims in a simulated rescue mission or a "rattlesnake" that snaps shut when someone crosses its field of vision. The choice of the scenario is subject to the age of the students – lower grades build a rattlesnake which is easier to build and to program while junior and senior classes go on a more complex rescue mission. Within this storyline, the four main tasks of the course are embedded: an introduction giving basic theoretical information, the construction phase, the programming process as well as the reflection or evaluation phase. The underlying didactical course concept focuses on own experiences made as a basis for all implicit learning processes. These processes primarily run playfully, practically and experimentally [3].

The school laboratory where the courses take place, however, is not located at school – it has been set up at RWTH Aachen University. This allows high school students to take a peek into the daily routine at University and is meant to facilitate the decision making process when it comes to choosing further steps after graduating from high school [5].

3 Enabling Higher Accessibility for Visually Impaired Students

3.1 Expert Design Workshops

In order to facilitate the process of redesigning the robotics course and reach a higher accessibility, researchers from RWTH Aachen University invited a team of interdisciplinary experts. In a series of expert design workshops, the roadmap of the redesign was created. The main goal of these workshops was to identify the key aspects of required adjustments in order to reach a distinctively higher level of accessibility.

In the course of the workshops, the participants gradually developed a grid of these requirements. In a first step they divided the course into its individual phases based on the established approach by Vieritz et al. [6]. They used the different phases of the course and analyzed the requirements and necessary adjustments for each individual part compared to those of the original course design. These phases are the introductory part, the construction phase, the programming phase as well as the phase for reflection. Combining their different experiences and testing single elements by simulating specifing eye disfunctions, the experts came to results in terms of requirements for each phase. These results are being presented and discussed in the chapters below divided into technical as well as didactical adjustments. At the end of chapter three, the developed grid gives a summarized overview of the results from the workshops.

3.2 Technical Requirements

Due to continuous research and rapid technical advancement, today, being visually impaired does not automatically exclude one from working on and with e.g. smaller objects or computers. It does, however, bring about specific technical requirements which have to be considered when designing a robotics course. According to the results of the design workshops, the identified requirements especially include auxiliary means which can be summed up as objects, software and computer settings. There are a lot of different eye dysfunctions which call for support by different objects. In order to increase accessibility these objects are e.g. magnifiers and common magnifying glasses. Other important objects for the different phases of the course are cameras and reading devices, printed handouts for every phase, additional lighting for the building process and sorting boxes for robot components.

In terms of software, screen readers such as JAWS or Dolphin, graphic programming using e.g. NXT-G [3] as well as textual programming using e.g. JBrick [7, 8] should be provided in the programming phase. Finally, the computers provided should allow for adjustments of graphic contrast on computer screens. These adjustments should also be possible on the provided work tables. Nevertheless, there is no "universal remedy" for increasing accessibility. In preparation of the course the teaching staff should therefore always acquaint themselves with the participants in order to be prepared for any special requirements the students might have.

3.3 *Didactical Adjustments*

Not every measure taken is helpful for every sort of handicap and not all changes can be made at once. In the presented case a fundamental distinction between different degrees of visual impairment up to sightlessness has been essential groundwork for further research and course development. Most advancements and adjustments have to be made gradually in order to reach full accessibility. This has proven to be a very helpful approach in the process of designing the new course. Some degrees of visual impairment, for example, are even contrary to one another [2], so there is a need for different technical as well as didactical approaches in one course to reduce or extinguish existing barriers for all participating students.

As a first result and requirement, printed manuals should be provided for the first three phases, the introductory part, the construction as well as the programming phase. This allows students to reread instructions at their individual pace.

Time has also proven to be one of the main but often underestimated factor [9]. Visually impaired students need to be given more time to work on their tasks in terms of reading instructions, following presentations as well as building and programming. The more severe the impairment is, the more time will be needed to finish a task. In addition to that, additional time needs to be invested in giving detailed information regarding the content of e.g. manuals or presentations, repeating this content, reflecting processes, practicing as well as post-processing. Practitioners from the workshop have come to the subjective conclusion that that the time necessary for a traditional course design should be at least multiplied by four after monitoring their own ability to work through the tasks of the class by wearing glasses that simulate an eye dysfunction. On an average it took them four times as long to finish the assigned tasks. Further research and evaluations of the course will have to prove whether that factor needs to be adjusted.

Another important adjustment relates to the teacher-student ratio. It has to be increased compared to traditional course designs which of course takes up additional time and resources on the teaching end. The required ratio can differ vastly as students have very diverse needs in terms of support. As we also know from

Phase	Content	Original Course Design and equipment	Technical Requirements for a barrier free course	Didactical Requirements for a barrier free course
Introduction	Theoretical Input	Power Point Presentation	- Laptops with screen readers - Magnifying glasses	- Detailed explanations and descriptions of what the slide shows - Repetition of content - Simple slide design with high contrast - Printed Manuals
Construction	Building of the robot	Unsorted boxes	- Sorting boxes - Magnifying glasses - Reading Device - Graphic contrast on work tables	- Pre-sorting of components - Room for extra time and practice - Continuous supervision and support - Printed construction manuals
Programming	Programming of the robot	Laptops	- Contrast settings - Screenreader (JAWS/Dolphin) - Extra lighting - Printed Manual instead of beamer - On-screen magnifier - Graphic programming	- Continuous supervision and support - Room for extra time and practice - Printed programming manuals
Reflexion	Reflecting the Processes and Outcomes			Room for extra time

Fig. 1 Results from the workshop: requirements for the new course design

Silva et al., even students without handicap perceive and process experiences in different preferred ways [10]. This has been confirmed also by the practitioners. Therefore, the supervisors need to provide a high level of flexibility regarding supervision and support throughout the course. Lastly the practitioners identified presorting the sorting boxes used in the construction phase as a helpful measure in the building process which does no longer exclude visually impaired students from the haptic and tangible experience of building a robot themselves.

Every course is highly influenced by diversity aspects and a thorough preparation and awareness of all possibilities and influences as well as a preanalysis of the expected target group of each course proves to be the key to a successful course design. Figure 1 sums up the results from the workshop in a grid.

4 Conclusion and Outlook

The paper has described the process of redesigning of a robotics course from an educational robotics laboratory. The redesign was performed in order to increase accessibility of the course for visually impaired students. The evaluation of an expert workshop has brought about a concept for the redesign which has been implemented and is currently being tested in a second run with various groups of visually impaired students. The developed grid of the workshop suggests that smaller as well as bigger adjustments to the designated phases of the lecture can lead to a higher level of accessibility. First anecdotal but enthusiastic feedback from the students leads to the gentle assumption that the applied changes suggested by the experts were successful.

Nevertheless, a huge part of the adjustments needs to be individually taken considering the needs and requirements that the specific dysfunctions of the target group bring about. At this point of research, there is no "one-fits-all"-solution to the challenge. As a consecutive step, evaluations of the designed courses will allow for a thorough analysis and serve the pursuit of continuous improvement. Additionally, it will be the key to future research. In order to broaden the range of accessibility, further research will have to focus on full accessibility also for blind students as well as other impairments such as hearing and e.g. physical disabilities.

References

1. S. Reuter, K. Yilanci, N. Nabil, M. Ossenkopf, F. Reinbacher, J. Schlierkamp, F. Weidler, J. Zagatta, D. Ewert, R. Vossen, S. Jeschke, Robotic education in the DLR_SCHOOL_LAB RWTH AACHEN. In: *Proceedings of the International Technology, Education and Development Conference INTED 2015*. 2015. In Process
2. S. Ludi, T. Reichlmayr, Developing inclusive outreach activities for students with visual impairments. In: *Proceedings of the 39th SIGCSE Technical Symposium on Computer Science Education 2008, USA*. 2008, pp. 439–443
3. A. Hansen, F. Hees, S. Jeschke, Hands on robotics. concept of a student laboratory on the basis of an experience-oriented learning model. In: *EDULEARN 2010, 5–7 July 2010*. IATED, Barcelona, Spain, 2010, Proceedings of the International Conference on Education and New Learning Technologies, pp. 6047–6057
4. S. Jeschke, F. Hees, N. Natho, O. Pfeiffer, A rescue robotics pbl course. In: *Proceedings of the ISCA 25th International Conference on Computers and their Applications (CATA), USA, 24–26 March 2010*. 2010, pp. 63–68

5. S. Jeschke, L. Knipping, M. Liebhardt, F. Muller, U. Vollmer, M. Wilke, X. Yan, Whats it like to be an engineer? robotics in academic engineering education. In: *Proceedings of the Canadian Conference on Electrical and Computer Engineering (CCECE), Niagara Falls, Canada, 4–7 May 2008*. 2008, pp. 941–946
6. H. Vieritz, D. Schilberg, S. Jeschke, Early accessibility evaluation in web application development. universal access in human-computer interaction. user and context diversity. In: *Proceedings Part II of the 7th International Conference, UAHCI 2013, held as Part of HCI International, Las Vegas, NV, USA, 21–26 July 2013*, *Lecture notes in computer science*, vol. 8010. Springer, Berlin, 2013, *Lecture notes in computer science*, vol. 8010, pp. 726–733
7. S. Ludi, R. T., The use of robotics to promote computing to pre-college students with visual impairments. ACM Transactions on Computing Education **11** (3), 2011
8. S. Ludi, M. Abadi, Y. Fujiki, P. Sankaran, S. Herzberg, Jbrick: Accessible lego mindstorm programming tool for users who are visually impaired. In: *Proceedings of the 12th International ACM SIGACCESS Conference on Computers and Accessibility, ASSETS 2010, Orlando, FL, USA, October 25–27, 2010*. 2010
9. M. Kabátová, L. Jaskova, P. Lecky, V. Lassakova, Robotic activities for visually impaired secondary school children. In: *Proceedings of 3rd International Workshop Teaching Robotics, Teaching with Robotics Integrating Robotics in School Curriculum*. 2012, pp. 22–31
10. D.L. Silva, L.D. Sabino, E.M. Adina, D.M. Lanuza, O.C. Baluyot, Transforming diverse learners through a brain-based 4mat cycle of learning. In: *Proceedings of the World Congress on Engineering and Computer Science 2011, Vol 1, WCECS 2011, October 19–21, San Francisco, USA*. 2011

Part IV
Professional Competency

Where Have all the Inventors Gone? Is There a Lack of Spirit of Research in Engineering Education Curricula?

Tobias Haertel, Claudius Terkowsky and Isa Jahnke

Abstract In political discussions and the research and innovation policy of the European Union, the topic "creativity" seems to become increasingly important. Innovation depends on good ideas, so the need of innovative solutions in a globalized world puts creativity in focus. Engineers, who embody the creative inventors and tinkerers more than any other occupation group, carry an important contribution (or even the societal responsibility) to solving current problems. However, engineering education has not been known to be particularly creative or to foster creativity. Based on the results of several interwoven research projects, this paper presents an approach to the nature of creativity in the context of higher education and a small pre-study about the actual activities to foster creativity in two engineering courses. The results indicate a lack of fostering creative learning to establish a spirit of research among students. For this reason, two examples of well elaborated didactical concepts are given, able to foster creativity in engineering education in adaptable dimensions.

Keywords Fostering Creativity in Higher Engineering Education · Higher Engineering Education Research · Remote Labs · Creativity Supporting Learning Scenarios · Curriculum Development

1 Introduction

The European Union has declared the year 2009 as the European Year of Creativity and Innovation. Facing tremendous problems, creativity and innovation were seen at the heart of the strategy to transform Europe into a knowledge-based society that is

T. Haertel (✉) · C. Terkowsky
Engineering Education Research Group, Center for Higher Education,
TU Dortmund University, Dortmund, Germany
e-mail: tobias.haertel@tu-dortmund.de

I. Jahnke
Director of Research for the Information Experience Lab with the School of Information Science and Learning Technologies, University of Missouri, Columbia, USA

Originally published in "Proceedings of the 15th International Conference on Interactive Collaborative Learning and 41st International Conference on Engineering Pedagogy", © IAOE 2012. Reprint by Springer International Publishing AG 2016, DOI 10.1007/978-3-319-46916-4_61

able to cope with ongoing and future problems. For example, new techniques to tackle climate change are urgently needed, new ideas on how to retain mobility of people, new concepts for energy production without fossil fuels. Engineers play an important role in addressing these challenges. Their ideas, their inventions, their creativity have brought Europe's prosperity, and it will depend on their inventions and creativity to ensure that progress in the future. According to Feisel & Rosa [1] an engineer should be able to "demonstrate appropriate levels of independent thought, creativity, and capability in real-world problem solving" (p. 127). But where, how, and when may s/he generate these competencies? The core of the questions raised here can also be found in a parable told by Hans-Jörg Bullinger, president of the Fraunhofer Gesellschaft, on the opening of the ball of the Association of German Engineers (VDI) 2005: "Our students found the lectures on the meaning and purpose of DIN standards only limited fun. In a nice irony they told the story that a mathematics student and a physics student had just met an engineering student. They could not agree on the volume of a golf ball. So everyone picked up the methodology which corresponded to his field. The mathematician measured the diameter and the indentations on the surface and began to count. The physicist put the ball in a full glass of water and determined the displacement of water. And what did the engineer? He looked in the DIN standard for golf balls." [2] In summary, the question is raised, in what way universities contribute to educate creative engineers nowadays. To find an answer to this question, two projects have been conducted. First, the German research project "Da Vinci – fostering creativity in higher education" started in 2008 (supported by the German Federal Ministry of Education and Research BMBF, 2008-2011). This project followed an interdisciplinary approach and was with the object of designing generic creativity supporting learning scenarios. A second project, the ELLI project ("excellent learning and teaching in engineering education", 2011-2016 funded by the German Federal Ministry of Education and Research "BMBF") still runs today and has the task to bring engineering students into focus and foster their creativity, based on the results of the Da Vinci project.

2 Fostering Creativity in Higher Education

Research on creativity in formal education hasn't led to an own discipline yet, but a lot of work on this topic from different disciplines has been done so far (e.g. psychology, philosophy, economics, pedagogy). The parallel activities from various disciplines are one reason for the immense heterogeneity of approaches that try to explain the nature of creativity, but even within the subjects contradictory attempts coexist [3, 4]. Choosing one of those approaches as a basis for fostering creativity at universities would always mean to exclude those students, teachers, and researchers who favor an opposite concept of creativity. Against this background, the existing research work on creativity can be used for inspiration, but it doesn't provide an expedient working basis. Regarding higher education, only few researchers have already tried to contextualize creativity. The effort of some British researchers has to be highlighted [5, 6]. They created the Imaginative Curriculum network, which carried out some sensible and helpful concepts of promoting creativity at universities. With quite an open understanding of creativity, they tried to include all individual perspectives on this issue. However, even their respectable attempt to explain creativity wasn't a proper science base for the DaVinci project. Due to the concept's openness, no concrete and practically manageable anchor points for fostering creativity in higher education were provided. For this reason, the DaVinci project was faced with the task of operationalizing the concept of creativity in higher education. In a qualitative study, 20 expert interviews were conducted. 10 interviewees were supposed to be experts in creativity, because they have won teaching awards, got a good review at 'my prof' ("meinprof.de"[1]) or offered a course about creativity. These experts came from different disciplines (for instance, computer science, science of economy, sociology, engineers). In contrast, the other 10 interviewees were teachers of pedagogy in order to find out how creativity is characterized in everyday teaching and whether there are disciplinary focuses. All interviews were transcribed, labeled, and ordered. The analysis results show a model of creativity in higher education (across all disciplines), which consists of 6 different facets (see Fig. 1) [7–10]:

1. Self-reflective learning – teachers help learners to break out of their receptive habitus and start to question any information given by the teacher. An internal dialogue takes place and knowledge becomes constructed rather than adopted.
2. Independent learning – teachers stop to determine the way students learn (different learning pathes, personalized learning). Instead, they support students to start for example to search for relevant literature on their own, they make their own decisions about structuring a text or they even find their own research questions and chose the adequate methods to answer it.
3. Curiosity and motivation – teachers support learners to be curious and to question the topics; this aspect relates to all measures that contribute to increased motivation, for instance the linking of a theoretical question to a practical example.
4. Learning by creating something – Teachers foster students to learn by creating a sort of product. Depending on the discipline, this might be a presentation, an interview, a questionnaire, a machine, a website, a computer program or similar. Students act like researchers.
5. Multi-perspective thinking – teachers create a learning environment where learners overcome the thinking within the limits of their disciplines or prejudiced thinking. They learn to look automatically from different points of view on an issue and they use thinking methods that prevent their brain from being "structurally lazy"[2].

---

[1] A German Internet portal where students can rate their teachers.

[2] According to [11] brains are used to work with mental patterns. The more successful such a pattern is, the stronger it becomes and the more often it is remembered and used again. Considering Spitzer's theory, for example most adults' brains have saved a very strong mental pattern for brushing their teeth. Regarding the brain, this is very helpful and effective, because those adults don't need to figure out each morning anew how to brush their teeth. Regarding creativity, this is obstructive, because those adults won't ever invent a new, maybe more effective way of brushing teeth.

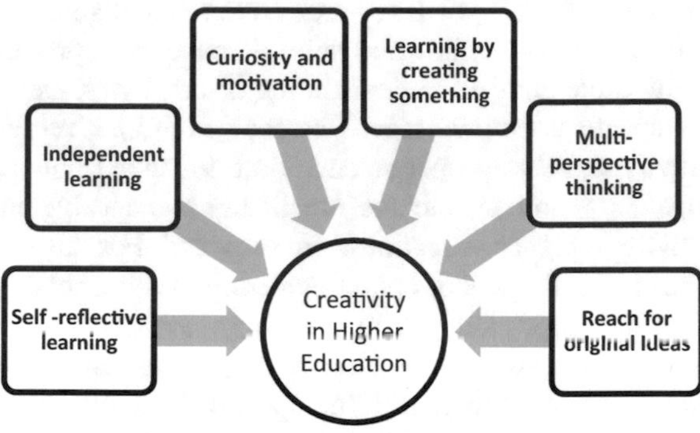

Fig. 1 6 facets of creativity in higher education

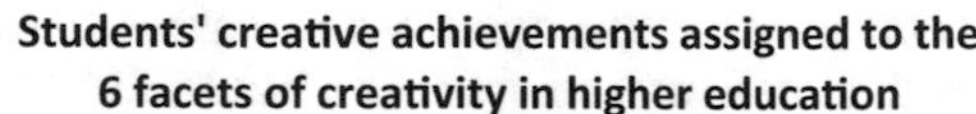

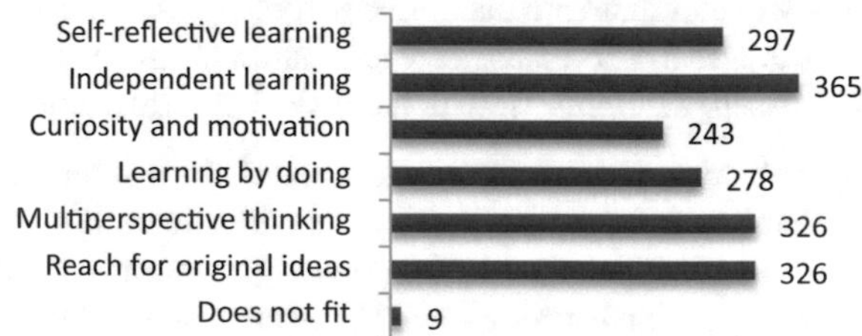

Fig. 2 Validation of the 6 facets of creativity in higher education

6. Reach for original ideas – teachers help learners to aim to get original, new ideas and prepare themselves to be as ready-to-receive as possible. Getting original ideas cannot be forced, but by the use of appropriate creative techniques and by creating a suitable environment (that allows making mistakes and expressing unconventional ideas without being laughed out or rejected), the reception of original ideas can be fostered.

To validate this model, an interdisciplinary online survey of teachers at three German universities[3] was conducted. Almost 300 teachers participated in the (non-representative) survey (n = 296) and gave 600 answers to the question "What is a creative achievement of your students?" (multiple answers possible). In the next step, they were asked to map their responses to one or more of the 6 facets. In addition, they had the opportunity to choose the option "does not fit" in case they thought that none of the 6 facets was suitable. Again, multiple answers were possible, and altogether 1.844 assignments were made (see Figs. 2 and 3). It is quite remarkable that the option "does not fit" was selected for nine answers only, which makes about 0, 5 % out of all answers. This result indicates that the 6 facets are actually able to comprise almost all of teachers' concepts of creativity in higher education. What is more, all six facets are about equally represented, which suggests that (across all

[3]The University Alliance Metropolis Ruhr: University of Duisburg-Essen, Ruhr-Universität Bochum and TU Dortmund University.

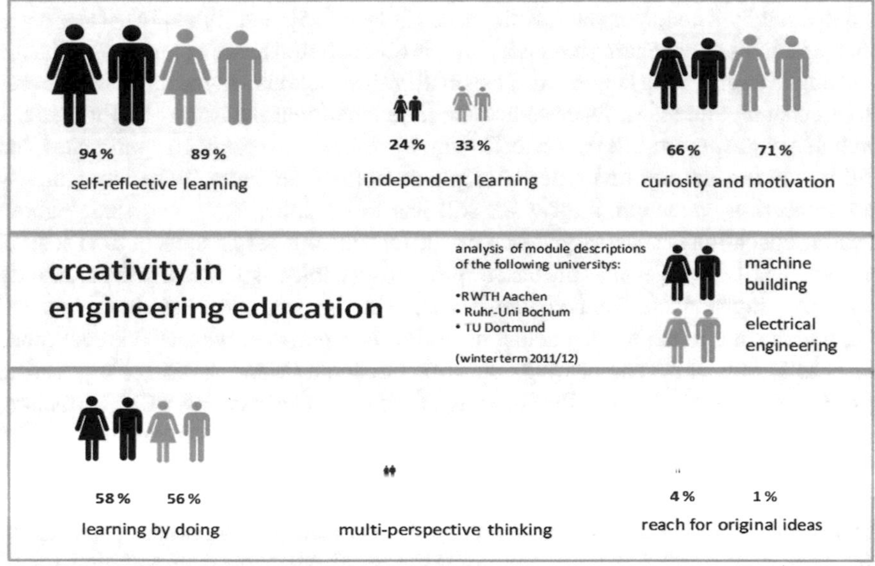

Fig. 3 Creativity in engineering education

disciplines) all 6 facets are equally important. There is no facet that really dominates the concept of creativity or that doesn't matter. As a result of the DaVinci project, this model of creativity in higher education has been used as a reflection tool for fostering creativity by many university teachers in several higher education workshops and has proven its usefulness.

3 Is there a Lack of Creativity in Higher Engineering Education Curicula??

Regarding creativity in the field of engineering education, some work has already been done: Thompson & Lordan [12] review essential findings and techniques in the scientific literature about creativity and try to transfer them to engineering education. Though, they don't substantiate the benefit of simply using creativity techniques in teaching courses, nor do they present appropriate learning scenarios. Cropley & Cropley [13] remarkably work on different theories about creativity from various disciplines, but even they are faced to the problem to define creativity and somehow remain stuck in the discussion about assessing creative efforts. Nevertheless, they describe an interesting learning scenario that guides students to reflect their own creativity. Byrge & Hansen [14] present the creative platform, a concept that focuses on confidence, concentration, motivation and diversified knowledge. Though, a concrete didactic scenario for engineering education is missing. Finally, such a scenario is delivered by Zhou, Holgaard, Kolmos & Nielsen [15], and Zhou [16]. They combine principles of enhancing creativity with problem based learning and project based learning in engineering education. They justify their actions very well, prove the creative benefits, and show concrete actionable educational measures. Unfortunately, such fine concepts are still very rare. To sum up, there a several good approaches, but still far too few or they are limited to single aspects of creativity. To foster creativity in engineering education, a lot work still has to be done. More knowledge about creative education is necessary. Based on the interdisciplinary results of the DaVinci project, thc ELLI project has the task of promoting subject-specifically the creativity of engineering students at universities. As a first step, a preliminary study was carried out, which should identify the actual need. For this purpose, we analyzed the module descriptions of six engineering education curricula (Manufacturing Engineering and Electrical and Electronic Engineering IT) of three German universities (Aachen, Bochum, Dortmund) in order to get to know which aspects of creativity are fostered in today's engineering education (see Fig. 3). The module description is only one small piece of the picture and we know this. But it is a start to analyze creativity. As a result, fostering the creativity-aspects 1 (self-reflective thinking), 3 (curiosity and motivation), and 4 (learning by doing) is highly developed in both courses of all three universities. With one exception these aspects have shares of over 50 %. On the other hand, the aspects 2 (independent learning), 5 (multi-perspective thinking) and 6 (reach for original ideas) can be found only in small proportions with percentages below 50 %, in aspects 5 and 6 with one exception even below 10 %. To sum up, these pre-analysis of the module descriptions shows that in the considered courses students were encouraged to think critically and self-reflective. They had to demonstrate motivation and commitment in their courses and they were trained to create something, to work practically. Independence, collaborative development of ideas and the exchange with other disciplines and for open-minded discussions, scenarios and experiments, however, were almost not required and promoted. Together with empirical experiences in engineering education a picture of diligent students is emerging, who rather work conscientiously on given tasks than finding new problems, questions and solutions on their own and in discussion with others. Also, the fact that in some of the courses the students were not free to choose the topic of their thesis reinforces this picture. Instead, they have to choose it out of a pool of given topics developed by the teachers. In this way, many learning processes that require creativity weren‘t done by the students, but by the teachers: the detection of relevant research questions, the deliberation whether an issue is workable, the creation of a structure and the assessment of eligible methods. Due to this, students aren't able to see the "big pictures" of their discipline, which is only seen by the teachers. They don't get in touch with the spirit of research: the (collaborative) reasoning about current issues in the community, setting up and discussing new (and sometimes as well risky) theories, the making of own decisions and seeking collegial advice. When students are able to see the big picture, they get a feeling about the value and importance of their work. Through these findings, the question rises whether this understanding of fostering creativity in engineering education is appropriate. However, students seem to have a different understanding of creativity. An interdisciplinary survey (n =

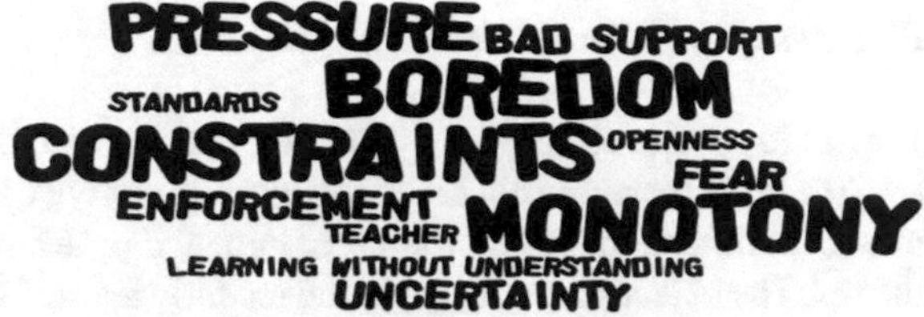

Fig. 4 Students' perspective on factors that hinder creativity

Fig. 5 Students' perspective of factors that foster creativity

320) at TU Dortmund University (Germany) shows that students regard "openness", "freedom", "stimulation", "inspiration" and "empowerment" as factors that promote their creativity (see Fig. 4). Pressure, constraints, boredom and monotony, on the other hand, hinder their creativity (see Fig. 5). At this stage of the investigation these results need to be treated very cautiously; they depend on a very small, arbitrary sample of only six courses from three different universities. Further studies must and will be done. Nevertheless, the study's results provide first hints about where to start when coming to fostering creativity in engineering education. In the following two approaches are described, which also take the spirit of research into account.

4 Two Examples of Best Practices

4.1 Fostering Creative Learning by iPad-Didactics

In a municipality in Denmark, seven K9 schools implemented iPads for all 200 teachers and 2,000 students from grade 0 to grade 9. In a first study, we conducted 15 classroom observations and 13 interviews with teachers. Here, we make one creative example visible that illustrated a typical iPad-Didactics case that can be easily transferred into engineering education [17].

4.1.1 A Creative Task for Science Learning

The case was a 9th grade physics classroom with a male teacher and 15 students (8 F; 7 M) from 8:00-9:30 a.m. in the morning. The main learning activity for the students was to design new experiments based on the prior knowledge the students gained from previous classes. The objective was to apply the recently learned knowledge and to show the teacher the learning outcome in the field of Sound, Light, Magnetism, Electricity and Chemistry. The teacher's instruction to the students was, "Please, show me something essential about sound or light, and create a new experiment". He also asked them to document their process of planning and conducting the experiment.

4.1.2 iPad-Based Group Learning

While some students gathered and built groups and started to work on the assignment, one group of two students were not sure how to start. The teacher thus created a new assignment for their personal needs but asking them to create a joint mindmap using the app Popplet. They brainstormed to collect their knowledge into one mindmap to identify their personal gap in knowledge. That gap then served as the starting point to plan the experiment.

The other groups had some ideas already and started with the experiment without making a mindmap. In one group there were up to 7 students who worked together, in other groups there were 3 students. As part of their experiment, the students used the Camera and Video recording features in the iPad, and took photos and made podcasts. They also podcast the preparation of the experiment in case the experiment would fail, to show the teacher what they had done till that point and to analyze why the experiment failed. iPads were thus used by the students to document the process of creating the experiment, in addition to the following ways in which they used them:

- Searching for information (in Google/Bing, Youtube etc.),
- the 'Textbook' app (an app that has a lot of textbooks),
- Pro Tuner (a tuning app),
- to upload their documentation (Dropbox app).

In this case, the students got the opportunity to reflect on their existing knowledge and to create new knowledge. The assessment was process-based and part of the learning process because the teacher could gauge how much the students had understood from previous lessons. The teacher said in the interview that followed, "How do I know when the students have learnt something? When they can apply it to the real world". He stressed that the students are able to check their theoretical knowledge by transferring this into concrete physical experiments. The teacher also immediately checked the results of the experiment in the class and gave feedback. Students then shared the results in Dropbox and got feedback from their peers.

4.1.3 Classroom Oberservations

One classroom observer described the feeling in the classroom "somewhat chaotic, but in a good way". He stated that, "some teachers would not have liked the informal way of doing teaching". However, the observers noted that all students were engaged in the task and looked genuine interested in their experiments. The main communication in this classroom was among the students. This corresponded to the teacher's statement during the interview that he strongly focused on "informal teaching", where he would rather be in the background and let the students experiment. He liked to foster a role change, where students become the person who teaches other students, and he acts like a process mentor who supports the personal learning needs of the students instead of telling them the facts. One example of this – during classroom observations – the teacher asked the students to first present their idea for a new experiment on a blank sheet of paper and so, the students themselves got a "aha" effect; they suddenly saw that the idea was not clear enough and why the experiment went wrong. The teacher stated that this assignment was based on his philosophy of students "learning by mistakes." He said his designs for learning are based on the idea that students "test their theories through experiments within a given field (e.g., sound, light) and translate it, to learn how it works in reality".

4.1.4 Creative Process-Product-Oriented Learning Outcomes

This case presented here represents active learning where there was a focus on action and a focus on students "to produce something." Read the digital didactical designs in detail in Jahnke & Kumar 2012 [17]. The teachers' design for teaching and their design for learning included active student participation, student engagement and student motivation by "doing" something. The students produced something and while doing so, they reflected and learned; the reflection research points to a positive relationship between being active and a deeper learning outcome [18].

4.1.5 Enabling the Spirit of Research

The didactical designs by the teachers were both process-oriented and product-oriented (see creativity facet 4, 'learning by creating something'). The teachers did not only focus on outcomes or exams/test only and did not expect students to reproduce the facts. Both the teachers had a learner-centered approach – they allowed their students to learn by making mistakes, they wanted to challenge their students, and yet, they scaffolded the learning process by providing feedback and personalizing the learning experience for students who struggled. This example illustrates how the iPad served as a "booster" to support the spirit of research in a way where learning is didactically designed as a process where the students produce something to stimulate communication and social exchange. The students acted like researchers; they planned an experiment and made it come true. In those stages, some students made mistakes and the experiment failed. But this is part of good research and the teacher supported his students to learn from their mistakes. Together, they analyzed what went wrong. Then, the students revised the plan and made a revised successful experiment. In all of our classroom observation cases, the iPad was integrated into a digital didactical design, in addition to active learning and a process-orientation, the didactical design included both teacher-student and student-student interaction and feedback. These elements together form the digital didactical approach that can foster a creative learning approach to establish a spirit of research among students.

4.2 *Fostering Creativity with Remote Labs: PeTEX – Platform for E-Learning and Telemetric Experimentation*

The didactical concept of the "Platform for E-Learning and Telemetric Experimentation" (PeTEX) is another example [19–27]: The PeTEX system is designed for the usage in higher education and for workplace learning. PeTEX combines a teleoperated experimentation platform (material testing, particularly forming, cutting, and joining) with a collaborative learning environment based on Moodle. It provides three different learning levels deploying three different didactic approaches, addressing three different problem types. The three levels correspond to three of the six facets of fostering creativity (see Table 1).

4.2.1 Three Consecutive Problem Levels to Foster Different Facets of Creativity

Beginner Level Learning with Interpolation Problems

Students in the beginner-level are guided through the learning platform and are asked to create predefined and expected order in a given complexity of elements and actions by identifying, assembling and executing all given elements and actions in the right order to solve the task, in the PeTEX case to correctly carry out predefined experiments. These predefined experiments consist of interpolation problems. According to [28–30] interpolation problems consist of three elements:

Table 1 Three consecutive learning levels, corresponding to the problem types and three facets of creativity

Learning levels	Didactic approach	Problem type	Creativity facet
1. level: beginner	Scripted learning paths	Interpolation problems	1. self-reflective learning
2. level: intermediate	Real world scenarios	Synthesis problems	4. learning by creating something
3. level: advanced	Research based learning	Dialectic problems	6. reach for original ideas

- a predefined starting point (1),
- a concrete terminal point (2), and
- a concrete and predefined solution process how to bridge the gap between starting point and terminal point (3).

The challenge of this kind of problem is to correctly fulfill a sufficient complex task according to the given and scripted path. It deals with recognizing of and acting in complexity: e.g. understanding the manual, identifying the relevant units of the real equipment introduced in the manual. The next step is to combine, assemble, and connect these elements in the right scripted technical and logical order in order to fulfill the pregiven task, and to produce the expected results.

The main creativity facet addressed by that kind of task is to break out of the receptive habitus and to start questioning the given information by transforming them into correct action.

Internediate Level Learning with Synthesis Problems

In the intermediate-level, learners have to transfer their knowledge to given real-world scenarios and are encouraged to perform their experiments in a self-directed way. According to [28–30] real world scenarios relate to synthesis problems which consist of three elements:

- a predefined starting point (1),
- a concrete terminal state (2), and
- no defined solution process to bridge the gap (3).

The challenge of this problem type is to find, to develop and to deploy a sufficient solution path to a given problem consisting of a presented starting point and an expected terminal point by applying divergent and convergent thinking to find and implement an appropriate solution for the given problem. The creative final product is the developed solution which is gained mostly "by doing" and the competencies the learner has gained with this kind of tasks are generated with "learning by doing" according to II.4 presented in this paper.

Advanced Level Learning with Dialectical Problems

Learners at the advanced level have to design own research questions and to develop the appropriate experiments. According to [28–30] those dialectical problems consist of

- no predefined starting point
- no predefined terminal point
- no predefined solution process

The challenge is to apply the developed knowledge, skills, and competencies of the learners to find and define novel and origin problems as research questions, defining a starting point, a final state, and the means for gaining it, like a concrete new product, prototype, theory, process, and so on.

4.2.2 Dealing with Increasing Complexity in PeTEX for Fostering "The Spirit of Research"

The more the students have worked with PeTEX, the more freedom they get to define their own research problems and to find the answers on their own. Furthermore, PeTEX provides collaboration, not only with other students (from other universities and even other countries), but also with lifelong learners. In summary, PeTEX offers an important contribution to foster the "spirit of research".

5 Radical Consequences and Challenging Open Questions

It remains unclear whether these points also play an important role from the perspective of the teachers and, furthermore, parts of the society:

- How to foster creativity in science and engineering education courses and curricula?
- How to train teachers efficiently and successfully in creativity fostering techniques?
- Are open experimentation and trying out new ideas, the search for the unknown new really important for a society in a globalized world economy?
- Does our economic society indeed need diligent professionals who execute given tasks instead of developing their own initiatives?
- Does our industry require graduates that are used to think multi-perspectively?
- What is the role of a new thinking culture?
- What wishes and visions do teachers, researchers, industry representatives, professional association representatives have with regard to the education of tomorrow's engineers and to their creativity and their "spirit of research"?
- What kind of education will be needed, if a society wants to bring up future inventors who are able to cope with the problem mentioned by the European Union?

These questions should soon be discussed in a broad social debate. Further studies on the impact of teaching creativity need to be done urgently.

References

1. L. Feisel, A. Rosa, The role of the laboratory in undergraduate engineering education. Journal of Engineering Education , 2005, pp. 121–130
2. VDI. Mitbrief0604. URL http://wiv.vdi-bezirksverein.de/mitbrief0604.htm
3. M. Dresler, T. Baudson, eds., *Kreativität. Beiträge aus den Natur- und Geisteswissenschaften*. S. Hirzel, Stuttgart, 2008
4. H. Lenk, *Creative ascents (Kreative Aufstiege: zur Philosophie und Psychologie der Kreativität)*. Suhrkamp, Frankfurt am Main, 2000
5. N. Jackson, Imagining a different world. In: *Developing Creativity in Higher Education*, ed. by N. Jackson, M. Oliver, M. Shaw, J. Wisdom, Routledge, London, 2006, pp. 1–9
6. P. Kleiman, Towards transformation: conceptions of creativity in higher education. Innovations in education and teaching tnternational **45** (3), 2008, pp. 209–217
7. T. Haertel, I. Jahnke, Wie kommt die kreativitätsförderung in die hochschullehre? Zeitschrift für Hochschulentwicklung **6** (3), 2011, pp. 238–245
8. T. Haertel, I. Jahnke, Kreativitätsförderung in der hochschullehre: ein 6-stufen-modell für alle fächer?! In: *Fachbezogene und fachübergreifende Hochschuldidaktik. Blickpunkt Hochschuldidaktik*, vol. 121, ed. by I. Jahnke, J. Wildt, W. Bertelsmann Verl., Bielefeld, 2011, pp. 135–146
9. I. Jahnke, T. Haertel, M. Winkler, Sechs facetten der kreativitätsförderung in der lehre – empirische erkenntnisse. In: *Der Bologna-Prozess aus Sicht der Hochschulforschung, Analysen und Impulse für die Praxis, Arbeitspapier*, vol. 148, ed. by S. Nickel, CHE gemeinnütziges Centrum für Hochschulentwicklung, Gütersloh, 2011, pp. 138–152
10. I. Jahnke, T. Haertel, Kreativitätsförderung in hochschulen - ein rahmenkonzept. Hochschulwesen **58** (3), 2010, pp. 88–96
11. M. Spitzer, *Geist im Netz: Modelle für Lernen, Denken und Handeln*. Spektrum, Heidelberg, u.a., 2000
12. G. Thompson, M. Lordan, A review of creativity principles applied to engineering design. Journal of Process Mechanical Engeneering **213** (1), 1999, pp. 17–31
13. D. Cropley, A. Cropley, Recognizing and fostering creativity in technological design education. Int J. Technol Des Educ **20**, 2010, pp. 345–358
14. C. Byrge, S. Hansen, The creative platform: A didactic for sharing and using knowledge in interdisciplinary and intercultural groups. In: *SEFI 2008 - Conference Proceedings*, 2008, p. 9
15. C. Zhou, J.E. Holgaard, A. Kolmos, J.D. Nielsen, Creativity development for engineering students: Cases of problem and project based learning. In: *Joint International IGIP-SEFI Annual Conference 2010*, Trnava,Slovakia, 22.09.2010
16. C. Zhou, Learning engineering knowledge and creativity by solving projects. International Journal of Engineering Pedagogy (iJEP) **2** (1), 2012, pp. 26–31
17. I. Jahnke, S. Kumar, ipad-didactics. didactical designs for ipad-classrooms: Experiences from danish schools and a swedish university. In: *The New Landscape of Mobile Learning: Redesigning Education in an App-based World*, ed. by C. Miller, A. Doering, Routledge, 2012
18. E. Chapman, Alternative approaches to assessing student engagement rates. Practical Assessment, Research & Evaluation **8** (13), 2003
19. C. [17] Chr. Pleul, I. Terkowsky, I. Jahnke, A.E. Tekkaya, Tele-operated laboratory experiments in engineering education – the uniaxial tensile test for material characterization in forming technology. In: *Using Remote Labs in Education. Two Little Ducks in Remote Experimentation*, ed. by J.G. Zubía, G.R. Alves, University of Deusto Bilbao, Spain, 2011, pp. 323–348
20. C. C. Terkowsky, C. Pleul, I. Jahnke, A.E. Tekkaya, Tele-operated laboratories for online production engineering education: Platform for e-learning and telemetric experimentation (petex). International journal of online engineering (iJOE). Special Issue: Educon **7**, 2011, pp. 37–43
21. C. Terkowsky, I. Jahnke, C. Pleul, A.E. Tekkaya, Platform for e-learning and telemetric experimentation (petex) - tele-operated laboratories for production engineering education. In: *Proceedings of the 2011 IEEE Global Engineering Education Conference (EDUCON) – "Learning Environments and Ecosystems in Engineering Education". IAOE*, 2011

22. C. Terkowsky, C. Pleul, I. Jahnke, A.E. Tekkaya, Platform for e-learning and tele-operative experimentation (petex) - hollistacally integrated laboratory experiments for manufacturing technology in engineering education. In: *Proceedings of SEFE Annual Conference, 1st World Engineering Education Flash Week*, ed. by J. Bernardino, J.C. Quadrado, Lissabon, Portugal, 2011, pp. 578–585
23. C. Terkowsky, I. Jahnke, C. Pleul, R. Licari, P. Johannssen, G. Buffa, M. Heiner, L. Fratini, E. Lo Valvo, M. Nicolescu, J. Wildt, A. Erman Tekkaya, Developing tele-operated laboratories for manufacturing engineering education. platform for e-learning and telemetric experimentation (petex). International Journal of Online Engineering (iJOE). IAOE. Special Issue: REV2010 6, 2010, pp. 60–70
24. C. Terkowsky, I. Jahnke, C. Pleul, R. Licari, P. Johannssen, G. Buffa, M. Heiner, L. Fratini, E. Lo Valvo, M. Nicolescu, J. Wildt, A. Erman Tekkaya, Developing tele-operated laboratories for manufacturing engineering education. platform for e-learning and telemetric experimentation (petex). In: *REV 2010 International Conference on Remote Engineering and Virtual Instrumentation, Stockholm, Sweden, Conference Proceedings. IAOE*, Vienna, 2010, pp. 97–107
25. C. Terkowsky, I. Jahnke, C. Pleul, D. May, T. Jungmann, A.E. Tekkaya, Petex@work. designing online engineering education. In: *Computer-Supported Collaborative Learning at the Workplace // Computer-supported collaborative learning at the workplace*, ed. by S.P. Goggins, I. Jahnke, V. Wulf, S.P. Goggins, I. Jahnke, V. Wulf, Computer-supported collaborative learning, Springer, New York, 2013, pp. 269–292
26. I. Jahnke, C. Terkowsky, C. Pleul, A.E. Tekkaya, Online learning with remote-configured experiments. In: *Interaktive Kulturen, DeLFI 2010 – 8. Tagung der Fachgruppe E-Learning der Gesellschaft für Informatik e.V*, ed. by M. Kerres, N. Ojstersek, U. Schroeder, U. Hoppe, 2010, pp. 265–277
27. D. May, C. Terkowsky, T. Haertel, C. Pleul, Using e-portfolios to support experiential learning and open the use of tele-operated laboratories for mobile devices. In: *REV2012 - Remote Engineering & Virtual Instrumentation, Bilbao, Spain, Conference Proceedings*, ed. by M.E. Auer, J. García Zubía, 2012
28. R.M. Rahn, *Vom Problem zur Lösung*. Heyne, 1990
29. D. Dörner, *Die Logik des Misslingens. Strategisches Denken in komplexen Situationen*. rororo, 2003
30. F. Vester, *Die Kunst vernetzt zu denken: Ideen und Werkzeuge für einen neuen Umgang mit Komplexität: Ideen und Werkzeuge für einen neuen Umgang mit Komplexität. Ein Bericht an den Club of Rome*. DTV, 2002

Developing Cultural Competency in Engineering Through Transnational Distance Learning

Stephanie Moore, Dominik May and Kari Wold

Abstract While cultural competency is a stated priority for engineering education in the United States, as emphasized by Outcome H in the ABET standards, it is often difficult to engage students in immersive international experiences that develop intercultural awareness. Undergraduate engineering students face packed curricula with little or no room for languages and an often unforgiving structure that puts them a year out of course sequences if they do travel for study abroad. In this case study, the authors examine how online education can be a transformational factor in this challenge. When designed to create interactive, engaging learning across nations, online education can support joint international experiences that develop cultural competency without requiring the time and expenses that are often a barrier for students. This online model could easily be scaled up to offer more students an international collaboration opportunity without institutional reliance on study abroad. This online transnational distance learning approach saves students and universities time and money, while accomplishing the intended professional competencies.

1 Introduction

With a growing awareness among many disciplines of the global, systemic nature of their professions, the spotlight has increasingly turned to transnational educational experiences that afford students the opportunities to experience other cultures or collaborate across borders as a means of preparing them for the work worlds that await them. Companies around the world expect future graduates to be able to work in an international environment and within international teams [1]. The past 10 years have seen greater attention to offering such transnational experiences for students

S. Moore (✉) · K. Wold
University of Virginia, Charlottesville, USA
e-mail: slm6un@virginia.edu

D. May
Engineering Education Research Group (EERG) at the Center for Higher Education (zhb), TU Dortmund University, Dortmund, Germany
e-mail: Dominik.may@tu-dortmund.de

Originally published in "Transnational Distance Learning and Building New Markets for Universities", © 2012. Reprint by Springer International Publishing AG 2016, DOI 10.1007/978-3-319-46916-4_62

in order to attract students from around the world as institutions are now largely seeking to establish connections with those from other countries in efforts to afford students new experiences for learning [2]. Designing and facilitating these experiences, however, introduces challenges that many instructors and their institutions do not fully anticipate, which could threaten individual and institutional commitment to these sorts of investments. Here we examine one such example created using the systematic process of instructional systems design as a way of providing the needed structure for the design of complex learning environments [3]. This case study highlights how a transnational collaborative instructional design process enabled cultural differences that appeared to be significant instructional constraints to be turned into instructional affordances.

In the discipline of engineering, recent modifications to expected outcomes for engineering students have included an emphasis on students' abilities to understand the impact of engineering solutions in global and societal contexts [4], thus providing broad impetus for developments in this area. Although articulated together, global and societal are two very different constructs that, while they intersect along some points, are largely two different types of awareness among engineering graduates. Understanding solutions in a global context often implies increased awareness of cultural differences, and therefore cultural considerations and implications that should be addressed during design, development, and implementation [5, 6]. In contrast, understanding the impact of solutions on a societal context often connotes more of an emphasis on social responsibility and attending concepts such as environmental impact, sustainability, safety, and other measures of societal impact. Through the collaborative design illustrated here, we seek to add the understanding of what it means to teach with a global perspective in addition to content and strategies in this particular course that already emphasizes the societal impact of engineering.

In examining the relationship between intercultural sensitivity and moral reasoning, Endicott, Bock, and Navarez [7] looked at intercultural development as a cognitive construct. They hypothesized the development of this construct is largely facilitated by multicultural experiences such as study-abroad and work-abroad experiences. While their research supported the relationship between these two types of reasoning and such experiences, we offer that providing such deep multicultural experiences for every student is prohibitively costly for any institution. Nevertheless the need to offer an engineering education that gives every student the fundamental knowledge and skills to function in the global workplace, where cultural appreciation and international collaboration have become basic skills, remains. We posit that distributed, transnational learning experiences, when based on a sound instructional design process, can facilitate the development of cultural competence in the absence of extensive study abroad opportunities, and serve as a stepping-stone experience when more immersive experiences are possible. The design of these environments requires two critical components:

1. Awareness of success factors and challenges for transnational collaboration
2. Employment of an instructional system design that takes intercultural and instructional parameters into consideration.

2 Background

2.1 Challenges to Transnational Education Collaborations

One of the major factors contributing to the success of transnational initiatives is governmental stability in a given partnering country. For example, shifts in government from a centralized governmental structure to a devolution in structure have hampered the funding for institutions in places like Indonesia, therefore crippling the country's efforts to implement transnational initiatives that would provide much-needed outside funding [8]. The Universitas Indonesia (UI) is one such institution that has encountered such setbacks in implementing transnational programs [9]. Specifically, corruption within the Indonesian government dissuades other countries from engaging in any form of transnational collaboration at UI [9]. It is important to emphasize international collaborations do exist at UI and in Indonesia; yet they are often between individuals that have connected through international exchanges. One engineer at UI, for instance, reported that his primary connection to international visits was through email or through yearly visits to institutions. While there are many connections among individuals, connections between institutions, which could add organizational strength, are all too infrequent, making partnerships unlikely.

In contrast, government stability promoted transnational collaborations among institutions in Australia and in China. For instance, the Chinese government's Decision on Further Educational Reform to Promote Quality Education in 1999 encouraged investment from the private sector that wished to attract students outside national borders. This initiative in turn stimulated investment in transnational programs [10]. In recent years the government has begun funding transnational programs in public institutions to attract outside educational funding and to boost global prestige. Currently, both the public and private sectors focus on establishing large research centers, as well as building laboratories that collaborate with other countries' universities. These innovations have greatly improved the breadth and depth of the country's transnational programs [10].

Other governmental factors hindering transnational programs involve desires to expand institutions' number of international students, instead of beginning or strengthening transnational programs. The government in India is one such case. Although its engineering institutions would like to develop transnational programs, institutional leadership views attracting international students as essential for its institutions' growth over promoting transnational initiatives [11]. Many universities within the UK also have the same goal of increasing their share in the international higher education market by 2020, keeping internationalization at home [12]. Their institutions have become economically competitive largely by developing what the government calls transnational mobile academics focused on travelling abroad instead of institutions developing transnational programs. However, it is important to note further cuts in governmental funding indicate a trend toward transnational education.

In contrast, other governments have adjusted their resources to focus on transnational students more so than international students. The Australian National University is one prime example. The institution has a longstanding reputation for attracting international students and resources from international institutions [9]. These programs, however, have relied on governmental funding of these international projects. As the government's funding has decreased, the Australian National University has sought to expand its transnational programs to stay a leader in all forms of international education. Governmental resources have also inadvertently promoted transnationalism in Malaysia as well [13]. Due to the lack of space in public institutions, the government now encourages the growth of private institutions, which then have heavily promoted transnational programs in order to attract students in a crowded market. The popularity of these programs is projected to greatly enhance the number and strength of these transnational programs in the next decade [13].

One last key factor in transnational initiatives regards accreditation. Currently, many accreditation models are regional or are not specific to one discipline, and in some countries accreditation is not a requirement for all universities. The same is true for academic disciplines. In one example, UNESCO's Global Forum on International Quality Assurance, Accreditation, and the Recognition of Qualifications provides guidance for cross-border accreditation, yet does not specify a discipline [14]. Ambiguity in this regard discourages institutions from instigating transnational education systems if their students will not be certified practitioners where they reside [15]. This has led the President of the European Network for Accreditation of Engineering Education, Augusti [16], to call for extensive transnational accreditation specific for engineering programs.

Some transnational accreditation collaborations specific to engineering do exist, however. For instance, the Accreditation Board for Engineering and Technology [17] is the largest accreditation body in engineering in the United States, and is well recognized by national and international accreditation boards through the Washington Accord Cross-Accreditation Agreement [18]. However, there are still numerous regional or national accreditation boards that are not part of this accord, making it difficult for transnational students to be confident their program will be accredited in their home countries [15]. For example, Europe has established the European Federation of National Engineering Association, yet the Engineering Council United Kingdom (ECUK) and the German Accreditation Council may also accredit engineers in Europe.

In our particular program, many of these barriers to transnational collaborations did not exist. Governments from both countries are very stable, and both institutions are under the Washington Accord [18]. Both institutions also host a variety of activities that range from transnational to international programs, so there were no institutional impediments to a truly trans-continental collaboration. Within that context, a focus on the design of an effective, engaging transnational learning environment could proceed-a process we viewed as a joint design rather than the one-way delivery of content from one country to another. To support that collaborative design, we adopted a formal process of instructional design and development to provide the framework for joint decision making.

2.2 *Instructional Design: A Formal Structure for Planning*

Starting around the 1950's, models began to emerge for the systematic improvement of instruction, deriving from work in the United States military sector [19], as well as others from Michigan State University and Oregon State University [3]. This work reflected detailed and complex models for how to systematically design instruction and improve it over time, drawing from a theoretical basis such as general systems theory. Over the years, a suite of models have been developed based on differing paradigms of thought such as behavioral, cognitive, and constructivist models. The general parts of any model of instructional design, however, remain largely the same (for a review of the models, see Gustafson & Branch, 2002) [3]. Those are Analysis, Design, Development, Implementation, and Evaluation (or ADDIE). For our purposes, we will go into one specific model in depth to explore how that aided in the design of an international learning experience, although it should be noted that there are other equally viable options. While other models could have been used, we selected the Morrison, Ross, and Kemp model [20] (2001), which is shown in Fig. 1, because the decision points this particular model emphasizes played a key role in the design of the course.

A systematic instructional design process maps out the different decision points and considerations that should go into the development of any effective learning environment. For transnational education, these models can be particularly helpful as they highlight the need to define learners' characteristics, expectations, and past experiences, as well as define institutional expectations. These models additionally spark discussion regarding constraints such as scheduling differences, time zones,

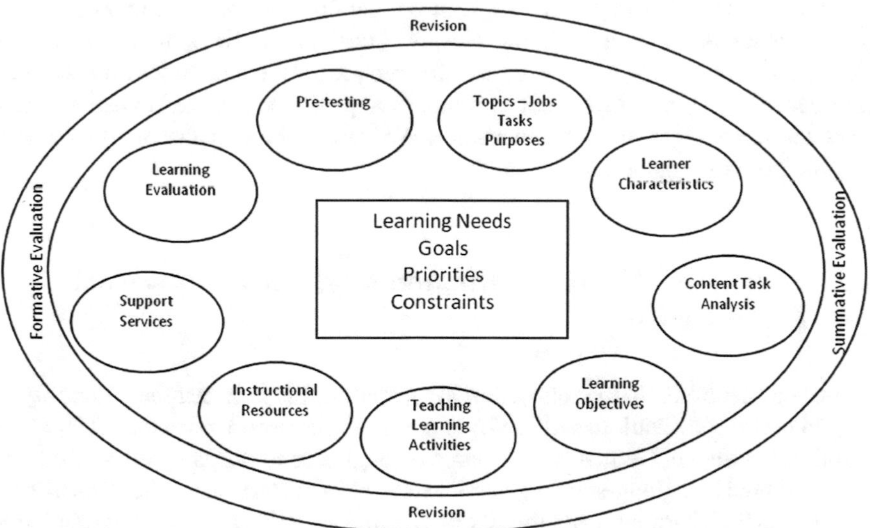

Fig. 1 The Morrison, Ross, and Kemp model of instructional design

technological infrastructures, and of course, the model highlights learning needs and goals. By working through certain details first, the shape of the course can be better developed as refined objectives are articulated and appropriated activities or strategies emerge from understanding the learners' needs and constraints.

As with most design processes, the actual process is not a linear flow as a two-dimensional visual suggests. Rather, these are the different tasks that an instructional designer undertakes to design an effective learning experience. While there is a temporal progression, there is need for dialogue between the steps. Of paramount importance initially are the learning needs, goals, priorities and constraints because any instructional design must revolve around defining these parameters. What may be more surprising in this depiction is that learning objectives come later in the process. As we shall see in the course design discussion, this makes sense because the objectives require input from task analysis, learner characteristics, and priorities in order for the objective to be accurate and useful.

3 Main Focus of the Chapter

Virtual collaborations provide an alternate path to international learning experiences that may overcome the barriers to sustainable programs. Many universities in the US have well developed infrastructures for online learning. Even these capabilities are presently limited to specific programs; they can be readily extended to international learning partnerships. Whereas traditional international learning opportunities rely on physically moving students, requiring a linear increase in cost versus the number of students, transnational distance learning can leverage distance and is therefore more sustainable. In the current stagnant economic environments, investments in infrastructure decrease are declining. However, curricula and support infrastructures can be maintained. However, these investments are only worthwhile if the learning experience for learners from both countries is a quality educational experience that provides outcomes in cultural competency. We explore the specifics of how we have designed such a program.

3.1 Case Study: Virginia-Dortmund Engineering Education Collaboration

In the summer of 2010, the foundation for a relationship between the University of Virginia (UVa) in Charlottesville, Virginia, and Technische Universität (TU) Dortmund in Dortmund, Germany, took shape through a serendipitous series of events. TU Dortmund had a long-standing relationship with the German Studies Department at UVa, but TU Dortmund and the city of Dortmund itself play a critical role in the area of engineering and industry within the German economy. Leadership at Dortmund sought to develop a stronger relationship with the School of Engineering and

Applied Sciences (SEAS) at UVa. UVa's SEAS was five years into the development and expansion of an online undergraduate program, called Engineers PRODUCED in Virginia (PRODUCED), and laying forward a vision for how to leverage this online infrastructure for international opportunities for students. From the meetings in the summer of 2010, the first step identified how to make this partnership a reality. That step was the delivery of a specific course from UVa to Dortmund that blended learners from both universities into one course. In the fall of 2011, a course that focused on the professional role of engineers in the design of global technological systems was broken up into two sections: one for traditional UVa students and the second as an international offering.

This specific course, "Science, Technology, and Contemporary Issues," was particularly well suited for international collaboration and addresses a lack in soft skills in the engineering curriculum. For instance, at an American Society of Mechanical Engineers Congress, industry leaders emphasized the need for international communication, preparation for global collaboration, and other "soft" skills be taught alongside technical skills in the engineering course sequence [21]. In seeking to address this gap, engineering students in SEAS at UVa have been required to take a three-course sequence throughout their degrees and complete a final paper prior to graduation, all of which are designed to develop student awareness and ability to plan for the broader impact of engineering endeavors on society and environment. This course, the first in the threecourse sequence, is required for all UVa students in their first year of studies. The general topic of the course is technology in society, but the course has different flavors and can be readily adapted. For example, the year prior to delivery of the international offering, the theme had been technology and democracy in which students examined the complex relationships between technologies and democratic political systems. They also participated in a simulation of a large engineering system, the levee system in New Orleans, to learn how to interact with various stakeholder groups.

For the fall of 2011, two versions of the course were offered: the traditional larger version of the course, designed specifically for on-campus students at UVa, and the online section designed for learners distributed around the world. The on-campus version had a new theme, engineering and the future. This time, the course emphasized how engineering reflects our visions of the future and how we can discuss the possible future implications of engineering developments such as nanotechnology, bio-engineering, and nuclear energy. In every section, regardless of the theme, students considered very similar general topics such as historical and cross-cultural examples of technological developments and the mutual relationship between technologies and societies that shaped each other. For the online section, however, the possibility emerged for a new theme, one that drew from both the technology-and-and the engineering-and-the-future themes. These themes were then woven together to produce a more global endeavor. Previously, the topic of "global perspective" had been a topic within the course, but faculty lamented the irony that exposure to this topic amounted to students sitting in a lecture hall looking at slides on the screen with definitions and concepts, but no actual interactions or activities that made this concept real for students.

Today, most engineering graduates work either into government jobs where their work has global implications, or into multinational corporations in which they must meet, plan, and work with peers around the world, often not in person and with the resources available to each at their own location [22, 23]. The engineering systems these students will work on are complex, operating on a vast scale that has implications for multiple countries. When a nuclear plant in one country malfunctions, the whole world experiences the consequences and policies and planning around the world respond to events in one country. According to technology ethicist Ian Barbour [24], as we scale up our technological systems, so too must we scale up our frameworks for ethics in decision making and planning to consider not just the technical implications but also the cultural, environmental, and global. This ability to consider larger, more complex systems and their long-term consequences is an increasingly critical skill for professional engineers and has been articulated as a priority outcome for engineering programs in the United States [21]. In fact, the generally accepted desirable attributes of engineering graduates articulated in the EC2000 standards have had a global impact already in "strengthen[ing] global linkages within the engineering profession and facilitate[ing] professional mobility" [25].

It is within this context of varying course themes that transnational distance learning became the unifying link. Two instructors, one from each institution, worked closely on the planning of the course starting with a week-long planning visit in Dortmund, Germany. This dedicated time proved critical to sorting out institutional differences and mapping out the overall schedules and structures for each institution to identify the shape of the actual course landscape before any instructional planning began. From there, joint objectives were articulated for the course. These objectives were for students to employ complex system thinking, discuss the role of human agency in planning desirable technological developments, and view engineering as a global endeavor that shapes our collective futures. Planning continued through bi-weekly online meetings until the start of the course with weekly reflection meetings during the course. The instructors met for a second week-long planning period in Germany to finalize details, test the technical platform between institutions and countries, and for personalized training on the systems by one instructor who had many years of experience in online teaching and who had used the learning platform extensively.

The instructors specifically emphasized cultural themes throughout the course, looking at a broad range of examples that played out differently in different cultures and reading cross-cultural comfort careful design in order to navigate through the logistical, structural, cultural, and assessment challenges [16, 26, 27]. They also recognized that designing this course would take more time and careful planning than is typical [28, 29]. Through collaborative design, challenges to transnational education turned what could have been a challenge into instructional opportunities. The rest of this chapter focuses on the design of the course's structure, sequencing, types of interaction and learning activities, resources, and student support to explore how we structured a transnational learning experience and leveraged challenges into opportunities.

3.2 Course Design

The design of international learning experiences was a complex challenge. Often the schedules, priorities, time differences, and differing curricular structures introduced competing constraints.

3.3 Scheduling

Our design process started with identifying these scheduling constraints first and mapping them out on a calendar to identify all the critical dates first. In the case of the partnership between TU Dortmund and UVa, the universities had completely different schedules: UVa's semester begins in late August and ends in December, whereas TU Dortmund's semester begins in October and ends in February. This meant that the fall UVa course started (and ended) 7 weeks earlier than the TU Dortmund schedule for students. This meant there would be 7 weeks of instruction delivered to the UVa students before the TU Dortmund students joined. Additionally, the UVa course is a 3-credit required course and is part of a three-course required sequence for all UVa engineering students. For the TU Dortmund students, the course was initially offered as a 2- to 3-credit elective in a structure where it is open to engineering and nonengineering students. This free-elective part of the curriculum is called "Studium Fundamentale" and offers the opportunity to the students to work on topics they are interested in and are not represented in the standard curriculum. So the students can get in contact with scientists who are not part of their own discipline and break out of their specialist discipline's scope of mind. It is meant as an initiation for an interdisciplinary and in the case of this particular course design for an international dialogue. The Studium Fundamentale offers the framework for a broader perspective on and understanding of defining and solving complex problems. Finally, the scheduling constraint of holidays and days off were considered-UVa has fall "Reading Days" during which students have a partial week off for studies as well as time off for Thanksgiving. TU Dortmund did not have any holidays that interrupted the class schedule.

3.3.1 Intercultural Learning Needs

At the outset, student needs were a priority for the course design. Time was spent with the students at each end discussing some of the cultural issues and differences within their own local learning communities. Because differences in emphasis, expectations, or interpretations can exist for the different participating university communities, colleagues suggested it could be beneficial to plan for "local" time when students meet only with others at their location. Consequently, the course included local conversation within the course structure. For the UVa students, there was content critical for them as part of the overall curricular requirements, but this content did not pertain to the TU Dortmund students. The UVa content was covered during the first 7 weeks of the course. The TU Dortmund students had different needs in preparing for the course, so TU's instructor offered a 3-day primer to the course prior to the first day of classes that students had to attend. This primer covered TU-specific emphases and covered some of the same content and examples from the first 7 weeks for UVa so that once students joined in the same live class session, they were all at the same point and ready for meaningful interactions. This offset in the institutions' schedules required careful sequencing of the topics to be covered, establishing when and at what point in the content the students should come together and what preparation was necessary for joint activities.

3.3.2 Goals

The final macro-structure considerations were the goals for global interactions and collaborations. In defining the goals, merely delivering content from one university to another would fail to meet basic objective for the partnership. The goal to expose students to different cultural considerations and to develop cross-culture collaboration skills made it necessary to have meaningful activities that get the students working jointly in teams on targeted projects. This goal was intimately related to the implications to the course design, from objectives to activities to technologies selection and content sequencing. Another consideration needed to support students working together in international teams was to establish learning environments that supported equal participation. One technique used was to emphasize examples that represented cross-cultural comparisons and considerations. With these considerations established, the detailed course design could begin by examining learner characteristics, establishing content sequencing, and selecting class readings that emphasized engineering concepts, cross-cultural considerations, and engineering as a global endeavor.

3.3.3 Learner Characteristics

After defining constraints, needs, priorities, and goals it was important to work out the special learner characteristics of German and United States students. To just bring these two groups together and adopt either an American or German course standard would miss the intent of the international collaboration. It was necessary to take into account the differences of both groups in order to design an internationalized course that reflected the characteristics of both, and not just one of them. As Fig. 2 shows, there were crucial differences between American and German students in language, culture, and learning expectations.

While the cultural backgrounds and language differences may seem more obvious, the differences in learning expectations proved to be very important. In the US, students are expected to attend every class session and come prepared, having read

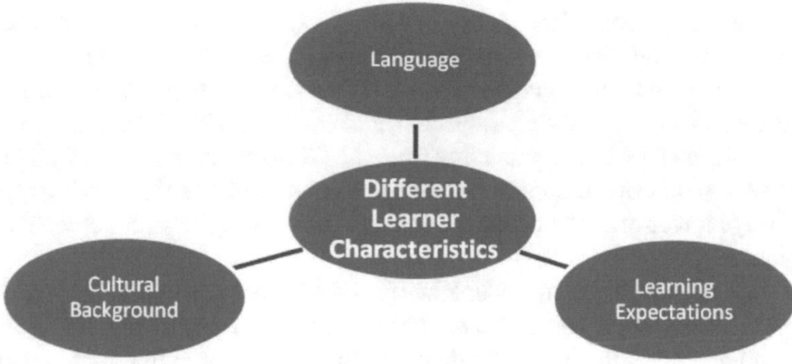

Fig. 2 Main differences between German and US students

in advance, and to participate. For this class in particular, the students are required to complete three papers, take three exams, and give at least two presentations. This is very different from the common European learning environment, in which students are not required to attend every session and are assessed primarily through one test at the end of the class. These differences in learner expectations dramatically influenced learner behavior.

In addition, the overall online course format could limit personal contact between the students groups, becoming a barrier to successful participation. As the details in the following design indicate, the course design addressed these differences. The content selection addressed differing cultural backgrounds and language, whereas the instructional activities were chosen according to the course goals, and student groups consisted of students with totally different learning styles.

3.3.4 Content Selection

While the majority of content selected came from previous readings and topics used in previous courses, selection of and sequencing of readings, based on the objectives and constraints were made. Some content was heavily American in its emphasis, and some of it was unique to UVa. That content was resequenced to be delivered early in the course before the TU students joined. Some content, not required for the TU students, felt to be important, was included into their three-day primer.

3.3.5 Language

The course language was English, which created a challenge for German students. Since English was a second language, students had to be made comfortable to speak English without being frightened of making mistakes. This was one of the German tutor's tasks to make clear that communication in a foreign language does mean speaking perfectly. Even if the ability to speak English is limited, it is still possible to communicate and interact in such a course concept. The, US students faced a similar challenge, having to grapple with texts that were written in German. Thus, for some readings German texts were selected, knowing that US students would have to translate and seek help in understanding from the German instructor and their German peers. This was of critical import, because it emphasized that the learning objective is not always lower-order text comprehension, but also higher-order objectives such as collaboration.

In an international market, students will need to be versed in different languages. In Chapter 7, Garman discusses the market for teaching Arabic online, and the need to be capable of speaking, understanding, reading, and writing in foreign languages. A major learning objective for the course was for the students to work together on clearly defining problems and interpreting the readings, so this reading selection began to function as strategy, not just content. Especially in the context of nuclear energy, it was important to consider the students' different cultural background. The view of nuclear energy is different in German and the United States. For this reason, the readings were selected to contain both viewpoints. Readings around the major class activity, the nuclear energy ethics commission, had to reflect multiple perspectives. Actual reports from both countries were selected as a way to compare and contrast the differences in approach.

3.3.6 Activities

The development of the course activities required they be in synch with the course concepts and learning expectations of German and United States universities. Whereas in American courses reading and preparing texts at home and discussing them during the course it normal, in Germany the most common course concept is the lecture, especially in engineering departments. The lecture is dominated by the professor and his presentation of scientific knowledge by mainly giving his lecture and using media like PowerPoint-slides. He is the acting person in the front, and students are expected to absorb his knowledge by listening, taking notes, and asking questions. Normally the lecture is accompanied by workshops or tutorials [30, 31]. For German students this transnational course was a very unusual pedagogical approach, requiring weekly readings and discussions. As a result, the pedagogy had to be adopted to meet German needs.

Typical class sessions included a mix of presentations by the lead instructor, and questioning of students regarding their understanding of the readings and discussions. The American instructor led most of these sessions. To move beyond content delivery into content engagement, class activities were sequenced (Fig. 3) to move students from learning about each other, to collaborative learning, and to participating in more authentic, immersive exercises.

The activities were explicitly geared towards scaffolding identification of cultural components of systems and team partnerships, which is important to understand the perspectives and influence of different stakeholders, or relevant social groups [32].

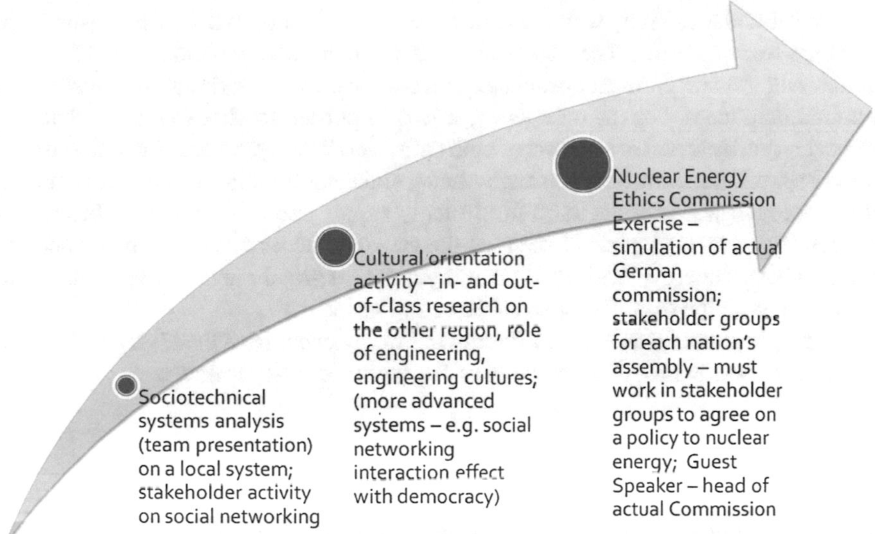

Fig. 3 Series of course activities designed to guide students to work in transnational teams

In this course, students worked in teams of two or groups to analyze socio-technical systems with local implications for their communities. Students demonstrated early on how well they understood and could identify those cultural aspects. The students also engaged in a series of group discussion and breakout group discussion activities. These discussion activities occurred at least once a week; some were whole class discussions, others were designed to break students into meaningful groups to discuss the topic from different perspectives. One example was an activity on stakeholder perspectives on the role of social networking technologies in Middle East peace protests. Students read or watched news reports from multiple sources and were then put into groups representing different stakeholders. Next the teams engaged in debates. A feature in the collaboration software made it easy to to mix student groups for various activities. These activities were crucial to get students comfortable working together talking to each other, even if they felt nervous about making mistakes, because this skill was essential for effective engagement in the final activity.

The second major activity was cultural orientation during which time they researched, read articles, and learned about the others' country, region, and university. For each group, the instructor from the other country facilitated that particular course session. Students had readings and Web tours to do outside of class, and in class they heard first-hand experiences and guest lectures on the role of engineering in the state. By the end, they were expected to understand more about the culture, the history, and the role of engineering in the other location, and how social cultures in a broad sense interact with engineering cultures [33].

A third student activity was a case study in which students worked in transnational distance-learning teams. The objective was for students to communicate and function across cultural and national borders to gain competence in working effectively in an international team. For the case study, it was important to choose a topic that was important and relevant to students of both countries. The topic needed to demonstrate the complex interaction between engineering, stakeholders and nations. In examining these needs, it was obvious that future energy supply could be one topic. In seeking to identify a precise topic, we selected the current discussions in Germany and the US on nuclear energy. Working on that topic gave a broad range of opportunities to connect it with the overall course subjects and goals.

The topic chosen was Nuclear Energy in Energy because of the high political and emotional sensitivities. A description of the topic summary follows.

Excursus: Nuclear Energy in Germany

Among the broad range of alternatives of future energy supply the nuclear power with its possibilities and its risks is the most discussed way of producing energy in nowadays. On the one hand nuclear power often is referred to as a clean and safe way of energy production. On the other hand especially the latter is more than controversial. Even if there are a lot of experts who say that the nuclear risks are manageable and-even more important-controllable, it appears as if reality shows the opposite. On April 26 th in 1986 in Chernobyl (today's Ukraine) one part of a nuclear power plant burned and exploded. This nuclear accident ended in a disaster. Large quantities of radioactive contamination were released into the atmosphere, which spread over much of Western USSR and Europe. This accident and its consequences dominated the discussion of peaceful use of nuclear power for years in the world and especially in Germany. At the beginning of the new century the German government decided that nuclear power is not controllable and too dangerous in order to produce electricity. They passed a law that regulated the way out of the nuclear power use by defining concrete residual terms for every nuclear power plant. Later on in 2010 a new German government from a different party modified the law by prolonging the residual terms. Nuclear power was seen as a necessary technology for power producing until the use of renewable energy would have been extended fully. This step-in German called "Laufzeitverlängeung"-led to a fundamental discussion in politics and society about the future energy supply. The happenings from March 2011 in Japan affected this discourse radically, not only in Germany but also in the whole world.

In light of the earthquake and tsunami in Japan and the impact of that natural disaster on the Fukushima Dai-Ichi Nuclear Plant, both Germany and the United States have initiated new public discussions on the presence and use of nuclear energy in each country. What's most interesting is the concrete reaction of both countries in answer to the catastrophe of Fukushima. It differs radically. While the US-President Obama expressed his confidence in nuclear power as a part of future energy production in a keynote speech at the end of March 2011, the German government shut down seven old nuclear power plants immediately and appointed an special ethics commission with different stake holders to debate on the future of energy supply in Germany and prepare a recommendation for opting out of the nuclear energy program. In addition, the government appointed a second committee with the task to test every nuclear power plant in Germany for its assault and disaster-hit security. Their report revealed that some of the nuclear power plants would not even abide the crash of a small plane. The ethics commission handed in their report in May 2011. Their recommendation for the future German energy supply was to opt out of the nuclear power use up to the year 2021. At the same time this proposal leads to the necessity of investment in alternative energy supply solutions-especially in renewable systems. During the summer of 2011, the German government worked on that recommendation and passed a law that says that Germany will opt out of nuclear energy within the next 10 years.

However nuclear energy is currently part of the energy infrastructure for both countries. As we will see, culture plays a major role in how these conversations evolve.

The pinnacle activity was the Nuclear Energy Ethics Commission Exercise. This activity connected a technical engineering topic-which energy supply and what are the side issues? Depending on how such an ethics commission is built and which stakeholders take part, concerns about the technical, future, social, religious, or financial aspects will vary. The word "ethics" in the commission's name underscores that the students have to think beyond just technical aspects.

For the Nuclear Energy Ethics Commission, students were split into two assemblies: a US assembly and a German assembly. Both assemblies had students from both countries that were carefully assigned to ensure equal distribution. As Fig. 4 shows, within each assembly students were put into stakeholder groups (5-6 stakeholder groups per assembly). The students had to work with the members of their own stakeholder groups to clearly identify their stakeholders' interests and craft a stakeholder statement they then presented to the rest of your assembly. After the stakeholder groups presented their positions, each assembly had to work between the stakeholder groups to develop a joint recommendation to the Commission on nuclear energy policy for their assembly's country.

This activity occurred over four class periods, and students were allowed to work together outside of class using the same collaboration software in self-organized group work. After the activity, there was a debriefing that included a guest speaker from one of the actual nuclear energy commissions join the students.

3.3.7 Instructional Resources

The online technology platform included synchronous tools that allowed students to collaborate synchronously and asynchronously. All had equal access to the communication and collaboration tools. The platform for this course was Elluminate Live! We tested Elluminate prior to delivery in both locations and from both campuses to

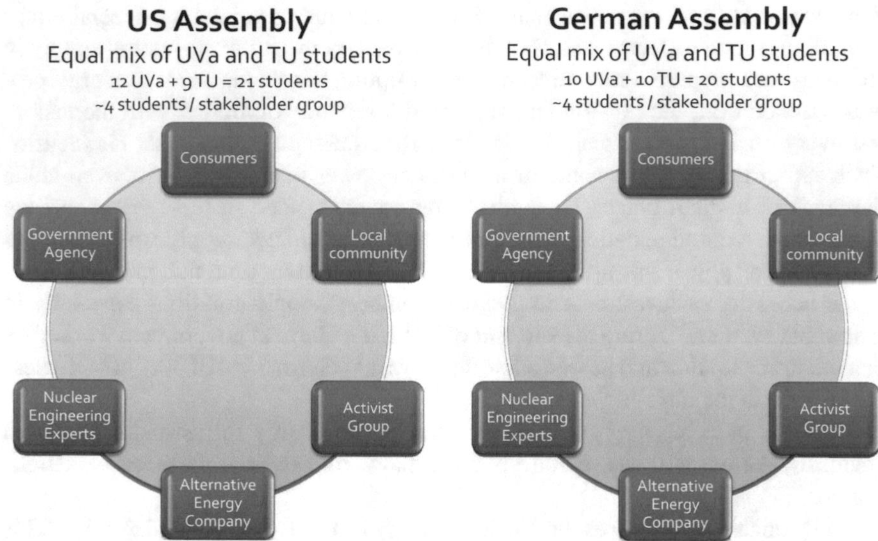

Fig. 4 Student assemblies represented stakeholders and a mix of UVa and TU students

ensure no firewall problems arose. As there were no problems with implementing Elluminate, the design was structured around the tool's features.

Unlike the videoconferencing limitations of a decade ago that required all students to be seated at one end of a room, looking at a camera at the other end, Elluminate allowed students to sign in from any computer that has the tools for video, desktop sharing, and whiteboard capabilities. For the class meetings, the instructor could moderate the environment; students could raise their hands, and post comments or questions. Elluminate also includes polling and quizzing features, as well as the ability to send students to breakout rooms. Within those breakout rooms, students have all the same tools and the instructors can go from room to room to check in on groups and talk with them. Class sessions can also be recorded so students can go back and revisit any day's topics. Students can also use this online environment to plan, discuss, share, and then give individual or joint presentations. They can create their own meetings outside of class time to meet as a group and self-organize their group projects, all using the same tools as the course. The instructors note that using the same software in and out of class significantly reduced the learning curve.

In addition to the live class environment, the course content, syllabus, and readings were loaded through a course shell using UVa's Learning Management System (LMS), Collab. UVa's Collab is an instance of the Sakai open-source environment. It includes a Resources area that house course documents, as well as assessments and quizzes, discussion board, and Wiki. The Elluminate sessions were also integrated into the course calendar in the course's Collab site, so the Collab site served as the main portal to all the material and sessions.

3.3.8 Support Services

The course provided out-of-class support for students. Instructors held online, live office hours on Elluminate Live!, and they were accessible by phone. This process was established to demonstrate that the instructors from both countries were "their" instructor. Office hours were scheduled at different days and times, and students were welcomed to contact either instructor during office hours. This was particularly crucial when students in the US had to read German texts, but it also gave students a relief valve at their respective ends to be able to go talk with an instructor at their home institution with questions that were unique to each context.

3.3.9 Learning and Performance Evaluation

The expectations each German and American student regarding the learning environment was quite different. This proved to be a significant challenge. For UVa students, the number of papers presentations, and exams were fixed because these were tied to specific standards and accreditation expectations. Given the German emphasis on the final exam, it was a concern that students from Germany would respond very negatively to the volume of deliverables and tests. The scheduling minimized this problem by enabling the UVa students to complete one of their papers and one exam before the German students joined the class; the second paper was due just after the collaboration began. Since the UVa students had extra assignments during the collaborative phase, the number of exams was reduced to just a midterm and a final. The third paper was required for all of the students with the option to make it a joint paper for all the German and United States students (using a wiki), if he determines collaboration would allay student concerns.

As for presentations, UVa students did one presentation prior to the German students joining. The second and third presentations were joint. All students participated in the final reflection activity at the end of class. The reflection was conducted during class time.

3.3.10 Assessment

Instructors from both countries developed the rubrics for evaluating the papers and grading the exams jointly; exams were jointly constructed. Grading was done independently. As a result, it was noted that the overall percentages and weighting for individual items varied for the UVa students vice the German students. The weighting was one specific way in which the marking differentiated the design for learners to accommodate differing expectations. Since little work has been done to date at this specific intersection of engineering, cultural competency and distance learning, assessment is key to continued development, improvement, and expansion of this course and of related initiatives.

One of the major challenges to assessment was a clear definition of what constitutes cultural competency, and how to measure it. In this study, we identified key concepts from literature on sociotechnical systems analysis and engineering cultures that we wanted students to apply in their presentations and papers and be able to identify, explain or use in their exams. For their presentations and papers on an analysis of sociotechnical systems, one of the major components of the grading rubric focused on their ability to identify cultural components of the system, provide accurate examples, and explain how those cultural aspects were influenced by or influenced the technology under consideration. Students who could explain an example of '"mutual shaping," which describes how society and technologies shape each other, demonstrated an advanced understanding of the role society plays in shaping the technologies we develop or implement.

During the cultural orientation activity, students were evaluated on their ability to explain cultural significances beyond the basic assumptions and provide a comparison or contrast to cultural aspects of their own countries. Throughout all activities and class discussions, on-going observations were made of student comments, extent of collaboration, discussion activities, and evaluation of the final products that student groups produced. Analysis of these artifacts is on-going based on the development of a coding schema that is a research project separate from the design process described in this chapter.

In addition to these performance-oriented assessments cited by Stiggins, Arter, Chappuis, & Chappuis [34], students were also evaluated through a midterm and final exam. The exams included objective questions and brief written responses that required students answer questions or write on "global perspective." Analyses of these instruments and their results are on-going.

4 Recommendations and Future Directions

Transnational distance learning is a cost-effective way for students to immerse themselves in an international learning experience virtually anywhere in the world, but does the cost effectiveness apply to universities? It was our experience in this project that transnational distance learning represented a substantial front-end investment in both time and money. This case study had many successes, and much of that was due to front-end planning, design and development that benefitted from an existing distance-learning infrastructure. This project was also strongly supported by administrators as a priority. These three aspects-existing infrastructure, administrative support, and front-end planning-should be carefully evaluated when an institution evaluating whether it is ready and able to go transnational. The administrative support made the pilot project possible, but will it be sustainable. Can the model be replicated? Will it become more sustainable as enrollment grows? All these questions must be answered.

Promising future directions of this sort of work include: (1) developing and measuring a suite of different approaches and activities to develop cultural competency, and (2) on-going work on how to measure and evaluate cultural competency. In the first instance, the activities and design solutions presented in this course reflect the one possible design for using online approaches for transnational learning. The variety of approaches from countries around the world suggests that a range of instructional activities may prove more or less effective in different contexts. In addition, this area could likely benefit from other types of instructional technologies, such as simulations or virtual worlds, in which students can see consequences for their decisions and manage a system to see how it evolves based on their collective decisions. Use of these technologies could also introduce measures and assessments beyond those explored in this chapter.

Regarding the development of cultural competency, much of the literature focuses on attitudinal or affective approaches, defined more as intercultural sensitivity. While this may play a role in defining how students develop cultural awareness, broadening the definition to the cognitive and performance domains could round out not only our understanding of the construct, but also provide us additional options for how we teach toward developing these important skills. Much work remains to be done in this area, suggesting a significant area for future research. And finally, if as Moore (2011) suggests cultural competency is part of professional ethics, then it is worth exploring whether the underdeveloped conative domain [35, 36] might provide a better structure and theoretical underpinning for defining cultural competency. As we explore this concept across domains of learning, we may develop a more systemic theory to the concept that it is one of the few areas that stretches across the cognitive, affective, performance, and conative.

5 Conclusion

For engineers, the rise in transnational distance learning programs marks a large shift toward international collaboration in a discipline that has historically been regionally or nationally focused. This national focus may be in part because engineering programs in the United States often put more weight on practical competence and hands-on learning over intercultural activities such as language programs, interactions with different cultures, or other international experiences [37]. A survey of 13,328 U.S. undergraduates studying engineering in 247 colleges and universities found, for instance, that engineers report the least amount of involvement in study abroad programs compared to undergraduates in other majors [37]. Instead, this same study found that these students were most heavily involved in field experiences or internships. The focus away from intercultural experiences may lead to engineering students graduating from institutions when these students have limited knowledge of how to address global problems and projects, which can lead to negative effects with global repercussions [38].

Many initiatives have sought to open the engineering curriculum in attempts to give students more opportunities for interactions with other cultures. However, the questions facing engineering institutions are largely to determine whether distance or virtual models of education are effective. Conflicting opinions regarding these methods seem to draw attention away from projects that reach across national borders. The reluctance of some engineering programs to branch into transnational distance learning may change as institutions see managing global projects are necessities for practicing engineers. For instance, practicing engineers in the engineering, procurement, and construction fields have found forming global virtual engineering teams both successful and needed; although Chen and Messner [5] cautioned that cultural differences such as how to appropriately communicate proved difficult without training. These ventures for engineers reinforce the need for transnational courses, such as the one illustrated here, to teach students skills they will need in their discipline. It is hoped institutions may create more transnational opportunities for students to enhance their awareness of the world around them, and therefore, enhance their awareness of how their disciplines may benefit from learning from those around the world.

As Finken, Provost and Vice Chancellor for Academic Affairs at the University of Wisconsin, stated, the path to internationalizing a university is a series of steps is "a thousand little things," not a single solution [39]. Similarly, this course is but one step for our students and our institutions towards global collaborations. Strategically, it provides a starting point for students, faculty, and institutions alike to develop an early stage of interactions on the path toward more immersive experiences and thus serves as a carefully designed experience for all students and a stepping stone for those who are able to travel for study abroad or other programs as part of an overall mosaic of transnational opportunities.

References

1. G.G. Hiller, S. Vogler-Lipp, eds., *Schlüsselqualifikation Interkulturelle Kompetenz an Hochschulen: Grundlagen, Konzepte, Methoden*. VS Verlag für Sozialwissenschaften/GWV Fachverlage, Wiesbaden, Wiesbaden, 2010
2. G. Helguero-Balcells, The bologna declaration agreement impact on us higher education: Recommendations for integration. International Journal of Learning **16**(10), 2009, pp. 241–251
3. K.L. Gustafson, R.M. Branch, *Survey of instructional development models*, 4th edn. Syracuse University, Syracuse, NY, 2002
4. Criteria for accrediting engineering programs, 2008. http://www.abet.org
5. C. Chen, J.I. Messner, A recommended practices system for a global virtual engineering team. Architectural Engineering & Design Management **6**(3), 2010, pp. 207–221
6. G. Dewey, J. Lucena, B. Moskal, R. Parkhurst, T. Bigley, C. Hays, S. Ruff, The globally competent engineer: Working effectively with people who define problems differently. Journal of Engineering Education, 2006, pp. 1–16
7. L. Endicott, T. Bock, D. Narvaez, Moral reasoning, intercultural development, and multicultural experiences: Relations and cognitive underpinnings. International Journal of Intercultural Relations **27**, 2003, pp. 403–419

8. A.R. Welch, Blurred vision? public and private higher education in indonesia. Higher Education **54**(5), 2007, pp. 665–687
9. S. Marginson, E. Sawir, University leaders' strategies in the global environment: A comparative study of universitas indonesia and the australian national university. Higher Education **52**(2), 2006, pp. 343–373
10. R. Yang, Transnational higher education in china: Contexts, characteristics and concerns. Australian Journal of Education **52**(3), 2008, pp. 272–286
11. S. Goel, Competency focused engineering education with reference to it-related disciplines: Is the indian system ready for transformation? Journal of Information Technology Education **5**(27–52), 2006
12. T. Kim, Transnational academic mobility, internationalization and interculturality in higher education. Intercultural Education **20**(5), 2009, pp. 395–405
13. M. Lee, Restructuring higher education in malaysia. Educational Research for Policy and Practice **3**(1), 2004, pp. 31–46
14. R.A. Skinner, The challenges of transnational online learning. Journal of Asynchronous Learning Networks **12**(2), 2008, pp. 83–89. http://sloanconsortium.org/publications/jaln_main
15. A. Patil, G. Codner, Accreditation of engineering education: Review, observations and proposal for global accreditation. European Journal of Engineering Education **32**(6), 2007, pp. 639–651
16. G. Augusti, Accreditation of engineering programmes: European perspectives and challenges in a global context. European Journal of Engineering Education **32**(3), 2007, pp. 273–283
17. Accreditation Board for Engineering and Technology, *Engineering criteria 2000*. ABET, Baltimore, MD, 1995
18. Washington accord. http://www.washingtonaccord.org/Washington-Accord/Accredited.cfm
19. L.C. Silvern, *Basic analysis*. Education and Training Consultants Company, Los Angeles, CA, 1965
20. G. Morrison, S. Ross, J. Kemp, *Designing effective instruction*, 3rd edn. John Wiley, New York, NY, 2001
21. J.C. Swearengen, S. Barnes, S. Coe, C. Reinhardt, K. Subramanian, Globalization and the undergraduate manufacturing engineering curriculum. Journal of Engineering Education **91**(2), 2002, pp. 255–261
22. C. Borri, E. Guberti, J. Melsa, International dimension in engineering education. European Journal of Engineering Education **32**(6), 2007, pp. 627–637
23. B.K. Jesiek, M. Borrego, K. Beddoes, Advancing global capacity for engineering education-research (agceer): Relating research to practice, policy, and industry. Journal of Engineering Education **99**(2), 2010, pp. 107–119
24. I. Barbour, *Ethics in the age of technology*. HarperOne, New York, NY, 1992
25. J.W. Prados, G.D. Peterson, L.R. Lattuca, Quality assurance of engineering education through accreditation: The impact of engineering criteria 2000 and its global influence. Journal of Engineering Education, 2005, pp. 165–184
26. J. Borrego, Roadmap for a successful transition to an online environment. Contemporary Issues in Education Research **3**(5), 2010, pp. 59–66
27. K.M. Passino, Educating the humanitarian engineer. Science and Engineering Ethics **15**(4), 2009, pp. 577–600
28. R.A. Perkins, Challenges and questions concerning "culturally-sensitive design". Tech-Trends: Linking Research & Practice to Improve Learning **52**(6), 2008, pp. 19–21
29. G. Samarawickrema, R. Benson, Helping academic staff to design electronic learning and teaching approaches. British Journal of Educational Technology **35**(5), 2004, pp. 659–662
30. K. Dumman, *Einsteigerhanduch Hochschullehre: Aus der praxis für die praxis: [Beginner's manual university teaching: From practice to practice.]*. Wiss. Buchges, Darmstadt, 2007
31. B. Szczyrba, Instruiueren, arrangieren, motivieren: Handlungsebenen professioneller lehre. In: *Neues Handbuch der Hochschullehre*, ed. by B. Brigitte, H.P. Voss, J. Wildt, Raabe Verlag, Berlin, 2006

32. T.J. Pinch, W.E. Bijker, The social construction of facts and artifacts: Or how the sociology of science and the sociology of technology might benefit each other. In: *The Social Construction of Technological Systems: New Directions in the Sociology and History of Technology*, ed. by W.E. Bijker, T.P. Hughes, T.J. & Pinch, The MIT Press, Cambridge, MA, 1987
33. G. Downey, J. Lucena, National identities in multinational worlds: Engineers and "engineering cultures". International Journal of Continuing Engineering Education and Lifelong Learning **15**(3–6), 2005, pp. 252–260
34. R.J. Stiggins, J. Arter, J. Chappuis, S. & Chappuis, *Classroom assessment for student learning: Doing it right – Using it well.* Assessment Training Institute, New York, NY, 2004
35. R.E. Snow, L. Corno, D. Jackson, Individual differences in affective and conative functions. In: *Handbook of Educational Psychology*, ed. by D.C. Berliner, R.C. Calfee, Macmillan, New York, NY, 1996, pp. 243–310
36. T.C. Reeves, Can educational research be both rigorous and relevant? Educational Designer **1**(4), 2011, pp. 1–24
37. G. Lichtenstein, A.C. McCormick, S.D. Sheppard, J. Puma, Comparing the undergraduate experience of engineers to all other majors: Significant differences are programmatic. Journal of Engineering Education **99**(4), 2010, pp. 305–317
38. B. Amadei, R. Sandekian, Model of integrating humanitarian development into engineering education. Journal of Professional Issues in Engineering Education and Practice **136**(2), 2010, pp. 84–92
39. K. Fisher. A wisconsin campus goes global, one step at a time. the chronicle of higher education., 2011. http://chronicle.com/article/A-Wisconsin-Campus-Goes/127523/

Constructive Alignment als didaktisches Konzept – Lehre planen in den Ingenieur- und Geisteswissenschaften

Britta Baumert and Dominik May

1 Einleitung

Stellen Sie sich vor (oder vielleicht müssen Sie es sich auch gar nicht vorstellen weil es genau so ist...), Sie sind eine junge Hochschulabsolventin oder ein junger Hochschulabsolvent. Im ersten Gespräch nach ihrer Anstellung sagt Ihr Professor dann folgendes: „Herzlich willkommen hier am Lehrstuhl. Wie Sie ja wissen, beginnt in der kommenden Woche unsere Vorlesung. Da Sie ja selbst noch sehr nah am Stoff sind, habe ich mich dazu entschlossen, von Ihnen die Übung leiten zu lassen. Die Unterlagen dazu finden Sie bei uns auf dem internen Laufwerk." Was ist jetzt tun? Mit etwas Glück liegen die Unterlagen tatsächlich auf dem Laufwerk und mit noch etwas mehr Glück sind diese sogar eine sehr gute Grundlage, um die Übung zu gestalten. Vielleicht trifft aber auch beides nicht zu und es ist ihnen überlassen, die Übung von Grund auf zu gestalten. Die Frage ist dann, welche Schritte zu tun sind, um diese Aufgabe bewältigen.

2 Constructive Alignment in der Theorie

Natürlich gibt es unzählige Ratgeber, Handbücher und Methodensammlungen, die dabei unterstützen können, gute Lehre sinnvoll zu gestalten. Doch muss man diese alle vorher durcharbeiten? Anstelle dessen soll im Folgenden mit dem *Constructive Alignment* ein grundlegendes Konzept zur Gestaltung von Lehre vorgestellt werden, welches unabhängig von Fachkulturen und -inhalten einsetzbar ist [1].

B. Baumert (✉)
Institut für Katholische Theologie, Universität Vechta,
Vechta, Germany
e-mail: britta.baumert@uni-vechta.de

D. May
Engineering Education Research Group, Center for Higher Education,
TU Dortmund University, Dortmund, Germany
e-mail: Dominik.may@tu-dortmund.de

Originally published in "Journal Hochschuldidaktik", Issue 1-2, Vol. #24, © 2013.
Reprint by Springer International Publishing AG 2016,
DOI 10.1007/978-3-319-46916-4_63

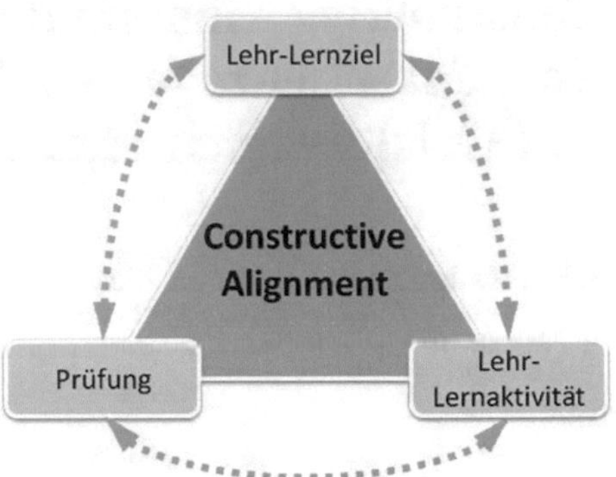

Abb. 1 Die drei Elemente des Constructive Alignment

Das *Constructive Alignment* wurde von Prof. John Biggs eingeführt. Es orientiert sich an drei Kernpunkten der Lehrgestaltung, da es die Lehr-Lernziele, die Lehr-Lernsituation und die Prüfung in einen Gesamtzusammenhang einordnet (siehe Abb. 1).

Kernaussage des Konzeptes ist, dass alle drei Kernpunkte voneinander abhängig sind und aufeinander abgestimmt sein müssen. Ist dies der Fall, ist die Lehrveranstaltung im Sinne des *Constructive Alignment* richtig gestaltet. Das bedeutet in der Praxis, dass die Lehr-Lernaktivität genau so gestaltet sein muss, dass die Studierenden die angestrebten Lehr-Lernziele auch erreichen können und dass die Prüfung auch genau das Erreichen dieser Ziele abprüft. Auch wenn das *Constructive Alignment nicht* vorschreibt an welchem Ende des Dreiecks Lehrende mit der Planung beginnen sollten, so bringt es Vorteile mit sich, bei der Planung mit den Lehr-Lernzielen zu beginnen. Sofern die Lehr-Lernziele richtig formuliert sind, geben sie vor, welche Lehr-Lernaktivität durchgeführt werden muss, um das Ziel zu erreichen. Beschreibt das Lernziel beispielsweise, dass die Studierenden nach der Veranstaltung in der Lage sein sollen, verschiedene Testmethoden beschreiben zu können und für einen Anwendungsfall die richtige auszuwählen, so ist klar, dass die Lehr-Lernaktivität die Beschreibung von Testmethoden und den Vorgang zur adäquaten Auswahl beinhalten muss. Letztlich muss dann in einem dritten Schritt die Prüfung ebenfalls genau diese Vorgänge beinhalten. Dabei ist egal, ob es sich um eine praktische Prüfung oder eine detaillierte schriftliche Prüfung handelt. Wichtig ist, dass den Studierenden die Möglichkeit gegeben wird, die Tätigkeiten in der Theorie oder Praxis abzubilden. Die reine Aufzählung unterschiedlicher Testmethoden wäre im Sinne des *Constructive Alignment* keine korrekt konzipierte Prüfungsleistung.

Nach dieser kurzen theoretischen Auseinandersetzung mit dem *Constructive Alignment* soll im Folgenden das Konzept anhand von zwei praktischen Lehreispielen aus den Fachbereichen Ingenieurwissenschaften und Theologie verdeutlicht werden.

3 Umsetzung des Constructive Alignment in den Ingenieurwissenschaften, ausgehend von den Lehr-Lernzielen

Das Tätigkeitsprofil von Ingenieurinnen und Ingenieuren hat sich in den letzten Jahren stark dahin gehend verändert, dass immer weitere Teile der täglichen Arbeit einen internationalen Kontext enthalten. Sei es die internationale Vermarktung eines Produktes, die weltweit verteilte Herstellung von Komponenten oder gar die Produktentwicklung, die auf mehrere internationale Standorte verteilt sein kann. Viele solcher Projekte werden dementsprechend in international besetzten Projektteams durchgeführt – teilweise sogar ohne dass die Teammitglieder sich persönlich treffen (vgl. [2]). Aus dieser Beobachtung heraus und basierend auf der Feststellung, dass Ingenieurstudierende heute wenig bis gar nicht auf eine derartige internationale Kollaboration vorbereitet werden, wurde im Jahr 2011 in Kooperation mit der University of Virginia in Charlottesville, USA (UVa) und der Technischen Universität Dortmund die internationale Onlinevorlesung „Als Ingenieur die Zukunft gestalten – Eine globale Herausforderung“ aufgebaut und bisher in zwei Durchgängen mit insgesamt über 80 Studierenden von beiden Universitäten durchgeführt (vgl. [3]). Ausgangpunkt für die Kursgestaltung sind die bereits angedeuteten Anforderungen aus der beruflichen Praxis von Ingenieurinnen und Ingenieuren, die entsprechend dem *Constructive Alignment* in Lehr-Lernziele umformuliert wurden. Diese Lehr-Lernziele werden den Studierenden zu Beginn der Veranstaltung detailliert vorgestellt. Nach der Veranstaltung sollen die Studierenden unter anderem in der Lage sein, …

1. … komplexe technische Systeme unter Berücksichtigung technischer, organisationaler und kultureller Aspekte mit Hilfe von konkreten Beispielen zu erläutern,
2. … im Rahmen der Ingenieurtätigkeit die globale Perspektive zu berücksichtigen und unterschiedliche Perspektiven zu vergleichen sowie abzuwägen,
3. … effektiv in internationalen Studierenden-Teams zusammenzuarbeiten und die Arbeit in Form von Präsentationen zu dokumentieren,
4. … mithilfe unterschiedlicher moderner Kommunikationsmöglichkeiten mit Ihren Kommilitonen im In- und Ausland zu kommunizieren sowie themengebunden zu kooperieren. (Dies ist eine beispielhafte Darstellung von lediglich vier der insgesamt sechs globalen Lehr-Lernzielen für diese Veranstaltung.)[1]

Aufbauend auf den Lehr-Lernzielen wurden die korrespondierenden Lehr- und Lernaktivitäten konzipiert. Dabei stellte sich immer wieder die Frage nach einer adäquaten Lern-Lernsituation, in der die Studierenden genau die Tätigkeit aus- bzw. einüben können, die das Lehr-Lernziel beschreibt. Hierbei ist es wichtig, die Aktivität der Studierenden im Fokus zu behalten, damit diese auch aktiv durchführen, was sie lernen sollen. In Anlehnung an Lernziel (1) bedeutet dies, dass die Studierenden eine Technologie anhand ihrer technischen, organisationalen und kulturellen Aspekte untersuchen müssen. Ein Vortrag durch die Dozentin oder den Dozenten

[1] vgl. [4]

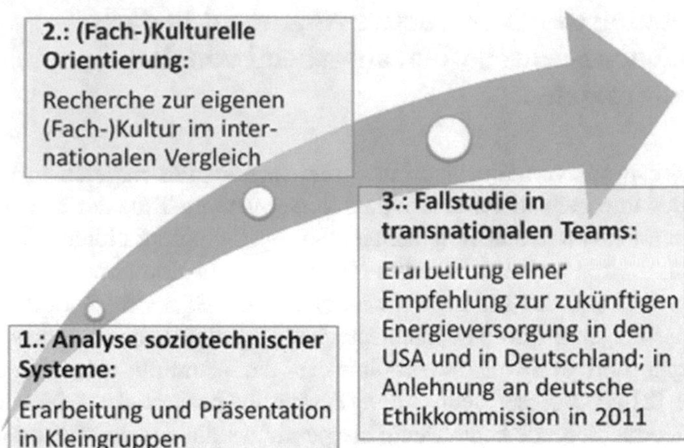

Abb. 2 Drei zentrale Lehr-Lernaktivitäten in der Veranstaltung „Als Ingenieur die Zukunft gestalten"

über die entsprechenden Inhalte würde im Sinne des *Constructive Alignment* den Lehr-Lernzielen nicht gerecht. In Anlehnung an die übrigen Lehr-Lernziele wurden für die beschriebene Lehrveranstaltung drei zentrale Lehr-Lernaktivitäten konzipiert, die aufeinander aufbauen und durch Vorträge der Dozierenden sowie Lektüreaufgaben für die Studierenden eingerahmt werden. Die Studierenden sollen zuerst in ihrer lokalen Gruppe eine Analyse sozio-technischer Systeme durchführen, daraufhin eine kulturelle Orientierung vornehmen und letztendlich in länderübergreifenden Teams eine Fallstudie gemeinsam bearbeiten (s. Abb. 2).

Während der ersten Lehreinheiten ist es Aufgabe der Studierenden in einem kleinen Team zusammenzuarbeiten und dabei technische, organisationale sowie kulturelle Aspekte von sozio-technischen Systemen einerseits zu identifizieren und andererseits ihre wechselseitige Beeinflussung zu analysieren. Die Studierenden sollen somit bereits frühzeitig die Kompetenz erlangen, technische, organisatorische aber auch kulturelle Aspekte bei Technologiebeispielen zu identifizieren, zu verstehen und in Verbindung zu bringen. Der zweite Veranstaltungsblock befasst sich inhaltlich mit der Orientierung der Studierenden in ihrer eigenen und der jeweils anderen (Fach-)Kultur. Dazu recherchieren sie und erstellen Präsentationen über das Land, die Stadt, den regionalen wirtschaftlichen Hintergrund und die Universität der jeweils anderen teilnehmenden Gruppe. Die dritte und komplexeste Lehr-Lernaktivität besteht aus einer Fallstudie. Die Studierenden sollen dabei in länderübergreifenden Teams eine Ethikkommission zu Zukunft der Energieversorgung bilden und eine gemeinsame Empfehlung zum Umgang mit Kernenergie in Deutschland und den USA erarbeiten. Es ist für die Studieren im Rahmen dieser Fallstudie somit notwendig, über ihre eigene Kulturgrenze hinaus und mit Hilfe moderner Kommunikationsmethoden zu kommunizieren und zu interagieren. Mit der damit verbundenen Kompetenzentwicklung zur Zusammenarbeit in internationalen Teams wird eines der Kernziele des Kurses umgesetzt. Entsprechend der unterschiedlichen Lehr-

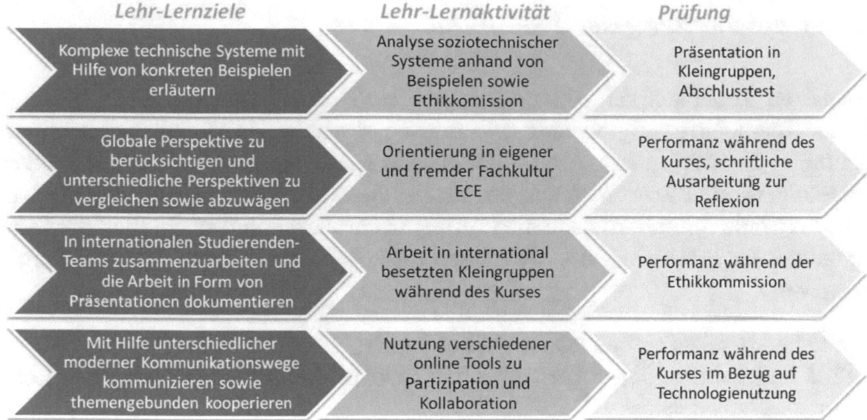

Abb. 3 Tabellarische Verdeutlichung des „Alignment" von Zielen, Aktivtäten und Prüfung

Lernziele und Lehr-Lernaktivitäten ist es ebenfalls notwendig, unterschiedliche Prüfungsmethoden anzuwenden. Ein Multiple-Choice-Test würde den Anforderungen des *Constructive Alignment* beispielsweise nicht gerecht. So zählen neben der aktiven Mitarbeit bereits während der Veranstaltung auch mehrere Gruppenpräsentationen, eine schriftliche Prüfung über die theoretischen Inhalte und eine schriftliche Ausarbeitung zur Reflexion zu den Prüfungsleistungen. Abbildung 3 fasst die erläuterten Kernbereiche der Lehrveranstaltung zusammen und verdeutlicht das Alignment von Lehr-Lernzielen, Lehr-Lernaktivitäten sowie Prüfungsleistungen.

4 Umsetzung des Constructive Alignment in der Lehrerbildung im Fach Theologie, ausgehend vom Modulhandbuch

Im Rahmen des neuen Lehrerausbildungsgesetzes von 2009 sind in allen Lehramts-Fächern neue Studienordnungen entstanden, die gemäß der Bologna-Reform auf dem Konzept der Kompetenzorientierung aufbauen. Im Zuge dessen gewinnen auch die fächerspezifischen Bestimmungen und die Modulhandbücher an Bedeutung, in denen die Kompetenzen für die einzelnen Module und Lehrveranstaltungen, aber auch die möglichen Prüfungsformen aufgeführt sind. Für die Lehr- und Prüfungsplanung im Sinne des Constructive Aligments kann nun die Auseinandersetzung mit den Modulhandbüchern besonders wertvoll sein, um bereits im Vorfeld der Planung zu prüfen, inwiefern Prüfungsform und die zu erwerbenden Kompetenzen aufeinander abgestimmt sind, bzw. inwiefern sie sich aufeinander beziehen lassen.

4.1 Untersuchung der Modulhandbücher

In unserem konkreten Fall soll am Beispiel des Theorie-Praxis-Moduls der katholischen Theologie gezeigt werden, wie aus den Vorgaben des Modulhandbuches ein stimmiges Lehr-Prüfungskonzept gemäß dem *Constructive Alignment* entwickelt werden kann. Hierfür werden ausgehend von den Modulhandbüchern zunächst die Lehr-Lernziele formuliert, dann die Prüfungen entwickelt und schließlich die Lehr-Lernaktivtäten hergeleitet. Bei der Lektüre des Modulhandbuches sind drei Fragen zu stellen:

- Welche Vorgaben enthält das Modulhandbuch?
- Sind im Modulhandbuch bereits Kompetenzen formuliert, aus denen sich Lehr-Lernziele ableiten lassen?
- Gibt es Vorgaben bezüglich der Prüfungsform?

Das Theorie-Praxis-Modul der katholischen Theologie besteht aus drei Lehrveranstaltungen:

1. Bibeldidaktik (Seminar / 3 Credits / 2 SWS)
2. Didaktik zu einem systematisch-theologischen Thema (Seminar / 3 Credits / 2 SWS)
3. Theorie-Praxis-Seminar (Seminar / 3 Credits / 2 SWS)

Da es sich um ein Modul handelt, das aus mehreren Lehrveranstaltungen besteht, ist darauf zu achten, welche Kompetenzen sich auf welche Lehrveranstaltung beziehen. Werden die einzelnen Lehrveranstaltungen von verschiedenen Lehrpersonen durchgeführt, ist es notwendig, dass sich die Lehrenden darüber austauschen, welche Kompetenzen in welcher Lehrveranstaltung erworben werden.

In den Modulhandbüchern sind in der Regel die im Modul zu behandelnden Lehrinhalte formuliert, die häufig nicht nur Aufschluss über die zu behandelnden Inhalte sondern auch Hinweise für die Lehr-Lernaktvität enthalten. In unserem Beispiel finden wir folgende Angaben:

1. **Vermittlung und Erprobung** verschiedener *bibeldidaktischer Ansätze*
2. **Erarbeitung der Struktur** *systematisch-theologischer Themen* anhand unterschiedlicher *(religions-)didaktischer Zugänge*
3. **Vermittlung** zwischen *Theorie und Praxis* im *schulischen Religionsunterricht*

Die fett gedruckten Elemente weisen bereits auf mögliche Lehr-Lernaktivitäten hin. Die kursiv gedruckten Elemente geben uns Aufschlüsse über die im Rahmen der Lehrveranstaltung zu behandelnden Inhalte.

In den Modulhandbüchern finden sich darüber hinaus weitere Hinweise indem zu entwickelnde Kompetenzen beschrieben werden, aus denen sich in der Regel die konkreten Lernziele ableiten lassen. In unserem Beispiel sind die Kompetenzen sogar bereits als Lernziele formuliert.

4.2 Kompetenzen

Nach dem Studium dieses Moduls sollen die Studierenden in der Lage sein,

(a) mithilfe verschiedener didaktischer Zugänge biblische Stoffe im Religionsunterricht zu erarbeiten,
(b) die Struktur systematisch-theologischer Themen so zu erarbeiten, dass für sie ein angemessener didaktischer Ansatz entwickelt werden kann,
(c) die Beziehung zwischen Theorie und Praxis im schulischen Religionsunterricht anhand konkreter Fallbeispiele zu erörtern.

4.3 Formulierung der Lehr-Lernziele

Welche Konsequenzen hat das nun für das Lehr-Prüfungskonzept? Zunächst einmal ist festzuhalten, dass die Lehr-Lernziele teilweise vorgegeben sind, jedoch erweitert und konkretisiert werden können. Die einzelnen Lehrveranstaltungen sollen sich gegenseitig ergänzen. Die Prüfungsformen sind vorgegeben. Die konkreten Lehr-Lernziele müssen so formuliert werden, dass sie sich zum einen aus den vorgegebenen Kompetenzen ergeben und zum anderen sowohl durch eine mündliche Prüfung als auch durch eine schriftliche Klausur prüfbar sind. Darüber hinaus ist darauf zu achten, dass die Lehr-Lernziele realistisch und durch die entsprechende Lehr-Lernaktivität zu erreichen sind.

Im Rahmen unseres Beispiels konzentrieren wir uns auf die Lehrveranstaltung „Didaktik zu einem systematisch-theologischen Thema". Ihr ordnen wir folgende Inhalte und Kompetenzen der Modulbeschreibung zu:

1. Erarbeitung der Struktur systematisch-theologischer Themen anhand unterschiedlicher (religions-)didaktischer Zugänge
2. Vermittlung zwischen Theorie und Praxis im schulischen Religionsunterricht
3. ein

1. **Erarbeitung der Struktur** *systematisch-theologischer Themen* anhand *unterschiedlicher (religions-)didaktischer Zugänge*
2. **Vermittlung** zwischen *Theorie und Praxis* im *schulischen Religionsunterricht*

(b) Nach dem Studium dieses Moduls sollen die Studierenden in der Lage sein, die Struktur systematisch-theologischer Themen so zu erarbeiten, dass für sie ein angemessener didaktischer Ansatz entwickelt werden kann.

Der erste Lehrinhalt sowie die erste Kompetenz werden der Lehrveranstaltung Bibeldidaktik zugeordnet. Der dritte Lehrinhalt ist Gegenstand sowohl der systematisch-theologischen Lehrveranstaltung als auch des Theorie-Praxis-Seminars. Die dritte Kompetenz wird ausschließlich dem Theorie-Praxis-Seminar zugeordnet. Dennoch bleiben die dort implizierten Lehr-Lernaktivitäten im Blick.

Da unsere Lehrveranstaltung jeweils ein konkretes systematisch-theologisches Thema zum Gegenstand hat, werden sich die folgenden Ausführungen auf das Thema Schöpfung beziehen.

Mögliche Lernziele für diese Lehrveranstaltung könnten wie folgt lauten:

- Die Studierenden können das didaktische Modell der „Elementarisierung" auf das Thema „Schöpfung" anwenden.
- Die Studierenden können Handlungskonzepte entwickeln, wie Unterrichtsmaterial fachgerecht und fachdidaktisch – entsprechend der im Seminar hergeleiteten Kriterien – im Religionsunterricht vermittelt werden kann.

4.4 Entwicklung der Prüfungsaufgaben

Mit Blick auf die Prüfung ist nun eine angemessene Aufgabenstellung für die Klausur und die mündliche Prüfung zu überlegen, die vergleichbar sind. Hinzu kommt die Aufnahme der Elemente aus den anderen Lehrveranstaltungen, die ja in die Modulprüfung einfließen müssen.

Eine mögliche Prüfungsaufgabe für die Klausur könnte wie folgt aussehen:

I. Analysieren Sie die Ihnen vorliegenden Schulbuchseiten in Hinblick auf ihren Einsatz im konfessionsgebundenen katholischen Religionsunterricht der Jahrgangsstufe 7 an einem Gymnasium unter besonderer Berücksichtigung des Elementarisierungsmodells und der Korrelation (75 %).
II. Konzipieren Sie für einen handlungsorientierten Religionsunterricht eigene Ansätze zum Einsatz dieser Materialien (25 %).

In Aufgabe zwei könnten nun auch die erworbenen Kenntnisse und Kompetenzen aus dem Theorie-Praxisseminar eingebracht werden. Für die mündliche Prüfung wäre folgende Aufgabenstellung denkbar:

I. Wählen Sie Unterrichtsmaterial für Ihre Schulform aus, das Sie nach dem Elementarisierungsmodell und dem Korrelationsmodell unter Berücksichtigung der Entwicklungsmodelle analysieren.
   i. Sie haben 20 Minuten Zeit, Ihre Analyse zu präsentieren.
   ii. Im Anschluss erfolgt eine 20-minütige Diskussion über Ihre Analyse.

4.5 Herleitung der Lehr-Lernaktivitäten

Aus diesen Prüfungsaufgaben ergibt sich in Zusammenhang mit den zuvor formulierten Lehr-Lernzielen die Lehr-Lernaktivität im Seminar. Denn die Lehrveranstaltung sollte so gestaltet werden, dass die Studierenden die Inhalte lernen und die Kompetenzen erwerben, die sie in der Prüfung benötigen. Die Lehr-Lernaktivität sollte sich an den Handlungen orientieren, die während der Prüfungsvorbereitung oder in der Prüfung relevant sind.

Für unser Beispiel ergeben sich dadurch folgende Lehr-Lernaktivitäten:

Die Studierenden sollen sich selbstständig Inhalte aneignen. Dementsprechend sollten im Seminar Texte gelesen und erarbeitet werden. Beurteilt wird in der mündlichen Prüfung die Diskussionskompetenz der Studierenden. Daher sollte auch im Seminar viel Raum für Diskussionen eingeplant werden. Diskutiert werden kann sowohl im Plenum, auf einem Podium oder in der Kleingruppe. In der mündlichen Prüfung müssen die Studierenden präsentieren. Also sollte es auch im Seminar Situationen geben, in denen die Studierenden präsentieren. Das kann in Form von Referaten, aber auch im Rahmen der Ergebnissicherung von Gruppenarbeiten etc. erfolgen. In beiden Prüfungsformen wird die Argumentationskompetenz geprüft. Das Argumentieren sollte daher im Seminar einen wichtigen Stellenwert einnehmen. Dabei sollten die Studierenden sowohl mündlich als auch schriftlich argumentieren üben. Kurze schriftliche Übungen können als Hausaufgabe oder im Seminar selbst eingefordert werden. Mündliches Argumentieren kann wie auch das Diskutieren in der Kleingruppe, auf dem Podium oder im Plenum erfolgen. Eine zentrale Aktivität in der Lehrveranstaltung sollte zudem das Analysieren von Schulbüchern und Unterrichtsmaterialien sein. Wenn die Studierenden in der Prüfung bzw. in der Prüfungsvorbereitung Unterrichtsmaterial analysieren müssen, ist es wichtig, dass sie auch zuvor bereits am konkreten Material gearbeitet haben.

5 Fazit

Anhand der beiden Praxisbeispiele konnte gezeigt werden, dass das *Constructive Alignment* unabhängig von Fach und Studiengang universal einsetzbar ist. Es hilft den Lehrenden und Prüfenden dabei, ihre Lehrplanung, Lehre und Prüfung so aufeinander abzustimmen, dass die Studierenden tatsächlich das lernen können, was sie lernen sollen und das Gelernte auch real Gegenstand der Prüfung ist. Das Modell eignet sich daher sowohl zur Lehrplanung als auch zur kritischen Reflexion der eigenen Lehr- und Prüfungspraxis. Darüber hinaus ist es besonders geeignet, um die Lehr- und Prüfungspraxis der jeweiligen Lehrstühle, Fächer und Studiengänge zu hinterfragen. Gerade mit Perspektive auf die Reakkreditierung der Studiengänge wäre das Constructive Alignment ein geeignetes Instrument zur Evaluierung der Prüfungsordnung.

Literaturverzeichnis

1. J. Biggs, C. Tang, *Teaching for quality learning at university. What the student does*. McGraw-Hill, Maidenhead, 2007
2. I.G. Barbour, *Ethics in the age of technology*. HarperOne, New York, 1992

3. S. Moore, D. May, K. Wold, Developing cultural competency in engineering through transnational distance learning. In: *Transnational Distance Learning and Building New Markets for Universities*, ed. by R. Hogan, IGI Global, Hershey (PA/USA), 2012, pp. 210–228
4. D. May, S. Moore, M. Eggeling, Transnationales kooperatives Lernen für Studierende der Ingenieurswissenschaften mittels Online-Lehrumgebungen zur Ausbildung interkultureller Kompetenz. In: *TeachING-LearnING.EU discussions – Innovationen für die Zukunft der Ingenieursaubildung*, ed. by A.E. Tekkaya, TeachING-LearnING.EU, Aachen, Bochum, Dortmund, 2013, pp. 233–241

FLExperimente als erster Schritt in der Laborausbildung – Wissenschaftliche Kompetenz aus dem Baukasten

Thorsten Jungmann and Philipp Ossenberg

1 Einleitung

Von Ingenieurinnen und Ingenieuren wird erwartet, dass sie Innovationen entwickeln und kreativ technische Probleme lösen. Um den Studierenden dies zu ermöglichen, ist ihre Ausbildung in Laboren ein wichtiger Bestandteil von Ingenieurstudiengängen. Dort lernen die angehenden Ingenieure u.a. den Umgang mit technischen Geräten auf dem Stand der Forschung. Als weitere Fähigkeiten werden in Laboren das Experimentieren, die methodische Suche nach Literatur und das Schreiben eines wissenschaftlichen Berichts erlernt. Studierende in den ersten Semestern besuchen Veranstaltungen wie Mathematik, Physik und Technische Mechanik, doch die wenigsten nutzen ihre dort erworbenen Fähigkeiten, um kreativ technische Probleme zu lösen, bis sie beginnen über ihrer Abschlussarbeit nachzudenken [1].

Das im Projekt ELLI (Exzellentes Lehren und Lernen in den Ingenieurwissenschaften) entwickelte Lehr-Lernformat FLExperiment soll neben der Fähigkeit zu experimentieren die Fähigkeit ausprägen, die zum wissenschaftlichen Arbeiten erforderlich sind. Ohne kostspielige Laborgeräte führen die Studierenden kleine Experimente mit einem flexiblen, professionellen Experimentierkasten durch (FLExperiment) und dokumentieren ihre Erkenntnisse in einem technischen Bericht nach wissenschaftlichen Standards. In diesem Beitrag wird das didaktische Konzept charakterisiert und beispielhaft ein FLExperiment aus dem Bereich der Kinetik vorgestellt. Das Format der FLExperimente haben die Autoren bereits an verschiedenen Stellen Veröffentlicht [2–4].

2 Didaktisches Konzept

Um sicherzustellen, dass die Studierenden die angestrebten Lernergebnisse (ILO) erreichen, wurde bei der Entwicklung der FLExperimente auf die Designmethode

---

T. Jungmann (✉) · P. Ossenberg
TU Dortmund University, Dortmund, Germany
e-mail: thorsten.jungmann@tu-dormund.de

Originally published in "HD MINT. MINTTENDRIN Lehre erleben. Tagungsband zum 1. HDMINT Symposium 2013", © 2013.
Reprint by Springer International Publishing AG 2016,
DOI 10.1007/978-3-319-46916-4_64

Constructive Alignment zurückgegriffen [5]. Das Design der FLExperimente verfolgt einen forschungsorientierten Ansatz, indem es den Lernzyklus nach Kolb mit dem Forschungszyklus nach Jungmann synchronisiert [6, 7].

2.1 Constructive Alignment

Die Designmethode Constructive Alignment basiert auf der Ausrichtung des Lehr-Lernprozesses und der Prüfungsform an den angestrebten Lernergebnissen (Intended Learning Outcomes). Bei der Umsetzung in ingenieurwissenschaftliche Curricula besteht die Herausforderung darin, die Lehr-Lernaktivität und die Prüfungsaktivität so zu gestalten, dass die aus den beruflichen und gesellschaftlichen Aufgabenstellungen hervorgehenden angestrebten Lernergebnisse (ILOs) durch die Studierenden erreicht und durch die Prüfung geprüft werden [8] (Abb. 1).

Um zu überprüfen, ob die Anforderungen des Constructive Alignment erreicht wurden, sollten die von Anderson und Krathwohl formulierten organisatorischen Fragen beim Entwerfen einer Lehrveranstaltung beantwortet werden:

1. „What is important for students to learn in the limited school and classroom time available? (the learning question)" [9]
   Diese Lerninhaltsfrage wird im Abschnitt 3.1 durch die Definition von drei angestrebten Lernergebnissen (ILOs) beantwortet. Die ILO sollen nach dem erfolgreichen Absolvieren mehrerer FLExperimente erreicht werden.
2. „How does one plan and deliver instruction that will result in high levels of learning for large numbers of students? (the instruction question)" [9]
   Die İnstruction Question" wird im Abschnitt 3.2 mit der Beschreibung der Aufgabe beantwortet, die die Studierenden zu lösen haben. Im Abschnitt 3.2 wird außerdem ein iterativer Prozess skizziert, den die Studierenden beim Experimentieren durchlaufen.

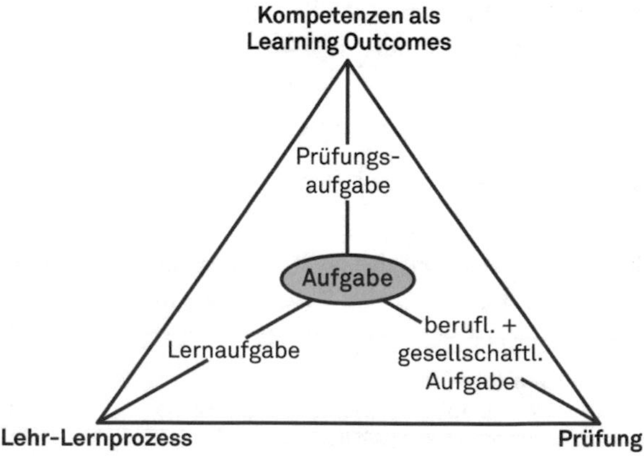

Abb. 1 Constructive Alignment

3. „How does one select or design assessment instruments and procedures that provide accurate information about how well students are learning? (the assessment question)“ [9]
   FLExperimente sind derzeit ein freiwilliges Angebot für die Studierenden. Aus diesem Grund benötigt das entwickelte Format derzeit keine formale Prüfung, die den Lernerfolg feststellt und eine Note generiert. Stattdessen erhalten die Studierenden durch geschulte Tutoren und wissenschaftliche Mitarbeiter ein qualitatives Feedback, das sowohl Rückmeldung über die Ergebnisse (wissenschaftlicher Bericht) als auch über den Prozess (wie die Ergebnisse entstanden sind) liefert. Das Feedback soll den Studierenden ihren Lernerfolg aufzeigen.
4. „How does one ensure that objectives, instruction, and assessment are consistent with one another? (the alignment question)“ [9]

Dies ist die dominante Frage während des gesamten Entwicklungsprozesses. Im Abschnitt 3.4 wird beschrieben, wie die Lehr-Lernaktivität und die Prüfungsaktivität auf die angestrebten Lernergebnisse abgestimmt sind.

2.2 *Forschendes Lernen*

Abbildung 2 zeigt Kolbs Lernzyklus synchronisiert mit einem von Jungmann entwickelten für die Ingenieurwissenschaften typischen Forschungszyklus [6, 7]. Das Erfahrungslernen nach Kolb ist ein „well accepted (...) efficient pedagogical model of Learning“ [10]. Das Erfahrungslernen wurde aus den Theorien von Lewin, Dewey und Piaget entwickelt. Der Lernzyklus in der Mitte von Abb. 2 entspricht weitestgehend dem von Kurt Lewin entwickelten Model of Action Research and Laboratory Training.

Nach Arnegger kann der Lernzyklus an jeder Position beginnen [11]. Für gewöhnlich starten Beschreibungen mit (1.) der Konkreten Erfahrung. In diesem Abschnitt des Lernzyklus werden die Studierenden durch eine Situation oder Erfahrung irritiert [12]. Dieser Zustand initiiert den Lernprozess des Erfahrungslernens. Im zweiten Schritt des Lernzyklus (2.) reflexives Beobachten wird die Erfahrung oder Situation durch die Studierenden reflektiert. Diese reflexive Beobachtung bildet die Grundlage für den (3.) Schritt der abstrakten Konzeptualisierung. In dieser Phase wird eine Theorie über die Erfahrung bzw. Situation gebildet, die im (4.) Schritt aktives Experimentieren durch Ausprobieren entweder bestätigt oder widerlegt wird. Der vierte Schritt erzeugt dann wiederum eine konkrete Erfahrung, d.h. der Lernzyklus beginnt von neuem.

Der in Abb. 2 gezeigte Forschungszyklus wurde von Jungmann [7] entwickelt, um Forschendes Lernen in der Ingenieurausbildung umzusetzen. Dabei folgte er dem Wilds Postulat, dass eine Synchronisation von Lern- und Forschungsprozess für jede Disziplin separat erfolgen sollte [12, 13].

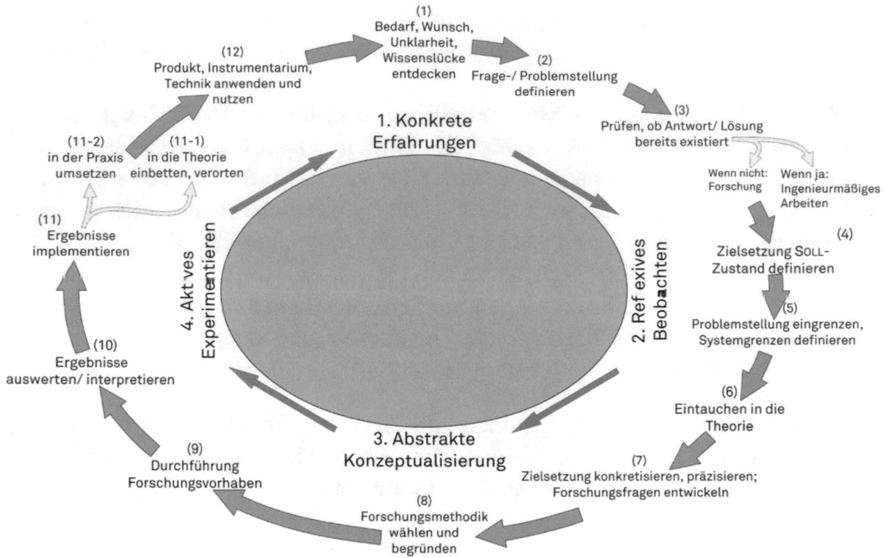

Abb. 2 Kolb's Lernzyklus synchronisiert mit dem Forschungszyklus nach Jungmann [7]

3 Lehr-/Lernformat: FLExperiment

In diesem Abschnitt wird das entwickelte Lehr-/Lernformat beschrieben. Es wird die Lernumgebung beschrieben, die angestrebten Lernergebnisse definiert, die Lehr-/Lernaktivität sowie die abschließende Assessment skizziert.

3.1 Beantwortung der Lerninhaltsfrage

Die angestrebten Lernergebnisse werden basierend auf dem folgenden Richtziel formuliert:

„The overall goal of engineering education is to prepare students to practice engineering and, in particular, to deal with the forces and materials of nature" [14]. Angestrebte Lernergebnisse (ILOs) definieren, was Studierende nach dem Lehr-/Lernprozess können sollten [5].

- ILO 1 Studierende führen Experimente erfolgreich durch.
- ILO 2 Studierende treffen Annahmen, verifizieren diese experimentell und berichtigen ihre Annahme.
- ILO 3 Studierende verfassen einen wissenschaftlichen Bericht.

Diese drei angestrebten Lernergebnisse sind als Grobziele für alle FLExperimente definiert. Zusätzliche, themenabhängige ILOs werden für jeden experimentellen Aufbau formuliert. Beispielsweise im Fall des in diesem Beitrag beschriebenen Aufbaus: Studierende sind mit der Energieerhaltung vertraut.

3.2 Beantwortung der „Instruction Question"

Die Studierenden erhalten die Aufgabe, einen wissenschaftlichen Bericht zu schreiben, der ihr Experiment inkl. des experimentellen Aufbaus, der Methoden, der verwendeten Materialien, der Forschungsergebnisse und der Schlussfolgerungen umfasst. Zu Beginn ihres Studiums erhalten die Studierenden Unterstützung in Form von Anleitungen für den experimentellen Aufbau und eine Hypothese, die es mithilfe des Experimentes zu bestätigen oder zu belegen gilt. Außerdem erhalten sie Leitfragen, die sie durch den Forschungsprozess führen sollen. Studierende in höheren Semestern müssen dann den Aufbau und die zu untersuchende Hypothese eigenständige erstellen.

Während Studierende ein FLExperiment durchführen, durchlaufen sie den in Abb. 3 gezeigten iterativen Prozess. Ausgehend von der gegebenen Hypothese führen die Studierenden ein erstes Experiment durch. Anschließend ist es erstmals Aufgabe der Studierenden, eine Annahme zu treffen bzw. einen Erklärungsansatz zu finden, der das Experiment mittels mathematischer und physikalischer Gleichungen beschreibt und eine Vorhersage über den Ausgang zulässt. Mithilfe der Annahme und dem Erklärungsansatz ist dann im nächsten Schritt die Beschreibung bzw. Vorhersage, wie das Experiment ausgeht, mittels einer Rechnung möglich. Auf eine Berechnung erfolgt dann die experimentelle Überprüfung, die im Regelfall scheitert und weitere Annahmen erforderlich macht. Nachdem die Studierenden durch mehrmaliges Anpassen ihrer Annahme, Rechnung und reales Verhalten im Experiment in Einklang gebracht haben, schreiben sie ihren wissenschaftlichen Bericht und ziehen dabei Schlussfolgerungen aus ihren Forschungsergebnissen.

Das FLExperiment, das in Abschnitt 4 beschrieben wird, startet mit einer gegebenen Hypothese. Diese entspricht im Lernzyklus nach Kolb [6] der abstrakten Konzep-

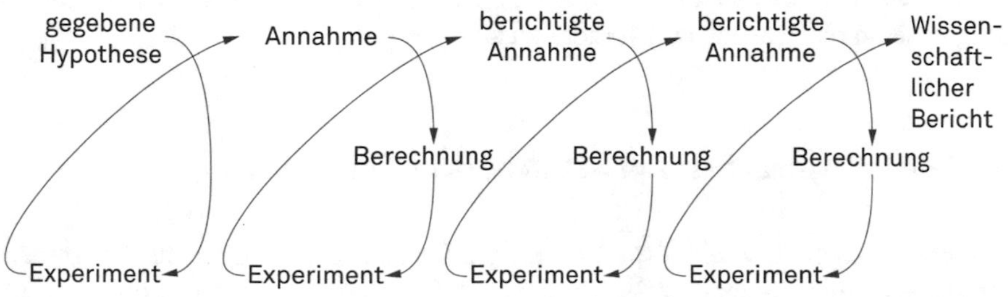

Abb. 3 Iterativer Prozess

tualisierung (3.), das darauf folgende erste Experiment dem aktiven Experimentieren (4.). Das erste Experiment dient dazu, die Studierenden zu irritieren und ihnen eine (1.) konkrete Erfahrung zu liefern. Während der konkreten Erfahrung haben die Studierenden die Gelegenheit, durch reflexives Beobachten (2.) und eine Abstrakte Konzeptualisierung einen Erklärungsansatz zu entwickeln bzw. eine Annahme aufzustellen, die im nächsten Schritt überprüft werden soll. Im iterativen Prozess folgt nach einer Rechnung das zweite Experiment und der Lernzyklus schließt sich und/oder wird fortgesetzt.

Während die Studierenden den iterativen Prozess durchlaufen und dabei dem Lernzyklus folgen, absolvieren sie ebenfalls Teile des Forschungszyklus.

Während des gesamten Prozesses können die Studierenden die Hilfe von Tutoren in Anspruch nehmen, die ausgebildet sind im Wissenschaftlichen Arbeiten und Erfahrung mit Vorgehen beim Experimentieren haben.

3.3 Beantwortung der „Assessment Question"

Beim derzeitigen Entwicklungsstand sind die FLExperimente ein zusätzliches Angebot für Studierende der Ingenieurwissenschaften und noch nicht in Curriculae integriert. Dem Constructive Alignment zufolge ist die Prüfung wichtiger Bestandteil des Lehr-/Lernprozesses. Anstatt einer strikten Bewertung erhalten die Studierenden ein qualitatives Feedback von einem wissenschaftlichen Mitarbeiter bezüglich ihres wissenschaftlichen Berichts. Wäre das Format FLExperiment in ein Curriculum integriert, würde bewertet werden welchen Lernfortschritt die Studierenden erreicht haben, dazu ist eine Operationalisierung der ILOs nötig.

3.4 Beantwortung der „Alignment Question"

Das erste, zweite und dritte angestrebte Lernergebnis wird durch die gegebene Aufgabe in der Lehr-/Lernaktivität umgesetzt. Das „Assessment" bzw. Feedback untersucht den Lernfortschritt, indem es den wissenschaftlichen Bericht überprüft. Die Studierenden werden während des Schreibprozesses unterstützt, sodass sie gut auf das „Assessment" (Feedback) vorbereitet sind.

4 Der Energieerhaltungssatz auf dem Prüfstand

In diesem Abschnitt wird beispielhaft ein FLExeriment mit möglichen Annahmen, Rechnungen und Ergebnissen beschrieben. Das FLExperiment „Der Looping", das erste ausgearbeitete Experiment ist aus dem Bereich der Technischen Mechanik. Es ermöglicht den Studierenden sich näher mit dem Prinzip der Energieerhaltung und

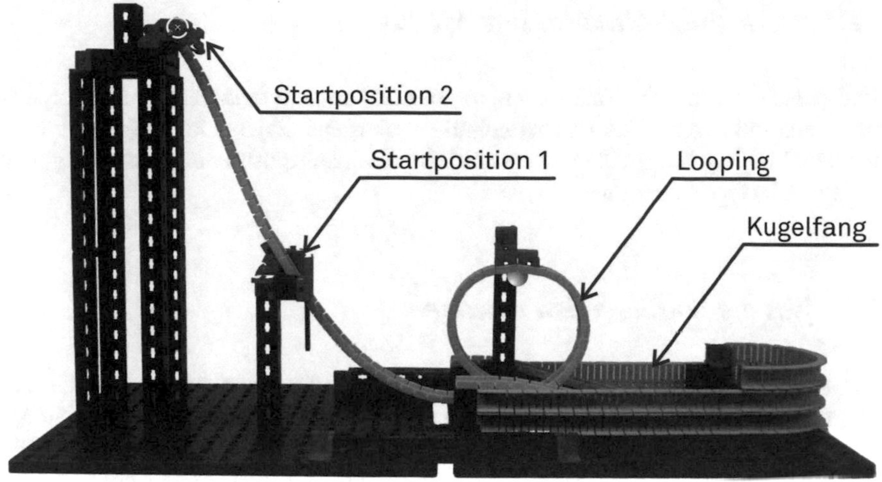

Abb. 4 Experimental set-up

deren Anwendung auseinandersetzen. Der experimentelle Aufbau ist in Abb. 4 zu sehen. Eine Stahlkugel kann von zwei festgelegten Startpositionen in einen Looping rollen. Abhängig von Starthöhe rollt die Stahlkugel entweder durch den Looping in den Kugelfang oder verhungert im Looping und rollt zurück.

4.1 Kick-off – Studierende arbeitsfähig machen

Wenn Studierende das FLExperiment „Der Looping" durchführen, sind der experimentelle Aufbau, eine zu überprüfende Hypothese und zwei Leitfragegen gegeben. Die Vorgaben sollen den Forschungsprozess in Gang setzten bzw. unterstützen.

Hypothese: Das reale Verhalten der Kugel beim Durchrollen des Loopings weicht von der mathematischen Modellvorstellung ab.

Anhand der folgenden zwei Leitfragen wird die Hypothese untersucht:

1. Welche Ausgangshöhe wird mindestens benötigt, damit eine Kugel der Masse m den Looping mit der gegebenen Höhe h_B durchrollt?
2. Welche Rolle spielt die Masse der Kugel?

Die Leitfragen können mit den zur Verfügung gestellten Ausrüstungen, Materialen und Methoden beantwortet werden.

4.2 Ausrüstung, Material und Methoden

Studierende erhalten die Materialien, um das Experiment aufzubauen. Das Experiment kann mit Teilen des Fischertechnik-Baukastens „Dynamic" aufgebaut werden. Der Looping verfügt über zwei unterschiedlich Startpositionen. Die beiliegende Stahlkugel wiegt 7 Gramm.

4.3 Was die Studierenden machen

Zu Beginn des FLExperiments wird der Looping von Studierenden mithilfe der zur Verfügung gestellten Anleitung aufgebaut. Dafür benötigen die Studierenden ungefähr 30 Minuten. Anschließend führen die Studierenden das erste Experiment durch, in dessen Folge sie für die Bearbeitung der nächsten Schritte weitere Informationen benötigen.

Die Aufgabe der Studierenden ist in diesem Schritt Informationen zu beschaffen, mit deren Hilfe sie die Hypothese bestätigen oder widerlegen können. In dieser Phase des FLExperimentes können die Studierenden die Universitätsbibliothek benutzen, um Literatur zu beschaffen. In diesem FLExperiment wird von Studierenden erwartet, dass sie Grundlagenliteratur finden und ihre Vorlesungsunterlagen aus den Bereichen Technische Mechanik, Physik und Mathematik nutzen.

4.4 Forschungsergebnis mit einem Experiment erzeugen

Die Studierenden durchlaufen mehrmals den iterativen Prozess. Während sie ihre Erkenntnisse einsetzen, stellen sie Abweichungen zwischen dem berechneten (vorhergesagtem) Verhalten und dem realen Verhalten im Experiment fest. Daraufhin müssen Annahmen angepasst und experimentell erneut überprüft werden. Dies erfolgt so lange, bis die Hypothese belegt oder widerlegt ist.

Im ersten Versuch könnten die Studierenden die Energieerhaltung nutzen, um die minimale Starthöhe rechnerisch zu ermitteln. Dazu kann die Annahme, die Kugel sei ein Massepunkt, hilfreich sein. In diesem Fall wären die kinetischen und potentiellen Energien auf einfache Weise zu betrachten. Zunächst muss jedoch festgestellt werden, mit welcher Geschwindigkeit die Kugel durch den Looping rollen muss, damit sie nicht runterfällt. Die minimale Geschwindigkeit kann mithilfe der Zentrifugalkraft ermittelt werden [15]:

$$F_z = m \cdot a_z = m \cdot \frac{\omega^2}{r} = m \cdot \frac{\omega^2 \cdot r^2}{r} = m \cdot \omega^2 \cdot r \tag{1}$$

Die Zentrifugalkraft wird im nächsten Schritt mit der Gewichtskraft gleichgesetzt. Wird die neue Gleichung dann nach v_{min} umgestellt, ergibt sich die minimale Geschwindigkeit:

$$v_{min} = \sqrt{g \cdot r} \tag{2}$$

Mit der minimalen Bahngeschwindigkeit der Kugel und dem Energieerhaltungssatz können die Studierenden dann die minimale Starthöhe h_{min} berechnen:

$$E_{pot,1} = E_{kin,2} + E_{pot,2} \Rightarrow h_{min} = 1,25 \cdot h_{Looping} \tag{3}$$

Die experimentelle Überprüfung des Ergebnisses scheitert. Die Stahlkugel kann den Looping nicht rollend passieren. Die Studierenden müssen nun ihre Annahmen anpassen.

Eine Möglichkeit, die bisher getroffenen Annahmen anzupassen, besteht in dem Versuch, das betrachtete System genauer zu beschreiben. Beispielsweise rollt die Kugel, und in den Energieerhaltungssatz ist ein rotatorischer Anteil einzubeziehen.

$$E_{pot,1} = E_{kin,trans,2} + E_{kin,rot,2} + E_{pot,2} \tag{4}$$

Um die Rotationsenergie der Kugel zu bestimmen, ist das Massenträgheitsmoment zu berücksichtigen, das mithilfe des folgenden Integrals bestimmt werden kann (oder durch Entnehmen einer Formel aus einer Tabelle) [16]:

$$J_x = \rho \int_V r^2 dV \Rightarrow h_{min} = 1,35 h_{Looping} \tag{5}$$

Mithilfe der Rotationsenergie kann ein neues h_{min} berechnet werden. Dieses weicht von der vorher berechneten minimalen Starthöhe ab. Auch die neue minimale Starthöhe wird experimentell überprüft. Das Ergebnis dieser Überprüfung: auch im zweiten Fall kann die Kugel nicht durch den Looping rollen.

Ein dritter Schritt sollte in jedem Fall die Reibung zwischen Kugel und Bahn mit einbeziehen. Während des letzten Schritts erfahren die Studierenden dann die Grenzen des Energieerhaltungssatzes, der „einfach“ nur in konservativen Systemen anwendbar ist.

Mithilfe einer weiteren Annahme über die Reibungsverluste könnte eine dritte minimale Starthöhe berechnet werden.

$$E_{pot,1} = E_{kin,trans,2} + E_{kin,rot,2} + E_{pot,2} + W_{fric} \tag{6}$$

Im Berufsalltag von Ingenieuren kommt es häufiger vor, dass sie bestimmte Einflussgrößen zunächst schätzen müssen. Nachdem die Studierenden die dritte Runde absolviert haben, können sie ihr Handeln reflektieren und ihre Ergebnisse strukturiert in einem wissenschaftlichen Bericht dokumentieren.

5 Fazit

Dieser Beitrag hat FLExperimente als Format Forschendn Lernen vorgestellt. Implementiert wird das Format derzeit in einer neuen Lernumgebung, der FLEx-Forschungswerkstatt für Studierende der Ingenieurwissenschaften. FLExperimente ermöglichen den Studierenden das Lösen von technischen Fragestellungen bereits zu Anfang ihres Studiums.

Durch FLExperimente kann Studierenden das Gesamtbild der Ingenieurwissenschaften in einer aktivierenden Lernatmosphäre näher gebracht werden. Auch kann die Fähigkeit zur Kreativität und Innovation bei den Studierenden ausgeprägt werden.

Literaturverzeichnis

1. T. Jungmann, Innovation und Kreativität durch forschendes Lernen. ATZagenda **1** (1), 2012, pp. 114–118
2. P. Ossenberg, T. Jungmann, Flexperiment. solving small technical problems by using mathematical and physical knowledge, skills and competencies in engineering education. In: *Proceedings of the 2013 IEEE Global Engineering Education Conference (EDUCON).* IEEE, 2013, pp. 597–601
3. P. Ossenberg, T. Jungmann, Experimentation in a research workshop: A peer-learning approach as a first step to scientific competence. Int. J. Eng. Ped **3** (S3), 2013, p. 27
4. T. Jungmann, P. Ossenberg, Scrutineering kinetics. engineering students put physical laws to the proof. In: *Innovationen für die Zukunft der Lehre in den Ingenieurwissenschaften*, ed. by A.E. Tekkaya, S. Jeschke, M. Petermann, D. May, N. Friese, C. Ernst, et al., TeachING-LearnING.EU discussions, Aachen, 2013
5. J.B. Biggs, C. Tang, *Teaching for quality learning at university. What the student does*, 3rd edn. McGraw-Hill, Maidenhead, 2007
6. D.A. Kolb. Experiential learning: experience as the source of learning and development, 1984. URL http://www.learningfromexperience.com/images/uploads/process-of-experiential-learning.pdf
7. T. Jungmann, Forschendes Lernen im Logistikstudium. Systematische Entwicklung, Implementierung und empirische Evaluation eines hochschuldidaktischen Modells am Beispiel des Projekt-managements. Dissertation, TU Dortmund, Herne, 2011. URL https://eldorado.tu-dortmund.de/bitstream/2003/28955/1/Dissertation.pdf
8. J. Wildt. Funktionen und Anforderungen von Prüfungen unter den Bedingungen neuer Lehrlernformen, 2009. URL http://www.hrk-bolog-na.de/bologna/de/download/dateien/Wildt_Funktion_Anforderung__Pruefungen.pdf
9. L.W. Anderson, D.R. Krathwohl, *A taxonomy for learning, teaching, and assessing. A revision of Bloom's taxonomy of educational objectives*, abridged ed. edn. Longman, New York, 2001
10. M. Abdulwahed, Z.K. Nagy, Applying kolb's experiential learning cycle for laboratory education. Journal of Engineering Education **98** (3), 2009, pp. 283–294. URL http://www.jee.org/2009/july/7.pdf
11. M. Arnegger. Begleittext zum Vortrag. Wie können soziale Kompetenzen vermittelt werden? Die Theorie des erfahrungsorientierten Lernens von David a. Kolb. Tagung "Vielfalt in der Fortbildung", 10.11.2010. URL http://www.xenos-berlin.de/attachments/article/199/Handout%20dieWille.pdf
12. J. Wildt, Guidelines for educators: 'From the sage on the stage to the guide at the side. In: *Neues Handbuch Hochschullehre*, vol. Griffmarke J 1.8, 2010

13. J. Wildt, Forschendes Lernen: Lernen im „Format“ der Forschung. Journal Hochschuldidaktik **20** (2), 2009, pp. 4–7
14. L.D. Feisel, A.J. Rosa, The role of the laboratory in undergraduate engineering education. Journal of Engineering Education **94**, 2005, pp. 121–130
15. P.A. Tipler, *Physics*, vol. Volume 1, 2nd edn. Worth Publishers, New York, NY., 1982
16. D. Gross, J. Schröder, W.A. Wall, *Technische Mechanik*, vol. Band 3: Kinetik, 11th edn. Springer, Berlin, Heidelberg, 2010

Recruiting the Right Engineering Students – Comparative Study of Common Approaches in German Higher Education

Thorsten Jungmann and Philipp Ossenberg

Zusammenfassung Excellent research requires excellent teaching, and excellent teaching requires excellent research. Recruiting the right students is a powerful measure to enhance the number of successful graduations and to reduce drop-out. The search for the right students is not limited to German Higher Engineering Education, but can be found in any higher educational system. This paper focuses on the most suitable approaches for German Universities to select the best engineering students when allocating study places for an education in the field of engineering. With recourse to pertinent quality factors we analyze the possibilities and limits of each approach and reflect on their efficacy in the mirror of pertinent quality factors.

Schlüsselwörter: Recruitment · Allocation of University Place · Reduction of Drop-Out

1 Positioning in the Student Lifecycle

Recruiting students is the first stage in the student lifecycle. The student lifecycle divides the course of studies into different, successive phases.

According to Schulmeister [1], this system includes Bachelor and Master, as well as Doctorate as the first phase of independent research. Also the transition from school to university and from university to the professional life is part of this cycle. According to Ebner [2] we differentiate between the perspective of students and the perspective of the universities. Viewed from the perspective of a student, the cycle is organized as follows (see Fig. 1): During the first phase, the orientation phase, the student gets an overview of all relevant university offers and then applies for one or more universities. After this application phase he enters the phase of actual study and afterwards moves on to the professional life as university graduate. From this phase he most likely enters the orientation phase again, which is the start of a new

T. Jungmann (✉) · P. Ossenberg
Center for Higher and Continuing Education Dortmund,
TU Dortmund University, Dortmund, Germany
e-mail: thorsten.jungmann@tu-dortmund.de

Originally published in "Proceedings of the 2013 IEEE Global Engineering Education Conference (EDUCON)", © 2013. Reprint by Springer International Publishing AG 2016, DOI 10.1007/978-3-319-46916-4_65

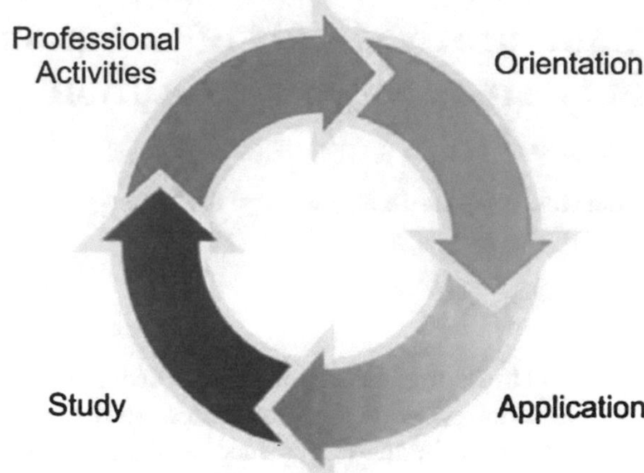

Fig. 1 Student lifecycle from the perspective of students

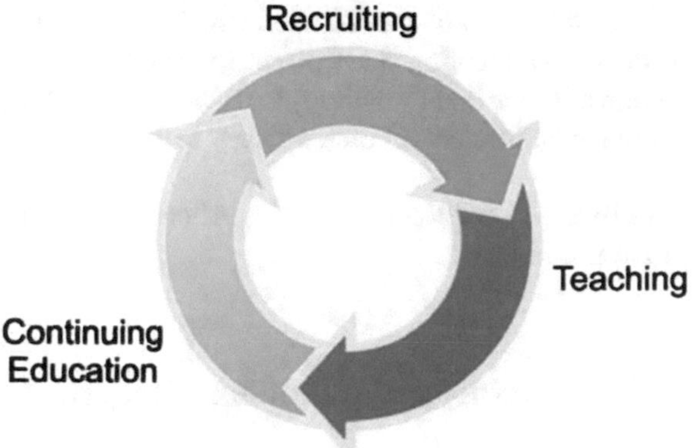

Fig. 2 Student lifecycle from the perspective of universities

cycle. At least this is what the idea of Life Long Learning suggests. When analyzing the matter for the perspective of universities (see Fig. 2), the student lifecycle will begin with the recruiting phase and the following teaching phase. After successful graduation this phase will lead to the phase of continuing education.

Study paths in the field of engineering sciences have a particular problem during the introductory phase, as one third of all registered students will quit their study during this stage [3]. Measures for increasing the number of successful graduates and to lower the chances of drop-outs in this field therefore concentrate on the search for the right students and the first introductory study phase, which means the time period between allocating the study places and the end of the second (and third respectively) semester [4]. The transition phase from school to university is also highly significant, as this phase is crucial in selecting the right field of study. The right field of study is that one particular field which promises the highest chance of graduation based on one's personal motivation and study-related aptitudes and inclination. Exemplary projects attending this matter are TeachING-LearnING.EU and ELLI (Exzellentes Lehren und Lernen in den Ingenieurwissenschaften), in which the Universities Dortmund, Bochum and Aachen cooperate for a higher standard in the engineering education. Furthermore, the image of the Engineering Profession plays an important role for recruiting students [5].

In contrast to the projects above, this paper does not focus on the possibilities of curricular, didactical or structural improvement of teaching or study conditions. This paper rather focuses on the instruments, which help universities to identify and recruit the students with the highest motivation, aptitudes and inclination for an education in engineering. Within this context and based on existing studies, the present text elaborates on the different instruments of student selection. With recourse to pertinent quality factors we analyze the possibilities and limits of each instrument. The paper closes with a recommendation for universities, on how to increase the number of successful graduations for their engineering study paths by recruiting the most suitable students.

2 Admission as a Selection Process

The recruiting phase in the student lifecycle stands for enrolment at a university and the path that leads here. At this point we would assume that the students' orientation phase has ended and they have made use of all relevant information and consulting offers and have subsequently identified the most suitable study path for themselves. Should the desired study program offer free admission, the future students can skip the phase of application and directly register at the university of their choice. If the university however imposes certain admission criteria or admission restrictions these will have to be considered before registering. At this point we first have to emphasize the difference between admission criteria and admission restrictions.

2.1 Admission Criteria vs. Admission Restrictions

Admission criteria have a long tradition for study programs like music, art or sports and have to be met before registration. In order to follow one of these particular study paths, applicants must meet a series of additional prerequisites and qualifications; sport studies, for example, will also test their students in practical exams.

Universities have the option to determine which additional requirements a future student must meet before registration. For example, the Faculty of Mechanical Engineering at the Technical University Darmstadt demands an average grade of 2.8 on the Abitur or a compensation in the subjects mathematics and physics (German grading system from 1.0 (excellent), 2.0 (good), 3.0 (satisfying), 4.0 (sufficient) to 5.0 (failed) and 6.0 (extremely failed)). As a basic principle, applicants who do not meet this requirement will not be admitted to this faculty [6] (compare practical example 5).

For medicine and any other study program that does not impose any particular requirements like athletic, artistic or other talents, admission restrictions will regulate the number of students. Such regulations are necessary at universities which have more study applicants than study places.

When allocating study places based on the numerus clausus, universities have the option to exert controlling influence for 60 % of all applicants through a selection process, according to the German Higher Education Framework Act (HRG – Hochschulrahmengesetz). The exact details of this selection process may be defined by the universities independently, based on §32 HRG [7]. This independence however may be restricted by laws and regulations on regional level.

2.2 *Selecting Individuals from a Group*

If individuals are to be selected form a larger group, four basic assumptions must be made in order to allow a selection [8]:

1. There are features that pave the way for the successful graduation of an applicant.
2. The presence of these features can be determined by certain selection procedures.
3. These features allow a prognosis for the academic success or the exercising of the intended profession.
4. The study-relevant abilities acquired during school or outside school are a part of these features.

The first step when planning a selection or allocating procedure is to identify the subject-specific features. Universities may highlight in this phase certain features relevant for a particular study profile. The matching between future students and study profile can be optimized in this phase.

2.3 *Quality Factors*

The selection procedures are subject to certain quality factors in order to ensure that they do not take place in arbitrary conditions; these criteria are described below, in line with the quality criteria applicable to exams and tests [9–13]:

Objectivity is synonym to factuality. In the context of a test, objectivity means that the result is independent from the examiner or two independent examiners will have the same evaluation result [14].

Reliability, or dependability, indicates to what extent the test results would be the same as previous if the test will be conducted repeatedly [14].

Validity allows a statement on whether the test measures those values which should be measured [14].

Fairness will be guaranteed if neither group of applicants is favored systematically [9].

Trainability indicates to what extent the test result may be influenced by practising the test subjects. If practice has no influence on the final result, the test is assumed to have a high level of validity. Targeted preparation for a test should not have any influence on the test result [10].

Economy is the relation between the cost, effort and benefit of a selection procedure [9].

Acceptance is provided, when every participant, including the public, accepts the process. Acceptance has a significant influence on "whether the right applicants respond and can be won over" [10].

3 Instruments for Study Place Allocation

Below we point out the instruments used by universities when allocating their study places. The structure of below paragraph is based on the classification for procedures according to Deidesheimer Kreis [8]. The listed techniques may be used for selecting future students for programs with particular requirements as well as allocating study places in programs with a specific numerus clausus.

3.1 School Grades

There are different methods for allocating study places based on school grades [8].

1. Course and subject selection
2. Reports or recommendations issued by the school
3. Ranking within the class
4. Average grade on the diploma
5. Grades for selected subjects

about (1) With this method only students who have enrolled for certain subjects during school are accepted for study. In Germany, not all schoolchildren have access to the entire range of subjects, however a bonus could be granted for enrolled subjects [8].

about (2) Universities may ask for reports or recommendation letters issued by the schools. This procedure however is not recommended, due to "lacking objectivity and comparability" [8].

about (3) The rank within a school class indicates how the performances of an applicant, expressed through school grades, can be assessed in comparison to his classmates. Ranking lists are not common practice in Germany, therefore rendering this method impracticable at present.

about (4) The most common method is to allocate study places based on the average grade on the high school diploma (Abitur). This method is used in the NC-procedure (cf. practical example 1). However, the average grade allows a general success prognosis, at best. Predictions for individual study programs will not be possible based only on the average grade [8].

about (5) A different approach is analyzing the grades for particular subjects. Grades for mathematics and physics have the highest validity for study programs in the area of engineering. According to Baron-Boldt this can be explained by the fact that these subjects require the highest level of abstract and cognitive abilities [15].

Another alternative is to do a weighting of specific individual subjects. When working with such a weighting, a value will be calculated based on the grades in the Abitur, which will then be used to determine the academic aptitude. This value may include all or only certain subjects with or without additional weighting.

Quality Factors

Points (1) to (3) should not be applied lightly in Germany, which is why following look at the quality factors will only refer to the average grade on the Leaving Certificate (4) and grades for selected subjects (5).

Average grades from the last school years are still considered the best individual means for predicting a successful course of studies, despite their detriments. Beside their diagnostic function, school grades provide a series of additional functions (for example pedagogical and selective). School grades therefore can only conditionally meet the requirement of objectivity. This has been a known fact for a long time. "Differences between evaluators" have first been evidenced in 1888 in England. Such differences are not only found in essay evaluations, but also in evaluations of mathematics tests [16].

A weak point of this procedure is therefore its low level of objectivity. A direct consequence of lacking objectivity is the lack of fairness and low reliability. In addition to that, the average grade allows only a general study prognosis, at best. Predictions for individual study programs will not be possible based on the average grade alone [8] (Table 1).

As the criteria for obtaining a study place lies in the past, this procedure cannot be trained. The lack of trainability is a positive feature.

From an economic point of view, the general availability of average grades speaks in favor of using these as selection criteria; integrating them into the selection procedure does not imply any additional expenses and does not generate additional costs [11].

Besides a good economy, acceptance for the selection procedure after school grades is also high. Of all the requirements, the average grade of the higher education entrance qualification has the highest predictive potential regarding validity. The grade for mathematics is of particular importance for engineering studies [17, 18].

3.2 Screening Tests

After analyzing past school performances as selection or allocation criteria in the previous paragraph, we will now dedicate this paragraph to describing tests as a

Table 1 – Practical example no. 1

Numerus Clausus – The local NC-Procedure
A numerus clausus will be used in cases where the number of applicants exceeds the number of study places. Numerus clausus is the lowest average Abitur grade admitted for registration at university. Pre-quotas allow applicants with other diplomas, different from the general higher education entrance qualification, to register for study. After exhausting the entire quota, the remaining study places will be allocated according to following procedure. The first step is generating three lists which will sort the applicants not eligible for registration in the pre-quota, according to various criteria. The first list is for the main ranking requirement, the average grade of the higher education entrance qualification (HEB – Hochschulzugangsberechtigung). The first 20% of the remaining places is allocated to the first applicants in the first list. Another 20% goes to applicants in the second list. This list is sorted firstly by waiting time as main criteria and afterwards by average grade of the HZB. Allocating the remaining 60% lies solely in the universities' responsibility, as these may now define the criteria for sorting the third list independently. One sorting criteria could be, for example, the results from an academic screening test. The Technical University Dortmund currently sorts the third list the same way they sort the first, so that admission is based solely on the average Abitur grade [16].

possible means of selecting the "right" students for engineering education. These evaluations can be divided into knowledge tests and aptitude tests. At first we will describe the knowledge tests which evaluate the school knowledge gained in the past, followed by a description of the study aptitude tests, which will diagnose the applicant's cognitive performance level. Together with aptitude tests they are then subject to the general quality factors for selection procedures.

3.2.1 Knowledge Tests – "Achievement Tests"

By successfully passing an achievement test, the applicants prove their basic knowledge in a field which can allow a successful study course. This way, the universities can define a minimum level of knowledge that can later be used as a basis during lectures. The knowledge level can be determined objectively and independent from school grades [11]. The knowledge evaluated with this method is expected to ensure the further study success.

Achievement tests can then again be divided into two subgroups, school subject-related and academic subject-related tests. School subject-related tests evaluate "school knowledge", another term would be "school performance tests". Academic subject-related tests evaluate the knowledge necessary for a successful study course. "Knowledge tests are widely spread and used as a selection instrument for university admission in the USA, Japan, in Belgium for engineering studies, China, Greece, Israel, South Korea and in Turkey" [8].

3.2.2 Aptitude Tests

Contrary to knowledge tests, aptitude tests do not evaluate knowledge, but the cognitive aptitudes which ensure a successful course of study [11].

According to Trost, there are general and specific aptitude tests. General aptitude tests evaluate a student's general aptitudes for study, whereas specific tests evaluate subject- or field-specific aptitudes [9].

The measuring range of general aptitude tests is targeted at the intellectual abilities that are fundamental to mastering an academic education, regardless of the desired field of study. This includes, for example, the ability to process complex information and draw correct conclusions from it [10].

A well-known representative of this type of test is the Scholastic Assessment Test (SAT) necessary for admission into a series of universities in the USA. The SAT is provided by the Educational Testing Service (ETS) residing in Princeton [10]. A detailed description of the SAT can be found in [19] and on the website of the [20].

Field-specific aptitude tests will measure those abilities of particular importance for successful study in a specific field or subject area (e.g. engineering sciences).

There are already numerous specific aptitude tests in German, developed by ITB Consulting or their predecessor, (…) partly for guidance and partly for selection purposes [10]. Aptitude tests in German are developed mainly by ITB Consulting GmbH, which offers licensing for task groups in aptitude tests for 2.500 – 4.000 € and will also evaluate the results [9].

Other alternatives for selecting students could be general personality tests and intelligence tests, whereby aptitude tests already have commonalities with intelligence tests (Tables 2 and 3).

Quality Factors

Standardized tests are characterized by a high level of objectivity, justified by the conditions during the test. Reliability is also provided in standardized types of tasks.

With objectivity and reliability provided, we can assume a high level of fairness.

Development, trial and implementation for aptitude tests are all time-consuming and cost-intensive.

However, once the tests have been developed and standardized, it is fairly economical to conduct and evaluate them and much less costly than conducting interviews,

Table 2 – Practical example no. 2

German National Academic Foundation (Studienstiftung des deutschen Volkes)
The German National Academic Foundation is, according to its own statement, »the largest and eldest organization for academically gifted« in Germany. Since 2010, pupils and students have the possibility to submit their application independently. The selection process begins with a general aptitude test developed and conducted by ITB-Consulting. A demo version of such a test can be found on the foundation's website. [19]

Table 3 – Practical example no. 3

Negative Test Experiences in Nürtingen
In the initial situation for a study about determining valid early indicators for the future study success, Wewel [21] describes following situation. After implementing an economics-related aptitude test for student selection it was noticed that the performance did not increase but showed a significant deterioration. Regionalization of the applicants was considered to be the reason for this. The testing procedure required the applicants to travel, which meant additional costs and probably discouraged applicants living in more remote regions from applying. These are two reasons why the aptitude test is no longer in use for student selection in Nürtingen [21].

for example [8]. Multiple-choice tests provide a very economical evaluation, as these tests allow mechanical evaluation.

According to the Deidesheimer Kreis [8], the predictive potential of knowledge tests is not as high as for Abitur grades, but still satisfactory. This is also evidenced by the meta analysis conducted by Hell [22] on the issue of validity in aptitude tests, which certifies a positive validity for these tests.

3.3 Foreign Language Exams

The previous chapter presented tests as a method for allocating study places. In this chapter we will describe a further means of testing, the foreign language exam. A foreign language exam will evaluate the degree to which the applicants master a foreign language (e.g. English). American universities require foreign students to successfully pass the TOEFL-Test [11].

Foreign languages have always played an important role in language disciplines and study programs with international orientation, as these require a corresponding minimum level of foreign language skills for successful studies in this field [11].

Quality Factors

Studies about quality factors in relation to student selection by foreign language exams could not be found. Therefore it is not known to what degree a passed exam in foreign languages could predict a successful study course. As these language exams are standardized testing procedures, the exam itself meets the quality requirements.

According to the Deidesheimer Kreis [8], foreign language exams as a study prerequisite are particularly relevant within those study programs, where mastering one or more foreign languages may guarantee future study or professional success.

3.4 Essays

In addition to the foreign language exams there is another instrument also related to the aptitude test. The essay is used as a procedure for selecting students, in the form of an application letter and as a written composition on a particular topic. The essay has been an integral part of selection procedures in the USA for a long time [8].

"Application letters, along with other subjective elements in the selection procedure, should give the best possible differentiated impression about the applicant and thus contribute to higher fairness for individual cases." [8]

Quality Factors

No available data regarding quality factors for essays as a procedure for student selection could be found. The requirement for objectivity cannot be met by this procedure (cf. paragraph A). This procedure is hardly economical, especially for high numbers of applicants.

3.5 Selection Interviews

Selection interviews are distinguished by the number of participants and by the level of structuring. Interviews allow the collection of objective, as well as subjective information about an applicant [8].

It is the only procedure that creates a direct contact between applicant and future professors. The candidates have a chance of presenting their entire personality with strengths and weaknesses. The university can thus compare their profile to the applicant and ensure a perfect fit between student and university [8].

Quality Factors

Selection interviews with applicants are extensive selection procedures, being very time-consuming for the professors, and therefore fail to be economic.

A selection interview is an extensive selection tool; it demands a lot of time from the professors [10]. Selection interviews with applicants are therefore not economical.

Findings from studies about objectivity and reliability do not allow a closing final conclusion on this matter. However, it can be assumed that selection interviews for determining aptitude can achieve only average objectivity and reliability levels, at best [10].

Even a combination with other procedures does not present an increase in accuracy [8].

Table 4 – Practical example no. 4

Admission to studies in Mechanical Engineering in Darmstadt The Faculty for Mechanical Engineering of the TU Darmstadt implements a two-stage procedure. At first students have to apply for studying Mechanical Engineering. Therefore they fill a biographical questionnaire TU Darmstadt admits all candidates to the Bachelor Program in Mechanical Engineering, who have an average Abitur grade of 2, 0 or above. These candidates must not pass the second stage. This approach is based on the assumption that a student with an average Abitur grade of 2, 0 or above is, on principle, eligible for studying mechanical engineering. The other students with worse average grades have to take part at a assessed interview with two university representatives. Using the interview result and the average Abitur grade another grade will be calculated. In case that this new grade is better than 2, 4 students are authorized to enroll. Candidates with an average grade below 2, 8 are not admitted to the Faculty of Mechanical Engineering at TU Darmstadt. [6]

3.6 *Combined Procedures*

Every previous paragraph to this point has presented self- contained procedures. To increase the prediction accuracy, above-named procedures may be combined with each other.

In the practical example 4 the first selection is based on school performance. If the applicant meets a series of further conditions he will then participate in a selection interview with the university's representatives. The selection interview could also be replaced by an aptitude test (Table 4).

Quality Factors

No data regarding quality factors is provided on this matter.

4 Conclusions

Based on the insight, that finding the "right" students for Engineering Education is a powerful measure to reduce drop- out and to enhance the number of successful graduations, we analyzed and assessed common approaches of selecting students and allocating study places with respect to efficacy, efficiency and applicability. The assessment was made by analysis of previously documented research findings, relevant scientific studies and pertinent quality factors for exams and tests.

Taking into consideration all assessed approaches, we conclude that none of the individual instruments, which are currently applied in practice, can provide reliable statements on aptitude or tendency for successful academic studies. Especially the widely used procedure to select students based on their Abitur grade appears to be rather unsuitable for selecting the right students for engineering studies, due to the lack of objectivity and predictive potential. Combined procedures, which compensate for the shortcomings of individual instruments with additional measures turn out to be more suitable.

Given that the ideal solution is currently not within sight, there is reason to conclude that the best approach available is a combination of (a) self-tests, which give students the opportunity to test their inclination in advance of their decision for or against an engineering study program, (b) school grades in Mathematics and Physics as preferably objective and reliable selection criteria, and (c) personal interviews or at least essays as a measure to test the motivation of applicants. This, of course, does not appear to be an economical set of instruments at the first glance. But given that this is common practice at those universities, that succeed to reduce their drop-out and enhance their graduation rate, it can be seen as the best possible approach to recruiting the "right" students for Engineering Education.

Literaturverzeichnis

1. R. Schulmeister, Der 'Student Lifecycle' als Organisationsprinzip für E-Learning. In: *E-university - update Bologna: Campus Innovation 2006*, ed. by R. Keil, Waxmann, Münster, 2007, pp. 229–261
2. L. Ebner, *Student Life Cycle: Optimierung der Verwaltungsabläufe*. 2011
3. U. Heublein, *Internationale Mobilität im Studium 2009: Wiederholungsuntersuchung zu studienbezogenen Aufenthalten deutscher Studierender in anderen Ländern*. Hannover, 2009
4. K. Weihe, *Studieneingangsphase: Einsichten und Empfehlungen des Fakultätentags Informatik*. 2010
5. H.O. Yurtseven, How does the image of engineering affect student recruitment and retention? Global Journal of Engineering Education **6** (1), 2002, pp. 17–23
6. Fachbereich Maschinenbau der Technischen Universität Darmstadt, *Satzung über die Eignungsfestellung für den Bachelor Studiengang Maschinenbau – Mechanical and Process Engeneering*. 2009
7. Hochschulrahmengesetz: Hrg, 2011
8. Deidesheimer Kreis, ed., *Hochschulzulassung und Studieneignungstests: Studienfeldbezogene Verfahren zur Feststellung der Eignung für Numerus-clausus- und andere Studiengänge*. Vandenhoeck & Ruprecht, Göttingen, 1997
9. G. Trost, *Deutsche und internationale Studierfähigkeitstests: Arten, Brauchbarkeit, Handhabung*. Bonn, 2003
10. G. Trost, K. Haase, *Hochschulzulassung: Auswahlmodelle für die Zukunft: Eine Entscheidungshilfe für die Hochschulen, Schriftenreihe der Landestiftung Baden-Württemberg*, vol. 6. Stifterverband für die Deutsche Wissenschaft, Essen, Stuttgart, 2005
11. C. Heine, K. Briedis, H.J. Didi, K. Haase, G. Trost, *Auswahl und Eignungsfeststellungsverfahren beim Hochschulzugang in Deutschland und ausgewählten Ländern. Eine Bestandsaufnahme, HISKurzinformation*, vol. A3. 2006
12. R. Dubs, *Besser schriftlich prüfen: Prüfungen valide und zuverlüssig durchführen*, vol. Griffmarke: H 5.1. 2006
13. J. Wildt, *Funktionen und Anforderungen von Prüfungen unter den Bedingungen neuer Lehr-Lernformen*. 2009
14. H. Schaub, K.G. Zenke, *Wörterbuch Pädagogik*. Dt. Taschenbuch-Verlag, Mänchen, 2007
15. J. Baron-Boldt, *Die Validität von Schulabschlußnoten für die Prognose von Ausbildungs- und Studienerfolg: Eine Metaanalyse nach dem Prinzip der Validitätsgeneralisierung*. Lang, Frankfurt am Main, 1989
16. Technische Universität Dortmund, TU Dortmund-NC-Verfahren, 2011

17. B. Hell, S. Trapmann, H. Schuler, Synopse der Hohenheimer Metaanalysen zur Prognostizierbarkeit des Studienerfolgs und Implikation für die Auswahl- und Beratungspraxis. In: *Studierendenauswahl und Studienentscheidung*, ed. by H. Schuler, B. Hell, Hogrefe, Göttingen, 2008, pp. 41–54
18. S. Trapmann, B. Hell, S. Weigand, H. Schuler, Die Validität von Schulnoten zur Vorhersage des Sudienerfolgs – eine Metaanalyse. Zeitschrift für Pädagogische Psychologie **21** (1), 2007, pp. 11–27
19. Studienstiftung des deutschen Volkes, 2011
20. Collegeboard, college admissions - sat - university & college search tool, 2012
21. M.C. Wewel, Ermittlung von validen Frühindikatoren für ein erfolgreiches Studium. Eine empirische Untersuchung zur Neukonzeption des Auswahlverfahrens bei der Zulassung. In: *Report – Beiträge zur Hochschuldidaktik, Studieneignung und Studierendenauswahl: Untersuchungen und Erfahrungsberichte*, vol. 42, ed. by M. Rentschler, H.P. Voss, Shaker, Aachen, 2008, pp. 13–41
22. B. Hell, S. Trapmann, H. Schuler, Eine Metaanalyse der Validität von fachspezifischen Studierfähigkeitstests im deutschsprachigen Raum. Empirische Pädagogik **21** (3), 2007, pp. 251–270

Das Studienkonzept „Querformat" – Förderung des interdisziplinären Dialogs

Tobias Berens and Ute Berbuir

Zusammenfassung Viele Herausforderungen des beruflichen Alltags von Ingenieuren, insbesondere Sicherheitsingenieuren, können nur im Miteinander verschiedener Akteure und in konstruktiver Ergänzung unterschiedlicher Lösungsansätze gemeistert werden. Zur Förderung der hierzu erforderlichen Fähigkeiten wurde ein Lehrveranstaltungsformat entwickelt, welches an Fragestellungen des Arbeits- und Gesundheitsschutzes den interdisziplinären Dialog praktiziert.

1 Hintergrund und Ziele des Lehrveranstaltungskonzeptes

Die Fähigkeit, über die Grenzen des eigenen fachlichen Hintergrunds hinaus zu sehen und sich immer wieder schnell in neue Prozesse und Kontexte hineindenken zu können, gilt als eine zentrale Schlüsselqualifikation im Arbeitsleben von Ingenieurinnen und Ingenieuren. Interdisziplinäre Teamarbeit zu stärken sollte daher essenzieller Bestandteil einer wissenschafts- und praxisorientierten akademischen Ingenieurausbildung sein.

Die Umsetzung dieser anspruchsvollen Aufgabe ist eines der Ziele des Projektes ELLI (Exzellentes Lehren und Lehren in den Ingenieurwissenschaften – gefördert durch das BMBF). In dessen Rahmen wurde das Lehrkonzept „Querformat" entwickelt und im Sommersemester 2013 an der Ruhr-Universität Bochum erprobt. Bei der Konzeption wurde dabei besonderer Wert darauf gelegt, dass das Lehrkonzept in die bestehenden Curricula eingebunden und mittelfristig an andere Hochschulstandorte übertragen werden kann.

Der Titel „Querformat" verdeutlicht dabei plakativ, dass es in der Lehrveranstaltung um einen Blick „quer" zu dem sonst üblichen, rein disziplinär geprägten Blick gehen soll. Interdisziplinarität „an sich", d.h. im Sinne einer wissenschaftstheoreti-

T. Berens (✉)
Berufsforschungs- und Beratungsinstitut für interdisziplinäre Technikgestaltung e.V., Deutschland, Germany
e-mail: tobias.berens@bit-bochum.de

U. Berbuir
Ruhr-Universität Bochum, Deutschland, Germany

Originally published in ""Querformat" - Förderung des interdisziplinären Dialogs, Sicherheitsingenieur, 3/2014", © 2013. Reprint by Springer International Publishing Switzerland 2016, DOI 10.1007/978-3-319-46916-4_66

schen Betrachtung, ist dabei nicht Gegenstand der Veranstaltung, sondern es geht darum, an einer realistischen Problemstellung das Zusammenwirken verschiedener Ansätze und Disziplinen zur Lösung eines Problems erkennbar und erfahrbar zu machen [1]. Das Lehrveranstaltungskonzept beinhaltet verschiedene Facetten eines interdisziplinären Dialogs bzw. interdisziplinärer Zusammenarbeit. Es werden unterschiedliche disziplinäre Blickwinkel auf eine Thematik dargestellt, wodurch sich die Studierenden in der Kleingruppenarbeit als Vertreter ihrer eigenen Disziplin erleben und in einen Dialog mit Vertretern anderer Fachgebiete kommen.

2 Lehr- und Lernziele

Ein Ziel der Lehrveranstaltung ist es, Wissen (im Sinne von Fachwissen) zu einem konkreten Themenfeld zu vermitteln. Neben diesem Wissenserwerb sollen die Studierenden lernen, interdisziplinäre Problemstellungen als solche zu erkennen und zu bearbeiten. Sie sollen Erfahrungen darin sammeln, Lösungsbeiträge zu entwickeln und diese in einen gemeinsamen, disziplinübergreifenden Lösungsansatz einzubringen. Es soll somit die Kommunikations- und Handlungsfähigkeit der Studierenden gefördert werden. Ein spezielles Anliegen im Hinblick auf interdisziplinäre Zusammenarbeit ist es, dass die Studierenden durch die Begegnung mit Vertretern anderer Disziplinen und die Zusammenarbeit mit ihnen zu einer positiven, im Sinne von wertschätzenden, Einstellung gegenüber Konzepten und Vertretern anderer Fachgebiete gelangen [2].

3 Das Konzept

3.1 Grundkonzept

Das Grundkonzept sieht dabei – ganz klassisch – die beiden Bestandteile Vorlesung und Übung vor, welche durch Exkursionen o.ä. ergänzt wird.

Die Vorlesung ist noch einmal in sich geteilt. In der sogenannten Basisvorlesung erhalten die Studierenden Grund- bzw. Überblickswissen zu einer Thematik. In der multidisziplinären Vertiefung werden bestimmte Aspekte des Themas durch Gastvorträge von Referenten aus verschiedenen Disziplinen näher beleuchtet. Diese unterschiedlichen Referenten bieten den Studierenden dabei sowohl Information, indem das Thema von verschiedenen Fachdisziplinen aus betrachtet und die Vielschichtigkeit der Themenstellung erkundet wird, als auch eine gewisse Inspiration, indem sie Einblick in Methoden und Inhalte ihrer Disziplinen geben.

Die Übung ist nach der Methode des Problem-based learning (PBL bzw. Problemorientierten Lernens PoL) organisiert. Kernelement dabei ist es, dass die Studierenden in strukturierter Form einen sogenannten Fall, d.h. eine realistische Problem-situation, die als textliche Beschreibung vorliegt, bearbeiten. Die Methode gibt die Struktur der Vorgehensweise in einem 7-Schritte-Verfahren vor. [3–5]

3.2 Wie läuft eine PBL-Übung ab?

Die PBL-Fallbearbeitung ist in einen zwei Wochen Rhythmus gegliedert. Zunächst bearbeiten die Studierenden in fachheterogen besetzten Kleingruppen einen Fall aus dem Themenbereich. Diese Fälle spiegeln Situationen wieder, die in der beruflichen Praxis so vorkommen können. Die Beschreibung der Fälle ist bewusst situativ gehalten, sodass es Teil des Bearbeitungsprozesses ist, nicht nur ein abstrakt formuliertes fachliches Problem zu betrachten, sondern „das“ Problem aus einer spezifische Situation heraus zu filtern.

Der erste grundlegende Schritt in der Bearbeitung ist es dann auch, dass die Gruppe ein gemeinsames Verständnis des Problems oder der Teilprobleme entwickelt, die in dem jeweiligen Fall stecken. Erst danach sammeln die Studierenden mögliche Gründe und Lösungsansätze, bewerten diese und überlegen, was ihnen an Information bzw. Wissen fehlt, um den Fall „zu lösen“ bzw. die beschriebene Situation zu bewerten und fundierte Handlungsempfehlungen abgeben zu können. Es werden Lernfragen formuliert, die von den Studierenden im Selbststudium bearbeitet werden.

Nach einer Woche treffen die Studierenden sich wieder, tragen ihre Rechercheergebnisse zusammen und kommen zu einer Lösung des Falles bzw. zu einer reflektierten Bewertung und der Formulierung von Handlungsempfehlungen. Die Ergebnisse werden den anderen Gruppen vorgestellt und diskutiert.

Diese Lehr- und Lernform (er)fordert von den Studierenden nicht nur die selbstständige Deduktion des Problems, sondern auch die Akzeptanz alternativer Handhabungen und Lösungswegen und genau hierin liegt eine der besonderen Herausforderungen in einer interdisziplinären Zusammenarbeit. Das PBL-Format mit seiner klaren Strukturierung in 7 Schritte bietet dabei einen Handlungsrahmen zur Zusammenarbeit, der als besonders geeignet erachtet wird, die Studierenden in einen konstruktiven Dialog zu bringen. Die Struktur leitet dazu an, sich einer Problemstellung offen zu nähern und verschiedene Perspektiven mit einzubeziehen, sodass gute Bedingungen für einen interdisziplinären Dialog geschaffen werden.

4 Die Pilotveranstaltung im Sommersemester 2013

Der erste Veranstaltungszyklus an der Ruhr-Universität Bochum lief unter der Überschrift: „Schutzbrille allein genügt nicht – interdisziplinäre Aspekte im Arbeitsschutz“. Das Thema Arbeits- und Gesundheitsschutz wurde gewählt, da es ein relevantes Thema ist, das im Prinzip jeden angeht. Es hat einen klaren und akzeptierten ingenieurwissenschaftlichen Bezug und birgt gleichzeitig viele interdisziplinäre

Ansätze in sich. Hinzu kommt, dass das Thema Arbeitsschutz in der Hochschullehre – gemessen an seiner universellen Bedeutung im Arbeitsleben – oftmals nur in sehr geringen Umfang behandelt wird und wenn, dann zumeist aus einem spezifischen Blickwinkel heraus und nicht in seiner multidisziplinären Breite.

Das Lehrveranstaltungsformat wurde im Rahmen des Projektes ELLI entwickelt und in Kooperation verschiedener Lehrstühle der Ruhr-Universität Bochum, der gemeinsamen Arbeitsstelle RUB/IGM sowie einem Vertreter der beruflichen Praxis umgesetzt. Der entsprechende Lehrauftrag wurde von einem Mitarbeiter des BIT (Berufsforschungs- und Beratungsinstitut für interdisziplinäre Technikgestaltung e.V.) übernommen, einer Einrichtung die im Kontext des Arbeits- und Gesundheitsschutzes mehr als 25 Jahre Erfahrung im interdisziplinären Dialog vorweisen kann. An der Pilotveranstaltung im Sommersemester 2013 haben 25 Studierende aus sieben verschiedenen Studiengängen teilgenommen. Jeweils vier bis fünf Studierende bildeten eine Arbeitsgruppe. Die Gruppen wurden von einem PBL-Tutor begleitet, wobei ein Tutor drei Gruppen betreut hat.

Fachinhaltliches Ziel der Veranstaltung bzw. der Basisvorlesung war es, den Studierenden ein solides Wissensfundament in Bezug auf das Thema Arbeits- und Gesundheitsschutz zu geben. Die Vorlesung orientierte sich an den Belangen des betrieblichen Arbeitsschutzes. Die Gefährdungsbeurteilung in ihren verschiedenen Facetten wurde als zentrales Element der gesetzlichen Arbeitsschutzanforderungen intensiv thematisiert. So wurde neben den rechtlichen Grundlagen, insbesondere die Verantwortung der Akteure diskutiert sowie Inhalte und Konzepte der konkreten Umsetzung vorgestellt. Ebenso wurde die Gefährdungsbeurteilung im Kontext eines ganzheitlichen Verständnisses von Arbeits- und Gesundheitsschutz thematisiert und Anknüpfungspunkte für unterschiedliche Fachdisziplinen aufgezeigt.

In den Vertiefungsvorlesungen wurden verschiedene Themenfelder erörtert. Dazu zählen z. B. „Die betriebliche Mitbestimmung im Rahmen des betrieblichen Arbeits- und Gesundheitsschutzes“, „Partizipatives Sicherheitsmanagement“ oder „Führung und Gesundheitsschutz im Wandel“. Dabei kamen Vertreter unterschiedlicher Branchen, Unternehmensgrößen und Disziplinen zu Wort: Sozialwissenschaften, Betriebswirtschaft, Arbeitswissenschaft, Pädagogik, Psychologie und Maschinenbau.

In den PBL-Fällen wurden ganz unterschiedliche Themen und Probleme beschrieben. Die Bandbreite reichte von Fällen, in denen konkrete Verbesserungsvorschläge für die Gestaltung von Arbeitsplätzen im Fokus des Bearbeitungsprozesses standen, bis hin zu Fällen, in denen es mehr um das Zusammenwirken von Akteuren, wie beispielsweise der Fachkraft für Arbeitssicherheit und des Betriebsarztes ging.

Um die Thematik noch praxisnäher darzustellen, wurde die Veranstaltung mit einer Betriebsbesichtigung in einem Kaltwalzwerk abgerundet. Die theoretisch vermittelten Konzepte zur Gefährdungsbeurteilung und zur Maschinensicherheit konnten vor Ort praktisch erfahren und mit den zuständigen Sicherheitsingenieuren und Fachkräften für Arbeitssicherheit diskutiert werden.

5 Erfahrungen

Insgesamt wurde die Veranstaltung gut angenommen. In der Evaluation wurde seitens der Studierenden insgesamt ein positives Fazit gezogen. Auch die Lehrenden ziehen eine sehr positive Bilanz des ersten Durchlaufs. Bestimmte Beobachtungen im Verlauf sowie verschiedene einzelne Punkte aus der Evaluation bieten wertvolle Ansätze für die Fortentwicklung des Konzeptes.

Die Ziele der Vorlesung hinsichtlich der Wissensvermittlung wurden voll erreicht. Hierfür sprechen sowohl die guten Klausurergebnisse als auch die Bewertungen in der Evaluation. In großer Übereinstimmung urteilten die Studierenden, dass sie in der Basisvorlesung solides Grundlagenwissen zum Thema Arbeitsschutz erwerben konnten. Auch bei der Frage, ob ihnen in den Vertiefungsvorlesungen die interdisziplinären Aspekte des Themas näher gebracht worden seien, sagten die meisten Studierenden, dass dies zutreffe.

Generell wurde der hohe Praxisbezug als positives Element hervorgehoben.

Die Zielerreichung hinsichtlich der weiteren Lehr- und Lernziele muss differenzierter betrachtet werden. In den PbL-Übungen zeigten die Studierenden, dass sie sich auf einen interdisziplinären Dialog einlassen und sich sehr konstruktiv um Lösungsansätze bemühen. Die Prüfungsfälle wurden von allen Gruppen mit großem Engagement bearbeitet, hervorragend aufbereitet und erfolgreich präsentiert. Die persönlichen Ausarbeitungen der Lernfragen des Prüfungsfalles zeigten eine breite Spanne an individuellen Leistungen. Generell wird das Lernen anhand realitätsnaher Problemstellungen als sehr motivierend empfunden, ebenso wie das eigenständige Erarbeiten eines Stoffes bzw. Lehrinhalts.

Die Studierenden hatten teilweise Schwierigkeiten mit der Bearbeitung der Fälle. Diese resultierten aus dem Wunsch, die eine „richtige“ Lösung zu finden, verbunden mit dem Anspruch, dass die Lehrenden alle für die Erarbeitung dieser einen richtigen Lösung notwendigen Informationen bereitstellen sollten bzw. dass es keine Vagheit in den Beschreibungen geben dürfe. Die Haltung mancher Studierenden war geprägt von der Vorstellung, dass es eine eindeutig definierte Aufgabenstellung geben müsse, zu der es auch „die eine Lösung“ gibt. Diesen Studierenden fiel es schwer, sich auf eine offenere Problemstellung einzulassen. Genau diese Bereitschaft sich einzulassen und alternative Handhabungen zu akzeptieren, ist eine Voraussetzung für interdisziplinäre Zusammenarbeit. Die PBL-Methode fordert genau dies ein und ist dementsprechend in besonderer Weise geeignet, interdisziplinäre Fragestellungen zu behandeln. Insgesamt wird die Methode von den Studierenden positiv bewertet und die Mehrheit wünscht sich mehr Lehrveranstaltungen in dieser Form.

6 Ausblick

Nach dem erfolgreichen ersten Durchlauf des Formates zeigt sich an folgenden Punkten Potenzial zur Weiterentwicklung: Die hauptsächliche Motivation, eine solche Lehrveranstaltung zu besuchen, ist das Interesse an der behandelten Thematik.

Daher sollten in der Zukunft solche Themen behandelt werden, die noch stärker in einer aktuellen gesellschaftlichen Diskussion sind, um für den nächsten Durchlauf, mehr Studierende aus geistes- und sozialwissenschaftlichen Studiengängen für die Lehrveranstaltung zu begeistern.

Es hat sich gezeigt, dass die PBL-Methode sehr gut geeignet ist, die Studierenden in einen Dialog zu bringen. Aufgrund der Fallkonstruktion, die immer Aspekte unterschiedlicher Disziplinen enthielt, waren die Studierenden herausgefordert, Themen zu betrachten, die nicht aus ihrem originären Fachgebiet/aus ihrer Disziplin stammen. Die strukturierte Vorgehensweise, die die PBL-Methode vorgibt, hat dabei einen formalen Rahmen geschaffen, dies zu bewältigen. Zukünftig sollte noch stärker darauf zu geachtet werden, dass die Fälle hinsichtlich ihrer Komplexität und fachlichen Anforderungen für die Gruppen gut bearbeitbar sind. Hinsichtlich der Methode sollten die Leitlinien der Begleitung weiter konkretisiert werden. Es gilt, eine gute Balance von Anleitung und Eigenständigkeit zu finden.

Insgesamt lässt sich festhalten, dass neben der Förderung des interdisziplinären Dialoges gleichsam der Wunsch, das Thema Arbeits- und Gesundheitsschutz mehr in den Fokus universitärer Ingenieursausbildung zu rücken, in diesem Pilotversuch erfüllt worden ist Die methodische Ausarbeitung der Lehrveranstaltung hat es ermöglicht, dieses Konzept auch für weitere Universitäten zur Verfügung zu stellen.

Literaturverzeichnis

1. M. Jungert, Was zwischen wem und warum eigentlich? In: *Interdisziplinarität – Theorie, Praxis, Probleme*, ed. by M. Jungert, E. Romfeld, T. Sukopp, U. Voigt, WGB, Darmstadt, 2010, pp. 1–12
2. G. Vollmer, Interdisziplinarität – unerlässlich, aber leider unmöglich?,. In: *Interdisziplinarität – Theorie, Praxis, Probleme*, ed. by M. Jungert, E. Romfeld, T. Sukopp, U. Voigt, WGB, Darmstadt, 2010, pp. 47–75
3. A. Weber, Problem-based learning – eine Lehr-und Lernform gehirngerechter und problemorientierter Didaktik. In: *Problembasiertes Lernen*, ed. by J. Zumbach, A. Weber, G. Olsowski, hep, Bern, 2007, pp. 15–32
4. F.G. Becker, V. Friske, Problemorientiertes Lehren und Lernen in der Betriebswirtschaftslehre: Entwicklung eines Moduls. In: *Problem Based Learning im Dialog*, ed. by M. Mair, G. Brezowar, G. Olsowski, J. Zumbach, Facultas, Wien, 2012, pp. 85–97
5. A. Slemeyer. Aktivierung von Studierenden durch Problemorientiertes Lernen: Homepage der Stabsstelle interne Fortbildung und Beratung. https://dbs-lin.rub.de/leh-reladen/problemorientiertes-lernen/akti-vierung-von-studierenden-durch-pro-blemorientiertes-lernen/

Landscape Format – A Course Concept to Stimulate Interdisciplinary Dialogue

Ute Berbuir, Tobias Berens, Marcus Petermann and Pia Wagner

Abstract Professional life of engineers is strongly formed by complex questions which include economical, ecological or social aspects. Therefore the ability to communicate with members of different disciplines is essential to cope with the multidisciplinary tasks. In the so called "landscape format courses" engineering students work together with students of the humanities. A theme of multidisciplinary relevance is illustrated in lectures by experts from different disciplines and intensively discussed in student working groups, arranged according to the "problem-based learning" method. This paper gives an account of work in progress and previous experiences.

Keywords Engineering Education · Social Implications of Technology · Interdisciplinary Approach · Problem Based Learning

1 Objective of an Interdisciplinary Dialogue

The competence to see beyond the limits of one's own professional background and the ability to become quickly acquainted with new processes and contexts, are considered to be key qualifications in later working life. Therefore it should be an essential part of engineering education to strengthen interdisciplinary teamwork. Against this background an appropriate course concept is developed in the framework of the project "Excellent Teaching and Learning in Engineering Science" (ELLI) which aims to promote the ability of students to do interdisciplinary teamwork. The teaching concept should be developed in such a kind that it can be integrated into existing curricula and also can be transferred to other universities.

U. Berbuir (✉) · M. Petermann · P. Wagner
Ruhr-Universität Bochum, Bochum, Germany
e-mail: Ute.Berbuir@uv.ruhr-uni-bochum.de

T. Berens
Berufsforschungs- und Beratungsinstitut für interdisziplinäre Technikgestaltung e.V., Bochum, Germany

Originally published in "Proceedings of Global Engineering Education Conference (EDUCON)", © 2014. Reprint by Springer International Publishing AG 2016, DOI 10.1007/978-3-319-46916-4_67

The appellation "Landscape format" illustrates that it is the aim of the course to take a look beyond the boundaries of disciplinary procedures and methods. Multi- or Interdisciplinarity "in itself", in the sense of epistemological consideration, is not the subject of the course. It is about how the interaction of different disciplines helps to solve problems. This kind of Interdisciplinarity can be denominated as "Composite Interdisciplinarity" [1].

The course concept will include different facets of an interdisciplinary dialogue and interdisciplinary cooperation. One facet is that different disciplinary perspectives on an issue are presented. Another facet is that the students see themselves as representatives of a discipline and come in a dialogue with representatives of other disciplines.

2 Presentation of the Course Concept

2.1 Educational Objectives

The aim of promoting interdisciplinary dialogue is further specified as follows: the students should acquire knowledge about an interdisciplinary issue or theme. In addition to the acquisition of this knowledge, they should learn to recognize interdisciplinary problems as such and learn how to deal with such problems. They should gain experience to develop approaches and to introduce them into a common, cross-disciplinary solution.

An open and appreciative attitude is regarded as a prerequisite for interdisciplinary collaboration [2]. Therefore it is a special objective that students get through the meeting with representatives of other disciplines and the cooperation with them in a positive and respectful attitude towards them and the concepts of their disciplines.

Over all, the communication skills and the professional abilities of the students should be developed and strengthened.

2.2 The Course Concept

The course concept consists of lectures and seminars which are supplemented by factory tours.

The lectures are divided into the parts "basics" and "exemplary insight". In the basic lecture, the students get an overview and fundamental knowledge of the subject. In the exemplary insight lecture a certain aspect is examined in detail by academics from various disciplines. The different speakers of the lecture series provide the students with information, as well as with inspiration: Information is given by regarding the topic from the viewpoints of various subject disciplines, discovering its contextual complexity. The academics also provide inspiration by giving insights into methods and contents of their respective disciplines while simultaneously embodying their individual professional biographies.

The lecture is accompanied by a Problem-Based Learning seminar. Problem-based learning (PBL) is known as an instructional learner-centered approach that empowers learners to conduct research, integrate theory and practice, and apply knowledge and skills to develop a viable solution to a defined problem [3]. The students work in heterogeneous assembled groups accompanied by a Tutor. The PBL-approach requires an autonomous deduction of the problem, as well as the acceptance of alternative handling. Exactly this point is a special challenge in an interdisciplinary cooperation. The PBL format, with the clearly structured seven step method, provides a framework for cooperation, which is considered particularly suitable to bring students in a constructive dialogue. The structure leads on to an open minded way to approximate a problem and to involve different perspectives.

The first PBL-step consists in the development of a common understanding within the group of the main problem of the case. Afterwards the group collects possible solutions, evaluate these, and reflect on the instruments, knowledge and information they lack in order to elaborate effective recommendations. Questions are formulated, which are distributed in the working groups and requires the students' individual work at home in order to deal with the questions by the next meeting. Here, the results of the single investigations are discussed in groups and lead to the formulation of final recommendations. The results of each group are presented and discussed with the others [4, 5].

3 The Pilot Event in the Summer Term 2013

In the summer term 2013, the first course took place and the leading theme was "Interdisciplinary aspects of occupational safety and health (OSH)". The subject OSH was chosen because it is an established topic with a clear and accepted engineering reference and simultaneously involves many interdisciplinary approaches in itself.

The basic lecture offered an overview of the topic of occupational safety and health. It focused on the risk assessment as the main element of legal requirement. In addition to the legal framework, the actors and the particular responsibilities in the field of OSH were explicated. Furthermore the way how to perform a risk assessment was illustrated. Afterwards the risk assessment was discussed in the context of a holistic understanding of OSH and the development of an integrative health management was outlined.

In the exemplary insight lecture representatives of the academic disciplines Social Sciences, Organizational Management, Mechanical Engineering, Pedagogics and Psychology contributed their expertise. Topics such as "The role of employee participation in the field of OSH" or "Cognition of endangerment" were treated.

The lecture and the seminar were held weekly. 25 students from 7 different degree programs participated, most of them having an engineering background. 3 Groups consisting of 4 students each worked parallel in a seminar assisted by one PBL-tutor.

A new case was handled every two weeks applying the PBL-method. The cases reflected situations that could occur in professional practice. A contextualization through a precise description of the situation was intentionally provided in order to force the students to filtrate the problem(s) from a close to reality situation instead of confronting them with an abstract formulated task.

4 Experiences

Generally students evaluated the course positively. Also the teachers drew a very positive balance. Particular observations concerning the course and single items that emerged from the evaluation offer an interesting response for the development of the course concept.

The lecture's learning goals were reached. Evidence for this can be found in the good results in the exam and in the evaluation. There is a consensus among the students that the basis lecture provided them with a firm knowledge of occupational safety and health. Similarly, most of them confirmed that thanks to the deepening lecture they got to know more about different valuable aspects related to the topic. The intensive practice orientation was generally seen as a positive element.

The accomplishment of further educational objectives has to be considered differently: In the PBL-seminar the students showed that they were able to get involved in an interdisciplinary dialogue and that they could constructively work on common solutions. All the cases were successfully handled and presented by all groups. The written elaborations of the different learning issues of the examination case, however, showed a wide range of personal results. Both, the group presentations and the written elaborations of the questions concerning the case were include in the final grade.

The practice oriented learning which based on the confrontation with problems close to reality and also the autonomous elaboration of learning contents were perceived as very motivating.

The student opinions of the "usefulness" of the PBL-method diverged. All response categories from "applies fully" up to "not true at all" were represented and the median was in the middle of the 5-part scale. In the evaluation students mentioned the necessity of being provided with even more instructions by PBL-exercises. Generally the students had problems with the vagueness respectively uncertainty of the cases. Many of the participants expressed the wish to find the "correct" solution, together with the claim, directed to the teacher, to dispose all the Information needed to solve the case. Acceptance of alternative handling – which is fundamental for the PBL-Method – led to irritation with some students.

In summary it can be said that the method was evaluated mostly positive and that the majority of participating students wants more courses of this kind.

References

1. M. Jungert, "Was zwischen wem und warum eigentlich? Grundsätzliche Fragen der Interdisziplinarität," in M. Jungert, E. Romfeld, T. Sukopp, U. Voigt, „Interdisziplinarität – Theorie, Praxis, Probleme," WGB Darmstadt, 2010, pp. 1–12.
2. G. Vollmer, "Interdisziplinarität – unerlässlich, aber leider unmöglich?," in M. Jungert, E. Romfeld, T. Sukopp, U. Voigt „Interdisziplinarität – Theorie, Praxis, Probleme," WGB Darmstadt, 2010, pp. 47–75.
3. J. R. Savery, "Overview of Problem-based Learning: Definitions and Distinctions." in Interdisciplinary Journal of Problem-based Learning, 1(1), 2006, pp. 9–20. Available at: http://dx.doi.org/10.7771/1541-5015.1002
4. A. Weber, „Problem-Based Learning – Eine Lehr-und Lernform gehirngerechter und problemorientierter Didaktik," in J. Zumbach, A. Weber, G. Olsowski, „Problembasiertes Lernen," hep Bern, 2007, pp. 15–32
5. F. G. Becker, V. Friske, „Problemorientiertes Lehren und Lernen in der Betriebswirtschaftslehre: Entwicklung eines Moduls," in M. Mair, G. Brezowar, G. Olsowski, J. Zumbach, „Problem Based Learning im Dialog," Facultas Wien, 2012, pp. 85–97.

Problemorientiertes Lernen

Ute Berbuir, Hille Lieverscheidt and Andreas Slemeyer

Zusammenfassung Von Hochschulabsolventinnen und -absolventen werden neben fundiertem Fachwissen in hohem Maße auch personale Kompetenzen gefordert. Selbstständigkeit und Eigenverantwortung gehören genauso dazu wie kommunikatives und kooperatives Verhalten. Durch problembasiertes Lernen werden solche Kompetenzen gefördert.

1 Einleitung

Die Methode des problemorientierten oder auch problembasierten Lernens und Lehrens (POL) nahm in den 1960er-Jahren ihren Anfang in der Medizinerausbildung in Kanada und ist mittlerweile weltweit erfolgreich im Einsatz. Sie ermöglicht selbstgesteuertes und kollaboratives Lernen. Sie zielt auf die Vermittlung von Kompetenzen und fördert den Praxisbezug des Studiums, indem Fallbeispiele zum Ausgangspunkt des Lernens gemacht und in kleinen Gruppen selbstständig bearbeitet werden.

In Deutschland wird POL vor allem in der Medizinerausbildung eingesetzt, obwohl Projekt- und Fallarbeit auch für andere Studiengänge typisch ist, etwa in den Ingenieur-, Wirtschafts- und Rechtswissenschaften. Als Argument gegen den Einsatz von POL wird häufig der Arbeitsaufwand angeführt, der als höher angenommen wird. Erfahrungen von Anwendern zeigen jedoch, dass der (Mehr-)Aufwand sich zu einem großen Teil auf die Einführungsphase bezieht und in der Regel durch höhere Motivation und Kreativität aufgewogen wird. Klar ist jedoch, dass Veränderungen in der Lehre nicht ohne Aufwand erreicht werden können.

Mit diesem Beitrag möchten wir Mut machen, POL einfach einmal auszuprobieren. Wir beschreiben die Grundprinzipien, lenken den Fokus auf wesentliche

U. Berbuir (✉) · H. Lieverscheidt
Ruhr-Universität Bochum, Bochum, Germany
e-mail: Ute.Berbuir@uv.ruhr-uni-bochum.de

A. Slemeyer
Technische Hochschule Mittelhessen, Gießen, Germany

Originally published in "Problemorientiertes Lernen, Raabe, duz Deutsche Universitätszeitung, 11", © 2014. Reprint by Springer International Publishing AG 2016, DOI 10.1007/978-3-319-46916-4_68

Erfolgsmerkmale, geben Tipps und Literaturhinweise zur Umsetzung und stellen drei konkrete Beispiele vor.

2 Die Prinzipien

Für POL gibt es keinen verbindlichen Standard, doch lassen sich Grundprinzipien nennen:

- Fallbeispiele aus dem Arbeitsalltag stehen am Anfang, das heißt vor der Wissensvermittlung.
- Fallbeispiele werden in Gruppen bearbeitet, die tutoriell begleitet werden.
- Die Bearbeitung erfolgt in definierten Schritten, den sogenannten „7-Steps": zunächst gemeinsame Fallanalyse (Schritte 1 bis 5), dann Selbststudienphase (Schritt 6) und zum Abschluss Rückkopplung der Ergebnisse in die Gruppe (Schritt 7).
- Die Studierenden entwickeln aus der Fallbearbeitung heraus eigenständig die Fragen, die sie im Selbststudium bearbeiten, und entscheiden somit selbst, was und wie sie lernen.
- Die Lehrenden agieren als Lernbegleiter.

Es gibt zahlreiche Literatur zum Einsatz von POL in verschiedenen Disziplinen (s. Punkt 6). Aus ihr lassen sich Anregungen zur Vorgehensweise und zur Formulierung der Fallbeispiele gewinnen.

3 Erfolgsmerkmale

Den Erfolg einer problemorientierten Lehrveranstaltung bestimmen im Wesentlichen

- die Qualität der Fallbeispiele. Sie sollten alltagsrelevant, herausfordernd und mit klarem Bezug zu den Lern- und Prüfungszielen sein.
- die Qualität der Lernbegleitung. Sie sollte Gruppenprozesse qualifiziert anleiten und den Teilnehmern konstruktives Feedback bieten.
- eine sinnvolle Einbettung. Die Platzierung von POL im Curriculum sollte stimmig, die Form des Leistungsnachweises angemessen (zum Beispiel in einem Lernportfolio, Fachgespräch etc.) sein.

Zur Entwicklung der Fallbeispiele empfiehlt es sich, Anregungen von Experten aus der beruflichen Praxis einzubeziehen. Ebenso ist eine Begleitung der Implementierung durch hochschuldidaktische Multiplikator/innen anzuraten. Die Form des Leistungsnachweises (z. B. durch Lernportfolio) sollte mit den Studierenden abgestimmt werden und justiziabel sein.

Im Folgenden werden drei Beispiele aus der Praxis vorgestellt. Während in den ersten beiden Beispielen vorhandene Module durch den Einsatz von POL umgestaltet werden, dient POL im dritten Beispiel als flächendeckendes Lernarrangement.

3.1 Beispiel 1: Problemorientiertes Lernen in einer Übung

Vor zwei Jahren ist die Methode POL in einer interdisziplinären Lehrveranstaltung mit Studierenden aus ingenieur- und gesellschaftswissenschaftlichen Studiengängen an der Ruhr-Universität Bochum eingeführt worden.

Die Umsetzung erfolgte innerhalb der bestehenden Curricula, wobei der Übungsteil einer fächerübergreifend angebotenen Lehrveranstaltung im POL-Format organisiert wurde. In dieser Struktur ist darauf zu achten, dass der Aufbau der Vorlesung beziehungsweise die Reihenfolge der Themen sinnvoll mit den Fällen abgestimmt wird. Die Bearbeitung der Fälle erfolgt in Gruppen von 4 oder 5 Personen. Ein/e POL-Tutor/in betreut dabei bis zu drei Kleingruppen. Die Lernerfolgskontrolle erfolgt anhand der Bearbeitung eines Prüfungsfalles. Bewertet wird die Präsentation des Gruppenergebnisses sowie die schriftliche Ausarbeitung zu einer individuellen Lernfrage.

Zeitliche Strukturierung

- Einführung der POL-Methode und Bearbeitung eines Probefalles (2 Übungstermine)
- Die Fallbearbeitung in den Übungen erfolgt im 2-Wochen Turnus:
- gemeinsame Fallanalyse, anschließend Selbststudienphase bis zur nächsten Übung.
- Zusammenführung der Ergebnisse in der Gruppe sowie Kurzdarstellung der Ergebnisse vor den anderen Gruppen mit gemeinsamer Diskussion und mit Feedback der Lehrenden.
- Abschluss: Bearbeitung des Prüfungsfalles (verlängerte Bearbeitungszeit 2 Wochen) mit anschließender Präsentation.

Bereitzustellende Ressourcen

In der Rolle eines Lernbegleiters geschulte Tutoren/innen, Arbeitsräume mit beweglichen Tischen, Moderationsmaterial.

3.2 Beispiel 2: Problemorientiertes Lernen in einem ingenieurwissenschaftlichem Modul

Die Erfahrungen stammen aus einer ingenieurwissenschaftlichen Lehrveranstaltung an der Technischen Hochschule Mittelhessen. Die Vorlesung und Übung zum Wahlpflichtfach Sensorik im Fachbereich Elektro- und Informationstechnik, angesiedelt im 5. beziehungsweise 6. Semester, wurde durch die Bearbeitung von POL-Fällen ersetzt. Die Studierenden arbeiten in Gruppen mit bis zu 5 Teilnehmer/innen.

Es wird anstelle einer Klausur ein Lernportfolio als Leistungsnachweis verwendet. Lerntexte, Literaturhinweise, Anleitungen und Formblätter werden auf der Lernplattform bereitgestellt.

Zeitliche Strukturierung

- Woche 1: Einführung in POL, Gruppenarbeit, Lernportfolio.
- Woche 2: Bearbeitung des 1. Falls in der Gruppe, Formulierung von Lernfragen, Verteilung der Rechercheaufträge, Selbststudium.
- Woche 3: Zusammentragen der Arbeitsergebnisse, Auswahl und Synthese für den Gruppenbericht.
- Woche 4: Präsentation der Gruppenberichte. Feedback an die Gruppen und Ergänzungen durch den Lernbegleiter. Beginn der Bearbeitung des 2. Falls in den Gruppen. Anmerkung: Da für die Bearbeitung eines Falls zwei Wochen benötigt werden, können bis zu fünf Fälle pro Semester eingebracht werden.
- Abschlusswoche: Fachgespräche mit den Gruppen, Evaluation.

Bereitzustellende Ressourcen Vergleiche Beispiel 1

3.3 Beispiel 3: Problemorientiertes Lernen in der medizinischen Ausbildung

Erfahrungen mit POL in der Medizinerausbildung werden an der Ruhr-Universität Bochum seit 13 Jahren in drei verschiedenen Curricula gesammelt.

Der neue integrierte Reformstudiengang (iRM) ist im Wintersemester 2013/2014 gestartet und löst die beiden vorausgegangenen Studiengänge ab. Er vereinigt die Vorzüge von Regel- und Modellstudiengang in einem themenbezogenen Curriculum in der Vorklinik, das durch problemorientiertes Lernen, systematische Untersuchungskurse, Anamnesetrainings und Übungen zur Arzt-Patient-Kommunikation ergänzt wird.

Die Studierenden bearbeiten in 10er-Gruppen Patientengeschichten aus dem klinischen Alltag. Im vorklinischen Studienabschnitt gibt es 14 Fälle verteilt auf vier Semester. In den Semestern 2 bis 4 gibt es pro thematischem Modul (etwa 4 Wochen „Herz/Kreislauf" oder „Atmung") einen POL-Fall. Dieser bildet die thematische Klammer für die vorklinischen Lehrinhalte des Moduls. Die POL-Fälle werden so ausgewählt, dass sowohl Anknüpfungspunkte zu den vorklinischen Inhalten zu finden sind, als auch Bezüge zu Anamnesegesprächen und zur Arzt-Patient-Interaktion.

Die in POL behandelten fachlichen Inhalte werden durch die Klausuren der beteiligten Fächer geprüft, in den POL-Sitzungen bekommen die Studierenden regelmäßiges Feedback zu ihren Schlüsselqualifikationen.

Zeitliche Strukturierung am Beispiel des thematischen Moduls „Spinale Sensorik" im 2. Semester:

- Woche 1: Vorlesungen, Übungen und Praktika zum Modulthema sowie POL-Fall „Ein Streifen macht noch keine Blutvergiftung"; Schritte 1 bis 5; Lernziele formulieren.

- Woche 2 und 3: Lernziele bearbeiten und durch begleitende Praktika und Übungen vertiefen.
- Woche 4: Schritt 7 des POL-Falles mit interaktiver Lernzielpräsentation der Gruppenmitglieder; Bearbeitung der Schritte 1 bis 5 des POL-Falles für das nächste Modul zum Thema „Periphere Motorik".

Bereitzustellende Ressourcen

Im Modellstudiengang Medizin wurden viele Innovationen probiert, evaluiert und weiter entwickelt, auf die die Fakultät heute zurückgreifen kann. Aus den Tutorentrainings der Anfangsphase hat sich ein Zertifikat Medizindidaktik entwickelt, das mittlerweile landes- und bundesweit vernetzt ist. Es gibt einen riesigen Pool von erprobten und evaluierten POL-Fällen, aus dem geschöpft werden kann. Im integrierten Reformstudiengang wurde POL curricular verankert, und es besteht die Herausforderung, das Format für 320 Studierende zu organisieren. Hierzu stimmen Fachvertreter/innen in extra Planungssitzungen das Curriculum untereinander ab. 64 Tutoren/innen betreuen die rund 32 Gruppen in zwei Jahrgängen der Vorklinik. Die Tutoren werden ausgebildet und in Austauschtreffen weiter begleitet. Die POL-Fälle werden passend zu den Lehrinhalten ausgewählt oder neu geschrieben. Kleingruppenräume müssen ausreichend zur Verfügung stehen, Lehrmaterialien wie Fallunterlagen, Infos zum Curriculum, Stundenpläne, Moderationsmaterial für die Gruppen müssen bereitgestellt werden.

4 Fazit

POL lebt vom Engagement der Beteiligten. Die Konstruktion der Fallbeispiele bietet für Lehrende eine anregende und kreative Herausforderung, sich mit dem eigenen Fach und der eigenen Lehre in neuer Weise auseinanderzusetzen. Der Lohn der Mühe sind motivierte Studierende, dic in einen lebendigen Dialog untereinander und mit den Lehrenden eintreten. POL fördert die Langzeitspeicherung von Zusammenhängen durch assoziatives Vernetzen und führt zu einer Verschiebung vom „surface level learning" zum „deep level learning". Die Methode birgt forschendes Lernen in sich und ist insofern eine ideale Vorbereitung für die zukünftigen beruflichen Herausforderungen unserer Absolventinnen und Absolventen.

4.1 Nutzen für Studierende

Beim problemorientierten Lernen können sich Studierende leichter austauschen und sich vernetzen. Sie trainieren ihre Konfliktlösungsfähigkeiten und können ihre Kommunikations- und Teamfähigkeit verbessern.

4.2 Nutzen für Lehrende

Beim problemorientierten Lernen übernehmen Studierende aktiv Verantwortung für ihren eigenen Lernprozess und Lernerfolg. Sie sind dadurch motivierter und bereit für einen lebendigen Dialog mit Lehrenden. POL bietet wegen der oftmals interdisziplinären Fragestellung Anregungen für Lehrende zum kollegialen Austausch mit Dozierenden aus anderen Fachbereichen.

5 Fünf Tipps zum Ausprobieren von POL

1. Planen Sie ausreichend Zeit für die Einführung der Methode ein
2. Vollziehen Sie Ihren Rollenwechsel zum Lernbegleiter ganz bewusst und machen Sie dies auch Ihren Studierenden gegenüber transparent.
3. Stimmen Sie die Fälle gezielt auf die Lernziele des Moduls ab und versuchen Sie, durch thematische Eingrenzung eine (zeitliche) Überforderung der Studierenden zu vermeiden.
4. Leiten Sie die Studierenden an, sich ihre eigenen Lernfortschritte bewusst zu machen, beispielsweise durch die Erstellung eines Lernportfolios.
5. Holen Sie sich gezielt das Feedback der Studierenden, um Verbesserungsmöglichkeiten für den nächsten Durchlauf zu fmden.

6 Tipps zum Weiterlesen

- Lieverscheidt, Hille: Zeitgemäß lehren und lernen in der medizinischen Ausbildung: Studierenden-orientiert, praxisnah, interaktiv. In: Curriculum Naturheilverfahren und Komplementärmedizin. Lehrinhalte und Medizindidaktik. KVC-Verlag Essen 2013, S. 5–62
- Slemeyer, Andreas: Problemorientiertes Lernen für eine Einzelveranstaltung: ein Fallbeispiel aus dem Ingenieurbereich. Neues Handbuch Hochschullehre, Lieferung 61, C 1.5, Raabe Verlag (2013)

- Strobel, Johannes; Barneveld, Angela van: When is PBL More Effective? Interdisciplinary Journal of Problem-based Learning, Vol. 3, Issue 1 (2009). http://docs.lib.purdue.edu/ijpbl/vol3/iss1/4/
- Savery, John R.: Overview of Problem-based Learning: Definitions and Distinctions. Interdisciplinary Journal of Problem-based Learning, Vol. 1, Issue 1 (2006). http://docs.lib.purdue.edu/ijpbl/vol1/iss1/3/

7 Materialien zum Download

- Die Ruhr-Universität Bochum hat für Lehrende eine ganze Reihe von Informationen zum Stichwort „Problemorientiertes Lernen" bereitgestellt. Die Materialien finden sich unter https://dbs-lin.rub.de/lehreladen/problemorientiertes-lernen/

Querformat – Ein Lehrveranstaltungskonzept zur Förderung des interdisziplinären Dialogs

Ute Berbuir, Marcus Petermann and Martina Schmohr

Zusammenfassung The professional life of engineers is strongly formed by complex questions which include economical, ecological or social aspects. Therefore, the ability to communicate with members of different disciplines is essential to cope with those multidisciplinary tasks. In the framework of the project „Excellent Teaching and Learning in Engineering Science" (ELLI), the course concept „landscape format" was developed which aims to promote the ability of students to do interdisciplinary teamwork. This paper gives an account of the conception as well as the experience of the pilot course and identifies prospects for further development.

1 Hintergrund und Ziele des Lehrveranstaltungskonzeptes

Die Fähigkeit, über die Grenzen des eigenen Faches hinaus zu sehen und sich immer wieder schnell in neue Prozesse und Kontexte hineindenken zu können, gilt als eine zentrale Schlüsselqualifikation im Arbeitsleben von Ingenieuren. Interdisziplinäre Teamarbeit zu stärken, sollte daher essenzieller Bestandteil einer wissenschafts- und praxisorientierten Ingenieurausbildung sein. Die Umsetzung dieser anspruchsvollen Aufgabe ist eines der Ziele des Projektes ELLI „Exzellentes Lehren und Lernen in den Ingenieurwissenschaften", gefördert innerhalb des Bund-Länder-Programms für bessere Studienbedingungen und mehr Qualität in der Lehre („Qualitätspakt Lehre"). In dessen Rahmen wurde das Lehrkonzept „Querformat" entwickelt und im Sommersemester 2013 an der Ruhr-Universität Bochum erprobt. Bei der Konzeption wurde besonderer Wert darauf gelegt, dass es in die bestehenden Fach-Curricula eingebunden und mittelfristig auch an andere Hochschulen übertragen werden kann.

Die Benennung des Konzepts als „Querformat" soll verdeutlichen, dass es in der Lehrveranstaltung um einen Blick quer zu dem sonst üblichen, rein disziplinär geprägten Blick gehen soll. Interdisziplinarität an sich, d.h. im Sinne einer wissenschaftstheoretischen Betrachtung, ist dabei nicht Gegenstand der Veranstaltung.

U. Berbuir (✉) · M. Petermann · M. Schmohr
Ruhr-Universität Bochum, Bochum, Germany
e-mail: Ute.Berbuir@uv.ruhr-uni-bochum.de

Originally published in "Schier, Carmen; Schwinger, Elke (Hg.), Interdisziplinarität und Transdisziplinarität als Herausforderung akademischer Bildung, Bielefeld, Transcript, S. 229-236", © 2014.
Reprint by Springer International Publishing AG 2016,
DOI 10.1007/978-3-319-46916-4_69

Vielmehr orientiert sich der Ansatz an einem Interdisziplinaritätsbegriff, der von Jungert (in Anlehnung an Heckhausen) folgendermaßen beschrieben wird:

„Zusammengesetzte Interdisziplinarität (Composite Interdisciplinarity): Heckhausen vergleicht diese Art von Interdisziplinarität mit einem Puzzle. Drängende praktische Probleme motivieren eine Zusammenarbeit verschiedener Disziplinen, etwa in der Friedensforschung oder auf dem Gebiet der Städteplanung, wo solch unterschiedliche Disziplinen wie Architektur, Ökonomie, Psychologie, Biologie und Ingenieurwissenschaften zusammenarbeiten. Dabei überlappen weder die Gegenstandsbereiche der jeweiligen Fächer ernsthaft noch deren Methoden oder theoretische Integrationsniveaus. Zusammengehalten wird dies eigentlich (auf theoretischer Ebene) unverbundene Fachensemble durch die Interdependenzen des komplexen, akuten Problembereichs, der die Einbeziehung aller Perspektiven erforderlich macht." [1]

Die Lehr-Lernziele des Lehrveranstaltungsformats „Querformat" adressieren Sach-, Sozial- und Selbstkompetenzen der Studierenden. Vor dem Hintergrund der Bedeutung des jeweiligen Themas als Bezugspunkt der Zusammenarbeit wird die inhaltliche, am Problem orientierte Behandlung in den Mittelpunkt gestellt. Es soll exemplarisch an praxisrelevanten Fragestellungen gearbeitet, Wissen vermittelt und Sachkompetenz aufgebaut werden. Bei den Sozialkompetenzen liegt der Schwerpunkt bei der Förderung der kommunikativen Fähigkeiten. Darüber hinaus soll auch die Selbstkompetenz, d.h. die Haltung der Studierenden zu fachübergreifender Zusammenarbeit gefördert bzw. deren Entwicklung angeregt werden. Auch wenn die folgenden Zitate und Inhalte vornehmlich aus Quellen stammen, die an Belangen interdisziplinärer Forschung orientiert sind, lässt sich daraus allgemein ableiten, dass die innere Haltung essenziell für das Funktionieren von interdisziplinärer Zusammenarbeit ist. Vollmer beispielsweise verweist dabei auf Rottländer und beschreibt

„[…] unter welchen Voraussetzungen interdisziplinäre Zusammenarbeit [Herv. i.O.] – trotzt aller Schwierigkeiten – möglich ist […]. Es sind drei innere Einstellungen: Erstens kann weder die Naturwissenschaft die Geisteswissenschaft beaufsichtigen oder ihr ihre Methoden aufzwingen noch umgekehrt. Zweitens muss man sich mit der Ausdrucksweise, der Terminologie und den Methoden der anderen Seite vertraut machen. Drittens muss man dem Partner aus der anderen Disziplin seine Erkenntnisse zunächst einmal glauben. Das heißt nicht, dass man alles vorbehaltlos schlucken muss: Kompetenzen können sich überschneiden, und man kann durchaus prüfen, wer seinen Kompetenzbereich überschritten hat, aber von den Thesen des Partners muss angenommen werden, dass sie auf vertretbare Weise gewonnen wurden." [2]

Auch Engler betont die innere Haltung und formuliert in einem Interview zum Thema „Chancen, Risiken und Grenzen einer breit angelegten wissenschaftlichen Zusammenarbeit" Folgendes: „Sie [die Interdisziplinarität] ist eine Geisteshaltung, das Endprodukt eines Sich-Öffnens gegenüber Fragestellungen, die nicht die alltäglichen eines Faches sind. Insofern ist sie auch das Produkt eines wachen und offenen Geistes." [3] In diesem Sinne ist es ein spezielles Anliegen des

Veranstaltungsformates, dass die Studierenden durch die Begegnung mit Vertretern anderer Disziplinen und durch die Zusammenarbeit mit ihnen zu einer wertschätzenden Einstellung gegenüber Konzepten und Vertretern anderer Fachgebiete gelangen.

2 Vorstellung des Lehrveranstaltungskonzeptes „Querformat"

Bei der Entwicklung des Lehrveranstaltungs- bzw. Modulkonzeptes wurde besonderer Wert darauf gelegt, dass es in die aktuellen Curricula eingebunden werden kann. Daher wurden bestehende Strukturelemente verwendet bzw. berücksichtigt, wozu sowohl der formale Aufbau des Moduls als auch der wöchentliche Veranstaltungsturnus gehört. „Querformat" läuft über ein Semester und besteht – formal gesehen – aus einer Vorlesung und einer Übung mit jeweils zwei Semesterwochenstunden. Die Leistungskontrolle beinhaltet die Prüfungselemente Klausur, Präsentation und eine schriftliche Ausarbeitung, die gewichtet in die Notengebung einfließen.

Die Vorlesung ist in sich geteilt. In rund der Hälfte der Termine, der sogenannten Basisvorlesung, erhalten die Studierenden Grund- bzw. Überblickswissen zu einer Thematik. In der multidisziplinären Vertiefung, die in gewisser Weise den Charakter einer Ringvorlesung hat, werden bestimmte Aspekte des Veranstaltungsthemas durch Gastvorträge von Referenten aus verschiedenen Disziplinen näher beleuchtet. Diese bieten den Studierenden dabei sowohl Information, um das Thema von verschiedenen Fachdisziplinen aus betrachten und die Vielschichtigkeit der Themenstellung erkunden zu können, als auch eine gewisse Inspiration, indem die Dozenten jeweils Einblick in Methoden und Inhalte ihrer Disziplinen geben.

Wie bereits erläutert, zielt die Veranstaltung auch auf Förderung und Entwicklung von Sozial- und Selbstkompetenzen der Studierenden. Daher wird ein Lehr-Lernformat benötigt, das Spielräume lässt, den Dialog der Beteiligten fördert und eine individuelle Auseinandersetzung mit der Fragestellung fordert. Vor diesem Hintergrund wurde für die Übung die Methode des problembasierten Lernens (PBL) ausgewählt. Diese, häufig auch problemorientiertes Lernen (PoL) oder problem-based learning genannt, ist eine etablierte Lehr-Lernform, die in den 1960er Jahren ihren Anfang in der Medizinausbildung in Kanada nahm und mittlerweile weltweit in der Hochschullehre erfolgreich im Einsatz ist. Sie gilt als eine Form gehirngerechter Didaktik, die selbstgesteuertes und kollaboratives Lernen ermöglicht und auf Kompetenzorientierung als auch transferwirksames Lernen zielt. Darüber hinaus fördert sie den Praxisbezug des Studiums, indem Fallbeispiele aus dem beruflichen Alltag zum Ausgangspunkt des Lernens gemacht und in kleinen Gruppen selbstständig bearbeitet werden [4, 5].

Die PBL-Fallbearbeitung in diesem aktuell entwickelten Lehrveranstaltungskonzept ist in einen vierzehntägigen Rhythmus gegliedert. Zunächst bearbeiten die Studierenden in fachheterogen besetzten Kleingruppen einen Fall aus dem Themenbereich. Diese Fälle spiegeln Situationen wieder, die in der beruflichen Praxis so vorkommen könnten. Sie beschreiben bewusst eine konkrete Situation, sodass es Teil des Bearbeitungsprozesses ist, nicht nur ein abstrakt formuliertes Problem zu betrachten, sondern dieses aus einer spezifischen Situation heraus zu filtern. Dementsprechend besteht der erste grundlegende Schritt in der Bearbeitung darin, dass die Gruppe ein gemeinsames Verständnis des Problems oder der Teilprobleme entwickelt, die in dem jeweiligen Fall stecken. Erst danach sammeln die Studierenden mögliche Gründe und Lösungsansätze, bewerten diese und überlegen, was Ihnen an Information bzw. Wissen fehlt, um den Fall zu lösen bzw. die beschriebene Situation zu bewerten und fundierte Handlungsempfehlungen abgeben zu können. Jede Gruppe formuliert Lernfragen, die von den einzelnen Mitgliedern im Selbststudium bearbeitet werden. Nach einer Woche treffen die Studierenden sich wieder, tragen ihre Rechercheergebnisse zusammen und kommen zu einer Lösung des Falles bzw. zur Formulierung von Handlungsempfehlungen. Die Ergebnisse werden den anderen Gruppen vorgestellt und diskutiert [6].

Diese Lehr- und Lernform erfordert von den Studierenden nicht nur die selbstständige Deduktion des Problems, sondern auch die Akzeptanz alternativer Handhabungen und Lösungswege. Insofern fordert und fördert das PBL-Format generell eine offene Haltung, die als eine Voraussetzung für das Gelingen von fachübergreifender Zusammenarbeit angesehen wird. Gleichzeitig bietet das PBL-Format mit seiner klaren Strukturierung in sieben Schritte einen Handlungsrahmen zur Zusammenarbeit, der als besonders geeignet erachtet wird, die Studierenden in einen konstruktiven Dialog zu bringen. Die Struktur leitet dazu an, sich einer Problemstellung offen zu nähern und verschiedene fachspezifische Perspektiven mit einzubeziehen, sodass gute Bedingungen für einen interdisziplinären Dialog geschaffen werden.

3 Die Pilotveranstaltung im Sommersemester 2013

Der erste Veranstaltungszyklus an der Ruhr-Universität Bochum lief unter der Überschrift: „Schutzbrille allein genügt nicht – interdisziplinäre Aspekte im Arbeitsschutz". Das Thema „Arbeits- und Gesundheitsschutz" wurde gewählt, da es ein relevantes Thema ist, das jeden Menschen angeht. Zudem hat es einen klaren und akzeptierten ingenieurwissenschaftlichen Bezug und birgt gleichzeitig viele interdisziplinäre Ansätze in sich. Die Pilotveranstaltung fand im Sommersemester 2013 statt und wurde im Rahmen des ELLI-Projektes in Kooperation verschiedener Lehrstühle der Ruhr-Universität Bochum, der gemeinsamen Arbeitsstelle RUB/IGM sowie Vertretern der beruflichen Praxis umgesetzt. Es nahmen 25 Studierende aus sieben verschiedenen Studiengängen teil, davon jedoch nur drei Studierende, die ein geistes- oder sozialwissenschaftliches Fach studierten. Jeweils vier bzw. fünf Studierende bildeten eine PBL-Arbeitsgruppe. Die Gruppen wurden von einem PBL-Tutor begleitet, wobei ein Tutor drei Gruppen betreute.

Fachinhaltliches Ziel der Veranstaltung bzw. der Basisvorlesung war es, den Studierenden ein solides Wissensfundament in Bezug auf das Thema zu geben. Die Vorlesung orientierte sich an den Belangen des betrieblichen Arbeitsschutzes.

Die Gefährdungsbeurteilung wurde als zentrales Element der gesetzlichen Arbeitsschutzanforderungen in der Vorlesung intensiv thematisiert und im Kontext eines ganzheitlichen Verständnisses von Arbeits- und Gesundheitsschutz dargestellt. Dabei wurde besonderer Wert darauf gelegt, Anknüpfungspunkte für unterschiedliche Fachdisziplinen aufzuzeigen. In den Vertiefungsvorlesungen wurden verschiedene Aspekte exemplarisch erörtert. Dazu zählen z.B. „Die betriebliche Mitbestimmung im Rahmen des betrieblichen Arbeits- und Gesundheitsschutzes", „Partizipatives Sicherheitsmanagement" oder „Führung und Gesundheitsschutz im Wandel". Dabei kamen Vertreter unterschiedlicher Branchen und Disziplinen wie der Sozialwissenschaft, Betriebswirtschaft, Arbeitswissenschaft, Pädagogik, Psychologie und Maschinenbau zu Wort. In den PBL-Fällen wurden ganz unterschiedliche Themen und Probleme beschrieben. Die Bandbreite reichte von Fällen, in denen konkrete Verbesserungsvorschläge für die Gestaltung von Arbeitsplätzen im Fokus des Bearbeitungsprozesses standen, bis hin zu Fällen, in denen es um gesellschaftliche Dimensionen des Arbeits- und Gesundheitsschutzes bzw. des demografischen Wandels und einer alter(n)sgerechten Arbeitsplatzgestaltung ging.

4 Erfahrungen

Die Veranstaltung wurde von den Studierenden gut angenommen und in der Evaluation insgesamt positiv bewertet. Auch die Lehrenden ziehen ein durchweg positives Fazit des ersten Durchlaufs. Einzelne Beobachtungen im Verlauf sowie spezifische Teilergebnisse der Evaluation bieten wertvolle Ansätze für die Fortentwicklung des Konzeptes. Im Folgenden werden die Evaluationsergebnisse und Beobachtungen in Bezug auf die jeweiligen Veranstaltungsteile und deren Zielerreichungsgrad diskutiert.

Die Ziele der Vorlesung hinsichtlich der Wissensvermittlung wurden voll erreicht. Hierfür sprechen sowohl die guten Klausurergebnisse als auch die Bewertungen in der Evaluation. In großer Übereinstimmung urteilten die Studierenden, dass sie in der Basisvorlesung solides Grundlagenwissen zum Thema Arbeits- und Gesundheitsschutz erwerben konnten. Auch bei der Frage, ob ihnen in den Vertiefungsvorlesungen die interdisziplinären Aspekte des Themas nähergebracht worden seien, äußerten die meisten Studierenden, dass dies zutreffe. Generell wurde der hohe Praxisbezug als positives Element hervorgehoben.

Die Zielerreichung der weitergehenden Lehr- und Lernziele bedarf einer differenzierten Betrachtung. In den PBL-Übungen zeigten die Studierenden, dass sie sich auf fachübergreifende Themen einlassen und sich sehr konstruktiv um Lösungsansätze bemühen. Die Prüfungsfälle wurden von allen Gruppen mit großem Engagement bearbeitet, hervorragend aufbereitet und erfolgreich präsentiert. Die persönlichen Ausarbeitungen der Lernfragen des Prüfungsfalles zeigten eine breite Spanne an individuellen Leistungen. Generell wird das Lernen anhand realitätsnaher Problemstellungen als sehr motivierend empfunden, ebenso wie das eigenständige Erarbeiten eines Stoffes bzw. Lehrinhalts. Insgesamt wird die PBL-Methode positiv bewertet und die Mehrheit wünscht sich mehr Lehrveranstaltungen in dieser Form. Für manche Studierende war jedoch die Bearbeitung der Fälle teilweise mit einer gewissen Verunsicherung verbunden. Trotz einer systematischen Einführung der PBL-Methode schien es für diese Studierenden schwer zu sein, die Fallbeschreibungen als exemplarischen Lernanlass zu sehen, d.h. zu akzeptieren, dass die Lösung des Falles nicht das eigentliche Ziel ist, sondern der Bearbeitungs- und Lernprozess. Die Haltung dieser Studierenden war geprägt von der Vorstellung, dass es „eigentlich" eine eindeutig definierte Aufgabenstellung geben müsse, zu der es auch nur die eine richtige Lösung geben könne. Diese Fokussierung auf die Lösung führte dementsprechend zu einer Erwartungshaltung, dass die Lehrenden alle für die Erarbeitung dieser einen richtigen Lösung notwendigen Informationen bereitstellen sollten und auch, dass es keine Vagheit in den Beschreibungen geben dürfe. Diese letztlich nicht erfüllte Erwartungshaltung führte bei diesen Studierenden zu einer gewissen Skepsis der Methode gegenüber. Es fiel diesen Studierenden schwer, sich auf eine offene Problemstellung einzulassen und unterschiedliche Herangehensweisen und Lösungsvarianten zu akzeptieren. Genau diese Bereitschaft aber charakterisiert eine Haltung, die, wie einleitend bereits erläutert, als unabdingbare Voraussetzung gelingender interdisziplinärer Zusammenarbeit gesehen wird. Die PBL-Methode fordert genau dies ein und ist dementsprechend in besonderer Weise geeignet, interdisziplinäre Fragestellungen zu behandeln. Insofern wird gerade in der Diskussion bzw. Reflexion dieser „Verunsicherung" ein Punkt erreicht, der die Studierenden herausfordert und zu einer persönlichen Entwicklung anregt. Im Rahmen der Veranstaltung wurden entsprechende Erfahrungsräume geschaffen und vielfältige Anstöße gegeben.

5 Ausblick für das Lehrveranstaltungsformat

Es hat sich gezeigt, dass die PBL-Methode sehr gut geeignet ist, die Studierenden in einen Dialog zu bringen. Aufgrund der Fallkonstruktion, die immer Aspekte unterschiedlicher Disziplinen enthielt, waren die Studierenden herausgefordert, Themen differenzierter zu betrachten, die nicht aus ihren originären Fachgebieten stammen. Die strukturierte Vorgehensweise, die die PBL-Methode vorgibt, hat dabei einen formalen Rahmen geschaffen, dies gut zu bewältigen. Gleichzeitig bergen die interdisziplinären, inhaltlich breit gefächerten Fälle die Gefahr einer Überforderung. Daher ist weiterhin besonders sorgfältig darauf zu achten, dass die Fälle hinsichtlich ihrer Komplexität und der fachlichen Anforderungen für die Gruppen gut bearbeitbar bleiben. Wichtig ist auch, ein inhaltliches Feedback zu geben und die Reflexion der Fallbearbeitung intensiv zu begleiten, sodass die Studierenden lernen, ihre eigene Arbeitsleistung einzuschätzen. Die Leitlinien der Begleitung durch Lehrende sollten weiter konkretisiert werden, um eine kontinuierlich gute Balance von Anleitung und Eigenständigkeit zu finden. Dies ist insbesondere für Studierende wichtig, die noch wenig Erfahrung mit offenen Lehr- und Lernformen haben.

Aus der Evaluation wurde deutlich, dass die hauptsächliche Motivation, eine solche Lehrveranstaltung zu besuchen, das Interesse an der behandelten Thematik ist. Daher sollten in der Zukunft solche Themen behandelt werden, die noch stärker in einer aktuellen gesellschaftlichen Diskussion sind, um für den nächsten Durchlauf mehr Studierende aus geistes- und sozialwissenschaftlichen Studiengängen für die Lehrveranstaltung zu gewinnen.

Literaturverzeichnis

1. M. Jungert, „Was zwischen wem und warum eigentlich?“: Grundsätzliche Fragen der Interdisziplinarität. In: *Interdisziplinarität*, ed. by M. Jungert, E. Romfeld, T. Sukopp, U. Voigt, WBG – Wissenschaftliche Buchgesellschaft, Darmstadt, 2010, pp. 1–12
2. G. Vollmer, Interdisziplinarität – unerlässlich, aber leider unmöglich? In: *Interdisziplinarität*, ed. by M. Jungert, E. Romfeld, T. Sukopp, U. Voigt, WBG – Wissenschaftliche Buchgesellschaft, Darmstadt, 2010, pp. 47–75
3. B. Engler, Eine Geisteshaltung, aber kein Fetisch. In: *Attempo! Forum der Universität Tübingen*, Tübingen, 2011, pp. 6–7
4. A. Siemeyer, Aktivierung von Studierenden durch problemorientiertes Lernen, 2016. https://dbs-lin.ruhr-uni-bochum.de/lehreladen/lehrformate-methoden/problemorientiertes-lernen/aktivierung-von-studierenden-durch-problemorientiertes-lernen/
5. A. Weber, Problem-Based Learning – eine Lehr- und Lernform gehirngerechter und problemorientierter Didaktik. In: *Problembasiertes Lernen: Konzepte, Werkzeuge und Fallbeispiele aus dem deutschsprachigen Raum*, ed. by J. Zumbach, A. Weber, G. Olsowski, Bern, 2007, pp. 15–32
6. G.F. Becker, V. Friske, Problemorientiertes Lehren und Lernen in der Betriebswirtschaftslehre: Entwicklung eines Moduls. In: *Problem Based Learning im Dialog*, ed. by M. Mair, G. Brezowar, Olsowski, G., Zumbach, J., Wien, 2012, pp. 85–97

Begriffsklärung zur Kompetenzorientierung

Thorsten Jungmann, Philipp Ossenberg and Sarah Wissemann

1 Einleitung

Die Forderung nach Kompetenzorientierung in Lehr- und Lernprozessen steht bei der Entwicklung und Akkreditierung von Studiengängen gleich welcher Art und Fachrichtung ganz weit oben auf der Agenda. Auch die Prüfungen sollen sich an den Kompetenzen orientieren, die im jeweiligen Modul in der Liste der angestrebten Lernergebnisse stehen. Leider wird der Begriff Kompetenz wenig einheitlich verwendet. Es besteht ein lebhafter theoretischer Diskurs um die Modellierung, Erfassung bzw. Messung von Kompetenzen, an dem sich die verschiedenen wissenschaftlichen Disziplinen beteiligen. Einflüsse aus dem europäischen und internationalen Diskurs tragen zur weiteren Diversifizierung der Begriffswelt bei. In der Folge sind die für die Studiengangentwicklung zentralen Begriffe wie Kompetenzen, Wissen, Kenntnisse (Knowledge), Fertigkeiten (Skills) und Fähigkeiten (Competences) uneinheitlich definiert. Dies irritiert und erschwert die Arbeit derjenigen, die z. B. natur- oder ingenieurwissenschaftliche Studiengänge entwickeln und nicht über das erziehungswissenschaftliche, psychologische, neurowissenschaftliche und soziologische Hintergrundwissen verfügen, das notwendig wäre, um die verschiedenen theoretischen Sichtweisen mitsamt ihren begrifflichen Abgrenzungen und Feinheiten nutzbringend in die Studiengangentwicklung einfließen lassen zu können.

Ziel dieses Beitrags ist, die stringente Verwendung des Kompetenzbegriffs in der Studiengangentwicklung zu ermöglichen. Dazu werden die verschiedenen Stränge des theoretischen Diskurses aufgegriffen und im Kontext der Studiengangentwicklung verortet. Anschließend wird ein Modell vorgestellt, das die stringente Verwendung des Kompetenzbegriffs im Kontext des Lehrens, Lernens und Prüfens an Hochschulen ermöglicht.

T. Jungmann (✉) · P. Ossenberg · S. Wissemann
TU Dortmund University, Jungmann Institut – Besser lehren, besser lernen.
Professor für Ingenieurwesen Direktor der FOM School of Engineering, Mitglied der Forschungsgruppe Ingenieurdidaktik, Dortmund, Germany
e-mail: tj@jungmann-institut.de

Originally published in "Die Neue Hochschule", © 2014.
Reprint by Springer International Publishing AG 2016,
DOI 10.1007/978-3-319-46916-4_70

2 Constructive Alignment als Gestaltungskonzept der Studiengangentwicklung

In der Studiengangentwicklung, insbesondere bei der kompetenzorientierten Gestaltung des Lehrens, Lernens und Prüfens, hat sich das Constructive Alignment als Gestaltungskonzept bewährt. Das von [1] beschriebene Vorgehen basiert auf der Ausrichtung der Prüfungsform und des Lehr-Lern-Prozesses an den beabsichtigten Lernergebnissen (intended learning outcomes). Das Constructive Alignment eignet sich für alle Ebenen der Studiengangentwicklung: von einzelnen Vorlesungsterminen über Module, ganze Studiengänge bis hin zu Studiengangfamilien als profilgebende Einheit von Fachbereichen und Fakultäten. Am Beginn der Entwicklung steht die Frage nach den Zielen: Wo soll es hingehen? Wozu sollen die Studierenden befähigt werden? Es schließt sich die Frage an, wie der Grad der Zielerreichung gemessen werden kann: Was können die Studierenden? In welchem Maße werden die beabsichtigten Lernergebnisse erreicht? Schließlich ist die Frage nach Didaktik und Methodik zu stellen: Auf welchem Weg wird die Lehrperson die Studierenden an das Ziel führen? Wie ist der Lehr-Lern-Prozess gestaltet? Welche Aktivitäten unternimmt die Lehrperson, um die Lernprozesse der Studierenden zu steuern und sie auf dem Weg zum Ziel zu unterstützen?

Am Beispiel eines Grundlagenmoduls im Studiengang Elektrotechnik besteht der erste Planungsschritt darin, die Kompetenzen zu definieren, über welche die Studierenden nach dem Besuch der zu diesem Modul gehörigen Lehrveranstaltungen verfügen sollen. Welche Kompetenzen das sind, richtet sich dem Constructive Alignment zufolge danach, welche Kompetenzen als Eingangsvoraussetzung für die nachfolgenden Module erforderlich sind. In der Konsequenz richten sich auch die Ziele der Module in der Studieneingangsphase an den Studiengangzielen aus. Soll das Studium beispielsweise dazu befähigen, wirtschaftliche und nachhaltige Lösungen für die Speicherung und bedarfsgerechte Verteilung von elektrischer Energie zu entwickeln, so gehört es zu den ersten beabsichtigten Lernergebnissen, die Verluste bei der Übertragung elektrischer Energie berechnen zu können. Die Feststellung, ob die beabsichtigten Lernergebnisse erreicht wurden, fällt umso leichter, je stärker bei der Definition der Ziele darauf geachtet wurde, sie spezifisch und messbar (operationalisierbar) zu formulieren. Es ist beispielsweise mit einer gestellten Rechenaufgabe einfach zu überprüfen, ob die Studierenden die Verlustleistung bei der Übertragung elektrischer Energie über eine Leitung bei gegebener Länge, gegebenem Querschnitt und Werkstoff berechnen können. Schwieriger ist festzustellen, ob sie über „grundlegendes Verständnis des spezifischen Leiterwiderstandes und dessen Wirkung auf die Wirtschaftlichkeit" verfügen. Das Beispiel verdeutlicht, dass durch die gewissenhafte kompetenzorientierte Formulierung von Modulzielen der Aufwand für die Entwicklung geeigneter Prüfungsaufgaben deutlich reduziert werden kann. Ist die zu den Zielen passende Prüfungsform definiert, kann der Lehr-Lern-Prozess so gestaltet werden, dass die erforderlichen Kompetenzen entwickelt werden können. Im Beispiel wäre dies durch die Herleitung und Einführung der entsprechenden Gleichung und das Üben verschiedener Anwendungsfälle zu erreichen.

Kompetenzen stehen im Mittelpunkt aller drei hier skizzierten Planungsschritte. Im Folgenden wird der Kompetenzbegriff zunächst aus verschiedenen theoretischen Perspektiven beleuchtet, um anschließend eine stringente begriffliche Basis für die weitere Diskussion über kompetenzorientiertes Lehren, Lernen und Prüfen zu bilden.

3 Kompetenzbegriff im theoretischen Diskurs

Van der Blij [2] definiert Kompetenz als die Fähigkeit, in einem gegebenen Kontext verantwortlich und angemessen zu handeln und dabei komplexes Wissen, Fertigkeiten und Einstellungen zu integrieren. Nach Weinert [3] sind Kompetenzen „die bei Individuen verfügbaren oder erlernbaren kognitiven Fähigkeiten und Fertigkeiten, bestimmte Probleme zu lösen, sowie die damit verbundenen motivationalen, volitionalen und sozialen Bereitschaften und Fähigkeiten, die Problemlösung in variablen Situationen erfolgreich und verantwortungsvoll nutzen zu können". Walzik [4] analysiert Weinerts Definition und extrahiert drei Dimensionen von Kompetenz: „Wissen („Fähigkeiten") ist kognitives Handeln. Es umfasst das Erkennen, Wissen und Verstehen (...). Werte und Einstellungen („Bereitschaften", „verantwortungsvoll") sind Grundhaltungen gegenüber Dingen, Situationen und Beziehungen zu sich und anderen Personen (...). Fertigkeiten zielen auf das handhabend-gestaltende Wirken im Umgang mit Dingen, anderen Personen und der eigenen Person ab."

Zusätzliche Unschärfe entsteht in Verbindung mit den Definitionen des Europäischen Qualifikationsrahmens für Lebenslanges Lernen (EQR). Im EQR ([5]) sind die Begriffe Kenntnisse und Fertigkeiten ergänzend als Grundlage für Kompetenz definiert: Kenntnisse sind „das Ergebnis der Verarbeitung von Information durch Lernen" und „bezeichnen die Gesamtheit der Fakten, Grundsätze, Theorien und Praxis". Darauf aufbauend sind Fertigkeiten „die Fähigkeit, Kenntnisse anzuwenden und Know-how einzusetzen, um Aufgaben auszuführen und Probleme zu lösen". Auf dieser Grundlage ist Kompetenz im EQR definiert als „die nachgewiesene Fähigkeit, Kenntnisse, Fertigkeiten sowie persönliche, soziale und methodische Fähigkeiten in Arbeits- oder Lernsituationen und für die berufliche und/oder persönliche Entwicklung zu nutzen". Am deutlichsten zeigt sich die begriffliche Unschärfe im Verständnis von Fertigkeiten. Ist eine Fertigkeit bei Walzik ein motorisches Handeln, das zwingend Aktionen der Extremitäten erfordert, kann eine Fertigkeit im EQR rein geistiger Natur sein, z. B. das Lösen einer Gleichung. Auch wird der Begriff Fähigkeiten teils mit Kompetenz gleichgesetzt, teils aber als ein Bestandteil von Kompetenz betrachtet. Neben der zuvor beschriebenen Differenzierung spannen zahlreiche Autoren (im Detail durchaus abweichend voneinander) die vier Kompetenzfelder Fach-, Methoden-, Sozial- und Selbstkompetenzen auf. Wildt [6] ergänzt noch die System- und die Organisationskompetenz. Regelmäßig wird die Fachkompetenz von den übrigen Kompetenzen abgegrenzt, die dann als Schlüsselkompetenzen oder fachübergreifende Kompetenzen zusammengefasst werden.

Abb. 1 xxx

Für das kompetenzorientierte Prüfen stellt sich die besondere Herausforderung, dass Kompetenz an sich nicht beobachtbar ist (vgl. [4] i. A. a. Chomsky). Um auf das Vorhandensein von Kompetenz zu schließen, muss Handeln sichtbar werden (Performanz). Performanz bezeichnet die „Art und Weise, in der sich ein Mensch angesichts einer konkreten Aufgabenstellung verhält und Lösungen sucht ..." ([7]). Unter Performanz wird ergänzend zum Begriff Kompetenz der aktuelle Einsatz von Fähigkeiten in bestimmten Situationen verstanden (vgl. [8]). Die Kompetenz von Studierenden zeigt sich in der Regel in der Prüfungsleistung. Der Blick auf Abbildung 1 verdeutlicht, dass die Verwendung des Kompetenzbegriffes in der Literatur noch vielfältiger und uneinheitlicher ist, als die hier skizzierten Perspektiven es vermuten lassen. Wie kann dieser Diskurs, der differenziert aus verschiedenen fachlichen Perspektiven geführt wird, in ein für die Studiengangentwicklung nutzbringendes Modell überführt werden?

4 Basismodell des Kompetenzbegriffes

Das Basismodell entsteht durch Abstraktion und Idealisierung. Demzufolge liegt der Fokus nicht auf inhaltlich vollständiger Repräsentation einzelner Theorien, sondern auf den für die Studiengangentwicklung bedeutsamsten Gemeinsamkeiten und Zusammenhängen der vorstehend skizzierten Definitionen und Theorien. Um die stringente Verwendung der Begriffe zu ermöglichen, werden Studiengänge, Module und Lehrveranstaltungen im Basismodell auf Lehr-Lern- und Prüfungsprozesse heruntergebrochen. Bei entsprechender Detaillierung können damit alle Lehr-Lern- und Prüfungsformen modelliert werden. Abbildung 2 veranschaulicht die Elemente dieses Modells und deren Zusammenhang, der im Folgenden beschrieben wird.

Im Modell tritt die Kompetenz in mehreren Stadien in Erscheinung. Sie setzt sich aus einzelnen, unterschiedlich weit entwickelten Fähigkeiten zusammen, die in ihrem Zusammenwirken die Kompetenz ausmachen. Fähigkeiten beschreiben, zu was eine Person in der Lage ist, also was jemand (tun) kann. Beim Einsatz

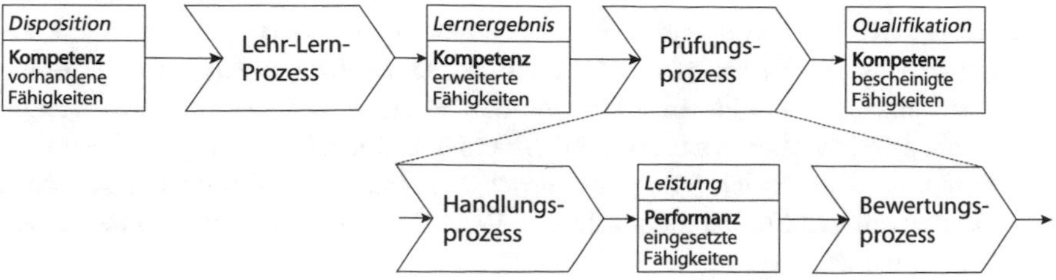

Abb. 2 xxx

der Fähigkeiten wirken Kenntnisse (Wissen) und Fertigkeiten (Können) zusammen. Kenntnisse beschreiben, was eine Person gelernt hat, also was sie weiß. Unter Fertigkeiten ist angewandtes Wissen zu verstehen, also Wissen, das sich in der Handlung zeigt. Hierbei ist sowohl kognitives als auch motorisches Handeln gemeint. Um die Stadien von Kompetenz zu unterscheiden, werden eindeutige Begriffe verwendet. Die Sammlung von Fähigkeiten, die Studierende in einen Lehr-Lern-Prozess einbringen wird als Disposition bezeichnet. Die erweiterte Sammlung von Fähigkeiten, die sich als Ergebnis des Lehr-Lern-Prozesses entwickelt, wird Lernergebnis genannt. Als Qualifikation wird im Modell die Sammlung von Fähigkeiten verstanden, die infolge einer bestandenen Prüfung bescheinigt werden. Im Handlungsprozess von Prüfungen setzen Studierende ihre Fähigkeiten ein. Es entsteht beobachtbares Verhalten. Im Prozess der Bewertung werden auf Grundlage der Leistung (Performanz) Rückschlüsse auf die Kompetenz gezogen, die an sich nicht sichtbar ist. Durch die Verwendung der vier Begriffe Disposition, Lernergebnis, Leistung und Qualifikation wird die Unterscheidung der verschiedenen Stadien der Kompetenz erst möglich. Durch die prozessorientierte Modellierung kann verdeutlicht werden, dass die Lernergebnisse eines vorgelagerten Lehr-Lern-Prozesses zentraler Bestandteil der Disposition zur Teilnahme an nachgelagerten Lehr-Lern-Prozessen sind. Zur Disposition zählen neben der Kompetenz auch Werte und Einstellungen und in besonderem Maße die Motivation.

5 Zusammenfassung

Im Studium sollen Studierende die Fähigkeiten entwickeln, die sie zur professionellen und kompetenten Teilhabe am Berufsleben benötigen. Um in der Studiengangentwicklung sicherzustellen, dass dieses Ziel erreicht wird, ist ein klarer Blick auf das kompetenzorientierte Lehren, Lernen und Prüfen von großer Bedeutung.

In diesem Beitrag werden die Begriffe, die in der einschlägigen Literatur wenig einheitlich verwendet werden, geklärt und geordnet. Das hier vorgestellte Basismodell ermöglicht die stringente Verwendung der Begriffe in der Debatte um die Kompetenzorientierung im Studium sowie bei der Entwicklung von Studiengängen.

Aspekte, die im vorliegenden Text nicht thematisiert werden, sind die Möglichkeiten zur Erfassung und Messung von Kompetenz, ebenso wie die Frage, wie die einzelnen Kompetenzfelder in der Planung von Lehrveranstaltungen metho-disch und didaktisch berücksichtigt werden können. Diese Fragen können idealerweise unter Berücksichtigung des curricularen und organisatorischen Rahmens sowie der jeweiligen Studierendenschaft in Workshops aufgegriffen und individuell beantwortet werden.

Literaturverzeichnis

1. J. Biggs, C. Tang, *Teaching for quality learning at university. What the student does*. SRHE and Open University Press, 2011
2. M. van der Blij, *Competentieprofielen over schillen en knoppen*. Stichting Digitale Universiteit, Utrecht, 2003
3. F.E. Weinert, *Leistungsmessung in Schulen*. Beltz, 2001
4. S. Walzik, *Kompetenzorientiert prüfen. Leistungsbewertungan der Hochschule in Theorieund Praxis*. Verlag Barbara Budrich, 2012
5. Europäisches Parlament, Europäischer Rat, Empfehlung des europäischen Parlaments und des Rates vom 23. April 2008 zur Einrichtung des europäischen Qualifikationsrahmens für lebenslanges Lernen. In: *Amtsblatt der Europäischen Union 51*, 2008, pp. 1–7
6. J. Wildt, Kompetenzen als "Learning Outcome". Journal Hochschuldidaktik **17** (1), 2006, pp. 6–9
7. H. Schaub, K.G. Zenke, *Wörterbuch Pädagogik*. Dt. Taschenbuch-Verlag, München, 2007
8. H.E. Tenorth, R. Tippelt, eds., *BELTZLexikon Pädagogik*. Beltz, Weinheim:, 2007

Research Workshop in Engineering Education – Draft of New Learning

Thorsten Jungmann and Philipp Ossenberg

Abstract At the beginning of their studies, first-year students learn basics in mathematics, physics and chemistry disintegrated from each other. Graduated engineers should be problem solvers and scientists, which demands for abilities that students can hardly learn in lectures. Recognizing the need of a learning format that covers these demands we developed a research workshop for students in engineering science based on previous work of Wildt, Schneider and Jungmann. With this development we implement research-based learning into engineering courses by synchronizing the students' learning process with a typical research process.

Keywords Engineering Education · Research Based Learning

1 Introduction

In this paper we describe a research workshop designed for students of engineering science. In our concept a research workshop is both a room and a combination of learning formats. We described the research workshop in an earlier paper as room which is "open for students to conduct research on their own projects, learning about the research process and make small experiments" [1]. We believe that students in engineering science should participate in research processes right from the beginning of their studies, especially in study programs which aim at educating engineering scientists rather than engineers.

2 Educational Design – Theoretical Perspective

In this chapter we describe the educational design. It is based on our educational belief, that it is our professional responsibility to make a sustainable and substantial contribution to our students' professional and personal development. The educational

T. Jungmann (✉) · P. Ossenberg
Center for Higher Education Dortmund, TU Dortmund University, Dortmund, Germany
e-mail: thorsten.jungmann@tu-dortmund.de

Originally published in "Proceedings of the 2014 IEEE Global Engineering Education Conference (EDUCON). "Engineering Education towards Openness and Sustainability", © IEEE 2014. Reprint by Springer International Publishing AG 2016, DOI 10.1007/978-3-319-46916-4_71 design is based on the principle of research-based learning. First we show how we systematically match the learning process with the research process of the respected subject. Second we present the peer-learning concept, which we applied to implement research-based learning into our learning format.

2.1 Research-Based Learning

Research-based learning as an educational concept in Higher Education goes back to the 1970s where the Conference of German Research Assistants postulated research-based learning as part of a scientific education performed by scientists in a science for a science-oriented profession [2]. Campbell [3] identifies practicing research in the mode of research-based learning as one of five "Engineering Education Themes" which are important for engineering education, especially in interdisciplinary subjects. According to Trempp [4] research-based learning means not only enabling student insight through research, but integrating students into the research process. Research-based learning is operationalized by formatting the learning processes in accordance with the research process in the respective subject, in other words through the synchronization of learning and research processes. Research-based learning in engineering education means, for example, defining the problem, elaborating the state of the art, developing the methodical design, preparing a model, carrying out the experiment, interpreting the findings in the context of methods and theory, and publishing the results. Students participate in selected phases or in the whole process. This selectiveness in implementation of research-based learning allows to adjust the didactical design according to the intended outcomes [5, 6].

Matching learning and research processes requires that both learning and research are modeled in circles that can be synchronized. We use Kolb's circle of experiential learning as an adequate model to describe the learning process [7]. The learning cycle (see Fig. 1) shows that learning takes place between feeling and thinking on the perception continuum, and between observing and doing on the processing continuum. The four steps of learning are consequently (1) concrete experience, (2) reflective observation, (3) abstract conceptualization, and (4) active experimentation. The research process can be modelled as shown in Fig. 2, according to Jungmann [6]. It starts with (1) a need or a knowledge gap (that someone perceives), followed by (2) a question or a first problem definition. Then – typical step in engineering – there is (3) a check of the same problem has been resolved before. If yes, the found method is applied to the problem, which leads to engineering as applied science. If no, the process continues and results in engineering science. (4) the objective is stated, (5) the problem is specified the systems limits are defined. (6) Theory is consulted before (7) the research questions are fine-tuned. Then (8) the research method is built up in dependence of the research questions. (9) The research is carried out, (10) results are interpreted and (11) implemented into practice and/or theory. If the findings are implemented into a new system or toolkit, this is (12) applied and used in technical practice. Once there is a new question or unclarity the cycle restarts.

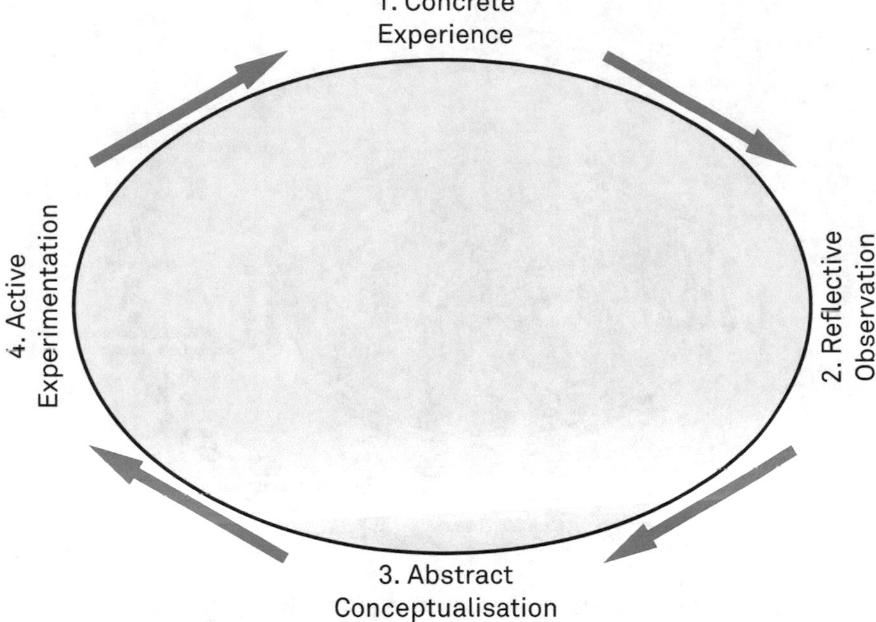

Fig. 1 Kolb's Learning Cycle

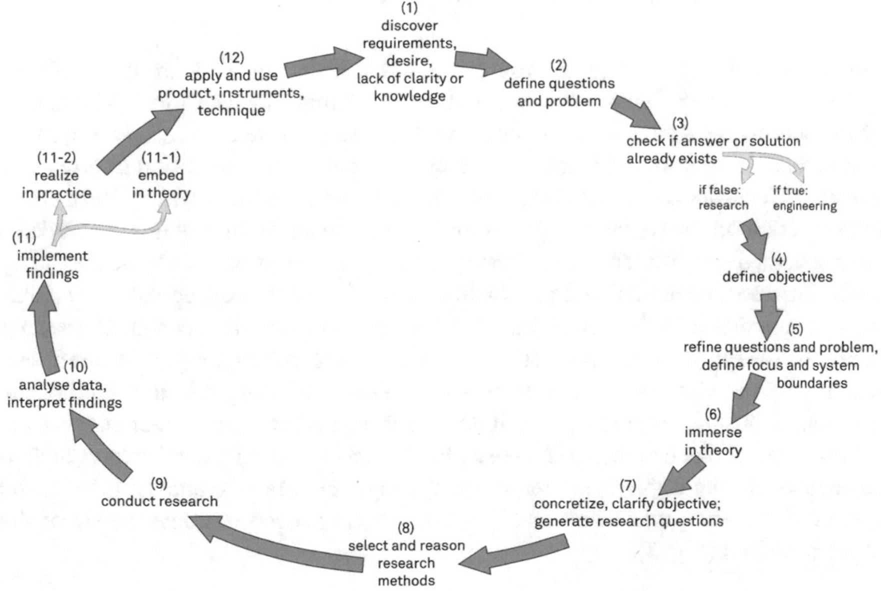

Fig. 2 Jungmann's Research Cycle [6]

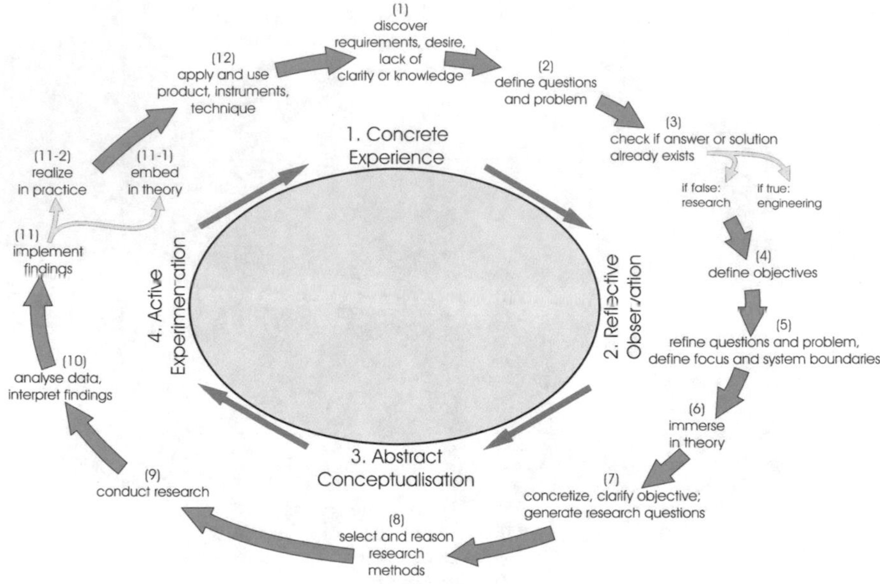

Fig. 3 Jungmann's Research Cycle synchronized with Kolb's Learning Cycle [6]

2.2 *Sychronizing Learning and Research*

We implement our concept of research-based learning by synchronizing Kolb's Learning Cycle with Jungmann's research cycle. In this way we follow Wildt's idea of synchronizing Kolb's learning cycle with a discipline-related research cycle to implement research-based learning. The cycle begins with perception of a need or question. Curiosity can be a driving aspect in both, research and learning. With growing reflection on the research object thinking supersedes feeling. After the students have specified the problem and set the system limits, the problem can be abstractly conceptualized and modelled. Immerse into theory fosters the conceptual understanding and supports both the development of research questions and design of research methods. Carrying out the projected research leads to activity, e.g. an experiment, which puts the abstract concept to the test. When analyzing and interpreting the resulting data students find out whether the subsequent concrete experience on the research object can be anticipated based on the abstract concept. Mistakes come to appearance during implementation of the findings into practice and/ or theory. The cycle restarts when unclarities and questions are perceived in consequence of the implementation (Fig. 3).

3 Research Workshop for Future Engineers – Practical Perspective

In this section we show how we put the theoretical idea into practice. We therefor describe the layout of our research workshop and the various learning formats.

3.1 Room

The layout of the research workshop is shown in Fig. 4. The room is divided into two sections by four rollable office cupboards (2). The front part is set-up for lecturing. It is equipped with a projector and a projection screen. There are four magnetic whiteboards (7) and a classical blackboard. A round table adjustable in height (5) makes it comfortable to present in front of the plenum. In the rear part students can work on exercises, develop ideas and learn in smaller groups. They can use flipcharts (1), pin boards (4), whiteboards (7) and professional facilitation equipment, which are stored in the office cupboards (2). There are also fifteen laptops for students use (10). Students find literature about research methods, presentation techniques and about scientific writing in book shelves (8). Last but not least students there are experimentation kits, which we use in tutored experimentation sessions.

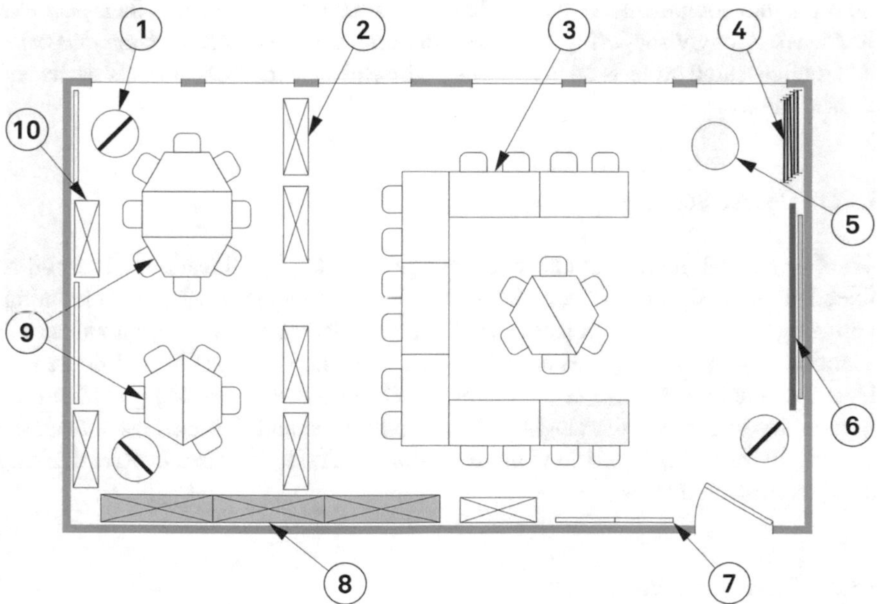

Fig. 4 Layout of the research workshop at TU Dortmund University

3.2 Organisation

The research workshop was initiated as a part of ELLI, a project funded by the German Federal Ministry of Education and Research (BMBF) aiming at improvement of teaching quality and study conditions in Engineering Education. The team consists of senior and junior scientists, who prepare the concepts and support the work, and students. These are trained to facilitate the other students learning processes in a peer-learning concept tailored to the needs of research-based learning.

3.3 Learning Formats

The equipment and staff, which we described in the sections above, support the students' learning and research processes in three different learning formats: opening hours, course sessions, and FLExperiments.

3.3.1 Opening Hours

During opening hours students can come to the workshop individually or in small groups to consult tutors, ask questions on problems in the fields of finding research questions, scientific writing, presenting data, using literature, etc. They can also use the room and equipment to work on their own research projects, e.g. their bachelor thesis. We strongly support the idea that students use the room to create and carry out small research projects on their owns in the earlier semesters, not only at the end of their studies.

3.3.2 Course Sessions

We offer several courses to enhance the student's key qualifications. The course objectives are designed to give students effective and efficient strategies for learning, managing their time and projects, preparing and giving decent presentations, and scientific writing. The courses are offered as workshop sessions from 3 hours to 2 days. The methods we use in the courses activate the students and give them the opportunity to practice the taught techniques to an extent that enables substantial learning success. The idea behind these sessions is also to initiate reflective thinking about the session themes.

3.3.3 FLExperiments

FLExperiment is an innovative learning format aiming at the students' ability to plan and perform experiments. We use professional experimentation kits to make models

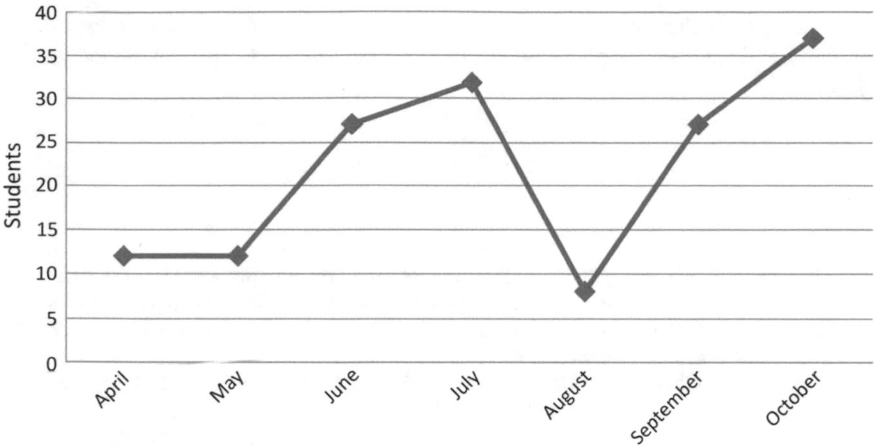

Fig. 5 Students per month

of real technical systems, which elucidate the relationship between technical models and real-world systems. FLExperiments are currently offered in addition to lectures, for example, in Physics and Technical Mechanics. With this new format we created a way of informal learning that is pending to be integrated into the curricula of the engineering courses [1, 8].

4 How do the Students Accept the Offer? – The Response

First we described the 'opening hours'. Figure 5 shows how many students have visited the research workshops from March to November 2013. At the beginning of our records only 12 students per month visited the opening hours. We ascribe the raise in June to an e-mail newsletter we send to our target group in May. After the downswing in August we started a campaign with flyers and another newsletter. In September and October the amount of visitors was increasing again. We assume that the fresh students make use of our offer to a greater extent than those from the higher semesters. They mainly use the research workshop to learn in groups and get into touch with engineering methods. In so far our goal of improving the study conditions is reached. The most frequented offers are the course sessions. Students evaluate the sessions very well.

5 Conclusion and Outlook

Our evaluation shows that the offers of the research workshop are well-accepted among students. What is still missing is acceptance by the engineering faculties. Currently, students carry out FLExperiments as an extracurricular voluntary activity.

Here there is further potential for curricular integration of the research workshop, which would enable us to give students credit points for taking part in the workshops offers.

References

1. P. Ossenberg, T. Jungmann, Flexperiment: Solving small technical problems by using mathematical and physical knowledge, skills and competencies in engnineering education. In: *Proceedings of the 2013 IEEE Global Engineering Education Conference (EDUCON)*. 2013, pp. 597–601
2. Forschendes lernen - wissenschaftliches prüfen
3. M.E. Campbell, Oh, know i got it! Journal of Engineering Education **88** (4), 1999, pp. 381–383. URL http://www.jee.org/1999/october/659.pdf
4. P. Trempp, Verknüpfung von lehre und forschung: Eine universitäre tradition als didaktische herausforderung. Beiträge zur Lehrerbildung **23** (3), 2005, pp. 339–348. URL http://www.afh.uzh.ch/schwerpunkte/universitaereDidaktik/BZL_2005_3_339-348.pdf
5. M. Healey, Linking research and teaching: Exploring disciplinary spaces and the role of inquiry-based learning. In: *Reshaping the university: New relationships between research, scholarship and teaching*, ed. by R. Barnett, McGraw-Hill/Open University Press, Maidenhead, 2005, pp. 67–78
6. J. Jungmann, Forschendes lernen im logistikstudium: Systematische entwicklung, implementierung und empirische evaluation eines hochschuldidaktischen modells am beispiel des projektmanagements. Dissertation, TU Dortmund, Herne, 2011
7. D.A. Kolb, *Experiential learning: experience as the source of learning and development*. Prentice Hall, Englewood Cliffs and NJ., 1984
8. P. Ossenberg, T. Jungmann, Experimentation in a research workshop: A peer-learning approach as a first step to scientific competence. Int. J. Eng. Ped **3** (S3), 2013, p. 27

ModellING Competences – Developing a Holistic Competence Model for Engineering Education

Dominik May and Philipp Ossendorf

Abstract The teachers' job at universities is to give students the chance to develop knowledge, skills and competences. In many cases this is summed up by a general discussion on competences. With this paper we combine various approaches to the term 'competence'. On the one hand we discuss competence development and integrate the discussion on taxonomies. On the other hand we present different alternatives to cluster the general term competence into different areas: for example a personal and a professional area. The combination of both approaches leads us to a newly and holistically defined competence model. With the help of this model it is possible to similarly describe existing courses with regard to the intended competence development and addressed competence areas. To verify our developed model we proof it by analyzing intended learning outcomes of a course and discuss the findings.

Keywords Engineering Education · Taxonimies · Competences · Competence Model

1 Introduction

The research presented in this paper is based on two initial questions:

- How can higher education courses be described with regards to the competence development and different areas?
- How can we make students able to work as professionals?

Our aim is to describe educational courses with regard to two different aspects. On the one hand we want to describe which competence areas in a course are in focus.

D. May (✉) · P. Ossendorf
Engineering Education Research Group, Center for Higher Education,
TU Dortmund University, Dortmund, Germany
e-mail: Dominik.may@tu-dortmund.de

Originally published in "Proceedings of 2014 International Conference on Interactive Collaborative Learning (ICL)", © IEEE 2014. Reprint by Springer International Publishing AG 2016, DOI 10.1007/978-3-319-46916-4_72

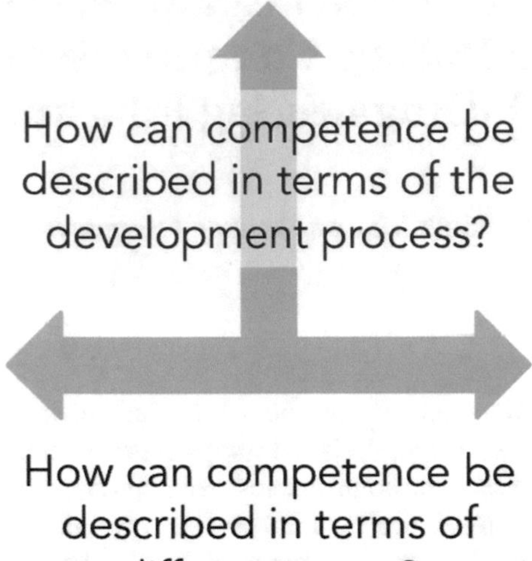

Fig. 1 Two dimensions of describing a course with regard to competence

On the other hand we want to describe which way the students have to go in order to develop the respective competences. In various disciplines exist different approaches to analyze and classify them. Hence, our focus lies on competence areas and the competence development process. Both perspectives can be visualized in one diagram by describing them with the use of horizontal and vertical dimensions (Fig. 1). Up to now these two dimensions of the term competence are more or less separated from each other and discoursed in two different lines of discussion. However in order to have a holistic view on courses and in order to fully describe these courses it is necessary to have both perspectives in mind. Therefore a competence model that integrates both approaches could be helpful.

2 Competence Development and Taxonomies

Before we look on various models to describe competences, we describe in this section how competences are developed. Therefore, we will have a look at two different perspectives. Firstly important terms will be explained and secondly different taxonomies will be displayed.

2.1 *The EQR Perspective*

The "European Qualification framework for Lifelong Learning" (EQR) is a tool to compare various European qualification levels [1]. With the "German Framework for

Lifelong Learning" (DQR) it has an equivalent for the German education system [2]. The idea underlying these frameworks is to define the different levels of education and training and make them comparable (from preschool to the doctoral degree). Each of these levels should be describable by knowledge, skills and competences. With this the EQR also includes an explicit competence model. However this model is less adequate for defining areas but it is very helpful to start into the discussion of competence development. Within the EQR explicit terms in context with the competence discussion are defined. These are the mentioned terms 'knowledge', 'skills' and 'competences'. These terms are of high importance for the understanding of competence development – means how competences are developed by individuals and what is needed as a basis. Even if these terms are frequently used in the discussion of education, there are uncountable definitions and explanations. For our work we will stick to the definition given by the European Parliament Council in context with the European Qualification Framework: "'knowledge' means the outcome of the assimilation of information through learning. Knowledge is the body of facts, principles, theories and practices that is related to a field of work or study. In the context of the European Qualifications Framework, knowledge is described as theoretical and/or factual; 'skills' means the ability to apply knowledge and use knowhow to complete tasks and solve problems. In the context of the European Qualifications Framework, skills are described as cognitive (involving the use of logical, intuitive and creative thinking) or practical (involving manual dexterity and the use of methods, materials, tools and instruments); 'competence' means the proven ability to use knowledge, skills and personal, social and/or methodological abilities, in work or study situations and in professional and personal development. In the context of the European Qualifications Framework, competence is described in terms of responsibility and autonomy" [1]. Taking these three definitions into account it becomes obvious that a competence development needs three different steps that are closely tied to each other. In order to describe skills it is necessary to use the term knowledge and further on it is necessary to use the terms knowledge and skills in order to describe competence. This observation will become important when we will be talking about competence development. A strong focus should be put on the last part of the "competence"-definition, that says "[…] competence is described in terms of responsibility and autonomy". Taking into account another description from van der Blij et al. [3], saying that competences is "the ability to act within a given context in a responsible and adequate way, while integrating complex knowledge, skills and attitudes" leads to the necessity to see competence within a situation. That means that competences necessarily have to be seen in strong connection to a given context or a situation that require a responsible and autonomous action. Only in such a situation it is possible to speak about competent action. However, we will come back to that later while describing competence development. Firstly another approach to competence development will be discussed.

2.2 *The Taxonomy Perspective*

We describe in this section the second approach we want to integrate in our holistic model with regard to competence development: Learning taxonomies. The term 'taxonomy' is borrowed from biology: "Taxonomy is the branch of biology that names and classifies species and groups them into broader categories" [4]. Taxonomies' main characteristic is that an upper category always includes all the bellows. Taxonomies in context of teaching and learning were introduced by the work of Bloom et al. in 1956 [5]. In this publication a group of examiners developed the first cognitive taxonomy to make exercises in exams comparable. This group also postulated the triad of the three different taxonomies: cognitive, affective and psychomotor. To integrate the taxonomy concept in our model we use Bloom's cognitive taxonomy revised by Anderson [6], Krathwohl's affective taxonomy [7] and Ferris's and Aziz's psychomotor extension of Bloom's taxonomy for our holistic model [8]. All three taxonomies consist of major types or levels and sub types or levels. We will focus on the major types/levels in order to keep the text from getting lost in details.

2.2.1 Cognitive Taxonomy by Anderson et al.

The most often cited cognitive taxonomy was published by Bloom et al. [4]. In 2001 it was replaced by a revised version, which addressed teachers to help them "clarify and communicate" what students should learn in their classes [6]. "In simplest terms, our revised framework is intended to help teachers teach, learners learn and assessors assess" [6]. To attain this a framework was developed divided in four dimensions of knowledge and six cognitive processes. This differentiation based on the assumption that an educational objective includes a cognitive process and a piece of knowledge.

Knowledge Dimension

Anderson et al. divide based on findings in the field of cognitive psychology that there are four knowledge dimensions: factual, conceptual, procedural and metacognitive knowledge. In their words: "Factual knowledge is knowledge of discrete, isolated content elements 'bits of information' (…). In contrast, Conceptual knowledge is knowledge of 'more complex, organized knowledge forms' (…). Procedural knowledge is 'knowledge of how to do something' (…). Finally, Metacognitive knowledge is 'knowledge about cognition in general as well as awareness of and knowledge about one's own cognition"' [6].

Cognitive Process Dimensions

Anderson et al. describe cognitive processes in six major types with all in all nineteen subtypes. The major types are: Remember, understand, apply, analyze, evaluate, and create. The cognitive process in which students "retrieve relevant knowledge from longterm memory" [6] is called Remember. If students are able to connect presented knowledge with existing concepts they reach the level Understand e.g. when students are "able to convert information from one representational form to another" [6]. If students are in a situation in which they have to "carry out or using a procedure" [6] the educational objective is part of the third major type called Apply. In this case students are able to apply procedures both to familiar and unfamiliar tasks. The cognitive process of Analyze enables students to split and examine complex issues in order to identify underlying principles. Subsequently the cognitive process Evaluate means making "judgments based on criteria and standards" [6]. The highest and sixth major type is called Create. "Objectives c1assified as Create have students make a new product by mentally reorganizing some elements or parts into a pattern or structure not c1early present before" [6].

2.2.2 Affektive Taxonomy by Krathwohl et al.

There are interactions between cognitive and affective processes so that students cannot act without the motivation to do it. Educational objectives presented by the affective taxonomy aim on "the development of interest or motivation" [7]. Finally students are able to build their own philosophy of life. The following taxonomy was published first by Krathwohl, Bloom and Masia in 1964. The affective domain starts with Receiving (Attending). Students on this level are inactive. The teacher's job is to draw students attentions to facts or phenomena [9]. At the second affective level, called Responding, students have to be active in terms of learning something. They direct their activities to facts or phenomena [9]. Valuing on the affective taxonomy's third level and means "that a thing, phenomenon, or behavior has worth" and is assessed [7]. This could be done by the students' individual process or they use "criterion of worth" received as a "social product" [7]. The level Organization and the process that organize values is important in situations in which multiple values are relevant. Three important steps need to be taken: (1) Creating a system of values by combining single values, (2) clarify the relationship of various values, and (3) prioritizing dominant values. The affective taxonomy's fifth level of is named Characterization by value or value complex. At this level the students' values "are organized into some kind of internally consistent system" [6]. In this context Krathwohl et al. caution that "formal education generally cannot reach this level (…)" [7].

2.2.3 Psychomotor Taxonomy by Ferris and Aziz

Taxonomies in the cognitive and affective domain described above are in most faculties generally accepted. In psychomotor domain exist many approaches to categorize levels of motor learning depending on field and audience. Ferris and Azis developed a taxonomy for educational objectives in engineering education for the psychomotor domain and published it in 2005 [8]. Students who reached the first level recognition of tools and materials are able to recognize names and safety information of tools as well as materials of their own profession. In various situation students are required to handle tools and or materials. On the second level handling tools and materials students learn methods for "holding, lifting, moving and setting down tools and materials" [10]. Building on this abilities students learn to "perform elementary tasks" [10] and reach the level basic operation of tools. Elementary in this case means that students indeed produce work pieces but with low complexity so that these can't be part of functional products. Students acquire the ability "to produce useful outcomes" by reaching the level competent operation of tools. Students are able to use a tool competent in other words "performing a range of tasks for which it was designed" [10]. On the Expert operation of tools level students "use tools to efficiently, effectively and safely perform tasks" [10]. Engineers are often faced with the task to create a work plan. Students who reached the planning work operations level are enabled to "to transform a product specification into the set of processes or tasks required to deliver the product or service" [10]. The highest level in this taxonomy is evaluation of outputs and planning means for improvement. "At this level the practitioner can look at a product and review it for quality of manufacture, identifying deficiencies and propose actions which would either correct or prevent the faults" [10].

2.3 *Summary and Evaluation for the Following*

So far we looked at the EQR's concept of competence development and different learning taxonomies to show how students' competence acquisition can be described. Both concepts use various levels, which have to be reached by the students stepbystep. The EQR carefully defines central terms and so it can be taken as an appropriate framework for the subsequent modeling. Learning taxonomies go more into detail, as they are appropriate for analyzing educational objectives. The cognitive and affective domains are accepted in various disciplines. The taxonomy in the psychomotor domain presented here is especially designed to formulate and analyze educational objectives in context of engineering education. The EQR's as well as the taxonomies' perspectives are compatible with each other. So we will combine them later on. In the following we will change the perspective from the vertical dimension to the horizontal dimension (Fig 1). By describing different competence models we will explain how competence can be clustered into different areas and which alternatives can be found in literature.

3 Competence Models

In literature can be found many different models describing competences. Most of them describe them by dividing the general view on competences in different clusters. For our discussion a short overview on these models is needed and will be given in the following.

3.1 *Threefield Matrix by Roth*

Especially in the discussion on vocational training a regularly mentioned model is the threefield model from Roth [11], who divides competences in the fields selfcompetence, socialcompetence, and subjectspecificcompetence as shown in Fig. 2 [12]. Selfcompetence in this context means the ability to act independently on basis of the own responsibility. Socialcompetence describes the ability to act, judge and take responsibility in social and political contexts. Finally subjectspecificcompetence is the ability to act and take responsibility in subject related situations. This model and most of the models based on Roth's definition include personal cognitive dispositions as well as affective and motivational aspects [12].

3.2 *Integrated 1x3x3model by the Standing Conference of the (German) Ministers of Education and Cultural Affairs*

Based on [11] other models have been developed in competence research. The standing conference of the (German) ministers of education and cultural affairs (KMK) regularly publishes recommendations for the development of framework curricula in vocational training. Even if these recommendation do not refer to higher education in the narrow sense the linked competence model is worthwhile to present it here. In former years the KMK defined the competence to act as the final aim for vocational training. This competence was divided on a lower level into professional competence, human competence and social competence. Based on these three competences the competence to learn and methodological competence can be developed [13]. Fifteen years later the KMK defined its understanding of competences slightly differently as

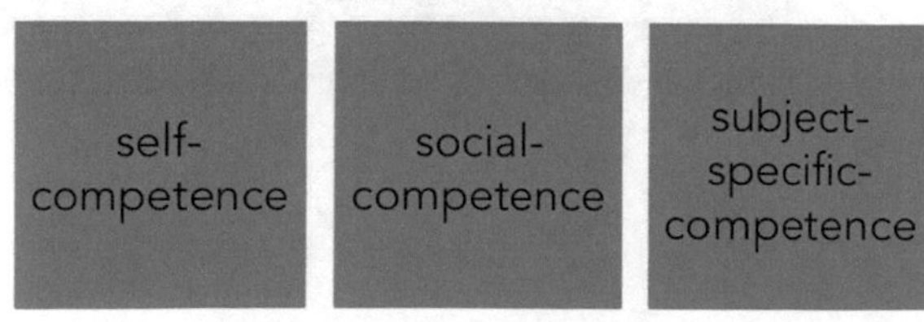

Fig. 2 Threefield competence matrix by Roth [11]

follows: The central aim of vocational school is to develop a comprehensive competence to act, which is defined as "the ability and the willingness of each individual to act adequately and […] responsibly in professional, social and private situations" [14]. Moreover the competence to act integrates professional competence, self competence and social competence. KMK defines the competences as follows [14]:

- Professional competence means willingness and ability to act on basis of professional knowledge and skills and solve tasks in a problemoriented, adequate, method driven as well as independent way and rate the result.
- Self competence describes (among others) the willingness and ability to rate personal chances, requirements as well as restrictions and develop the own talents and personal future plan. This competence includes autonomy, critical faculties, selfconfidence, reliability, sense of responsibility and sense of duty.
- Social competence means the willingness and ability to develop and live in social relationships, have a sense for sympathy and antipathy, and interact with others in a rational and responsible way.

Components of each of these three competences on a lower level are method competence (the willingness and ability to work on tasks and problems in a focused and planed way), the communication competence (the willingness and ability to understand and shape communicative situations by realizing, understanding and presenting the personal and the other side's intentions), and learning competence (the willingness and ability to understand, rate and structure new information and connections independently or in a team by developing learning strategies and use them for lifelong learning). This understanding of the relation between different areas of competence leads to the model presented in Fig. 3.

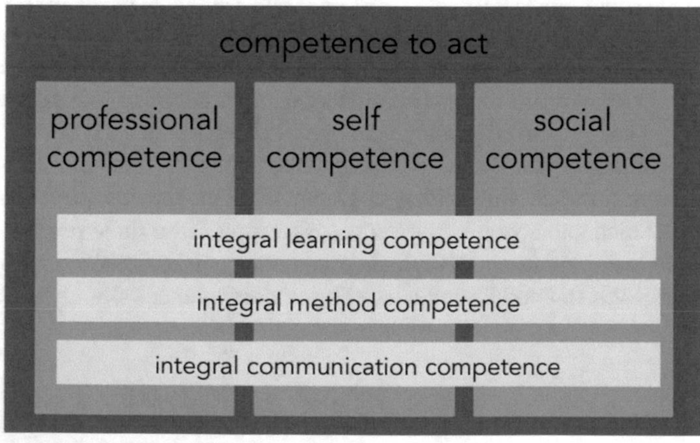

Fig. 3 Competence model based on the KMK [14]

3.3 *Fourfield Matrix by Wildt or Erpenbeck*

Other competence models do not see a threefield matrix but extend this view by adding a fourth field. Even these models differ slightly from each other (e.g. [15, 16]). Hence only two will be given in the following. Wildt defines the four fields as follows (own translation): professionalcompetence, socialcompetence, selfcompetence, and methodcompetence [16]. The integration of these four competences leads on a metalevel to personal action competence [16]. Even if this grid is closely related to [11] and an often used one, in this context it might to lead to a problem. Looking for example on engineering methods and its successful usage, it is not very clear in which competence field it belongs. Does the usage of professional engineering methods rather show professional competence than method competence or is the other way around the case? Because of this, a slightly different shaped grid may be more practical to use in this case. Erpenbeck et al. divide the different fields of competences as follows (see Fig. 4; [15]):

- Professional and method competences mean the ability to solve subjectspecific problems creatively by using subjectspecific knowledge, skills, tools and methods. To rate knowledge and to develop methods further are also aspects of these competences.
- Personal competences mean the ability to act in a selfreflective and independent way. It integrates aspects like being able to evaluate yourself and your own actions, to generate a productive attitude as well as values, to unfold own abilities, to develop the own person creatively, and to learn.
- Social and communication competences mean the ability to act in a cooperative and communicative way, which includes the capability to act within a team and develop plans, activities or solutions together.

Fig. 4 Fourfield competecne model by [15]

- Activity and implementation competences mean the ability to act actively and in an organized way in order to achieve goals and/or to put plans into action (on your own or in a team). Thus these competences describe the ability to perform successfully by integrating own emotions, motivation, skills and personal experience and all the other three competences.

3.4 Four Column Structure in the German Qualification Framework for Lifelong Learning

Whereas the EQR's approach to the competence definition is more related to the development based on the definition of the terms knowledge, skills and competences, the German equivalent (German Qualification Framework for Lifelong Learning; DQR) takes these three terms and puts them into a holistic model [2]. Even if the DQR's principal aim is not the understanding of competences but the ability to compare and rate different qualification levels the structure presented there includes a model, which should be included here. The DQR model is divided into four columns, which are divided into two different competence areas: Professional competence and personal competence. Professional competence in this context includes the knowledge and skills. Moreover it describes the willingness and the ability to independently work on tasks and problems in a professionally adequate and methoddriven way and rate the results. Hence knowledge and skills form the two columns of professional competence. By looking at the definition it becomes obvious that the wording is very similar to the one in II.B, which is not surprising as the work on the DQR is also supported by the KMK. Personal competence includes the two columns social competence and independence. It describes the willingness and the ability to evolve and independently as well as responsibly shape the own live with regard to the respective social, cultural and professional context. Again the similarities to the definition from III.B are easy to see. As the similarities are clear it is not necessary to define social

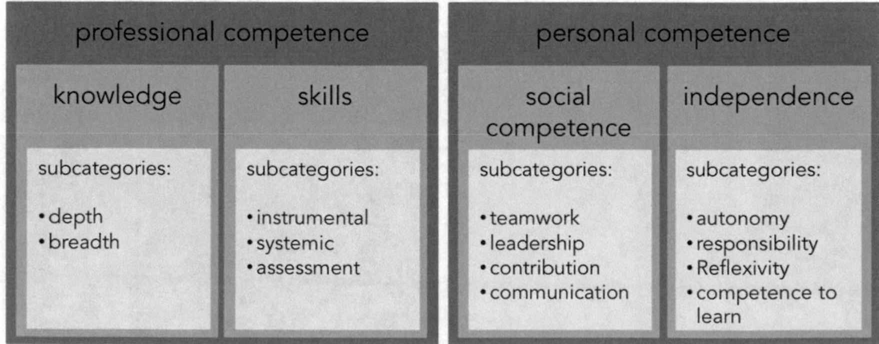

Fig. 5 Four competence columns in the DQR [2]′

competence and independence at this point again. Within these four columns the DQR differentiates into several subcategories in order to explain knowledge, skills, social competence, and independence more in detail. This leads to a general model with three levels building on each other (Fig. 5).

3.5 Summary and Evaluation for the Following

The comparison of several different models and approaches to categorize competence clearly shows that most of them have somc strong similarities but as well show differences. The main difference can be seen in the different number of competence areas shown by the different models. Whereas the last example from the DQR only includes two main areas the most if the others needed three or more areas in order to describe the general competence to act. However in this observation there can be seen a similarity, too. All of the models show clearly that there is strong need for clustering different competences to areas. It seems to be apparent that at least two main areas can be identified. On the one hand there is the area of professional competences, however they are constructed on a lower lever. This area is related to the profession, which means that differences in the competency requirements from profession to profession are mainly expressed in this area. On the other hand there can be identified the area of personal competences, which is not that much related to the profession. The focus in this context is in the individual and its connection to the environment. Even in the fourfield competence models this differentiation can be seen. Nevertheless all the competence areas in the different models are defined very broadly as they are usable for all different fields of specialization and serve as general concept. So it is necessary to fill the universal concept with concrete engineering content in order to make them usable for curriculum development and put a little bit more "meat on the bones".

4 Combining the two Approaches to a Holistic Model

In the last two sections different approaches to the term competence were presented. As these approaches do include different perspectives and different understandings of competences it is time to sum up and make all the prior explanations usable for the new model's development. Therefore we will refer back to the introduction and the twodimensional understanding of the term competence.

4.1 Connections to Existing Models

Firstly for the vertical dimension (competence development) the discussion from II will be used in order to integrate the process of competence development into the model. This makes clear that competence development heavily relies on knowledge and skills. Secondly in III the question regarding different competence areas was focused. That represents the horizontal dimension. The discussion led to the observation that in general two to four different areas of competences can be identified. However this is not enough for defining a new holistic competence model. Hence, we will borrow different aspects of the different approaches and views for the modeling process and combine them to a new model. In the following we will explain the taken issues:

1. Several models defined a general competence to act as the final aim. This broadly determined competence consists on an underlying structure of a set of other competences.
2. Even if several competence areas or clusters can be identified, the most obvious differentiation can be seen in contrasting competences with regard to the profession and competences with regard to the individual or it's interaction with others. Hence professional and personal competence areas can be defined. We will call these areas 'competence dimensions', as these are the most general clusters.
3. The personal competence dimension subsequently can be split into selfcompetence and social competence. Selfcompetence is called personal competence in other models (e.g. [15]) and tackles individual aspects of a person. Social competence puts the acting person into contact with others and therefore deals with communication and interaction in its different forms.
4. As indicated by the EQR competence development needs knowledge and skills. Hence we argue that talking about the models vertical perspective knowledge is the basis. Skills are built on that and competence finally integrates both.
5. This threestep structure reflects the taxonomy model, too. Each of the presented taxonomy does include different levels or types building on one another. That means that each level or type in each of the taxonomies can be assigned either to knowledge, skills or competences.

These five fundamental assumptions build the basis for our newly and holistically defined competence model. They were complemented by personal considerations, which sometimes enhance and sometimes even contradict existing models. Before we show the new model, and explain it more in detail a last note has to be made. Even if we use the DQR and the EQR for our considerations on the competence model, the documents' general idea and its aims were something different. They should represent a tool to compare several qualification levels. However these levels add a third dimension to our model. Whereas the Xaxis describes the competence areas and the Yaxis represents the competence development, the Zaxis describes the educational levels (see Fig. 6).

4.2 X-Axis: Competence Areas

All these considerations lead to the new holistic model in Fig. 6. As indicated above the general educational aim underlying this model is the development of the competence to act in explicit situations. This competence is divided into a personal and a professional competence dimension. Within these two dimensions several areas are defined. Here we go beyond existing models and change the areas indicated by the existing models. Whereas the personal dimension can still be divided into an area with regard to individual and interpersonal aspects (others call this social competence), the professional dimension is divided into a more general area with regard to a discipline and into a field specific area. In the case of engineering for example this would mean that the general area of engineering needs explicit competences (for example in mathematics or physics) and the specific field of mechanical engineering adds a more specific collection of competences, that are not required in all engineering disciplines. However both areas include professional as well as methodological competences. The idea behind this differentiation into two dimensions and four areas is that a modular design principle can be applied. The understanding based on such a principle gives us the opportunity to define competence requirements from the general to the detailed reading the model from left to right. We argue that the personal competence dimension can be defined more or less independently from any profession, because individual as well as interpersonal competences are required in

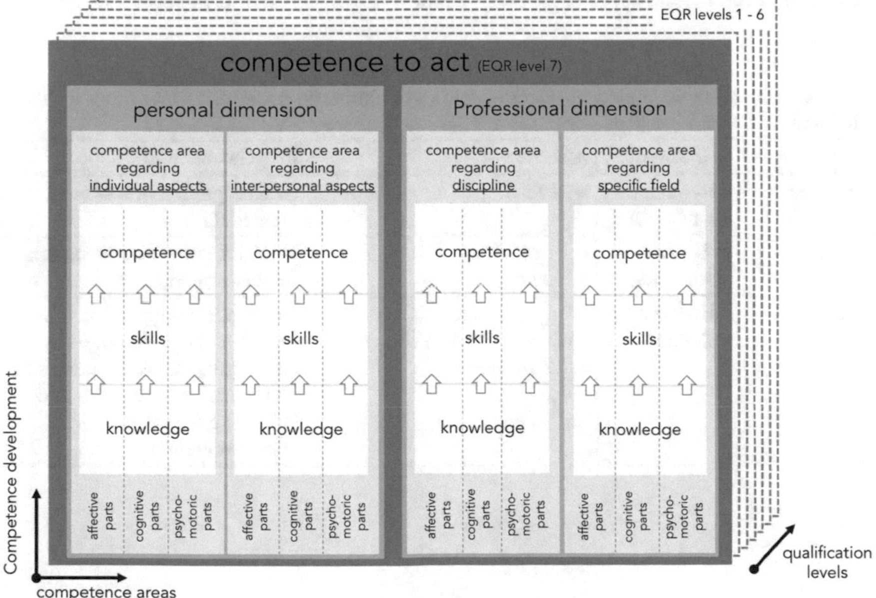

Fig. 6 Holistically defined competence model for desrcibing higher education courses

nearly any profession. Narrowing the focus means to look at a specific discipline and going even more into detail means to have a look at a specific field within this discipline. That implies that an existing set of competences can be easily adapted to a new field (as long as it is in the same discipline) by just striking out the appropriate competences for the one field and adding the needed competences for the new field.

4.3 Y-Axis: Competence Development

On the Yaxis the competence development is described. In this context the new model is very much oriented to the sources presented in II. We argue that the basis for competence development is knowledge. Skills are developed on a higher level and take knowledge into account. On the highest level competences are developed, which again refer to skills and knowledge. This model's new aspects in this context are the three taxonomies that are integrated into all the competence areas. This means that competence development in each of the areas shows links to all three taxonomies. It is obvious that these links are more or less strong, depending on the taxonomy and the area. For example the personal competence dimension will show stronger linkages to the affective taxonomy whereas the professional dimension will show more cognitive and psychomotoric parts. A next step for this model was to connect the taxonomies' levels with the steps from knowledge to skills and competences. This was implemented by a simple approach, taking the taxonomies' types or levels and assigning them to knowledge, skills or competence. The result is shown in Table 1.

Table 1 Integration of cognitive, affective and psychomotoric taxonomies into the knowledge, skills and competence system

	Cognitive	Affective	Psycho-motoric
Knowledge	• Remember • Understand	• Receiving	• Recognition of tools and materials
Skills	• Apply • Analyze	• Responding • Valuing	• Handling tools and materials • Basic operations
Competence	• Evaluate • Create	• Organization • Characterization by a value or value complex	• Competent operation of tools • Expert operation of tools • Planning of work operations • Evaluation of outputs and planning means for improvement

4.4 Z-Axis: Qualification Level

As shown in Fig. 6 the different qualification levels are represented by the ZAxis. In this paper's discussion we focus the 7th level as we are talking about higher education at universities up to the master level. At this point a clarification should be made. The model does indicate the different knowledge, skills and competences that have to be gained on the presented qualification level. It is not the model's aim to represent all the knowledge, skills and competences an individual developed throughout its qualification on different levels. That means that basic competences like reading, writing or calculating are not necessarily to be shown here, as these competences are gained on lower levels. So we are taking rather an institutional perspective than the individual's perspective with this approach. Above we explain our holistic model concerning competence development in higher education. However we started with the aim of describing higher education courses with regard to competence areas and competence development. That this is possible with the help of the designed model will be shown in the following section. With the FLExperiments a higher education course is classified with the model's help.

5 Analyzing FLExperiments – The Holistic Model on Trial

In order to proof the developed model we analyze in this section a new learning format implemented at the TU Dortmund University. The format is called FLExperiments and aims on competence development in context of laboratory education [17]. FLExperiments are integrated in the learning location research workshop, which was launched in 2011 [18]. The research workshop's general aim is to implement research based learning in engineering education. Therefore the authors developed various learning scenarios like the FLExperiments or several other courses to improve the students' key qualifications. Last semester we started to integrate courses into the curricular of engineering programs. For each of the formats and courses in the research workshop the teaching staff carefully formulates indented learning outcomes to define the courses' aims from the students' perspective. The following three intended learning outcomes were formulated to describe the FLExperiments' objectives:

- "ILO 1: Students acquire skills to successfully perform experiments.
- ILO 2: Students are able to create assumptions, verify them by experimentation and to refine their assumption based on their findings.
- ILO 3: Students are able to write a scientific report." [17]

These learning outcomes will be classified in the following with the new model's help in order to find out which competence dimension and area are addressed. First of all we want to put ILO 1 in more concrete terms in order to make it manageable for our context. The revised ILO 1 would be: 'Students are able to perform experiments successfully by using various skills'. Now we classify ILO 1 as learning outcome that addresses the 'professional' dimension and 'discipline' area. In both wordings the keyword 'skills' is used. Hence, we classify ILO 1 on the 'skills' level. What can be said about taxonomies finally? – We see ILO 1 only in the 'psychomotor' domain. In the taxonomy of Ferris and Azis ILO 1 reaches the level 'competent operation of tools'. Moreover we assign the activities helping students to reach the second intended learning outcome (ILO 2) to the 'professional' dimension, too. It shows no connection to a specific field so the competence area is 'discipline'. For ILO2 the students have to use their knowledge and experiences to create assumptions and to find a way to verify them. The keyword 'create' indicates to the level 'competence'. From the taxonomies' perspective ILO 2 can be classified in the 'cognitive' domain. Writing "scientific reports" from ILO 3 is not a unique characteristic of engineering science in the narrow sense. That alone would mean that ILO 3 has to be seen in the 'personal' dimension. Although writing could be a part of general communication, ILO 3 in this example aims at discipline specific abilities. So we classify it as part of the 'professional' dimension and in addition to that as part of the 'discipline' competence area. Not only for students writing under the consideration of scientific standards is a challenge. That means scientific writing is a complex task, which requires various skills and special knowledge. Hence, we assign ILO 3 to the level 'competence'. Interesting in this particular context is that we can classify ILO 3 in all three taxonomy domains. Firstly, in the cognitive domain the highest type 'create' is addressed. Secondly, in the affective domain students have to reach the level 'characterization by value or value complex'. Otherwise it could happen that they ignore parts of scientific standards like not presenting other's work as the own. Thirdly, writing a scientific report is more efficient if the writer is able to effectively use the keyboard by using ten fingers. Hence, in the psychomotor domain it would be adequate to reach the skill level and with this show the taxonomy level 'competent operation of tools'. As a conclusion it can be said that all intended learning outcomes are in the 'professional' dimension and in the competence area regarding 'discipline'. That is not coincidental as the research workshop and in particular the FLExperiments do not aim on a specific target group like electrical or mechanical engineering students but rather the whole engineering science discipline. In this section we showed that the developed holistic model can be used in order to classify intended learning outcomes. With this it was possible to describe a course concept with regard to the intended competence development and the addressed competence areas. That means that we were able to proof the accuracy of the developed holistic model in a small scale. Doing so for other courses will be the next step for the future. This leads us to our final conclusions and prospective future work.

6 Future Work and Conclusion

In this paper we present an approach to combine different theoretical perspectives to the competence discussion. This leads to a newly defined holistic competence model, which represents competence areas as well as the competence development process. We have recognized parts of the discussion about the term competence and about taxonomies as integrative part of the discussion and build up a model that takes into account both. To check the plausibility of the developed model we used it to analyze the intended learning outcomes of a selfdeveloped course. It was shown that were able to classify each of the course's intended learning outcomes with the help of the model. That means that the new model has successfully passed the first verification. Next steps can be divided in ongoing verification of the developed model and further development. Both aspects are for the authors highly relevant. The additional verification means to use the model for the description of more courses. Our aim was to define a holistic model, which can describe nearly every course. That our model is able to do so is not verified yet but this will be focused in the future. Based on this verification and the gained experiences the further development can be carried out. In the long term it will be interesting to check if the model shows even more possibilities for application. For example from our perspective it might be helpful to not only describe courses but to examine the courses in terms of needed previous knowledge, skills of competences. For example if a teacher uses the model and notices that all the connected learning outcomes are on the 'competence' level it is important to doublecheck if the students already show the knowledge and skills necessary to develop the addressed competences. If not this would be an important topic for improving the course. However, we see many possibilities to use the newly defined model for higher education course design. Especially the integration of the taxonomies' perspective is new in this context. In the future we will go on with the model's development and specification.

References

1. European Parliament and Council, Recommendation of the european parliament and of the council of 23 april 2008 on the establishment of the european qualifications framework for lifelong learning (text with eea relevance): (text with eea relevance). Official Journal of the European Union **51** (C111), 2008, pp. 1–7. http://eur-lex.europa.eu/legal-content/EN/ALL/?uri=CELEX:32008H0506
2. Handbuch zum deutschen qualifikationsrahmen struktur - zuordnungen - verfahren - zuständigkeiten
3. M. van der Blij, J. Boon, H. van Lieshout, H. Schafer, H. Schrijen, *Competentieprofielen: over schillen en knoppen*. e-Comopetence profiles. 2002
4. N.A. Campbell, *Biology: Concepts & connections*, 6th edn. Pearson Cummings, San Francisco, 2009
5. B.S. Bloom, M.D. Engelhart, W.H. Hill, E.J. Furst, D.R. Krathwohl, *Taxonomy of Educational Objectives: The Classification of Educational Goals.*, vol. Handbook I: Cognitive Domain. 1956

6. L.W. Anderson, D.R. Krathwohl, *A taxonomy for learning, teaching, and assessing: A revision of Bloom's taxonomy of educational objectives*. Longman, New York, 2001
7. D.R. Krathwohl, B.S. Bloom, B.B. Masia, *Taxonomy of Educational Objectives: The Classification of Educational Goals*, vol. Handbook II: Affevtive Domain. 1965
8. T.L. Ferris, S. Aziz. A psychomotor skills extension to bloom's taxonomy of education objectives for engineering education: Exploring innovation in education and research, 2005. http://slo.sbcc.edu/wp-content/uploads/bloom-psychomotor.pdf
9. J. Schofnegger, H. Zöpfl, *Affektive Ziele*. Ehrenwirth, München, 1978
10. T.L. Ferris, Bloom's taxonomy of educational objectives: A psychomotor skills extension for engineering and science education. International Journal of Engineering Education **26** (3), 2010, pp. 699–707
11. H. Roth, *Pädagogische Antrhopologie: Entwicklung und Erziehung*, vol. II, 1st edn. Schroedel Verlag, Hannover, 1971
12. R. Nickolaus, S. Seeber, Berufliche kompetenzen: Modellierungen und diagnostische verfahren. In: *Handbuch Berufspädagogische Diagnostik*, ed. by A. Frey, U. Lissmann, B. Schwarz, Beltz, Weinheim, Basel, pp. 155–180
13. Handreichung für die erarbeitung von rahmenplänen der kmk für den berufsbezogenen unterricht in der berufsschule und ihre abstimmung mit ausbildungsordnungen des bundes für anerkannte ausbildungsberuf
14. Handreichung für die erarbeitung von rahmenlehrplänen der kultusministerkonferenz für den berufsbezogenen unterricht in der berufsschule und ihre abstimmung mit ausbildungsordnungen des bundes für anerkannte ausbildungsberufe
15. J. Erpenbeck, L. Rosenstiel, eds., *Handbuch Kompetenzmessung: Erkennen, verstehen und bewerten von Kompetenzen in der betrieblichen, pädagogischen und psychologischen Praxis*, 2nd edn. Schäffer-Poeschel, Stuttgart, 2007
16. Wildt, Kompetenzen als 'learning outcomes. Journal Hochschuldidaktik **17** (1), 2006, pp. 6–9. http://www.hdz.tu-dortmund.de/fileadmin/JournalHD/2006/Journal_HD_2006_1.pdf
17. P. Ossenberg, T. Jungmann, Experimentation in a research workshop: A peer-learning approach as a first step to scientific competence. International Journal of Engineering Pedagogy (iJEP) **3** (3), 2013, pp. 27–31
18. T. Jungmann, P. Ossenberg, Research workshop in engineering education draft of new learning. In: *2014 IEEE Global Engineering Education Conference (EDUCON)*. 2014, pp. 83–87

The Globally Competent Engineer – What Different Stakeholders Say About Educating Engineers for a Globalized World

Dominik May and Erman Tekkaya

Abstract The world is becoming more and more globalized. That has an important impact on the working environment of engineers. Designing, producing and distributing products either in international companies, in international working-teams or at least for international markets is part of today's economic reality. Even if goods have been sold or purchased all over the globe for centuries now, today's markets have been grown closer together than ever. This is the potential working environment engineering students are facing. In contrast to that engineering education still is very local oriented. The number of international courses or even programs in which the students can train to act in intercultural settings is small in comparison to the amount of engineering courses at universities. Changing this imbalance means developing possibilities to train globally competent engineers in local educational systems. An essential step in this work is to define the globally competent engineer itself. What are competences or attributes graduates need in order to act successfully in international contexts? Answering this question the paper takes three sequential steps. First of all the term "intercultural competence" will be discussed. In a second step a literature research on attributes of globally acting engineers will be shown. Finally official documents of four accreditation agencies for engineering programs will be analyzed with focus on international aspects. By doing so four central areas of intercultural competence that are related to engineering can be identified: (International) communication, understanding of the engineering profession in a global context, (international) teamwork and ethical reasoning. In the conclusion all three steps will be connected and a catalogue of four central competences will be discussed.

Keywords International Engineering Education · Accreditation Agencies · Globally Competent Engineer

D. May (✉)
Engineering Education Research Group, Center for Higher Education,
TU Dortmund University, Dortmund, Germany
e-mail: Dominik.may@tu-dortmund.de

E. Tekkaya
Institute of Forming Technology and Lightweight Construction (IUL),
TU Dortmund University, Dortmund, Germany

Originally published in "Proceedings of 2014 International Conference on Interactive Collaborative Learning (ICL) as part of 2014 World Engineering Education Forum: Engineering Education for a Global Community", © IEEE 2014. Reprint by Springer International Publishing AG 2016, DOI 10.1007/978-3-319-46916-4_73

1 Introduction

Without a doubt the world's economy is becoming more and more globalized. In nearly every sector the purchasing, producing and selling of products is worldwide connected. Quite often complex technical products are composed of parts, which are produced in completely different areas of world - that counts for smart phones as well as for airplanes [1]. In fact it is not only the case that components are produced all over the world. The more important aspect is that design and production teams that are working on these components are coming from all over the world and have to work successfully together. This fact of course has implications for the engineering profession. Engineers are increasingly facing working contexts that are globally connected and their colleagues in project teams may be distributed all over the world. In order to be able to act in such environments, engineers must show global competency. In other words, globally competent engineers are needed. In contrast to this, engineering education is still very local oriented. Except for those, who are studying or working for some time abroad, engineering students are mainly focused on their home university's environment for their entire course of studies. Even if there are examples for international or transnational courses in higher engineering education in Germany, these cases remain exceptions (e.g. [2, 3]). That seems to be surprising looking at the situation in the economy explained above.

Facing a more and more globalized economy without preparing students to work in such an environment will lead to a problem on the long term. It represents an imbalance between the educational system and the labor market. The central question in this context is:

- "What does it mean in detail to educate engineering students to globally competent engineers?"

In order to answers this question many perspectives and stakeholders in the educational system have to be taken into account (Fig. 1). Besides the industry and the science sector as prospective employers, at least the students, the teachers, politics, accreditation agencies and finally education research have to be taken into consideration. It is obvious that discussing all those stakeholders and their perspectives is too much for just one paper. Hence, this paper will concentrate on three perspectives: Intercultural competence research, education research, and accreditation agencies.

Having a look at the intercultural competence research by explaining a definition and a model of intercultural competence will be the first step (see Section 2).

The second step in this paper will be to have a special look at educational research with focus on intercultural competence in engineering. There are already publications answering the question concerning attributes of globally competent engineers. However they mainly present lists of different attributes. These lists differ from publication to publication. That is why an analysis of these lists for differences and similarities is necessary. At this point employers will be taken into account insofar as their perspective is included in the presented lists (see Section 3.1).

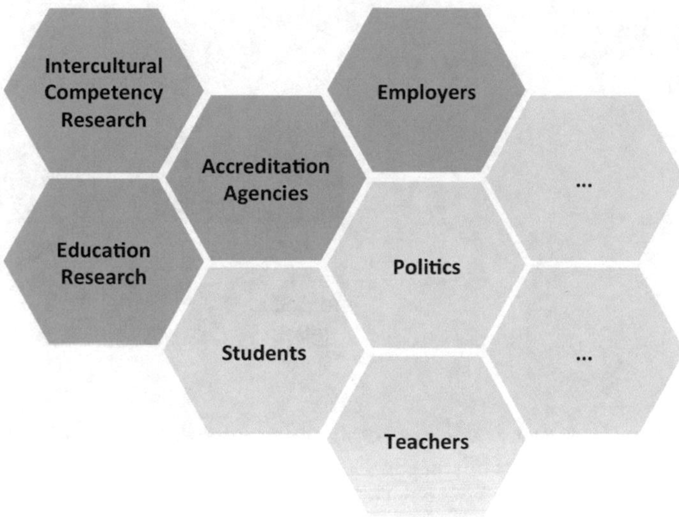

Fig. 1 Stakeholders with different perspectives on engineering education

In order go even deeper into the discussion national accreditation agencies will be focused in a third step. They are playing an important role in the higher education sector. As accreditation agencies are working on a national level, it is an important step to look into the official documents of multiple agencies all around the world and search for their criteria for adequate engineering education with regard to interculturality. This paper will have a look on four different accreditation agencies' documents for engineering education. As the authors are working at a German university the ASIIN – the biggest agency for engineering curricula in Germany – will be in focus. With the ABET from the USA another agency, that is mentioned very often in Anglo-American literature on engineering education, will be taken into account. Locking at Japan and China will enrich their views. Expressed criteria for adequate engineering education will be inspected for statements on international aspects (see Section 3.2).

This paper's final part will bring the three perspectives together. For the methodological perspective these steps mean to proceed from the general to the detail. On the most detailed level clusters for globally competent engineers will be worked out and checked against the background of the less detailed levels above. Findings from this analysis will be summarized and conclusions for future research on globally competent engineers will be drawn at the end.

2 Intercultural Competence

Many different disciplines are involved into the discussion on intercultural competence. Pedagogues or psychologists discuss it from their perspectives and even economists take it into account by asking for its importance for international eco-

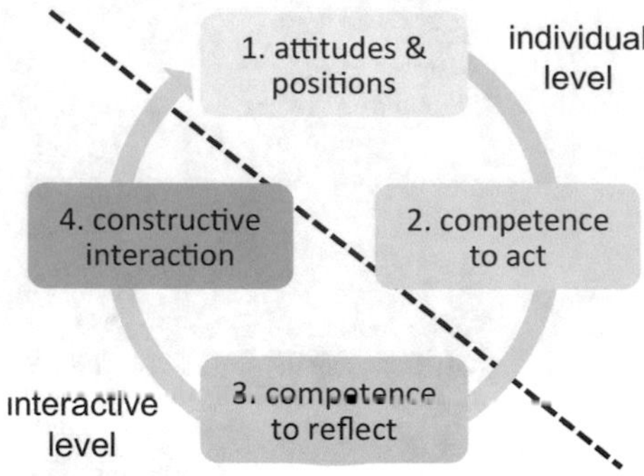

Fig. 2 Learning model for intercultural competence following [5]

nomic relations. However, it will not be possible to present the full discussion in its diversity and complexity in this paper. Nevertheless a working definition is mandatory at this point in order to have a basis for the discussion to build on. The German Bertelsman Stiftung defines intercultural competence on basis of [4] as follows:

"Intercultural competence describes the competence to effectively and adequately interact in intercultural situations on basis of explicit attitudes as well as the special ability to act and reflect" (own translation, [5]).

This definition shows two crucial aspects. Firstly the final aim of intercultural competence is being able to act in explicit situations, which show any aspect of interculturality. That means for the understanding of intercultural competence that it always has something to do with interaction and communication with other people. Following the definition this has to be done in an effective and adequate way, whatever this may be in the respective situation. Secondly this competence bases on personal attitudes, interaction and reflection. Building on this [5] defines a learning model for intercultural competence in a four-step circle (Fig. 2).

These four steps finally lead to the development of intercultural competence. [5] explains the different steps more in detail as follows:

- Attitudes & Positions
  - Appreciation of diversity
  - Ambiguity tolerance
- Competence to act
  - Extensive cultural knowledge
  - Communication skills
  - Conflict-solving skills
- Competence to Reflect
  - Relativization of reference framework
  - Ability for empathy

- Constructive interaction
  - Avoidance of the violation of rules
  - Achievement of objectiveays

In addition to that, these four steps can be split into two different areas. Attitudes and positions as well as the competence to act can be seen on an individual level. Here the individual person is in focus and it is made clear that the person has to show a personal volitional basis, basic knowledge, and central skills in order to move into the next steps. The last two steps can be seen on an interactive level. Here the person is not longer on its own but in contact with others. Nevertheless even these steps have different focuses. Whereas the competence to reflect aims on an internal effect the constructive interaction aims on an external effect. The internal effect can lead to a re-evaluation of the own and the others' way of thinking and acting. The external effect is the final aim of the whole process: the achievement of a common goal in the respective situation.

Summing up and saving the important aspects for the following it can be said that intercultural competence has four central facets building on each other: (1) Personal motivation and attitude, (2) cultural knowledge and the ability to communicate and to solve problems are needed as a basis. This should be enriched by (3) the competence to reflect (internal effect). All these points are necessary for (4) constructive interaction (external effect). These four facets represent an individual and an interactive level.

3 Globally Competent Engineers in Literature and Accreditation

In the following the discussion on intercultural competence will be focused more on engineering. Therefore literature on globally competent engineers will be presented first (Section 3.1). This will be expanded by the presentation of different accreditation agencies and their official documents on engineering competences. From these lists competences in context with intercultural collaboration will be extracted (Section 3.2).

3.1 Globally Competent Engineers in Education Research Literature

In order to compare the intercultural competence definition with existing work on globally competent engineers, two examples from literature will be taken into account and presented first. [1] worked out an overview of several different sources that all worked out lists with competences or needed attributes to work on global contexts.

One of them is [6], who developed a list of all in all thirteen characteristics. Following him the globally competent engineer

1. can appreciate other cultures;
2. is able to communicate across cultures;
3. is familiar with the history, government, and economic systems of several target countries;
4. speaks a second language at a conversational level;
5. speaks a second language at a professional/ technical level;
6. is proficient working in or directing a team of ethnic and cultural diversity;
7. can effectively deal with ethical issues arising from cultural or national differences;
8. understands cultural differences relating to product design, manufacture, and use;
9. has an understanding of the connectedness of the world and the workings of the global economy;
10. understands implications of cultural differences on how engineering tasks might be approached;
11. has some exposure to international aspects of topics such as supply chain management, intellectual property, liability and risk, and business practices;
12. has had a chance to practice engineering in a global context, whether through an international internship, a service-learning opportunity, a virtual global engineering project, or some other form of experience;
13. views himself/herself as citizens of the world, as well as of a particular company.

This list is the result of his work, in which he investigated the meaning and different attributes of global competency. By asking university faculty and administrators as well as managers in industry [6] identified the attributes 1, 6, 2, 12, and 7 as the most important.

In addition to that list the catalogue worked out by [7] will be displayed in the following. They argue, that showing global competency means to show

- awareness of global political and societal issues;
- understanding of cross and multicultural issues;
- understanding of the globalized nature of engineering education;
- knowledge of the international labor market and workplace imperatives;
- understanding of the international business, economy and world market;
- competency in applying engineering solutions/applications in a global context.

Even more lists like these exist in literature (see e.g. [8, 9]) but for this paper the two given examples should be enough. Combining the five most important attributes from [6] with the full list from [7] and striking out duplications leads to the following list of ten central attributes. The globally competent engineer. . .

(a) . . .is aware of and understands ethical aspects in cross-cultural and multicultural issues arising from national differences, and he is able to effectively deal with that.
(b) . . .is able to appreciate other cultures.
(c) . . .is able to be a member or even a leader of a ethnically or culturally diverse team.
(d) . . .communicates effectively across different nationalities and cultures.
(e) . . .is aware of global political and societal issues.
(f) . . .understands the globalized nature of engineering education.
(g) . . .has knowledge of the international labor market and workplace imperatives.
(h) . . .understands the international business, economy and world market.
(i) . . .is competent in applying engineering solutions/applications in a global context.
(j) (. . .is experienced in working in global contexts whether through an international internship, a service-learning opportunity, a virtual global engineering project, or some other form of experience.)

The competence under (j) is put in brackets, as this is not an attribute in the narrow sense. It is more a way to gain the other attributes or improve the own intercultural competences.

Up to now two different perspectives on the discussion about globally competent engineers have been taken into account. With the accreditation agencies and their official documents a third viewpoint will be integrated.

3.2 *Globally Competent Engineers in Accredition Agency Documents*

Several agencies all over the world are involved in the accreditation and assessment of engineering curricula. They represent an independent third party authority, which accredits higher education curricula. Within the accreditation process the agencies define on the one hand how engineering curricula should be designed and on the other hand what intended learning outcomes and the graduates' competences should be. The way they are involved into the curricula design process and the model they are following during the accreditation process differ from country to country. This fact consequently leads to a problem in terms of comparability of engineering programs when talking about engineering education on a global level. Research results published in 2007, for example, claim that there "[. . .] is an urgent need for a systematic global model of engineering accreditation that can be used to assess global professional skills and attributes of engineering graduates" [10]. In addition to that it is difficult to judge from a distance the different agencies' importance and their role in the local higher education system. Hence, the following should be understood as a collection of different international views on engineering education.

However it is worthwhile looking at the different agencies' documents on a content analyses level. Hence, competences they claim to be important for future engineers in a global context will be compared. With the ASIIN, the ABET, the CEEAA, and JABEE all in all four examples will be explained shortly - mainly in order to show their general requirements concerning engineering education's learning outcomes. The aim at this point is to see, if and how globalization can be found in the accreditation agencies' requirements for engineering education. Therefore the agencies' accreditation documents available on their homepages are analyzed and aspects that can be traced back to any kind of global cooperation, interculturality, etc. are marked (in italics).

3.2.1 Germany

ASIIN is the German accreditation agency working on engineering education programs. As their criteria are pretty similar to the ones defined by the European Accreditation of Engineering Programs [11] this is the only example taken into account from Europe. ASIIN defines for engineering practice on a master level all in all twenty-three outcomes and divides them into six different areas [12]:
Area Knowledge and Understanding:

- broad knowledge and deep understanding of mathematical and natural sciences and engineering principles as well as its interdisciplinary extensions and application-oriented knowledge about areas of specialty
- a critical awareness of new findings in the discipline

Area Engineering Methods:

- ability to analyze and solve problems, which are unusual or incomplete and have competitive specifications, scientifically and/or application-oriented
- ability to frame complex problems in new or evolving areas of the own discipline
- ability to apply innovative methods in the problem solving process and generate new methods

Area Engineering related Development and Design:

- ability to generate new concepts and solutions confronted with unusual questions with respect to other disciplines
- ability to use the own creativity in order to develop new products, processes and methods
- ability to use engineering judgment in order to work with complex, technical imprecise or incomplete information

Area Scrutinizing and Evaluation:

- ability to identify, find, and gather information
- ability to plan and implement analytic, exemplary or experimental investigations
- ability to evaluate data critically and draw conclusions

- ability to examine and judge on new or upcoming technologies in the own field of specialization

Area Engineering Practice:

- ability to classify knowledge from different areas, combine that systematically in order to put that into practice and handle complexity
- ability to become acquainted with something new in fast and systematic way
- ability to rate methods and its limitations
- ability to reflect systematically on non-technical effects of the engineering work and incorporate that into acting in a responsible way

Area Interdisciplinary Competences:

- have an understanding of health- and security-related as well as legal consequences of engineering practice, do reflect on implications of engineering solutions for society and ecology, and commit themselves to follow professional engineering ethics as well as norms
- know and understand project-management as well as economic methods (for example risk management) and know their limitations
- recognize the necessity of self-directed as well as life-long learning and are able to do so
- ability to work effectively on your own or in a team and act as a team coordinator if necessary
- ability to use different methods in order to communicate effectively with engineering colleagues from and the general public
- are able to work and communicate confidently in national and international contexts
- are able to act as leader of teams, composed of different disciplines and levels

3.2.2 United States

The ABET document "Criteria for Accrediting Engineering Programs" [13] from 2012 is very often cited in literature on engineering competences and learning outcomes. Within this document the American accreditation association working on engineering education defines eleven outcomes for engineering programs. Following the ABET, engineering students should show the following [13]:

(a) an ability to apply knowledge of mathematics, science, and engineering
(b) an ability to design and conduct experiments, as well as to analyze and interpret data
(c) an ability to design a system, component, or process to meet desired needs within realistic constraints such as economic, environmental, social, political, ethical, health and safety, manufacturability, and sustainability
(d) an ability to function on multidisciplinary teams
(e) an ability to identify, formulate, and solve engineering problems

(f) an understanding of professional and ethical responsibility
(g) an ability to communicate effectively
(h) the broad education necessary to understand the impact of engineering solutions in a global, economic, environmental, and societal context
(i) a recognition of the need for, and an ability to engage in life-long learning
(j) a knowledge of contemporary issues
(k) an ability to use the techniques, skills, and modern engineering tools necessary for engineering practice.

3.2.3 China

On the homepage of the China Engineering Education Accreditation Association (CEEAA) can be found outcomes for engineering graduates. Talking about program accreditation, the CEEAA states, that "the programs must demonstrate through evaluation that their graduates achieve the following" [14] ten central competences:

1. a knowledge of humanities and an understanding of social, professional and ethical responsibility;
2. a knowledge of mathematics, natural sciences and business management necessary for engineering practice.
3. a knowledge of basic engineering and the discipline; an experience of systematic engineering practice; a knowledge of contemporary issues and development trend of the discipline;
4. an ability to design and conduct experiments, as well as to analyze and interpret data;
5. an ability to innovate; an attitude and awareness of innovation; an ability to synthesize theories and techniques to design a system and process within the economic, environmental, legal, safety, health and ethical constraints;
6. an ability to apply basic methods of literature and document searching, and use modern information technologies to get relevant information;
7. a knowledge of policies, laws and regulations on the production, design, research and development, environment protection and sustainable development related to the profession and industry; a recognition of the impact of engineering to the external world and society;
8. an ability to manage, communicate and function in teams;
9. an understanding of lifelong learning; an ability to learn and adjust to development.
10. an global vision; an ability to communicate, compete and cooperate in the multicultural context.

3.2.4 Japan

The Japanese Accreditation Board for Engineering Education defines in its document "JABEE Common Criteria for Accreditation of Professional Education Programs" under criterion 1 all in all nine learning outcomes for engineering education [15]:

(a) An ability of multidimensional thinking with knowledge from global perspective
(b) An ability of understanding of effects and impact of professional activities on society and nature, and of professionals' social responsibility
(c) Knowledge of and ability to apply mathematics and natural sciences
(d) Knowledge of the related professional fields, and ability to apply
(e) Design ability to respond to requirements of the society by utilizing various sciences, technologies and information
(f) Communication skills including logical writing, presentation and debating
(g) An ability of independent and life-long learning
(h) An ability to manage and accomplish tasks systematically under given constraints
(i) An ability to work in a team

In this chapter two different perspectives on engineering education are displayed. In the following these perspectives will be discussed, separately in a first step and in context with each other in a second step.

4 Comparing the Accreditation Boards' Lists with Literature on Globally Competent Engineers

In Section 3.2 the different competences, outcomes or attributes for engineering education defined by several accreditation agencies are displayed. The aim was to investigate if globalization or international working contexts are a topic openly addressed by the agencies. Firstly it can be stated that in each of these lists only one of the items directly addresses any international aspect by naming it. Nevertheless there can be identified – if directly or indirectly – competences that are important in an international context.

4.1 Communication

Even if the four compared agencies do show a global perspective they mainly do not explicitly express what should be understood by the outcomes. What, as an example, is meant by being able to work and communicate confidently in national and international contexts (ASIIN)? Especially the term confidently leaves open what this competency really means. The Chinese example does not really help in this context but it goes a little bit more into detail by not only addressing communication in general but also asking for the ability to compete and cooperate in international contexts.

The problem instead remains the same. What does it mean to be able to compete and cooperate? The fact that these terms are antonyms makes this competence even more difficult to understand.

4.2 Global Interconnections

The ABET expression under (h) (to understand the impact of engineering solutions in a global [. . .] context) also tackles global interconnections, although in this case communication is not addressed. The competence for effective communication is addressed by outcome (g) and is demanded in a general understanding. Coming back to (h), it can be stated instead, that the globalized economy does not only require international communication aspects but also leads to the necessity to see the engineers' work itself and the deliverables in a global context. The JABEE shows the same approach by not talking about international communication but demanding for the ability of multidimensional thinking with knowledge from global perspective. These requirements express that an engineer in a globalized world must be able to take multiple perspectives — that also includes global perspectives – into consideration and see its own profession globally connected.

Apart from those two competence areas that openly express the international interconnection of engineering work, there can be identified two more areas. Even if they do not openly address international aspects they also have to be taken into account when talking about globally competent engineers. In the following it will be explained why they must be understood in the same context.

4.3 (International) Teamwork

A competency that is expressed by each of the four agencies is being able to work in (multidisciplinary) teams. Even if this does not necessarily mean that a team must be build of team-members from different nations and cultures, this increasingly will be the case in the future [1]. Considering the reasoning that the engineers' work is more and more internationally connected – which in fact means that teams are distributed globally – the "teamwork-competency" has to be taken into account. Moreover teamwork has to be understood in context with the (international) communication competency explained above. Cooperation in teams always has something to do with personal interaction and by this with communication. Hence teamwork should not be missing, in order to work out a list of competences for globally competent engineers.

4.4 Ethical Reasoning

The fourth competence takes the same approach like "understanding of the engineering profession in a global context". Each of the accreditation agencies shown here are talking about ethical aspects. Some of them (see the German, American and Chinese example) are even using the term "ethics" itself. JABEE uses other words but express the same point by demanding "an ability of understanding of effects and impact of professional activities on society and nature, and of professionals' social responsibility". Such competences express that engineers should not only have the technical aspects of their solutions in mind but should also think about global ethical implications in terms of society, nature and security.

This of course must be listed under competences for globally competent engineers as ethical reasoning hardly stops at national boarders. Quite the contrary is the case. As soon as technical products or solutions are produced, sold or distributed around the world ethical reasoning must be carried out with global aspects in mind.

Even if the results are not overwhelming four different competence areas could be clearly identified by the investigation on the agencies' documents. Having this list in mind and compare it with the one worked out in III.A it becomes obvious that all four competence areas for globally competent engineers can also be found in the literature, even if different words are used there. Moreover all of the aspects from the list in III.A can be allocated to the four areas:
Communication

- communicates effectively across different nationalities and cultures

Global interconnections

- is competent in applying engineering solutions/applications in a global context
- understands the international business, economy and world market
- has knowledge of the international labor market and workplace imperatives
- understands the globalized nature of engineering education
- (is experienced in working in global contexts whether through an international internship, a service-learning opportunity, a virtual global engineering project, or some other form of experience)
- is aware of global political and societal issues;

(International) teamwork

- is able to be a member or even a leader of a ethnically or culturally diverse team

Ethical reasoning

- is aware of and understands ethical aspects in cross-cultural and multicultural issues arising from national differences, and he is able to effectively deal with that;
- is able to appreciate other cultures

That supports the view that these four areas can be seen as the main important areas in which global competencies can be described. In the following this paper's results will be summed up and brought into context with the general discussion on intercultural competence. Moreover future research tasks will be displayed.

5 Conclusion and Future Work

In order to draw conclusions for this research the questions raised in I will be used and answered. Honestly accreditation agencies do not say much about international aspects in engineering education. On basis of four chosen agencies and looking at their official documents for accreditation, it can be shown that there are only some exceptions. Nevertheless four different competence areas could be identified. Hence, these attributes can be seen as the focus aspects of internationality for the accreditation agencies.

In order to enrich this research and work out even more different aspects of global competency, literature on global engineering was taken into account and compared with the four attributes. This comparison showed that each of the four identified competence areas above are considered in literature. The scientific discussion in literature instead does provide even longer catalogues of competences for globally competent engineers but all of these competence items can be allocated to one of the identified areas. Hence, these four areas can be supported by attributes of globally competent engineers in engineering education literature. Based on these results the authors finally argue that globally competent engineers have at least to be able to show competences in four different areas (Fig. 3).

Tying this result back to the discussion on the learning model for intercultural competences makes clear that a comparison is difficult to do. The discussions, even if both basically aim on intercultural or global competence, take place on totally different levels. However, there can be seen some loose similarities. Just as the model from II shows that intercultural competence has two parts (the individual and

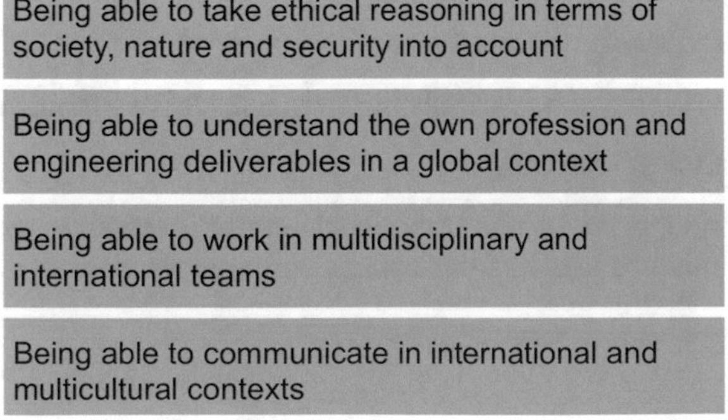

Fig. 3 Areas global competent engineers have to show competences

Being able to take ethical reasoning in terms of society, nature and security into account

Being able to understand the own profession and engineering deliverables in a global context

Being able to work in multidisciplinary and international teams

Being able to communicate in international and multicultural contexts the interactive level), the same can be seen in the four competence areas from Fig. 3. Whereas the first two are pointing to internal processes (like reasoning or the ability to understand) the third and the fourth area clearly take interaction into focus. Going more into detail these areas seem to describe the step two and three in the model (competence to act and competence to reflect) by filling these steps with concrete content. Nevertheless both discussions are still independent. One is focusing on the learning process or better on the intercultural competence development and the other is focusing on different areas of intercultural competence. Hence, the question raised above has to be answered (at the moment) in two separate answers. It will be the task for the future to broaden this research and to meaningfully combine both discussions. Moreover it will be necessary to work out a model that goes beyond just naming the competences but as well describes the interconnections between them and the needed development process. Only by doing this it can be worked out a holistic model on educating globally competent engineers looking at the development process itself and the respective areas. This model would represent a complete picture of the – in future needed – globally competent engineer

References

1. S.A. Rajala, Beyond 2020: Preparing engineers for the future. In: *Proceedings IEEE*, vol. 100. 2012, vol. 100, pp. 1376–1383
2. S. Moore, D. May, K. Wold, Developing cultural competency in engineering through transnational distance learning. In: *Transnational Distance Learning and Building New Markets for Universities*, ed. by R. Hogan, IGI Global, Hershey (PA/USA), 2012
3. G. Schuh, T. Potente, R. Varandani, C. Witthohn, ipodia-innovative, international, interactive higher education. In: *TeachING-LearnING.EU discussions - Innovationen für die Zukunft der Lehre in den Ingenieurwissenschaften*, ed. by A.E. Tekkaya, S. Jeschke, M. Petermann, D. May, N. Friese, C. Ernst, S. Lenz, K. Müller, K. Schuster. 2013
4. D.K. Deardorff, The identification and assessment of intercultural competence as a student outcome of internationalization at institutions of higher education in the united states: Unpublished dissertation. Ph.D. thesis, North Carolina State University, Raleigh, NC, 2004
5. Bertelsmann Stiftung, *Interkulturelle Kompetenz - Schlüsselkompetenz des 21. Jahrhunderts?: Thesenpapier der Bertelsmann Stiftung auf Basis der Interkulturellen-Kompetenz-Modelle von Dr. Darla K. Deardorf.* Gütersloh, 2006
6. A. Parkinson, The rationale for developing global competence. Online Journal for Global Engineering Education **4** (2), 2009. URL http://digitalcommons.uri.edu/cgi/viewcontent.cgi?article=1018&context=ojgee
7. A. Patil, C.S. Nair, Codner G., Global accreditation for the global engineering attributes: A way forward. In: *Proceedings of the 2008 AaeE Conference*. 2008
8. A.D. Chan, J. Fishbein, A global engineer fort he global community. The Journal of Policy Engagement **1** (2), 2009, pp. 4–9
9. L.G. Brown. Asee - global engineer project 2013. URL http://www.sefi.be/?p=2850
10. A. Patil, Codner G., Accreditation of engineering education: review, observations and proposal for global accreditation. European Journal of Engineering Education **32** (6), 2007, pp. 639–651. URL http://www.tandfonline.com/doi/citedby/10.1080/03043790701520594#tabModule
11. Eur-ace: Framework standards for the accreditation if engineering programmes, 2008. URL http://www.enaee.eu/wp-content/uploads/2012/01/EUR-ACE_Framework-Standards_2008-11-0511.pdf

12. Fachspezifisch ergänzende hinweise zur akkreditierung von bachelor- und masterstudiengängen des maschinenbaus, der verfahrenstechnik und des chemieingenieurwesens, 2011. URL http://www.asiin-ev.de/media/feh/ASIIN_FEH_01_Maschinenbau_und_Verfahrenstechnik_2011-12-09.pdf
13. Abet, criteria for accrediting engineering programs: Effective for reviews during the 2013-2014 accreditation cycle, 2012. URL http://www.abet.org/uploadedFiles/Accreditation/Accreditation_Step_by_Step/Accreditation_Documents/Current/2013_-_2014/eac-criteria-2013-2014.pdf
14. Accreditation criteria: General criteria. URL http://www.ceeaa.org.cn/criteriaG_en.html
15. Japanese Accreditation Board for Engineering Education. Jabee common criteria for accreditation of professional education programs, 2012. URL http://www.jabee.org/english/examination_accreditation/documents/

What Should They Learn? – A Short Comparison Between Different Areas of Competence and Accreditation Boards' Criteria for Engineering Education

Dominik May and Claudius Terkowsky

Abstract The question "What should engineering students learn for being successful engineers?" is and always was a driver for intense discussions about curriculum development in engineering education. Contributions to this question differ between various types of education institutions and organizations, various fields of specialization, and even various countries. Such differences make it necessary, that a framework, which describes the students' intended learning outcomes in engineering education programs, must be designed openly to represent engineering education in general and in the same way accurately enough to answer the question above. Therefore this work-in-progress-paper firstly discusses a general model of areas of competence, secondly looks at different accreditation boards' criteria for engineering education, thirdly combines the boards' criteria with the general areas of competences and fourthly derives conclusions for engineering education in laboratories.

Keywords Engineering Education · Accreditation Board · Areas of Competence · Learning Outcomes · Framework · Curriculum Development · Laboratories

1 Introduction

In many papers, publications and conferences all over the world the question concerning what engineering students should learn during their study is discussed intensively. Hence it might be questionable if another paper on this topic is necessary. Most of these publications written by teachers, education experts, or experts for engineering curricula represent the academic perspective. The industry - that means in this context the companies that hire the vast majority of the engineering graduates - has another sight on this topic. These companies constantly argue that the graduates coming from the university are not "ready" for working in industry because of a

D. May (✉) · C. Terkowsky
Engineering Education Research Group, Center for Higher Education,
TU Dortmund University, Dortmund, Germany
e-mail: Dominik.may@tu-dortmund.de

Originally published in "Proceedings of EDUCON2014 - IEEE Global Engineering Education Conference: "Engineering Education towards Openness and Sustainability", © 2014. Reprint by Springer International Publishing AG 2016, DOI 10.1007/978-3-319-46916-4_74

severe lack of competences (only two examples from Germany [1, 2]). Hence the companies have to invest in the graduates in form of trainings etc. first, before they can work successfully in a professional environment. Looking a little bit closer on [1, 2] it becomes obvious that technical knowledge and professional competence seldom is the origin for the industry's complaints. Lacks of competences mainly are identified in these fields of competences that are often described as key competences, which are necessary to perform successfully in a working environment but are not directly understood as professional competences in a technical understanding. Looking on the broad discussion on the one hand an the ongoing complaints on the other hand, it is obvious that the discussion on intended learning outcomes in engineering education has to be carried on. Even if higher engineering education, especially in Germany, does not understand itself as a training center for industry, nevertheless the gap between supply and demand should be reduced. Doing so a broader context might help. Hence a first step will be made in this paper as several criteria for modern engineering education defined by two different accreditation boards are summed up and brought into relation with a general understanding of competences.

2 What Competence Research Contributes to the Discussion

The discussion about competences can be divided into two main parts. On the one hand it is discussed, what competences 'are', how competences are defined, how they are developed, and how they differ from other concepts like knowledge, skills or qualifications [3–5]. On the other hand - and that is the focus for this paper - there is the question concerning different fields of competences. Without a doubt there are many different competences, which have to be developed by the students at university. The question is, how they can be clustered in order to handle them and make this cluster useful for curriculum development.

In competence research four main fields of competences have been identified [6, 7]. These fields are (own translation): professional-competence, social-competence, self-competence, and methodological-competence. The integration of these four competences leads on a meta-level to personal action competence [7]. Even if this grid is a quite logical and often used one, in this context it might to lead to a problem. Looking for example on engineering methods and its successful usage, it is not very clear in which competence field it belongs. Does the usage of professional engineering methods rather show professional competence than methodological competence or is the other way around the case? Because of this, a slightly different shaped grid is more practical to use in this case. [4] divides the different fields of competences as follows:

Professional and methodological competences mean the ability to solve subject-specific problems creatively by using subject-specific knowledge, skills, tools and methods. To rate knowledge and to develop methods further are also aspects of these competences.

Personal competences mean the ability to act in a self-reflective and independent way. It integrates aspects like being able to evaluate yourself and your own actions, to generate a productive attitude as well as values, to unfold own abilities, to develop the own person creatively, and to learn.

Social and communication competences mean the ability to act in a co-operative and communicative way, which includes the capability to act within a team and develop plans, activities or solutions together.

Activity and implementation competences mean the ability to act actively and in an organized way in order to achieve goals and/or to put plans into action (on your own or in a team). Thus these competences describe the ability to perform successfully by integrating own emotions, motivation, skills and personal experience and all the other three competences.

These four areas of competence (or the slightly different shaped grid, like shown above) are defined very broadly as they are usable for all different fields of specialization and serve as general concept. So it is necessary to fill this universal concept with concrete engineering content in order to make them usable for curriculum development and put a little bit more "meat on the bones".

3 What Accreditation Boards Say

There are multiple sources that curriculum developers can use in order to define what engineering students should learn during their studies. Beside the industry and the engineering science sector itself, accreditation agencies are an important stakeholder in this process, as they finally judge, if a course of studies is officially approved or not. Hence looking at what they say about the engineering degree course's outcome and connected criteria is interesting and an important step to define what engineering students should learn. Even if it is obvious that the development of such criteria is a back and forth process between different stakeholders and not a one-way procedure, in which the agencies act completely decoupled from practice, this paper will focus on the agencies' official documents and take them basis for further discussion.

Nearly every country has its own accreditation agencies. So there has to be made a choice first. Within this paper two different agencies will be taken into focus: The Accreditation Board for Engineering and Technology (ABET; USA; www.abet.org) and the Accreditation Agency for degree programs in Engineering, Computer Science, Natural Science, Mathematic and Teaching Qualification (ASIIN; Germany; www.asiin.de) which refers mainly to the European Network for Accreditation of Engineering Education (ENAEE; Europe; www.enaee.eu). ASIIN was chosen because it is the most important agency for engineering programs in Germany. Just looking at Germany, however, would lead to a very narrow perspective. Hence the German view will be complemented by an American perspective. Especially the ABET criteria [8] are pretty often cited and discussed in the engineering education field [9–11], so they should be taken into account here, too. In the following it will be given a short overview on what the agencies say about students outcomes in engineering education in order to prepare Section 4, in which that will be brought in context with the areas of competence from Section 2.

The ASIIN defined in addition to their general criteria for program accreditation [12] requirements for bachelor and master programs with a focus on engineering practice or engineering science in mechanical engineering in the following areas [13] (own translation):

- Knowledge and Understanding
- Engineering Methods
- Engineering related Development and Design
- Scrutinizing and Evaluation
- Engineering Practice
- Interdisciplinary Competences

These requirements are pretty much identical to the ones defined in 2008 by the ENAEE Administrative Council approved under the "Framework Standards for the Accreditation of Engineering Programmes [sic]" (EUR-ACE) [14]. Below these areas concrete outcomes are defined.

In 2012 the ABET defined all in all eleven (a - k) student outcomes for engineering education [8]. These criteria are:

(a) An ability to apply knowledge of mathematics, science, and engineering
(b) An ability to design and conduct experiments, as well as to analyze and interpret data
(c) An ability to design a system, component, or process to meet desired needs within realistic constraints such as economic, environmental, social, political, ethical, health and safety, manufacturability, and sustainability
(d) An ability to function on multidisciplinary teams
(e) An ability to identify, formulate, and solve engineering problems
(f) An understanding of professional and ethical responsibility
(g) An ability to communicate effectively
(h) The broad education necessary to understand the impact of engineering solutions in a global, economic, environmental, and societal context
(i) A recognition of the need for, and an ability to engage in life-long learning
(j) A knowledge of contemporary issues
(k) An ability to use the techniques, skills, and modern engineering tools necessary for engineering practice

Whereas the areas defined by the ASIIN and EUR-ACE show a high level of conformity on a general level, the ABET criteria are framed on another level of detail. In order to compare all three of them it is necessary to go more in detail and look at the requirements ASIIN and EUR-ACE explain in the different areas. This will be done in the following, accompanied by the next step, which is the classification in the four areas of competences.

4 How to Accreditation Requirements fit into the Four Fields of Competences

The four areas of competences outlined in 2 will be taken at this point to have a closer look on the requirements for respectively outcomes in engineering programs defined by the three accreditation agencies ASIIN/ENAEE [13] and ABET [8]. In the following, these will be assigned to the four areas of competences. (Note: Looking at the ASIIN's explanation, only requirements for master degrees with a focus on engineering practice and engineering science in mechanical engineering are taken into account)

4.1 Professional and Methodological Competence

4.1.1 ASIIN

Area Knowledge and Understanding:

- broad knowledge and deep understanding of mathematical and natural sciences and engineering principles as well as its interdisciplinary extensions and application-oriented knowledge about areas of specialty
- a critical awareness of new findings in the discipline

Area Engineering Methods:

- ability to analyze and solve problems, which are unusual or incomplete and have competitive specifications, scientifically and/or application-oriented
- ability to frame complex problems in new or evolving areas of the own discipline
- ability to apply innovative methods in the problem solving process and generate new methods

Area Engineering related Development and Design:

- ability to generate new concepts and solutions confronted with unusual questions with respect to other disciplines
- ability to use the own creativity in order to develop new products, processes and methods
- ability to use engineering judgment in order to work with complex, technical imprecise or incomplete information

Area Scrutinizing and Evaluation:

- ability to classify knowledge from different areas, combine that systematically in order to put that into practice and handle complexity
- ability to become acquainted with something new in fast and systematic way
- ability to rate methods and its limitations

- ability to reflect systematically on non-technical effects of the engineering work and incorporate that into acting in a responsible way

Area Interdisciplinary Competences:

- have an understanding of health- and security-related as well as legal consequences of engineering practice, do reflect on implications of engineering solutions for society and ecology, and commit themselves to follow professional engineering ethics as well as norms
- know and understand project-management as well as economic methods (for example risk management) and know their limitations

4.1.2 ABET

(a) An ability to apply knowledge of mathematics, science, and engineering
(b) An ability to design and conduct experiments, as well as to analyze and interpret data
(c) An ability to design a system, component, or process to meet desired needs within realistic constraints such as economic, environmental, social, political, ethical, health and safety, manufacturability, and sustainability
(e) An ability to identify, formulate, and solve engineering problems
(h) The broad education necessary to understand the impact of engineering solutions in a global, economic, environmental, and societal context
(k) An ability to use the techniques, skills, and modern engineering tools necessary for engineering practice

4.2 Personal Competences

4.2.1 ASIIN

Area: Interdisciplinary Competences:

- recognize the necessity of self-directed as well as life-long learning and are able to do so
- ability to work effectively on your own or in a team and act as a team coordinator if necessary

4.2.2 ABET

(i) A recognition of the need for, and an ability to engage in life-long learning
(f) An understanding of professional and ethical responsibility
(j) A knowledge of contemporary issues

4.3 Social and Communication Competences

4.3.1 ASIIN

- ability to use different methods in order to communicate effectively with engineering colleagues from and the general public
- are able to work and communicate confidently in national and international contexts
- are able to act as leader of teams, composed of different disciplines and levels

4.3.2 ABET

(d) An ability to function on multidisciplinary teams
(g) An ability to communicate effectively

4.4 Conclusion

Going into detail, it is not very surprising that the vast amount of aspects in the accreditation agencies' requirements are concerning technical or professional engineering and methodological aspects. Even some of the so-called "transferable skills" or "interdisciplinary competences" are not really transferable to other disciplines as the headline may indicate. The competence "have an understanding of health- and security-related as well as legal consequences of engineering practice, do reflect on implications of engineering solutions for society and ecology, and commit themselves to follow professional engineering ethics as well as norms" (interdisciplinary competences; ASIIN) for example, can hardly be identified as a real transferable competence even if it expresses the necessity for engineers to not only focus in the own work but put it into relation to other disciplines.

Very surprising instead is the aspect that none of the different requirements and defined outcomes can be mapped to thc fourth field of competence "Activity and Implementation competence" — or at least cannot easily be mapped to this field. At a later point it should be examined why this is the case. Is this for example because these competences should not be gained during studies? Another explanation could be that these competences are highly integrative and only describable by explaining others (for example more technical competences). This short analysis already leads us the last part of this paper.

5 What that means for Education in Laboratories

5.1 ABET: Discussing and Defining the Fundamental Objectives of Engineering Instructional Laboratories

In 2002, ABET held a colloquy to the query of what are, in broad terms, the accurate goals of a laboratory experience in Undergraduate Engineering Education. Throughout that talk the participants agreed to outline the Instructional Laboratory Experience as "personal interaction with equipment/tools leading to the accumulation of knowledge and skills required in a practice-oriented profession" [15]. The following are the comprehensive set of learning objectives for the engineering laboratory developed and agreed by the participants of the colloquy [15, 16]. All objectives start with the following: "By completing the laboratories in the engineering undergraduate curriculum, you will be able to...."

1. *Instrumentation.* ...apply appropriate sensors, instrumentation, and/or software tools to make measurements of physical quantities.
2. *Models.* ...identify the strengths and limitations of theoretical models as predictors of real-world behaviors. This may include evaluating whether a theory adequately describes a physical event and establishing or validating a relationship between measured data and underlying physical principles.
3. *Experiment.* ...devise an experimental approach, specify appropriate equipment and procedures, implement these procedures, and interpret the resulting data to characterize an engineering material, component, or system.
4. *Data Analysis.* ...demonstrate the ability to collect, analyze, and interpret data, and to form and support conclusions. Make order of magnitude judgments and use measurement unit systems and conversions.
5. *Design.* ...build, or assemble a part, product, or system, including using specific methodologies, equipment, or materials; meeting client requirements; developing system specifications from requirements; and testing and debugging a prototype, system, or process using appropriate tools to satisfy requirements.
6. *Learn from Failure.* ...identify unsuccessful outcomes due to faulty equipment, parts, code, construction, process, or design, and then re-engineer effective solutions.
7. *Creativity.* ...demonstrate appropriate levels of independent thought, creativity, and capability in real-world problem solving.
8. *Psychomotor.* ...demonstrate competence in selection, modification, and operation of appropriate engineering tools and resources.
9. *Safety.* ...identify health, safety, and environmental issues related to technological processes and activities, and deal with them responsibly.
10. *Communication.* ...communicate effectively about laboratory work with a specific audience, both orally and in writing, at levels ranging from executive summaries to comprehensive technical reports.

11. *Teamwork.* . . .work effectively in teams, including structure individual and joint accountability; assign roles, responsibilities, and tasks; monitor progress; meet deadlines; and integrate individual contributions into a final deliverable.
12. *Ethics in the Laboratory.* . . .behave with highest ethical standards, including reporting information objectively and interacting with integrity.
13. *Sensory Awareness.* . . .use the human senses to gather information and to make sound engineering judgments in formulating conclusions about real-world problems.

[15] concluded, that "engineering instructional laboratories provide a fertile field for educational research in the future. While it is always interesting and rewarding to develop new laboratory experiments and experiences, future research should be aimed at developing a more thorough understanding of this critical component of the undergraduate" experience. Moreover, [16] demands a "further understanding of the fundamental objectives of instructional laboratories: While the ABET/Sloan colloquy produced a useful list of objectives, these need to be ‚calibrated' by comparison to objectives currently in use and by developing an understanding of the objectives on a disciplinary basis. Activities might include a discipline-specific survey of faculty or an analysis of proposals received by funding agencies such as the National Science Foundation".

5.2 ASIIN

Yet, the ABET discussion on "The Fundamental Objectives of Engineering Instructional Laboratories" neither has been officially echoed in Germany's accreditation orders for Engineering Education curricula, nor was there a separate broader discussion on that topic. Until now, there have been very lonely efforts in Germany to awake enthusiasm and a broader debate on laboratory learning objectives in engineering education (see e.g., [16–22]).

6 Future Plans

This is a work-in-progress paper that explains just a first step of future work. Hence it mainly shows descriptive work by showing how engineering requirements and defined learning outcomes can be integrated into a general four-field-grid of competences. This paper combines the view of two different accreditation agencies and their explanations concerning engineering education and what graduates should have learned during studies. It is shown that the field of activity and implementation competences cannot be identified directly in the agencies documents. This might be because of the highly integrative characteristics of this field of competence or using the four general competences simply is not appropriate in this context. Another competence grid for example is given by [23] using the areas technical, professional and global competence. Especially the latter might be a very interesting field of competences as the world's markets more and more globalize and that leads undeniably to special competence requirements (as seen, some of them are integrated in the explained agencies' documents already). So a new field of competence, which especially deals with some kind of global competence, could be important to define and examine.

Going more into detail and refining the work explained in this paper will mainly drive future work. The final aim is to generate an adequate grid in which all intended learning outcomes for engineering programs are integrated. Using such a grid can help in the future curriculum developers to define new programs and set focus areas. A next step will be to broaden the view even more, which means going more into detail what industry expects from graduates. Looking at the accreditation agencies on the one hand and at industry demand on the other and define differences as well as similarities will be a very interesting step in the future.

References

1. K.H. Minks, Kompetenzen für den arbeitsmarkt:: Was wird vermittelt, was vermisst? In: *Bachelor- und Master-Ingenieure:. Welche Kompetenzen verlangt der Arbeitsmarkt*, ed. by Stifterverband für die Deutsche Wissenschaft, Positionen, Essen, 2004, pp. 32–40
2. F.S. [1] Becker, *Herausforderungen für Elektroingenieure/innen: Entwicklungen im Arbeitsumfeld, Erwartungen von Personalverantwortlichen: Tipps für Berufsstart und Karriere*. Frankfurt am Main, 2012
3. G.P. Bunk, Kompetenzvermittlung in der beruflichen aus- und weiterbildung in deutschland. Europäische Zeitschrift Berufsbildung (1), 1994, pp. 9–15
4. J. Erpenbeck, L. Rosenstiel, *Handbuch Kompetenzmessung: Erkennen, verstehen und bewerten von Kompetenzen in der betrieblichen, pädagogischen und psychologischen Praxis*, 2nd edn. Schäffer-Poeschel, Stuttgart, 2007
5. M. Bernien, Anforderungen an eine qualitative und quantitative darstellung der beruflichen kompetenzentwicklung. In: *Berufliche Weiterbildung in der Transformation - Fakten und Visionen, 1997*, vol. Kompetenzentwicklung, Waxmann, Münster, 1997, pp. 17–84
6. J. Erpenbeck, W. Heyse, *Die Kompetenzbiographien. Strategien der Kompetenzentwicklung*. Münster, 1999
7. J. Wildt, Kompetenzen als "learning outcome. Journal Hochschuldidaktik **17** (1), 2006, pp. 6–9
8. Criteria for accrediting engineering programs: Effective for reviews during the 2013-2014 accreditation cycle, 2012
9. P. Blumenthal, U. Grothus, Developing global competence in engineering students:: U.s. and german approaches. The Online Journal for Global Engineering Education **3** (2), 2008
10. G. Heitmann, Elemente einer qualitativ hochwertigen und vergleichbaren europäischen ingenieurausbildung. In: *Bachelor- und Master-Ingenieure:. Welche Kompetenzen verlangt der Arbeitsmarkt*, ed. by Stifterverband für die Deutsche Wissenschaft, Positionen, Essen, 2004, pp. 18–23
11. S. Moore, D. May, K. Wold, Developing cultural competency in engineering through transnational distance learning. In: *Transnational Distance Learning and Building New Markets for Universities*, ed. by R. Hogan, IGI Global, Hershey (PA/USA), 2012

12. Allgemeine kriterien für die akkreditierung von studiengängen: Ingenieurwissenschaften, informatik, architektur, naturwissenschaften, mathematik und ihre kombinationen mit anderen fachgebieten, 2012
13. Fachspezifisch ergänzende hinweise zur akkreditierung von bachelor- und masterstudiengängen des maschinenbaus, der verfahrenstechnik und des chemieingenieurwesens, 2011
14. Eur-ace framework standards for the accreditation of engineering programmes, 2008. URL http://www.enaee.eu/eur-ace-system/eur-ace-framework-standards
15. L.D. Feisel, G.D. Peterson, A colloquy on learning objectives for engineering education laboratories. In: *Proceedings of the 2002 American Society for Engineering Education Annual Conference & Exposition Copyright*. 2002
16. H.G. Bruchmüller, A. Haug, *Labordidaktik für Hochschulen – Eine Einführung zum Praxisorientierten Projekt-Labor*, *Schriftenreihe report*, vol. 40. Leuchtturm-Verlag, 2001
17. A. Haug, *Labordidaktik in der Ingenieurausbildung*. VDE-Verlag GmbH, Berlin, 1980
18. URL http://www.acatech.de/?id=1841
19. C. Terkowsky, C. Pleul, I. Jahnke, A.E. Tekkaya, Tele-operated laboratories for production engineering education - platform for e-learning and telemetric experimentation (petex). International Journal of Online Engineering (iJOE). Special Issue EDUCON 2011 **7** (Issue S1), 2011, pp. 37–43
20. C. Terkowsky, T. Haertel, E. Bielski, D. May, Creativity@school: Mobile learning environments involving remote labs ande-portfolios. a conceptual framework to foster the inquiring mind in secondary stem education. In: *IT Innovative Practices inSecondary Schools: Remote Experiments*, ed. by J.C. Zubía, O. Dziabenko, Bilbao, Spain, 2013, pp. 255–280
21. C. Terkowsky, I. Jahnke, C. Pleul, D. May, T. Jungmann, A.E. Tekkaya, Petex@work. designing cscl@work for online engineering education. In: *Computer-Supported Collaborative Learning at the Workplace - CSCL@Work, Computer-Supported Collaborative Learning Series*, vol. 14, ed. by S.P. Goggins, I. Jahnke, V. Wulf, Springer, 2013, pp. 269–292
22. C. Terkowsky, T. Haertel, E. Bielski, D. May, Bringing the inquiring mind back into the labs. a conceptual framework to foster the creative attitude in higher engineering education. In: *Proceedings of EDUCON2014 – IEEE Global Engineering Education Conference*. 2014
23. Y. Chang, D. Atkinson, E.D. Hirleman, International research and engineering education: Impacts and best practices. Online J. Global Eng. Educ **4** (2), 2009

On Learning Objectives and Learning Activities to Foster Creativity in the Engineering Lab

Claudius Terkowsky and Tobias Heartel

Abstract Creativity involves coming up with something novel, something different. Up-to-date laboratory learning approaches in combination with inventive ICT can offer an immense variety of novel opportunities for experimentation and learning in the modes of creative inquiry. Fostering and encouraging creative laboratory learning in engineering education may not only animate what is learned but also includes the chance to tighten students' understanding and creative self-efficacy. The presented conceptual framework proposes a learning space based on portable devices in combination with an e-portfolio system. The aligned teaching and learning approaches aim at facilitating and fostering creative laboratory learning in engineering education. To this end, this article features six different scaffolding tasks to design learning objectives and activities for fostering creativity in the lab. It illustrates how the proposed personal learning environments might enhance or substitute formal classroom activities and laboratory work in order to achieve sophisticated learning objectives.

Keywords Creative Learning · Laboratory Learning · Online Labs · Personal Learning Environments · Mobile Learning · Engineering Education

1 Introduction

According to [1] creativity involves coming up with something novel, something different. While creativity mainly deals with generating of ideas and novel solutions to problems, engineering is more concerned with bringing forth technological solutions to the task at hand, e.g. by designing and developing artefacts, processes, models, systems and services. But [2] stresses: "If creativity is so central to engineering, why is it not an obvious part of the engineering curriculum at every university?" Hitting the same line, [3] studied findings on fostering creativity in engineering education,

C. Terkowsky (✉) · T. Heartel
Engineering Education Research Group, Center for Higher Education Dortmund,
TU Dortmund University, Vogelpothsweg 78, 44227 Dortmund, Germany
e-mail: claudius.terkowsky@tu-dortmund.de

Originally published in "Proceedings of 2014 International Conference on Interactive Collaborative Learning (ICL)", © IEEE 2014.
Reprint by Springer International Publishing AG 2016,
DOI 10.1007/978-3-319-46916-4_75

and concluded that there is insignificant provision for creative students, until yet. To foster the inquiring minds in engineering education, creative laboratory learning may not only animate what is learned but also involves the opportunity to consolidate students' understanding as well as creative self-efficacy [4–7]. Findings on the virtue of laboratory work delineate that involving students in authentic investigation can encourage them to:

- develop their own comprehension of scientific theories, models, and concepts [8]
- develop knowledge schemes and solve engineering tasks at hand [9, 10]
- experience science as scientists [11]
- cultivate positive mindsets on science and engineering [12]
- develop critical thinking as well as decision-making abilities [13], and
- creatively nurture their own research questions [4–7]

"Unfortunately, science laboratory materials and exercises usually provided to teachers of science, K-16, are still centred on traditional methods of the past decades" [14]. This work is a subtask of the collaborative project ELLI–Excellent Teaching and Learning in Engineering Education, which is funded by the German Ministry of Research and Education between 2011-2016. As opposed to this, laboratories based on research and inquiry flip teachers' roles into inviting students

- to frame problems themselves as well as
- to relate their findings to preceding work
- to indicate the intention of their proposed investigation
- to specify the task at hand
- to project outcomes (deemed possible or impossible)
- to determine problem-solving approaches, and
- to lastly accomplish the study [15, 16]

Beyond this, the utilisation of mobile devices can intensify creative thinking processes since original ideas usually emerge unexpectedly [17]. Having the mobile device on hand gives the opportunity to the user to capture these ideas by making notes, audio recordings, or logging first artefacts and actions by camera for enabling future access and continuative improvement [18]. Four contemporary tides in ICT-based teaching and learning development with the potential to support creative engineering education are distinguishable: personal learning environments (1), flipped or inverted classrooms (2), portable devices (3), and online labs (4).

1. Personal learning environments (PLE) are "educational technology which responds to the way people are using technology for learning and which allows them to shape their own learning spaces themselves, to form and join communities and to create, consume, remix, and share material" [19]. PLEs "imply redrawing the balance between institutional learning and learning in the wider world" [20], and thus provide more responsibility and more independence for learners.
2. Flipped or inverted learning turns the focus of the class on the students instead of the teacher. With inverted learning, students can absorb the material online as

homework, or wherever and whenever they want, and then practice what they've learned with guidance from the teacher, when they need it. This novel learning style increases students' engagement and achievement, and provides to all forms of personalized learning [21].
3. Portable devices are perhaps the most increasing kind of technology for informal learning. The propagation of portable devices like tablet PCs and smart phones offers a distinguished capability to nurture new ways of informal and creative learning–anytime and anywhere [22].
4. The practice of remote and virtual experimentation [23–25] can be delivered to the learner by integrating them into technology enhanced and didactically aligned collaborative online learning systems [26] like cloud-based PLE.

This raises three crucial questions:

- How can engineering education foster (and be fostered by) creativity and creative engineering in general? How can students be enabled to gain the chance of executing lab work in a more creative mode in particular?
- Which might be the best manner for learners to capture and document their creative learning process documentation and to write their laboratory reports, even exams? How can teachers guide through these processes?
- How can this be nurtured by PLEs, portable devices and online experimentation facilities?

In this regard, the proposed solution is a personal learning environment based on portable devices assimilating online experimentation test beds and an e-portfolio system to expedite and cultivate creative learning in science and engineering studies.

2 Creative Learning with PLEs and Online Labs

2.1 E-Portfolios as Personal Learning Environments

E-Portfolio-based PLE software, e.g., 'Mahara' can conveniently be combined with an LCMS based on Moodle. The integrating application 'Mahoodle' assimilates properties and functions of the teacher-led LCMS 'Moodle' and the learner-led system 'Mahara' for e-portfolios. This mesh-up can be regarded as "a facility for an individual [or a group] to access, aggregate, configure and manipulate digital artefacts of their ongoing learning experiences" [27].

2.2 Changing in and out of Interaction with Portable Technology

Especially smart phones and portable computers open up a wide-ranging diversity of situations for creative inquiry regardless of time location and ambiance. As students are permanently on the move and changing in and out of interaction with technology, spare time periods can be spontaneously utilized for learning and working with e-portfolio software and the related laboratory equipment. This can be initiated virtually anywhere, regardless of most conventional software limitations, once access to a respective learning environment had been obtained: a far-reaching feature which is widely spread throughout cloud computing [28, 29].

2.3 Interactive Web Labs

Many programs are going to include Web labs into their education for enhancing the efficacy of rare equipment as well as to share it with other institutions and locations [23–25, 30–33]. According to [7], in Web labs students can be engaged in:

- "blended and online learning scenarios regardless of time location, as far as Internet access is available
- learning activities which take longer than a typical class meeting time
- multi-part assignments which require students to use equipment for several short periods over the span of one week or longer
- socio-technically enhanced opportunities for student collaboration
- building up their own knowledge schemes using tele-operated equipment provided by the online labs" [7]

Even risky experiments which may be too dangerous for human interactions or experiments with precious equipment can be conducted in a completely virtual fashion.

2.4 Digital Documentation of Learning Processes

By establishing personal portfolios, e.g. a digital field diary, learners are allowed for documenting their specific learning and research developments [34, 35]. As stated in [7], students will learn to be able to:

- "arrange all data and information they would choose to collect or share with others in different orders
- present experiments and their results or to show photos from the experimental set-up
- write notes and reflections on their experiments during their research-based learning processes
- explain their research results and thoughts to themselves and others
- collect ideas in creative moments, and to organize and improve them whenever desired

- support collaboration by allowing other learners and teachers to have access to their e-portfolios
- prepare, write and revise the lab report as a living document drawing on their learner-generated multimedia content" [7]

These considered learning objectives and related learning activities are significant and essential to nurture the students' personal creative learning cycles [36]. Above, the e-portfolio as a digital field diary and personal learning documentation can always offer guidance and reflection in topics of students' own inquiry [7, 37–39]. Beyond that, other eligible persons or groups are able to view the collection within the portfolio. It can be stated that the e-portfolio is not only a versatile instrument for both individual learning documentation and learning-related reflection processes but also an especially valuable tool for collaborative communication. Finally, course instructors get the opportunity to guide and appraise the students' outcomes by observing, intervening, considering and reviewing their learners' e-portfolio work [7].

2.5 *Learning Objectives and Aktivities to Foster Creativity in Higher Engineering Education*

According to [40], learning objectives "are statements, written from the students' perspective, indicating the level of understanding and performance they are expected to achieve as a result of engaging in the teaching and learning experience". In addition to that, learning activities describe the actual related students' actions that indicate the cognitive (and psychomotor and affective) engagement they are expected to mature and finally master. Whilst general learning objectives and related activities for the engineering lab are well established by [41], creativity still remains marginally addressed in this context. To this effect – foster creativity in the context of education – a model of six facets for defining, formulating, stimulating, and analysing learning objectives and activities is established by [42–46].

2.6 *The Six Facets of Creative Learning Objectives and Activities*

According to [7], these six facets are:

1. Students (are able to) develop self-reflective learning skills: Learners are relieved from the constraints formed by their receptive habits and empowered to question information given by the teacher. An internal dialogue takes place and knowledge becomes "constructed" rather than "adopted".
2. Students (are able to) mature independent learning skills: Teachers stop to restrain the way students learn. Instead, students are free to discover relevant literature on

their own and, for example, to make their own guidelines about structuring a text or even to find their own research questions and to choose appropriate methods of answering them.
3. Students (are able to) nurture curiosity and motivation: This aspect is related to all measures that contribute to increased motivation, for instance, linking a theoretical question to a practical example or representation.
4. Students (are able to) learn from "learning by doing" tasks: Students learn by creating a kind of "product", their respective artefact. Depending on the discipline, this might be a presentation, an interview, a questionnaire, a machine, a website, a computer program or similar. Students are guided into the role of "real" researchers.
5. Students (are able to) develop multi-perspective thinking skills: Learners overcome thinking within the limits of their respective disciplines or prejudiced thinking. Along with that, they learn to consider an issue from different points of view and to use thinking methods which prevent their brains from being trapped in ordinary disciplinary paths.
6. Students (are able to) reach for original ideas: Learners are facilitated to embrace new, original ideas and to prepare themselves to be as ready-to-receive as possible. Although the acquirement of original ideas cannot be forced, the reception of original ideas can be fostered by applying appropriate creative techniques and by creating a suitable environment (allowing students to make mistakes and to express unconventional ideas without being laughed at or rejected).

Hereafter, six different model tasks to scaffold creative learning in the engineering lab are presented in order to explicate how personal learning environments containing the combination of Web labs, e-portfolios and portable devices can improve formal laboratory learning.

2.7 *Model Tasks to Scaffold Creative Laboratory Learning Objectives and Aktivities*

The model tasks are established on the six creativity facets framework.

2.7.1 Students (are able to) Develop Self-reflective Learning Skills by Evolving Critical Thinking with Technological Pitfalls

In order to nurture critical awareness of technological pitfalls, a misleading environment can be deployed to introduce a critical approach to given or default information. In order to achieve this goal, teachers can create tasks which are impossible to solve, or can provide false information that leads to seemingly correct but erroneous results of the experiments. In both cases, the students will be irritated while performing the experiment as well as challenged to find the cause. A common engineering topic is surface coating in the field of micro technology. In order to foster self-reflective learning, teachers could ask their students to produce a surface coating for a substrate that is suitable for frying eggs. To stimulate a critical attitude towards given information, teachers can provide students with faulty instructions for developing and producing such non-stick coatings. The result could be a surface that would melt under high temperature and intermingle with the fried egg (have a look that students do not start to eat it). In this case, students would have to verify the given information as well as to clarify the mistake and reproduce the surface coating which fulfils demand. The achieved learning outcome would comprise (besides technical knowledge about surface coatings and their) that students mistrust and reflect given information rather than simply receiving it, as they can experience the consequences of applying misleading information without putting it into question.

2.7.2 Students (are able to) Develop Independent Learning Skills by Improving Self-reliance and Self-confidence Towards Technical Issues

Online lab work combined with PLEs make it possible to let students learn more autonomously. For instance, PLEs permit searching for information, planning and running experiments as well as learning independent decision making. Advanced materials are another common topic in higher engineering education. Instead of simply handing students pre-selected information about the characteristics of selected advanced materials, teachers could ask them a relatively open question, for instance: "Which material is the best choice to attach an iPhone to a mirror in a fashion (in order to be able to watch it during the morning toilet) that it cannot be separated manually but by use of a second material without any damage of both?" Students are asked to collect the necessary information for themselves in the Internet rather than in textbooks. In order to give assistance, teachers have to carefully monitor their students' learning progress. If they notice difficulties or confusion among their students, teachers will have to help them "to help themselves" without providing completed solutions which will only lead to well-known solutions. For this purpose, PLEs are valuable tools that enable teachers to monitor their students' learning steps and provide feedback to them, while students on the system's other end can ask for help at any time. As a reward for using this learning scenario, students will not only incorporate a wide range of aspects about advanced materials but will also learn to plan small research projects, as well as putting them into practice and feeling responsible for them as well as their individual learning successes. In the long run, performing small learning tasks on their own will improve their self-reliance and self-confidence towards technical issues.

2.7.3 Students (are able to) Nurture Curiosity and Motivation by Intertwining Technical Issues with Students' Real-World Experiences

The question of rendering subject matter more interesting for students in order to tackle their curiosity and motivation is always worth asking. A renowned methodology to increase students' motivation is to apply practical questions and tasks. Therefore, combining practical issues taken directly from the students' daily environment as well as online experiments or simulations might be the silver bullet of situated experimental learning. For example, a lecture in electronics engineering about the characteristics of transistors could be very boring. But it can easily be spiced up with a small homework as a preparation for the transistor lecture: Ask the students to spend one day completely without the use of transistors. Collect their experiences in the lecture and try to find transistors they used even if they thought were absent (it is not that easy to raise awareness of the ubiquitous use of transistors). By noticing the relevance of transistors for modern life, students might become more interested in lab works on for example the temperature limits of those transistors keeping their life civilized. To this end, the use of online labs may have several advantages regarding students' creativity: they are readily available throughout the day, so students can pursue experiments whenever they have a good idea and therefore do not have to wait for the next lab lesson to test their personal ideas. They can conduct it whenever they feel creative and wherever they are, all they need is simply their online mobile device.

2.7.4 Students (are able to) Learn in the Mode of "Learn by Doing" by Designing and Building Functional Models

Since experimental learning as a means of scaffolding always inherits some sort of product to be created, this learning method appears suitable in order to achieve this goal as it might also strengthen the students' awareness of their creative potential. Besides, exposing the learning outcome to a larger interest and external assessments are valuable factors in tackling students' intrinsic motivation. For this purpose, the PLEs can be used to render the readily documented learning process accessible for externals. For example, in aerospace engineering students are expected to learn the fundamentals of rheology. Instead of just presenting them the contents of a course book in a lecture, students could be asked to develop the most effective paper plane by assistance of several materials which can be found in the laboratory. By doing so, students should parenthetically be enabled to determine the drag coefficient of different designs. Moreover, by using online simulations of wind tunnels to try several varieties of their paper plane, students get involved into increasing its range by reducing its drag in an informal, playful manner. Furthermore, they would have to find the proper material for realizing the best design with regard to ductility and – of course – weight. At this point, with emphasis on rheology, different aspects of aerospace engineering would be mentioned and the coherences would become visible. Students would be enabled to see a bigger part of the picture while applying and thus internalizing the necessary professional knowledge. Finally, the inventors of the most effective paper plane (is it really the one with the lowest cd value?) would receive a prize.

2.7.5 Students (are able to) Evolve Multi-perspective Thinking by Overcoming Cognitive Barriers

This scenario can raise students' capability to reconsider their questions from varying viewpoints: envisaging a student who had executed a lab task that generated odd results or does not know how to construe them appropriately. While writing his e-portfolio as documentation for the teacher's evaluation, s/he could start the "creative-help app", a wizard application within the PLE intended to help him develop different perspectives on the same problem: The student is asked to do a (mental) headstand following the question: "What else could I do to get the wrong results from experimenting?" If this should not suffice, the student will be asked to describe his experimental design and assumptions in a way that a ten-year-old could understand it. If those methods which are fairly close to the problem still cannot help him, the "creative-help app" will suggest a force-fit technique by showing a picture that does not have anything in common with a problem and asking the student to find relationships between the picture and his experiment. To depart from the beaten track can help students to look at their problem from completely different perspectives. This may result in unconventional or provocative ideas at first sight, but rethinking the obviously unsuitable solutions sometimes leads to the one really good idea which would not have appeared without making the detour [7, 46, 47].

2.7.6 Students (are able to) Reach for Original Ideas by Breaking all Rules and Posing One's Own Questions

The admission for "breaking all rules and posing one's own questions" is the actual challenge in creativity education. There are several options available to arrange for learner-led inquiries, investigation, discovery and reasoning. However, serious "Breaking all Rules and Posing one's Own Questions" may also create a lot of severe resistance from colleagues, superiors, institutions and others who still need to be involved and convinced to transform an idea into innovation. At this point, it is up to the reader to design a "Reach for Original Ideas by Breaking all Rules and Posing one's Own Questions" learning scenario for the lab course.

3 Conclusion

The integration of online labs and e-portfolios into mobile personal learning environments can establish unique scenarios to nurture students' creative performing in engineering education. Permitting students to learn this self-governing mode already integrates fostering their creativity with regard to the introduced and adapted model of six facets for defining, formulating, stimulating, and analysing learning objectives and activities to cultivate creativity in engineering education. This is one crucial approach to engage students in high-level learning outcomes and by this means developing the basis of fundamental domain-specific and generic competences for their successful participation in the prospective world of work. Moreover, the depicted methods can nurture attitudes like curiosity, creative self-efficacy, agency and responsibility. If students are empowered to evolve their own research questions, to choose suitable experimentation designs and finally to perform the experiment, they will be able to develop a spirit of inquiry and research. This spirit is one important premise for developing original ideas. And, generating of original ideas may the prerequisite for bringing forth technological solutions by designing and developing artefacts, processes, models, systems and services – or in short: engineering.

References

1. D. H. Cropley, Creativity in engineering. In: *Multidisciplinary Contributions to the Science of Creative Thinking*, ed. by G. E. Corazza, S. Agnoli, Springer, 2014
2. K. Kazerounian, S. Foley, Barriers to creativity in engineering education: A study of instructors and students perceptions. Journal of Mechanical Design, 2007, p. 761
3. D. H. Cropley, A. J. Cropley, Engineering creativity: A systems concept of functional creativity. In: *Faces of the Muse: How People Think, Work and Act Creatively in Diverse Domains*, ed. by J. C. Kaufman, J. Baer, Lawrence Erlbaum, Hillsdale, New York, 2005, pp. 169–185
4. C. Terkowsky, I. Jahnke, C. Pleul, D. May, T. Jungmann, A.E. Tekkaya, Petex@work. designing cscl@work for online engineering education. In: *Computer-Supported Collaborative Learning at the Workplace - CSCL@Work, Computer-Supported Collaborative Learning Series*, vol. 14, ed. by S. P. Goggins, I. Jahnke, V. Wulf, Springer, 2013, pp. 269–292
5. C. Terkowsky, I. Jahnke, C. Pleul, A. E. Tekkaya, Platform for e-learning and telemetric experimentation (petex) - tele-operated laboratories for production engineering education. In: *Proceedings of the 2011 IEEE Global Engineering Education Conference (EDUCON) Ĺearning Environments and Ecosystems in Engineering Education. IAOE*, ed. by M. E. Auer, Y. Al-Zouvi, E. Tovar. 2011, pp. 491–497
6. T. Haertel, C. Terkowsky, D. May, C. Pleul, Entwicklung von remote-labs zum erfahrungsbasierten lernen. In: *Themenheft Kompetenzen, Kompetenzorientierung und Employability in der Hochschule (Teil 2), Zeitschrift für Hochschulentwicklung ZFHE*, ed. by N. Schaper, T. Schlömer, M. Paechter, 2013, pp. 79–87
7. C. Terkowsky, T. Haertel, E. Bielski, D. May, Bringing the inquiring mind back into the labs. a conceptual framework to foster the creative attitude in higher engineering education. In: *Proceedings of EDUCON2014 – IEEE Global Engineering Education Conference: Ëngineering Education towards Openness and Sustainability*. April 3-5, 2014
8. G. Roehrig, A. Lulie, M. Edwards, Versatile vee maps. The Science Teacher **68**, 2001, pp. 28–31

9. J. German, S. Haskins, S. Auls, Analysis of nine high school biology laboratory manuals: promoting scientific inquiry. Journal of Research in Science Teaching **33** (5), 1996, pp. 475–499
10. D.M. C. Terkowsky, T. Haertel, C. Pleul, Experiential learning with remote labs and e-portfolios - integrating tele-operated experiments into personal learning environments. International Journal of Online Engineering (iJOE) **9** (1), 2013, pp. 12–20. doi:10.3991/ijoe.v9i1.2364
11. A. Okebukola, B. Ogunniyi, Cooperative, competitive and individualistic science laboratory interaction patterns effects on students' achievement and acquisition of practical skills. Journal of Reserach in Science Teaching **21** (9), 1984, pp. 875–884
12. R. Lehrer, L. Schauble, A. Petrosino, Reconsidering the role of experiment in science education. In: *Designing for science : implications from everyday, classroom, and professional settings*, ed. by K. D. Crowley, C. D. Schunn, T. Okada, Lawrence Erlbaum, Mahwah, NJ, 2001, pp. 251–277
13. F. Abd-El-Khalik, S. BouJaoude, R. Duschl, N.G. Lederman, R. Mamlok-Naaman, A. Hofstein, Inquiry in science education: International perspectives. Science Education **88**, 2004, pp. 397–419
14. D. W. Sunal, C. S. Sunal, C. Sundberg, E. L. Wright, The importance of laboratory work and technology in science teaching. In: *The Impact of the Laboratory and Technology on Learning and Teaching Science K-16*, ed. by D. W. Sunal, E.L. Wright, C. Sundberg, IAP-Information Age Publishing, 2008, pp. 1–28
15. P. Tamir, How are laboratories used. Journal of Research and Science Teaching **14** (9), 1977, pp. 311–316
16. C. W. Keys, Revitalizing instruction in scientific genres: Connecting knowledge production with writing to learn in science. Science Education **83** (2), 1999, pp. 115–130
17. T. Haertel, I. Jahnke, Wie kommt die kreativitätsförderung in die hochschullehre. Zeitschrift für Hochschulentwicklung **6** (3), 2011, pp. 238–245
18. D. May, C. Terkowsky, T. Haertel, C. Pleul, The laboratory in your hand - making remote laboratories accessible through mobile devices. In: *Proceedings of the 2013 IEEE Global Engineering Education Conference (EDUCON), S̈ynergy from Classic and Future Engineering Education.* March 13-15, 2013. IEEE, 2013, pp. 335–344
19. G. Attwell, Personal learning environments - the future of elearning? eLearning Papers **2**, 2007, p. 5
20. M. v. Harmelen, Personal learning environments. In: *Proceedings of the 6th International Conference on Advanced Learning Technologies (ICALT'06).* 2006
21. J. Bergmann, A. Sams, *Flip Your Classroom: Reach Every Student in Every Class Every Day*. ISTE, Washington, D. C., 2012
22. E. Scanlon, A. Jones, J. Waycott, Mobile technologies: Prospects for their use in informal science learning (portable learning: experiences with mobile devices). Journal of Interactive Media in Education, 2005
23. J. García Zubía, G. R. Alves, eds., *Using Remote Labs in Education. Two Little Ducks in Remote Experimentation. Engineering, no. 8.* University of Deusto, Bilbao, Spain, 2011
24. A. K. M. Azad, M. E. Auer, V. J. Harward, eds., *Internet Accessible Remote Laboratories: Scalable E-Learning Tools for Engineering and Science Disciplines*. Engineering Science Reference, 2012
25. G. Zubía, O. Dziabenko, eds., *IT Innovative Practices in Secondary Schools: Remote Experiments*. University of Deusto Bilbao, 2013
26. M. Faßler, C. Terkowsky, *Urban Fictions. Die Zukunft des Städtischen*. W. Fink, 2006
27. R. Lubensky. The present and future of personal learning environments (ple), 2012. http://www.deliberations.com.au/2006/12/present-and-future-of-personal-learning.html
28. D. May, C. Terkowsky, T. Haertel, C. Pleul, Using e-portfolios to support experiential learning and open the use of tele-operated laboratories for mobile devices. REV2012 - Remote Engineering & Virtual Instrumentation, Bilbao, Spain, Conference Proceedings , 2012, pp. 172–180. doi:10.1109/REV.2012.6293126

29. S. P. Rochadel, J. B. Silva, T. D. Luz, G. R. Alves, Utilization of remote experimentation in mobile devices for education. International Journal of Interactive Mobile Technologies (iJIM) **6** (3), 2012, pp. 42–47
30. C. Terkowsky, I. Jahnke, C. Pleul, R. Licari, P. Johannssen, G. Buffa, M. Heiner, L. Fratini, E. Lo Valvo, M. Nicolescu, J. Wildt, A. E. Tekkaya, Developing tele-operated laboratories for manufacturing engineering education. platform for e-learning and telemetric experimentation (petex). International Journal of Online Engineering (iJOE) **6** (Secial Issue 1: REV2010, Vienna, IAOE), 2010, pp. 60–70. doi:10.3991/ijoe.v6s1.1378
31. C. Terkowsky, T. Haertel, Fostering the creative attitude with remote lab learning environments: An essay on the spirit of research in engineering education. International Journal of Online Engineering (iJOE) **9** (5), 2013, pp. 13–20. doi:10.3991/ijoe.v9iS5.2750
32. C. Terkowsky, T. Pleul, I. Jahnke, A. E. Tekkaya, Tele-operated laboratories for online production engineering education - platform for e-learning and telemetric experimentation (petex). International Journal of Online Engineering (iJOE) **7** (Special Issue: Educon 2011, Vienna, IAOE), 2011, pp. 37–43. doi:10.3991/ijoe.v7iS1.1725
33. C. Pleul, C. Terkowsky, I. Jahnke, A. E. Tekkaya, Tele-operated laboratory experiments in engineering education - the uniaxial tensile test for material characterization in forming technology. In: *Using Remote Labs in Education. Two Little Ducks in Remote Experimentation. Engineering, no. 8*, ed. by J. García Zubía, G. R. Alves, University of Deusto, Bilbao, Spain, 2011, pp. 323–348
34. T. Meyer et al, ed., *Kontrolle und Selbstkontrolle – Zur Ambivalenz von E-Portfolios in Bildungsprozessen.* VS Verlag für Sozialwissenschaften – Springer Fachmedien Wiesbaden GmbH, Wiesbaden, 2011
35. E. Reichert, Das e-portfolio - eine mediale technologie zur herstellung von kontrolle und selbstkontrolle. In: *Kontrolle und Selbstkontrolle – Zur Ambivalenz von E-Portfolios in Bildungsprozessen*, ed. by T. Meyer et al, VS Verlag für Sozialwissenschaften – Springer Fachmedien Wiesbaden GmbH, Wiesbaden, 2011, pp. 89–108
36. T. Haertel, C. Terkowsky, I. Jahnke, Where have all the inventors gone? is there a lack of spirit of research in engineering education. In: *15th International Conference on Interactive Collaborative Learning and 41st International Conference on Engineering Pedagogy*. 2012
37. C. Pleul, C. Terkowsky, I. Jahnke, A. E. Tekkaya, Platform for e-learning and telemetric experimentation - holistically integrated laboratory experiments for engineering education in manufacturing technology. In: *Proceedings WEE2011. 1st World Engineering Education Flash Week Lisbon, Portugal, SEFI - European Society for Engineering Education*, ed. by J. Bernadino, J. C. Quadrato. 2011, pp. 578–585
38. C. Terkowsky, C. Pleul, I. Jahnke, A. E. Tekkaya, Petex: Platform for elearning and telemetric experimentation. In: *Praxiseinblicke Forschendes Lernen, TeachING.LearnING.EU*, ed. by U. Bach, T. Jungmann, K. Müller, Aachen, Dortmund, Bochum, 2011, pp. 28–31
39. D. May, C. Terkowsky, T. Haertel, C. Pleul, Bringing remote labs and mobile learning together. International Journal of Interactive Mobile Technologies (iJIM) **7** (3), 2013, pp. 54–62
40. J. Biggs, C. Tang, *Teaching for Quality Learning at Univerfsity. What the Student Does.* Society for Research into Higher Education & Open University Press, McGraw Hill, 2009
41. L. D. Feisel, A. J. Rosa, The role of the laboratory in undergraduate engineering education. Journal of Engineering Education , 2005, pp. 121–130
42. I. Jahnke, T. Haertel, Kreativitätsförderung in hochschulen - ein rahmenkonzept. Das Hochschulwesen **58**, 2010, pp. 88–96
43. T. Haertel, I. Jahnke, Kreativitätsförderung in der hochschullehre: ein 6-stufen-modell für alle fächer?! In: *Fachbezogene und fachübergreifende Hochschuldidaktik. Blickpunkt Hochschuldidaktik, Band 121*, ed. by I. Jahnke, J. Wildt, Bertelsmann Verl., 2011, pp. 135–146
44. I. Jahnke, T. Haertel, M. Winkler, Sechs facetten der kreativitätsförderung in der lehre – empirische erkenntnisse. In: *Der Bologna-Prozess aus Sicht der Hochschulforschung, Analysen und Impulse für die Praxis*, ed. by S. Nickel, Arbeitspapier Nr, CHE gemeinnütziges Centrum für Hochschulentwicklung, 2011, pp. 138–152

45. T. Haertel, C. Terkowsky, Where have all the inventors gone? the lack of spirit of research in engineering education. In: *Proceedings of the 2012 Conference on Modern Materials, Technics and Technologies in Mechanical Engineering. The Ministry of Higher and Secondary Specialized Education (MHSSE) of the Republic of Uzbekistan*. 2012, pp. 507–512
46. C. Terkowsky, T. Haertel, Where have all the inventors gone? the neglected spirit of research in engineering education curricula. In: *Proceedings of the 2012 Conference on Actual Problems of Development of Light Industry in Uzbekistan on the Basis of Innovations. The Ministry of Higher and Secondary Specialized Education (MHSSE) of the Republic of Uzbekistan and The Tashkent Institute of Textile and Light Industry (TITLI)*. 2012, pp. 5–8
47. C. Terkowsky, T. Haertel, E. Bielski, D. May, Creativity@school: Mobile learning environments involving remote labs and e-portfolios. a conceptual framework to foster the inquiring mind in secondary stem education. In: *IT Innovative Practices in Secondary Schools: Remote Experiments*, ed. by J. García Zubía, O. Dziabenk, University of Deusto, Bilbao, Spain, 2013, pp. 255–280

Bringing the Inquiring Mind Back into the Labs – A Conceptual Framework to Foster the Creative Attitude in Higher Engineering Education

Claudius Terkowsky, Tobias Haertel, Emanuel Bielski and Dominik May

Abstract Contemporary laboratory learning in combination with state-of-the-art ICT can provide a vast variety of novel opportunities for creative experimentation and inquiry learning. Fostering and encouraging creative laboratory learning in engineering education may not only breathe life into what is learned but also involves the chance to consolidate students' understanding as well as creative self-efficacy. The presented conceptual framework proposes an individual learning environment based on mobile technology in combination with an e-portfolio system to facilitate and foster creative laboratory learning in engineering education. Since new ideas largely appear spontaneously, the application of mobile devices aims at boosting creative thinking processes. Having the mobile device handy permits the user to get hold on at least a quick note of an idea or to record it, as well as storing observed artefacts and activities for further elaboration. This article also features six different task-based scaffolding scenarios to foster creativity in the lab in order to illustrate how the proposed personal learning environments might enrich formal classroom activities and laboratory work to attain high-level learning outcomes.

Keywords Creative Learning · Laboratory Learning · Online Labs · Personal Learning Environments · Mobile Learning · Engineering Education

1 Introduction

Research on the effects of laboratory activities outlines that involving students in authentic inquiry helps them to:

- construct their own understanding of scientific concepts [1]
- construct knowledge schemes and solve problems [2, 3]
- practice science as scientists [4]

C. Terkowsky (✉) · T. Haertel · E. Bielski · D. May
Engineering Education Research Group, Center for Higher Education,
TU Dortmund University, Vogelpothsweg 78, 44227 Dortmund, Germany
e-mail: claudius.terkowsky@tu-dortmund.de

Originally published in "Proceedings of EDUCON2014 - IEEE Global Engineering Education Conference: "Engineering Education towards Openness and Sustainability",

DOI 10.1007/978-3-319-46916-4_76

- nurture positive attitudes on science [5]
- evolve critical thinking as well as decision-making skills [6], and
- creatively develop their own research questions [7–9]

"Unfortunately, science laboratory materials and exercises usually provided to teachers of science, K-16, are still centred on traditional methods of the past decades" [10].

In contrast to this, inquiry-based laboratories flip teachers' roles into asking students to formulate problems themselves as well as to relate their investigation to previous work, to state the purpose of the investigation, to identify the problem, to predict possible results, to specify approaches and resolution procedures and to finally perform the investigation [11, 12]. Moreover, the application of mobile devices can boost creative thinking processes as new ideas arise spontaneously in most cases [13]. Having the mobile device handy enables the user to grasp these ideas by making a note, to record it or to store first artefacts and activities to attain later access for further development [14].

Three concurrent trends in ICT-based learning technology development to support creative engineering education are recognizable throughout the latest scientific publications: personal learning environments (1), portable devices (2) and online labs (3)

- Personal learning environments (PLE) are "educational technology which responds to the way people are using technology for learning and which allows them to shape their own learning spaces themselves, to form and join communities and to create, consume, remix, and share material" [15]. PLEs provide more responsibility and more independence for learners. They "imply redrawing the balance between institutional learning and learning in the wider world" [16].
- Personal or mobile devices are perhaps the most rapidly growing category of technology for informal learning environments. The increasing diffusion of portable devices such as tablet PCs, Personal Digital Assistants (PDAs) and mobile phones offers a valuable potential to support new ways of self-directed, informal and creative learning- anytime and anywhere [17].
- The experience of online labs [18–20] can be delivered to the learner by technically and didactically integrating them into collaborative learning systems like cloud-based PLE.

This raises three essential topics in this context:

- How can creativity foster and enhance engineering education? How can students be enabled to gain the chance of conveying experimentation in the mode of a creative learner-driven inquiry?
- Which might be the best way for students to document their creative learning processes and to write their laboratory reports, even exams? How can teachers guide through these processes?
- How can this be fostered by personal mobile learning environments, portable devices and online labs?

In this respect, innovative ICTs provide novel, manifold opportunities to support students in their personal learning process documentations to develop scientific competencies on the basis of learner-generated multimedia content of laboratory experiences, field work and documentation.

The proposed solution is a personal learning environment based on mobile technology which integrates online labs and an e-portfolio system to facilitate and nurture creative science and technology learning.

2 Creative Learning with PLEs and Online Labs

2.1 E-Portfolios as Personal Learning Environments

E-portfolios are accessible online and provide a collection of digital, interactive and multimedial data e.g. texts, tables, photos, videos, audio etc. E-Portfolio-based PLE software, e.g. "Mahara" can conveniently be combined with an LCMS based on Moodle. The integrating PLR application "Mahoodle" intertwines properties and functions of the teacher-led LCMS Moodle and the learner-led system Mahara for e-portfolios. This mesh-up can be regarded as "a facility for an individual [or a group] to access, aggregate, configure and manipulate digital artefacts of their ongoing learning experiences" [21].

In addition, portable devices open up an extensive variety of novel occasions for creative inquiry regardless of time, space and situation. Since students are always on the move and changing in and out of interaction with technology, periods of spare time can be utilized for learning and working with e-portfolio software and the related laboratory equipment. This can be initiated virtually anywhere, regardless of most conventional software limitations [22, 23] once access to a respective learning environment had been obtained: a far-reaching feature which is widely spread throughout cloud computing.

2.2 Interactive Labs in the Web

Many programs are going to incorporate online labs into their instruction for extending the effectiveness of scarce resources as well as to share equipment with other institutions and locations [8, 18, 19, 24–27]. Students can be engaged in:

- blended and online learning scenarios regardless of time location, as far as Internet access is available
- learning activities which take longer than a typical class meeting time
- multi-part assignments which require students to use equipment for several short periods over the span of one week or longer
- socio-technically enhanced opportunities for student collaboration

- building up their own knowledge schemes using tele-operated equipment provided by the online labs

Even risky experiments which may be to hazardous for human interactions or with precious equipment can be conducted in a completely virtual fashion.

2.3 Learning Process Documentation

By creating and designing personal portfolios like a multimedia field diary, learners are empowered to document their own learning and research processes [28, 29]. They get the chance to:

- arrange all data and information they would choose to collect or share with others in different orders
- present experiments and their results or to show photos from the experimental set-up
- write notes and reflections on their experiments during their research-based learning processes
- explain their research results and thoughts to themselves and others
- collect ideas in creative moments, and to organize and improve them whenever desired
- support collaboration by allowing other learners and teachers to have access to their e-portfolios
- prepare, write and revise the lab report as a living document drawing on their learner-generated multimedia content

This reflection on learning processes and outcomes is an important aspect required to support the students' personal creative learning cycles [30]. Especially for learners, the e-portfolio as a field diary and personal learning documentation can always provide guidance and a checkpoint in topics of their own inquiry [31–33].

In the same way, teachers have the option to evaluate the actions of learners by observing, considering and reviewing their students' e-portfolios. Since other selectable persons or groups are able to view the collection within the portfolio, it can be stated that the e-portfolio is not only a versatile instrument for both individual learning documentation and learning-related reflection processes but also an especially valuable tool for collaborative communication.

2.4 Learning Objectives to Foster Creativity in Higher Engineering Education

A model of six facets for fostering creativity in the context of education for analysing, defining and stimulating new learning objectives and learning activities is delivered by [34–37]. These six facets are:

1. Developing self-reflective learning skills: Learners are relieved from the constraints formed by their receptive habits and empowered to question information given by the teacher. An internal dialogue takes place and knowledge becomes "constructed" rather than "adopted".
2. Developing independent learning skills: Teachers stop to restrain the way students learn. Instead, students are free to discover relevant literature on their own and, for example, to make their own guidelines about structuring a text or even to find their own research questions and to choose appropriate methods of answering them.
3. Enhancing curiosity and motivation: This aspect is related to all measures that contribute to increased motivation, for instance, linking a theoretical question to a practical example or representation.
4. Learning by doing: Students learn by creating a kind of "product", their respective artefact. Depending on the discipline, this might be a presentation, an interview, a questionnaire, a machine, a website, a computer program or similar. Students are guided into the role of "real" researchers.
5. Evolving multi-perspective thinking: Learners overcome thinking within the limits of their respective disciplines or prejudiced thinking. Along with that, they learn to consider an issue from different points of view and to use thinking methods which prevent their brains from being trapped in ordinary disciplinary paths.
6. Reaching for original ideas: Learners are facilitated to embrace new, original ideas and to prepare themselves to be as ready-to-receive as possible. Although the acquirement of original ideas cannot be forced, the reception of original ideas can be fostered by applying appropriate creative techniques and by creating a suitable environment (allowing students to make mistakes and to express unconventional ideas without being laughed at or rejected).

In the following, six different task-based scaffolding scenarios are presented in order to explain how personal learning environments featuring the combination of labs (online and offline), e-portfolios and mobile devices can enrich formal classroom activities and laboratory work.

2.5 *Scaffolding Scenarios for Creative Laboratory Learning*

The scenarios are founded and premised on the model of the six creativity facets as mentioned above.

2.5.1 Developing Self-reflective Learning Skills: Evolving Critical Thinking with Technological Pitfalls

In order to raise critical awareness of technological pitfalls, a misleading environment can be utilized to make students familiar with a critical approach to information. In order to achieve this goal, teachers can assign tasks to their students which are impossible to solve, or can provide false information that leads to seemingly correct but erroneous results of the experiments. In both cases, the students will be irritated while performing the experiment as well as challenged to find the cause.

A common engineering topic is surface coating in the field of micro technology. In order to foster self-reflective learning, teachers could ask their students to produce a surface coating for a substrate that is suitable for frying eggs. To stimulate a critical attitude towards given information, teachers can provide students with faulty instructions for developing and producing such non-stick coatings. The result could be a surface that would melt under high temperature and intermingle with the fried egg (have a look that students do not start to eat it). In this case, students would have to verify the given information as well as to clarify the mistake and reproduce the surface coating which fulfils demand.

The achieved learning outcome would comprise (besides technical knowledge about surface coatings and their behaviour) that students mistrust and reflect given information rather than simply receiving it, as they can experience the consequences of applying misleading information without putting it into question.

2.5.2 Developing Independent Learning Skills: Improving Self-reliance and Self-confidence Towards Technical Issues

Online experiments in combination with PLEs make it possible to let students learn more independently. They enable them to search for information, to plan and to conduct experiments as well as to make decisions on their own.

Advanced materials are another common topic in higher engineering education. Instead of simply handing students pre-selected information about the characteristics of selected advanced materials, teachers could ask them a relatively open question, for instance: "Which material is the best choice to attach an iPhone to a mirror in a fashion (in order to be able to watch it during the morning toilet) that it cannot be separated manually but by use of a second material without any damage of both?" Students are asked to collect the necessary information for themselves in the Internet rather than in textbooks.

In order to give assistance, teachers have to carefully monitor their students' learning progress. If they notice difficulties or confusion among their students, teachers will have to help them "to help themselves" without providing completed solutions which will only lead to well-known solutions. For this purpose, PLEs are valuable tools that enable teachers to monitor their students' learning steps and provide feedback to them, while students on the system's other end can ask for help at any time.

As a reward for using this learning scenario, students will not only incorporate a wide range of aspects about advanced materials but will also learn to plan small research projects, as well as putting them into practice and feeling responsible for them as well as their individual learning successes. In the long run, performing small learning tasks on their own will improve their self-reliance and self-confidence towards technical issues.

2.5.3 Tackling Curiosity and Motivation by Intertwining Technical Issues with Students' Real-World Experiences

Learning with experiments in combination with real-world questions are a suitable combination to support students' motivation to learn, compared to sitting in a monotone classroom and being restrained to receive information presented by a teacher without active multi-directional participation. Nevertheless, the question of rendering subject matter more interesting for students in order to tackle their curiosity and motivation is always worth asking.

A well-known approach towards increasing students' motivation is to apply practical questions and tasks. Therefore, combining practical issues taken directly from the students' daily environment as well as online experiments or simulations might be the silver bullet of situated experimental learning.

For example, a lecture in electronics engineering about the characteristics of transistors could be very boring. But it can easily be spiced up with a small homework as a preparation for the transistor lecture: Ask the students to spend one day completely without the use of transistors. Collect their experiences in the lecture and try to find transistors they used even if they thought were absent (it is not that easy to raise awareness of the ubiquitous use of transistors). By noticing the relevance of transistors for modern life, students might become more interested in lab works on for example the temperature limits of those transistors keeping their life civilized.

To this end, the use of online labs may have several advantages regarding students' creativity: they are readily available throughout the day, so students can pursue experiments whenever they have a good idea and therefore do not have to wait for the next lab lesson to test their personal ideas. They can conduct it whenever they feel creative and wherever they are, all they need is simply their online mobile device.

2.5.4 Learning by Doing: Designing and Building Functional Models

Since experimental learning as a means of scaffolding always inherits some sort of product to be created, this learning method appears suitable in order to achieve this goal as it might also strengthen the students' awareness of their creative potential. Besides, exposing the learning outcome to a larger interest and external assessments are valuable factors in tackling students' intrinsic motivation. For this purpose, the PLEs can be used to render the readily documented learning process accessible for externals.

For example, in aerospace engineering students are expected to learn the fundamentals of rheology. Instead of just presenting them the contents of a course book in a lecture, students could be asked to develop the most effective paper plane by assistance of several materials which can be found in the laboratory. By doing so, students should parenthetically be enabled to determine the drag coefficient of different designs. Moreover, by using online simulations of wind tunnels to try several varieties of their paper plane, students get involved into increasing its range by reducing its drag in an informal, playful manner. Furthermore, they would have to find the proper material for realizing the best design with regard to ductility and – of course – weight.

At this point, with emphasis on rheology, different aspects of aerospace engineering would be mentioned and the coherences would become visible. Students would be enabled to see a bigger part of the picture while applying and thus internalizing the necessary professional knowledge.

Finally, the inventors of the most effective paper plane (is it really the one with the lowest cd value?) would receive a prize.

2.5.5 Evolving Multi-perspective Thinking: Overcoming Cognitive Barriers

This scenario can foster students' ability to re-think about their questions from independent perspectives: imagining a student who had performed an experiment assigned to him by the teacher, he/she contingently does not know why the experiment did not show the expected results or does not know how to interpret them correctly. He asks himself why the experiment did not work according to plan but cannot find a satisfactory answer. While writing his e-portfolio as documentation for the teacher's evaluation, he could start the "creative-help app", a wizard application within the PLE intended to help him develop different perspectives on the same problem:

- The student is asked to do a (mental) headstand following the question: "What else could I do to get the wrong results from experimenting?"
- If this should not suffice, the student will be asked to describe his experimental design and assumptions in a way that a ten-year-old could understand it.
- If those methods which are fairly close to the problem still cannot help him, the "creative-help app" will suggest a force-fit technique by showing a picture that does not have anything in common with a problem and asking the student to find relationships between the picture and his experiment.

This set of methods can help students to leave the well-trodden paths and forces them to look at their problem from completely different perspectives. This often results in unconventional or provocative ideas, but rethinking the obviously unsuitable solutions sometimes leads to the one really good idea which would not have appeared without making the detour [30, 38].

2.5.6 Reaching for Original Ideas: Breaking all Rules and Posing one's own Questions

The permission for "breaking all rules and posing one's own questions" is the supreme discipline in creativity education. There are several options available to arrange for learner-led inquiries, investigation, discovery and reasoning.

At this point, it is up to the reader to design a "Reach for Original Ideas by Breaking all Rules and Posing one's Own Questions" learning scenario for the lab course. However, serious "Breaking all Rules and Posing one's Own Questions" may also create a lot of severe resistance from colleagues, superiors, institutions and others who still need to be involved and convinced to transform an idea into innovation. See [38] for an empathic and very special example of facet 6.

3 Conclusion

In this paper the authors explained how the integration of online labs and e-portfolios into mobile personal learning environments can offer novel opportunities to foster students' creative learning in engineering education. Allowing students to learn this self-confident, independent way already incorporates fostering their creativity with respect to the six facets of creativity in education. Each (online) experiment is some sort of a product (facet 4, learning by doing), is supposed to increase students' curiosity and motivation (facet 3, compared to just sitting in a monotone classroom listening to the teacher) and fosters their self-reliance (facet 2, by doing something on their own).

This is one essential way for students to reach high-level learning outcomes and thereby developing the basis of fundamental competences for their future academic or professional life, as well as tackling attitudes like curiosity, creative self-efficacy, agency and responsibility. If students are empowered to evolve their own research questions, to choose suitable experimentation designs and finally to perform the experiment, they will be able to develop a "spirit of inquiry and research" [30]. This spirit is one important premise for developing original ideas.

References

1. G. Roehrig, A. Lulie, M. Edwards, Versatile vee maps. The Science Teacher **68**, 2001, pp. 28–31
2. J. German, S. Haskins, S. Auls, Analysis of nine high school biology laboratory manuals: promoting scientific inquiry. Analysis of nine high school biology laboratory manuals: promoting scientific inquiry **33** (5), 1996, pp. 475–499
3. C. Terkowsky, D. May, T. Haertel, C. Pleul, Experiential learning with remote labs and e-portfolios - integrating tele-operated experiments into personal learning environments. International Journal of Online Engineering **9** (1), 2013, pp. 12–20. URL http://dx.doi.org/10.3991/ijoe.v9i1.2364
4. A. Okebukola, B. Ogunniyi, Cooperative, competitive and individualistic science laboratory interaction patterns effects on students' achievement and acquisition of practical skills. Journal of Reserach in Science Teaching **21** (9), 1984, pp. 875–884
5. R. Lehrer, L. Schauble, A. Petrosino, Reconsidering the role of experiment in science education. In: *Designing for science*, ed. by K.D. Crowley, C.D. Schunn, T. Okada, Lawrence Erlbaum Associates, Mahwah, N.J., 2001, pp. 251–277

6. F. Abd-El-Khalik, S. BouJaoude, R. Duschl, N.G. Lederman, R. Mamlok-Naaman, A. Hofstein, Inquiry in science education: International perspectives. Science Education **88** (3), 2004, pp. 397–419
7. C. Terkowsky, T. Haertel, Fostering the creative attitude with remote lab learning environments: An essay on the spirit of research in engineering education. International Journal of Online Engineering **9** (Special Issue 5 "EDUCON2013"), 2013, pp. 13–20
8. C. Terkowsky, I. Jahnke, C. Pleul, R. Licari, V. Johannssen, G. Buffa, M. Heiner, L. Fratini, E. Lo Valvo, M. Nicolescu, J. Wildt, A.E. Tekkaya, Developing tele-operated laboratories for manufacturing engineering education. platform for e-learning and telemetric experimentation (petex). International Journal of Online Engineering **6** (Special Issue 1: REV2010), 2010, pp. 60–70
9. T. Haertel, C. Terkowsky, D. May, C. Pleul, Entwicklung von remote labs zum erfahrungs basierten lernen. Zeitschrift für Hochschulentwicklung. Themenheft Kompetenzen, Kompetenzorientierung und Employability in der Hochschule (Teil 2) **8** (1), 2013, pp. 79–87
10. D.W. Sunal, C.S. Sunal, C. Sundberg, E.L. Wright, The importance of laboratory work and technology in science teaching. In: *The Impact of the Laboratory and Technology on Learning and Teaching Science K-16*, ed. by D.W. Sunal, E. Wright, C. Sundberg, IAP-Information Age Publishing, 2008, pp. 1–28
11. P. Tamir, How are laboratories used? Journal of Research and Science Teaching **14** (9), 1977, pp. 311–316
12. C.W. Keys, Revitalizing instruction in scientific genres: Connecting knowledge production with writing to learn in science. Science Education **83** (2), 1999, pp. 115–130
13. T. Haertel, I. Jahnke, Wie kommt die kreativitätsförderung in die hochschullehre? Zeitschrift für Hochschulentwicklung **6** (3), 2011, pp. 238–245
14. D. May, C. Terkowsky, T. Haertel, C. Pleul, The laboratory in your hand - making remote laboratories accessible through mobile devices. In: *Proceedings of the 2013 IEEE Global Engineering Education Conference (EDUCON)*. IEEE, 2013, pp. 335–344
15. G. Attwell, Personal learning environments - the future of elearning? eLearning Papers **2**, 2007, p. 5
16. M.v. Harmelen, Personal learning environments. In: *Proceedings of the 6th International Conference on Advanced Learning Technologies (ICALT'06)*. 2006
17. E. Scanlon, A. Jones, J. Waycott, Mobile technologies: Prospects for their use in informal science learning (portable learning: experiences with mobile devices). Journal of Interactive Media in Education (Special Issue), 2005. URL http://jime.open.ac.uk/article/2005-25/303
18. J. García Zubía, G.R. Alves, eds., *Using Remote Labs in Education. Two Little Ducks in Remote Experimentation, Engineering*, vol. 8. University of Deusto, Bilbao, Spain, 2011
19. A. Azad, M.E. Auer, V.J. Harward, eds., *Internet Accessible Remote Laboratories: Scalable E-Learning Tools for Engineering and Science Disciplines*. Engineering Science Reference, 2012
20. J. García Zubía, O. Dziabenko, eds., *IT Innovative Practices in Secondary Schools: Remote Experiments*. University of Deusto, Bilbao, Spain, 2013
21. R. Lubensky. The present and future of personal learning environments (ple), 2012. URL http://www.deliberations.com.au/2006/12/present-and-future-of-personal-learning.html
22. D. May, C. Terkowsky, T. Haertel, C. Pleul, Using e-portfolios to support experiential learning and open the use of tele-operated laboratories for mobile devices. In: *REV2012 - Remote Engineering & Virtual Instrumentation*. Bilbao, Spain, 2012, pp. 172–180
23. W. Rochadel, S.P. Silva, J.B. Silva, T.D. Luz, G.R. Alves, Utilization of remote experimentation in mobile devices for education. International Journal of Interactive Mobile Technologies **6** (3), 2012, pp. 42–47
24. J.G. Zubía, O. Dziabenko, eds., *IT Innovative Practices in Secondary Schools: Remote Experiments*. Bilbao, Spain, 2013
25. C. Terkowsky, I. Jahnke, C. Pleul, D. May, T. Jungmann, A.E. Tekkaya, Petex@work. designing cscl@work for online engineering education. In: *Computer-Supported Collaborative Learning at the Workplace - CSCL@Work, Computer-Supported Collaborative Learning Series*, vol. 14, ed. by S.P. Goggins, I. Jahnke, V. Wulf, Springer, 2013, pp. 269–292

26. C. Terkowsky, C. Pleul, I. Jahnke, A.E. Tekkaya, Tele-operated laboratories for online production engineering education. platform for e-learning and telemetric experimentation (petex). International Journal of Online Engineering (Vol.7 Special Issue: Educon 2011), 2011, pp. 37–43
27. C. Pleul, C. Terkowsky, I. Jahnke, A.E. Tekkaya, Tele-operated laboratory experiments in engineering education - the uniaxial tensile test for material characterization in forming technology. In: *Using Remote Labs in Education. Two Little Ducks in Remote Experimentation, Engineering*, vol. 8, ed. by J. García Zubía, G.R. Alves, University of Deusto, Bilbao, Spain, 2011, pp. 323–348
28. G. Reinmann, S. Sippel, Königsweg oder sackgasse? – e-portfolios für das forschende lernen. In: *Kontrolle und Selbstkontrolle – Zur Ambivalenz von E-Portfolios in Bildungsprozessen*, ed. by T. Meyer, et al., VS Verlag für Sozialwissenschaften – Springer Fachmedien Wiesbaden GmbH, Wiesbaden, 2011
29. R. Reichert, Das e-portfolio - eine mediale technologie zur herstellung von kontrolle und selbstkontrolle. In: *Kontrolle und Selbstkontrolle – Zur Ambivalenz von E-Portfolios in Bildungsprozessen*, ed. by T. Meyer, et al., VS Verlag für Sozialwissenschaften – Springer Fachmedien Wiesbaden GmbH, Wiesbaden, 2011
30. C. Terkowsky, T. Haertel, Where have all the inventors gone? the neglected spirit of research in engineering education curricula. In: *Proceedings of the 2012 Conference on Actual Problems of Development of Light Industry in Uzbekistan on the Basis of Innovations*. 2012, pp. 5–8
31. C. Pleul, C. Terkowsky, I. Jahnke, A.E. Tekkaya, Platform for e-learning and telemetric experimentation - holistically integrated laboratory experiments for engineering education in manufacturing technology. In: *Proceedings WEE2011. 1st World Engineering Education Flash Week Lisbon*, ed. by J. Bernadino, J.C. Quadrato. SEFI - European Society for Engineering Education, Portugal, 2011, pp. 578–585
32. C. Terkowsky, C. Pleul, I. Jahnke, A.E. Tekkaya, Petex: Platform for elearning and telemetric experimentation. In: *Praxiseinblicke Forschendes Lernen, TeachING.LearnING.EU*, ed. by U. Bach, T. Jungmann., K. Müller, Aachen, Dortmund, Bochum, 2011, pp. 28–31
33. D. May, C. Terkowsky, T. Haertel, C. Pleul, Bringing remote labs and mobile learning together. International Journal of Interactive Mobile Technologies **7** (3), 2013, pp. 54–62
34. I. Jahnke, T. Haertel, Kreativitätsförderung in hochschulen - ein rahmenkonzept. Das Hochschulwesen **58**, 2010, pp. 88–96
35. T. Haertel, I. Jahnke, Kreativitätsförderung in der hochschullehre: ein 6-stufen-modell für alle fächer?! In: *Fachbezogene und fachübergreifende Hochschuldidaktik. Blickpunkt Hochschuldidaktik*, vol. 121, ed. by I. Jahnke, J. Wildt, Bertelsmann Verlag, 2011, pp. 135–146
36. I. Jahnke, T. Haertel, M. Winkler, Sechs facetten der kreativitätsförderung in der lehre – empirische erkenntnisse. In: *Der Bologna-Prozess aus Sicht der Hochschulforschung, Analysen und Impulse für die Praxis*, ed. by S. Nickel, 2011, pp. 138–152
37. T. Haertel, C. Terkowsky, Where have all the inventors gone? the lack of spirit of research in engineering education. In: *Proceedings of the 2012 Conference on Modern Materials, Technics and Technologies in Mechanical Engineering*. Andijan Area, Andijan City, Uzbekistan, 2012, pp. 507–512
38. C. Terkowsky, T. Haertel, E. Bielski, D. May, Creativity@school: Mobile learning environments involving remote labs and e-portfolios. a conceptual framework to foster the inquiring mind in secondary stem education. In: *IT Innovative Practices in Secondary Schools: Remote Experiments*, ed. by J. García Zubía, O. Dziabenko, University of Deusto, Bilbao, Spain, 2013, pp. 255–280

Creative Students Need Creative Teachers – Fostering the Creativity of Teachers: A Blind Spot in Higher Engineering Education?

Tobias Haertel, Claudius Terkowsky and Monika Radtke

Abstract Fostering students' creativity is one indicator for good teaching and learning in higher education. Recently, several approaches for fostering creativity in higher education have been developed. In Germany and other countries, innovative learning and teaching concepts that allow students to unfold their creativity have been put into practice. However, further education trainings for teachers have shown that teachers' creativity has to be promoted in order to foster students' creativity. The implementation of innovative learning scenarios is a creative process itself, and teachers need the courage to overcome fears that emerge in creative situations. In this paper, the connection between creativity and courage is shown and used to analyse the experiences that have been made with the task "do something unusual" in further education trainings.

Keywords Creativity in Higher Education · Teachers Creativity · Innovation in Teaching and Learning

1 Introduction

The term "creativity" is not recorded in the German "Charta guter Lehre" [11] and not even in the otherwise substantial keyword collection of the German "Qualitätszirkel" [11]. Nevertheless many works in higher education deal with creativity with the aim to make a contribution to good teaching and learning practices [1, 3, 9, 10, 14, 17]. The implementation of teaching and learning scenarios that foster students' creativity requires itself an act of creativity for the teacher. Longstanding experience in higher educational trainings on creativity shows how challenging this can be for teachers. They usually tempt to prefer incremental changes across from radical innovations [7]. According to this it is important to foster teachers' creativity (besides students'

T. Haertel (✉) · C. Terkowsky · M. Radtke
Engineering Education Research Group, Center for Higher Education,
TU Dortmund University, Dortmund, Germany
e-mail: tobias.haertel@tu-dortmund.de

Originally published in ?Proceedings of: World Engineering Education Forum 2015: 18th International Conference on Interactive Collaborative Learning and 43rd International Conference on Engineering Pedagogy", © 2014. Reprint by Springer International Publishing AG 2016, DOI 10.1007/978-3-319-46916-4_77

creativity) but not in terms of creating new ideas but with the "courage to create" [16]. The relation between creativity and courage will be discussed in the following. Subsequently a practical exercise will be presented that deals with this concrete context and that was realized with teachers and students several times.

2 Creativity and Courage

As a reason why teachers have troubles with the realization of innovative teaching and learning scenarios was identified, that "failing" is considered as harmful in the professional socialization of scientists and associated with nonprofessional work and the waste of resources [7]. This corresponds with the findings of [21]. With a view to the waste of creative potential in higher education he reports from a failure of universities and describes them as a place where innovations, the searching for new ideas and thinking against the norms are not only unpopular but rather get systematically prevented [21]. Teachers need to overcome this obstacle to take the risk of "failing" and to leave the norms of the higher educational system behind. [2] describe all creative actions as "navigating in open systems". Routines and traditions give people the feeling of safety, they allow a safe navigation along known expectations, and who remains on this paths doesn't have to fear criticism or negative feedback: "Laws, rules, regulations, conventions and taboos reduce complexity and create a relatively safe frame for a peaceful, organized life and coexistence." [2] speak of "closed systems" that bring about inner safety. However creativity means to leave this closed system and to strike off into a new closed system. When somebody is creative, he leaves the well-known routines, traditions and norms for a while. For the instance of the transition they are in an "open system" in which they do not know the expectations they are faced with. They cannot know which feedback they get (positive or negative) for their creative actions — if they get criticized or honored: "In the moment of the transition from one closed system into another one appears a temporary open system, an exposed position that contains risks and accordingly is frightening."

Children already learn the ability to orientate themselves towards the expectations of others very early in the age between three and a half and four years. From then children are able to capture the mental conditions of others, to predict their behaving and feelings and to distill predictions for their actions [5, 19, 20]. That gives the premise to behave prosocial towards others, which is normally sanctioned positively and leads to stronger acceptance from others [19]. In [15] prosocial behaviour is influenced by 10 basic values of all human cultures identified [18] (Table 1). Thereby values like self-direction and conformity are in conflict with each other. Transferred to higher education the assumption can be made, that in earlier academically development conformity was stronger fostered than self-direction. [4] shows for the discipline law how students get socialized to conformity through the pressure to perform and through exam anxiety: "Successful candidates for an exam stand out through little failure anxiety, endurance, conflict avoiding socialization, readiness to assimilate

Table 1 [18] Ten basic values and their operational definitions (cited from [15])

Value	Operational definition
Benevolence	Preserving and enhancing the welfare of those with whom one is in frequent personal contact
Universalism	Understanding, appreciation, tolerance, and protection for the welfare of all people and for nature
Self-direction	Independent thought and action make own decisions, create, and innovate
Stimulation	Excitement, novelty, adventure, and risks in life
Hedonism	Have a good time, have fun, and do things that bring gratification for oneself
Achievement	Personal success through demonstrating competence, seeking admiration, and impressing others
Power	Wealth, control, or dominance over people and resources
Security	Safety and stability for oneself and one's country
Conformity	Restraint of actions and intentions that could upset or harm others and/or violate social norms and rules
Tradition	Respect, commitment, and acceptance of the customs and ideas that one's culture or religion provides

and confidence. In such a way described successful candidates identify themselves easily with the privileges and the authority to exert power from the position they are going to occupy after their exams." In the professional life of academics people are precariously occupied in the social middle class and so in a group that reacts on the fears of the pressure to perform, overextension and social decline with the "return of conformity" [13].

Summarized can be said that teachers have to face their fears in open systems if they want to realize an innovative teaching method to foster their students' creativity.

3 Do Something Unusual

On this occasion [16] reports about the "courage to create" being necessary. For him the courage to overcome personal fears belongs to creativity. In his work about the correlation between courage and workplace [12] defines - referring to [6] - the term courage as "acting intentionally in the face of risks, threats, or obstacles in the pursuit of morally worthy goals". According to [8] courage can be learned slowly step by step. Inspired by that a special training was integrated into higher engineering educational workshops for teachers ("Through the barricades", "Rage against the machine") and also into students' courses ("Studying creative", "Creativity in engineering education") at different universities. These trainings first of all demonstrate the meaning of courage for creativity and secondly – repeated regularly with small increases - want to offer an approach to act more out of courage and to leave conformity.

The participants get the same task in all trainings: Do something unusual! Not in the sense of a test of courage but as an encouragement to overcome challenging difficult social situations: Leave the comforting and safe terrains of normative regulative social routines and traditions, go against something that you for yourself have declared as an interacting norm.

After setting the task participants always ask what is valid as a norm transgression and how spectacular it should be. It can be seen that norms are understood and sensed in a different, subjective way. What builds a norm transgression for one person, is completely "normal" and inside the norms for another one. That is why it is important that every participant is able to find a solution for this task along his very own norms All the same task every participant would lead to a comparability of mastering it but it would also contain the danger of demanding too little or – even worse – too much of individual participants. So all participants should find a solution that corresponds with their individual situation and requires some courage but doesn't overwhelm them. No solution is too unspectacular, no solution is wrong, as long as it gets realized anyway.

Giving this task there are some observations that can be done regularly:

- Teachers often face this task with (sometimes very strong) resistance. In workshops with engineering teachers the task is set as a homework between the first and second workshop day. Homework as such seems to cause discomfort for many of the teachers. Getting faced with the point that they expect the same from their students and that the task takes just a little time (with some creativity it can be solved by the way), they come to a substantial rejection. As experience teaches a majority rejects the task because its sense is not seen or the teachers do not consider it as important or they do not have any ideas how to solve it. Summarized giving this task can contain a danger for the relationship between the workshop moderation and the participants even though the rest of the day was successful and it often ends up with the open announcement of the participants that they are not going to do this.
- Students face this task (that has to be fulfilled until the next date) mainly with curiosity but they also mention some concerns that their ideas to solve the task could be too unspectacular for the other participants. The task gets accepted by a majority after repeating for several times that no idea is too unspectacular and no one should ask too much of himself or herself and at least the task is no competition but the possibility to deal with each ones individual willingness to creativity.

The different understandings of teachers and students are visible in the processing of the task:

- As said before teachers in the majority do not solve the task or at least not in the sense it should be. Popular and in some way understandable creative is the solution "For me it is unusual not to fulfill given tasks, that's why I didn't do this homework". Many teachers also solve the problem intentionally wrong, e.g. "Normally I never indicate while driving a car, so this time I did so. That is unusual for me.", "Since a long time my wife and I think about euthanizing our old and very ill dog but never found the courage to do so. Now yesterday we've done that as part of the homework." or "I normally never listen to my best friend on the phone. Yesterday I did so. This was unusual for me and it wasn't worth it.". But there also are always teachers who take the homework serious and solve it correctly. A positive example is a teacher from Berlin who first also rejected the task completely but then decided to buy a rose and to give it as a present to the first person that sits down beside her in the subway on her way to the second workshop day. She didn't feel comfortable enough with the appearance of the first person but the second person was an old man to whom she gave the rose and wished him a nice day. She found out that the man was on his way home after a hospital stay for weeks because of a cancer therapy.

- Students just refuse the homework very rarely. Mostly in their reports an accurate examination of the problem can be observed: In the majority the students find a personal benefit and search for a solution that fits to their individual context despite their fears of the uncertain situation. The solutions vary quite strong because of the diverse structure of the student groups. For example one student was active in a Death-Metal group. In their meetings everyone wears black clothes. To solve the task she went there with usual street wear. Another student opened a beer and drank it in a seminar and still another one played songs in the pedestrian mall and collected money for that. One student decided to take off his shoes and walked the long way home without wearing them, while another student went shopping in his Pajama. To give two more examples, one student decided to sit down in the middle of a small and highly frequented passage inside the library, and yet another student dialed a phone number she did not know to overcome her fear of talking to strangers. Also some students do not solve the task correctly. For example one student did a bungee jump. But in the group of students these are exceptions.

In the following discussion the participants should describe the feelings they had before while and after solving their tasks. A typical process with individual variances can be observed: With the development of the idea and the decision to realize it there comes a stage of uneasiness that becomes stronger and stronger and reaches its climax right before the actual realization. In the situation itself the uneasiness changes into a concentration on the situation. The own actions and the reactions of the surrounding are observed in many details. After the realization there comes a time of relaxation and a feeling of proudness to have stood the challenge.

4 Findings and Discussion

The concrete reasons why engineering teachers struggle so much more with the task "Do something unusual" than students do is not known. But the observation corresponds with the argumentation that teachers have spent more time on the subject-specific and professional socialization at their university and so have internalized the norms through positive feedback of conformity which complicates going against the established routines. Although the academic system with its main purpose of developing knew knowledge depends on creativity and non-conformity (like e.g. art), it can be seen how difficult it is for teachers to leave their usual ways of conformity and to do something unusual. Higher educational trainings that only focus on fostering students' creativity are not enough here. On this occasion trainings to foster students' creativity should always be linked to trainings to foster teacher's creativity. Creative teachers then cannot just foster their students' creativity in a better way. With their escape of the closed system in economy they are able to see and name its deficits. Only thereby the conditions for changes are created.

References

1. H.-K. Adriansen, How criticality affects students' creativity. In: *Teaching creativity – creativity in teaching*, ed. by C. Nygaard, N. Courtney, C. Holtham, Libri Publishing, Oxfordshire, 2010, pp. 65–84
2. U. Bertram, W. Preißing, *Navigieren im offenen System: Unternehmensführung ist ein künstlerischer Prozess*. Container Verl., Leonberg, 2007
3. E. Brodin, L. Frick, Conceptualizing and encouraging critical creativity in doctoral education. International Journal for Researcher Development **2** (2), 2011, pp. 133–151
4. L. Dammann, Sozialisation durch prüfungsangst und leistungsdruck: Wirkung und funktuon des ersten juristischen staatsexamen. Forum Recht. Zwischen Wir und Ich: Europäische Idee und nationale Interessen (2), 2006, pp. 60–64
5. J. A. Fodor, A theory of the cild's theory of mind. Cognition **44**, 1992, pp. 282–296
6. N. H. Goud, Courage: Its nature and development. Journal of Humanistic Counseling, Education & Development **44**, 2005, pp. 102–116
7. T. Haertel, C. Terkowsky, P. Ossenberg, Kreativität in der hochschullehre: Was geht? In: *Lernen, Lehren, Beraten auf Augenhöhe*, ed. by J. Tosic, Hochschule Niederrhein, University of Applied Sciences, 2015, pp. 46–53
8. K. Hoffmann, *"Die Logik des Mutes: Dein Mutmacher bist Du selbst,"* 2nd edn. Springer, 2013
9. N. Jackson, M. Oliver, M. Shaw, J. Wisdom, eds., *Developing Creativity in Higher Education: An imaginative curriculum*. Routledge, London, 2006
10. I. Jahnke, T. Haertel, Kreativitätsförderung in hochschulen – ein rahmenkonzept. Hochschulwesen **3**, 2010, pp. 88–96
11. B. Jorzik, *Charta guter Lehre. Grundsätze und Leitlinien für eine bessere Lehrkultur*. Essen, 2013
12. M. Koerner, Courage as identity work: accounts of workplace courage. Academy of Management Journal **75** (5), 2014, pp. 3–93
13. C. Koppetsch, *Die Wiederkehr der Konformität. Streifzüge durch die gefährdete Mitte*. Campus-Verl., 2013
14. S. Lange, Learning through creative conversations. In: *Teaching creativity – creativity in teaching*, ed. by C. Nygaard, N. Courtney, C. Holtham, Libri Publishing, Oxfordshire, 2010, pp. 173–188
15. J.-E. Lönnqvist, M. Verkasalo, P. Wichardt, G. Walkowitz, Personal values and prosocial behaviour in strategic interactions: Distinguishing value-expressive from value-ambivalent behaviours. European Journal of Social Psychology, Eur. J. Soc. Psychol **43**, 2013, pp. 554–569
16. R. May, *The courage to create*. Nortin, rev. ed., New York, 1994
17. A. Raiker, Creativity and reflection: some theoretical perspectives arising from practice. In: *Teaching creativity – creativity in teaching*, ed. by C. Nygaard, N. Courtney, C. Holtham, Hochschule Niederrhein, University of Applied Sciences, Oxfordshire, 2010, pp. 121–138

18. S. H. Schwartz, Universals in the content and structure of values: Theory and empirical tests in 20 countries. In: *Advances in experimental social psychology*, ed. by M. Zanna, Academic Press., vol. 25, New York, 1992, pp. 1–65
19. E. Seifermann, H. M. Buhl, Soziale kognitionen, sozialverhalten und akzeptanz durch gleichaltrige bei kindern im vorschulalter. Diskurs Kindheits- und Jugendforschung **3**, 2012, pp. 321–332
20. V. Slaughter, A. Gopnik, Conceptual coherence in the child's theory of mind: Training children to understand belief. Child Development **67**, 1996, pp. 2967–2988
21. Wolf Wagner, *Tatort Universität: Vom Versagen deutscher Hochschulen und ihrer Rettung*. Klett-Cotta, Stuttgart, 2010

Entrepreneurship and Gender in Higher Engineering Education in Germany

Dominik May, Bengue Hosch Dayican, Liudvika Leisyte, Karsten Lensing, Lisa Sigl and Claudius Terkowsky

Keywords Entrepreneurship · Gender · Content Analysis · Engineering Curricula

1 Introduction

In the past years European economic and employment policies increasingly underline the strategically important role of higher education institutions (HEIs) in boosting Europe's innovation potential through supplying highly skilled labour [1]. Two key priorities are particularly emphasized: The first is to embed entrepreneurship into higher education curricula in order to further develop the knowledge triangle that integrates education, research, and innovation with each other [2]. This particularly counts for applied disciplines such as engineering [3]. The second priority is to enforce gender equality in labour force participation and increase representation of women in skilled employment [4] which is again most evident in engineering by incorporating gender issues in teaching plans and creating more awareness for gender balance in labour markets [5]. These objectives have subsequently been incorporated into national and regional policies. In Germany, funding for knowledge and technology transfer for economic and societal applications is already a central instrument in policy strategies of federal and state level ministries of education and science [6, 7]. The question is in how far these political demands are reflected in higher engineering education practice. In this study we focus on engineering curricula of nine leading technical universities to understand how entrepreneurship and gender studies have been incorporated in engineering programmes. Hence, this research brings together entrepreneurial and gender research with research on higher engineering education.

D. May (✉) · B. Hosch Dayican · L. Leisyte · K. Lensing · L. Sigl · C. Terkowsky
Engineering Education Research Group, Center for Higher Education,
TU Dortmund University, Dortmund, Germany
e-mail: dominik.may@tu-dortmund.de

Originally published in "Proceedings of SEFI Annual Conference 2015", © 2015.
Reprint by Springer International Publishing AG 2016,
DOI 10.1007/978-3-319-46916-4_78

2 Entrepreneurship and Entrepreneurship Education

The expression entrepreneurship initially meant the course of undertaking a new task [8]. Today, the term entrepreneurship largely refers to risk taking in producing innovations, for example, through creating new companies [9]. Entrepreneurship is a subsuming term of particular expertise related to business skills, beyond those of engineering skills. Entrepreneurs have multiple tasks: "bringing the first product to the market and of building and financing a new organization. [...] Entrepreneurship is a highrisk, highpotential reward activity. In modern society, engineers are increasingly expected to move to positions of leadership and to take on additional roles as entrepreneurs. [...] In many regions, entrepreneurship is a significant source of new jobs and economic growth, and is strongly incentivised by governments and universities [...]" [8].

Entrepreneurship education in general can be described and defined as the means and approaches used to teach students to start novel businesses and run such businesses successfully [10]. Or as Aulets puts it: "Preparation for entrepreneurship, that is, the starting of a new company, involves unique competencies that can be learned" [11]. Entrepreneurial action includes breeding and identifying ideas with the potential to be developed into goods or services finding success in the market [12]. Students in engineering entrepreneurship programs generate competencies for teamwork, effective communication, independent thinking, understanding business basics, design for end users, and openended problem solving [3]. Even though engineering entrepreneurship education is an emerging topic, its provision to engineering students seems not to be widespread at universities worldwide [13]. The inclusion of entrepreneurship, creativity and innovation in engineering education curricula demands a change in mind set and disposition on the part of teaching staff to partake in, or at least tolerate modifications in the engineering syllabus [3].

During the 1990s, quite a number of engineering schools established novel education programs, putting emphasis on engineering design skills and introduced aspects of social sciences into the syllabus of engineering design. These additions include

- science and technology studies
- user research with ethnographical methodology, and
- entrepreneurship and marketing development [8]

However, in most engineering education programs, these novel topics ended up as a addons without sufficient integration into engineering and sciences subjects, not contributing further to the disciplinary emptiness in engineering education syllabuses [14].

Recently, entrepreneurship education has emerged in different fashion in various engineering schools in Europe, Asia and the U.S. and engineering schools have reacted to a range of challenges with changing their views on entrepreneurship in engineering education:

- entrepreneurship is seen as an additional competence: engineering students should learn in dedicated courses making them able to sustain their predominately technologydriven perspective on innovation [15]
- entrepreneurship is provided through management and business courses especially for engineers: Management and business courses have been redesigned and adapted especially to the needs of engineers based on the idea of markets and economic processes creating the selection mechanisms that determine which technologies will survive [16]
- entrepreneurship is seen as critical thinking in engineering: entrepreneurship as critique and the capacity to provide significant problem solving for a society facing a series of new challenges that range from a
  - restructuring of industrial mass production,
  - globalization of trade and technology, and an
  - increased embedding of technology in social activities [17]

One example to address these additional skills of entrepreneurship is the CDIO-Syllabus 2.0. It includes a section with the following topics: company founding, formulation, leadership, and organization; business plan development; company capitalization and finances; innovative product marketing; conceiving products and services around new technologies; the innovation system, networks, infrastructure, and services; building the team and initiating engineering processes (the engineering process according to the CDIO approach: conceiving, designing, implementing, and operating); managing intellectual property [8].

3 The Gender Gap in Engineering Research and Entrepreneurship

Entrepreneurship in engineering is an intersection of two gendered professional cultures and therewith particularly prone to a gender gap. It is a known fact that engineering is amongst the academic fields where the underrepresentation of women is most striking. While overall, women represent 37 % of grade B academic staff and 20 % of overall academic staff, figures for engineering are 23 % and 11 % [18]. Gender diversity in engineering organizations still had limited success and respective policy initiatives have met resistance, hostility or indifference by the managers as well as women engineers. It seems to require a cultural change in engineering environments, particularly also on the educational part of professional training [5]. Additionally, academic entrepreneurship has been discussed as being less attractive to women; amongst others because they were less likely to have received training in business and management or a lower preference for being selfemployed [19].

Many attempts have been taken to overcome the gender gap in STEM fields by attracting and retaining more female students. One of the prominent approaches is to attract and retain more women in the STEM professions by making the curricula of natural science and engineering education more gender inclusive [20]. An inclusive curriculum can be incorporated into engineering programs in several ways, varying from designing the courses in a way that is responsive to the needs of students from different gender groups to introducing additional courses aiming at teaching gender and diversity competences to students. This implies that professional training for engineering students should include topics such as social justice, ethics, gender equality, and mechanisms of inclusion and exclusion as well as providing gender sensitivity examples and role models [21]. Universities in several countries have taken action in this regard to implement gender inclusiveness in engineering education. In Germany, there have been attempts to combine gender studies with technological sciences, for instance at the University of Hamburg by integrating a gender studies module into the Mathematics, Informatics and Natural Sciences Faculty [22]. In this new module, courses on gender and natural science have been offered at the introductory, advanced and research levels which, as later evaluations have shown, were hardly attended by students majoring in science and engineering studies. This demonstrated that there is a need to implement gender related courses directly within the faculties of science and engineering.

With respect to the gap in entrepreneurship in the STEM fields, literature has suggested different explanations including gaps in enrolment, gaps in female faculty mentorship, gaps in seniority/experience, funding and training gaps and job satisfaction priorities. Besides challenges within professional cultures, previous research also suggests that graduate training (environments) and postdoctoral training have relative importance in explaining the gender gap in STEM entrepreneurship [23]. Still however, gender aware entrepreneurship training is limited [24] and recent studies have identified an ongoing need for a contemporary image of women's entrepreneurship in Germany [25], which could introduce cultural change through awareness raising in general through gender sensitive curriculum in general which will be more likely to instil the interest in entrepreneurship among female students in entrepreneurial opportunities as well as raise their confidence to engage in highrisk entrepreneurship in technology.

4 Research Question and Methodology

In the above context, the aim of this paper is to explore the extent to which gender and entrepreneurship training are integrated in engineering training at universities in Germany and in this way the universities are responding to the policy imperatives at European and national levels. Curriculum descriptions serve as a data source for this study. We assume that if entrepreneurship and gender are taught at the universities it must be visible in the official curricula descriptions. We pose two research questions:

- In how far are entrepreneurship and gender manifested in curricula of selected German technical universities based on the number of modules tackling these topics? and

- How prominent are the topics entrepreneurship and gender represented within the identified modules?

The methodology for this research is based on content analysis, which may be broadly defined as "any systematic reduction of a flow of text (or other symbols) to a standard set of statistically manipulable symbols representing the presence, the intensity, or the frequency of some characteristics relevant to social science" [26]. Amongst a variety of analysis techniques, one could distinguish between the basic categories of qualitative versus quantitative, or thematic versus relational types of content analysis. For addressing our research question the most relevant type appears to be the thematic content analysis (TCA), which "aims at an assessment of the (frequency of the) presence of specified themes, issues, actors, states of affairs, words or ideas in the texts or visuals to be analysed" [27]. In order to perform TCA one has to operationalize concepts using predetermined keywords, whereupon the frequency distribution of the keywords demonstrates whether the themes, issues or actors appear more or less frequently in the analysed data source [27]. In this paper we will make use of this merely descriptive procedure, and limit ourselves to generating descriptive inferences from the data at hand by focusing on frequency lists of selected keywords that represent entrepreneurship and gender diversity to address our research questions. Below we introduce our database and analysis procedure more in detail.

We examined the mechanical engineering curricula for bachelor and master level study programs (published between 2011 and 2014) from nine leading German technical universities (TU 9) as the main data source of this paper. In line with the TCA procedure, the curricula were analysed in terms of the appearance of entrepreneurship and gender as an explicit topic within the curriculum. The focus was put on mechanical engineering curricula, as this can be seen as the core discipline within the engineering field. Furthermore, especially in context with the gender topic the underrepresentation of female students in mechanical engineering is widely documented and discussed. Hence, this discipline is of special interest. For operationalizing entrepreneurship we used the terms (and its German equivalents) "entrepreneurship" (Unternehmertum) itself, "innovation" (Innovation), "venture"/"venturing" (Unternehmung), and "business" (Business/Unternehmen). Gender was operationalized by looking for the terms "gender" (Geschlecht), "diversity" (Diversität), and "inclusiveness" (Inklusivität). These keywords were generated by selecting the most frequently occurring words in literature on entrepreneurship and gender in engineering education to address the respective topics. They were then used as search terms to determine how frequent the topics of entrepreneurship and gender appear in the curricula. The universities and the examined documents can be seen in Table 1.

All in all we examined 1211 different module descriptions of mechanical engineering programmes at bachelor level and 2068 at master level at these nine universities. The addition of these two figures in order to find out the total number of examined modules is not expedient as some of the modules appear on bachelor as well as on master level. In general the module descriptions at the studied universities can be split into two to three main sections: The front page and the table of contents, the main body with the modules' descriptions, and finally in some cases there can be found an

Table 1 The analysed universities and documents. Source: the documents were downloaded from the official universities' homepage

University	Document title			
Course of studies	**Level (Bsc/M)**	**Release Date**	**Pages**	**Modules**
Technische Universität München	Modulhandbuch 17 400 Maschinenwesen;			
	Modulhandbuch 16 401 Maschinenwesen			
Maschinenwesen	Bachelor	20.01.14	679	312
	Master	20.01.14	957	259
RWTH Aachen University	Prüfungsordnung für den Bachelorstudiengang Maschinenbau der RWTH, Anlage 1: Modulkatalog			
	Prüfungsordnung für den Masterstudiengang Maschinenbau der RWTH, Anlage 1: Modulkatalog			
Maschinenbau	Bachelor	16.03.12	319	115
	Master	31.03.11	369	188
Technische Universität Berlin	Modulkatalog Bachelor Maschinenbau SoSe 2013;			
	Modulkatalog Master Maschinenbau SoSe 2013			
Maschinenbau	Bachelor	Summerterm 2013	215	102
	Master	Summerterm 2013	279	115
Karlsruhe Institute of Technology	Modulhandbuch BSc Maschinenbau (Langfassung);			
	Modulhandbuch MSc Maschinenbau (Langfassung)			
Maschinenbau	Bachelor	01.10.13	478	310
	Master	24.10.12	587	357
Leibniz Universität Hannover	Modulkatalog zur PO2010 – Studienführer für den Studiengang Maschinenbau			
Maschinenbau	Bachelor	Winter 2012/2013	113	51
	Master	Winter 2012/2013	113	267
Technische Universität Braunschweig	Modulhandbuch: Bachelor Maschinenbau (BPO 2012);			
	Modulhandbuch: Master Maschinenbau			
Maschinenbau	Bachelor	2012	337	117
	Master	24.08.12	875	336
Technische Universität Darmstadt	Formblatt Modulbeschreibungen FB16			
Maschinenbau	Bachelor	March 2007	114	57
	Master	March 2007	334	50
Technische Universität Dresden	Studienordnung für den Bachelor-Studiengang Maschinenbau Anlage 1: Modulbeschreibungen für den Diplom-Aufbaustudiengang Maschinenbau;			
	Studienordnung für den Diplom-Aufbaustudiengang Maschinenbau Anlage 1: Modulbeschreibungen für den Diplom-Aufbaustudiengang Maschinenbau			
Maschinenbau	Bachelor	17.07.13	97	76
	Master	09.03.12	96	90
University of Stuttgart	Modulhandbuch Studiengang Bachelor of Science Maschinenbau;			
	Modulhandbuch Studiengang Master of Science Maschinenbau			
Maschinenbau	Bachelor	25.03.13	157	71
	Master	27.03.12	690	406

additional index or further explanations at the document's end. For the analysis we used only the part in the middle section. This means for the word counting that if a module has one of the search terms directly in its title, it appears at least three times in the document; once in the table of contents, once in the module description and once in the final index. For our research we counted the explicit term only once in this case in order to prevent distortion. Moreover, we counted it as often as it appeared within the module description, even if it appeared several times in one module. If a module for example had the word "entrepreneurship" in the title and additionally eight times in the description we counted the term nine times. In addition to that, we differentiated between thematic relevant usage of terms in context with entrepreneurship and gender and thematic irrelevant usage. An example can be explained by looking at the term "business". If the word business was used in context with, let us say, business plan writing we counted it, as this shows an obvious connection with entrepreneurship. In contrast we did not count it if, for example, the word business was used in context with basic accounting. From our perspective in this case the term is not used in context with entrepreneurship in the narrow meaning. Building on this methodology a word counting was done on terms that are linked to entrepreneurship and gender. The results will be explained in the following.

5 Findings

As the topic entrepreneurship and gender were examined separately we present the findings one by one. In the first section we present data analysis in terms of word counts. We show how many modules tackle the examined topics. Further we study the identified modules and examine in which contexts entrepreneurship and gender are taught. Here we differentiate between modules, which have one of the topics as a main subject in its learning outcomes. On the other hand entrepreneurship and/or gender can be addressed in modules as one topic among others. This differentiation is important to rate the importance of entrepreneurship and gender within the curriculum. For us the underlying assumption is, that if these topics are of high importance they are represented by modules with a strong focus on them.

5.1 Entrepreneurship

To understand the usage of entrepreneurship in engineering curricula we searched the terms "entrepreneurship", "innovation", "venture/venturing", and "business". The search showed varied frequencies of different terms as shown in Table 2.

As the terms were handled separately, the data in Table 2 is not yet adjusted in terms of double appearance of search terms. That means that if in one module description "entrepreneurship" and "innovation" appeared, this module was counted twice. Taking a closer look and deleting the modules that counted twice, has revealed that in total 32 modules at bachelor level and 67 modules at master level used the search terms in the title and/or in the description. Hence, these modules could be identified as modules that include the entrepreneurship topic. However, these results show clearly that the entrepreneurship topic (represented by the terms above) is not very much represented in the module descriptions. As we looked at 1211 module descriptions at bachelor level the 32 found modules just make up a proportion of

Table 2 Frequencies of terms in studied modules: entrepreneurship

Search terms	Quantity of term appearance in documents	Quantity of modules using terms in titles and/or in description	
		Bachelor level	Master level
Entrepreneurship	55	5	8
Innovation	208	26	54
Venture/Venturing	4	0	2
Business	26	4	10

Table 3 Results of the second step TCA: entrepreneurship

Topic	Quantity of modules having a strong focus on the entrepreneurship topic	
	Bachelor level	Master level
Entrepreneurship	10/1211 =0,82%	15/2068 =0,73%

2.6 %. At master level this proportion is with 3.2 % of 2068 modules slightly higher but still pretty low.

A second step for our data analysis was to study the importance of the entrepreneurship topic within the identified modules. It aimed to understand if this topic is in focus of the modules or is just one topic out of several others. For us this can serve as an additional evidence of the importance of entrepreneurship in engineering higher education. Thus we furthered the thematic content analysis of the identified modules by taking a second step in which we manually coded the content description and rated the importance of entrepreneurship as a subject within the module between "strong focus" and "one topic among others". Taking the figures from above and filtering out the modules with a strong focus shows that at bachelor level only 10 and at master level only 15 modules remained. This finally leads to the proportion of 0.82 % (bachelor) and 0.73 % (master), which again is pretty low (See Table 3).

5.2 Gender

Just as for the terms on entrepreneurship the modules were examined in order to find the search terms on gender. These terms were "gender", "diversity", and "inclusiveness". The results are much more disappointing than for entrepreneurship (See Table 4).

The findings show that the gender topic is barely represented in the module descriptions as seen in Table 4. The term "gender" appears twice and the term "diversity" only appears once in all of the module descriptions. Talking about the proportion of modules tackling the gender topic within the description the figures are insignificant low: 0.17 % of the modules at bachelor level do show a connection to gender as a subject and none at the master level there could not be found even one.

Table 4 Frequencies of terms in studied modules: gender

Terms	Quantity of term appearance in documents	Quantity of modules using terms in titles and/or in description	
		Bachelor level	Master level
Gender	2	1	0
Diversity	1	1	0
Inclusiveness	0	0	0

Table 5 Results of the second step TCA: gender

Topic	Quantity of modules having a strong focus on the gender topic	
	Bachelor level	Master level
Gender	1/1211=0,08%	-

Making the same second step for gender as we did for the entrepreneurship topic and finding out the modules that have gender as a main focus, only one module on bachelor level remains (proportion: 0.8 %) (Table 5).

5.3 *Conclusion and Future Work*

Based on the performed thematic content analyses of modules of mechanical engineering curricular at German technical universities we showed that the entrepreneurship topic is represented in most of the examined curricula, even if the way how it is represented differs significantly. On the one hand we find educational modules that fully address the entrepreneurship topic. On the other hand there are quite a number of modules that include it only in one or two parts of the full course. However, the results are not overwhelming. Looking at the proportion of modules on entrepreneurship in the curricula description clearly shows that not even 5 % of the modules address this topic in any way, neither at bachelor nor at master level. Identifying the modules that have entrepreneurship in focus this proportion even declines to 1 %. These results differ significantly with the results on the gender topic. Only a very small amount of modules could be identified that include gender topic in module descriptions. Only in two of the modules we found the terms "gender" or "diversity" in its description and only one out of these two provides gender sensitivity training. In addition, we did not find any module where both entrepreneurship and gender topics were included in module description. Hence these two topics are still being kept separate in mechanical engineering curricula at German technical universities. Thus, the policy imperatives to integrate entrepreneurship training in engineering university education have been to some extent implemented, while we cannot see this for the inclusion of gender awareness training.

Thus it is still questionable how the policy demands should be fulfilled if the situation stays as it is. It can be assumed that if entrepreneurship and gender are not taught to a wider range at the universities, the future graduates will not acquire entrepreneurship and gender sensitivity skills. Furthermore, the fact that gender as a topic barely appears within the module description can be a reason for the on going underrepresentation of female students in engineering programs.

For further research we will go on researching these official program documents. As they were partly updated during the last year we will look into the new documents and check if and in how far the situation may have changed. Furthermore we will study in depth the examples where we identified the presence of entrepreneurship and gender topics. Specifically, we will examine how entrepreneurship and gender are taught, what types of learning outcomes are defined, what are the methodological approaches used in teaching these course and finally what are institutional contexts where these courses are provided. These will be the guiding questions for the future in order to identify best practice examples and to provide recommendations for teaching and learning designs in mechanical engineering. From our perspective these are necessary steps in order to improve engineering education and in this way to meet political as well as social demands of the future.

References

1. European Institute of Innovation & Technology, *Catalyzing Innovation in the Knowledge Triangle. Practices from the EIT Knowledge and Innovation Communities*. Publications Office of the European Union, Luxembourg, 2012
2. The Council of the European Union, Conclusions of the council and of the representatives of the governments of the member states, meeting within the council, of 26 november 2009 on developing the role of education in a fully-functioning knowledge triangle. Official Journal of the European Union **302**, 2009, pp. 3–5
3. B. T, T. Seelig, S. Sheppart, P. Weilerstein, Entrepreneurship: Its role in engineering education. The Bridge – Linking Engineering and Society **43** (2), 2013, pp. 35–40
4. European Commission, *Europe 2020. A strategy for smart, sustainable and inclusive growth*. European Commission, Brussels, 2010
5. R. Sharp, S. Franzway, J. Mills, J. Jill, Flawed policy, failed politics? challenging the sexual politics on managing diversity in engineering organization. Gender, Work and Organization **19** (6), 2012, pp. 555–572
6. Bundesministerium für Bildung und Forschung. Die neue hightech-strategie. innovationen für deutschland, 2014. URL http://www.bmbf.de/pub/HTS_Broschure_barrierefrei.pdf
7. Ministerium für Innovation, Wissenschaft und Forschung des Landes Nordrhein-Westfalen. Forschungsstrategie fortschritt nrw. forschung und innovation für nachhaltige entwicklung 2013 – 2020, 2013. URL http://www.wissenschaft.nrw.de/mediathek/broschueren/
8. E. F. Crawley, J. Malmquist, S. Östlund, D. R. Brodeur, K. Edström, F Edward, *Rethinking Engineering Education - The CDIO Approach*, 2nd edn. Springer, Cham, Heidelberg, New york, Dordrecht, London., 2014
9. H. C. Menzel, I. Aalito, U. Aalito, On the way to creativity: Engineering as intraprenuership in organizations. Technovation **27**, 2007, pp. 732–743
10. L. C. Tung, *The Impact of Entrepreneurship Education on Entrepreneural Intention of Engineering Students*. PhD thesis, City University of Hong Kong, 2011

11. B. Aulet, *Disciplined entrepreneurship: 24 steps to a successful startup*. Wiley, Hoboken, 2013
12. T. B. Ward, Cognition, creativity, and entrepreneurship. Journal of Business Venturing **19**, 2004
13. . February 20-22, Entrepreneurship education in engineering: A literature review, and an integrated embedment proposal. In: *Recent Advances in Educational Methods, Proceedings of the 10th International Conference on Engineering Education (EDUCATION '13)*. February 20-22, 2013, pp. 106–111
14. G. Downey, Are engineers losing control of technology? from 'problem solving' to 'problem definition and solution' in engineering education. Chemical Engineering Research and Design **83**, 2005, pp. 583–595
15. D. Goldberg, *The entrepreneurial engineer*. Wiley, New York, 2006
16. ASME. Vision 2028 for mechanical engineering, 2008. URL http://www.files.asme.org
17. U. Jorgenswilen, S. Brodersen, H. Lindegaard, P. Boelskifte, Foundations for a new type of design engineers: Experiences from dtu. In: *Proceedings of the ICED 2011 Conference*. August 15-18, 2011
18. She figures, *Gender in Research and Innovation. Statistics and Indicators*. European Commission, 2012
19. I. Verheul, R. Thurik, I. Grilo, P. van der Zwan, Explaining preferences and actual involvement in self-employment: gender and the entrepreneurial personality. Journal of Economic Psychology (33), 2012, pp. 325–341
20. J. E. Mills, M. E. Ayre, J. Gill, Perceptions and understanding of gender inclusive curriculum in engineering education. In: *SEFI 36th annual conference proceedings, Quality assessment, employability and innovation*. 2008, pp. 1–10
21. A. Béraud, A. S. Godfroy, J. Michel, eds., *GIEE 2011: Gender and Interdisciplinary Education for Engineers. Formation Interdisciplinaire des Ingénieurs et Problème du Genre*. Sense Publishers, Rotterdam, 2012
22. H. Götschel, Gender and science studies competence for students in engineering, natural sciences, and science education. the project "degendering science" at the university of hamburg, germany. In: *GIEE 2011: Gender and Interdisciplinary Education for Engineers. Formation Interdisciplinaire des Ingénieurs et Problème du Genre*, ed. by A. Béraud, A. S. Godfroy, J. Michel, Sense Publishers, Rotterdam, 2012, pp. 101–114
23. M. E. Blume-Kohout. Understanding the gender gap in stem fields entrepreneurship, sba office for advocacy, 2014. URL https://www.sba.gov/sites/default/files/Gender%20Gap%20in%20STEM%20Fields_0.pdf
24. I. Treanor, Entrepreneurship education: exploring the gender dimension: A gender and enterprise network, hea sponsored, discussion workshop. International Journal of Gender and Entrepreneurship **4** (2), 2012, pp. 206–210
25. K. Ettl, F. Welter, Gender, context and entrepreneurial learning. International Journal of Gender and Entrepreneurship **2** (2), 2010, pp. 108–129
26. G. Shapiro, J. Markoff, *Revolutionary Demands. A Content Analysis of the Cahiers de Doléances of 1789*. Stanford University Press, Stanford, CA, 1998
27. P. Pennings, H. Keman, J . Kleinnijenhuis, *Doing Research in Political Science*, 2nd edn. Sage Publications, London, 2006

Printed in the United States
By Bookmasters